Path to College Mathematics

Elayn Martin-Gay

University of New Orleans

Pearson

330 Hudson Street, NY NY 10013

Editorial Director: *Chris Hoag*
Editor in Chief: *Michael Hirsch*
Acquisitions Editor: *Mary Beckwith*
Editorial Assistant: *Alison Oehmen*
Content Producer: *Lauren Morse*
Managing Producer: *Scott Disanno*
VP, Product Marketing, Business, Mathematics, Statistics, Engineering, Computer Science, Foundations: *Roxanne McCarley*
Media Producer: *Audra Walsh*
Executive Content Manager, MathXL: *Eric Gregg*
MathXL QA Supervisor, TestGen: *Mary Durnwald*
Senior Author Support/Technology Specialist: *Joe Vetere*
Rights and Permissions Project Manager: *Gina Cheselka*
Manufacturing Buyer: *Carol Melville, LSC Communications*
Associate Director of Design: *Blair Brown*
Cover Design: *Studio Montage*
Text Art: *Scientific Illustrators*
Project Management, Interior Design, Answer Art, and Composition: *Integra*
Cover Image: *Summer tree. Abstract paper nature background: iStock/Getty Images; Abstract color mosaic background. Red vector background: Strizh/Shutterstock*

Acknowledgements of third party content appear on page **xxii**, which constitutes an extension of this copyright page.

Library of Congress Cataloging-in-Publication Data
This title has been cataloged with the Library of Congress.

1 16

Pearson

ISBN 10: 0-13-465440-4
ISBN 13: 978-0-13-465440-9

This book is dedicated to students everywhere—and we should all be students. After all, is there anyone among us who really knows too much? Take that hint and continue to learn something new every day of your life.

Best of wishes from a fellow student: Elayn Martin-Gay

Contents

Preface

I have written *Path to College Mathematics* to prepare non-STEM and STEM students for success in future math courses. The text is designed to be flexible and to provide the necessary topics and skills while helping students to accelerate to their next course and career goals.

For Students Who Choose a Non-STEM Path

Upon successful completion of a course using this text, students are prepared to enter a variety of mathematics college courses—including a liberal arts mathematics course, an introduction to statistics course, and a finite mathematics course. Preparation for these courses occurs through coverage in the following key areas:

- An abundance of real-life applications including those with fractions, decimals, percents and conversions
- A brief introduction to probability and statistics, including counting methods and permutations and combinations
- An introduction and continual review of reading histograms and frequency distributions, as well as other visual presentation of data

For Students Who Choose a STEM Path

Upon successful completion of a course using this text, students are prepared to continue with STEM-level mathematics. Preparation for this path occurs by the following:

- An abundance of algebraic concepts, including linear and quadratic equations, the rectangular coordinate system, functions, introduction to linear regressions, and work with exponents and scientific notation—just to name a few concepts

Special effort has been made to ensure students will be able to solve real-life applications, develop critical thinking and reasoning skills, and encourage conceptual understanding.

Pedagogical Features

Problem-Solving Process This is formally introduced in Chapter 2 with a four-step process that is integrated throughout the text. The four steps are **Understand, Translate, Solve,** and **Interpret.** The repeated use of these steps in a variety of examples shows their wide applicability. Reinforcing the steps can increase students' comfort level and confidence in tackling problems.

Examples Detailed, step-by-step examples are included. Many examples reflect real life. Additional support is provided in the annotated examples.

Practice Exercises Throughout the text, each worked-out example has a parallel Practice Exercise. These invite students to be actively involved in the learning process. Students should try each Practice Exercise after finishing the corresponding example. Learning by doing will help students grasp ideas before moving on to other concepts. Answers to Practice Exercises are provided in the back of the text.

Helpful Hints Helpful Hints contain practical advice on applying mathematical concepts. Strategically placed where students are most likely to need immediate reinforcement, Helpful Hints help students avoid common trouble areas and mistakes.

Concept Checks This feature allows students to gauge their grasp of an idea as it is being presented in the text. Concept Checks stress conceptual understanding at the point of use and help suppress misconceived notions before they start. Answers appear at the bottom of the page.

Exercise Sets The exercises sets were carefully developed to ensure that even- and odd-numbered exercises are paired. The exercise sets also include several special types of exercises. These are:

- **Vocabulary, Readiness & Video Check Questions** These exercises appear at the beginning of the section exercise sets and are also available for assignment in MyMathLab. The **Vocabulary** exercises are fill-in-the-blank using a word bank. The exercises center on the key vocabulary and concepts in the section. The **Readiness** exercises center on a student's understanding of a concept that is necessary in order to continue to the exercise set. The **Video Check Questions** are included in every section for every learning objective. These exercises are a great way to assess whether students have viewed and understood the key concepts presented in the Lecture videos.
- **Mixed Practice Exercises** Found in the section exercise sets, these exercises require students to determine the problem type and strategy needed to solve it just as they would need to do on a test.
- **Guided Applications** These exercises appear in many sections throughout the text. These applications prompt students on how to set up the problem and get started with the solution process. These guided exercises will help students prepare to solve application exercises on their own.
- **Translating Exercises** Many exercise sets contain translating exercises to help students develop and maintain their ability to translate English to algebra.
- **Concept Extensions** These exercises are included in the section exercise sets to provide further opportunities for critical thinking and application of the concepts.
- **Review and Preview** Found in the section exercise sets, these exercises help students retain concepts and skills from earlier sections as well as prepare them for upcoming material.
- **Writing Exercises** ✎ These exercises occur in almost every exercise set and require students to provide a written response to explain concepts or justify their thinking.

Integrated Reviews A unique, mid-chapter exercise set that helps students assimilate new skills and concepts that they have learned separately over several sections. These reviews provide yet another opportunity for students to work with mixed exercises as they master the topics.

End-of-Chapter Material offers students support for review and retention of the topics, skills, and concepts learned in the chapter. This material includes:

- **Vocabulary Check** Provides an opportunity for students to become more familiar with the use of mathematical terms as they strengthen their verbal skills. These appear at the end of each chapter before the Chapter Highlights. Vocabulary, Readiness & Video Check exercises provide practice at the section level.
- **Chapter Highlights** Found at the end of every chapter, these contain key definitions and concepts with examples to help students understand and retain what they have learned and help them organize their notes and study for tests.
- **Chapter Review** The end of every chapter contains a comprehensive review of the topics introduced in the chapter. The Chapter Review offers exercises keyed to every section in the chapter as well as Mixed Review exercises that are not keyed to sections.
- **Getting Ready for the Test** can be found before each Chapter Test. These exercises increase student success by helping students prepare for their chapter test. The purpose of these exercises is to check students' conceptual understanding of the topics in the chapter as well as common student errors. It is suggested that students complete and check these exercises before taking a practice Chapter Test. All Getting Ready for the Test exercises are either multiple-choice or matching, and all answers can be found in the answer section.

Video Solutions of all Getting Ready for the Test exercises can be found in MyMathLab. These video solutions contain brief explanations and reminders of material in the chapter. Where applicable, incorrect choices contain explanations.

Getting Ready for the Test exercise numbers marked in blue indicate that the question is available in **Learning Catalytics.** lc

- **Chapter Test and Chapter Test Prep Videos** The Chapter Test is structured to include exercises that involve common student errors. The Chapter Test Prep Videos give students instant author access to a step-by-step video solution of each exercise on the chapter test.
- **Cumulative Review** Follows every chapter in the text (except Chapters R and 1). Each odd-numbered exercise contained in the Cumulative Review is an earlier worked example in the text that is referenced in the back of the book along with the answer.

Exercise Set Resource Icons Located at the opening of each exercise set, these icons remind students of the resources available for extra practice and support:

MyMathLab®

See Student Resource descriptions page xi for details on the individual resources available.

Exercise Icons These icons facilitate the assignment of specialized exercises and let students know what resources can support them.

- Video icon: exercise worked in MyMathLab.
- △ Triangle icon: identifies exercises involving geometric concepts.
- Pencil icon: indicates a written response is needed.
- Calculator icon: optional exercises intended to be solved using a scientific or graphing calculator.

Optional: Calculator Exploration Boxes and Calculator Exercises The optional Calculator Explorations provide keystrokes and exercises at appropriate places to give students an opportunity to become familiar with these tools. Section exercises that are best completed by using a calculator are identified by for ease of assignment.

Media Program and Key resources

- **The Martin-Gay MyMathLab** course provides extensive exercise coverage, including assignable video check questions and a complete, text-specific video program. There are section lectures videos for every section, which students can also access at the specific objective level; Getting Ready for the Test video solutions; Student Success Tips videos; and an extensive number of watch clips at the exercise level to help students while doing homework in MathXL.
- **Interactive Lecture Series**, featuring your text author Elayn Martin-Gay, provides students with active learning at their own pace. The videos offer the following resources and more:

  A complete lecture for each section of the text highlights key examples and exercises from the text. Pop-ups reinforce key terms, definitions, and concepts.

  An interface with menu navigation features allows students to quickly find and focus on the examples and exercises they need to review.

  Interactive Concept Check exercises measure students' understanding of key concepts and common trouble spots.

New Student Success Tips Videos

Student Success Tips Videos are 3- to 5-minute video segments designed to be daily reminders to students to continue practicing and maintaining good organizational and study habits. They are organized in three categories and are available in MyMathLab. The categories are:

1. Success Tips that apply to any course in college in general, such as Time Management.
2. Success Tips that apply to any mathematics course. One example is based on understanding that mathematics is a course that requires homework to be completed in a timely fashion.
3. Section- or content-specific Success Tips to help students avoid common mistakes or to better understand concepts that often prove challenging. One example of this type of tip is how to apply the order of operations to simplify an expression.

- **The Lecture Series** also includes the following resources for test prep:

  Getting Ready for the Test Videos

  Video Solutions of all exercises can be found in MyMathLab. These video solutions contain brief explanations and reminders of material in the chapter. Where applicable, incorrect choices contain explanations.

  Getting Ready for the Test exercise numbers marked in blue indicate that the question is available in **Learning Catalytics.** lc

  The Chapter Test Prep Videos help students during their most teachable moment—when they are preparing for a test. This innovation provides step-by-step solutions for the exercises found in each Chapter Test. The chapter test prep videos are also available on YouTube™. The videos are captioned in English and Spanish.

- **Learning Catalytics** is an interactive student response tool that uses students' smartphones, tablets, or laptops to engage them in more sophisticated tasks and thinking. Generate class discussion, guide your lecture, and promote peer-to-peer learning with real-time analytics. Accessible through MyMathLab, instructors can use Learning Catalytics to:
  - Pose a variety of open-ended questions that help your students develop critical thinking skills.
  - Monitor responses to find out where students are struggling.
  - Use real-time data to adjust your instructional strategy and try other ways of engaging your students during class.
  - Manage student interactions by automatically grouping students for discussion, teamwork, and peer-to-peer learning.

  For *Path to College Mathematics*, Getting Ready for the Test exercises marked in blue are available in Learning Catalytics. To search for the questions in Learning Catalytics, search the question library for **MGPath** and the chapter number. For example, search **MGPath4** for questions from Chapter 4.

Key Resources

There are two key resources to support students in mastering necessary learning objectives and concepts: the Video Organizer and the Key Concept Activity Lab Workbook.

- **The Video Organizer** is designed to help students take notes and work practice exercises while watching the Interactive Lecture Series videos (available in MyMathLab). All content in the Video Organizer is presented in the same order as it is presented in the videos, making it easy for students to create a course notebook and build good study habits.

- Covers all of the video examples in order.
- Provides ample space for students to write down key definitions and properties.
- Includes Play and Pause button icons to prompt students to follow along with the author for some exercises while they try others on their own.

The Video Organizer is available in a loose-leaf, notebook-ready format. It is also available for download in MyMathLab.

- **Key Concept Activity Lab Workbook** includes Extension Exercises, Exploration Activities, Conceptual Exercises, and Group Activities. These activities are a great way to engage students in conceptual projects and exploration as well as group work.

Student and Instructor Resources

STUDENT RESOURCES

Lecture Series Videos	Video Organizer	Student Solutions Manual
Provides students with active learning at their pace. The videos offer: • A complete lecture for each text section. The interface allows easy navigation to examples and exercises students need to review. • Interactive Concept Check exercises • Student Success Tips Videos • Getting Ready for the Test Videos • Chapter Test Prep Videos Available in MyMathLab	Designed to help students take notes and work practice exercises while watching the Interactive Lecture Series videos. • Covers all of the video examples in order. • Provides ample space for students to write down key definitions and rules. • Includes Play and Pause button icons to prompt students to follow along with the author for some exercises while they try others on their own. Available in loose-leaf, notebook-ready format and in MyMathLab.	Provides completely worked-out solutions to the odd-numbered section exercises; all exercises in the Integrated Reviews, Chapter Reviews, Chapter Tests, and Cumulative Reviews. **Key Concept Activity Lab Workbook** includes Extension Exercises, Exploration Activities, Conceptual Exercises, and Group Activities.

INSTRUCTOR RESOURCES

Annotated Instructor's Edition	Instructor's Resource Manual with Tests and Mini-Lectures
Contains all the content found in the student edition, plus the following: • Classroom example paired to each example • Answers to exercises on the same text page • Teaching Tips throughout the text, placed at key points • Video Answer Section	• Mini-lectures for each text section • Additional Practice worksheets for each section • Several forms of test per chapter—free response and multiple choice • Answers to all items **Instructor's Solutions Manual** **TestGen®** (Available for download from the IRC)
Instructor-to-Instructor Videos—available in the Instructor Resources section of the MyMathLab course.	**Online Resources** **MyMathLab®** (access code required) **MathXL®** (access code required)

Get the Most Out of MyMathLab®

MyMathLab is the leading online homework, tutorial, and assessment program for teaching and learning mathematics, built around Pearson's best-selling content. MyMathLab helps students and instructors improve results; it provides engaging experiences and personalized learning for each student so learning can happen in any environment. Plus, it offers flexible and time-saving course management features to allow instructors to easily manage their classes while remaining in complete control, regardless of course format.

Preparedness

One of the biggest challenges in many mathematics courses is making sure students are adequately prepared with the prerequisite skills needed to successfully complete their course work. MyMathLab offers a variety of content and course options to support students with just-in-time remediation and key-concept review.

- **MyMathLab with Integrated Review**—available for Developmental Mathematics through Calculus—can be used for just-in-time prerequisite review or co-requisite courses. These courses provide videos on review topics, along with pre-made, assignable skills-check quizzes and personalized review homework assignments.
- In recent years many new course models have emerged, as institutions "redesign" to help improve retention and results. At Pearson, we're focused on creating solutions tailored to support your plans and programs. In addition to the new Integrated Review courses, we offer new All-in-One solutions, non-STEM pathways, and STEM-track options.

Used by more than 37 million students worldwide, MyMathLab delivers consistent, measurable gains in student learning outcomes, retention, and subsequent course success.

Acknowledgments

Many people helped me develop this text, and I will attempt to thank some of them here. Cindy Trimble was *invaluable* for contributing to the overall accuracy of the text. Elaine Page was *invaluable* for her many suggestions and contributions during the development and writing of this text. Gina Linko and Lauren Morse provided guidance throughout the production process.

A very special thank you goes to my editor, Mary Beckwith, for being there 24/7/365, as my students say. Last, my thanks to the staff at Pearson for all their support: Michael Hirsch, Jennifer Edwards, Roxanne McCarley, Chris Hoag, and Paul Corey.

I would like to thank the following reviewers for their input and suggestions:

Sheila Anderson, *Housatonic Community College*
Tom Blackburn, *Northeastern Illinois University*
Gail Burkett, *Palm Beach State College*
James Butterbach, *Joliet Junior College*
Laura Dyer, *Southwestern Illinois College*
Sharon Edgemon, *Bakersfield College*
Hope Essien, *Olive-Harvey College*
Randa Kress, *Idaho State University*
Ted Lai, *Hudson Community College*
Nicole Lang, *North Hennepin Community College*
Lee LaRue, *Paris Junior College*
Jeri Lee, *Des Moines Area Community College*
Jean McArthur, *Joliet Junior College*
Michael Montano, *Riverside Community College*
Lisa J. Music, *Big Sandy Community and Technical College*
Linda Padilla, *Joliet Junior College*
Scott Perkins, *Lake Sumter Community College*
Marilyn Platt, *Gaston College*
Sandy Spears, *Jefferson Community College*
Ping Charlene Tintera, *Texas A & M University*
Jane Wampler, *Housatonic Community College*
Peter Zimmer, *West Chester University*

I would also like to thank the following dedicated group of instructors who participated in our focus groups, Martin-Gay Summits, and our design review for the series. Their feedback and insights have helped to strengthen this edition of the text. These instructors include:

Billie Anderson, *Tyler Junior College*
Cedric Atkins, *Mott Community College*
Andrea Barnett, *Tri-County Technical College*
Lois Beardon, *Schoolcraft College*
Michelle Beerman, *Pasco Hernando Community College*
Laurel Berry, *Bryant & Stratton College*
John Beyers, *University of Maryland*
Bob Brown, *Community College of Baltimore County–Essex*
Lisa Brown, *Community College of Baltimore County–Essex*
NeKeith Brown, *Richland College*
Sue Brown, *Guilford Technical Community College*
Gail Burkett, *Palm Beach State College*
Cheryl Cantwell, *Seminole Community College*
Jackie Cohen, *Augusta State College*
Julie Dewan, *Mohawk Valley Community College*
Janice Ervin, *Central Piedmont Community College*
Richard Fielding, *Southwestern College*
Cindy Gaddis, *Tyler Junior College*
Nita Graham, *St. Louis Community College*

Pauline Hall, *Iowa State College*
Elizabeth Hamman, *Cypress College*
Pat Hussey, *Triton College*
Dorothy Johnson, *Lorain County Community College*
Sonya Johnson, *Central Piedmont Community College*
Irene Jones, *Fullerton College*
Paul Jones, *University of Cincinnati*
Kathy Kopelousos, *Lewis and Clark Community College*
Nancy Lange, *Inver Hills Community College*
Judy Langer, *Westchester Community College*
Lisa Lindloff, *McLinnan Community College*
Sandy Lofstock, *St. Petersburg College*
Kathy Lovelle, *Westchester Community College*
Jamie Malek, *Florida State College*
Jean McArthur, *Joliet Junior College*
Kevin McCandless, *Evergreen Valley College*
Daniel Miller, *Niagara County Community College*
Marica Molle, *Metropolitan Community College*
Carol Murphy, *San Diego Miramar College*
Greg Nguyen, *Fullerton College*
Eric Ollila, *Jackson Community College*
Linda Padilla, *Joliet Junior College*
Davidson Pierre, *State College of Florida*
Marilyn Platt, *Gaston College*
Susan Poss, *Spartanburg Community College*
Natalie Rivera, *Estrella Mountain Community College*
Judy Roane, *Pearl River Community College*
Claudinna Rowley, *Montgomery Community College, Rockville*
Ena Salter, *Manatee Community College*
Carole Shapero, *Oakton Community College*
Janet Sibol, *Hillsborough Community College*
Anne Smallen, *Mohawk Valley Community College*
Barbara Stoner, *Reading Area Community College*
Jennifer Strehler, *Oakton Community College*
Ellen Stutes, *Louisiana State University Eunice*
Tanomo Taguchi, *Fullerton College*
MaryAnn Tuerk, *Elsin Community College*
Gwen Turbeville, *J. Sargeant Reynolds Community College*
Walter Wang, *Baruch College*
Leigh Ann Wheeler, *Greenville Technical Community College*
Valerie Wright, *Central Piedmont Community College*

A special thank you to those students who participated in our design review: Katherine Browne, Mike Bulfin, Nancy Canipe, Ashley Carpenter, Jeff Chojnachi, Roxanne Davis, Mike Dieter, Amy Dombrowski, Kay Herring, Todd Jaycox, Kaleena Levan, Matt Montgomery, Tony Plese, Abigail Polkinghorn, Harley Price, Eli Robinson, Avery Rosen, Robyn Schott, Cynthia Thomas, and Sherry Ward.

Elayn Martin-Gay

About the Author

Elayn Martin-Gay has taught mathematics at the University of New Orleans for more than 25 years. Her numerous teaching awards include the local University Alumni Association's Award for Excellence in Teaching and Outstanding Developmental Educator at University of New Orleans, presented by the Louisiana Association of Developmental Educators.

Prior to writing textbooks, Elayn Martin-Gay developed an acclaimed series of lecture videos to support developmental mathematics students in their quest for success. These highly successful videos originally served as the foundation material for her texts. Today, the videos are specific to each book in the Martin-Gay series. The author has also created Chapter Test Prep videos to help students during their most "teachable moment"—as they prepare for a test—along with instructor-to-instructor videos that provide teaching tips, hints, and suggestions for each developmental mathematics course, including basic mathematics, prealgebra, beginning algebra, and intermediate algebra. Her most recent innovations are the Algebra Prep Apps for the iPhone and iPod Touch. These apps embrace the different learning styles, schedules, and paces of students and provide them with quality math tutoring.

Elayn is the author of numerous published textbooks as well as multimedia interactive mathematics, all specializing in developmental mathematics courses. She has participated as an author across the broadest range of educational materials: textbooks, videos, tutorial software, and courseware. This offers an opportunity of various combinations for an integrated teaching and learning package offering great consistency for the student.

Applications Index

F

G

Photo Credits

Cover: Strizh/Shutterstock; iStock/Getty Images

Chapter R

Page R-1 William87/Fotolia

Page R-19 Pressmaster/Shutterstock

Page R-23 Clay Gay

Page R-48 Abdesign/iStock/Getty Images

Page R-49 Tranger/iStock/Getty Images

Chapter 1

Page 6 Maciej Noskowski/Getty Images

Page 11 Asya M/Fotolia

Page 22 Speedfighter/Fotolia

Chapter 2

Page 56 Umberto Shtanzman/Shutterstock

Page 73 Iofoto/Fotlia

Page 74 (top) Apdesign/Shutterstock; (bottom) Galyna Andrushko/Shutterstock

Page 75 Phovoir/Shutterstock

Page 87 Michael Reynoids/EPA/Newscom

Page 88 Jason Maeh/Shutterstock

Page 89 EdBockStock/Shutterstock

Page 93 06photo/Shutterstock

Page 94 Ollirg/Fotolia

Page 97 narvf/Fotolia

Page 98 Tyler Olson/Shutterstock

Page 99 Photoncatcher/iStock/Getty Images

Page 100 LuckyImages/Fotolia

Page 107 (top) Gillmar/Shutterstock; (bottom) Ayzek/Shutterstock

Page 116 Vstock/UpperCut Images/Getty Images

Page 126 Orhan Cam/Shutterstock

Chapter 3

Page 135 Alin Brotea/Shutterstock

Page 141 Michaeljung/Shutterstock

Page 144 (top) Snvv/Shutterstock; (bottom) Artem Kursin/Shutterstock

Page 146 (top) Dotshock/Shutterstock; (bottom) Dmitry Kalinovsky/Shutterstock

Page 147 BlueSkyImage/Shutterstock

Page 157 Gburba/Fotolia

Page 165 BlueSkyImages/Fotolia

Page 166 (top) Recreativeworks/Shutterstock; (bottom) NetPhotos/Alamy Stock Photo

Page 171 GoodMood Photo/Shutterstock

Page 174 Moonrise/Fotolia

Page 175 Pashin Georgiy/Shutterstock

Page 189 yauhen0409/Fotolia

Page 220 Naypong/Shutterstock

Chapter 4

Page 274 (top) Henry Czauderna/Fotolia; (bottom) Elenarts/Fotolia

Page 279 Jabiru/Fotolia

Page 280 (top) Luc Novovitch/Alamy Stock Photo; (bottom) Byron Moore/Fotolia

Chapter 5

Page 284 Wavebreakmedia/Shutterstock

Page 310 (left) Kojihirano/Fotolia; (right, top) Dirk Vonten/Fotolia; (right, bottom) E.O./Shutterstock

Page 346 (top) Benng/Shutterstock; (bottom) Zoe Schumacher/Fotolia

Page 353 (top, left) Jejim/Shutterstock; (top, right) An Qi/Alamy Stock Photo; (bottom) Jabiru/Fotolia

Page 356 (top) Gennadiy Poznyakov/Fotolia; (bottom) Staff/Getty Images

Chapter 6

Page 385 (top) Asa Gauen/Alamy Stock Photo; (bottom) Donyanedomam/Fotolia

Page 399 Frankix/Fotolia

Page 402 (top) Jorg Hackemann/Fotolia; (bottom) Shuttoz/Fotolia

Page 406 Daphot75/Fotolia

Page 414 (top) ATM2003/Shutterstock; (bottom) Worker/Shutterstock

Page 423 (left) Bill/Fotolia; (right) Vuktopua/Fotolia

Page 436 (left) Huaxiadragon/Fotolia; (right) Galina Barskaya/Shutterstock

Page 437 Dave/Fotolia

Chapter 7

Page 453 Leekris/Fotolia

Page 464 WavebreakMedia/Shutterstock

Page 466 (top) VIktor/Fotolia; (bottom) Monkey Business Images/Shutterstock

Page 481 Monkey Business/Fotolia

Chapter 8

Page 506 (top) Ultrashock/Shutterstock; (bottom) Robraine/Shutterstock

Page 511 Wrangler/Fotolia

Page 512 Peterz/Fotolia

Appendix

Page 531 Rachel Youdelman/Pearson, Inc.

Page 540 Dragonlmages/Fotolia

Page 548 Cherkas/Shutterstock

Page 549 Joggie Botman/Fotolia

CHAPTER R
Prealgebra Review

CHECK YOUR PROGRESS

This optional review chapter covers basic topics and skills from prealgebra, such as fractions, decimals, percents, and reading graphs. Knowledge of these topics is needed for success in college mathematics.

What Majors Do College Freshmen Choose?

The following graph is called a circle graph or a pie chart. Each sector (shaped like a piece of pie) shows the fraction of entering college freshmen who choose to major in each discipline shown. In Exercise Set R.2, Exercises 65–70, we show this same circle graph, but in 3-D design. We simplify some of the fractions in it and also study sector size versus fraction value.

College Freshman Majors

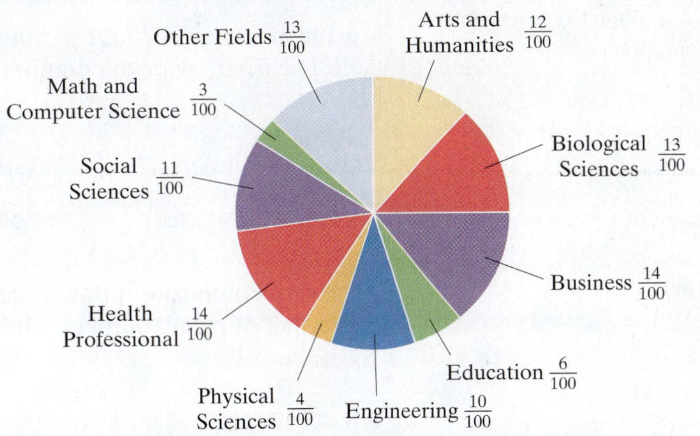

Source: The Higher Education Research Institute

R.1 Fractions and Mixed Numbers

OBJECTIVES

1 Write Fractions in Simplest Form.

2 Multiply and Divide Fractions.

3 Add and Subtract Fractions.

4 Perform Operations on Mixed Numbers.

OBJECTIVE 1

Writing Fractions in Simplest Form

A quotient of two numbers such as $\frac{2}{9}$ is called a **fraction.** The parts of a fraction are:

Fraction bar $\rightarrow \frac{2}{9}$ $\leftarrow$ Numerator, $\leftarrow$ Denominator

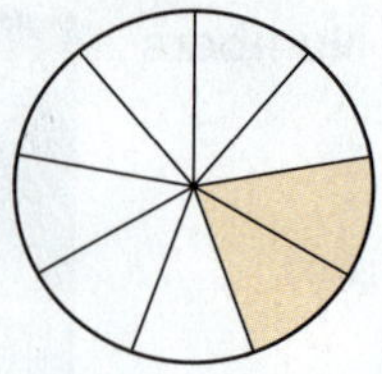

$\frac{2}{9}$ of the circle is shaded.

A fraction may be used to refer to part of a whole. For example, $\frac{2}{9}$ of the circle above is shaded. The denominator 9 tells us how many equal parts the whole circle is divided into, and the numerator 2 tells us how many equal parts are shaded.

To simplify fractions, we can factor the numerator and the denominator. In the statement $3 \cdot 5 = 15$, 3 and 5 are called **factors** and 15 is the **product.** (The raised dot symbol indicates multiplication.)

$$\underset{\text{factor}}{\underset{\uparrow}{3}} \quad \cdot \quad \underset{\text{factor}}{\underset{\uparrow}{5}} \quad = \quad \underset{\text{product}}{\underset{\uparrow}{15}}$$

To **factor** 15 means to write it as a product. The number 15 can be factored as $3 \cdot 5$ or as $1 \cdot 15$.

A fraction is said to be **simplified** or in **lowest terms** when the numerator and the denominator have no factors in common other than 1. For example, the fraction $\frac{5}{11}$ is in lowest terms since 5 and 11 have no common factors other than 1.

To help us simplify fractions, we write the numerator and the denominator as products of **prime numbers.**

Helpful Hint

The natural number 1 is neither prime nor composite.

Prime Number and Composite Number

A **prime number** is a natural number, other than 1, whose only factors are 1 and itself. The first few prime numbers are

2, 3, 5, 7, 11, 13, 17, 19, 23, 29, and so on.

A natural number, other than 1, that is not a prime number is called a **composite number.**

Every composite number can be written as a product of prime numbers. We call this product of prime numbers the **prime factorization** of the composite number.

EXAMPLE 1 Write each of the following numbers as a product of primes.

a. 40 **b.** 63

Solution

a. First, write 40 as the product of any two whole numbers other than 1.

$$40 = 4 \cdot 10$$

Next, factor each of these numbers. Continue this process until all of the factors are prime numbers.

$$\begin{aligned} 40 &= 4 \cdot 10 \\ &= 2 \cdot 2 \cdot 2 \cdot 5 \end{aligned}$$

All the factors are now prime numbers. Then 40 written as a product of primes is

$$40 = 2 \cdot 2 \cdot 2 \cdot 5$$

b. $$\begin{aligned} 63 &= 9 \cdot 7 \\ &= 3 \cdot 3 \cdot 7 \end{aligned}$$ □

PRACTICE 1 Write each of the following numbers as a product of primes.

a. 36 **b.** 200 ■

To use prime factors to write a fraction in lowest terms (or simplified form), apply the fundamental principle of fractions.

Fundamental Principle of Fractions

If $\frac{a}{b}$ is a fraction and c is a nonzero real number, then

$$\frac{a \cdot c}{b \cdot c} = \frac{a}{b}$$

To understand why this is true, we use the fact that since c is not zero, $\frac{c}{c} = 1$.

$$\frac{a \cdot c}{b \cdot c} = \frac{a}{b} \cdot \frac{c}{c} = \frac{a}{b} \cdot 1 = \frac{a}{b}$$

We will call this process dividing out the common factor of c.

EXAMPLE 2 Simplify each fraction (write it in lowest terms).

a. $\frac{42}{49}$ **b.** $\frac{11}{27}$ **c.** $\frac{88}{20}$

Solution

a. Write the numerator and the denominator as products of primes; then apply the fundamental principle to the common factor 7.

$$\frac{42}{49} = \frac{2 \cdot 3 \cdot 7}{7 \cdot 7} = \frac{2 \cdot 3}{7} \cdot \frac{7}{7} = \frac{2 \cdot 3}{7} = \frac{6}{7}$$

b. $\frac{11}{27} = \frac{11}{3 \cdot 3 \cdot 3}$

There are no common factors other than 1, so $\frac{11}{27}$ is already in simplest form.

c. $\frac{88}{20} = \frac{2 \cdot 2 \cdot 2 \cdot 11}{2 \cdot 2 \cdot 5} = \frac{2}{2} \cdot \frac{2}{2} \cdot \frac{2 \cdot 11}{5} = \frac{22}{5}$ □

PRACTICE 2 Simplify each fraction (write it in lowest terms).

a. $\frac{63}{72}$ **b.** $\frac{64}{12}$ **c.** $\frac{7}{25}$ ■

✔ CONCEPT CHECK

Explain the error in the following steps.

a. $\frac{15}{55} = \frac{15}{55} = \frac{1}{5}$ (crossed out: $\frac{1\,5}{5\,5}$)

b. $\frac{6}{7} = \frac{5+1}{5+2} = \frac{1}{2}$ (crossed out)

OBJECTIVE

2 Multiplying and Dividing Fractions

To multiply two fractions, multiply numerator times numerator to obtain the numerator of the product; multiply denominator times denominator to obtain the denominator of the product.

Multiplying Fractions

$$\frac{a}{b} \cdot \frac{c}{d} = \frac{a \cdot c}{b \cdot d} \quad \text{if } b \neq 0 \text{ and } d \neq 0$$

Helpful Hint

Remember that division by 0 is undefined.

EXAMPLE 3 Multiply $\frac{2}{15}$ and $\frac{5}{13}$. Simplify the product if possible.

Solution

$$\frac{2}{15} \cdot \frac{5}{13} = \frac{2 \cdot 5}{15 \cdot 13} \quad \begin{array}{l}\text{Multiply numerators.}\\ \text{Multiply denominators.}\end{array}$$

Next, simplify the product by dividing the numerator and the denominator by any common factors.

$$= \frac{2 \cdot \cancel{5}^{1}}{3 \cdot \cancel{5}_{1} \cdot 13}$$

$$= \frac{2}{39}$$ □

PRACTICE

3 Multiply $\frac{3}{8}$ and $\frac{7}{9}$. Simplify the product if possible. ■

Before dividing fractions, we first define **reciprocals.** Two fractions are reciprocals of each other if their product is 1.

For example:

The reciprocal of $\frac{2}{3}$ is $\frac{3}{2}$ because $\frac{2}{3} \cdot \frac{3}{2} = \frac{6}{6} = 1$.

The reciprocal of 5 is $\frac{1}{5}$ because $5 \cdot \frac{1}{5} = \frac{5}{1} \cdot \frac{1}{5} = \frac{5}{5} = 1$.

To divide fractions, multiply the first fraction by the reciprocal of the second fraction.

Dividing Fractions

$$\frac{a}{b} \div \frac{c}{d} = \frac{a}{b} \cdot \frac{d}{c}, \quad \text{if } b \neq 0, d \neq 0, \text{ and } c \neq 0$$

EXAMPLE 4 Divide. Simplify all quotients if possible.

a. $\frac{4}{5} \div \frac{5}{16}$ **b.** $\frac{7}{10} \div 14$ **c.** $\frac{3}{8} \div \frac{3}{10}$

Solution

a. $\frac{4}{5} \div \frac{5}{16} = \frac{4}{5} \cdot \frac{16}{5} = \frac{4 \cdot 16}{5 \cdot 5} = \frac{64}{25}$ The numerator and denominator have no common factors.

Answers to Concept Check:
answers may vary

b. $\dfrac{7}{10} \div 14 = \dfrac{7}{10} \div \dfrac{14}{1} = \dfrac{7}{10} \cdot \dfrac{1}{14} = \dfrac{\overset{1}{\cancel{7}} \cdot 1}{2 \cdot 5 \cdot 2 \cdot \underset{1}{\cancel{7}}} = \dfrac{1}{20}$

c. $\dfrac{3}{8} \div \dfrac{3}{10} = \dfrac{3}{8} \cdot \dfrac{10}{3} = \dfrac{\overset{1}{\cancel{3}} \cdot \overset{1}{\cancel{2}} \cdot 5}{\underset{1}{\cancel{2}} \cdot 2 \cdot 2 \cdot \underset{1}{\cancel{3}}} = \dfrac{5}{4}$ □

PRACTICE 4 Divide. Simplify all quotients if possible.

a. $\dfrac{3}{4} \div \dfrac{4}{9}$ **b.** $\dfrac{5}{12} \div 15$ **c.** $\dfrac{7}{6} \div \dfrac{7}{15}$ ■

OBJECTIVE 3

Adding and Subtracting Fractions

To add or subtract fractions with the same denominator, combine numerators and place the sum or difference over the common denominator.

Adding and Subtracting Fractions with the Same Denominator

$$\frac{a}{b} + \frac{c}{b} = \frac{a + c}{b}, \quad \text{if } b \neq 0$$

$$\frac{a}{b} - \frac{c}{b} = \frac{a - c}{b}, \quad \text{if } b \neq 0$$

EXAMPLE 5 Add or subtract as indicated. Simplify each result if possible.

a. $\dfrac{2}{7} + \dfrac{4}{7}$ **b.** $\dfrac{3}{10} + \dfrac{2}{10}$ **c.** $\dfrac{9}{7} - \dfrac{2}{7}$ **d.** $\dfrac{5}{3} - \dfrac{1}{3}$

Solution

a. $\dfrac{2}{7} + \dfrac{4}{7} = \dfrac{2 + 4}{7} = \dfrac{6}{7}$

b. $\dfrac{3}{10} + \dfrac{2}{10} = \dfrac{3 + 2}{10} = \dfrac{5}{10} = \dfrac{\overset{1}{\cancel{5}}}{2 \cdot \underset{1}{\cancel{5}}} = \dfrac{1}{2}$

c. $\dfrac{9}{7} - \dfrac{2}{7} = \dfrac{9 - 2}{7} = \dfrac{7}{7} = 1$

d. $\dfrac{5}{3} - \dfrac{1}{3} = \dfrac{5 - 1}{3} = \dfrac{4}{3}$ □

PRACTICE 5 Add or subtract as indicated. Simplify each result if possible.

a. $\dfrac{8}{5} - \dfrac{3}{5}$ **b.** $\dfrac{8}{5} - \dfrac{2}{5}$ **c.** $\dfrac{3}{5} + \dfrac{1}{5}$ **d.** $\dfrac{5}{12} + \dfrac{1}{12}$ ■

Whole

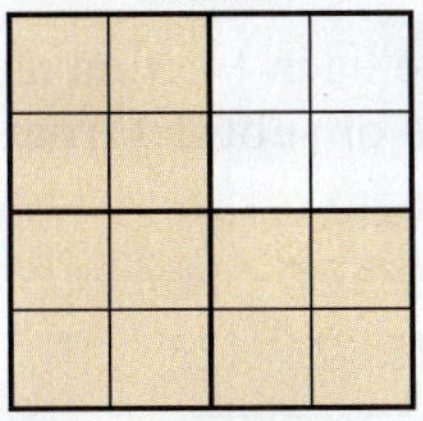

$\dfrac{3}{4} = \dfrac{12}{16}$

To add or subtract fractions without the same denominator, first write the fractions as equivalent fractions with a common denominator. **Equivalent fractions** are fractions that represent the same quantity. For example,

$$\frac{3}{4} \text{ and } \frac{12}{16} \text{ are equivalent fractions}$$

since they represent the same portion of a whole, as the diagram shows. Count the larger squares, and the shaded portion is $\dfrac{3}{4}$. Count the smaller squares, and the shaded portion is $\dfrac{12}{16}$. Thus, $\dfrac{3}{4} = \dfrac{12}{16}$.

We can write equivalent fractions by multiplying a given fraction by 1, as shown in the next example. Multiplying a fraction by 1 does not change the value of the fraction.

EXAMPLE 6 Write $\frac{2}{5}$ as an equivalent fraction with a denominator of 20.

Solution Since $5 \cdot 4 = 20$, multiply the fraction by $\frac{4}{4}$. Multiplying by $\frac{4}{4} = 1$ does not change the value of the fraction.

Multiply by $\frac{4}{4}$ or 1.

$$\frac{2}{5} = \frac{2}{5} \cdot \frac{4}{4} = \frac{2 \cdot 4}{2 \cdot 4} = \frac{8}{20}$$

Thus, $\frac{2}{5} = \frac{8}{20}$. □

PRACTICE 6 Write $\frac{2}{3}$ as an equivalent fraction with a denominator of 21.

To add or subtract with different denominators, we first write the fractions as **equivalent fractions** with the same denominator. We use the smallest or **least common denominator, or LCD.** (The LCD is the same as the least common multiple of the denominators.)

EXAMPLE 7 Add or subtract as indicated. Write each answer in simplest form.

a. $\frac{2}{5} + \frac{1}{4}$ **b.** $\frac{19}{6} - \frac{23}{12}$ **c.** $\frac{1}{2} + \frac{17}{22} - \frac{2}{11}$

Solution

a. Fractions must have a common denominator before they can be added or subtracted. Since 20 is the smallest number that both 5 and 4 divide into evenly, 20 is the **least common denominator** (LCD). Write both fractions as equivalent fractions with denominators of 20. Since

$$\frac{2}{5} \cdot \frac{4}{4} = \frac{2 \cdot 4}{5 \cdot 4} = \frac{8}{20} \quad \text{and} \quad \frac{1}{4} \cdot \frac{5}{5} = \frac{1 \cdot 5}{4 \cdot 5} = \frac{5}{20}$$

then

$$\frac{2}{5} + \frac{1}{4} = \frac{8}{20} + \frac{5}{20} = \frac{13}{20}$$

b. The LCD is 12. We write both fractions as equivalent fractions with denominators of 12.

$$\frac{19}{6} - \frac{23}{12} = \frac{38}{12} - \frac{23}{12}$$

$$= \frac{15}{12} = \frac{\overset{1}{\cancel{3}} \cdot 5}{2 \cdot 2 \cdot \underset{1}{\cancel{3}}} = \frac{5}{4}$$

c. The LCD for denominators 2, 22, and 11 is 22. First, write each fraction as an equivalent fraction with a denominator of 22. Then add or subtract from left to right.

$$\frac{1}{2} = \frac{1}{2} \cdot \frac{11}{11} = \frac{11}{22}, \quad \frac{17}{22} = \frac{17}{22}, \quad \text{and} \quad \frac{2}{11} = \frac{2}{11} \cdot \frac{2}{2} = \frac{4}{22}$$

Then

$$\frac{1}{2} + \frac{17}{22} - \frac{2}{11} = \frac{11}{22} + \frac{17}{22} - \frac{4}{22} = \frac{24}{22} = \frac{12}{11}$$ □

PRACTICE 7 Add or subtract as indicated. Write answers in simplest form.

a. $\frac{5}{11} + \frac{1}{7}$ **b.** $\frac{5}{21} - \frac{1}{6}$ **c.** $\frac{1}{3} + \frac{29}{30} - \frac{4}{5}$

OBJECTIVE
4 Performing Operations on Mixed Numbers

Helpful Hint
Proper Fraction:
numerator < denominator
Improper fraction:
numerator ≥ denominator

Recall that a mixed number contains a whole number part and a fraction part.

To multiply or divide mixed numbers, first write each mixed number as an improper fraction. To recall how this is done, let's write $3\frac{1}{5}$ as an improper fraction.

$$3\frac{1}{5} = 3 + \frac{1}{5} = \frac{15}{5} + \frac{1}{5} = \frac{16}{5}$$

Because of the steps above, notice that we can use a shortcut process for writing a mixed number as an improper fraction.

$$3\frac{1}{5} = \frac{5 \cdot 3 + 1}{5} = \frac{16}{5}$$

EXAMPLE 8 Divide: $2\frac{1}{8} \div 1\frac{2}{3}$

Solution First write each mixed number as an improper fraction.

$$2\frac{1}{8} = \frac{8 \cdot 2 + 1}{8} = \frac{17}{8}; \quad 1\frac{2}{3} = \frac{3 \cdot 1 + 2}{3} = \frac{5}{3}$$

Now divide as usual.

$$2\frac{1}{8} \div 1\frac{2}{3} = \frac{17}{8} \div \frac{5}{3} = \frac{17}{8} \cdot \frac{3}{5} = \frac{51}{40}$$

The fraction $\frac{51}{40}$ is improper. To write it as an equivalent mixed number, remember that the fraction bar means division and divide.

$$\begin{array}{r} 1\frac{11}{40} \\ 40\overline{)51} \\ \underline{-40} \\ 11 \end{array}$$

Thus, the quotient is $\frac{51}{40}$ or $1\frac{11}{40}$. □

PRACTICE
8 Multiply: $5\frac{1}{6} \cdot 4\frac{2}{5}$

As a general rule, if the original exercise contains mixed numbers, write the result as a mixed number if possible.

When adding or subtracting mixed numbers, you might want to use the following method.

EXAMPLE 9 Subtract: $50\frac{1}{6} - 38\frac{1}{3}$

Solution

$$\begin{array}{rcrcr} 50\frac{1}{6} & = & 50\frac{1}{6} & = & 49\frac{7}{6} \\ \underline{-38\frac{1}{3}} & = & \underline{-38\frac{2}{6}} & = & \underline{-38\frac{2}{6}} \\ & & & & 11\frac{5}{6} \end{array}$$

$$50\frac{1}{6} = 49 + 1 + \frac{1}{6} = 49\frac{7}{6}$$

□

PRACTICE
9 Subtract: $76\frac{1}{12} - 35\frac{1}{4}$

Vocabulary, Readiness & Video Check

Use the choices below to fill in each blank. Choices may be used more than once or not at all.

simplified	reciprocals	equivalent	denominator
product	factors	fraction	numerator

1. A quotient of two numbers, such as $\frac{5}{8}$, is called a(n) ____________.

2. In the fraction $\frac{3}{11}$, the number 3 is called the ____________ and the number 11 is called the ____________.

3. To factor a number means to write it as a(n) ____________.

4. A fraction is said to be ____________ when the numerator and the denominator have no common factors other than 1.

5. In $7 \cdot 3 = 21$, the numbers 7 and 3 are called ____________ and the number 21 is called the ____________.

6. The fractions $\frac{2}{9}$ and $\frac{9}{2}$ are called ____________.

7. Fractions that represent the same quantity are called ____________ fractions.

Martin-Gay Interactive Videos

See Video R.1

Watch the section lecture video and answer the following questions.

OBJECTIVE 1 **8.** What is the common factor in the numerator and denominator of Example 1? What principle is used to simplify this fraction?

OBJECTIVE 2 **9.** During the solving of Example 3, what two things change in the first step?

OBJECTIVE 3 **10.** What is the first step needed in order to subtract the fractions in Example 6 and why?

OBJECTIVE 4 **11.** For Example 7, why is the sum not left as $4\frac{7}{6}$?

R.1 Exercise Set MyMathLab®

Represent the shaded part of each geometric figure by a fraction.

1.

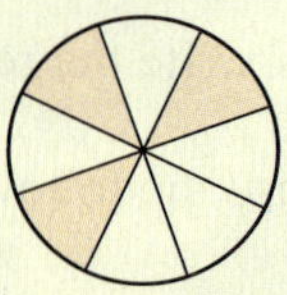

2.

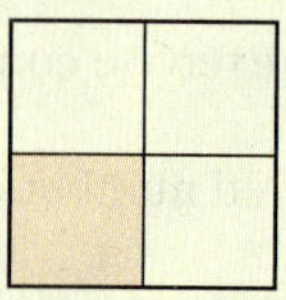

3.

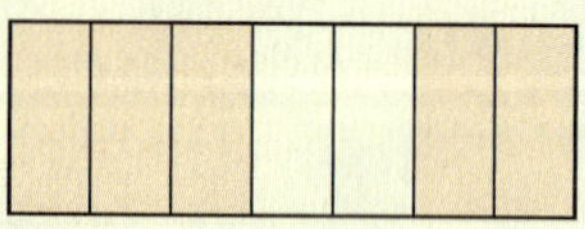

4.

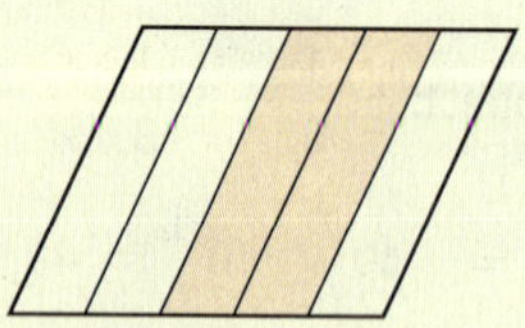

Write each number as a product of primes. See Example 1.

5. 33 **6.** 60 **7.** 98

8. 27 **9.** 20 **10.** 56

11. 75 **12.** 32 **13.** 45

14. 24

Simplify each fraction (write it in lowest terms). See Example 2.

15. $\frac{2}{4}$ **16.** $\frac{3}{6}$ **17.** $\frac{10}{15}$

18. $\frac{15}{20}$ **19.** $\frac{3}{7}$ **20.** $\frac{5}{9}$

21. $\frac{18}{30}$ **22.** $\frac{42}{45}$ **23.** $\frac{120}{244}$

24. $\frac{360}{700}$

Multiply or divide as indicated. Simplify the answer if possible. See Examples 3 and 4.

25. $\frac{1}{2} \cdot \frac{3}{4}$ **26.** $\frac{7}{11} \cdot \frac{3}{5}$ **27.** $\frac{2}{3} \cdot \frac{3}{4}$

28. $\frac{7}{8} \cdot \frac{3}{21}$ **29.** $\frac{1}{2} \div \frac{7}{12}$ **30.** $\frac{7}{12} \div \frac{1}{2}$

31. $\frac{3}{4} \div \frac{1}{20}$ **32.** $\frac{3}{5} \div \frac{9}{10}$ **33.** $\frac{7}{10} \cdot \frac{5}{21}$

34. $\frac{3}{35} \cdot \frac{10}{63}$ **35.** $\frac{25}{9} \cdot \frac{1}{3}$ **36.** $\frac{1}{4} \cdot \frac{19}{6}$

Recall that the **area** *of a plane figure is a measure of the amount of surface of the figure. Find the area of each figure below. (The area of a rectangle is the product of its length and width. The area of a triangle is $\frac{1}{2}$ the product of its base and height.)*

△ **37.**

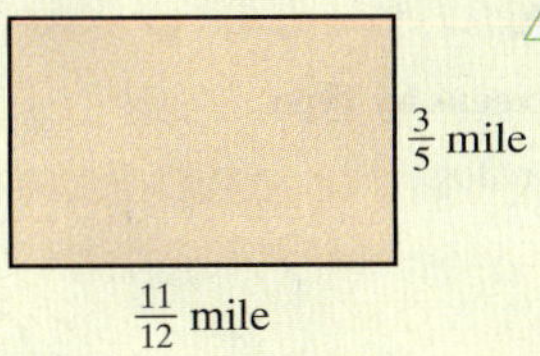

△ **38.**

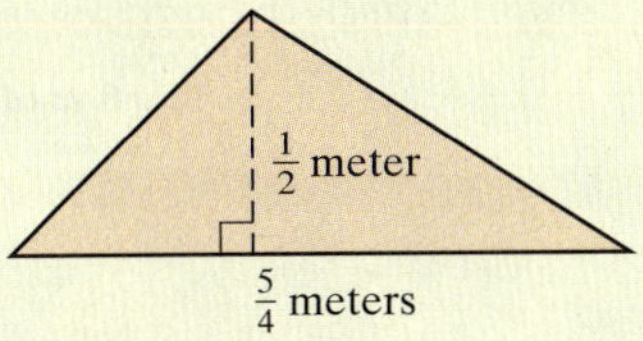

△ **39.**

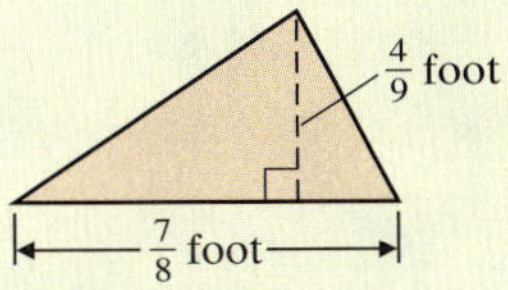

△ **40.**

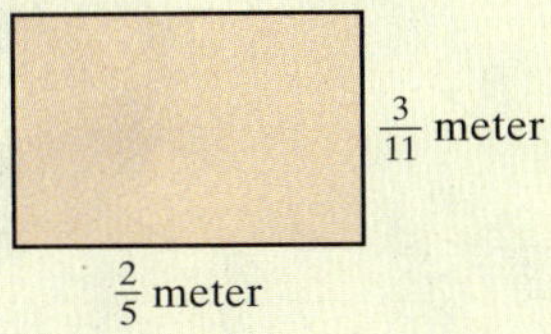

Add or subtract as indicated. Write the answer in simplest form. See Example 5.

41. $\frac{4}{5} - \frac{1}{5}$ **42.** $\frac{6}{7} - \frac{1}{7}$

43. $\frac{4}{5} + \frac{1}{5}$ **44.** $\frac{6}{7} + \frac{1}{7}$

45. $\frac{17}{21} - \frac{10}{21}$ **46.** $\frac{18}{35} - \frac{11}{35}$

47. $\frac{23}{105} + \frac{4}{105}$ **48.** $\frac{13}{132} + \frac{35}{132}$

Write each fraction as an equivalent fraction with the given denominator. See Example 6.

49. $\frac{7}{10}$ with a denominator of 30

50. $\frac{2}{3}$ with a denominator of 9

51. $\frac{2}{9}$ with a denominator of 18

52. $\frac{8}{7}$ with a denominator of 56

53. $\frac{4}{5}$ with a denominator of 20

54. $\frac{4}{5}$ with a denominator of 25

Add or subtract as indicated. Write the answer in simplest form. See Example 7.

55. $\frac{2}{3} + \frac{3}{7}$ **56.** $\frac{3}{4} + \frac{1}{6}$

57. $\frac{4}{15} - \frac{1}{12}$ **58.** $\frac{11}{12} - \frac{1}{16}$

59. $\frac{5}{22} - \frac{5}{33}$ **60.** $\frac{7}{10} - \frac{8}{15}$

61. $\frac{12}{5} - 1$ **62.** $2 - \frac{3}{8}$

Each circle in Exercises 63–68 represents a whole, or 1. Use subtraction to determine the unknown part of the circle.

63.

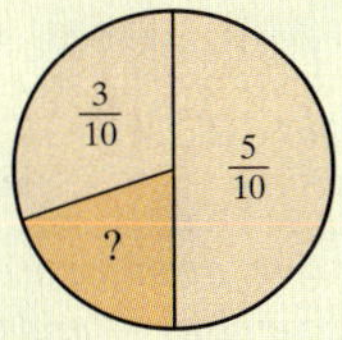

64.

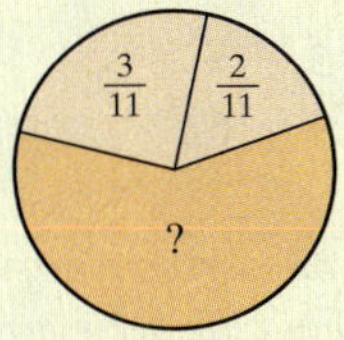

65.

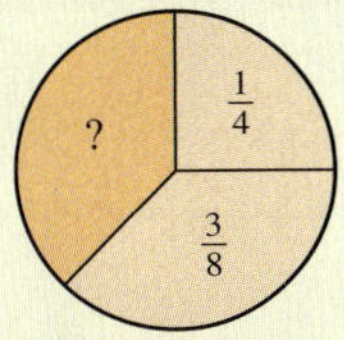

66.

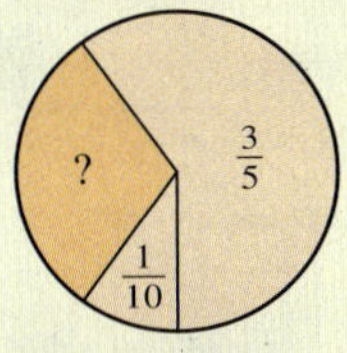

67.

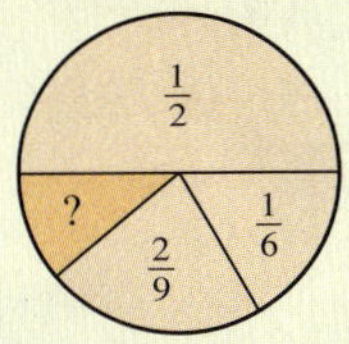

68.

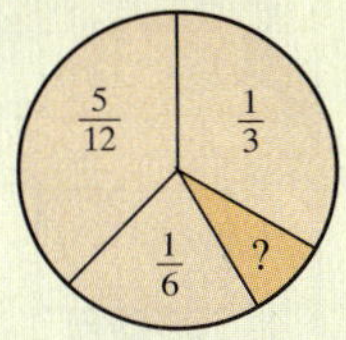

Perform the indicated operations. See Examples 8 and 9.

69. $5\frac{1}{9} \cdot 3\frac{2}{3}$ **70.** $2\frac{3}{4} \cdot 1\frac{7}{8}$

71. $8\frac{3}{5} \div 2\frac{9}{10}$ **72.** $1\frac{7}{8} \div 3\frac{8}{9}$

73. $17\frac{2}{5} + 30\frac{2}{3}$ **74.** $26\frac{11}{20} + 40\frac{7}{10}$

75. $8\frac{11}{12} - 1\frac{5}{6}$ **76.** $4\frac{7}{8} - 2\frac{3}{16}$

MIXED PRACTICE

Perform the following operations. Write answers in simplest form.

77. $\frac{10}{21} + \frac{5}{21}$ **78.** $\frac{11}{35} + \frac{3}{35}$

79. $\frac{10}{3} - \frac{5}{21}$ **80.** $\frac{11}{7} - \frac{3}{35}$

81. $\frac{2}{3} \cdot \frac{3}{5}$ **82.** $\frac{3}{4} \cdot \frac{7}{12}$

83. $\frac{2}{3} \div \frac{3}{5}$ **84.** $\frac{3}{4} \div \frac{7}{12}$

85. $5 + \frac{2}{3}$ **86.** $7 + \frac{1}{10}$

87. $7\frac{2}{5} \div \frac{1}{5}$ **88.** $9\frac{5}{6} \div \frac{1}{6}$

89. $\frac{1}{2} - \frac{14}{33}$ **90.** $\frac{7}{15} - \frac{7}{25}$

91. $\frac{23}{105} - \frac{2}{105}$

92. $\frac{57}{132} - \frac{13}{132}$

93. $1\frac{1}{2} + 3\frac{2}{3}$

94. $2\frac{3}{5} + 4\frac{7}{10}$

95. $\frac{2}{3} - \frac{5}{9} + \frac{5}{6}$

96. $\frac{8}{11} - \frac{1}{4} + \frac{1}{2}$

Recall that the ***perimeter*** *of a plane figure is the total distance around the figure. Find the perimeter of each figure in Exercises 97 and 98.*

△ **97.**

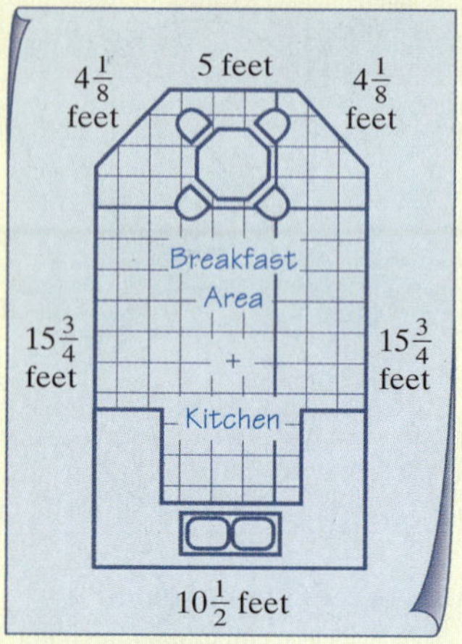

△ **98.**

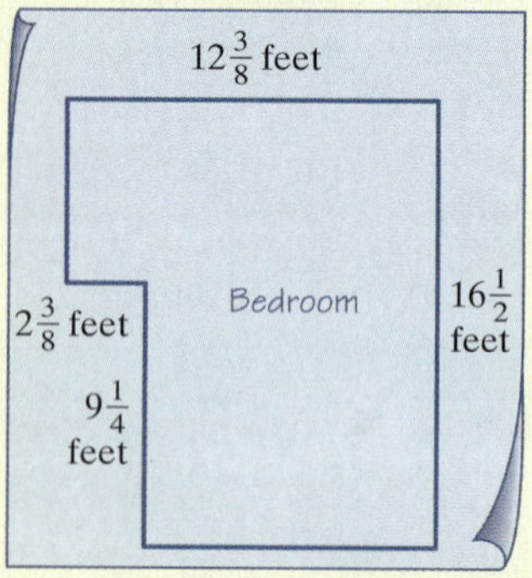

99. In your own words, explain how to add two fractions with different denominators.

100. In your own words, explain how to multiply two fractions.

The following trail chart is given to visitors at the Lakeview Forest Preserve. Use the chart for Exercises 101 and 102.

Trail Name	*Distance (miles)*
Robin Path	$3\frac{1}{2}$
Red Falls	$5\frac{1}{2}$
Green Way	$2\frac{1}{8}$
Autumn Walk	$1\frac{3}{4}$

101. How much longer is Red Falls Trail than Green Way Trail?

102. Find the total distance traveled by someone who hiked along all four trails.

CONCEPT EXTENSIONS

The graph shown is called a circle graph or a pie chart. Use the graph to answer Exercises 103 through 106.

Fraction of U.S. Screens by Type

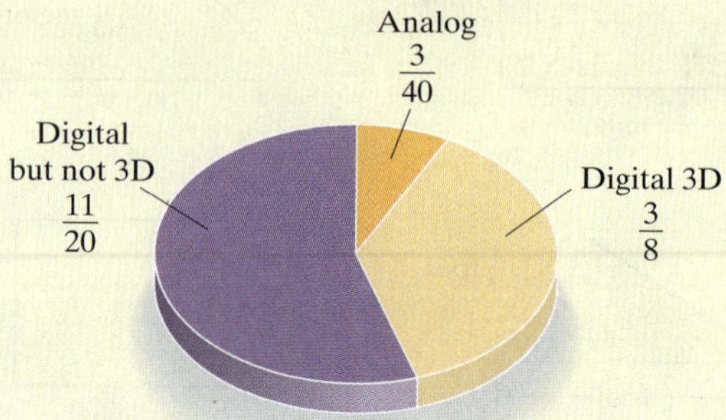

Data from Motion Picture Association of America

103. What fraction of U.S. movie screens are analog?

104. What fraction of U.S. movie screens are digital but not 3-D?

105. What fraction of U.S. movie screens are digital?

106. What fraction of U.S. movie screens are analog or digital 3-D?

For Exercises 107 through 110, determine whether the work is correct or incorrect. If incorrect, find the error and correct. See the Concept Check in this section.

107. $\frac{12}{24} = \frac{2 + 4 + 6}{2 + 4 + 6 + 12} = \frac{1}{12}$

108. $\frac{30}{60} = \frac{2 \cdot 3 \cdot 5}{2 \cdot 2 \cdot 3 \cdot 5} = \frac{1}{2}$

109. $\frac{2}{7} + \frac{9}{7} = \frac{11}{14}$

110. $\frac{16}{28} = \frac{2 \cdot 5 + 6 \cdot 1}{2 \cdot 5 + 6 \cdot 3} = \frac{1}{3}$

R.2 Decimals

OBJECTIVES

1 Write Decimals as Fractions.

2 Add, Subtract, Multiply, and Divide Decimals.

3 Round Decimals to a Given Decimal Place.

4 Write Fractions as Decimals.

OBJECTIVE 1

Writing Decimals as Fractions

Like fractional notation, **decimal notation** is used to denote a part of a whole. Below is a **place value chart** that shows the value of each place.

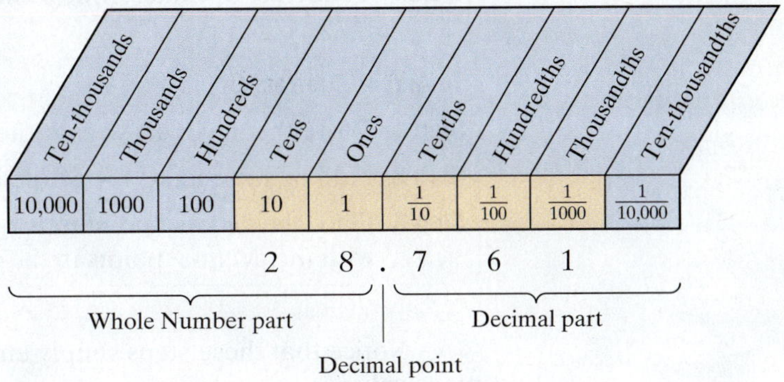

✔ CONCEPT CHECK

Fill in the blank: In the number 52.634, the 3 is in the _____ place.

a. tens **b.** ones **c.** tenths
d. hundredths **e.** thousandths

The next chart shows decimals written as fractions.

Decimal Form	Fractional Form
0.1	$\frac{1}{10}$
tenths	
0.07	$\frac{7}{100}$
hundredths	
2.31	$\frac{231}{100}$
hundredths	
0.9862	$\frac{9862}{10,000}$
ten-thousandths	

To write a decimal as a fraction, use place values.

EXAMPLES Write each decimal as a fraction. Do not simplify.

1. $\underbrace{0.37}_{\text{2 decimal places}} = \frac{37}{\underbrace{100}_{\text{2 zeros}}}$

2. $\underbrace{1.3}_{\text{1 decimal place}} = \frac{13}{\underbrace{10}_{\text{1 zero}}}$

3. $\underbrace{2.649}_{\text{3 decimal places}} = \frac{2649}{\underbrace{1000}_{\text{3 zeros}}}$

Answer to Concept Check:
d

(Continued on next page)

PRACTICE 1–3 Write each decimal as a fraction. Do not simplify.

1. 0.27 **2.** 5.1 **3.** 7.685

OBJECTIVE 2

Adding, Subtracting, Multiplying, and Dividing Decimals

To **add** or **subtract** decimals, follow the steps below.

To Add or Subtract Decimals

Step 1. Write the decimals so that the decimal points line up vertically.

Step 2. Add or subtract as for whole numbers.

Step 3. Place the decimal point in the sum or difference so that it lines up vertically with the decimal points in the problem.

Notice that these steps simply ensure that we add or subtract digits with the same place value.

EXAMPLE 4 Add.

a. 5.87 + 23.279 + 0.003 **b.** 7 + 0.23 + 0.6

Solution

a.
$$\begin{array}{r} 5.87 \\ 23.279 \\ +\ 0.003 \\ \hline 29.152 \end{array}$$

b.
$$\begin{array}{r} 7. \\ 0.23 \\ +0.6 \\ \hline 7.83 \end{array}$$

PRACTICE 4 Add.

a. 7.19 + 19.782 + 1.006 **b.** 12 + 0.79 + 0.03

EXAMPLE 5 Subtract.

a. 32.15 − 11.237 **b.** 70 − 0.48

Solution

a.
$$\begin{array}{r} \scriptstyle 1\ 11\ 4\ 10 \\ 3\not{2}.\not{1}\not{5}\not{0} \\ -\ 1\ 1.\ 2\ 3\ 7 \\ \hline 2\ 0.\ 9\ 1\ 3 \end{array}$$

b.
$$\begin{array}{r} \scriptstyle 6\ 9\ \ 9\ 10 \\ \not{7}\not{0}.\not{0}\not{0} \\ -\ \ 0.\ 4\ 8 \\ \hline 6\ 9.\ 5\ 2 \end{array}$$

PRACTICE 5 Subtract.

a. 84.23 − 26.982 **b.** 90 − 0.19

Now let's study the following product of decimals. Notice the pattern in the decimal points.

$$0.03 \times 0.6 = \frac{3}{100} \times \frac{6}{10} = \frac{18}{1000} \quad \text{or} \quad 0.018$$

0.03: 2 decimal places; 0.6: 1 decimal place; 0.018: 3 decimal places

In general, to **multiply** decimals, follow the steps below.

To Multiply Decimals

Step 1. Multiply the decimals as though they are whole numbers.

Step 2. The decimal point in the product is placed so that the number of decimal places in the product is equal to the **sum** of the number of decimal places in the factors.

EXAMPLE 6 Multiply.

a. 0.072 × 3.5 **b.** 0.17 × 0.02

Solution

a.

$$\begin{array}{r l} 0.072 & \text{3 decimal places} \\ \times\ \ 3.5 & \text{1 decimal place} \\ \hline 360 & \\ 216\ \ & \\ \hline 0.2520 & \text{4 decimal places} \end{array}$$

Helpful Hint

The product 0.2520 may also be written as 0.252.

b.

$$\begin{array}{r l} 0.17 & \text{2 decimal places} \\ \times\ 0.02 & \text{2 decimal places} \\ \hline 0.0034 & \text{4 decimal places} \end{array}$$

□

PRACTICE 6 Multiply.

a. 0.31 × 4.6 **b.** 1.26 × 0.03

To divide a decimal by a whole number using long division, we place the decimal point in the quotient directly above the decimal point in the dividend. For example,

$$\begin{array}{r} 2.47 \\ 3\overline{)7.41} \\ \underline{-6} \\ 1\,4 \\ \underline{-1\,2} \\ 21 \\ \underline{-21} \\ 0 \end{array}$$

To check, see that 2.47 × 3 = 7.41

Helpful Hint

Don't forget the names of the numbers in a division problem.

$$\text{divisor}\overline{)\text{dividend}}^{\ \text{quotient}}$$

In general, to **divide** decimals, use the steps below.

To Divide Decimals

Step 1. Move the decimal point in the divisor to the right until the divisor is a whole number.

Step 2. Move the decimal point in the dividend to the right the **same number of places** as the decimal point was moved in Step 1.

Step 3. Divide. The decimal point in the quotient is directly over the moved decimal point in the dividend.

EXAMPLE 7 Divide.

a. 9.46 ÷ 0.04 **b.** 31.5 ÷ 0.007

Solution

a.

$$\begin{array}{r} 236.5 \\ 004.\overline{)946.0} \\ -8 \\ \hline 14 \\ -12 \\ \hline 26 \\ -24 \\ \hline 20 \\ -20 \\ \hline 0 \end{array}$$

A zero is inserted to continue dividing.

b.

$$\begin{array}{r} 4500. \\ 0007.\overline{)31500.} \\ -28 \\ \hline 35 \\ -35 \\ \hline 0 \end{array}$$

Zeros are inserted in order to move the decimal point three places to the right.

PRACTICE 7 Divide.

a. 21.75 ÷ 0.5 **b.** 15.6 ÷ 0.006

OBJECTIVE 3

Rounding Decimals

We **round** the decimal part of a decimal number in nearly the same way as we round whole numbers. The only difference is that we drop digits to the right of the rounding place, instead of replacing these digits by 0s. For example,

24.954 rounded to the nearest hundredth is 24.95
(↑ hundredths place)

> **To Round Decimals to a Place Value to the Right of the Decimal Point**
>
> **Step 1.** Locate the digit to the right of the given place value.
>
> **Step 2.**
> - If this digit is 5 or greater, add 1 to the digit in the given place value and drop all digits to its right.
> - If this digit is less than 5, drop all digits to the right of the given place.

EXAMPLE 8 Round 7.8265 to the nearest hundredth.

Solution 7.8265 (hundredths place: 2)

Step 1. Locate the digit to the right of the hundredths place.
Step 2. This digit is 5 or greater, so we add 1 to the hundredths place digit and drop all digits to its right.

Thus, 7.8265 rounded to the nearest hundredth is 7.83.

PRACTICE 8 Round 12.9187 to the nearest hundredth.

EXAMPLE 9 Round 19.329 to the nearest tenth.

Solution 19.329 (tenths place: 3)

Step 1. Locate the digit to the right of the tenths place.
Step 2. This digit is less than 5, so we drop this digit and all digits to its right.

Thus, 19.329 rounded to the nearest tenth is 19.3.

PRACTICE 9 Round 245.348 to the nearest tenth.

OBJECTIVE 4

Writing Fractions as Decimals

To write fractions as decimals, interpret the fraction bar as division and find the quotient.

To Write a Fraction as a Decimal

Divide the numerator by the denominator.

EXAMPLE 10 Write $\frac{1}{4}$ as a decimal.

Solution

$$\begin{array}{r} 0.25 \\ 4\overline{)1.00} \\ \underline{-8} \\ 20 \\ \underline{-20} \\ 0 \end{array} \qquad \frac{1}{4} = 0.25$$

□

PRACTICE 10 Write $\frac{2}{5}$ as a decimal.

■

EXAMPLE 11 Write $\frac{2}{3}$ as a decimal.

Solution

$$\begin{array}{r} 0.666 \\ 3\overline{)2.000} \\ \underline{-1\,8} \\ 20 \\ \underline{-18} \\ 20 \\ \underline{-18} \\ 2 \end{array}$$

This division pattern will continue so that $\frac{2}{3} = 0.6666\ldots$.

A bar can be placed over the digit 6 to indicate that it repeats. We call this a **repeating decimal.**

$$\frac{2}{3} = 0.666\ldots = 0.\overline{6}$$

□

PRACTICE 11 Write $\frac{5}{6}$ as a decimal.

■

We can also write a decimal approximation for $\frac{2}{3}$. For example, $\frac{2}{3}$ rounded to the nearest hundredth is 0.67. This can be written as $\frac{2}{3} \approx 0.67$. The $\approx$ sign means "is approximately equal to."

✔ CONCEPT CHECK

The notation $0.5\overline{2}$ is the same as

a. $\frac{52}{100}$ b. $\frac{52\ldots}{100}$ c. $0.52222222\ldots$

Answer to Concept Check:
c

EXAMPLE 12 Write $\frac{22}{7}$ as a decimal. Round to the nearest hundredth.

Solution

$$\begin{array}{r} 3.142 \\ 7\overline{)22.000} \\ \underline{-21} \\ 1\,0 \\ \underline{-7} \\ 30 \\ \underline{-28} \\ 20 \\ \underline{-14} \\ 6 \end{array} \approx 3.14$$

If rounding to the nearest hundredth, carry the division process out to one more decimal place, the thousandths place.

The fraction $\frac{22}{7}$ in decimal form is approximately 3.14. (The fraction $\frac{22}{7}$ is an approximation for π.) □

PRACTICE 12 Write $\frac{1}{9}$ as a decimal. Round to the nearest thousandth. ■

Vocabulary, Readiness & Video Check

Fill in each blank with one of the choices listed below. Choices may be used more than once or not at all.

vertically	sum	numerator
decimal	denominator	

1. Like fractional notation, ____________ notation is used to denote a part of a whole.
2. To write fractions as decimals, divide the ____________ by the ____________.
3. To add or subtract decimals, write the decimals so that the decimal points line up ____________.
4. When multiplying decimals, the decimal point in the product is placed so that the number of decimal places in the product is equal to the ____________ of the number of decimal places in the factors.

Martin-Gay Interactive Videos

See Video R.2

Watch the section lecture video and answer the following questions.

OBJECTIVE 1
5. From Example 1, how does reading a decimal number correctly help us write it as an equivalent fraction?

OBJECTIVE 2
6. From Examples 3 and 4, complete each statement with "do" or "do not." When adding or subtracting decimal numbers, we ______ line up decimal points. When multiplying decimal numbers, we ______ need to line up decimal points.

OBJECTIVE 3
7. From the lecture before Example 7, explain the difference between rounding whole numbers and rounding decimal numbers to a place to the right of the decimal point.

OBJECTIVE 4
8. Complete this statement based on the lecture before Example 8. To write a fraction as a decimal, we divide the ______ by the ______.

R.2 Exercise Set MyMathLab®

Write each decimal as a fraction. Do not simplify. See Examples 1 through 3.

1. 0.6
2. 0.9
3. 1.86
4. 7.23
5. 0.114
6. 0.239
7. 123.1
8. 892.7

Add or subtract as indicated. See Examples 4 and 5.

9. $5.7 + 1.13$
10. $2.31 + 6.4$
11. $24.6 + 2.39 + 0.0678$
12. $32.4 + 1.58 + 0.0934$
13. $8.8 - 2.3$
14. $7.6 - 2.1$
15. $18 - 2.78$
16. $28 - 3.31$

Multiply or divide as indicated. See Examples 6 and 7.

17. $\begin{array}{r} 0.2 \\ \times\ 0.6 \\ \hline \end{array}$
18. $\begin{array}{r} 0.7 \\ \times\ 0.9 \\ \hline \end{array}$
19. $\begin{array}{r} 0.063 \\ \times\ \ 4.2 \\ \hline \end{array}$
20. $\begin{array}{r} 0.079 \\ \times\ \ 3.6 \\ \hline \end{array}$
21. $5\overline{)8.4}$
22. $2\overline{)11.7}$
23. $0.82\overline{)4.756}$
24. $0.92\overline{)3.312}$

MIXED PRACTICE

Perform the indicated operation. See Examples 4 through 7.

25. $\begin{array}{r} 45.02 \\ 3.006 \\ +\ 8.405 \\ \hline \end{array}$
26. $\begin{array}{r} 65.0028 \\ 5.0903 \\ +\ 6.9 \\ \hline \end{array}$
27. $\begin{array}{r} 6.75 \\ \times\ \ 10 \\ \hline \end{array}$
28. $\begin{array}{r} 8.91 \\ \times\ 100 \\ \hline \end{array}$
29. $0.6\overline{)42}$
30. $0.9\overline{)36}$
31. $\begin{array}{r} 654.9 \\ -\ 56.67 \\ \hline \end{array}$
32. $\begin{array}{r} 863.2 \\ -\ 39.45 \\ \hline \end{array}$
33. $\begin{array}{r} 5.62 \\ \times\ 7.7 \\ \hline \end{array}$
34. $\begin{array}{r} 8.03 \\ \times\ 5.5 \\ \hline \end{array}$
35. $0.063\overline{)52.92}$
36. $0.054\overline{)51.84}$
37. $\begin{array}{r} 16.003 \\ \times\ 5.31 \\ \hline \end{array}$
38. $\begin{array}{r} 31.006 \\ \times\ \ 3.71 \\ \hline \end{array}$

Round each decimal to the given place value. See Examples 8 and 9.

39. 0.57, nearest tenth
40. 0.75, nearest tenth
41. 0.234, nearest hundredth
42. 0.452, nearest hundredth
43. 0.5945, nearest thousandth
44. 63.4529, nearest thousandth
45. 98,207.23, nearest tenth
46. 68,936.543, nearest tenth
47. 12.347, nearest hundredth
48. 42.9878, nearest thousandth

Write each fraction as a decimal. If the decimal is a repeating decimal, write using the bar notation and then round to the nearest hundredth. See Examples 10 through 12.

49. $\frac{3}{4}$
50. $\frac{9}{25}$
51. $\frac{1}{3}$
52. $\frac{7}{9}$
53. $\frac{7}{16}$
54. $\frac{5}{8}$
55. $\frac{6}{11}$
56. $\frac{1}{6}$
57. $\frac{29}{6}$
58. $\frac{34}{9}$

CONCEPT EXTENSIONS

Solve. See the Concept Checks in this section.

59. In the number 3.659, identify the place value of the
    a. 6 **b.** 9 **c.** 3
60. The notation $0.\overline{67}$ is the same as
    a. 0.6777 . . . **b.** 0.67666 . . . **c.** 0.6767 . . .
61. In your own words, describe how to multiply decimal numbers.
62. In your own words, describe how to add or subtract decimal numbers.

The chart shows the average number of pounds of various dairy products consumed by each U.S. citizen. Use this chart for Exercises 63 and 64. (Source: Dairy Information Center)

Dairy Product	*Pounds*
Fluid Milk	213.4
Cheese	30.8
Butter	4.4

63. How much more fluid milk products than cheese products does the average U.S. citizen consume?
64. What is the total amount of these milk products consumed by the average U.S. citizen annually?

The following graph is called a circle graph or pie chart. Each sector (shaped like a piece of pie) shows the fraction of entering college freshmen who choose to major in each discipline shown. The whole circle represents the entire class of college freshmen. Use this graph to answer Exercises 65 through 70. Write each fraction answer in simplest form.

College Freshmen Majors

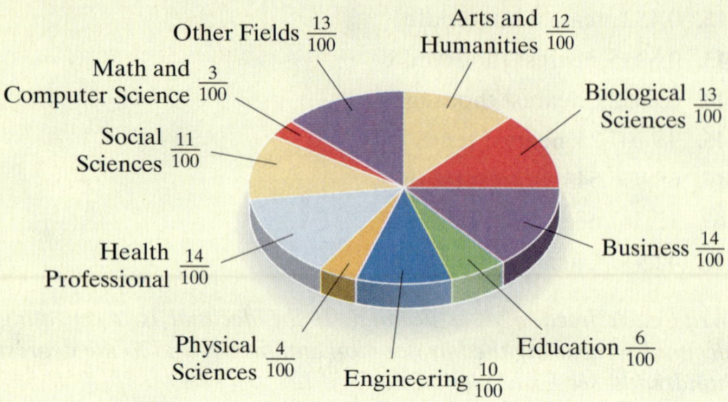

Source: The Higher Education Research Institute

65. What fraction of entering college freshmen plan to major in education?

66. What fraction of entering college freshmen plan to major in engineering?

67. Write your answer to Exercise 65 as a decimal.

68. Write your answer to Exercise 66 as a decimal.

69. Why is the Business sector the same size as the Health Professional sector?

70. Why is the Physical Sciences sector smaller than the Business sector?

R.3 Percents, Decimals, and Fractions

OBJECTIVES

1 Understand Percent.

2 Write Percents as Decimals or Fractions.

3 Write Decimals or Fractions as Percents.

4 Solve Applications with Percents, Decimals, and Fractions.

OBJECTIVE

1 Understanding Percent

The word **percent** comes from the Latin phrase *per centum,* which means **"per 100."** For example, 53% (53 percent) means 53 per 100. In the square below, 53 of the 100 squares are shaded. Thus, 53% of the figure is shaded.

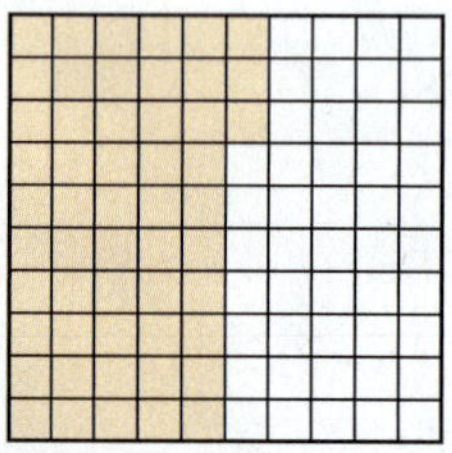

53 of 100 squares are shaded or 53% is shaded.

Since 53% means 53 per 100, 53% is the ratio of 53 to 100, or $\frac{53}{100}$.

$$53\% = \frac{53}{100}$$

Also,

$$7\% = \frac{7}{100} \quad \text{7 parts per 100 parts}$$

$$73\% = \frac{73}{100} \quad \text{73 parts per 100 parts}$$

$$109\% = \frac{109}{100} \quad \text{109 parts per 100 parts}$$

Percent

Percent means **per one hundred.** The "%" symbol is used to denote percent.

Percent is used in a variety of everyday situations. For example,

- 88.5% of the U.S. population uses the Internet.
- The store is having a 20%-off sale.
- For the past two years, the enrollment in community colleges increased 15%.
- The South is the home of 49% of all frequent paintball participants.
- 55% of the cinema screens in the U.S. are digital non-3-D.

EXAMPLE 1 In a survey of 100 people, 17 people drive blue cars. What percent of the people drive blue cars?

Solution Since 17 people out of 100 drive blue cars, the fraction is $\frac{17}{100}$. Then

$$\frac{17}{100} = 17\% \qquad \square$$

PRACTICE 1 Of 100 students in a club, 27 are freshmen. What percent of the students are freshmen? ■

EXAMPLE 2 45 out of every 100 young college graduates return home to live (at least temporarily). What percent of college graduates is this? (*Source:* Independent Insurance Agents of America)

Solution

$$\frac{45}{100} = 45\% \qquad \square$$

PRACTICE 2 31 out of 100 college instructors are in their forties. What percent of these instructors are in their forties? ■

OBJECTIVE 2 Writing Percents as Decimals or Fractions

Since percent means "per hundred," we have that

$$1\% = \frac{1}{100} = 0.01$$

In other words, the percent symbol means "per hundred" or, equivalently, "$\frac{1}{100}$" or "0.01." Thus

$$\text{Write 87\% as a fraction: } 87\% = 87 \times \frac{1}{100} = \frac{87}{100}$$

or

$$\text{Write 87\% as a decimal: } 87\% = 87 \times (0.01) = 0.87$$

Results are the same.

Of course, we know that the end results are the same; that is,

$$\frac{87}{100} = 0.87$$

The above gives us two options for converting percents. We can replace the percent symbol, %, by $\frac{1}{100}$ or by 0.01 and then multiply.

For consistency, when we

- convert from a percent to a *decimal,* we will drop the % symbol and multiply by 0.01.
- convert from a percent to a *fraction,* we will drop the % symbol and multiply by $\frac{1}{100}$.

Let's practice writing percents as decimals, then writing percents as fractions.

Writing a Percent as a Decimal

Replace the percent symbol with its decimal equivalent, 0.01; then multiply.

$$43\% = 43(0.01) = 0.43$$

Helpful Hint

If it helps, think of writing a percent as a decimal by

Percent → Remove the % symbol and move the decimal point 2 places to the left. → Decimal

EXAMPLES Write each percent as a decimal.

3. $23\% = 23(0.01) = 0.23$ Replace the percent symbol with 0.01. Then multiply.

4. $4.6\% = 4.6(0.01) = 0.046$ Replace the percent symbol with 0.01. Then multiply.

5. $190\% = 190(0.01) = 1.90$ or 1.9

6. $0.74\% = 0.74(0.01) = 0.0074$

7. $100\% = 100(0.01) = 1.00$ or 1

Helpful Hint

We just learned that 100% = 1.

PRACTICE 3–7 Write each percent as a decimal.

3. 49% **4.** 3.1% **5.** 175%

6. 0.46% **7.** 600%

✔ CONCEPT CHECK

Why is it incorrect to write the percent 0.033% as 3.3 in decimal form?

Now let's write percents as fractions.

Writing a Percent as a Fraction

Replace the percent symbol with its fraction equivalent, $\frac{1}{100}$; then multiply. Don't forget to simplify the fraction if possible.

$$43\% = 43 \cdot \frac{1}{100} = \frac{43}{100}$$

Answer to Concept Check:
To write a percent as a decimal, the decimal point should be moved two places to the left, not to the right. So the correct answer is 0.00033.

EXAMPLES Write each percent as a fraction or mixed number in simplest form.

8. $40\% = 40 \cdot \frac{1}{100} = \frac{40}{100} = \frac{2 \cdot \overset{1}{\cancel{20}}}{5 \cdot \underset{1}{\cancel{20}}} = \frac{2}{5}$

9. $1.9\% = 1.9 \cdot \frac{1}{100} = \frac{1.9}{100}$. We don't want the numerator of the fraction to contain a decimal, so we multiply by 1 in the form of $\frac{10}{10}$.

$$= \frac{1.9}{100} \cdot \frac{10}{10} = \frac{1.9 \cdot 10}{100 \cdot 10} = \frac{19}{1000}$$

10. $125\% = 125 \cdot \frac{1}{100} = \frac{125}{100} = \frac{5 \cdot \overset{1}{\cancel{25}}}{4 \cdot \underset{1}{\cancel{25}}} = \frac{5}{4}$ or $1\frac{1}{4}$

11. $33\frac{1}{3}\% = 33\frac{1}{3} \cdot \frac{1}{100} = \frac{100}{3} \cdot \frac{1}{100} = \frac{\overset{1}{\cancel{100}} \cdot 1}{3 \cdot \underset{1}{\cancel{100}}} = \frac{1}{3}$

Write as an improper fraction.

12. $100\% = 100 \cdot \frac{1}{100} = \frac{100}{100} = 1$

Helpful Hint
Just as in Example 7, we confirm that 100% = 1.

PRACTICE 8–12 Write each percent as a fraction or mixed number in simplest form.

8. 50% **9.** 2.3% **10.** 150%

11. $66\frac{2}{3}\%$ **12.** 12%

OBJECTIVE 3

Writing Decimals or Fractions as Percents

To write a decimal or fraction as a percent, we use the result of Examples 7 and 12. In these examples, we found that 1 = 100%.

Write 0.38 as a percent: $0.38 = 0.38(1) = 0.38(100\%) = 38.\%$

Write $\frac{1}{5}$ as a percent: $\frac{1}{5} = \frac{1}{5}(1) = \frac{1}{5} \cdot 100\% = \frac{100}{5}\% = 20\%$

First, let's practice writing decimals as percents.

Writing a Decimal as a Percent

Multiply by 1 in the form of 100%.

$$0.27 = 0.27(100\%) = 27.\%$$

Helpful Hint

If it helps, think of writing a decimal as a percent by reversing the steps in the Helpful Hint on page R-20.

Percent ← | Move the decimal point 2 places to the right and attach a % symbol. | ← Decimal

EXAMPLES Write each decimal as a percent.

13. $0.65 = 0.65(100\%) = 65.\%$ or 65% Multiply by 100%.

14. $1.25 = 1.25(100\%) = 125.\%$ or 125%

15. $0.012 = 0.012(100\%) = 001.2\%$ or 1.2%

16. $0.6 = 0.6(100\%) = 060.\%$ or 60%

Helpful Hint
A zero was inserted as a placeholder.

PRACTICE 13–16 Write each decimal as a percent.

13. 0.14 **14.** 1.75 **15.** 0.057 **16.** 0.5

✔ CONCEPT CHECK
Why is it incorrect to write the decimal 0.0345 as 34.5% in percent form?

Now let's write fractions as percents.

Writing a Fraction as a Percent

Multiply by 1 in the form of 100%.

$$\frac{1}{8} = \frac{1}{8} \cdot 100\% = \frac{1}{8} \cdot \frac{100}{1}\% = \frac{100}{8}\% = 12\frac{1}{2}\% \quad \text{or} \quad 12.5\%$$

Helpful Hint

From Examples 7 and 12, we know that

$$100\% = 1$$

Recall that when we multiply a number by 1, we are not changing the value of that number. This means that when we multiply a number by 100%, we are not changing its value but rather writing the number as an equivalent percent.

EXAMPLES Write each fraction or mixed number as a percent.

17. $\frac{7}{20} = \frac{7}{20} \cdot 100\% = \frac{7}{20} \cdot \frac{100}{1}\% = \frac{700}{20}\% = 35\%$

18. $\frac{2}{3} = \frac{2}{3} \cdot 100\% = \frac{2}{3} \cdot \frac{100}{1}\% = \frac{200}{3}\% = 66\frac{2}{3}\%$

19. $2\frac{1}{4} = \frac{9}{4} \cdot 100\% = \frac{9}{4} \cdot \frac{100}{1}\% = \frac{900}{4}\% = 225\%$

Helpful Hint
$\frac{200}{3} = 66.\overline{6}$. Thus, another way to write $\frac{200}{3}\%$ is $66.\overline{6}\%$.

PRACTICE 17–19 Write each fraction or mixed number as a percent.

17. $\frac{3}{25}$ **18.** $\frac{9}{40}$ **19.** $5\frac{1}{2}$

Answer to Concept Check:
To change a decimal to a percent, multiply by 100%, or move the decimal point *only* two places to the right. So the correct answer is 3.45%.

✔ CONCEPT CHECK
Which digit in the percent 76.4582% represents

a. A tenth percent?
b. A thousandth percent?
c. A hundredth percent?
d. A ten percent?

Answers to Concept Check:
a. 4 **b.** 8 **c.** 5 **d.** 7

EXAMPLE 20 Write $\frac{1}{12}$ as a percent. Round to the nearest hundredth percent.

Solution

$$\frac{1}{12} = \frac{1}{12} \cdot 100\% = \frac{1}{12} \cdot \frac{100\%}{1} = \frac{100}{12}\% \approx 8.33\%$$

("approximately")

$$\begin{array}{r} 8.333 \approx 8.33 \\ 12\overline{)100.000} \\ \underline{-96} \\ 4\,0 \\ \underline{-3\,6} \\ 40 \\ \underline{-36} \\ 40 \\ \underline{-36} \\ 4 \end{array}$$

Thus, $\frac{1}{12}$ is approximately 8.33%. □

PRACTICE 20 Write $\frac{3}{17}$ as a percent. Round to the nearest hundredth percent.

OBJECTIVE 4

Solving Applications with Percents, Decimals, and Fractions

Let's summarize what we have learned so far about percents, decimals, and fractions:

Summary of Converting Percents, Decimals, and Fractions

- *To write a percent as a decimal,* replace the % symbol with its decimal equivalent, 0.01; then multiply.
- *To write a percent as a fraction,* replace the % symbol with its fraction equivalent, $\frac{1}{100}$; then multiply.
- *To write a decimal or fraction as a percent,* multiply by 100%.

If we let x represent a number, below we summarize using symbols.

Write a percent as a decimal:	Write a percent as a fraction:	Write a number as a percent:
$x\% = x(0.01)$	$x\% = x \cdot \frac{1}{100}$	$x = x \cdot 100\%$

EXAMPLE 21 In the last ten years, automobile thefts in the continental United States have decreased 39.8%. Write this percent as a decimal and as a fraction. (*Source:* The American Automobile Manufacturers Association)

Solution

(Continued on next page)

As a decimal: $39.8\% = 39.8(0.01) = 0.398$

As a fraction: $39.8\% = 39.8 \cdot \frac{1}{100} = \frac{39.8}{100} = \frac{39.8}{100} \cdot \frac{10}{10} = \frac{398}{1000} = \frac{\overset{1}{\cancel{2}} \cdot 199}{\underset{1}{\cancel{2}} \cdot 500} = \frac{199}{500}$.

Thus, 39.8% written as a decimal is 0.398, and written as a fraction is $\frac{199}{500}$. □

PRACTICE 21 A family decides to spend no more than 27.5% of its monthly income on rent. Write 27.5% as a decimal and as a fraction. ■

EXAMPLE 22 An advertisement for a stereo system reads "$\frac{1}{4}$ off." What percent off is this?

Solution Write $\frac{1}{4}$ as a percent.

$$\frac{1}{4} = \frac{1}{4} \cdot 100\% = \frac{1}{4} \cdot \frac{100\%}{1} = \frac{100}{4}\% = 25\%$$

Thus, "$\frac{1}{4}$ off" is the same as "25% off." □

PRACTICE 22 Provincetown's budget for waste disposal increased by $1\frac{3}{4}$ times over the budget from last year. What percent increase is this? ■

Note: It is helpful to know a few basic percent conversions. See Appendix A for a handy reference of percent, decimal, and fraction equivalencies.

There is also a table in Appendix A about how to find common percents of a number.

Vocabulary, Readiness & Video Check

Use the choices below to fill in each blank. Some choices may be used more than once.

$\frac{1}{100}$ 0.01 100% percent

1. ________ means "per hundred."

2. ________ = 1.

3. The % symbol is read as ________.

4. To write a decimal or a fraction as a percent, multiply by 1 in the form of ________.

5. To write a percent as a *decimal,* drop the % symbol and multiply by ________.

6. To write a percent as a *fraction,* drop the % symbol and multiply by ________.

Martin-Gay Interactive Videos

See Video R.3

Watch the section lecture video and answer the following questions.

OBJECTIVE 1

7. From the lecture before Example 1, what is the most important thing to remember about percent?

OBJECTIVE 2

8. From Example 4, what is the percent equivalent of 1?

OBJECTIVE 3

9. Complete this statement based on Example 9: Multiplying by 100% is the same as multiplying by ____.

OBJECTIVE 4

10. From Example 15, what is the main difference between writing a percent as a decimal and writing a percent as a fraction?

R.3 Exercise Set MyMathLab®

Solve. See Examples 1 and 2.

1. In a survey of 100 college students, 96 use the Internet. What percent use the Internet?

2. A basketball player makes 81 out of 100 attempted free throws. What percent of free throws are made?

One hundred adults were asked to name their favorite sport, and the results are shown in the circle graph.

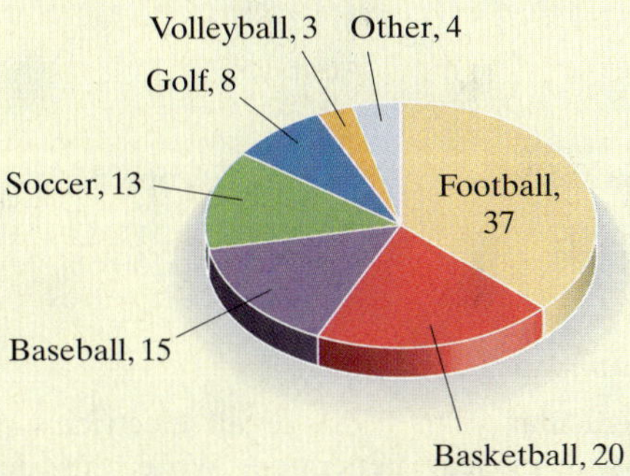

3. What sport was preferred by most adults? What percent preferred this sport?

4. What sport was preferred by the least number of adults? What percent preferred this sport?

5. What percent of adults preferred football or soccer?

6. What percent of adults preferred basketball or baseball?

Write each percent as a decimal. See Examples 3 through 7.

7. 41% **8.** 62% **9.** 6%

10. 3% **11.** 100% **12.** 136%

13. 73.6% **14.** 45.7% **15.** 2.8%

16. 1.4% **17.** 0.6% **18.** 0.9%

19. 300% **20.** 500% **21.** 32.58%

22. 72.18%

Write each percent as a fraction or mixed number in simplest form. See Examples 8 through 12.

23. 8% **24.** 22% **25.** 4%

26. 2% **27.** 4.5% **28.** 7.5%

29. 175% **30.** 275% **31.** 6.25%

32. 8.75% **33.** $10\frac{1}{3}\%$ **34.** $7\frac{3}{4}\%$

35. $22\frac{3}{8}\%$ **36.** $21\frac{7}{8}\%$

Write each decimal as a percent. See Examples 13 through 16.

37. 0.22 **38.** 0.44 **39.** 0.006

40. 0.008 **41.** 5.3 **42.** 2.7

43. 0.056 **44.** 0.019 **45.** 0.2228

46. 0.1115 **47.** 3.00 **48.** 9.00

49. 0.7 **50.** 0.8

Write each fraction or mixed number as a percent. See Examples 17 through 19.

51. $\frac{7}{10}$ **52.** $\frac{9}{10}$ **53.** $\frac{4}{5}$

54. $\frac{2}{5}$ **55.** $\frac{34}{50}$ **56.** $\frac{41}{50}$

57. $\frac{3}{8}$ **58.** $\frac{5}{16}$ **59.** $\frac{1}{3}$

60. $\frac{5}{6}$ **61.** $4\frac{1}{2}$ **62.** $6\frac{1}{5}$

63. $1\frac{9}{10}$ **64.** $2\frac{7}{10}$

Write each fraction as a percent. Round to the nearest hundredth percent. See Example 20.

65. $\frac{9}{11}$ **66.** $\frac{11}{12}$

67. $\frac{4}{15}$ **68.** $\frac{10}{11}$

MIXED PRACTICE

Complete each table. See Examples 1 through 20.

69.

Percent	*Decimal*	*Fraction*
60%		
	0.235	
		$\frac{4}{5}$
$33\frac{1}{3}\%$		
		$\frac{7}{8}$
7.5%		

70.

Percent	*Decimal*	*Fraction*
	0.525	
		$\frac{3}{4}$
$66\frac{1}{2}\%$		
		$\frac{5}{6}$
100%		
		$\frac{7}{50}$

71.

Percent	*Decimal*	*Fraction*
200%		
	2.8	
705%		
		$4\frac{27}{50}$

72.

Percent	*Decimal*	*Fraction*
800%		
	3.2	
608%		
		$9\frac{13}{50}$

Write each percent as a decimal and a fraction. See Examples 21 and 22.

73. People take aspirin for a variety of reasons. The most common use of aspirin is to prevent heart disease, accounting for 38% of all aspirin use. (*Source:* Bayer Market Research)

74. The 2020 goal for adult cigarette smokers is 12%. (*Source:* Centers for Disease Control and Prevention)

75. In the United States recently, 35.8% of households had no landlines, just cell phones. (*Source:* CTIA—The Wireless Association)

76. Japan exports 80.5% of all motorcycles manufactured there. (*Source:* Japan Automobile Manufacturers Association)

77. Approximately 91% of all eighth-grade students in public schools recently reported that they use a computer at home. (*Source:* National Center for Education Statistics)

78. Approximately 11% of the American population is not covered by health insurance. (*Source:* U.S. Census Current Population Study)

In Exercises 79 through 82, write the percent from the circle graph as a decimal and a fraction.

World Population by Continent

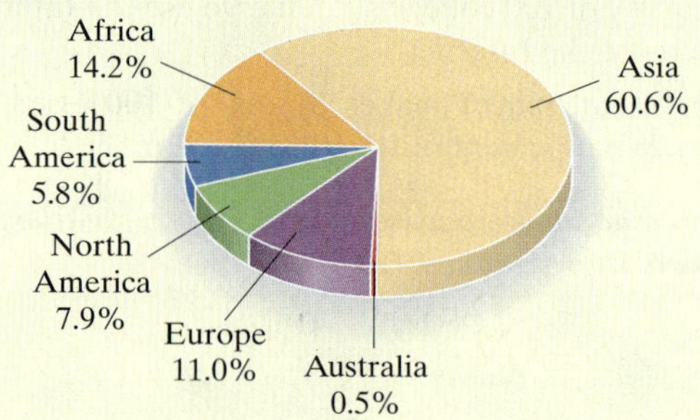

79. Australia: 0.5%

80. Europe: 11%

81. Africa: 14.2%

82. Asia: 60.6%

Solve. See Examples 21 and 22.

83. In a particular year, 0.781 of all electricity produced in France was nuclear generated. Write this decimal as a percent.

84. The United States' share of the total world motor vehicle production is 0.123. Write this decimal as a percent. (*Source:* OICA)

85. The mirrors on the Hubble Space Telescope are able to lock onto a target without deviating more than $\frac{7}{1000}$ of an arc-second. Write this fraction as a percent. (*Source:* NASA)

86. In a particular year, $\frac{1}{4}$ of light trucks sold in the United States were white. Write this fraction as a percent.

87. The U.S. Postal Service handles 47% of the world's mail. Write this percent as a decimal. (*Source:* U.S. Postal Service)

88. In the United States, 97% of households contain a television. Write this percent as a fraction. (*Source:* Nielsen Media)

CONCEPT EXTENSIONS

Solve. See the Concept Checks in this section.

89. Given the percent 52.8647%, round as indicated.

a. Round to the nearest tenth percent.

b. Round to the nearest hundredth percent.

90. Given the percent 0.5269%, round as indicated.

a. Round to the nearest tenth percent.

b. Round to the nearest hundredth percent.

91. Which of the following are correct?

a. 6.5% = 0.65

b. 7.8% = 0.078

c. 120% = 0.12

d. 0.35% = 0.0035

92. Which of the following are correct?

a. 0.231 = 23.1%

b. 5.12 = 0.0512%

c. 3.2 = 320%

d. 0.0175 = 0.175%

Recall that 1 = 100%. This means that 1 whole is 100%. Use this for Exercises 93 and 94. (Source: Some Body by Dr. Pete Rowen*)*

93. The four blood types are A, B, O, and AB. (Each blood type can also be further classified as Rh-positive or Rh-negative depending upon whether your blood contains protein or not.) Given the percent blood types for people in the United States below, calculate the percent of the U.S. population with AB blood type.

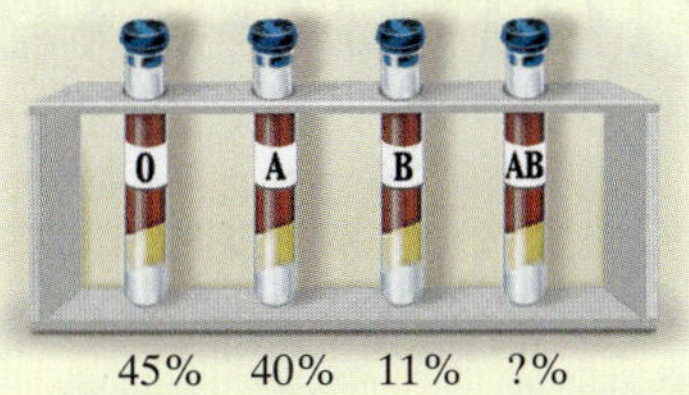

94. The top four components of bone are below. Find the missing percent.

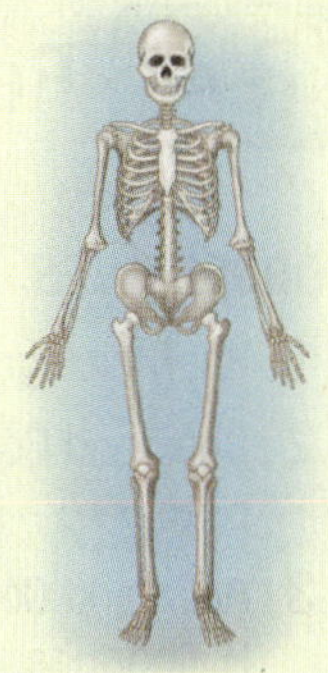

1. Minerals—45%

2. Living tissue—30%

3. Water—20%

4. Other—?

What percent of the figure is shaded?

95. 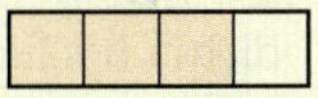

96. 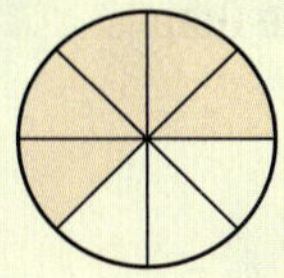

Fill in the blanks.

97. A fraction written as a percent is greater than 100% when the numerator is ________ than the denominator. (greater/less)

98. A decimal written as a percent is less than 100% when the decimal is ________ than 1. (greater/less)

Write each fraction as a decimal and then write each decimal as a percent. Round the decimal to three decimal places (nearest thousandth) and the percent to the nearest tenth percent.

99. $\frac{21}{79}$

100. $\frac{56}{102}$

The bar graph shows the predicted fastest-growing occupations by percent that require an associate degree or more education. Use this graph for Exercises 101 through 104.

Fastest-Growing Occupations 2010–2020 (projected)

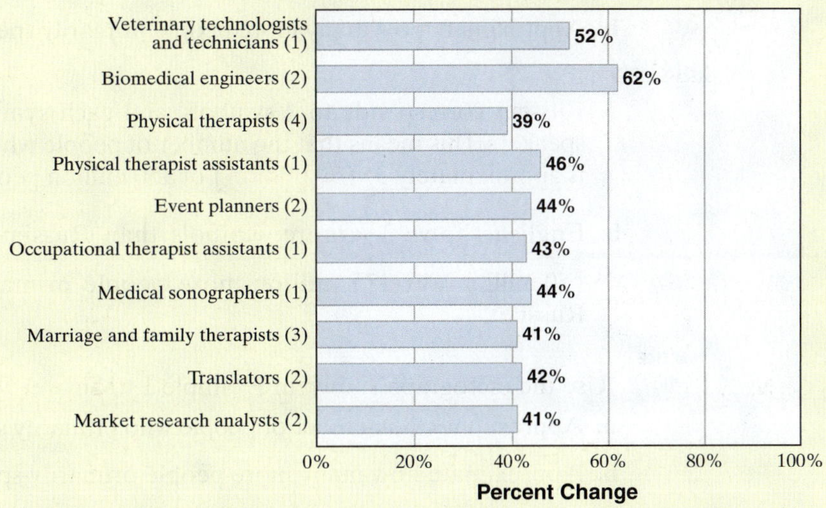

(*Note:* (1) = associate degree; (2) = bachelor degree; (3) = master degree; (4) = doctoral or professional degree)

Source: Bureau of Labor Statistics

101. What occupation is predicted to be the fastest growing?

102. What occupation is predicted to be the second fastest growing?

103. Write the percent change for physical therapists as a decimal.

104. Write the percent change for event planners as a decimal.

105. In your own words, explain how to write a percent as a decimal.

106. In your own words, explain how to write a decimal as a percent.

R.4 Reading Pictographs, and Bar, Line, and Circle Graphs

OBJECTIVES

1. Read Pictographs.
2. Read and Construct Bar Graphs.
3. Read and Construct Histograms.
4. Read Line Graphs.
5. Read Circle Graphs.

Often data are presented visually in a graph. In this section, we practice reading several kinds of graphs including pictographs, bar graphs, and line graphs.

OBJECTIVE

1 Reading Pictographs

A **pictograph** such as the one below is a graph in which pictures or symbols are used. This type of graph contains a key that explains the meaning of the symbol used. An advantage of using a pictograph to display information is that comparisons can easily be made. A disadvantage of using a pictograph is that it is often hard to tell what fractional part of a symbol is shown. For example, in the pictograph below, Arabic shows a part of a symbol, but it's hard to read with any accuracy what fractional part of a symbol is shown.

EXAMPLE 1 Calculating Languages Spoken

The following pictograph shows the top eight most-spoken (primary) languages. Use this pictograph to answer the questions.

Top 8 Most-Spoken (Primary) Languages

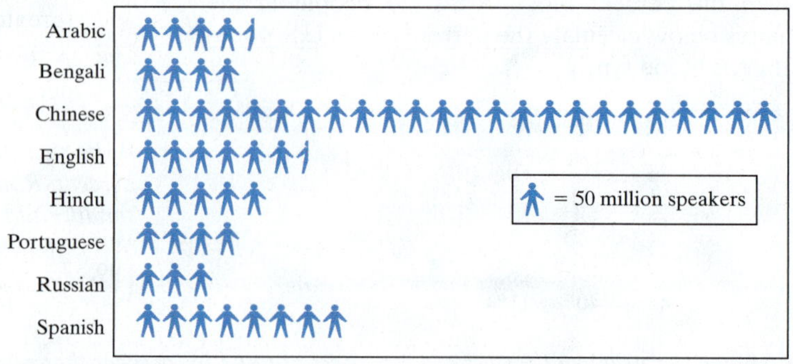

Source: www.ethnologue.com

a. Approximate the number of people who primarily speak Russian.

b. Approximate how many more people primarily speak English than Russian.

Solution

a. Russian corresponds to 3 symbols, and each symbol represents 50 million speakers. This means that the number of people who primarily speak Russian is approximately $3 \cdot (50 \text{ million})$ or 150 million people.

b. English shows $3\frac{1}{2}$ more symbols than Russian. This means that $3\frac{1}{2} \cdot (50 \text{ million})$ or 175 million more people primarily speak English than Russian. □

PRACTICE

1 Use the pictograph shown in Example 1 to answer the following questions:

a. Approximate the number of people who primarily speak Spanish.

b. Approximate how many more people primarily speak Spanish than Arabic. ■

OBJECTIVE

2 Reading and Constructing Bar Graphs

Another way to visually present data is with a **bar graph.** Bar graphs can appear with vertical bars or horizontal bars. We now practice reading the height or length of the bars contained in a bar graph. An advantage to using bar graphs is that a scale is usually included for greater accuracy. Care must be taken when reading bar graphs, as well as other types of graphs—they may be misleading, as shown later in this section.

EXAMPLE 2 Finding the Number of Endangered Species

The following bar graph shows the number of endangered species in the United States in a recent year. Use this graph to answer the questions.

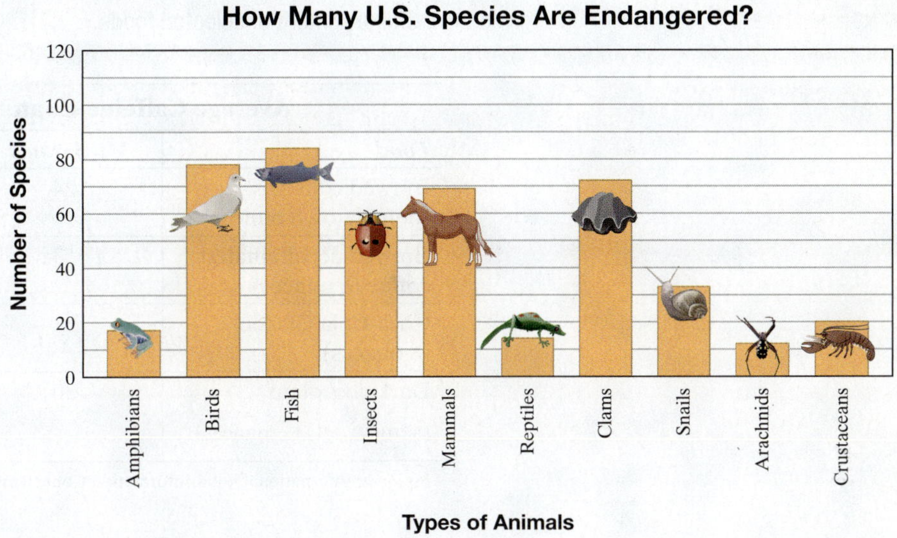

Source: U.S. Fish and Wildlife Service

a. Approximate the number of endangered species that are clams.

b. Which category has the most endangered species?

Solution

a. To approximate the number of endangered species that are clams, we go to the top of the bar that represents clams. From the top of this bar, we move horizontally to the left until the scale is reached. We read the height of the bar on the scale as approximately 72. There are approximately 72 clam species that are endangered, as shown.

b. The most endangered species is represented by the tallest (longest) bar. The tallest bar corresponds to fish.

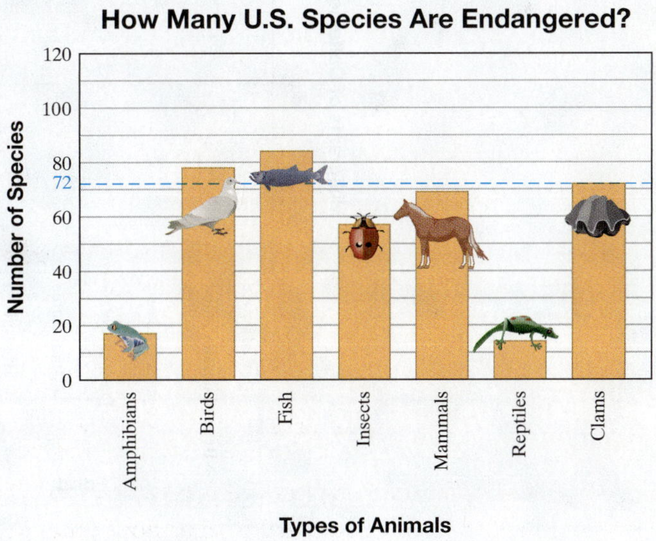

Source: U.S. Fish and Wildlife Service

□

(Continued on next page)

PRACTICE 2 Use the bar graph in Example 2 to answer the following questions:

a. Approximate the number of endangered species that are birds.

b. Which category shows the fewest endangered species?

Next, we practice constructing a bar graph.

EXAMPLE 3 Draw a vertical bar graph using the information in the table below, which gives the caffeine content of selected foods.

Average Caffeine Content of Selected Foods

Food	*Milligrams*	*Food*	*Milligrams*
Brewed coffee (percolator, 8 ounces)	124	Instant coffee (8 ounces)	104
Brewed decaffeinated coffee (8 ounces)	3	Brewed tea (U.S. brands, 8 ounces)	64
Coca-Cola Classic (8 ounces)	31	Pibb Xtra (8 ounces)	27
Dark chocolate (semisweet, $1\frac{1}{2}$ ounces)	30	Milk chocolate (8 ounces)	9

(*Sources:* International Food Information Council and the Coca-Cola Company)

Solution We draw and label a vertical line and a horizontal line as shown below on the left. These lines are also called axes. We place the different food categories along the horizontal axis. Along the vertical axis, we place a scale.

There are many choices of scales that would be appropriate. Notice that the milligrams range from a low of 3 to a high of 124. From this information, we use a scale that starts at 0 and then shows multiples of 20 so that the scale is not too cluttered. The scale stops at 140, the smallest multiple of 20 that will allow all milligrams to be graphed. It may also be helpful to draw horizontal lines along the scale markings to help draw the vertical bars at the correct height. The finished bar graph is shown below on the right.

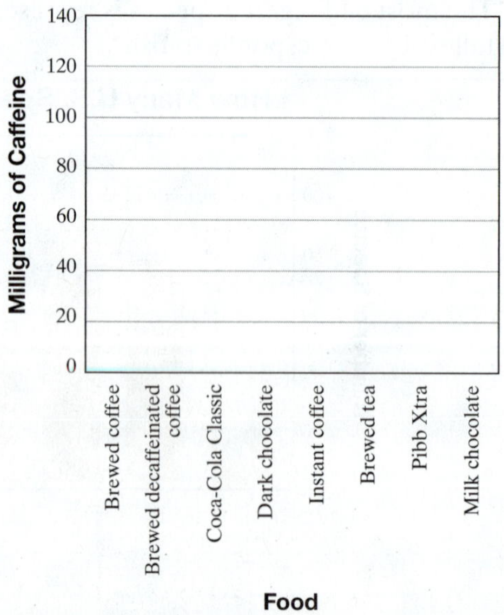

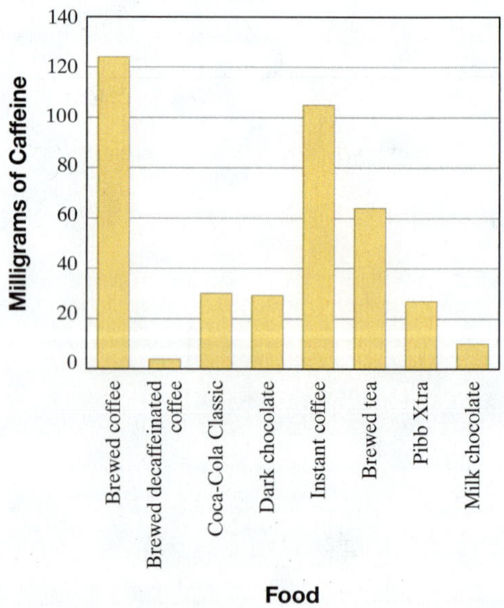

PRACTICE 3 Draw a vertical bar graph using the information in the table about electoral votes for selected states.

Total Electoral Votes by Selected States

State	*Electoral Votes*
Texas	38
California	55
Florida	29
Nebraska	5
Indiana	11
Georgia	16

(*Source: World Almanac 2016*)

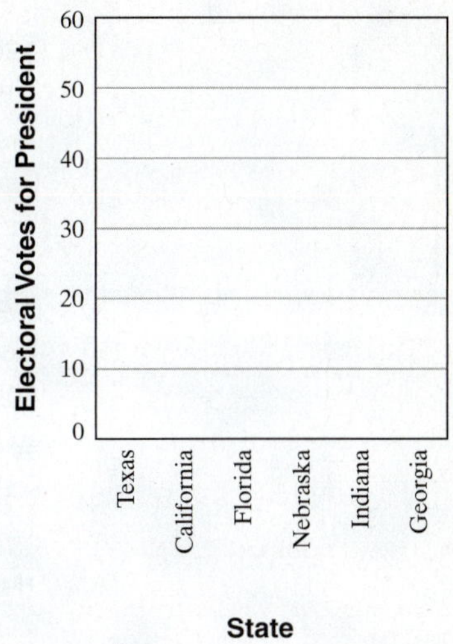

As mentioned previously, graphs can be misleading. Both graphs below show the same information, but with different scales. Special care should be taken when forming conclusions from the appearance of a graph.

Notice the ⚡ symbol on each vertical scale on the graphs below. This symbol alerts us that numbers are missing from that scale

Which Graph is Misleading?

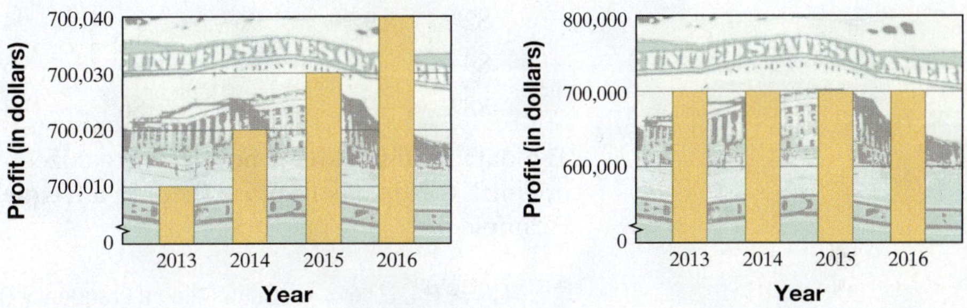

Are profits shown in the graphs above greatly increasing, or are they remaining about the same? Can you understand why the graph to the left is misleading?

OBJECTIVE 3

Reading and Constructing Histograms

Suppose that the test scores of 36 students are summarized in the table below:

Student Scores	*Frequency (Number of Students)*
40–49	1
50–59	3
60–69	2
70–79	10
80–89	12
90–99	8

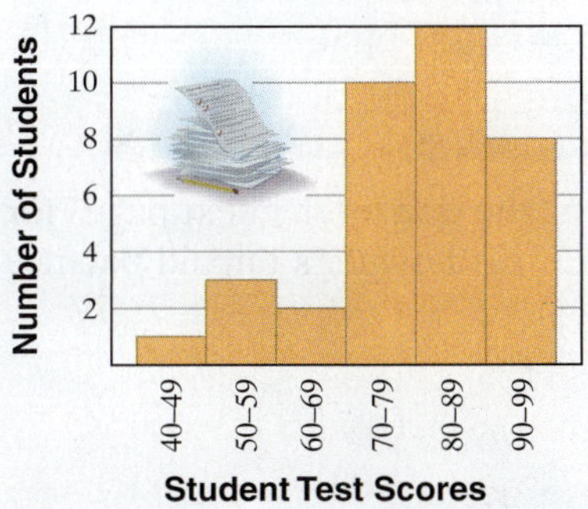

The results in the table can be displayed in a histogram. (See the histogram to the left.) A **histogram** is a special bar graph. The width of each bar represents a range of numbers called a **class interval.** The height of each bar corresponds to how many times a number in the class interval occurs and is called the **class frequency.** The bars in a histogram lie side by side with no space between them.

EXAMPLE 4 Reading a Histogram on Student Test Scores

Use the histogram on the previous page to determine how many students scored 50–59 on the test.

Solution We find the bar representing 50–59. The height of this bar is 3, which means 3 students scored 50–59 on the test. □

PRACTICE 4 Use the histogram on the previous page to determine how many students scored 80–89 on the test. ■

EXAMPLE 5 Reading a Histogram on Student Test Scores

Use the histogram on the previous page to determine how many students scored 80 or above on the test.

Solution We see that two different bars fit this description. There are 12 students who scored 80–89 and 8 students who scored 90–99. The sum of these two categories is 12 + 8 or 20 students. Thus, 20 students scored 80 or above on the test. □

PRACTICE 5 Use the histogram on the previous page to determine how many students scored less than 80 on the test. ■

Now we will look at a way to construct histograms.

The daily high temperatures for 1 month in New Orleans, Louisiana, are recorded in the following list:

85°	90°	95°	89°	88°	94°
87°	90°	95°	92°	95°	94°
82°	92°	96°	91°	94°	92°
89°	89°	90°	93°	95°	91°
88°	90°	88°	86°	93°	89°

The data in this list have not been organized and can be hard to interpret. One way to organize the data is to place them in a **frequency distribution table.** We will do this in Example 6.

EXAMPLE 6 Completing a Frequency Distribution on Temperature

Complete the frequency distribution table for the preceding temperature data.

Solution Go through the data and place a tally mark in the second column of the table next to the class interval. Then count the tally marks and write each total in the third column of the table.

Class Intervals (Temperatures)	*Tally*	*Class Frequency (Number of Days)*
82°–84°	|	1
85°–87°	|||	3
88°–90°	𝍸 𝍸 |	11
91°–93°	𝍸 ||	7
94°–96°	𝍸 |||	8

□

PRACTICE 6 Complete a frequency distribution table (such as the one on the next page) for the data below. Each number represents a credit card owner's unpaid balance for one month.

0	53	89	125
265	161	37	76
62	201	136	42

Class Intervals (Credit Card Balances)	*Tally*	*Class Frequency (Number of Months)*
\$0–\$49	____	____
\$50–\$99	____	____
\$100–\$149	____	____
\$150–\$199	____	____
\$200–\$249	____	____
\$250–\$299	____	____

EXAMPLE 7 Constructing a Histogram

Construct a histogram from the frequency distribution table in Example 6.

Solution

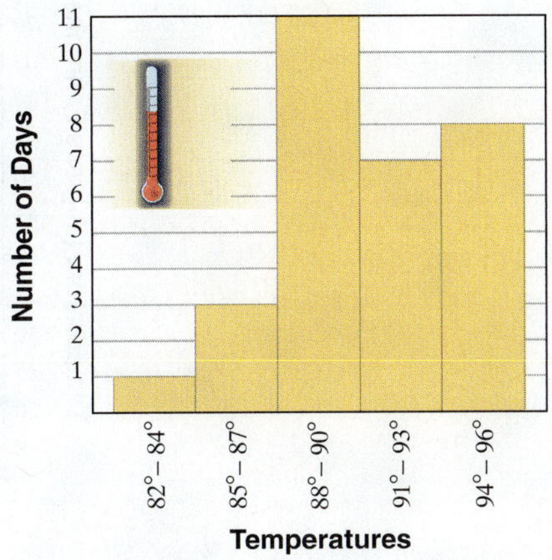

PRACTICE 7 Construct a histogram from the frequency distribution table for Practice 6.

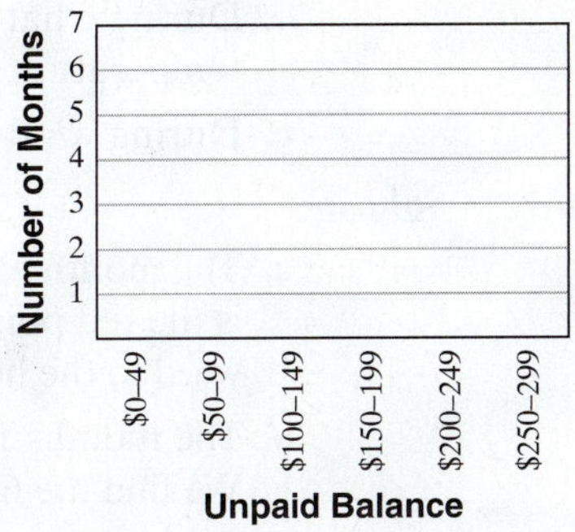

✔ CONCEPT CHECK

Which of the following sets of data is better suited to representation by a histogram? Explain.

Set 1		**Set 2**	
Grade on Final	***# of Students***	***Section Number***	***Avg. Grade on Final***
51–60	12	150	78
61–70	18	151	83
71–80	29	152	87
81–90	23	153	73
91–100	25		

Answer to Concept Check:
Set 1; the grades are arranged in ranges of scores.

OBJECTIVE

4 Reading Line Graphs

Another common way to display information with a graph is by using a **line graph.** An advantage of a line graph is that it can be used to visualize relationships between two quantities. A line graph can also be very useful in showing changes over time.

EXAMPLE 8 Reading Temperatures from a Line Graph

The following line graph shows the average daily temperature for each month in Omaha, Nebraska. Use this graph to answer the questions below.

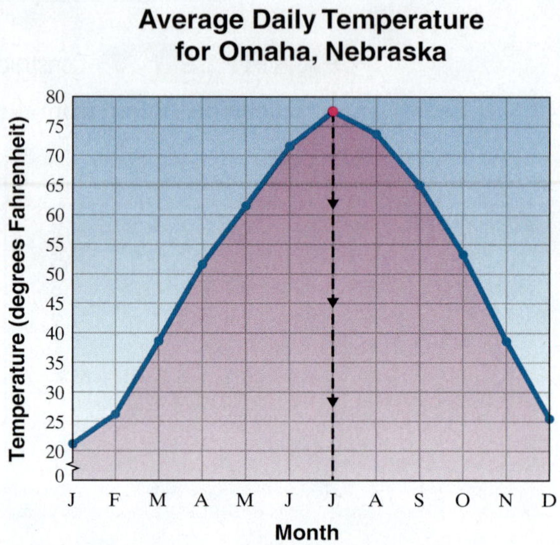

Source: National Climatic Data Center

a. During what month is the average daily temperature the highest?

b. During what month, from July through December, is the average daily temperature 65°F?

c. During what months is the average daily temperature less than 30°F?

Solution

a. The month with the highest temperature corresponds to the highest point. This is the red point shown on the graph above. We follow this highest point downward to the horizontal month scale and see that this point corresponds to July.

b. The months July through December correspond to the right side of the graph. We find the 65°F mark on the vertical temperature scale and move to the right until a point on the right side of the graph is reached. From that point, we move downward to the horizontal month scale and read the corresponding month. During the month of September, the average daily temperature is 65°F.

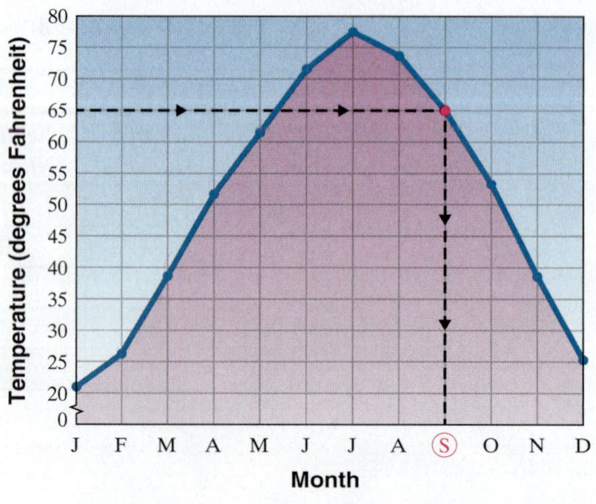

Source: National Climatic Data Center

c. To see what months the temperature is less than 30°F, we find what months correspond to points that fall below the 30°F mark on the vertical scale. These months are January, February, and December.

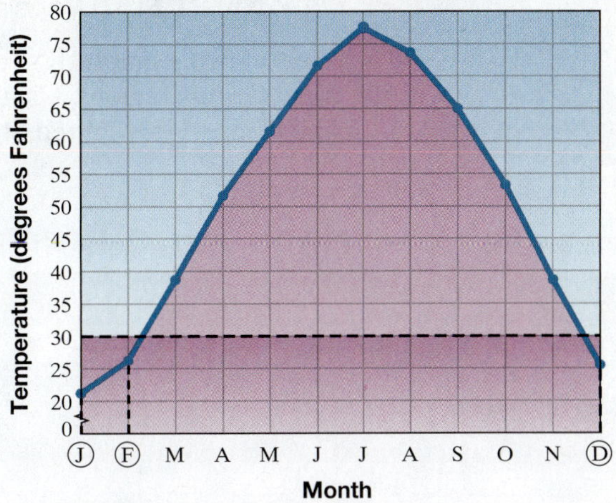

Source: National Climatic Data Center □

PRACTICE 8 Use the temperature graph in Example 8 to answer the following questions:

a. During what month is the average daily temperature the lowest?

b. During what month is the average daily temperature 25°F?

c. During what months is the average daily temperature greater than 70°F? ■

OBJECTIVE 5

Reading Circle Graphs

In Exercise Set R.3, the following **circle graph** was shown. This particular graph shows the favorite sport for 100 adults.

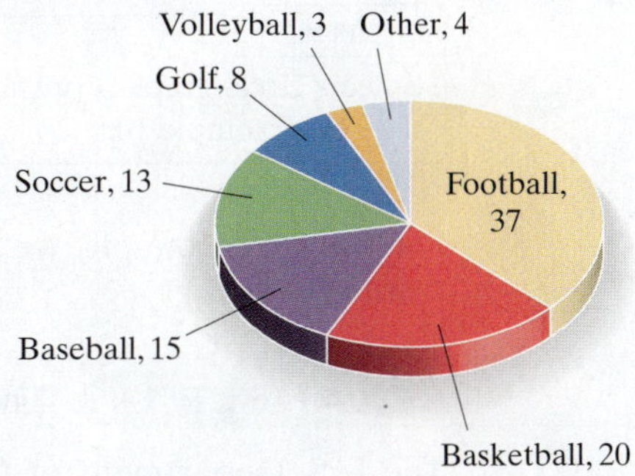

Each sector of the graph (shaped like a piece of pie) shows a category and the relative size of the category. In other words, the most popular sport is football, and it is represented by the largest sector.

EXAMPLE 9 Find the fraction of adults preferring basketball to total adults. Write the fraction in simplest form.

Solution The fraction is

$$\frac{\text{people preferring basketball}}{\text{total adults}} = \frac{20}{100} = \frac{1}{5}$$ □

PRACTICE 9 Find the fraction of adults preferring golf to total adults. Write the fraction in simplest form. ■

A circle graph is often used to show percents in different categories, with the whole circle representing 100%.

EXAMPLE 10 Using a Circle Graph

The following graph shows the percent of visitors to the United States in a recent year from various regions. Using the circle graph shown, determine the percent of visitors who came to the United States from Mexico or Canada.

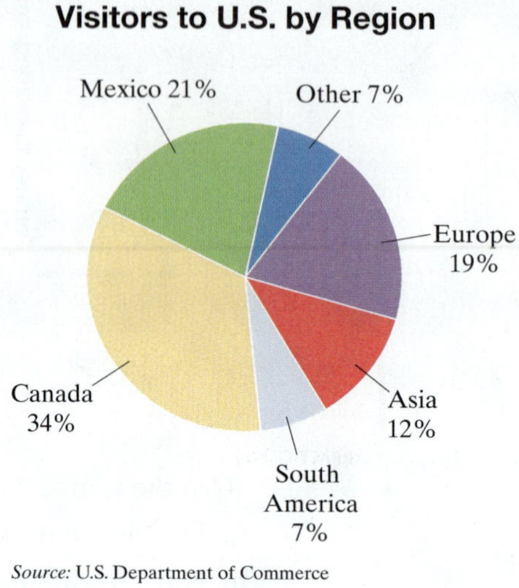

Solution To find this percent, we add the percents corresponding to Mexico and Canada. The percent of visitors to the United States that came from Mexico or Canada is

$$34\% + 21\% = 55\%$$

□

PRACTICE 10 Using the circle graph shown in Example 10, determine the percent of visitors to the United States that came from Europe, Asia, or South America. ■

> **Helpful Hint**
> Since a circle graph represents a whole, the percents should add to 100% or 1. Notice this is true for Example 10.

For the next example, we recall that in mathematics, the word *of* usually translates to "multiplication."

EXAMPLE 11 Finding Percent of Population

The U.S. Department of Commerce forecasts 90 million international visitors to the United States in 2020. Use the circle graph in Example 10 and predict the number of tourists that might be from Europe.

Solution To find 19% "of" 90 million, remember that *of* means "to multiply."

$$\begin{aligned} \text{visitors (from Europe)} &= \boxed{19\%}\ \boxed{\text{of}}\ \boxed{90\text{ million}} \\ \text{visitors} &= 0.19 \cdot 90{,}000{,}000 \\ &= 0.19(90{,}000{,}000) \\ &= 17{,}100{,}000 \end{aligned}$$

Thus, 17,100,000 tourists might come from Europe in 2020. □

PRACTICE 11 Use the information in Example 11 and the circle graph from Example 10 to predict the number of tourists from Mexico in 2020. ■

CONCEPT CHECK

Can the following data be represented by a circle graph? Why or why not?

Responses to the Question "In Which Activities Are You Involved?"	
Intramural sports	60%
On-campus job	42%
Fraternity/sorority	27%
Academic clubs	21%
Music programs	14%

Answer to Concept Check:
no; the percents add up to more than 100%

Vocabulary, Readiness & Video Check

Fill in each blank with one of the choices below.

pictograph	bar	class frequency	sector	1
histogram	line	class interval	circle	100

1. A ____________ graph presents data using vertical or horizontal bars.

2. A ____________ is a graph in which pictures or symbols are used to visually present data.

3. A ____________ graph displays information with a line that connects data points.

4. A ____________ is a special bar graph in which the width of each bar represents a ____________ and the height of each bar represents the ____________.

5. In a ____________ graph, each section (shaped like a piece of pie) shows a category and the relative size of the category.

6. A circle graph contains pie-shaped sections, each called a ____________.

7. If a circle graph has percent labels, the percents should add to ____________%.

8. If a circle graph has fraction labels, the fractions should add to ____________.

Martin-Gay Interactive Videos

See Video R.4

Watch the section lecture video and answer the following questions.

OBJECTIVE 1 **9.** From the pictograph in Example 1, how would you approximate the number of wildfires for any given year?

OBJECTIVE 2 **10.** What is one advantage of displaying data in a bar graph?

OBJECTIVE 3 **11.** Complete this statement based on the lecture before Example 6: A histogram is a special kind of ____.

OBJECTIVE 4 **12.** From the line graph in Examples 10–13, what year averaged the greatest number of goals per game average and what was this average?

OBJECTIVE 5 **13.** From Example 16, when a circle graph shows different parts or percents of some whole category, what is the sum of the percents in the whole circle graph?

R.4 Exercise Set MyMathLab®

The following pictograph shows the number of acres devoted to wheat production in selected states. Use this graph to answer Exercises 1 through 8. See Example 1. (Source: U.S. Department of Agriculture)

Annual Wheat Acreage in Selected Top States

1. Which state plants the greatest quantity of acreage in wheat?
2. Which of the states shown plant the least amount of wheat acreage?
3. Approximate the number of acres of wheat planted in Oklahoma.
4. Approximate the number of acres of wheat planted in Kansas.
5. Which state plants about 6,000,000 acres of wheat?
6. Which state plants about 2,000,000 acres of wheat?
7. Which two states together plant about the same acreage of wheat as North Dakota?
8. Which two states together plant about the same acreage of wheat as Kansas?

The following pictograph shows the average number of wildfires in the United States between 2009 and 2015. Use this graph to answer Exercises 9 through 16. See Example 1. (Source: National Interagency Fire Center)

Wildfires in the United States

9. Approximate the number of wildfires in 2013.
10. Approximately how many wildfires were there in 2012?
11. Which year, of the years shown, had the most wildfires?
12. In what years were the number of wildfires greater than 72,000?
13. What was the amount of decrease in wildfires from 2012 to 2013?
14. What was the amount of decrease in wildfires from 2010 to 2012?
15. What was the average annual number of wildfires from 2012 to 2014? (*Hint:* How do you calculate an average?)
16. Give a possible explanation for the number of wildfires in 2013.

The National Weather Service has exacting definitions for hurricanes; they are tropical storms with winds in excess of 74 mph. The following bar graph shows the number of hurricanes, by month, that have made landfall on the mainland United States between 1851 and 2016. Use this graph to answer Exercises 17 through 22. See Example 2. (Source: National Weather Service: National Hurricane Center)

Hurricanes Making Landfall in the United States, by Month, 1851–2016

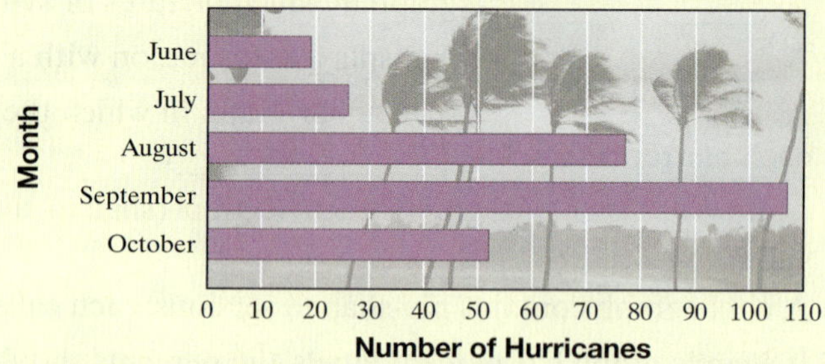

17. In which month did the most hurricanes make landfall in the United States?
18. In which month did the fewest hurricanes make landfall in the United States?
19. Approximate the number of hurricanes that made landfall in the United States during the month of August.
20. Approximate the number of hurricanes that made landfall in the United States in September.
21. In 2008 alone, two hurricanes made landfall during the month of August. What fraction of all the 77 hurricanes that made landfall during August is this?
22. In 2007, only one hurricane made landfall on the United States during the entire season, in the month of September. If there have been 107 hurricanes to make landfall in the month of September since 1851, approximately what percent of these arrived in 2007?

The following horizontal bar graph shows a recent population of the world's largest cities (including their suburbs). Use this graph to answer Exercises 23 through 28. See Example 2. (Source: Dept. of Economic and Social Affairs)

World's Largest Cities
(including Suburbs)

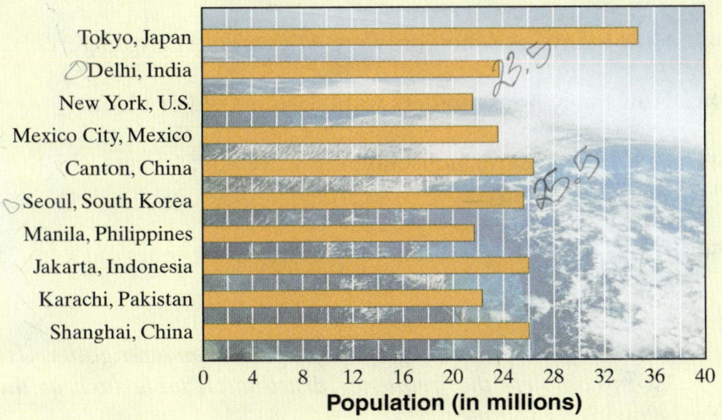

23. Name the city with the largest population, and estimate its population.

24. Name the cities whose population is between 19 million and 22 million.

25. Name the city in the United States with the largest population, and estimate its population.

26. Name the two cities that have approximately the same population.

27. How much larger (in terms of population) is Seoul, South Korea, than Delhi, India?

28. How much larger (in terms of population) is, Shanghai, China, than Mexico City, Mexico?

Use the information given to draw a vertical bar graph. Clearly label the bars. See Example 3.

29.

Fiber Content of Selected Foods

Food	*Grams of Total Fiber*
Kidney beans $\left(\frac{1}{2}\text{ c}\right)$	4.5
Oatmeal, cooked $\left(\frac{3}{4}\text{ c}\right)$	3.0
Peanut butter, chunky (2 tbsp)	1.5
Popcorn (1 c)	1.0
Potato, baked with skin (1 med)	4.0
Whole wheat bread (1 slice)	2.5

(*Sources:* American Dietetic Association and National Center for Nutrition and Dietetics)

Fiber Content of Selected Foods

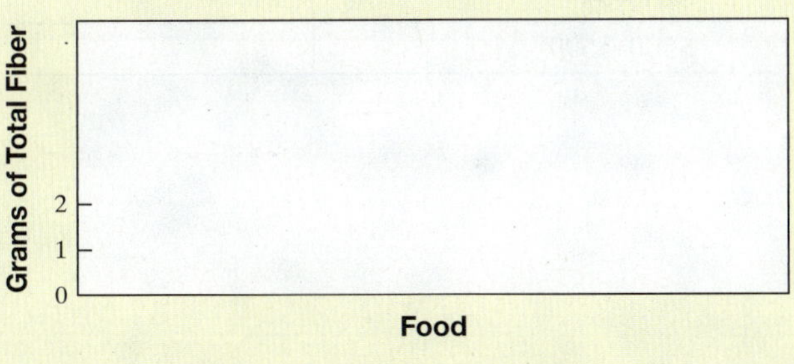

30.

U.S. Annual Food Sales

Year	*Sales in Billions of Dollars*
2000	834
2005	1056
2010	1139
2015	1512

(*Source:* U.S. Department of Agriculture)

U.S. Annual Food Sales

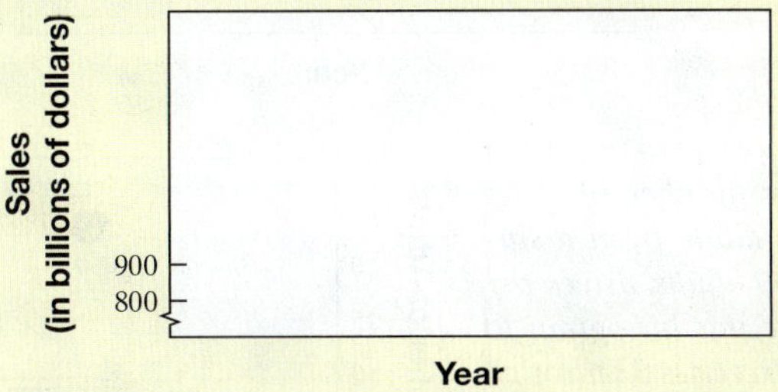

31.

Best-Selling Albums of All Time (U.S. Sales)

Album	*Estimated Sales (in millions)*
Pink Floyd: *The Wall* (1979)	23
Michael Jackson: *Thriller* (1982)	29
Billy Joel: *Greatest Hits Volumes I - II* (1985)	23
Eagles: *Their Greatest Hits* (1976)	29
Led Zeppelin: *Led Zeppelin IV* (1971)	23

(*Source:* Recording Industry Association of America)

Best-Selling Albums of All Time
(U.S. sales)

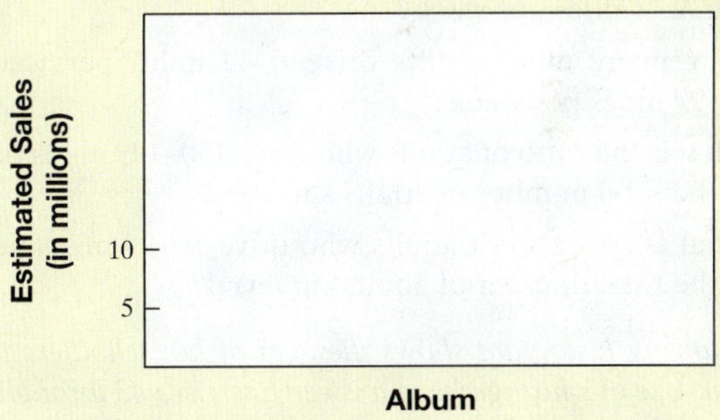

32.

Worldwide Commercial Space Launch Projections

Year	*Number of Commercial Launches*
2018	15
2019	12
2020	10
2021	11
2022	10

Source: Federal Aviation Administration

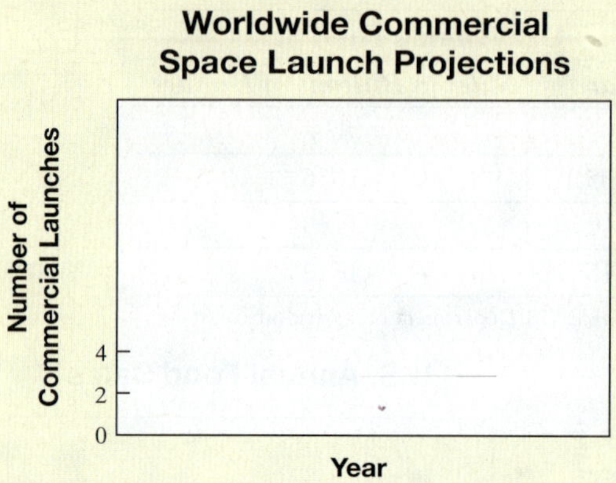

The histogram to the right shows the number of miles that each adult, from a survey of 100 adults, drives per week. Use this histogram to answer Exercises 33 through 42. See Examples 4 and 5.

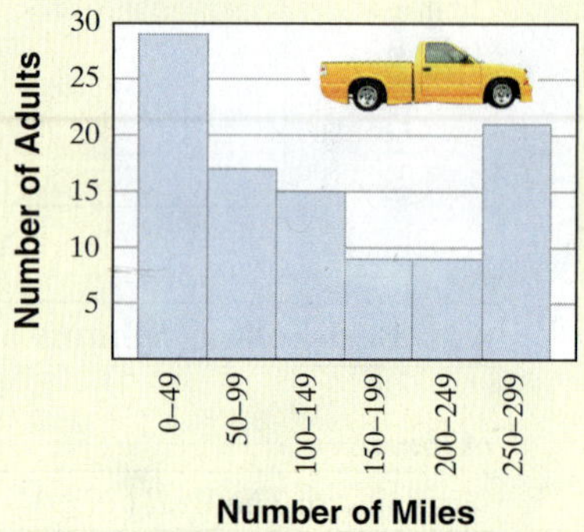

33. How many adults drive 100–149 miles per week?

34. How many adults drive 200–249 miles per week?

35. How many adults drive fewer than 150 miles per week?

36. How many adults drive 200 miles or more per week?

37. How many adults drive 100–199 miles per week?

38. How many adults drive 150–249 miles per week?

39. How many more adults drive 250–299 miles per week than 200–249 miles per week?

40. How many more adults drive 0–49 miles per week than 50–99 miles per week?

41. What is the ratio of adults who drive 150–199 miles per week to the total number of adults surveyed?

42. What is the ratio of adults who drive 50–99 miles per week to the total number of adults surveyed?

The following histogram shows the ages of householders for a recent year. Use this histogram to answer Exercises 43 through 50. For Exercises 45 through 50, estimate to the nearest whole million. See Examples 4 and 5.

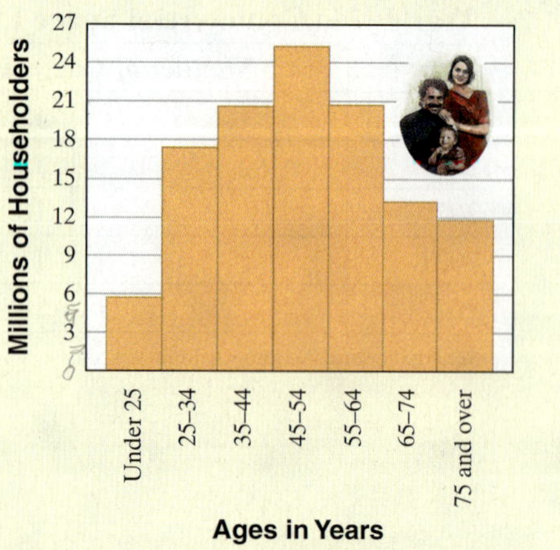

Source: U.S. Bureau of the Census, *Current Population Reports*

43. The most householders were in what age range?

44. The fewest number of householders were in what age range?

45. How many householders were 55–64 years old?

46. How many householders were 35–44 years old?

47. How many householders were 44 years old or younger?

48. How many householders were 55 years old or older?

49. How many more householders were 45–54 years old than 55–64 years old?

50. How many more householders were 45–54 years old than 75 and over?

The following list shows the golf scores for an amateur golfer. Use this list to complete the frequency distribution table such as the one below. See Example 6.

78	84	91	93	97
97	95	85	95	96
101	89	92	89	100

	Class Intervals (Scores)	Tally	Class Frequency (Number of Games)
51.	70–79		
52.	80–89		
53.	90–99		
54.	100–109		

Twenty-five people in a survey were asked to give their current checking account balances. Use the balances shown in the following list to complete the frequency distribution table to the right. See Example 6.

$53	$105	$162	$443	$109
$468	$47	$259	$316	$228
$207	$357	$15	$301	$75
$86	$77	$512	$219	$100
$192	$288	$352	$166	$292

	Class Intervals (Account Balances)	Tally	Class Frequency (Number of People)
55.	$0–$99		
56.	$100–$199		
57.	$200–$299		
58.	$300–$399		
59.	$400–$499		
60.	$500–$599		

61. Use the frequency distribution table from Exercises **51** through **54** to construct a histogram. See Example 7.

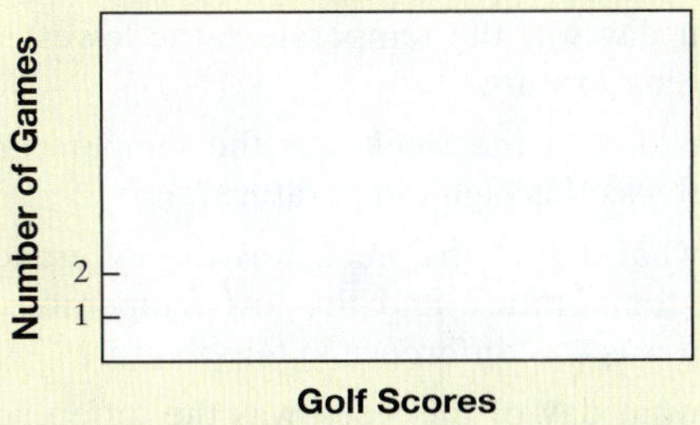

62. Use the frequency distribution table from Exercises **55** through **60** to construct a histogram. See Example 7.

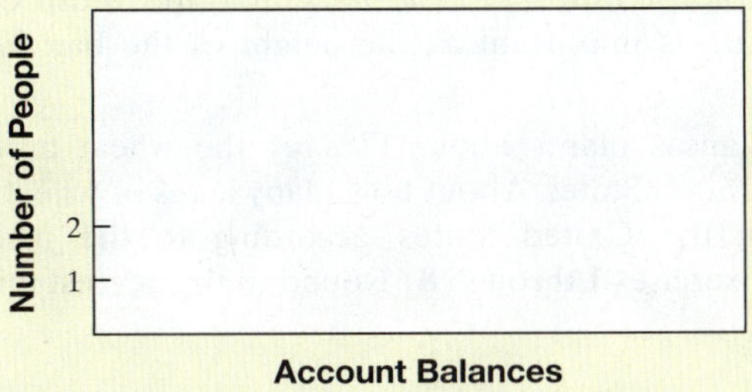

Beach Soccer World Cup is now held every two years. The following line graph shows the World Cup goals per game average for beach soccer during the years shown. Use this graph to answer Exercises 63 through 70. See Example 8.

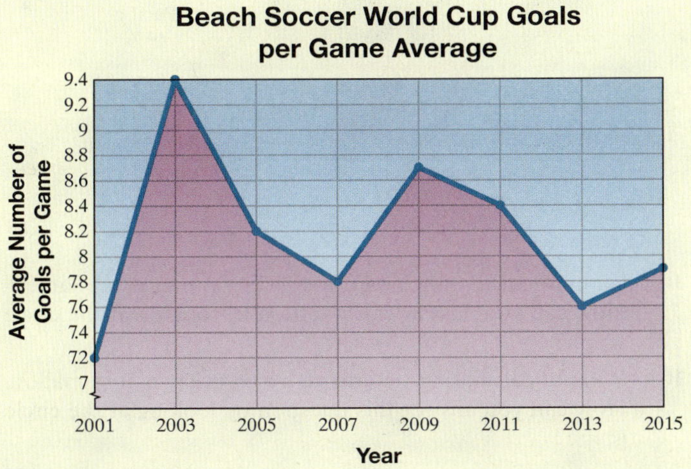

Source: FIFA.com

63. Find the average number of goals per game in 2015.

64. Find the average number of goals per game in 2013.

65. During what year shown was the average number of goals per game the highest?

66. During what year shown was the average number of goals per game the lowest?

67. From 2007 to 2009, did the average number of goals per game increase or decrease?

68. From 2009 to 2011, did the average number of goals per game increase or decrease?

69. During what year(s) shown were the average goals per game less than 8?

70. During what year(s) shown were the average goals per game greater than 8?

The following circle graph is a result of surveying 700 college students. They were asked where they live while attending college. Use this graph to answer Exercises 71 through 76. Write all ratios as fractions in simplest form. See Example 9.

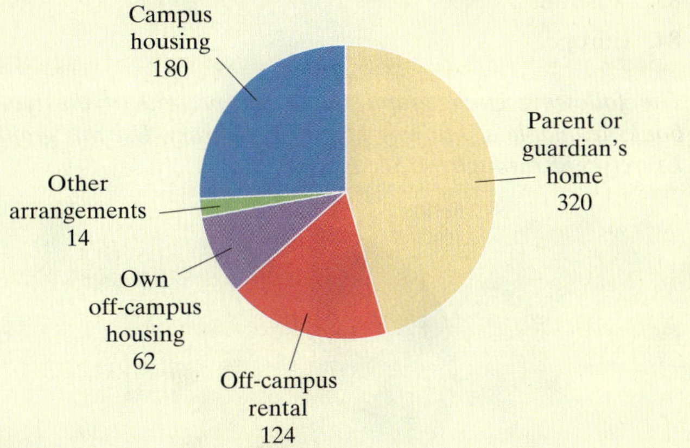

71. Where do most of these college students live?

72. Besides the category "Other arrangements," where do the fewest of these college students live?

73. Find the ratio (fraction) of students living in campus housing to total students.

74. Find the ratio (fraction) of students living in off-campus rentals to total students.

75. Find the ratio (fraction) of students living in campus housing to students living in a parent or guardian's home.

76. Find the ratio (fraction) of students living in off-campus rentals to students living in a parent or guardian's home.

The following circle graph shows the percent of the land area of the continents of Earth. Use this graph for Exercises 77 through 80. See Example 10.

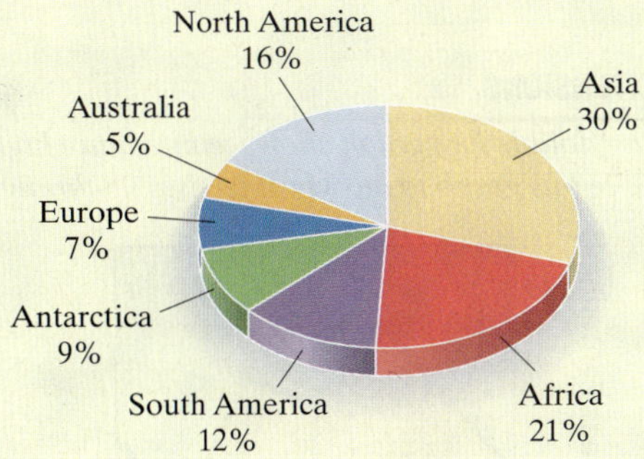

Source: National Geographic Society

77. Which continent is the largest?

78. Which continent is the smallest?

79. What percent of the land on Earth is accounted for by Asia and Europe together?

80. What percent of the land on Earth is accounted for by North and South America?

The total amount of land from the continents is approximately 57,000,000 square miles. Use the graph to find the area of the continents given in Exercises 81 through 84. See Example 11.

81. Asia

82. South America

83. Australia

84. Europe

The following circle graph shows the percent of the types of books available at Midway Memorial Library. Use this graph for Exercises 85 through 94. See Example 10.

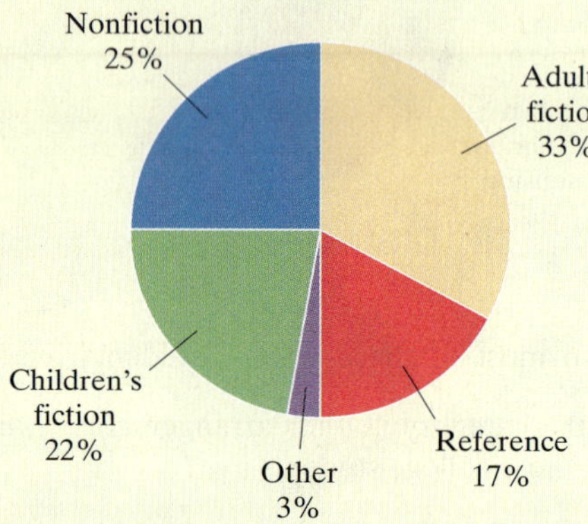

85. What percent of books are classified as some type of fiction?

86. What percent of books are nonfiction or reference?

87. What is the second-largest category of books?

88. What is the third-largest category of books?

If this library has 125,600 books, find how many books are in each category given in Exercises 89 through 94. See Example 11.

89. Nonfiction

90. Reference

91. Children's fiction

92. Adult's fiction

93. Reference or other

94. Nonfiction or other

CONCEPT EXTENSIONS

The following double line graph shows temperature highs and lows for a week. Use this graph to answer Exercises 95 through 100.

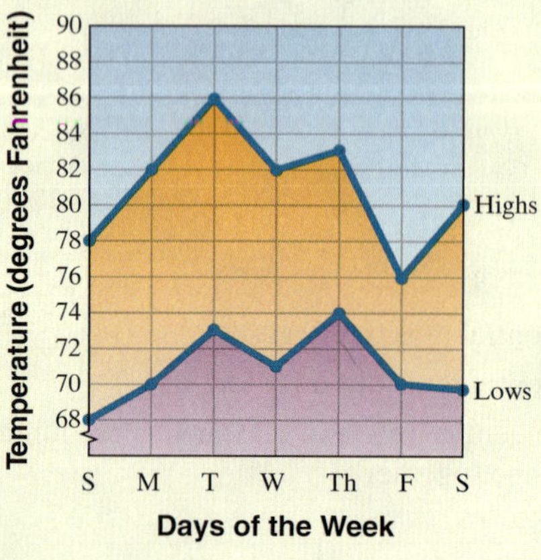

95. What was the high temperature reading on Thursday?

96. What was the low temperature reading on Thursday?

97. What day was the temperature the lowest? What was this low temperature?

98. What day of the week was the temperature the highest? What was this high temperature?

99. On what day of the week was the difference between the high temperature and the low temperature the greatest? What was this difference in temperature?

100. On what day of the week was the difference between the high temperature and the low temperature the least? What was this difference in temperature?

101. True or false? With a bar graph, the width of the bar is just as important as the height of the bar. Explain your answer.

102. Kansas plants about 17% of the wheat acreage in the United States. About how many acres of wheat are planted in the United States, according to the pictograph for Exercises **1** through **8**? Round to the nearest million acre.

The following circle graph shows the relative sizes of the great oceans.

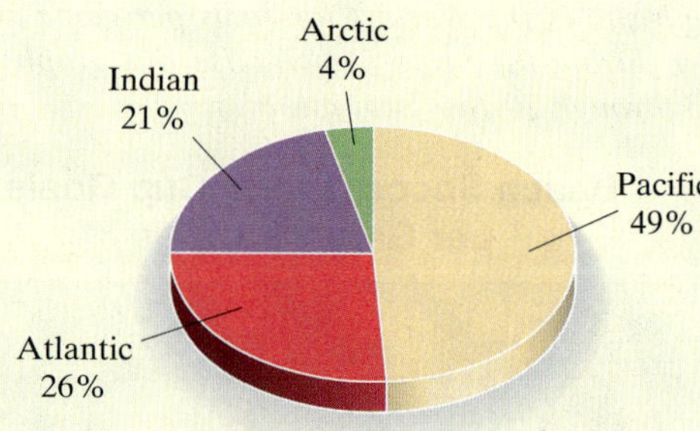

Source: Philip's World Atlas

103. Without calculating, determine which ocean is the largest. How can you answer this question by looking at the circle graph?

104. Without calculating, determine which ocean is the smallest. How can you answer this question by looking at the circle graph?

These oceans together make up 264,489,800 square kilometers of Earth's surface. Find the square kilometers for each ocean.

105. Pacific Ocean

106. Atlantic Ocean

107. Indian Ocean

108. Arctic Ocean

The following circle graph summarizes the results of a survey of online spending in America. Let's use these results to make predictions about the online spending behavior of a community of 2800 Internet users age 18 and over. Use this graph for Exercises 109 through 114. Round to the nearest whole. (Note: Because of rounding, these percents do not have a sum of 100%.)

Online Spending per Month

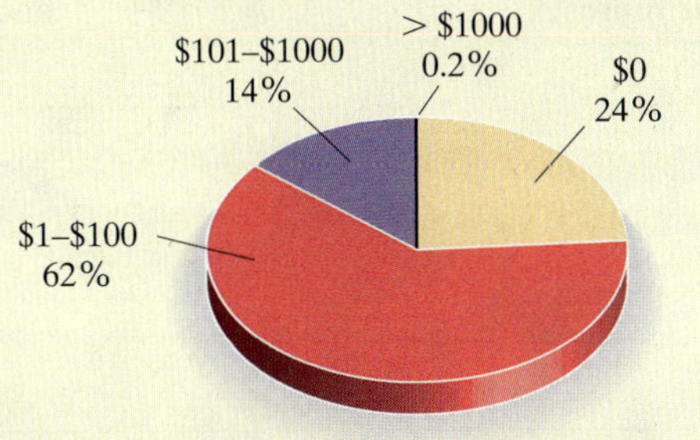

Source: The Digital Future Report, 2013

109. How many of the survey respondents said that they spend $0 online each month?

110. How many of the survey repondents said that they spend $1–$100 online each month?

111. How many of the survey respondents said that they spend $0 to $100 online each month?

112. How many of the survey respondents said that they spend $1 to $1000 online each month?

113. Find the ratio of *number* of respondents who spend $0 online to *number* of respondents who spend $1–$100 online. Write the ratio as a fraction. Simplify the fraction if possible.

114. Find the ratio of *percent* of respondents who spend $101–$1000 online to *percent* of those who spend $1–$100. Write the ratio as a fraction with integers in the numerator and denominator. Simplify the fraction if possible.

See the Concept Checks in this section.

115. Can the data below be represented by a circle graph? Why or why not?

Responses to the Question "What Classes Are You Taking?"	
Math	80%
English	72%
History	37%
Biology	21%
Chemistry	14%

116. True or false? The smaller a sector in a circle graph, the smaller the percent of the total it represents. Explain why.

Chapter R Vocabulary Check

Fill in each blank with one of the words or phrases listed below.

mixed number	factor	class frequency	percent	of	$\frac{1}{100}$	circle	bar
histogram	composite number	class interval	simplified	100%	0.01	line	pictograph
prime number	equivalent						

1. To ________________ means to write as a product.

2. A(n) ________________ is a natural number greater than 1 that is not prime.

3. Fractions that represent the same portion of a whole are called ________________ fractions.

4. A(n) ________________ is a natural number greater than 1 whose only factors are 1 and itself.

5. A fraction is ________________ when the numerator and the denominator have no factors in common other than 1.

6. A(n) ________________ contains a whole number part and a fraction part.

7. In a mathematical statement, ________________ usually means "multiplication."

8. ________________ means "per hundred."

9. To write a decimal or fraction as a percent, multiply by ________________

10. The decimal equivalent of the % symbol is ________________.

11. The fraction equivalent of the % symbol is ________________.

12. A(n) ________________ graph presents data using vertical or horizontal bars.

13. A(n) ________________ is a graph in which pictures or symbols are used to visually present data.

14. A(n) ________________ graph displays information with a line that connects data points.

15. In a(n) ________________ graph, each section (shaped like a piece of pie) shows a category and the relative size of the category.

16. A(n) ________________ is a special bar graph in which the width of each bar represents a(n) ________________ and the height of each bar represents the ________________.

Chapter R Highlights

DEFINITIONS AND CONCEPTS	EXAMPLES
Section R.1 Fractions and Mixed Numbers	
A quotient of two integers is called a **fraction.** The **numerator** of a fraction is the top number. The **denominator** of a fraction is the bottom number.	$\frac{13 \leftarrow \text{numerator}}{17 \leftarrow \text{denominator}}$
If $a \cdot b = c$, then a and b are **factors** and c is the **product.**	$\underset{\text{factor}}{7} \cdot \underset{\text{factor}}{9} = \underset{\text{product}}{63}$
A fraction is in **lowest terms** or **simplest form** when the numerator and the denominator have no factors in common other than 1.	$\frac{13}{17}$ is in simplest form.
To write a fraction in simplest form, factor the numerator and the denominator; then apply the fundamental principle.	Write in simplest form. $\frac{6}{14} = \frac{2 \cdot 3}{2 \cdot 7} = \frac{3}{7}$
Two fractions are **reciprocals** if their product is 1. The reciprocal of $\frac{a}{b}$ is $\frac{b}{a}$.	The reciprocal of $\frac{6}{25}$ is $\frac{25}{6}$.
To multiply fractions, numerator times numerator is the numerator of the product and denominator times denominator is the denominator of the product.	Perform the indicated operations. $\frac{2}{5} \cdot \frac{3}{7} = \frac{6}{35}$
To divide fractions, multiply the first fraction by the reciprocal of the second fraction.	$\frac{5}{9} \div \frac{2}{7} = \frac{5}{9} \cdot \frac{7}{2} = \frac{35}{18}$
To add fractions with the same denominator, add the numerators and place the sum over the common denominator.	$\frac{5}{11} + \frac{3}{11} = \frac{8}{11}$
To subtract fractions with the same denominator, subtract the numerators and place the difference over the common denominator.	$\frac{13}{15} - \frac{3}{15} = \frac{10}{15} = \frac{2}{3}$
Fractions that represent the same quantity are called **equivalent fractions.**	$\frac{1}{5} = \frac{1 \cdot 4}{5 \cdot 4} = \frac{4}{20}$ $\frac{1}{5}$ and $\frac{4}{20}$ are equivalent fractions.

DEFINITIONS AND CONCEPTS	EXAMPLES

Section R.2 Decimals

To write decimals as fractions, use place values.

$$0.11 = \frac{11}{100}$$

To Add or Subtract Decimals

Step 1. Write the decimals so that the decimal points line up vertically.

Step 2. Add or subtract as for whole numbers.

Step 3. Place the decimal point in the sum or difference so that it lines up vertically with the decimal points in the problem.

Subtract: $2.8 - 1.04$ Add: $25 + 0.02$

$$\begin{array}{r} \overset{7\,10}{2.\not{8}\not{0}} \\ -1.0\,4 \\ \hline 1.7\,6 \end{array} \qquad \begin{array}{r} 25. \\ +\ 0.02 \\ \hline 25.02 \end{array}$$

To Multiply Decimals

Step 1. Multiply the decimals as though they are whole numbers.

Step 2. The decimal point in the product is placed so that the number of decimal places in the product is equal to the **sum** of the number of decimal places in the factors.

Multiply: 1.48×5.9

$$\begin{array}{r} 1.4\,8 \\ \times \quad 5.9 \\ \hline 1\,3\,3\,2 \\ 7\,4\,0 \\ \hline 8.7\,3\,2 \end{array} \begin{array}{l} \leftarrow 2 \text{ decimal places} \\ \leftarrow 1 \text{ decimal place} \\ \\ \\ \leftarrow 3 \text{ decimal places} \end{array}$$

To Divide Decimals

Step 1. Move the decimal point in the divisor to the right until the divisor is a whole number.

Step 2. Move the decimal point in the dividend to the right the **same number of places** as the decimal point was moved in Step 1.

Step 3. Divide. The decimal point in the quotient is directly over the moved decimal point in the dividend.

Divide: $1.118 \div 2.6$

$$\begin{array}{r} 0.43 \\ 26.\overline{)11.18} \\ -10\,4 \\ \hline 78 \\ -78 \\ \hline 0 \end{array}$$

To write fractions as decimals, divide the numerator by the denominator.

Write $\frac{3}{8}$ as a decimal.

$$\begin{array}{r} 0.375 \\ 8\overline{)3.000} \\ -2\,4 \\ \hline 60 \\ -56 \\ \hline 40 \\ -40 \\ \hline 0 \end{array}$$

Section R.3 Percents, Decimals, and Fractions

Percent means "per hundred." The % symbol denotes percent.

$51\% = \frac{51}{100}$ 51 per 100

$7\% = \frac{7}{100}$ 7 per 100

(continued)

DEFINITIONS AND CONCEPTS	EXAMPLES
Section R.3 Percents, Decimals, and Fractions (continued)	
To write a percent as a decimal, replace the % symbol with its decimal equivalent, 0.01, and multiply.	$32\% = 32(0.01) = 0.32$
To write a decimal as a percent, multiply by 100%.	$0.08 = 0.08(100\%) = 08.\% = 8\%$
To write a percent as a fraction, replace the % symbol with its fraction equivalent, $\frac{1}{100}$, and multiply.	$25\% = 25 \cdot \frac{1}{100} = \frac{25}{100} = \frac{\overset{1}{\cancel{25}}}{4 \cdot \underset{1}{\cancel{25}}} = \frac{1}{4}$
To write a fraction as a percent, multiply by 100%.	$\frac{1}{6} = \frac{1}{6} \cdot 100\% = \frac{1}{6} \cdot \frac{100}{1}\% = \frac{100}{6}\% = 16\frac{2}{3}\%$

Section R.4 Reading Pictographs, and Bar, Line, and Circle Graphs

A **pictograph** is a graph in which pictures or symbols are used to visually present data.

A **line graph** displays information with a line that connects data points.

A **bar graph** presents data using vertical or horizontal bars.

The bar graph on the right shows the number of acres of corn harvested in a recent year for leading states.

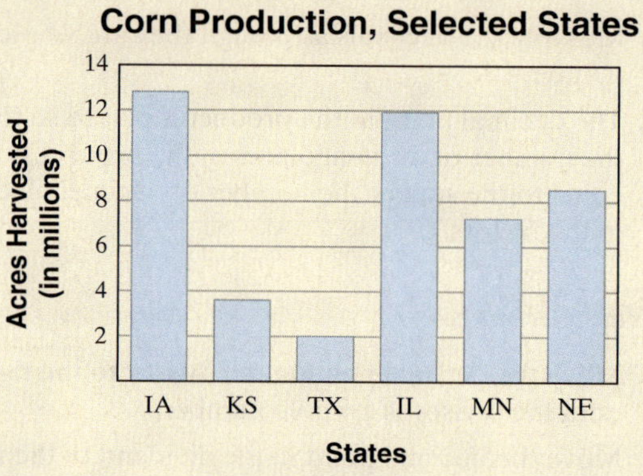

Source: U.S. Department of Agriculture

1. Approximately how many acres of corn were harvested in Iowa?

   12,800,000 acres

2. About how many more acres of corn were harvested in Illinois than Nebraska?

$$\begin{array}{r} 12 \text{ million} \\ -\ 8.5 \text{ million} \\ \hline 3.5 \text{ million} \end{array} \quad \text{or } 3{,}500{,}000 \text{ acres}$$

A **histogram** is a special bar graph in which the width of each bar represents a **class interval** and the height of each bar represents the **class frequency.** The histogram on the right shows student quiz scores.

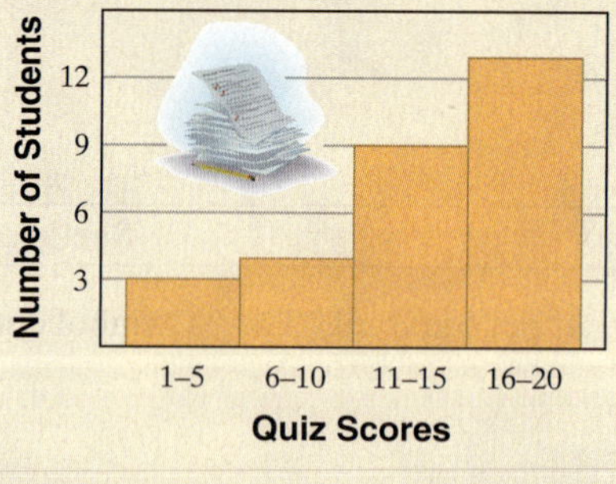

1. How many students received a score of 6–10?

   4 students

2. How many students received a score of 11–20?

   $9 + 13 = 22$ students

DEFINITIONS AND CONCEPTS	**EXAMPLES**

Section R.4 Reading Pictographs, and Bar, Line, and Circle Graphs (continued)

In a **circle graph,** each section (shaped like a piece of pie) shows a category and the relative size of the category.

The circle graph on the right classifies tornadoes by wind speed.

Tornado Wind Speeds

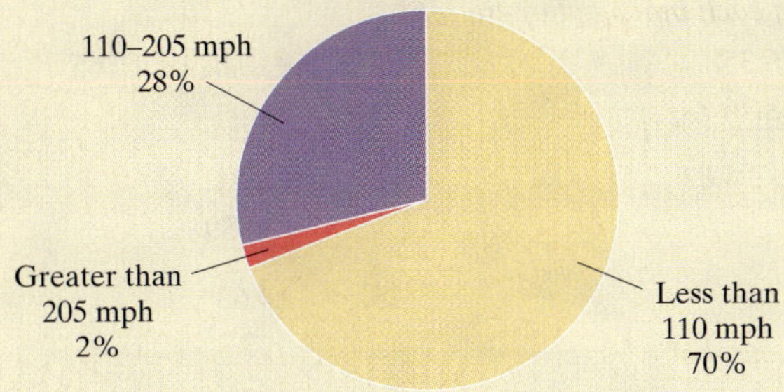

Source: National Oceanic and Atmospheric Administration

1. What percent of tornadoes have wind speeds of 110 mph or greater?

$$28\% + 2\% = 30\%$$

2. If there were 939 tornadoes in the United States in 2012, how many of these might we expect to have had wind speeds less than 110 mph? Find 70% of 939.

$$70\%(939) = 0.70(939) = 657.3 \approx 657$$

Around 657 tornadoes would be expected to have had wind speeds of less than 110 mph.

Chapter R Review

(R.1) Write the number as a product of prime factors.

1. 36 **2.** 120

Perform the indicated operations. Write results in lowest terms.

3. $\frac{8}{15} \cdot \frac{27}{30}$ **4.** $\frac{7}{8} \div \frac{21}{32}$ **5.** $\frac{7}{15} + \frac{5}{6}$

6. $\frac{3}{4} - \frac{3}{20}$ **7.** $2\frac{3}{4} + 6\frac{5}{8}$ **8.** $7\frac{1}{6} - 2\frac{2}{3}$

9. $5 \div \frac{1}{3}$ **10.** $2 \cdot 8\frac{3}{4}$

Each circle represents a whole, or 1. Determine the unknown part of the circle.

11.

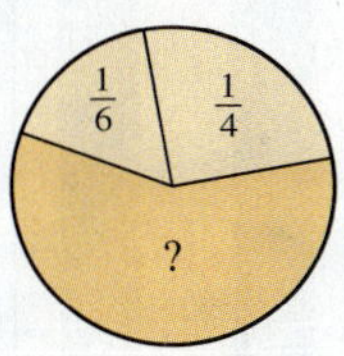

12.

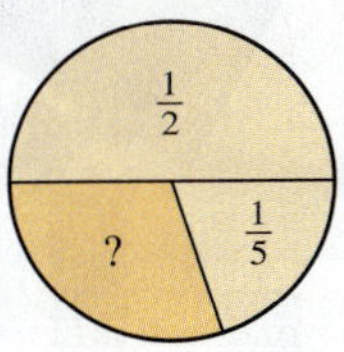

Find the area and the perimeter of each figure.

△ **13.**

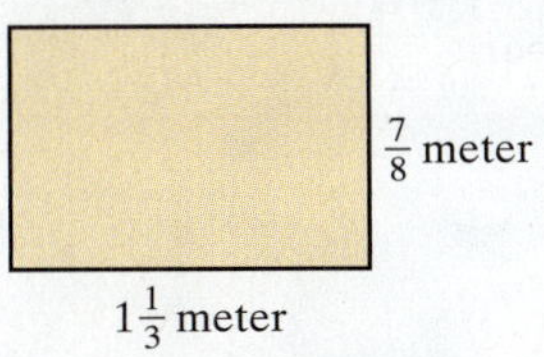

△ **14.**

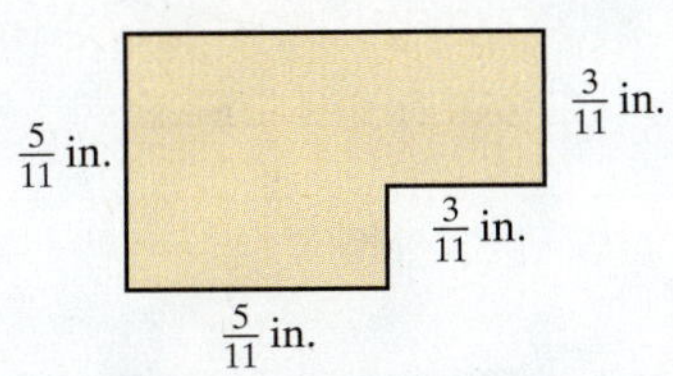

Octuplets were born in the U.S. in 2009. The following chart gives the octuplets' birthweights. The babies are listed in order of birth.

Baby	*Gender*	*Birthweight (pounds)*
Baby A	boy	$2\frac{1}{2}$
Baby B	girl	$2\frac{1}{8}$
Baby C	boy	$3\frac{1}{16}$
Baby D	girl	$2\frac{3}{16}$
Baby E	boy	$1\frac{3}{4}$
Baby F	boy	$2\frac{9}{16}$
Baby G	boy	$1\frac{13}{16}$
Baby H	boy	$2\frac{7}{16}$

15. What was the total weight of the boy octuplets?

16. What was the total weight of the girl octuplets?

17. Find the combined weight of all eight octuplets.

18. Which baby weighed the most?

19. Which baby weighed the least?

20. How much more did the heaviest baby weigh than the lightest baby?

(R.2) *Write each decimal as a fraction. Do not simplify.*

21. 1.81 **22.** 0.035

Perform each indicated operation.

23. $\begin{array}{r} 76.358 \\ +18.76 \\ \hline \end{array}$ **24.** $35 + 0.02 + 1.765$

25. $18 - 4.62$ **26.** $\begin{array}{r} 804.062 \\ -112.489 \\ \hline \end{array}$

27. $\begin{array}{r} 7.6 \\ \times\ 12 \\ \hline \end{array}$ **28.** $\begin{array}{r} 14.63 \\ \times\ \ 3.2 \\ \hline \end{array}$

29. $27\overline{)772.2}$ **30.** $0.06\overline{)13.8}$

Round each decimal to the given place value.

31. 0.7652, nearest hundredth

32. 25.6293, nearest tenth

Write each fraction as a decimal. If the decimal is a repeating decimal, write it using the bar notation and then round to the nearest thousandth.

33. $\frac{1}{2}$ **34.** $\frac{3}{8}$

35. $\frac{4}{11}$ **36.** $\frac{5}{6}$

(R.3) *Solve.*

37. In a survey of 100 adults, 37 preferred pepperoni on their pizzas. What percent preferred pepperoni?

38. A basketball player made 77 out 100 attempted free throws. What percent of free throws was made?

Write each percent as a decimal.

39. 26% **40.** 75%

41. 3.5% **42.** 1.5%

43. 275% **44.** 400%

45. 47.85% **46.** 85.34%

Write each decimal as a percent.

47. 1.6 **48.** 0.055

49. 0.076 **50.** 0.085

51. 0.71 **52.** 0.65

53. 6 **54.** 9

Write each percent as a fraction or mixed number in simplest form.

55. 7% **56.** 15%

57. 25% **58.** 8.5%

59. 10.2% **60.** $16\frac{2}{3}\%$

61. $33\frac{1}{3}\%$ **62.** 110%

Write each fraction or mixed number as a percent.

63. $\frac{2}{5}$ **64.** $\frac{7}{10}$

65. $\frac{7}{12}$ **66.** $1\frac{2}{3}$

67. $1\frac{1}{4}$ **68.** $\frac{3}{5}$

69. $\frac{1}{16}$ **70.** $\frac{5}{8}$

(R.4) *The following pictograph shows the number of new homes constructed for one year, by region. Use this graph to answer Exercises 71 through 76.*

New Home Construction for One Year by Region

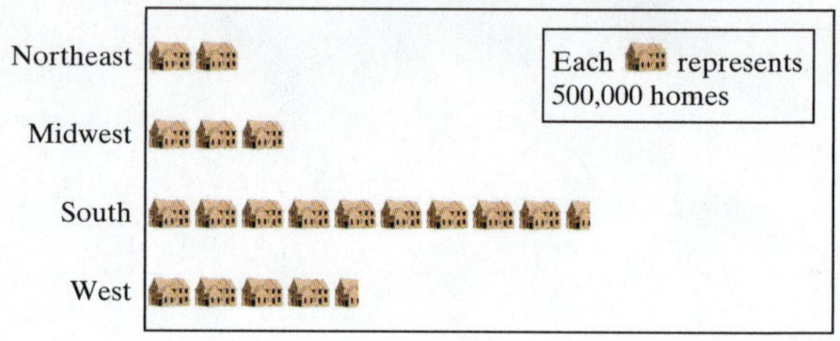

Source: U.S. Census Bureau

71. How many new homes were constructed in the West during the given year?

72. How many new homes were constructed in the South during the given year?

73. Which region had the most new homes constructed?

74. Which region had the fewest new homes constructed?

75. Which region(s) had 2,000,000 or more new homes constructed?

76. Which region(s) had fewer than 2,000,000 new homes constructed?

The following bar graph shows the percent of persons age 25 or over who completed four or more years of college. Use this graph to answer Exercises 77 through 80.

Four or More Years of College by Persons Age 25 or Over

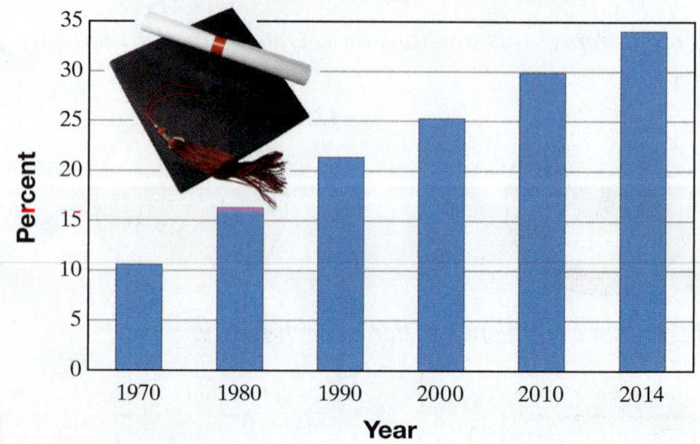

Source: U.S. Census Bureau

77. Approximate the percent of persons who had completed four or more years of college by 2010.

78. What year shown had the greatest percent of persons completing four or more years of college?

79. What years shown had 20% or more of persons completing four or more years of college?

80. Describe any patterns you notice in this graph.

The following line graph shows the total number of Olympic medals awarded during the Summer Olympics since 1992. Use this graph to answer Exercises 81 through 86.

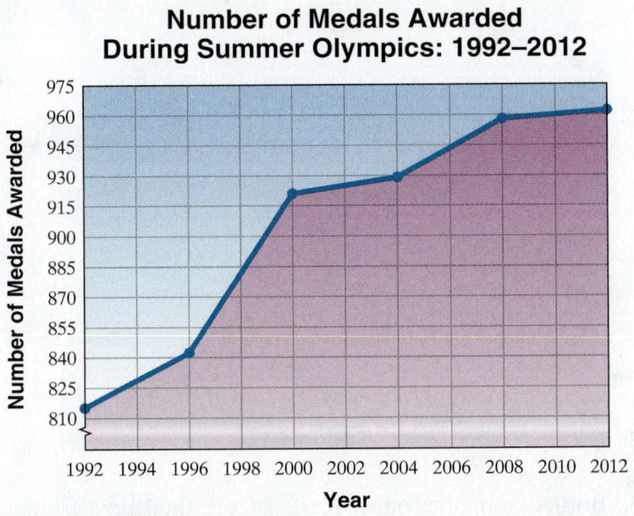

Source: International Olympic Committee

81. Approximate the number of medals awarded during the Summer Olympics of 2012.

82. Approximate the number of medals awarded during the Summer Olympics of 2000.

83. Approximate the number of medals awarded during the Summer Olympics of 2004.

84. Approximate the number of medals awarded during the Summer Olympics of 1992.

85. How many more medals were awarded at the Summer Olympics of 1996 than at the Summer Olympics of 1992?

86. How many more medals were awarded at the Summer Olympics of 2012 than at the Summer Olympics of 1992?

The following histogram shows the hours worked per week by the employees of Southern Star Furniture. Use this histogram to answer Exercises 87 through 90.

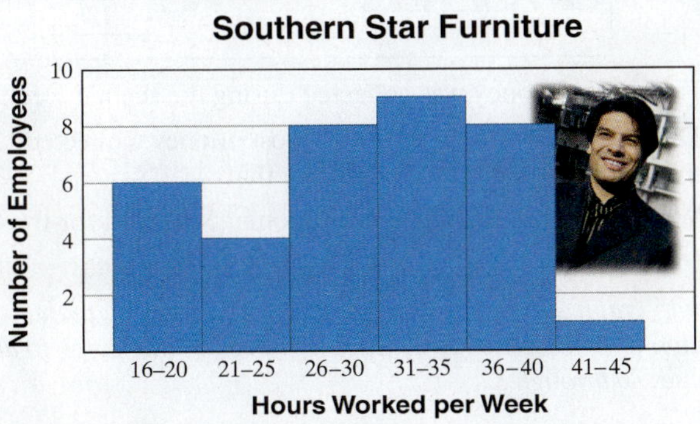

87. How many employees work 41–45 hours per week?

88. How many employees work 21–25 hours per week?

89. How many employees work 30 hours or less per week?

90. How many employees work 36 hours or more per week?

Following is a list of monthly record high temperatures for New Orleans, Louisiana. Use this list to complete the frequency distribution table below.

83	96	101	92
85	100	92	102
89	101	87	84

	Class Intervals (Temperatures)	*Tally*	*Class Frequency (Number of Months)*
91.	80°–89°		
92.	90°–99°		
93.	100°–109°		

94. Use the table from Exercises 91–93 to draw a histogram.

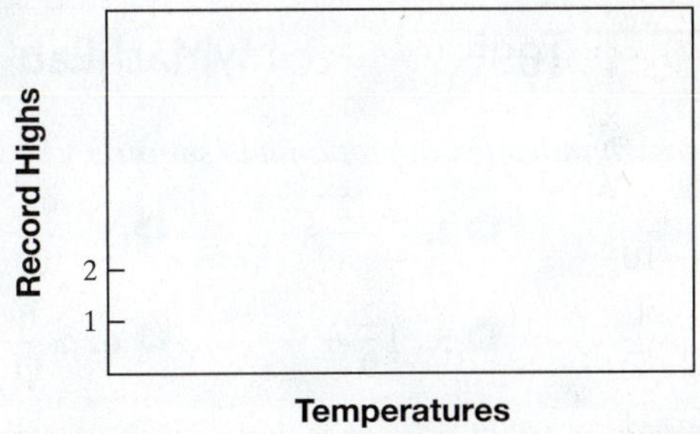

The following circle graph shows a family's $4000 monthly budget. Use this graph to answer Exercises 95 through 100. Write all ratios as fractions in simplest form.

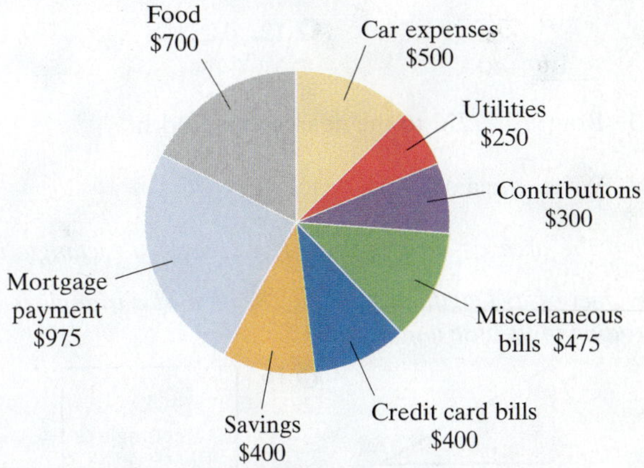

95. What is the largest budget item?

96. What is the smallest budget item?

97. How much money is budgeted for the mortgage payment and utilities?

98. How much money is budgeted for savings and contributions?

99. Find the ratio (fraction) of the mortage payment to the total monthly budget.

100. Find the ratio (fraction) of food to the total monthly budget.

In a recent year, there were approximately 62 buildings 200 meters or taller completed in the world. The following circle graph shows the percent of these buildings by region. Use this graph to answer Exercises 101 through 104. Round each answer to the nearest whole.

Percent of Tall Buildings Completed
(200 Meters or Taller by Continent)

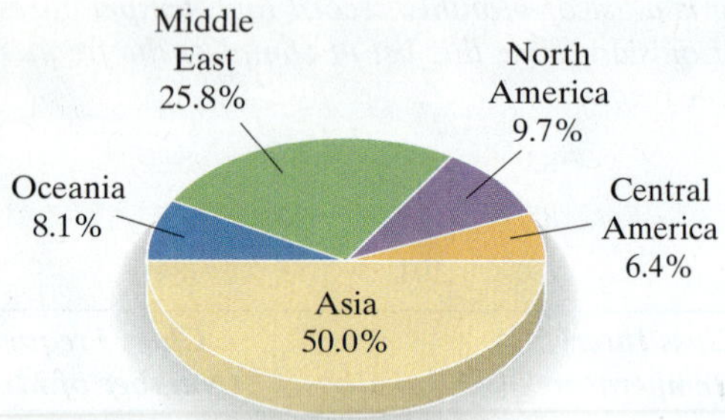

Source: Council on Tall Buildings and Urban Habitats

101. How many completed tall buildings were located in Asia?

102. How many completed tall buildings were located in North America?

103. How many completed tall buildings were located in Oceania?

104. How many completed tall buildings were located in the Middle East?

Chapter R Test

MyMathLab® YouTube

Perform each indicated operation and simplify.

1. $\frac{5}{8} + \frac{7}{10}$ **2.** $\frac{2}{3} \cdot \frac{27}{49}$ **3.** $\frac{9}{10} \div 18$

4. $\frac{8}{9} - \frac{1}{12}$ **5.** $1\frac{2}{9} + 3\frac{2}{3}$ **6.** $5\frac{6}{11} - 3\frac{7}{22}$

7. $6\frac{7}{8} \div \frac{1}{8}$ **8.** $2\frac{1}{10} \cdot 6\frac{1}{2}$

Perform each indicated operation.

9. $43 + 0.21 + 1.9$ **10.** $123.6 - 57.72$

11. $\begin{array}{r} 7.93 \\ \times\ 1.6 \\ \hline \end{array}$ **12.** $0.25\overline{)80}$

13. Round 23.7272 to the nearest hundredth.

14. Write $\frac{7}{8}$ as a decimal.

Find the area of each figure. (The area of a rectangle is the product of its length and width. The area of a triangle is $\frac{1}{2}$ the product of its base and height.)

15. △

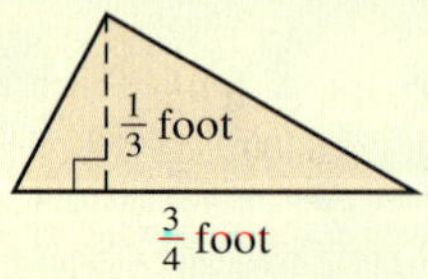

16. △

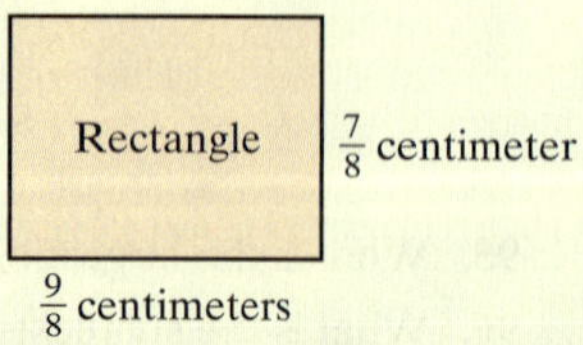

Write each percent as a decimal.

17. 85% **18.** 0.6%

Write each decimal as a percent.

19. 0.056 **20.** 6.1

Write each percent as a fraction or a mixed number in simplest form.

21. 120% **22.** 38.5%

Write each fraction or mixed number as a percent.

23. $\frac{3}{8}$ **24.** $1\frac{3}{4}$

25. Bottled water accounts for $\frac{1}{5}$ of the total noncarbonated beverage category in convenience stores. Write $\frac{1}{5}$ as a percent. (*Source:* Grocery Manufacturers of America)

26. In small firms in the United States, 64% of full-time employees receive medical insurance benefits. Write 64% as a fraction. (*Source:* U.S. Bureau of Labor Statistics)

The following pictograph shows the money collected each week from a wrapping paper fundraiser. Use this graph to answer Exercises 27 through 29.

Weekly Wrapping Paper Sales

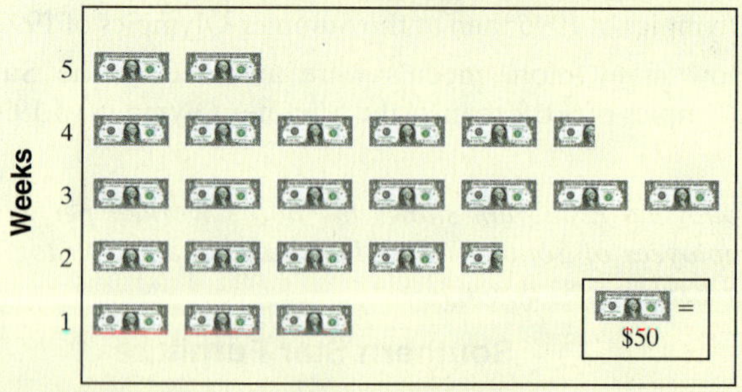

27. How much money was collected during the second week?

28. During which week was the most money collected? How much money was collected during that week?

29. What was the total amount of money collected for the fundraiser?

The bar graph on the next page shows the normal monthly precipitation in centimeters for Chicago, Illinois. Use this graph to answer Exercises 30 through 32.

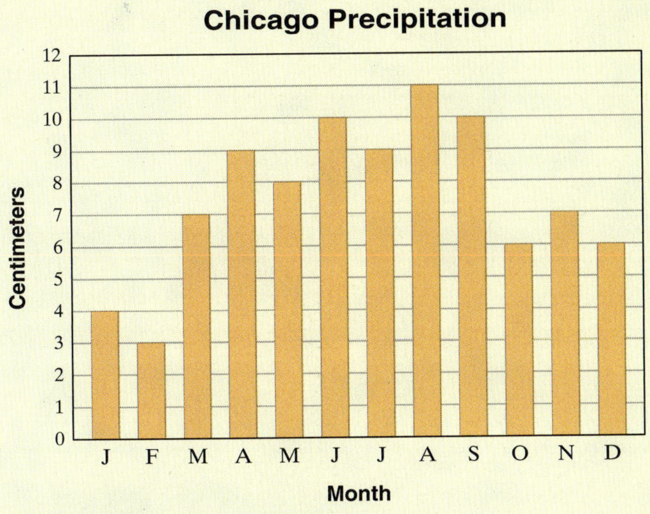

Source: U.S. National Oceanic and Atmospheric Administration, *Climatography of the United States*, No. 81

30. During which month(s) does Chicago normally have more than 9 centimeters of precipitation?

31. During which month does Chicago normally have the least amount of precipitation? How much precipitation occurs during that month?

32. During which month(s) does 7 centimeters of precipitation normally occur?

The result of a survey of 200 people is shown in the following circle graph. Each person was asked to tell his or her favorite type of music. Use this graph to answer Exercises 33 and 34.

33. Find the ratio (fraction) of those who prefer rock music to the total number surveyed.

34. Find the ratio (fraction) of those who prefer country music to those who prefer jazz.

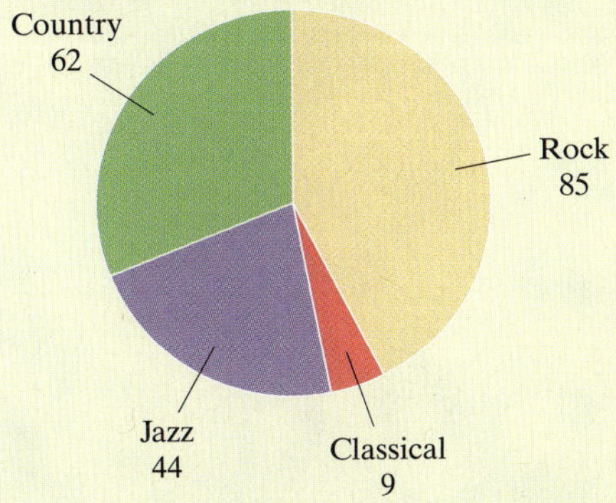

The following circle graph shows the age distribution of the population of the United States in 2015. Suppose there were 326 million people in the United States in 2015. Use the graph to find how many people are expected to be in the age groups given.

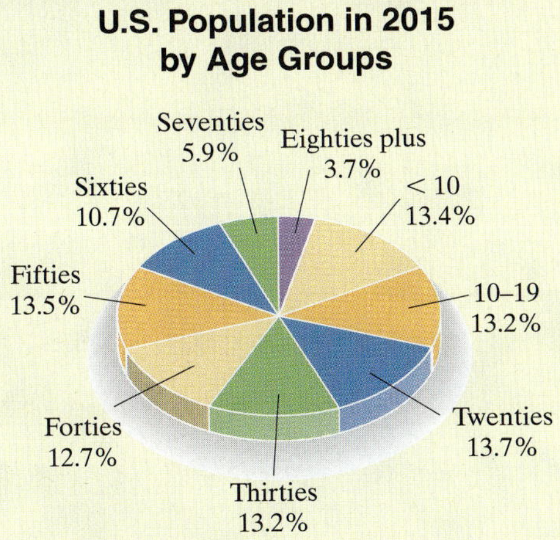

Source: U.S. Census Bureau

35. Twenties (Round to nearest whole million.)

36. Eighties plus (Round to nearest whole million.)

A professor measures the heights of the students in her class. The results are shown in the following histogram. Use this histogram to answer Exercises 37 and 38.

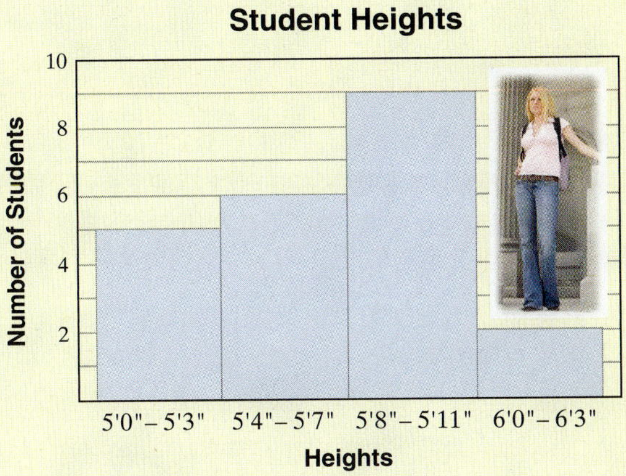

37. How many students are 5′8″–5′11″ tall?

38. How many students are 5′7″ or shorter?

39. The history test scores of 25 students are shown below. Use these scores to complete the frequency distribution table.

70	86	81	65	92
43	72	85	69	97
82	51	75	50	68
88	83	85	77	99
77	63	59	84	90

Class Intervals (Scores)	*Tally*	*Class Frequency (Number of Students)*
40–49		
50–59		
60–69		
70–79		
80–89		
90–99		

40. Use the results of Exercise 39 to draw a histogram.

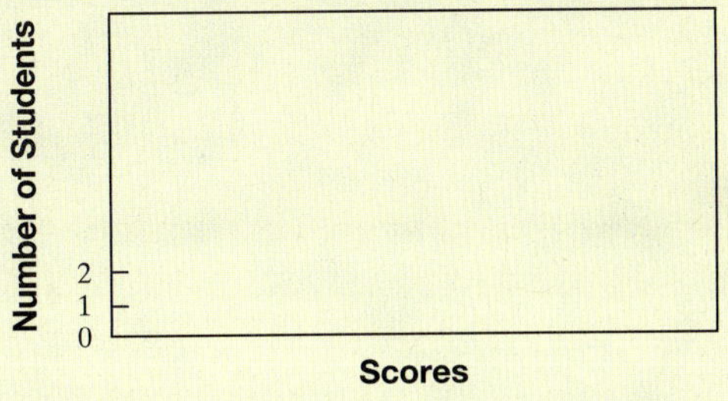

CHAPTER 1

Review of Real Numbers

1.1 Study Skill Tips for Success in Mathematics

1.2 Symbols and Sets of Numbers

1.3 Exponents, Order of Operations, Variable Expressions, and Equations

Integrated Review—Order of Operations

1.4 Operations on Real Numbers and Evaluating Square Roots

1.5 Properties of Real Numbers

CHECK YOUR PROGRESS

Vocabulary Check
Chapter Highlights
Chapter Review
Getting Ready for the Test
Chapter Test

In this chapter, we review the basic symbols and words—the language—of arithmetic and introduce using variables in place of numbers.

A Selection of Resources for Success in This Mathematics Course

Textbook

Instructor

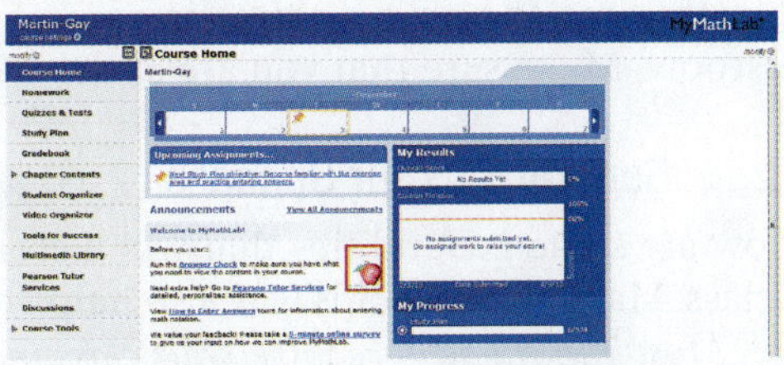
MyMathLab and MathXL

Video Organizer

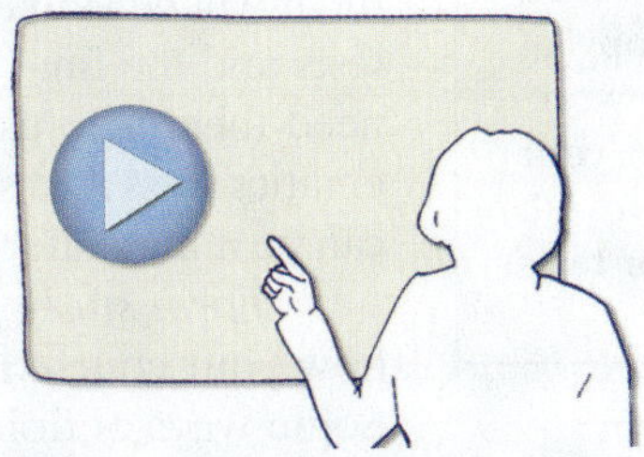
Interactive Lecture Series

For more information about the resources illustrated above, read Section 1.1.

1.1 Study Skill Tips for Success in Mathematics

OBJECTIVES

1 Get Ready for This Course.

2 Understand Some General Tips for Success.

3 Know How to Use This Text.

4 Know How to Use Text Resources.

5 Get Help as Soon as You Need It.

6 Learn How to Prepare for and Take an Exam.

7 Develop Good Time Management.

Before reading Section 1.1, you might want to ask yourself a few questions.

1. When you took your last math course, were you organized? Were your notes and materials from that course easy to find, or were they disorganized and hard to find—if you saved them at all?
2. Were you satisfied—really satisfied—with your performance in that course? In other words, do you feel that your outcome represented your best effort?

If the answer is "no" to these questions, then it is time to make a change. Changing to or resuming good study skill habits is not a process you can start and stop as you please. It is something that you must remember and practice each and every day. To begin, continue reading this section.

OBJECTIVE 1 Getting Ready for This Course

Now that you have decided to take this course, remember that a *positive attitude* will make all the difference in the world. Your belief that you can succeed is just as important as your commitment to this course. Make sure you are ready for this course by having the time and positive attitude that it takes to succeed.

Make sure that you are familiar with the way that this course is being taught. Is it a traditional course, in which you have a printed textbook and meet with an instructor? Is it taught totally online, and your textbook is electronic and you e-mail your instructor? Or is your course structured somewhere in between these two methods? (Not all of the tips that follow will apply to all forms of instruction.)

Also make sure that you have scheduled your math course for a time that will give you the best chance for success. For example, if you are also working, you may want to check with your employer to make sure that your work hours will not conflict with your course schedule.

On the day of your first class period, double-check your schedule and allow yourself extra time to arrive on time in case of traffic problems or difficulty locating your classroom. Make sure that you are aware of and bring all necessary class materials.

OBJECTIVE 2 General Tips for Success

Below are some general tips that will increase your chance for success in a mathematics class. Many of these tips will also help you in other courses you may be taking.

> **Helpful Hint**
> **MyMathLab® and MathXL®**
> When assignments are turned in online, keep a hard copy of your complete written work. You will need to refer to your written work to be able to ask questions and to study for tests later.

Most important! Organize your class materials. In the next couple pages, many ideas will be presented to help you organize your class materials—notes, any handouts, completed homework, previous tests, etc. In general, you MUST have these materials organized. All of them will be valuable references throughout your course and when studying for upcoming tests and the final exam. One way to make sure you can locate these materials when you need them is to use a three-ring binder. This binder should be used solely for your mathematics class and should be brought to each and every class or lab. This way, any material can be immediately inserted in a section of this binder and will be there when you need it.

Form study groups and/or exchange names and e-mail addresses. Depending on how your course is taught, you may want to keep in contact with your fellow students. Some ways of doing this are to form a study group—whether in person or through the Internet. Also, you may want to ask if anyone is interested in exchanging e-mail addresses or another form of contact.

Choose to attend all class periods. If possible, sit near the front of the classroom. This way, you will see and hear the presentation better. It may also be easier for you to participate in classroom activities.

> **Helpful Hint**
> **MyMathLab® and MathXL®**
> If you are doing your homework online, you can work and re-work those exercises that you struggle with until you master them. Try working through all the assigned exercises twice before the due date.

Do your homework. You've probably heard the phrase "practice makes perfect" in relation to music and sports. It also applies to mathematics. You will find that the more time you spend solving mathematics exercises, the easier the process becomes. Be sure to schedule enough time to complete your assignments before the due date assigned by your instructor.

> **Helpful Hint**
>
> **MyMathLab® and MathXL®**
> If you are completing your homework online, it's important to work each exercise on paper before submitting the answer. That way, you can check your work and follow your steps to find and correct any mistakes.

Check your work. Review the steps you took while working a problem. Learn to check your answers in the original exercises. You may also compare your answers with the "Answers to Selected Exercises" section in the back of the book. If you have made a mistake, try to figure out what went wrong. Then correct your mistake. If you can't find what went wrong, **don't** erase your work or throw it away. Show your work to your instructor, a tutor in a math lab, or a classmate. It is easier for someone to find where you had trouble if he or she looks at your original work.

Learn from your mistakes and be patient with yourself. Everyone, even your instructor, makes mistakes. (That definitely includes me—Elayn Martin-Gay.) Use your errors to learn and to become a better math student. The key is finding and understanding your errors.

Was your mistake a careless one, or did you make it because you can't read your own math writing? If so, try to work more slowly or write more neatly and make a conscious effort to carefully check your work.

Did you make a mistake because you don't understand a concept? Take the time to review the concept or ask questions to better understand it.

Did you skip too many steps? Skipping steps or trying to do too many steps mentally may lead to preventable mistakes.

Know how to get help if you need it. It's all right to ask for help. In fact, it's a good idea to ask for help whenever there is something that you don't understand. Make sure you know when your instructor has office hours and how to find his or her office. Find out whether math tutoring services are available on your campus. Check on the hours, location, and requirements of the tutoring service.

Don't be afraid to ask questions. You are not the only person in class with questions. Other students are normally grateful that someone has spoken up.

> **Helpful Hint**
>
> **MyMathLab® and MathXL®**
> Be aware of assignments and due dates set by your instructor. Don't wait until the last minute to submit work online.

Turn in assignments on time. This way, you can be sure that you will not lose points for being late. Show every step of a problem and be neat and organized. Also be sure that you understand which problems are assigned for homework. If allowed, you can always double-check the assignment with another student in your class.

OBJECTIVE

3 Knowing and Using Your Text

Flip through the pages of this text or view the e-text pages on a computer screen. Start noticing examples, exercise sets, end-of-chapter material, and so on. Every text is organized in some manner. Learn the way this text is organized by reading about and then finding an example in your text of each type of resource listed below. Finding and using these resources throughout your course will increase your chance of success.

- *Practice Exercises.* Each example in every section has a parallel Practice exercise. As you read a section, try each Practice exercise after you've finished the corresponding example. This "learn-by-doing" approach will help you grasp ideas before you move on to other concepts. Answers are at the back of the text.
- *Symbols at the Beginning of an Exercise Set.* If you need help with a particular section, the symbols listed at the beginning of each exercise set will remind you of the numerous resources available.
- *Objectives.* The main section of exercises in each exercise set is referenced by an example(s). There is also often a section of exercises entitled "Mixed Practice," which is referenced by two or more examples or sections. These are mixed exercises written to prepare you for your next exam. Use all of this referencing if you have trouble completing an assignment from the exercise set.
- *Icons (Symbols).* Make sure that you understand the meaning of the icons that are beside many exercises. ▶ tells you that the corresponding exercise may be viewed on the video segment that corresponds to that section. ✎ tells you that this exercise is a writing exercise in which you should answer in complete sentences. △ tells you that the exercise involves geometry. ▦ tells you that this exercise is worked more efficiently with the aid of a calculator. Also, a feature called Graphing Calculator Explorations may be found before select exercise sets.

- *Integrated Reviews.* Found in the middle of each chapter, these reviews offer you a chance to practice—in one place—the many concepts that you have learned separately over several sections.
- *End-of-Chapter Opportunities.* There are many opportunities at the end of each chapter to help you understand the concepts of the chapter.

  Vocabulary Checks contain key vocabulary terms introduced in the chapter.

  Chapter Highlights contain chapter summaries and examples.

  Chapter Reviews contain review exercises. The first part is organized section by section and the second part contains a set of mixed exercises.

  Getting Ready for the Tests contain conceptual exercises written to prepare students for chapter test directions as well as mixed sections of exercises.

  Chapter Tests are sample tests to help you prepare for an exam. The Chapter Test Prep Videos found in the Interactive Lecture Series, MyMathLab, and YouTube provide the video solution to each question on each Chapter Test.

  Cumulative Reviews start at Chapter 2 and are reviews consisting of material from the beginning of the book to the end of that particular chapter.
- *Student Resources in Your Textbook.* You will find a **Student Resources** section at the back of this textbook. It contains the following to help you study and prepare for tests:

  Study Skills Builders contain study skills advice. To increase your chance for success in the course, read these study tips and answer the questions.
- *Resources to Check Your Work.* The **Answers to Selected Exercises** section provides answers to all odd-numbered section exercises and odd-numbered review exercises and all integrated review, getting ready for the test and chapter test exercises.

OBJECTIVE

4 Knowing and Using Video and Notebook Organizer Resources

Video Resources

Below is a list of video resources that are all made by me—the author of your text, Elayn Martin-Gay. By making these videos, I can be sure that the methods presented are consistent with those in the text.

- *Interactive DVD Lecture Series.* Exercises marked with a ▶ are fully worked out by the author on the DVDs and within MyMathLab. The lecture series provides approximately 20 minutes of instruction per section and is organized by Objective.
- *Chapter Test Prep Videos.* These videos provide solutions to all of the Chapter Test exercises worked out by the author. They can be found in MyMathLab, the Interactive Lecture series, and YouTube. This supplement is very helpful before a test or exam.
- *Student Success Tips.* These video segments are about 3 minutes long and are daily reminders to help you continue practicing and maintaining good organizational and study habits.
- *Getting Ready for the Chapter Test Videos.* These video segments provide solutions to each question. These videos can be found within MyMathLab and the Interactive Lecture Series.

Helpful Hint

MyMathLab®

In MyMathLab, you have access to the following video resources:

- Lecture Videos for each section
- Getting Ready for the Chapter Test
- Chapter Test Prep Videos

Use these videos provided by the author to prepare for class, review, and study for tests.

Notebook Organizer Resource

This resource is in three-ring notebook-ready form. It is to be inserted in a three-ring binder and completed. This resource is numbered according to the sections in your text to which they refer.

- *Video Organizer.* This organizer is closely tied to the Interactive Lecture (Video) Series. Each section should be completed while watching the lecture video on the same section. Once completed, you will have a set of notes to accompany the Lecture (Video) Series section by section.

OBJECTIVE

5 Getting Help

If you have trouble completing assignments or understanding the mathematics, get help as soon as you need it! This tip is presented as an objective on its own because it is so important. In mathematics, usually the material presented in one section builds on your understanding of the previous section. This means that if you don't understand the concepts covered during a class period, there is a good chance that you will not understand the concepts covered during the next class period. If this happens to you, get help as soon as you can.

Where can you get help? Many suggestions have been made in this section on where to get help, and now it is up to you to get it. Try your instructor, a tutoring center, or a math lab, or you may want to form a study group with fellow classmates. If you do decide to see your instructor or go to a tutoring center, make sure that you have a neat notebook and are ready with your questions.

> **Helpful Hint**
>
> **MyMathLab® and MathXL®**
>
> - Use the **Help Me Solve This** button to get step-by-step help for the exercise you are working. You will need to work an additional exercise of the same type before you can get credit for having worked it correctly.
> - Use the **Video** button to view a video clip of the author working a similar exercise.

OBJECTIVE

6 Preparing For and Taking an Exam

Make sure that you allow yourself plenty of time to prepare for a test. If you think that you are a little "math anxious," it may be that you are not preparing for a test in a way that will ensure success. The way that you prepare for a test in mathematics is important. To prepare for a test:

> **Helpful Hint**
>
> **MyMathLab® and MathXL®**
>
> Review your written work for previous assignments. Then, go back and re-work previous assignments. Open a previous assignment, and click **Similar Exercise** to generate new exercises. Re-work the exercises until you fully understand them and can work them without help features.

1. Review your previous homework assignments.
2. Review any notes from class and section-level quizzes you have taken. (If this is a final exam, also review chapter tests you have taken.)
3. Review concepts and definitions by reading the Chapter Highlights at the end of each chapter.
4. Practice working out exercises by completing the Chapter Review and Getting Ready for the Test found at the end of each chapter. (If this is a final exam, go through a Cumulative Review. There is one found at the end of each chapter except Chapter 1. Choose the review found at the end of the latest chapter that you have covered in your course.) *Don't stop here!*
5. It is important that you place yourself in conditions similar to test conditions to find out how you will perform. In other words, as soon as you feel that you know the material, get a few blank sheets of paper and take a sample test. There is a Chapter Test available at the end of each chapter, or you can work selected problems from the Chapter Review. Your instructor may also provide you with a review sheet. During this sample test, do not use your notes or your textbook. Then check your sample test. If your sample test is the Chapter Test in the text, don't forget that the video solutions are in MyMathLab, the Interactive Lecture Series, and YouTube. If you are not satisfied with the results, study the areas that you are weak in and try again.
6. On the day of the test, allow yourself plenty of time to arrive where you will be taking your exam.

When taking your test:

1. Read the directions on the test carefully.
2. Read each problem carefully as you take the test. Make sure that you answer the question asked.
3. Watch your time and pace yourself so that you can attempt each problem on your test.
4. If you have time, check your work and answers.
5. Do not turn your test in early. If you have extra time, spend it double-checking your work.

OBJECTIVE

7 Managing Your Time

As a college student, you know the demands that classes, homework, work, and family place on your time. Some days you probably wonder how you'll ever get everything done. One key to managing your time is developing a schedule. Here are some hints for making a schedule:

1. Make a list of all your weekly commitments for the term. Include classes, work, regular meetings, extracurricular activities, etc. You may also find it helpful to list such things as laundry, regular workouts, grocery shopping, etc.
2. Next, estimate the time needed for each item on the list. Also make a note of how often you will need to do each item. Don't forget to include time estimates for the reading, studying, and homework you do outside of your classes. You may want to ask your instructor for help estimating the time needed.
3. In the exercise set that follows, you are asked to block out a typical week on the schedule grid given. Start with items with fixed time slots like classes and work.
4. Next, include the items on your list with flexible time slots. Think carefully about how best to schedule items such as study time.
5. Don't fill up every time slot on the schedule. Remember that you need to allow time for eating, sleeping, and relaxing! You should also allow a little extra time in case some items take longer than planned.
6. If you find that your weekly schedule is too full for you to handle, you may need to make some changes in your workload, classload, or other areas of your life. You may want to talk to your advisor, manager or supervisor at work, or someone in your college's academic counseling center for help with such decisions.

1.1 Exercise Set MyMathLab®

1. What is your instructor's name?
2. What are your instructor's office location and office hours?
3. What is the best way to contact your instructor?
4. Do you have the name and contact information of at least one other student in class?
5. Will your instructor allow you to use a calculator in this class?
6. Why is it important that you write step-by-step solutions to homework exercises and keep a hard copy of all work submitted?
7. Is there a tutoring service available on campus? If so, what are its hours? What services are available?
8. Have you attempted this course before? If so, write down ways that you might improve your chances of success during this next attempt.
9. List some steps that you can take if you begin having trouble understanding the material or completing an assignment. If you are completing your homework in MyMathLab® and MathXL®, list the resources you can use for help.
10. How many hours of studying does your instructor advise for each hour of instruction?
11. What does the ╲ icon in this text mean?
12. What does the △ icon in this text mean?
13. What does the ▶ icon in this text mean?
14. What are Practice exercises?
15. When might be the best time to work a Practice exercise?
16. Where are the answers to Practice exercises?
17. What answers are contained in this text and where are they?
18. What are Study Skills Builders and where are they?
19. What and where are Integrated Reviews?
20. How many times is it suggested that you work through the homework exercises in MathXL® before the submission deadline?
21. How far in advance of the assigned due date is it suggested that homework be submitted online? Why?
22. Chapter Highlights are found at the end of each chapter. Find the Chapter 1 Highlights and explain how you might use it and how it might be helpful.

23. Chapter Reviews are found at the end of each chapter. Find the Chapter 1 Review and explain how you might use it and how it might be useful.

24. Chapter Tests are at the end of each chapter. Find the Chapter 1 Test and explain how you might use it and how it might be helpful when preparing for an exam on Chapter 1. Include how the Chapter Test Prep Videos may help. If you are working in MyMathLab® and MathXL®, how can you use previous homework assignments to study?

25. Getting Ready for the Tests are at the end of each chapter. Find the Chapter 1 Getting Ready for the Test and explain how you might use it and how it might be helpful when preparing for an exam on Chapter 1.

26. What is the Video Organizer? Explain the contents and how it might be used.

27. Explain how the Video Organizer can help you when watching a lecture video.

28. Read or reread Objective 7 and fill out the schedule grid below.

	Monday	*Tuesday*	*Wednesday*	*Thursday*	*Friday*	*Saturday*	*Sunday*
1:00 a.m.							
2:00 a.m.							
3:00 a.m.							
4:00 a.m.							
5:00 a.m.							
6:00 a.m.							
7:00 a.m.							
8:00 a.m.							
9:00 a.m.							
10:00 a.m.							
11:00 a.m.							
Noon							
1:00 p.m.							
2:00 p.m.							
3:00 p.m.							
4:00 p.m.							
5:00 p.m.							
6:00 p.m.							
7:00 p.m.							
8:00 p.m.							
9:00 p.m.							
10:00 p.m.							
11:00 p.m.							
Midnight							

1.2 Symbols and Sets of Numbers

OBJECTIVES

1 Use a Number Line to Order Numbers.

2 Translate Sentences into Mathematical Statements.

3 Identify Natural Numbers, Whole Numbers, Integers, Rational Numbers, Irrational Numbers, and Real Numbers.

4 Find the Absolute Value of a Real Number.

5 Find the Opposite of a Real Number.

OBJECTIVE

1 Using a Number Line to Order Numbers

We begin with a review of the set of natural numbers and the set of whole numbers and how we use symbols to compare these numbers. A **set** is a collection of objects, each of which is called a **member** or **element** of the set. A pair of brace symbols { } encloses the list of elements and is translated as "the set of" or "the set containing."

Natural Numbers

The set of **natural numbers** is {1, 2, 3, 4, 5, 6, . . .}.

Whole Numbers

The set of **whole numbers** is {0, 1, 2, 3, 4, . . .}.

Helpful Hint

The three dots (an ellipsis) means that the list continues in the same manner indefinitely.

These numbers can be pictured on a **number line.** We will use number lines often to help us visualize distance and relationships between numbers.

To draw a number line, first draw a line. Choose a point on the line and label it 0. This point is also called the **origin.** To the right of 0, label any other point 1. Being careful to use the same distance as from 0 to 1, mark off equally spaced distances. Label these points 2, 3, 4, 5, and so on. Since the whole numbers continue indefinitely, it is not possible to show every whole number on this number line. The arrow at the right end of the line indicates that the pattern continues indefinitely. (See the number line drawn in the margin.)

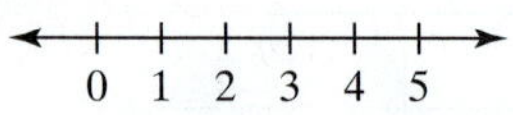

A Number Line

Picturing whole numbers on a number line helps us see the order of the numbers. Symbols can be used to describe concisely in writing the order that we see.

The **equal symbol** = means "is equal to."

The symbol ≠ means "is not equal to."

These symbols may be used to form a **mathematical statement.** The statement might be true or it might be false. The two statements below are both true.

$2 = 2$ states that "two is equal to two."

$2 \neq 6$ states that "two is not equal to six."

If two numbers are not equal, one number is larger than the other.
The symbol > means "is greater than."
The symbol < means "is less than." For example,

$3 < 5$ states that "three is less than five."

$2 > 0$ states that "two is greater than zero."

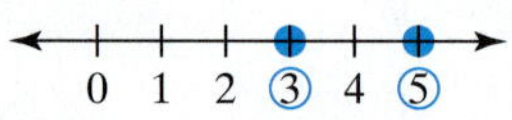

$3 < 5$

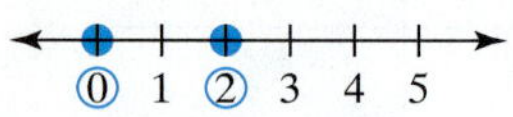

$2 > 0$ or $0 < 2$

On a number line, we see that a number **to the right of** another number is **larger.** Similarly, a number **to the left of** another number is smaller. For example, 3 is to the left of 5 on a number line, which means that 3 is less than 5, or $3 < 5$. Similarly, 2 is to the right of 0 on a number line, which means 2 is greater than 0, or $2 > 0$. Since 0 is to the left of 2, we can also say that 0 is less than 2, or $0 < 2$.

The symbols ≠, <, and > are called **inequality symbols.**

Helpful Hint

Notice that $2 > 0$ has exactly the same meaning as $0 < 2$. Switching the order of the numbers and reversing the direction of the inequality symbol does not change the meaning of the statement.

$3 < 5$ has the same meaning as $5 > 3$.

Also notice that, when the statement is true, the inequality arrow points to the smaller number.

EXAMPLE 1 Insert $<$, $>$, or $=$ in the space between each pair of numbers to make each statement true

a. 2 3 **b.** 7 4 **c.** 72 27

Solution

a. $2 < 3$ since 2 is to the left of 3 on a number line.

b. $7 > 4$ since 7 is to the right of 4 on a number line.

c. $72 > 27$ since 72 is to the right of 27 on a number line. □

PRACTICE 1 Insert $<$, $>$, or $=$ in the space between each pair of numbers to make each statement true.

a. 5 8 **b.** 6 4 **c.** 16 82 ■

Two other symbols are used to compare numbers.
The symbol $\leq$ means "is less than or equal to."
The symbol $\geq$ means "is greater than or equal to." For example,

$7 \leq 10$ states that "seven is less than or equal to ten."

This statement is true since $7 < 10$ is true. If either $7 < 10$ or $7 = 10$ is true, then $7 \leq 10$ is true.

$3 \geq 3$ states that "three is greater than or equal to three."

This statement is true since $3 = 3$ is true. If either $3 > 3$ or $3 = 3$ is true, then $3 \geq 3$ is true.

The statement $6 \geq 10$ is false since neither $6 > 10$ nor $6 = 10$ is true. The symbols $\leq$ and $\geq$ are also called **inequality symbols.**

EXAMPLE 2 Tell whether each statement is true or false.

a. $8 \geq 8$ **b.** $8 \leq 8$ **c.** $23 \leq 0$ **d.** $23 \geq 0$

Solution

a. True. Since $8 = 8$ is true, then $8 \geq 8$ is true.

b. True. Since $8 = 8$ is true, then $8 \leq 8$ is true.

c. False. Since neither $23 < 0$ nor $23 = 0$ is true, then $23 \leq 0$ is false.

d. True. Since $23 > 0$ is true, then $23 \geq 0$ is true. □

PRACTICE 2 Tell whether each statement is true or false.

a. $9 \geq 3$ **b.** $3 \geq 8$ **c.** $25 \leq 25$ **d.** $4 \leq 14$ ■

OBJECTIVE 2

Translating Sentences

Now, let's use the symbols discussed to translate sentences into mathematical statements.

EXAMPLE 3 Translate each sentence into a mathematical statement.

a. Nine is less than or equal to eleven.

b. Eight is greater than one.

c. Three is not equal to four.

(Continued on next page)

Solution

a. nine → 9; is less than or equal to → $\leq$; eleven → 11

b. eight → 8; is greater than → $>$; one → 1

c. three → 3; is not equal to → $\neq$; four → 4

□

PRACTICE

3 Translate each sentence into a mathematical statement.

a. Three is less than eight.

b. Fifteen is greater than or equal to nine.

c. Six is not equal to seven.

OBJECTIVE

3 Identifying Common Sets of Numbers

Whole numbers are not sufficient to describe many situations in the real world. For example, quantities less than zero must sometimes be represented, such as temperatures less than 0 degrees. The number line below shows numbers less than zero, or negative numbers. Also noted are numbers greater than zero, or positive numbers. The number 0 is neither positive nor negative.

> **Helpful Hint**
> 0 is neither a positive number nor a negative number.

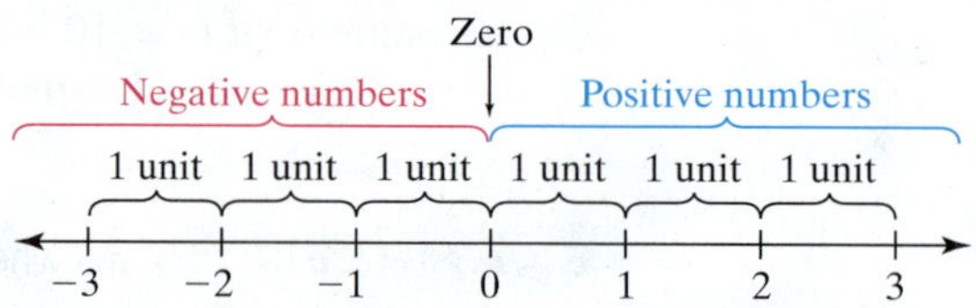

Notice that numbers less than 0 are to the left of 0 and are labeled $-1, -2, -3$, and so on. A "$-$" sign, such as the one in -1, tells us that the number is to the left of 0 on a number line. In words, -1 is read "negative one." A "$+$" sign or no sign tells us that a number lies to the right of 0 on a number line. For example, 3 and $+3$ both mean positive three.

A number is **graphed** on a number line by shading the point on the number line that corresponds to the number. Below we graph natural numbers and whole numbers, and we introduce and graph another set of numbers called integers.

> **Helpful Hint**
> A new set of numbers called integers are defined and graphed.

Natural numbers: $\{1, 2, 3, \ldots\}$

Whole numbers: $\{0, 1, 2, 3, \ldots\}$

Integers: $\{\ldots, -3, -2, -1, 0, 1, 2, 3, \ldots\}$

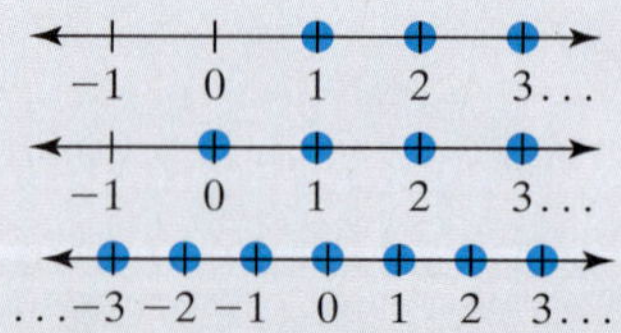

Each listing of three dots, ..., is called an **ellipsis** and means to continue in the same pattern.

EXAMPLE 4 Use an integer to express the number in the following. "Pole of Inaccessibility, Antarctica, is the coldest location in the world, with an average annual temperature of 72 degrees below zero." (*Source: The Guinness Book of Records*)

Solution The integer -72 represents 72 degrees below zero. □

PRACTICE 4 Use an integer to express the number in the following. The elevation of Laguna Salada in Mexico is 10 meters below sea level. (*Source: The World Almanac*) ■

A set can also be written in **set builder notation,** which describes the members of a set but does not list them. The following set is written in set builder notation.

$$\{x \mid x \text{ is a natural number less than } 3\}$$

The set of all numbers x — such that — x is a natural number less than 3

We can use set builder notation to describe three other common sets of numbers.

Identifying Numbers

Real numbers: $\{x \mid x \text{ corresponds to a point on a number line}\}$

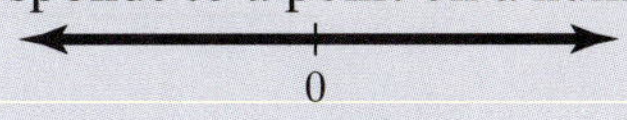

Rational numbers: $\left\{\frac{a}{b} \middle| a \text{ and } b \text{ are integers and } b \neq 0\right\}$

Irrational numbers: $\{x \mid x \text{ is a real number and } x \text{ is not a rational number}\}$

Helpful Hint

Notice from the definition that all real numbers are either rational or irrational.

Every rational number can be written as a decimal that either repeats in a pattern or terminates. For example,

Rational Numbers

$$7 = \frac{7}{1} \qquad \frac{5}{4} = 1.25$$

$$\frac{1}{2} = 0.5 \qquad \frac{2}{3} = 0.6666666\ldots = 0.\overline{6}$$

An irrational number written as a decimal neither terminates nor repeats. For example, π and $\sqrt{2}$ are irrational numbers. Their decimal form neither terminates nor repeats. Decimal approximations of each are below:

Irrational Numbers

$$\pi \approx 3.141592\ldots \qquad \sqrt{2} \approx 1.414213\ldots$$

Notice that every integer is also a rational number since each integer can be written as the quotient of itself and 1:

$$3 = \frac{3}{1}, \quad 0 = \frac{0}{1}, \quad -8 = \frac{-8}{1}$$

Not every rational number, however, is an integer. The rational number $\frac{2}{3}$, for example, is not an integer. Some square roots are rational numbers and some are irrational numbers. For example, $\sqrt{2}$, $\sqrt{3}$, and $\sqrt{7}$ are irrational numbers while $\sqrt{25}$ is a rational number because $\sqrt{25} = 5 = \frac{5}{1}$. The set of rational numbers together with the set of irrational numbers make up the set of real numbers. To help you make the distinction between rational and irrational numbers, here are a few examples of each.

Real Numbers			
Rational Numbers			***Irrational Numbers***
Examples		***Equivalent Quotient of Integers,*** $\frac{a}{b}$	***Examples***
$-\frac{2}{3}$	=	$\frac{-2}{3}$ or $\frac{2}{-3}$	$\sqrt{5}$
$\sqrt{25}$	=	$\frac{5}{1}$	$\frac{\sqrt{6}}{7}$
5	=	$\frac{5}{1}$	$-\sqrt{3}$
0	=	$\frac{0}{1}$	π
1.2	=	$\frac{12}{10}$	$\frac{2}{\sqrt{3}}$
$3\frac{7}{8}$	=	$\frac{31}{8}$	

Some rational and irrational numbers are graphed below.

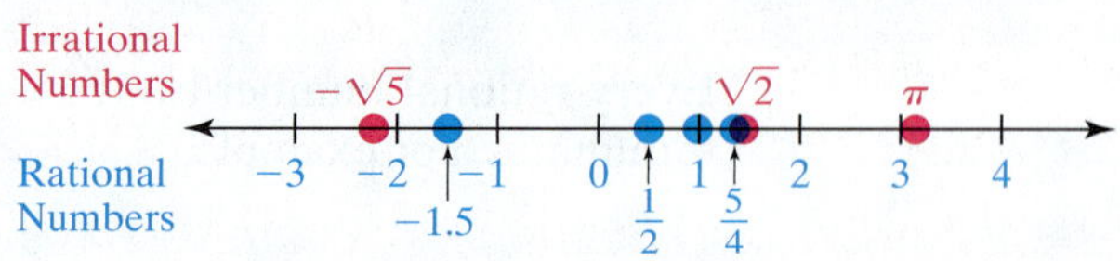

Helpful Hint

From our previous definitions, we have that

Every real number is either

a rational number or an irrational number

Recall that real numbers can be positive, negative, or 0. Numbers to the left of 0 are called **negative numbers;** numbers to the right of 0 are called **positive numbers.** Positive and negative numbers are also called **signed numbers.**

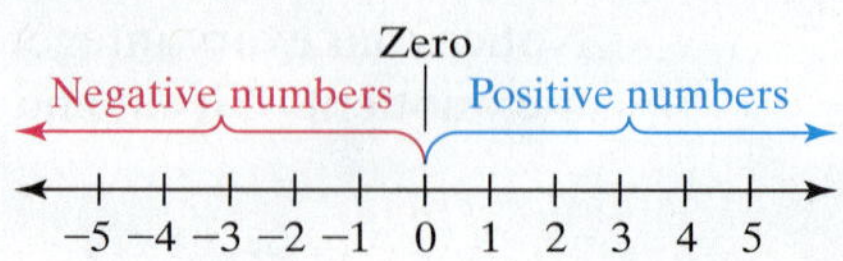

Several different sets of numbers have been discussed in this section. The following diagram shows the relationships among these sets of real numbers.

Common Sets of Numbers

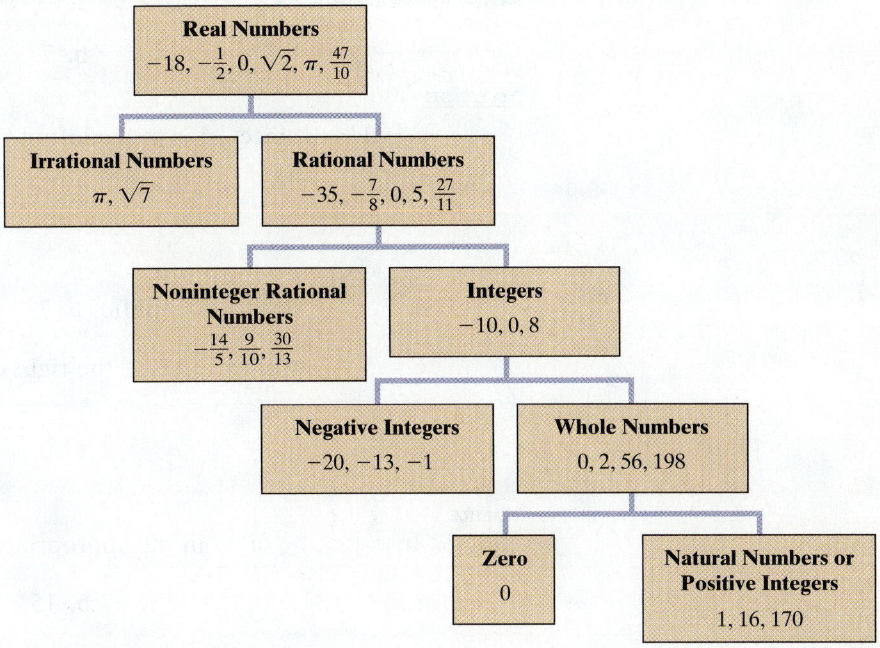

EXAMPLE 5 Given the set $\left\{-2, 0, \frac{1}{4}, -1.5, 112, -3, 11, \sqrt{2}\right\}$, list the numbers in this set that belong to the set of:

a. Natural numbers **b.** Whole numbers
c. Integers **d.** Rational numbers
e. Irrational numbers **f.** Real numbers

Solution

a. The natural numbers are 11 and 112.

b. The whole numbers are 0, 11, and 112.

c. The integers are −3, −2, 0, 11, and 112.

d. Recall that integers are rational numbers also. The rational numbers are $-3, -2, -1.5, 0, \frac{1}{4}$, 11, and 112.

e. The irrational number is $\sqrt{2}$.

f. The real numbers are all numbers in the given set. □

PRACTICE 5 Given the set $\left\{25, \frac{7}{3}, -15, -\frac{3}{4}, \sqrt{5}, -3.7, 8.8, -99\right\}$, list the numbers in this set that belong to the set of:

a. Natural numbers **b.** Whole numbers
c. Integers **d.** Rational numbers
e. Irrational numbers **f.** Real numbers ■

We now extend the meaning and use of inequality symbols such as $<$ and $>$ to all real numbers.

Order Property for Real Numbers

For any two real numbers a and b, a is less than b if a is to the left of b on a number line.

$a < b$ or also $b > a$

EXAMPLE 6 Insert <, >, or = in the appropriate space to make each statement true.

a. $-1 \quad 0$ **b.** $7 \quad \frac{14}{2}$ **c.** $-5 \quad -6$

Solution

a. $-1 < 0$ since -1 is to the left of 0 on a number line.

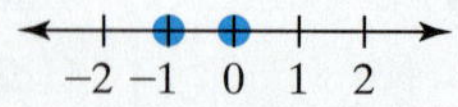

$-1 < 0$

b. $7 = \frac{14}{2}$ since $\frac{14}{2}$ simplifies to 7.

c. $-5 > -6$ since -5 is to the right of -6 on a number line.

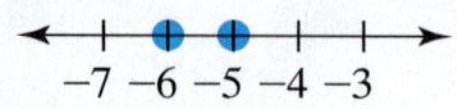

$-5 > -6$

PRACTICE 6 Insert <, >, or = in the appropriate space to make each statement true.

a. $0 \quad 3$ **b.** $15 \quad -5$ **c.** $3 \quad \frac{12}{4}$

OBJECTIVE 4

Finding the Absolute Value of a Real Number

A number line also helps us visualize the distance between numbers. The distance between a real number a and 0 is given a special name called the **absolute value** of a. "The absolute value of a" is written in symbols as $|a|$.

> **Absolute Value**
>
> The absolute value of a real number a, denoted by $|a|$, is the distance between a and 0 on a number line.

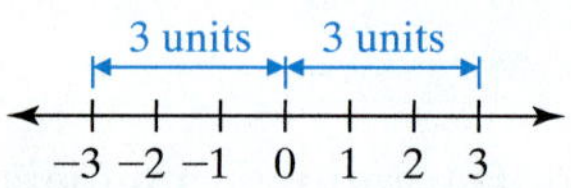

For example, $|3| = 3$ and $|-3| = 3$ since both 3 and -3 are a distance of 3 units from 0 on a number line.

> **Helpful Hint**
>
> Since $|a|$ is a distance, $|a|$ is always either positive or 0, never negative. That is, **for any real number a, $|a| \geq 0$.**

EXAMPLE 7 Find the absolute value of each number.

a. $|4|$ **b.** $|-5|$ **c.** $|0|$ **d.** $\left|-\frac{1}{2}\right|$ **e.** $|5.6|$

Solution

a. $|4| = 4$ since 4 is 4 units from 0 on a number line.

b. $|-5| = 5$ since -5 is 5 units from 0 on a number line.

c. $|0| = 0$ since 0 is 0 units from 0 on a number line.

d. $\left|-\frac{1}{2}\right| = \frac{1}{2}$ since $-\frac{1}{2}$ is $\frac{1}{2}$ unit from 0 on a number line.

e. $|5.6| = 5.6$ since 5.6 is 5.6 units from 0 on a number line.

PRACTICE 7 Find the absolute value of each number.

a. $|-8|$ **b.** $|9|$ **c.** $|-2.5|$ **d.** $\left|\frac{5}{11}\right|$ **e.** $|\sqrt{3}|$

EXAMPLE 8 Insert $<, >$, or $=$ in the appropriate space to make each statement true.

a. $|0| \quad 2$ **b.** $|-5| \quad 5$ **c.** $|-3| \quad |-2|$ **d.** $|5| \quad |6|$ **e.** $|-7| \quad |6|$

Solution

a. $|0| < 2$ since $|0| = 0$ and $0 < 2$. **b.** $|-5| = 5$ since $5 = 5$.

c. $|-3| > |-2|$ since $3 > 2$. **d.** $|5| < |6|$ since $5 < 6$.

e. $|-7| > |6|$ since $7 > 6$. □

PRACTICE 8 Insert $<, >$, or $=$ in the appropriate space to make each statement true.

a. $|8| \quad |-8|$ **b.** $|-3| \quad 0$ **c.** $|-7| \quad |-11|$ **d.** $|3| \quad |2|$ **e.** $|0| \quad |-4|$ ■

OBJECTIVE 5

Finding the Opposite of a Number

A number line can also help us visualize opposites. Two numbers that are the same distance from 0 on a number line but are on opposite sides of 0 are called **opposites.**

See the definition illustrated on the number lines below.

> **Helpful Hint**
> The opposite of 0 is 0.

The opposite of 6.5 is -6.5

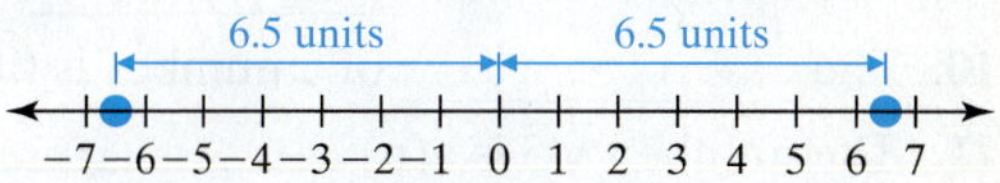

The opposite of $\frac{2}{3}$ is $-\frac{2}{3}$.

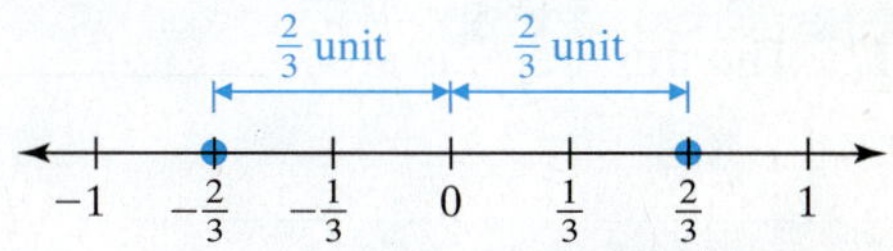

The opposite of -4 is 4.

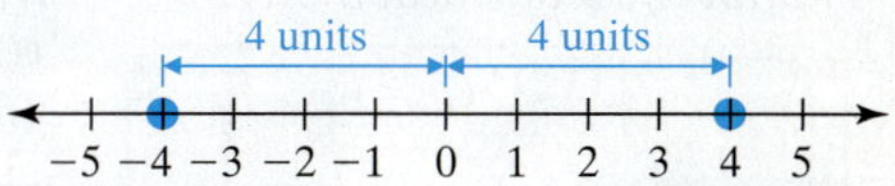

> **Helpful Hint**
> In Section 1.5, we learn that another name for opposite is additive inverse.

> **Opposite**
> The opposite of a number a is the number $-a$.

Above, we state that the opposite of a number a is $-a$. This means that the opposite of -4 is $-(-4)$. But from the number line above, the opposite of -4 is 4. This means that $-(-4) = 4$, and in general, we have the following property.

> **Double Negative Property**
> For every real number a, $-(-a) = a$.

EXAMPLE 9 Write the opposite of each number.

a. 8 **b.** $\frac{1}{5}$ **c.** -9.6

Solution

a. The opposite of 8 is -8.

b. The opposite of $\frac{1}{5}$ is $-\frac{1}{5}$.

c. The opposite of -9.6 is $-(-9.6) = 9.6$. □

PRACTICE 9 Write the opposite of each number.

a. 5.4 **b.** $-\frac{3}{5}$ **c.** 18 ■

Vocabulary, Readiness & Video Check

Use the choices below to fill in each blank. Some choices will be used more than once and some not at all.

real	natural	whole	irrational	a	absolute value
$\lvert b \rvert$	inequality	integers	rational	$-a$	

1. The ______________ numbers are $\{0, 1, 2, 3, 4, \ldots\}$.

2. The ______________ numbers are $\{1, 2, 3, 4, 5, \ldots\}$.

3. The symbols $\neq$, $\leq$, and $>$ are called ______________ symbols.

4. The ______________ are $\{\ldots, -3, -2, -1, 0, 1, 2, 3, \ldots\}$.

5. The ______________ numbers are {all numbers that correspond to points on a number line}.

6. The ______________ numbers are $\left\{\frac{a}{b} \middle| a \text{ and } b \text{ are integers}, b \neq 0\right\}$.

7. The ______________ numbers are {nonrational numbers that correspond to points on a number line}.

8. The distance between a number b and 0 on a number line is ______________.

9. The opposite of a is ______________.

10. The ______________ of a number is that number's distance from 0 on a number line.

11. The number $\sqrt{5}$ is a(n) ______________ number.

12. The number $\frac{5}{7}$ is a(n) ______________ number.

Martin-Gay Interactive Videos

See Video 1.2

Watch the section lecture video and answer the following questions.

OBJECTIVE 1 — **13.** In Example 2, why is the symbol $<$ inserted between the two numbers?

OBJECTIVE 2 — **14.** Write the sentence given in Example 4 and translate it to a mathematical statement, using symbols.

OBJECTIVE 3 — **15.** Which sets of numbers does the number in Example 6 belong to? Why is this number not an irrational number?

OBJECTIVE 4 — **16.** Complete this statement based on the lecture given before Example 8. The _____ of a real number a, denoted by $|a|$, is the distance between a and 0 on a number line.

OBJECTIVE 5 — **17.** Based on the lecture before Example 10, the description of opposite using a number line is given. Explain the difference between the absolute value of a number and the opposite of a number using the number line descriptions.

1.2 Exercise Set MyMathLab®

Insert $<$, $>$, or $=$ in the appropriate space to make the statement true. See Examples 1, 4 and 6.

1. 7 3

2. 9 15

3. 6.26 6.26

4. 2.13 1.13

5. 0 7

6. 20 0

7. −2 2

8. −4 −6

9. The freezing point of water is 32° Fahrenheit. The boiling point of water is 212° Fahrenheit. Write an inequality statement using $<$ or $>$ comparing the numbers 32 and 212.

10. The freezing point of water is 0° Celsius. The boiling point of water is 100° Celsius. Write an inequality statement using $<$ or $>$ comparing the numbers 0 and 100.

△ **11.** An angle measuring 30° is shown and an angle measuring 45° is shown. Use the inequality symbol $\leq$ or $\geq$ to write a statement comparing the numbers 30 and 45.

△ **12.** The sum of the measures of the angles of a triangle is 180°. The sum of the measures of the angles of a parallelogram is 360°. Use the inequality symbol $\leq$ or $\geq$ to write a statement comparing the numbers 360 and 180.

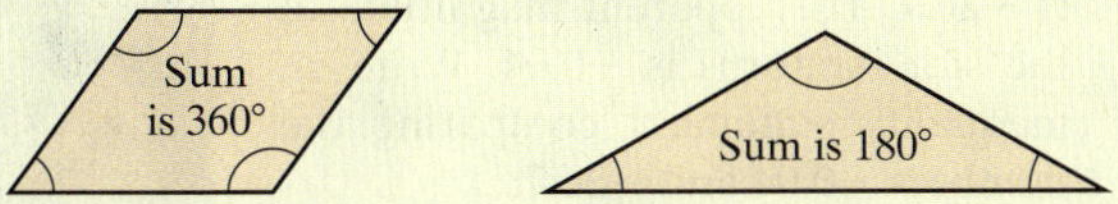

Are the following statements true or false? See Examples 2 and 6.

13. $11 \leq 11$ **14.** $4 \geq 7$
15. $10 > 11$ **16.** $17 > 16$
17. $3 + 8 \geq 3(8)$ **18.** $8 \cdot 8 \leq 8 \cdot 7$
19. $9 > 0$ **20.** $4 < 7$
21. $-6 > -2$ **22.** $0 < -15$

TRANSLATING

Write each sentence as a mathematical statement. See Example 3.

23. Eight is less than twelve.
24. Fifteen is greater than five.
25. Five is greater than or equal to four.
26. Negative ten is less than or equal to thirty-seven.
27. Fifteen is not equal to negative two.
28. Negative seven is not equal to seven.

Use integers to represent the values in each statement. See Example 4.

29. The highest elevation in California is Mt. Whitney, with an altitude of 14,494 feet. The lowest elevation in California is Death Valley, with an altitude of 282 feet below sea level. (*Source:* U.S. Geological Survey)

30. Driskill Mountain, in Louisiana, has an altitude of 535 feet. New Orleans, Louisiana, lies 8 feet below sea level. (*Source:* U.S. Geological Survey)

31. The number of graduate students at the University of Texas at Austin is 28,000 fewer than the number of undergraduate students. (*Source:* University of Texas at Austin)

32. The number of students admitted to the class of 2014 at UCLA is 80,784 fewer students than the number that had applied. (*Source:* UCLA)

33. Aaron Miller deposited \$350 in his savings account. He later withdrew \$126.

34. Aris Peña was deep-sea diving. During her dive, she ascended 30 feet and later descended 50 feet.

Tell which set or sets each number belongs to: natural numbers, whole numbers, integers, rational numbers, irrational numbers, and real numbers. See Example 5.

35. 0 **36.** $\frac{1}{4}$
37. -2 **38.** $-\frac{1}{2}$
39. 6 **40.** 5
41. $\frac{2}{3}$ **42.** $\sqrt{3}$
43. $-\sqrt{5}$ **44.** $-1\frac{5}{9}$

Tell whether each statement is true or false.

45. Every rational number is also an integer.
46. Every negative number is also a rational number.
47. Every natural number is positive.
48. Every rational number is also a real number.
49. 0 is a real number.
50. Every real number is also a rational number.
51. Every whole number is an integer.
52. $\frac{1}{2}$ is an integer.
53. A number can be both rational and irrational.
54. Every whole number is positive.

Insert $<$, $>$, or $=$ in the appropriate space to make a true statement. See Examples 6 through 8.

55. $-10 \quad -100$ **56.** $-200 \quad -20$
57. $32 \quad 5.2$ **58.** $7.1 \quad -7$
59. $\frac{18}{3} \quad \frac{24}{3}$ **60.** $\frac{8}{2} \quad \frac{12}{3}$
61. $-51 \quad -50$ **62.** $|-20| \quad -200$
63. $|-5| \quad -4$ **64.** $0 \quad |0|$
65. $|-1| \quad |1|$ **66.** $\left|\frac{2}{5}\right| \quad \left|-\frac{2}{5}\right|$
67. $|-2| \quad |-3|$ **68.** $-500 \quad |-50|$
69. $|0| \quad |-8|$ **70.** $|-12| \quad \frac{24}{2}$

Write the opposite of each number. See Example 9.

71. -6.2 **72.** -7.8
73. $\frac{4}{7}$ **74.** $\frac{9}{5}$
75. $-\frac{2}{3}$ **76.** $-\frac{14}{3}$
77. 0 **78.** 10.3

CONCEPT EXTENSIONS

Recall that the graph below is called a bar graph. This particular bar graph shows cranberry production from the top five cranberry-producing states.

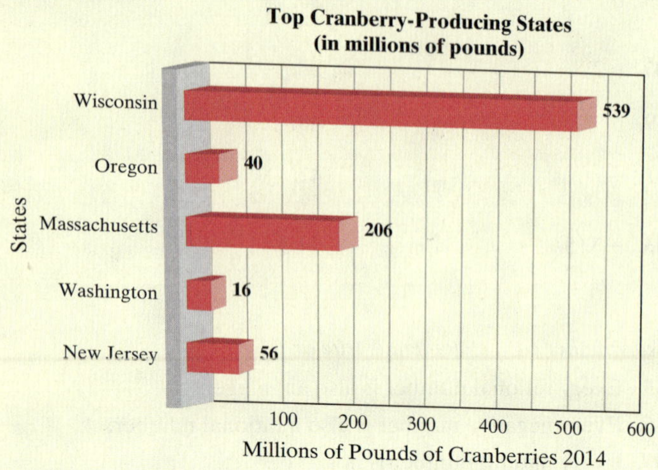

Data from National Agricultural Statistics Service

79. Write an inequality comparing the 2014 cranberry production in Oregon with the 2014 cranberry production in Washington.

80. Write an inequality comparing the 2014 cranberry production in Massachusetts with the 2014 cranberry production in Wisconsin.

81. Determine the difference between the 2014 cranberry production in Washington and the 2014 cranberry production in New Jersey.

82. According to the bar graph, which two states were the closest in terms of millions of pounds in 2014 cranberry crops?

This bar graph shows the number of people admitted into the Baseball Hall of Fame since its founding. Each bar represents a decade, and the height of the bar represents the number of Hall of Famers admitted in that decade.

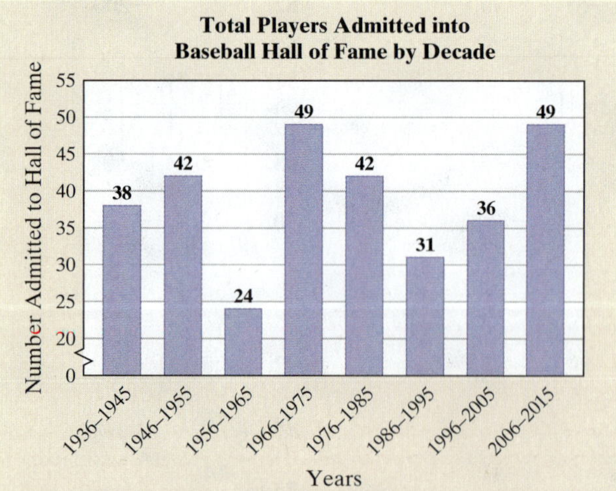

(*Note:* The symbol means that some numbers are missing. Along the vertical data line, notice the numbers between 0 and 20 are missing or not shown.)

Source: BASEBALL-Reference.com

83. In which decade(s) was the number of players admitted the greatest?

84. What was the greatest number of players admitted shown?

85. In which decade(s) was the number of players admitted greater than 40?

86. In which decade(s) was the number of players admitted fewer than 30?

87. Write an inequality statement comparing the number of players admitted in 1936–1945 and in 1966–1975.

88. Do you notice any trends shown by this bar graph?

The apparent magnitude of a star is the measure of its brightness as seen by someone on Earth. The smaller the apparent magnitude, the brighter the star. Use the apparent magnitudes in the table below to answer Exercises 89 through 94.

89. The apparent magnitude of the Sun is -26.7. The apparent magnitude of the star Arcturus is -0.04. Write an inequality statement comparing the numbers -0.04 and -26.7.

Star	*Apparent Magnitude*	*Star*	*Apparent Magnitude*
Arcturus	−0.04	Spica	0.98
Sirius	−1.46	Rigel	0.12
Vega	0.03	Regulus	1.35
Antares	0.96	Canopus	−0.72
Sun	−26.7	Hadar	0.61

(Data from *Norton's Star Atlas and Reference Handbook*, 20th Edition, edited by Ian Ridpath. © 2004 Pearson Education, Inc.)

90. The apparent magnitude of Antares is 0.96. The apparent magnitude of Spica is 0.98. Write an inequality statement comparing the numbers 0.96 and 0.98.

91. Which is brighter, the Sun or Arcturus?

92. Which is dimmer, Antares or Spica?

93. Which star listed is the brightest?

94. Which star listed is the dimmest?

Rewrite the following inequalities so that the inequality symbol points in the opposite direction and the resulting statement has the same meaning as the given one.

95. $25 \geq 20$

96. $-13 \leq 13$

97. $0 < 6$

98. $75 > 73$

99. $-10 > -12$

100. $-4 < -2$

101. In your own words, explain how to find the absolute value of a number.

102. Explain why $-(-2)$ and $-|-2|$ simplify to different numbers.

1.3 Exponents, Order of Operations, Variable Expressions, and Equations

OBJECTIVES

1. Define and Use Exponents and the Order of Operations.
2. Evaluate Algebraic Expressions, Given Replacement Values for Variables.
3. Determine Whether a Number Is a Solution of a Given Equation.
4. Translate Phrases into Expressions and Sentences into Statements.

OBJECTIVE

1 Using Exponents and the Order of Operations

Frequently in algebra, products occur that contain repeated multiplication of the same factor. For example, the volume of a cube whose sides each measure 2 centimeters is $(2 \cdot 2 \cdot 2)$ cubic centimeters. We may use **exponential notation** to write such products in a more compact form. For example,

$$2 \cdot 2 \cdot 2 \quad \textit{may be written as} \quad 2^3.$$

The 2 in 2^3 is called the **base;** it is the repeated factor. The 3 in 2^3 is called the **exponent** and is the number of times the base is used as a factor. The expression 2^3 is called an **exponential expression.**

exponent

base $2^3 = 2 \cdot 2 \cdot 2 = 8$

2 is a factor 3 times

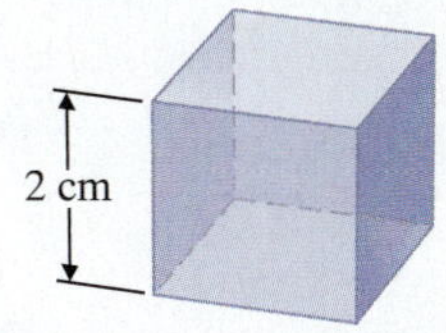

Volume is $(2 \cdot 2 \cdot 2)$ cubic centimeters.

EXAMPLE 1 Evaluate the following.

a. 3^2 [read as "3 squared" or as "3 to the second power"]
b. 5^3 [read as "5 cubed" or as "5 to the third power"]
c. 2^4 [read as "2 to the fourth power"]
d. 7^1 **e.** $\left(\frac{3}{7}\right)^2$

Solution

a. $3^2 = 3 \cdot 3 = 9$ **b.** $5^3 = 5 \cdot 5 \cdot 5 = 125$
c. $2^4 = 2 \cdot 2 \cdot 2 \cdot 2 = 16$ **d.** $7^1 = 7$
e. $\left(\frac{3}{7}\right)^2 = \left(\frac{3}{7}\right)\left(\frac{3}{7}\right) = \frac{9}{49}$ □

PRACTICE

1 Evaluate.

a. 1^3 **b.** 5^2 **c.** $\left(\frac{1}{10}\right)^2$ **d.** 9^1 **e.** $\left(\frac{2}{5}\right)^3$ ■

Helpful Hint

$2^3 \neq 2 \cdot 3$ since 2^3 indicates repeated **multiplication** of the same factor.

$$2^3 = 2 \cdot 2 \cdot 2 = 8, \text{ whereas } 2 \cdot 3 = 6.$$

Using symbols for mathematical operations is a great convenience. However, the more operation symbols present in an expression, the more careful we must be when performing the indicated operation. For example, in the expression $2 + 3 \cdot 7$, do we add first or multiply first? To eliminate confusion, **grouping symbols** are used. Examples of grouping symbols are parentheses ( ), brackets [ ], braces { }, and the fraction bar. If we wish $2 + 3 \cdot 7$ to be simplified by adding first, we enclose $2 + 3$ in parentheses.

$$(2 + 3) \cdot 7 = 5 \cdot 7 = 35$$

If we wish to multiply first, $3 \cdot 7$ may be enclosed in parentheses.

$$2 + (3 \cdot 7) = 2 + 21 = 23$$

To eliminate confusion when no grouping symbols are present, use the following agreed-upon order of operations.

Order of Operations

Simplify expressions using the order below.

If grouping symbols such as parentheses are present, simplify expressions within those first, starting with the innermost set. If fraction bars are present, simplify the numerator and the denominator separately.

1. Evaluate exponential expressions or absolute values in order from left to right.

2. Multiply or divide in order from left to right.

3. Add or subtract in order from left to right.

Now simplify $2 + 3 \cdot 7$. There are no grouping symbols and no exponents, so we multiply and then add.

$$2 + 3 \cdot 7 = 2 + 21 \quad \text{Multiply.}$$
$$= 23 \quad \text{Add.}$$

EXAMPLE 2 Simplify each expression.

a. $6 \div 3 + 5^2$ **b.** $20 \div 5 \cdot 4$ **c.** $\dfrac{2(12 + 3)}{|-15|}$ **d.** $3 \cdot 4^2$ **e.** $\dfrac{3}{2} \cdot \dfrac{1}{2} - \dfrac{1}{2}$

Solution

a. Evaluate 5^2 first.

$$6 \div 3 + 5^2 = 6 \div 3 + 25$$

Next divide, then add.

$$= 2 + 25 \quad \text{Divide.}$$
$$= 27 \quad \text{Add.}$$

b. $20 \div 5 \cdot 4 = 4 \cdot 4$

$= 16$

Helpful Hint

Remember to multiply or divide in order from left to right.

c. First, simplify the numerator and the denominator separately.

$$\frac{2(12 + 3)}{|-15|} = \frac{2(15)}{15} \quad \text{Simplify numerator and denominator separately.}$$
$$= \frac{30}{15}$$
$$= 2 \quad \text{Simplify.}$$

d. In this example, only the 4 is squared. The factor of 3 is not part of the base because no grouping symbol includes it as part of the base.

$$3 \cdot 4^2 = 3 \cdot 16 \quad \text{Evaluate the exponential expression.}$$
$$= 48 \quad \text{Multiply.}$$

e. The order of operations applies to operations with fractions in exactly the same way as it applies to operations with whole numbers.

$$\frac{3}{2} \cdot \frac{1}{2} - \frac{1}{2} = \frac{3}{4} - \frac{1}{2} \quad \text{Multiply.}$$
$$= \frac{3}{4} - \frac{2}{4} \quad \text{The least common denominator is 4.}$$
$$= \frac{1}{4} \quad \text{Subtract.}$$

□

PRACTICE 2 Simplify each expression.

a. $6 + 3 \cdot 9$ **b.** $4^3 \div 8 + 3$ **c.** $\left(\frac{2}{3}\right)^2 \cdot |-8|$

d. $\dfrac{9(14 - 6)}{|-2|}$ **e.** $\frac{7}{4} \cdot \frac{1}{4} - \frac{1}{4}$

Helpful Hint

Be careful when evaluating exponential expressions. In $3 \cdot 4^2$, the exponent 2 applies only to the base 4. In $(3 \cdot 4)^2$, the parentheses are a grouping symbol, so the exponent 2 applies to the product $3 \cdot 4$. Thus, we multiply first.

$$3 \cdot 4^2 = 3 \cdot 16 = 48 \qquad (3 \cdot 4)^2 = (12)^2 = 144$$

Expressions that include many grouping symbols can be confusing. When simplifying these expressions, keep in mind that grouping symbols separate the expression into distinct parts. Each is then simplified separately.

EXAMPLE 3 Simplify: $\dfrac{3 + |4 - 3| + 2^2}{6 - 3}$

Solution The fraction bar serves as a grouping symbol and separates the numerator and denominator. Simplify each separately. Also, the absolute value bars here serve as a grouping symbol. We begin in the numerator by simplifying within the absolute value bars.

$$\frac{3 + |4 - 3| + 2^2}{6 - 3} = \frac{3 + |1| + 2^2}{6 - 3}$$ Simplify the expression inside the absolute value bars.

$$= \frac{3 + 1 + 2^2}{3}$$ Find the absolute value and simplify the denominator.

$$= \frac{3 + 1 + 4}{3}$$ Evaluate the exponential expression.

$$= \frac{8}{3}$$ Simplify the numerator. □

PRACTICE 3 Simplify: $\dfrac{6^2 - 5}{3 + |6 - 5| \cdot 8}$

EXAMPLE 4 Simplify: $3[4 + 2(10 - 1)]$

Solution Notice that both parentheses and brackets are used as grouping symbols. Start with the innermost set of grouping symbols.

Helpful Hint

Be sure to follow order of operations and resist the temptation to incorrectly add 4 and 2 first.

$3[4 + 2(10 - 1)] = 3[4 + 2(9)]$ Simplify the expression in parentheses.

$= 3[4 + 18]$ Multiply.

$= 3[22]$ Add.

$= 66$ Multiply. □

PRACTICE 4 Simplify: $4[25 - 3(5 + 3)]$

EXAMPLE 5 Simplify: $\dfrac{8 + 2 \cdot 3}{2^2 - 1}$

Solution

$$\frac{8 + 2 \cdot 3}{2^2 - 1} = \frac{8 + 6}{4 - 1} = \frac{14}{3}$$ □

PRACTICE 5 Simplify: $\dfrac{36 \div 9 + 5}{5^2 - 3}$

OBJECTIVE

2 Evaluating Algebraic Expressions

In algebra, we use symbols, usually letters such as x, y, or z, to represent unknown numbers. A symbol that is used to represent a number is called a **variable.** An **algebraic expression** is a collection of numbers, variables, operation symbols, and grouping symbols. For example,

$$2x, \quad -3, \quad 2x + 10, \quad 5(p^2 + 1), \quad \text{and} \quad \frac{3y^2 - 6y + 1}{5}$$

are algebraic expressions.

Expression	*Meaning*
$2x$	$2 \cdot x$
$5(p^2 + 1)$	$5 \cdot (p^2 + 1)$
$3y^2$	$3 \cdot y^2$
xy	$x \cdot y$

If we give a specific value to a variable, we can **evaluate an algebraic expression.** To evaluate an algebraic expression means to find its numerical value once we know the values of the variables.

Algebraic expressions are often used in problem solving. For example, the expression

$$16t^2$$

gives the distance in feet (neglecting air resistance) that an object will fall in t seconds.

EXAMPLE 6 Evaluate each expression if $x = 3$ and $y = 2$.

a. $2x - y$ **b.** $\frac{3x}{2y}$ **c.** $\frac{x}{y} + \frac{y}{2}$ **d.** $x^2 - y^2$

Solution

a. Replace x with 3 and y with 2.

$$2x - y = 2(3) - 2 \quad \text{Let } x = 3 \text{ and } y = 2.$$
$$= 6 - 2 \quad \text{Multiply.}$$
$$= 4 \quad \text{Subtract.}$$

b. $\frac{3x}{2y} = \frac{3 \cdot 3}{2 \cdot 2} = \frac{9}{4}$ Let $x = 3$ and $y = 2$.

c. Replace x with 3 and y with 2. Then simplify.

$$\frac{x}{y} + \frac{y}{2} = \frac{3}{2} + \frac{2}{2} = \frac{5}{2}$$

d. Replace x with 3 and y with 2.

$$x^2 - y^2 = 3^2 - 2^2 = 9 - 4 = 5$$

□

PRACTICE

6 Evaluate each expression if $x = 2$ and $y = 5$.

a. $2x + y$ **b.** $\frac{4x}{3y}$ **c.** $\frac{3}{x} + \frac{x}{y}$ **d.** $x^3 + y^2$

■

OBJECTIVE

3 Determining Whether a Number Is a Solution of an Equation

Many times, a problem-solving situation is modeled by an equation. An **equation** is a mathematical statement that two expressions have equal value. An equal sign "=" is used to equate the two expressions. For example,

$$3 + 2 = 5, \quad 7x = 35, \quad \frac{2(x - 1)}{3} = 0, \text{ and } I = PRT \text{ are all equations.}$$

Helpful Hint

An equation contains an equal sign "=". An algebraic expression does not.

✔ CONCEPT CHECK

Which of the following are equations? Which are expressions?

a. $5x = 8$ **b.** $5x - 8$ **c.** $12y + 3x$ **d.** $12y = 3x$

When an equation contains a variable, deciding which values of the variable make the equation a true statement is called **solving** the equation for the variable. A **solution** of an equation is a value for the variable that makes the equation true. For example, 3 is a solution of the equation $x + 4 = 7$ because if x is replaced with 3, the statement is true.

$$x + 4 = 7$$
$$\downarrow$$
$$3 + 4 = 7 \quad \text{Replace } x \text{ with 3.}$$
$$7 = 7 \quad \text{True}$$

Similarly, 1 is not a solution of the equation $x + 4 = 7$ because $1 + 4 = 7$ is **not** a true statement.

EXAMPLE 7 Decide whether 2 is a solution of $3x + 10 = 8x$.

Solution Replace x with 2 and see if a true statement results.

$$3x + 10 = 8x \quad \text{Original equation}$$
$$3(2) + 10 \stackrel{?}{=} 8(2) \quad \text{Replace } x \text{ with 2.}$$
$$6 + 10 \stackrel{?}{=} 16 \quad \text{Simplify each side.}$$
$$16 = 16 \quad \text{True}$$

Since we arrived at a true statement after replacing x with 2 and simplifying both sides of the equation, 2 is a solution of the equation. □

PRACTICE 7 Decide whether 4 is a solution of $9x - 6 = 7x$. ■

OBJECTIVE 4

Translating Phrases to Expressions and Sentences to Statements

Now that we know how to represent an unknown number by a variable, let's practice translating phrases into algebraic expressions and sentences into statements. Oftentimes, solving problems requires the ability to translate word phrases and sentences into symbols. Below is a list of some key words and phrases to help us translate.

Helpful Hint

Order matters when subtracting and dividing, so be especially careful with these translations.

Addition (+)	***Subtraction*** (−)	***Multiplication*** (·)	***Division*** (÷)	***Equality*** (=)
Sum	Difference of	Product	Quotient	Equals
Plus	Minus	Times	Divide	Gives
Added to	Subtracted from	Multiply	Into	Is/Was/Should be
More than	Less than	Twice	Ratio	Yields
Increased by	Decreased by	Of	Divided by	Amounts to
Total	Less			Represents/Is the same as

Answers to Concept Check:
equations: a, d; expressions: b, c.

EXAMPLE 8 Write an algebraic expression that represents each phrase. Let the variable x represent the unknown number.

a. The sum of a number and 3 **b.** The product of 3 and a number

c. Twice a number **d.** 10 decreased by a number

e. 5 times a number, increased by 7

Solution

a. $x + 3$ since "sum" means to add

b. $3 \cdot x$ and $3x$ are both ways to denote the product of 3 and x

c. $2 \cdot x$ or $2x$

d. $10 - x$ because "decreased by" means to subtract

e. $\underbrace{5x}_{\text{5 times a number}} + 7$

□

PRACTICE 8 Write an algebraic expression that represents each phrase. Let the variable x represent the unknown number.

a. Six times a number **b.** A number decreased by 8

c. The product of a number and 9 **d.** Two times a number, plus 3

e. The sum of 7 and a number

■

Helpful Hint

Make sure you understand the difference when translating phrases containing "decreased by," "subtracted from," and "less than."

Phrase	*Translation*	
A number decreased by 10	$x - 10$	Notice the order.
A number subtracted from 10	$10 - x$	
10 less than a number	$x - 10$	
A number less 10	$x - 10$	

Now let's practice translating sentences into equations.

EXAMPLE 9 Write each sentence as an equation or inequality. Let x represent the unknown number.

a. The quotient of 15 and a number is 4.

b. Three subtracted from 12 is a number.

c. Four times a number, added to 17, is not equal to 21.

d. Triple a number is less than 48.

Solution

a. In words: the quotient of 15 and a number | is | 4

↓ ↓ ↓

Translate: $\frac{15}{x} = 4$

b. In words: three subtracted **from** 12 | is | a number

↓ ↓ ↓

Translate: $12 - 3 = x$

Care must be taken when the operation is subtraction. The expression $3 - 12$ would be incorrect. Notice that $3 - 12 \neq 12 - 3$.

c. In words: four times a number → added to → 17 → is not equal to → 21

Translate: $4x \quad + \quad 17 \quad \neq \quad 21$

d. In words: triple a number → is less than → 48

Translate: $3x \quad < \quad 48$

PRACTICE 9 Write each sentence as an equation or inequality. Let x represent the unknown number.

a. A number increased by 7 is equal to 13.

b. Two less than a number is 11.

c. Double a number, added to 9, is not equal to 25.

d. Five times 11 is greater than or equal to an unknown number.

Graphing Calculator Explorations

Exponents

To evaluate exponential expressions on a calculator, find the key marked [y^x] or [^]. To evaluate, for example, 3^5, press the following keys: [3] [y^x] [5] [=] or [3] [^] [5] [=] (or [ENTER]).

The display should read [243] or [3^5 243]

Order of Operations

Some calculators follow the order of operations, and others do not. To see whether your calculator has the order of operations built in, use your calculator to find $2 + 3 \cdot 4$. To do this, press the following sequence of keys:

[2] [+] [3] [×] [4] [=] (or [ENTER]).

The correct answer is 14 because the order of operations is to multiply before we add. If the calculator displays [14], then it has the order of operations built in.

Even if the order of operations is built in, parentheses must sometimes be inserted. For example, to simplify $\frac{5}{12 - 7}$, press the keys

[5] [÷] [(] [1] [2] [−] [7] [)] [=] (or [ENTER]).

The display should read [1] or [5/(12 − 7) 1]

Use a calculator to evaluate each expression.

1. 5^4 **2.** 7^4

3. 9^5 **4.** 8^6

5. $2(20 - 5)$ **6.** $3(14 - 7) + 21$

7. $24(862 - 455) + 89$ **8.** $99 + (401 + 962)$

9. $\dfrac{4623 + 129}{36 - 34}$ **10.** $\dfrac{956 - 452}{89 - 86}$

Vocabulary, Readiness & Video Check

Use the choices below to fill in each blank.

equation	variable	base	grouping
expression	solution	solving	exponent

1. In the expression 5^2, the 5 is called the __________ and the 2 is called the __________.

2. The symbols ( ), [ ], and { } are examples of __________ symbols.

3. A symbol that is used to represent a number is called a(n) __________.

4. A collection of numbers, variables, operation symbols, and grouping symbols is called a(n) __________.

5. A mathematical statement that two expressions are equal is called a(n) __________.

6. A value for the variable that makes an equation a true statement is called a(n) __________.

7. Deciding what values of a variable make an equation a true statement is called __________ the equation.

Martin-Gay Interactive Videos

See Video 1.3

Watch the section lecture video and answer the following questions.

OBJECTIVE 1

8. In Example 3 and the lecture before, what is the main point made about the order of operations?

OBJECTIVE 2

9. What happens with the replacement value for z in Example 6 and why?

OBJECTIVE 3

10. Is the value 0 a solution of the equation given in Example 9? How is this determined?

OBJECTIVE 4

11. Earlier in this video the point was made that equations have =, while expressions do not. In the lecture before Example 10, translating from English to math is discussed and another difference between expressions and equations is explained. What is it?

1.3 Exercise Set MyMathLab®

Evaluate. See Example 1.

1. 3^5 **2.** 2^5

3. 3^3 **4.** 4^4

5. 1^5 **6.** 1^8

7. 5^1 **8.** 8^1

9. 7^2 **10.** 9^2

11. $\left(\frac{2}{3}\right)^4$ **12.** $\left(\frac{6}{11}\right)^2$

13. $\left(\frac{1}{5}\right)^3$ **14.** $\left(\frac{1}{2}\right)^5$

15. $(1.2)^2$ **16.** $(1.5)^2$

17. $(0.04)^3$ **18.** $(0.03)^3$

MIXED PRACTICE

Simplify each expression. See Examples 2 through 5.

19. $5 + 6 \cdot 2$
20. $8 + 5 \cdot 3$
21. $4 \cdot 8 - 6 \cdot 2$
22. $12 \cdot 5 - 3 \cdot 6$
23. $2(8 - 3)$
24. $5(6 - 2)$
25. $2 + (5 - 2) + 4^2$
26. $6 - 2 \cdot 2 + 2^5$
27. $5 \cdot 3^2$
28. $2 \cdot 5^2$
29. $\frac{1}{4} \cdot \frac{2}{3} - \frac{1}{6}$
30. $\frac{3}{4} \cdot \frac{1}{2} + \frac{2}{3}$
31. $2[5 + 2(8 - 3)]$
32. $3[4 + 3(6 - 4)]$
33. $\frac{19 - 3 \cdot 5}{6 - 4}$
34. $\frac{4 \cdot 3 + 2}{4 + 3 \cdot 2}$
35. $\frac{|6 - 2| + 3}{8 + 2 \cdot 5}$
36. $\frac{15 - |3 - 1|}{12 - 3 \cdot 2}$
37. $\frac{3 + 3(5 + 3)}{3^2 + 1}$
38. $\frac{3 + 6(8 - 5)}{4^2 + 2}$
39. $\frac{6 + |8 - 2| + 3^2}{18 - 3}$
40. $\frac{16 + |13 - 5| + 4^2}{17 - 5}$
41. $2 + 3[10(4 \cdot 5 - 16) - 30]$
42. $3 + 4[8(5 \cdot 5 - 20) - 39]$
43. $\left(\frac{2}{3}\right)^3 + \frac{1}{9} + \frac{1}{3} \cdot \frac{4}{3}$
44. $\left(\frac{3}{8}\right)^2 + \frac{1}{4} + \frac{1}{8} \cdot \frac{3}{2}$

For Exercises 45 and 46, match each expression in the first column with its value in the second column.

45.

a. $(6 + 2) \cdot (5 + 3)$	19
b. $(6 + 2) \cdot 5 + 3$	22
c. $6 + 2 \cdot 5 + 3$	64
d. $6 + 2 \cdot (5 + 3)$	43

46.

a. $(1 + 4) \cdot 6 - 3$	15
b. $1 + 4 \cdot (6 - 3)$	13
c. $1 + 4 \cdot 6 - 3$	27
d. $(1 + 4) \cdot (6 - 3)$	22

Evaluate each expression when $x = 1, y = 3,$ and $z = 5$. See Example 6.

47. $3y$
48. $4x$
49. $\frac{z}{5x}$
50. $\frac{y}{2z}$
51. $3x - 2$
52. $6y - 8$
53. $|2x + 3y|$
54. $|5z - 2y|$
55. $xy + z$
56. $yz - x$
57. $5y^2$
58. $2z^2$

Evaluate each expression when $x = 12, y = 8,$ and $z = 4$. See Example 6.

59. $\frac{x}{z} + 3y$
60. $\frac{y}{z} + 8x$
61. $x^2 - 3y + x$
62. $y^2 - 3x + y$
63. $\frac{x^2 + z}{y^2 + 2z}$
64. $\frac{y^2 + x}{x^2 + 3y}$

Neglecting air resistance, the expression $16t^2$ gives the distance in feet an object will fall in t seconds.

65. Complete the chart below. To evaluate $16t^2$, remember to first find t^2, then multiply by 16.

Time t (in seconds)	*Distance $16t^2$ (in feet)*
1	
2	
3	
4	

66. Does an object fall the same distance *during* each second? Why or why not? (See Exercise 65.)

Decide whether the given number is a solution of the given equation. See Example 7.

67. Is 5 a solution of $3x + 30 = 9x$?
68. Is 6 a solution of $2x + 7 = 3x$?
69. Is 0 a solution of $2x + 6 = 5x - 1$?
70. Is 2 a solution of $4x + 2 = x + 8$?
71. Is 8 a solution of $2x - 5 = 5$?
72. Is 6 a solution of $3x - 10 = 8$?
73. Is 2 a solution of $x + 6 = x + 6$?
74. Is 10 a solution of $x + 6 = x + 6$?
75. Is 0 a solution of $x = 5x + 15$?
76. Is 1 a solution of $4 = 1 - x$?

TRANSLATING

Write each phrase as an algebraic expression. Let x represent the unknown number. See Example 8.

77. Fifteen more than a number
78. A number increased by 9
79. Five subtracted from a number
80. Five decreased by a number
81. The ratio of a number and 4
82. The quotient of a number and 9
83. Three times a number, increased by 22
84. Twice a number, decreased by 72

TRANSLATING

Write each sentence as an equation or inequality. Use x to represent any unknown number. See Example 9.

85. One increased by two equals the quotient of nine and three.
86. Four subtracted from eight is equal to two squared.
87. Three is not equal to four divided by two.
88. The difference of sixteen and four is greater than ten.
89. The sum of 5 and a number is 20.

90. Seven subtracted from a number is 0.

91. The product of 7.6 and a number is 17.

92. 9.1 times a number equals 4.

93. Thirteen minus three times a number is 13.

94. Eight added to twice a number is 42.

CONCEPT EXTENSIONS

Fill in each blank with one of the following:

add subtract multiply divide

95. To simplify the expression $1 + 3 \cdot 6$, first ________.

96. To simplify the expression $(1 + 3) \cdot 6$, first ________.

97. To simplify the expression $(20 - 4) \cdot 2$, first ________.

98. To simplify the expression $20 - 4 \div 2$, first ________.

99. Are parentheses necessary in the expression $2 + (3 \cdot 5)$? Explain your answer.

100. Are parentheses necessary in the expression $(2 + 3) \cdot 5$? Explain your answer.

△ *Recall that perimeter measures the distance around a plane figure and area measures the amount of surface of a plane figure. The expression* ***2l* + *2w*** *gives the* ***perimeter*** *of the* ***rectangle*** *below (measured in units), and the expression* ***lw*** *gives its* ***area*** *(measured in square units). Complete the chart below for the given lengths and widths. Be sure to include units.*

	Length: l	Width: w	Perimeter of Rectangle: $2l + 2w$	Area of Rectangle: lw
101.	4 in.	3 in.		
102.	6 in.	1 in.		
103.	5.3 in.	1.7 in.		
104.	4.6 in.	2.4 in.		

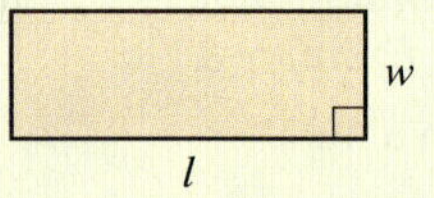

105. Study the perimeters and areas found in the chart on the previous page. Do you notice any trends?

106. In your own words, explain the difference between an expression and an equation.

107. Insert one set of parentheses so that the following expression simplifies to 32.

$$20 - 4 \cdot 4 \div 2$$

108. Insert one set of parentheses so that the following expression simplifies to 28.

$$2 \cdot 5 + 3^2$$

Determine whether each is an expression or an equation. See the Concept Check in this section.

109. **a.** $5x + 6$

b. $2a = 7$

c. $3a + 2 = 9$

d. $4x + 3y - 8z$

e. $5^2 - 2(6 - 2)$

110. **a.** $3x^2 - 26$

b. $3x^2 - 26 = 1$

c. $2x - 5 = 7x - 5$

d. $9y + x - 8$

e. $3^2 - 4(5 - 3)$

△ **111.** Why is 4^3 usually read as "four cubed"? (*Hint:* What is the volume of the **cube** below?)

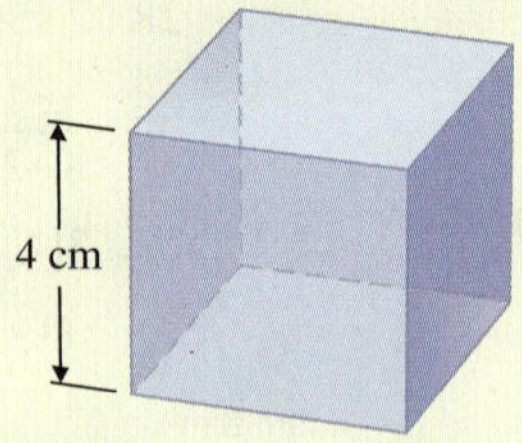

△ **112.** Why is 8^2 usually read as "eight squared"? (*Hint:* What is the area of the **square** below?)

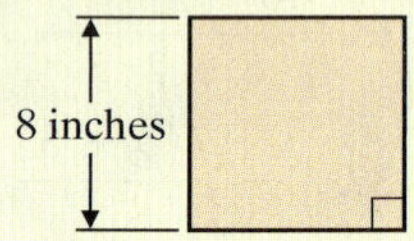

113. Write any expression, using 3 or more numbers, that simplifies to 11.

114. Write any expression, using 4 or more numbers, that simplifies to 7.

△ **115.** Recall that the area of a figure is the total enclosed surface of the figure. Area is measured in square units. The expression lw represents the area of a rectangle when l is its length and w is its width. Find the area of the following rectangular shaped lot.

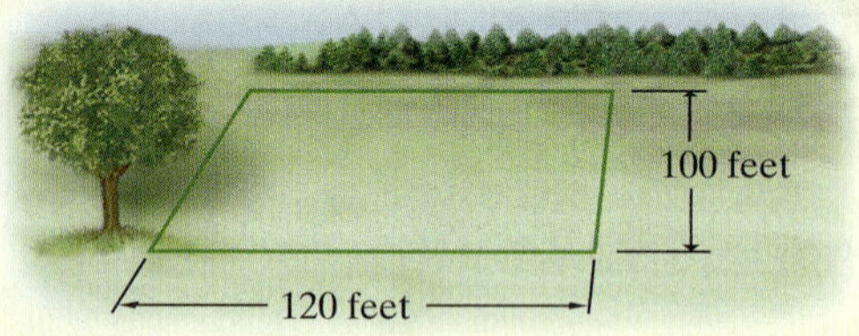

△ **116.** A trapezoid is a four-sided figure with exactly one pair of parallel sides. The expression $\frac{1}{2}h(B + b)$ represents its area, when B and b are the lengths of the two parallel sides and h is the height between these sides. Find the area if $B = 15$ inches, $b = 7$ inches, and $h = 5$ inches.

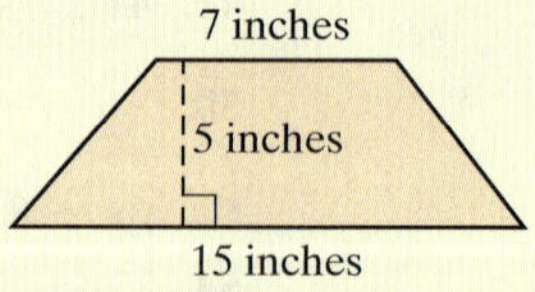

117. The expression $\frac{d}{t}$ represents the average speed in miles per hour if a distance of d miles is traveled in t hours. Find the speed to the nearest whole number if the distance between Dallas, Texas, and Kaw City, Oklahoma, is 432 miles, and it takes Peter Callac 8.5 hours to drive the distance.

118. The expression $\frac{I}{PT}$ represents the rate of interest being charged if a loan of P dollars for T years required I dollars in interest to be paid. Find the interest rate if a \$650 loan for 3 years to buy a used IBM personal computer requires \$126.75 in interest to be paid.

Integrated Review Order of Operations

Sections 1.2–1.3

Find the value of each expression when $x = 1, y = 3$, and $z = 4$.

1. z^2

2. y^3

3. $\frac{4x - z}{2y}$

4. $x(y + 2z)$

Perform the indicated operations.

5. $6 + 3 \cdot 5$

6. $4 + 7 \cdot 2$

7. $\frac{13}{2 - 2}$

8. $12^2 - 10^2$

9. $12 - 3^2$

10. $4 + 5(6 - 2)$

11. $9 + 2[(10 - 8)^2 + (3)^2]$

12. $8 - 6 + [4(2) + 2(5)]$

Answer the following with positive, negative, or 0.

13. The opposite of a positive number is a _______ number.

14. $-|-8|$ evaluates to a _______ number.

15. The absolute value of a negative number is a _______ number.

16. The absolute value of zero is _______.

17. The reciprocal of a positive number is a _______ number.

18. The opposite of 0 is _______.

19. The absolute value of a positive number is a _______ number.

20. The opposite of a negative number is a _______ number.

Fill in the chart:

	Number	*Opposite*	*Absolute Value*
21.	$\frac{1}{7}$		
22.	$-\frac{12}{5}$		
23.		-3	
24.		$\frac{9}{11}$	

1.4 Operations on Real Numbers and Evaluating Square Roots

OBJECTIVES

1. Add and Subtract Real Numbers.
2. Multiply and Divide Real Numbers.
3. Evaluate Expressions Containing Exponents.
4. Find Square Roots of Numbers.
5. Use the Order of Operations.
6. Evaluate Algebraic Expressions.

OBJECTIVE

1 Adding and Subtracting Real Numbers

When solving problems, we often have to add real numbers. For example, if the New Orleans Saints lose 5 yards in one play, then lose another 7 yards in the next play, their total loss may be described by $-5 + (-7)$.

The addition of two real numbers may be summarized by the following.

Adding Real Numbers

1. To add two numbers with the *same sign*, add their absolute values and attach their common sign.
2. To add two numbers with *different signs*, subtract the smaller absolute value from the larger absolute value and attach the sign of the number with the larger absolute value.

For example, to add $-5 + (-7)$, first add their absolute values.

$$|-5| = 5, \quad |-7| = 7, \quad \text{and} \quad 5 + 7 = 12$$

Next, attach their common negative sign.

$$-5 + (-7) = -12$$

(This represents a total loss of 12 yards for the New Orleans Saints in the example above.)

To find $(-4) + 3$, first subtract their absolute values.

$$|-4| = 4, \quad |3| = 3, \quad \text{and} \quad 4 - 3 = 1$$

Next, attach the sign of the number with the larger absolute value.

$$(-4) + 3 = -1$$

EXAMPLE 1 Add.

a. $-3 + (-11)$ b. $3 + (-7)$ c. $-10 + 15$

d. $-8.3 + (-1.9)$ e. $-\frac{1}{4} + \frac{1}{2}$ f. $-\frac{2}{3} + \frac{3}{7}$

Solution

a. $-3 + (-11) = -(3 + 11) = -14$ ← Add their absolute values: $3 + 11 = 14$; Use their common sign.

b. $3 + (-7) = -4$

c. $-10 + 15 = 5$

d. $-8.3 + (-1.9) = -10.2$

e. $-\frac{1}{4} + \frac{1}{2} = -\frac{1}{4} + \frac{1 \cdot 2}{2 \cdot 2} = -\frac{1}{4} + \frac{2}{4} = \frac{1}{4}$

f. $-\frac{2}{3} + \frac{3}{7} = -\frac{14}{21} + \frac{9}{21} = -\frac{5}{21}$ □

PRACTICE 1 Add.

a. $-6 + (-2)$ b. $5 + (-8)$ c. $-4 + 9$

d. $-3.2 + (-4.9)$ e. $-\frac{3}{5} + \frac{2}{3}$ f. $-\frac{5}{11} + \frac{3}{22}$ ■

Subtraction of two real numbers may be defined in terms of addition.

Subtracting Real Numbers

If a and b are real numbers,

$$a - b = a + (-b)$$

In other words, to subtract a real number, we add its opposite.

EXAMPLE 2 Subtract.

a. $2 - 8$ b. $-8 - (-1)$ c. $-11 - 5$ d. $10.7 - (-9.8)$

e. $-\frac{2}{3} - \frac{1}{2}$ f. $1 - 0.06$ g. Subtract 7 from 4.

Solution

a. $2 - 8 = 2 + (-8) = -6$ (Add the opposite) b. $-8 - (-1) = -8 + (1) = -7$ (Add the opposite)

c. $-11 - 5 = -11 + (-5) = -16$ d. $10.7 - (-9.8) = 10.7 + 9.8 = 20.5$

e. $-\frac{2}{3} - \frac{1}{2} = -\frac{2 \cdot 2}{3 \cdot 2} - \frac{1 \cdot 3}{2 \cdot 3} = -\frac{4}{6} + \left(-\frac{3}{6}\right) = -\frac{7}{6}$

f. $1 - 0.06 = 1 + (-0.06) = 0.94$ g. $4 - 7 = 4 + (-7) = -3$ □

Helpful Hint

Order is important when subtracting. "Subtract 7 from 4" translates to the order $4 - 7$ only.

PRACTICE 2 Subtract.

a. $3 - 11$ b. $-6 - (-3)$ c. $-7 - 5$ d. $4.2 - (-3.5)$

e. $-\frac{5}{7} - \frac{1}{3}$ f. $3 - 1.2$ g. Subtract 9 from 2. ■

To add or subtract three or more real numbers, add or subtract from left to right.

EXAMPLE 3 Simplify the following expressions.

a. $11 + 2 - 7$ **b.** $-5 - 4 + 2$

Solution

a. $11 + 2 - 7 = 13 - 7 = 6$ **b.** $-5 - 4 + 2 = -9 + 2 = -7$ □

PRACTICE

3 Simplify the following expressions.

a. $13 + 5 - 6$ **b.** $-6 - 2 + 4$ ■

OBJECTIVE

2 Multiplying and Dividing Real Numbers

To discover sign patterns when you multiply real numbers, recall that multiplication by a positive integer is the same as repeated addition. For example,

$$3(2) = 2 + 2 + 2 = 6$$
$$3(-2) = (-2) + (-2) + (-2) = -6$$

Notice here that $3(-2) = -6$. This illustrates that the product of two numbers with different signs is negative. We summarize sign patterns for multiplying any two real numbers as follows.

> **Multiplying Two Real Numbers**
>
> The product of two numbers with the *same* sign is positive.
>
> The product of two numbers with *different* signs is negative.

Also recall that the product of zero and any real number is zero.

$$0 \cdot a = 0$$

> **Product Property of 0**
>
> $0 \cdot a = 0$ Also $a \cdot 0 = 0$

EXAMPLE 4 Multiply.

a. $(-8)(-1)$ **b.** $-2\left(\frac{1}{6}\right)$ **c.** $-1.2(0.3)$ **d.** $0(-11)$

e. $\frac{1}{5}\left(-\frac{10}{11}\right)$ **f.** $(7)(1)(-2)(-3)$ **g.** $8(-2)(0)$

Solution

a. Since the signs of the two numbers are the same, the product is positive. Thus $(-8)(-1) = +8$, or 8.

b. Since the signs of the two numbers are different or unlike, the product is negative. Thus $-2\left(\frac{1}{6}\right) = -\frac{2}{6} = -\frac{1}{3}$.

c. $-1.2(0.3) = -0.36$

d. $0(-11) = 0$

e. $\frac{1}{5}\left(-\frac{10}{11}\right) = -\frac{10}{55} = -\frac{2}{11}$

f. To multiply three or more real numbers, you may multiply from left to right.

$$\begin{aligned}(7)(1)(-2)(-3) &= 7(-2)(-3)\\ &= -14(-3)\\ &= 42\end{aligned}$$

g. Since zero is a factor, the product is zero.

$$(8)(-2)(0) = 0$$ □

(Continued on next page)

PRACTICE 4 Multiply.

a. $(-5)(3)$ **b.** $(-7)\left(-\frac{1}{14}\right)$ **c.** $5.1(-2)$ **d.** $14(0)$

e. $\left(-\frac{1}{4}\right)\left(\frac{8}{13}\right)$ **f.** $6(-1)(-2)(3)$ **g.** $5(-2.3)$

Helpful Hint

The following sign patterns may be helpful when we are multiplying.

1. An odd number of negative factors gives a negative product.
2. An even number of negative factors gives a positive product.

Recall that $\frac{8}{4} = 2$ because $2 \cdot 4 = 8$. Likewise, $\frac{8}{-4} = -2$ because $(-2)(-4) = 8$. Also, $\frac{-8}{4} = -2$ because $(-2)4 = -8$, and $\frac{-8}{-4} = 2$ because $2(-4) = -8$. From these examples, we can see that the sign patterns for division are the same as for multiplication.

Dividing Two Real Numbers

The quotient of two numbers with the *same* sign is positive.

The quotient of two numbers with *different* signs is negative.

Recall from your knowledge of fractions that division by a nonzero real number b is the same as multiplication by $\frac{1}{b}$. In other words,

$$\frac{a}{b} = a \div b = a \cdot \frac{1}{b}$$

This means that to simplify $\frac{a}{b}$, we can divide by b or multiply by $\frac{1}{b}$. Recall from R.1 that the nonzero numbers b and $\frac{1}{b}$ are called **reciprocals.**

Notice that b *must* be a nonzero number. We do not define division by 0. For example, $5 \div 0$, or $\frac{5}{0}$, is undefined. To see why, recall that if $5 \div 0 = n$, a number, then $n \cdot 0 = 5$. This is not possible since $n \cdot 0 = 0$ for any number n and is never 5. Thus far, we have learned that we cannot divide 5 or any other nonzero number by 0.

Can we divide 0 by 0? By the same reasoning, if $0 \div 0 = n$, a number, then $n \cdot 0 = 0$. This is true for any number n so that the quotient $0 \div 0$ would not be a single number. To avoid this, we say that

Helpful Hint

If $n \neq 0$, $\frac{0}{n} = 0$ and $\frac{n}{0}$ is undefined.

Division by 0 is undefined.

EXAMPLE 5 Divide.

a. $\frac{20}{-4}$ **b.** $\frac{-9}{-3}$ **c.** $-\frac{3}{8} \div 3$ **d.** $\frac{-40}{10}$ **e.** $\frac{-1}{10} \div \frac{-2}{5}$ **f.** $\frac{8}{0}$

Solution

a. Since the signs are different or unlike, the quotient is negative and $\frac{20}{-4} = -5$.

b. Since the signs are the same, the quotient is positive and $\frac{-9}{-3} = 3$.

c. $-\frac{3}{8} \div 3 = -\frac{3}{8} \cdot \frac{1}{3} = -\frac{1}{8}$ **d.** $\frac{-40}{10} = -4$

e. $\frac{-1}{10} \div \frac{-2}{5} = -\frac{1}{10} \cdot -\frac{5}{2} = \frac{1}{4}$ **f.** $\frac{8}{0}$ is undefined. □

PRACTICE 5 Divide.

a. $\frac{-16}{8}$ **b.** $\frac{-15}{-3}$ **c.** $-\frac{2}{3} \div 4$

d. $\frac{54}{-9}$ **e.** $-\frac{1}{12} \div \left(-\frac{3}{4}\right)$ **f.** $\frac{0}{-7}$ ■

With sign rules for division, we can understand why the positioning of the negative sign in a fraction does not change the value of the fraction. For example,

$$\frac{-12}{3} = -4, \quad \frac{12}{-3} = -4, \quad \text{and} \quad -\frac{12}{3} = -4$$

Since all the fractions equal -4, we can say that

$$\frac{-12}{3} = \frac{12}{-3} = -\frac{12}{3}$$

In general, the following holds true.

> If a and b are real numbers and $b \neq 0$, then $\frac{a}{-b} = \frac{-a}{b} = -\frac{a}{b}$.

OBJECTIVE 3

Evaluating Expressions Containing Exponents

Now that we know how to perform operations on negative numbers, let's review exponents and evaluate exponential expressions with negative bases.

Exponents

If a is a real number and n is a natural number, then the ***n*th power of *a*,** or ***a* raised to the *n*th power,** written as a^n, is the product of n factors, each of which is a.

$$a^n = \underbrace{a \cdot a \cdot a \cdot a \cdot \cdots \cdot a}_{a \text{ is a factor } n \text{ times.}}$$

(exponent: n; base: a)

Recall that it is not necessary to write an exponent of 1. For example, 3 is assumed to be 3^1.

EXAMPLE 6 Evaluate each expression.

a. 3^2 **b.** $\left(\frac{1}{2}\right)^4$ **c.** -5^2

d. $(-5)^2$ **e.** -5^3 **f.** $(-5)^3$

Solution

a. $3^2 = 3 \cdot 3 = 9$ **b.** $\left(\frac{1}{2}\right)^4 = \left(\frac{1}{2}\right)\left(\frac{1}{2}\right)\left(\frac{1}{2}\right)\left(\frac{1}{2}\right) = \frac{1}{16}$

c. $-5^2 = -(5 \cdot 5) = -25$ **d.** $(-5)^2 = (-5)(-5) = 25$

e. $-5^3 = -(5 \cdot 5 \cdot 5) = -125$ **f.** $(-5)^3 = (-5)(-5)(-5) = -125$ □

PRACTICE 6 Evaluate each expression.

a. 2^3 **b.** $\left(\frac{1}{3}\right)^2$ **c.** -12^2

d. $(-12)^2$ **e.** -4^3 **f.** $(-4)^3$ ■

✔ CONCEPT CHECK

When $(-8.2)^7$ is evaluated, will the value be positive or negative? How can you tell without making any calculations?

Helpful Hint

Be very careful when simplifying expressions such as -5^2 and $(-5)^2$.

$$-5^2 = -(5 \cdot 5) = -25 \quad \text{and} \quad (-5)^2 = (-5)(-5) = 25$$

Without parentheses, the base to square is 5, not −5.

OBJECTIVE

4 Finding Square Roots of Numbers

The opposite of squaring a number is taking the **square root** of a number. For example, since the square of 4, or 4^2, is 16, we say that a square root of 16 is 4. The notation $\sqrt{a}$ denotes the **positive,** or **principal, square root** of a nonnegative number a. We then have in symbols that $\sqrt{16} = 4$. The negative square root of 16 is written $-\sqrt{16} = -4$. The square root of a negative number such as $\sqrt{-16}$ is not a real number. Why? There is no real number that, when squared, gives a negative number.

EXAMPLE 7 Find the square roots.

a. $\sqrt{9}$ **b.** $\sqrt{25}$ **c.** $\sqrt{\frac{1}{4}}$ **d.** $-\sqrt{36}$ **e.** $\sqrt{-36}$

Solution

a. $\sqrt{9} = 3$ since 3 is positive and $3^2 = 9$.

b. $\sqrt{25} = 5$ since $5^2 = 25$.

c. $\sqrt{\frac{1}{4}} = \frac{1}{2}$ since $\left(\frac{1}{2}\right)^2 = \frac{1}{4}$.

d. $-\sqrt{36} = -6$

e. $\sqrt{-36}$ is not a real number. □

PRACTICE

7 Find the square roots.

a. $\sqrt{49}$ **b.** $\sqrt{\frac{1}{16}}$ **c.** $-\sqrt{64}$ **d.** $\sqrt{-64}$ **e.** $\sqrt{100}$ ■

OBJECTIVE

5 Using the Order of Operations

To practice performing operations on negative numbers, let's once again simplify expressions. The same order of operations from the previous section is listed.

Order of Operations

Simplify expressions using the order that follows.

If grouping symbols such as parentheses are present, simplify expressions within those first, starting with the innermost set. If fraction bars are present, simplify the numerator and denominator separately.

1. Evaluate exponential expressions, roots, or absolute values in order from left to right.
2. Multiply or divide in order from left to right.
3. Add or subtract in order from left to right.

Answer to Concept Check:
negative; the exponent is an odd number

Helpful Hint

Fraction bars, radical signs, and absolute value bars can sometimes be used as grouping symbols. For example,

	Fraction Bar	*Radical Sign*	*Absolute Value Bars*
Grouping Symbol	$\frac{-1-7}{6-11}$	$\sqrt{15+1}$	$\lvert -7.2-\sqrt{4}\rvert$
Not Grouping Symbol	$-\frac{8}{9}$	$\sqrt{9}$	$\lvert -3.2\rvert$

EXAMPLE 8 Simplify.

a. $20 \div 2 \cdot 10$ **b.** $1 + 2(1-4)^2$ **c.** $\frac{|-2|^3+1}{-7-\sqrt{4}}$

Solution

a. Be careful! Here, we multiply or divide in order from left to right. Thus, divide, then multiply.

$$20 \div 2 \cdot 10 = 10 \cdot 10 = 100$$

b. Remember the order of operations so that you are *not* tempted to add 1 and 2 first.

$$\begin{aligned} 1+2(1-4)^2 &= 1+2(-3)^2 && \text{Simplify inside grouping symbols first.}\\ &= 1+2(9) && \text{Write } (-3)^2 \text{ as 9.}\\ &= 1+18 && \text{Multiply.}\\ &= 19 && \text{Add.} \end{aligned}$$

c. Simplify the numerator and the denominator separately; then divide.

$$\begin{aligned} \frac{|-2|^3+1}{-7-\sqrt{4}} &= \frac{2^3+1}{-7-2} && \text{Write } |-2| \text{ as 2 and } \sqrt{4} \text{ as 2.}\\ &= \frac{8+1}{-9} && \text{Write } 2^3 \text{ as 8.}\\ &= \frac{9}{-9} = -1 && \text{Simplify the numerator, then divide.} \end{aligned}$$

□

PRACTICE 8 Simplify.

a. $14 - 3 \cdot 4$ **b.** $3(5-8)^2$ **c.** $\frac{|-5|^2+4}{\sqrt{4}-3}$ ■

Besides parentheses, other symbols used for grouping expressions are brackets [ ] and braces { }. These other grouping symbols are commonly used when we group expressions that already contain parentheses.

EXAMPLE 9 Simplify: $3 - [(4-6) + 2(5-9)]$

Solution

$$\begin{aligned} 3-[(4-6)+2(5-9)] &= 3-[-2+2(-4)] && \text{Simplify within the innermost sets of parentheses.}\\ &= 3-[-2+(-8)]\\ &= 3-[-10]\\ &= 13 \end{aligned}$$

□

Helpful Hint

When grouping symbols occur within grouping symbols, remember to perform operations on the innermost set first.

PRACTICE 9 Simplify: $5 - [(3-5) + 6(2-4)]$ ■

EXAMPLE 10 Simplify: $\dfrac{-5\sqrt{30-5}+(-2)^2}{4^2+|7-10|}$

Solution Here, the fraction bar, radical sign, and absolute value bars serve as grouping symbols. Thus, we simplify within the radical sign and absolute value bars first, remembering to calculate above and below the fraction bar separately.

$$\frac{-5\sqrt{30-5}+(-2)^2}{4^2+|7-10|}=\frac{-5\sqrt{25}+(-2)^2}{4^2+|-3|}=\frac{-5\cdot 5+4}{16+3}=\frac{-25+4}{16+3}$$

$$=\frac{-21}{19}\text{ or }-\frac{21}{19}$$

□

PRACTICE 10 Simplify: $\dfrac{-2\sqrt{12+4}-(-3)^2}{6^2+|1-9|}$

✔ **CONCEPT CHECK**

True or false? If two people use the order of operations to simplify a numerical expression and neither makes a calculation error, it is not possible that they each obtain a different result. Explain.

OBJECTIVE

6 Evaluating Algebraic Expressions

Recall from Section 1.3 that an algebraic expression is formed by numbers and variables connected by the operations of addition, subtraction, multiplication, division, raising to powers, and/or taking roots. Also, if numbers are substituted for the variables in an algebraic expression and the operations performed, the result is called **the value of the expression** for the given replacement values. This entire process is called **evaluating an expression.**

EXAMPLE 11 Evaluate each expression when $x = 4$ and $y = -3$.

a. $3x - 7y$ **b.** $-2y^2$ **c.** $\dfrac{\sqrt{x}}{y} - \dfrac{y}{x}$

Solution For each expression, replace x with 4 and y with -3.

a. $3x - 7y = 3\cdot 4 - 7(-3)$ Let $x = 4$ and $y = -3$.

$= 12 - (-21)$ Multiply.

$= 12 + 21$ Write as an addition.

$= 33$ Add.

b. $-2y^2 = -2(-3)^2$ Let $y = -3$.

$= -2(9)$ Write $(-3)^2$ as 9.

$= -18$ Multiply.

Helpful Hint
In $-2(-3)^2$, the exponent 2 goes with the base of -3 only.

c. $\dfrac{\sqrt{x}}{y} - \dfrac{y}{x} = \dfrac{\sqrt{4}}{-3} - \dfrac{-3}{4}$

$= -\dfrac{2}{3} + \dfrac{3}{4}$ Write $\sqrt{4}$ as 2.

$= -\dfrac{2}{3}\cdot\dfrac{4}{4} + \dfrac{3}{4}\cdot\dfrac{3}{3}$ The LCD is 12.

$= -\dfrac{8}{12} + \dfrac{9}{12}$ Write each fraction with a denominator of 12.

$= \dfrac{1}{12}$ Add. □

Answer to Concept Check:
true; answers may vary

PRACTICE

11 Evaluate each expression when $x = 16$ and $y = -5$

a. $2x - 7y$ **b.** $-4y^2$ **c.** $\frac{\sqrt{x}}{y} - \frac{y}{x}$

Note: Sometimes variables such as x_1 and x_2 will be used in this book. The small 1 and 2 are called **subscripts.** The variable x_1 can be read as "x sub 1," and the variable x_2 can be read as "x sub 2." The important thing to remember is that they are two different variables. For example, if $x_1 = -5$ and $x_2 = 7$, then

$$x_1 - x_2 = -5 - 7 = -12.$$

EXAMPLE 12 The algebraic expression $\frac{5(x - 32)}{9}$ represents the equivalent temperature in degrees Celsius when x is the temperature in degrees Fahrenheit. Complete the following table by evaluating this expression at the given values of x.

Degrees Fahrenheit	x	−4	10	32
Degrees Celsius	$\frac{5(x-32)}{9}$			

Solution To complete the table, evaluate $\frac{5(x - 32)}{9}$ at each given replacement value.

When $x = -4$,

$$\frac{5(x-32)}{9} = \frac{5(-4-32)}{9} = \frac{5(-36)}{9} = -20$$

When $x = 10$,

$$\frac{5(x-32)}{9} = \frac{5(10-32)}{9} = \frac{5(-22)}{9} = -\frac{110}{9} \text{ or } -12\frac{2}{9}$$

When $x = 32$,

$$\frac{5(x-32)}{9} = \frac{5(32-32)}{9} = \frac{5 \cdot 0}{9} = 0$$

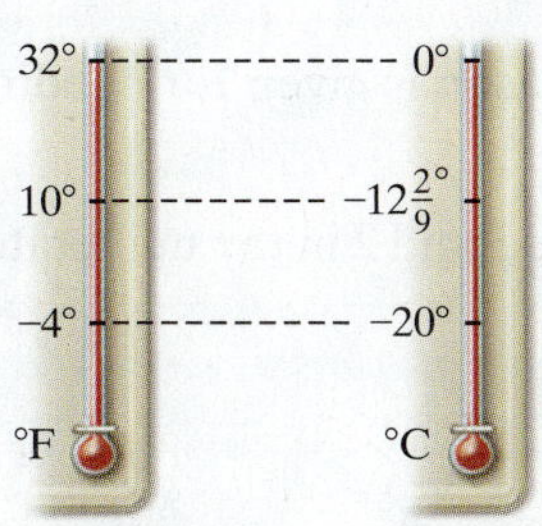

Degrees Fahrenheit Degrees Celsius

The completed table is

Degrees Fahrenheit	x	−4	10	32
Degrees Celsius	$\frac{5(x-32)}{9}$	−20	$-\frac{110}{9}$ or $-12\frac{2}{9}$	0

Thus, −4°F is equivalent to −20°C, 10°F is equivalent to $-\frac{110°}{9}$ C, and 32°F is equivalent to 0°C.

PRACTICE

12 The algebraic expression $\frac{9}{5}x + 32$ represents the equivalent temperature in degrees Fahrenheit when x is the temperature in degrees Celsius. Complete the following table by evaluating this expression at the given values of x.

Degrees Celsius	x	−5	10	25
Degrees Fahrenheit	$\frac{9}{5}x + 32$			

Vocabulary, Readiness & Video Check

Use the choices below to fill in each blank. Some choices may be used more than once and some used not at all.

exponent	undefined	base	1	$\frac{-a}{-b}$	$\frac{a}{b}$
square root	reciprocal	0	9	$\frac{-a}{b}$	$\frac{a}{-b}$

1. $0 \cdot a =$ ________.
2. $\frac{0}{4}$ simplifies to ________ while $\frac{4}{0}$ is ________.
3. The ________ of the nonzero number b is $\frac{1}{b}$.
4. The fraction $-\frac{a}{b} =$ ________ $=$ ________.
5. A(n) ________ is a shorthand notation for repeated multiplication of the same number.
6. In $(-5)^2$, the 2 is the ________ and the -5 is the ________.
7. The opposite of squaring a number is taking the ________ of a number.
8. Using order of operations, $9 \div 3 \cdot 3 =$ ________.

Martin-Gay Interactive Videos

See Video 1.4

Watch the section lecture video and answer the following questions.

OBJECTIVE 1
9. From Example 2, we learn that subtraction of two real numbers may be defined in terms of what operation?

OBJECTIVE 2
10. From Examples 3 and 4, explain the significance of the signs of numbers when multiplying and dividing two real numbers.

OBJECTIVE 3
11. Explain the significance of the use of parentheses when comparing Examples 7 and 8.

OBJECTIVE 4
12. From Example 9, what does the $\sqrt{\ }$ notation mean?

OBJECTIVE 5
13. In the lecture before Example 12, what reason is given for needing the order of operations?

OBJECTIVE 6
14. In Example 15, why can you not first add the 3 and 2 in the numerator?

1.4 Exercise Set MyMathLab®

Add or subtract as indicated. See Examples 1 through 3.

1. $-3 + 8$
2. $12 + (-7)$
3. $-14 + (-10)$
4. $-5 + (-9)$
5. $8 + (-7)$
6. $6 + (-4)$
7. $-14 + 2$
8. $-10 + 5$
9. $-2 + (-3)$
10. $-7 + (-4)$
11. $-16 + 16$
12. $23 + (-23)$
13. $-6 - 4$
14. $-12 - 8$
15. $4 - 9$
16. $8 - 11$
17. $16 - (-3)$
18. $12 - (-5)$
19. $-6 - (-11)$
20. $-4 - (-16)$
21. $-4.3 - 6.7$
22. $-8.2 - (-6.6)$
23. $13 - 17$
24. $15 - (-1)$
25. $\frac{11}{15} - \left(-\frac{3}{5}\right)$
26. $\frac{7}{10} - \frac{4}{5}$
27. $19 - 10 - 11$
28. $-13 - 4 + 9$
29. $-\frac{4}{5} - \left(-\frac{3}{10}\right)$
30. $-\frac{5}{2} - \left(-\frac{2}{3}\right)$

TRANSLATING

Translate each phrase to an expression and simplify.

31. Subtract −5 from 8.
32. Subtract 3 from −2.
33. Subtract 8 from 7.
34. Subtract 9 from −4.
35. Decrease −8 by 15.
36. Decrease 11 by −14.

Multiply or divide as indicated. See Examples 4 and 5.

37. $-5 \cdot 12$
38. $-3 \cdot 8$
39. $2(-1)$
40. $7(-4)$
41. $-5(-10)$
42. $-6(-11)$
43. $-17 \cdot 0$
44. $-5 \cdot 0$
45. $\frac{-9}{3}$
46. $\frac{-20}{5}$
47. $\frac{-12}{-4}$
48. $\frac{-36}{-6}$
49. $\frac{0}{-2}$
50. $\frac{-2}{0}$
51. $3\left(-\frac{1}{18}\right)$
52. $5\left(-\frac{1}{50}\right)$
53. $(-0.7)(-0.8)$
54. $(-0.9)(-0.5)$
55. $9.1 \div (-1.3)$
56. $22.5 \div (-2.5)$
57. $-4(-2)(-1)$
58. $-5(-3)(-2)$

Evaluate each expression. See Example 6.

59. -7^2
60. $(-7)^2$
61. $(-6)^2$
62. -6^2
63. $(-2)^3$
64. -2^3
65. $\left(-\frac{1}{3}\right)^3$
66. $\left(-\frac{1}{2}\right)^4$

Find the following roots. See Example 7.

67. $\sqrt{49}$
68. $\sqrt{81}$
69. $-\sqrt{\frac{4}{9}}$
70. $-\sqrt{\frac{4}{25}}$
71. $\sqrt{64}$
72. $\sqrt{121}$
73. $\sqrt{16}$
74. $\sqrt{1}$
75. $\sqrt{-100}$
76. $\sqrt{-25}$

MIXED PRACTICE

Simplify each expression. Write Exercises 97 and 98 as decimals. See Examples 1 through 10.

77. $3(5-7)^4$
78. $7(3-8)^2$
79. $-3^2 + 2^3$
80. $-5^2 - 2^4$
81. $\frac{3.1 - (-1.4)}{-0.5}$
82. $\frac{4.2 - (-8.2)}{-0.4}$
83. $(-3)^2 + 2^3$
84. $(-15)^2 - 2^4$
85. $-8 \div 4 \cdot 2$
86. $-20 \div 5 \cdot 4$
87. $-8\left(-\frac{3}{4}\right) - 8$
88. $-10\left(-\frac{2}{5}\right) - 10$
89. $\frac{8 + (-4)^2}{4 - 12}$
90. $\frac{6 + (-2)^2}{4 - 9}$
91. $\frac{-3 - 5^2}{2(-7)}$
92. $\frac{-2 - 4^2}{3(-6)}$
93. $\frac{6 - 2(-3)}{4 - 3(-2)}$
94. $\frac{8 - 3(-2)}{2 - 5(-4)}$
95. $\frac{|5-9| + |10-15|}{|2(-3)|}$
96. $\frac{|-3+6| + |-2+7|}{|-2 \cdot 2|}$
97. $\frac{(3 - \sqrt{9}) - (-5 - 1.3)}{-3}$
98. $\frac{-\sqrt{16} - (6 - 2.4)}{-2}$
99. $\frac{-4\sqrt{80+1} + (-4)^2}{3^3 + |-2(3)|}$
100. $\frac{(-2)^4 + 3\sqrt{120-20}}{4^3 + |5(-1)|}$

If $x = -5$ and $y = -3$, evaluate each expression. See Example 11.

101. $3x + 2y$
102. $4x + 5y$
103. $2x^2 - y^2$
104. $x^2 - 2y^2$
105. $x^3 + 3y$
106. $y^3 + 3x$
107. $\frac{2x - 5}{y - 2}$
108. $\frac{2y - 12}{x - 4}$
109. $\frac{-3 - y}{x - 4}$
110. $\frac{4 - 2x}{y + 3}$

Evaluate each expression when $x = 9$ and $y = -2$. See Example 11.

111. $\frac{\sqrt{x}}{y} - \frac{y}{x}$
112. $\frac{y}{2x} - \frac{\sqrt{x}}{3y}$
113. $\frac{3 + 2|x - y|}{x + 2y}$
114. $\frac{5 + 2|y - x|}{x + 6y}$

See Example 12.

△ **115.** The algebraic expression $8 + 2y$ represents the perimeter of a rectangle with width 4 and length y.

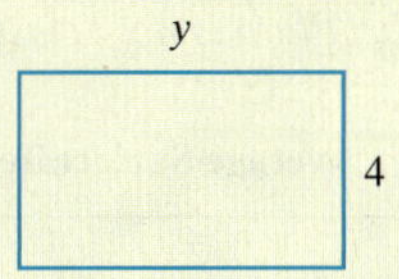

a. Complete the table that follows by evaluating this expression at the given values of y.

Length	y	5	7	10	100
Perimeter	$8 + 2y$				

b. Use the results of the table in part **a** to answer the following question. As the width of a rectangle remains the same and the length increases, does the perimeter increase or decrease? Explain how you arrived at your answer.

△ **116.** The algebraic expression πr^2 represents the area of a circle with radius r.

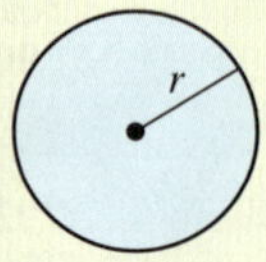

a. Complete the table below by evaluating this expression at given values of r. (Use 3.14 for π.)

Radius	r	2	3	7	10
Area	πr^2				

b. As the radius of a circle increases (see the table), does its area increase or decrease? Explain your answer.

117. The algebraic expression $\frac{100x + 5000}{x}$ represents the cost per bookshelf (in dollars) of producing x bookshelves.

a. Complete the table below.

Number of Bookshelves	x	10	100	1000
Cost per Bookshelf	$\frac{100x + 5000}{x}$			

b. As the number of bookshelves manufactured increases (see the table), does the cost per bookshelf increase or decrease? Why do you think that this is so?

118. If c is degrees Celsius, the algebraic expression $1.8c + 32$ represents the equivalent temperature in degrees Fahrenheit.

a. Complete the table below.

Degrees Celsius	c	−10	0	50
Degrees Fahrenheit	$1.8c + 32$			

b. As degrees Celsius increase (see the table), do degrees Fahrenheit increase or decrease?

CONCEPT EXTENSIONS

Study the bar graph below showing the average surface temperatures of planets. Use Exercises 119 and 120 to complete the planet temperatures on the graph. (Pluto is now classified as a dwarf planet.)

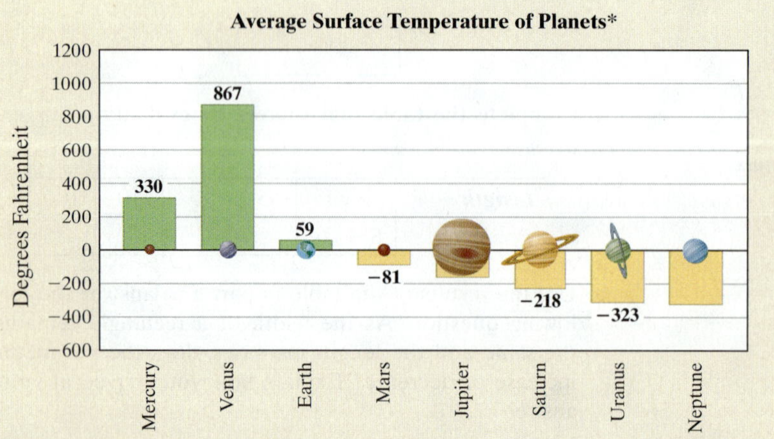

*For some planets, the temperature given is the temperature where the atmosphere pressure equals 1 Earth atmosphere; data from *The World Almanac*

119. The surface temperature of Jupiter is twice the temperature of Mars. Find this temperature.

120. The surface temperature of Neptune is equal to the temperature of Mercury divided by −1. Find this temperature.

Complementary angles have a sum of 90° and supplementary angles have a sum of 180°. Find each unknown complementary or supplementary angle.

△ **121.**

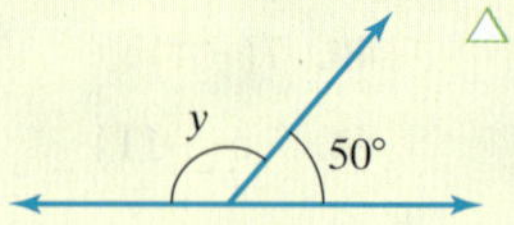

△ **122.**

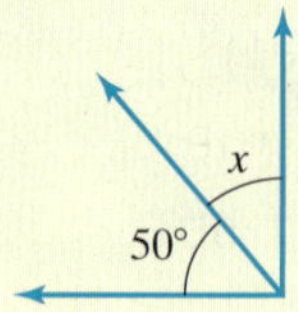

△ **123.**

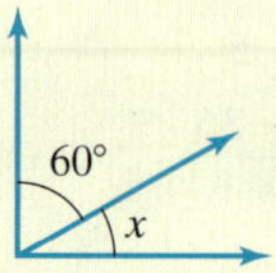

△ **124.**

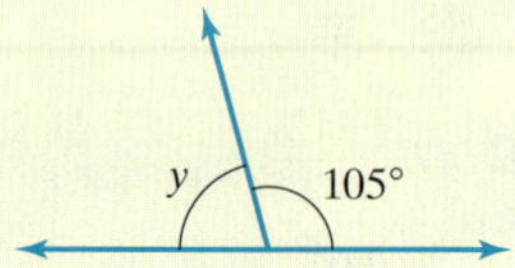

125. Explain why the product of an even number of negative numbers is a positive number.

126. If a and b are any real numbers, is the statement $a \cdot b = b \cdot a$ always true? Why or why not?

Each calculation below is incorrect. Find the error and correct it.

127. $7 + (-10) \stackrel{?}{=} 17$

128. $-4 + 14 \stackrel{?}{=} -18$

129. $-10 + (-12) \stackrel{?}{=} -120$

130. $-15 + (-17) \stackrel{?}{=} 32$

If a is a positive number and b is a negative number, fill in the blanks with the words positive or negative.

131. $-a$ is ________.

132. $-b$ is ________.

133. $a + a$ is ________.

134. $b + b$ is ________.

For Exercises 135 through 138, determine whether each statement is true or false.

135. The sum of two negative numbers is always a negative number.

136. The sum of two positive numbers is always a positive number.

137. The sum of a positive number and a negative number is always a negative number.

138. The sum of zero and a negative number is always a negative number.

If q is a negative number, r is a negative number, and t is a positive number, determine whether each expression simplifies to a positive or negative number. If it is not possible to determine, state so.

139. $q + t$

140. $t + r$

141. $t(q + r)$

142. $r(q - t)$

1.5 Properties of Real Numbers

OBJECTIVES

1 Use the Commutative and Associative Properties.

2 Use the Distributive Property.

3 Use the Identity and Inverse Properties.

OBJECTIVE 1

Using the Commutative and Associative Properties

In this section, we give names to properties of real numbers with which we are already familiar. Throughout this section, the variables a, b, and c represent real numbers.

We know that order does not matter when adding numbers. For example, we know that $7 + 5$ is the same as $5 + 7$. This property is given a special name—the **commutative property of addition.** We also know that order does not matter when multiplying numbers. For example, we know that $-5(6) = 6(-5)$. This property means that multiplication is commutative also and is called the **commutative property of multiplication.**

Commutative Properties

Addition: $a + b = b + a$

Multiplication: $a \cdot b = b \cdot a$

These properties state that the *order* in which any two real numbers are added or multiplied does not change their sum or product. For example, if we let $a = 3$ and $b = 5$, then the commutative properties guarantee that

$$3 + 5 = 5 + 3 \quad \text{and} \quad 3 \cdot 5 = 5 \cdot 3$$

Helpful Hint

Is subtraction also commutative? Try an example. Does $3 - 2 = 2 - 3$? **No!** The left side of this statement equals 1; the right side equals -1. There is no commutative property of subtraction. Similarly, there is no commutative property for division. For example, $10 \div 2$ does not equal $2 \div 10$.

EXAMPLE 1 Use a commutative property to complete each statement.

a. $x + 5 =$ ______ **b.** $3 \cdot x =$ ______

Solution

a. $x + 5 = 5 + x$ By the commutative property of addition

b. $3 \cdot x = x \cdot 3$ By the commutative property of multiplication

PRACTICE 1 Use a commutative property to complete each statement.

a. $x \cdot 8 =$ ______ **b.** $x + 17 =$ ______

✔ CONCEPT CHECK

Which of the following pairs of actions are commutative?

a. "raking the leaves" and "bagging the leaves"

b. "putting on your left glove" and "putting on your right glove"

c. "putting on your coat" and "putting on your shirt"

d. "reading a novel" and "reading a newspaper"

Let's now discuss grouping numbers. We know that when we add three numbers, the way in which they are grouped or associated does not change their sum. For example, we know that $2 + (3 + 4) = 2 + 7 = 9$. This result is the same if we group the numbers differently. In other words, $(2 + 3) + 4 = 5 + 4 = 9$ also. Thus, $2 + (3 + 4) = (2 + 3) + 4$. This property is called the **associative property of addition.**

We also know that changing the grouping of numbers when multiplying does not change their product. For example, $2 \cdot (3 \cdot 4) = (2 \cdot 3) \cdot 4$ (check it). This is the **associative property of multiplication.**

Answers to Concept Check:
b, d

Associative Properties

Addition:	$(a + b) + c = a + (b + c)$
Multiplication:	$(a \cdot b) \cdot c = a \cdot (b \cdot c)$

These properties state that the way in which three numbers are *grouped* does not change their sum or their product.

EXAMPLE 2 Use an associative property to complete each statement.

a. $5 + (4 + 6) =$ ______

b. $(-1 \cdot 2) \cdot 5 =$ ______

Solution

a. $5 + (4 + 6) = (5 + 4) + 6$ By the associative property of addition

b. $(-1 \cdot 2) \cdot 5 = -1 \cdot (2 \cdot 5)$ By the associative property of multiplication □

PRACTICE 2 Use an associative property to complete each statement.

a. $(2 + 9) + 7 =$ ______

b. $-4 \cdot (2 \cdot 7) =$ ______ ■

Helpful Hint

Remember the difference between the commutative properties and the associative properties. The commutative properties have to do with the *order* of numbers, and the associative properties have to do with the *grouping* of numbers.

Let's now illustrate how these properties can help us simplify expressions.

EXAMPLE 3 Simplify each expression.

a. $10 + (x + 12)$ **b.** $-3(7x)$

Solution

a. $10 + (x + 12) = 10 + (12 + x)$ By the commutative property of addition

$= (10 + 12) + x$ By the associative property of addition

$= 22 + x$ Add.

b. $-3(7x) = (-3 \cdot 7)x$ By the associative property of multiplication

$= -21x$ Multiply. □

PRACTICE 3 Simplify each expression.

a. $(5 + x) + 9$ **b.** $5(-6x)$ ■

OBJECTIVE 2

Using the Distributive Property

The **distributive property of multiplication over addition** is used repeatedly throughout algebra. It is useful because it allows us to write a product as a sum or a sum as a product.

We know that $7(2 + 4) = 7(6) = 42$. Compare that with $7(2) + 7(4) = 14 + 28 = 42$. Since both original expressions equal 42, they must equal each other, or

$$7(2 + 4) = 7(2) + 7(4)$$

This is an example of the distributive property. The product on the left side of the equal sign is equal to the sum on the right side. We can think of the 7 as being distributed to each number inside the parentheses.

Distributive Property of Multiplication Over Addition

$$a(b + c) = ab + ac$$

Since multiplication is commutative, this property can also be written as

$$(b + c)a = ba + ca$$

The distributive property can also be extended to more than two numbers inside the parentheses. For example,

$$\begin{aligned} 3(x + y + z) &= 3(x) + 3(y) + 3(z) \\ &= 3x + 3y + 3z \end{aligned}$$

Since we define subtraction in terms of addition, the distributive property is also true for subtraction. For example,

$$\begin{aligned} 2(x - y) &= 2(x) - 2(y) \\ &= 2x - 2y \end{aligned}$$

EXAMPLE 4 Use the distributive property to write each expression without parentheses. Then simplify if possible.

a. $2(x + y)$ **b.** $-5(-3 + 2z)$ **c.** $5(x + 3y - z)$

d. $-1(2 - y)$ **e.** $-(3 + x - w)$ **f.** $\frac{1}{2}(6x + 14) + 10$

Solution

a. $$\begin{aligned} 2(x + y) &= 2 \cdot x + 2 \cdot y \\ &= 2x + 2y \end{aligned}$$

b. $$\begin{aligned} -5(-3 + 2z) &= -5(-3) + (-5)(2z) \\ &= 15 - 10z \end{aligned}$$

c. $$\begin{aligned} 5(x + 3y - z) &= 5(x) + 5(3y) - 5(z) \\ &= 5x + 15y - 5z \end{aligned}$$

d. $$\begin{aligned} -1(2 - y) &= (-1)(2) - (-1)(y) \\ &= -2 + y \end{aligned}$$

Helpful Hint

Notice in part (**e**) that $-(3 + x - w)$ is first rewritten as $-1(3 + x - w)$.

e. $$\begin{aligned} -(3 + x - w) &= -1(3 + x - w) \\ &= (-1)(3) + (-1)(x) - (-1)(w) \\ &= -3 - x + w \end{aligned}$$

f. $\frac{1}{2}(6x + 14) + 10 = \frac{1}{2}(6x) + \frac{1}{2}(14) + 10$ Apply the distributive property.

$= 3x + 7 + 10$ Multiply.

$= 3x + 17$ Add. □

(Continued on next page)

PRACTICE 4 Use the distributive property to write each expression without parentheses. Then simplify if possible.

a. $5(x - y)$ **b.** $-6(4 + 2t)$

c. $2(3x - 4y - z)$ **d.** $(3 - y) \cdot (-1)$

e. $-(x - 7 + 2s)$ **f.** $\frac{1}{2}(2x + 4) + 9$

We can use the distributive property in reverse to write a sum as a product.

EXAMPLE 5 Use the distributive property to write each sum as a product.

a. $8 \cdot 2 + 8 \cdot x$ **b.** $7s + 7t$

Solution

a. $8 \cdot 2 + 8 \cdot x = 8(2 + x)$ **b.** $7s + 7t = 7(s + t)$

PRACTICE 5 Use the distributive property to write each sum as a product.

a. $5 \cdot w + 5 \cdot 3$ **b.** $9w + 9z$

OBJECTIVE

3 Using the Identity and Inverse Properties

Next, we look at the **identity properties.**

The number 0 is called the identity for addition because when 0 is added to any real number, the result is the same real number. In other words, the *identity* of the real number is not changed.

The number 1 is called the identity for multiplication because when a real number is multiplied by 1, the result is the same real number. In other words, the *identity* of the real number is not changed.

Identities for Addition and Multiplication

0 is the identity element for addition.

$$a + 0 = a \quad \text{and} \quad 0 + a = a$$

1 is the identity element for multiplication.

$$a \cdot 1 = a \quad \text{and} \quad 1 \cdot a = a$$

Notice that 0 is the *only* number that can be added to any real number with the result that the sum is the same real number. Also, 1 is the *only* number that can be multiplied by any real number with the result that the product is the same real number.

Opposites were introduced in Section 1.2. Another name for an opposite is an **additive inverse.** Two numbers are called additive inverses or opposites if their sum is 0. The additive inverse or opposite of 6 is -6 because $6 + (-6) = 0$. The additive inverse or opposite of -5 is 5 because $-5 + 5 = 0$.

Reciprocals or **multiplicative inverses** were introduced in Section R.1. Two nonzero numbers are called reciprocals or multiplicative inverses if their product is 1. The reciprocal or multiplicative inverse of $\frac{2}{3}$ is $\frac{3}{2}$ because $\frac{2}{3} \cdot \frac{3}{2} = 1$. Likewise, the reciprocal of -5 is $-\frac{1}{5}$ because $-5\left(-\frac{1}{5}\right) = 1$.

✔ CONCEPT CHECK

Which of the following, $1, -\frac{10}{3}, \frac{3}{10}, 0, \frac{10}{3}, -\frac{3}{10}$, is the

a. opposite of $-\frac{3}{10}$? **b.** reciprocal of $-\frac{3}{10}$?

Additive or Multiplicative Inverses

The numbers a and $-a$ are additive inverses or opposites of each other because their sum is 0; that is,

$$a + (-a) = 0$$

The numbers b and $\frac{1}{b}$ (for $b \neq 0$) are reciprocals or multiplicative inverses of each other because their product is 1; that is,

$$b \cdot \frac{1}{b} = 1$$

EXAMPLE 6 Name the property or properties illustrated by each true statement.

Solution

a. $3 \cdot y = y \cdot 3$ — commutative property of multiplication (order changed)

b. $(x + 7) + 9 = x + (7 + 9)$ — associative property of addition (grouping changed)

c. $(b + 0) + 3 = b + 3$ — identity element for addition

d. $0.2 \cdot (z \cdot 5) = 0.2 \cdot (5 \cdot z)$ — commutative property of multiplication (order changed)

e. $-2 \cdot \left(-\frac{1}{2}\right) = 1$ — multiplicative inverse property

f. $-2 + 2 = 0$ — additive inverse property

g. $-6 \cdot (y \cdot 2) = (-6 \cdot 2) \cdot y$ — commutative and associative properties of multiplication (order and grouping changed) □

PRACTICE 6 Name the property or properties illustrated by each true statement.

a. $(7 \cdot 3x) \cdot 4 = (3x \cdot 7) \cdot 4$ — commutative property of multiplication

b. $6 + (3 + y) = (6 + 3) + y$ — associative property of addition

c. $8 + (t + 0) = 8 + t$ — identity element for addition

d. $-\frac{3}{4} \cdot \left(-\frac{4}{3}\right) = 1$ — multiplicative inverse property

e. $(2 + x) + 5 = 5 + (2 + x)$ — commutative property of addition

f. $3 + (-3) = 0$ — additive inverse property

g. $(-3b) \cdot 7 = (-3 \cdot 7) \cdot b$ — commutative and associative properties of multiplication

Answers to Concept Check:

a. $\frac{3}{10}$ **b.** $-\frac{10}{3}$

Vocabulary, Readiness & Video Check

Use the choices below to fill in each blank.

distributive property	associative property of multiplication	commutative property of addition
opposites or additive inverses	associative property of addition	
reciprocals or multiplicative inverses	commutative property of multiplication	

1. $x + 5 = 5 + x$ is a true statement by the ______________________.

2. $x \cdot 5 = 5 \cdot x$ is a true statement by the ______________________.

3. $3(y + 6) = 3 \cdot y + 3 \cdot 6$ is true by the ________________.

4. $2 \cdot (x \cdot y) = (2 \cdot x) \cdot y$ is a true statement by the ______________________.

5. $x + (7 + y) = (x + 7) + y$ is a true statement by the ______________________.

6. The numbers $-\frac{2}{3}$ and $-\frac{3}{2}$ are called ______________________.

7. The numbers $-\frac{2}{3}$ and $\frac{2}{3}$ are called ____________________.

Martin-Gay Interactive Videos

See Video 1.5

Watch the section lecture video and answer the following questions.

OBJECTIVE 1

8. The commutative properties are discussed in Examples 1 and 2 and the associative properties are discussed in Examples 3–7. What's the one word used again and again to describe the commutative property? The associative property?

OBJECTIVE 2

9. In Example 10, what point is made about the term 2?

OBJECTIVE 3

10. Complete these statements based on the lecture given before Example 12.

- The identity element for addition is _____ because if we add _____ to any real number, the result is that real number.
- The identity element for multiplication is ______ because any real number times _____ gives a result of that original real number.

1.5 Exercise Set MyMathLab®

Use a commutative property to complete each statement. See Example 1.

1. $x + 16 =$ ______ **2.** $4 + y =$ ______

3. $-4 \cdot y =$ ______ **4.** $-2 \cdot x =$ ______

5. $xy =$ ______ **6.** $ab =$ ______

7. $2x + 13 =$ ______ **8.** $19 + 3y =$ ______

Use an associative property to complete each statement. See Example 2.

9. $(xy) \cdot z =$ ______ **10.** $3 \cdot (xy) =$ ______

11. $2 + (a + b) =$ ________ **12.** $(y + 4) + z =$ ________

13. $4 \cdot (ab) =$ ______ **14.** $(-3y) \cdot z =$ ________

15. $(a + b) + c =$ ________

16. $6 + (r + s) =$ ________

Use the commutative and associative properties to simplify each expression. See Example 3.

17. $8 + (9 + b)$ **18.** $(r + 3) + 11$

19. $4(6y)$ **20.** $2(42x)$

21. $\frac{1}{5}(5y)$ **22.** $\frac{1}{8}(8z)$

23. $(13 + a) + 13$ **24.** $7 + (x + 4)$

25. $-9(8x)$ **26.** $-3(12y)$

27. $\frac{3}{4}\left(\frac{4}{3}s\right)$ **28.** $\frac{2}{7}\left(\frac{7}{2}r\right)$

29. $\frac{2}{3} + \left(\frac{4}{3} + x\right)$ **30.** $\frac{7}{9} + \left(\frac{2}{9} + y\right)$

Use the distributive property to write each expression without parentheses. Then simplify the result. See Example 4.

31. $4(x + y)$
32. $7(a + b)$
33. $9(x - 6)$
34. $11(y - 4)$
35. $2(3x + 5)$
36. $5(7 + 8y)$
37. $7(4x - 3)$
38. $3(8x - 1)$
39. $3(6 + x)$
40. $2(x + 5)$
41. $-2(y - z)$
42. $-3(z - y)$
43. $-7(3y + 5)$
44. $-5(2r + 11)$
45. $5(x + 4m + 2)$
46. $8(3y + z - 6)$
47. $-4(1 - 2m + n)$
48. $-4(4 + 2p + 5q)$
49. $-(5x + 2)$
50. $-(9r + 5)$
51. $-(r - 3 - 7p)$
52. $-(q - 2 + 6r)$
53. $\frac{1}{2}(6x + 8)$
54. $\frac{1}{4}(4x - 2)$
55. $-\frac{1}{3}(3x - 9y)$
56. $-\frac{1}{5}(10a - 25b)$
57. $3(2r + 5) - 7$
58. $10(4s + 6) - 40$
59. $-9(4x + 8) + 2$
60. $-11(5x + 3) + 10$
61. $-4(4x + 5) - 5$
62. $-6(2x + 1) - 1$

Use the distributive property to write each sum as a product. See Example 5.

63. $4 \cdot 1 + 4 \cdot y$
64. $14 \cdot z + 14 \cdot 5$
65. $11x + 11y$
66. $9a + 9b$
67. $(-1) \cdot 5 + (-1) \cdot x$
68. $(-3)a + (-3)b$
69. $30a + 30b$
70. $25x + 25y$

Name the property or properties illustrated by each true statement. See Example 6.

71. $3 \cdot 5 = 5 \cdot 3$
72. $4(3 + 8) = 4 \cdot 3 + 4 \cdot 8$
73. $2 + (x + 5) = (2 + x) + 5$
74. $(x + 9) + 3 = (9 + x) + 3$
75. $9(3 + 7) = 9 \cdot 3 + 9 \cdot 7$
76. $1 \cdot 9 = 9$
77. $(4 \cdot y) \cdot 9 = 4 \cdot (y \cdot 9)$
78. $6 \cdot \frac{1}{6} = 1$
79. $0 + 6 = 6$
80. $(a + 9) + 6 = a + (9 + 6)$
81. $-4(y + 7) = -4 \cdot y + (-4) \cdot 7$
82. $(11 + r) + 8 = (r + 11) + 8$
83. $-4 \cdot (8 \cdot 3) = (8 \cdot -4) \cdot 3$
84. $r + 0 = r$

CONCEPT EXTENSIONS

Fill in the table with the opposite (additive inverse) and the reciprocal (multiplicative inverse). Assume that the value of each expression is not 0.

	Expression	Opposite	Reciprocal
85.	8		
86.	$-\frac{2}{3}$		
87.	x		
88.	$4y$		
89.			$\frac{1}{2x}$
90.		$7x$	

Decide whether each statement is true or false. See the second Concept Check in this section.

91. The opposite of $-\frac{a}{2}$ is $-\frac{2}{a}$.

92. The reciprocal of $-\frac{a}{2}$ is $\frac{a}{2}$.

Determine which pairs of actions are commutative. See the first Concept Check in this section.

93. "taking a test" and "studying for the test"
94. "putting on your shoes" and "putting on your socks"
95. "putting on your left shoe" and "putting on your right shoe"
96. "reading the sports section" and "reading the comics section"
97. "mowing the lawn" and "trimming the hedges"
98. "baking a cake" and "eating the cake"
99. "dialing a number" and "turning on the cell phone"
100. "feeding the dog" and "feeding the cat"

Name the property illustrated by each step.

101. a. $\triangle + (\square + \bigcirc) = (\square + \bigcirc) + \triangle$

b. $= (\bigcirc + \square) + \triangle$

c. $= \bigcirc + (\square + \triangle)$

102. **a.** $(x + y) + z = x + (y + z)$

b. $= (y + z) + x$

c. $= (z + y) + x$

103. Explain why 0 is called the identity element for addition.

104. Explain why 1 is called the identity element for multiplication.

105. Write an example that shows that division is not commutative.

106. Write an example that shows that subtraction is not commutative.

Chapter 1 Vocabulary Check

Fill in each blank with one of the words or phrases listed below.

set	inequality symbols	opposites	absolute value
grouping symbols	exponent	base	reciprocals
variable	equation	solution	

1. The symbols $\neq$, $<$, and $>$ are called ______________.

2. A mathematical statement that two expressions are equal is called a(n) ______________.

3. The ______________ of a number is the distance between that number and 0 on a number line.

4. A symbol used to represent a number is called a(n) ______________.

5. Two numbers that are the same distance from 0 but lie on opposite sides of 0 are called ______________.

6. A(n) ______________ of an equation is a value for the variable that makes the equation a true statement.

7. Two numbers whose product is 1 are called ______________.

8. In 2^3, the 2 is called the ______________ and the 3 is called the ______________.

9. Parentheses and brackets are examples of ______________.

10. A(n) ______________ is a collection of objects.

Chapter 1 Highlights

DEFINITIONS AND CONCEPTS	EXAMPLES
Section 1.2 Symbols and Sets of Numbers	
A **set** is a collection of objects, called **elements,** enclosed in braces.	$\{a, c, e\}$
Natural Numbers: $\{1, 2, 3, 4, \dots\}$ **Whole Numbers:** $\{0, 1, 2, 3, 4, \dots\}$ **Integers:** $\{\dots, -3, -2, -1, 0, 1, 2, 3, \dots\}$ **Rational Numbers:** {real numbers that can be expressed as a quotient of integers} **Irrational Numbers:** {real numbers that cannot be expressed as a quotient of integers} **Real Numbers:** {all numbers that correspond to a point on a number line}	Given the set $\left\{-3.4, \sqrt{3}, 0, \frac{2}{3}, 5, -4\right\}$, list the numbers that belong to the set of Natural numbers: 5 Whole numbers: 0, 5 Integers: $-4, 0, 5$ Rational numbers: $-4, -3.4, 0, \frac{2}{3}, 5$ Irrational numbers: $\sqrt{3}$ Real numbers: $-4, -3.4, 0, \frac{2}{3}, \sqrt{3}, 5$

DEFINITIONS AND CONCEPTS | **EXAMPLES**

Section 1.2 Symbols and Sets of Numbers (continued)

A line used to picture numbers is called a **number line.**

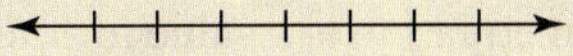

The **absolute value** of a real number a, denoted by $|a|$, is the distance between a and 0 on a number line.

$|5| = 5 \quad |0| = 0 \quad |-2| = 2$

Symbols: $=$ is equal to

$\neq$ is not equal to

$>$ is greater than

$<$ is less than

$\leq$ is less than or equal to

$\geq$ is greater than or equal to

$$-7 = -7$$
$$3 \neq -3$$
$$4 > 1$$
$$1 < 4$$
$$6 \leq 6$$
$$18 \geq -\frac{1}{3}$$

Order Property for Real Numbers

For any two real numbers a and b, a is less than b if a is to the left of b on a number line.

$-3 < 0 \quad 0 > -3 \quad 0 < 2.5 \quad 2.5 > 0$

The **opposite** of a number a is the number $-a$.

The opposite of 5 is −5. The opposite of −11 is 11.

Section 1.3 Exponents, Order of Operations, Variable Expressions, and Equations

The expression a^n is an **exponential expression.** The number a is called the **base;** it is the repeated factor. The number n is called the **exponent;** it is the number of times that the base is a factor.

$$4^3 = 4 \cdot 4 \cdot 4 = 64$$
$$7^2 = 7 \cdot 7 = 49$$

Order of Operations

Simplify expressions in the following order. If grouping symbols are present, simplify expressions within those first, starting with the innermost set. Also, simplify the numerator and the denominator of a fraction separately.

1. Evaluate exponential expressions or absolute values in order from left to right.
2. Multiply or divide in order from left to right.
3. Add or subtract in order from left to right.

$$\frac{8^2 + 5(7 - 3)}{3 \cdot 7} = \frac{8^2 + 5(4)}{21}$$
$$= \frac{64 + 5(4)}{21}$$
$$= \frac{64 + 20}{21}$$
$$= \frac{84}{21}$$
$$= 4$$

A symbol used to represent a number is called a **variable.**

Examples of variables are:

$$q, x, z$$

An **algebraic expression** is a collection of numbers, variables, operation symbols, and grouping symbols.

Examples of algebraic expressions are:

$$5x, 2(y - 6), \frac{q^2 - 3q + 1}{6}$$

To evaluate an algebraic expression containing a variable, substitute a given number for the variable and simplify.

Evaluate $x^2 - y^2$ if $x = 5$ and $y = 3$.

$$x^2 - y^2 = (5)^2 - (3)^2$$
$$= 25 - 9$$
$$= 16$$

A mathematical statement that two expressions are equal is called an **equation.**

Examples of equations are:

$$3x - 9 = 20$$
$$A = \pi r^2$$

(continued)

DEFINITIONS AND CONCEPTS | **EXAMPLES**

Section 1.3 Exponents, Order of Operations, Variable Expressions, and Equations (continued)

A **solution** of an equation is a value for the variable that makes the equation a true statement.

Determine whether 4 is a solution of $5x + 7 = 27$.

$$5x + 7 = 27$$
$$5(4) + 7 \stackrel{?}{=} 27$$
$$20 + 7 \stackrel{?}{=} 27$$
$$27 = 27 \quad \text{True}$$

4 is a solution.

Section 1.4 Operations on Real Numbers and Evaluating Square Roots

Adding Real Numbers

1. To add two numbers with the same sign, add their absolute values and attach their common sign.
2. To add two numbers with different signs, subtract the smaller absolute value from the larger absolute value and attach the sign of the number with the larger absolute value.

$$\frac{2}{7} + \frac{1}{7} = \frac{3}{7}$$
$$-5 + (-2.6) = -7.6$$
$$-18 + 6 = -12$$
$$20.8 + (-10.2) = 10.6$$

Subtracting Real Numbers

$$a - b = a + (-b)$$

$$18 - 21 = 18 + (-21) = -3$$

Multiplying and Dividing Real Numbers

The product or quotient of two numbers with the same sign is positive.

$$(-8)(-4) = 32 \qquad \frac{-8}{-4} = 2$$
$$8 \cdot 4 = 32 \qquad \frac{8}{4} = 2$$

The product or quotient of two numbers with different signs is negative.

$$-17 \cdot 2 = -34 \qquad \frac{-14}{2} = -7$$
$$4(-1.6) = -6.4 \qquad \frac{22}{-2} = -11$$

A natural number **exponent** is a shorthand notation for repeated multiplication of the same factor.

$$3^4 = 3 \cdot 3 \cdot 3 \cdot 3 = 81$$

The notation $\sqrt{a}$ denotes the **positive,** or **principal, square root** of a nonnegative number a.

$$\sqrt{a} = b \text{ if } b^2 = a \text{ and } b \text{ is positive.}$$

$$\sqrt{49} = 7$$

Order of Operations

Simplify expressions using the order that follows. If grouping symbols such as parentheses are present, simplify expressions within those first, starting with the innermost set. If fraction bars are present, simplify the numerator and denominator separately.

1. Evaluate exponential expressions, roots, or absolute values in order from left to right.
2. Multiply or divide in order from left to right.
3. Add or subtract in order from left to right.

Simplify $\dfrac{42 - 2(3^2 - \sqrt{16})}{-8}$.

$$\frac{42 - 2(3^2 - \sqrt{16})}{-8} = \frac{42 - 2(9 - 4)}{-8}$$
$$= \frac{42 - 2(5)}{-8}$$
$$= \frac{42 - 10}{-8}$$
$$= \frac{32}{-8} = -4$$

The algebraic expression $2\pi r$ represents the circumference of (distance around) a circle of radius r.

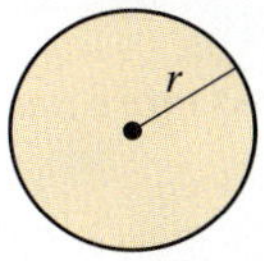

85. Complete the table below by evaluating the expression at the given values of r. (Use 3.14 for π.)

Radius	r	1	10	100
Circumference	$2\pi r$			

86. As the radius of a circle increases, does the circumference of the circle increase or decrease?

(1.5) *Name the property illustrated.*

87. $-6 + 5 = 5 + (-6)$

88. $6 \cdot 1 = 6$

89. $3(8 - 5) = 3 \cdot 8 - 3 \cdot 5$

90. $4 + (-4) = 0$

91. $2 + (3 + 9) = (2 + 3) + 9$

92. $2 \cdot 8 = 8 \cdot 2$

93. $6(8 + 5) = 6 \cdot 8 + 6 \cdot 5$

94. $(3 \cdot 8) \cdot 4 = 3 \cdot (8 \cdot 4)$

95. $4 \cdot \frac{1}{4} = 1$

96. $8 + 0 = 8$

Use the distributive property to write each expression without parentheses.

97. $5(y - 2)$

98. $-3(z + y)$

99. $-(7 - x + 4z)$

100. $\frac{1}{2}(6z - 10)$

101. $-4(3x + 5) - 7$

102. $-8(2y + 9) - 1$

MIXED REVIEW

Insert <, >, or = in the space between each pair of numbers.

103. $-|-11| \quad |11.4|$

104. $-1\frac{1}{2} \quad -2\frac{1}{2}$

Perform the indicated operations.

105. $-7.2 + (-8.1)$

106. $14 - 20$

107. $4(-20)$

108. $\frac{-20}{4}$

109. $-\frac{4}{5}\left(\frac{5}{16}\right)$

110. $-0.5(-0.3)$

111. $8 \div 2 \cdot 4$

112. $(-2)^4$

113. $\frac{-3 - 2(-9)}{-15 - 3(-4)}$

114. $5 + 2[(7 - 5)^2 + (1 - 3)]$

115. $-\frac{5}{8} \div \frac{3}{4}$

116. $\frac{-15 + (-4)^2 + |-9|}{10 - 2 \cdot 5}$

Complete the table.

	Number	***Opposite of Number***	***Reciprocal of Number***
117.	$-\frac{3}{4}$		
118.		-5	

The bar graph shows the U.S. life expectancy at birth for females born in the years shown. Use the graph to calculate the increase in life expectancy over each ten-year period shown. (The first row has been completed for you.)

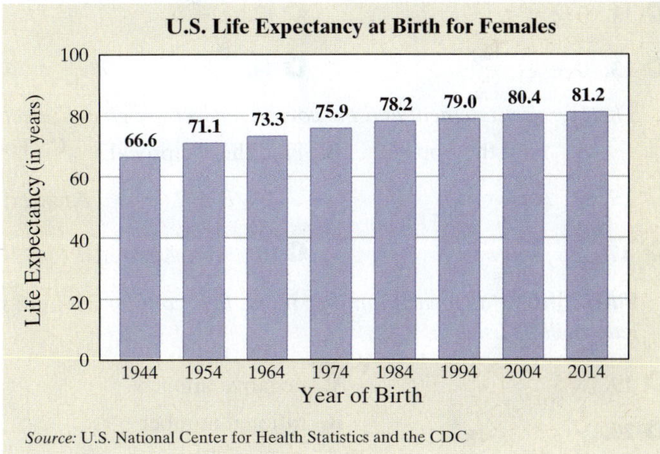

	Year	***Increase in Life Expectancy (in years) from 10 Years Earlier***
	1954	4.5
119.	1964	
120.	1974	
121.	1984	
122.	1994	
123.	2004	
124.	2014	

Chapter 1 Getting Ready for the Test

1c

All the exercises below are **Multiple Choice.** *Choose the correct letter(s). Also, letters may be used more than once.*

Select the given operation between the two numbers.

1. For $-5 + (-3)$, the operation is
   A. addition **B.** subtraction **C.** multiplication **D.** division
2. For $-5(-3)$, the operation is
   A. addition **B.** subtraction **C.** multiplication **D.** division

Identify each as an
A. equation or an **B.** expression

3. $6x + 2 + 4x - 10$
4. $6x + 2 = 4x - 10$
5. $-2(x - 1) = 12$
6. $-7\left(x + \frac{1}{2}\right) - 22$

For the exercises below, a and b are negative numbers. State whether each expression simplifies to
A. positive number **B.** negative number **C.** 0 **D.** not possible to determine

7. $a + b$
8. $a \cdot b$
9. $\frac{a}{b}$
10. $a - 0$
11. $0 \cdot b$
12. $a - b$
13. $0 + b$
14. $\frac{0}{a}$

The exercise statement and the correct answer are given. Select the correct directions.
A. Find the opposite. **B.** Find the reciprocal. **C.** Evaluate or simplify.

15. 5 Answer: $\frac{1}{5}$
16. $3 + 2(-8)$ Answer: -13
17. 2^3 Answer: 8
18. -7 Answer: 7

MATCHING *Match each number in the first column with the set of numbers that it is an element of in the second column. Not every letter will be used.*

19. $\sqrt{5}$
20. $\frac{7}{8}$
21. 0
22. -12

A. negative integer
B. rational number
C. irrational number
D. whole number
E. natural number

Chapter 1 Test

MyMathLab® YouTube

Translate the statement into symbols.

1. The absolute value of negative seven is greater than five.
2. The sum of nine and five is greater than or equal to four.

Simplify the expression.

3. $-13 + 8$
4. $-13 - (-2)$
5. $12 \div 4 \cdot 3 - 6 \cdot 2$
6. $(13)(-3)$
7. $(-6)(-2)$
8. $\frac{|-16|}{-8}$
9. $\frac{-8}{0}$
10. $\frac{|-6| + 2}{5 - 6}$
11. $5 - 12 \div 3(2)$
12. $5^2 - 3^4$
13. $(4 - 9)^3 - |-4 - 6|^2$
14. $3(-4)^2 - 80$
15. $6[5 + 2(3 - 8) - 3]$
16. $\frac{-12 + 3 \cdot 8}{4}$
17. $\frac{(4 - \sqrt{16}) - (-7 - 20)}{-2(1 - 4)^2}$

Insert <, >, or = in the appropriate space to make each of the following statements true.

18. $-3 \quad -7$
19. $4 \quad -8$
20. $2 \quad |-3|$
21. $|-2| \quad -1 - (-3)$
22. In the state of Massachusetts, there are 2221 licensed child care centers and 10,993 licensed home-based child care providers. Write an inequality statement comparing the numbers 2221 and 10,993. (*Source:* Children's Foundation)

23. Given $\left\{-5, -1, 0, \frac{1}{4}, 1, 7, 11.6, \sqrt{7}, 3\pi\right\}$, list the numbers in this set that also belong to the set of:

a. Natural numbers

b. Whole numbers

c. Integers

d. Rational numbers

e. Irrational numbers

f. Real numbers

If $x = 6, y = -2$, and $z = -3$, evaluate each expression.

24. $x^2 + y^2$

25. $x + yz$

26. $2 + 3x - y$

27. $\frac{y + z - 1}{x}$

Identify the property illustrated by each expression.

28. $8 + (9 + 3) = (8 + 9) + 3$

29. $6 \cdot 8 = 8 \cdot 6$

30. $-6(2 + 4) = -6 \cdot 2 + (-6) \cdot 4$

31. $\frac{1}{6}(6) = 1$

32. Find the opposite of -9.

33. Find the reciprocal of $-\frac{1}{3}$.

The New Orleans Saints were 22 yards from the goal when the following series of gains and losses occurred.

	Gains and Losses in Yards
First Down	5
Second Down	−10
Third Down	−2
Fourth Down	29

34. During which down did the greatest loss of yardage occur?

35. Was a touchdown scored?

36. The temperature at the Winter Olympics was a frigid 14 degrees below zero in the morning, but by noon it had risen 31 degrees. What was the temperature at noon?

37. A health insurance provider had net incomes of \$356 million, \$460 million, and −\$166 million in 3 consecutive years. What was the health insurance provider's total net income for these three years?

38. A stockbroker decided to sell 280 shares of stock, which decreased in value by \$1.50 per share yesterday. How much money did she lose?

CHAPTER

2 Equations, Inequalities, and Problem Solving

CHECK YOUR PROGRESS

Much of mathematics relates to deciding which statements are true and which are false. For example, the statement $x + 7 = 15$ is an equation stating that the sum $x + 7$ has the same value as 15. Is this statement true or false? It is false for some values of x and true for just one value of x, namely 8. Our purpose in this chapter is to learn ways of deciding which values make an equation or an inequality true.

What Are iOS and Android Systems?

Smartphones, tablets, and iPads are everywhere. What are the operating systems behind these devices? Two such systems are iOS and Android. iOS is a mobile operating system developed by Apple, Inc., for Apple hardware, and Android is a mobile operating system developed by Google, primarily for touchscreen devices such as smartphones and tablet computers. The average American now spends 162 minutes, that is, 2 hours and 42 minutes, daily on one (or more) of these devices. This is an increase of 4 minutes a day from last year, and it seems that this amount of time will only continue to grow.

In Section 2.5, Exercises 57 through 60, we will explore how this time is spent.

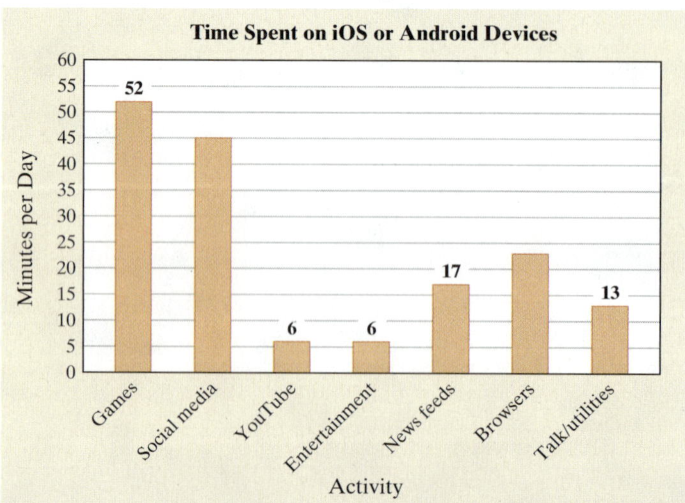

2.1 Simplifying Algebraic Expressions

OBJECTIVES

1 Identify Terms, Like Terms, and Unlike Terms.

2 Combine Like Terms.

3 Use the Distributive Property to Remove Parentheses.

4 Write Word Phrases as Algebraic Expressions.

As we explore in this section, an expression such as $3x + 2x$ is not as simple as possible because—even without replacing x by a value—we can perform the indicated addition.

OBJECTIVE

1 Identifying Terms, Like Terms, and Unlike Terms

Before we practice simplifying expressions, some new language of algebra is presented. A **term** is a number or the product of a number and variables raised to powers.

Terms

$$-y, \quad 2x^3, \quad -5, \quad 3xz^2, \quad \frac{2}{y}, \quad 0.8z$$

The **numerical coefficient** (sometimes also simply called the **coefficient**) of a term is the numerical factor. The numerical coefficient of $3x$ is 3. Recall that $3x$ means $3 \cdot x$.

Term	*Numerical Coefficient*
$3x$	3
$\frac{y^3}{5}$	$\frac{1}{5}$ since $\frac{y^3}{5}$ means $\frac{1}{5} \cdot y^3$
$0.7ab^3c^5$	0.7
z	1
$-y$	-1
-5	-5

Helpful Hint

The term $-y$ means $-1y$ and thus has a numerical coefficient of -1.
The term z means $1z$ and thus has a numerical coefficient of 1.

EXAMPLE 1 Identify the numerical coefficient of each term.

a. $-3y$ **b.** $22z^4$ **c.** y **d.** $-x$ **e.** $\frac{x}{7}$

Solution

a. The numerical coefficient of $-3y$ is -3.

b. The numerical coefficient of $22z^4$ is 22.

c. The numerical coefficient of y is 1, since y is $1y$.

d. The numerical coefficient of $-x$ is -1, since $-x$ is $-1x$.

e. The numerical coefficient of $\frac{x}{7}$ is $\frac{1}{7}$, since $\frac{x}{7}$ means $\frac{1}{7} \cdot x$.

PRACTICE

1 Identify the numerical coefficient of each term.

a. t **b.** $-7x$ **c.** $-\frac{w}{5}$ **d.** $43x^4$ **e.** $-b$

Terms with the same variables raised to exactly the same powers are called **like terms.** Terms that aren't like terms are called **unlike terms.**

Like Terms	*Unlike Terms*	*Reason*
$3x, 2x$	$5x, 5x^2$	Why? Same variable x but different powers x and x^2
$-6x^2y, 2x^2y, 4x^2y$	$7y, 3z, 8x^2$	Why? Different variables
$2ab^2c^3, ac^3b^2$	$6abc^3, 6ab^2$	Why? Different variables and different powers

Helpful Hint

In like terms, each variable and its exponent must match exactly, but these factors don't need to be in the same order.

$2x^2y$ and $3yx^2$ are like terms.

EXAMPLE 2 Determine whether the terms are like or unlike.

a. $2x, 3x^2$ **b.** $4x^2y, x^2y, -2x^2y$ **c.** $-2yz, -3zy$ **d.** $-x^4, x^4$

Solution

a. Unlike terms, since the exponents on x are not the same.

b. Like terms, since each variable and its exponent match.

c. Like terms, since $zy = yz$ by the commutative property.

d. Like terms. □

PRACTICE 2 Determine whether the terms are like or unlike.

a. $-4xy, 5yx$ **b.** $5q, -3q^2$

c. $3ab^2, -2ab^2, 43ab^2$ **d.** $y^5, \frac{y^5}{2}$ ■

OBJECTIVE 2

Combining Like Terms

An algebraic expression containing the sum or difference of like terms can be simplified by applying the distributive property. For example, by the distributive property, we rewrite the sum of the like terms $3x + 2x$ as

$$3x + 2x = (3 + 2)x = 5x$$

Also,

$$-y^2 + 5y^2 = -1y^2 + 5y^2 = (-1 + 5)y^2 = 4y^2$$

Simplifying the sum or difference of like terms is called **combining like terms.**

EXAMPLE 3 Simplify each expression by combining like terms.

a. $7x - 3x$ **b.** $10y^2 + y^2$

c. $8x^2 + 2x - 3x$ **d.** $9n^2 - 5n^2 + n^2$

Solution

a. $7x - 3x = (7 - 3)x = 4x$

b. $10y^2 + y^2 = 10y^2 + 1y^2 = (10 + 1)y^2 = 11y^2$

c. $8x^2 + 2x - 3x = 8x^2 + (2 - 3)x = 8x^2 - x$

d. $9n^2 - 5n^2 + n^2 = (9 - 5 + 1)n^2 = 5n^2$ □

PRACTICE 3 Simplify each expression by combining like terms.

a. $-3y + 11y$ **b.** $4x^2 + x^2$

c. $5x - 3x^2 + 8x^2$ **d.** $20y^2 + 2y^2 - y^2$ ■

The previous example suggests the following:

Combining Like Terms

To **combine like terms,** add the numerical coefficients and multiply the result by the common variable factors.

EXAMPLE 4 Simplify each expression by combining like terms.

a. $2x + 3x + 5 + 2$ **b.** $-5a - 3 + a + 2$ **c.** $4y - 3y^2$

d. $2.3x + 5x - 6$ **e.** $-\frac{1}{2}b + b$

Solution Use the distributive property to combine like terms.

a. $2x + 3x + 5 + 2 = (2 + 3)x + (5 + 2)$
$= 5x + 7$

b. $-5a - 3 + a + 2 = -5a + 1a + (-3 + 2)$
$= (-5 + 1)a + (-3 + 2)$
$= -4a - 1$

c. $4y - 3y^2$ These two terms cannot be combined because they are unlike terms.

d. $2.3x + 5x - 6 = (2.3 + 5)x - 6$
$= 7.3x - 6$

e. $-\frac{1}{2}b + b = -\frac{1}{2}b + 1b = \left(-\frac{1}{2} + 1\right)b = \frac{1}{2}b$ □

PRACTICE 4 Use the distributive property to combine like terms.

a. $3y + 8y - 7 + 2$ **b.** $6x - 3 - x - 3$ **c.** $\frac{3}{4}t - t$

d. $9y + 3.2y + 10 + 3$ **e.** $5z - 3z^4$ ■

OBJECTIVE

3 Using the Distributive Property

Simplifying expressions makes frequent use of the distributive property to also remove parentheses.

It may be helpful to study the examples below.

$$+(3a + 2) = +1(3a + 2) = +1(3a) + (+1)(2) = 3a + 2$$

(+ means +1)

$$-(3a + 2) = -1(3a + 2) = -1(3a) + (-1)(2) = -3a - 2$$

(− means −1)

EXAMPLE 5 Find each product by using the distributive property to remove parentheses.

a. $5(3x + 2)$ **b.** $-2(y + 0.3z - 1)$ **c.** $-(9x + y - 2z + 6)$

Solution

a. $5(3x + 2) = 5 \cdot 3x + 5 \cdot 2$ Apply the distributive property.
$= 15x + 10$ Multiply.

(Continued on next page)

b. $-2(y + 0.3z - 1) = -2(y) + (-2)(0.3z) + (-2)(-1)$ Apply the distributive property.
$= -2y - 0.6z + 2$ Multiply.

c. $-(9x + y - 2z + 6) = -1(9x + y - 2z + 6)$ Distribute −1 over each term.
$= -1(9x) - 1(y) - 1(-2z) - 1(6)$
$= -9x - y + 2z - 6$ □

PRACTICE 5 Find each product by using the distributive property to remove parentheses.

a. $3(2x - 7)$ **b.** $-5(x - 0.5z - 5)$
c. $-(2x - y + z - 2)$ ■

Helpful Hint

If a "−" sign precedes parentheses, the sign of each term inside the parentheses is changed when the distributive property is applied to remove parentheses.

Examples:

$-(2x + 1) = -2x - 1$ $-(-5x + y - z) = 5x - y + z$
$-(x - 2y) = -x + 2y$ $-(-3x - 4y - 1) = 3x + 4y + 1$

When simplifying an expression containing parentheses, we often use the distributive property in both directions—first to remove parentheses and then again to combine any like terms.

EXAMPLE 6 Simplify each expression.

a. $3(2x - 5) + 1$ **b.** $-2(4x + 7) - (3x - 1)$ **c.** $9 + 3(4x - 10)$

Solution

a. $3(2x - 5) + 1 = 6x - 15 + 1$ Apply the distributive property.
$= 6x - 14$ Combine like terms.

b. $-2(4x + 7) - (3x - 1) = -8x - 14 - 3x + 1$ Apply the distributive property.
$= -11x - 13$ Combine like terms.

c. $9 + 3(4x - 10) = 9 + 12x - 30$ Apply the distributive property.
$= -21 + 12x$ Combine like terms.
or $12x - 21$ □

Helpful Hint

Don't forget to use the distributive property and multiply before adding or subtracting like terms.

PRACTICE 6 Simplify each expression.

a. $4(9x + 1) + 6$ **b.** $-7(2x - 1) - (6 - 3x)$ **c.** $8 - 5(6x + 5)$ ■

EXAMPLE 7 Write the phrase below as an algebraic expression. Then simplify if possible.

"Subtract $4x - 2$ from $2x - 3$."

Solution "Subtract $4x - 2$ **from** $2x - 3$" translates to $(2x - 3) - (4x - 2)$. Next, simplify the algebraic expression.

$(2x - 3) - (4x - 2) = 2x - 3 - 4x + 2$ Apply the distributive property.
$= -2x - 1$ Combine like terms. □

PRACTICE 7 Write the phrase below as an algebraic expression. Then simplify if possible.

"Subtract $7x - 1$ from $2x + 3$."

OBJECTIVE 4

Writing Word Phrases as Algebraic Expressions

Next, we practice writing word phrases as algebraic expressions.

EXAMPLE 8 Write the following phrases as algebraic expressions and simplify if possible. Let x represent the unknown number.

a. Twice a number, plus 6

b. The difference of a number and 4, divided by 7

c. Five added to triple the sum of a number and 1

d. The sum of twice a number, 3 times the number, and 5 times the number

Solution

a.

In words:	twice a number	plus	6
	↓	↓	↓
Translate:	$2x$	$+$	6

b.

In words:	the difference of a number and 4	divided by	7
	↓	↓	↓
Translate:	$(x - 4)$	$\div$	7 or $\frac{x-4}{7}$

c.

In words:	five	added to	triple	the sum of a number and 1
	↓	↓	↓	↓
Translate:	5	$+$	$3\cdot$	$(x + 1)$

Next, we simplify this expression.

$$5 + 3(x + 1) = 5 + 3x + 3 \quad \text{Use the distributive property.}$$
$$= 8 + 3x \quad \text{Combine like terms.}$$

d. The phrase "the sum of" means that we add.

In words:	twice a number	added to	3 times the number	added to	5 times the number
	↓	↓	↓	↓	↓
Translate:	$2x$	$+$	$3x$	$+$	$5x$

Now let's simplify.

$$2x + 3x + 5x = 10x \quad \text{Combine like terms.}$$

PRACTICE 8 Write the following phrases as algebraic expressions and simplify if possible. Let x represent the unknown number.

a. Three added to double a number

b. Six subtracted from the sum of 5 and a number

c. Two times the sum of 3 and a number, increased by 4

d. The sum of a number, half the number, and 5 times the number

Vocabulary, Readiness & Video Check

Use the choices below to fill in each blank. Some choices may be used more than once.

like	numerical coefficient	term	distributive
unlike	combine like terms	expression	

1. $23y^2 + 10y - 6$ is called a(n) ________________ while $23y^2$, $10y$, and -6 are each called a(n) ________________.
2. To simplify $x + 4x$, we ________________.
3. The term y has an understood ________________ of 1.
4. The terms $7z$ and $7y$ are ________________ terms and the terms $7z$ and $-z$ are ________________ terms.
5. For the term $-\frac{1}{2}xy^2$, the number $-\frac{1}{2}$ is the ________________.
6. $5(3x - y)$ equals $15x - 5y$ by the ________________ property.

Martin-Gay Interactive Videos

See Video 2.1

Watch the section lecture video and answer the following questions.

OBJECTIVE 1

7. Example 7 shows two terms with exactly the same variables. Why are these terms not considered like terms?

OBJECTIVE 2

8. Example 8 shows us that when combining like terms, we are actually applying what property?

OBJECTIVE 3

9. The expression in Example 11 shows a minus sign before parentheses. When using the distributive property to multiply and remove parentheses, what number are we actually distributing to each term within the parentheses?

OBJECTIVE 4

10. Write the phrase given in Example 14, translate it into an algebraic expression, then simplify it. Why are we able to simplify it?

2.1 Exercise Set MyMathLab®

Identify the numerical coefficient of each term. See Example 1.

1. $-7y$
2. $3x$
3. x
4. $-y$
5. $17x^2y$
6. $1.2xyz$

Indicate whether the terms in each list are like or unlike. See Example 2.

7. $5y, -y$
8. $-2x^2y, 6xy$
9. $2z, 3z^2$
10. $ab^2, -7ab^2$
11. $8wz, \frac{1}{7}zw$
12. $7.4p^3q^2, 6.2p^3q^2r$

Simplify each expression by combining any like terms. See Examples 3 and 4.

13. $7y + 8y$
14. $3x + 2x$
15. $8w - w + 6w$
16. $c - 7c + 2c$
17. $3b - 5 - 10b - 4$
18. $6g + 5 - 3g - 7$
19. $m - 4m + 2m - 6$
20. $a + 3a - 2 - 7a$
21. $5g - 3 - 5 - 5g$
22. $8p + 4 - 8p - 15$
23. $6.2x - 4 + x - 1.2$
24. $7.9y - 0.7 - y + 0.2$
25. $6x - 5x + x - 3 + 2x$
26. $8h + 13h - 6 + 7h - h$
27. $7x^2 + 8x^2 - 10x^2$
28. $8x^3 + x^3 - 11x^3$
29. $6x + 0.5 - 4.3x - 0.4x + 3$
30. $0.4y - 6.7 + y - 0.3 - 2.6y$

Simplify each expression. First use the distributive property to remove any parentheses. See Examples 5 and 6.

31. $5(y - 4)$
32. $7(r - 3)$
33. $-2(x + 2)$
34. $-4(y + 6)$
35. $7(d - 3) + 10$
36. $9(z + 7) - 15$
37. $-5(2x - 3y + 6)$
38. $-2(4x - 3z - 1)$

39. $-(3x - 2y + 1)$

40. $-(y + 5z - 7)$

41. $5(x + 2) - (3x - 4)$

42. $4(2x - 3) - 2(x + 1)$

Write each of the following as an algebraic expression. Simplify if possible. See Example 7.

43. Add $6x + 7$ to $4x - 10$.

44. Add $3y - 5$ to $y + 16$.

45. Subtract $7x + 1$ from $3x - 8$.

46. Subtract $4x - 7$ from $12 + x$.

47. Subtract $5m - 6$ from $m - 9$.

48. Subtract $m - 3$ from $2m - 6$.

MIXED PRACTICE

Simplify each expression. See Examples 3 through 7.

49. $2k - k - 6$

50. $7c - 8 - c$

51. $-9x + 4x + 18 - 10x$

52. $5y - 14 + 7y - 20y$

53. $-4(3y - 4) + 12y$

54. $-3(2x + 5) - 6x$

55. $3(2x - 5) - 5(x - 4)$

56. $2(6x - 1) - (x - 7)$

57. $-2(3x - 4) + 7x - 6$

58. $8y - 2 - 3(y + 4)$

59. $5k - (3k - 10)$

60. $-11c - (4 - 2c)$

61. Subtract $6x - 1$ from $3x + 4$

62. Subtract $4 + 3y$ from $8 - 5y$

63. $3.4m - 4 - 3.4m - 7$

64. $2.8w - 0.9 - 0.5 - 2.8w$

65. $\frac{1}{3}(7y - 1) + \frac{1}{6}(4y + 7)$

66. $\frac{1}{5}(9y + 2) + \frac{1}{10}(2y - 1)$

67. $2 + 4(6x - 6)$

68. $8 + 4(3x - 4)$

69. $0.5(m + 2) + 0.4m$

70. $0.2(k + 8) - 0.1k$

71. $10 - 3(2x + 3y)$

72. $14 - 11(5m + 3n)$

73. $6(3x - 6) - 2(x + 1) - 17x$

74. $7(2x + 5) - 4(x + 2) - 20x$

75. $\frac{1}{2}(12x - 4) - (x + 5)$

76. $\frac{1}{3}(9x - 6) - (x - 2)$

TRANSLATING

Write each phrase as an algebraic expression and simplify if possible. Let x represent the unknown number. See Examples 7 and 8.

77. Twice a number, decreased by four

78. The difference of a number and two, divided by five

79. Seven added to double a number

80. Eight more than triple a number

81. Three-fourths of a number, increased by twelve

82. Eleven, increased by two-thirds of a number

83. The sum of 5 times a number and -2, added to 7 times the number

84. The sum of 3 times a number and 10, subtracted from 9 times the number

85. Eight times the sum of a number and six

86. Six times the difference of a number and five

87. Double a number, minus the sum of the number and ten

88. Half a number, minus the product of the number and eight

89. The sum of 2, three times a number, -9, and four times the number

90. The sum of twice a number, -1, five times the number, and -12

REVIEW AND PREVIEW

Recall that a histogram is a special bar graph. The width of each bar represents a range of numbers (class intervals), and the height of each bar (class frequency) tells us how many times a number in the class interval occurs. Use the histogram below to answer the exercises. (See Section R.4.)

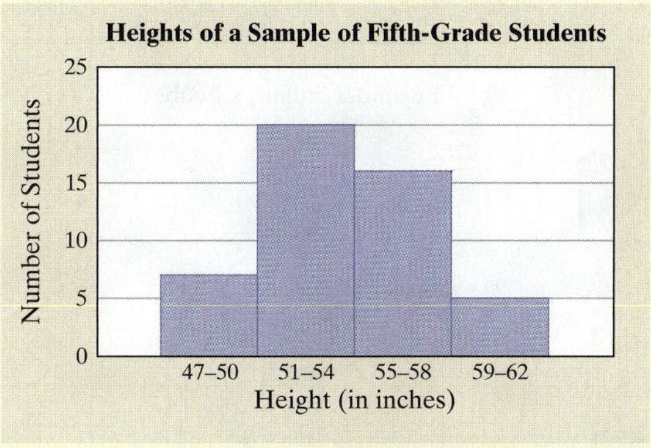

91. Most students are in what height range?

92. The fewest number of students are in what height range?

93. How many students are 55 inches or taller?

94. How many students are 54 inches or shorter?

95. What is the total number of students sampled in this graph?

96. How many students are 51 inches or taller?

CONCEPT EXTENSIONS

△ **97.** Recall that the perimeter of a figure is the total distance around the figure. Given the following rectangle, express the perimeter as an algebraic expression containing the variable x.

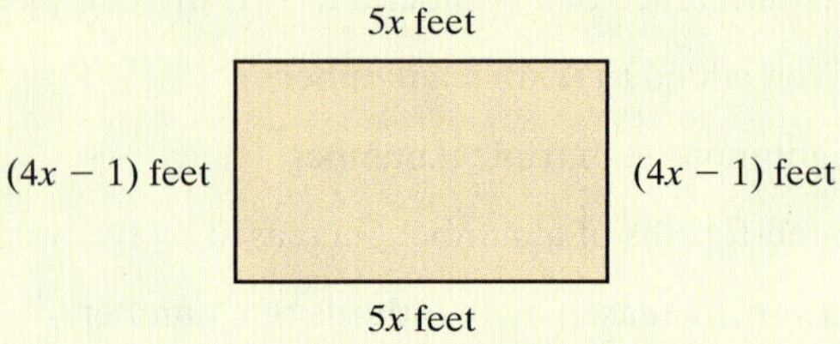

△ **98.** Given the following triangle, express its perimeter as an algebraic expression containing the variable x.

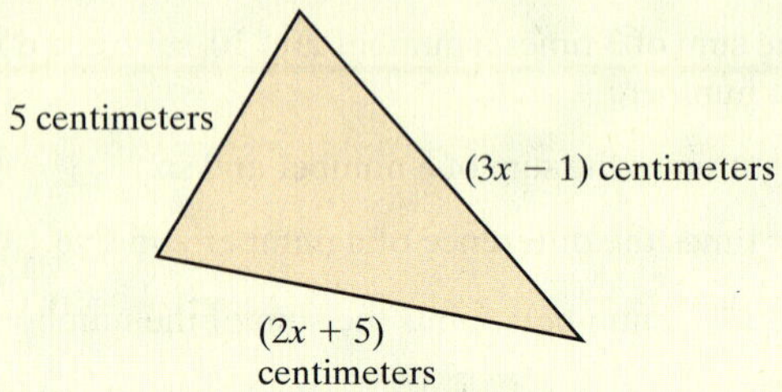

Given the following two rules, determine whether each scale in Exercises 99 through 102 is balanced.

Rule 1:

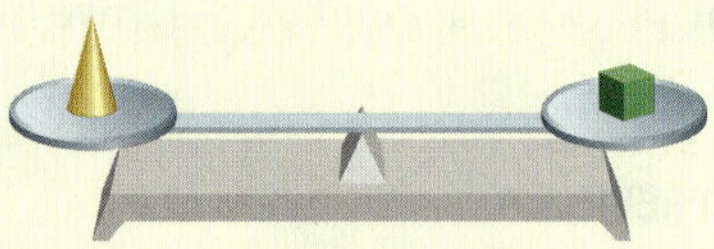

1 cone balances 1 cube

Rule 2:

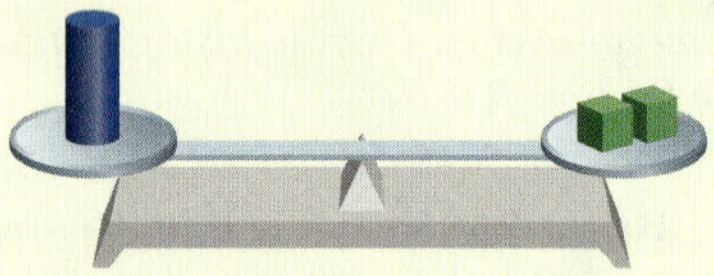

1 cylinder balances 2 cubes

99.

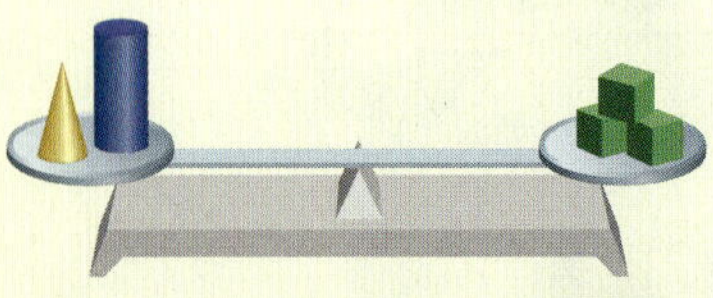

100.

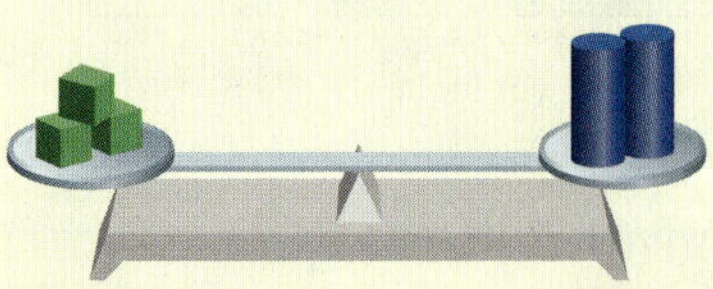

101.

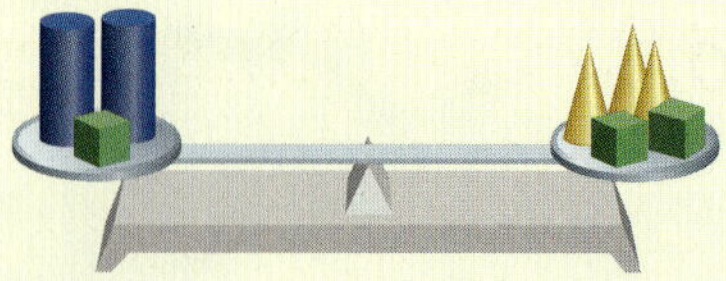

102.

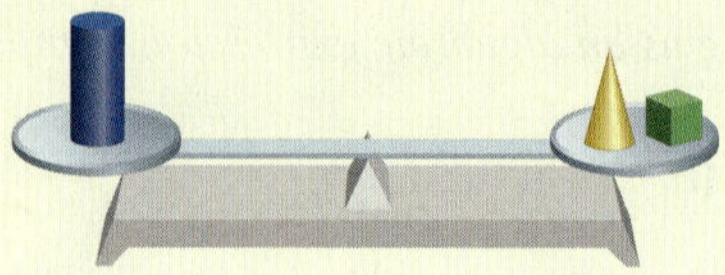

Write each algebraic expression described.

103. Write an expression with 4 terms that simplifies to $3x - 4$.

104. Write an expression of the form ________ (_____ + _____) whose product is $6x + 24$.

105. To convert from feet to inches, we multiply by 12. For example, the number of inches in 2 feet is $12 \cdot 2$ inches. If one board has a length of $(x + 2)$ *feet* and a second board has a length of $(3x - 1)$ *inches,* express their total length in inches as an algebraic expression.

106. The value of 7 nickels is $5 \cdot 7$ cents. Likewise, the value of x nickels is $5x$ cents. If the money box in a drink machine contains x *nickels,* $3x$ *dimes,* and $(30x - 1)$ *quarters*, express their total value in cents as an algebraic expression.

107. In your own words, explain how to combine like terms.

108. Do like terms always contain the same numerical coefficients? Explain your answer.

For Exercises 109 through 114, see the example below.

Example

Simplify: $-3xy + 2x^2y - (2xy - 1)$

Solution

$$-3xy + 2x^2y - (2xy - 1)$$
$$= -3xy + 2x^2y - 2xy + 1 = -5xy + 2x^2y + 1$$

Simplify each expression.

109. $5b^2c^3 + 8b^3c^2 - 7b^3c^2$

110. $4m^4p^2 + m^4p^2 - 5m^2p^4$

111. $3x - (2x^2 - 6x) + 7x^2$

112. $9y^2 - (6xy^2 - 5y^2) - 8xy^2$

113. $-(2x^2y + 3z) + 3z - 5x^2y$

114. $-(7c^3d - 8c) - 5c - 4c^3d$

2.2 The Addition and Multiplication Properties of Equality

OBJECTIVES

1 Define Linear Equations and Use the Addition Property of Equality to Solve Linear Equations.

2 Use the Multiplication Property of Equality to Solve Linear Equations.

3 Use Both Properties of Equality to Solve Linear Equations.

4 Write Word Phrases as Algebraic Expressions.

OBJECTIVE

1 Defining Linear Equations and Using the Addition Property

Recall from Section 1.3 that an equation is a statement that two expressions have the same value. Also, a value of the variable that makes an equation a true statement is called a solution or root of the equation. The process of finding the solution of an equation is called **solving** the equation for the variable. In this section we concentrate on solving **linear equations** in one variable.

Linear Equation in One Variable

A linear equation in one variable can be written in the form

$$ax + b = c$$

where a, b, and c are real numbers and $a \neq 0$.

Evaluating a linear equation for a given value of the variable, as we did in Section 1.3, can tell us whether that value is a solution, but we can't rely on evaluating an equation as our method of solving it.

Instead, to solve a linear equation in x, we write a series of simpler equations, all *equivalent* to the original equation, so that the final equation has the form

$$\mathbf{x = number} \qquad \text{or} \qquad \mathbf{number = x}$$

Equivalent equations are equations that have the same solution. This means that the "number" above is the solution of the original equation.

The first property of equality that helps us write simpler equivalent equations is the **addition property of equality.**

Addition Property of Equality

If a, b, and c are real numbers, then

$$a = b \qquad \text{and} \qquad a + c = b + c$$

are equivalent equations.

This property guarantees that adding the same number to both sides of an equation does not change the solution of the equation. Since subtraction is defined in terms of addition, we may also **subtract the same number from both sides** without changing the solution.

A good way to picture a true equation is as a balanced scale. Since it is balanced, each side of the scale weighs the same amount.

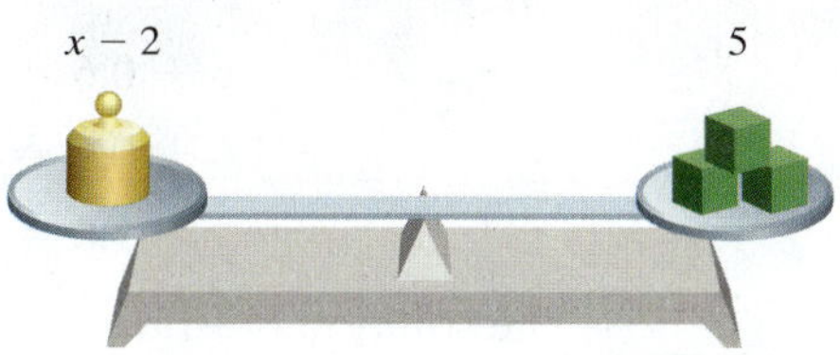

If the same weight is added to or subtracted from each side, the scale remains balanced.

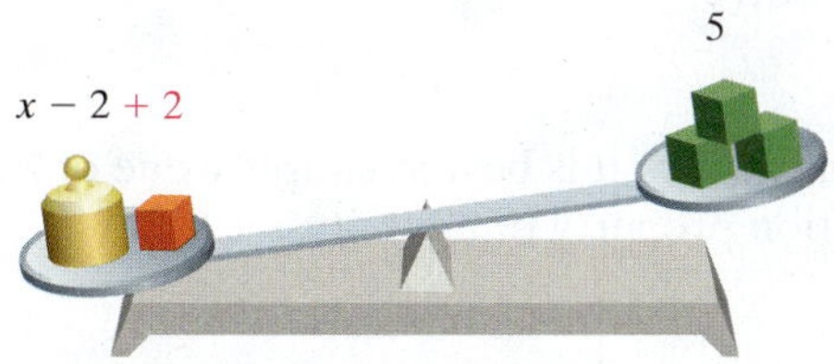

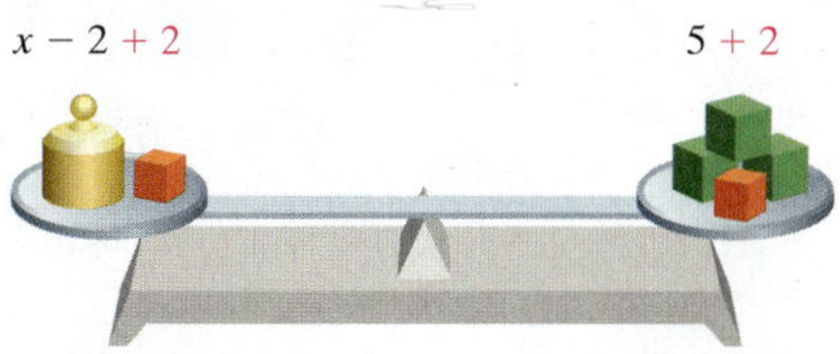

We use the addition property of equality to write equivalent equations until the variable is by itself on one side of the equation, and the equation looks like "x = number" or "number = x."

EXAMPLE 1 Solve: $x - 7 = 10$ for x

Solution To solve for x, we want x alone on one side of the equation. To do this, we add 7 to both sides of the equation.

$$x - 7 = 10$$
$$x - 7 + 7 = 10 + 7 \quad \text{Add 7 to both sides.}$$
$$x = 17 \quad \text{Simplify.}$$

The solution of the equation $x = 17$ is obviously 17. Since we are writing equivalent equations, the solution of the equation $x - 7 = 10$ is also 17.

Check: To check, replace x with 17 in the original equation.

$$x - 7 = 10$$
$$17 - 7 \stackrel{?}{=} 10 \quad \text{Replace } x \text{ with 17 in the original equation.}$$
$$10 = 10 \quad \text{True}$$

Since the statement is true, 17 is the solution. □

PRACTICE 1 Solve: $x + 3 = -5$ for x ■

✔ CONCEPT CHECK
Use the addition property to fill in the blanks so that the middle equation simplifies to the last equation.

$$x - 5 = 3$$
$$x - 5 + \_\_ = 3 + \_\_$$
$$x = 8$$

EXAMPLE 2 Solve: $y + 0.6 = -1.0$ for y

Solution To get y alone on one side of the equation, subtract 0.6 from both sides of the equation.

$$y + 0.6 = -1.0$$
$$y + 0.6 - 0.6 = -1.0 - 0.6 \quad \text{Subtract 0.6 from both sides.}$$
$$y = -1.6 \quad \text{Combine like terms.}$$

Check: To check the proposed solution, −1.6, replace y with −1.6 in the original equation.

$$y + 0.6 = -1.0$$
$$-1.6 + 0.6 \stackrel{?}{=} -1.0 \quad \text{Replace } y \text{ with } -1.6 \text{ in the original equation.}$$
$$-1.0 = -1.0 \quad \text{True}$$

The solution is −1.6. □

PRACTICE 2 Solve: $y - 0.3 = -2.1$ for y ■

Answer to Concept Check:
5; 5

Many times, it is best to simplify one or both sides of an equation before applying the addition property of equality.

EXAMPLE 3 Solve: $2x + 3x - 5 + 7 = 10x + 3 - 6x - 4$

Solution First we simplify both sides of the equation.

$$2x + 3x - 5 + 7 = 10x + 3 - 6x - 4$$
$$5x + 2 = 4x - 1 \quad \text{Combine like terms on each side of the equation.}$$

Next, we want all terms with a variable on one side of the equation and all numbers on the other side.

$$5x + 2 - 4x = 4x - 1 - 4x \quad \text{Subtract } 4x \text{ from both sides.}$$
$$x + 2 = -1 \quad \text{Combine like terms.}$$
$$x + 2 - 2 = -1 - 2 \quad \text{Subtract 2 from both sides to get } x \text{ alone.}$$
$$x = -3 \quad \text{Combine like terms.}$$

Check:
$$2x + 3x - 5 + 7 = 10x + 3 - 6x - 4 \quad \text{Original equation}$$
$$2(-3) + 3(-3) - 5 + 7 \stackrel{?}{=} 10(-3) + 3 - 6(-3) - 4 \quad \text{Replace } x \text{ with } -3.$$
$$-6 - 9 - 5 + 7 \stackrel{?}{=} -30 + 3 + 18 - 4 \quad \text{Multiply.}$$
$$-13 = -13 \quad \text{True}$$

The solution is -3. □

PRACTICE 3 Solve: $8x - 5x - 3 + 9 = x + x + 3 - 7$ ■

If an equation contains parentheses, we use the distributive property to remove them.

EXAMPLE 4 Solve: $7 = -5(2a - 1) - (-11a + 6)$

Solution

$$7 = -5(2a - 1) - (-11a + 6)$$
$$7 = -10a + 5 + 11a - 6 \quad \text{Apply the distributive property.}$$
$$7 = a - 1 \quad \text{Combine like terms.}$$
$$7 + 1 = a - 1 + 1 \quad \text{Add 1 to both sides to get } a \text{ alone.}$$
$$8 = a \quad \text{Combine like terms.}$$

Check to see that 8 is the solution. □

PRACTICE 4 Solve: $2 = 4(2a - 3) - (7a + 4)$ ■

> **Helpful Hint**
> We may solve an equation so that the variable is alone on either side of the equation. For example, $8 = a$ is equivalent to $a = 8$.

When solving equations, we may sometimes encounter an equation such as

$$-x = 5.$$

This equation is not solved for x because x is not isolated. One way to solve this equation for x is to recall that

"$-$" can be read as "the opposite of."

We can read the equation $-x = 5$ then as "the opposite of $x = 5$." If the opposite of x is 5, this means that x is the opposite of 5 or -5.

In summary,

$$-x = 5 \quad \text{and} \quad x = -5$$

are equivalent equations and $x = -5$ is solved for x.

OBJECTIVE

2 Using the Multiplication Property

As useful as the addition property of equality is, it cannot help us solve every type of linear equation in one variable. For example, adding or subtracting a value on both sides of the equation does not help solve

$$\frac{5}{2}x = 15.$$

Instead, we apply another important property of equality, the **multiplication property of equality.**

Multiplication Property of Equality

If a, b, and c are real numbers and $c \neq 0$, then

$$a = b \qquad \text{and} \qquad ac = bc$$

are equivalent equations.

This property guarantees that multiplying both sides of an equation by the same nonzero number does not change the solution of the equation. Since division is defined in terms of multiplication, we may also **divide both sides of the equation by the same nonzero number** without changing the solution.

EXAMPLE 5 Solve: $\frac{5}{2}x = 15$

Solution To get x alone, multiply both sides of the equation by the reciprocal of $\frac{5}{2}$, which is $\frac{2}{5}$.

$$\frac{5}{2}x = 15$$

$$\frac{2}{5} \cdot \frac{5}{2}x = \frac{2}{5} \cdot 15 \quad \text{Multiply both sides by } \frac{2}{5}.$$

$$\left(\frac{2}{5} \cdot \frac{5}{2}\right)x = \frac{2}{5} \cdot 15 \quad \text{Apply the associative property.}$$

$$1x = 6 \quad \text{Simplify.}$$

Helpful Hint

Don't forget to multiply *both sides* by $\frac{2}{5}$.

or

$$x = 6$$

Check: Replace x with 6 in the original equation.

$$\frac{5}{2}x = 15 \quad \text{Original equation}$$

$$\frac{5}{2}(6) \stackrel{?}{=} 15 \quad \text{Replace } x \text{ with 6.}$$

$$15 = 15 \quad \text{True}$$

The solution is 6. □

PRACTICE

5 Solve: $\frac{4}{5}x = 16$

In the equation $\frac{5}{2}x = 15$, $\frac{5}{2}$ is the coefficient of x. When the coefficient of x is a *fraction,* we will get x alone by multiplying by the reciprocal. When the coefficient of x is an integer or a decimal, it is usually more convenient to divide both sides by the coefficient. (Dividing by a number is, of course, the same as multiplying by the reciprocal of the number.)

EXAMPLE 6 Solve: $-3x = 33$

Solution Recall that $-3x$ means $-3 \cdot x$. To get x alone, we divide both sides by the coefficient of x, that is, -3.

$$-3x = 33$$

$$\frac{-3x}{-3} = \frac{33}{-3} \quad \text{Divide both sides by } -3.$$

$$1x = -11 \quad \text{Simplify.}$$

$$x = -11$$

Check:

$$-3x = 33 \quad \text{Original equation}$$

$$-3(-11) \stackrel{?}{=} 33 \quad \text{Replace } x \text{ with } -11.$$

$$33 = 33 \quad \text{True}$$

The solution is -11. □

PRACTICE 6 Solve: $8x = -96$ ■

EXAMPLE 7 Solve: $\frac{y}{7} = 20$

Solution Recall that $\frac{y}{7} = \frac{1}{7}y$. To get y alone, we multiply both sides of the equation by 7, the reciprocal of $\frac{1}{7}$.

$$\frac{y}{7} = 20$$

$$\frac{1}{7}y = 20$$

$$7 \cdot \frac{1}{7}y = 7 \cdot 20 \quad \text{Multiply both sides by 7.}$$

$$1y = 140 \quad \text{Simplify.}$$

$$y = 140$$

Check:

$$\frac{y}{7} = 20 \quad \text{Original equation}$$

$$\frac{140}{7} \stackrel{?}{=} 20 \quad \text{Replace } y \text{ with } 140.$$

$$20 = 20 \quad \text{True}$$

The solution is 140. □

PRACTICE 7 Solve: $\frac{x}{5} = 13$ ■

OBJECTIVE 3 Using Both the Addition and Multiplication Properties

Next, we practice solving equations using both properties.

EXAMPLE 8 Solve: $12a - 8a = 10 + 2a - 13 - 7$

Solution First, simplify both sides of the equation by combining like terms.

$$12a - 8a = 10 + 2a - 13 - 7$$
$$4a = 2a - 10 \quad \text{Combine like terms.}$$

To get all terms containing a variable on one side, subtract $2a$ from both sides.

$$4a - 2a = 2a - 10 - 2a \quad \text{Subtract } 2a \text{ from both sides.}$$
$$2a = -10 \quad \text{Simplify.}$$
$$\frac{2a}{2} = \frac{-10}{2} \quad \text{Divide both sides by 2.}$$
$$a = -5 \quad \text{Simplify.}$$

Check: Check by replacing a with -5 in the original equation. The solution is -5. □

PRACTICE 8 Solve: $6b - 11b = 18 + 2b - 6 + 9$ ■

OBJECTIVE 4

Writing Word Phrases as Algebraic Expressions

Next, we practice writing word phrases as algebraic expressions.

EXAMPLE 9

a. The sum of two numbers is 8. If one number is 3, find the other number.

b. The sum of two numbers is 8. If one number is x, write an expression representing the other number.

c. An 8-foot board is cut into two pieces. If one piece is x feet, express the length of the other piece in terms of x.

Solution

a. If the sum of two numbers is 8 and one number is 3, we find the other number by subtracting 3 from 8. The other number is $8 - 3$ or 5.

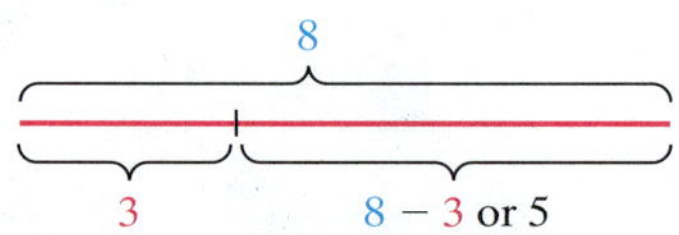

b. If the sum of two numbers is 8 and one number is x, we find the other number by subtracting x from 8. The other number is represented by $8 - x$.

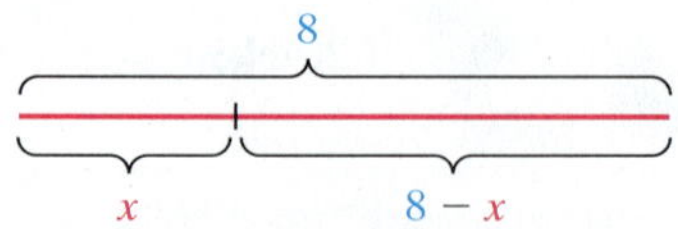

c. If an 8-foot board is cut into two pieces and one piece is x feet, we find the other length by subtracting x from 8. The other piece is $(8 - x)$ feet.

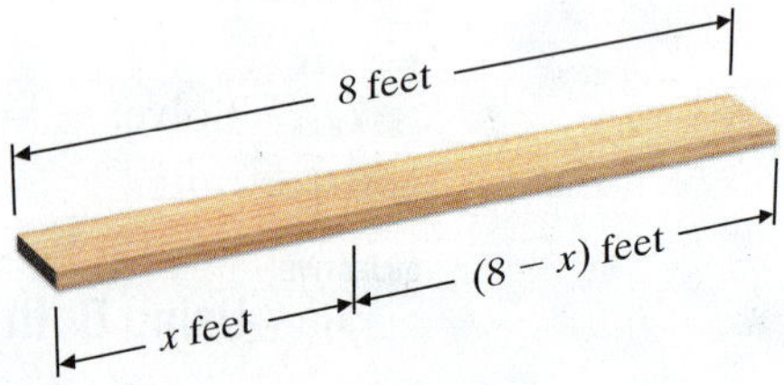

□

PRACTICE
9
a. The sum of two numbers is 9. If one number is 2, find the other number.
b. The sum of two numbers is 9. If one number is x, write an expression representing the other number.
c. A 9-foot rope is cut into two pieces. If one piece is x feet, express the length of the other piece in terms of x.

EXAMPLE 10 If x is the first of three consecutive integers, express the sum of the three integers in terms of x. Simplify if possible.

Solution An example of three consecutive integers is

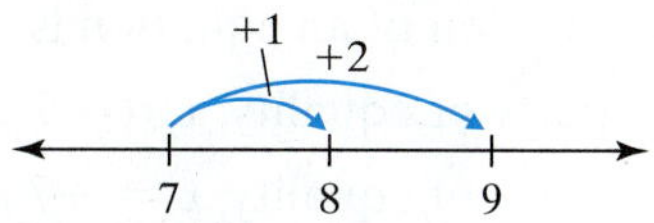

The second consecutive integer is always 1 more than the first, and the third consecutive integer is 2 more than the first. If x is the first of three consecutive integers, the three consecutive integers are

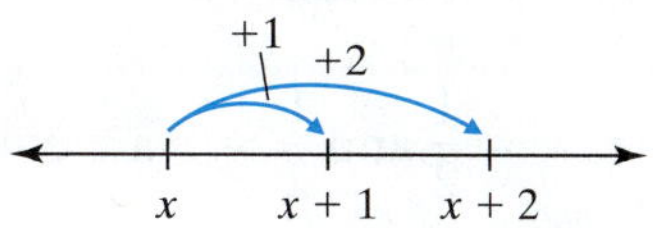

Their sum is

In words:	first integer	+	second integer	+	third integer
	↓		↓		↓
Translate:	x	+	$(x + 1)$	+	$(x + 2)$

which simplifies to $3x + 3$.

PRACTICE
10 If x is the first of three consecutive *even* integers, express their sum in terms of x.

Below are examples of consecutive even and odd integers.

Consecutive even integers:

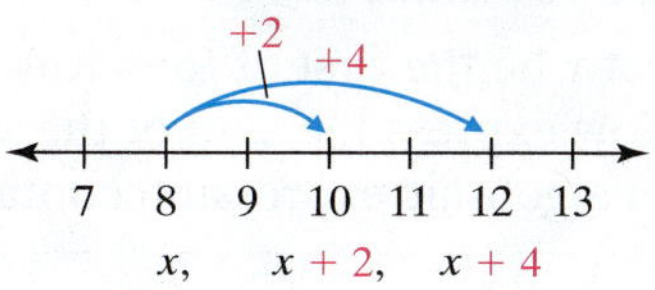

Consecutive odd integers:

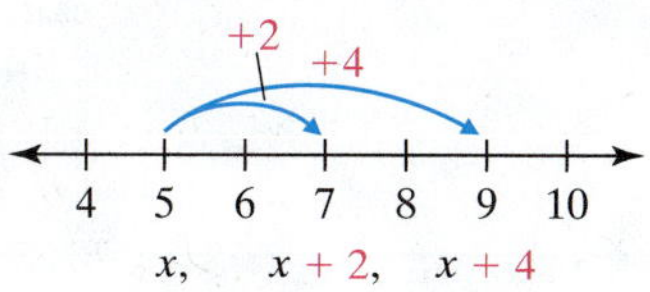

Helpful Hint

If x is an odd integer, then $x + 2$ is the next odd integer. This 2 simply means that odd integers are always 2 units from each other. (The same is true for even integers. They are always 2 units from each other.)

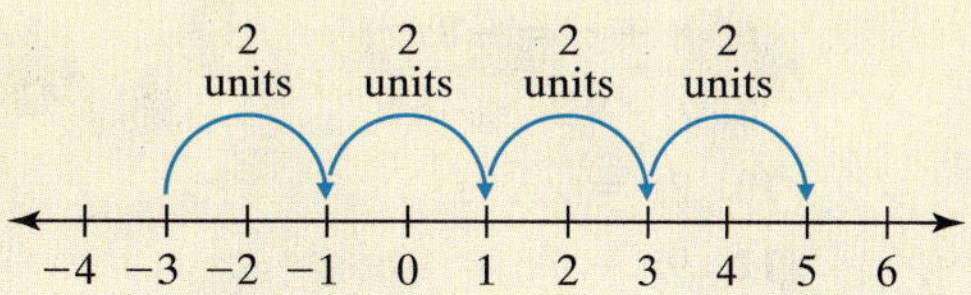

Vocabulary, Readiness & Video Check

Use the choices below to fill in each blank. Some choices may be used more than once.

addition	solving	expression	true	multiplication
equivalent	equation	solution	false	

1. The difference between an equation and an expression is that a(n) ____________ contains an equal sign, whereas an ____________ does not.

2. ____________ equations are equations that have the same solution.

3. A value of the variable that makes an equation a true statement is called a(n) ____________ of the equation.

4. The process of finding the solution of an equation is called ____________ the equation for the variable.

5. By the ____________ property of equality, $x = -2$ and $x + 10 = -2 + 10$ are equivalent equations.

6. By the ____________ property of equality, $x = -7$ and $x - 5 = -7 - 5$ are equivalent equations.

7. By the ____________ property of equality, $y = \frac{1}{2}$ and $5 \cdot y = 5 \cdot \frac{1}{2}$ are equivalent equations.

8. By the ____________ property of equality, $9x = -63$ and $\frac{9x}{9} = \frac{-63}{9}$ are equivalent equations.

9. True or false: The equations $x = \frac{1}{2}$ and $\frac{1}{2} = x$ are equivalent equations. ____________

10. True or false: The equations $\frac{z}{4} = 10$ and $4 \cdot \frac{z}{4} = 10$ are equivalent equations. ____________

Martin-Gay Interactive Videos

See Video 2.2

Watch the section lecture video and answer the following questions.

OBJECTIVE 1

11. Complete this statement based on the lecture given before Example 1. The addition property of equality means that if we have an equation, we can add the same real number to _____ of the equation and have an equivalent equation.

OBJECTIVE 2

12. Complete this statement based on the lecture given before Example 4. We can multiply both sides of an equation by _____ nonzero number and have an equivalent equation.

OBJECTIVE 3

13. Both the addition and multiplication properties of equality are used to solve Examples 6 and 7. In each of these examples, what property is applied first? What property is applied last? What conclusion, if any, can you make?

OBJECTIVE 4

14. Let x be the first of four consecutive integers, as in Example 10. Now express the sum of the second integer and the fourth integer as an algebraic expression containing x.

2.2 Exercise Set MyMathLab®

Solve each equation. Check each solution. See Examples 1 and 2.

1. $x + 7 = 10$

2. $x + 14 = 25$

3. $x - 2 = -4$

4. $y - 9 = 1$

5. $3 + x = -11$

6. $8 + z = -8$

7. $r - 8.6 = -8.1$

8. $t - 9.2 = -6.8$

9. $8x = 7x - 3$

10. $2x = x - 5$

11. $5b - 0.7 = 6b$

12. $9x + 5.5 = 10x$

13. $7x - 3 = 6x$

14. $18x - 9 = 19x$

Solve each equation. See Examples 3 and 4.

15. $2x + x - 6 = 2x + 5$

16. $7y - 2 = 2y + 4y + 2$

17. $3t - t - 7 = t - 7$

18. $4c + 8 - c = 8 + 2c$

19. $7x + 2x = 8x - 3$

20. $3n + 2n = 7 + 4n$

21. $-2(x + 1) + 3x = 14$

22. $10 = 8(3y - 4) - 23y + 20$

Solve each equation. See Example 6.

23. $-5x = 20$

24. $-7x = -49$

25. $3x = 0$

26. $-2x = 0$

27. $-x = -12$

28. $-y = 8$

29. $3x + 2x = 50$

30. $-y + 4y = 33$

Solve each equation. See Examples 5 and 7.

31. $\frac{2}{3}x = -8$

32. $\frac{3}{4}n = -15$

33. $\frac{1}{6}d = \frac{1}{2}$

34. $\frac{1}{8}v = \frac{1}{4}$

35. $\frac{a}{-2} = 1$

36. $\frac{d}{15} = 2$

37. $\frac{k}{7} = 0$

38. $\frac{f}{-5} = 0$

39. In your own words, explain the addition property of equality.

40. In your own words, explain the multiplication property of equality.

MIXED PRACTICE

Solve each equation. Check each solution. See Examples 1 through 8.

41. $2x - 4 = 16$

42. $3x - 1 = 26$

43. $-x + 2 = 22$

44. $-x + 4 = -24$

45. $6a + 3 = 3$

46. $8t + 5 = 5$

47. $6x + 10 = -20$

48. $-10y + 15 = 5$

49. $5 - 0.3k = 5$

50. $2 + 0.4p = 2$

51. $-2x + \frac{1}{2} = \frac{7}{2}$

52. $-3n - \frac{1}{3} = \frac{8}{3}$

53. $\frac{x}{3} + 2 = -5$

54. $\frac{b}{4} - 1 = -7$

55. $10 = 2x - 1$

56. $12 = 3j - 4$

57. $6z - 8 - z + 3 = 0$

58. $4a + 1 + a - 11 = 0$

59. $10 - 3x - 6 - 9x = 7$

60. $12x + 30 + 8x - 6 = 10$

61. $\frac{5}{6}x = 10$

62. $-\frac{3}{4}x = 9$

63. $1 = 0.3x - 0.5x - 5$

64. $19 = 0.4x - 0.9x - 6$

65. $z - 5z = 7z - 9 - z$

66. $t - 6t = -13 + t - 3t$

67. $0.4x - 0.6x - 5 = 1$

68. $0.1x - 0.6x - 6 = 19$

69. $6 - 2x + 8 = 10$

70. $-5 - 6y + 6 = 19$

71. $-3a + 6 + 5a = 7a - 8a$

72. $4b - 8 - b = 10b - 3b$

73. $20 = -3(2x + 1) + 7x$

74. $-3 = -5(4x + 3) + 21x$

See Example 9.

75. Two numbers have a sum of 20. If one number is p, express the other number in terms of p.

76. Two numbers have a sum of 13. If one number is y, express the other number in terms of y.

77. A 10-foot board is cut into two pieces. If one piece is x feet long, express the other length in terms of x.

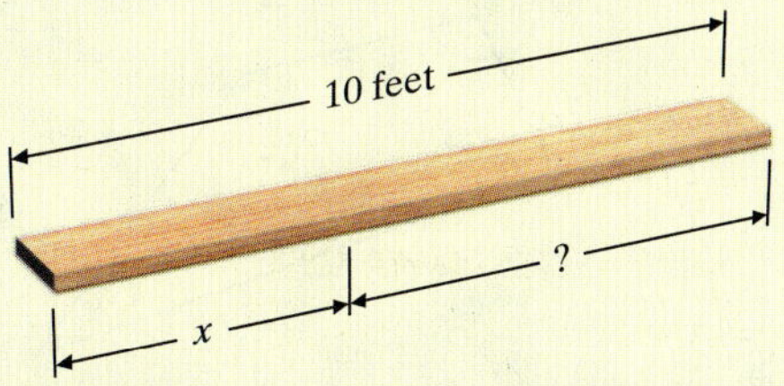

78. A 5-foot piece of string is cut into two pieces. If one piece is x feet long, express the other length in terms of x.

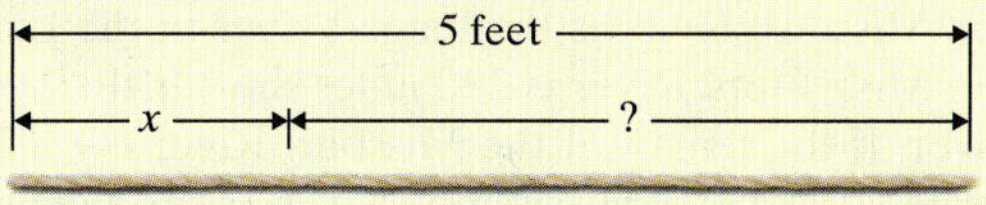

79. Two angles are *supplementary* if their sum is 180°. If one angle measures $x°$, express the measure of its supplement in terms of x.

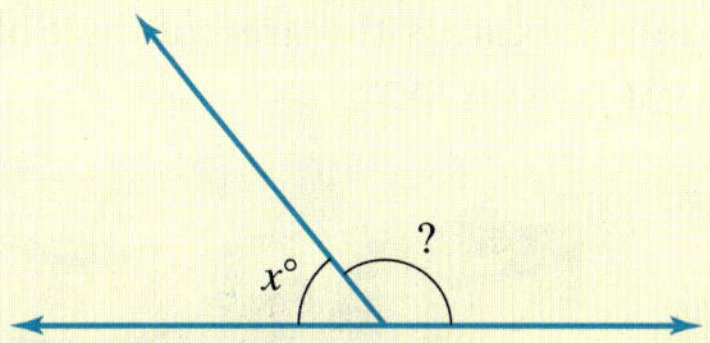

80. Two angles are *complementary* if their sum is 90°. If one angle measures $x°$, express the measure of its complement in terms of x.

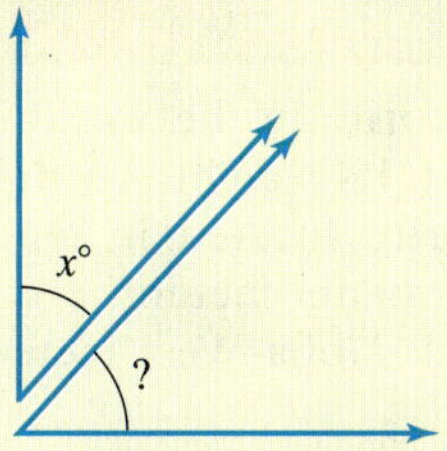

81. In a mayoral election, April Catarella received 284 more votes than Charles Pecot. If Charles received n votes, how many votes did April receive?

82. The length of the top of a computer desk is $1\frac{1}{2}$ feet longer than its width. If its width measures m feet, express its length as an algebraic expression in m.

83. The Verrazano-Narrows Bridge in New York City is the longest suspension bridge in North America. The Golden Gate Bridge in San Francisco is 60 feet shorter than the Verrazano-Narrows Bridge. If the length of the Verrazano-Narrows Bridge is m feet, express the length of the Golden Gate Bridge as an algebraic expression in m. (*Source: World Almanac*, 2000).

84. The longest interstate highway in the U.S. is I-90, which connects Seattle, Washington, and Boston, Massachusetts. The second longest interstate highway, I-80 (connecting San Francisco, California, and Teaneck, New Jersey), is 178.5 miles shorter than I-90. If the length of I-80 is m miles, express the length of I-90 as an algebraic expression in m. (*Source:* U.S. Department of Transportation–Federal Highway Administration)

85. In a recent year, the number of graduate students at the University of Texas at Austin was approximately 28,000 fewer than the number of undergraduate students. If the number of undergraduate students was n, how many graduate students attend UT Austin? (*Source:* University of Texas at Austin)

86. The Missouri River is the longest river in the United States. The Mississippi River is 200 miles shorter than the Missouri River. If the length of the Missouri River is r miles, express the length of the Mississippi River as an algebraic expression in r. (*Source:* U.S. Geological Survey)

87. The area of the Sahara Desert in Africa is 7 times the area of the Gobi Desert in Asia. If the area of the Gobi Desert is x square miles, express the area of the Sahara Desert as an algebraic expression in x.

88. The largest meteorite in the world is the Hoba West located in Namibia. Its weight is 3 times the weight of the Armanty meteorite located in Outer Mongolia. If the weight of the Armanty meteorite is y kilograms, express the weight of the Hoba West meteorite as an algebraic expression in y.

Write each algebraic expression described. Simplify if possible. See Example 10.

89. If x represents the first of two consecutive odd integers, express the sum of the two integers in terms of x.

90. If x is the first of four consecutive even integers, write their sum as an algebraic expression in x.

91. If x is the first of four consecutive integers, express the sum of the first integer and the third integer as an algebraic expression containing the variable x.

92. If x is the first of two consecutive integers, express the sum of 20 and the second consecutive integer as an algebraic expression containing the variable x.

93. Classrooms on one side of the science building are all numbered with consecutive even integers. If the first room on this side of the building is numbered x, write an expression in x for the sum of five classroom numbers in a row. Then simplify this expression.

94. Two sides of a quadrilateral have the same length, x, while the other two sides have the same length, both being the next consecutive odd integer. Write the sum of these lengths. Then simplify this expression.

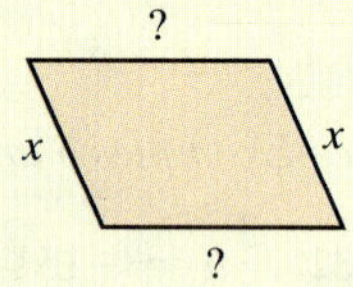

REVIEW AND PREVIEW

Simplify each expression. See Section 2.1.

95. $5x + 2(x - 6)$

96. $-7y + 2y - 3(y + 1)$

97. $-(x - 1) + x$

98. $-(3a - 3) + 2a - 6$

The circle graph below shows each continent/region's percent of Earth's land. Use this graph for Exercises 7 through 10. See Section R.4.

Continent/Region Land Area (percent of Earth's Land)

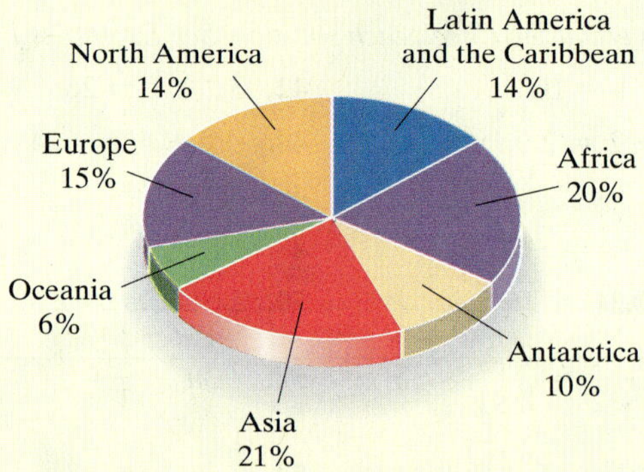

99. What percent of land is not included in North America?

100. What percent of land is in Asia, Antarctica, or Africa?

101. If Earth's land area is 56.4 million square miles, find the land area of Europe.

102. If Earth's land area is 56.4 million square miles, find the land area of Africa.

CONCEPT EXTENSIONS

△ **103.** The sum of the angles of a triangle is 180°. If one angle of a triangle measures $x°$ and a second angle measures $(2x + 7)°$, express the measure of the third angle in terms of x. Simplify the expression.

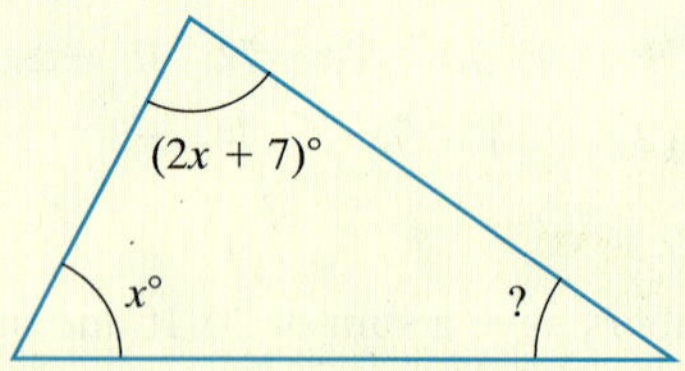

△ **104.** A quadrilateral is a four-sided figure like the one shown below whose angle sum is 360°. If one angle measures $x°$, a second angle measures $3x°$, and a third angle measures $5x°$, express the measure of the fourth angle in terms of x. Simplify the expression.

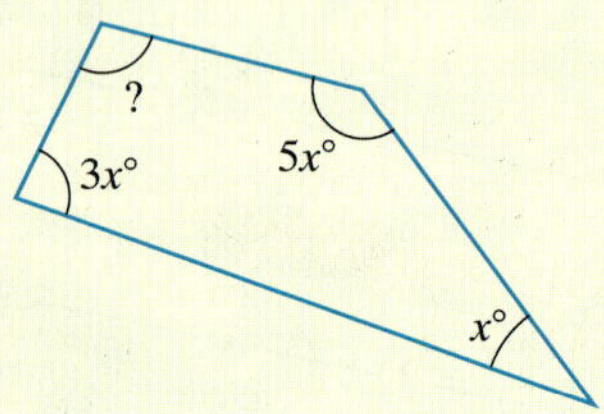

105. Write two terms whose sum is $-3x$.

106. Write four terms whose sum is $2y - 6$.

Use the addition property to fill in numbers between the parentheses so that the middle equation simplifies to the last equation. See the Concept Check in this section.

107.
$$x - 4 = -9$$
$$x - 4 + (\quad) = -9 + (\quad)$$
$$x = -5$$

108.
$$a + 9 = 15$$
$$a + 9 + (\quad) = 15 + (\quad)$$
$$a = 6$$

Fill in the blanks with numbers of your choice so that each equation has the given solution. Note: *Each blank may be replaced with a different number.*

109. _____ $+ x =$ _____; Solution: -3

110. $x -$ _____ $=$ _____; Solution: -10

111. Let $x = 1$ and then $x = 2$ in the equation $x + 5 = x + 6$. Is either number a solution? How many solutions do you think this equation has? Explain your answer.

112. Let $x = 1$ and then $x = 2$ in the equation $x + 3 = x + 3$. Is either number a solution? How many solutions do you think this equation has? Explain your answer.

Fill in the blank with a number so that each equation has the given solution.

113. $6x =$ _____; solution: -8

114. _____ $x = 10$; solution: $\frac{1}{2}$

115. A licensed nurse practitioner is instructed to give a patient 2100 milligrams of an antibiotic over a period of 36 hours. If the antibiotic is to be given every 4 hours starting immediately, how much antibiotic should be given in each dose? To answer this question, solve the equation $9x = 2100$.

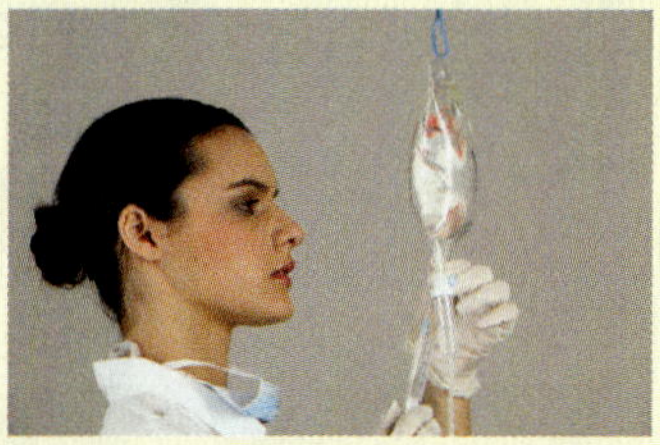

116. Suppose you are a pharmacist and a customer asks you the following question. His child is to receive 13.5 milliliters of a nausea medicine over a period of 54 hours. If the nausea medicine is to be administered every 6 hours starting immediately, how much medicine should be given in each dose?

Use a calculator to determine whether the given value is a solution of the given equation.

117. $8.13 + 5.85y = 20.05y - 8.91; y = 1.2$

118. $3(a + 4.6) = 5a + 2.5; a = 6.3$

Solve each equation.

119. $-3.6x = 10.62$

120. $4.95y = -31.185$

121. $7x - 5.06 = -4.92$

122. $0.06y + 2.63 = 2.5562$

2.3 Solving Linear Equations

OBJECTIVES

1. Apply a General Strategy for Solving a Linear Equation.
2. Solve Equations Containing Fractions.
3. Solve Equations Containing Decimals.
4. Recognize Identities and Equations with No Solution.

OBJECTIVE

1 Applying a General Strategy for Solving a Linear Equation

We now present a general strategy for solving linear equations. One new piece of strategy is a suggestion to "clear an equation of fractions" as a first step. Doing so makes the equation less tedious, since operating on integers is usually more convenient than operating on fractions.

Solving Linear Equations in One Variable

Step 1. Multiply on both sides by the least common denominator (LCD) to clear the equation of fractions if they occur.

Step 2. Use the distributive property to remove parentheses if they occur.

Step 3. Simplify each side of the equation by combining like terms.

Step 4. Get all variable terms on one side and all numbers on the other side by using the addition property of equality.

Step 5. Get the variable alone by using the multiplication property of equality.

Step 6. Check the solution by substituting it into the original equation.

EXAMPLE 1 Solve: $4(2x - 3) + 7 = 3x + 5$

Solution There are no fractions, so we begin with Step 2.

$$4(2x - 3) + 7 = 3x + 5$$

Step 2. $8x - 12 + 7 = 3x + 5$ Apply the distributive property.

Step 3. $8x - 5 = 3x + 5$ Combine like terms.

Step 4. Get all variable terms on the same side of the equation by subtracting $3x$ from both sides, then adding 5 to both sides.

$8x - 5 - 3x = 3x + 5 - 3x$ Subtract $3x$ from both sides.

$5x - 5 = 5$ Simplify.

$5x - 5 + 5 = 5 + 5$ Add 5 to both sides.

$5x = 10$ Simplify.

Step 5. Use the multiplication property of equality to get x alone.

$\frac{5x}{5} = \frac{10}{5}$ Divide both sides by 5.

$x = 2$ Simplify.

Step 6. Check.

Helpful Hint
When checking solutions, remember to use the original written equation.

$4(2x - 3) + 7 = 3x + 5$ Original equation

$4[2(2) - 3] + 7 \stackrel{?}{=} 3(2) + 5$ Replace x with 2.

$4(4 - 3) + 7 \stackrel{?}{=} 6 + 5$

$4(1) + 7 \stackrel{?}{=} 11$

$4 + 7 \stackrel{?}{=} 11$

$11 = 11$ True

The solution is 2 or the solution set is $\{2\}$. □

PRACTICE
1 Solve: $2(4a - 9) + 3 = 5a - 6$ ■

EXAMPLE 2 Solve: $8(2 - t) = -5t$

Solution First, we apply the distributive property.

$$8(2 - t) = -5t$$

Step 2. $16 - 8t = -5t$ Use the distributive property.

Step 4. $16 - 8t + 8t = -5t + 8t$ To get variable terms on one side, add $8t$ to both sides.

$16 = 3t$ Combine like terms.

Step 5. $\frac{16}{3} = \frac{3t}{3}$ Divide both sides by 3.

$\frac{16}{3} = t$ Simplify.

Step 6. Check.

$8(2 - t) = -5t$ Original equation

$8\left(2 - \frac{16}{3}\right) \stackrel{?}{=} -5\left(\frac{16}{3}\right)$ Replace t with $\frac{16}{3}$.

$8\left(\frac{6}{3} - \frac{16}{3}\right) \stackrel{?}{=} -\frac{80}{3}$ The LCD is 3.

$$8\left(-\frac{10}{3}\right) \stackrel{?}{=} -\frac{80}{3} \quad \text{Subtract fractions.}$$

$$-\frac{80}{3} = -\frac{80}{3} \quad \text{True}$$

The solution is $\frac{16}{3}$. □

PRACTICE 2 Solve: $7(x - 3) = -6x$ ■

OBJECTIVE 2

Solving Equations Containing Fractions

If an equation contains fractions, we can clear the equation of fractions by multiplying both sides by the LCD of all denominators. By doing this, we avoid working with time-consuming fractions.

EXAMPLE 3 Solve: $\frac{x}{2} - 1 = \frac{2}{3}x - 3$

Solution We begin by clearing fractions. To do this, we multiply both sides of the equation by the LCD of 2 and 3, which is 6.

$$\frac{x}{2} - 1 = \frac{2}{3}x - 3$$

Step 1. $6\left(\frac{x}{2} - 1\right) = 6\left(\frac{2}{3}x - 3\right)$ Multiply both sides by the LCD, 6.

Step 2. $6\left(\frac{x}{2}\right) - 6(1) = 6\left(\frac{2}{3}x\right) - 6(3)$ Apply the distributive property.

$3x - 6 = 4x - 18$ Simplify.

> **Helpful Hint**
> Don't forget to multiply *each* term by the LCD.

There are no longer grouping symbols and no like terms on either side of the equation, so we continue with Step 4.

$$3x - 6 = 4x - 18$$

Step 4. $3x - 6 - 3x = 4x - 18 - 3x$ To get variable terms on one side, subtract $3x$ from both sides.

$-6 = x - 18$ Simplify.

$-6 + 18 = x - 18 + 18$ Add 18 to both sides.

$12 = x$ Simplify.

Step 5. The variable is now alone, so there is no need to apply the multiplication property of equality.

Step 6. Check.

$\frac{x}{2} - 1 = \frac{2}{3}x - 3$ Original equation

$\frac{12}{2} - 1 \stackrel{?}{=} \frac{2}{3} \cdot 12 - 3$ Replace x with 12.

$6 - 1 \stackrel{?}{=} 8 - 3$ Simplify.

$5 = 5$ True

The solution is 12. □

PRACTICE 3 Solve: $\frac{3}{5}x - 2 = \frac{2}{3}x - 1$ ■

EXAMPLE 4 Solve: $\frac{2(a+3)}{3} = 6a + 2$

Solution We clear the equation of fractions first.

$$\frac{2(a+3)}{3} = 6a + 2$$

Step 1. $3 \cdot \frac{2(a+3)}{3} = 3(6a + 2)$ Clear the fraction by multiplying both sides by the LCD, 3.

$$2(a+3) = 3(6a+2)$$

Step 2. Next, we use the distributive property and remove parentheses.

$2a + 6 = 18a + 6$ Apply the distributive property.

Step 4. $2a + 6 - 6 = 18a + 6 - 6$ Subtract 6 from both sides.

$$2a = 18a$$

$2a - 18a = 18a - 18a$ Subtract $18a$ from both sides.

$$-16a = 0$$

Step 5. $\frac{-16a}{-16} = \frac{0}{-16}$ Divide both sides by -16.

$a = 0$ Write the fraction in simplest form.

Step 6. To check, replace a with 0 in the original equation. The solution is 0. □

PRACTICE 4 Solve: $\frac{4(y+3)}{3} = 5y - 7$ ■

> **Helpful Hint**
> Remember: When solving an equation, it makes no difference on which side of the equation variable terms lie. Just make sure that constant terms (number terms) lie on the other side.

OBJECTIVE

3 Solving Equations Containing Decimals

When solving a problem about money, you may need to solve an equation containing decimals. If you choose, you may multiply to clear the equation of decimals.

EXAMPLE 5 Solve: $0.25x + 0.10(x - 3) = 0.05(22)$

Solution First we clear this equation of decimals by multiplying both sides of the equation by 100. Recall that multiplying a decimal number by 100 has the effect of moving the decimal point 2 places to the right.

$$0.25x + 0.10(x - 3) = 0.05(22)$$

> **Helpful Hint**
> By the distributive property, 0.10 is multiplied by x and -3. Thus to multiply each term here by 100, we only need to multiply 0.10 by 100.

Step 1. $0.25x + 0.10(x - 3) = 0.05(22)$ Multiply both sides by 100.

$$25x + 10(x - 3) = 5(22)$$

Step 2. $25x + 10x - 30 = 110$ Apply the distributive property.

Step 3. $35x - 30 = 110$ Combine like terms.

Step 4. $35x - 30 + 30 = 110 + 30$ Add 30 to both sides.

$35x = 140$ Combine like terms.

Step 5. $$\frac{35x}{35} = \frac{140}{35} \quad \text{Divide both sides by 35.}$$
$$x = 4$$

Step 6. To check, replace x with 4 in the original equation. The solution is 4. □

PRACTICE 5 Solve: $0.35x + 0.09(x + 4) = 0.03(12)$ ■

OBJECTIVE 4

Recognizing Identities and Equations with No Solution

So far, each equation that we have solved has had a single solution. However, not every equation in one variable has a single solution. Some equations have no solution, while others have an infinite number of solutions. For example,

$$x + 5 = x + 7$$

has no solution since no matter which **real number** we replace x with, the equation is false.

real number + 5 = **same** real number + 7 **FALSE**

On the other hand,

$$x + 6 = x + 6$$

has infinitely many solutions since x can be replaced by any real number and the equation is always true.

real number + 6 = **same** real number + 6 **TRUE**

The equation $x + 6 = x + 6$ is called an **identity.** The next two examples illustrate special equations like these.

EXAMPLE 6 Solve: $-2(x - 5) + 10 = -3(x + 2) + x$

Solution

$$\begin{aligned} -2(x - 5) + 10 &= -3(x + 2) + x \\ -2x + 10 + 10 &= -3x - 6 + x && \text{Apply the distributive property on both sides.} \\ -2x + 20 &= -2x - 6 && \text{Combine like terms.} \\ -2x + 20 + 2x &= -2x - 6 + 2x && \text{Add } 2x \text{ to both sides.} \\ 20 &= -6 && \text{Combine like terms.} \end{aligned}$$

The final equation contains no variable terms, and there is no value for x that makes $20 = -6$ a true equation. We conclude that there is **no solution** to this equation. In set notation, we can indicate that there is no solution with the empty set, { }, or use the empty set or null set symbol, ∅. In this chapter, we will simply write *no solution*. □

PRACTICE 6 Solve: $4(x + 4) - x = 2(x + 11) + x$ ■

EXAMPLE 7 Solve: $3(x - 4) = 3x - 12$

Solution

$$\begin{aligned} 3(x - 4) &= 3x - 12 \\ 3x - 12 &= 3x - 12 && \text{Apply the distributive property.} \end{aligned}$$

(Continued on next page)

The left side of the equation is now identical to the right side. Every real number may be substituted for x and a true statement will result. We arrive at the same conclusion if we continue.

$$3x - 12 = 3x - 12$$
$$3x - 12 + 12 = 3x - 12 + 12 \quad \text{Add 12 to both sides.}$$
$$3x = 3x \quad \text{Combine like terms.}$$
$$3x - 3x = 3x - 3x \quad \text{Subtract } 3x \text{ from both sides.}$$
$$0 = 0$$

Again, one side of the equation is identical to the other side. Thus, $3(x - 4) = 3x - 12$ is an **identity** and **all real numbers** are solutions. In set notation, this is {all real numbers}. □

PRACTICE

7 Solve: $12x - 18 = 9(x - 2) + 3x$ ■

✔ CONCEPT CHECK

Suppose you have simplified several equations and obtain the following results. What can you conclude about the solutions to the original equation?

a. $7 = 7$ **b.** $x = 0$ **c.** $7 = -4$

Graphing Calculator Explorations

Checking Equations

We can use a calculator to check possible solutions of equations. To do this, replace the variable by the possible solution and evaluate both sides of the equation separately.

Equation: $3x - 4 = 2(x + 6)$ *Solution:* $x = 16$

$$3x - 4 = 2(x + 6) \quad \text{Original equation}$$
$$3(16) - 4 \stackrel{?}{=} 2(16 + 6) \quad \text{Replace } x \text{ with 16.}$$

Now evaluate each side with your calculator.

Evaluate left side:

[3] [×] [16] [−] [4] then [=] or [ENTER] Display: [44] or [3*16 − 4 / 44]

Evaluate right side:

[2] [(] [16] [+] [6] [)] then [=] or [ENTER] Display: [44] or [2(16 + 6) / 44]

Since the left side equals the right side, the solution checks.

Use a calculator to check the possible solutions to each equation.

1. $2x = 48 + 6x;\ x = -12$

2. $-3x - 7 = 3x - 1;\ x = -1$

3. $5x - 2.6 = 2(x + 0.8);\ x = 4.4$

4. $-1.6x - 3.9 = -6.9x - 25.6;\ x = 5$

5. $\frac{564x}{4} = 200x - 11(649);\ x = 121$

6. $20(x - 39) = 5x - 432;\ x = 23.2$

Answers to Concept Check:

a. Every real number is a solution.
b. The solution is 0.
c. There is no solution.

Vocabulary, Readiness & Video Check

Throughout algebra, it is important to be able to identify equations and expressions.

Remember,
- an equation contains an equal sign and
- an expression does not.

Also,
- we solve equations and
- we simplify or perform operations on expressions.

Identify each as an equation or an expression.

1. $x = -7$ ________

2. $x - 7$ ________

3. $4y - 6 + 9y + 1$ ________

4. $4y - 6 = 9y + 1$ ________

5. $\frac{1}{x} - \frac{x-1}{8}$ ________

6. $\frac{1}{x} - \frac{x-1}{8} = 6$ ________

7. $0.1x + 9 = 0.2x$ ________

8. $0.1x^2 + 9y - 0.2x^2$ ________

Martin-Gay Interactive Videos

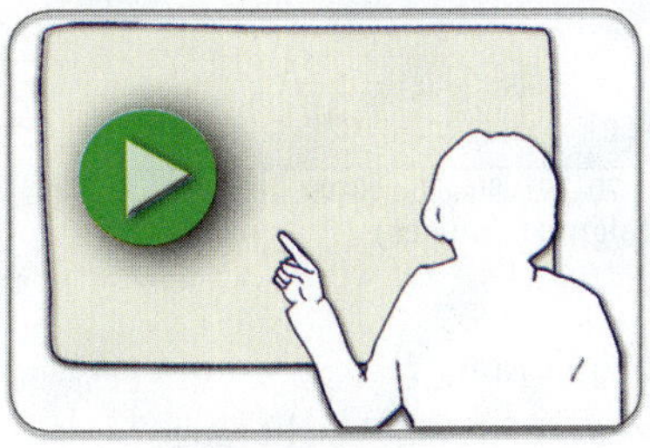

See Video 2.3

Watch the section lecture video and answer the following questions.

OBJECTIVE 1

9. The general strategy for solving linear equations in one variable is discussed after Example 1. How many properties are mentioned in this strategy and what are they?

OBJECTIVE 2

10. In the first step for solving Example 2, both sides of the equation are being multiplied by the LCD. Why is the distributive property mentioned?

OBJECTIVE 3

11. In Example 3, why is the number of decimal places in each term of the equation important?

OBJECTIVE 4

12. Complete each statement based on Examples 4 and 5.
When solving an equation and all variable terms subtract out:
a. If you have a true statement, then the equation has ____ solution(s).
b. If you have a false statement, then the equation has ____ solution(s).

2.3 Exercise Set MyMathLab®

Solve each equation. See Examples 1 and 2.

1. $-4y + 10 = -2(3y + 1)$

2. $-3x + 1 = -2(4x + 2)$

3. $15x - 8 = 10 + 9x$

4. $15x - 5 = 7 + 12x$

5. $-2(3x - 4) = 2x$

6. $-(5x - 10) = 5x$

7. $5(2x - 1) - 2(3x) = 1$

8. $3(2 - 5x) + 4(6x) = 12$

9. $-6(x - 3) - 26 = -8$

10. $-4(n - 4) - 23 = -7$

11. $8 - 2(a + 1) = 9 + a$

12. $5 - 6(2 + b) = b - 14$

13. $4x + 3 = -3 + 2x + 14$

14. $6y - 8 = -6 + 3y + 13$

15. $-2y - 10 = 5y + 18$

16. $-7n + 5 = 8n - 10$

Solve each equation. See Examples 3 through 5.

17. $\frac{2}{3}x + \frac{4}{3} = -\frac{2}{3}$

18. $\frac{4}{5}x - \frac{8}{5} = -\frac{16}{5}$

19. $\frac{3}{4}x - \frac{1}{2} = 1$

20. $\frac{2}{9}x - \frac{1}{3} = 1$

21. $0.50x + 0.15(70) = 35.5$

22. $0.40x + 0.06(30) = 9.8$

23. $\frac{2(x + 1)}{4} = 3x - 2$

24. $\frac{3(y + 3)}{5} = 2y + 6$

25. $x + \frac{7}{6} = 2x - \frac{7}{6}$

26. $\frac{5}{2}x - 1 = x + \frac{1}{4}$

27. $0.12(y - 6) + 0.06y = 0.08y - 0.70$

28. $0.60(z - 300) + 0.05z = 0.70z - 205$

Solve each equation. See Examples 6 and 7.

29. $4(3x + 2) = 12x + 8$

30. $14x + 7 = 7(2x + 1)$

31. $\frac{x}{4} + 1 = \frac{x}{4}$

32. $\frac{x}{3} - 2 = \frac{x}{3}$

33. $3x - 7 = 3(x + 1)$

34. $2(x - 5) = 2x + 10$

35. $-2(6x - 5) + 4 = -12x + 14$

36. $-5(4y - 3) + 2 = -20y + 17$

MIXED PRACTICE

Solve. See Examples 1 through 7.

37. $\frac{6(3 - z)}{5} = -z$

38. $\frac{4(5 - w)}{3} = -w$

39. $-3(2t - 5) + 2t = 5t - 4$

40. $-(4a - 7) - 5a = 10 + a$

41. $5y + 2(y - 6) = 4(y + 1) - 2$

42. $9x + 3(x - 4) = 10(x - 5) + 7$

43. $\frac{3(x - 5)}{2} = \frac{2(x + 5)}{3}$

44. $\frac{5(x - 1)}{4} = \frac{3(x + 1)}{2}$

45. $0.7x - 2.3 = 0.5$

46. $0.9x - 4.1 = 0.4$

47. $5x - 5 = 2(x + 1) + 3x - 7$

48. $3(2x - 1) + 5 = 6x + 2$

49. $4(2n + 1) = 3(6n + 3) + 1$

50. $4(4y + 2) = 2(1 + 6y) + 8$

51. $x + \frac{5}{4} = \frac{3}{4}x$

52. $\frac{7}{8}x + \frac{1}{4} = \frac{3}{4}x$

53. $\frac{x}{2} - 1 = \frac{x}{5} + 2$

54. $\frac{x}{5} - 7 = \frac{x}{3} - 5$

55. $2(x + 3) - 5 = 5x - 3(1 + x)$

56. $4(2 + x) + 1 = 7x - 3(x - 2)$

57. $0.06 - 0.01(x + 1) = -0.02(2 - x)$

58. $-0.01(5x + 4) = 0.04 - 0.01(x + 4)$

59. $\frac{9}{2} + \frac{5}{2}y = 2y - 4$

60. $3 - \frac{1}{2}x = 5x - 8$

61. $-2y - 10 = 5y + 18$

62. $7n + 5 = 10n - 10$

63. $0.6x - 0.1 = 0.5x + 0.2$

64. $0.2x - 0.1 = 0.6x - 2.1$

65. $0.02(6t - 3) = 0.12(t - 2) + 0.18$

66. $0.03(2m + 7) = 0.06(5 + m) - 0.09$

REVIEW AND PREVIEW

Recall that a histogram is a special bar graph. The width of each bar represents a range of numbers (class intervals), and the height of each bar (class frequency) tells us how many times a number in the class interval occurs. Use the histogram below to answer the exercises. (See Section R.4.)

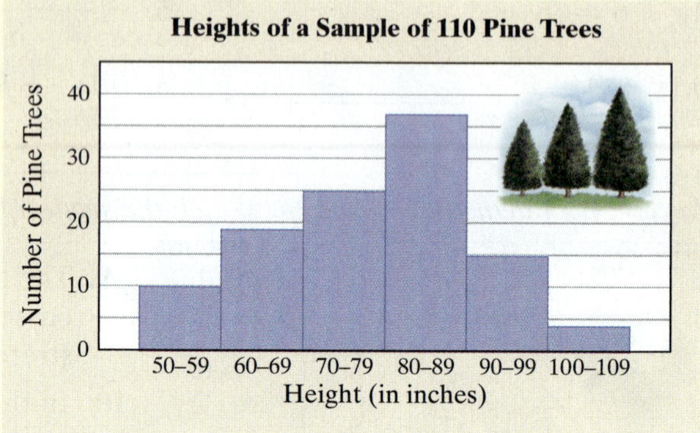

67. Most trees are in what height range?

68. The fewest number of trees are in what height range?

69. How many trees have a height of 69 inches or less?

70. How many trees have a height of 90 inches or more?

71. What fraction of the 110 trees sampled have a height of 70–79 inches?

72. What fraction of the 110 trees sampled have a height of 50–59 inches?

See Section 2.1.

73. A plot of land is in the shape of a triangle. If one side is x meters, a second side is $(2x - 3)$ meters, and a third side is $(3x - 5)$ meters, express the perimeter of the lot as a simplified expression in x.

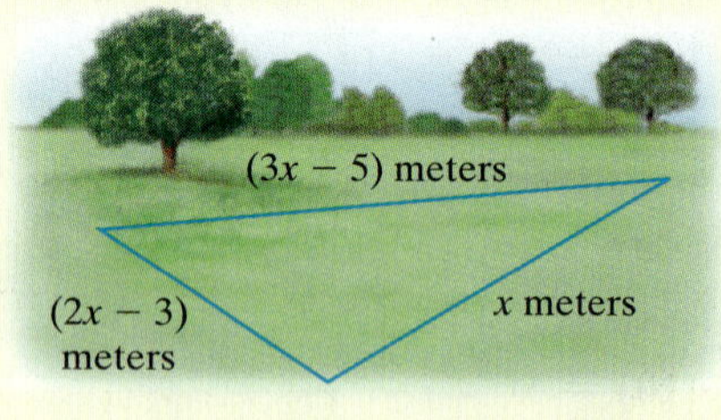

74. A portion of a board has length x feet. The other part has length $(7x - 9)$ feet. Express the total length of the board as a simplified expression in x.

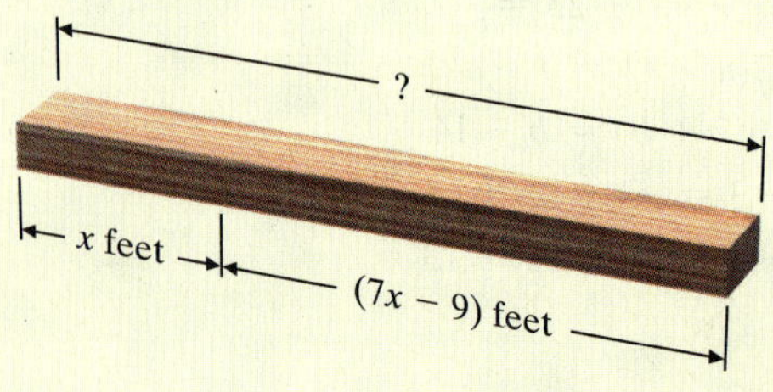

CONCEPT EXTENSIONS

See the Concept Check in this section.

75. **a.** Solve: $x + 3 = x + 3$ for x

b. If you simplify an equation and get $0 = 0$, what can you conclude about the solution(s) of the original equation?

c. On your own, construct an equation for which every real number is a solution.

76. **a.** Solve: $x + 3 = x + 5$ for x

b. If you simplify an equation and get $3 = 5$, what can you conclude about the solution(s) of the original equation?

c. On your own, construct an equation that has no solution.

For Exercises 77 through 82, match each equation in the first column with its solution in the second column. Items in the second column may be used more than once.

77. $5x + 1 = 5x + 1$

78. $3x + 1 = 3x + 2$

79. $2x - 6x - 10 = -4x + 3 - 10$

80. $x - 11x - 3 = -10x - 1 - 2$

81. $9x - 20 = 8x - 20$

82. $-x + 15 = x + 15$

A. all real numbers

B. no solution

C. 0

83. Explain the difference between simplifying an expression and solving an equation.

84. On your own, write an expression and then an equation. Label each.

For Exercises 85 and 86, **a.** *Write an equation for perimeter.* **b.** *Solve the equation in part (a).* **c.** *Find the length of each side.*

△ **85.** The perimeter of a geometric figure is the sum of the lengths of its sides. The perimeter of the following pentagon (five-sided figure) is 28 centimeters.

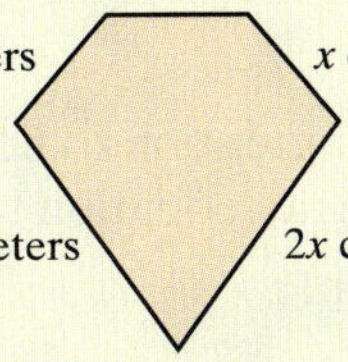

△ **86.** The perimeter of the following triangle is 35 meters.

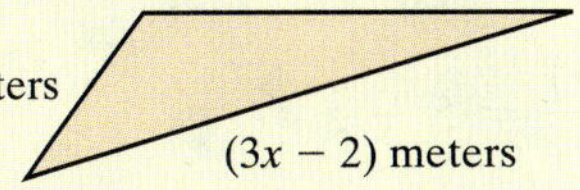

Fill in the blanks with numbers of your choice so that each equation has the given solution. Note: Each blank may be replaced by a different number.

87. $x +$ ____ $= 2x -$ ____; solution: 9

88. $-5x -$ ____ $=$ ____; solution: 2

Solve.

89. $1000(7x - 10) = 50(412 + 100x)$

90. $1000(x + 40) = 100(16 + 7x)$

91. $0.035x + 5.112 = 0.010x + 5.107$

92. $0.127x - 2.685 = 0.027x - 2.38$

For Exercises 93 through 96, see the example below.

Example

Solve: $t(t + 4) = t^2 - 2t + 6$

Solution

$$\begin{aligned} t(t + 4) &= t^2 - 2t + 6 \\ t^2 + 4t &= t^2 - 2t + 6 \\ t^2 + 4t - t^2 &= t^2 - 2t + 6 - t^2 \\ 4t &= -2t + 6 \\ 4t + 2t &= -2t + 6 + 2t \\ 6t &= 6 \\ t &= 1 \end{aligned}$$

Solve each equation.

93. $x(x - 3) = x^2 + 5x + 7$

94. $t^2 - 6t = t(8 + t)$

95. $2z(z + 6) = 2z^2 + 12z - 8$

96. $y^2 - 4y + 10 = y(y - 5)$

Integrated Review Solving Linear Equations

Sections 2.1–2.3

Solve. Feel free to use the steps given in Section 2.3.

1. $x - 10 = -4$

2. $y + 14 = -3$

3. $9y = 108$

4. $-3x = 78$

5. $-6x + 7 = 25$

6. $5y - 42 = -47$

7. $\frac{2}{3}x = 9$

8. $\frac{4}{5}z = 10$

9. $\frac{r}{-4} = -2$

10. $\frac{y}{-8} = 8$

11. $6 - 2x + 8 = 10$

12. $-5 - 6y + 6 = 19$

13. $2x - 7 = 2x - 27$

14. $3 + 8y = 8y - 2$

15. $-3a + 6 + 5a = 7a - 8a$

16. $4b - 8 - b = 10b - 3b$

17. $-\frac{2}{3}x = \frac{5}{9}$

18. $-\frac{3}{8}y = -\frac{1}{16}$

19. $10 = -6n + 16$

20. $-5 = -2m + 7$

21. $3(5c - 1) - 2 = 13c + 3$

22. $4(3t + 4) - 20 = 3 + 5t$

23. $\frac{2(z + 3)}{3} = 5 - z$

24. $\frac{3(w + 2)}{4} = 2w + 3$

25. $-2(2x - 5) = -3x + 7 - x + 3$

26. $-4(5x - 2) = -12x + 4 - 8x + 4$

27. $0.02(6t - 3) = 0.04(t - 2) + 0.02$

28. $0.03(m + 7) = 0.02(5 - m) + 0.03$

29. $-3y = \frac{4(y - 1)}{5}$

30. $-4x = \frac{5(1 - x)}{6}$

31. $\frac{5}{3}x - \frac{7}{3} = x$

32. $\frac{7}{5}n + \frac{3}{5} = -n$

33. $9(3x - 1) = -4 + 49$

34. $12(2x + 1) = -6 + 66$

35. $\frac{1}{10}(3x - 7) = \frac{3}{10}x + 5$

36. $\frac{1}{7}(2x - 5) = \frac{2}{7}x + 1$

37. $5 + 2(3x - 6) = -4(6x - 7)$

38. $3 + 5(2x - 4) = -7(5x + 2)$

2.4 An Introduction to Problem Solving

OBJECTIVES

1 Solve Problems Involving Direct Translations.

2 Solve Problems Involving Relationships Among Unknown Quantities.

3 Solve Problems Involving Consecutive Integers.

OBJECTIVE 1 Solving Direct Translation Problems

In previous sections, we practiced writing word phrases and sentences as algebraic expressions and equations to help prepare for problem solving. We now use these translations to help write equations that model a problem. The problem-solving steps given next may be helpful.

General Strategy for Problem Solving

1. UNDERSTAND the problem. During this step, become comfortable with the problem. Some ways of doing this are to:

Read and reread the problem.

Choose a variable to represent the unknown.

Construct a drawing whenever possible.

Propose a solution and check. Pay careful attention to how you check your proposed solution. This will help when writing an equation to model the problem.

2. TRANSLATE the problem into an equation.

3. SOLVE the equation.

4. INTERPRET the results: *Check* the proposed solution in the stated problem and state your conclusion.

Much of problem solving involves a direct translation from a sentence to an equation.

EXAMPLE 1 Finding an Unknown Number

Twice a number, added to seven, is the same as three subtracted from the number. Find the number.

Solution Translate the sentence into an equation and solve.

In words:	twice a number	added to	seven	is the same as	three subtracted from the number
	↓	↓	↓	↓	↓
Translate:	$2x$	$+$	7	$=$	$x - 3$

Helpful Hint

Order matters when subtracting (and dividing), so be especially careful with these translations.

To solve, begin by subtracting x from both sides to get all variable terms on one side.

$$2x + 7 = x - 3$$
$$2x + 7 - x = x - 3 - x \quad \text{Subtract } x \text{ from both sides.}$$
$$x + 7 = -3 \quad \text{Combine like terms.}$$
$$x + 7 - 7 = -3 - 7 \quad \text{Subtract 7 from both sides.}$$
$$x = -10 \quad \text{Combine like terms.}$$

Check the solution in the problem as it was originally stated. To do so, replace "number" in the sentence with −10. Twice "−10" added to 7 is the same as 3 subtracted from "−10."

$$2(-10) + 7 = -10 - 3$$
$$-13 = -13$$

The unknown number is −10. □

PRACTICE 1 Three times a number, minus 6, is the same as two times the number, plus 3. Find the number. ■

Helpful Hint

When checking solutions, go back to the original stated problem, rather than to your equation, in case errors have been made in translating into an equation.

EXAMPLE 2 Finding an Unknown Number

Twice the sum of a number and 4 is the same as four times the number, decreased by 12. Find the number.

Solution

1. UNDERSTAND. Read and reread the problem. If we let

   x = the unknown number, then

   "the sum of a number and 4" translates to "$x + 4$" and "four times the number" translates to "$4x$."

2. TRANSLATE.

twice	the sum of a number and 4	is the same as	four times the number	decreased by	12
↓	↓	↓	↓	↓	↓
2	$(x + 4)$	=	$4x$	−	12

3. SOLVE.

$$2(x + 4) = 4x - 12$$
$$2x + 8 = 4x - 12 \quad \text{Apply the distributive property.}$$
$$2x + 8 - 4x = 4x - 12 - 4x \quad \text{Subtract } 4x \text{ from both sides.}$$
$$-2x + 8 = -12$$
$$-2x + 8 - 8 = -12 - 8 \quad \text{Subtract 8 from both sides.}$$
$$-2x = -20$$
$$\frac{-2x}{-2} = \frac{-20}{-2} \quad \text{Divide both sides by } -2.$$
$$x = 10$$

4. INTERPRET.

Check: Check this solution in the problem as it was originally stated. To do so, replace "number" with 10. Twice the sum of "10" and 4 is 28, which is the same as 4 times "10" decreased by 12.

State: The number is 10. □

(Continued on next page)

PRACTICE 2 Three times a number, decreased by 4, is the same as double the difference of the number and 1.

OBJECTIVE 2

Solving Problems Involving Relationships Among Unknown Quantities

The next three examples have to do with relationships among unknown quantities.

EXAMPLE 3 **Finding the Length of a Board**

Balsa wood sticks are commonly used for building models (for example, bridge models). A 48-inch balsa wood stick is to be cut into two pieces so that the longer piece is 3 times the shorter. Find the length of each piece.

Solution

1. UNDERSTAND the problem. To do so, read and reread the problem. You may also want to propose a solution. For example, if 10 inches represents the length of the shorter piece, then $3(10) = 30$ inches is the length of the longer piece, since it is 3 times the length of the shorter piece. This guess gives a total stick length of 10 inches + 30 inches = 40 inches, too short. However, the purpose of proposing a solution is not to guess correctly but to help understand the problem better and how to model it.

   Since the length of the longer piece is given in terms of the length of the shorter piece, let's let

$$x = \text{length of shorter piece; then}$$
$$3x = \text{length of longer piece}$$

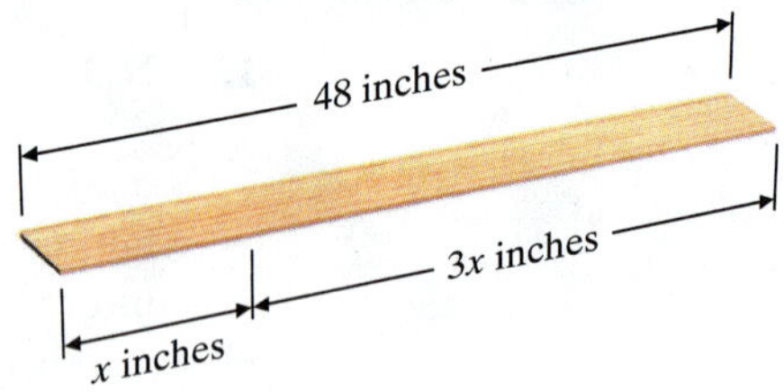

2. TRANSLATE the problem. First, we write the equation in words.

length of shorter piece	added to	length of longer piece	equals	total length of stick
↓	↓	↓	↓	↓
x	$+$	$3x$	$=$	48

3. SOLVE.

$$x + 3x = 48$$
$$4x = 48 \quad \text{Combine like terms.}$$
$$\frac{4x}{4} = \frac{48}{4} \quad \text{Divide both sides by 4.}$$
$$x = 12$$

4. INTERPRET.

Check: Check the solution in the stated problem. If the shorter piece of stick is 12 inches, the longer piece is $3 \cdot (12 \text{ inches}) = 36$ inches, and the sum of the two pieces is 12 inches + 36 inches = 48 inches.

State: The shorter piece of balsa wood is 12 inches, and the longer piece of balsa wood is 36 inches.

Helpful Hint
Make sure that units are included in your answer, if appropriate.

PRACTICE 3 A 45-inch board is to be cut into two pieces so that the longer piece is 4 times the shorter. Find the length of each piece.

EXAMPLE 4 **Finding the Number of Republican and Democratic Representatives**

The 114th Congress began on January 3, 2015, and had a total of 435 Democrat and Republican representatives. There were 59 fewer Democratic representatives than Republican. Find the number of representatives from each party. (*Source:* Congress.gov)

Solution

1. UNDERSTAND the problem. Read and reread the problem. Let's suppose that there were 200 Republican representatives. Since there were 59 fewer Democrats than Republicans, there must have been $200 - 59 = 141$ Democrats. The total number of Republicans and Democrats was then $200 + 141 = 341$. This is incorrect since the total should be 435, but we now have a better understanding of the problem.

In general, if we let

$$x = \text{number of Republicans, then}$$
$$x - 59 = \text{number of Democrats}$$

2. TRANSLATE the problem. First, we write the equation in words.

number of Republicans	added to	number of Democrats	equals	435
↓	↓	↓	↓	↓
x	$+$	$(x - 59)$	$=$	435

3. SOLVE.

$$x + (x - 59) = 435$$
$$2x - 59 = 435 \quad \text{Combine like terms.}$$
$$2x - 59 + 59 = 435 + 59 \quad \text{Add 59 to both sides.}$$
$$2x = 494$$
$$\frac{2x}{2} = \frac{494}{2} \quad \text{Divide both sides by 2.}$$
$$x = 247$$

4. INTERPRET.

Check: If there were 247 Republican representatives, then there were $247 - 59 = 188$ Democratic representatives. The total number of representatives was then $247 + 188 = 435$. The results check.

State: There were 247 Republican and 188 Democratic representatives at the beginning of the 114th Congress. □

PRACTICE 4 In 2015, there were 7 fewer Democratic State Governors than Republican State Governors. Find the number of State Governors from each party. Alaska had an independent governor, so use a total of 49, representing the other 49 states. (*Source:* National Conference of State Legislatures). ■

△ **EXAMPLE 5** **Finding Angle Measures**

If the two walls of the Vietnam Veterans Memorial in Washington, D.C., were connected, an isosceles triangle would be formed. The measure of the third angle is 97.5° more than the measure of either of the other two equal angles. Find the measure of the third angle. (*Source:* National Park Service)

(Continued on next page)

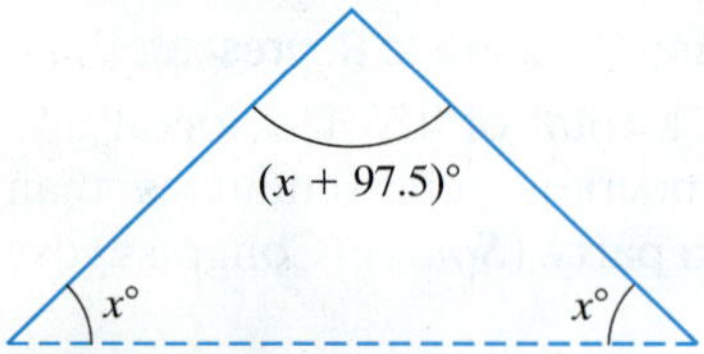

Solution

1. UNDERSTAND. Read and reread the problem. We then draw a diagram (recall that an isosceles triangle has two angles with the same measure) and let

x = degree measure of one angle

x = degree measure of the second equal angle

$x + 97.5$ = degree measure of the third angle

2. TRANSLATE. Recall that the sum of the measures of the angles of a triangle equals 180.

measure of first angle		measure of second angle		measure of third angle	equals	180
↓		↓		↓	↓	↓
x	$+$	x		$+ (x + 97.5)$	$=$	180

3. SOLVE.

$$x + x + (x + 97.5) = 180$$

$$3x + 97.5 = 180 \quad \text{Combine like terms.}$$

$$3x + 97.5 - 97.5 = 180 - 97.5 \quad \text{Subtract 97.5 from both sides.}$$

$$3x = 82.5$$

$$\frac{3x}{3} = \frac{82.5}{3} \quad \text{Divide both sides by 3.}$$

$$x = 27.5$$

4. INTERPRET.

Check: If $x = 27.5$, then the measure of the third angle is $x + 97.5 = 125$. The sum of the angles is then $27.5 + 27.5 + 125 = 180$, the correct sum.

State: The third angle measures 125°.* □

PRACTICE

5 The second angle of a triangle measures three times as large as the first. If the third angle measures 55° more than the first, find the measures of all three angles. ■

OBJECTIVE

3 Solving Consecutive Integer Problems

The next example has to do with consecutive integers. Recall what we have learned thus far about these integers.

	Example	*General Representation*
Consecutive Integers	11, 12, 13 (+1, +1)	Let x be an integer. $x,\ x + 1, x + 2$ (+1, +1)
Consecutive Even Integers	38, 40, 42 (+2, +2)	Let x be an **even** integer. $x,\ x + 2, x + 4$ (+2, +2)
Consecutive Odd Integers	57, 59, 61 (+2, +2)	Let x be an **odd** integer. $x,\ x + 2, x + 4$ (+2, +2)

*The two walls actually meet at an angle of 125 degrees 12 minutes. The measurement of 97.5° given in the problem is an approximation.

EXAMPLE 6 Some states have a single area code for the entire state. Two such states have area codes that are consecutive odd integers. If the sum of these integers is 1208, find the two area codes. (*Source:* North American Numbering Plan Administration)

Solution

1. UNDERSTAND. Read and reread the problem. If we let

$$x = \text{the first odd integer, then}$$
$$x + 2 = \text{the next odd integer}$$

Helpful Hint

Remember, the 2 here means that odd integers are 2 units apart, for example, the odd integers 13 and $13 + 2 = 15$.

2. TRANSLATE.

first odd integer	the sum of	next odd integer	is	1208
↓	↓	↓	↓	↓
x	$+$	$(x + 2)$	$=$	1208

3. SOLVE.

$$\begin{aligned} x + x + 2 &= 1208 \\ 2x + 2 &= 1208 \\ 2x + 2 - 2 &= 1208 - 2 \\ 2x &= 1206 \\ \frac{2x}{2} &= \frac{1206}{2} \\ x &= 603 \end{aligned}$$

4. INTERPRET.

Check: If $x = 603$, then the next odd integer $x + 2 = 603 + 2 = 605$. Notice their sum, $603 + 605 = 1208$, as needed.

State: The area codes are 603 and 605.

Note: New Hampshire's area code is 603 and South Dakota's area code is 605. □

PRACTICE 6 The sum of three consecutive even integers is 144. Find the integers. ■

Vocabulary, Readiness & Video Check

Fill in the table.

1.	A number: → x	Double the number:	→ Double the number, decreased by 31:
2.	A number: → x	Three times the number:	→ Three times the number, increased by 17:
3.	A number: → x	The sum of the number and 5:	→ Twice the sum of the number and 5:
4.	A number: → x	The difference of the number and 11:	→ Seven times the difference of the number and 11:
5.	A number: → y	The difference of 20 and the number:	→ The difference of 20 and the number, divided by 3:
6.	A number: → y	The sum of -10 and the number:	→ The sum of -10 and the number, divided by 9:

Martin-Gay Interactive Videos

See Video 2.4

Watch the section lecture video and answer the following questions.

OBJECTIVE 1

7. At the end of Example 1, where are you told is the best place to check the solution of an application problem?

OBJECTIVE 2

8. The solution of the equation for Example 3 is $x = 43$. Why is this not the solution to the application?

OBJECTIVE 3

9. What are two things that should be checked to make sure the solution of Example 4 is correct?

2.4 Exercise Set MyMathLab®

TRANSLATING

Write each of the following as an equation. Then solve. See Examples 1 and 2.

1. The sum of six times a number, and 1, is equal to five times the number. Find the number.

2. The difference of three times a number, and 1, is the same as twice the number. Find the number.

3. Three times a number, minus 6, is equal to two times the number, plus 8. Find the number.

4. The sum of 4 times a number, and -2, is equal to the sum of 5 times the number, and -2. Find the number.

5. Twice the difference of a number and 8 is equal to three times the sum of the number and 3. Find the number.

6. Five times the sum of a number and -1 is the same as 6 times the difference of the number and 5. Find the number.

7. Twice the sum of -2 and a number is the same as the number decreased by $\frac{1}{2}$. Find the number.

8. If the difference of a number and four is doubled, the result is $\frac{1}{4}$ less than the number. Find the number.

Solve. For Exercises 9 and 10, the solutions have been started for you. See Examples 3 through 5.

9. A 25-inch piece of steel is cut into three pieces so that the second piece is twice as long as the first piece, and the third piece is one inch more than five times the length of the first piece. Find the lengths of the pieces.

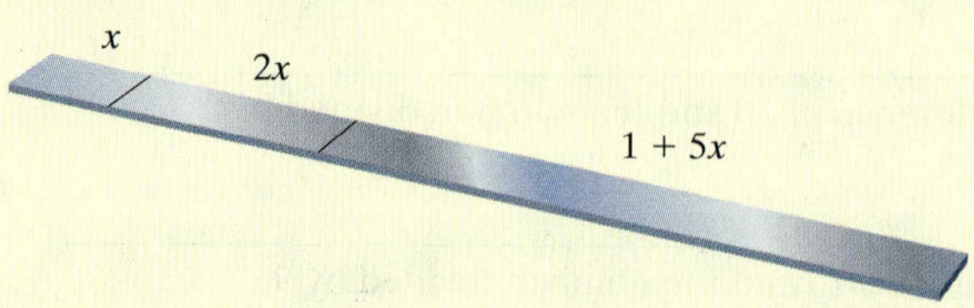

Start the solution:

1. UNDERSTAND the problem. Reread it as many times as needed.
2. TRANSLATE into an equation. (Fill in the blanks below.)

total length of steel	equals	length of first piece	plus	length of second piece	plus	length of third piece
↓	↓	↓	↓	↓	↓	↓
25	=	____	+	____	+	____

Finish with:

3. SOLVE and 4. INTERPRET

10. A 46-foot piece of rope is cut into three pieces so that the second piece is three times as long as the first piece, and the third piece is two feet more than seven times the length of the first piece. Find the lengths of the pieces.

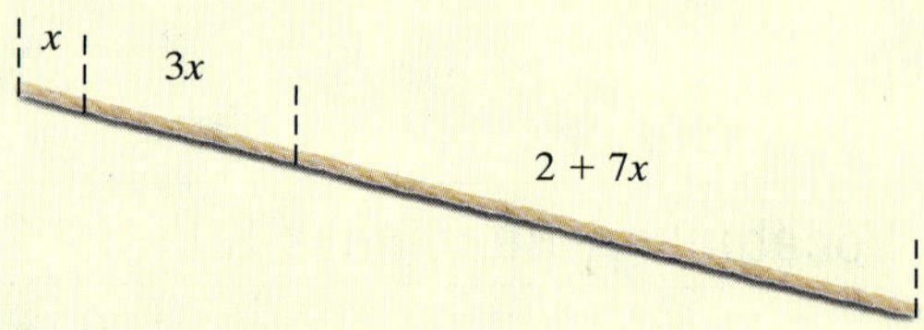

Start the solution:

1. UNDERSTAND the problem. Reread it as many times as needed.
2. TRANSLATE into an equation. (Fill in the blanks below.)

total length of rope	equals	length of first piece	plus	length of second piece	plus	length of third piece
↓	↓	↓	↓	↓	↓	↓
46	=	____	+	____	+	____

Finish with:

3. SOLVE and 4. INTERPRET

11. A 40-inch board is to be cut into three pieces so that the second piece is twice as long as the first piece, and the third piece is 5 times as long as the first piece. If x represents the length of the first piece, find the lengths of all three pieces.

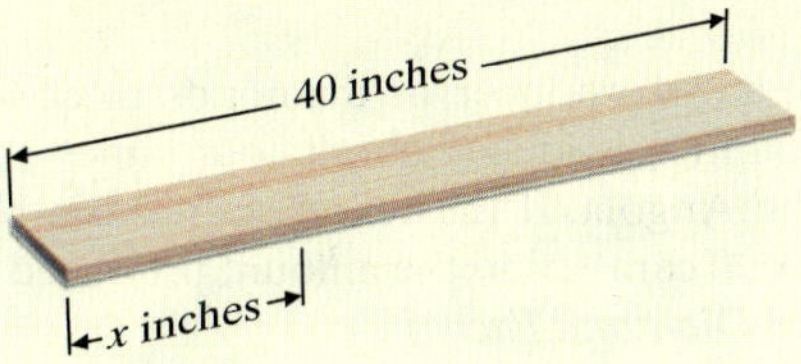

12. A 21-foot beam is to be cut into three pieces so that the second and the third piece are each 3 times the length of the first piece. If x represents the length of the shorter piece, find the lengths of all three pieces.

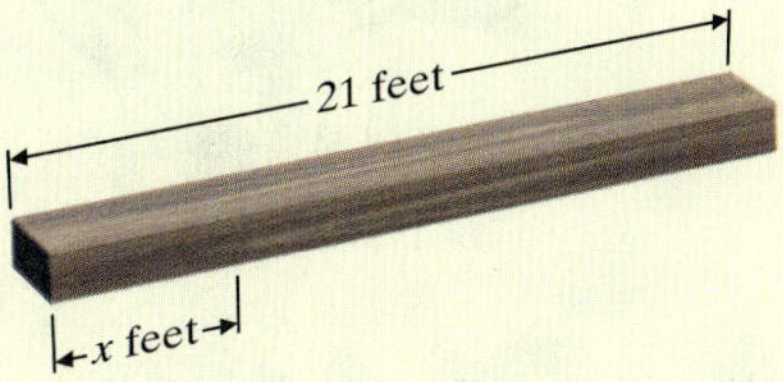

For Exercises 13 and 14, use each table to find the value of x. Then write a sentence to explain, in words, the meaning and the value of x. (Source: IHS Screen Digest*)*

13. The fastest growing type of movie screen in 2014 was 3D.

Type of Screen	*Number of Screens*
3D	x
Non-3D	10,973 more than x
Total screens	43,265

14. The vast majority of movie screens in the United States in 2014 were digital.

Type of Screen	*Number of Screens*
Analog	x
Digital	39,771 more than x
Total screens	43,265

△ **15.** The flag of Equatorial Guinea contains an isosceles triangle. (Recall that an isosceles triangle contains two angles with the same measure.) If the measure of the third angle of the triangle is 30° more than twice the measure of either of the other two angles, find the measure of each angle of the triangle. (*Hint:* Recall that the sum of the measures of the angles of a triangle is 180°.)

△ **16.** The flag of Brazil contains a parallelogram. One angle of the parallelogram is 15° less than twice the measure of the angle next to it. Find the measure of each angle. (*Hint:* Recall that opposite angles of a parallelogram have the same measure and that the sum of the measures of the angles is 360°.)

Solve. See Example 6. For Exercises 17 through 24, fill in the table. Most of the first row has been completed for you.

	First Integer →	*Next Integers*			→ *Indicated Sum*
17. Three consecutive integers:	Integer: x	$x + 1$	$x + 2$		Sum of the three consecutive integers, simplified:
18. Three consecutive integers:	Integer: x				Sum of the second and third consecutive integers, simplified:
19. Three consecutive even integers:	Even integer: x				Sum of the first and third even consecutive integers, simplified:
20. Three consecutive odd integers:	Odd integer: x				Sum of the three consecutive odd integers, simplified:
21. Four consecutive integers:	Integer: x				Sum of the four consecutive integers, simplified:
22. Four consecutive integers:	Integer: x				Sum of the first and fourth consecutive integers, simplified:
23. Three consecutive odd integers:	Odd integer: x				Sum of the second and third consecutive odd integers, simplified:
24. Three consecutive even integers:	Even integer: x				Sum of the three consecutive even integers, simplified:

25. The left and right page numbers of an open book are two consecutive integers whose sum is 469. Find these page numbers.

26. The room numbers of two adjacent classrooms are two consecutive even numbers. If their sum is 654, find the classroom numbers.

27. To make an international telephone call, you need the code for the country you are calling. The codes for Belgium, France, and Spain are three consecutive integers whose sum is 99. Find the code for each country. (*Source: The World Almanac and Book of Facts*, 2007)

28. To make an international telephone call, you need the code for the country you are calling. The codes for Mali Republic, Côte d'Ivoire, and Niger are three consecutive odd integers whose sum is 675. Find the code for each country.

MIXED PRACTICE

Solve. See Examples 1 through 6.

29. The area of the Sahara Desert is 7 times the area of the Gobi Desert. If the sum of their areas is 4,000,000 square miles, find the area of each desert.

30. The largest meteorite in the world is the Hoba West, located in Namibia. Its weight is 3 times the weight of the Armanty meteorite, located in Outer Mongolia. If the sum of their weights is 88 tons, find the weight of each.

31. A 17-foot piece of string is cut into two pieces so that the longer piece is 2 feet longer than twice the length of the shorter piece. Find the lengths of both pieces.

32. A 25-foot wire is to be cut so that the longer piece is one foot longer than 5 times the length of the shorter piece. Find the length of each piece.

33. Five times a number, subtracted from ten, is triple the number. Find the number.

34. Nine is equal to ten subtracted from double a number. Find the number.

35. The greatest producer of diamonds in carats is Botswana. This country produces about four times the amount produced in Angola. If the total produced in both countries is 40,000,000 carats, find the amount produced in each country. (*Source: Diamond Facts.*)

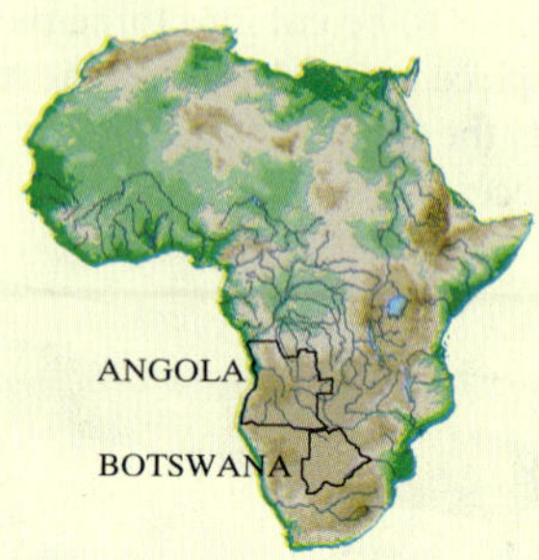

36. Beetles have the greatest number of different species. There are twenty times the number of beetle species as grasshopper species, and the total number of species for both is 420,000. Find the number of species for each type of insect.

△ **37.** The measures of the angles of a triangle are 3 consecutive even integers. Find the measure of each angle.

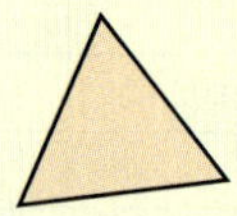

△ **38.** A quadrilateral is a polygon with 4 sides. The sum of the measures of the 4 angles in a quadrilateral is 360°. If the measures of the angles of a quadrilateral are consecutive odd integers, find the measures.

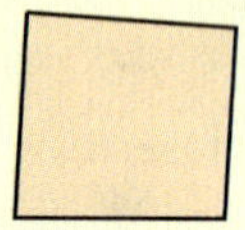

39. For the 2014 Winter Olympics, the total number of medals won by athletes from each of the countries of Netherlands, Canada, and Norway are three consecutive integers whose sum is 75. Find the number of medals for each country.

40. The code to unlock a student's combination lock happens to be three consecutive odd integers whose sum is 51. Find the integers.

41. If the sum of a number and five is tripled, the result is one less than twice the number. Find the number.

42. Twice the sum of a number and six equals three times the sum of the number and four. Find the number.

43. Two angles are supplementary if their sum is 180°. A larger angle measures eight degrees more than three times the measure of a smaller angle. If x represents the measure of the smaller angle and these two angles are supplementary, find the measure of each angle.

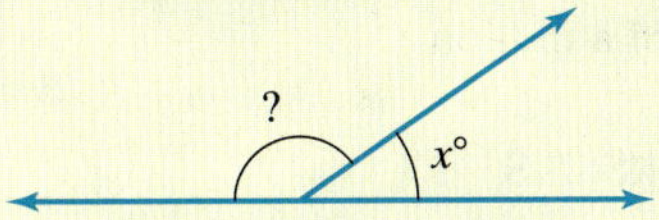

44. Two angles are complementary if their sum is 90°. A larger angle measures three degrees less than twice the measure of a smaller angle. If x represents the measure of the smaller angle and these two angles are complementary, find the measure of each angle.

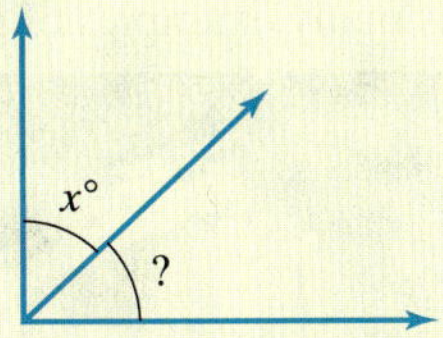

45. If the quotient of a number and 4 is added to $\frac{1}{2}$, the result is $\frac{3}{4}$. Find the number.

46. The sum of $\frac{1}{5}$ and twice a number is equal to $\frac{4}{5}$ subtracted from three times the number. Find the number.

47. The sum of $\frac{2}{3}$ and four times a number is equal to $\frac{5}{6}$ subtracted from five times the number. Find the number.

48. If $\frac{3}{4}$ is added to three times a number, the result is $\frac{1}{2}$ subtracted from twice the number. Find the number.

49. Currently, the two fastest trains in the world are both Chinese, the Shanghai Maglev and the Harmony CRH. The sum of their fastest speeds is 503 miles per hour. If the maximum speed of the Maglev is 31 miles per hour faster than the speed of the Harmony, find the speeds of each. (*Source:* Railway-technology.com)

50. The Pentagon is the world's largest office building in terms of floor space. It has three times the amount of floor space as the Empire State Building. If the total floor space for these two buildings is approximately 8700 thousand square feet, find the floor space of each building.

51. One-third of a number is five-sixths. Find the number.

52. Seven-eighths of a number is one-half. Find the number.

53. The number of counties in California and the number of counties in Montana are consecutive even integers whose sum is 114. If California has more counties than Montana, how many counties does each state have? (*Source: The World Almanac and Book of Facts*)

54. A student is building a bookcase with stepped shelves for her dorm room. She buys a 48-inch board and wants to cut the board into three pieces with lengths equal to three consecutive even integers. Find the three board lengths.

55. A geodesic dome, based on the design by Buckminster Fuller, is composed of two types of triangular panels. One of these is an isosceles triangle. In one geodesic dome, the measure of the third angle is 76.5° more than the measure of either of the two equal angles. Find the measure of the third angle. (*Source:* Buckminster Fuller Institute)

56. The measures of the angles of a particular triangle are such that the second and third angles are each four times larger than the smallest angle. Find the measures of the angles of this triangle.

57. A 30-foot piece of siding is cut into three pieces so that the second piece is four times as long as the first piece and the third piece is five times as long as the first piece. If x represents the length of the first piece, find the lengths of all three pieces.

58. A 48-foot-long piece of cable wire is to be cut into three pieces so that the second piece is five times as long as the first piece and the third piece is six times as long as the first piece. If x represents the length of the first piece, find the lengths of all three pieces.

The graph below shows the best-selling video games for 2014. Use this graph for Exercises 59 through 64.

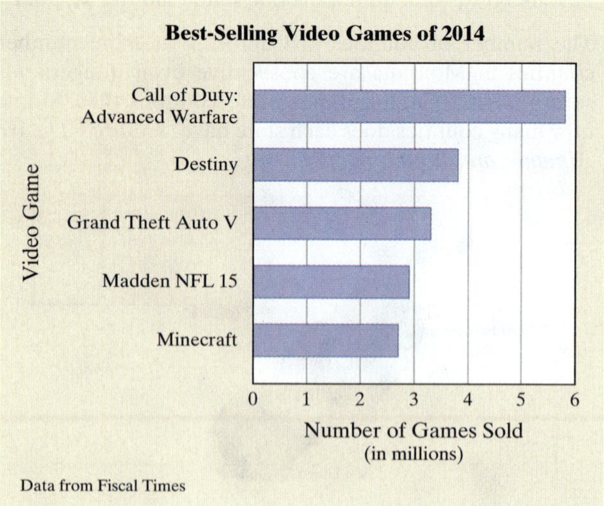

59. Which video game was the best-selling game of 2014?

60. Which video games sold between 3 and 5 million copies in 2014?

61. Madden NFL 15 and Destiny together sold 6.7 million copies in 2014. Destiny sold 0.9 million copies more than Madden NFL 15. Find the number of sales of each video game.

62. Grand Theft Auto V and Minecraft together sold 6 million copies in 2014. Grand Theft Auto V sold 0.6 million more copies than Minecraft. Find the number of sales of each video game.

Compare the lengths of the bars in the graph with your results for the exercises below. Are your answers reasonable?

63. Exercise 61 **64.** Exercise 62

REVIEW AND PREVIEW

The circle graph below shows the breakdown of total pets owned in the United States. Use this graph to answer the questions. See Section R.4.

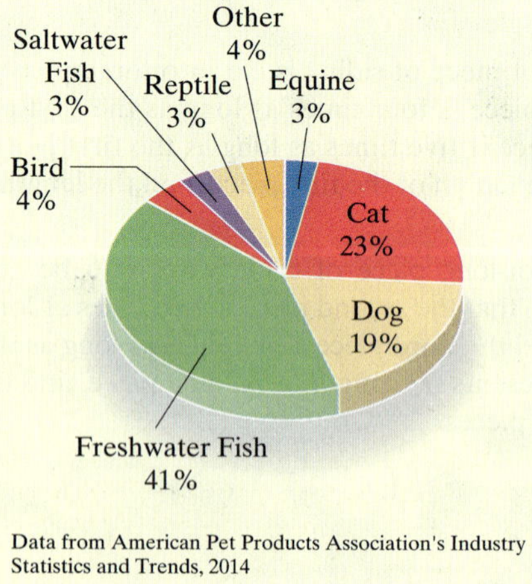

65. What percent of pets owned in the United States are cats or dogs?

66. What percent of pets owned in the United States are not birds?

67. In 2014, 396.12 million pets were owned in the United States. How many of these were cats? (Round to the nearest tenth of a million.)

68. In 2014, 396.12 million pets were owned in the United States. How many of these were dogs? (Round to the nearest tenth of a million.)

CONCEPT EXTENSIONS

△ **69.** A golden rectangle is a rectangle whose length is approximately 1.6 times its width. The early Greeks thought that a rectangle with these dimensions was the most pleasing to the eye, and examples of the golden rectangle are found in many early works of art. For example, the Parthenon in Athens contains many examples of golden rectangles.

Mike Hallahan would like to plant a rectangular garden in the shape of a golden rectangle. If he has 78 feet of fencing available, find the dimensions of the garden.

△ **70.** Dr. Dorothy Smith gave the students in her geometry class at the University of New Orleans the following question. Is it possible to construct a triangle such that the second angle of the triangle has a measure that is twice the measure of the first angle and the measure of the third angle is 5 times the measure of the first? If so, find the measure of each angle. (*Hint:* Recall that the sum of the measures of the angles of a triangle is 180°.)

71. Only male crickets chirp. They chirp at different rates depending on their species and the temperature of their environment. Suppose a certain species is currently chirping at a rate of 90 chirps per minute. At this rate, how many chirps occur in one hour? In one 24-hour day? In one year?

72. The human eye blinks once every 5 seconds on average. How many times does the average eye blink in one hour? In one 16-hour day while awake? In one year while awake?

73. In your own words, explain why a solution of a word problem should be checked using the original wording of the problem and not the equation written from the wording.

74. Give an example of how you recently solved a problem using mathematics.

Recall from Exercise 69 that a golden rectangle is a rectangle whose length is approximately 1.6 times its width.

△ **75.** It is thought that for about 75% of adults, a rectangle in the shape of the golden rectangle is the most pleasing to the eye. Draw three rectangles, one in the shape of the golden rectangle, and poll your class. Do the results agree with the percentage given above?

△ **76.** Examples of golden rectangles can be found today in architecture and manufacturing packaging. Find an example of a golden rectangle in your home. A few suggestions: the front face of a book, the floor of a room, the front of a box of food.

For Exercises 77 and 78, measure the dimensions of each rectangle and decide which one best approximates the shape of a golden rectangle.

△ **77.**

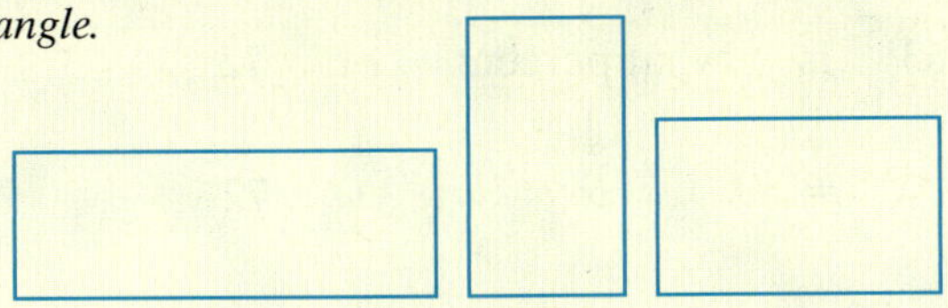

(a) **(b)** **(c)**

△ **78.**

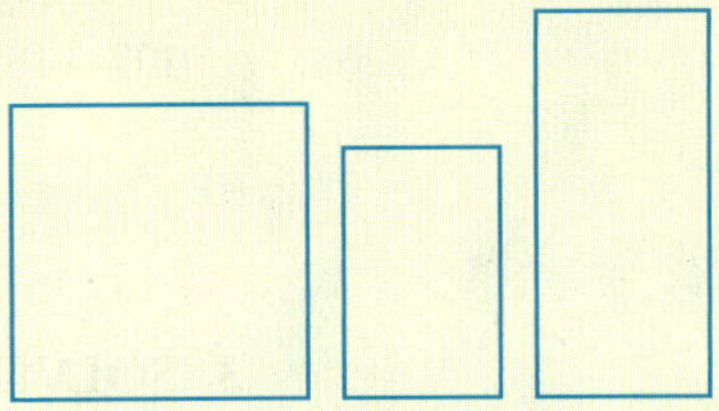

(a) **(b)** **(c)**

2.5 Percent and Finance Problem Solving

OBJECTIVES

1 Solve Percent Equations.

2 Solve Discount and Markup Problems.

3 Solve Percent of Increase and Percent of Decrease Problems.

4 Solve Simple and Compound Interest Problems.

This section is devoted to solving problems in the categories listed. The same problem-solving steps used in the previous section are also followed in this section. They are listed below for review.

General Strategy for Problem Solving

1. UNDERSTAND the problem. During this step, become comfortable with the problem. Some ways of doing this are as follows:

   Read and reread the problem.

   Choose a variable to represent the unknown.

   Construct a drawing whenever possible.

   Propose a solution and check. Pay careful attention to how you check your proposed solution. This will help when writing an equation to model the problem.
2. TRANSLATE the problem into an equation.
3. SOLVE the equation.
4. INTERPRET the results: *Check* the proposed solution in the stated problem and *state* your conclusion.

OBJECTIVE

1 Solving Percent Equations

Many of today's statistics are given in terms of percent: a basketball player's free throw percent, current interest rates, stock market trends, and nutrition labeling, just to name a few. In this section, we first explore percent, percent equations, and applications involving percents and finance. See Chapter R if a further review of percents is needed.

EXAMPLE 1 The number 63 is what percent of 72?

Solution

1. UNDERSTAND. Read and reread the problem. Next, let's suppose that the percent is 80%. To check, we find 80% of 72.

$$80\% \text{ of } 72 = 0.80(72) = 57.6$$

This is close but not 63. At this point, though, we have a better understanding of the problem, we know the correct answer is close to and greater than 80%, and we know how to check our proposed solution later.

Let $x =$ the unknown percent.

(Continued on next page)

2. TRANSLATE. Recall that "is" means "equals" and "of" signifies multiplying. Let's translate the sentence directly.

the number 63	is	what percent	of	72
↓	↓	↓	↓	↓
63	=	x	·	72

3. SOLVE.

$$63 = 72x$$

$$0.875 = x \quad \text{Divide both sides by 72.}$$

$$87.5\% = x \quad \text{Write as a percent.}$$

4. INTERPRET.

Check: Verify that 87.5% of 72 is 63.

State: The number 63 is 87.5% of 72. □

PRACTICE 1 The number 35 is what percent of 56? ■

EXAMPLE 2 The number 120 is 15% of what number?

Solution

1. UNDERSTAND. Read and reread the problem.

Let $x =$ the unknown number.

2. TRANSLATE.

the number 120	is	15%	of	what number
↓	↓	↓	↓	↓
120	=	15%	·	x

3. SOLVE.

$$120 = 0.15x \quad \text{Write 15\% as 0.15.}$$

$$800 = x \quad \text{Divide both sides by 0.15.}$$

4. INTERPRET.

Check: Check the proposed solution by finding 15% of 800 and verifying that the result is 120.

State: Thus, 120 is 15% of 800. □

PRACTICE 2 The number 198 is 55% of what number? ■

OBJECTIVE 2

Solving Discount and Markup Problems

Percents are frequently used in finance. In this objective, we study discounting and markup. The next example has to do with discounting the price of a cell phone.

EXAMPLE 3 Cell Phones Unlimited recently reduced the price of a \$140 phone by 20%. What are the discount and the new price?

Solution

1. UNDERSTAND. Read and reread the problem. Make sure you understand the meaning of the word *discount*. Discount is the amount of money by which the cost of an item has been decreased. To find the discount, we simply find 20% of \$140. In other words, we have the formulas,

discount = percent · original price   Then

new price = original price − discount

2, 3. TRANSLATE and SOLVE.

$$\begin{aligned} \text{discount} &= \text{percent} \cdot \text{original price} \\ &= 20\% \cdot \$140 \\ &= 0.20 \cdot \$140 \\ &= \$28 \end{aligned}$$

Thus, the discount in price is \$28.

$$\begin{aligned} \text{new price} &= \text{original price} - \text{discount} \\ &= \$140 - \$28 \\ &= \$112 \end{aligned}$$

4. INTERPRET.

Check: Check your calculations in the formulas; see also whether our results are reasonable. They are.

State: The discount in price is \$28, and the new price is \$112. □

PRACTICE 3 A used treadmill, originally purchased for \$480, was sold at a garage sale at a discount of 85% of the original price. What were the discount and the new price? ■

A concept similar to discount is markup. What is the difference between the two? A **discount** is subtracted from the original price while a **markup** is added to the original price. For markups,

markup = percent · original price

new price = original price + markup

Helpful Hint

Discounts are subtracted from the original price, while markups are added.

Common markup applications are sales tax and calculating tips. The next example has to do with sales tax. In other words, the markup is the sales tax.

EXAMPLE 4 Find the markup (sales tax) and the new price on the purchase of an \$85.50 atlas in a city where the sales tax rate is 7.5%. Round the sales tax to the nearest cent.

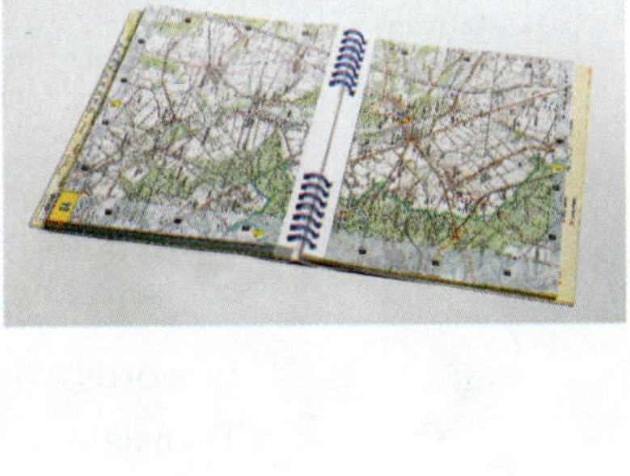

(Continued on next page)

Solution The purchase price is \$85.50 and the tax rate is 7.5%.

$$\text{markup (sales tax)} = \text{percent} \cdot \text{original price}$$

$$\begin{aligned} \text{sales tax} &= 7.5\% \cdot \$85.50 \\ &= 0.075 \cdot \$85.5 && \text{Write 7.5\% as a decimal.} \\ &\approx \$6.41 && \text{Rounded to the nearest cent} \end{aligned}$$

Thus, the sales tax is \$6.41. Next find the new price.

$$\text{new price} = \text{original price} + \text{sales tax}$$

$$\begin{aligned} \text{new price} &= \$85.50 + \$6.41 \\ &= \$91.91 \end{aligned}$$

The sales tax on \$85.50 is \$6.41, and the new price is \$91.91. □

PRACTICE 4 Find the markup (sales tax) and the new price on the purchase of a \$129.99 graphing calculator in a city where the sales tax is 9.5%. Round the sales tax to the nearest cent.

OBJECTIVE 3

Solving Percent of Increase and Percent of Decrease Problems

Percent of increase or percent of decrease is a common way to describe how some measurement has increased or decreased. For example, crime increased by 8%, teachers received a 5.5% increase in salary, or a company decreased its number of employees by 10%. The next example is a review of percent of increase.

EXAMPLE 5 Calculating the Percent Increase of Attending College

The tuition and fees cost of attending a public four-year college for in-state students rose from \$4020 in 1996 to \$9410 in 2015. Find the percent of increase. Round to the nearest tenth of a percent. (*Source:* The College Board)

Solution

1. UNDERSTAND. Read and reread the problem. Notice that the new tuition, \$9410, is over double the old tuition of \$4020. Because of that, we know that the percent of increase is over 100%. To see this, let's guess that the percent of increase is 100%. To check, we find 100% of \$4020 to find the *increase* in cost. Then we add this increase to \$4020 to find the *new cost*. In other words, 100%(\$4020) = 1.00(\$4020) = \$4020, the *increase* in cost. The *new cost* would be old cost + increase = \$4020 + \$4020 = \$8040, close to the actual new cost of \$9410. We now know that the increase is close to, but more than, 100% and we know how to check our proposed solution.

   Let x = the percent of increase.

2. TRANSLATE. First, find the **increase** and then the **percent of increase.** The increase in cost is found by:

In words:	amount of increase	=	new cost	−	original cost
Translate:	increase	=	\$9410	−	\$4020
		=	\$5390		

   Next, find the percent of increase. The percent of increase or percent of decrease is always a percent of the original number or, in this case, the old cost.

In words:	increase	is	what percent	of	original cost
Translate:	\$5390	=	x	·	\$4020

3. SOLVE.

$$5390 = 4020x$$
$$1.341 \approx x \qquad \text{Divide both sides by 4020 and round to 3 decimal places.}$$
$$134.1\% \approx x \qquad \text{Write as a percent.}$$

Note: After the amount of increase is found, the following may be used:

$$\text{percent of increase} = \frac{\text{amount of increase}}{\text{original amount}}$$

Thus,

$$x = \frac{\$5390}{\$4020} \approx 1.341 = 134.1\%$$

4. INTERPRET.

Check: Check the proposed solution

State: The percent of increase in cost was approximately 134.1%. □

PRACTICE 5 The tuition and fees cost of attending a public two-year college for in-state students rose from $1900 in 1996 to $3435 in 2015. Find the percent of increase. Round to the nearest tenth of a percent. (*Source:* The College Board) ■

Percent of decrease is found using a method similar to that in Example 5. First find the amount of decrease, then determine what percent of the original or first amount is that decrease. Each time, the below may be used:

Percent of Increase of Decrease

$$\text{percent of increase} = \frac{\text{amount of increase}}{\text{original amount}} \qquad \text{percent of decrease} = \frac{\text{amount of decrease}}{\text{original amount}}$$

Each time, write the quotient as a percent.

Read the next example carefully. For Example 5, we were asked to find percent of increase. In Example 6, we are given the percent of increase and asked to find the number before the increase.

EXAMPLE 6 Most of the movie screens globally project analog film, but the number of cinemas using digital is increasing. Find the number of digital screens worldwide in 2012 if, after a 43% increase, the number in 2014 was 127,689. Round to the nearest whole number. (*Source:* Motion Picture Association of America)

Solution

1. UNDERSTAND. Read and reread the problem. Let's guess a solution and see how we would check our guess. If the number of digital screens worldwide in 2012 was 80,000, we would see if 80,000 plus the increase is 127,689; that is,

$$80{,}000 + 43\%(80{,}000) = 80{,}000 + 0.43(80{,}000) = 80{,}000 + 34{,}400 = 114{,}400$$

Since 114,400 is too small, we know that our guess of 80,000 is too small. We also have a better understanding of the problem. Let

$$x = \text{number of digital screens in 2012.}$$

2. TRANSLATE. To translate into an equation, we remember that

In words:	number of digital screens in 2012	plus	increase	equals	number of digital screens in 2014
	↓	↓	↓	↓	↓
Translate:	x	$+$	$0.43x$	$=$	127,689

(Continued on next page)

3. SOLVE.

$$1.43x = 127{,}689 \quad \text{Add like terms.}$$
$$x = \frac{127{,}689}{1.43}$$
$$x \approx 89{,}293$$

4. INTERPRET.

Check: Recall that x represents the number of digital screens worldwide in 2012. If this number is approximately 89,293, let's see if 89,293 plus the increase is close to 127,689. (We use the word "close" because it is rounded.)

$$89{,}293 + 43\%(89{,}293) = 89{,}293 + 0.43(89{,}293) = 1.43(89{,}293) = 127{,}688.99$$

which is close to 127,689.

State: There were approximately 89,293 digital screens worldwide in 2012. □

PRACTICE 6 The fastest-growing sector of digital theater screens is 3D. Find the number of digital 3D screens in the United States and Canada in 2012 if, after a 9.6% increase, the number in 2014 was 16,146. Round to the nearest whole. (*Source:* MPAA) ■

OBJECTIVE 4

Solving Simple and Compound Interest Problems

Interest is money charged for using other people's money. When you borrow money, you pay interest. When you loan or invest money, you earn interest. The money borrowed, loaned, or invested is called the **principal amount,** or simply **principal.** Interest is normally stated in terms of a percent of the principal for a given period of time. The **interest rate** is the percent used in computing the interest. Unless stated otherwise, *the rate is understood to be per year.* When the interest is computed on the original principal, it is called **simple interest.** Simple interest is calculated using the following:

Simple Interest

$$\text{Simple Interest} = \text{Principal} \cdot \text{Rate} \cdot \text{Time}$$
$$I = P \cdot R \cdot T$$

where the rate is understood to be per year and time is in years.

EXAMPLE 7 Find the simple interest after 2 years on \$500 at an interest rate of 12%.

Solution In this example, $P = \$500$, $R = 12\%$, and $T = 2$ years. Replace the variables with values in the formula $I = PRT$.

$$I = P \cdot R \cdot T$$
$$I = \$500 \cdot 12\% \cdot 2 \quad \text{Let } P = \$500, R = 12\%, \text{ and } T = 2.$$
$$= \$500 \cdot (0.12) \cdot 2 \quad \text{Write 12\% as a decimal.}$$
$$= \$120 \quad \text{Multiply.}$$

The simple interest is \$120. □

PRACTICE 7 Find the simple interest after 3 years on \$650 at an interest rate of 15%. ■

If time is not given in years, we need to convert the given time to years.

EXAMPLE 8 A recent college graduate borrowed \$2400 at 10% simple interest for 8 months to buy a used Toyota Corolla. Find the simple interest he paid.

Solution Since there are 12 months in a year, we first find what part of a year 8 months is.

$$8 \text{ months} = \frac{8}{12} \text{ year} = \frac{2}{3} \text{ year}$$

Now we find the simple interest.

$$\begin{aligned} I &= P \cdot R \cdot T \\ &= \$2400 \cdot (0.10) \cdot \frac{2}{3} \quad \text{Let } P = \$2400, R = 10\% \text{ or } 0.10, \text{ and } T = \frac{2}{3}. \\ &= \$160 \end{aligned}$$

The interest on his loan is \$160. □

PRACTICE 8 A new employee needed a short-term loan for moving expenses. The employee borrowed \$1500 at 8% simple interest for 4 months. Find the simple interest paid. ■

✔ CONCEPT CHECK
Suppose in Example 8 you had obtained an answer of \$16,000. How would you know that you had made a mistake in this problem?

Answer to Concept Check:
\$16,000 is too much interest.

When money is borrowed, the borrower pays the original amount borrowed, or the principal, as well as the interest. When money is invested, the investor receives the original amount invested, or the principal, as well as the interest. In either case, the **total amount** is the sum of the principal and the interest.

Finding the Total Amount of a Loan or Investment

total amount (paid or received) = principal + interest

EXAMPLE 9 Find the total amount paid for the loan in Example 8.

Solution We add the interest to the principal.

$$\begin{aligned} \text{total amount} &= \text{principal} + \text{interest} \\ &\downarrow \qquad\quad \downarrow \qquad\quad \downarrow \\ \text{total amount} &= \$2400 \quad + \quad \$160 \\ &= \$2560 \end{aligned}$$

The total amount paid is \$2560 □

PRACTICE 9 Find the total amount paid for the loan in Practice 8. ■

✔ CONCEPT CHECK
Which investment would earn more interest: an amount of money invested at 8% interest for 2 years, or the same amount of money invested at 8% for 3 years? Explain.

Answer to Concept Check:
8% for 3 years. Since the interest rate is the same, the longer you keep the money invested, the more interest you earn.

Recall that simple interest depends on the original principal only. Another type of interest is compound interest. **Compound interest** is computed not only on the principal, but also on the interest already earned in previous compounding periods. Compound interest is used more often than simple interest.

Let's see how compound interest differs from simple interest. Suppose that \$2000 is invested at 7% interest **compounded annually** for 3 years. This means that interest is added to the principal at the end of each year and that next year's interest is computed on this new amount. In this section, we round dollar amounts to the nearest cent.

	Amount at Beginning of Year	*Principal*	·	*Rate*	·	*Time*	= *Interest*	*Amount at End of Year*
1st year	\$2000	\$2000	·	0.07	·	1	= \$140	\$2000 + 140 = \$2140
2nd year	\$2140	\$2140	·	0.07	·	1	= \$149.80	\$2140 + 149.80 = \$2289.80
3rd year	\$2289.80	\$2289.80	·	0.07	·	1	= \$160.29	\$2289.80 + 160.29 = \$2450.09

The compound interest earned can be found by

$$\begin{array}{ccccc} \text{total amount} & - & \text{original principal} & = & \text{compound interest} \\ \downarrow & & \downarrow & & \downarrow \\ \$2450.09 & - & \$2000 & = & \$450.09 \end{array}$$

The simple interest earned would have been

$$\begin{array}{ccccccc} \text{principal} & \cdot & \text{rate} & \cdot & \text{time} & = & \text{interest} \\ \downarrow & & \downarrow & & \downarrow & & \downarrow \\ \$2000 & \cdot & 0.07 & \cdot & 3 & = & \$420 \end{array}$$

Since compound interest earns "interest on interest," compound interest earns more than simple interest.

Computing compound interest using the method above can be tedious. We can use a calculator and the compound interest formula below to compute compound interest more quickly.

Compound Interest Formula

The total amount A in an account is given by

$$A = P\left(1 + \frac{r}{n}\right)^{n \cdot t}$$

where P is the principal, r is the interest rate written as a decimal, t is the length of time in years, and n is the number of times compounded per year.

EXAMPLE 10 \$1800 is invested at 2% interest compounded annually. Find the total amount after 3 years.

Solution "Compounded annually" means 1 time a year, so $n = 1$. Also, $P = \$1800$, $r = 2\% = 0.02$, and $t = 3$ years.

$$\begin{aligned} A &= P\left(1 + \frac{r}{n}\right)^{n \cdot t} \\ &= 1800\left(1 + \frac{0.02}{1}\right)^{1 \cdot 3} \\ &= 1800(1.02)^3 \\ &\approx 1910.17 \quad \text{Round to 2 decimal places.} \end{aligned}$$

Helpful Hint

Remember order of operations. **First** evaluate $(1.02)^3$, then multiply by 1800.

The total amount at the end of 3 years is \$1910.17. □

PRACTICE 10 $2500 is invested at 3% interest compounded annually. Find the total amount after 5 years.

EXAMPLE 11 $4000 is invested at 5.3% compounded quarterly for 10 years. Find the total amount at the end of 10 years.

Solution "Compounded quarterly" means 4 times a year, so $n = 4$. Also, $P = \$4000$, $r = 5.3\% = 0.053$, and $t = 10$ years.

$$A = P\left(1 + \frac{r}{n}\right)^{n \cdot t}$$

$$= 4000\left(1 + \frac{0.053}{4}\right)^{4 \cdot 10}$$

$$= 4000(1.01325)^{40}$$

$$\approx 6772.12$$

The total amount after 10 years is $6772.12.

PRACTICE 11 $6800 is invested at 3.2% compounded semiannually for 4 years. Find the total amount after 4 years.

Vocabulary, Readiness & Video Check

Tell whether the percent labels in the circle graphs are correct. Recall that percents in a circle graph have a sum of 100%.

1.

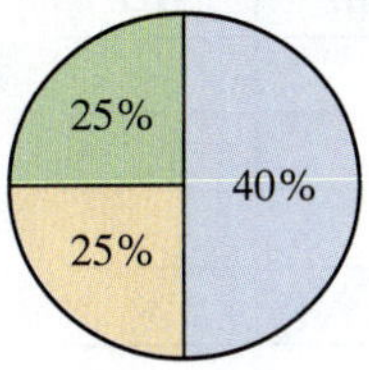

2.

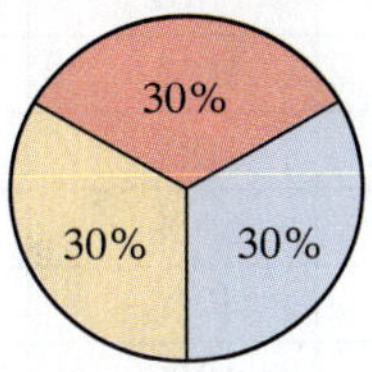

3.

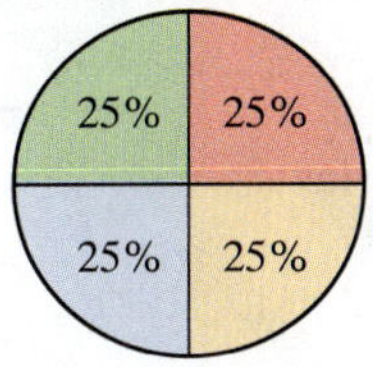

4.

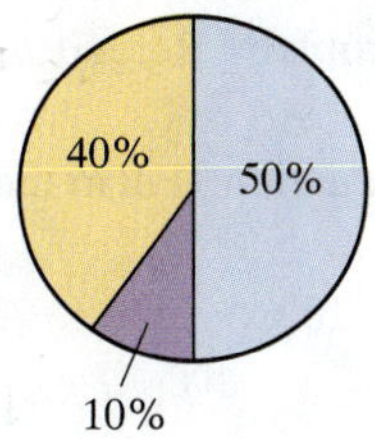

Use the choices below to fill in each blank. Some choices may be used more than once.

total amount   simple   principal amount   compound

5. To calculate ______________ interest, use $I = P \cdot R \cdot T$.

6. To calculate ______________ interest, use $A = P\left(1 + \frac{r}{n}\right)^{n \cdot t}$.

7. ______________ interest is computed not only on the original principal, but also on interest already earned in previous compounding periods.

8. When interest is computed on the original principal only, it is called ______________ interest.

9. ______________ (paid or received) = principal + interest.

10. The ______________ is the money borrowed, loaned, or invested.

Martin-Gay Interactive Videos

See Video 2.5

Watch the section lecture video and answer the following questions.

OBJECTIVE 1

11. Answer these questions based on how Example 2 was translated into an equation.
a. What does "is" translate to?
b. What does "of" translate to?
c. How do you write a percent as an equivalent decimal?

OBJECTIVE 2

12. At the end of Example 3 you are told that the process for finding discount is *almost* the same as finding markup.
a. How is discount similar?
b. How does discount differ?

OBJECTIVE 3

13. According to Example 4, what amount must you find before you can find a percent of increase in price? How do you find this amount?

OBJECTIVE 4

14. Complete this statement based on the lecture before Example 6: Simple interest is charged on the _____ only.

15. In Example 7, how often is the interest compounded, and what number does this translate to in the formula?

2.5 Exercise Set MyMathLab®

Find each number described. For Exercises 1 and 2, the solutions have been started for you. See Examples 1 and 2.

1. What number is 16% of 70?

Start the solution:

1. UNDERSTAND the problem. Reread it as many times as needed.
2. TRANSLATE into an equation. (Fill in the blanks below.)

What number	is	16%	of	70
↓	↓	↓	↓	↓
x	___	0.16	___	70

Finish with:

3. SOLVE and 4. INTERPRET

2. What number is 88% of 1000?

Start the solution:

1. UNDERSTAND the problem. Reread it as many times as needed.
2. TRANSLATE into an equation. (Fill in the blanks below.)

What number	is	88%	of	1000
↓	↓	↓	↓	↓
x	___	0.88	___	1000

Finish with:

3. SOLVE and 4. INTERPRET

3. The number 28.6 is what percent of 52?

4. The number 87.2 is what percent of 436?

5. The number 45 is 25% of what number?

6. The number 126 is 35% of what number?

Find the amount of discount and the sale price. See Example 3.

	Original Price	Discount Rate	Amount of Discount	Sale Price
7.	\$89	10%		
8.	\$74	20%		
9.	\$196.50	50%		
10.	\$110.60	40%		
11.	\$21,700	15%		
12.	\$17,800	12%		

Find the sales tax and new price. See Example 4.

	Purchase Price	Tax Rate	Sales Tax	New Price
13.	\$305	9%		
14.	\$243	8%		
15.	\$56	5.5%		
16.	\$65	8.4%		

Find the amount of increase and the percent of increase. See Example 5.

	Original Amount	New Amount	Amount of Increase	Percent of Increase
17.	50	80		
18.	8	12		
19.	65	117		
20.	68	170		

Find the amount of decrease and the percent of decrease. See Example 5.

	Original Amount	New Amount	Amount of Decrease	Percent of Decrease
21.	8	6		
22.	25	20		
23.	160	40		
24.	200	162		

Solve. See Example 6.

25. Find the original price of a pair of shoes if the sale price is \$78 after a 25% discount.

26. Find the original price of a popular pair of shoes if the increased price is \$80 after a 25% increase.

27. Find last year's salary if, after a 4% pay raise, this year's salary is \$44,200.

28. Find last year's salary if after a 3% pay raise, this year's salary is \$55,620.

Find the simple interest and the total amount. See Examples 7 through 9.

	Principal	Rate	Time	Simple Interest	Total Amount
29.	\$160	11.5%	4 years		
30.	\$950	12.5%	5 years		
31.	\$5000	10%	$1\frac{1}{2}$ years		
32.	\$1500	14%	$2\frac{1}{4}$ years		
33.	\$375	18%	6 months		
34.	\$775	15%	8 months		
35.	\$2500	16%	21 months		
36.	\$1000	10%	18 months		

Find the total amount. See Examples 10 and 11.

	Principal	Time	Rate	Compounding	Total Amount
37.	\$600	5 years	9%	quarterly	
38.	\$10,000	15 years	4%	daily	
39.	\$1200	20 years	11%	annually	
40.	\$5800	1 year	7%	semiannually	
41.	\$500	4 years	6%	quarterly	
42.	\$2500	19 years	5%	daily	

MIXED PRACTICE

Solve. If needed, round money amounts to two decimal places and all other amounts to one decimal place. See Examples 1 through 6.

43. Find 23% of 20.

44. Find 140% of 86.

45. The number 40 is 80% of what number?

46. The number 56.25 is 45% of what number?

47. The number 144 is what percent of 480?

48. The number 42 is what percent of 35?

49. A used automobile dealership recently reduced the price of a used compact car by 8%. If the price of the car before the discount was \$18,500, find the discount and the new price.

50. A music store is advertising a 25%-off sale on all new releases. Find the discount and the sale price of a newly released CD that regularly sells for \$12.50.

51. A birthday celebration meal is \$40.50 including tax. Find the total cost if a 15% tip is added to the given cost.

52. A retirement dinner for two is \$65.40 including tax. Find the total cost if a 20% tip is added to the given cost.

Use the graph below for Exercises 53 and 54.

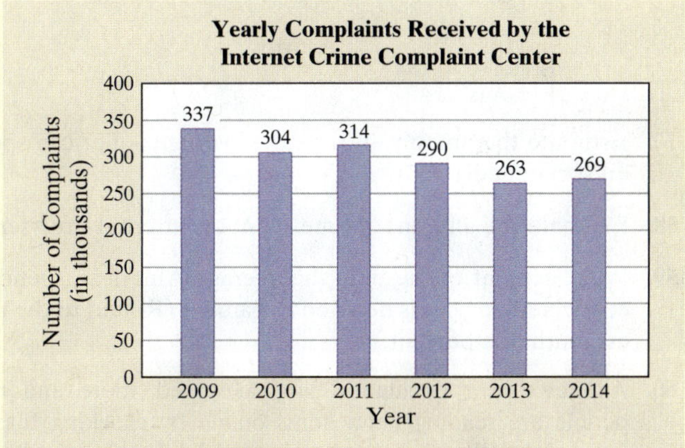

Data from Internet Crime Complaint Center (www.ic3.gov)

53. The number of Internet-crime complaints decreased from 2012 to 2013. Find the percent of decrease. Round to the nearest tenth of a percent.

54. The number of Internet-crime complaints decreased from 2011 to 2012. Find the percent of decrease. Round to the nearest tenth of a percent.

△ **55.** By decreasing each dimension by 1 unit, the area of a rectangle decreased from 40 square feet (on the left) to 28 square feet (on the right). Find the percent of decrease in area.

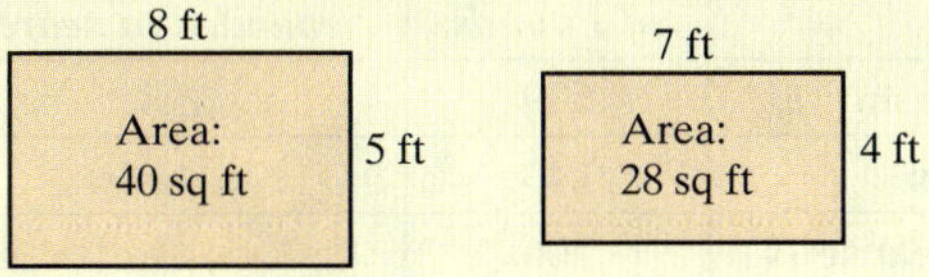

△ **56.** By decreasing the length of the side by one unit, the area of a square decreased from 100 square meters to 81 square meters. Find the percent of decrease in area.

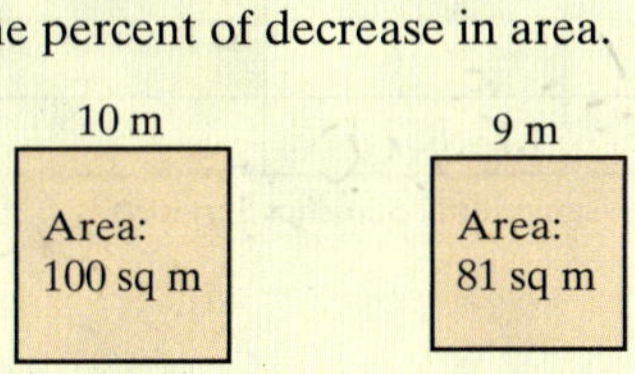

The average American spends 162 minutes per day on his or her mobile device. The graph below is a result of mobile measurement research on how Americans spend their time daily on their connected iOS or Android devices. Use this graph for Exercises 57 through 60.

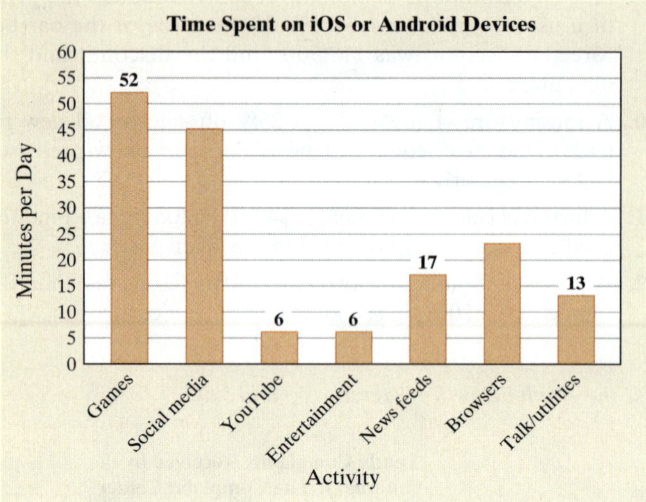

Data from Flurry Analytics

57. Estimate the amount of time the average American spends on social media daily.

58. Estimate the amount of time spent on Internet browsers.

59. What percent of the time the average American spends on connected devices is devoted to games? (Round to the nearest tenth of a percent.)

60. As newspaper circulation has decreased, more and more people are reading news items online by checking Internet news sites, television station sites, and other news-collecting sites. What percent of the average American's online time is spent following news? (Round to the nearest tenth of a percent.)

For Exercises 61 and 62, fill in the percent column in each table. Each table contains a worked-out example.

61.

Top Cranberry-Producing States (in millions of pounds)

State	*Millions of Pounds*	*Percent of Total (rounded to nearest percent)*
Wisconsin	539	
Oregon	40	
Massachusetts	206	
New Jersey	56	Example: $\frac{56}{857} \approx 7\%$
Washington	16	
Total	857	

Data from National Agricultural Statistics Service

62.

Worldwide Unit Case Volume for Coca-Cola (in billions of cases)

World Region	*Case Volume*	*Percent of Total (rounded to nearest percent)*
North America	5.9	
Latin America	8.2	
Europe	3.9	Example: $\frac{3.9}{28.2} \approx 14\%$
Eurasia and Africa	4.3	
Pacific	5.9	
Total	28.2	

Data from Coca-Cola Company

63. Nordstrom advertised a 25%-off sale. If a London Fog coat originally sold for \$256, find the decrease in price and the new price.

64. A gasoline station decreased the price of a \$0.95 cola by 15%. Find the decrease in price and the new price.

65. In 2010, Americans consumed approximately 11.1 pounds of romaine and leaf lettuce per capita. By 2013, consumption had dropped to 10.3 pounds per capita. Find the percent of decrease. Round to the nearest tenth of a percent. (*Source:* USDA)

66. Due to arid conditions in lettuce-growing areas at the end of 2014, the cost of growing and shipping iceberg lettuce to stores was about \$1.19. In certain areas, consumers purchased lettuce for about \$1.49 a head. Find the percent of increase. Round to the nearest tenth of a percent. (*Source:* USDA)

67. The number of registered vehicles on the road in the United States is constantly increasing. In 2013, there were approximately 256 million registered vehicles. This represents an increase of 2% over 2006. How many registered vehicles were there in the United States in 2006? Round to the nearest million. (*Source:* Federal Highway Safety Administration)

68. A student at the University of New Orleans makes money by buying and selling used cars. Charles bought a used car and later sold it for a 20% profit. If he sold it for \$4680, how much did Charles pay for the car?

△ 69. By doubling each dimension, the area of a parallelogram increased from 36 square centimeters to 144 square centimeters. Find the percent of increase in area.

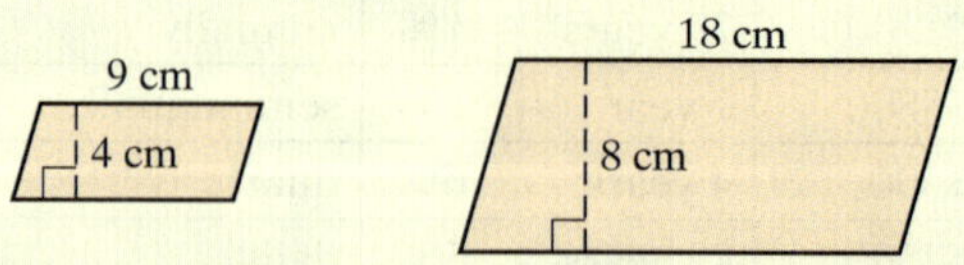

△ 70. By doubling each dimension, the area of a triangle increased from 6 square miles to 24 square miles. Find the percent of increase in area.

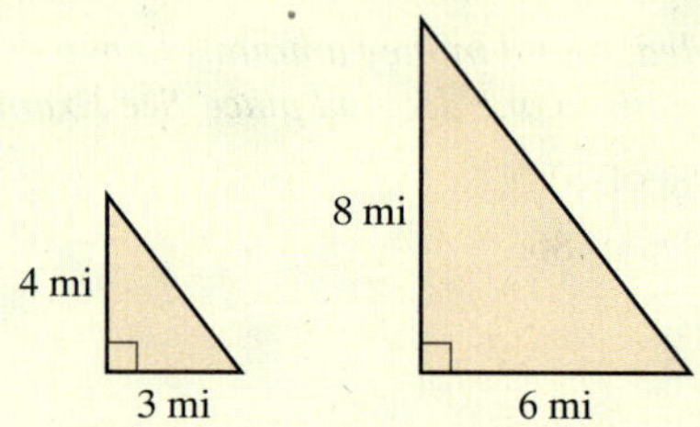

71. The number of farms in the United States is decreasing. In 1982 there were approximately 2.5 million farms, while in 2014, there were only approximately 2.1 million farms. Find the percent of decrease in the number of farms. (*Source:* USDA agricultural census)

72. During the 2007–2008 term, the Supreme court made 72 decisions, while during the 2013–2014 term they made 88 decisions. Find the percent of increase in the number of decisions. Round to the nearest tenth of a percent. (*Source:* supremecourt.gov)

73. A company recently downsized its number of employees by 35%. If there are still 78 employees, how many employees were there prior to the layoffs?

74. The average number of children born to each U.S. woman has decreased by 44% since 1920. If this average is now 1.9, find the average in 1920. Round to the nearest tenth.

Solve. See Examples 7 through 9.

75. A company borrows \$162,500 for 5 years at a simple interest rate of 12.5%. Find the interest paid on the loan and the total amount paid back.

76. \$265,000 is borrowed to buy a house. If the simple interest rate on the 30-year loan is 8.25%, find the interest paid on the loan and the total amount paid back.

77. A money market fund advertises a simple interest rate of 9%. Find the total amount received on an investment of \$5000 for 15 months.

78. The Real Service Company takes out a 270-day (9-month) short-term, simple interest loan of \$4500 to finance the purchase of some new equipment. If the interest rate is 14%, find the total amount that the company pays back.

Find the total amount in each compound interest account. See Examples 10 and 11.

79. \$6150 is compounded semiannually at a rate of 14% for 15 years.

80. \$2060 is compounded annually at a rate of 15% for 10 years.

81. \$1560 is compounded daily at a rate of 8% for 5 years.

82. \$1450 is compounded quarterly at a rate of 10% for 15 years.

83. \$10,000 is compounded semiannually at a rate of 9% for 20 years.

84. \$3500 is compounded daily at a rate of 8% for 10 years.

85. \$2000 is compounded annually at a rate of 8% for 5 years.

86. \$2000 is compounded semiannually at a rate of 8% for 5 years.

87. \$2000 is compounded quarterly at a rate of 8% for 5 years.

88. \$2000 is compounded daily at a rate of 8% for 5 years.

89. Scoville units are used to measure the hotness of a pepper. Measuring 577 thousand Scoville units, the "Red Savina" habañero pepper was known as the hottest chili pepper. That changed with the discovery of the Naga Jolokia pepper from India. It measures 48% hotter than the habañero. Find the measure of the Naga Jolokia pepper. Round to the nearest thousand units.

90. As of this writing, the women's record for throwing a disc (like a heavy Frisbee) was set by Valarie Jenkins of the United States in 2008. Her throw was 148.00 meters. The men's world record was set by Christian Sandstrom of Sweden in 2002. His throw was 68.9% farther than Valarie's. Find the distance of his throw. Round to the nearest meter. (*Source:* World Flying Disc Federation)

REVIEW AND PREVIEW

Place <, >, or = in the appropriate space to make each a true statement. See Sections 1.2, 1.3, and 1.4.

91. $-5 \quad -7$

92. $\frac{12}{3} \quad 2^2$

93. $|-5| \quad -(-5)$

94. $-3^3 \quad (-3)^3$

95. $(-3)^2 \quad -3^2$

96. $|-2| \quad -|-2|$

CONCEPT EXTENSIONS

97. Must the percents in a circle graph have a sum of 100%? Why or why not?

98. a. Can an item be marked up by more than 100%? Why or why not?

b. Can an item be discounted by more than 100%? Why or why not?

Tipping *One very useful application of percent is mentally calculating a tip. Recall that to find 10% of a number, simply move the decimal point one place to the left. To find 20% of a number, just double 10% of the number. To find 15% of a number, find 10% and then add to that number half of the 10% amount. Mentally fill in the chart below. To do so, start by rounding the bill amount to the nearest dollar.*

Tipping Chart

	Bill Amount	*10%*	*15%*	*20%*
99.	\$40.21			
100.	\$15.89			
101.	\$72.17			
102.	\$9.33			

103. Suppose that the original price of a shirt is \$50. Which is better, a 60% discount or a discount of 30% followed by a discount of 35% of the reduced price? Explain your answer.

104. Which is better, a 30% discount followed by an additional 25% off or a 20% discount followed by an additional 40% off? To see, suppose an item costs \$100 and calculate each discounted price. Explain your answer.

Standardized nutrition labels like the one below have been displayed on food items since 1994. The percent column on the right shows the percent of daily values (based on a 2000-calorie diet) shown at the bottom of the label. For example, a serving of this food contains 4 grams of total fat and the recommended daily fat based on a 2000-calorie diet is less than 65 grams of fat. This means that $\frac{4}{65}$ *or approximately 6% (as shown) of your daily recommended fat is taken in by eating a serving of this food. Use this nutrition label to answer Exercises 105 through 107.*

Nutrition Facts

Serving Size	18 Crackers (31g)
Servings Per Container	About 9
Amount Per Serving	
Calories 130	Calories from Fat 35
	% Daily Value*
Total Fat 4g	**6%**
Saturated Fat 0.5g	**3%**
Polyunsaturated Fat 0g	
Monounsaturated Fat 1.5g	
Cholesterol 0mg	**0%**
Sodium 230mg	*x*
Total Carbohydrate 23g	*y*
Dietary Fiber 2g	**8%**
Sugars 3g	
Protein 2g	
Vitamin A 0% •	Vitamin C 0%
Calcium 2% •	Iron 6%

* Percent Daily Values are based on a 2,000 calorie diet. Your daily values may be higher or lower depending on your calorie needs.

	Calories	2,000	2,500
Total Fat	Less than	65g	80g
Sat. Fat	Less than	20g	25g
Cholesterol	Less than	300mg	300mg
Sodium	Less than	2400mg	2400mg
Total Carbohydrate		300g	375g
Dietary Fiber		25g	30g

105. Based on a 2000-calorie diet, what percent of daily value of sodium is contained in a serving of this food? In other words, find x in the label. (Round to the nearest tenth of a percent.)

106. Based on a 2000-calorie diet, what percent of daily value of total carbohydrate is contained in a serving of this food? In other words, find y in the label. (Round to the nearest tenth of a percent.)

107. Notice on the nutrition label that one serving of this food contains 130 calories and 35 of these calories are from fat. Find the percent of calories from fat. (Round to the nearest tenth of a percent.) It is recommended that no more than 30% of calorie intake come from fat. Does this food satisfy this recommendation?

Use the nutrition label below to answer Exercises 108 and 109.

NUTRITIONAL INFORMATION PER SERVING	
Serving Size: 9.8 oz	**Servings Per Container: 1**
Calories 280	Polyunsaturated Fat 1g
Protein 12g	Saturated Fat 3g
Carbohydrate 45g	Cholesterol 20mg
Fat 6g	Sodium 520mg
Percent of Calories from Fat....?	Potassium 220mg

108. If fat contains approximately 9 calories per gram, find the percent of calories from fat in one serving of this food. (Round to the nearest tenth of a percent.)

109. If protein contains approximately 4 calories per gram, find the percent of calories from protein from one serving of this food. (Round to the nearest tenth of a percent.)

110. Find a food that contains more than 30% of its calories per serving from fat. Analyze the nutrition label and verify that the percents shown are correct.

2.6 Solving Linear Inequalities

OBJECTIVES

1 Define Linear Inequality in One Variable, Graph Solution Sets on a Number Line, and Use Interval Notation.

2 Solve Linear Inequalities.

3 Solve Compound Inequalities.

4 Solve Inequality Applications.

OBJECTIVE

1 Graphing Solution Sets to Linear Inequalities and Using Interval Notation

In Chapter 1, we reviewed these inequality symbols and their meanings:

$<$ means "is less than" $\leq$ means "is less than or equal to"

$>$ means "is greater than" $\geq$ means "is greater than or equal to"

Equations	*Inequalities*
$x = 3$	$x \leq 3$
$5n - 6 = 14$	$5n - 6 < 14$
$12 = 7 - 3y$	$12 \geq 7 - 3y$
$\frac{x}{4} - 6 = 1$	$\frac{x}{4} - 6 > 1$

A linear inequality is similar to a linear equation except that the equality symbol is replaced with an inequality symbol.

> **Linear Inequality in One Variable**
>
> A **linear inequality in one variable** is an inequality that can be written in the form
>
> $$ax + b < c$$
>
> where a, b, and c are real numbers and a is not 0.

This definition and all other definitions, properties, and steps in this section also hold true for the inequality symbols $>$, $\geq$, and $\leq$.

A **solution of an inequality** is a value of the variable that makes the inequality a true statement. The solution set is the set of all solutions. For the inequality $x < 3$, replacing x with any number less than 3, that is, to the left of 3 on a number line, makes the resulting inequality true. This means that any number less than 3 is a solution of the inequality $x < 3$.

Since there are infinitely many such numbers, we cannot list all the solutions of the inequality. We *can* use set notation and write

$$\{x \mid x < 3\}.$$

Recall that this is read: $\{x$ ↑ the set of all x; $|$ ↑ such that; $x < 3\}$ ↑ x is less than 3.

We can also picture the solutions on a number line. If we use open/closed-circle notation, the graph of $\{x|x < 3\}$ looks like the following.

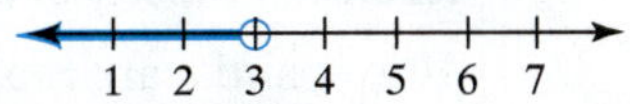

In this text, a convenient notation, called **interval notation,** will be used to write solution sets of inequalities. To help us understand this notation, a different graphing notation will be used. Instead of an open circle, we use a parenthesis; instead of a closed circle, we use a bracket. With this new notation, the graph of $\{x|x < 3\}$ now looks like

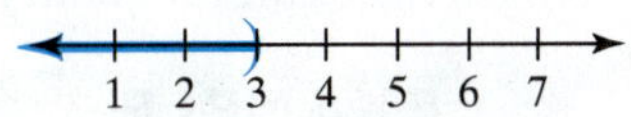

and can be represented in interval notation as $(-\infty, 3)$. The symbol $-\infty$, read as "negative infinity," does not indicate a number but does indicate that the shaded arrow to the left never ends. In other words, the interval $(-\infty, 3)$ includes *all* numbers less than 3.

Picturing the solutions of an inequality on a number line is called **graphing** the solutions or graphing the inequality, and the picture is called the **graph** of the inequality.

To graph $\{x|x \leq 3\}$ or simply $x \leq 3$, shade the numbers to the left of 3 and place a bracket at 3 on the number line as shown below. The bracket indicates that 3 **is** a solution: 3 **is** less than or equal to 3. In interval notation, we write $(-\infty, 3]$.

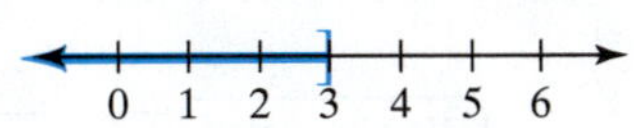

Helpful Hint

When writing an inequality in interval notation, it may be easier to graph the inequality first, then write it in interval notation. To help, think of the number line as approaching $-\infty$ to the left and $+\infty$ or ∞ to the right. Then simply write the interval notation by following your shading from left to right.

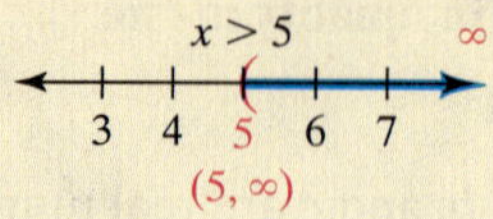

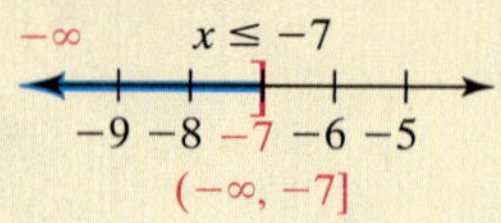

EXAMPLE 1 Graph $x \geq -1$ Then write the solutions in interval notation.

Solution We place a bracket at -1 since the inequality symbol is $\geq$ and -1 is greater than or equal to -1. Then we shade to the right of -1.

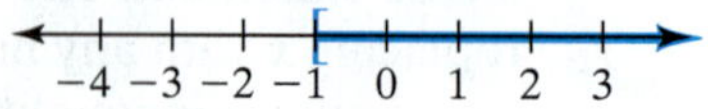

In interval notation, this is $[-1, \infty)$. □

PRACTICE
1 Graph $x < 5$. Then write the solutions in interval notation. ■

OBJECTIVE
2 Solving Linear Inequalities

When solutions of a linear inequality are not immediately obvious, they are found through a process similar to the one used to solve a linear equation. Our goal is to get the variable alone, and we use properties of inequality similar to properties of equality.

Addition Property of Inequality

If a, b, and c are real numbers, then

$$a < b \quad \text{and} \quad a + c < b + c$$

are equivalent inequalities.

This property also holds true for subtracting values, since subtraction is defined in terms of addition. In other words, adding or subtracting the same quantity from both sides of an inequality does not change the solution of the inequality.

EXAMPLE 2 Solve $x + 4 \leq -6$ for x. Graph the solution set and write it in interval notation.

Solution To solve for x, subtract 4 from both sides of the inequality.

$x + 4 \leq -6$ — Original inequality

$x + 4 - 4 \leq -6 - 4$ — Subtract 4 from both sides.

$x \leq -10$ — Simplify.

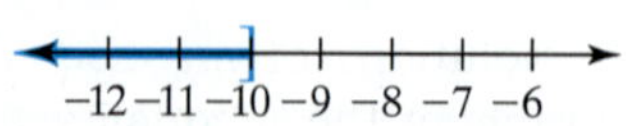

The solution set is $(-\infty, -10]$. □

PRACTICE
2 Solve $x + 11 \geq 6$. Graph the solution set and write it in interval notation. ■

Helpful Hint

Notice that any number less than or equal to -10 is a solution to $x \leq -10$. For example, solutions include

$$-10, \ -200, \ -11\frac{1}{2}, \ -7\pi, \ -\sqrt{130}, \ -50.3$$

An important difference between linear equations and linear inequalities is shown when we multiply or divide both sides of an inequality by a nonzero real number. For example, start with the true statement $6 < 8$ and multiply both sides by 2. As we see below, the resulting inequality is also true.

$$6 < 8 \quad \text{True}$$
$$2(6) < 2(8) \quad \text{Multiply both sides by 2.}$$
$$12 < 16 \quad \text{True}$$

But if we start with the same true statement $6 < 8$ and multiply both sides by -2, the resulting inequality is not a true statement.

$$6 < 8 \quad \text{True}$$
$$-2(6) < -2(8) \quad \text{Multiply both sides by } -2.$$
$$-12 < -16 \quad \text{False}$$

Notice, however, that if we reverse the direction of the inequality symbol, the resulting inequality is true.

$$-12 < -16 \quad \text{False}$$
$$-12 > -16 \quad \text{True}$$

This demonstrates the multiplication property of inequality.

Multiplication Property of Inequality

1. If a, b, and c are real numbers, and c is **positive,** then

$$a < b \quad \text{and} \quad ac < bc$$

are equivalent inequalities.

2. If a, b, and c are real numbers, and c is **negative,** then

$$a < b \quad \text{and} \quad ac > bc$$

are equivalent inequalities.

Because division is defined in terms of multiplication, this property also holds true when dividing both sides of an inequality by a nonzero number. If we multiply or divide both sides of an inequality by a negative number, **the direction of the inequality symbol must be reversed for the inequalities to remain equivalent.**

Helpful Hint

Whenever both sides of an inequality are multiplied or divided by a negative number, the direction of the inequality symbol **must be** reversed to form an equivalent inequality.

EXAMPLE 3 Solve $-2x \le -4$. Graph the solution set and write it in interval notation.

Solution Remember to reverse the direction of the inequality symbol when dividing by a negative number.

Helpful Hint

Don't forget to reverse the direction of the inequality symbol.

$$-2x \le -4$$
$$\frac{-2x}{-2} \ge \frac{-4}{-2} \quad \text{Divide both sides by } -2 \text{ and reverse the direction of the inequality symbol.}$$
$$x \ge 2 \quad \text{Simplify.}$$

The solution set $[2, \infty)$ is graphed as shown.

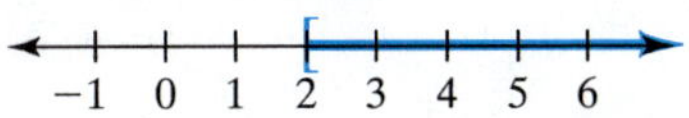

□

PRACTICE 3 Solve $-5x \ge -15$. Graph the solution set and write it in interval notation. ■

EXAMPLE 4 Solve $2x < -4$. Graph the solution set and write it in interval notation.

Solution

Helpful Hint

Do not reverse the inequality symbol.

$$2x < -4$$

$$\frac{2x}{2} < \frac{-4}{2}$$ Divide both sides by 2. Do not reverse the direction of the inequality symbol.

$$x < -2$$ Simplify.

The solution set $(-\infty, -2)$ is graphed as shown.

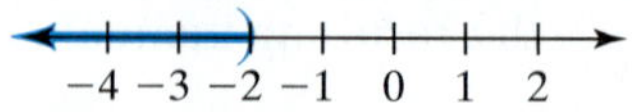

PRACTICE

4 Solve $3x > -9$. Graph the solution set and write it in interval notation.

✔ **CONCEPT CHECK**

Fill in the blank with $<$, $>$, $\le$, or $\ge$.

a. Since $-8 < -4$, then $3(-8)$_____$3(-4)$.

b. Since $5 \ge -2$, then $\frac{5}{-7}$ _____ $\frac{-2}{-7}$.

c. If $a < b$, then $2a$_____$2b$.

d. If $a \ge b$, then $\frac{a}{-3}$ _____ $\frac{b}{-3}$.

The following steps may be helpful when solving inequalities. Notice that these steps are similar to the ones given in Section 2.4 for solving equations.

Solving Linear Inequalities in One Variable

Step 1. Clear the inequality of fractions by multiplying both sides of the inequality by the least common denominator (LCD) of all fractions in the inequality.

Step 2. Remove grouping symbols such as parentheses by using the distributive property.

Step 3. Simplify each side of the inequality by combining like terms.

Step 4. Write the inequality with variable terms on one side and numbers on the other side by using the addition property of inequality.

Step 5. Get the variable alone by using the multiplication property of inequality.

Helpful Hint

Don't forget that if both sides of an inequality are multiplied or divided by a negative number, the direction of the inequality symbol must be reversed.

EXAMPLE 5 Solve $-4x + 7 \ge -9$. Graph the solution set and write it in interval notation.

Solution

$$-4x + 7 \ge -9$$

$$-4x + 7 - 7 \ge -9 - 7$$ Subtract 7 from both sides.

$$-4x \ge -16$$ Simplify.

$$\frac{-4x}{-4} \le \frac{-16}{-4}$$ Divide both sides by -4 and reverse the direction of the inequality symbol.

$$x \le 4$$ Simplify.

Answers to Concept Check:

a. $<$ **b.** $\le$ **c.** $<$ **d.** $\le$

The solution set $(-\infty, 4]$ is graphed as shown.

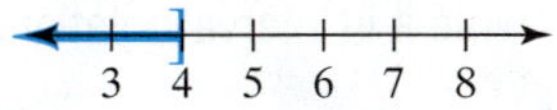

□

PRACTICE 5 Solve $45 - 7x \le -4$. Graph the solution set and write it in interval notation. ■

EXAMPLE 6 Solve $2x + 7 \le x - 11$. Graph the solution set and write it in interval notation.

Solution

$$2x + 7 \le x - 11$$

$$2x + 7 - x \le x - 11 - x \quad \text{Subtract } x \text{ from both sides.}$$

$$x + 7 \le -11 \quad \text{Combine like terms.}$$

$$x + 7 - 7 \le -11 - 7 \quad \text{Subtract 7 from both sides.}$$

$$x \le -18 \quad \text{Combine like terms.}$$

The graph of the solution set $(-\infty, -18]$ is shown.

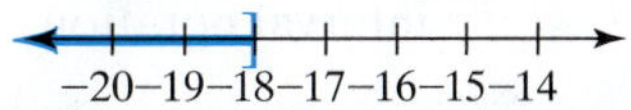

□

PRACTICE 6 Solve $3x + 20 \le 2x + 13$. Graph the solution set and write it in interval notation. ■

EXAMPLE 7 Solve $-5x + 7 < 2(x - 3)$. Graph the solution set and write it in interval notation.

Solution

$$-5x + 7 < 2(x - 3)$$

$$-5x + 7 < 2x - 6 \quad \text{Apply the distributive property.}$$

$$-5x + 7 - 2x < 2x - 6 - 2x \quad \text{Subtract } 2x \text{ from both sides.}$$

$$-7x + 7 < -6 \quad \text{Combine like terms.}$$

$$-7x + 7 - 7 < -6 - 7 \quad \text{Subtract 7 from both sides.}$$

$$-7x < -13 \quad \text{Combine like terms.}$$

$$\frac{-7x}{-7} > \frac{-13}{-7} \quad \text{Divide both sides by } -7 \text{ and reverse the direction of the inequality symbol.}$$

$$x > \frac{13}{7} \quad \text{Simplify.}$$

The graph of the solution set $\left(\frac{13}{7}, \infty\right)$ is shown.

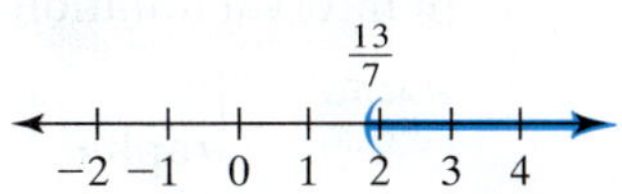

□

PRACTICE 7 Solve $6 - 5x > 3(x - 4)$. Graph the solution set and write it in interval notation. ■

EXAMPLE 8 Solve $2(x - 3) - 5 \leq 3(x + 2) - 18$. Graph the solution set and write it in interval notation.

Solution

$$2(x - 3) - 5 \leq 3(x + 2) - 18$$

$$2x - 6 - 5 \leq 3x + 6 - 18 \quad \text{Apply the distributive property.}$$

$$2x - 11 \leq 3x - 12 \quad \text{Combine like terms.}$$

$$-x - 11 \leq -12 \quad \text{Subtract } 3x \text{ from both sides.}$$

$$-x \leq -1 \quad \text{Add 11 to both sides.}$$

$$\frac{-x}{-1} \geq \frac{-1}{-1} \quad \text{Divide both sides by } -1 \text{ and reverse the direction of the inequality symbol.}$$

$$x \geq 1 \quad \text{Simplify.}$$

The graph of the solution set $[1, \infty)$ is shown.

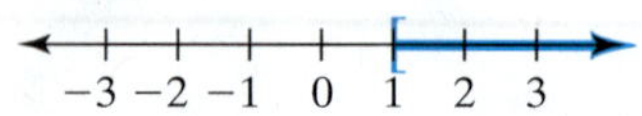

□

PRACTICE

8 Solve $3(x - 4) - 5 \leq 5(x - 1) - 12$. Graph the solution set and write it in interval notation.

OBJECTIVE

3 Solving Compound Inequalities

Inequalities containing one inequality symbol are called **simple inequalities,** while inequalities containing two inequality symbols are called **compound inequalities.** A compound inequality is really two simple inequalities in one. The compound inequality

$$3 < x < 5 \quad \text{means} \quad 3 < x \textbf{ and } x < 5$$

This can be read "x is greater than 3 and less than 5."

A solution of a compound inequality is a value that is a solution of both of the simple inequalities that make up the compound inequality. For example,

$$4\frac{1}{2} \text{ is a solution of } 3 < x < 5 \text{ since } 3 < 4\frac{1}{2} \textbf{ and } 4\frac{1}{2} < 5.$$

To graph $3 < x < 5$, place parentheses at both 3 and 5 and shade between.

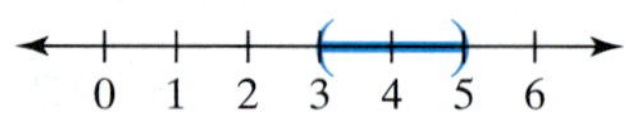

EXAMPLE 9 Graph $2 < x \leq 4$. Write the solutions in interval notation.

Solution Graph all numbers greater than 2 and less than or equal to 4. Place a parenthesis at 2, a bracket at 4, and shade between.

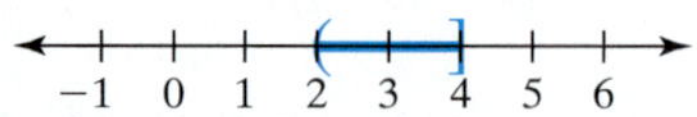

In interval notation, this is $(2, 4]$. □

PRACTICE

9 Graph $-3 \leq x < 1$. Write the solutions in interval notation.

When we solve a simple inequality, we isolate the variable on one side of the inequality. When we solve a compound inequality, we isolate the variable in the middle part of the inequality. Also, when solving a compound inequality, we must perform the same operation to all **three** parts of the inequality: left, middle, and right.

EXAMPLE 10 Solve $-1 \le 2x - 3 < 5$. Graph the solution set and write it in interval notation.

Solution

$$-1 \le 2x - 3 < 5$$

$$-1 + 3 \le 2x - 3 + 3 < 5 + 3 \quad \text{Add 3 to all three parts.}$$

$$2 \le 2x < 8 \quad \text{Combine like terms.}$$

$$\frac{2}{2} \le \frac{2x}{2} < \frac{8}{2} \quad \text{Divide all three parts by 2.}$$

$$1 \le x < 4 \quad \text{Simplify.}$$

The graph of the solution set $[1, 4)$ is shown.

□

PRACTICE 10 Solve $-4 < 3x + 2 \le 8$. Graph the solution set and write it in interval notation. ■

EXAMPLE 11 Solve $3 \le \frac{3x}{2} + 4 \le 5$. Graph the solution set and write it in interval notation.

Solution

$$3 \le \frac{3x}{2} + 4 \le 5$$

$$2(3) \le 2\left(\frac{3x}{2} + 4\right) \le 2(5) \quad \text{Multiply all three parts by 2 to clear the fraction.}$$

$$6 \le 3x + 8 \le 10 \quad \text{Distribute.}$$

$$-2 \le 3x \le 2 \quad \text{Subtract 8 from all three parts.}$$

$$\frac{-2}{3} \le \frac{3x}{3} \le \frac{2}{3} \quad \text{Divide all three parts by 3.}$$

$$-\frac{2}{3} \le x \le \frac{2}{3} \quad \text{Simplify.}$$

The graph of the solution set $\left[-\frac{2}{3}, \frac{2}{3}\right]$ is shown.

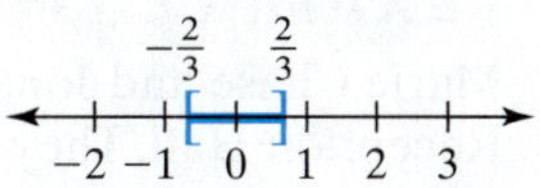

□

PRACTICE 11 Solve $1 < \frac{3}{4}x + 5 < 6$. Graph the solution set and write it in interval notation. ■

OBJECTIVE 4

Solving Inequality Applications

Problems containing words such as "at least," "at most," "between," "no more than," and "no less than" usually indicate that an inequality should be solved instead of an equation. In solving applications involving linear inequalities, use the same procedure you use to solve applications involving linear equations.

Some Inequality Translations			
$\geq$	$\leq$	$<$	$>$
at least	at most	is less than	is greater than
no less than	no more than		

EXAMPLE 12 12 subtracted from 3 times a number is less than 21. Find all numbers that make this statement true.

Solution

1. UNDERSTAND. Read and reread the problem. This is a direct translation problem, and let's let

$$x = \text{the unknown number.}$$

2. TRANSLATE.

12	subtracted from	three times a number	is less than	21
$3x$	$-$	12	$<$	21

3. SOLVE.

$$3x - 12 < 21$$

$$3x < 33 \quad \text{Add 12 to both sides.}$$

$$\frac{3x}{3} < \frac{33}{3} \quad \text{Divide both sides by 3 and do not reverse the direction of the inequality symbol.}$$

$$x < 11 \quad \text{Simplify.}$$

4. INTERPRET.

Check: Check the translation; then let's choose a number less than 11 to see if it checks. For example, let's check 10. 12 subtracted from 3 times 10 is 12 subtracted from 30, or 18. Since 18 is less than 21, the number 10 checks.

State: All numbers less than 11 make the original statement true. □

PRACTICE 12 Twice a number, subtracted from 35, is greater than 15. Find all numbers that make this true. ■

EXAMPLE 13 **Staying Within Budget**

Marie Chase and Jonathan Edwards are having their wedding reception at the Gallery Reception Hall. They may spend at most \$2000 for the reception. If the reception hall charges a \$100 cleanup fee plus \$36 per person, find the greatest number of people that they can invite and still stay within their budget.

Solution

1. UNDERSTAND. Read and reread the problem. Next, guess a solution. If 40 people attend the reception, the cost is $\$100 + \$36(40) = \$100 + \$1440 = \$1540$. Let

$$x = \text{the number of people who attend the reception.}$$

2. TRANSLATE.

In words:	cleanup fee	+	cost per person	must be less than or equal to	2000
Tlanslate:	100	+	$36x$	$\leq$	2000

3. SOLVE.

$$
\begin{aligned}
100 + 36x &\le 2000 \\
36x &\le 1900 \quad \text{Subtract 100 from both sides.} \\
x &\le 52\frac{7}{9} \quad \text{Divide both sides by 36.}
\end{aligned}
$$

4. INTERPRET.

Check: Since x represents the number of people, we round down to the nearest whole, or 52. Notice that if 52 people attend, the cost is

$\$100 + \$36(52) = \$1972$. If 53 people attend, the cost is

$\$100 + \$36(53) = \$2008$, which is more than the given 2000.

State: Marie Chase and Jonathan Edwards can invite at most 52 people to the reception. □

PRACTICE 13 Kasonga is eager to begin his education at his local community college. He has budgeted \$1500 for college this semester. His local college charges a \$300 matriculation fee and costs an average of \$375 for tuition, fees, and books for each three-credit course. Find the greatest number of classes Kasonga can afford to take this semester. ■

Vocabulary, Readiness & Video Check

Use the choices below to fill in each blank. Choices may be used more than once.

expression inequality equation

1. $6x - 7(x + 9)$ __________

2. $6x = 7(x + 9)$ __________

3. $6x < 7(x + 9)$ __________

4. $5y - 2 \ge -38$ __________

Decide which number listed is not a solution to each given inequality.

5. $x \ge -3$; $-3, 0, -5, \pi$ ________

6. $x < 6$; $-6, |-6|, 0, -3.2$ ________

Martin-Gay Interactive Videos

See Video 2.6

Watch the section lecture video and answer the following questions.

OBJECTIVE 1

7. Using Example 1 from the video as a reference, explain the connection between the graph of an inequality and interval notation.

OBJECTIVE 2

8. The steps for solving a linear inequality in one variable are discussed in the lecture before Example 6. Why are you told to be very careful when you use Step 5?

OBJECTIVE 3

9. For Example 8, explain how the solving would change if the compound inequality simplified to $0 < -3x < 14$ instead of $0 < 3x < 14$.

OBJECTIVE 4

10. What is the phrase in Example 9 that tells you to translate to an *inequality*? What does this phrase translate to?

2.6 Exercise Set MyMathLab®

Graph each set of numbers given in interval notation. Then write an inequality statement in x describing the numbers graphed.

1. $[2, \infty)$
2. $(-3, \infty)$
3. $(-\infty, -5)$
4. $(-\infty, 4]$

Graph each inequality on a number line. Then write the solutions in interval notation. See Example 1.

5. $x \le -1$
6. $y < 0$
7. $x < \frac{1}{2}$
8. $z < -\frac{2}{3}$
9. $y \ge 5$
10. $x > 3$

Solve each inequality. Graph the solution set and write it in interval notation. See Examples 2 through 4.

11. $2x < -6$
12. $3x > -9$
13. $x - 2 \ge -7$
14. $x + 4 \le 1$
15. $-8x \le 16$
16. $-5x < 20$

Solve each inequality. Graph the solution set and write it in interval notation. See Examples 5 and 6.

17. $3x - 5 > 2x - 8$
18. $3 - 7x \ge 10 - 8x$
19. $4x - 1 \le 5x - 2x$
20. $7x + 3 < 9x - 3x$

Solve each inequality. Graph the solution set and write it in interval notation. See Examples 7 and 8.

21. $x - 7 < 3(x + 1)$
22. $3x + 9 \le 5(x - 1)$
23. $-6x + 2 \ge 2(5 - x)$
24. $-7x + 4 > 3(4 - x)$
25. $4(3x - 1) \le 5(2x - 4)$
26. $3(5x - 4) \le 4(3x - 2)$
27. $3(x + 2) - 6 > -2(x - 3) + 14$
28. $7(x - 2) + x \le -4(5 - x) - 12$

MIXED PRACTICE

Solve the following inequalities. Graph each solution set and write it in interval notation.

29. $-2x \le -40$
30. $-7x > 21$
31. $-9 + x > 7$
32. $y - 4 \le 1$
33. $3x - 7 < 6x + 2$
34. $2x - 1 \ge 4x - 5$
35. $5x - 7x \ge x + 2$
36. $4 - x < 8x + 2x$
37. $\frac{3}{4}x > 2$
38. $\frac{5}{6}x \ge -8$
39. $3(x - 5) < 2(2x - 1)$
40. $5(x + 4) < 4(2x + 3)$
41. $4(2x + 1) < 4$
42. $6(2 - x) \ge 12$
43. $-5x + 4 \ge -4(x - 1)$
44. $-6x + 2 < -3(x + 4)$
45. $-2(x - 4) - 3x < -(4x + 1) + 2x$
46. $-5(1 - x) + x \le -(6 - 2x) + 6$
47. $\frac{1}{4}(x + 4) < \frac{1}{5}(2x + 3)$
48. $\frac{1}{3}(3x - 1) < \frac{1}{2}(x + 4)$

Graph each inequality. Then write the solutions in interval notation. See Example 9.

49. $-1 < x < 3$
50. $2 \le y \le 3$
51. $0 \le y < 2$
52. $-1 \le x \le 4$

Solve each inequality. Graph the solution set and write it in interval notation. See Examples 10 and 11.

53. $-3 < 3x < 6$
54. $-5 < 2x < -2$
55. $2 \le 3x - 10 \le 5$
56. $4 \le 5x - 6 \le 19$
57. $-4 < 2(x - 3) \le 4$
58. $0 < 4(x + 5) \le 8$
59. $-2 < 3x - 5 < 7$
60. $1 < 4 + 2x \le 7$
61. $-6 < 3(x - 2) \le 8$
62. $-5 \le 2(x + 4) < 8$

Solve the following. For Exercises 65 and 66, the solutions have been started for you. See Examples 12 and 13.

63. Six more than twice a number is greater than negative fourteen. Find all numbers that make this statement true.

64. One more than five times a number is less than or equal to ten. Find all such numbers.

△ **65.** The perimeter of a rectangle is to be no greater than 100 centimeters and the width must be 15 centimeters. Find the maximum length of the rectangle.

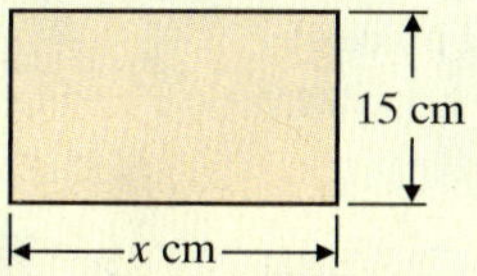

Start the solution:

1. UNDERSTAND the problem. Reread it as many times as needed.
2. TRANSLATE into an equation. (Fill in the blanks below.)

the perimeter of the rectangle	is no greater than	100
↓	↓	↓
$x + 15 + x + 15$	______	100

Finish with:

3. SOLVE and 4. INTERPRET

△ **66.** One side of a triangle is three times as long as another side, and the third side is 12 inches long. If the perimeter can be no longer than 32 inches, find the maximum lengths of the other two sides.

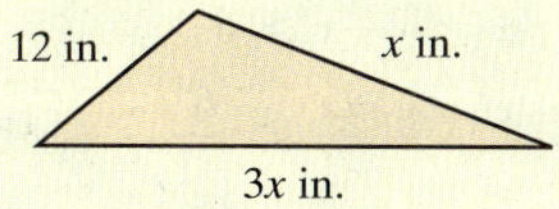

Start the solution:

1. UNDERSTAND the problem. Reread it as many times as needed.
2. TRANSLATE into an equation. (Fill in the blanks below.)

the perimeter of the rectangle	is no greater than	32
↓	↓	↓
$12 + 3x + x$	______	32

Finish with:

3. SOLVE and 4. INTERPRET

67. Ben Holladay bowled 146 and 201 in his first two games. What must he bowl in his third game to have an average of at least 180? (*Hint:* The average of a list of numbers is their sum divided by the number of numbers in the list.)

68. On an NBA team, the two forwards measure 6′8″ and 6′6″ tall and the two guards measure 6′0″ and 5′9″ tall. How tall should the center be if they wish to have a starting team average height of at least 6′5″?

69. Dennis and Nancy Wood are celebrating their 30th wedding anniversary by having a reception at Tiffany Oaks reception hall. They have budgeted $3000 for their reception. If the reception hall charges a $50.00 cleanup fee plus $34 per person, find the greatest number of people that they may invite and still stay within their budget.

70. A surprise retirement party is being planned for Pratap Puri. A total of $860 has been collected for the event, which is to be held at a local reception hall. This reception hall charges a cleanup fee of $40 and $15 per person for drinks and light snacks. Find the greatest number of people that may be invited and still stay within the $860 budget.

71. A 150-pound person uses 5.8 calories per minute when walking at a speed of 4 mph. How long must a person walk at this speed to use at least 200 calories? Round up to the next minute. (*Source:* Home & Garden Bulletin No. 72)

72. A 170-pound person uses 5.3 calories per minute when bicycling at a speed of 5.5 mph. How long must a person ride a bike at this speed to use at least 200 calories? Round up to the next minute. (*Source:* Same as Exercise 71)

73. Twice a number, increased by one, is between negative five and seven. Find all such numbers.

74. Half a number, decreased by four, is between two and three. Find all such numbers.

REVIEW AND PREVIEW

Use the bar graph below to answer Exercises 75 through 78. See Section R.4. ***Note:*** *Some of these players are still competing; thus, their total prize money may increase.*

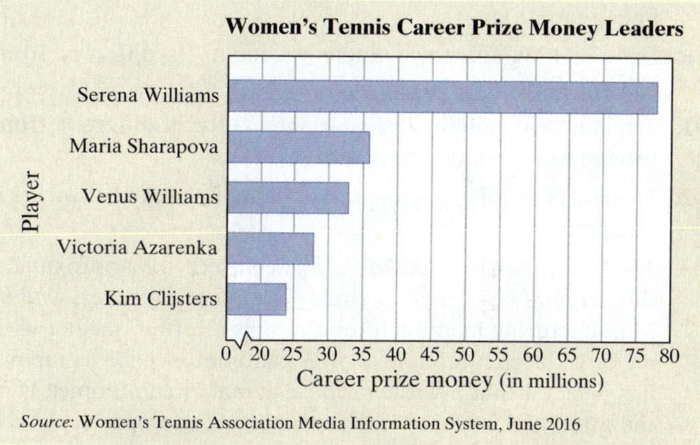

75. Which women's tennis player has earned the most prize money in her career?

76. Which women tennis player on the graph has earned the least prize money in her career.

77. What is the approximate spread in lifetime prize money between Victoria Azarenka and Maria Sharapova?

78. Estimate how much more prize money Venus Williams has earned in her career than Kim Clijsters.

79. During her career, Victoria Azarenka won 29 doubles and singles tournaments. Assume her prize money of $28 million was earned only for tournament wins. Find her average prize money for tournament win.

80. During her career, Kim Clijsters won 56 doubles and singles tournaments. Assume her prize money of $24 million was earned only for tournament wins, find her average prize money per tournament win.

CONCEPT EXTENSIONS

Fill in the box with $<$, $>$, $\leq$, or $\geq$. See the Concept Check in this section.

81. Since $3 < 5$, then $3(-4) \square 5(-4)$.

82. If $m \leq n$, then $2m \square 2n$.

83. If $m \leq n$, then $-2m \square -2n$.

84. If $-x < y$, then $x \square -y$.

85. When solving an inequality, when must you reverse the direction of the inequality symbol?

86. If both sides of the inequality $-3x < 30$ are divided by -3, do you reverse the direction of the inequality symbol? Why or why not?

Solve.

87. Eric Daly has scores of 75, 83, and 85 on his history tests. Use an inequality to find the scores he can make on his final exam to receive a B in the class. The final exam counts as **two** tests, and a B is received if the final course average is greater than or equal to 80.

88. Maria Lipco has scores of 85, 95, and 92 on her algebra tests. Use an inequality to find the scores she can make on her final exam to receive an A in the course. The final exam counts as **three** tests, and an A is received if the final course average is greater than or equal to 90. Round to one decimal place.

89. Explain how solving a linear inequality is similar to solving a linear equation.

90. Explain how solving a linear inequality is different from solving a linear equation.

91. Explain how solving a linear inequality is different from solving a compound inequality.

92. Explain how solving a linear inequality is similar to solving a compound inequality.

93. The formula $C = 3.14d$ can be used to approximate the circumference of a circle given its diameter. Waldo Manufacturing manufactures and sells a certain washer with an outside circumference of 3 centimeters. The company has decided that a washer whose actual circumference is in the interval $2.9 \leq C \leq 3.1$ centimeters is acceptable. Use a compound inequality and find the corresponding interval for diameters of these washers. (Round to three decimal places.)

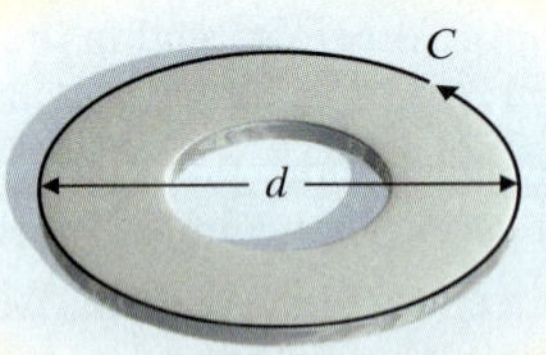

94. A company manufactures plastic Easter eggs that open. The company has determined that if the circumference of the opening of each part of the egg is in the interval $118 \leq C \leq 122$ millimeters, the eggs will open and close comfortably. Use a compound inequality and find the corresponding interval for diameters of these openings. (Round to two decimal places.)

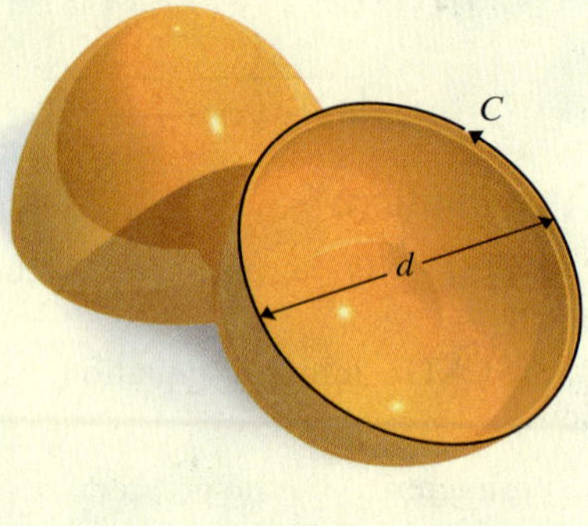

For Exercises 95 through 98, see the example below.

Solve $x(x - 6) > x^2 - 5x + 6$. Graph the solution set and write it in interval notation.

Solution

$$x(x - 6) > x^2 - 5x + 6$$
$$x^2 - 6x > x^2 - 5x + 6$$
$$x^2 - 6x - x^2 > x^2 - 5x + 6 - x^2$$
$$-6x > -5x + 6$$
$$-x > 6$$
$$\frac{-x}{-1} < \frac{6}{-1}$$
$$x < -6$$

The solution set $(-\infty, -6)$ is graphed as shown.

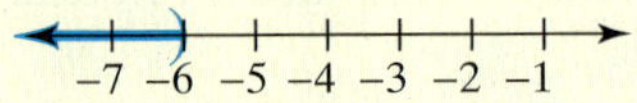

Solve each inequality. Graph the solution set and write it in interval notation.

95. $x(x + 4) > x^2 - 2x + 6$

96. $x(x - 3) \geq x^2 - 5x - 8$

97. $x^2 + 6x - 10 < x(x - 10)$

98. $x^2 - 4x + 8 < x(x + 8)$

Chapter 2 Vocabulary Check

Fill in each blank with one of the words or phrases listed below.

like terms	numerical coefficient	is	linear inequality in one variable	no solution	percent of increase
equivalent equations	0	of	compound inequalities	reversed	percent of decrease
linear equation in one variable	unlike terms	percent	all real numbers	the same	compound interest

1. Terms with the same variables raised to exactly the same powers are called ______________.
2. If terms are not like terms, they are ______________.
3. A(n) ______________ can be written in the form $ax + b = c$.
4. A(n) ______________ can be written in the form $ax + b < c$ (or $>$, $\leq$, $\geq$).
5. Inequalities containing two inequality symbols are called ______________.
6. The solution(s) to the equation $2x + 5 = x + 5$ is/are ______________.
7. The ______________ of a term is its numerical factor.
8. Equations that have the same solution are called ______________.
9. The solution(s) to the equation $x + 5 = x + 5$ is/are ______________.
10. The solution(s) to the equation $x + 5 = x + 4$ is/are ______________.
11. If both sides of an inequality are multiplied or divided by the same positive number, the direction of the inequality symbol is ______________.
12. If both sides of an inequality are multiplied or divided by the same negative number, the direction of the inequality symbol is ______________.
13. In a mathematical statement, ______________ usually means "multiplication."
14. In a mathematical statement, ______________ means "equal."
15. ______________ means "per hundred."
16. ______________ is computed not only on the principal, but also on interest already earned in previous compounding periods.
17. ______________ $= \dfrac{\text{amount of decrease}}{\text{original amount}}$.
18. ______________ $= \dfrac{\text{amount of increase}}{\text{original amount}}$.

Chapter 2 Highlights

DEFINITIONS AND CONCEPTS	EXAMPLES
Section 2.1 Simplifying Algebraic Expressions	
The **numerical coefficient** of a **term** is its numerical factor.	***Term*** — ***Numerical Coefficient*** $-7y$ — -7 x — 1 $\frac{1}{5}a^2b$ — $\frac{1}{5}$
Terms with the same variables raised to exactly the same powers are **like terms.**	***Like Terms*** — ***Terms Unlike Terms*** $12x, -x$ — $3y, 3y^2$ $-2xy, 5yx$ — $7a^2b, -2ab^2$
To combine like terms, add the numerical coefficients and multiply the result by the common variable factor.	$9y + 3y = 12y$ $-4z^2 + 5z^2 - 6z^2 = -5z^2$
To remove parentheses, apply the distributive property.	$-4(x + 7) + 10(3x - 1)$ $= -4x - 28 + 30x - 10$ $= 26x - 38$

DEFINITIONS AND CONCEPTS	EXAMPLES
Section 2.2 The Addition and Multiplication Properties of Equality	
A **linear equation in one variable** can be written in the form $ax + b = c$ where $a, b,$ and c are real numbers and $a \neq 0$.	***Linear Equations*** $-3x + 7 = 2$ $3(x - 1) = -8(x + 5) + 4$
Equivalent equations are equations that have the same solution.	$x - 7 = 10$ and $x = 17$ are equivalent equations.
Addition Property of Equality Adding the same number to or subtracting the same number from both sides of an equation does not change its solution.	$y + 9 = 3$ $y + 9 - 9 = 3 - 9$ $y = -6$
Multiplication Property of Equality Multiplying both sides or dividing both sides of an equation by the same nonzero number does not change its solution.	$\frac{2}{3}a = 18$ $\frac{3}{2}\left(\frac{2}{3}a\right) = \frac{3}{2}(18)$ $a = 27$
Section 2.3 Solving Linear Equations	
To Solve Linear Equations	Solve: $\frac{5(-2x + 9)}{6} + 3 = \frac{1}{2}$
1. Clear the equation of fractions.	**1.** $6 \cdot \frac{5(-2x + 9)}{6} + 6 \cdot 3 = 6 \cdot \frac{1}{2}$ $5(-2x + 9) + 18 = 3$
2. Remove any grouping symbols such as parentheses.	**2.** $-10x + 45 + 18 = 3$ Distributive property
3. Simplify each side by combining like terms.	**3.** $-10x + 63 = 3$ Combine like terms.
4. Write variable terms on one side and numbers on the other side using the addition property of equality.	**4.** $-10x + 63 - 63 = 3 - 63$ Subtract 63. $-10x = -60$
5. Get the variable alone using the multiplication property of equality.	**5.** $\frac{-10x}{-10} = \frac{-60}{-10}$ Divide by -10. $x = 6$
6. Check by substituting in the original equation.	**6.** $\frac{5(-2x + 9)}{6} + 3 = \frac{1}{2}$ $\frac{5(-2 \cdot 6 + 9)}{6} + 3 \stackrel{?}{=} \frac{1}{2}$ $\frac{5(-3)}{6} + 3 \stackrel{?}{=} \frac{1}{2}$ $-\frac{5}{2} + \frac{6}{2} \stackrel{?}{=} \frac{1}{2}$ $\frac{1}{2} = \frac{1}{2}$ True

DEFINITIONS AND CONCEPTS	EXAMPLES
Section 2.4 An Introduction to Problem Solving	
Problem-Solving Steps	The height of the Hudson volcano in Chile is twice the height of the Kiska volcano in the Aleutian Islands. If the sum of their heights is 12,870 feet, find the height of each.
1. UNDERSTAND the problem.	**1.** Read and reread the problem. Guess a solution and check your guess. Let x be the height of the Kiska volcano. Then $2x$ is the height of the Hudson volcano. x (Kiska) $2x$ (Hudson)
2. TRANSLATE the problem.	**2.** In words: height of Kiska ↓ added to ↓ height of Hudson ↓ is ↓ 12,870 Translate: $x + 2x = 12{,}870$
3. SOLVE.	**3.** $x + 2x = 12{,}870$ $3x = 12{,}870$ $x = 4290$
4. INTERPRET the results.	**4.** *Check:* If x is 4290, then $2x$ is 2(4290) or 8580. Their sum is 4290 + 8580 or 12,870, the required amount. *State:* The Kiska volcano is 4290 feet tall and the Hudson volcano is 8580 feet tall.
Section 2.5 Percent and Finance Problem Solving	
Use the same problem-solving steps to solve a problem containing percents.	32% of what number is 36.8?
1. UNDERSTAND.	**1.** Read and reread. Propose a solution and check. Let x = the unknown number.
2. TRANSLATE.	**2.** 32% ↓ of ↓ what number ↓ is ↓ 36.8 ↓ $32\% \cdot x = 36.8$
3. SOLVE.	**3.** *Solve:* $32\% \cdot x = 36.8$ $0.32x = 36.8$ $\frac{0.32x}{0.32} = \frac{36.8}{0.32}$ Divide by 0.32. $x = 115$ Simplify.
4. INTERPRET.	**4.** *Check, then state:* 32% of 115 is 36.8.
Simple Interest interest = principal · rate · time $I = P \cdot R \cdot T$ where the rate is understood to be per year, unless told otherwise.	Find the simple interest after 3 years on \$800 at an interest rate of 5%. interest ↓ = principal ↓ · rate ↓ · time ↓ interest = \$800 · 5% · 3 = \$800 · 0.05 · 3 Write 5% as 0.05. = \$120 Multiply. The interest is \$120.

(*continued*)

DEFINITIONS AND CONCEPTS	EXAMPLES

Section 2.5 Percent and Finance Problem Solving (continued)

Compound Interest Formula

The total amount A in an account is

$$A = P\left(1 + \frac{r}{n}\right)^{n\cdot t}$$

where P is the principal, r is the interest rate written as a decimal, t is time in years, and n is the number of times compounded per year.

\$800 is invested at 5% compounded quarterly for 10 years. Find the total amount at the end of 10 years. "Compounded quarterly" means $n = 4$. Also, $P = \$800$, $r = 5\% = 0.05$, and $t = 10$ years.

$$\begin{aligned} A &= P\left(1 + \frac{r}{n}\right)^{n\cdot t} \\ &= 800\left(1 + \frac{0.05}{4}\right)^{4\cdot 10} \\ &= 800(1.0125)^{40} \\ &\approx 1314.90 \end{aligned}$$

The total amount is \$1314.90.

Section 2.6 Solving Linear Inequalities

A **linear inequality in one variable** is an inequality that can be written in one of the forms:

$$ax + b < c \qquad ax + b \le c$$
$$ax + b > c \qquad ax + b \ge c$$

where a, b, and c are real numbers and a is not 0.

Linear Inequalities

$$2x + 3 < 6 \qquad 5(x - 6) \ge 10$$
$$\frac{x - 2}{5} > \frac{5x + 7}{2} \qquad \frac{-(x + 8)}{9} \le \frac{-2x}{11}$$

Addition Property of Inequality

Adding the same number to or subtracting the same number from both sides of an inequality does not change the solutions.

$$y + 4 \le -1$$
$$y + 4 - 4 \le -1 - 4 \quad \text{Subtract 4.}$$
$$y \le -5$$

$(-\infty, -5]$ 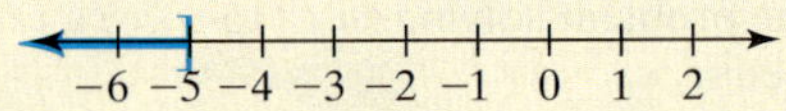

Multiplication Property of Inequality

Multiplying or dividing both sides of an inequality by the same positive number does not change its solutions.

$$\frac{1}{3}x > -2$$
$$3\left(\frac{1}{3}x\right) > 3\cdot -2 \quad \text{Multiply by 3.}$$
$$x > -6 \quad (-6, \infty)$$

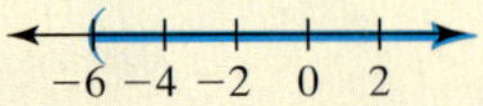

Multiplying or dividing both sides of an inequality by the same **negative number and reversing the direction of the inequality symbol** does not change its solutions.

$$-2x \le 4$$
$$\frac{-2x}{-2} \ge \frac{4}{-2} \quad \text{Divide by } -2\text{, reverse inequality symbol.}$$
$$x \ge -2 \quad [-2, \infty)$$

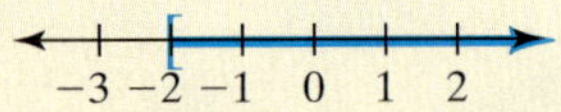

To Solve Linear Inequalities

Solve: $3(x + 2) \le -2 + 8$

1. Clear the equation of fractions.
2. Remove grouping symbols.
3. Simplify each side by combining like terms.
4. Write variable terms on one side and numbers on the other side, using the addition property of inequality.
5. Get the variable alone, using the multiplication property of inequality.

1. No fractions to clear. $3(x + 2) \le -2 + 8$
2. $3x + 6 \le -2 + 8$ Distributive property
3. $3x + 6 \le 6$ Combine like terms.
4. $3x + 6 - 6 \le 6 - 6$ Subtract 6.
   $3x \le 0$
5. $\frac{3x}{3} \le \frac{0}{3}$ Divide by 3.
   $x \le 0 \quad (-\infty, 0]$

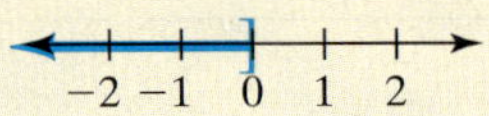

DEFINITIONS AND CONCEPTS	EXAMPLES
Section 2.6 Solving Linear Inequalities (continued)	
Inequalities containing two inequality symbols are called **compound inequalities.**	***Compound Inequalities*** $-2 < x < 6$ $5 \le 3(x-6) < \frac{20}{3}$
To solve a compound inequality, isolate the variable in the middle part of the inequality. Perform the same operation to all three parts of the inequality: left, middle, right.	Solve: $-2 < 3x + 1 < 7$ $-2 - 1 < 3x + 1 - 1 < 7 - 1$ Subtract 1. $-3 < 3x < 6$ $\frac{-3}{3} < \frac{3x}{3} < \frac{6}{3}$ Divide by 3. $-1 < x < 2$ $(-1, 2)$ 

Chapter 2 Review

(2.1) Simplify the following expressions.

1. $5x - x + 2x$
2. $0.2z - 4.6x - 7.4z$
3. $\frac{1}{2}x + 3 + \frac{7}{2}x - 5$
4. $\frac{4}{5}y + 1 + \frac{6}{5}y + 2$
5. $2(n-4) + n - 10$
6. $3(w+2) - (12 - w)$
7. Subtract $7x - 2$ from $x + 5$.
8. Subtract $1.4y - 3$ from $y - 0.7$.

Write each of the following as algebraic expressions.

9. Three times a number decreased by 7
10. Twice the sum of a number and 2.8 added to 3 times the number

(2.2) Solve each equation.

11. $8x + 4 = 9x$
12. $5y - 3 = 6y$
13. $\frac{2}{7}x + \frac{5}{7}x = 6$
14. $3x - 5 = 4x + 1$
15. $2x - 6 = x - 6$
16. $4(x+3) = 3(1+x)$
17. $6(3+n) = 5(n-1)$
18. $5(2+x) - 3(3x+2) = -5(x-6) + 2$

Use the addition property to fill in the blanks so that the middle equation simplifies to the last equation.

19. $x - 5 = 3$
$x - 5 + \_\_ = 3 + \_\_$
$x = 8$

20. $x + 9 = -2$
$x + 9 - \_\_ = -2 - \_\_$
$x = -11$

Choose the correct algebraic expression.

21. The sum of two numbers is 10. If one number is x, express the other number in terms of x.
    a. $x - 10$ **b.** $10 - x$
    c. $10 + x$ **d.** $10x$

22. Mandy is 5 inches taller than Melissa. If x inches represents the height of Mandy, express Melissa's height in terms of x.
    a. $x - 5$ **b.** $5 - x$
    c. $5 + x$ **d.** $5x$

△ 23. If one angle measures $x°$, express the measure of its complement in terms of x.
    a. $(180 - x)°$ **b.** $(90 - x)°$
    c. $(x - 180)°$ **d.** $(x - 90)°$

△ 24. If one angle measures $(x + 5)°$, express the measure of its supplement in terms of x.
    a. $(185 + x)°$
    b. $(95 + x)°$
    c. $(175 - x)°$
    d. $(x - 170)°$

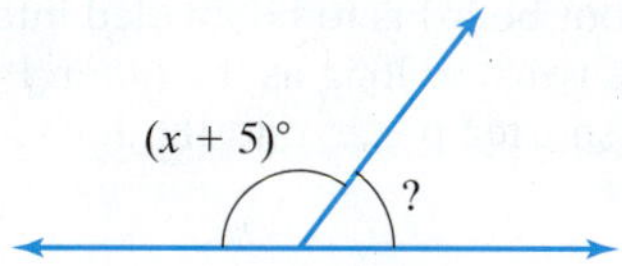

25. $\frac{3}{4}x = -9$
26. $\frac{x}{6} = \frac{2}{3}$
27. $-5x = 0$
28. $-y = 7$
29. $0.2x = 0.15$
30. $\frac{-x}{3} = 1$
31. $-3x + 1 = 19$
32. $5x + 25 = 20$
33. $7(x-1) + 9 = 5x$
34. $7x - 6 = 5x - 3$
35. $-5x + \frac{3}{7} = \frac{10}{7}$
36. $5x + x = 9 + 4x - 1 + 6$
37. Write the sum of three consecutive integers as an expression in x. Let x be the first integer.
38. Write the sum of the first and fourth of four consecutive even integers. Let x be the first even integer.

(2.3) Solve each equation.

39. $\frac{5}{3}x + 4 = \frac{2}{3}x$

40. $\frac{7}{8}x + 1 = \frac{5}{8}x$

41. $-(5x + 1) = -7x + 3$

42. $-4(2x + 1) = -5x + 5$

43. $-6(2x - 5) = -3(9 + 4x)$

44. $3(8y - 1) = 6(5 + 4y)$

45. $\frac{3(2 - z)}{5} = z$

46. $\frac{4(n + 2)}{5} = -n$

47. $0.5(2n - 3) - 0.1 = 0.4(6 + 2n)$

48. $-9 - 5a = 3(6a - 1)$

49. $\frac{5(c + 1)}{6} = 2c - 3$

50. $\frac{2(8 - a)}{3} = 4 - 4a$

51. $200(70x - 3560) = -179(150x - 19{,}300)$

52. $1.72y - 0.04y = 0.42$

(2.4) Solve each of the following.

53. The height of the Washington Monument is 50.5 inches more than 10 times the length of a side of its square base. If the sum of these two dimensions is 7327 inches, find the height of the Washington Monument. (*Source:* National Park Service)

54. A 12-foot board is to be divided into two pieces so that one piece is twice as long as the other. If x represents the length of the shorter piece, find the length of each piece.

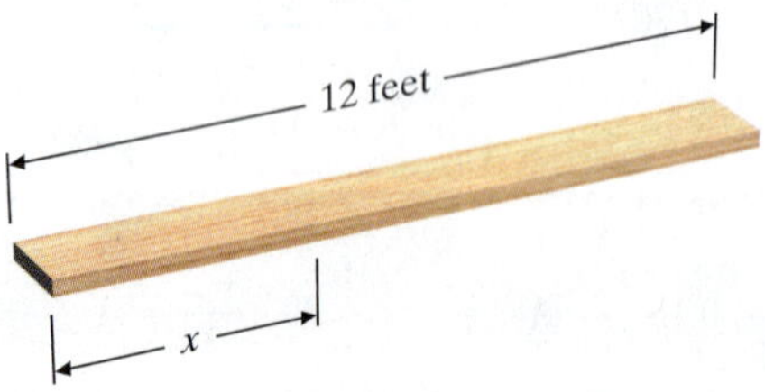

55. In 2015, Target made the decision to close all its Canadian retail stores. Before this decision, Target operated 1926 stores. The number of Target U.S. stores was 69 less than 14 times the number of Target Canada stores. How many Target stores were located in each country?

56. Find three consecutive integers whose sum is −114.

57. The quotient of a number and 3 is the same as the difference of the number and two. Find the number.

58. Double the sum of a number and 6 is the opposite of the number. Find the number.

(2.5) Find each of the following.

59. The number 9 is what percent of 45?

60. The number 59.5 is what percent of 85?

61. The number 137.5 is 125% of what number?

62. The number 768 is 60% of what number?

63. The price of a small diamond ring was recently increased by 11%. If the ring originally cost $1900, find the markup and the new price of the ring.

64. The ACT Assessment is a college entrance exam taken by about 57% of college-bound students. The national average score was 20.7 in 1993 and rose to 21.0 in 2014. Find the percent of increase. (Round to the nearest hundredth of a percent.)

The graph below shows the percents of cell phone users who have engaged in various behaviors while driving and talking on their cell phones. Use this graph to answer Exercises 65 through 68.

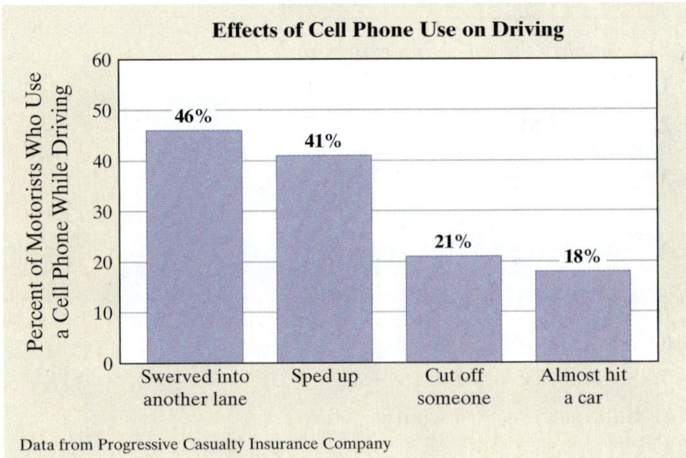

65. What percent of motorists who use a cell phone while driving have almost hit another car?

66. What is the most common effect of cell phone use on driving?

Suppose that a cell-phone service has an estimated 4600 customers who use their cell phones while driving. Use this information for Exercises 67 and 68.

67. How many of these customers would you expect to have cut someone off while driving and talking on their cell phones?

68. How many of these customers would you expect to have sped up while driving and talking on their cell phones?

69. If a price decreases from $250 to $170, what is the percent of decrease?

70. Find the original price of a DVD if the sale price is $19.20 after a 20% discount.

Solve.

71. If the sales tax rate is 9.5%, what is the new price charged for a $250 coat?

72. Find the sales tax paid on a \$25.50 purchase if the sales tax rate is 8.5%. Round to the nearest cent.

73. A \$3000 mink coat is on sale for 30% off. Find the discount and the new price.

74. A \$90 calculator is on sale for 10% off. Find the discount and the new price.

75. Find the simple interest due on \$4000 loaned for 4 months with a 12% interest rate.

76. Find the simple interest due on \$6500 loaned for 3 months with a 20% interest rate.

77. Find the total amount in an account if \$5500 is compounded annually at 12% for 15 years.

78. Find the total amount in an account if \$6000 is compounded semiannually at 11% for 10 years.

79. Find the total amount in an account if \$100 is compounded quarterly at 12% for 5 years.

80. Find the total amount in an account if \$1000 is compounded quarterly at 18% for 20 years.

(2.6) *Solve and graph the solutions of each of the following inequalities.*

81. $x > 0$

82. $x \le -2$

83. $0.5 \le y < 1.5$

84. $-1 < x < 1$

85. $-3x > 12$

86. $-2x \ge -20$

87. $x + 4 \ge 6x - 16$

88. $5x - 7 > 8x + 5$

89. $-3 < 4x - 1 < 2$

90. $2 \le 3x - 4 < 6$

91. $4(2x - 5) \le 5x - 1$

92. $-2(x - 5) > 2(3x - 2)$

93. Tina earns \$175 per week plus a 5% commission on all her sales. Find the minimum amount of sales to ensure that she earns at least \$300 per week.

94. Ellen Catarella shot rounds of 76, 82, and 79 golfing. What must she shoot on her next round so that her average will be below 80?

MIXED REVIEW

Solve each equation.

95. $6x + 2x - 1 = 5x + 11$

96. $2(3y - 4) = 6 + 7y$

97. $4(3 - a) - (6a + 9) = -12a$

98. $\frac{x}{3} - 2 = 5$

99. $2(y + 5) = 2y + 10$

100. $7x - 3x + 2 = 2(2x - 1)$

Solve.

101. The sum of six and twice a number is equal to seven less than the number. Find the number.

102. A 23-inch piece of string is to be cut into two pieces so that the length of the longer piece is three more than four times the shorter piece. If x represents the length of the shorter piece, find the lengths of both pieces.

103. The number 72 is 45% of what number?

104. A company recently increased its number of employees from 235 to 282. Find the percent of increase.

105. A dinette set sells for \$568.00. If the sales tax rate is 8.75%, find the new price of the dinette set.

106. The original price of a video game is \$23.00. It is on sale for 15% off. What is the amount of the discount?

107. Find the total amount due on a loan of \$5500 for 9 years at 12.5% simple interest.

108. Find the total amount due on a 6-month loan of \$1400 at a simple interest rate of 13%.

Solve each inequality. Graph the solution set.

109. $4x - 7 > 3x + 2$

110. $-5x < 20$

111. $-3(1 + 2x) + x \ge -(3 - x)$

Chapter 2 Getting Ready for the Test

1c

MULTIPLE CHOICE *Exercises 1–4 below are given. Choose the best directions (choice **A**, **B**, **C**, or **D**) below for each exercise.*

A. Solve for x. **B.** Simplify. **C.** Identify the numerical coefficient. **D.** Are these like or unlike terms?

1. Given: $-3x^2$

2. Given: $4x - 5 = 2x + 3$

3. Given: $5x^2$ and $4x$

4. Given: $4x - 5 + 2x + 3$

MULTIPLE CHOICE

5. Subtracting $100z$ from $8m$ translates to

A. $100z - 8m$ **B.** $8m - 100z$ **C.** $-800zm$ **D.** $92zm$

6. Subtracting $7x - 1$ from $9y$ translates to:

A. $7x - 1 - 9y$ **B.** $9y - 7x - 1$ **C.** $9y - (7x - 1)$ **D.** $7x - 1 - (9y)$

MATCHING *Match each equation in the first column with its solution in the second column. Items in the second column may be used more than once.*

7. $7x + 6 = 7x + 9$ — A. all real numbers
8. $2y - 5 = 2y - 5$ — B. no solution
9. $11x - 13 = 10x - 13$ — C. the solution is 0
10. $x + 15 = -x + 15$

MULTIPLE CHOICE

11. To solve $5(3x - 2) = -(x + 20)$, we first use the distributive property and remove parentheses by multiplying. Once this is done, the equation is

    A. $15x - 2 = -x + 20$ **B.** $15x - 10 = -x - 20$ **C.** $15x - 10 = -x + 20$ **D.** $15x - 7 = -x - 20$

12. To solve $\frac{8x}{3} + 1 = \frac{x - 2}{10}$ we multiply through by the LCD, 30. Once this is done, the simplified equation is

    A. $80x + 1 = 3x - 6$ **B.** $80x + 6 = 3x - 6$ **C.** $8x + 1 = x - 2$ **D.** $80x + 30 = 3x - 6$

Chapter 2 Test

MyMathLab® YouTube

Simplify each of the following expressions.

1. $2y - 6 - y - 4$
2. $2.7x + 6.1 + 3.2x - 4.9$
3. $4(x - 2) - 3(2x - 6)$
4. $7 + 2(5y - 3)$

Solve each of the following equations.

5. $-\frac{4}{5}x = 4$
6. $4(n - 5) = -(4 - 2n)$
7. $5y - 7 + y = -(y + 3y)$
8. $4z + 1 - z = 1 + z$
9. $\frac{2(x + 6)}{3} = x - 5$
10. $\frac{1}{2} - x + \frac{3}{2} = x - 4$
11. $-0.3(x - 4) + x = 0.5(3 - x)$
12. $-4(a + 1) - 3a = -7(2a - 3)$
13. $-2(x - 3) = x + 5 - 3x$

Solve each of the following inequalities. Graph each solution set and write it in interval notation.

14. $3x - 5 \geq 7x + 3$
15. $x + 6 > 4x - 6$
16. $-2 < 3x + 1 < 8$
17. $\frac{2(5x + 1)}{3} > 2$

Solve each of the following applications.

18. A number increased by two-thirds of the number is 35. Find the number.

19. A rectangular deck is to be built so that the width and length are two consecutive even integers and the perimeter is 252 feet. Find the dimensions of the deck.

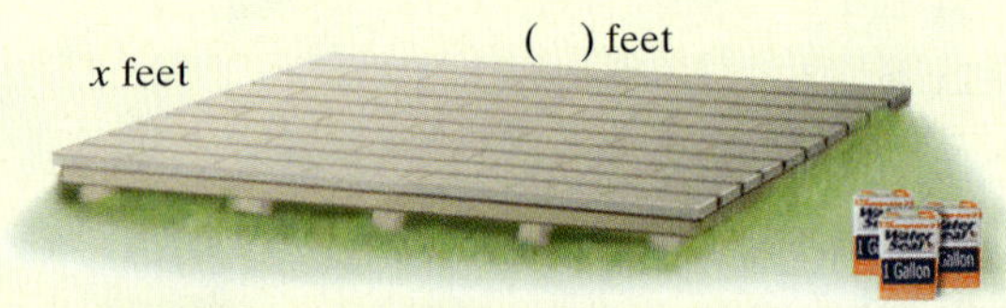

20. Some states have a single area code for the entire state. Two such states have area codes where one is double the other. If the sum of these integers is 1203, find the two area codes. (*Source:* North American Numbering Plan Administration)

The following graph shows the breakdown of tornadoes occurring in the United States by strength. The corresponding Fujita Tornado Scale categories are shown in parentheses. Use this graph to answer Exercise 24.

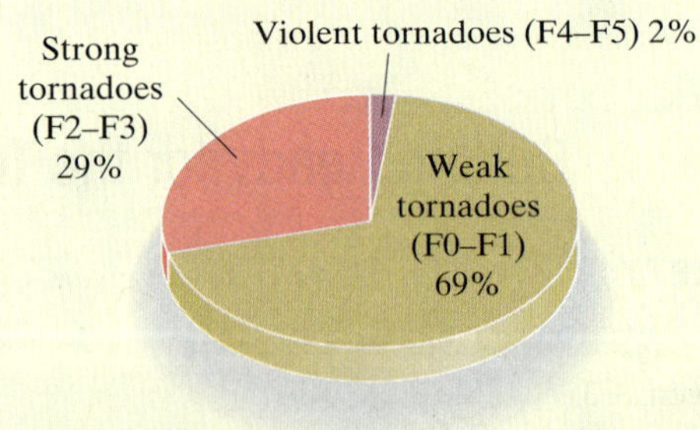

Data from National Climatic Data Center

21. According to the National Climatic Data Center, in an average year, about 800 tornadoes are reported in the United States. How many of these would you expect to be classified as "weak" tornadoes?

22. The number 72 is what percent of 180?

23. The number of employees of a company decreased from 225 to 189. Find this percent of decrease.

Solve. If necessary, round percents to the nearest tenth, dollar amounts to the nearest cent, and all other numbers to the nearest whole.

24. If the local sales tax rate is 8.25%, find the new price charged for a stereo system priced at \$354.

25. A town's population increased from 25,200 to 26,460. Find the percent of increase.

26. A \$120 framed picture is on sale for 15% off. Find the discount and the sale price.

27. Find the simple interest earned on \$2000 saved for $3\frac{1}{2}$ years at an interest rate of 9.25%.

28. \$1365 is compounded annually at 8%. Find the total amount in the account after 5 years.

29. A couple borrowed \$400 from a bank at 13.5% simple interest for 6 months for car repairs. Find the total amount due the bank at the end of the 6-month period.

Chapter 2 Cumulative Review

1. Given the set $\left\{-2, 0, \frac{1}{4}, -1.5, 112, -3, 11, \sqrt{2}\right\}$, list the numbers in this set that belong to the set of:

a. Natural numbers
b. Whole numbers
c. Integers
d. Rational numbers
e. Irrational numbers
f. Real numbers

2. Given the set $\left\{7, 2, -\frac{1}{5}, 0, \sqrt{3}, -185, 8\right\}$, list the numbers in this set that belong to the set of:

a. Natural numbers
b. Whole numbers
c. Integers
d. Rational numbers
e. Irrational numbers
f. Real numbers

3. Find the absolute value of each number.

a. $|4|$ **b.** $|-5|$
c. $|0|$ **d.** $\left|-\frac{1}{2}\right|$
e. $|5.6|$

4. Find the absolute value of each number.

a. $|5|$ **b.** $|-8|$
c. $\left|-\frac{2}{3}\right|$

5. Write each of the following numbers as a product of primes.

a. 40 **b.** 63

6. Write each number as a product of primes.

a. 44 **b.** 90

7. Write $\frac{2}{5}$ as an equivalent fraction with a denominator of 20.

8. Write $\frac{2}{3}$ as an equivalent fraction with a denominator of 24.

9. Simplify: $3[4 + 2(10 - 1)]$

10. Simplify: $5[16 - 4(2 + 1)]$

11. Decide whether 2 is a solution of $3x + 10 = 8x$.

12. Decide whether 3 is a solution of $5x - 2 = 4x$.

Add.

13. $-3 + (-11)$

14. $-2 + (-8)$

15. $-10 + 15$

16. $-3 + 10$

17. Subtract.

a. $2 - 8$ **b.** $-8 - (-1)$
c. $10.7 - (-9.8)$ **d.** $\frac{2}{3} - \frac{1}{2}$

18. Subtract

a. $-2.7 - 8.4$ **b.** $-\frac{4}{5} - \left(-\frac{3}{5}\right)$
c. $\frac{1}{4} - \left(-\frac{1}{2}\right)$

19. Simplify.

a. $11 + 2 - 7$ **b.** $-5 - 4 + 2$

20. Simplify.

a. $10 + 3 - 20$ **b.** $-9 - 3 + 15$

△ **21.** Find each unknown complementary or supplementary angle. Recall that complementary angles have a sum of 90° and supplementary angles have a sum of 180°.

a.

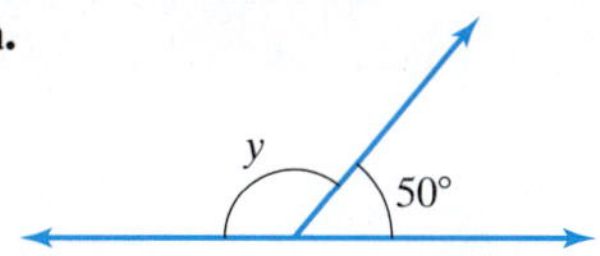

b.

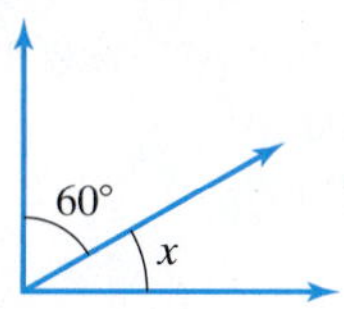

△ **22.** Find each unknown complementary or supplementary angle.

a.

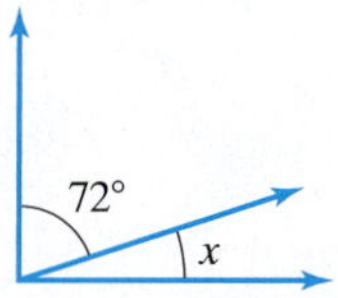

b.

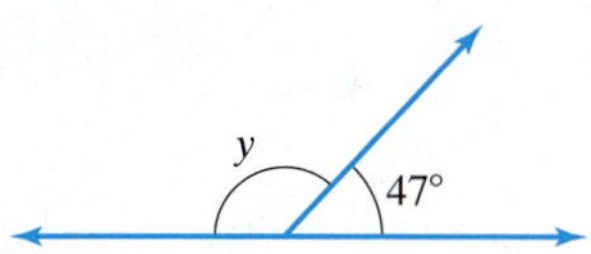

23. Find each product.

a. $(-8)(-1)$ **b.** $-1.2(0.3)$

c. $\frac{1}{5}\left(-\frac{10}{11}\right)$

24. Find each product.

a. $(4.5)(-0.08)$

b. $-\frac{3}{4}\cdot\left(-\frac{8}{17}\right)$

25. Divide.

a. $\frac{20}{-4}$ **b.** $\frac{-9}{-3}$

c. $\frac{-1}{10} \div \frac{-2}{5}$ **d.** $-\frac{3}{8} \div 3$

26. Divide.

a. $\frac{-32}{8}$ **b.** $\frac{-108}{-12}$

c. $\frac{-5}{7} \div \left(\frac{-9}{2}\right)$

27. Use a commutative property to complete each statement.

a. $x + 5 =$ _____

b. $3 \cdot x =$ _____

28. Use a commutative property to complete each statement.

a. $y + 1 =$ _____

b. $y \cdot 4 =$ _____

29. Use the distributive property to write each sum as a product.

a. $8 \cdot 2 + 8 \cdot x$

b. $7s + 7t$

30. Use the distributive property to write each sum as a product.

a. $4 \cdot y + 4 \cdot \frac{1}{3}$

b. $0.10x + 0.10y$

31. Subtract $4x - 2$ from $2x - 3$.

32. Subtract $10x + 3$ from $-5x + 1$.

Solve.

33. $y + 0.6 = -1.0$

34. $\frac{5}{6} + x = \frac{2}{3}$

35. $7 = -5(2a - 1) - (-11a + 6)$

36. $-3x + 1 - (-4x - 6) = 10$

37. $\frac{y}{7} = 20$

38. $\frac{x}{4} = 18$

39. $4(2x - 3) + 7 = 3x + 5$

40. $6x + 5 = 4(x + 4) - 1$

41. Twice the sum of a number and 4 is the same as four times the number, decreased by 12. Find the number.

42. A number increased by 4 is the same as 3 times the number decreased by 8. Find the number.

43. Solve $x + 4 \le -6$ for x. Graph the solution set and write it in interval notation.

44. Solve $x - 3 > 2$ for x. Graph the solution set and write it in interval notation.

CHAPTER 3

Graphing and Systems in Two Variables

In the previous chapter, we learned to solve and graph the solutions of linear equations and inequalities in one variable. Now we define and present techniques for solving and graphing linear equations and inequalities in two variables.

International Tourist Arrivals Forecast for 2020–2030 (numbers shown in millions)

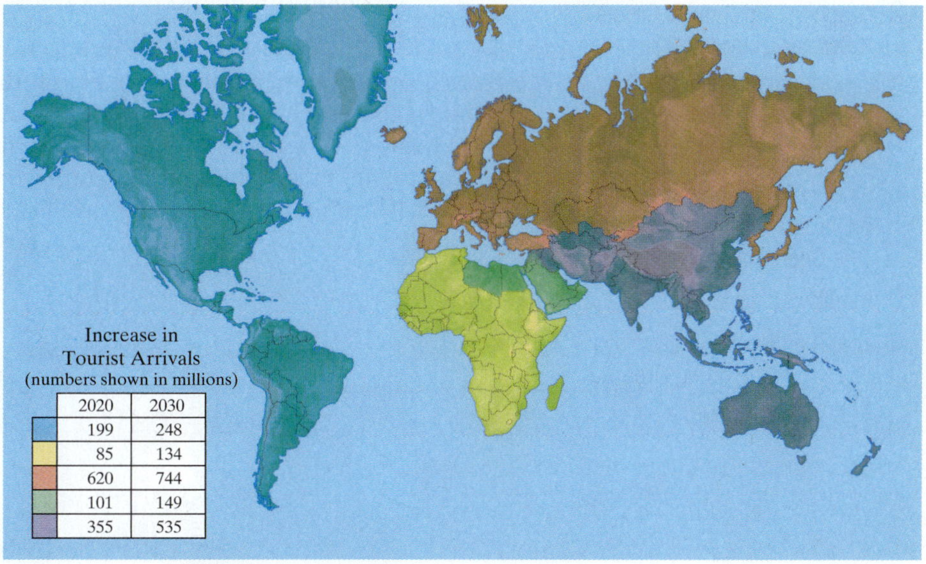

What Is Tourism Toward 2030?

Tourism 2020 Vision is the World Tourism Organization's long-term forecast of world tourism through 2020. *Tourism Towards 2030* is its new program title for longer-term forecasts to 2030. The bar graph below shows the forecast for number of tourists, which is extremely important as these numbers greatly affect a country's economy. In Section 3.4, we study this graph further.

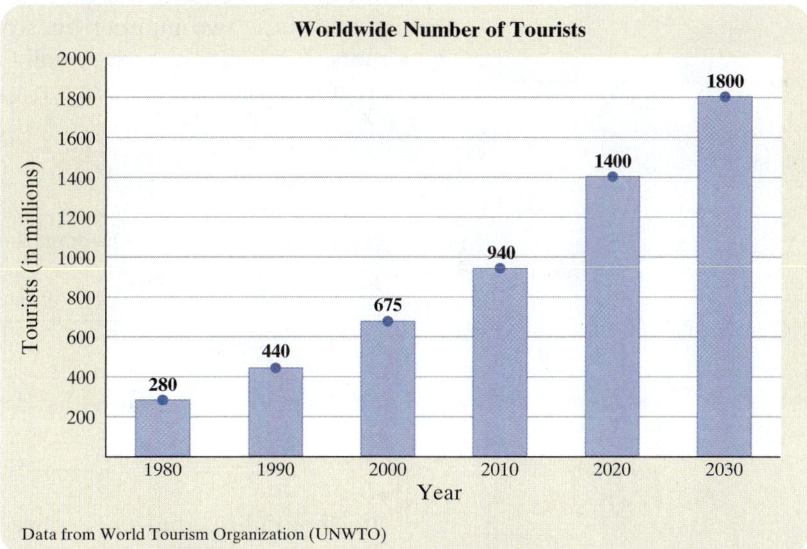

3.1 Scatter Plots and Graphing Linear Equations

OBJECTIVES

1 Define the Rectangular Coordinate System and Plot Ordered Pairs of Numbers.

2 Graph Paired Data to Create a Scatter Diagram.

3 Determine Whether an Ordered Pair Is a Solution of an Equation in Two Variables.

4 Find the Missing Coordinate of an Ordered Pair Solution, Given One Coordinate of the Pair.

5 Identify Linear Equations.

6 Graph a Linear Equation by Finding and Plotting Ordered Pair Solutions.

In today's world, where the exchange of information must be fast and entertaining, graphs are becoming increasingly popular. They provide a quick way of making comparisons, drawing conclusions, and approximating quantities. (For a review of reading graphs, see Section R.4.)

OBJECTIVE

1 Defining the Rectangular Coordinate System and Plotting Ordered Pairs of Numbers

Notice in the broken-line graph below that two numbers are associated with each point of the graph. For example, 15 minutes after lighting a cigarette, the pulse rate is 80 beats per minute. If we agree to write the time first and the pulse rate second, we can say there is a point on the graph corresponding to the **ordered pair** of numbers (15, 80). A few more ordered pairs are listed alongside their corresponding points.

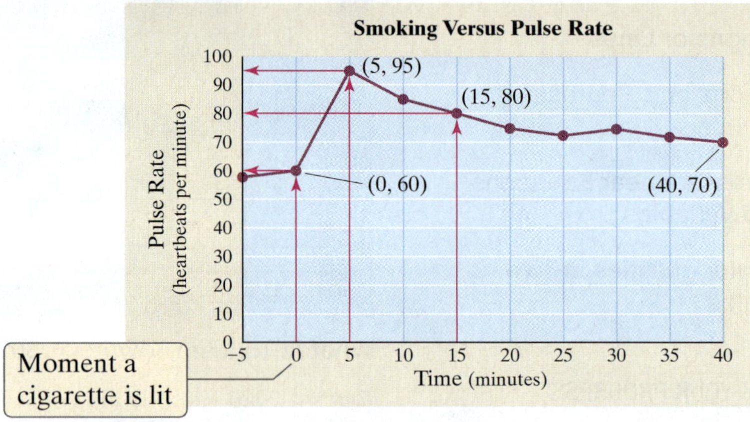

In general, we use this same ordered pair idea to describe the location of a point in a plane (such as a piece of paper). We start with a horizontal and a vertical axis. Each axis is a number line and, for the sake of consistency, we construct our axes to intersect at the 0 coordinate of both. This point of intersection is called the **origin.** Notice that these two number lines or axes divide the plane into four regions called **quadrants.** The quadrants are usually numbered with Roman numerals as shown. The axes are not considered to be in any quadrant.

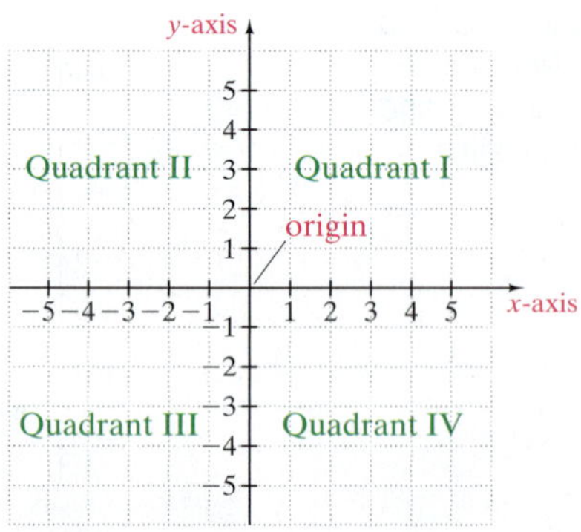

It is helpful to label axes, so we label the horizontal axis the ***x*-axis** and the vertical axis the ***y*-axis.** We call the system described above the **rectangular coordinate system.**

Just as with the pulse rate graph, we can then describe the locations of points by ordered pairs of numbers. We list the horizontal ***x*-axis** measurement first and the vertical ***y*-axis** measurement second.

To plot or graph the point corresponding to the ordered pair

$$(a, b)$$

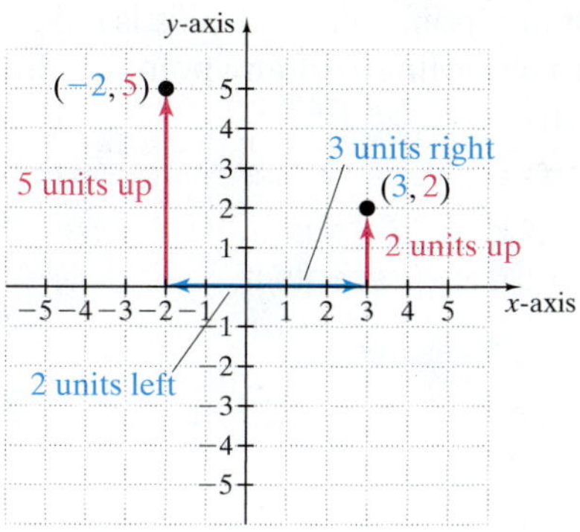

we start at the origin. We then move a units left or right (right if a is positive, left if a is negative). From there, we move b units up or down (up if b is positive, down if b is negative). For example, to plot the point corresponding to the ordered pair (3, 2), we start at the origin, move 3 units right and from there move 2 units up. (See the figure to the left.) The x-value, 3, is called the ***x*-coordinate** and the y-value, 2, is called the ***y*-coordinate.** From now on, we will call the point with coordinates (3, 2) simply the point (3, 2). The point (−2, 5) is graphed to the left also.

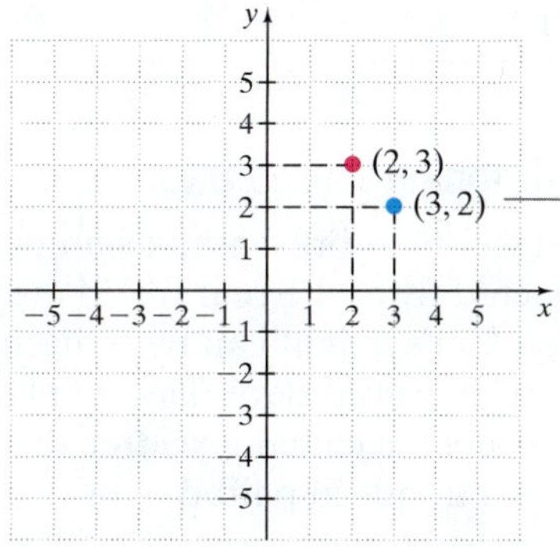

Helpful Hint

Order matters when coordinates are listed. Notice that (2, 3) and (3, 2) describe the location of two different points of the plane.

✔ CONCEPT CHECK

Is the graph of the point (−5, 1) in the same location as the graph of the point (1, −5)? Explain.

Helpful Hint

Don't forget that **each ordered pair corresponds to exactly one point in the plane and that each point in the plane corresponds to exactly one ordered pair.**

EXAMPLE 1 On a single coordinate system, plot each ordered pair. State in which quadrant, if any, each point lies.

a. (5, 3) **b.** (−5, 3) **c.** (−2, −4) **d.** (1, −2) **e.** (0, 0)

f. (0, 2) **g.** (−5, 0) **h.** $\left(0, -5\frac{1}{2}\right)$ **i.** $\left(4\frac{2}{3}, -3\right)$

Solution

a. Point (5, 3) lies in quadrant I.

b. Point (−5, 3) lies in quadrant II.

c. Point (−2, −4) lies in quadrant III.

d. Point (1, −2) lies in quadrant IV.

e.–h. Points (0, 0), (0, 2), (−5, 0), and $\left(0, -5\frac{1}{2}\right)$ lie on axes, so they are not in any quadrant.

i. Point $\left(4\frac{2}{3}, -3\right)$ lies in quadrant IV.

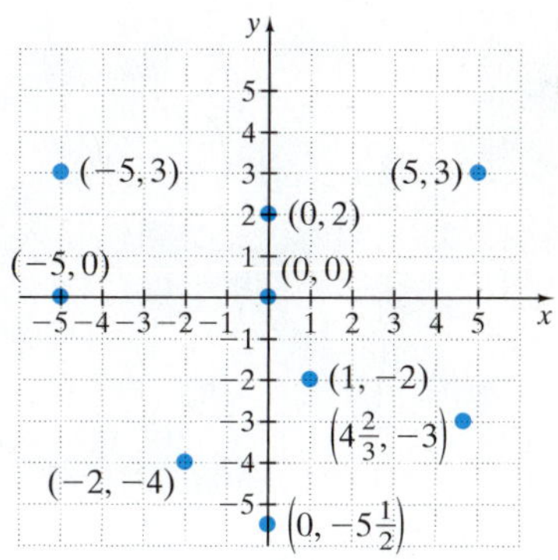

PRACTICE 1 On a single coordinate system, plot each ordered pair. State in which quadrant, or on which axis, each point lies.

a. (4, −3) **b.** (−3, 5) **c.** (0, 4) **d.** (−6, 1)

e. (−2, 0) **f.** (5, 5) **g.** $\left(3\frac{1}{2}, 1\frac{1}{2}\right)$ **h.** (−4, −5)

Answer to Concept Check:
The graph of point (−5, 1) lies in quadrant II and the graph of point (1, −5) lies in quadrant IV. They are *not* in the same location.

From Example 1, notice that the y-coordinate of any point on the x-axis is 0. For example, the point $(-5, 0)$ lies on the x-axis. Also, the x-coordinate of any point on the y-axis is 0. For example, the point $(0, 2)$ lies on the y-axis.

✔ CONCEPT CHECK

For each description of a point in the rectangular coordinate system, write an ordered pair that represents it.

a. Point A is located three units to the left of the y-axis and five units above the x-axis.

b. Point B is located six units below the origin.

OBJECTIVE

2 Graphing Paired Data

Data that can be represented as an ordered pair is called **paired data.** Many types of data collected from the real world are paired data. For instance, the annual measurement of a child's height can be written as an ordered pair of the form (year, height in inches) and is paired data. The graph of paired data as points in the rectangular coordinate system is called a **scatter diagram.** Scatter diagrams can be used to look for patterns and trends in paired data.

EXAMPLE 2 The table gives the annual net sales (in billions of dollars) for Target stores for the years shown. (*Source:* Corporate.target.com)

Year	*Target Net Sales (in billions of dollars)*
2009	65
2010	67
2011	68
2012	72
2013	71
2014	73
2015	74

a. Write this paired data as a set of ordered pairs of the form (year, sales in billions of dollars).

b. Create a scatter diagram of the paired data.

c. What trend in the paired data does the scatter diagram show?

Solution

a. The ordered pairs are $(2009, 65)$, $(2010, 67)$, $(2011, 68)$, $(2012, 72)$, $(2013, 71)$, $(2014, 73)$, $(2015, 74)$.

b. We begin by plotting the ordered pairs. Because the x-coordinate in each ordered pair is a year, we label the x-axis "Year" and mark the horizontal axis with the years given. Then we label the y-axis or vertical axis "Net Sales (in billions of dollars)." In this case we can mark the vertical axis in multiples of 2. Since no net sale is less than 64, we use the notation ⫮ to skip to 64, then proceed by multiples of 2.

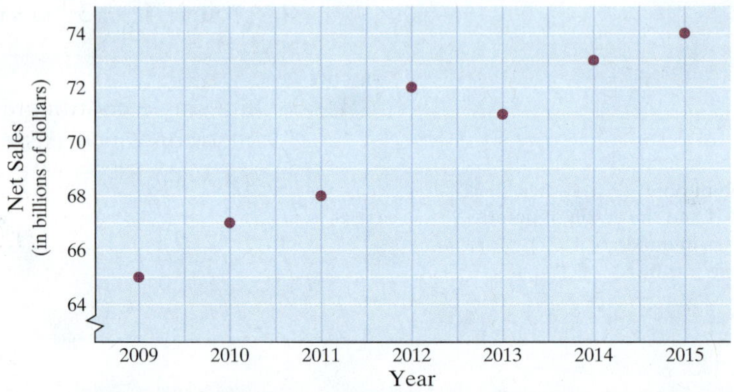

Answers to Concept Check:

a. $(-3, 5)$ **b.** $(0, -6)$

c. The scatter diagram shows that Target net sales were increasing over the years 2009–2012 and 2013–2015. □

PRACTICE
2 The table gives the approximate annual number of wildfires (in thousands) that have occurred in the United States for the years shown. (*Source:* National Interagency Fire Center)

Year	*Wildfires (in thousands)*
2008	79
2009	79
2010	72
2011	74
2012	68
2013	48
2014	63
2015	68

a. Write this paired data as a set of ordered pairs of the form (year, number of wildfires in thousands).
b. Create a scatter diagram of the paired data. ■

OBJECTIVE
3 Determining Whether an Ordered Pair Is a Solution

Let's see how we can use ordered pairs to record solutions of equations containing two variables. An equation in one variable such as $x + 1 = 5$ has one solution, which is 4: The number 4 is the value of the variable x that makes the equation true.

An equation in two variables, such as $2x + y = 8$, has solutions consisting of two values, one for x and one for y. For example, $x = 3$ and $y = 2$ is a solution of $2x + y = 8$ because, if x is replaced with 3 and y with 2, we get a true statement.

$$2x + y = 8$$
$$2(3) + 2 = 8$$
$$8 = 8 \quad \text{True}$$

The solution $x = 3$ and $y = 2$ can be written as (3, 2), an **ordered pair** of numbers. The first number, 3, is the x-value and the second number, 2, is the y-value.

In general, an ordered pair is a **solution** of an equation in two variables if replacing the variables by the values of the ordered pair results in a true statement.

EXAMPLE 3 Determine whether each ordered pair is a solution of the equation $x - 2y = 6$.

a. (6, 0) **b.** (0, 3) **c.** $\left(1, -\frac{5}{2}\right)$

Solution

a. Let $x = 6$ and $y = 0$ in the equation $x - 2y = 6$.

$$x - 2y = 6$$
$$6 - 2(0) = 6 \quad \text{Replace } x \text{ with 6 and } y \text{ with 0.}$$
$$6 - 0 = 6 \quad \text{Simplify.}$$
$$6 = 6 \quad \text{True}$$

(6, 0) is a solution, since $6 = 6$ is a true statement.

(Continued on next page)

b. Let $x = 0$ and $y = 3$.

$$\begin{aligned} x - 2y &= 6 \\ 0 - 2(3) &= 6 \quad \text{Replace } x \text{ with 0 and } y \text{ with 3.} \\ 0 - 6 &= 6 \\ -6 &= 6 \quad \text{False} \end{aligned}$$

(0, 3) is *not* a solution, since $-6 = 6$ is a false statement.

c. Let $x = 1$ and $y = -\frac{5}{2}$ in the equation.

$$\begin{aligned} x - 2y &= 6 \\ 1 - 2\left(-\frac{5}{2}\right) &= 6 \quad \text{Replace } x \text{ with 1 and } y \text{ with } -\frac{5}{2}. \\ 1 + 5 &= 6 \\ 6 &= 6 \quad \text{True} \end{aligned}$$

$\left(1, -\frac{5}{2}\right)$ is a solution, since $6 = 6$ is a true statement. □

PRACTICE 3 Determine whether each ordered pair is a solution of the equation $x + 3y = 6$.

a. (3, 1) **b.** (6, 0) **c.** $\left(-2, \frac{2}{3}\right)$ ■

OBJECTIVE 4

Completing Ordered Pair Solutions

If one value of an ordered pair solution of an equation is known, the other value can be determined. To find the unknown value, replace one variable in the equation by its known value. Doing so results in an equation with just one variable that can be solved for the variable using the methods of Chapter 2.

EXAMPLE 4 Complete the following ordered pair solutions for the equation $3x + y = 12$.

a. (0,) **b.** (, 6) **c.** (−1,)

Solution

a. In the ordered pair (0,), the x-value is 0. Let $x = 0$ in the equation and solve for y.

$$\begin{aligned} 3x + y &= 12 \\ 3(0) + y &= 12 \quad \text{Replace } x \text{ with 0.} \\ 0 + y &= 12 \\ y &= 12 \end{aligned}$$

The completed ordered pair is (0, 12).

b. In the ordered pair (, 6), the y-value is 6. Let $y = 6$ in the equation and solve for x.

$$\begin{aligned} 3x + y &= 12 \\ 3x + 6 &= 12 \quad \text{Replace } y \text{ with 6.} \\ 3x &= 6 \quad \text{Subtract 6 from both sides.} \\ x &= 2 \quad \text{Divide both sides by 3.} \end{aligned}$$

The ordered pair is (2, 6).

c. In the ordered pair $(-1, \quad)$, the x-value is -1. Let $x = -1$ in the equation and solve for y.

$$\begin{aligned} 3x + y &= 12 \\ 3(-1) + y &= 12 \quad \text{Replace } x \text{ with } -1. \\ -3 + y &= 12 \\ y &= 15 \quad \text{Add 3 to both sides.} \end{aligned}$$

The ordered pair is $(-1, 15)$. □

PRACTICE 4 Complete the following ordered pair solutions for the equation $2x - y = 8$.

a. $(0, \quad)$ **b.** $(\quad, 4)$ **c.** $(-3, \quad)$ ■

Solutions of equations in two variables can also be recorded in a **table of values,** as shown in the next example.

EXAMPLE 5 Complete the table for the equation

$$y = \frac{1}{2}x - 5.$$

	x	y
a.	−2	
b.	0	
c.		0

Solution

a. Let $x = -2$.

$$\begin{aligned} y &= \frac{1}{2}x - 5 \\ y &= \frac{1}{2}(-2) - 5 \\ y &= -1 - 5 \\ y &= -6 \end{aligned}$$

b. Let $x = 0$.

$$\begin{aligned} y &= \frac{1}{2}x - 5 \\ y &= \frac{1}{2}(0) - 5 \\ y &= 0 - 5 \\ y &= -5 \end{aligned}$$

c. Let $y = 0$.

$$\begin{aligned} y &= \frac{1}{2}x - 5 \\ 0 &= \frac{1}{2}x - 5 \quad \text{Now, solve for } x. \\ 5 &= \frac{1}{2}x \quad \text{Add 5.} \\ 10 &= x \quad \text{Multiply by 2.} \end{aligned}$$

Ordered pairs: **a.** $(-2, -6)$, **b.** $(0, -5)$, **c.** $(10, 0)$

The completed table is

	x	y
a.	−2	−6
b.	0	−5
c.	10	0

□

PRACTICE 5 Complete the table for the equation $y = \frac{1}{5}x - 2$.

	x	y
a.	−10	
b.	0	
c.		0

■

OBJECTIVE 5

Identifying Linear Equations

In the previous examples, we found that equations in two variables may have more than one solution. For example, both $(6, 0)$ and $(2, -2)$ are solutions of the equation $x - 2y = 6$. In fact, this equation has an infinite number of solutions. Other solutions include $(0, -3)$, $(4, -1)$, and $(-2, -4)$. If we graph these solutions, notice that a pattern appears.

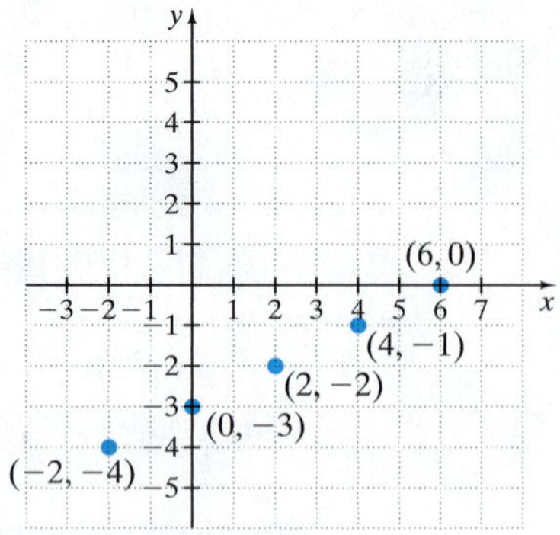

These solutions all appear to lie on the same line, which has been filled in below. It can be shown that every ordered pair solution of the equation corresponds to a point on this line, and every point on this line corresponds to an ordered pair solution. Thus, we say that this line is the **graph of the equation** $x - 2y = 6$.

Helpful Hint

Notice that we can only show part of a line on a graph. The arrowheads on each end of the line remind us that the line actually extends indefinitely in both directions.

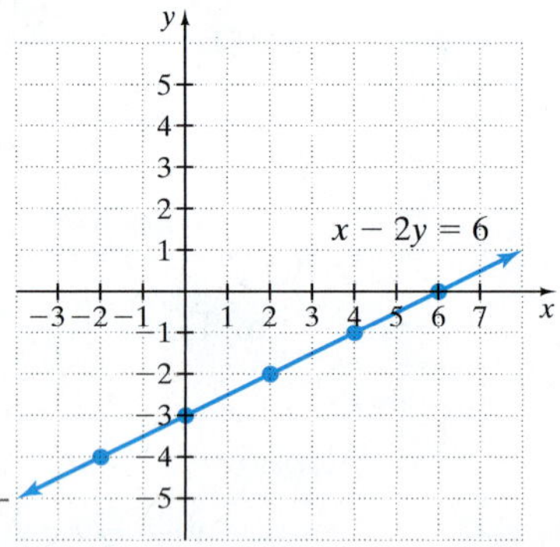

The equation $x - 2y = 6$ is called a **linear equation in two variables** and **the graph of every linear equation in two variables is a line.**

Linear Equation in Two Variables

A linear equation in two variables is an equation that can be written in the form

$$Ax + By = C$$

where A, B, and C are real numbers and A and B are not both 0. **The graph of a linear equation in two variables is a straight line.**

The form $Ax + By = C$ is called **standard form**.

Helpful Hint

Notice in the standard form $Ax + By = C$, the understood exponent on both x and y is 1.

Examples of Linear Equations in Two Variables

$2x + y = 8$ (Standard Form) $\quad -2x = 7y \quad y = \frac{1}{3}x + 2 \quad y = 7$

Before we graph linear equations in two variables, let's practice identifying these equations.

EXAMPLE 6 Determine whether each equation is a linear equation in two variables.

a. $x - 1.5y = -1.6$ **b.** $y = -2x$ **c.** $x + y^2 = 9$ **d.** $x = 5$

Solution

a. This is a linear equation in two variables because it is written in the form $Ax + By = C$ with $A = 1$, $B = -1.5$, and $C = -1.6$.

b. This is a linear equation in two variables because it can be written in the form $Ax + By = C$.

$$y = -2x$$
$$2x + y = 0 \quad \text{Add } 2x \text{ to both sides.}$$

c. This is *not* a linear equation in two variables because y is squared.

d. This is a linear equation in two variables because it can be written in the form $Ax + By = C$.

$$x = 5$$
$$x + 0y = 5 \quad \text{Add } 0 \cdot y.$$

□

PRACTICE 6 Determine whether each equation is a linear equation in two variables.

a. $3x + 2.7y = -5.3$ **b.** $x^2 + y = 8$

c. $y = 12$ **d.** $5x = -3y$

OBJECTIVE 6

Graphing Linear Equations by Plotting Ordered Pair Solutions

From geometry, we know that a straight line is determined by just two points. Graphing a linear equation in two variables, then, requires that we find just two of its infinitely many solutions. Once we do so, we plot the solution points and draw the line connecting the points. Usually, we find a third solution as well, as a check.

EXAMPLE 7 Graph the linear equation $-5x + 3y = 15$.

Solution Find three ordered pair solutions of $-5x + 3y = 15$.

Let $x = 0$.	**Let $y = 0$.**	**Let $x = -2$.**
$-5x + 3y = 15$	$-5x + 3y = 15$	$-5x + 3y = 15$
$-5 \cdot 0 + 3y = 15$	$-5x + 3 \cdot 0 = 15$	$-5(-2) + 3y = 15$
$0 + 3y = 15$	$-5x + 0 = 15$	$10 + 3y = 15$
$3y = 15$	$-5x = 15$	$3y = 5$
$y = 5$	$x = -3$	$y = \frac{5}{3}$

The ordered pairs are $(0, 5)$, $(-3, 0)$, and $\left(-2, \frac{5}{3}\right)$. The graph of $-5x + 3y = 15$ is the line through the three points.

x	y
0	5
−3	0
−2	$\frac{5}{3} = 1\frac{2}{3}$

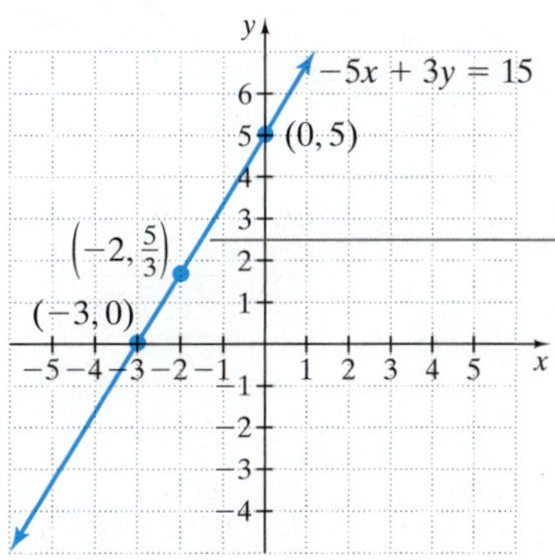

Helpful Hint

All three points should fall on the same straight line. If not, check your ordered pair solutions for a mistake.

□

PRACTICE 7 Graph the linear equation $3x - 4y = 12$.

EXAMPLE 8 Graph the linear equation $y = 3x$.

Solution To graph this linear equation, we find three ordered pair solutions. Since this equation is solved for y, choose three x-values.

If $x = 2, y = 3 \cdot 2 = 6$.
If $x = 0, y = 3 \cdot 0 = 0$.
If $x = -1, y = 3 \cdot -1 = -3$.

x	y
2	6
0	0
−1	−3

Next, graph the ordered pair solutions listed in the table above and draw a line through the plotted points as shown below. The line is the graph of $y = 3x$. Every point on the graph represents an ordered pair solution of the equation and every ordered pair solution is a point on this line.

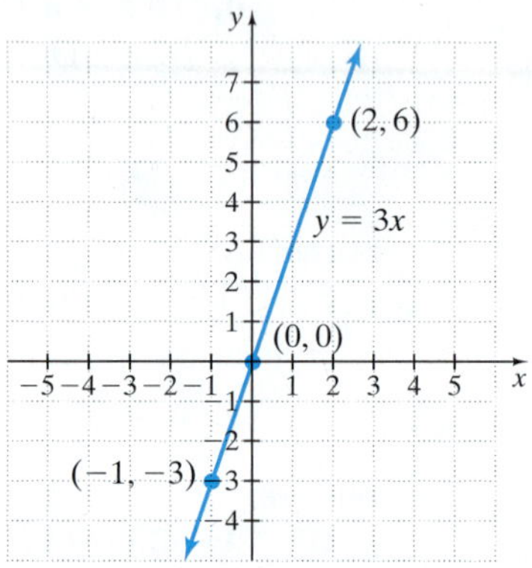

□

PRACTICE
8 Graph the linear equation $y = -2x$. ■

To prepare for functions and linear regression later in this chapter, we will concentrate on linear equations solved for y.

> **Helpful Hint**
> When graphing a linear equation in two variables, if it is
> - solved for y, it may be easier to find ordered pair solutions by choosing x-values.

EXAMPLE 9 Graph the linear equation $y = -\frac{1}{3}x + 2$.

Solution Find three ordered pair solutions, graph the solutions, and draw a line through the plotted solutions. To avoid fractions, choose x-values that are multiples of 3 to substitute in the equation. When a multiple of 3 is multiplied by $-\frac{1}{3}$, the result is an integer. See the calculations below used to fill in the table.

If $x = 6$, then $y = -\frac{1}{3} \cdot 6 + 2 = -2 + 2 = 0$

If $x = 0$, then $y = -\frac{1}{3} \cdot 0 + 2 = 0 + 2 = 2$

If $x = -3$, then $y = -\frac{1}{3} \cdot -3 + 2 = 1 + 2 = 3$

x	y
6	0
0	2
−3	3

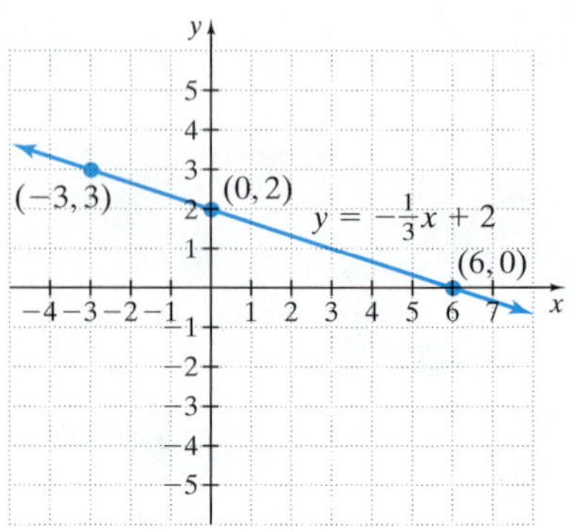

□

PRACTICE
9 Graph the linear equation $y = \frac{1}{2}x + 3$. ■

Let's compare the graphs in Examples 8 and 9. The graph of $y = 3x$ tilts upward (as we follow the line from left to right) and the graph of $y = -\frac{1}{3}x + 2$ tilts downward (as we follow the line from left to right). We will learn more about the tilt, or slope, of a line in Section 3.2.

Notice that the graph of $y = -\frac{1}{3}x + 2$ crosses the y-axis at 2. This happens because when $x = 0$, $y = -\frac{1}{3}x + 2$ becomes $y = -\frac{1}{3} \cdot 0 + 2 = 2$. The graph contains the point $(0, 2)$, which is on the y-axis.

In general, if a linear equation in two variables is solved for y, we say that it is written in the form $y = mx + b$. The graph of this equation contains the point $(0, b)$ because when $x = 0$, $y = mx + b$ is $y = m \cdot 0 + b = b$.

The graph of $y = mx + b$ crosses the y-axis at $(0, b)$.

We will review this again in Section 3.2.

Linear equations are often used to model real data as seen in the next example.

EXAMPLE 10 Estimating the Number of Registered Nurses

The occupation expected to have the most employment growth in the next few years is registered nurse. The number of people y (in thousands) employed as registered nurses in the United States can be estimated by the linear equation $y = 58.1x + 2619$, where x is the number of years after the year 2008. (*Source:* U.S. Bureau of Labor Statistics)

a. Graph the equation.

b. Use the graph to predict the number of registered nurses in the year 2018.

Solution

a. To graph $y = 58.1x + 2619$, choose x-values and substitute in the equation.

If $x = 0$, then $y = 58.1(0) + 2619 = 2619$.
If $x = 2$, then $y = 58.1(2) + 2619 = 2735.2$.
If $x = 5$, then $y = 58.1(5) + 2619 = 2909.5$.

x	y
0	2619
2	2735.2
5	2909.5

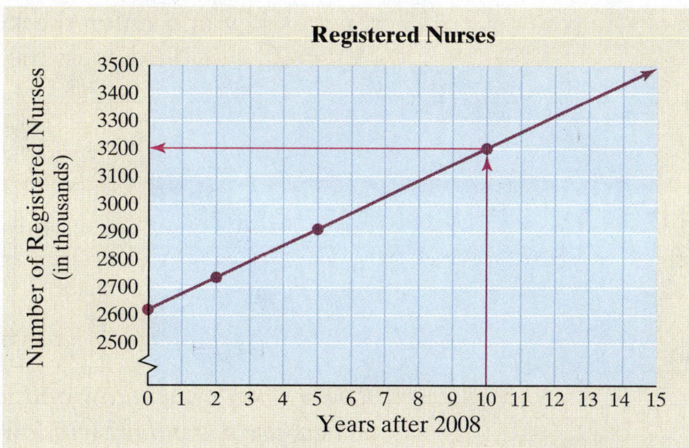

b. To use the graph to *predict* the number of registered nurses in the year 2018, we need to find the y-coordinate that corresponds to $x = 10$. (10 years after 2008 is the year 2018.) To do so, find 10 on the x-axis. Move vertically upward to the

(Continued on next page)

graphed line and then horizontally to the left. We approximate the number on the y-axis to be 3200. Thus, in the year 2018, we predict that there will be 3200 thousand registered nurses. (The actual value, using 10 for x, is 3200.) □

PRACTICE 10 One of the occupations expected to have a large growth in employment in the next few years is computer software application engineers. The number of people y (in thousands) employed as computer software application engineers in the United States can be estimated by the linear equation $y = 17.5x + 515$, where x is the number of years after 2008. (*Source:* Based on data from the Bureau of Labor Statistics)

a. Graph the equation.

b. Use the graph to predict the number of computer software application engineers in the year 2020. ■

Helpful Hint

Make sure you understand that models are mathematical approximations of the data for the known years. (For example, see the model in Example 10.) Any number of unknown factors can affect future years, so be cautious when using models to predict.

Graphing Calculator Explorations

In this section, we begin an optional study of graphing calculators and graphing software packages for computers. These graphers use the same point plotting technique that was introduced in this section. The advantage of this graphing technology is, of course, that graphing calculators and computers can find and plot ordered pair solutions much faster than we can. Note, however, that the features described in these boxes may not be available on all graphing calculators.

The rectangular screen where a portion of the rectangular coordinate system is displayed is called a **window.** We call it a **standard window** for graphing when both the x- and y-axes show coordinates between -10 and 10. This information is often displayed in the window menu on a graphing calculator as

Xmin = −10
Xmax = 10
Xscl = 1 The scale on the x-axis is one unit per tick mark.
Ymin = −10
Ymax = 10
Yscl = 1 The scale on the y-axis is one unit per tick mark.

To use a graphing calculator to graph the equation $y = 2x + 3$, press the [Y =] key and enter the keystrokes [2] [x] [+] [3]. The top row should now read $Y_1 = 2x + 3$. Next press the [GRAPH] key, and the display should look like this:

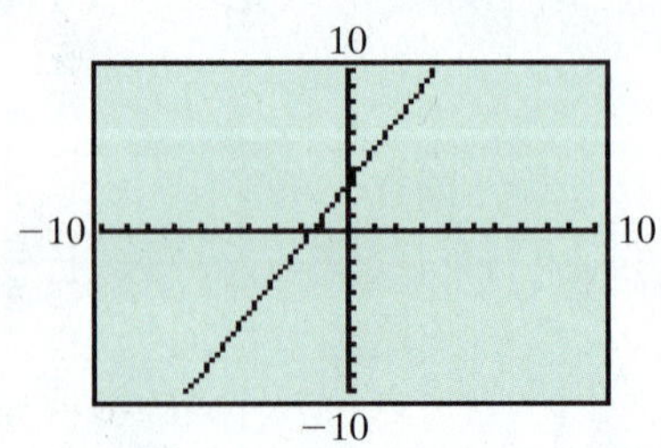

Use a standard window and graph the following linear equations. (Unless otherwise stated, use a standard window when graphing.)

1. $y = -3x + 7$ **2.** $y = -x + 5$ **3.** $y = 2.5x - 7.9$

4. $y = -1.3x + 5.2$ **5.** $y = -\frac{3}{10}x + \frac{32}{5}$ **6.** $y = \frac{2}{9}x - \frac{22}{3}$

Vocabulary, Readiness & Video Check

Use the choices below to fill in each blank. The exercises below all have to do with the rectangular coordinate system.

origin	x-coordinate	x-axis	standard	four	linear
quadrants	y-coordinate	y-axis	solution	one	

1. The horizontal axis is called the __________ and the vertical axis is called the __________.

2. The intersection of the horizontal axis and the vertical axis is a point called the __________.

3. The axes divide the plane into regions called __________. There are __________ of these regions.

4. In the ordered pair of numbers $(-2, 5)$, the number -2 is called the __________ and the number 5 is called the __________.

5. Each ordered pair of numbers corresponds to __________ point in the plane.

6. An ordered pair is a(n) __________ of an equation in two variables if replacing the variables by the coordinates of the ordered pair results in a true statement.

7. An equation that can be written in the form $Ax + By = C$ is called a(n) __________ equation in two variables.

8. The form $Ax + By = C$ is called __________ form.

Martin-Gay Interactive Videos

See Video 3.1

Watch the section lecture video and answer the following questions.

OBJECTIVE 1

9. Several points are plotted in Examples 1–6. Where do you always start when plotting a point? How does the 1st coordinate tell you to move? How does the 2nd coordinate tell you to move?

OBJECTIVE 2

10. In the lecture before Examples 7 and 8, what connection is made between data and graphing?

OBJECTIVE 3

11. An ordered pair is a solution of an equation if, when the variables are replaced with their values, a true statement results. In Example 9, three ordered pairs are tested. What are the last two points to be tested? What lesson can be learned by the results of testing these two points and why?

OBJECTIVE 4

12. In Example 10, when one variable of a linear equation in two variables is replaced by a replacement value, what type of equation results?

OBJECTIVE 5

13. Exponents aren't mentioned in the definition of a linear equation in two variables. However, in determining whether Example 13 is a linear equation in two variables, the exponents or powers on the variables are discussed. Explain.

OBJECTIVE 6

14. In the lecture before Example 15, it's mentioned that you need only two points to determine a line. Why then are three ordered pair solutions found in Examples 15–17?

15. What does a graphed line represent as discussed at the end of Examples 15 and 17?

3.1 Exercise Set MyMathLab®

Plot each ordered pair. State in which quadrant or on which axis each point lies. See Example 1.

1. a. $(1, 5)$ **b.** $(-5, -2)$
c. $(-3, 0)$ **d.** $(0, -1)$
e. $(2, -4)$ **f.** $\left(-1, 4\frac{1}{2}\right)$
g. $(3.7, 2.2)$ **h.** $\left(\frac{1}{2}, -3\right)$

2. a. $(2, 4)$ **b.** $(0, 2)$
c. $(-2, 1)$ **d.** $(-3, -3)$
e. $\left(3\frac{3}{4}, 0\right)$ **f.** $(5, -4)$
g. $(-3.4, 4.8)$ **h.** $\left(\frac{1}{3}, -5\right)$

Find the x- and y-coordinates of each labeled point. See Example 1.

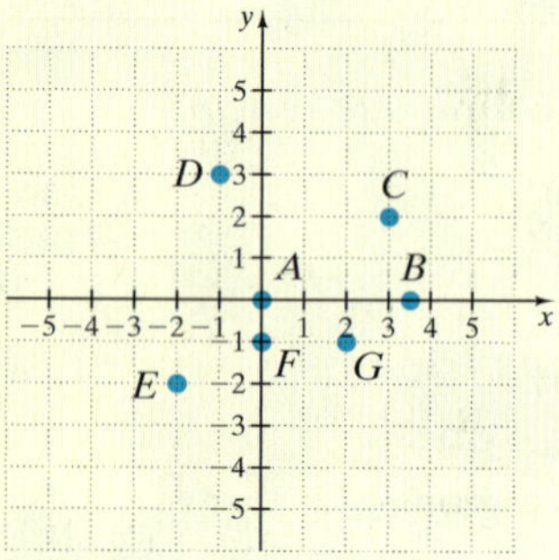

3. A
4. B
5. C
6. D
7. E
8. F
9. G

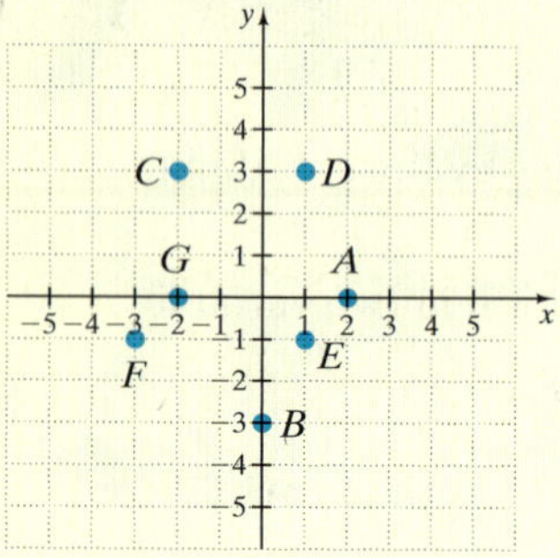

10. A
11. B
12. C
13. D
14. E
15. F
16. G

Solve. See Example 2.

17. The table shows the worldwide box office (in billions of dollars) for the movie industry during the years shown. (*Source:* Motion Picture Association of America)

Year	*Box Office (in billions of dollars)*
2011	32.6
2012	34.7
2013	35.9
2014	36.4
2015	38.3

a. Write this paired data as a set of ordered pairs of the form (year, box office).

b. In your own words, write the meaning of the ordered pair (2015, 38.3).

c. Create a scatter diagram of the paired data. Be sure to label the axes appropriately.

d. What trend in the paired data does the scatter diagram show?

18. The table shows the amount of money (in billions of dollars) Americans spent on their pets for the years shown. (*Source:* American Pet Products Manufacturers Association)

Year	*Pet-Related Expenditures (in billions of dollars)*
2012	53.3
2013	55.7
2014	58.0
2015	60.3
2016	62.8

a. Write this paired data as a set of ordered pairs of the form (year, pet-related expenditures).

b. In your own words, write the meaning of the ordered pair (2014, 58.0).

c. Create a scatter diagram of the paired data. Be sure to label the axes appropriately.

d. What trend in the paired data does the scatter diagram show?

19. The table shows the distance from the equator (in miles) and the average annual snowfall (in inches) for each of eight selected U.S. cities. (*Source:* National Climatic Data Center, Wake Forest University Albatross Project)

City	*Distance from Equator (in miles)*	*Average Annual Snowfall (in inches)*
1. Atlanta, GA	2313	2
2. Austin, TX	2085	1
3. Baltimore, MD	2711	21
4. Chicago, IL	2869	39
5. Detroit, MI	2920	42
6. Juneau, AK	4038	99
7. Miami, FL	1783	0
8. Winston-Salem, NC	2493	9

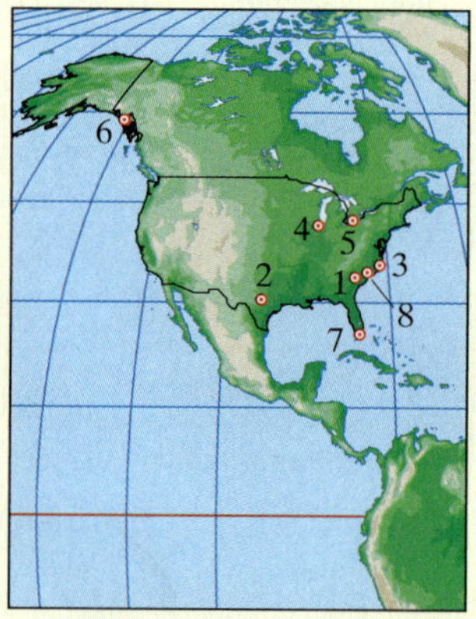

a. Write this paired data as a set of ordered pairs of the form (distance from equator, average annual snowfall).

b. Create a scatter diagram of the paired data. Be sure to label the axes appropriately.

c. What trend in the paired data does the scatter diagram show?

20. The table shows the average farm size (in acres) in the United States during the years shown. (*Source:* National Agricultural Statistics Service)

Year	*Average Farm Size (in acres)*
2009	418
2010	418
2011	429
2012	433
2013	435
2014	437

a. Write this paired data as a set of ordered pairs of the form (year, average farm size).

b. Create a scatter diagram of the paired data. Be sure to label the axes appropriately.

Determine whether each ordered pair is a solution of the given linear equation. See Example 3.

21. $2x + y = 7$; $(3, 1)$, $(7, 0)$, $(0, 7)$

22. $3x + y = 8$; $(2, 3)$, $(0, 8)$, $(8, 0)$

23. $x = -\frac{1}{3}y$; $(0, 0)$, $(3, -9)$

24. $y = -\frac{1}{2}x$; $(0, 0)$, $(4, 2)$

Complete each ordered pair so that it is a solution of the given linear equation. See Examples 4 and 5.

25. $x - 4y = 4$; $(\quad, -2)$, $(4, \quad)$

26. $x - 5y = -1$; $(\quad, -2)$, $(4, \quad)$

27. $y = \frac{1}{4}x - 3$; $(-8, \quad)$, $(\quad, 1)$

28. $y = \frac{1}{5}x - 2$; $(-10, \quad)$, $(\quad, 1)$

Complete the table of ordered pairs for each linear equation. See Examples 4 and 5.

29. $y = -7x$

x	y
0	
−1	
	2

30. $y = -9x$

x	y
	0
−3	
	2

31. $y = \frac{1}{2}x$

x	y
0	
−6	
	1

32. $y = \frac{1}{3}x$

x	y
0	
−6	
	1

33. $y = 2x - 12$

x	y
0	
	−2
3	

34. $y = 5x + 10$

x	y
	0
	5
0	

35. $2x + 7y = 5$

x	y
0	
	0
	1

36. $x - 6y = 3$

x	y
0	
1	
	−1

Determine whether each equation is a linear equation in two variables. See Example 6.

37. $-x = 3y + 10$

38. $y = x - 15$

39. $x = y$

40. $x = y^3$

41. $x^2 + 2y = 0$

42. $0.01x - 0.2y = 8.8$

MIXED PRACTICE

Graph each linear equation. See Examples 7 through 9.

43. $x + y = 1$

44. $x + y = 7$

45. $x - y = -2$

46. $-x + y = 6$

47. $x - 2y = 6$

48. $-x + 5y = 5$

49. $y = 6x + 3$

50. $y = -2x + 7$

51. $y = x$

52. $y = -x$

53. $x = -3y$

54. $x = -5y$

55. $x + 3y = 9$

56. $2x + y = 2$

57. $y = \frac{1}{2}x + 2$

58. $y = \frac{1}{4}x + 3$

59. $3x - 2y = 12$

60. $2x - 7y = 14$

61. $y = -3.5x + 4$

62. $y = -1.5x - 3$

Complete the table of ordered pairs for each equation. Then plot the ordered pair solutions.

63. $x = -5y$

x	y
	0
	1
10	

64. $y = -3x$

x	y
0	
−2	
	9

65. $y = \frac{1}{3}x + 2$

x	y
0	
−3	
	0

66. $y = \frac{1}{2}x + 3$

x	y
0	
−4	
	0

Solve. See Example 10.

67. The cost in dollars y of producing x computer desks is given by $y = 80x + 5000$.

a. Complete the table.

x	100	200	300
y			

b. Find the number of computer desks that can be produced for \$8600. (*Hint:* Find x when $y = 8600$.)

68. The hourly wage y of an employee at a certain production company is given by $y = 0.25x + 9$ where x is the number of units produced by the employee in an hour.

a. Complete the table.

x	0	1	5	10
y				

b. Find the number of units that an employee must produce each hour to earn an hourly wage of \$12.25. (*Hint:* Find x when $y = 12.25$.)

69. The average annual cinema admission price y (in dollars) from 2010 through 2014 is given by $y = 0.09x + 7.85$. In this equation, x represents the number of years after 2010. (*Source:* Motion Picture Association of America)

a. Complete the table.

x	0	2	4
y			

b. Find the year in which the average cinema admission price was approximately \$8.10.

(*Hint:* Find x when $y = 8.10$ and round to the nearest whole number.)

c. Use the given equation to predict when the cinema admission price might be \$10.00. (Use the hint for part **b.**)

d. In your own words, write the meaning of the ordered pair (1, 7.94).

70. The average amount of money y spent per person per year on music from iTunes from 2008 to 2013 can be approximated by $y = -5.74x + 38.05$. In this equation, x represents the number of years since 2008. (*Source:* billboard.com)

a. Complete the table.

x	1	3	5
y			

b. Find the year in which the yearly average amount spent on music in iTunes was approximately \$26.00. (*Hint:* Find x when $y = 26$ and round to the nearest whole number.)

The graph below shows the number of U.S. Walmart stores for each year. Use this graph to answer Exercises 71 through 74.

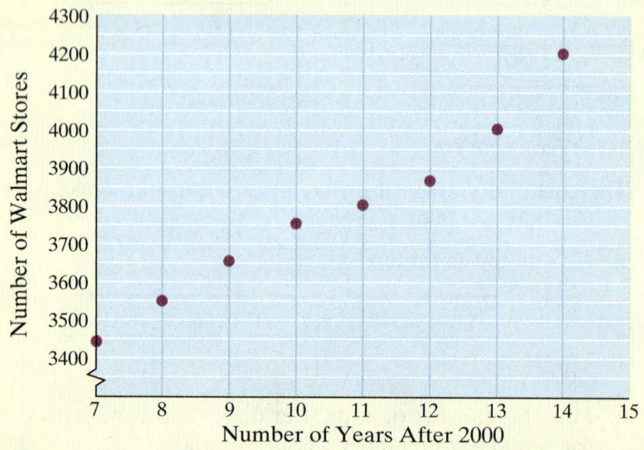

Data from Walmart

71. The ordered pair (14, 4203) is a point of the graph. Write a sentence describing the meaning of this ordered pair.

72. The ordered pair (12, 3868) is a point of the graph. Write a sentence describing the meaning of this ordered pair.

73. Estimate the increase in Walmart stores for years 8, 9, and 10.

74. Use a straightedge or ruler and this graph to predict the number of Walmart stores in the year 2018.

75. Describe what is similar about the coordinates of points whose graph lies on the x-axis.

76. Describe what is similar about the coordinates of points whose graph lies on the y-axis.

Solve. See Example 10.

77. Jogging is one of the few sports that has been consistently increasing over the past few years. The number of people jogging (in millions) from the years 2000 to 2009 is given by the equation $y = x + 23$, where x is the number of years after 2000. (*Source:* Based on data from the National Sporting Goods Association)

a. Use this equation or a graph of it to complete the ordered pair (8,   ).

b. Write a sentence explaining the meaning of the answer to part **a.**

c. If this trend continues, how many joggers will there be in 2017?

78. The revenue y (in billions of dollars) for Home Depot stores during the years 2010 through 2014 is given by the equation $y = 3.2x + 65.2$, where x is the number of years after 2010. (*Source:* Based on data from Home Depot stores)

a. Use this equation or a graph of it to complete the ordered pair (3,   ).

b. Write a sentence explaining the meaning of the answer to part **a.**

c. If this trend continues, predict the revenue for Home Depot stores for the year 2018.

79. One American rite of passage is a driver's license. The number of people y (in millions) who have a driver's license can be estimated by the linear equation $y = 2.2x + 190$, where x is the number of years after 2000. (*Source:* Federal Highway Administration)

a. Use this equation to complete the ordered pair (12,   ).

b. Write a sentence explaining the meaning of the ordered pair in part **a.**

c. If this trend continues, predict the number of people with driver's licenses in 2020.

80. The percent of U.S. households y with at least one computer can be approximated by the linear equation $y = 2.4x + 51$, where x is the number of years since 2000. (*Source:* Pew Research)

a. Use the equation to complete the ordered pair (10,).

b. Write a sentence explaining the meaning of the ordered pair found in part **a**.

c. If this trend continues, predict the percent of U.S. households that have at least one computer in 2018.

d. Explain any issues with your answer to part **c**.

REVIEW AND PREVIEW

The following bar graph shows the top 10 tourist destinations and the number of tourists that visit each destination per year forecasted for 2020. Use this graph to answer Exercises 81 through 86. See Section R.4.

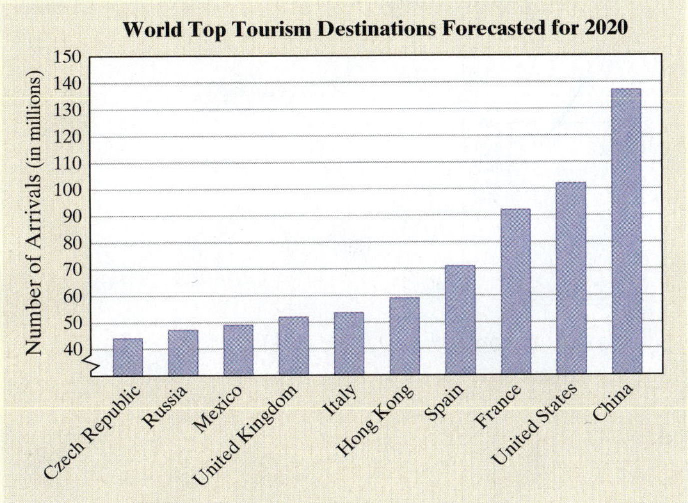

Data from UNWTO

81. Which location shown is predicted to be the most popular tourist destination?

82. Which location shown is predicted to be the least popular tourist destination?

83. Which locations shown are predicted to have more than 70 million tourists per year?

84. Which locations shown are predicted to have more than 100 million tourists per year?

85. Estimate the predicted number of tourists per year whose destination is Italy.

86. Estimate the predicted number of tourists per year whose destination is Mexico.

CONCEPT EXTENSIONS

Answer each exercise with true or false.

87. Point $(-1, 5)$ lies in quadrant IV.

88. Point $(3, 0)$ lies on the y-axis.

89. For the point $\left(-\frac{1}{2}, 1.5\right)$, the first value, $-\frac{1}{2}$, is the x-coordinate and the second value, 1.5, is the y-coordinate.

90. The ordered pair $\left(2, \frac{2}{3}\right)$ is a solution of $2x - 3y = 6$.

For Exercises 91 through 95, fill in each blank with "0," "positive," or "negative." For Exercises 96 and 97, fill in each blank with "x" or "y."

	Point	*Location*
91.	(________, ________)	quadrant III
92.	(________, ________)	quadrant I
93.	(________, ________)	quadrant IV
94.	(________, ________)	quadrant II
95.	(________, ________)	origin
96.	(number, 0)	_ -axis
97.	(0, number)	_ -axis

98. Give an example of an ordered pair whose location is in (or on)

a. quadrant I **b.** quadrant II
c. quadrant III **d.** quadrant IV
e. x-axis **f.** y-axis

Solve. See a Concept Check in this section.

99. Is the graph of $(3, 0)$ in the same location as the graph of $(0, 3)$? Explain why or why not.

100. Give the coordinates of a point such that if the coordinates are reversed, their location is the same.

101. In general, what points can have coordinates reversed and still have the same location?

102. In your own words, describe how to plot or graph an ordered pair of numbers.

103. Graph the nonlinear equation $y = x^2$ by completing the table shown. Plot the ordered pairs and connect them with a smooth curve.

x	y
0	
1	
-1	
2	
-2	

104. Graph the nonlinear equation $y = |x|$ by completing the table shown. Plot the ordered pairs and connect them. This curve is "V" shaped.

x	y
0	
1	
-1	
2	
-2	

3.2 Intercepts, Slope, and Rate of Change

OBJECTIVES

1. Graph a Linear Equation by Finding and Plotting Intercepts.
2. Find the Slope of a Line Given Two Points of the Line.
3. Find the Slope of a Line Given Its Equation.
4. Graph and Find the Slopes of Horizontal and Vertical Lines.
5. Interpret Slope as a Rate of Change.
6. Extension—Compare the Slopes of Parallel and Perpendicular Lines.

OBJECTIVE 1

Graphing a Linear Equation by Using Intercepts

In this section, we graph linear equations in two variables by identifying intercepts. For example, the graph of $y = 4x - 8$ is shown on the right. Notice that this graph crosses the y-axis at the point $(0, -8)$. This point is called the **y-intercept.** Likewise, the graph crosses the x-axis at $(2, 0)$, and this point is called the **x-intercept.**

The intercepts are $(2, 0)$ and $(0, -8)$.

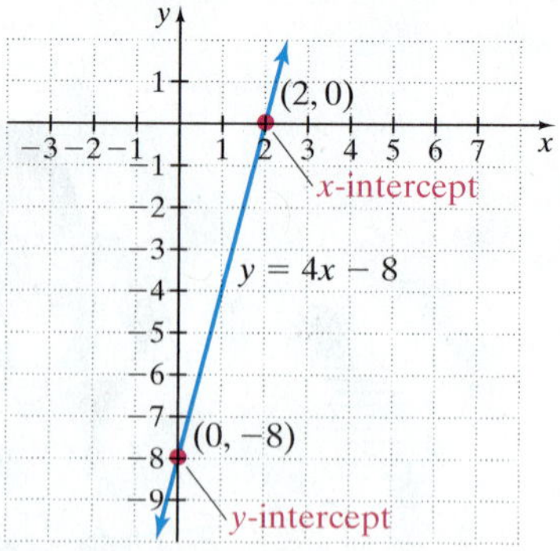

Helpful Hint

If a graph crosses the x-axis at $(-3, 0)$ and the y-axis at $(0, 7)$, then

$$\underbrace{(-3, 0)}_{x\text{-intercept}} \qquad \underbrace{(0, 7)}_{y\text{-intercept}}$$

Notice that for the y-intercept, the x-value is 0 and for the x-intercept, the y-value is 0.
Note: Sometimes in mathematics, you may see just the number -3 stated as the x-intercept, and 7 stated as the y-intercept.

EXAMPLE 1 Identify the x- and y-intercepts.

a.

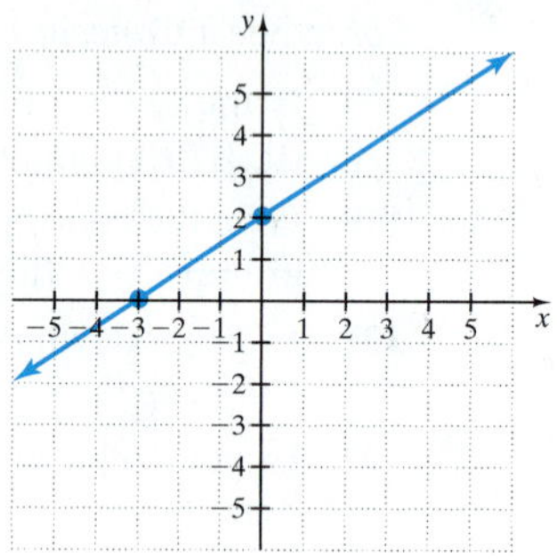

Solution

x-intercept: $(-3, 0)$
y-intercept: $(0, 2)$

b.

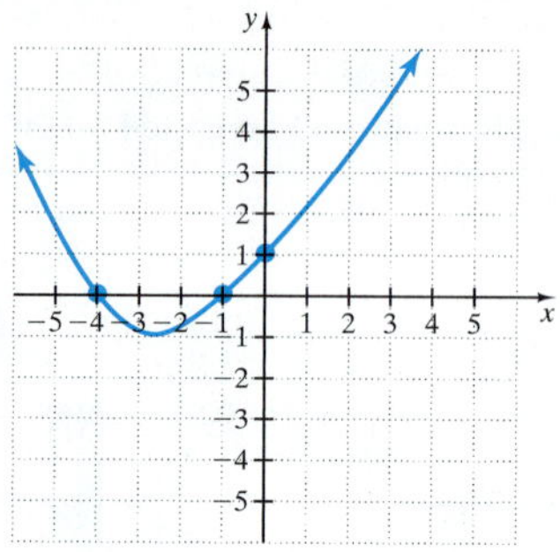

Solution

x-intercepts: $(-4, 0)$, $(-1, 0)$
y-intercept: $(0, 1)$

Helpful Hint

Notice that any time $(0, 0)$ is a point of a graph, then it is an x-intercept and a y-intercept. Why? It is the only point that lies on both axes.

c.

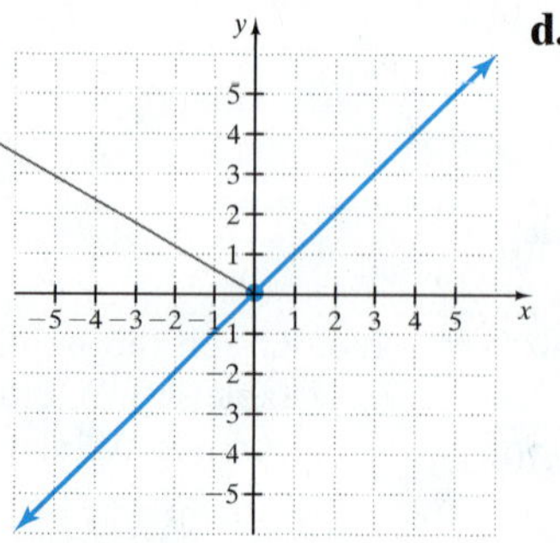

Solution

x-intercept: $(0, 0)$
y-intercept: $(0, 0)$

d.

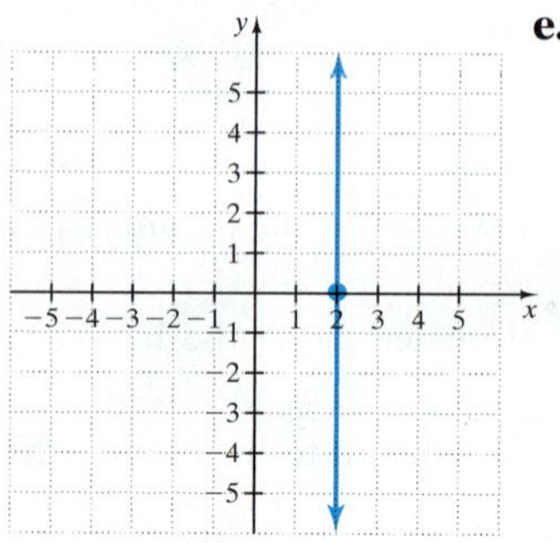

Solution

x-intercept: $(2, 0)$
y-intercept: none

e.

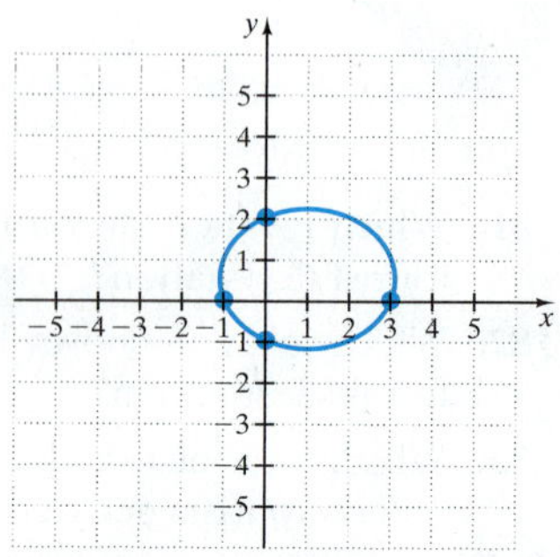

Solution

x-intercepts: $(-1, 0)$, $(3, 0)$
y-intercepts: $(0, 2)$, $(0, -1)$

□

PRACTICE 1 Identify the x- and y-intercepts.

a.

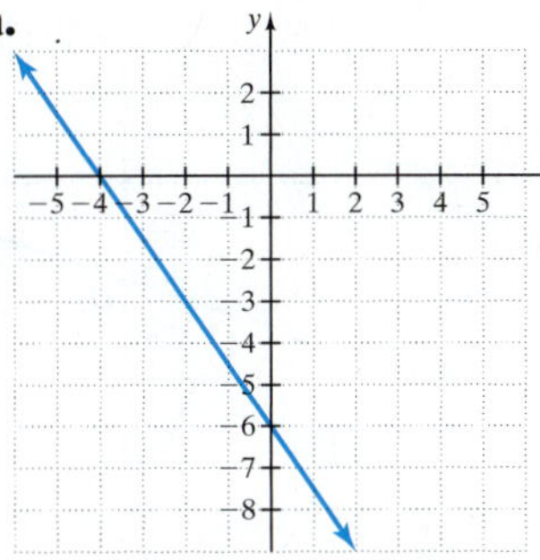

b.

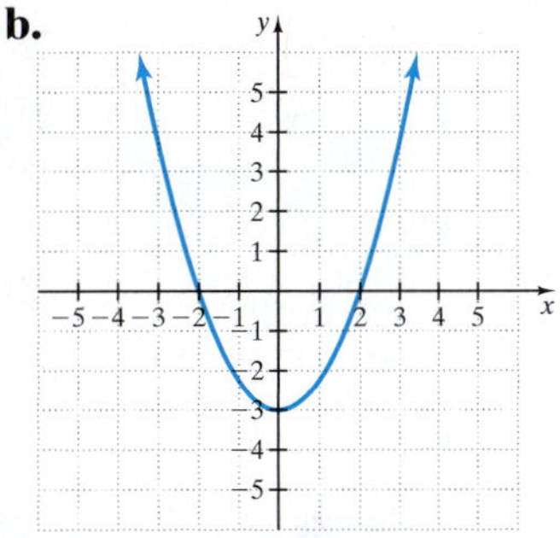

c.

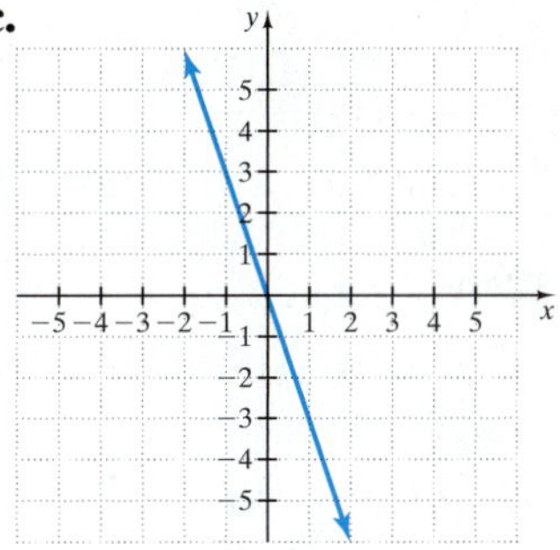

d.

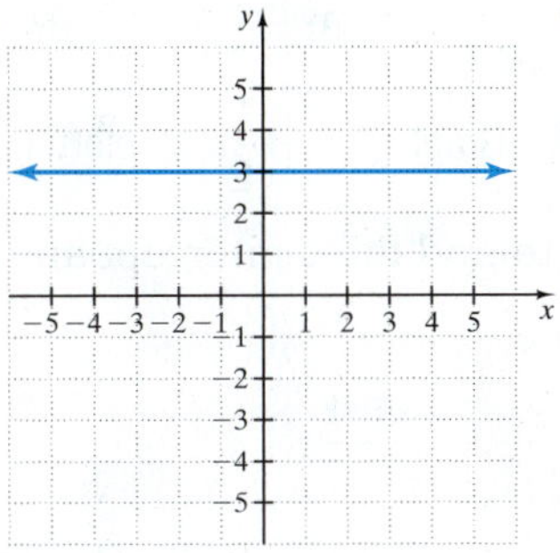

e.

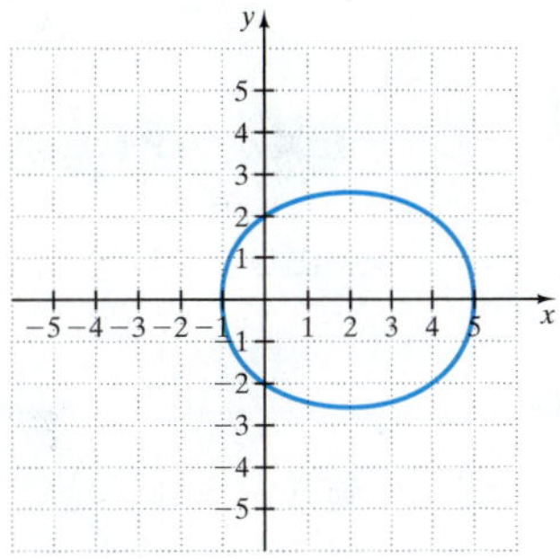

Given the equation of a line, intercepts are usually easy to find since one coordinate is 0.

One way to find the y-intercept of a line, given its equation, is to let $x = 0$, since a point on the y-axis has an x-coordinate of 0. To find the x-intercept of a line, let $y = 0$, since a point on the x-axis has a y-coordinate of 0.

Finding x- and y-intercepts

To find the x-intercept, let $y = 0$ and solve for x.

To find the y-intercept, let $x = 0$ and solve for y.

EXAMPLE 2 Graph $x - 3y = 6$ by finding and plotting intercepts.

Solution Let $y = 0$ to find the x-intercept and let $x = 0$ to find the y-intercept.

Let $y = 0$	Let $x = 0$
$x - 3y = 6$	$x - 3y = 6$
$x - 3(0) = 6$	$0 - 3y = 6$
$x - 0 = 6$	$-3y = 6$
$x = 6$	$y = -2$

The x-intercept is $(6, 0)$ and the y-intercept is $(0, -2)$. We find a third ordered pair solution to check our work. If we let $y = -1$, then $x = 3$. Plot the points $(6, 0)$, $(0, -2)$, and $(3, -1)$. The graph of $x - 3y = 6$ is the line drawn through these points, as shown.

(Continued on next page)

x	y
6	0
0	−2
3	−1

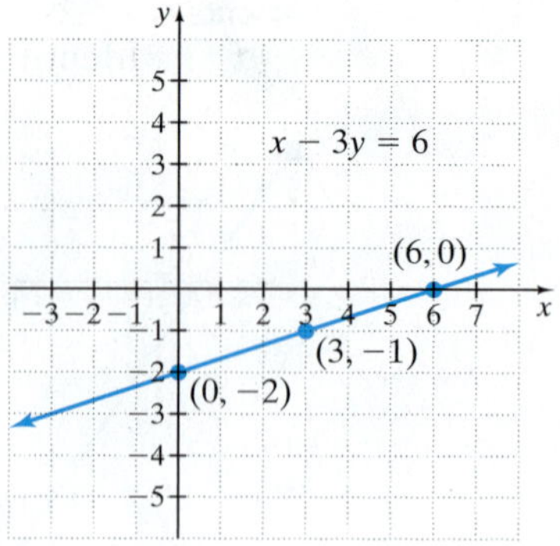

PRACTICE 2 Graph $x + 2y = -4$ by finding and plotting intercepts.

EXAMPLE 3 Graph $y = -\frac{1}{2}x$ by plotting intercepts.

Solution Let $y = 0$ to find the x-intercept and $x = 0$ to find the y-intercept.

Let $y = 0$	Let $x = 0$
$y = -\frac{1}{2}x$	$y = -\frac{1}{2}x$
$0 = -\frac{1}{2}x$	$y = -\frac{1}{2}(0)$
$0 = x$	$y = 0$

Both the x-intercept and y-intercept are (0, 0). In other words, when $x = 0$, then $y = 0$, which gives the ordered pair (0, 0). Also, when $y = 0$, then $x = 0$, which gives the same ordered pair (0, 0). This happens when the graph passes through the origin. Since two points are needed to determine a line, we must find at least one more ordered pair that satisfies $y = -\frac{1}{2}x$. We will let $x = 2$ to find a second ordered pair solution and let $x = -2$ as a checkpoint.

Let $x = 2$	Let $x = -2$
$y = -\frac{1}{2}(2)$	$y = -\frac{1}{2}(-2)$
$y = -1$	$y = 1$

The ordered pairs are (0, 0), (2, −1), and (−2, 1). Plot these points to graph $y = -\frac{1}{2}x$.

x	y
0	0
2	−1
−2	1

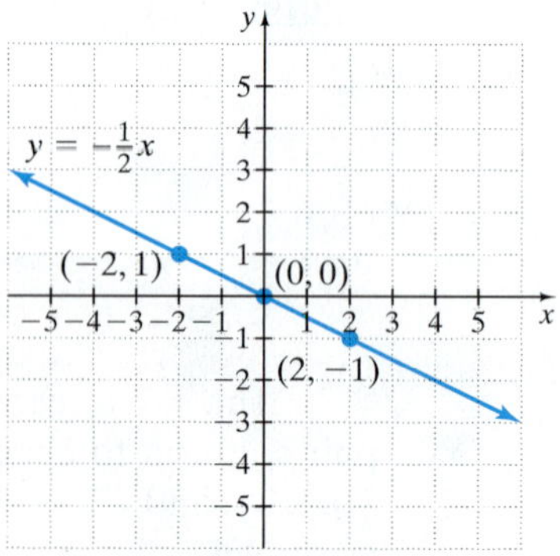

PRACTICE 3 Graph $y = \frac{1}{3}x$ by plotting intercepts.

OBJECTIVE

2 Finding the Slope of a Line Given Two Points of the Line

You have probably noticed by now that a key feature of a line is its slant or steepness. In mathematics, the slant or steepness of a line is formally known as its **slope.** We measure the slope of a line by the ratio of vertical change to the corresponding horizontal change as we move along the line.

On the line below, for example, suppose that we begin at the point (1, 2) and move to the point (4, 6). The vertical change is the change in y-coordinates: $6 - 2$ or 4 units. The corresponding horizontal change is the change in x-coordinates: $4 - 1 = 3$ units. The ratio of these changes is

$$\text{slope} = \frac{\text{change in } y \text{ (vertical change)}}{\text{change in } x \text{ (horizontal change)}} = \frac{4}{3}$$

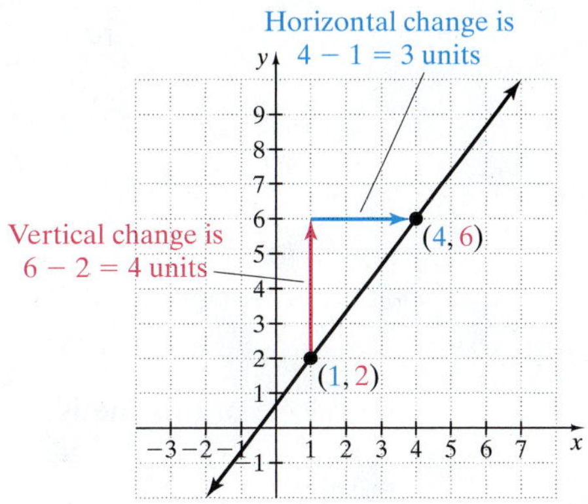

The slope of this line, then, is $\frac{4}{3}$. This means that for every 4 units of change in y-coordinates, there is a corresponding change of 3 units in x-coordinates.

Helpful Hint

It makes no difference what two points of a line are chosen to find its slope. The slope of a line is the same everywhere on the line.

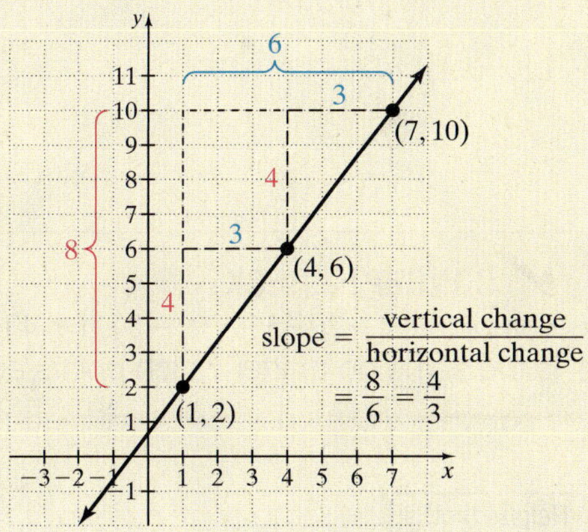

To find the slope of a line, then, choose any two points of the line. Label the x-coordinates of the two points x_1 and x_2 (read "x sub one" and "x sub two"), and label the corresponding y-coordinates y_1 and y_2.

The vertical change or **rise** between these points is the difference in the y-coordinates: $y_2 - y_1$. The horizontal change or **run** between the points is the difference of the x-coordinates: $x_2 - x_1$. The slope of the line is the ratio of $y_2 - y_1$ to $x_2 - x_1$, and we traditionally use the letter m to denote slope: $m = \frac{y_2 - y_1}{x_2 - x_1}$.

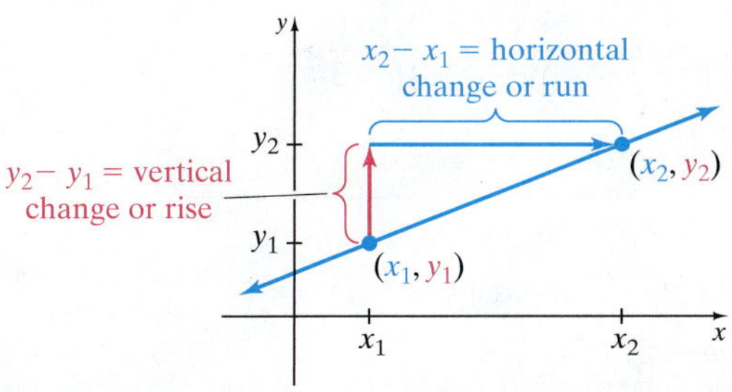

Slope of a Line

The slope m of the line containing the points (x_1, y_1) and (x_2, y_2) is given by

$$m = \frac{\text{rise}}{\text{run}} = \frac{\text{change in } y}{\text{change in } x} = \frac{y_2 - y_1}{x_2 - x_1}, \quad \text{as long as } x_2 \neq x_1$$

EXAMPLE 4 Find the slope of the line through $(-1, 5)$ and $(2, -3)$. Graph the line.

Solution If we let (x_1, y_1) be $(-1, 5)$, then $x_1 = -1$ and $y_1 = 5$. Also, let (x_2, y_2) be $(2, -3)$ so that $x_2 = 2$ and $y_2 = -3$. Then, by the definition of slope,

$$m = \frac{y_2 - y_1}{x_2 - x_1} = \frac{-3 - 5}{2 - (-1)} = \frac{-8}{3} = -\frac{8}{3}$$

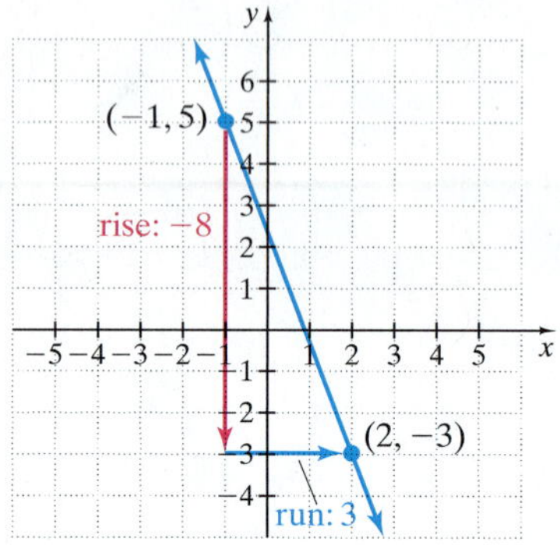

The slope of the line is $-\frac{8}{3}$. □

PRACTICE 4 Find the slope of the line through $(-4, 11)$ and $(2, 5)$. ■

Helpful Hint

When finding slope, it makes no difference which point is identified as (x_1, y_1) and which is identified as (x_2, y_2). Just remember that whatever y-value is first in the numerator, its corresponding x-value must be first in the denominator. Another way to calculate the slope in Example 4 is:

$$m = \frac{y_2 - y_1}{x_2 - x_1} = \frac{5 - (-3)}{-1 - 2} = \frac{8}{-3} \text{ or } -\frac{8}{3} \quad \leftarrow \text{Same slope as found in Example 4.}$$

✔ CONCEPT CHECK

The points $(-2, -5)$, $(0, -2)$, $(4, 4)$, and $(10, 13)$ all lie on the same line. Work with a partner and verify that the slope is the same no matter which points are used to find slope.

EXAMPLE 5 Find the slope of the line through $(-1, -2)$ and $(2, 4)$. Graph the line.

Solution Let (x_1, y_1) be $(2, 4)$ and (x_2, y_2) be $(-1, -2)$.

$$m = \frac{y_2 - y_1}{x_2 - x_1} = \frac{-2 - 4}{-1 - 2} = \frac{-6}{-3} = 2$$

(-2: y-value; -1: corresponding x-value)

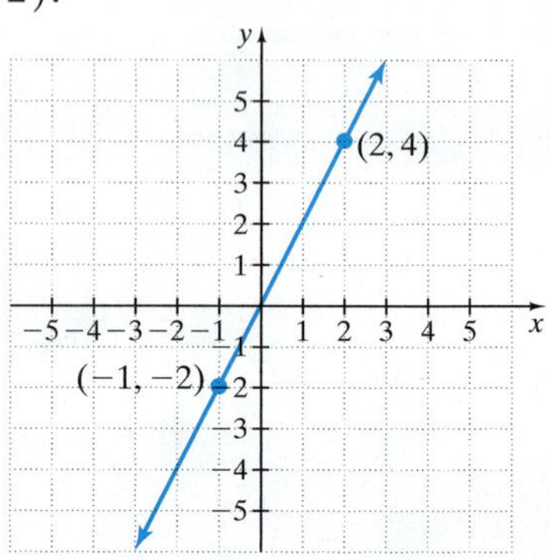

□

Helpful Hint

The slope for Example 5 is the same if we let (x_1, y_1) be $(-1, -2)$ and (x_2, y_2) be $(2, 4)$.

$$m = \frac{4 - (-2)}{2 - (-1)} = \frac{6}{3} = 2$$

(4: y-value; 2: corresponding x-value)

Answer to Concept Check:
$m = \frac{3}{2}$

PRACTICE
5 Find the slope of the line through $(-3, -1)$ and $(3, 1)$.

✔ CONCEPT CHECK

What is wrong with the following slope calculation for the points $(3, 5)$ and $(-2, 6)$?

$$m = \frac{5 - 6}{-2 - 3} = \frac{-1}{-5} = \frac{1}{5}$$

Notice that the slope of the line in Example 4 is negative, whereas the slope of the line in Example 5 is positive. Let your eye follow the line with negative slope from left to right and notice that the line "goes down." Following the line with positive slope from left to right, notice that the line "goes up." This is true in general.

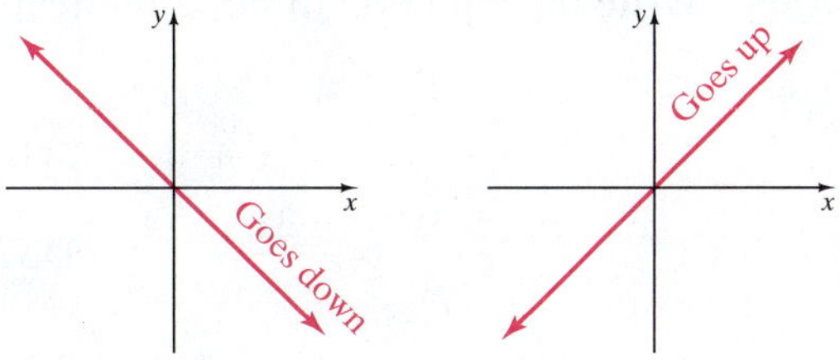

Negative slope   Positive slope

Helpful Hint

To decide whether a line "goes up" or "goes down," always follow the line from left to right.

OBJECTIVE

3 Finding the Slope of a Line Given Its Equation

As we have seen, the slope of a line is defined by two points on the line. Thus, if we know the equation of a line, we can find its slope by finding two of its points. For example, let's find the slope of the line

$$y = 3x + 2$$

To find two points, we can choose two values for x and substitute to find corresponding y-values. If $x = 0$, for example, $y = 3 \cdot 0 + 2$ or $y = 2$. If $x = 1$, $y = 3 \cdot 1 + 2$ or $y = 5$. This gives the ordered pairs $(0, 2)$ and $(1, 5)$. Using the definition for slope, we have

$$m = \frac{5 - 2}{1 - 0} = \frac{3}{1} = 3 \quad \text{The slope is 3.}$$

Notice that the slope, 3, is the same as the coefficient of x in the equation $y = 3x + 2$.

Also, recall from Section 3.1 that the graph of an equation of the form $y = mx + b$ has y-intercept $(0, b)$.

This means that the y-intercept of the graph of $y = 3x + 2$ is $(0, 2)$. This is true in general and the form $y = mx + b$ is appropriately called the **slope-intercept form.**

$$y = mx + b$$

↑ slope ↑ y-intercept $(0, b)$

Slope-Intercept Form

When a linear equation in two variables is written in slope-intercept form,

$$y = mx + b$$

m is the slope of the line and $(0, b)$ is the y-intercept of the line.

Answer to Concept Check:
The order in which the x- and y-values are used must be the same.

$$m = \frac{5 - 6}{3 - (-2)} = \frac{-1}{5} = -\frac{1}{5}$$

EXAMPLE 6 Find the slope and y-intercept of the line whose equation is $y = \frac{3}{4}x + 6$

Solution The equation is in slope-intercept form, $y = mx + b$.

$$y = \frac{3}{4}x + 6$$

The coefficient of x, $\frac{3}{4}$, is the slope and the constant term, 6, is the y-value of the y-intercept, (0, 6). □

PRACTICE 6 Find the slope and y-intercept of the line whose equation is $y = \frac{2}{3}x - 2$. ■

EXAMPLE 7 Find the slope and the y-intercept of the line whose equation is $3x - 4y = 4$.

Solution Write the equation in slope-intercept form by solving for y.

$$3x - 4y = 4$$

$$-4y = -3x + 4 \quad \text{Subtract } 3x \text{ from both sides.}$$

$$\frac{-4y}{-4} = \frac{-3x}{-4} + \frac{4}{-4} \quad \text{Divide both sides by } -4.$$

$$y = \frac{3}{4}x - 1 \quad \text{Simplify.}$$

The coefficient of x, $\frac{3}{4}$, is the slope, and the y-intercept is $(0, -1)$. □

PRACTICE 7 Find the slope and the y-intercept of the line whose equation is $5x + 2y = 8$. ■

OBJECTIVE 4

Graphing and Finding Slopes of Horizontal and Vertical Lines

Recall that if a line tilts upward from left to right, its slope is positive. If a line tilts downward from left to right, its slope is negative. Let's now graph and find the slopes of two special lines, horizontal and vertical lines.

EXAMPLE 8 Graph: $y = -3$. Then find the slope of the graph.

Solution The equation $y = -3$ can be written as $0x + y = -3$, so it is a linear equation in two variables. For any x-value chosen, y is -3. If we choose 4, 1, and -2 as x-values, the ordered pair solutions are $(4, -3)$, $(1, -3)$, and $(-2, -3)$. Use these ordered pairs to graph $y = -3$.

x	y
4	−3
1	−3
−2	−3

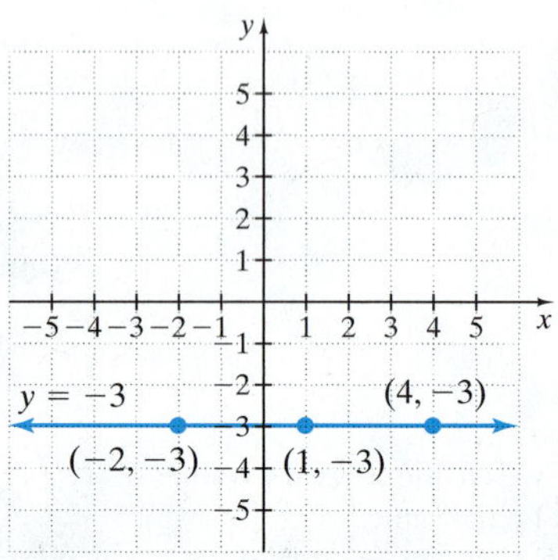

Slope of the graph:
Let's use points $(4, -3)$ and $(1, -3)$.

$$m = \frac{y_2 - y_1}{x_2 - x_1} = \frac{-3 - (-3)}{1 - 4} = \frac{0}{-3} = 0$$

The graph is a horizontal line with y-intercept $(0, -3)$ and no x-intercept. This horizontal line has a slope of 0. □

PRACTICE

8 Graph: $y = 2$ ■

Horizontal Lines

The graph of $y = c$, where c is a real number, is a horizontal line with y-intercept $(0, c)$.

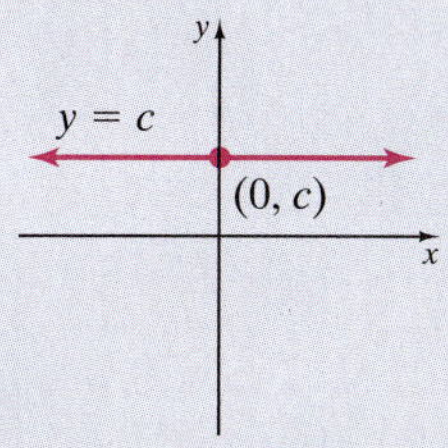

Slope:
All horizontal lines have a slope of 0.

Any two points of a horizontal line will have the same y-values. This means that the y-values will always have a difference of 0 for all horizontal lines. Thus, **all horizontal lines have a slope 0.**

EXAMPLE 9 Graph: $x = 2$. Then find the slope of the graph.

Solution The equation $x = 2$ is a linear equation in two variables because it can be written as $x + 0y = 2$. For any y-value chosen, notice that x is 2. No other value for x satisfies $x + 0y = 2$. Any ordered pair whose x-coordinate is 2 is a solution of $x + 0y = 2$. We will use the ordered pair solutions $(2, 3)$, $(2, 0)$, and $(2, -3)$ to graph $x = 2$.

x	y
2	3
2	0
2	−3

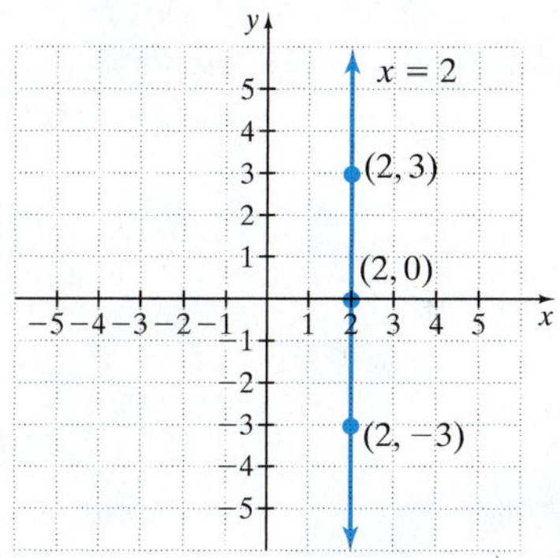

Slope of the graph:
Let's use points $(2, 3)$ and $(2, 0)$.

$$m = \frac{y_2 - y_1}{x_2 - x_1} = \frac{0 - 3}{2 - 2} = \frac{-3}{0} \text{ or undefined}$$

The graph is a vertical line with x-intercept $(2, 0)$. Note that this graph has no y-intercept because x is never 0. This vertical line has undefined slope. □

PRACTICE

9 Graph: $x = -2$ ■

Vertical Lines

The graph of $x = c$, where c is a real number, is a vertical line with x-intercept $(c, 0)$.

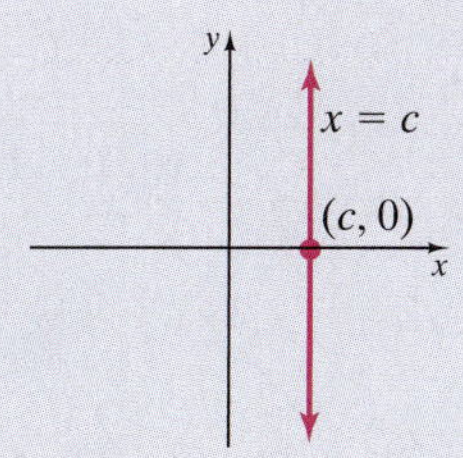

Slope:
All vertical lines have undefined slope.

Any two points of a vertical line will have the same x-values. This means that the x-values will always have a difference of 0 for all vertical lines. Thus **all vertical lines have undefined slope.**

> **Helpful Hint**
>
> Slope of 0 and undefined slope are not the same. Vertical lines have undefined slope or no slope, while horizontal lines have a slope of 0.

Here is a general review of slope.

Summary of Slope

Slope m of the line through (x_1, y_1) and (x_2, y_2) is given by the equation $m = \dfrac{y_2 - y_1}{x_2 - x_1}$.

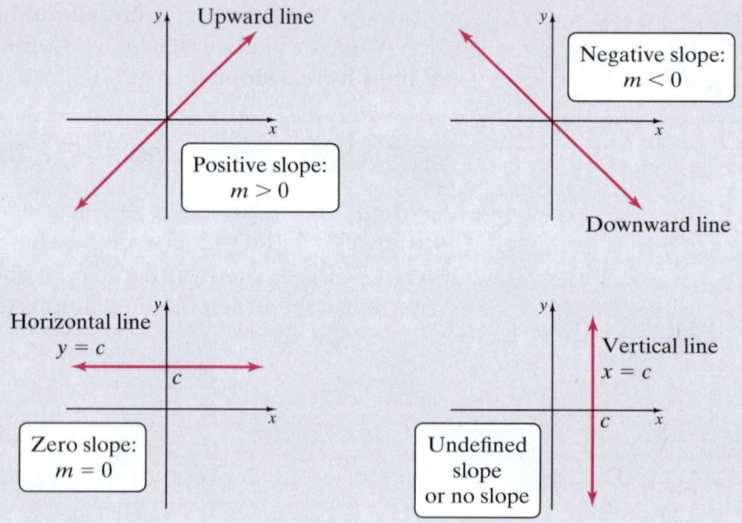

OBJECTIVE

5 Slope as a Rate of Change

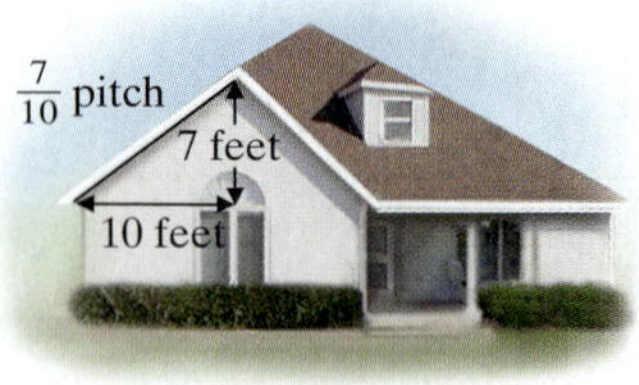

Slope can also be interpreted as a rate of change. In other words, slope tells us how fast y is changing with respect to x. To see this, let's look at a few of the many real-world applications of slope. For example, the pitch of a roof, used by builders and architects, is its slope. The pitch of the roof on the left is $\dfrac{7}{10}\left(\dfrac{\text{rise}}{\text{run}}\right)$. This means that the roof rises vertically 7 feet for every horizontal 10 feet. The rate of change for the roof is 7 vertical feet (y) per 10 horizontal feet (x).

The grade of a road is its slope written as a percent. A 7% grade, as shown below, means that the road rises (or falls) 7 feet for every horizontal 100 feet. $\left(\text{Recall that } 7\% = \dfrac{7}{100}.\right)$ Here, the slope of $\dfrac{7}{100}$ gives us the rate of change. The road rises (in our diagram from left to right) 7 vertical feet (y) for every 100 horizontal feet (x).

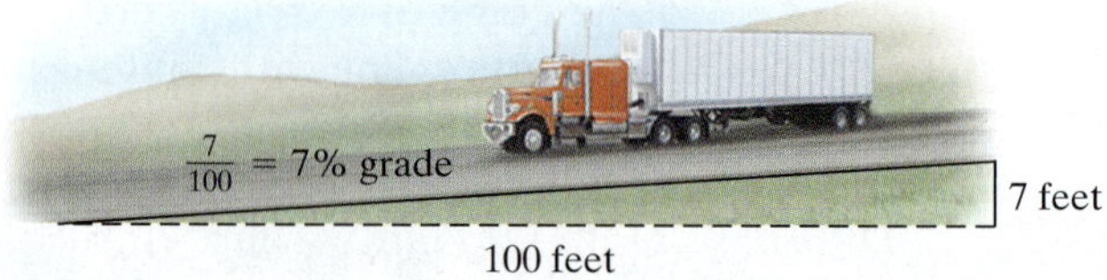

EXAMPLE 10 **Finding the Grade of a Road**

At one part of the road to the summit of Pikes Peak, the road rises at a rate of 15 vertical feet for a horizontal distance of 250 feet. Find the grade of the road.

Solution Recall that the grade of a road is its slope written as a percent.

$$\text{grade} = \frac{\text{rise}}{\text{run}} = \frac{15}{250} = 0.06 = 6\%$$

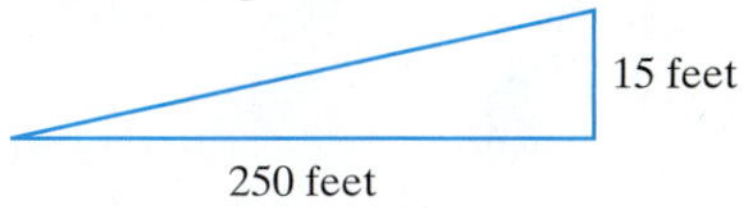

The grade is 6%. □

PRACTICE 10 One part of the Mt. Washington (New Hampshire) cog railway rises about 1794 feet over a horizontal distance of 7176 feet. Find the grade of this part of the railway. ■

EXAMPLE 11 **Finding the Slope of a Line**

The following graph shows annual food and drink sales y (in billions of dollars) for year x.

a. Find the slope of the line and attach the proper units for the rate of change.

b. Then write a sentence explaining the meaning of the slope for this application.

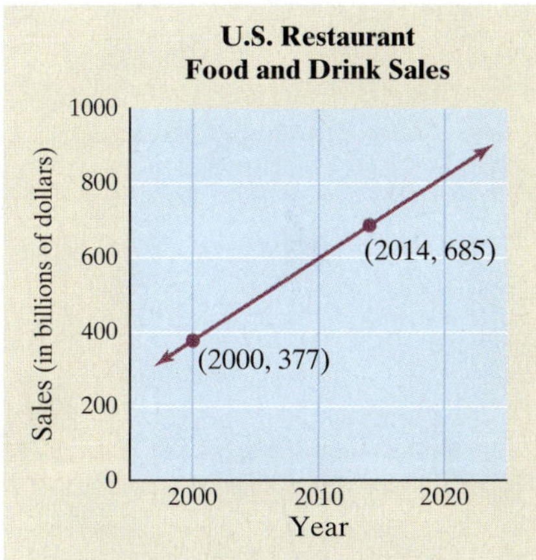

Data from National Restaurant Association

Solution

a. Use (2000, 377) and (2014, 685) to calculate slope.

$$m = \frac{685 - 377}{2014 - 2000} = \frac{308}{14} = \frac{22 \text{ billion dollars}}{1 \text{ year}}$$

b. This means that the rate of change of restaurant food and drink sales increases by 22 billion dollars every 1 year, or \$22 billion per year. □

PRACTICE 11 The following graph shows the cost y (in dollars) of having laundry done at the Wash-n-Fold, where x is the number of pounds of laundry.

a. Find the slope of the line, and attach the proper units for the rate of change.

b. Then write a sentence explaining the meaning of the slope for this application.

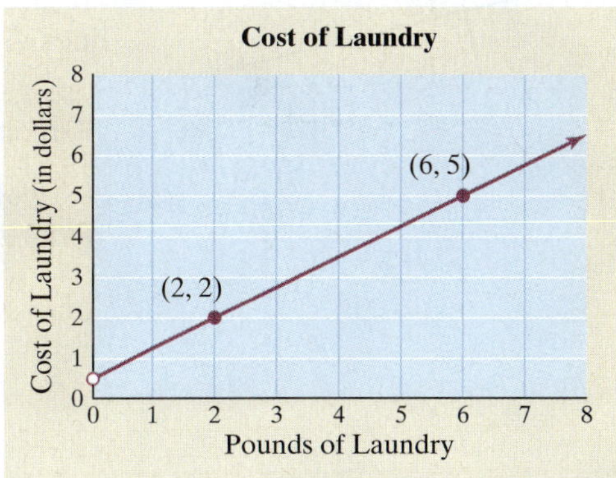

■

OBJECTIVE 6

Extension—Slopes of Parallel and Perpendicular Lines

Two lines in the same plane are **parallel** if they do not intersect. Slopes of lines can help us determine whether lines are parallel. Parallel lines have the same steepness, so it follows that they have the same slope.

For example, the graphs of

$$y = -2x + 4$$

and

$$y = -2x - 3$$

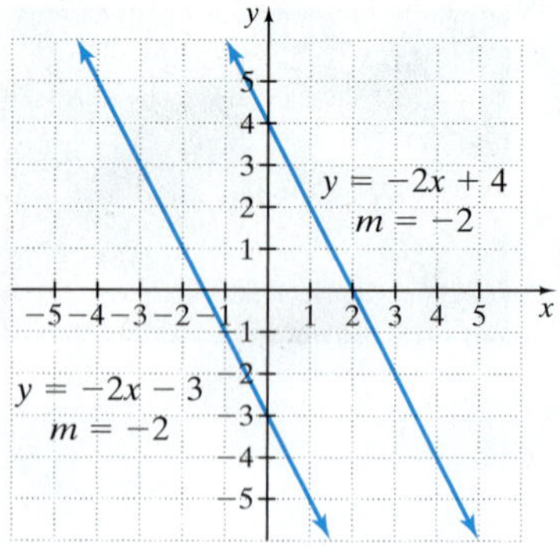

are shown. These lines have the same slope, -2. They also have different y-intercepts, so the lines are distinct and parallel. (If the y-intercepts were the same also, the lines would be the same.)

Parallel Lines

Nonvertical parallel lines have the same slope and different y-intercepts.

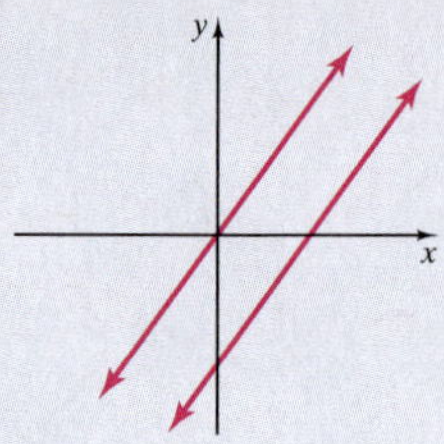

Two lines are **perpendicular** if they lie in the same plane and meet at a 90° (right) angle. How do the slopes of perpendicular lines compare? The product of the slopes of two perpendicular lines is -1.

For example, the graphs of

$$y = 4x + 1$$

and

$$y = -\frac{1}{4}x - 3$$

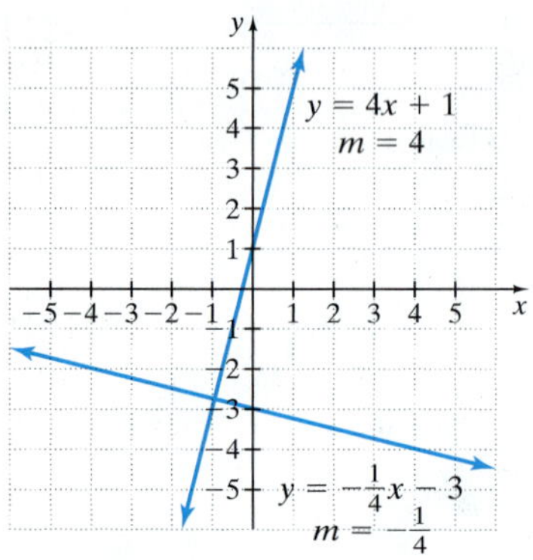

are shown. The slopes of the lines are 4 and $-\frac{1}{4}$. Their product is $4\left(-\frac{1}{4}\right) = -1$, so the lines are perpendicular.

Perpendicular Lines

If the product of the slopes of two lines is -1, then the lines are perpendicular.

(Two nonvertical lines are perpendicular if the slope of one is the negative reciprocal of the slope of the other.)

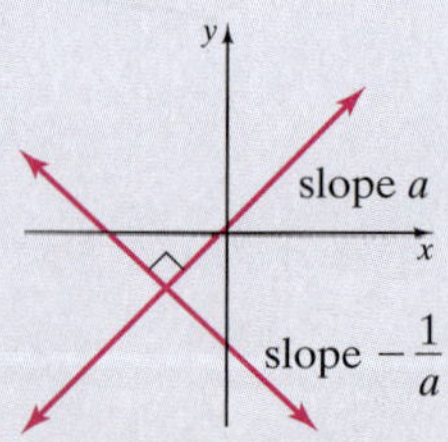

Helpful Hint

Here are examples of numbers that are negative (opposite) reciprocals.

Number	*Negative Reciprocal*	*Their Product Is* −1.
$\frac{2}{3}$	$-\frac{3}{2}$	$\frac{2}{3} \cdot -\frac{3}{2} = -\frac{6}{6} = -1$
-5 or $-\frac{5}{1}$	$\frac{1}{5}$	$-5 \cdot \frac{1}{5} = -\frac{5}{5} = -1$

> **Helpful Hint**
>
> Here are a few important facts about vertical and horizontal lines.
>
> - Two distinct vertical lines are parallel.
> - Two distinct horizontal lines are parallel.
> - A horizontal line and a vertical line are always perpendicular.

EXAMPLE 12 Determine whether each pair of lines is parallel, perpendicular, or neither.

a. $y = -\frac{1}{5}x + 1$
$2x + 10y = 3$

b. $x + y = 3$
$-x + y = 4$

c. $3x + y = 5$
$2x + 3y = 6$

Solution

a. The slope of the line $y = -\frac{1}{5}x + 1$ is $-\frac{1}{5}$. We find the slope of the second line by solving its equation for y.

$$2x + 10y = 3$$
$$10y = -2x + 3 \quad \text{Subtract } 2x \text{ from both sides.}$$
$$y = \frac{-2}{10}x + \frac{3}{10} \quad \text{Divide both sides by 10.}$$
$$y = -\frac{1}{5}x + \frac{3}{10} \quad \text{Simplify.}$$

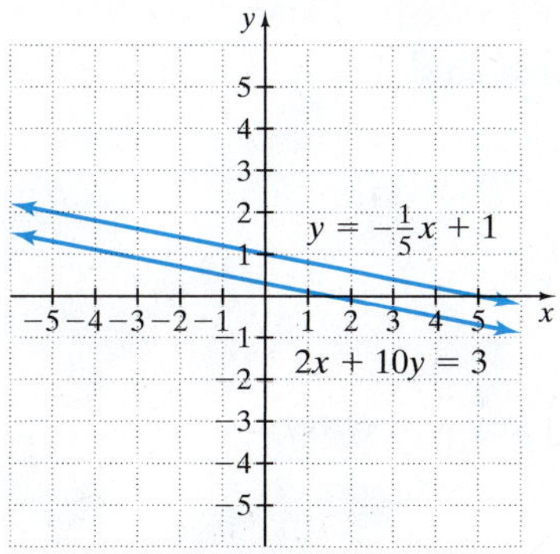

The slope of this line is $-\frac{1}{5}$ also. Since the lines have the same slope and different y-intercepts, they are parallel, as shown in the margin.

b. To find each slope, we solve each equation for y.

$$x + y = 3 \qquad\qquad -x + y = 4$$
$$y = -x + 3 \qquad\qquad y = x + 4$$

The slope is -1. The slope is 1.

The slopes are not the same, so the lines are not parallel. Next we check the product of the slopes: $(-1)(1) = -1$. Since the product is -1, the lines are perpendicular, as shown in the figure.

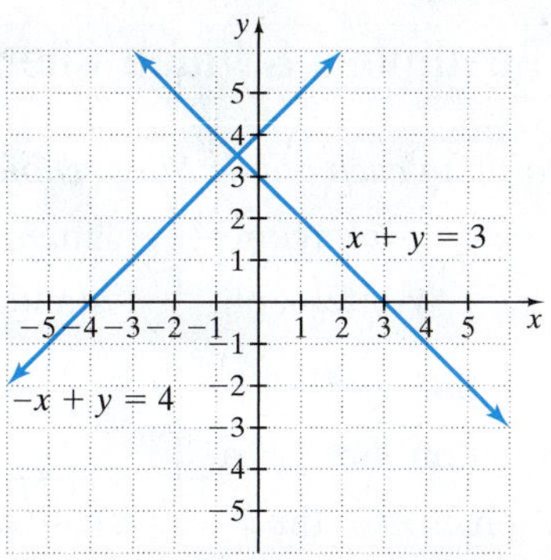

c. We solve each equation for y to find each slope. The slopes are -3 and $-\frac{2}{3}$. The slopes are not the same and their product is not -1. Thus, the lines are neither parallel nor perpendicular. □

(Continued on next page)

Answers to Concept Check:
a. any two lines with $m = 3$ and y-intercept not $\left(0, \frac{1}{2}\right)$
b. any two lines with $m = -\frac{1}{3}$

PRACTICE 12 Determine whether each pair of lines is parallel, perpendicular, or neither.

a. $y = -5x + 1$
$x - 5y = 10$

b. $x + y = 11$
$2x + y = 11$

c. $2x + 3y = 21$
$6y = -4x - 2$

✔ CONCEPT CHECK

Consider the line $-6x + 2y = 1$.

a. Write the equations of two lines parallel to this line.

b. Write the equations of two lines perpendicular to this line.

Graphing Calculator Explorations

You may have noticed that to use the [Y =] key on a grapher to graph an equation, the equation must be solved for y. For example, to graph $2x + 3y = 7$, we solve this equation for y.

$$2x + 3y = 7$$

$$3y = -2x + 7 \quad \text{Subtract } 2x \text{ from both sides.}$$

$$\frac{3y}{3} = -\frac{2x}{3} + \frac{7}{3} \quad \text{Divide both sides by 3.}$$

$$y = -\frac{2}{3}x + \frac{7}{3} \quad \text{Simplify.}$$

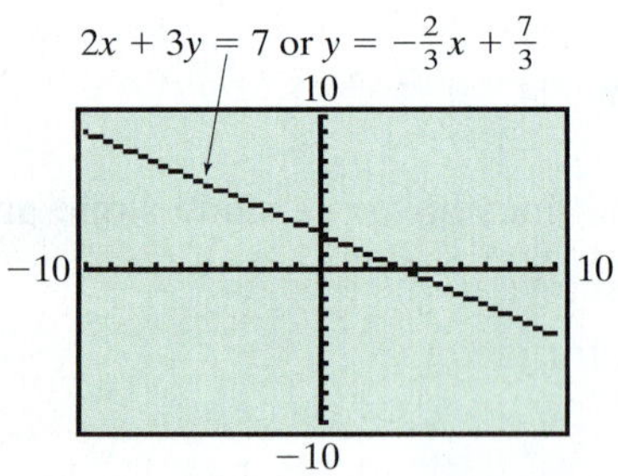

To graph $2x + 3y = 7$ or $y = -\frac{2}{3}x + \frac{7}{3}$, press the [Y =] key and enter

$$Y_1 = -\frac{2}{3}x + \frac{7}{3}$$

Graph each linear equation.

1. $x = 3.78y$

2. $-2.61y = x$

3. $3x + 7y = 21$

4. $-4x + 6y = 21$

5. $-2.2x + 6.8y = 15.5$

6. $5.9x - 0.8y = -10.4$

Vocabulary, Readiness & Video Check

Use the choices below to fill in each blank. Some choices may be used more than once or not at all.

x	vertical	x-intercept	linear	m	0	positive	undefined
y	horizontal	y-intercept	standard	b	slope	negative	

1. The graph of the equation $y = -1$ is a __________ line.

2. The graph of the equation $x = 5$ is a __________ line.

3. A point where a graph crosses the y-axis is called a(n) __________.

4. A point where a graph crosses the x-axis is called a(n) __________.

5. Given an equation of a line, to find the x-intercept (if there is one), let __________ = 0 and solve for __________.

6. Given an equation of a line, to find the y-intercept (if there is one), let __________ = 0 and solve for __________.

7. The measure of the steepness or tilt of a line is called __________.

8. If an equation is written in the form $y = mx + b$, the value of the letter __________ is the value of the slope of the graph.

9. The slope of a horizontal line is __________.

10. The slope of a vertical line is __________.

11. If the graph of a line moves upward from left to right, the line has __________ slope.

12. If the graph of a line moves downward from left to right, the line has __________ slope.

13. Given two points of a line, slope $= \dfrac{\text{change in } \_\_}{\text{change in } \_\_}$.

Martin-Gay Interactive Videos

See Video 3.2

Watch the section lecture video and answer the following questions.

OBJECTIVE 1

14. At the end of Example 2, patterns are discussed. What reason is given for why x-intercepts have y-values of 0? For why y-intercepts have x-values of 0?

15. In Example 3, the goal is to use the x- and y-intercepts to graph a line. Yet once the two intercepts are found, a third point is also found before the line is graphed. Why do you think this practice of finding a third point is continued?

OBJECTIVE 2

16. What important point is made during Example 5 having to do with the order of the points in the slope formula?

OBJECTIVE 3

17. From Example 9, how do you write an equation in "slope-intercept form"? Once the equation is in slope-intercept form, how do you identify the slope?

OBJECTIVE 4

18. From Examples 11 and 12, what can you say about the coefficient of x when the equation of a horizontal line is written as $Ax + By = C$? What about the coefficient of y when the equation of a vertical line is written as $Ax + By = C$?

19. In the lecture after Example 14, different slopes are summarized. What is the difference between zero slope and undefined slope? What does "no slope" mean?

OBJECTIVE 5

20. Writing the slope as a rate of change in Example 15 gave real-life meaning to the slope. What step in the general strategy for problem solving does this correspond to?

OBJECTIVE 6

21. From Example 17, what form of the equations is best to determine if two lines are parallel or perpendicular? Why?

3.2 Exercise Set MyMathLab®

Identify the intercepts. See Example 1.

1.

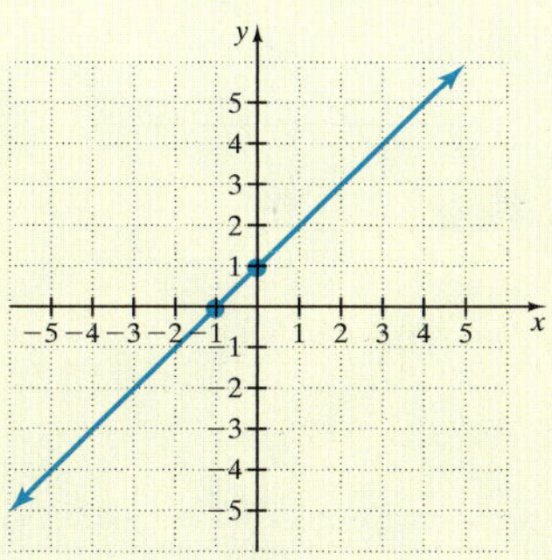

2.

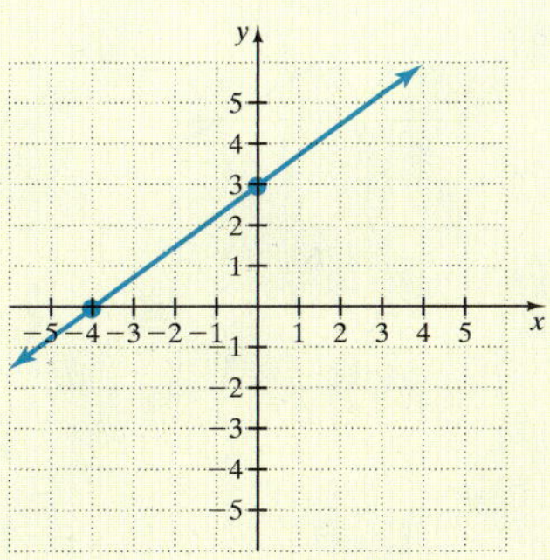

3.

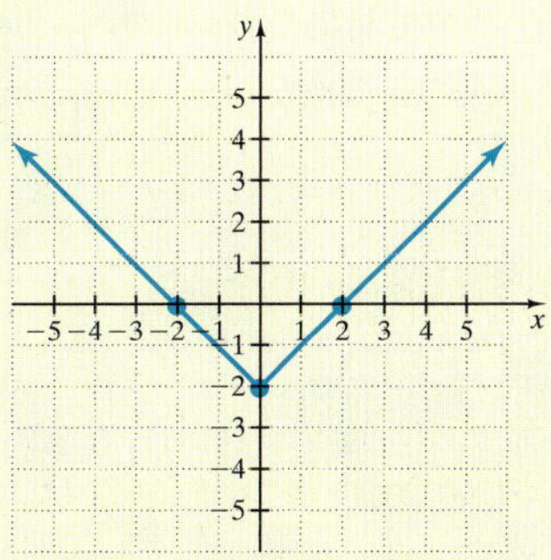

4.

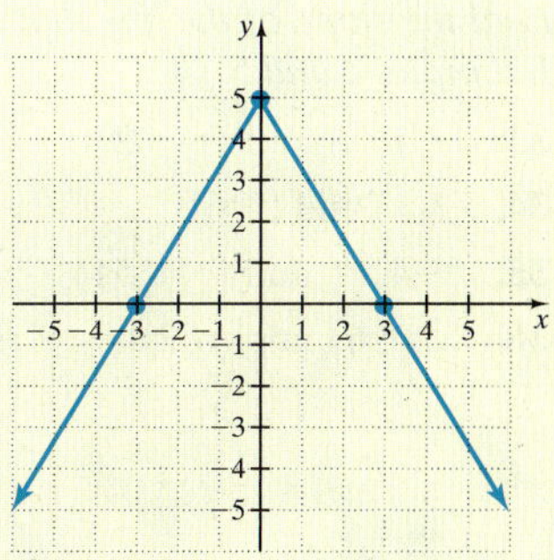

5.

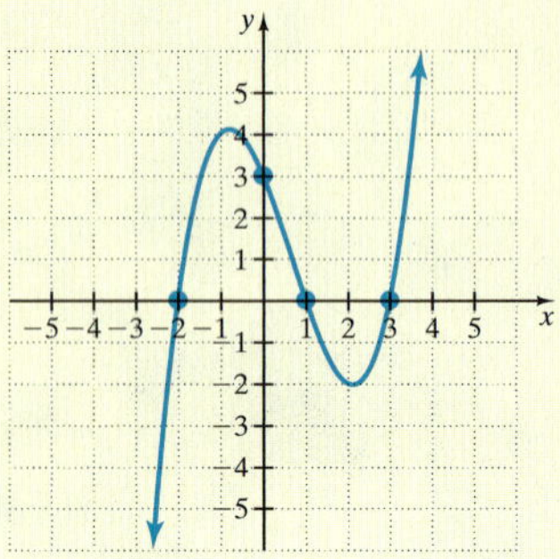

6.

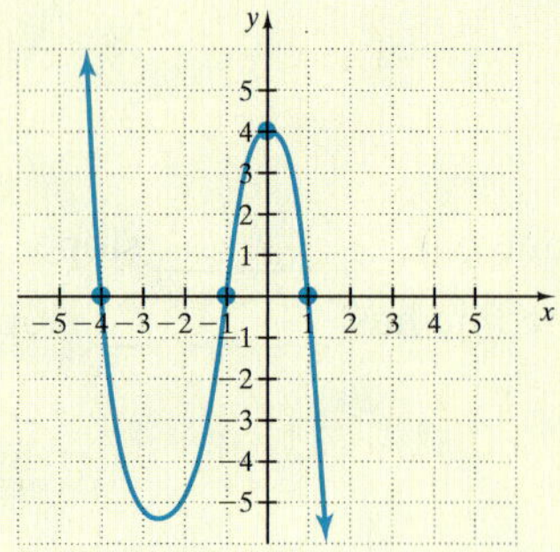

7.

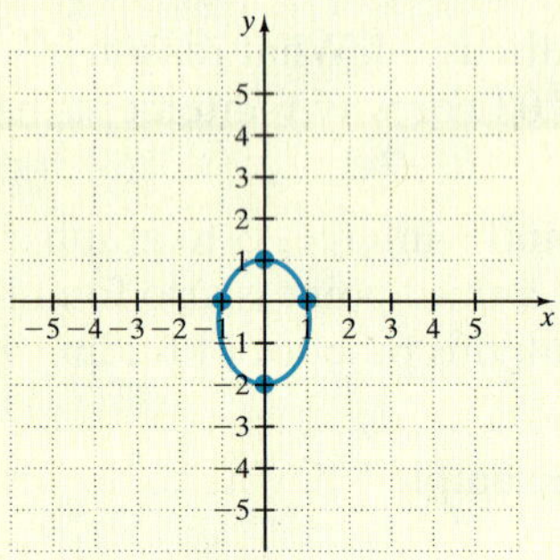

8.

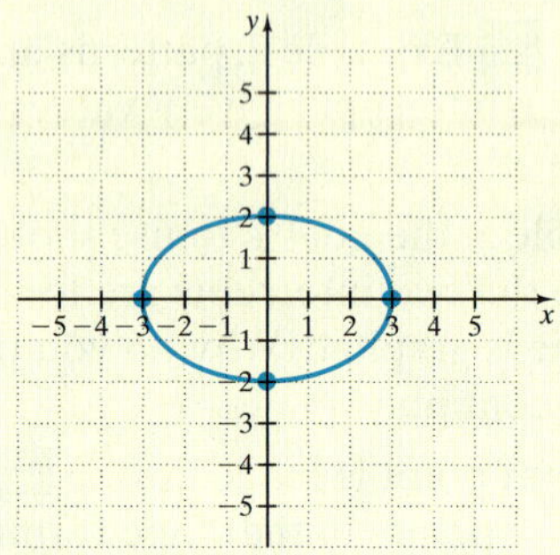

Solve. See Example 1.

9. What is the greatest number of intercepts for a line?

10. What is the least number of intercepts for a line?

11. What is the least number of intercepts for a circle?

12. What is the greatest number of intercepts for a circle?

Graph each linear equation by finding and plotting its intercepts. See Examples 2 and 3.

13. $x - y = 3$

14. $x - y = -4$

15. $x = 5y$

16. $x = 2y$

17. $-x + 2y = 6$

18. $x - 2y = -8$

19. $2x - 4y = 8$

20. $2x + 3y = 6$

21. $y = 2x$

22. $y = -2x$

23. $y = 3x + 6$

24. $y = 2x + 10$

Graph each linear equation and find its slope. See Examples 8 and 9.

25. $x = -1$

26. $y = 5$

27. $y = 0$

28. $x = 0$

29. $y = -7$

30. $x = 2$

31. $x + 3 = 0$

32. $y - 6 = 0$

Find the slope of the line that passes through the given points. See Examples 4 and 5.

33. $(-1, 5)$ and $(6, -2)$

34. $(3, 1)$ and $(2, 6)$

35. $(-4, 3)$ and $(-4, 5)$

36. $(6, -6)$ and $(6, 2)$

37. $(-2, 8)$ and $(1, 6)$

38. $(4, -3)$ and $(2, 2)$

39. $(5, 1)$ and $(-2, 1)$

40. $(0, 13)$ and $(-4, 13)$

Find the slope of each line if it exists. See Examples 4 and 5.

41.

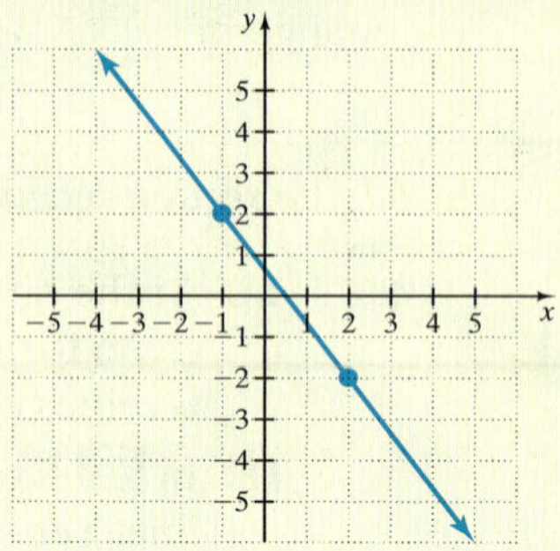

42.

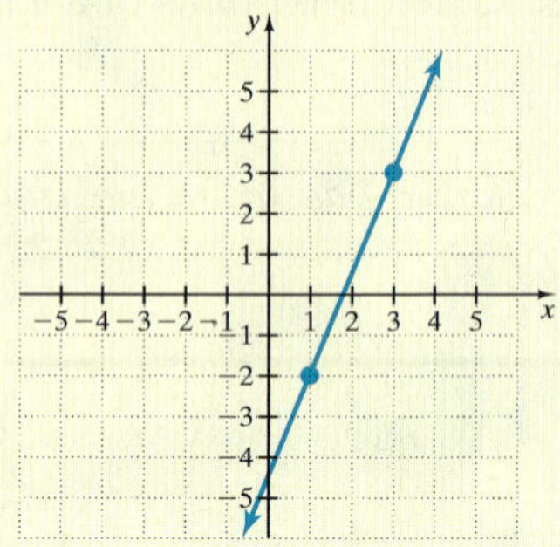

43.

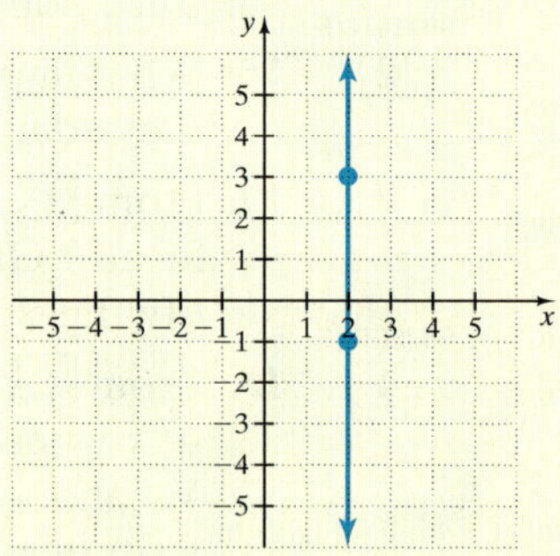

44.

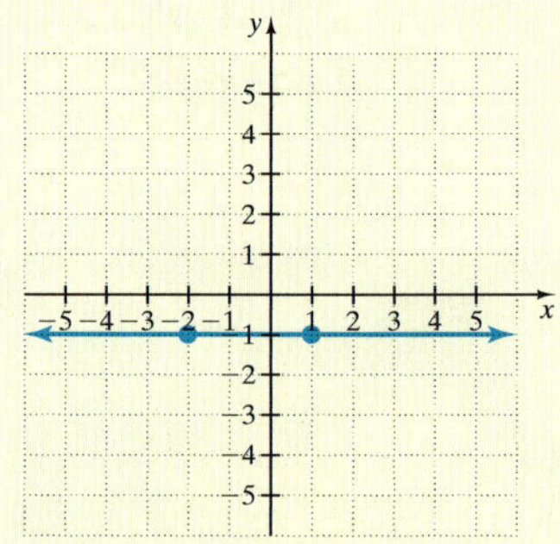

45.

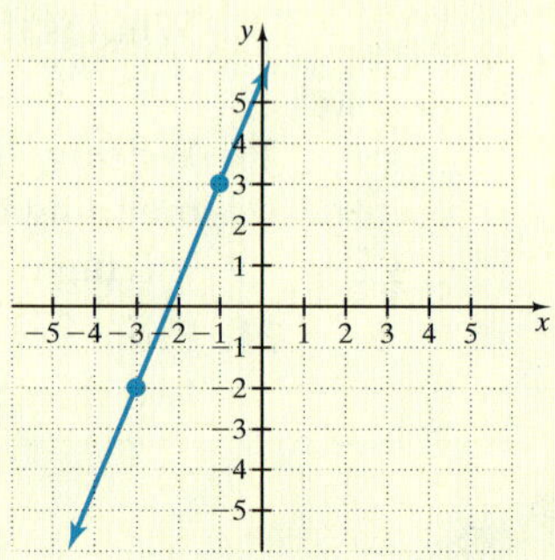

46.

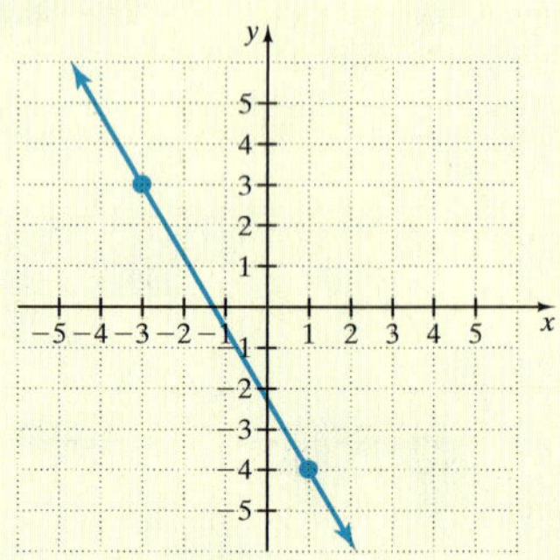

State whether the slope of the line is positive, negative, 0, or undefined. See the box on p. 156.

47.

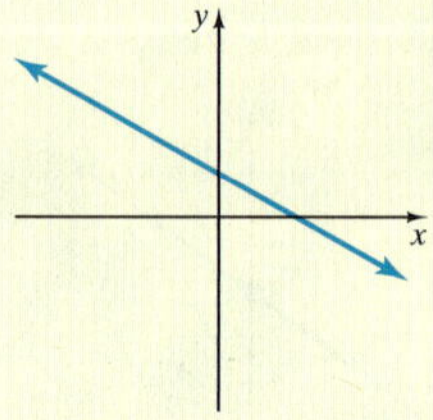

48.

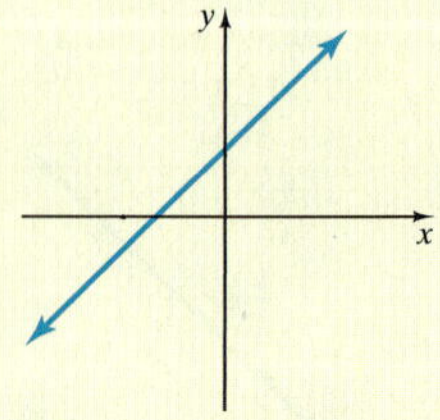

49.

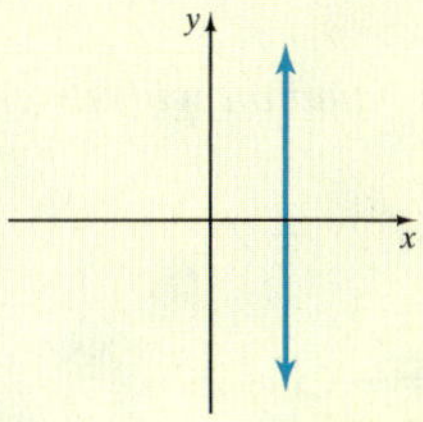

50.

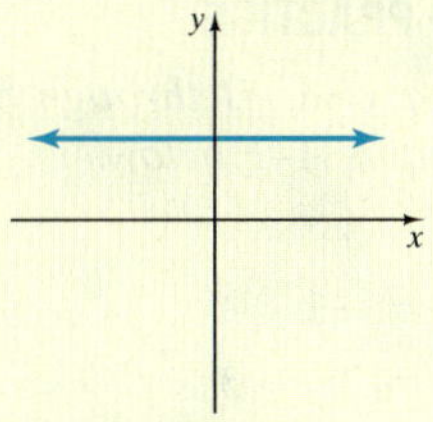

In Exercises 51 through 56, match each line with its slope. See Examples 4, 5, 8 and 9 and the box on p. 156.

A. $m = 0$ **B.** undefined slope **C.** $m = 3$

D. $m = 1$ **E.** $m = -\frac{1}{2}$ **F.** $m = -\frac{3}{4}$

51.

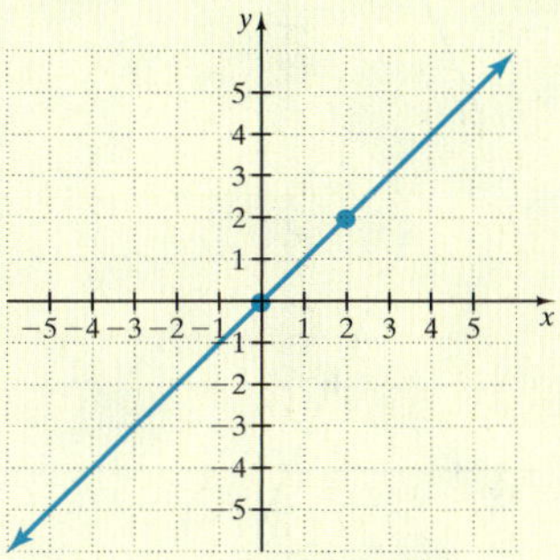

52.

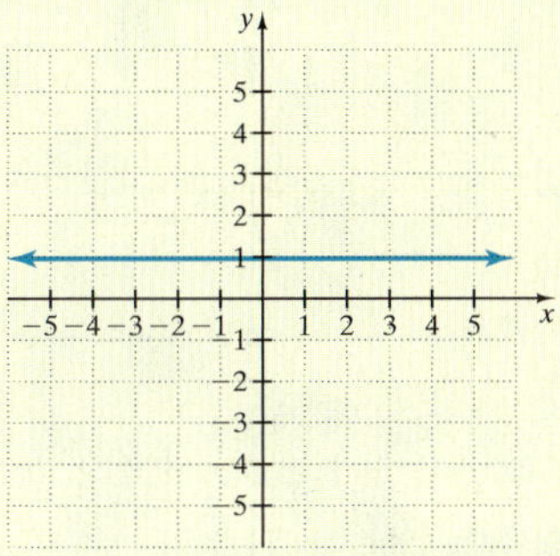

53.

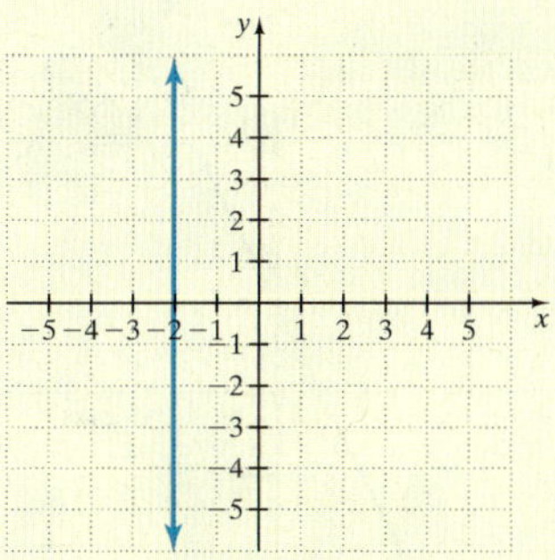

54.

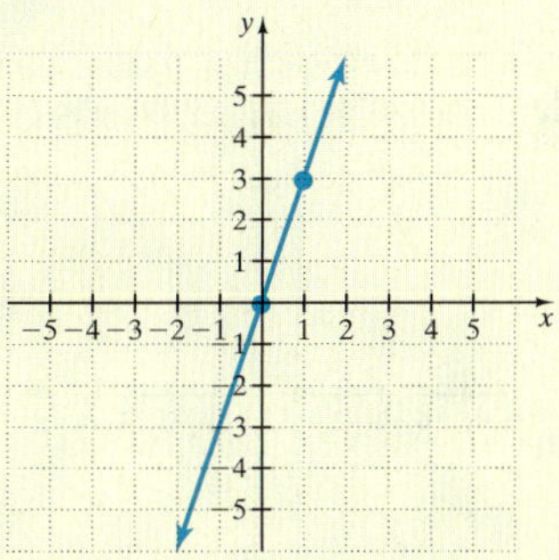

55.

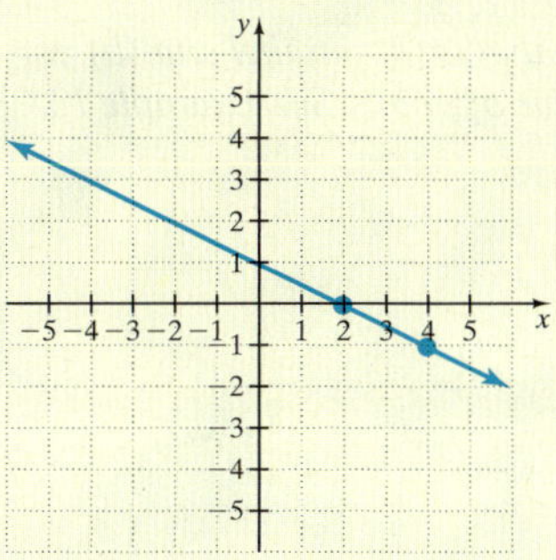

56.

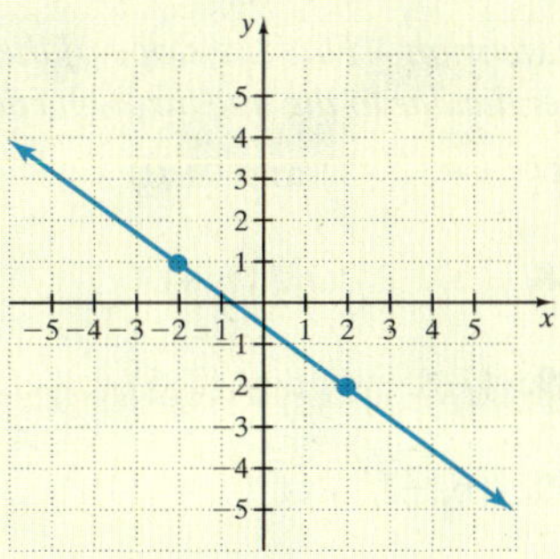

MIXED PRACTICE

Find the slope of each line. See Examples 6 through 9.

57. $y = 5x - 2$

58. $y = -2x + 6$

59. $y = -0.3x + 2.5$

60. $y = -7.6x - 0.1$

61. $2x + y = 7$

62. $-5x + y = 10$

63. $2x - 3y = 10$

64. $3x - 5y = 1$

65. $x = 1$

66. $y = -2$

67. $y = -3$

68. $x = 5$

69. $-3x - 4y = 6$

70. $-4x - 7y = 9$

The pitch of a roof is its slope. Find the pitch of each roof shown. See Example 10. (Note: Pitch of a roof is a positive value.)

71.

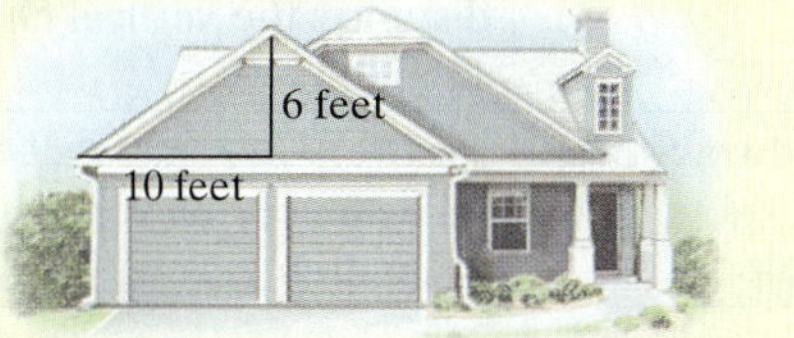

72.

The grade of a road is its slope written as a percent. Find the grade of each road shown. See Example 10. (Note: Grade of a road is a positive value.)

73.

74.

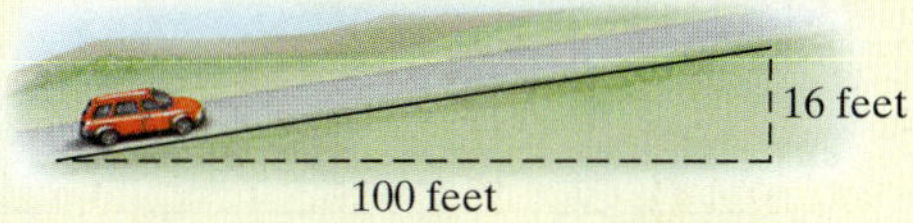

75. One of Japan's superconducting "bullet" trains is researched and tested at the Yamanashi Maglev Test Line near Otsuki City. The steepest section of the track has a rise of 2580 meters for a horizontal distance of 6450 meters. What is the grade of this section of track? (*Source:* Japan Railways Central Co.)

76. Professional plumbers suggest that a sewer pipe should rise 0.25 inch for every horizontal foot. Find the recommended slope for a sewer pipe. Round to the nearest hundredth.

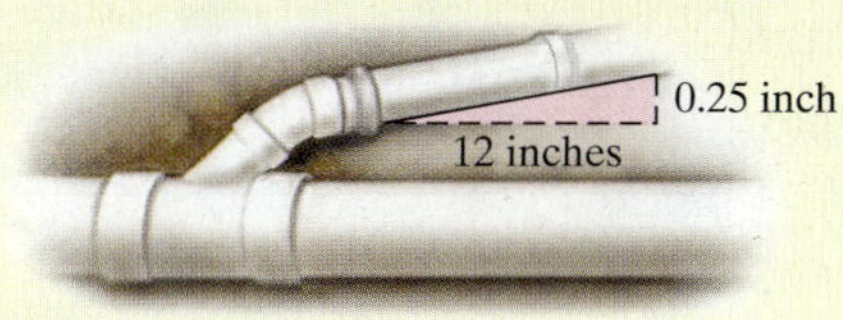

77. There has been controversy over the past few years about the world's steepest street. The *Guinness Book of Records* actually listed Baldwin Street, in Dunedin, New Zealand, as the world's steepest street, but Canton Avenue in the Pittsburgh neighborhood of Beechview may be steeper. Calculate each grade to the nearest percent.

		Grade (%)
Canton Avenue	for every 30 meters of horizontal distance, the vertical change is 11 meters	
Baldwin Street	for every 2.86 meters of horizontal distance, the vertical change is 1 meter	

78. According to federal regulations, a wheelchair ramp should rise no more than 1 foot for a horizontal distance of 12 feet. Write the slope as a grade. Round to the nearest tenth of a percent.

For Exercises 79 and 80, find the slope of each line and write a sentence explaining the meaning of the slope as a rate of change. Don't forget to attach the proper units. See Example 11.

79. This graph approximates the number of U.S. households that have computers y (in millions) for year x. (*Source:* U.S. census and statistics)

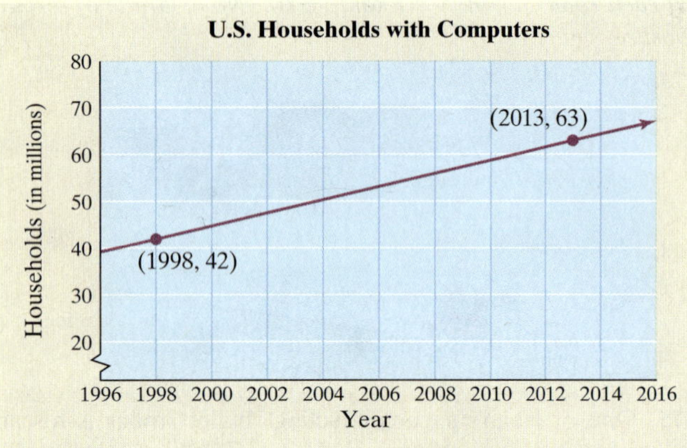

80. The graph approximates the amount of money y (in billions of dollars) spent worldwide on leisure travel and tourism for year x. (*Source:* World Travel and Tourism Council)

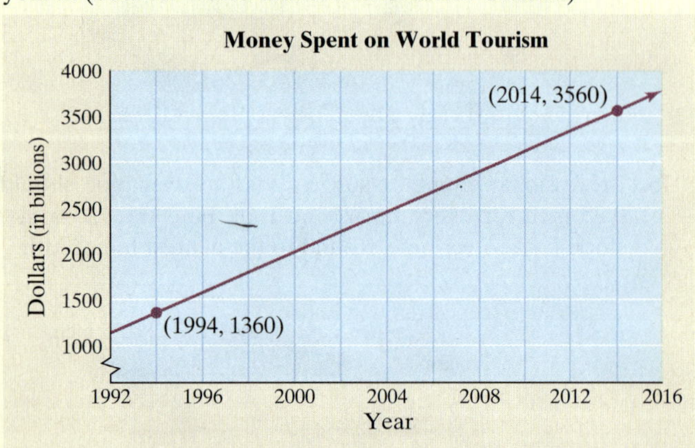

MIXED PRACTICE

For Exercises 81 through 86, match each equation with its graph. See graphs A–F below.

81. $y = 3$ **82.** $y = 2x + 2$

83. $x = -1$ **84.** $x = 3$

85. $y = 2x + 3$ **86.** $y = -2x$

A.

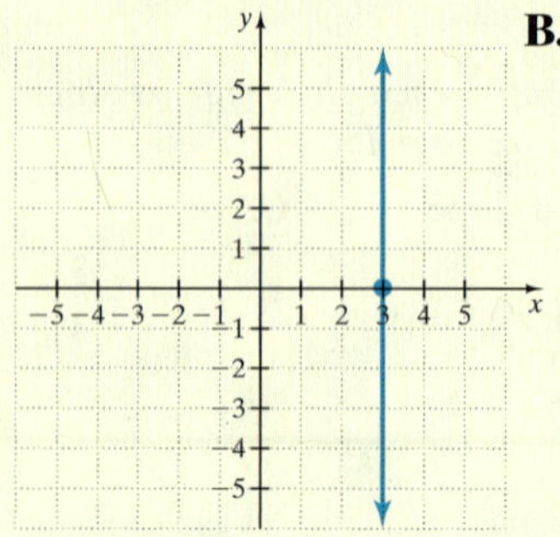

B.

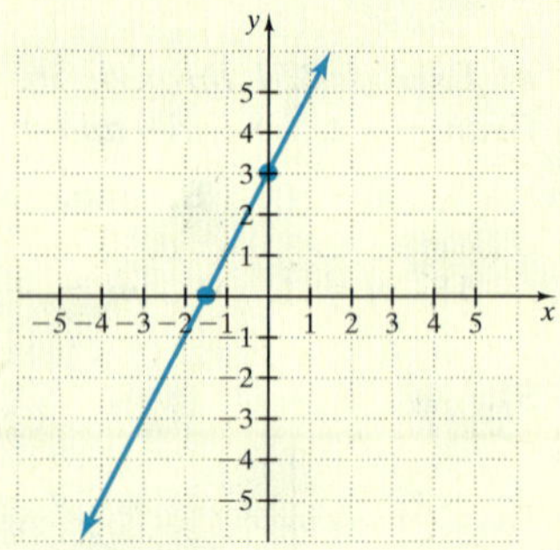

C.

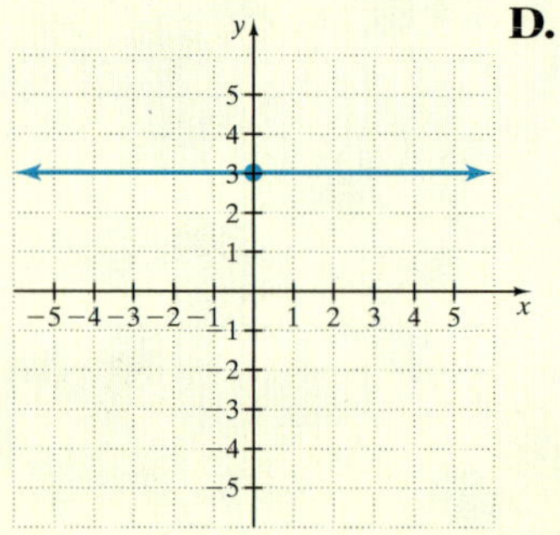

D.

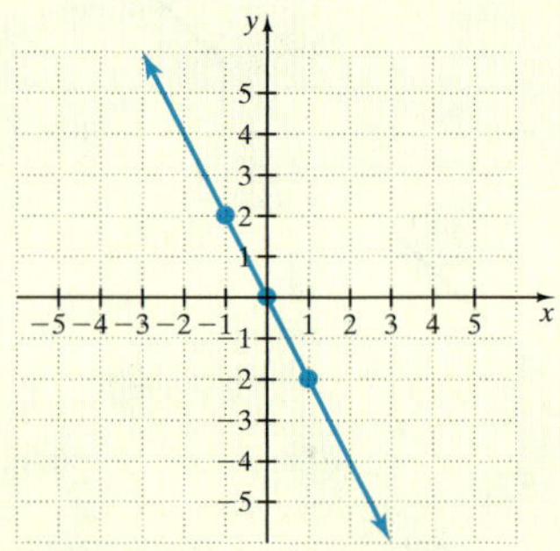

E.

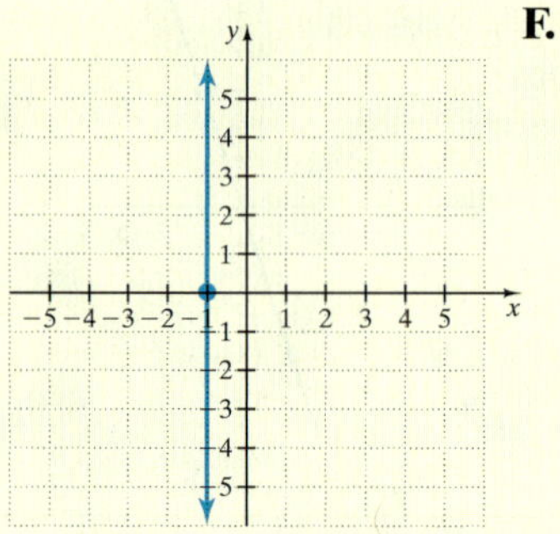

F.

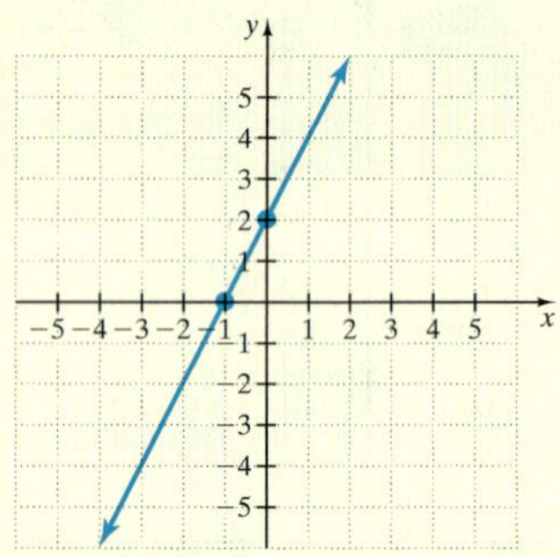

△ *Extension—Find the slope of the line that is (a) parallel and (b) perpendicular to the line through each pair of points. See Example 12.*

87. $(-3, -3)$ and $(0, 0)$

88. $(6, -2)$ and $(1, 4)$

89. $(-8, -4)$ and $(3, 5)$

90. $(6, -1)$ and $(-4, -10)$

△ *Extension—Determine whether each pair of lines is parallel, perpendicular, or neither. See Example 12.*

91. $y = \frac{2}{9}x + 3$

$y = -\frac{2}{9}x$

92. $y = \frac{1}{5}x + 20$

$y = -\frac{1}{5}x$

93. $6x = 5y + 1$
$-12x + 10y = 1$

94. $-x + 2y = -2$
$2x = 4y + 3$

95. $6 + 4x = 3y$
$3x + 4y = 8$

96. $10 + 3x = 5y$
$5x + 3y = 1$

REVIEW AND PREVIEW

The line graph below shows the number of students per teacher in U.S. public elementary and secondary schools. Use this graph for Exercises 97 through 102. See Section R.4.

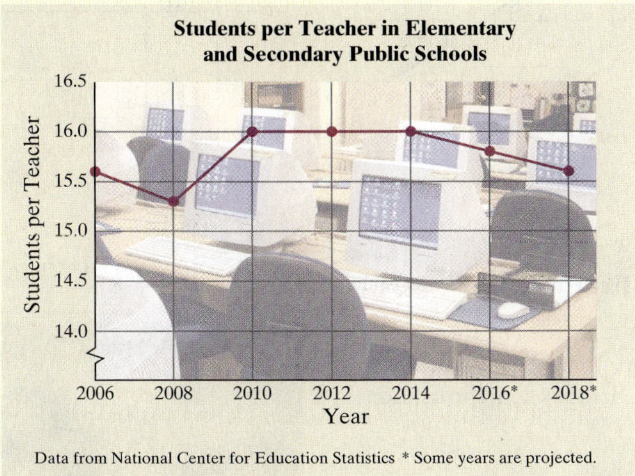

97. Approximate the number of students per teacher predicted in 2016.

98. Approximate the number of students per teacher predicted in 2018.

99. Between what years shown did the greatest increase in number of students per teacher occur?

100. What was the first year shown that the number of students per teacher fell below 15.5?

101. During what period was the student per teacher number at 16?

102. Discuss any trends shown by this line graph.

CONCEPT EXTENSIONS

Answer the following true or false.

103. All lines have an x-intercept *and* a y-intercept.

104. The graph of $y = 4x$ contains the point $(0, 0)$.

105. The graph of $x + y = 5$ has an x-intercept of $(5, 0)$ and a y-intercept of $(0, 5)$.

106. The graph of $y = 5x$ contains the point $(5, 1)$.

Solve.

107. Since 2009, the number of analog theater screens has been on the decline in the U.S. The number of analog movie screens y each year can be estimated by the equation $y = -7000x + 31{,}800$, where x represents the number of years since 2009. (*Source:* MPAA)

a. Find the x-intercept of this equation. Round to the nearest tenth.

b. What does this x-intercept mean?

c. Use part (b) to comment on your opinion of the limitations of using equations to model real data.

108. The price of admission to a movie theater has been steadily increasing. The price of regular admission y (in dollars) to a movie theater may be represented by the equation $y = 0.24x + 5.28$, where x is the number of years after 2000. (*Source:* Based on data from Motion Picture Association of America)

a. Find the x-intercept of this equation.

b. What does this x-intercept mean?

c. Use part (b) to comment on your opinion of the limitations of using equations to model real data.

Solve.

109. Find x so that the pitch of the roof is $\frac{1}{3}$.

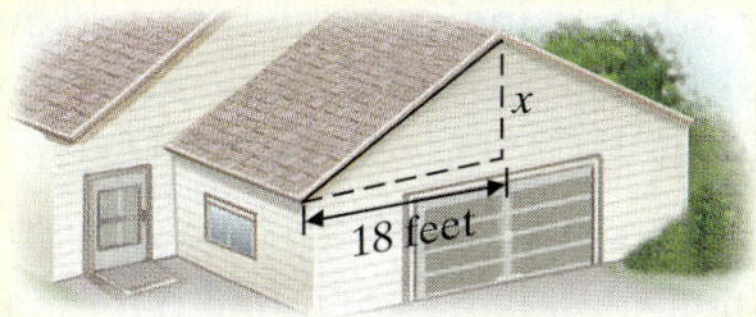

110. Find x so that the pitch of the roof is $\frac{2}{5}$.

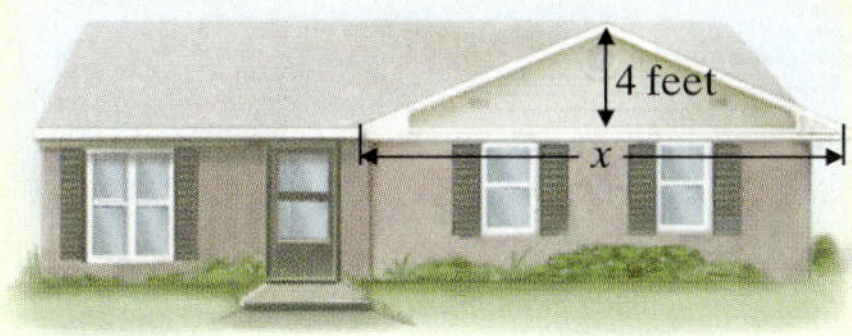

Integrated Review Summary on Slope and Graphing Linear Equations

Sections 3.1–3.2

Find the slope of each line.

1.

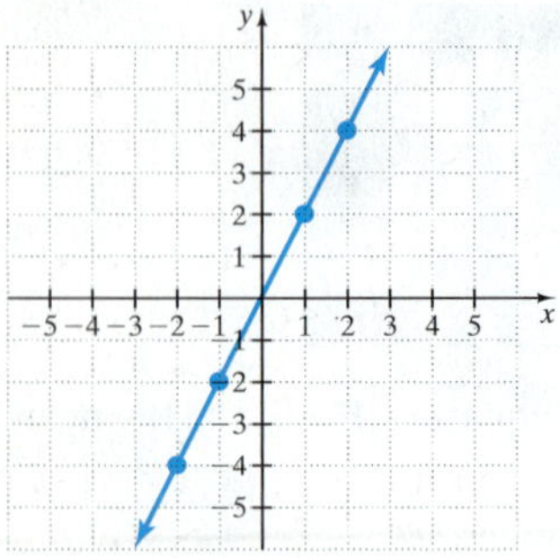

2.

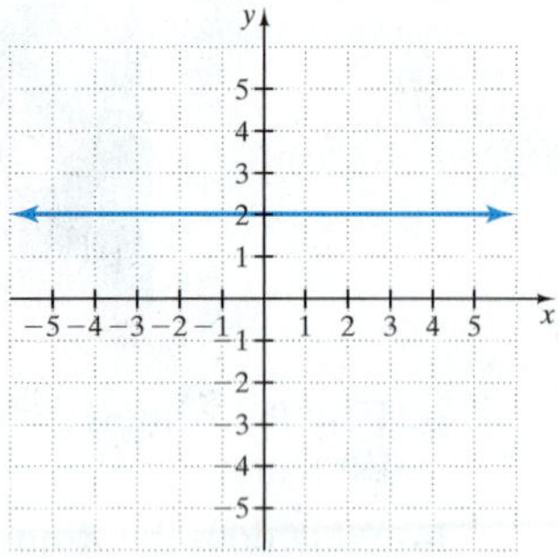

3.

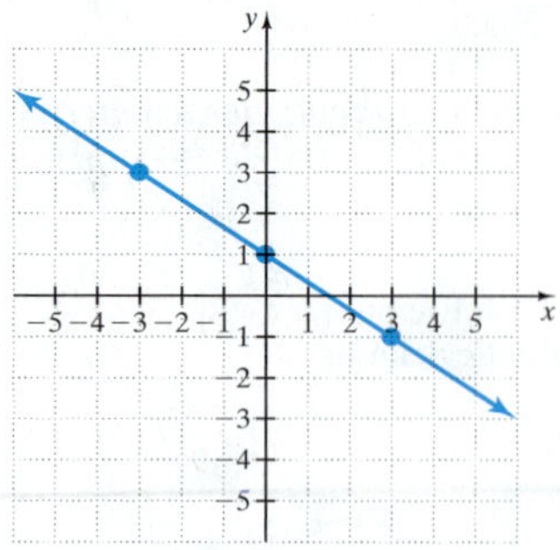

4.

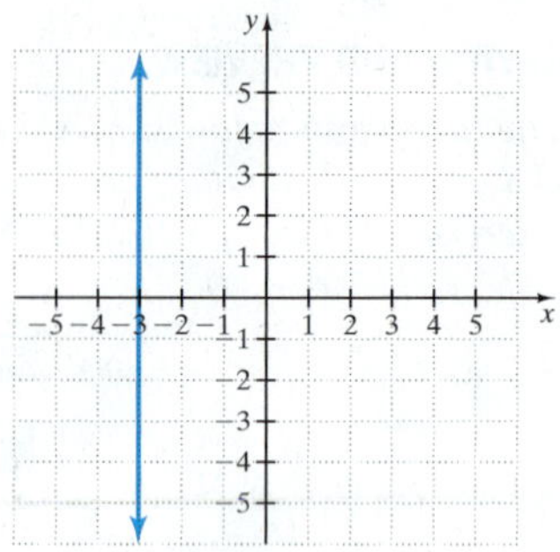

Graph each linear equation.

5. $y = -2x$ **6.** $x + y = 3$ **7.** $x = -1$ **8.** $y = 4$

9. $x - 2y = 6$ **10.** $y = 3x + 2$ **11.** $5x + 3y = 15$ **12.** $2x - 4y = 8$

13. In the years 2002 through 2013 the number of bridges on public roads (in thousands) in the United States can be modeled by the linear equation $y = 4.09x + 490$, where x is the number of years after 2002 and y is the number of bridges (in thousands). (*Source:* U.S. Dept. of Transportation)

a. Find the y-intercept of the line.

b. Write a sentence explaining the meaning of this intercept.

c. Find the slope of this line.

d. Write a sentence explaining the meaning of the slope as a rate of change.

14. Online advertising is a means of promoting products and services using the Internet. The revenue y (in billions of dollars) for online advertising for the years 2009 through 2014 is given by $y = 5.3x + 22.7$, where x is the number of years after 2009.

a. Use this equation to complete the ordered pair (4,)

b. Write a sentence explaining the meaning of the answer to part (a).

Extension—Determine whether the lines through the points are parallel, perpendicular, or neither.

15. $y = -\frac{1}{5}x + \frac{1}{3}$

$3x = -15y$

16. $x - y = \frac{1}{2}$

$3x - y = \frac{1}{2}$

3.3 Equations of Lines

OBJECTIVES

1. Use the Slope-Intercept Form to Graph a Linear Equation.
2. Use the Slope-Intercept Form to Write an Equation of a Line.
3. Use the Point-Slope Form to Find an Equation of a Line Given Its Slope and a Point on the Line.
4. Use the Point-Slope Form to Find an Equation of a Line Given Two Points on the Line.
5. Find Equations of Vertical and Horizontal Lines.
6. Use the Point-Slope Form to Solve Problems.

Recall that the form $y = mx + b$ is appropriately called the *slope-intercept form* of a linear equation.

(m: slope; b: y-intercept is $(0, b)$)

Slope-Intercept Form

When a linear equation in two variables is written in **slope-intercept form,**

$$y = mx + b$$

(m: slope; $(0, b)$, y-intercept)

then m is the slope of the line and $(0, b)$ is the y-intercept of the line.

OBJECTIVE

1 Using the Slope-Intercept Form to Graph an Equation

We can use the slope-intercept form of the equation of a line to graph a linear equation.

EXAMPLE 1 Use the slope-intercept form to graph the equation

$$y = \frac{3}{5}x - 2.$$

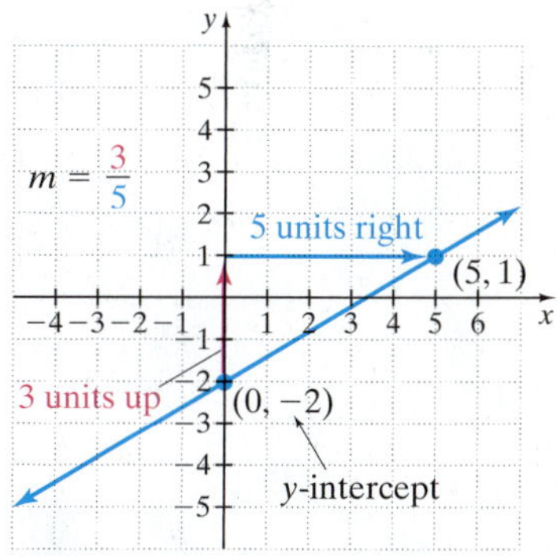

Solution Since the equation $y = \frac{3}{5}x - 2$ is written in slope-intercept form $y = mx + b$, the slope of its graph is $\frac{3}{5}$ and the y-intercept is $(0, -2)$. To graph this equation, we begin by plotting the point $(0, -2)$. From this point, we can find another point of the graph by using the slope $\frac{3}{5}$ and recalling that slope is $\frac{\text{rise}}{\text{run}}$. We start at the y-intercept and move 3 units up since the numerator of the slope is 3; then we move 5 units to the right since the denominator of the slope is 5. We stop at the point $(5, 1)$. The line through $(0, -2)$ and $(5, 1)$ is the graph of $y = \frac{3}{5}x - 2$. □

PRACTICE

1 Graph: $y = \frac{2}{3}x - 5$ ■

EXAMPLE 2 Use the slope-intercept form to graph the equation $4x + y = 1$.

Solution First we write the given equation in slope-intercept form.

$$4x + y = 1$$
$$y = -4x + 1$$

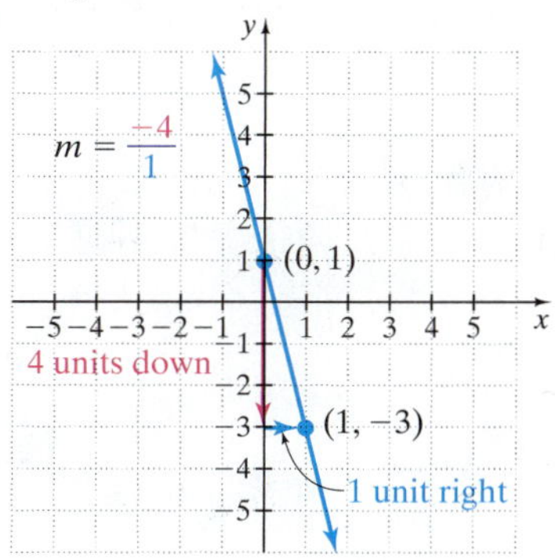

The graph of this equation will have slope -4 and y-intercept $(0, 1)$. To graph this line, we first plot the point $(0, 1)$. To find another point of the graph, we use the slope -4, which can be written as $\frac{-4}{1}$ $\left(\frac{4}{-1}\text{ could also be used}\right)$. We start at the point $(0, 1)$ and move 4 units down (since the numerator of the slope is -4) and then 1 unit to the right (since the denominator of the slope is 1).

Helpful Hint

In Example 2, if we interpret the slope of -4 as $\frac{4}{-1}$, we arrive at $(-1, 5)$ for a second point. Notice that this point is also on the line.

(Continued on next page)

We arrive at the point $(1, -3)$. The line through $(0, 1)$ and $(1, -3)$ is the graph of $4x + y = 1$. □

PRACTICE 2 Use the slope-intercept form to graph the equation $3x - y = 2$. ■

OBJECTIVE 2

Using the Slope-Intercept Form to Write an Equation

The slope-intercept form can also be used to write an equation of a line when we know its slope and y-intercept.

EXAMPLE 3 Find an equation of the line with y-intercept $(0, -3)$ and slope of $\frac{1}{4}$.

Solution We are given the slope and the y-intercept. We let $m = \frac{1}{4}$ and $b = -3$ and write the equation in slope-intercept form, $y = mx + b$.

$$y = mx + b$$

$$y = \frac{1}{4}x + (-3) \quad \text{Let } m = \frac{1}{4} \text{ and } b = -3.$$

$$y = \frac{1}{4}x - 3 \quad \text{Simplify.}$$

□

PRACTICE 3 Find an equation of the line with y-intercept $(0, 7)$ and slope of $\frac{1}{2}$. ■

OBJECTIVE 3

Writing an Equation Given Slope and a Point

Thus far, we have seen that we can write an equation of a line if we know its slope and y-intercept. We can also write an equation of a line if we know its slope and any point on the line. To see how we do this, let m represent slope and (x_1, y_1) represent the point on the line. Then if (x, y) is any other point of the line, we have that

$$\frac{y - y_1}{x - x_1} = m$$

$$y - y_1 = m(x - x_1) \quad \text{Multiply both sides by } (x - x_1).$$

↑ slope

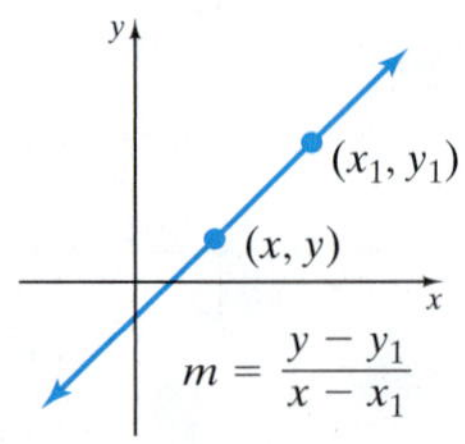

This is the *point-slope form* of the equation of a line.

Point-Slope Form of the Equation of a Line

The **point-slope form** of the equation of a line is

$$y - y_1 = m(x - x_1)$$

slope: m; (x_1, y_1) point on the line

where m is the slope of the line and (x_1, y_1) is a point on the line.

EXAMPLE 4 Find an equation of the line with slope -2 that passes through $(-1, 5)$. Write the equation in slope-intercept form, $y = mx + b$, and in standard form, $Ax + By = C$.

Solution Since the slope and a point on the line are given, we use point-slope form $y - y_1 = m(x - x_1)$ to write the equation. Let $m = -2$ and $(-1, 5) = (x_1, y_1)$.

$$
\begin{aligned}
y - y_1 &= m(x - x_1) \\
y - 5 &= -2[x - (-1)] && \text{Let } m = -2 \text{ and } (x_1, y_1) = (-1, 5). \\
y - 5 &= -2(x + 1) && \text{Simplify.} \\
y - 5 &= -2x - 2 && \text{Use the distributive property.}
\end{aligned}
$$

To write the equation in slope-intercept form, $y = mx + b$, we simply solve the equation for y. To do this, we add 5 to both sides.

$$
\begin{aligned}
y - 5 &= -2x - 2 \\
y &= -2x + 3 && \text{Slope-intercept form.} \\
2x + y &= 3 && \text{Add } 2x \text{ to both sides and we have standard form.}
\end{aligned}
$$

□

PRACTICE 4 Find an equation of the line passing through (2, 3) with slope 4. Write the equation in slope-intercept form, $y = mx + b$, and in standard form, $Ax + By = C$. ■

OBJECTIVE

4 Writing an Equation Given Two Points

We can also find an equation of a line when we are given any two points of the line.

EXAMPLE 5 Find an equation of the line through $(2, 5)$ and $(-3, 4)$. Write the equation in slope-intercept form, $y = mx + b$.

Solution First, use the two given points to find the slope of the line.

$$m = \frac{4 - 5}{-3 - 2} = \frac{-1}{-5} = \frac{1}{5}$$

Next we use the slope $\frac{1}{5}$ and either one of the given points to write the equation in point-slope form. We use (2, 5). Let $x_1 = 2$, $y_1 = 5$, and $m = \frac{1}{5}$.

$$
\begin{aligned}
y - y_1 &= m(x - x_1) && \text{Use point-slope form.} \\
y - 5 &= \frac{1}{5}(x - 2) && \text{Let } x_1 = 2,\ y_1 = 5, \text{ and } m = \frac{1}{5}. \\
5(y - 5) &= 5 \cdot \frac{1}{5}(x - 2) && \text{Multiply both sides by 5 to clear fractions.} \\
5y - 25 &= x - 2 && \text{Use the distributive property and simplify.}
\end{aligned}
$$

To write the equation in slope-intercept form, $y = mx + b$, we simply solve the equation for y.

$$
\begin{aligned}
5y &= x + 23 && \text{Add 25 to both sides.} \\
\frac{5y}{5} &= \frac{x}{5} + \frac{23}{5} && \text{Divide both sides by 5.} \\
y &= \frac{1}{5}x + \frac{23}{5} && \text{Simplify.}
\end{aligned}
$$

□

PRACTICE 5 Find an equation of the line through $(-1, 6)$ and (3, 1). Write the equation in slope-intercept form. ■

Helpful Hint

In this section, we concentrate on slope-intercept form to prepare for the next section—Functions and Regression Lines.

OBJECTIVE

5 Finding Equations of Vertical and Horizontal Lines

Recall from Section 3.2 that:

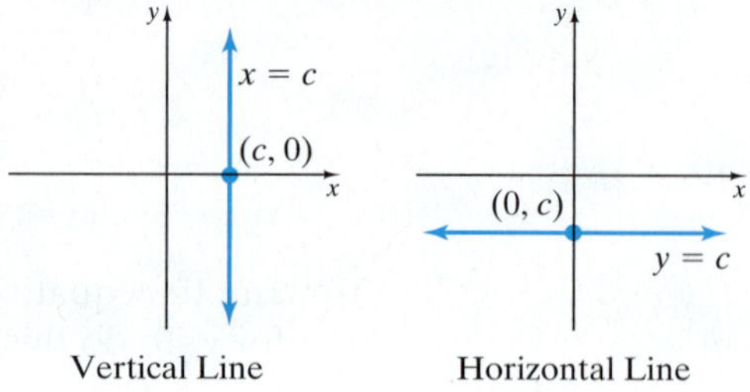

EXAMPLE 6 Find an equation of the vertical line through $(-1, 5)$.

Solution The equation of a vertical line can be written in the form $x = c$, so an equation for the vertical line passing through $(-1, 5)$ is $x = -1$.

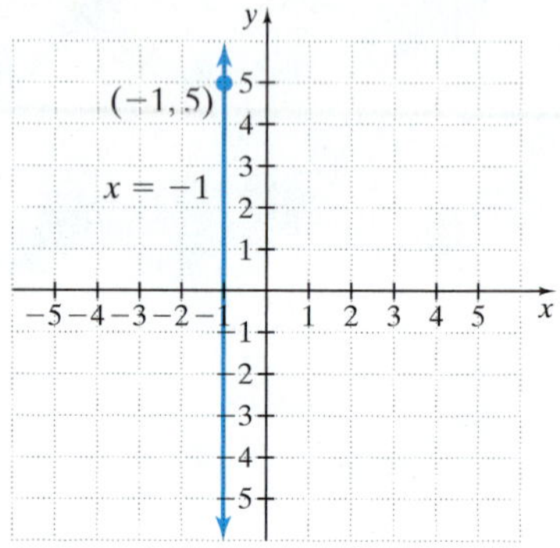

□

PRACTICE

6 Find an equation of the vertical line through $(3, -2)$. ■

EXAMPLE 7 Find an equation of the line parallel to the line $y = 5$ and passing through $(-2, -3)$.

Solution Since the graph of $y = 5$ is a horizontal line, any line parallel to it is also horizontal. The equation of a horizontal line can be written in the form $y = c$. An equation for the horizontal line passing through

$$(-2, -3) \text{ is } y = -3.$$

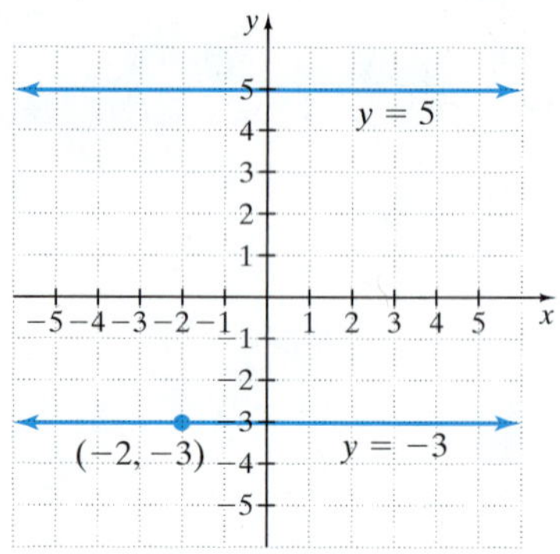

□

PRACTICE

7 Find an equation of the line parallel to the line $y = -2$ and passing through $(4, 3)$. ■

OBJECTIVE

6 Using the Point-Slope Form to Solve Problems

Problems occurring in many fields can be modeled by linear equations in two variables. The next example is from the field of marketing and shows how consumer demand of a product depends on the price of the product.

EXAMPLE 8 **Predicting the Sales of T-Shirts**

A web-based T-shirt company has learned that by pricing a clearance-sale T-shirt at \$6, sales will reach 2000 T-shirts per day. Raising the price to \$8 will cause the sales to fall to 1500 T-shirts per day.

a. Assume that the relationship between sales price and number of T-shirts sold is linear and write an equation describing this relationship. Write the equation in slope-intercept form.

b. Predict the daily sales of T-shirts if the price is \$7.50.

Solution

a. First, use the given information and write two ordered pairs. Ordered pairs will be in the form (sales price, number sold) so that our ordered pairs are (6, 2000) and (8, 1500). Use the point-slope form to write an equation. To do so, we find the slope of the line that contains these points.

$$m = \frac{2000 - 1500}{6 - 8} = \frac{500}{-2} = -250$$

Next, use the slope and either one of the points to write the equation in point-slope form. We use (6, 2000).

$$y - y_1 = m(x - x_1) \quad \text{Use point-slope form.}$$
$$y - 2000 = -250(x - 6) \quad \text{Let } x_1 = 6,\ y_1 = 2000, \text{ and } m = -250.$$
$$y - 2000 = -250x + 1500 \quad \text{Use the distributive property.}$$
$$y = -250x + 3500 \quad \text{Write in slope-intercept form.}$$

b. To predict the sales if the price is \$7.50, we find y when $x = 7.50$.

$$y = -250x + 3500$$
$$y = -250(7.50) + 3500 \quad \text{Let } x = 7.50.$$
$$y = -1875 + 3500$$
$$y = 1625$$

If the price is \$7.50, sales will reach 1625 T-shirts per day. □

PRACTICE 8 The new Camelot condos were selling at a rate of 30 per month when they were priced at \$150,000 each. Lowering the price to \$120,000 caused the sales to rise to 50 condos per month.

a. Assume that the relationship between number of condos sold and price is linear, and write an equation describing this relationship. Write the equation in slope-intercept form. Use ordered pairs of the form (number sold, sales price).

b. How should the condos be priced if the developer wishes to sell 60 condos per month? ■

The preceding example may also be solved by using ordered pairs of the form (number sold, sales price).

Forms of Linear Equations

$Ax + By = C$	**Standard form** of a linear equation. A and B are not both 0.
$y = mx + b$	**Slope-intercept form** of a linear equation. The slope is m and the y-intercept is $(0, b)$.
$y - y_1 = m(x - x_1)$	**Point-slope form** of a linear equation. The slope is m and (x_1, y_1) is a point on the line.
$y = c$	**Horizontal line** The slope is 0 and the y-intercept is $(0, c)$.
$x = c$	**Vertical line** The slope is undefined and the x-intercept is $(c, 0)$.

Parallel and Perpendicular Lines

Nonvertical parallel lines have the same slope.
The product of the slopes of two nonvertical perpendicular lines is -1.

Graphing Calculator Explorations

A grapher is a very useful tool for discovering patterns. To discover the change in the graph of a linear equation caused by a change in slope, try the following. Use a standard window and graph a linear equation in the form $y = mx + b$. Recall that the graph of such an equation will have slope m and y-intercept b.

First, graph $y = x + 3$. To do so, press the [Y =] key and enter $Y_1 = x + 3$. Notice that this graph has slope 1 and that the y-intercept is 3. Next, on the same set of axes, graph $y = 2x + 3$ and $y = 3x + 3$ by pressing [Y =] and entering $Y_2 = 2x + 3$ and $Y_3 = 3x + 3$.

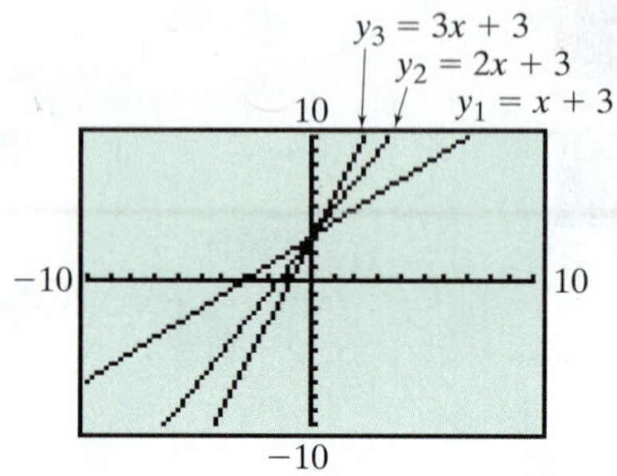

Notice the difference in the graph of each equation as the slope changes from 1 to 2 to 3. How would the graph of $y = 5x + 3$ appear? To see the change in the graph caused by a change in negative slope, try graphing $y = -x + 3$, $y = -2x + 3$, and $y = -3x + 3$ on the same set of axes.

Use a grapher to graph the following equations. For each exercise, graph the first equation and use its graph to predict the appearance of the other equations. Then graph the other equations on the same set of axes and check your prediction.

1. $y = x; y = 6x, y = -6x$

2. $y = -x; y = -5x, y = -10x$

3. $y = \frac{1}{2}x + 2; y = \frac{3}{4}x + 2, y = x + 2$

4. $y = x + 1; y = \frac{5}{4}x + 1, y = \frac{5}{2}x + 1$

5. $y = -7x + 5; y = 7x + 5$

6. $y = 3x - 1; y = -3x - 1$

Vocabulary, Readiness & Video Check

Use the choices below to fill in each blank. Some choices may be used more than once and some not at all.

b	(y_1, x_1)	point-slope	vertical	standard
m	(x_1, y_1)	slope-intercept	horizontal	

1. The form $y = mx + b$ is called ____________ form. When a linear equation in two variables is written in this form, ____________ is the slope of its graph and (0, ____________) is its y-intercept.

2. The form $y - y_1 = m(x - x_1)$ is called ____________ form. When a linear equation in two variables is written in this form, ____________ is the slope of its graph and ____________ is a point on the graph.

Martin-Gay Interactive Videos

See Video 3.3

Watch the section lecture video and answer the following questions.

OBJECTIVE 1

3. We can use the slope-intercept form to graph a line. Complete these statements based on Example 1. Start by graphing the _______. Find another point by applying the slope to this point—rewrite the slope as a _______ if necessary.

OBJECTIVE 2

4. In Example 3, what is the y-intercept?

OBJECTIVE 3

5. In Example 4 we use the point-slope form to find the equation of a line given the slope and a point. How do we then write this equation in standard form?

OBJECTIVE 4

6. The lecture before Example 5 discusses how to find the equation of a line given two points. Is there any circumstance when you might want to use the slope-intercept form to find the equation of a line given two points? If so, when?

OBJECTIVE 5

7. Solve Examples 6 and 7 again, this time using the point $(-2, -3)$ in each exercise.

OBJECTIVE 6

8. From Example 8, we are told to use ordered pairs of the form (time, speed). Explain why it is important to keep track of how you define your ordered pairs and/or define your variables.

3.3 Exercise Set MyMathLab®

Use the slope-intercept form to graph each equation. See Examples 1 and 2.

1. $y = 2x + 1$
2. $y = -4x - 1$
3. $y = \frac{2}{3}x + 5$
4. $y = \frac{1}{4}x - 3$
5. $y = -5x$
6. $y = -6x$
7. $4x + y = 6$
8. $-3x + y = 2$
9. $4x - 7y = -14$
10. $3x - 4y = 4$
11. $x = \frac{5}{4}y$
12. $x = \frac{3}{2}y$

Write an equation of the line with each given slope, m, and y-intercept, (0, b). See Example 3.

13. $m = 5, b = 3$
14. $m = -3, b = -3$
15. $m = -4, b = -\frac{1}{6}$
16. $m = 2, b = \frac{3}{4}$
17. $m = \frac{2}{3}, b = 0$
18. $m = -\frac{4}{5}, b = 0$
19. $m = 0, b = -8$
20. $m = 0, b = -2$
21. $m = -\frac{1}{5}, b = \frac{1}{9}$
22. $m = \frac{1}{2}, b = -\frac{1}{3}$

Find an equation of each line with the given slope that passes through the given point. Write the equation in slope-intercept form, $y = mx + b$. See Example 4.

23. $m = 6;\ (2, 2)$
24. $m = 4;\ (1, 3)$
25. $m = -8;\ (-1, -5)$
26. $m = -2;\ (-11, -12)$
27. $m = \frac{3}{2};\ (5, -6)$
28. $m = \frac{2}{3};\ (-8, 9)$
29. $m = -\frac{1}{2};\ (-3, 0)$
30. $m = -\frac{1}{5};\ (4, 0)$

Find an equation of the line passing through each pair of points. Write the equation in slope-intercept form, $y = mx + b$. See Example 5.

31. $(3, 2)$ and $(5, 6)$
32. $(6, 2)$ and $(8, 8)$
33. $(-1, 3)$ and $(-2, -5)$
34. $(-4, 0)$ and $(6, -1)$
35. $(2, 3)$ and $(-1, -1)$
36. $(7, 10)$ and $(-1, -1)$
37. $(0, 0)$ and $\left(-\frac{1}{8}, \frac{1}{13}\right)$
38. $(0, 0)$ and $\left(-\frac{1}{2}, \frac{1}{3}\right)$

Find an equation of each line. See Example 6.

39. Vertical line through $(0, 2)$
40. Horizontal line through $(1, 4)$
41. Horizontal line through $(-1, 3)$
42. Vertical line through $(-1, 3)$
43. Vertical line through $\left(-\frac{7}{3}, -\frac{2}{5}\right)$
44. Horizontal line through $\left(\frac{2}{7}, 0\right)$

Find an equation of each line. See Example 7.

45. △ Parallel to $y = 5$, through $(1, 2)$
46. △ Perpendicular to $y = 5$, through $(1, 2)$
47. △ Perpendicular to $x = -3$, through $(-2, 5)$
48. △ Parallel to $y = -4$, through $(0, -3)$
49. △ Parallel to $x = 0$, through $(6, -8)$
50. △ Perpendicular to $x = 7$, through $(-5, 0)$

MIXED PRACTICE

See Examples 3 through 7. Find an equation of each line described. Write each equation in slope-intercept form (solved for y), when possible.

51. With slope $-\frac{1}{2}$, through $\left(0, \frac{5}{3}\right)$

52. With slope $\frac{5}{7}$, through $(0, -3)$

53. Through $(10, 7)$ and $(7, 10)$

54. Through $(5, -6)$ and $(-6, 5)$

55. With undefined slope, through $\left(-\frac{3}{4}, 1\right)$

56. With slope 0, through $(6.7, 12.1)$

57. Slope 1, through $(-7, 9)$

58. Slope 5, through $(6, -8)$

59. Slope -5, y-intercept $(0, 7)$

60. Slope -2; y-intercept $(0, -4)$

61. Through $(6, 7)$, parallel to the x-axis

62. Through $(1, -5)$, parallel to the y-axis

63. Through $(2, 3)$ and $(0, 0)$

64. Through $(4, 7)$ and $(0, 0)$

65. Through $(-2, -3)$, perpendicular to the y-axis

66. Through $(0, 12)$, perpendicular to the x-axis

67. Slope $-\frac{4}{7}$, through $(-1, -2)$

68. Slope $-\frac{3}{5}$, through $(4, 4)$

Solve. Assume each exercise describes a linear relationship. Write the equations in slope-intercept form. See Example 8.

69. A rock is dropped from the top of a 400-foot cliff. After 1 second, the rock is traveling 32 feet per second. After 3 seconds, the rock is traveling 96 feet per second.

a. Assume that the relationship between time and speed is linear and write an equation describing this relationship. Use ordered pairs of the form (time, speed).

b. Use this equation to determine the speed of the rock 4 seconds after it was dropped.

70. A Hawaiian fruit company is studying the sales of a pineapple sauce to see if this product is to be continued. At the end of its first year, profits on this product amounted to \$30,000. At the end of the fourth year, profits were \$66,000.

a. Assume that the relationship between years on the market and profit is linear and write an equation describing this relationship. Use ordered pairs of the form (years on the market, profit).

b. Use this equation to predict the profit at the end of 7 years.

71. Sales of automobiles worldwide continues to grow as emerging economies provide their citizens more purchasing power. In 2011, there were 66 thousand new car sales worldwide, and in 2014, this number increased to 72 thousand cars. (*Source:* statista)

a. Write an equation describing the relationship between time and the number of new automobiles sold worldwide. Use ordered pairs of the form (years past 2011, number of automobiles sold).

b. Use this equation to predict the number of new car sales worldwide in the year 2018.

72. In 2004, the restaurant industry in the United States employed about 11.9 million people. In 2014, this number increased to 13.5 million employees. (*Source:* National Restaurant Association)

a. Write an equation describing the relationship between time and the number of restaurant industry employees. Use ordered pairs of the form (years past 2004, number of restaurant employees in millions).

b. Use this equation to predict the number of restaurant industry employees in 2018. Round to the nearest tenth of a million.

73. In 2006, the U.S. population (or persons) per square mile of land area was 85. In 2014, the persons per square mile was 89.48.

a. Write an equation describing the relationship between year and persons per square mile. Use ordered pairs of the form (years past 2006, persons per square mile).

b. Use this equation to predict the persons per square mile in 2018. Round to the nearest tenth of a person.

74. Dunkin' Donuts started as a New England brand coffee and donut shop, but has expanded recently throughout the world. In 2011, there were 10,083 stores worldwide. By 2014, this had grown to 10,884. (*Source:* Dunkin' Donuts)

a. Write an equation describing the relationship between year and number of Dunkin' Donuts stores. Use ordered pairs of the form (years past 2011, numbers of Dunkin' Donuts stores).

b. Use this equation to predict the number of Dunkin' Donuts stores in 2018.

75. It has been said that newspapers are disappearing, replaced by various electronic media. In 2003, newspaper circulation (the number of copies distributed in a day) was about 55 million. In 2013, this had dropped to 44 million. (*Source:* Newspaper Association of America).

a. Write an equation describing the relationship between time and circulation. Use ordered pairs of the form (years past 2003, number of newspapers circulated in millions).

b. Use this equation to predict newspaper circulation in 2018.

76. A certain chain of book stores is slowly closing down stores. Suppose that in 2006 there were 3991 stores and in 2010 there were 3200 stores.

a. Write an equation describing the relationship between time and number of store locations. Use ordered pairs of the form (years past 2006, number of stores).

b. Use this equation to predict the number of stores in 2018.

77. The Pool Fun Company has learned that, by pricing a newly released Fun Noodle at \$3, sales will reach 10,000 Fun Noodles per day during the summer. Raising the price to \$5 will cause sales to fall to 8000 Fun Noodles per day.

a. Assume that the relationship between price and number of Fun Noodles sold is linear and write an equation describing this relationship. Use ordered pairs of the form (price, number sold).

b. Predict the daily sales of Fun Noodles if the price is \$3.50.

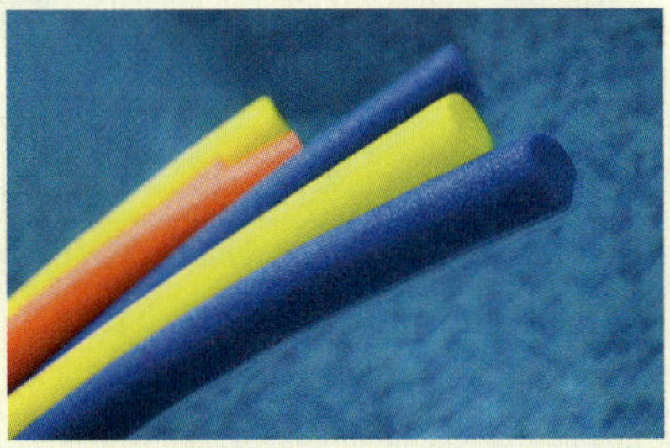

78. The value of a building bought in 2000 may be depreciated (or decreased) as time passes for income tax purposes. Seven years after the building was bought, this value was \$225,000 and 12 years after it was bought, this value was \$195,000.

a. If the relationship between number of years past 2000 and the depreciated value of the building is linear, write an equation describing this relationship. Use ordered pairs of the form (years past 2000, value of building).

b. Use this equation to estimate the depreciated value of the building in 2018.

REVIEW AND PREVIEW

This histogram shows the population of moviegoers in the United States and Canada by age. Use this graph to answer Exercises 79–86. See Section R.4.

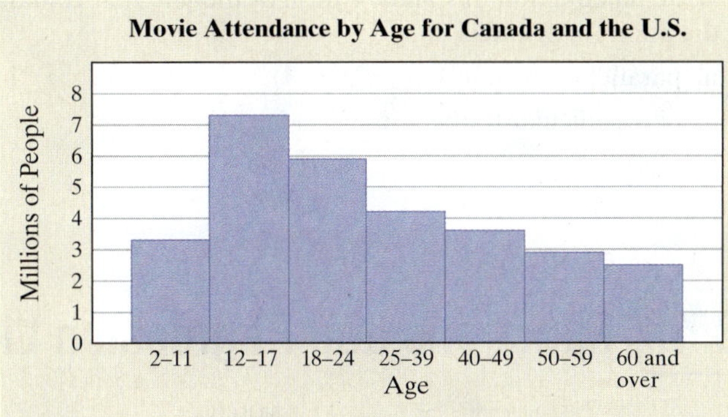

Source: Motion Picture Association of America

79. What age group has the highest moviegoer population?

80. What age group has the lowest moviegoer population?

81. Estimate the number of moviegoers age 18–24.

82. Estimate the number of moviegoers age 60 and over.

83. Estimate the number of moviegoers age 50 and over.

84. Estimate the number of moviegoers age 12–24.

85. Are the class intervals in this histogram the same?

86. Which class interval contains the smallest age group?

CONCEPT EXTENSIONS

For Exercises 87 through 90, identify the form that the linear equation in two variables is written in. For Exercises 91 and 92, identify the appearance of the graph of the equation.

87. $y - 7 = 4(x + 3)$; ______________ form

88. $5x - 9y = 11$; ______________ form

89. $y = \frac{3}{4}x - \frac{1}{3}$; ______________ form

90. $y + 2 = -\frac{1}{3}(x - 2)$; ______________ form

91. $y = \frac{1}{2}$; ______________ line

92. $x = -17$; ______________ line

93. Given the equation of a nonvertical line, explain how to find the slope without finding two points on the line.

94. Given two points on a nonvertical line, explain how to use the point-slope form to find an equation of the line.

△ **95.** Write an equation in standard form of the line that contains the point $(-1, 2)$ and is

a. parallel to the line $y = 3x - 1$.

b. perpendicular to the line $y = 3x - 1$.

△ **96.** Write an equation in standard form of the line that contains the point $(4, 0)$ and is

a. parallel to the line $y = -2x + 3$.

b. perpendicular to the line $y = -2x + 3$.

△ **97.** Write an equation in standard form of the line that contains the point $(3, -5)$ and is

a. parallel to the line $3x + 2y = 7$.

b. perpendicular to the line $3x + 2y = 7$.

△ **98.** Write an equation in standard form of the line that contains the point $(-2, 4)$ and is

a. parallel to the line $x + 3y = 6$.

b. perpendicular to the line $x + 3y = 6$.

3.4 Functions and Regression Lines

OBJECTIVES

1 Identify Relations, Domains, and Ranges.

2 Identify Functions.

3 Use the Vertical Line Test.

4 Use Function Notation.

5 Use Linear Regression.

OBJECTIVE 1 Identifying Relations, Domains, and Ranges

In previous sections, we have discussed the relationships between two quantities. For example, the relationship between the length of the side of a square x and its area y is described by the equation $y = x^2$. Ordered pairs can be used to write down solutions of this equation. For example, (2, 4) is a solution of $y = x^2$, and this notation tells us that the x-value 2 is related to the y-value 4 for this equation. In other words, when the length of the side of a square is 2 units, its area is 4 square units.

Examples of Relationships Between Two Quantities

Area of Square: $y = x^2$

Some Ordered Pairs

x	y
2	4
5	25
7	49
12	144

Equation of Line: $y = x + 2$

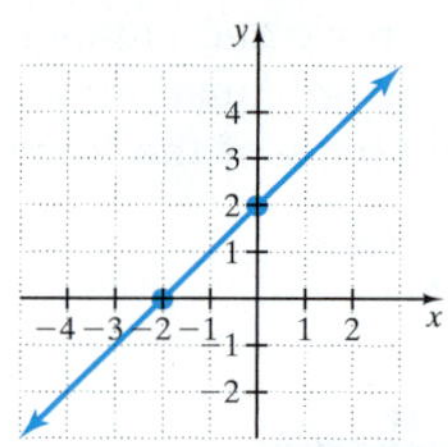

Some Ordered Pairs

x	y
−3	−1
0	2
2	4
9	11

Internet Advertising Revenue

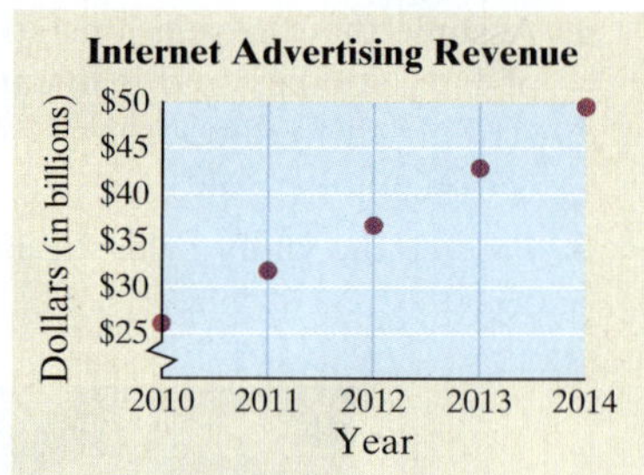

Ordered Pairs

Year	*Billions of Dollars*
2010	26
2011	31.7
2012	36.6
2013	42.8
2014	49.5

(*Source:* PricewaterhouseCoopers LLP)

A set of ordered pairs is called a **relation.** The set of all x-coordinates is called the **domain** of a relation, and the set of all y-coordinates is called the **range** of a relation. Equations such as $y = x^2$ are also called relations since equations in two variables define a set of ordered pair solutions.

EXAMPLE 1 Find the domain and the range of the relation $\{(0, 2), (3, 3), (-1, 0), (3, -2)\}$.

Solution The domain is the set of all x-values or $\{-1, 0, 3\}$, and the range is the set of all y-values, or $\{-2, 0, 2, 3\}$. □

PRACTICE 1 Find the domain and the range of the relation $\{(1, 3), (5, 0), (0, -2), (5, 4)\}$. ■

OBJECTIVE 2

Identifying Functions

Some relations are also functions.

> **Function**
>
> A function is a set of ordered pairs that assigns to each x-value exactly one y-value.

In other words, a function cannot have two ordered pairs with the same x-coordinate but different y-coordinates.

EXAMPLE 2 Determine whether each relation is also a function.

a. $\{(-1, 1), (2, 3), (7, 3), (8, 6)\}$ **b.** $\{(0, -2), (1, 5), (0, 3), (7, 7)\}$

Solution

a. Although the ordered pairs (2, 3) and (7, 3) have the same y-value, each x-value is assigned to only one y-value, so this set of ordered pairs is a function.

b. The x-value 0 is assigned to two y-values, -2 and 3, so this set of ordered pairs is not a function. □

PRACTICE 2 Determine whether each relation is also a function.

a. $\{(4, 1), (3, -2), (8, 5), (-5, 3)\}$ **b.** $\{(1, 2), (-4, 3), (0, 8), (1, 4)\}$ ■

Relations and functions can be described by a graph of their ordered pairs.

EXAMPLE 3 Determine whether each graph is the graph of a function.

a. **b.**

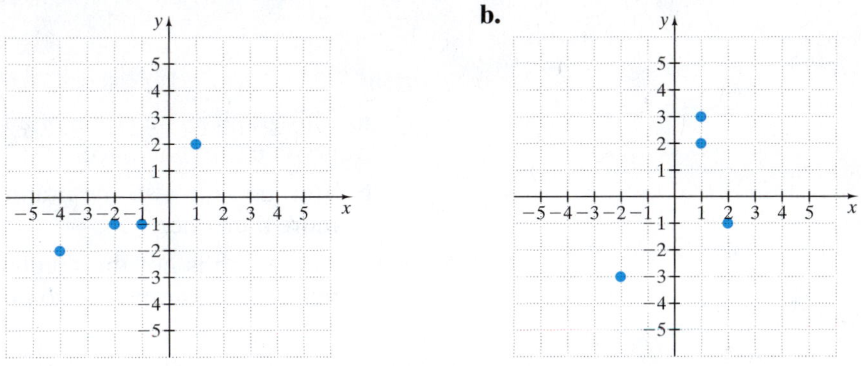

Solution

a. This is the graph of the relation $\{(-4, -2), (-2, -1), (-1, -1), (1, 2)\}$. Each x-coordinate has exactly one y-coordinate, so this is the graph of a function.

b. This is the graph of the relation $\{(-2, -3), (1, 2), (1, 3), (2, -1)\}$. The x-coordinate 1 is paired with two y-coordinates, 2 and 3, so this is not the graph of a function. □

(Continued on next page)

PRACTICE

3 Determine whether each graph is the graph of a function.

a.

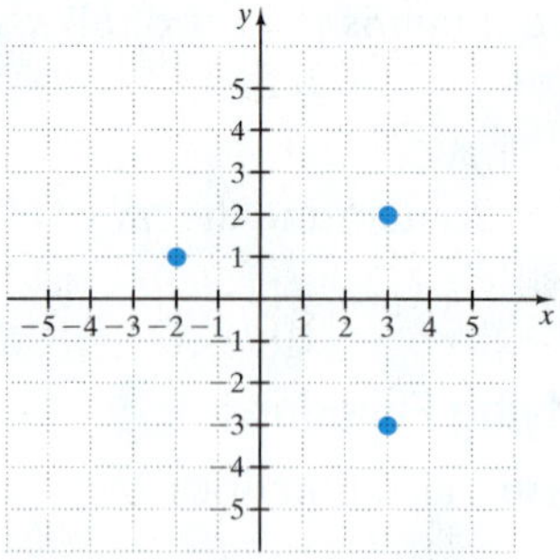

b.

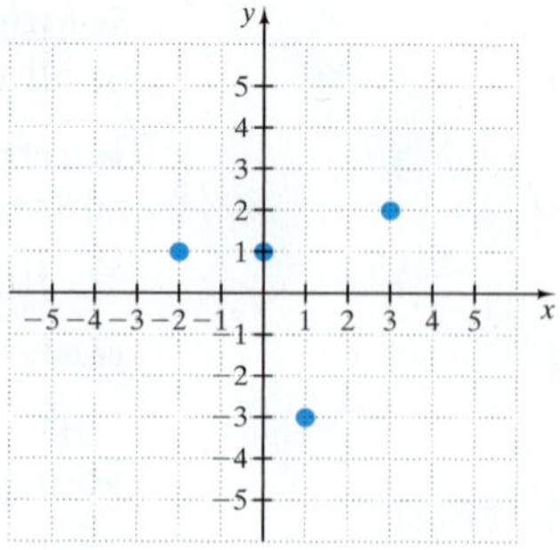

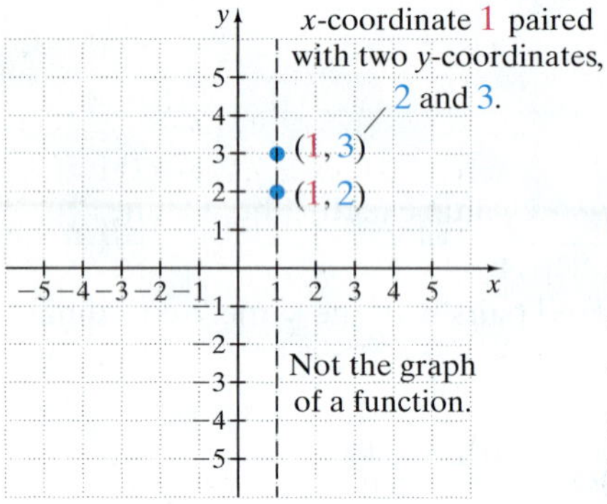

OBJECTIVE

3 Using the Vertical Line Test

The graph in Example 3(b) was not the graph of a function because the x-coordinate 1 was paired with two y-coordinates, 2 and 3. Notice that when an x-coordinate is paired with more than one y-coordinate, a vertical line can be drawn that will intersect the graph at more than one point. We can use this fact to determine whether a relation is also a function. We call this the **vertical line test.**

Vertical Line Test

If a vertical line can be drawn so that it intersects a graph more than once, the graph is not the graph of a function.

This vertical line test works for all types of graphs on the rectangular coordinate system.

EXAMPLE 4 Use the vertical line test to determine whether each graph is the graph of a function.

a.

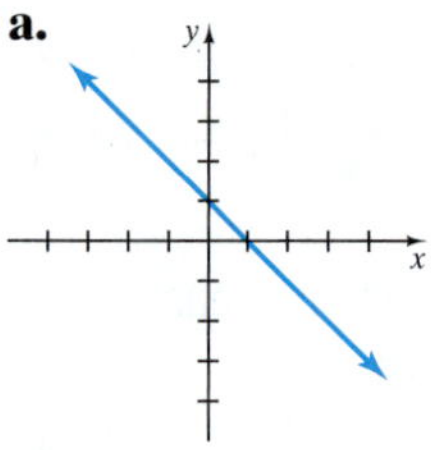

b.

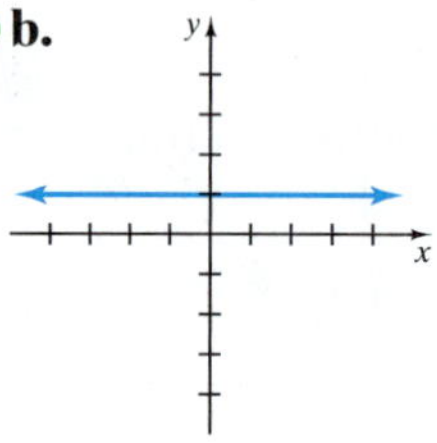

c.

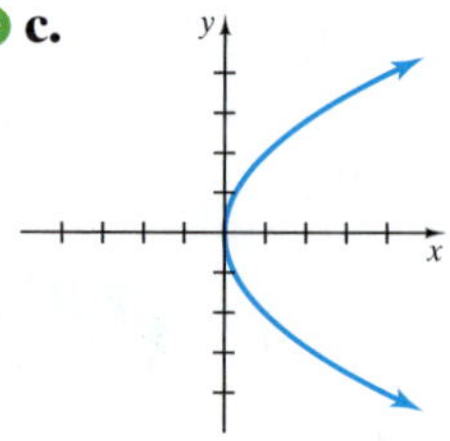

d.

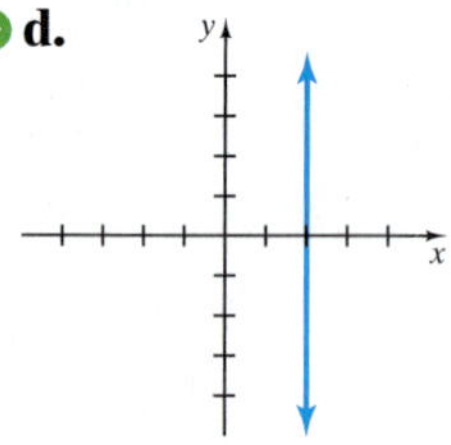

Solution

a. This graph is the graph of a function since no vertical line will intersect this graph more than once.

b. This graph is also the graph of a function; no vertical line will intersect it more than once.

c. This graph is not the graph of a function. Vertical lines can be drawn that intersect the graph in two points. An example of one is shown.

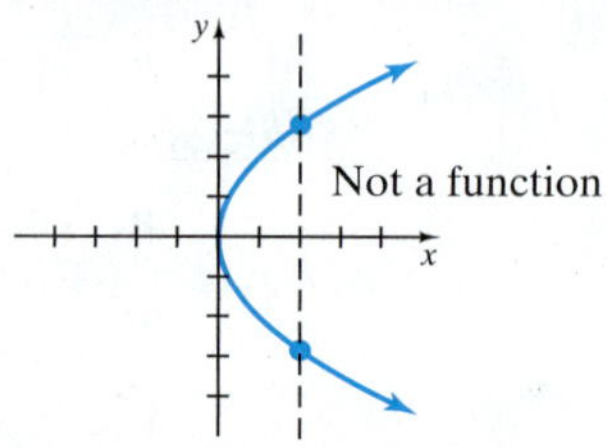

d. This graph is not the graph of a function. A vertical line can be drawn that intersects this line at every point.

PRACTICE 4 Use the vertical line test to determine whether each graph is the graph of a function.

a.

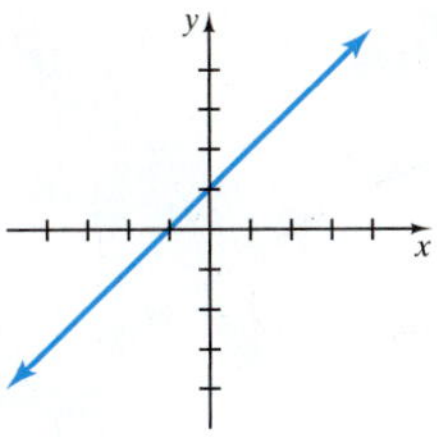

b.

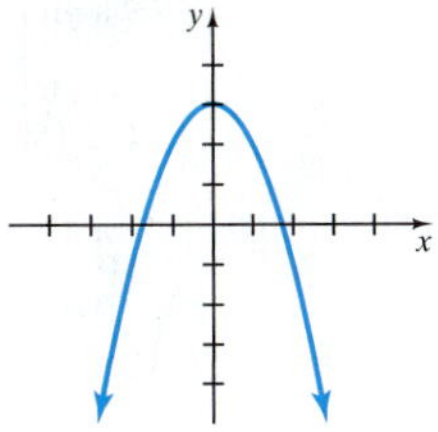

c.

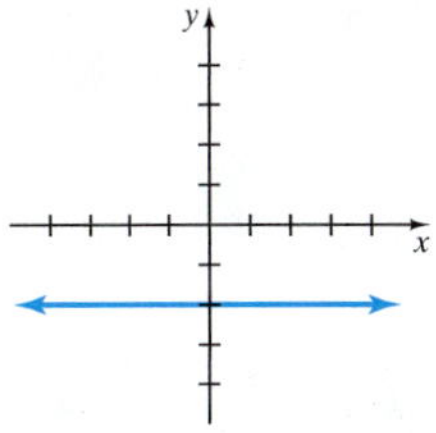

d.

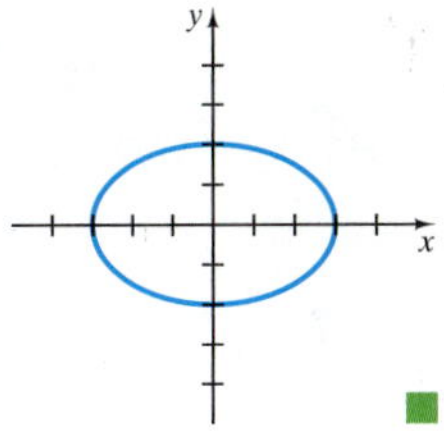

■

Recall that the graph of a linear equation is a line, and a line that is not vertical will pass the vertical line test. **Thus, all linear equations are functions except those of the form $x = c$, which are vertical lines.**

EXAMPLE 5 Decide whether the equation describes a function.

a. $y = x$ **b.** $y = 2x + 1$ **c.** $y = 5$ **d.** $x = -1$

Solution **a, b,** and **c** are functions because their graphs are nonvertical lines. **d** is not a function because its graph is a vertical line. □

PRACTICE 5 Decide whether the equation describes a function.

a. $y = 2x$ **b.** $y = -3x - 1$ **c.** $y = 8$ **d.** $x = 2$ ■

Examples of functions can often be found in magazines, newspapers, books, and other printed material in the form of tables or graphs such as that in Example 6.

EXAMPLE 6 The graph shows the sunrise time for Indianapolis, Indiana, for the year. Use this graph to answer the questions.

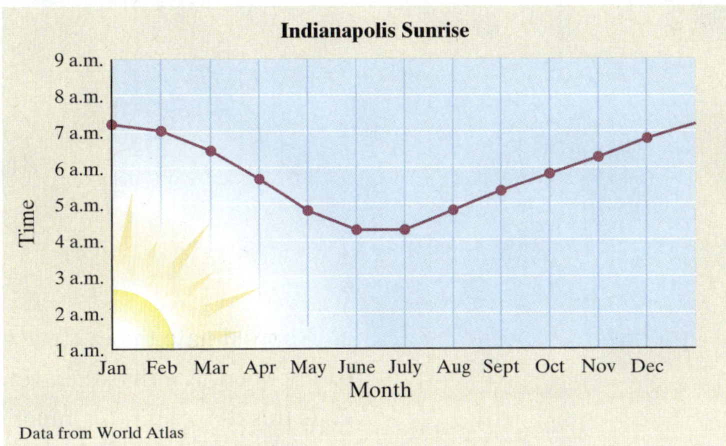

a. Approximate the time of sunrise on February 1.

b. Approximately when does the sun rise at 5 a.m.?

c. Is this the graph of a function?

Solution

a. To approximate the time of sunrise on February 1, we find the mark on the horizontal axis that corresponds to February 1. From this mark, we move vertically upward until the graph is reached. From that point on the graph,

(Continued on next page)

we move horizontally to the left until the vertical axis is reached. The vertical axis there reads 7 a.m.

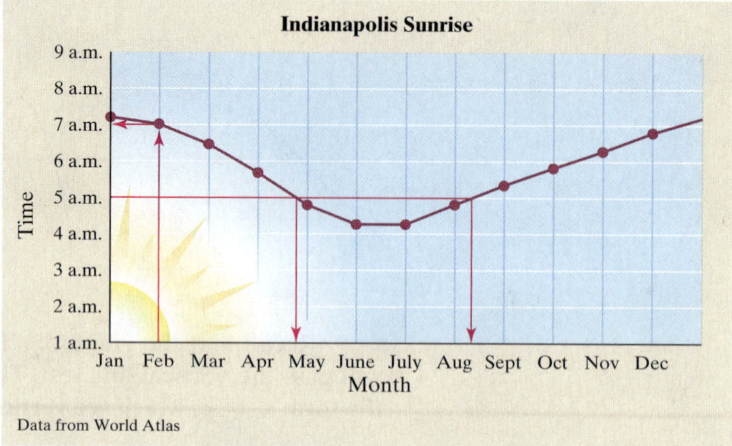

b. To approximate when the sun rises at 5 a.m., we find 5 a.m. on the time axis and move horizontally to the right. Notice that we will reach the graph twice, corresponding to two dates for which the sun rises at 5 a.m. We follow both points on the graph vertically downward until the horizontal axis is reached. The sun rises at 5 a.m. at approximately the end of the month of April and near the middle of the month of August.

c. The graph is the graph of a function since it passes the vertical line test. In other words, for every day of the year in Indianapolis, there is exactly one sunrise time. □

PRACTICE

6 The graph shows the average monthly temperature for Chicago, Illinois, for the year. Use this graph to answer the questions.

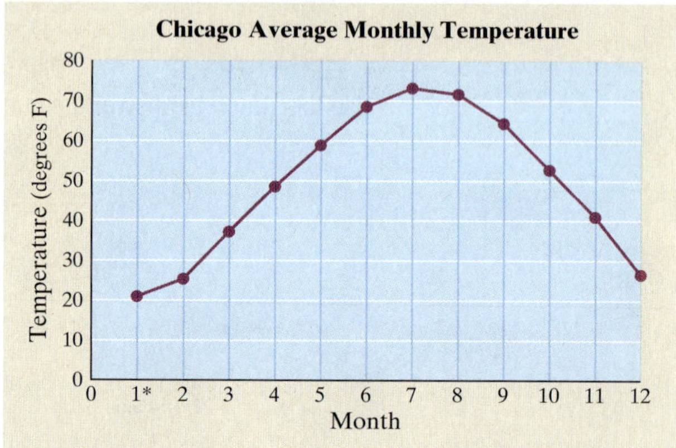

* (1 is Jan.; 12 is Dec.)

a. Approximate the average monthly temperature for June.

b. For what month is the average monthly temperature 25°?

c. Is this the graph of a function? ■

OBJECTIVE

4 Using Function Notation

The graph of the linear equation $y = 2x + 1$ passes the vertical line test, so we say that $y = 2x + 1$ is a function. In other words, $y = 2x + 1$ gives us a rule for writing ordered pairs where every x-coordinate is paired with one and only one y-coordinate.

The variable y is a function of the variable x. For each value of x, there is only one value of y. Thus, we say the variable x is the **independent variable** because any value in the domain can be assigned to x. The variable y is the **dependent variable** because its value depends on x.

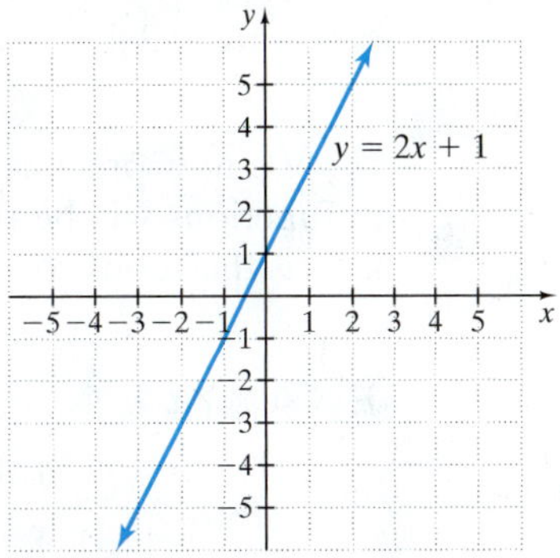

We often use letters such as f, g, and h to name functions. For example, the symbol $f(x)$ means *function of* x and is read "f of x." This notation is called **function notation.** The equation $y = 2x + 1$ can be written as $f(x) = 2x + 1$ using function notation, and these equations mean the same thing. In other words, $y = f(x)$.

The notation $f(1)$ means to replace x with 1 and find the resulting y or function value. Since

$$f(x) = 2x + 1$$

then

$$\begin{aligned} f(1) &= 2(1) + 1 \\ &= 3 \end{aligned}$$

This means that, when $x = 1$, y or $f(x) = 3$, and we have the ordered pair $(1, 3)$. Now let's find $f(2)$, $f(0)$, and $f(-1)$.

$$\begin{aligned} f(x) &= 2x + 1 \\ f(2) &= 2(2) + 1 \\ &= 4 + 1 \\ &= 5 \end{aligned} \qquad \begin{aligned} f(x) &= 2x + 1 \\ f(0) &= 2(0) + 1 \\ &= 0 + 1 \\ &= 1 \end{aligned} \qquad \begin{aligned} f(x) &= 2x + 1 \\ f(-1) &= 2(-1) + 1 \\ &= -2 + 1 \\ &= -1 \end{aligned}$$

Ordered Pair: $(2, 5)$ $\qquad (0, 1)$ $\qquad (-1, -1)$

Helpful Hint

Note that, for example, if $f(2) = 5$, the corresponding ordered pair is $(2, 5)$.

Helpful Hint

Note that $f(x)$ is a special symbol in mathematics used to denote a function. The symbol $f(x)$ is read "f of x." It does **not** mean $f \cdot x$ (f times x).

EXAMPLE 7 Given $g(x) = x^2 - 3$, find the following. Then write the corresponding ordered pairs generated.

a. $g(2)$ **b.** $g(-2)$ **c.** $g(0)$

Solution

a.
$$\begin{aligned} g(x) &= x^2 - 3 \\ g(2) &= 2^2 - 3 \\ &= 4 - 3 \\ &= 1 \end{aligned}$$

b.
$$\begin{aligned} g(x) &= x^2 - 3 \\ g(-2) &= (-2)^2 - 3 \\ &= 4 - 3 \\ &= 1 \end{aligned}$$

c.
$$\begin{aligned} g(x) &= x^2 - 3 \\ g(0) &= 0^2 - 3 \\ &= 0 - 3 \\ &= -3 \end{aligned}$$

Ordered Pairs:	$g(2) = 1$ gives $(2, 1)$	$g(-2) = 1$ gives $(-2, 1)$	$g(0) = -3$ gives $(0, -3)$

□

(Continued on next page)

PRACTICE 7 Given $h(x) = x^2 + 5$, find the following. Then write the corresponding ordered pairs generated.

a. $h(2)$ **b.** $h(-5)$ **c.** $h(0)$

We now practice finding the domain and the range of a function. The domains of our functions will be the set of all possible real numbers that x can be replaced by. The range is the set of corresponding y-values.

EXAMPLE 8 Find the domain of each function.

a. $g(x) = \frac{1}{x}$ **b.** $f(x) = 2x + 1$

Solution

a. Recall that we cannot divide by 0, so the domain of $g(x)$ is the set of all real numbers except 0. In interval notation, we can write $(-\infty, 0) \cup (0, \infty)$.

b. In this function, x can be any real number. The domain of $f(x)$ is the set of all real numbers, or $(-\infty, \infty)$ in interval notation.

PRACTICE 8 Find the domain of each function.

a. $h(x) = 6x + 3$ **b.** $f(x) = \frac{1}{x^2}$

CONCEPT CHECK
Suppose that the value of f is -7 when the function is evaluated at 2. Write this situation in function notation.

EXAMPLE 9 Find the domain and the range of each function graphed. Use interval notation.

a.

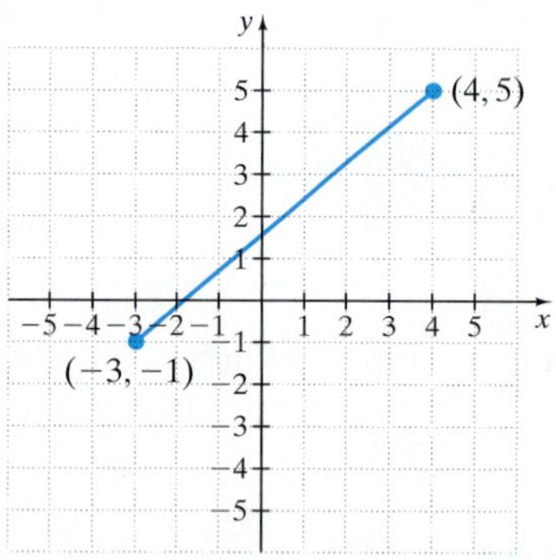

b.

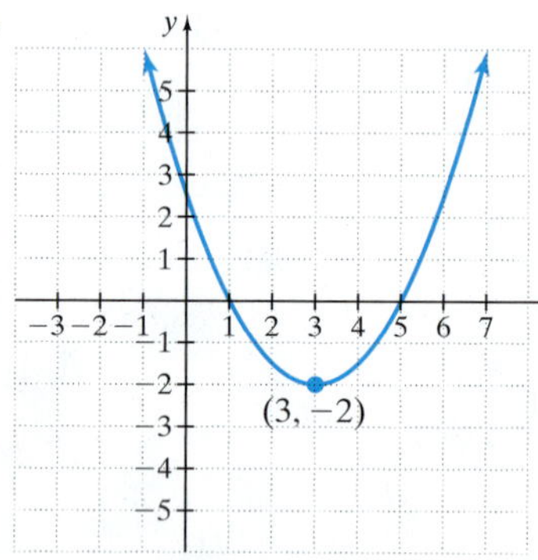

Solution

a.

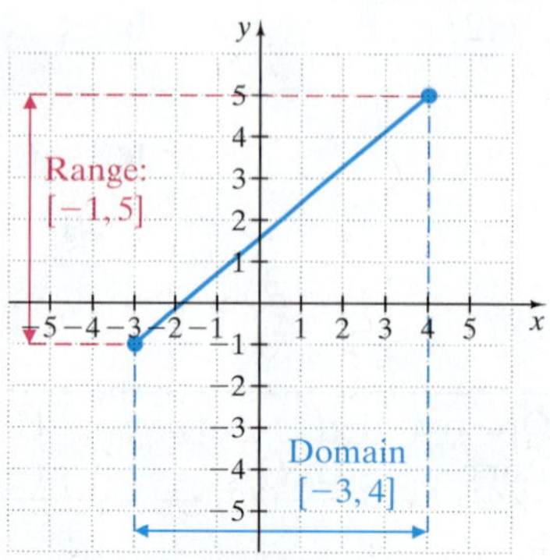

b.

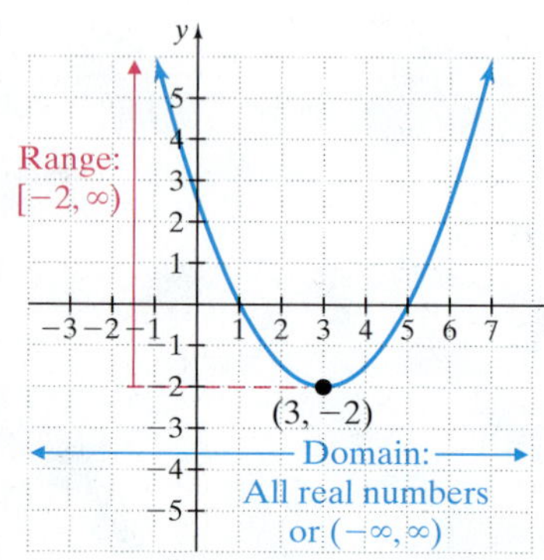

Answer to Concept Check: $f(2) = -7$

PRACTICE 9 Find the domain and the range of each function graphed. Use interval notation.

a.

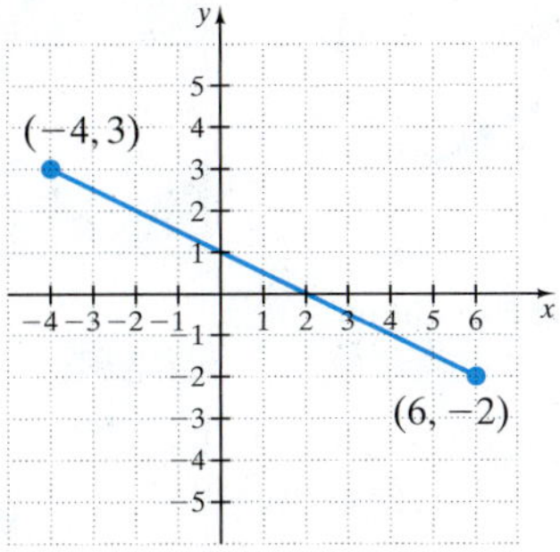

b.

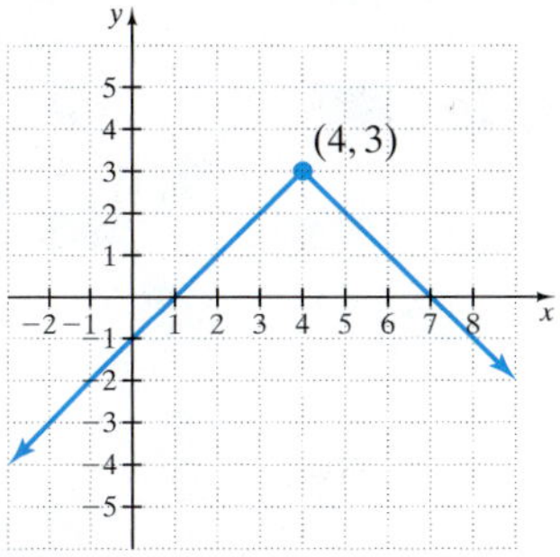

OBJECTIVE 5

Using Regression Lines

Let's suppose a company has been experiencing a steady increase in profit for the past five years, as shown below in the company's scatter plot. A scatter plot such as the one below is a visual way to notice if there is a relationship or correlation between two variables in the plot—in this case, time and profit.

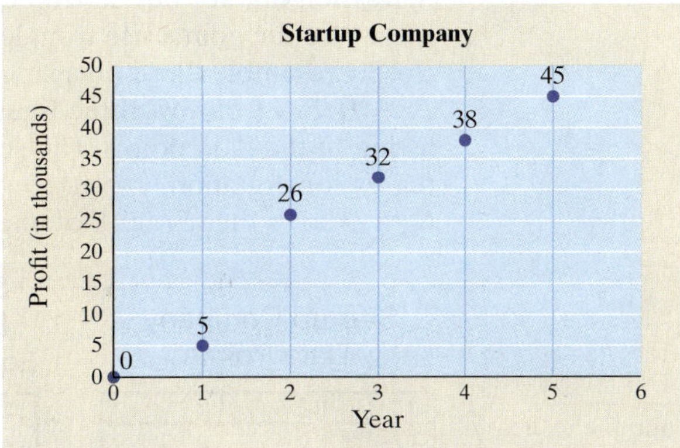

Notice that if the points in the scatter plot are connected, the result would resemble a line. In order to plan for the future, let's model the scatter plot of the company's profits with a line that "best fits" their data points. We call this "line of best fit" a **linear regression.** Simply stated, a regression line is a line that best resembles the data points.

> **Helpful Hint**
>
> There are other models besides linear, such as quadratic, cubic, and exponential. In this chapter, we will concentrate on regression lines, or linear models, only.

It is helpful to know how well a line approximates or fits the data. We measure this relationship by using a number that we call a correlation coefficient, which we denote by r. This correlation coefficient, r, is a number between and including -1 and 1. On the next page we show examples of different values of r, the correlation coefficient. If $r = 1$, we have a **perfect positive correlation.** This means that all the points lie on the regression line and this line rises from left to right. If $r = -1$, we have a **perfect negative correlation.** This means that all the data points lie on the regression line and this line falls from left to right.

The easiest way to obtain the equation of a regression line and the corresponding correlation coefficient r is by using a statistical program, such as one found in a graphing calculator. These calculators usually contain a least squares program that performs the calculations for us.

Examples of Correlation Coefficients, r

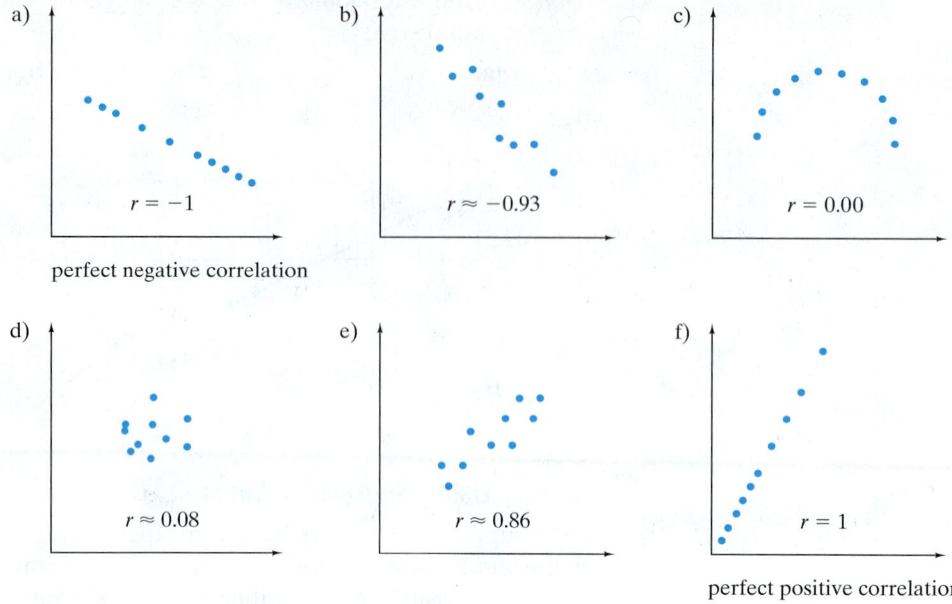

After studying these examples of correlation coefficients, we can surmise that the regression line for our startup company should have a positive correlation coefficient, r, because the points rise from left to right and r should be close to 1 because the points closely resemble those on a single straight line.

Below we show three figures from a graphing calculator. The first figure shows a listing of the data points. The second figure shows the values for a and b in the linear regression equation, $y = ax + b$, as well as the r-value of approximately 0.97. The last figure shows the scatter plot and the regression line for our startup company.

Helpful Hint

Using $y = ax + b$ and the values of a and b in the middle figure, the equation of the linear regression line is approximately

$$y = 9.43x + 0.76$$

Startup Company: Data Points

Startup Company: Linear Regression Equation and $r \approx 0.97$

Startup Company: Scatter Plot and Regression Line

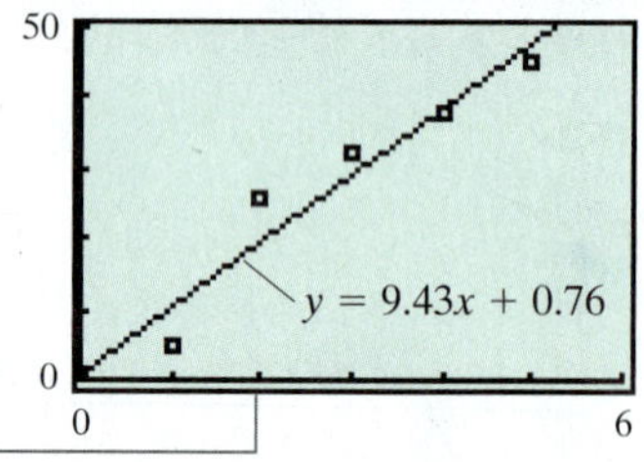

The graph below is the Chapter 3 Opener shown as a scatter plot. The data point heights are labeled, as well as the approximate regression line and the r-value calculated on a graphing calculator.

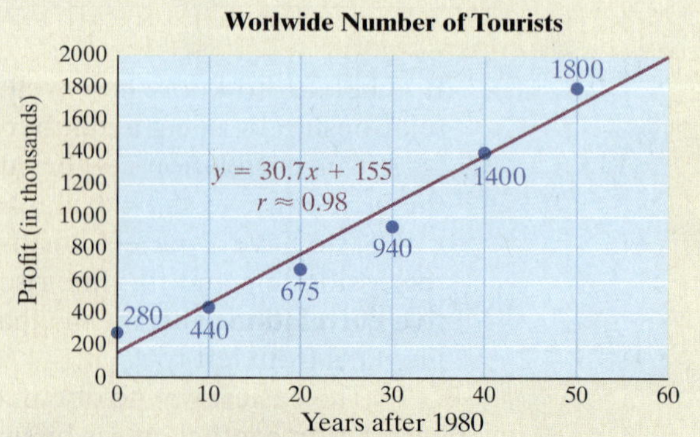

Note: "Years after 1980" means that we let $x = 0$ correspond to the year 1980. Thus, we use $x = 10$ for the year 1990, $x = 20$ for the year 2000, and so on.

Vocabulary, Readiness & Video Check

Use the choices below to fill in each blank. Some choices may not be used.

$x = c$	horizontal	domain	relation
$y = c$	vertical	range	function

1. A set of ordered pairs is called a(n) __________.

2. A set of ordered pairs that assigns to each x-value exactly one y-value is called a(n) __________.

3. The set of all y-coordinates of a relation is called the __________.

4. The set of all x-coordinates of a relation is called the __________.

5. All linear equations are functions except those whose graphs are __________ lines.

6. All linear equations are functions except those whose equations are of the form __________.

Martin-Gay Interactive Videos

See Video 3.4

Watch the section lecture video and answer the following questions.

OBJECTIVE 1

7. In the lecture before Example 1, relations are discussed. Why can an equation in two variables define a relation?

OBJECTIVE 2

8. Based on Examples 2 and 3, can a set of ordered pairs with no repeated x-values, but with repeated y-values be a function? For example: $\{(0, 4), (-3, 4), (2, 4)\}$.

OBJECTIVE 3

9. After reviewing Example 8, explain why the vertical line test works.

OBJECTIVE 4

10. In Example 10, three function values were found and their corresponding ordered pairs were written. For example, $f(0) = 2$ corresponds to $(0, 2)$. Write the other two function values found and their corresponding ordered pairs.

OBJECTIVE 5

11. In Example 14, how do we know that the correlation coefficient r is positive?

3.4 Exercise Set MyMathLab®

Find the domain and the range of each relation. See Example 1.

1. $\{(2, 4), (0, 0), (-7, 10), (10, -7)\}$

2. $\{(3, -6), (1, 4), (-2, -2)\}$

3. $\{(0, -2), (1, -2), (5, -2)\}$

4. $\{(5, 0), (5, -3), (5, 4), (5, 3)\}$

Determine whether each relation is also a function. See Example 2.

5. $\{(1, 1), (2, 2), (-3, -3), (0, 0)\}$

6. $\{(11, 6), (-1, -2), (0, 0), (3, -2)\}$

7. $\{(-1, 0), (-1, 6), (-1, 8)\}$

8. $\{(1, 2), (3, 2), (1, 4)\}$

MIXED PRACTICE

Determine whether each graph is the graph of a function. See Examples 3 and 4.

9.

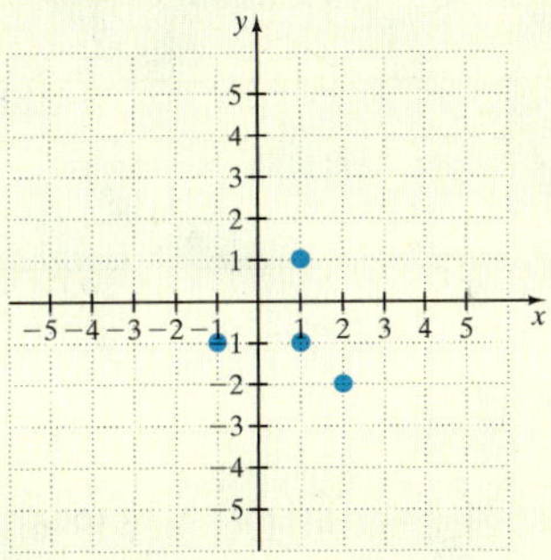

10.

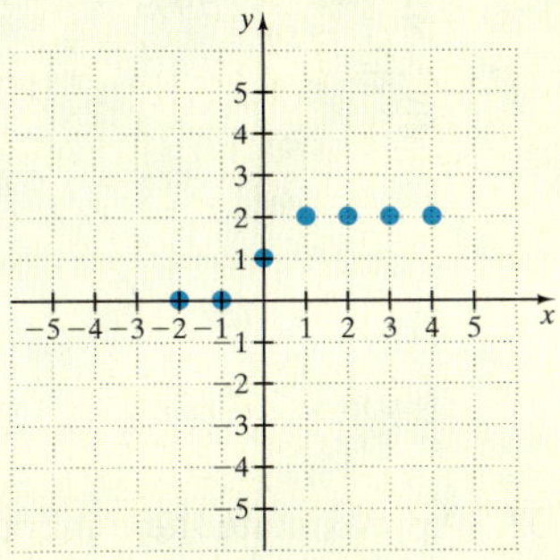

11.

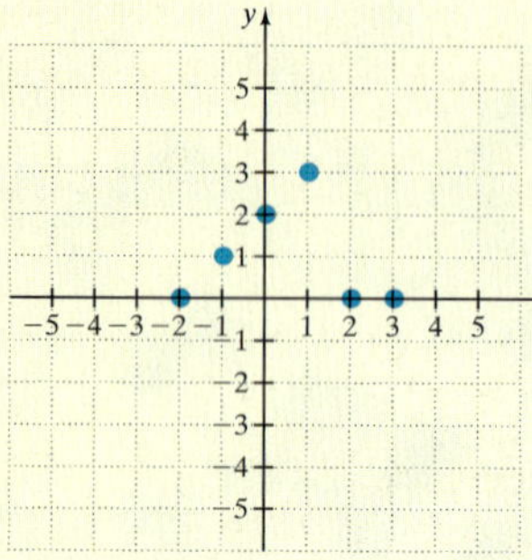

12.

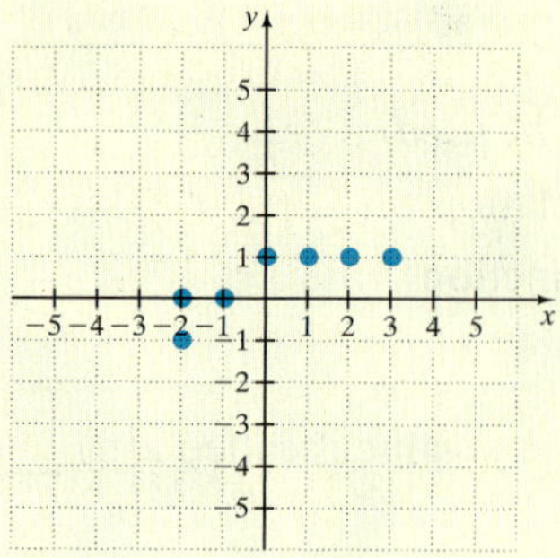

13.

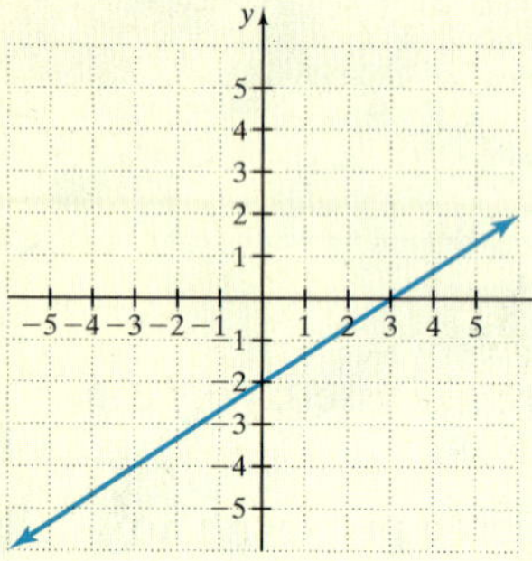

14.

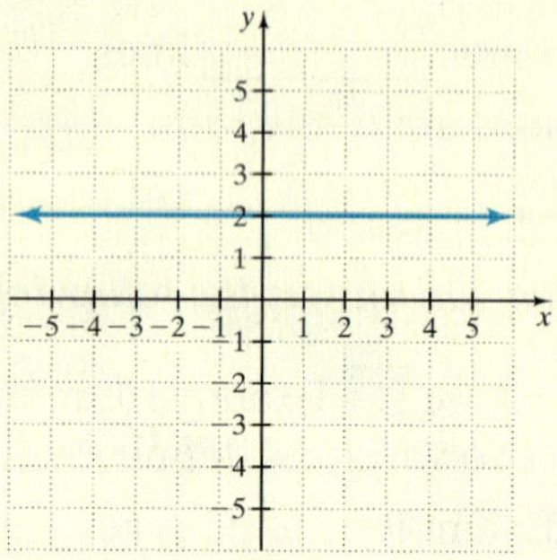

15.

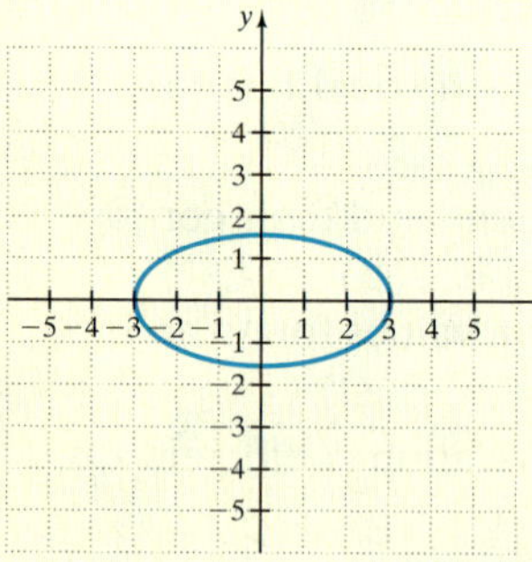

16.

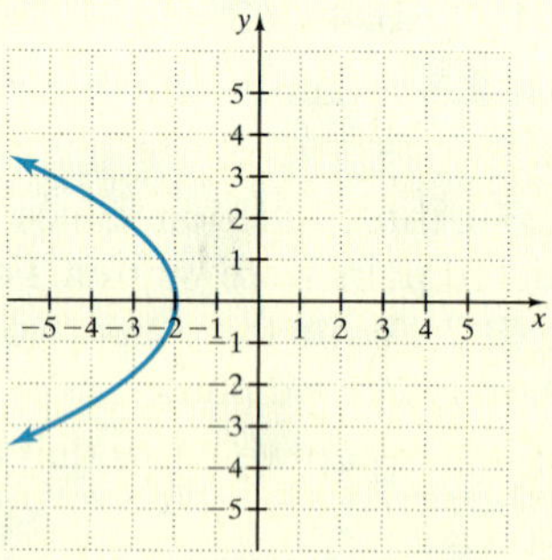

Decide whether the equation describes a function. See Example 5.

17. $y = x + 1$

18. $y = x - 1$

19. $y - x = 7$

20. $2x - 3y = 9$

21. $y = 6$

22. $x = 3$

23. $x = -2$

24. $y = -9$

25. $x = y^2$

26. $y = x^2 - 3$

The graph shows the sunset times for Seward, Alaska. Use this graph to answer Exercises 27 through 32. See Example 6.

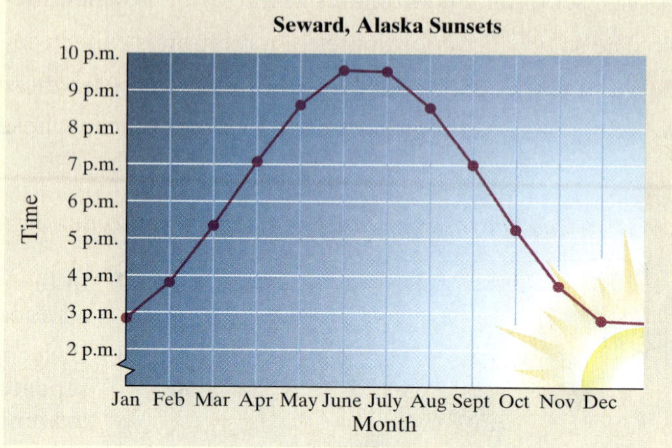

27. Approximate the time of sunset on June 1.

28. Approximate the time of sunset on November 1.

29. Approximate when the sunset is at 3 p.m.

30. Approximate when the sunset is at 9 p.m.

31. Is this graph the graph of a function? Why or why not?

32. Do you think a graph of sunset times for any location will always be a function? Why or why not?

This graph shows the U.S. hourly minimum wage for each year shown. Use this graph to answer Exercises 33 through 38. See Example 6.

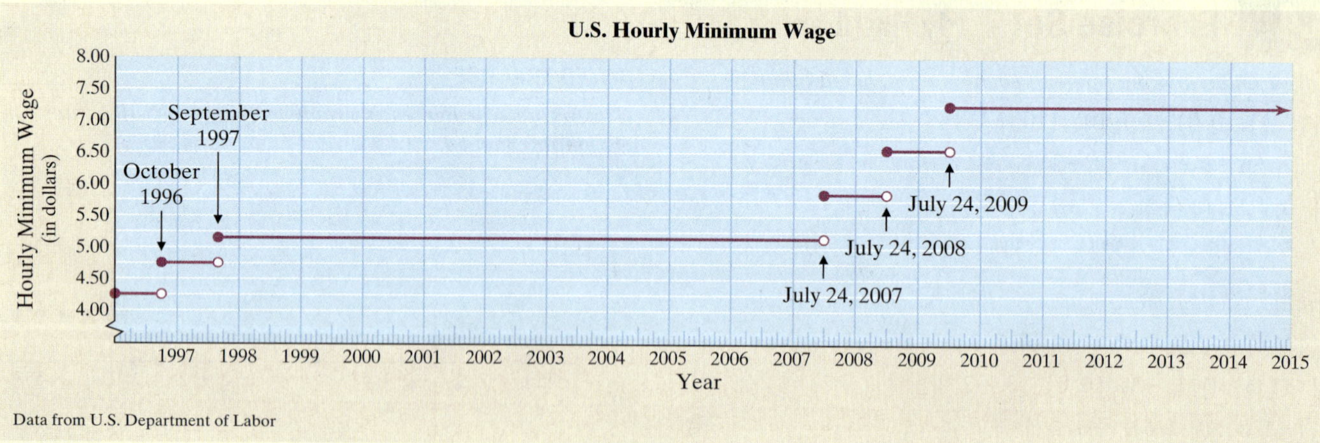

33. Approximate the minimum wage before October 1996.

34. Approximate the minimum wage in 2006.

35. Approximate the year when the minimum wage increased to over \$7.00 per hour.

36. According to the graph, what hourly wage was in effect for the greatest number of years?

37. Is this graph the graph of a function? Why or why not?

38. Do you think that a similar graph of your hourly wage on January 1 of every year (whether you are working or not) will be the graph of a function? Why or why not?

This graph shows the cost of mailing a large envelope through the U.S. Postal Service by weight at this time. Use this graph to answer Exercises 39 through 44.

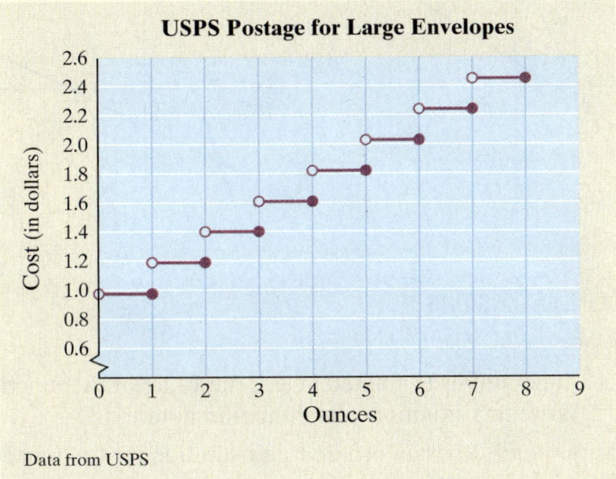

39. Approximate the postage to mail a large envelope weighing more than 4 ounces but not more than 5 ounces.

40. Approximate the postage to mail a large envelope weighing more than 7 ounces but not more than 8 ounces.

41. Give the weight of a large envelope that costs about $2.00 to mail.

42. If you have $1.50, what is the weight of the largest envelope you can mail for that money?

43. Is this graph a function? Why or why not?

44. Do you think that a similar graph of postage to mail a first-class letter will be the graph of a function? Why or why not?

Find $f(-2)$, $f(0)$, and $f(3)$ for each function. See Example 7.

45. $f(x) = 2x - 5$ **46.** $f(x) = 3 - 7x$

47. $f(x) = x^2 + 2$ **48.** $f(x) = x^2 - 4$

49. $f(x) = 3x$ **50.** $f(x) = -3x$

51. $f(x) = |x|$ **52.** $f(x) = |2 - x|$

Find $h(-1)$, $h(0)$, and $h(4)$ for each function. See Example 7.

53. $h(x) = -5x$ **54.** $h(x) = -3x$

55. $h(x) = 2x^2 + 3$ **56.** $h(x) = 3x^2$

For each given function value, write a corresponding ordered pair.

57. $f(3) = 6$ **58.** $f(7) = -2$

59. $g(0) = -\frac{1}{2}$ **60.** $g(0) = -\frac{7}{8}$

61. $h(-2) = 9$ **62.** $h(-10) = 1$

Find the domain of each function. See Example 8.

63. $f(x) = 3x - 7$ **64.** $g(x) = 5 - 2x$

65. $h(x) = \frac{1}{x + 5}$ **66.** $f(x) = \frac{1}{x - 6}$

67. $g(x) = |x + 1|$ **68.** $h(x) = |2x|$

Find the domain and the range of each relation graphed. See Example 9.

69.

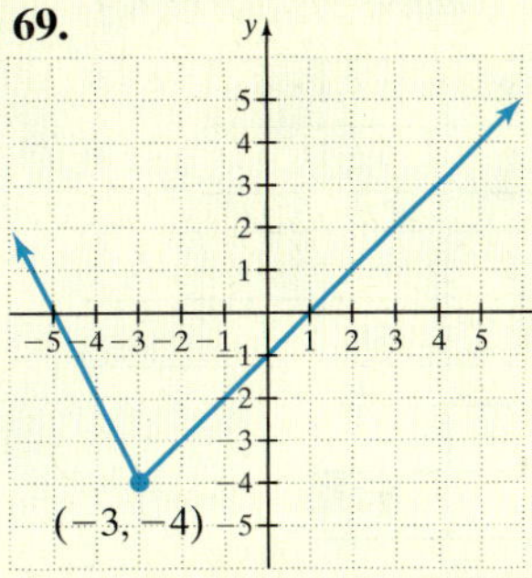

70.

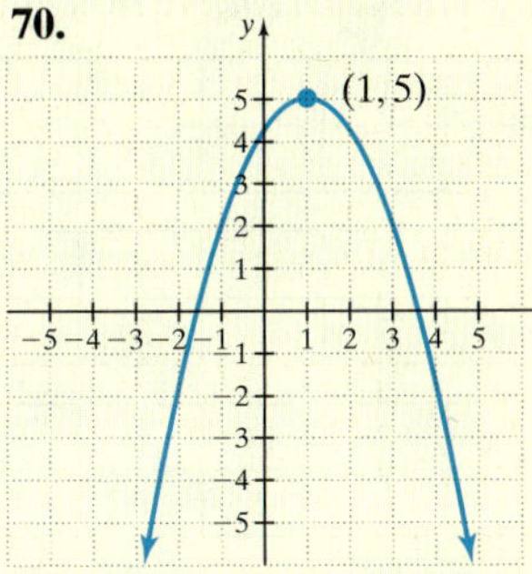

71.

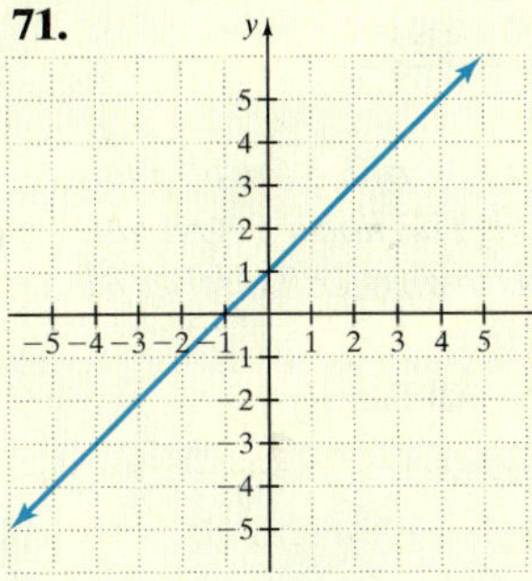

72.

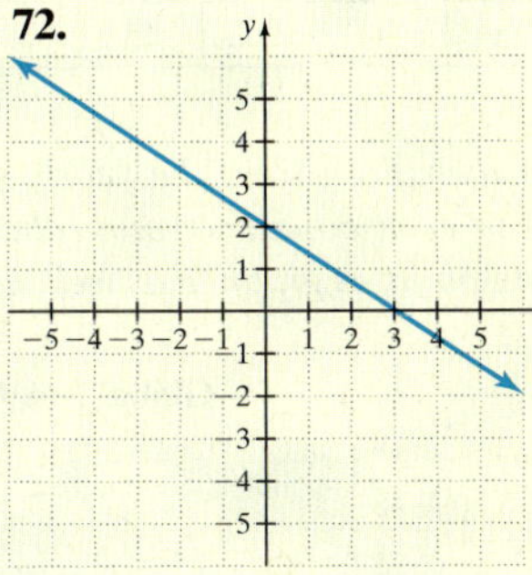

73.

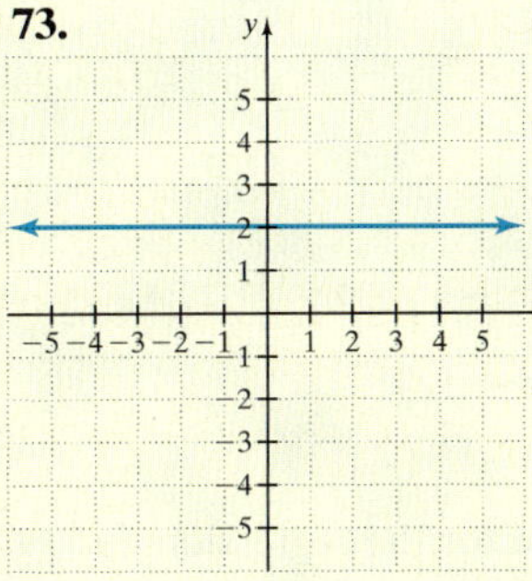

74.

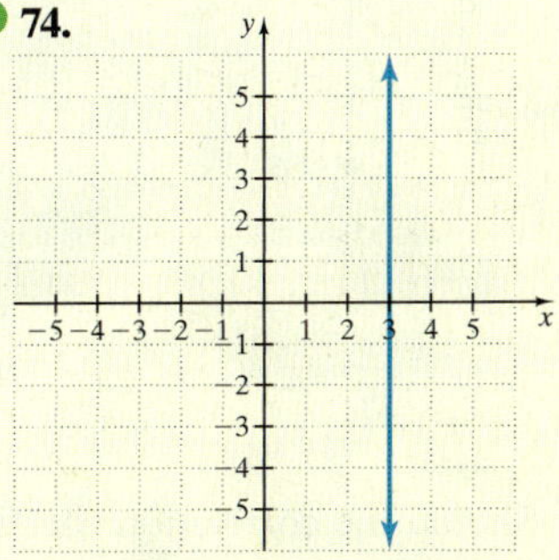

Use the graph of f below to answer Exercises 75 through 80.

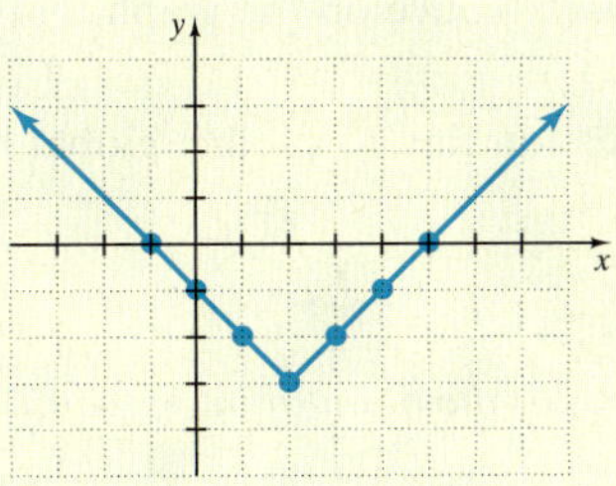

75. Complete the ordered pair solution for f. (0,)

76. Complete the ordered pair solution for f. (3,)

77. $f(0) =$ ____?

78. $f(3) =$ ____?

79. If $f(x) = 0$, find the value(s) of x.

80. If $f(x) = -1$, find the value(s) of x.

The examples of correlation coefficients, r, in this section are labeled a.–f. on page 184. For Exercises 81–84, draw a scatter plot for the data visually and match each to the closest correlation coefficient, a–f.

81.

x	1	3	4	5	8	7	9
y	1	4	4.5	5	3	4	1

82.

x	1	3	4	3	2	1	3	5
y	3	1	3	5	2.5	4	3	2

83.

x	2	14	6	8	10	4
y	8	2	6	5	4	7

84.

x	3	8	5	7	1	4
y	9	24	15	21	3	12

The scatter plot shows the yearly global box office for all films released in each country around the world. The linear regression for the data points is graphed as well. Use this to answer Exercises 85–88.

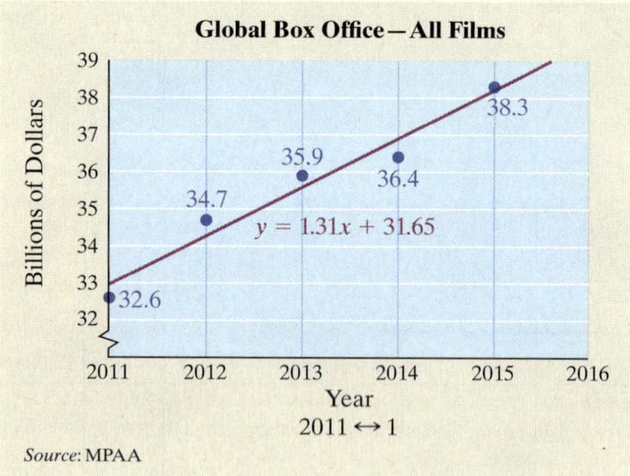

Source: MPAA

85. Should the correlation coefficient for the regression line be positive or negative?

86. What is the slope of the regression line?

87. Use the regression line to predict global box office for the year 2020. (*Hint:* Notice on the graph that $x = 1$ represents the year 2011.)

88. Use the regression line to predict global box office for the year 2022.

REVIEW AND PREVIEW

Find the coordinates of the point of intersection. See Section 3.1.

89.

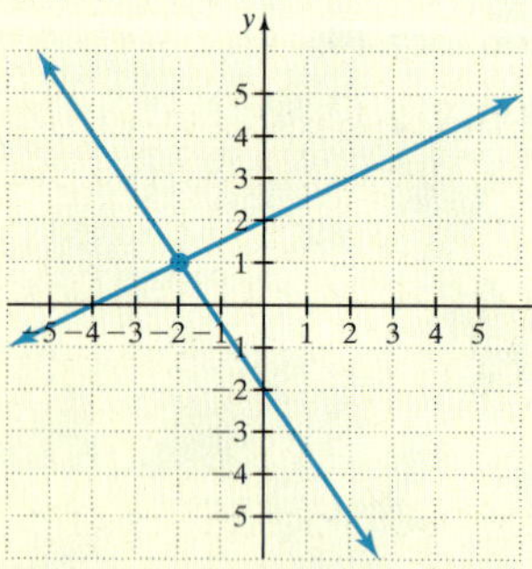

90.

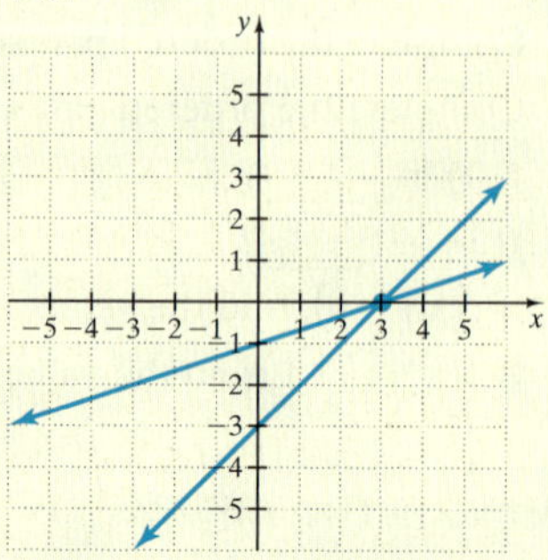

91.

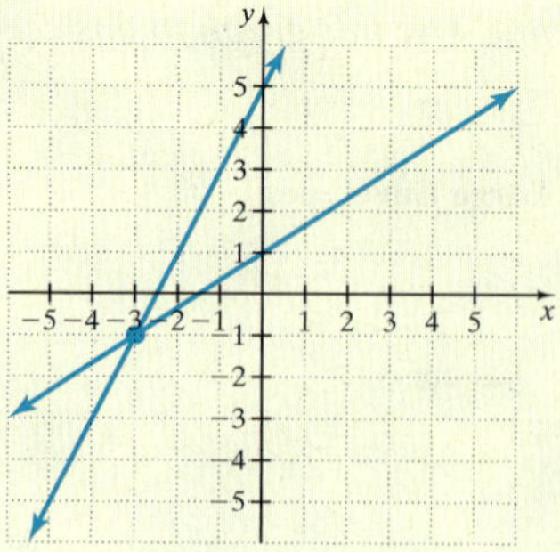

92.

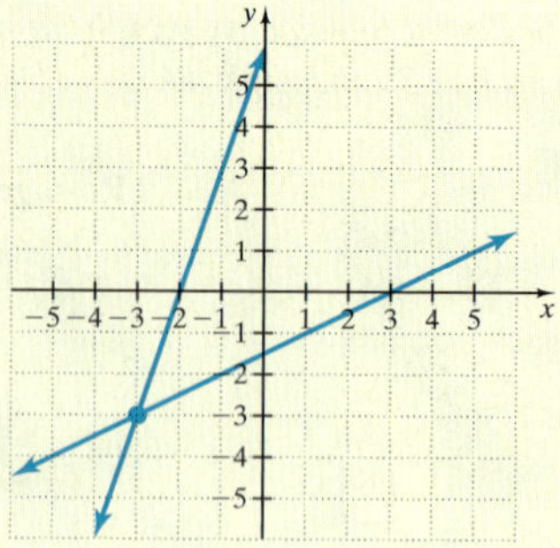

CONCEPT EXTENSIONS

Solve. See the Concept Check in this section.

93. If a function f is evaluated at -5, the value of the function is 12. Write this situation using function notation.

94. Suppose $(9, 20)$ is an ordered-pair solution for the function g. Write this situation using function notation.

The graph of the function f is below. Use this graph to answer Exercises 95 through 98.

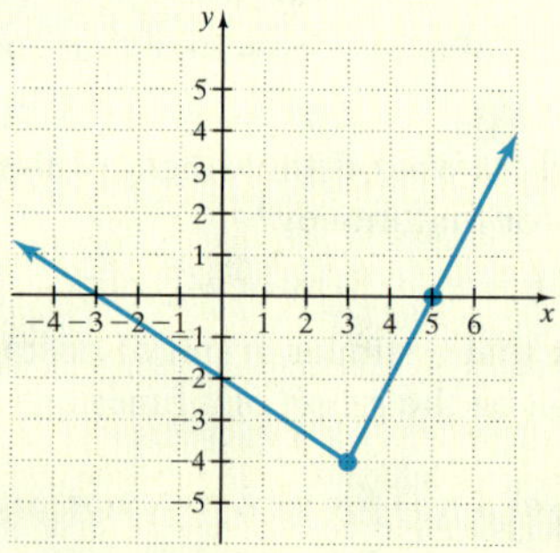

95. Write the coordinates of the lowest point of the graph.

96. Write the answer to Exercise 95 in function notation.

97. An x-intercept of this graph is $(5, 0)$. Write this using function notation.

98. Write the other x-intercept of this graph (see Exercise 97) using function notation.

99. Forensic scientists use the function

$$H(x) = 2.59x + 47.24$$

to estimate the height of a woman in centimeters given the length x of her femur bone.

a. Estimate the height of a woman whose femur measures 46 centimeters.

b. Estimate the height of a woman whose femur measures 39 centimeters.

100. The dosage in milligrams D of Ivermectin, a heartworm preventive for a dog who weighs x pounds, is given by the function

$$D(x) = \frac{136}{25}x$$

a. Find the proper dosage for a dog that weighs 35 pounds.

b. Find the proper dosage for a dog that weighs 70 pounds.

101. In your own words, define **(a)** function; **(b)** domain; **(c)** range.

102. Explain the vertical line test and how it is used.

103. Since $y = x + 7$ is a function, rewrite the equation using function notation.

104. Since $y = 3$ is a function, rewrite this equation using function notation.

See the example below for Exercises 105 through 108.

Example

If $f(x) = x^2 + 2x + 1$, find $f(\pi)$.

Solution:

$$f(x) = x^2 + 2x + 1$$
$$f(\pi) = \pi^2 + 2\pi + 1$$

Given the following functions, find the indicated values.

105. $f(x) = 2x + 7$

a. $f(2)$ **b.** $f(a)$

106. $g(x) = -3x + 12$

a. $g(5)$ **b.** $g(r)$

107. $h(x) = x^2 + 7$

a. $h(3)$ **b.** $h(a)$

108. $f(x) = x^2 - 12$

a. $f(12)$ **b.** $f(a)$

3.5 Systems of Linear Equations in Two Variables

OBJECTIVES

1. Determine Whether an Ordered Pair Is a Solution of a System of Two Linear Equations.
2. Solve a System by Graphing.
3. Solve a System by Substitution.
4. Solve a System by Elimination.
5. Solve Cost and Revenue Problems That Can Be Modeled by a System of Two Linear Equations.

An important problem that often occurs in the fields of business and economics concerns the concepts of revenue and cost. For example, suppose that a small manufacturing company begins to manufacture and sell iPad protective cases. The revenue of a company is the company's income from selling these cases, and the cost is the amount of money that a company spends to manufacture these cases. The following coordinate system shows the graphs of revenue and cost for the protective cases.

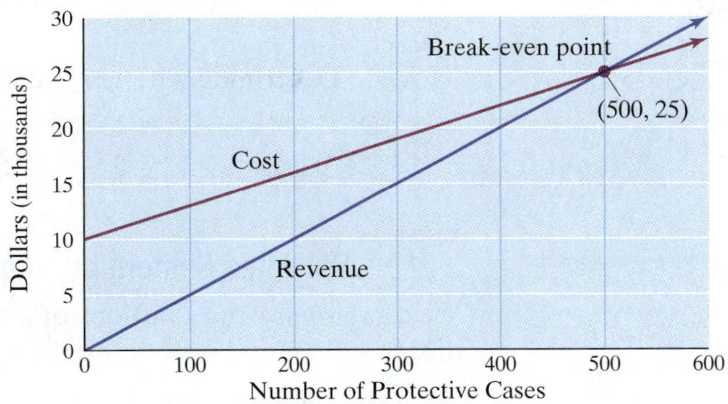

These lines intersect at the point $(500, 25)$. This means that when 500 protective cases are manufactured and sold, both cost and revenue are \$25,000. In business, this point of intersection is called the **break-even point.** Notice that for x-values (cases sold) less than 500, the cost graph is above the revenue graph, meaning that cost of manufacturing is greater than revenue, and so the company is losing money. For x-values (cases sold) greater than 500, the revenue graph is above the cost graph, meaning that revenue is greater than cost, and so the company is making money.

Recall from Section 3.1 that each line is the graph of some linear equation in two variables. Both equations together form a **system of equations.** The common point of intersection is called the **solution of the system.** Some examples of systems of linear equations in two variables are

Systems of Linear Equations in Two Variables

$$\begin{cases} x - 2y = -7 \\ 3x + y = 0 \end{cases} \qquad \begin{cases} x = 5 \\ x + \dfrac{y}{2} = 9 \end{cases} \qquad \begin{cases} x - 3 = 2y + 6 \\ y = 1 \end{cases}$$

OBJECTIVE

1 Determining Whether an Ordered Pair Is a Solution

Recall that a solution of an equation in two variables is an ordered pair (x, y) that makes the equation true. A **solution of a system** of two equations in two variables is an ordered pair (x, y) that makes both equations true.

EXAMPLE 1 Determine whether the given ordered pair is a solution of the system.

a. $\begin{cases} -x + y = 2 \\ 2x - y = -3 \end{cases}$ $(-1, 1)$ **b.** $\begin{cases} 5x + 3y = -1 \\ x - y = 1 \end{cases}$ $(-2, 3)$

Solution

a. We replace x with -1 and y with 1 in each equation.

$-x + y = 2$ First equation
$-(-1) + (1) \stackrel{?}{=} 2$ Let $x = -1$ and $y = 1$.
$1 + 1 \stackrel{?}{=} 2$
$2 = 2$ True

$2x - y = -3$ Second equation
$2(-1) - (1) \stackrel{?}{=} -3$ Let $x = -1$ and $y = 1$.
$-2 - 1 \stackrel{?}{=} -3$
$-3 = -3$ True

Since $(-1, 1)$ makes *both* equations true, it is a solution.

b. We replace x with -2 and y with 3 in each equation.

$5x + 3y = -1$ First equation
$5(-2) + 3(3) \stackrel{?}{=} -1$ Let $x = -2$ and $y = 3$.
$-10 + 9 \stackrel{?}{=} -1$
$-1 = -1$ True

$x - y = 1$ Second equation
$(-2) - (3) \stackrel{?}{=} 1$ Let $x = -2$ and $y = 3$.
$-5 = 1$ False

Since the ordered pair $(-2, 3)$ does not make *both* equations true, it is not a solution of the system. □

PRACTICE

1 Determine whether the given ordered pair is a solution of the system.

a. $\begin{cases} -x - 4y = 1 \\ 2x + y = 5 \end{cases}$ $(3, -1)$ **b.** $\begin{cases} 4x + y = -4 \\ -x + 3y = 8 \end{cases}$ $(-2, 4)$ ■

OBJECTIVE

2 Solving a System by Graphing

We can *estimate* the solutions of a system by graphing each equation on the same coordinate system and estimating the coordinates of any point of intersection.

EXAMPLE 2 Solve the system by graphing.

$$\begin{cases} x + y = 2 \\ 3x - y = -2 \end{cases}$$

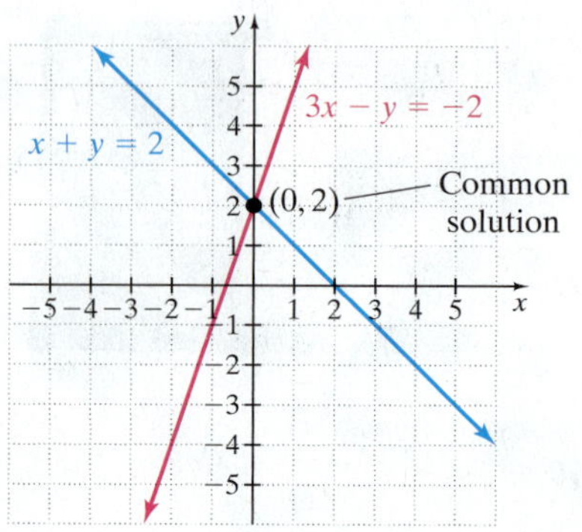

Solution First we graph the linear equations on the same rectangular coordinate system. These lines intersect at one point as shown. The coordinates of the point of intersection appear to be $(0, 2)$. We check this estimated solution by replacing x with 0 and y with 2 in *both* equations.

$x + y = 2$	First equation	$3x - y = -2$	Second equation
$0 + 2 \stackrel{?}{=} 2$	Let $x = 0$ and $y = 2$.	$3(0) - 2 \stackrel{?}{=} -2$	Let $x = 0$ and $y = 2$.
$2 = 2$	True	$-2 = -2$	True

Helpful Hint
Using set notation, we may write the solution set as $\{(0, 2)\}$.

The ordered pair $(0, 2)$ is the solution of the system. A system that has at least one solution, such as this one, is said to be **consistent.** □

PRACTICE 2 Solve the system by graphing. $\begin{cases} y = 5x \\ 2x + y = 7 \end{cases}$ ■

Helpful Hint
Reading values from graphs may not be accurate. Until a proposed solution is checked in both equations of the system, we can only assume that we have *estimated* a solution.

In Example 2, we have a **consistent system.** To review, a system that has at least one solution is said to be consistent.

Later, we will talk about **dependent equations.** For now, we define an **independent equation** to be an equation in a system of equations that cannot be algebraically derived from any other equation in the system.

Thus:

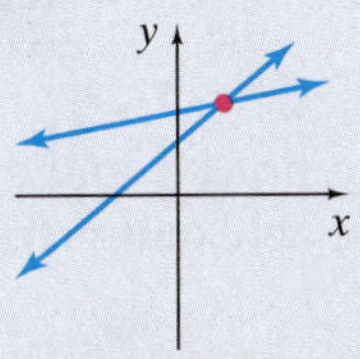

One solution

Consistent System: The system has at least one solution.

Independent Equations: Each equation in the system cannot be algebraically derived from the other. (Note: Their graphs are **not** the same.)

EXAMPLE 3 Solve the system by graphing.

$$\begin{cases} x - 2y = 4 \\ x = 2y \end{cases}$$

Solution We graph each linear equation.

The lines appear to be parallel. To be sure, let's write each equation in slope-intercept form, $y = mx + b$. To do so, we solve for y.

$x - 2y = 4$	First equation
$-2y = -x + 4$	Subtract x from both sides.
$y = \frac{1}{2}x - 2$	Divide both sides by -2
$x = 2y$	Second equation
$\frac{1}{2}x = y$	Divide both sides by 2.
$y = \frac{1}{2}x$	

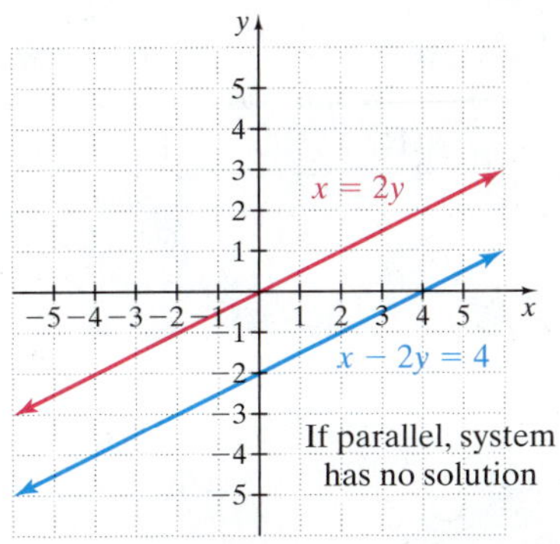

The graphs of these equations have the same slope, $\frac{1}{2}$, but different y-intercepts, so these lines are parallel. Therefore, the system has no solution since the equations have no common solution (there are no intersection points). A system that has no solution is said to be **inconsistent.** □

(Continued on next page)

PRACTICE 3 Solve the system by graphing. $\begin{cases} y = \frac{3}{4}x + 1 \\ 3x - 4y = 12 \end{cases}$ ■

In Example 3, we have an **inconsistent system.** To review, a system that has no solution is said to be inconsistent.

Let's now talk about the equations in this system. Each equation in this system cannot be algebraically derived from the other, so each equation is independent of the other. Thus:

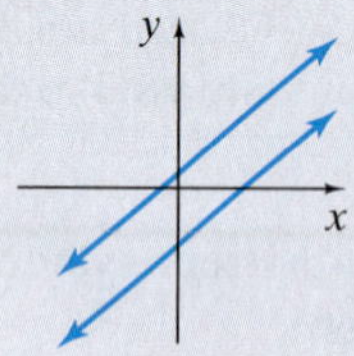

No solution

Inconsistent System: The system has no solution.

Independent Equations: Each equation in the system cannot be algebraically derived from the other. (Note: Their graphs are **not** the same.)

Helpful Hint

- If a system of equations has *at least one solution,* the system is *consistent.*
- If a system of equations has *no solution,* the system is *inconsistent.*

The pairs of equations in Examples 2 and 3 are called independent because their graphs differ. In Example 4, we see an example of dependent equations.

EXAMPLE 4 Solve the system by graphing.

$$\begin{cases} 2x + 4y = 10 \\ x + 2y = 5 \end{cases}$$

Solution We graph each linear equation. We see that the graphs of the equations are the same line. To confirm this, notice that if both sides of the second equation are multiplied by 2, the result is the first equation. This means that the equations have identical solutions. Any ordered pair solution of one equation satisfies the other equation also. These equations are said to be **dependent equations.** The solution set of the system is $\{(x, y) | x + 2y = 5\}$ or, equivalently, $\{(x, y) | 2x + 4y = 10\}$ since the lines describe identical ordered pairs. Written the second way, the solution set is read "the set of all ordered pairs (x, y), such that $2x + 4y = 10$." There is an infinite number of solutions to this system.

□

PRACTICE 4 Solve the system by graphing. $\begin{cases} 3x - 2y = 4 \\ -9x + 6y = -12 \end{cases}$ ■

In Example 4, we have a **consistent system.** To review, a system that has at least one solution is said to be consistent. In fact, the system in Example 4 has an infinite number of solutions.

Let's now define **dependent equations.** We define a **dependent equation** to be an equation in a system of equations that can be algebraically derived from another equation in the system.

Thus:

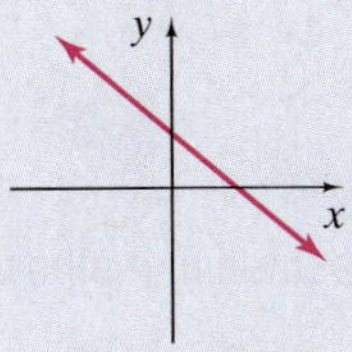

Infinite number of solutions

Consistent System: The system has at least one solution.

Dependent Equations: An equation in a system of equations can be algebraically derived from another. (Note: Their graphs are the same.)

Helpful Hint

- If the graphs of two equations *differ,* they are *independent* equations.
- If the graphs of two equations are the *same,* they are *dependent* equations.

CONCEPT CHECK

The equations in the system are dependent and the system has an infinite number of solutions. Which ordered pairs below are solutions?

$$\begin{cases} -x + 3y = 4 \\ 2x + 8 = 6y \end{cases}$$

a. $(4, 0)$ **b.** $(-4, 0)$ **c.** $(-1, 1)$

We can summarize the information discovered in Examples 2 through 4 as follows.

Possible Solutions to Systems of Two Linear Equations

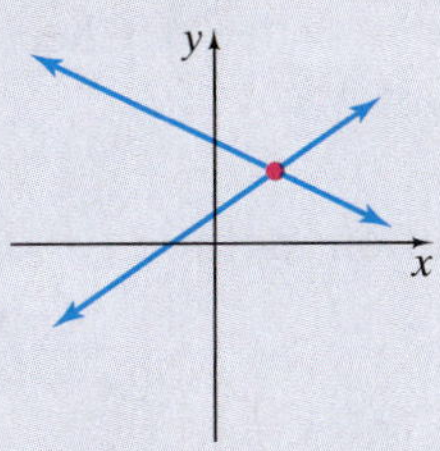

One solution: consistent system; independent equations

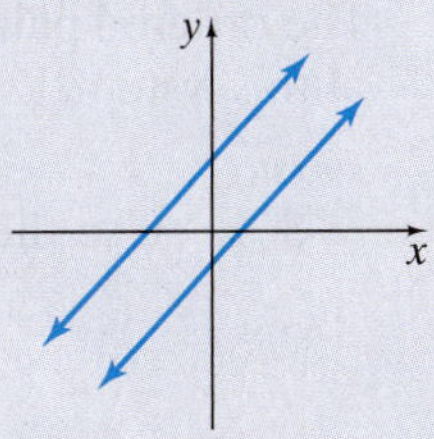

No solution: inconsistent system; independent equations

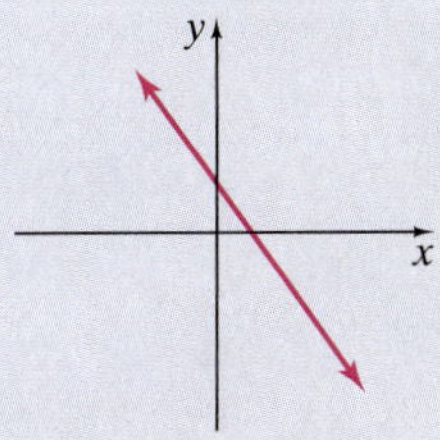

Infinite number of solutions: consistent system; dependent equations

CONCEPT CHECK

How can you tell just by looking at the following system that it has no solution?

$$\begin{cases} y = 3x + 5 \\ y = 3x - 7 \end{cases}$$

How can you tell just by looking at the following system that it has infinitely many solutions?

$$\begin{cases} x + y = 5 \\ 2x + 2y = 10 \end{cases}$$

Answer to Concept Check: b, c

Answer to Concept Check: answers may vary

OBJECTIVE

3 Solving a System by Substitution

Graphing the equations of a system by hand is often a good method of finding approximate solutions of a system, but it is not a reliable method of finding exact solutions of a system. We turn instead to two algebraic methods of solving systems. We use the first method, the **substitution method,** to solve the system

$$\begin{cases} 2x + 4y = -6 & \text{First equation} \\ x = 2y - 5 & \text{Second equation} \end{cases}$$

EXAMPLE 5 Use the substitution method to solve the system.

$$\begin{cases} 2x + 4y = -6 & \text{First equation} \\ x = 2y - 5 & \text{Second equation} \end{cases}$$

Solution In the second equation, we are told that x is equal to $2y - 5$. Since they are equal, we can *substitute* $2y - 5$ for x in the first equation. This will give us an equation in one variable, which we can solve for y.

$$2x + 4y = -6 \quad \text{First equation}$$

$$2(2y - 5) + 4y = -6 \quad \text{Substitute } 2y - 5 \text{ for } x.$$

$$4y - 10 + 4y = -6$$

$$8y = 4$$

$$y = \frac{4}{8} = \frac{1}{2} \quad \text{Solve for } y.$$

The y-coordinate of the solution is $\frac{1}{2}$. To find the x-coordinate, we replace y with $\frac{1}{2}$ in the second equation, $x = 2y - 5$.

$$x = 2y - 5$$

$$x = 2\left(\frac{1}{2}\right) - 5 = 1 - 5 = -4$$

The ordered pair solution is $\left(-4, \frac{1}{2}\right)$. Check to see that $\left(-4, \frac{1}{2}\right)$ satisfies both equations of the system. □

PRACTICE

5 Use the substitution method to solve the system.

$$\begin{cases} y = 4x + 7 \\ 2x + y = 4 \end{cases}$$

■

Solving a System of Two Equations Using the Substitution Method

Step 1. Solve one of the equations for one of its variables.

Step 2. Substitute the expression for the variable found in Step 1 into the other equation.

Step 3. Find the value of one variable by solving the equation from Step 2.

Step 4. Find the value of the other variable by substituting the value found in Step 3 into the equation from Step 1.

Step 5. Check the ordered pair solution in *both* original equations.

✔ **CONCEPT CHECK**

As you solve the system $\begin{cases} 2x + y = -5 \\ x - y = 5 \end{cases}$ you find that $y = -5$. Is this the solution of the system?

Answer to Concept Check:
No, the solution will be an ordered pair.

EXAMPLE 6 Solve the system:

$$\begin{cases} 7x - 3y = -14 \\ -3x + y = 6 \end{cases}$$

Solution To avoid introducing fractions, we will solve the second equation for y.

$$-3x + y = 6 \qquad \text{Second equation}$$
$$y = 3x + 6$$

Next, substitute $3x + 6$ for y in the first equation.

$$7x - 3y = -14 \qquad \text{First equation}$$
$$7x - 3(3x + 6) = -14 \qquad \text{Let } y = 3x + 6.$$
$$7x - 9x - 18 = -14 \qquad \text{Use the distributive property.}$$
$$-2x - 18 = -14 \qquad \text{Simplify.}$$
$$-2x = 4 \qquad \text{Add 18 to both sides.}$$
$$x = -2 \qquad \text{Divide both sides by } -2.$$

To find the corresponding y-value, substitute -2 for x in the equation $y = 3x + 6$. Then $y = 3(-2) + 6$ or $y = 0$. The solution of the system is $(-2, 0)$. Check this solution in both equations of the system. □

PRACTICE 6 Solve the system:

$$\begin{cases} 5x + 3y = -9 \\ -2x + y = 8 \end{cases}$$

Helpful Hint

When solving a system of equations by the substitution method, begin by solving an equation for one of its variables. If possible, solve for a variable that has a coefficient of 1 or -1. This way, we avoid working with time-consuming fractions.

✔ CONCEPT CHECK

To avoid fractions, which of the equations below would you use to solve for x?

a. $3x - 4y = 15$ **b.** $14 - 3y = 8x$ **c.** $7y + x = 12$

OBJECTIVE 4

Solving a System by Elimination

The **elimination method,** or **addition method,** is a second algebraic technique for solving systems of equations. For this method, we rely on a version of the addition property of equality, which states that "equals added to equals are equal."

If $A = B$ and $C = D$ then $A + C = B + D$.

EXAMPLE 7 Use the elimination method to solve the system.

$$\begin{cases} x - 5y = -12 & \text{First equation} \\ -x + y = 4 & \text{Second equation} \end{cases}$$

Solution Since the left side of each equation is equal to the right side, we add equal quantities by adding the left sides of the equations and the right sides of the equations. This sum gives us an equation in one variable, y, which we can solve for y.

(Continued on next page)

Answer to Concept Check:
c

$$\begin{array}{rll} x - 5y &= -12 & \text{First equation} \\ -x + y &= 4 & \text{Second equation} \\ \hline -4y &= -8 & \text{Add} \\ y &= 2 & \text{Solve for } y. \end{array}$$

The y-coordinate of the solution is 2. To find the corresponding x-coordinate, we replace y with 2 in either original equation of the system. Let's use the second equation.

$$\begin{array}{rll} -x + y &= 4 & \text{Second equation} \\ -x + 2 &= 4 & \text{Let } y = 2. \\ -x &= 2 & \\ x &= -2 & \end{array}$$

The ordered pair solution is $(-2, 2)$. Check to see that $(-2, 2)$ satisfies both equations of the system. □

PRACTICE 7 Use the elimination method to solve the system.

$$\begin{cases} 3x - y = 5 \\ 5x + y = 11 \end{cases}$$

The steps below summarize the elimination method.

Solving a System of Two Linear Equations Using the Elimination Method

Step 1. Rewrite each equation in standard form, $Ax + By = C$.

Step 2. If necessary, multiply one or both equations by some nonzero number so that the coefficients of a variable are opposites of each other.

Step 3. Add the equations.

Step 4. Find the value of one variable by solving the equation from Step 3.

Step 5. Find the value of the second variable by substituting the value found in Step 4 into either original equation.

Step 6. Check the proposed ordered pair solution in *both* original equations.

EXAMPLE 8 Use the elimination method to solve the system.

$$\begin{cases} 3x - 2y = 10 \\ 4x - 3y = 15 \end{cases}$$

Solution If we add the two equations, the sum will still be an equation in two variables. Notice, however, that we can eliminate y when the equations are added if we multiply both sides of the first equation by 3 and both sides of the second equation by -2. Then

$$\begin{cases} 3(3x - 2y) = 3(10) \\ -2(4x - 3y) = -2(15) \end{cases} \quad \text{simplifies to} \quad \begin{cases} 9x - 6y = 30 \\ -8x + 6y = -30 \end{cases}$$

Next we add the left sides and add the right sides.

$$\begin{array}{rl} 9x - 6y &= 30 \\ -8x + 6y &= -30 \\ \hline x \qquad &= 0 \end{array}$$

To find y, we let $x = 0$ in either equation of the system.

$$\begin{array}{rll} 3x - 2y &= 10 & \text{First equation} \\ 3(0) - 2y &= 10 & \text{Let } x = 0. \\ -2y &= 10 & \\ y &= -5 & \end{array}$$

The ordered pair solution is $(0, -5)$. Check to see that $(0, -5)$ satisfies both equations of the system. □

PRACTICE 8 Use the elimination method to solve the system.

$$\begin{cases} 3x - 2y = -6 \\ 4x + 5y = -8 \end{cases}$$

■

EXAMPLE 9 Use the elimination method to solve the system.

$$\begin{cases} 3x + \dfrac{y}{2} = 2 \\ 6x + y = 5 \end{cases}$$

Solution If we multiply both sides of the first equation by -2, the coefficients of x in the two equations will be opposites. Then

$$\begin{cases} -2\left(3x + \dfrac{y}{2}\right) = -2(2) \\ 6x + y = 5 \end{cases} \quad \text{simplifies to} \quad \begin{cases} -6x - y = -4 \\ 6x + y = 5 \end{cases}$$

Now we can add the left sides and add the right sides.

$$\begin{aligned} -6x - y &= -4 \\ 6x + y &= 5 \\ \hline 0 &= 1 \end{aligned} \quad \text{False}$$

The resulting equation, $0 = 1$, is false for all values of y or x. Thus, the system has no solution. The solution set is $\{\ \}$ or $\varnothing$. This system is inconsistent, and the graphs of the equations are parallel lines. □

PRACTICE 9 Use the elimination method to solve the system.

$$\begin{cases} 8x + y = 6 \\ 2x + \dfrac{y}{4} = -2 \end{cases}$$

■

EXAMPLE 10 Use the elimination method to solve the system.

$$\begin{cases} -5x - 3y = 9 \\ 10x + 6y = -18 \end{cases}$$

Solution To eliminate x when the equations are added, we multiply both sides of the first equation by 2. Then

$$\begin{cases} 2(-5x - 3y) = 2(9) \\ 10x + 6y = -18 \end{cases} \quad \text{simplifies to} \quad \begin{cases} -10x - 6y = 18 \\ 10x + 6y = -18 \end{cases}$$

Next we add the equations.

$$\begin{aligned} -10x - 6y &= 18 \\ 10x + 6y &= -18 \\ \hline 0 &= 0 \end{aligned}$$

The resulting equation, $0 = 0$, is true for all possible values of y or x. Notice in the original system that if both sides of the first equation are multiplied by -2, the result is the second equation. This means that the two equations are equivalent. They have the same solution set and there is an infinite number of solutions. Thus, the equations of this system are dependent, and the solution set of the system is

$$\{(x, y) \mid -5x - 3y = 9\} \quad \text{or, equivalently,} \quad \{(x, y) \mid 10x + 6y = -18\}.$$ □

(Continued on next page)

PRACTICE 10 Use the elimination method to solve the system.

$$\begin{cases} -3x + 2y = -1 \\ 9x - 6y = 3 \end{cases}$$

Helpful Hint

Remember that not all ordered pairs are solutions of the system in Example 10, only the infinite number of ordered pairs that satisfy $-5x - 3y = 9$ or, equivalently, $10x + 6y = -18$.

OBJECTIVE

5 Solving Problems with Cost and Revenue Functions

Recall that businesses are often computing cost and revenue functions or equations to predict sales, to determine whether prices need to be adjusted, and to see whether the company is making or losing money. Recall also that the value at which revenue equals cost is called the break-even point. When revenue is less than cost, the company is losing money; when revenue is greater than cost, the company is making money.

EXAMPLE 11 **Finding a Break-Even Point**

A manufacturing company recently purchased \$3000 worth of new equipment to offer new personalized stationery to its customers. The cost of producing a package of personalized stationery is \$3.00, and it is sold for \$5.50. Find the number of packages that must be sold for the company to break even.

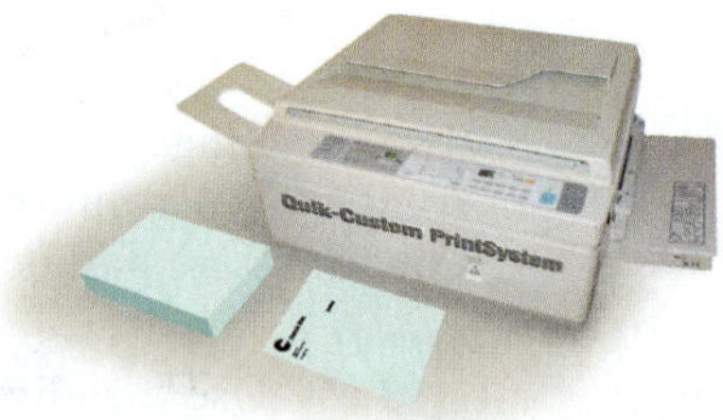

Solution

1. UNDERSTAND. Read and reread the problem. Notice that the cost to the company will include a one-time cost of \$3000 for the equipment and then \$3.00 per package produced. The revenue will be \$5.50 per package sold.

   To model this problem, we will let

   x = number of packages of personalized stationery

   $C(x)$ = total cost of producing x packages of stationery

   $R(x)$ = total revenue from selling x packages of stationery

2. TRANSLATE. The revenue equation is

In words:	revenue for selling x packages of stationery	=	price per package	·	number of packages
	↓		↓		↓
Translate:	$R(x)$	=	5.5	·	x

The cost equation is

In words:	cost for producing x packages of stationery	=	cost per package	·	number of packages	+	cost for equipment
	↓		↓		↓		↓
Translate:	$C(x)$	=	3	·	x	+	3000

Since the break-even point is when $R(x) = C(x)$, we solve the equation

$$5.5x = 3x + 3000$$

3. SOLVE.

$$5.5x = 3x + 3000$$
$$2.5x = 3000 \quad \text{Subtract } 3x \text{ from both sides.}$$
$$x = 1200 \quad \text{Divide both sides by 2.5.}$$

4. INTERPRET.

Check: To see whether the break-even point occurs when 1200 packages are produced and sold, see if revenue equals cost when $x = 1200$. When $x = 1200$, $R(x) = 5.5x = 5.5(1200) = 6600$ and $C(x) = 3x + 3000 = 3(1200) + 3000 = 6600$. Since $R(1200) = C(1200) = 6600$, the break-even point is 1200.

State: The company must sell 1200 packages of stationery to break even. The graph of this system is shown.

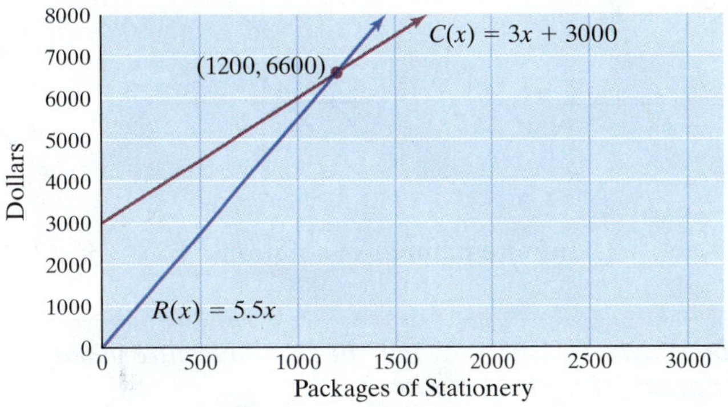

□

PRACTICE 11 An online-only electronics firm recently purchased $3000 worth of new equipment to create shock-proof packaging for its products. The cost of producing one shock-proof package is $2.50, and the firm charges the customer $4.50 for the packaging. Find the number of packages that must be sold for the company to break even. ■

Graphing Calculator Explorations

A graphing calculator may be used to approximate solutions of systems of equations by graphing each equation on the same set of axes and approximating any points of intersection. For example, approximate the solution of the system

$$\begin{cases} y = -2.6x + 5.6 \\ y = 4.3x - 4.9 \end{cases}$$

First use a standard window and graph both equations on a single screen.

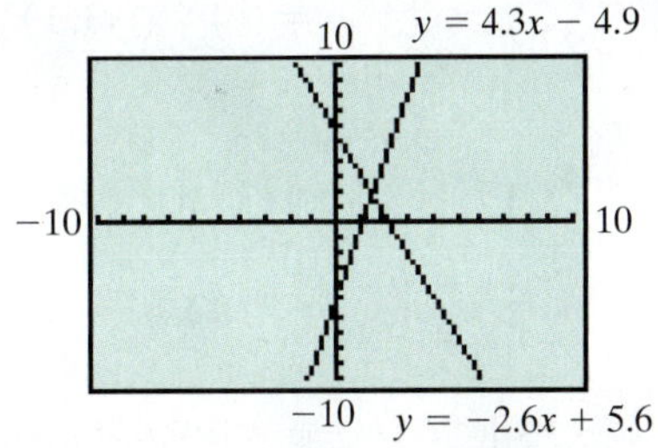

The two lines intersect. To approximate the point of intersection, trace to the point of intersection and use an Intersect feature of the graphing calculator or a Zoom In feature.

Using either method, we find that the approximate point of intersection is (1.52, 1.64).

Solve each system of equations. Approximate the solutions to two decimal places.

1. $y = -1.65x + 3.65$
   $y = 4.56x - 9.44$
2. $y = 7.61x + 3.48$
   $y = -1.26x - 6.43$
3. $2.33x - 4.72y = 10.61$
   $5.86x + 6.22y = -8.89$
4. $-7.89x - 5.68y = 3.26$
   $-3.65x + 4.98y = 11.77$

Vocabulary, Readiness & Video Check

Match each graph with the solution of the corresponding system.

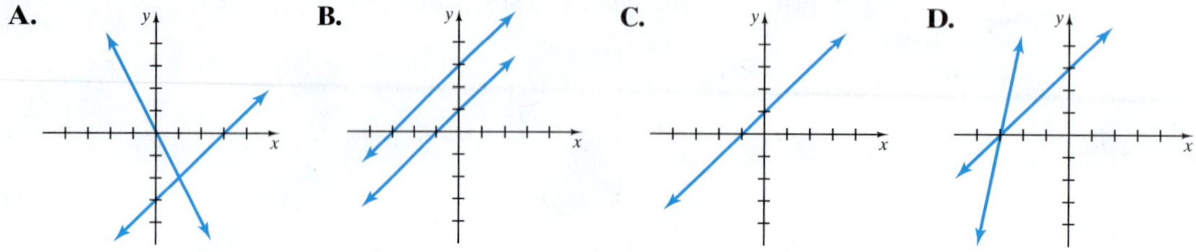

1. no solution **2.** Infinite number of solutions **3.** $(1, -2)$ **4.** $(-3, 0)$

Martin-Gay Interactive Videos

See Video 3.5

Watch the section lecture video and answer the following questions.

OBJECTIVE 1
5. In Example 1, the ordered pair is a solution of the first equation of the system. Why is this not enough to determine if the ordered pair is a solution of the system?

OBJECTIVE 2
6. From Example 2, what potential drawbacks does the graphing method have?

OBJECTIVE 3
7. The system in Example 3 needs one of its equations solved for a variable as a first step. What important point is then emphasized?

OBJECTIVE 4
8. Why is Step 2 in Solving a System by Elimination skipped in Example 5?

OBJECTIVE 5
9. Based on Example 6, explain the meaning of a break-even point. How do you find the break-even point algebraically?

3.5 Exercise Set MyMathLab®

Determine whether each given ordered pair is a solution of each system. See Example 1.

1. $\begin{cases} x - y = 3 \\ 2x - 4y = 8 \end{cases}$ $(2, -1)$

2. $\begin{cases} x - y = -4 \\ 2x + 10y = 4 \end{cases}$ $(-3, 1)$

3. $\begin{cases} 2x - 3y = -9 \\ 4x + 2y = -2 \end{cases}$ $(3, 5)$

4. $\begin{cases} 2x - 5y = -2 \\ 3x + 4y = 4 \end{cases}$ $(4, 2)$

5. $\begin{cases} y = -5x \\ x = -2 \end{cases}$ $(-2, 10)$

6. $\begin{cases} y = 6 \\ x = -2y \end{cases}$ $(-12, 6)$

7. $\begin{cases} 3x + 7y = -19 \\ -6x = 5y + 8 \end{cases}$ $\left(\frac{2}{3}, -3\right)$

8. $\begin{cases} 4x + 5y = -7 \\ -8x = 3y - 1 \end{cases}$ $\left(\frac{3}{4}, -2\right)$

Solve each system by graphing. See Examples 2 through 4.

9. $\begin{cases} x + y = 1 \\ x - 2y = 4 \end{cases}$

10. $\begin{cases} 2x - y = 8 \\ x + 3y = 11 \end{cases}$

11. $\begin{cases} 2y - 4x = 0 \\ x + 2y = 5 \end{cases}$

12. $\begin{cases} 4x - y = 6 \\ x - y = 0 \end{cases}$

13. $\begin{cases} 3x - y = 4 \\ 6x - 2y = 4 \end{cases}$

14. $\begin{cases} -x + 3y = 6 \\ 3x - 9y = 9 \end{cases}$

Solve each system of equations by the substitution method. See Examples 5 and 6.

15. $\begin{cases} x + y = 10 \\ y = 4x \end{cases}$

16. $\begin{cases} 5x + 2y = -17 \\ x = 3y \end{cases}$

17. $\begin{cases} 4x - y = 9 \\ 2x + 3y = -27 \end{cases}$

18. $\begin{cases} 3x - y = 6 \\ -4x + 2y = -8 \end{cases}$

19. $\begin{cases} 2x + 3y = -1 \\ 3x - y = 4 \end{cases}$

20. $\begin{cases} 2x + y = -5 \\ 5x + 2y = -8 \end{cases}$

21. $\begin{cases} \frac{x}{3} + y = \frac{4}{3} \\ -x + 2y = 11 \end{cases}$

22. $\begin{cases} \frac{x}{8} - \frac{y}{2} = 1 \\ \frac{x}{3} - y = 2 \end{cases}$

Solve each system of equations by the elimination method. See Examples 7 through 10.

23. $\begin{cases} -x + 2y = 0 \\ x + 2y = 5 \end{cases}$

24. $\begin{cases} -2x + 3y = 0 \\ 2x + 6y = 3 \end{cases}$

25. $\begin{cases} 5x + 2y = 1 \\ x - 3y = 7 \end{cases}$

26. $\begin{cases} 6x - y = -5 \\ 4x - 2y = 6 \end{cases}$

27. $\begin{cases} \frac{3}{4}x + \frac{5}{2}y = 11 \\ \frac{1}{16}x - \frac{3}{4}y = -1 \end{cases}$

28. $\begin{cases} \frac{2}{3}x + \frac{1}{4}y = -\frac{3}{2} \\ \frac{1}{2}x - \frac{1}{4}y = -2 \end{cases}$

29. $\begin{cases} 3x - 5y = 11 \\ 2x - 6y = 2 \end{cases}$

30. $\begin{cases} 6x - 3y = -3 \\ 4x + 5y = -9 \end{cases}$

31. $\begin{cases} x - 2y = 4 \\ 2x - 4y = 4 \end{cases}$

32. $\begin{cases} -x + 3y = 6 \\ 3x - 9y = 9 \end{cases}$

33. $\begin{cases} 3x + y = 1 \\ 2y = 2 - 6x \end{cases}$

34. $\begin{cases} y = 2x - 5 \\ 8x - 4y = 20 \end{cases}$

MIXED PRACTICE

Solve each system of equations.

35. $\begin{cases} 2x + 5y = 8 \\ 6x + y = 10 \end{cases}$

36. $\begin{cases} x - 4y = -5 \\ -3x - 8y = 0 \end{cases}$

37. $\begin{cases} 2x + 3y = 1 \\ x - 2y = 4 \end{cases}$

38. $\begin{cases} -2x + y = -8 \\ x + 3y = 11 \end{cases}$

39. $\begin{cases} \frac{1}{3}x + y = \frac{4}{3} \\ -\frac{1}{4}x - \frac{1}{2}y = -\frac{1}{4} \end{cases}$

40. $\begin{cases} \frac{3}{4}x - \frac{1}{2}y = -\frac{1}{2} \\ x + y = -\frac{3}{2} \end{cases}$

41. $\begin{cases} 2x + 6y = 8 \\ 3x + 9y = 12 \end{cases}$

42. $\begin{cases} x = 3y - 1 \\ 2x - 6y = -2 \end{cases}$

43. $\begin{cases} 4x + 2y = 5 \\ 2x + y = -1 \end{cases}$

44. $\begin{cases} 3x + 6y = 15 \\ 2x + 4y = 3 \end{cases}$

45. $\begin{cases} 10y - 2x = 1 \\ 5y = 4 - 6x \end{cases}$

46. $\begin{cases} 3x + 4y = 0 \\ 7x = 3y \end{cases}$

47. $\begin{cases} 5x - 2y = 27 \\ -3x + 5y = 18 \end{cases}$

48. $\begin{cases} 3x + 4y = 2 \\ 2x + 5y = -1 \end{cases}$

49. $\begin{cases} x = 3y + 2 \\ 5x - 15y = 10 \end{cases}$

50. $\begin{cases} y = \frac{1}{7}x + 3 \\ x - 7y = -21 \end{cases}$

51. $\begin{cases} 2x - y = -1 \\ y = -2x \end{cases}$

52. $\begin{cases} x = \frac{1}{5}y \\ x - y = -4 \end{cases}$

53. $\begin{cases} 2x = 6 \\ y = 5 - x \end{cases}$

54. $\begin{cases} x = 3y + 4 \\ -y = 5 \end{cases}$

55. $\begin{cases} \frac{x + 5}{2} = \frac{6 - 4y}{3} \\ \frac{3x}{5} = \frac{21 - 7y}{10} \end{cases}$

56. $\begin{cases} \frac{y}{5} = \frac{8 - x}{2} \\ x = \frac{2y - 8}{3} \end{cases}$

57. $\begin{cases} 4x - 7y = 7 \\ 12x - 21y = 24 \end{cases}$

58. $\begin{cases} 2x - 5y = 12 \\ -4x + 10y = 20 \end{cases}$

59. $\begin{cases} \frac{2}{3}x - \frac{3}{4}y = -1 \\ -\frac{1}{6}x + \frac{3}{8}y = 1 \end{cases}$

60. $\begin{cases} \frac{1}{2}x - \frac{1}{3}y = -3 \\ \frac{1}{8}x + \frac{1}{6}y = 0 \end{cases}$

61. $\begin{cases} 0.7x - 0.2y = -1.6 \\ 0.2x - y = -1.4 \end{cases}$

62. $\begin{cases} -0.7x + 0.6y = 1.3 \\ 0.5x - 0.3y = -0.8 \end{cases}$

63. $\begin{cases} 4x - 1.5y = 10.2 \\ 2x + 7.8y = -25.68 \end{cases}$

64. $\begin{cases} x - 3y = -5.3 \\ 6.3x + 6y = 3.96 \end{cases}$

Solve. See Example 11. For Exercises 65 and 66, the solutions have been started for you.

65. One number is two more than a second number. Twice the first is 4 less than 3 times the second. Find the numbers.

Start the solution:

1. UNDERSTAND the problem. Since we are looking for two numbers, let

x = one number and

y = second number.

2. TRANSLATE. Since we have assigned two variables, we will translate the facts into two equations. (Fill in the blanks.)

First equation:

In words:	One number	is	two	more than	second number
	↓	↓			
Translate:	x	=	____	____	____

Second equation:

In words:	Twice the first number	is	4	less than	3 times the second number
	↓	↓			
Translate:	$2x$	=	____	____	____

Finish with:

3. SOLVE the system and
4. INTERPRET the results.

66. Three times one number minus a second is 8, and the sum of the numbers is 12. Find the numbers.

Start the solution:

1. UNDERSTAND the problem. Since we are looking for two numbers, let

x = one number and

y = second number.

2. TRANSLATE. Since we have assigned two variables, we will translate the facts into two equations. (Fill in the blanks.)

First equation:

In words:	Three times one number	minus	a second number	is	8
	↓	↓	↓	↓	↓
Translate:	$3x$	____	____	=	8

Second equation:

In words:	The sum of the numbers		is	12
	↓		↓	↓
Translate:	x	+ ____	____	12

Finish with:

3. SOLVE the system and
4. INTERPRET the results.

Given the cost function $C(x)$ and the revenue function $R(x)$, find the number of units x that must be sold to break even.

67. $C(x) = 30x + 10{,}000$ $R(x) = 46x$

68. $C(x) = 12x + 15{,}000$ $R(x) = 32x$

69. $C(x) = 1.2x + 1500$ $R(x) = 1.7x$

70. $C(x) = 0.8x + 900$ $R(x) = 2x$

71. $C(x) = 75x + 160{,}000$ $R(x) = 200x$

72. $C(x) = 105x + 70{,}000$ $R(x) = 245x$

73. The planning department of Abstract Office Supplies has been asked to determine whether the company should introduce a new computer desk next year. The department estimates that \$6000 of new manufacturing equipment will need to be purchased and that the cost of constructing each desk will be \$200. The department also estimates that the revenue from each desk will be \$450.

a. Determine the revenue function $R(x)$ from the sale of x desks.

b. Determine the cost function $C(x)$ for manufacturing x desks.

c. Find the break-even point.

74. Baskets, Inc., is planning to introduce a new woven basket. The company estimates that \$500 worth of new equipment will be needed to manufacture this new type of basket and that it will cost \$15 per basket to manufacture. The company also estimates that the revenue from each basket will be \$31.

a. Determine the revenue function $R(x)$ from the sale of x baskets.

b. Determine the cost function $C(x)$ for manufacturing x baskets.

c. Find the break-even point. Round up to the next whole basket.

REVIEW AND PREVIEW

White gold is made by mixing (yellow) gold with other metals, as shown on the circle graph below. Use this graph to answer Exercises 75 through 82. See Section R.4.

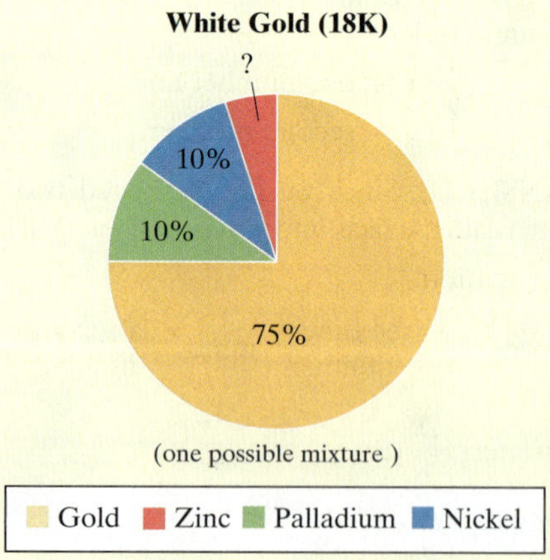

75. What percent of white gold is zinc?

76. What percent of white gold is not gold?

77. What fraction of white gold is nickel?

78. What fraction of white gold is palladium?

79. A 20-gram necklace of white gold contains how much zinc?

80. A 20-gram necklace of white gold contains how much nickel?

81. A 20-gram necklace of white gold contains how much gold?

82. A 20-gram necklace of white gold contains how much palladium?

CONCEPT EXTENSIONS

Without graphing, determine whether each system has one solution, no solution, or an infinite number of solutions. See the second Concept Check in this section.

83. $\begin{cases} y = 2x - 5 \\ y = 2x + 1 \end{cases}$

84. $\begin{cases} y = 3x - \dfrac{1}{2} \\ y = -2x + \dfrac{1}{5} \end{cases}$

85. $\begin{cases} x + y = 3 \\ 5x + 5y = 15 \end{cases}$

86. $\begin{cases} y = 5x - 2 \\ y = -\dfrac{1}{5}x - 2 \end{cases}$

87. Can a system consisting of two linear equations have exactly two solutions? Explain why or why not.

88. Suppose the graph of the equations in a system of two equations in two variables consists of a circle and a line. Discuss the possible number of solutions for this system.

The concept of supply and demand is used often in business. In general, as the unit price of a commodity increases, the demand for that commodity decreases. Also, as a commodity's unit price increases, the manufacturer normally increases the supply. The point where supply is equal to demand is called the equilibrium point. The following shows the graph of a demand equation and the graph of a supply equation for previously rented DVDs. The x-axis represents the number of DVDs in thousands, and the y-axis represents the cost of a DVD. Use this graph to answer Exercises 89 through 92.

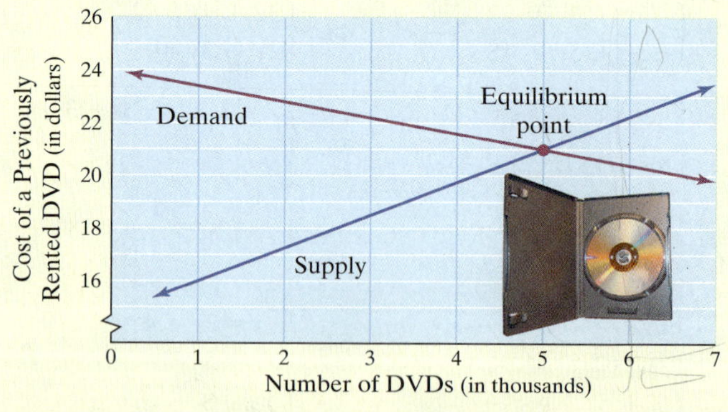

89. Find the number of DVDs and the price per DVD when supply equals demand.

90. When x is between 3 and 4, is supply greater than demand or is demand greater than supply?

91. When x is greater than 6, is supply greater than demand or is demand greater than supply?

92. For what x-values are the y-values corresponding to the supply equation greater than the y-values corresponding to the demand equation?

The revenue equation for a certain brand of toothpaste is $y = 2.5x$, where x is the number of tubes of toothpaste sold and y is the total income for selling x tubes. The cost equation is $y = 0.9x + 3000$, where x is the number of tubes of toothpaste manufactured and y is the cost of producing x tubes. The following set of axes shows the graph of the cost and revenue equations. Use this graph for Exercises 93 through 98.

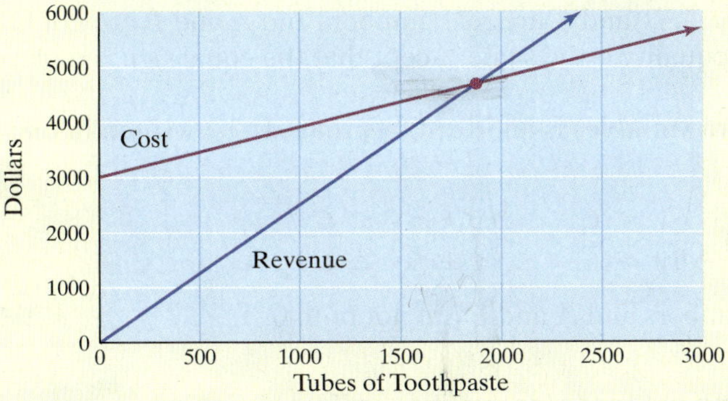

93. Find the coordinates of the point of intersection, or break-even point, by solving the system

$$\begin{cases} y = 2.5x \\ y = 0.9x + 3000 \end{cases}$$

94. Explain the meaning of the x-value of the point of intersection.

95. If the company sells 2000 tubes of toothpaste, does the company make money or lose money?

96. If the company sells 1000 tubes of toothpaste, does the company make money or lose money?

97. For what x-values will the company make a profit? (*Hint:* For what x-values is the revenue graph "higher" than the cost graph?)

98. For what x-values will the company lose money? (*Hint:* For what x-values is the revenue graph "lower" than the cost graph?)

99. Write a system of two linear equations in x and y that has the ordered pair solution (2, 5).

100. Which method would you use to solve the system?

$$\begin{cases} 5x - 2y = 6 \\ 2x + 3y = 5 \end{cases}$$

Explain your choice.

101. The number of mobile phone subscriptions has been growing steadily since 2000, while the number of landline accounts has been decreasing. The number per hundred people y of mobile phone subscriptions in the United States can be modeled by the linear equation $y = 3.9x + 44.1$. The number per hundred people y of landline phone subscriptions in the United States can be modeled by $y = -1.8x + 67.3$. In both models, $x = 0$ represents the year 2000. (*Source:* World Telecommunications/ICT Indicators Database)

a. What does the slope of each equation tell you about the patterns of mobile phone and landline subscriptions?

b. Solve this system of equations. Round your final results to the nearest whole numbers.

c. Explain the meaning of your answer to part **b**.

102. The amount of U.S. federal government income y (in trillions of dollars), for fiscal year x from 2012 to 2015 ($x = 0$ represents 2012), can be modeled by the linear equation $y = 0.35x + 5.04$. The amount of U.S. federal government expenditures y (in trillions of dollars) for the same period can be modeled by $y = 0.03x + 6.03$. (*Source:* Based on data from U.S. Department of the Treasury)

a. What does the slope of each equation tell you about the patterns of U.S. federal government income and expenditures?

b. Solve this system of equations. Round your final results to the nearest whole numbers.

c. Did expenditures ever equal income during the period from 2012 to 2015? If so, in what year?

Solve each system. To do so, you may want to let $a = \frac{1}{x}$ (if x is in the denominator) and let $b = \frac{1}{y}$ (if y is in the denominator.)

103. $\begin{cases} \frac{1}{x} + y = 12 \\ \frac{3}{x} - y = 4 \end{cases}$

104. $\begin{cases} x + \frac{2}{y} = 7 \\ 3x + \frac{3}{y} = 6 \end{cases}$

105. $\begin{cases} \frac{1}{x} + \frac{1}{y} = 5 \\ \frac{1}{x} - \frac{1}{y} = 1 \end{cases}$

106. $\begin{cases} \frac{2}{x} + \frac{3}{y} = 5 \\ \frac{5}{x} - \frac{3}{y} = 2 \end{cases}$

107. $\begin{cases} \frac{2}{x} - \frac{4}{y} = 5 \\ \frac{1}{x} - \frac{2}{y} = \frac{3}{2} \end{cases}$

108. $\begin{cases} \frac{2}{x} + \frac{3}{y} = -1 \\ \frac{3}{x} - \frac{2}{y} = 18 \end{cases}$

109. $\begin{cases} \frac{3}{x} - \frac{2}{y} = -18 \\ \frac{2}{x} + \frac{3}{y} = 1 \end{cases}$

110. $\begin{cases} \frac{5}{x} + \frac{7}{y} = 1 \\ -\frac{10}{x} - \frac{14}{y} = 0 \end{cases}$

3.6 Linear Inequalities in Two Variables

OBJECTIVES

1 Determine Whether an Ordered Pair Is a Solution of a Linear Inequality in Two Variables.

2 Graph a Linear Inequality in Two Variables.

3 Solve a System of Linear Inequalities.

In Objective 3, we continue our work with systems by solving systems of linear inequalities. Before that objective, we first need to learn to graph a single linear inequality in two variables.

Recall that a linear equation in two variables is an equation that can be written in the form $Ax + By = C$ where A, B, and C are real numbers and A and B are not both 0. The definition of a linear inequality is the same except that the equal sign is replaced with an inequality symbol.

A **linear inequality in two variables** is an inequality that can be written in one of the forms:

$$Ax + By < C \qquad Ax + By \le C$$
$$Ax + By > C \qquad Ax + By \ge C$$

where A, B, and C are real numbers and A and B are not both 0.

OBJECTIVE

1 Determining Solutions of Linear Inequalities in Two Variables

Just as for linear equations in x and y, an ordered pair is a **solution** of an inequality in x and y if replacing the variables by coordinates of the ordered pair results in a true statement.

EXAMPLE 1 Determine whether each ordered pair is a solution of the inequality $2x - y < 6$.

a. $(5, -1)$ **b.** $(2, 7)$

Solution

a. We replace x with 5 and y with -1 and see if a true statement results.

$$2x - y < 6$$
$$2(5) - (-1) \stackrel{?}{<} 6 \quad \text{Replace } x \text{ with 5 and } y \text{ with } -1.$$
$$10 + 1 \stackrel{?}{<} 6$$
$$11 < 6 \quad \text{False}$$

The ordered pair $(5, -1)$ is not a solution since $11 < 6$ is a false statement.

b. We replace x with 2 and y with 7 and see if a true statement results.

$$2x - y < 6$$
$$2(2) - (7) \stackrel{?}{<} 6 \quad \text{Replace } x \text{ with 2 and } y \text{ with 7.}$$
$$4 - 7 \stackrel{?}{<} 6$$
$$-3 < 6 \quad \text{True}$$

The ordered pair $(2, 7)$ is a solution since $-3 < 6$ is a true statement. □

PRACTICE

1 Determine whether each ordered pair is a solution of $x - 4y > 8$.

a. $(-3, 2)$ **b.** $(9, 0)$ ■

OBJECTIVE

2 Graphing Linear Inequalities in Two Variables

The linear equation $x - y = 1$ is graphed on the top of the next page. Recall that all points on the line correspond to ordered pairs that satisfy the equation $x - y = 1$.

Notice the line defined by $x - y = 1$ divides the rectangular coordinate system plane into 2 sides. All points on one side of the line satisfy the inequality $x - y < 1$ and all points on the other side satisfy the inequality $x - y > 1$. The next graph shows a few examples of this.

Check	$x - y < 1$
$(1, 3)$	$1 - 3 < 1$ True
$(-2, 1)$	$-2 - 1 < 1$ True
$(-4, -4)$	$-4 - (-4) < 1$ True

Check	$x - y > 1$
$(4, 1)$	$4 - 1 > 1$ True
$(2, -2)$	$2 - (-2) > 1$ True
$(0, -4)$	$0 - (-4) > 1$ True

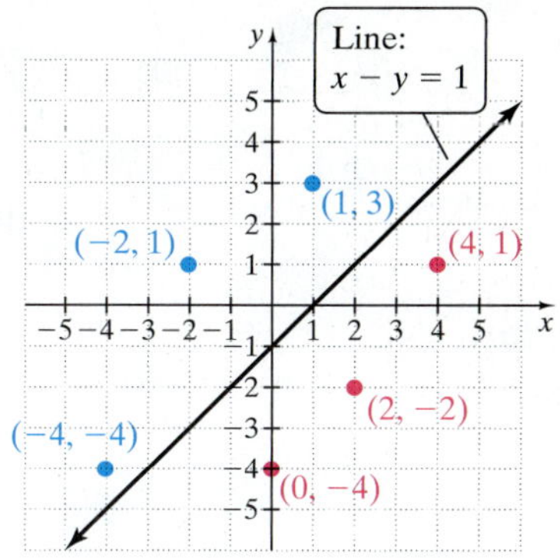

The graph of $x - y < 1$ is the region shaded blue, and the graph of $x - y > 1$ is the region shaded red below.

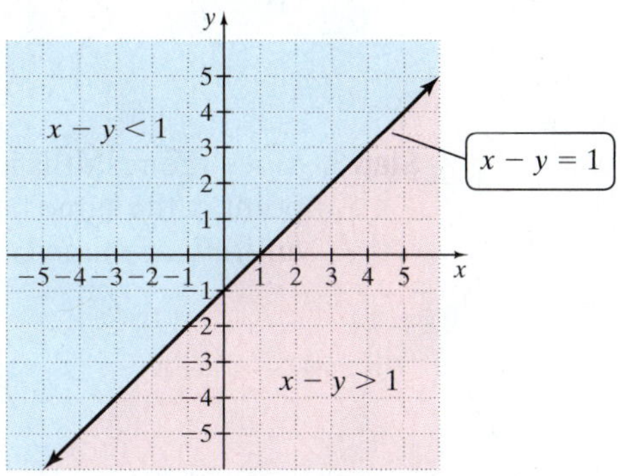

The region to the left of the line and the region to the right of the line are called **half-planes.** Every line divides the plane (similar to a sheet of paper extending indefinitely in all directions) into two half-planes; the line is called the **boundary.**

Recall that the inequality $x - y \leq 1$ means

$$x - y = 1 \quad \text{or} \quad x - y < 1$$

Thus, the graph of $x - y \leq 1$ is the half-plane $x - y < 1$ along with the boundary line $x - y = 1$.

Graphing a Linear Inequality in Two Variables

Step 1. Graph the boundary line found by replacing the inequality sign with an equal sign. If the inequality sign is $>$ or $<$, graph a dashed boundary line (indicating that the points on the line are not solutions of the inequality). If the inequality sign is $\geq$ or $\leq$, graph a solid boundary line (indicating that the points on the line are solutions of the inequality).

Step 2. Choose a point, *not* on the boundary line, as a test point. Substitute the coordinates of this test point into the *original* inequality.

Step 3. If a true statement is obtained in Step 2, shade the half-plane that contains the test point. If a false statement is obtained, shade the half-plane that does not contain the test point.

EXAMPLE 2 Graph: $x + y < 7$

Solution

Step 1. First we graph the boundary line by graphing the equation $x + y = 7$. We graph this boundary as a dashed line because the inequality sign is $<$, and thus the points on the line are not solutions of the inequality $x + y < 7$.

(Continued on next page)

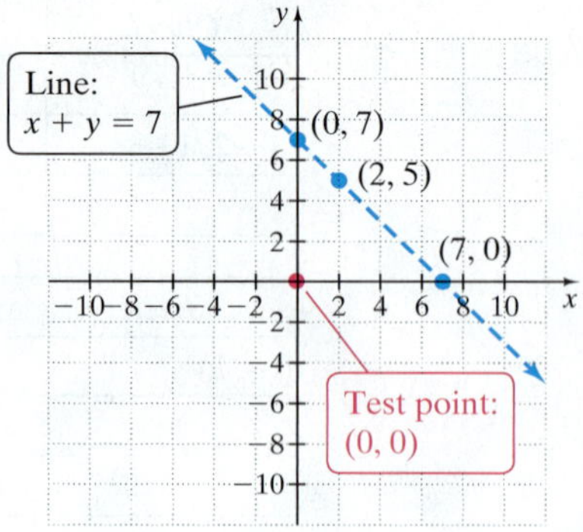

Step 2. Next, we choose a test point, being careful not to choose a point on the boundary line. We choose $(0, 0)$. Substitute the coordinates of $(0, 0)$ into $x + y < 7$.

$$x + y < 7 \quad \text{Original inequality}$$
$$0 + 0 \overset{?}{<} 7 \quad \text{Replace } x \text{ with 0 and } y \text{ with 0.}$$
$$0 < 7 \quad \text{True}$$

Step 3. Since the result is a true statement, $(0, 0)$ is a solution of $x + y < 7$, and every point in the same half-plane as $(0, 0)$ is also a solution. To indicate this, shade the entire half-plane containing $(0, 0)$, as shown.

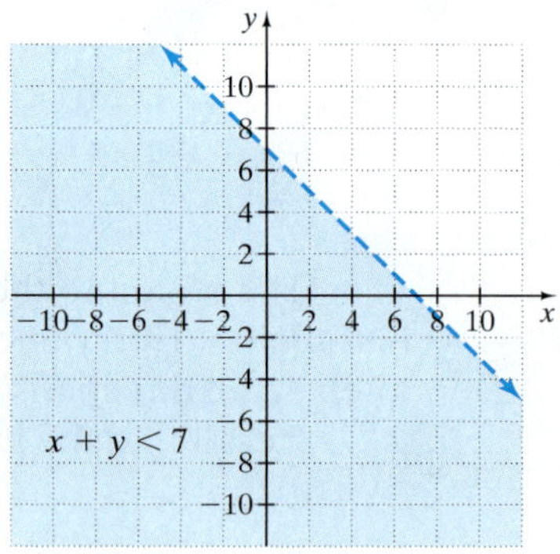

Graph of $x + y < 7$

PRACTICE 2 Graph: $x + y > 5$

✔ CONCEPT CHECK

Determine whether $(0, 0)$ is included in the graph of

a. $y \geq 2x + 3$ **b.** $x < 7$ **c.** $2x - 3y < 6$

EXAMPLE 3 Graph: $2x - y \geq 3$

Solution

Step 1. We graph the boundary line by graphing $2x - y = 3$. We draw this line as a solid line because the inequality sign is $\geq$, and thus the points on the line are solutions of $2x - y \geq 3$.

Step 2. Once again, $(0, 0)$ is a convenient test point since it is not on the boundary line. We substitute 0 for x and 0 for y into the original inequality.

$$2x - y \geq 3$$
$$2(0) - 0 \overset{?}{\geq} 3 \quad \text{Let } x = 0 \text{ and } y = 0.$$
$$0 \geq 3 \quad \text{False}$$

Answers to Concept Check:
a. no **b.** yes **c.** yes

Step 3. Since the statement is false, no point in the half-plane containing (0, 0) is a solution. Therefore, we shade the half-plane that does not contain (0, 0). Every point in the shaded half-plane and every point on the boundary line is a solution of $2x - y \geq 3$.

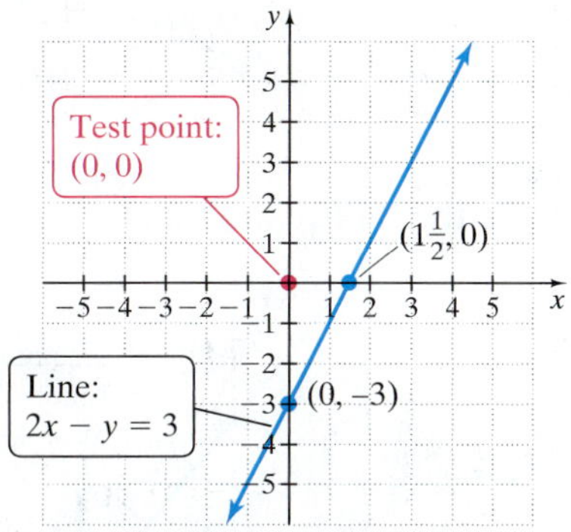

Step 1 and Step 2 on page 206

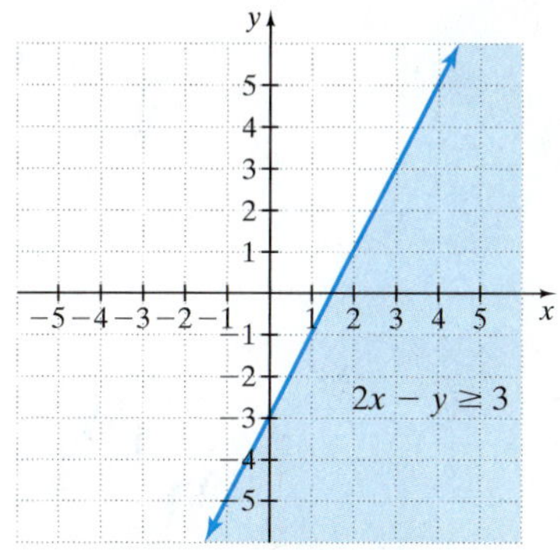

Graph of $2x - y \geq 3$ □

PRACTICE 3 Graph: $3x - y \geq 4$ ■

Helpful Hint

When graphing an inequality, make sure the test point is substituted into the **original inequality.** For Example 3, we substituted the test point (0, 0) into the **original inequality** $2x - y \geq 3$, *not* $2x - y = 3$.

EXAMPLE 4 Graph: $x > 2y$

Solution

Step 1. We find the boundary line by graphing $x = 2y$. The boundary line is a dashed line since the inequality symbol is $>$.

Step 2. We cannot use (0, 0) as a test point because it is a point on the boundary line. We choose instead (0, 2).

$$x > 2y$$
$$0 \stackrel{?}{>} 2(2) \quad \text{Let } x = 0 \text{ and } y = 2.$$
$$0 > 4 \quad \text{False}$$

Step 3. Since the statement is false, we shade the half-plane that does not contain the test point (0, 2), as shown.

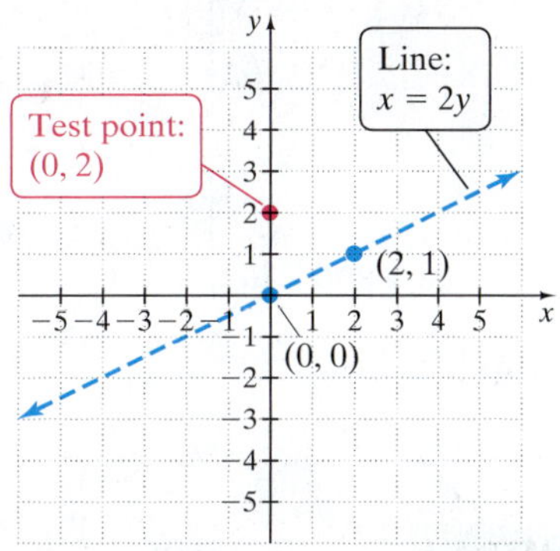

Step 1 and Step 2 above

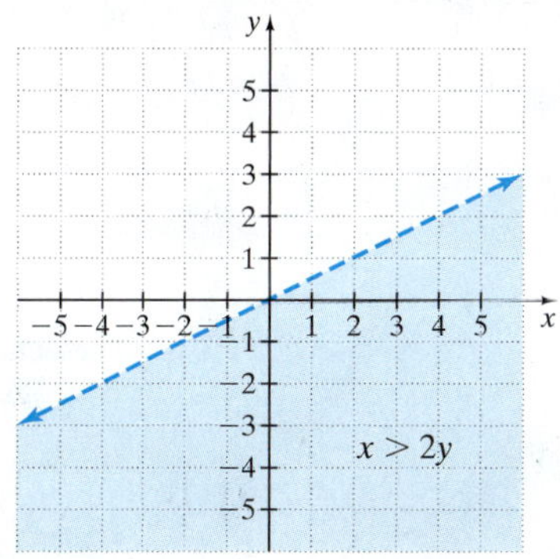

Graph of $x > 2y$ □

PRACTICE 4 Graph: $x > 3y$ ■

EXAMPLE 5 Graph: $5x + 4y \leq 20$

Solution We graph the solid boundary line $5x + 4y = 20$ and choose $(0, 0)$ as the test point.

$$5x + 4y \leq 20$$
$$5(0) + 4(0) \stackrel{?}{\leq} 20 \quad \text{Let } x = 0 \text{ and } y = 0.$$
$$0 \leq 20 \quad \text{True}$$

We shade the half-plane that contains $(0, 0)$, as shown.

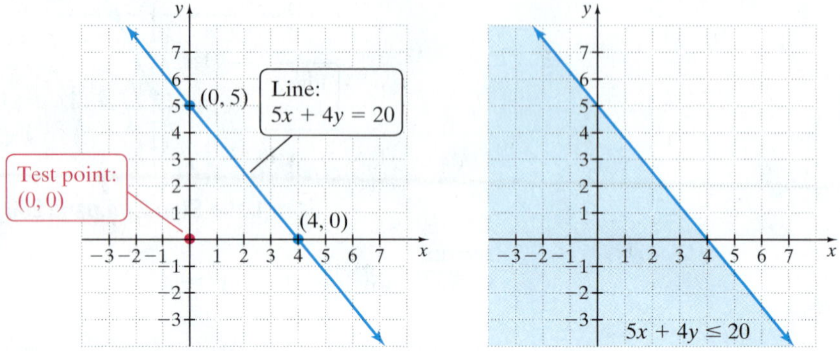

Steps 1 and 2 to graph $5x + 4y \leq 20$ **Graph of $5x + 4y \leq 20$** □

PRACTICE 5 Graph: $3x + 4y \geq 12$ ■

EXAMPLE 6 Graph: $y > 3$

Solution We graph the dashed boundary line $y = 3$ and choose $(0, 0)$ as the test point. (Recall that the graph of $y = 3$ is a horizontal line with y-intercept 3.)

$$y > 3$$
$$0 \stackrel{?}{>} 3 \quad \text{Let } y = 0.$$
$$0 > 3 \quad \text{False}$$

We shade the half-plane that does not contain $(0, 0)$, as shown.

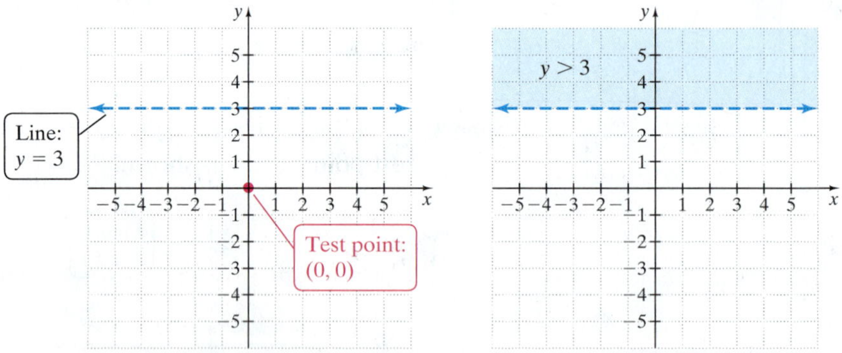

Steps 1 and 2 to graph $y > 3$ **Graph of $y > 3$** □

PRACTICE 6 Graph: $x > 3$ ■

EXAMPLE 7 Graph: $y \leq \frac{2}{3}x - 4$

Solution Graph the solid boundary line $y = \frac{2}{3}x - 4$. This equation is in slope-intercept form with slope $\frac{2}{3}$ and y-intercept -4.

We use this information to graph the line. Then we choose (0, 0) as our test point.

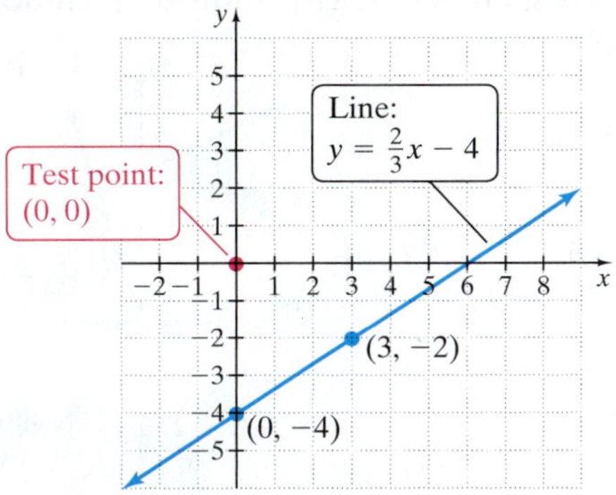

Check the test point.

$y \leq \frac{2}{3}x - 4$

$0 \overset{?}{\leq} \frac{2}{3} \cdot 0 - 4$

$0 \leq -4$ False

Since false, we shade the half-plane that does not contain (0, 0), as shown.

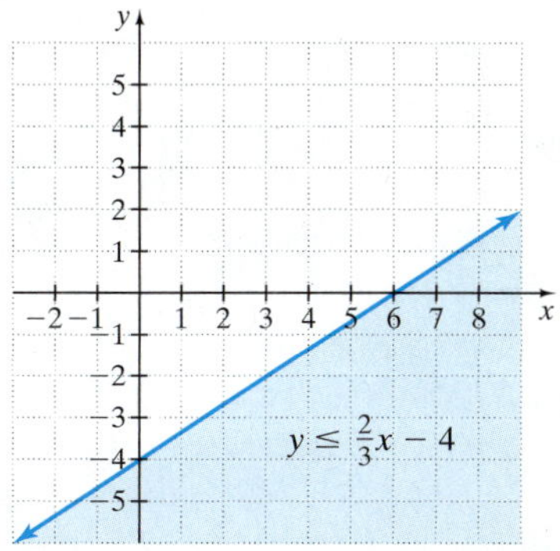

Steps 1 and 2 to graph $y \leq \frac{2}{3}x - 4$

Graph of $y \leq \frac{2}{3}x - 4$ □

PRACTICE 7 Graph: $y < \frac{3}{4}x - 1$ ■

OBJECTIVE 3

Solving Systems of Linear Inequalities

Just as two linear equations make a system of linear equations, two linear inequalities make a **system of linear inequalities.** Systems of inequalities are very important in a process called linear programming. Many businesses use linear programming to find the most profitable way to use limited resources such as employees, machines, or buildings.

A **solution of a system of linear inequalities** is an ordered pair that satisfies each inequality in the system. The set of all such ordered pairs is the solution set of the system. Graphing this set gives us a picture of the solution set. We can graph a system of inequalities by graphing each inequality in the system and identifying the region of overlap.

EXAMPLE 8 Graph the solution of the system: $\begin{cases} 3x \geq y \\ x + 2y \leq 8 \end{cases}$

Solution We begin by graphing each inequality on the same set of axes. The graph of the solution of the system is the region contained in the graphs of both inequalities. It is their intersection.

First, graph $3x \geq y$. The boundary line is the graph of $3x = y$. Sketch a solid boundary line since the inequality $3x \geq y$ means $3x > y$ or $3x = y$. The test point (1, 0) satisfies the inequality, so shade the half-plane that includes (1, 0).

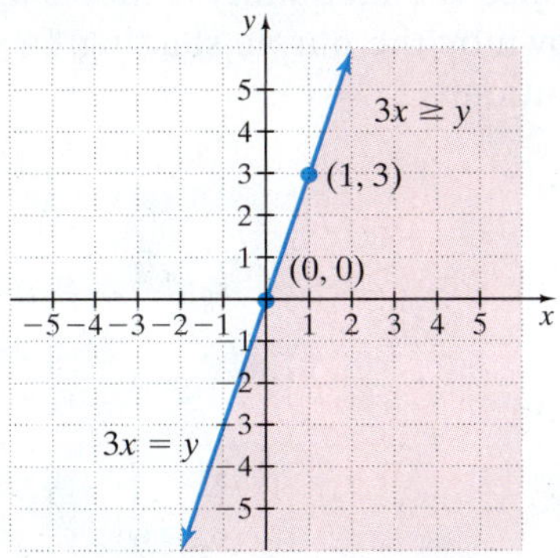

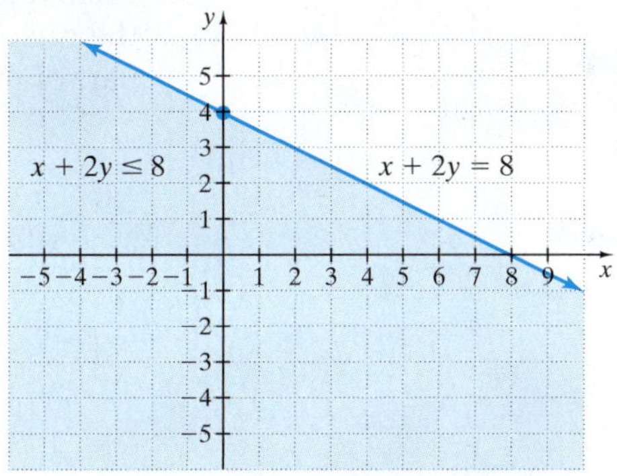

Next, graph $x + 2y \leq 8$ on the same set of axes. (For clarity, the graph of $x + 2y \leq 8$ is shown on a separate set of axes.) Sketch a solid boundary line $x + 2y = 8$. The test point (0, 0) satisfies the inequality $x + 2y \leq 8$, so shade the half-plane that includes (0, 0).

(Continued on next page)

An ordered pair solution of the system must satisfy both inequalities. These solutions are points that lie in both shaded regions. The solution of the system is the purple shaded region as seen below. This solution includes parts of both boundary lines.

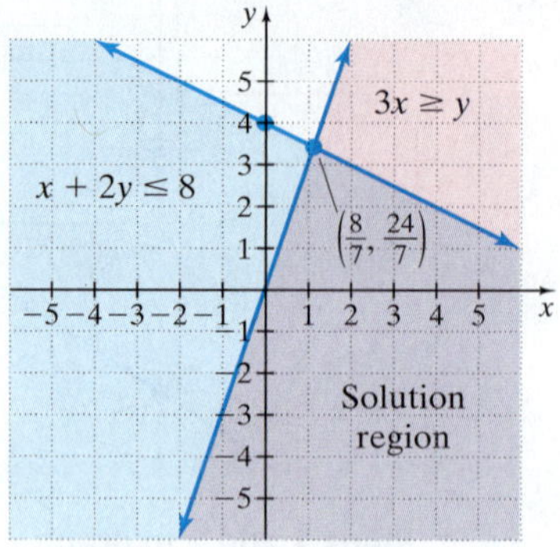

PRACTICE 8 Graph the solution of the system: $\begin{cases} 4x \le y \\ x + 3y \ge 9 \end{cases}$

In linear programming, it is sometimes necessary to find the coordinates of the **corner point:** the point at which the two boundary lines intersect. To find the point of intersection, solve the related linear system

$$\begin{cases} 3x = y \\ x + 2y = 8 \end{cases}$$

by the substitution method or the addition method. The lines intersect at $\left(\frac{8}{7}, \frac{24}{7}\right)$, the corner point of the graph.

Graphing the Solution of a System of Linear Inequalities

Step 1. Graph each inequality in the system on the same set of axes.

Step 2. The solutions of the system are the points common to the graphs of all the inequalities in the system.

EXAMPLE 9 Graph the solution of the system: $\begin{cases} x - y < 2 \\ x + 2y > -1 \end{cases}$

Solution Graph both inequalities on the same set of axes. Both boundary lines are dashed lines since the inequality symbols are < and >. The solution of the system is the region shown by the purple shading. In this example, the boundary lines are not a part of the solution.

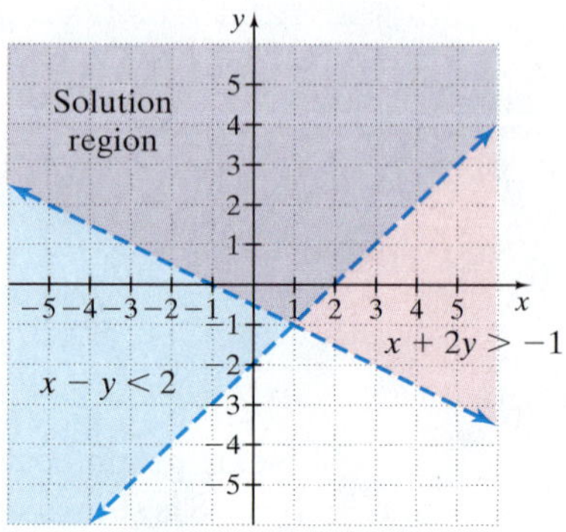

PRACTICE 9 Graph the solution of the system: $\begin{cases} x - y > 4 \\ x + 3y < -4 \end{cases}$

EXAMPLE 10 Graph the solution of the system: $\begin{cases} -3x + 4y < 12 \\ x \geq 2 \end{cases}$

Solution Graph both inequalities on the same set of axes.

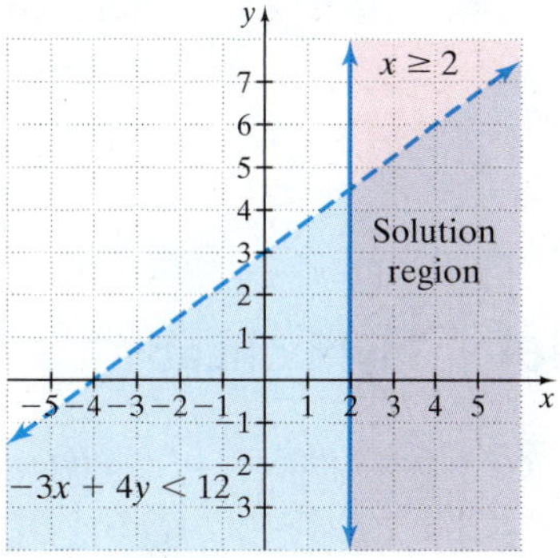

The solution of the system is the purple shaded region, including a portion of the line $x = 2$.

PRACTICE 10 Graph the solution of the system: $\begin{cases} y \leq 6 \\ -2x + 5y > 10 \end{cases}$

Vocabulary, Readiness & Video Check

Use the choices below to fill in each blank. Some choices may be used more than once and some not at all.

true	$x < 3$	$y < 3$	half-planes	yes
false	$x \leq 3$	$y \leq 3$	linear inequality in two variables	no

1. The statement $5x - 6y < 7$ is an example of a(n) ______________.
2. A boundary line divides a plane into two regions called ______________.
3. True or false: The graph of $5x - 6y < 7$ includes its corresponding boundary line. ______________
4. True or false: When graphing a linear inequality, to determine which side of the boundary line to shade, choose a point *not* on the boundary line. ______________
5. True or false: The boundary line for the inequality $5x - 6y < 7$ is the graph of $5x - 6y = 7$. ______________
6. The graph of ______________ is

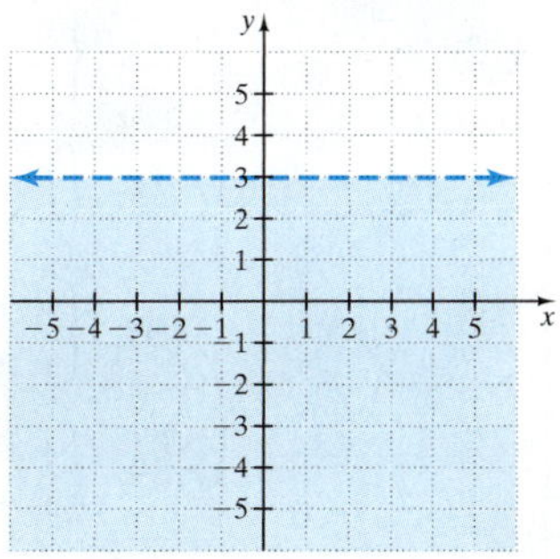

Martin-Gay Interactive Videos

See Video 3.6

Watch the section lecture video and answer the following questions.

OBJECTIVE 1

7. From Example 1, how do you determine whether an ordered pair in x and y is a solution of an inequality in x and y?

OBJECTIVE 2

8. From Example 3, how do you find the equation of the boundary line? How do you determine if the points on the boundary line are solutions of the inequality?

OBJECTIVE 3

9. In Example 4, did the graph of the first inequality of the system limit where we could choose the test point for the second inequality? Why or why not?

3.6 Exercise Set MyMathLab®

Determine which ordered pairs given are solutions of the linear inequality in two variables. See Example 1.

1. $3x - 5y \le -4$; $(-1, -1)$, $(4, 0)$
2. $2x + y \ge 10$; $(-1, -4)$, $(5, 0)$
3. $x < -y$; $(0, 2)$, $(-5, 1)$
4. $y > 3x$; $(0, 0)$, $(-1, -4)$

MIXED PRACTICE

Graph each inequality. See Examples 2 through 7.

5. $x + y \le 1$
6. $x + y \ge -2$
7. $2x + y > -4$
8. $x + 3y \le 3$
9. $x + 6y \le -6$
10. $7x + y > -14$
11. $2x + 5y > -10$
12. $5x + 2y \le 10$
13. $2x + 7y > 5$
14. $3x + 5y \le -2$
15. $x - 2y \ge 3$
16. $4x + y \le 2$
17. $4x + y < 8$
18. $9x + 2y \ge -9$
19. $y \ge 2x$
20. $x < 5y$
21. $x \ge 0$
22. $y \le 0$
23. $y \le -3$
24. $x > -\frac{2}{3}$
25. $2x - 7y > 0$
26. $5x + 2y \le 0$

In Exercises 27 through 32, match each graph with its inequality.

A. $x > 2$ **B.** $y < 2$ **C.** $y < 2x$

D. $y \le -3x$ **E.** $2x + 3y < 6$ **F.** $3x + 2y > 6$

27.

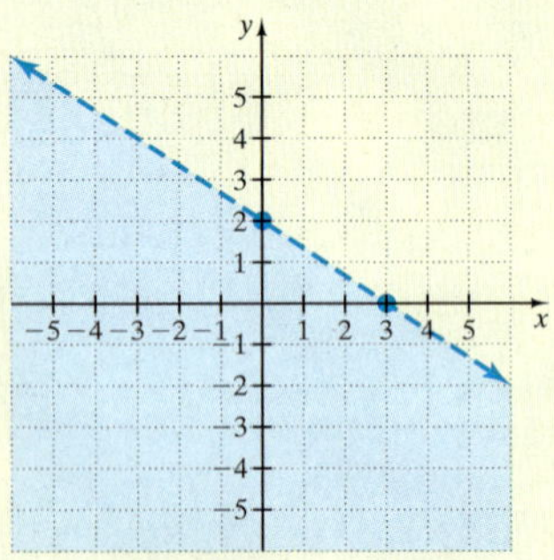

28.

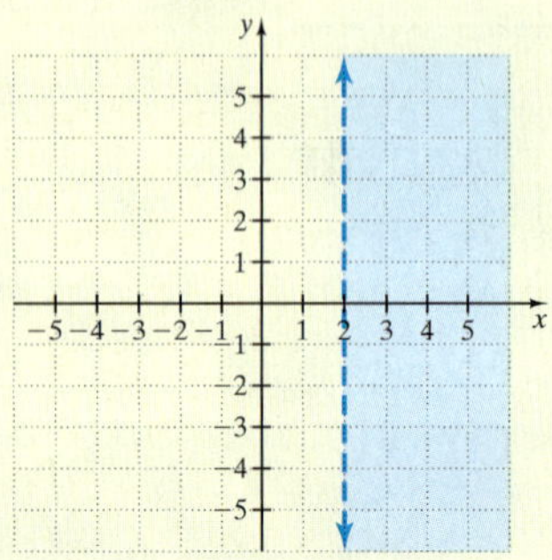

29.

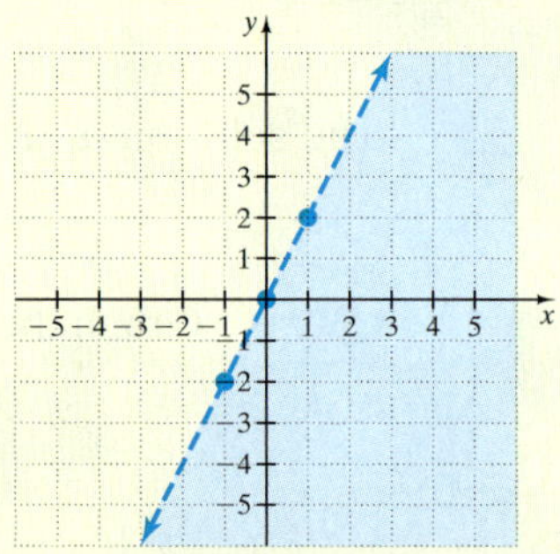

30.

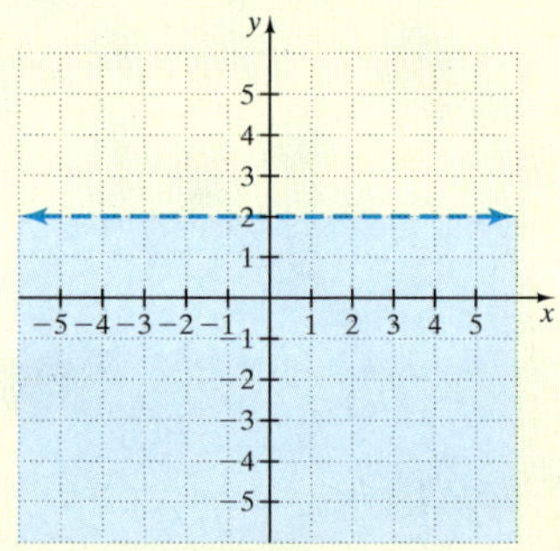

31.

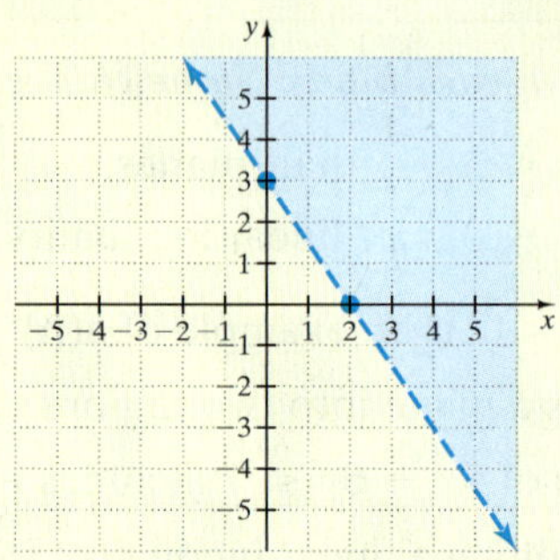

32.

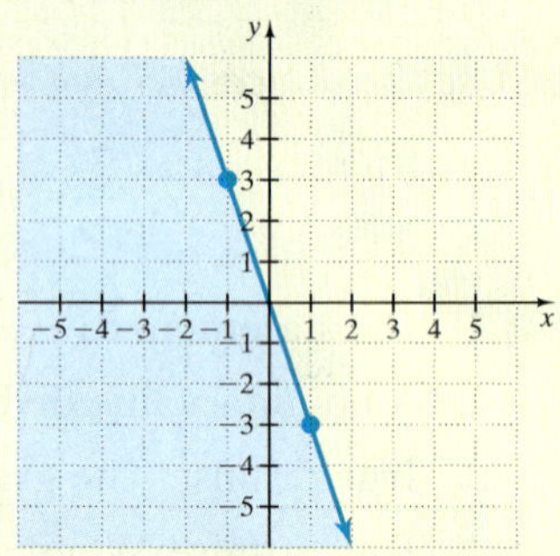

MIXED PRACTICE

Graph the solution of each system of linear inequalities. See Examples 8 through 10.

33. $\begin{cases} y \ge x + 1 \\ y \ge 3 - x \end{cases}$

34. $\begin{cases} y \ge x - 3 \\ y \ge -1 - x \end{cases}$

35. $\begin{cases} y < 3x - 4 \\ y \le x + 2 \end{cases}$

36. $\begin{cases} y \le 2x + 1 \\ y > x + 2 \end{cases}$

37. $\begin{cases} y \le -2x - 2 \\ y \ge x + 4 \end{cases}$

38. $\begin{cases} y \le 2x + 4 \\ y \ge -x - 5 \end{cases}$

39. $\begin{cases} x \ge 3y \\ x + 3y \le 6 \end{cases}$

40. $\begin{cases} -2x < y \\ x + 2y < 3 \end{cases}$

41. $\begin{cases} 3x - 4y \ge -6 \\ 2x + y \le 7 \end{cases}$

42. $\begin{cases} 4x - y \ge -2 \\ 2x + 3y \le -8 \end{cases}$

43. $\begin{cases} x \le 2 \\ y \ge -3 \end{cases}$

44. $\begin{cases} x \ge -3 \\ y \ge -2 \end{cases}$

45. $\begin{cases} y \ge 1 \\ x < -3 \end{cases}$

46. $\begin{cases} y > 2 \\ x \ge -1 \end{cases}$

47. $\begin{cases} 2x + 3y < -8 \\ x \ge -4 \end{cases}$

48. $\begin{cases} 3x + 2y \le 6 \\ x < 2 \end{cases}$

49. $\begin{cases} y \ge \frac{1}{2}x + 2 \\ y \le \frac{1}{2}x - 3 \end{cases}$

50. $\begin{cases} y \ge -\frac{3}{2}x + 3 \\ y < -\frac{3}{2}x + 6 \end{cases}$

For each system of inequalities, choose the corresponding graph.

51. $\begin{cases} y < 5 \\ x > 3 \end{cases}$

52. $\begin{cases} y > 5 \\ x < 3 \end{cases}$

53. $\begin{cases} y \le 5 \\ x < 3 \end{cases}$

54. $\begin{cases} y > 5 \\ x \ge 3 \end{cases}$

A.

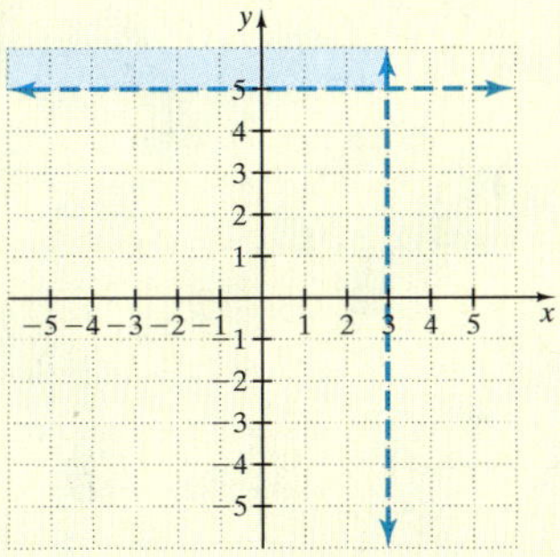

B.

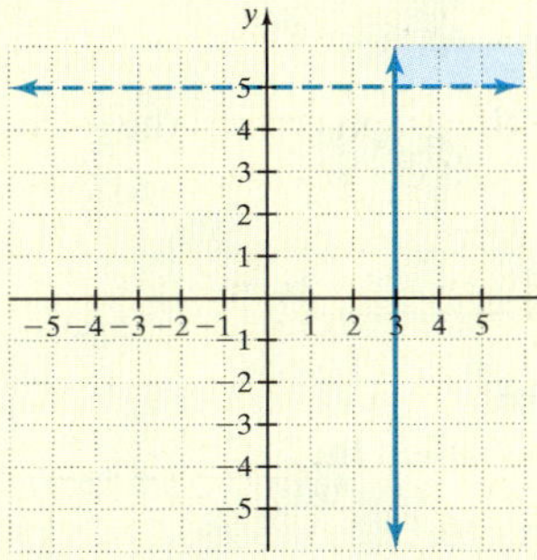

C.

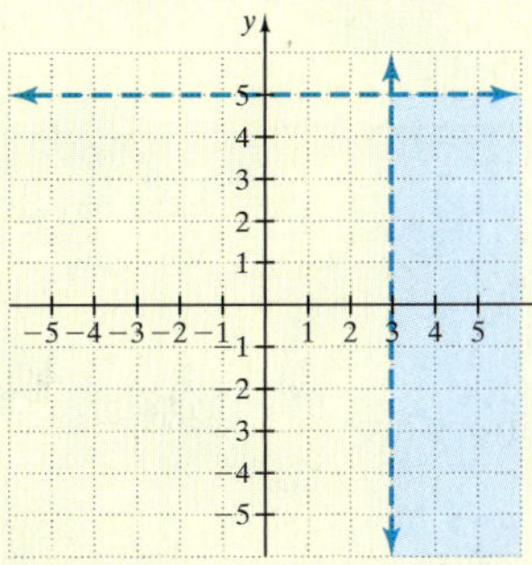

D.

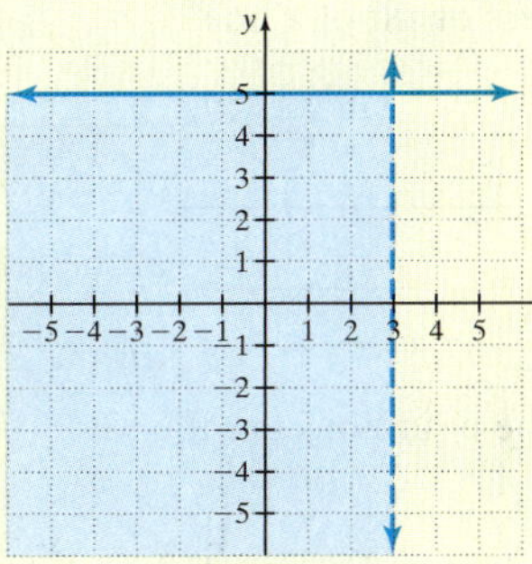

REVIEW AND PREVIEW

Use the histogram to answer Exercises 55–62. See Section R.4.

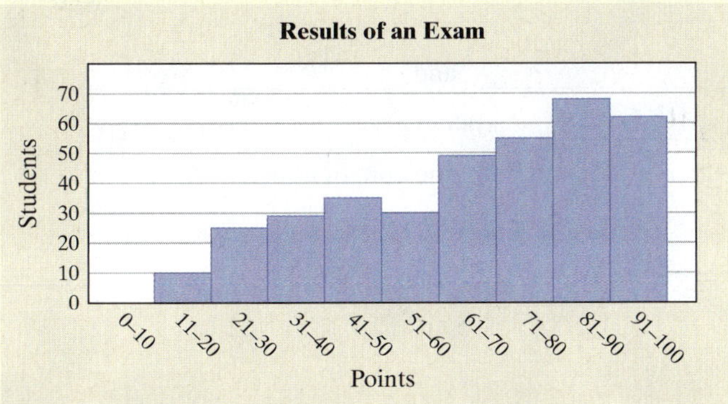

55. The greatest number of students scored in what interval?

56. The fewest number of students scored in what interval?

57. How many students scored in the class interval 21–30?

58. How many students scored in the class interval 71–80?

A total of 363 students took the exam described by the histogram. Use this for Exercises 59–62. Round each to the nearest tenth of a percent.

59. What percent of students scored 51–60?

60. What percent of students scored 0–10?

61. What percent of students scored 71 or higher?

62. What percent of students scored 40 or less?

CONCEPT EXTENSIONS

Determine whether (1, 1) *is included in each graph. See the Concept Check in this section.*

63. $3x + 4y < 8$

64. $y > 5x$

65. $y \ge -\frac{1}{2}x$

66. $x > 3$

67. Explain how to decide which region to shade to show the solution region of the following system.

$$\begin{cases} x \ge 3 \\ y \ge -2 \end{cases}$$

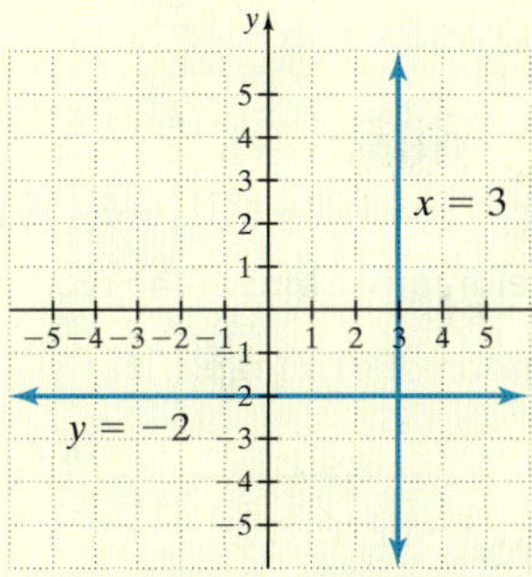

68. Describe the location of the solution region of the system

$$\begin{cases} x > 0 \\ y > 0. \end{cases}$$

69. Graph the solution. $\begin{cases} 2x - y \le 6 \\ x \ge 3 \\ y > 2 \end{cases}$

70. Graph the solution. $\begin{cases} x + y < 5 \\ y < 2x \\ x \ge 0 \\ y \ge 0 \end{cases}$

71. Explain why a point on the boundary line should not be chosen as the test point.

72. Describe the graph of a linear inequality.

73. The price for a taxi cab in a small city is \$2.50 per mile, x, while traveling, and \$0.25 every minute, y, while waiting. If you have \$20 to spend on a cab ride, the inequality

$$2.5x + 0.25y \le 20$$

represents your situation. Graph this inequality in the first quadrant only.

74. A word processor charges \$22 per hour, x, for typing a first draft, and \$15 per hour, y, for making changes and editing to create a second draft. If you need a document created (first and second drafts) and have \$100, the inequality

$$22x + 15y \le 100$$

represents your situation. Graph the inequality in the first quadrant only.

Chapter 3 Vocabulary Check

Fill in each blank with one of the words or phrases listed below. Choices may be used more than once or not at all.

relation	function	domain	range	standard
y-axis	x-axis	solution	linear	slope
x-intercept	y-intercept	y	x	system of linear equations
consistent	independent	dependent	slope-intercept	inconsistent
substitution	addition	system of linear inequalities	point-slope	

1. An ordered pair is a(n) ______________ of an equation in two variables if replacing the variables by the coordinates of the ordered pair results in a true statement.
2. The vertical number line in the rectangular coordinate system is called the ______________.
3. A(n) ______________ equation can be written in the form $Ax + By = C$.
4. A(n) ______________ is a point of the graph where the graph crosses the x-axis.
5. The form $Ax + By = C$ is called ______________ form.
6. A(n) ______________ is a point of the graph where the graph crosses the y-axis.
7. The equation $y = 7x - 5$ is written in ______________ form.
8. The equation $y + 1 = 7(x - 2)$ is written in ______________ form.
9. To find an x-intercept of a graph, let ______________ $= 0$.
10. The horizontal number line in the rectangular coordinate system is called the ______________.
11. To find a y-intercept of a graph, let ______________ $= 0$.
12. The ______________ of a line measures the steepness or tilt of a line.
13. A set of ordered pairs that assigns to each x-value exactly one y-value is called a(n) ______________.
14. The set of all x-coordinates of a relation is called the ______________ of the relation.
15. The set of all y-coordinates of a relation is called the ______________ of the relation.
16. A set of ordered pairs is called a(n) ______________.
17. In a system of linear equations in two variables, if the graphs of the equations are the same, the equations are ______________ equations.
18. Two or more linear equations are called a(n) ______________.
19. A system of equations that has at least one solution is called a(n) ______________ system.
20. A(n) ______________ of a system of two equations in two variables is an ordered pair of numbers that is a solution of both equations in the system.
21. Two algebraic methods for solving systems of equations are ______________ and ______________.
22. A system of equations that has no solution is called a(n) ______________ system.
23. In a system of linear equations in two variables, if the graphs of the equations are different, the equations are ______________ equations.
24. Two or more linear inequalities are called a(n) ______________.

Chapter 3 Highlights

DEFINITIONS AND CONCEPTS	EXAMPLES

Section 3.1 Scatter Plots and Graphing Linear Equations

The **rectangular coordinate system** consists of a plane and a vertical and a horizontal number line intersecting at their 0 coordinates. The vertical number line is called the ***y*-axis** and the horizontal number line is called the ***x*-axis.** The point of intersection of the axes is called the **origin.**

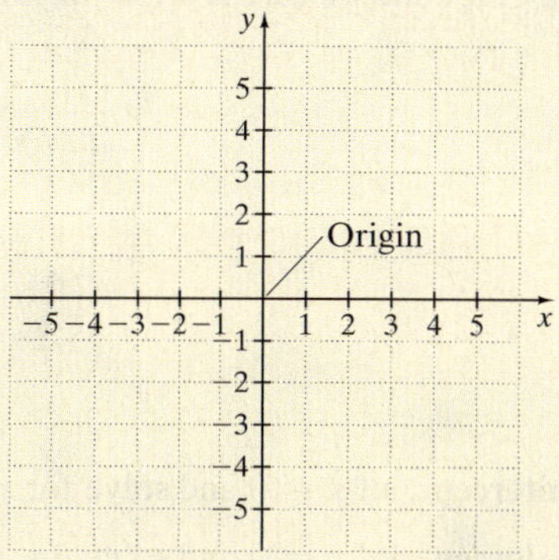

To **plot** or **graph** an ordered pair means to find its corresponding point on a rectangular coordinate system.

To plot or graph an ordered pair such as $(3, -2)$, start at the origin. Move 3 units to the right and, from there, 2 units down.

To plot or graph $(-3, 4)$, start at the origin. Move 3 units to the left and, from there, 4 units up.

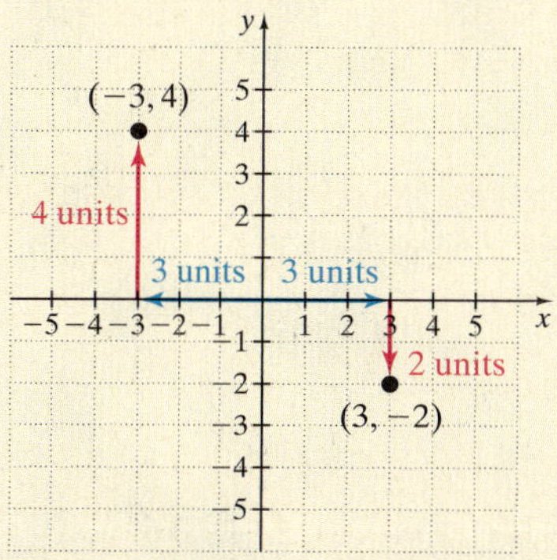

An ordered pair is a **solution** of an equation in two variables if replacing the variables by the coordinates of the ordered pair results in a true statement.

Determine whether $(-1, 5)$ is a solution of $2x + 3y = 13$.

$$2x + 3y = 13$$
$$2(-1) + 3 \cdot 5 = 13 \quad \text{Let } x = -1, y = 5$$
$$-2 + 15 = 13$$
$$13 = 13 \quad \text{True}$$

A **linear equation in two variables** is an equation that can be written in the form $Ax + By = C$ where A and B are not both 0. The form $Ax + By = C$ is called **standard form.**

Linear Equations

$$3x + 2y = -6 \qquad x = -5$$
$$y = 3 \qquad y = -x + 10$$

$x + y = 10$ is in standard form.

To graph a linear equation in two variables, find three ordered pair solutions. Plot the solution points and draw the line connecting the points.

Graph $x - 2y = 5$.

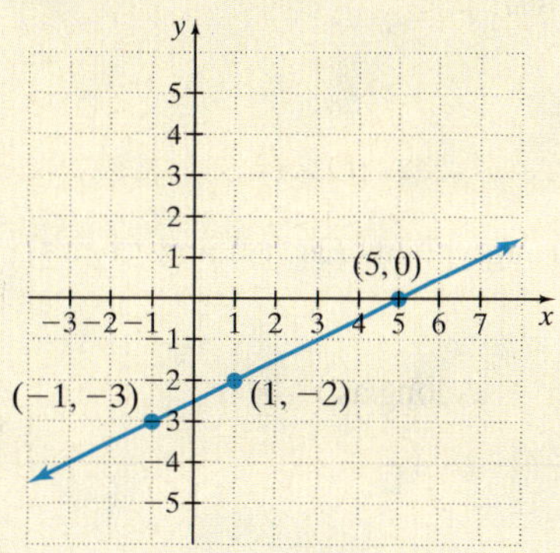

x	y
5	0
1	−2
−1	−3

DEFINITIONS AND CONCEPTS | **EXAMPLES**

Section 3.2 Intercepts, Slope, and Rate of Change

An **intercept** of a graph is a point where the graph intersects an axis. If a graph intersects the x-axis at a, then $(a, 0)$ is the **x-intercept.** If a graph intersects the y-axis at b, then $(0, b)$ is the **y-intercept.**

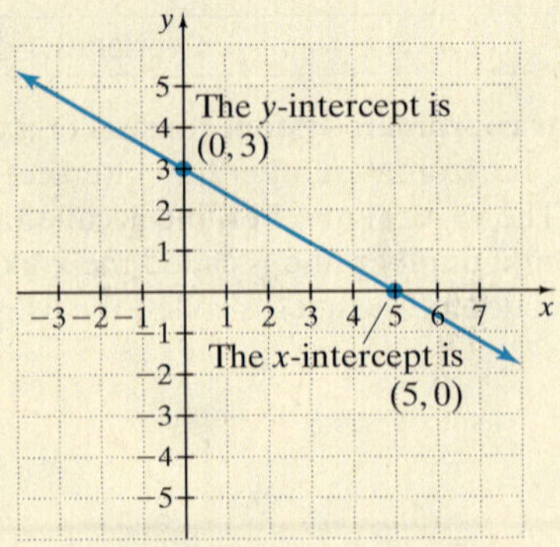

To find the x-intercept, let $y = 0$ and solve for x.

To find the y-intercept, let $x = 0$ and solve for y.

Graph $2x - 5y = -10$ by finding intercepts.

If $y = 0$, then

$$2x - 5 \cdot 0 = -10$$
$$2x = -10$$
$$\frac{2x}{2} = \frac{-10}{2}$$
$$x = -5$$

If $x = 0$, then

$$2 \cdot 0 - 5y = -10$$
$$-5y = -10$$
$$\frac{-5y}{-5} = \frac{-10}{-5}$$
$$y = 2$$

The x-intercept is $(-5, 0)$. The y-intercept is $(0, 2)$.

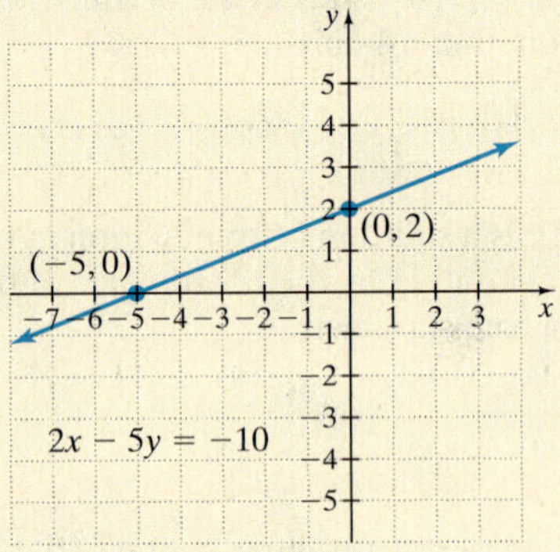

The graph of $x = c$ is a vertical line with x-intercept $(c, 0)$. The graph of $y = c$ is a horizontal line with y-intercept $(0, c)$.

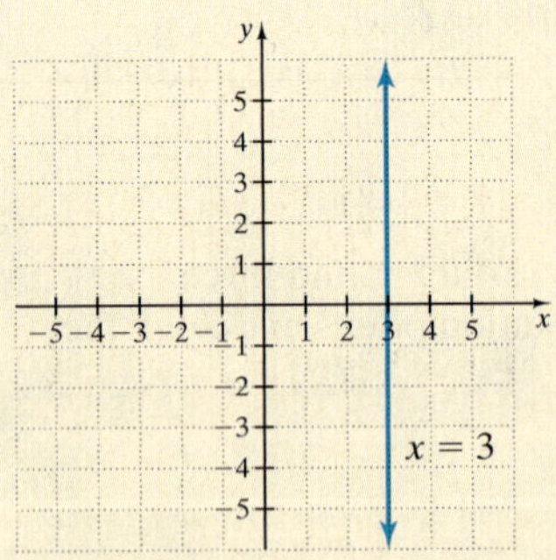

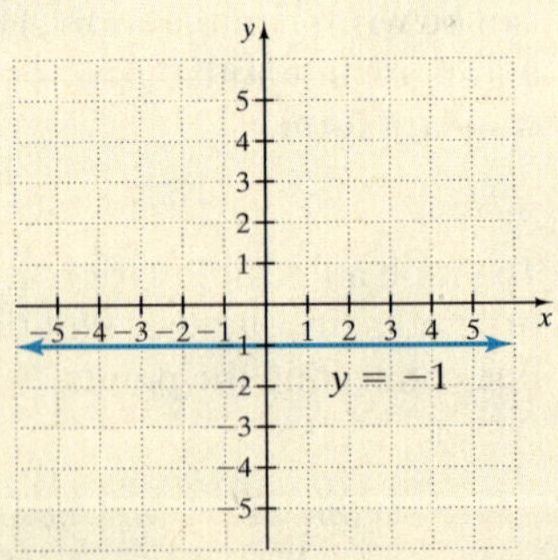

The **slope m** of the line through points (x_1, y_1) and (x_2, y_2) is given by

$$m = \frac{y_2 - y_1}{x_2 - x_1} \quad \text{as long as } x_2 \neq x_1$$

The slope of the line through points $(-1, 6)$ and $(-5, 8)$ is

$$m = \frac{y_2 - y_1}{x_2 - x_1} = \frac{8 - 6}{-5 - (-1)} = \frac{2}{-4} = -\frac{1}{2}$$

DEFINITIONS AND CONCEPTS | **EXAMPLES**

Section 3.2 Intercepts, Slope, and Rate of Change (Continued)

A horizontal line has slope 0.
The slope of a vertical line is undefined.

The slope of the line $y = -5$ is 0.
The line $x = 3$ has undefined slope.

Nonvertical parallel lines have the same slope.

Two nonvertical lines are perpendicular if the slope of one is the negative reciprocal of the slope of the other.

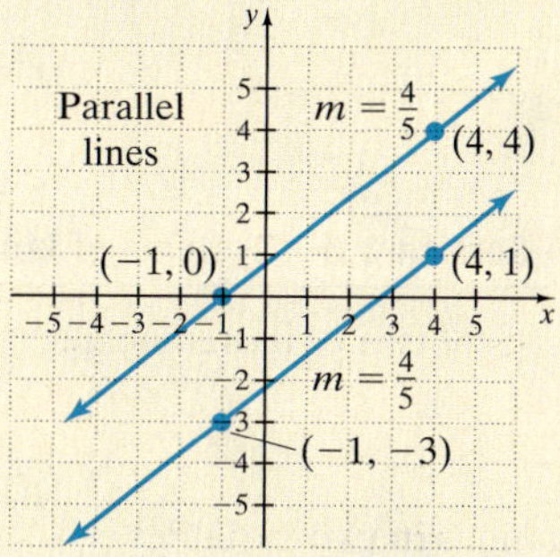

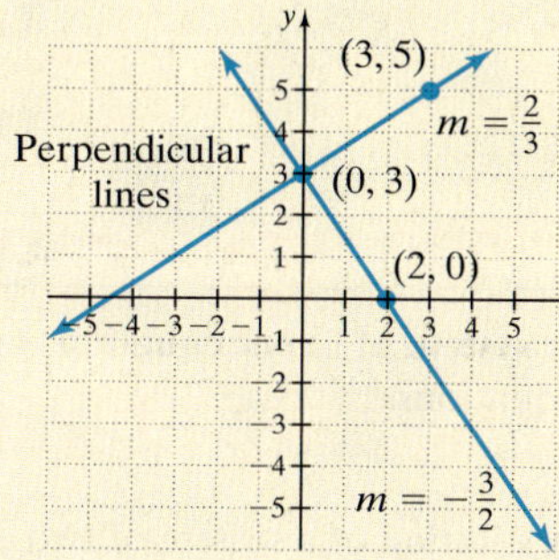

Section 3.3 Equations of Lines

Slope-Intercept Form

$$y = mx + b$$

m is the slope of the line.
$(0, b)$ is the y-intercept.

Find the slope and the y-intercept of the line whose equation is $2x + 3y = 6$.

Solve for y:

$$2x + 3y = 6$$
$$3y = -2x + 6 \quad \text{Subtract } 2x.$$
$$y = -\frac{2}{3}x + 2 \quad \text{Divide by 3.}$$

The slope of the line is $-\frac{2}{3}$ and the y-intercept is $(0, 2)$.

Find an equation of the line with slope 3 and y-intercept $(0, -1)$.
The equation is $y = 3x - 1$.

Point-Slope Form

$$y - y_1 = m(x - x_1)$$

m is the slope.
(x_1, y_1) is a point on the line.

Find an equation of the line with slope $\frac{3}{4}$ that contains the point $(-1, 5)$.

$$y - 5 = \frac{3}{4}[x - (-1)]$$
$$4(y - 5) = 3(x + 1) \quad \text{Multiply by 4.}$$
$$4y - 20 = 3x + 3 \quad \text{Distribute.}$$
$$-3x + 4y = 23 \quad \text{Subtract } 3x \text{ and add 20.}$$

Section 3.4 Functions and Regression Lines

A set of ordered pairs is a **relation.** The set of all x-coordinates is called the **domain** of the relation, and the set of all y-coordinates is called the **range** of the relation.

A **function** is a set of ordered pairs that assigns to each x-value exactly one y-value.

The domain of the relation $\{(0, 5), (2, 5), (4, 5), (5, -2)\}$ is $\{0, 2, 4, 5\}$. The range is $\{-2, 5\}$.

Vertical Line Test

If a vertical line can be drawn so that it intersects a graph more than once, the graph is not the graph of a function.

Which are graphs of functions?

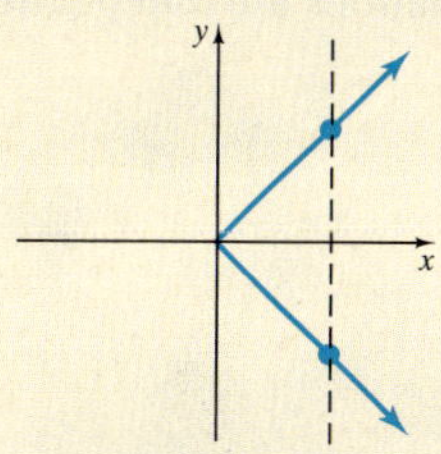

This graph is not the graph of a function.

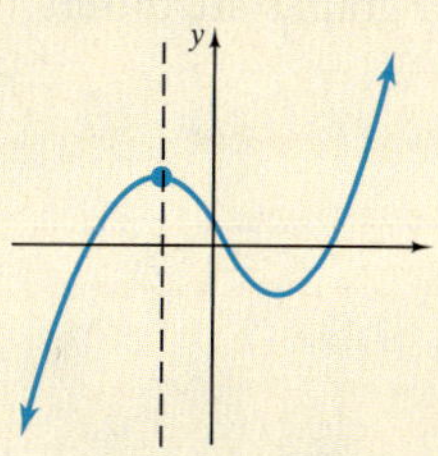

This graph is the graph of a function.

(continued)

DEFINITIONS AND CONCEPTS	EXAMPLES

Section 3.4 Functions and Regression Lines (Continued)

The symbol $f(x)$ means **function of x.** This notation is called **function** notation.

If $f(x) = 2x^2 + 6x - 1$, find $f(3)$.

$$\begin{aligned} f(3) &= 2(3)^2 + 6 \cdot 3 - 1 \\ &= 2 \cdot 9 + 18 - 1 \\ &= 18 + 18 - 1 \\ &= 35 \end{aligned}$$

Section 3.5 Systems of Linear Equations in Two Variables

A **system of linear equations** consists of two or more linear equations.

$$\begin{cases} x - 3y = 6 \\ y = \frac{1}{2}x \end{cases}$$

A **solution** of a system of two equations in two variables is an ordered pair (x, y) that makes both equations true.

Determine whether $(2, -5)$ is a solution of the system.

$$\begin{cases} x + y = -3 \\ 2x - 3y = 19 \end{cases}$$

Replace x with 2 and y with -5 in both equations.

$$\begin{aligned} x + y &= -3 \\ 2 + (-5) &\stackrel{?}{=} -3 \\ -3 &= -3 \quad \text{True} \end{aligned}$$

$$\begin{aligned} 2x - 3y &= 19 \\ 2(2) - 3(-5) &\stackrel{?}{=} 19 \\ 4 + 15 &\stackrel{?}{=} 19 \\ 19 &= 19 \quad \text{True} \end{aligned}$$

$(2, -5)$ is a solution of the system.

Geometrically, a solution of a system in two variables is a point of intersection of the graphs of the equations.

Solve by graphing: $\begin{cases} y = 2x - 1 \\ x + 2y = 13 \end{cases}$

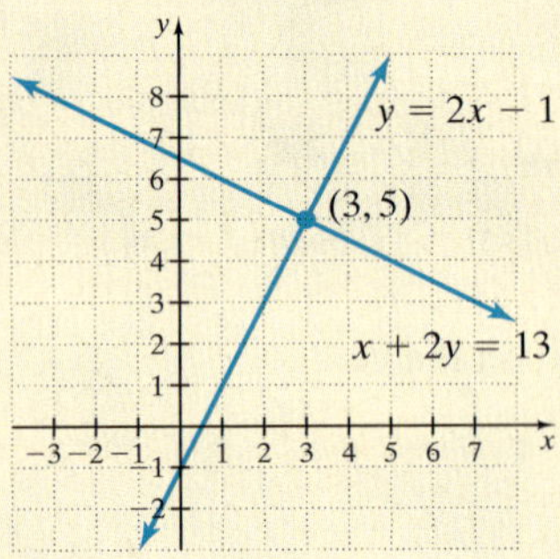

A system of equations with at least one solution is a **consistent system.** A system that has no solution is an **inconsistent system.**

If the graphs of two linear equations are identical, the equations are **dependent.**

If their graphs are different, the equations are **independent.**

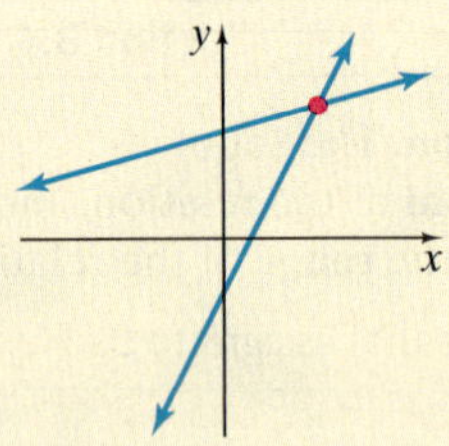

One solution:
Independent equations
Consistent system

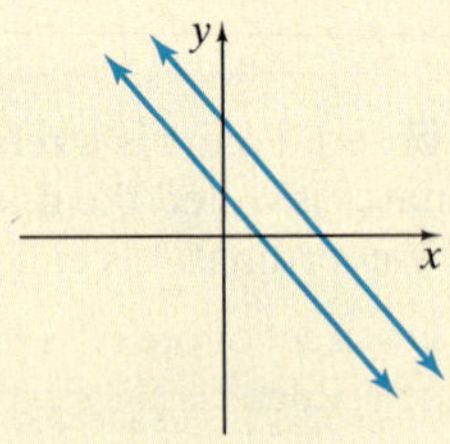

No solution:
Independent equations
Inconsistent system

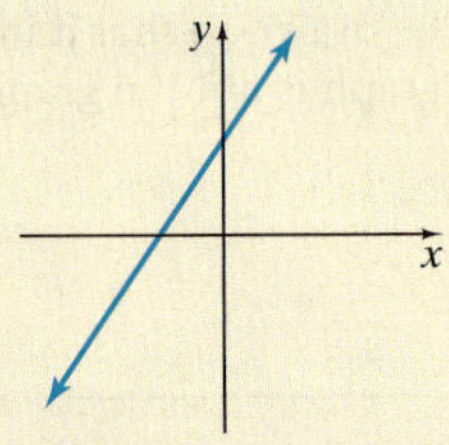

Infinite number of solutions:
Dependent equations
Consistent system

DEFINITIONS AND CONCEPTS	EXAMPLES
Section 3.5 Systems of Linear Equations in Two Variables (Continued)	
Solving a System of Linear Equations by the Substitution Method **Step 1.** Solve one equation for a variable. **Step 2.** Substitute the expression for the variable into the other equation. **Step 3.** Solve the equation from Step 2 to find the value of one variable. **Step 4.** Substitute the value from Step 3 in either original equation to find the value of the other variable. **Step 5.** Check the solution in both original equations.	Solve by substitution. $\begin{cases} y = x + 2 \\ 3x - 2y = -5 \end{cases}$ Since the first equation is solved for y, substitute $x + 2$ for y in the second equation. $3x - 2y = -5$ Second equation $3x - 2(x + 2) = -5$ Let $y = x + 2$. $3x - 2x - 4 = -5$ $x - 4 = -5$ Simplify. $x = -1$ Add 4. To find y, let $x = -1$ in $y = x + 2$, so $y = -1 + 2 = 1$. The solution $(-1, 1)$ checks.
Solving a System of Linear Equations by the Elimination Method **Step 1.** Rewrite each equation in standard form, $Ax + By = C$. **Step 2.** Multiply one or both equations by a nonzero number so that the coefficients of a variable are opposites. **Step 3.** Add the equations. **Step 4.** Find the value of the remaining variable by solving the resulting equation. **Step 5.** Substitute the value from Step 4 into either original equation to find the value of the other variable. **Step 6.** Check the solution in both original equations.	Solve by elimination. $\begin{cases} x - 3y = -3 \\ -2x + y = 6 \end{cases}$ Multiply both sides of the first equation by 2. $2x - 6y = -6$ $-2x + y = 6$ $-5y = 0$ Add. $y = 0$ Divide by -5. To find x, let $y = 0$ in an original equation. $x - 3y = -3$ $x - 3 \cdot 0 = -3$ $x = -3$ The solution $(-3, 0)$ checks.
Section 3.6 Linear Inequalities in Two Variables	
A **linear inequality in two variables** is an inequality that can be written in one of the forms: $Ax + By < C \quad Ax + By \le C$ $Ax + By > C \quad Ax + By \ge C$ Where A, B, and C are real numbers and A and B are not both 0.	***Linear Inequalities*** $2x - 5y < 6 \quad x \ge -5$ $y > -8x \quad y \le 2$
To graph a linear inequality: **1.** Graph the boundary line by graphing the related equation. Draw the line solid if the inequality symbol is $\le$ or $\ge$. Draw the line dashed if the inequality symbol is $<$ or $>$. **2.** Choose a test point not on the line. Substitute its coordinates into the original inequality. **3.** If the resulting inequality is true, shade the half-plane that contains the test point. If the inequality is not true, shade the half-plane that does not contain the test point.	Graph $2x - y \le 4$. **1.** Graph $2x - y = 4$. Draw a solid line because the inequality symbol is $\le$. **2.** Check the test point $(0, 0)$ in the inequality $2x - y \le 4$. $2 \cdot 0 - 0 \overset{?}{\le} 4$ Let $x = 0$ and $y = 0$. $0 \le 4$ True **3.** The inequality is true, so we shade the half-plane containing $(0, 0)$. 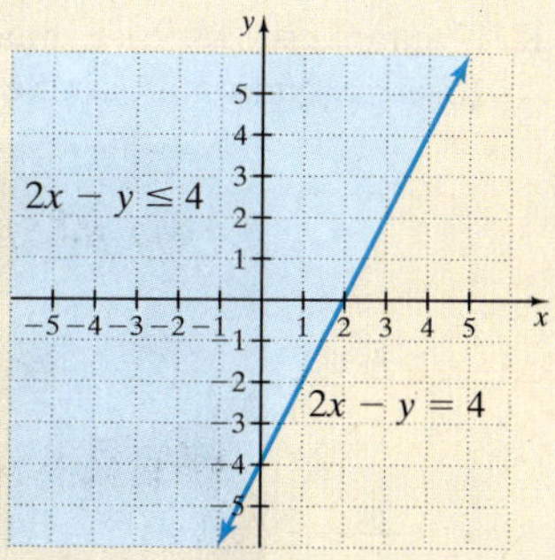

(continued)

DEFINITIONS AND CONCEPTS	EXAMPLES
Section 3.6 Linear Inequalities in Two Variables (Continued)	
A system of linear inequalities consists of two or more linear inequalities.	System of Linear Inequalities $\begin{cases} x - y \geq 3 \\ y \leq -2x \end{cases}$
To graph a system of inequalities, graph each inequality in the system. The overlapping region is the solution of the system.	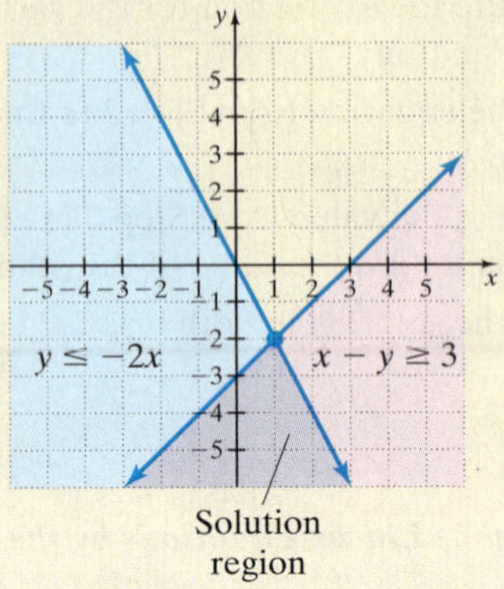

Chapter 3 Review

(3.1) Plot the following ordered pairs on a Cartesian coordinate system.

1. $(-7, 0)$
2. $\left(0, 4\frac{4}{5}\right)$
3. $(-2, -5)$
4. $(1, -3)$
5. $(0.7, 0.7)$
6. $(-6, 4)$
7. A local office supply store uses quantity pricing. The table shows the price per box of #10 security envelopes for different numbers of envelopes in a box purchased.

Price per Box of Envelopes (in dollars)	*Number of Envelopes in Box*
5.00	50
8.50	100
20.00	250
27.00	500

   a. Write each paired data as an ordered pair of the form (price per box of envelopes, number of envelopes in box).

   b. Create a scatter diagram of the paired data. Be sure to label the axes appropriately.

8. The table shows the annual overnight stays in national parks. (*Source:* National Park Service)

Year	*Overnight Stays in National Parks (in millions)*
2008	13.9
2009	14.6
2010	14.6
2011	14.0
2012	14.3
2013	13.5
2014	14.1

   a. Write each paired data as an ordered pair of the form (year, number of overnight stays).

   b. Create a scatter diagram of the paired data. Be sure to label the axes properly.

Determine whether each ordered pair is a solution of the given equation.

9. $7x - 8y = 56$; $(0, 56)$, $(8, 0)$
10. $-2x + 5y = 10$; $(-5, 0)$, $(1, 1)$
11. $x = 13$; $(13, 5)$, $(13, 13)$
12. $y = 2$; $(7, 2)$, $(2, 7)$

Complete the ordered pairs so that each is a solution of the given equation.

13. $-2 + y = 6x$; $(7, \quad)$
14. $y = 3x + 5$; $\left(\quad, -8\right)$

Complete the table of values for each given equation; then plot the ordered pairs. Use a single coordinate system for each exercise.

15. $9 = -3x + 4y$

x	y
	0
	3
9	

16. $x = 2y$

x	y
	0
	5
	−5

The cost in dollars of producing x compact disc holders is given by $y = 5x + 2000$. Use this equation for Exercises 17 and 18.

17. Complete the following table.

x	y
1	
100	
1000	

18. Find the number of compact disc holders that can be produced for $6430.

Graph each linear equation.

19. $x - y = 1$

20. $x + y = 6$

21. $x - 3y = 12$

22. $5x - y = -8$

23. $x = 3y$

24. $y = -2x$

25. $2x - 3y = 6$

26. $4x - 3y = 12$

(3.2) *Identify the intercepts.*

27.

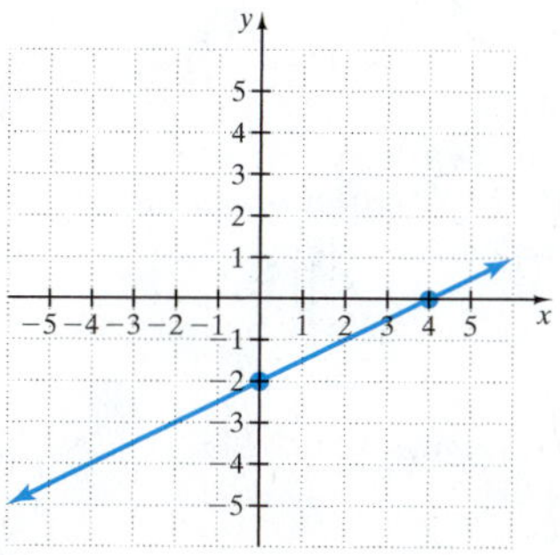

28.

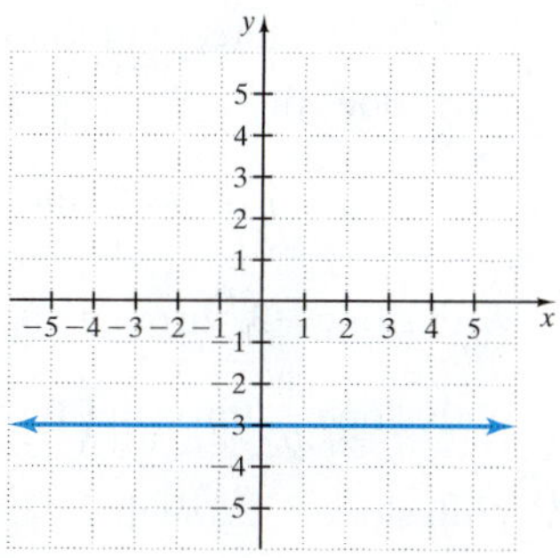

29.

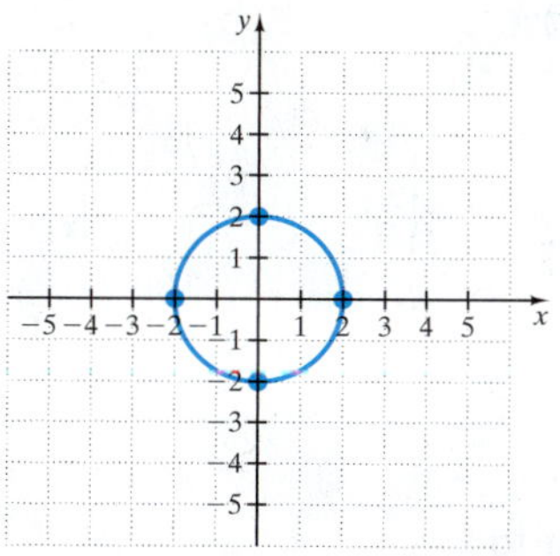

30.

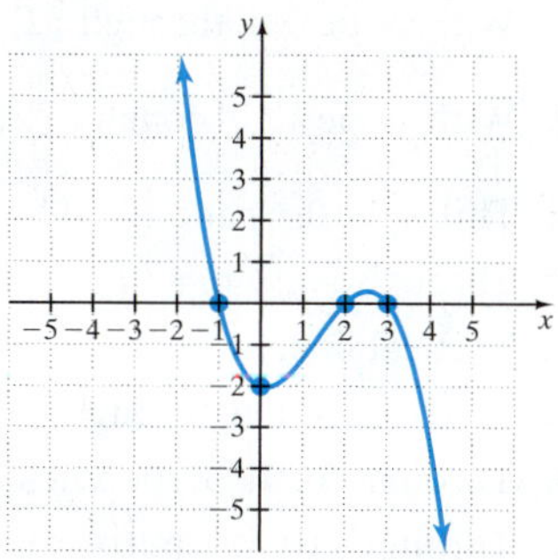

Graph each linear equation by finding its intercepts.

31. $x - 3y = 12$

32. $-4x + y = 8$

33. $y = -3$

34. $x = 5$

35. $y = -3x$

36. $x = 5y$

37. $x - 2 = 0$

38. $y + 6 = 0$

Find the slope of each line.

39.

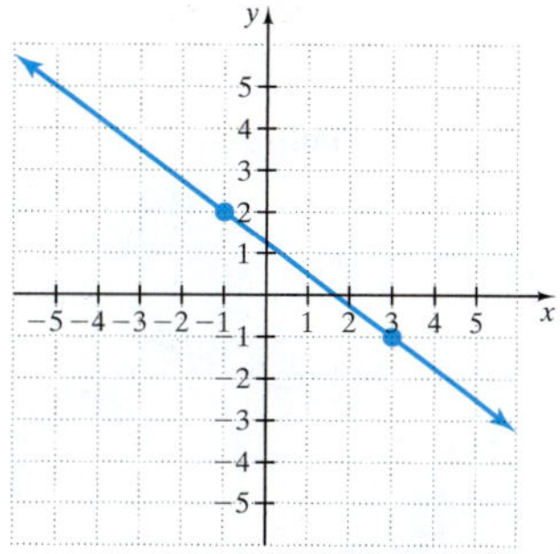

40.

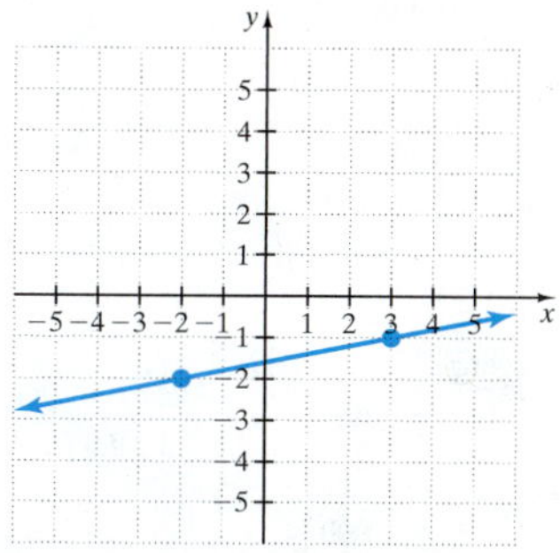

For Exercises 41 through 44, match each slope with its line.

A.

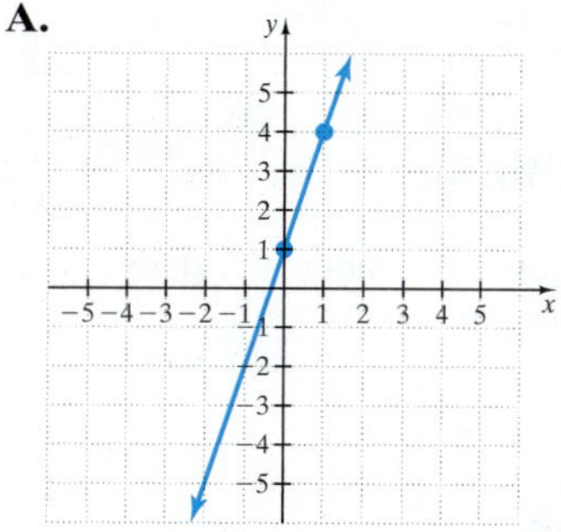

B.

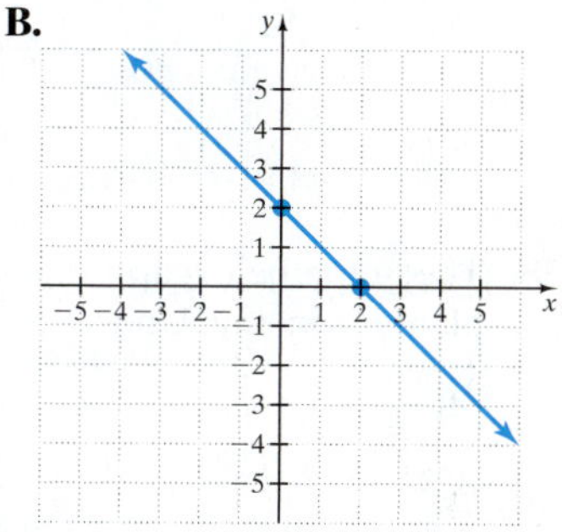

C.

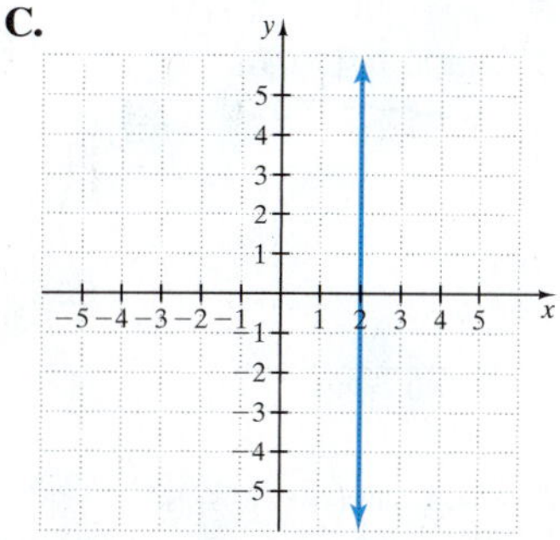

D.

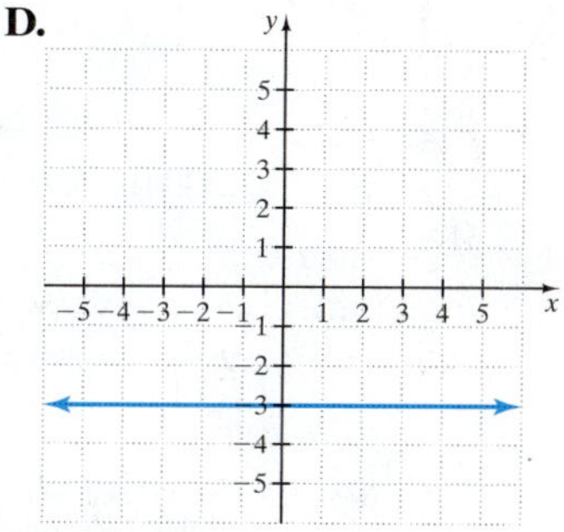

41. $m = 0$

42. $m = -1$

43. undefined slope

44. $m = 3$

Find the slope of the line that goes through the given points.

45. $(2, 5)$ and $(6, 8)$

46. $(4, 7)$ and $(1, 2)$

47. $(1, 3)$ and $(-2, -9)$

48. $(-4, 1)$ and $(3, -6)$

Find the slope of each line.

49. $y = 3x + 7$

50. $x - 2y = 4$

51. $y = -2$

52. $x = 0$

Extension–Determine whether each pair of lines is parallel, perpendicular, or neither.

53. $x - y = -6$
$x + y = 3$

54. $3x + y = 7$
$-3x - y = 10$

55. $y = 4x + \frac{1}{2}$
$4x + 2y = 1$

56. $x = 4$
$y = -2$

Find the slope of each line. Then write a sentence explaining the meaning of the slope as a rate of change. Don't forget to attach the proper units.

57. The graph below approximates the number of U.S. college students (in thousands) earning an associate's degree for each year x.

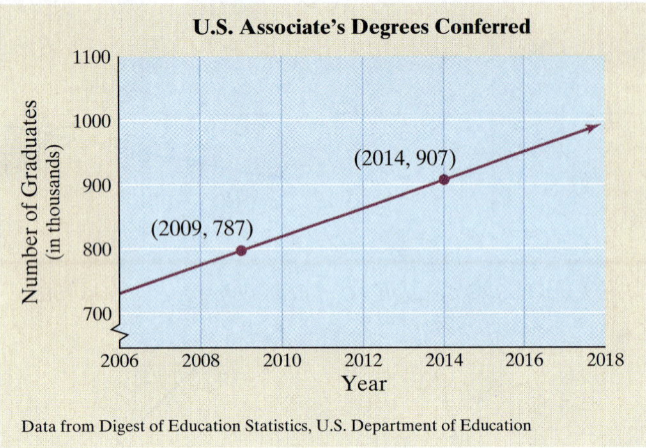

58. The graph below approximates the number of kidney transplants y in the United States for year x.

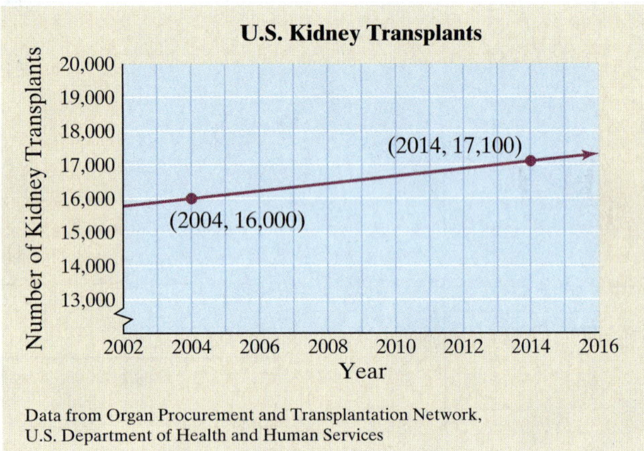

(3.3) *Determine the slope and the y-intercept of the graph of each equation.*

59. $3x + y = 7$

60. $x - 6y = -1$

61. $y = 2$

62. $x = -5$

Use the slope-intercept form to graph each equation.

63. $y = 3x - 1$

64. $y = -3x$

65. $5x - 3y = 15$

66. $-x + 2y = 8$

Write an equation of each line in slope-intercept form.

67. slope -5; y-intercept $\left(0, \frac{1}{2}\right)$

68. slope $\frac{2}{3}$; y-intercept $(0, 6)$

For Exercises 69 through 72, match each equation with its graph.

A.

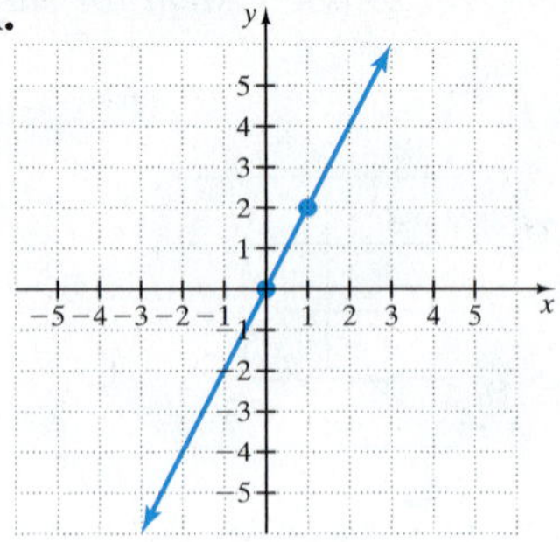

B.

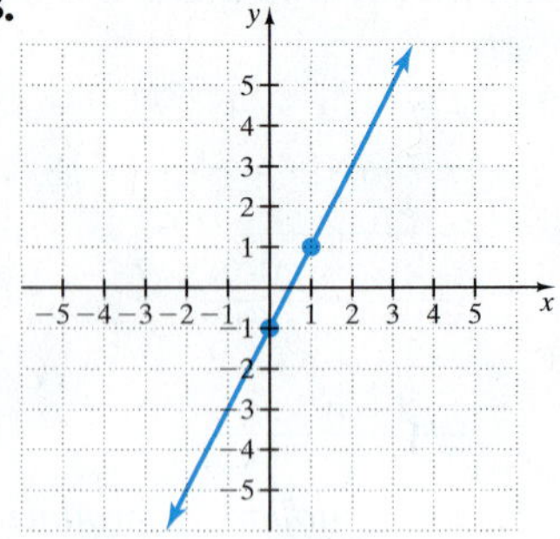

C.

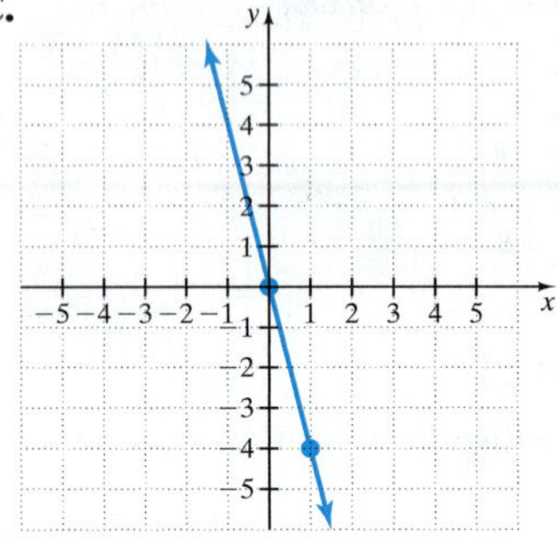

D.

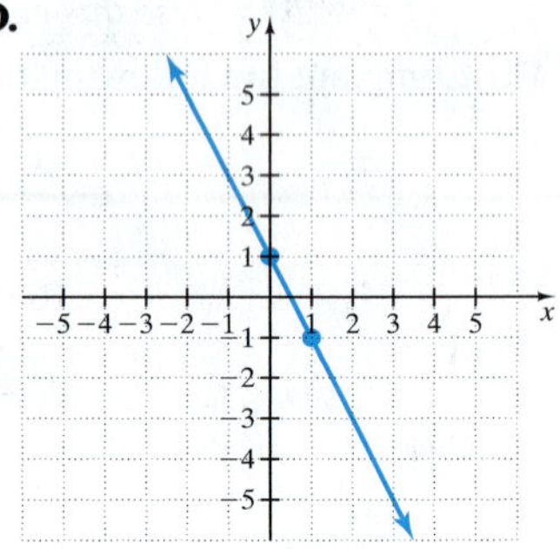

69. $y = -4x$

70. $y = 2x$

71. $y = 2x - 1$

72. $y = -2x + 1$

College is getting more expensive every year. The average cost for tuition and fees at a public two-year college y from 1995 through 2011 can be approximated by the linear equation $y = 56x + 1859$, where x is the number of years after 1995. Use this information for Exercises 73 and 74. (Source: The College Board: Trends in College Pricing 2010)

73. Find the y-intercept of this equation.

74. What does the y-intercept mean?

Write an equation of each line. Write each equation in standard form, $Ax + By = C$, or $x = c$ or $y = c$ form.

75. With slope -3, through $(0, -5)$

76. With slope $\frac{1}{2}$, through $\left(0, -\frac{7}{2}\right)$

77. With slope 0, through $(-2, -3)$

78. With slope 0, through the origin

79. With slope -6, through $(2, -1)$

80. With slope 12, through $\left(\frac{1}{2}, 5\right)$

81. Through $(0, 6)$ and $(6, 0)$

82. Through $(0, -4)$ and $(-8, 0)$

83. Vertical line, through $(5, 7)$

84. Horizontal line, through $(-6, 8)$

85. Through $(6, 0)$, perpendicular to $y = 8$

86. Through $(10, 12)$, perpendicular to $x = -2$

(3.4) *Determine whether each of the following are functions.*

87. $\{(7, 1), (7, 5), (2, 6)\}$

88. $\{(0, -1), (5, -1), (2, 2)\}$

89. $7x - 6y = 1$

90. $y = 7$

91. $x = 2$

92. $y = x^3$

93.

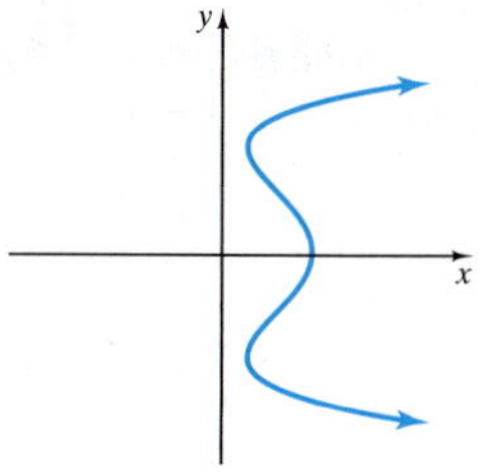

94.

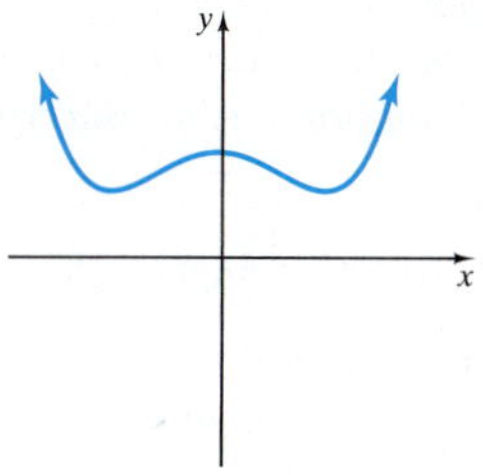

Given the following functions, find the indicated function values.

95. Given $f(x) = -2x + 6$, find

a. $f(0)$ **b.** $f(-2)$ **c.** $f\left(\frac{1}{2}\right)$

96. Given $h(x) = -5 - 3x$, find

a. $h(2)$ **b.** $h(-3)$ **c.** $h(0)$

97. Given $g(x) = x^2 + 12x$, find

a. $g(3)$ **b.** $g(-5)$ **c.** $g(0)$

98. Given $h(x) = 6 - |x|$, find

a. $h(-1)$ **b.** $h(1)$ **c.** $h(-4)$

Find the domain of each function.

99. $f(x) = 2x + 7$

100. $g(x) = \dfrac{7}{x - 2}$

Find the domain and the range of each relation graphed.

101.

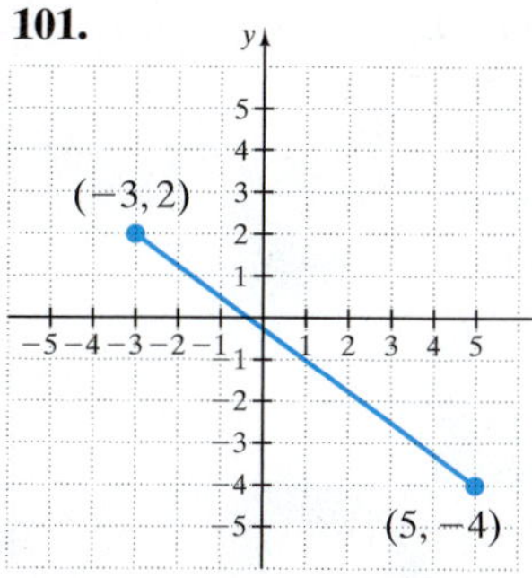

102.

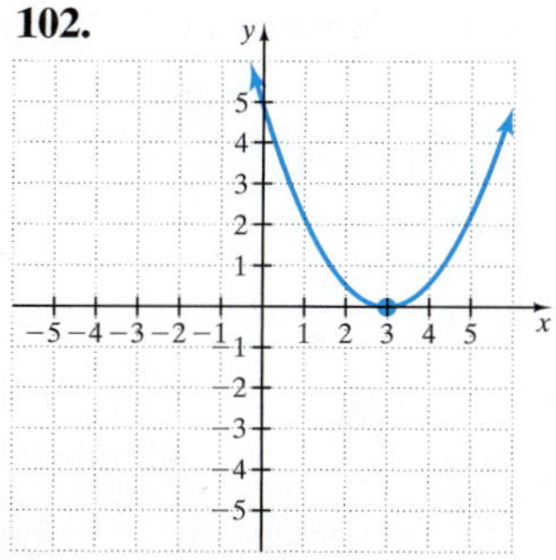

103.

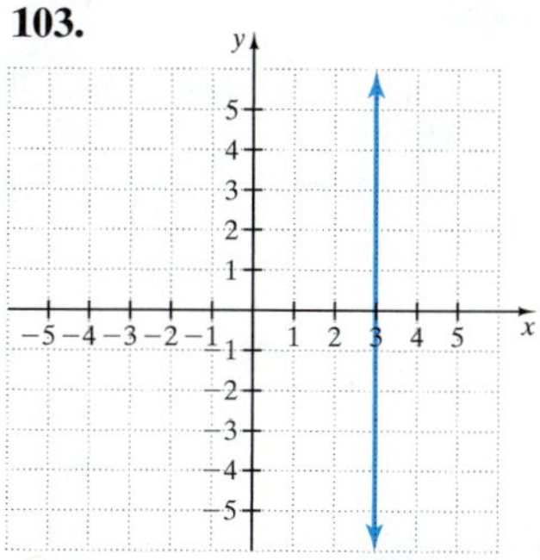

104.

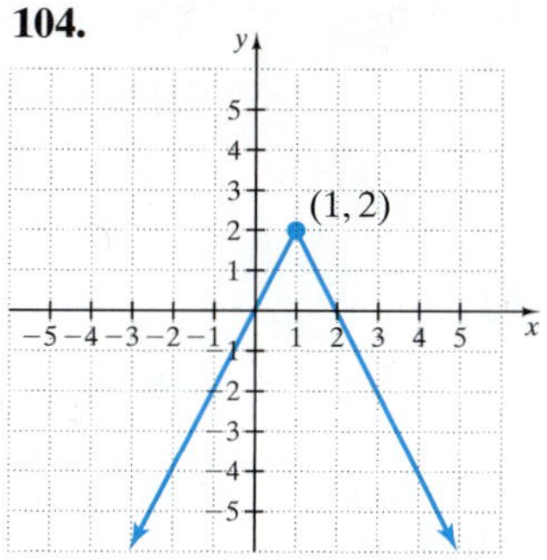

***(3.5)** Determine whether any of the following ordered pairs satisfy the system of linear equations.*

105. $\begin{cases} 2x - 3y = 12 \\ 3x + 4y = 1 \end{cases}$

a. $(12, 4)$ **b.** $(3, -2)$ **c.** $(-3, 6)$

106. $\begin{cases} 5x - 6y = 18 \\ 2y - x = -4 \end{cases}$

a. $(-6, -8)$ **b.** $\left(3, \frac{5}{2}\right)$ **c.** $\left(3, -\frac{1}{2}\right)$

Solve each system of equations by graphing.

107. $\begin{cases} x + y = 5 \\ x - y = 1 \end{cases}$

108. $\begin{cases} x + y = 3 \\ x - y = -1 \end{cases}$

109. $\begin{cases} 2x + y = 5 \\ x = -3y \end{cases}$

110. $\begin{cases} 3x + y = -2 \\ y = -5x \end{cases}$

111. $\begin{cases} y = 3x \\ -6x + 2y = 6 \end{cases}$

112. $\begin{cases} x - 2y = 2 \\ -2x + 4y = -4 \end{cases}$

Solve each system of equations by the substitution method.

113. $\begin{cases} y = 2x + 6 \\ 3x - 2y = -11 \end{cases}$

114. $\begin{cases} y = 3x - 7 \\ 2x - 3y = 7 \end{cases}$

115. $\begin{cases} x + 3y = -3 \\ 2x + y = 4 \end{cases}$

116. $\begin{cases} 3x + y = 11 \\ x + 2y = 12 \end{cases}$

117. $\begin{cases} 4y = 2x + 6 \\ x - 2y = -3 \end{cases}$

118. $\begin{cases} 9x = 6y + 3 \\ 6x - 4y = 2 \end{cases}$

119. $\begin{cases} x + y = 6 \\ y = -x - 4 \end{cases}$

120. $\begin{cases} -3x + y = 6 \\ y = 3x + 2 \end{cases}$

Solve each system of equations by the addition or elimination method.

121. $\begin{cases} 2x + 3y = -6 \\ x - 3y = -12 \end{cases}$

122. $\begin{cases} 4x + y = 15 \\ -4x + 3y = -19 \end{cases}$

123. $\begin{cases} 2x - 3y = -15 \\ x + 4y = 31 \end{cases}$

124. $\begin{cases} x - 5y = -22 \\ 4x + 3y = 4 \end{cases}$

***(3.6)** Graph each inequality.*

125. $5x + 4y < 20$

126. $x + 3y > 4$

127. $y \geq -7$

128. $y \leq -4$

129. $-x \leq y$

130. $x \geq -y$

Graph the solutions of the following systems of linear inequalities.

131. $\begin{cases} y \geq 2x - 3 \\ y \leq -2x + 1 \end{cases}$

132. $\begin{cases} y \leq -3x - 3 \\ y \leq 2x + 7 \end{cases}$

133. $\begin{cases} -3x + 2y > -1 \\ y < -2 \end{cases}$

134. $\begin{cases} -2x + 3y > -7 \\ x \geq -2 \end{cases}$

MIXED REVIEW

Complete the table of values for each given equation.

135. $2x - 5y = 9$

x	y
	1
2	
	−3

136. $x = -3y$

x	y
0	
	1
6	

Find the intercepts for each equation.

137. $2x - 3y = 6$

138. $-5x + y = 10$

Graph each linear equation.

139. $x - 5y = 10$

140. $x + y = 4$

141. $y = -4x$

142. $2x + 3y = -6$

143. $x = 3$

144. $y = -2$

Find the slope of the line that passes through each pair of points.

145. $(3, -5)$ and $(-4, 2)$

146. $(1, 3)$ and $(-6, -8)$

Find the slope of each line.

147.

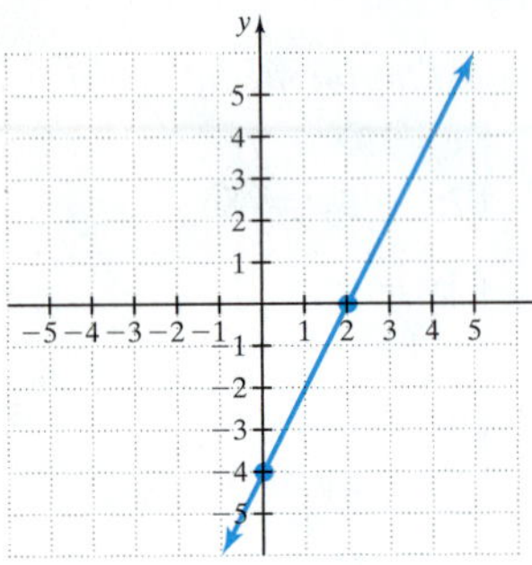

148.

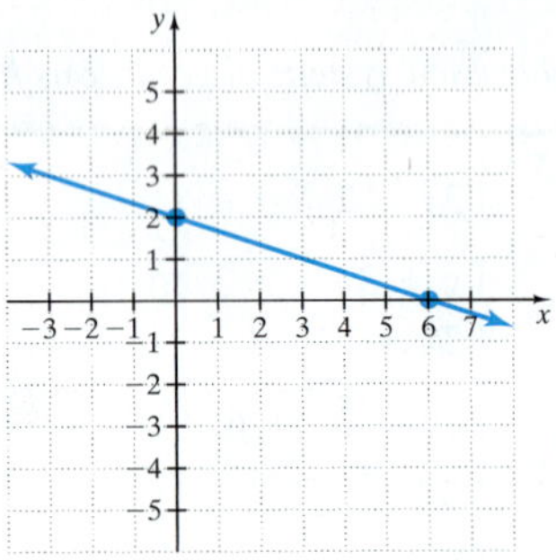

Determine the slope and y-intercept of the graph of each equation.

149. $-2x + 3y = -15$

150. $6x + y - 2 = 0$

Write an equation of the line with the given slope that passes through the given point. Write the equation in the form $y = mx + b$.

151. $m = -5$; $(3, -7)$

152. $m = 3$; $(0, 6)$

Write an equation of the line passing through each pair of points. Write the equation in the form $y = mx + b$.

153. $(-3, 9)$ and $(-2, 5)$

154. $(3, 1)$ and $(5, -9)$

Solve each system of equations.

155. $\begin{cases} x + 4y = 11 \\ 5x - 9y = -3 \end{cases}$

156. $\begin{cases} x + 9y = 16 \\ 3x - 8y = 13 \end{cases}$

157. $\begin{cases} y = -2x \\ 4x + 7y = -15 \end{cases}$

158. $\begin{cases} 3y = 2x + 15 \\ -2x + 3y = 21 \end{cases}$

159. $\begin{cases} x - 3y = -11 \\ 4x + 5y = -10 \end{cases}$

160. $\begin{cases} -x - 15y = 44 \\ 2x + 3y = 20 \end{cases}$

Graph each inequality.

161. $x + 6y < 6$

162. $x + y > -2$

lc

Chapter 3 Getting Ready for the Test

MULTIPLE CHOICE *For Exercises 1 and 2, choose the ordered pair that is NOT a solution of the linear equation.*

1. $x - y = 5$

A. $(7, 2)$ **B.** $(0, -5)$ **C.** $(-2, 3)$ **D.** $(-2, -7)$

2. $y = 4$

A. $(4, 0)$ **B.** $(0, 4)$ **C.** $(2, 4)$ **D.** $(100, 4)$

3. What is the most and then the fewest number of intercepts a line may have?

A. most: 2; fewest: 1 **B.** most: infinite number; fewest: 1 **C.** most: 2; fewest: 0
D. most: infinite number; fewest; 0

4. Choose the linear equation:

A. $\sqrt{x} - 3y = 7$ **B.** $2x = 6^2$ **C.** $4x^3 + 6y^3 = 5^3$ **D.** $y = |x|$

5. **MULTIPLE CHOICE** *The ordered pair $(-1, 2)$ is a solution to what system?*

A. $\begin{cases} 5x - y = -7 \\ x - y = 3 \end{cases}$ **B.** $\begin{cases} 3x - y = -5 \\ x + y = 1 \end{cases}$

C. $\begin{cases} x = 2 \\ x + y = 1 \end{cases}$ **D.** $\begin{cases} y = -1 \\ x + y = -3 \end{cases}$

MULTIPLE CHOICE *For Exercises 6 and 7, choose the best answer.*

6. An ordered pair solution for the function $f(x)$ is $(0, 5)$. This solution using function notation is:

A. $f(5) = 0$ **B.** $f(5) = f(0)$ **C.** $f(0) = 5$ **D.** $0 = 5$

7. Given: $(2, 3)$ and $(0, 9)$. Final Answer: $y = -3x + 9$. Select the correct directions:

A. Find the slope of the line through the two points.

B. Find an equation of the line through the two points. Write the equation in standard form.

C. Find an equation of the line through the two points. Write the equation in slope-intercept form.

MULTIPLE CHOICE *For Exercises 8–11, use the graph to fill in each blank using the choices below.*

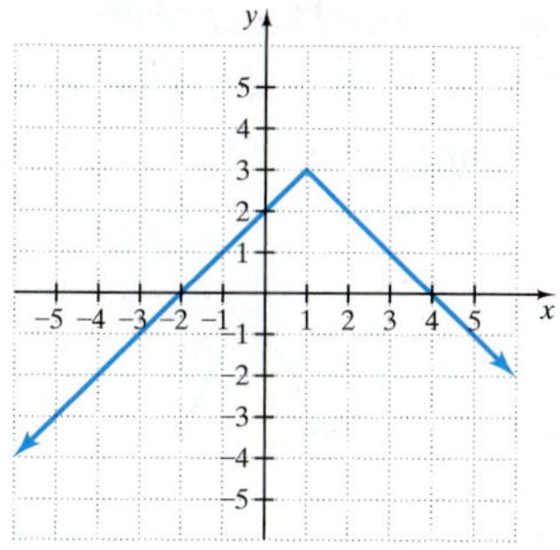

A. -2 **B.** 2 **C.** 4 **D.** 0 **E.** 3

8. $f(0) =$ ____________.

9. $f(4) =$ ____________.

10. If $f(x) = 0$, then $x =$ ____________ or $x =$ ____________.

11. $f(1) =$ ____________.

12. MULTIPLE CHOICE *When solving a system of two linear equations in two variables, all variables subtract out and the resulting equation is $0 = 5$. What does this mean?*

A. the solution is $(0, 5)$ **B.** the system has an infinite number of solutions **C.** the system has no solution

MATCHING *Match each system with its solution. Letter choices may be used more than once or not at all.*

13. $\begin{cases} y = 5x + 2 \\ y = -5x + 2 \end{cases}$

14. $\begin{cases} y = 4x + 2 \\ 8x - 2y = -4 \end{cases}$

15. $\begin{cases} y = \frac{1}{2}x - 3 \\ y = \frac{1}{2}x + 7 \end{cases}$

16. $\begin{cases} y = 6x \\ y = -\frac{1}{6}x \end{cases}$

A. no solution
B. one solution
C. two solutions
D. an infinite number of solutions

MULTIPLE CHOICE *Choose the correct choice for Exercises 17 and 18. The system for these exercises is:*

$$\begin{cases} 5x - y = -8 \\ 2x + 3y = 1 \end{cases}$$

17. When solving, if we decide to multiply the first equation above by 3, the result of the first equation is:

A. $15x - 3y = -8$ **B.** $6x + 9y = 1$ **C.** $6x + 9y = 3$ **D.** $15x - 3y = -24$

18. When solving, if we decide to multiply the second equation above by -5, the result of the second equation is:

A. $-10x - 15y = 1$ **B.** $-25x + 5y = 40$ **C.** $-10x - 15y = -5$ **D.** $-25x + 5y = -8$

Chapter 3 Test

MyMathLab® YouTube

Graph the following.

1. $y = \frac{1}{2}x$

2. $2x + y = 8$

3. $5x - 7y = 10$

4. $y = -1$

For Exercises 5 through 8, find the slopes of the lines.

5.

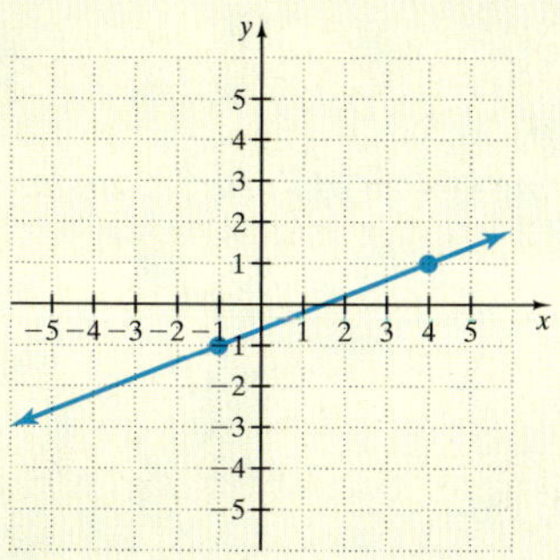

6.

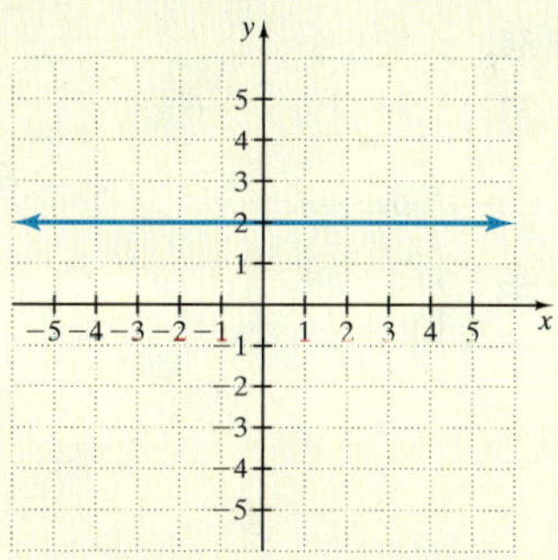

7. Through $(6, -5)$ and $(-1, 2)$

8. $x = 6$

9. Determine the slope and the y-intercept of the graph of $7x - 3y = 2$.

Find equations of the following lines. For Exercises 10 and 11, write the equation in the form $Ax + By = C$. For Exercises 12 and 13, write the equation in the form $y = mx + b$.

10. With slope of $-\frac{1}{4}$, through $(2, 2)$

11. Through the origin and $(6, -7)$

12. Through $(2, -5)$ and $(1, 3)$

13. With slope $\frac{1}{8}$ and y-intercept $(0, 12)$

Determine whether each graph is the graph of a function.

14.

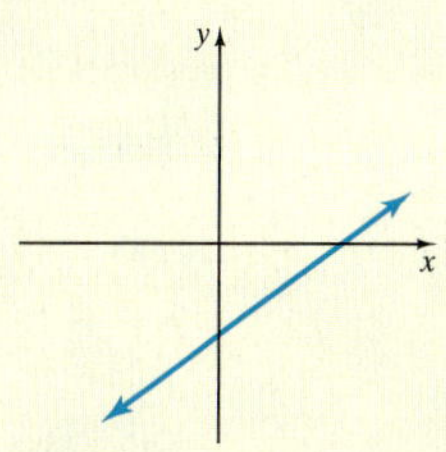

15.

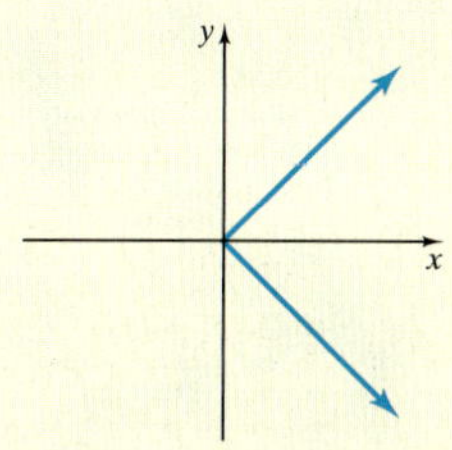

Given the following function, find the indicated function values.

16. $h(x) = x^3 - x$

a. $h(-1)$ **b.** $h(0)$ **c.** $h(4)$

For Exercises 17 and 18,

a. *Identify the x- and y-intercepts.*

b. *Find the domain and the range of each function graphed.*

17.

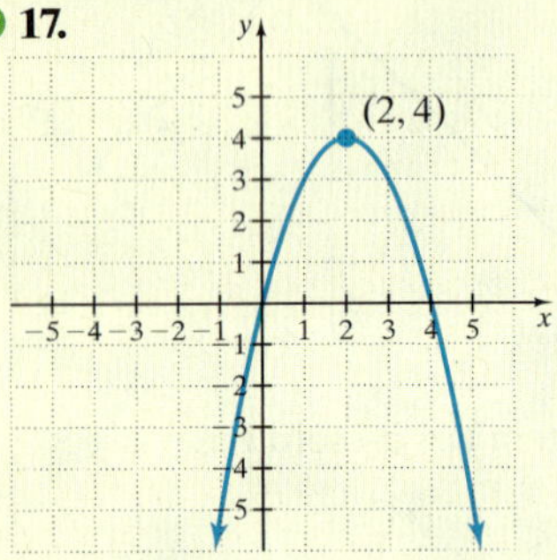

18.

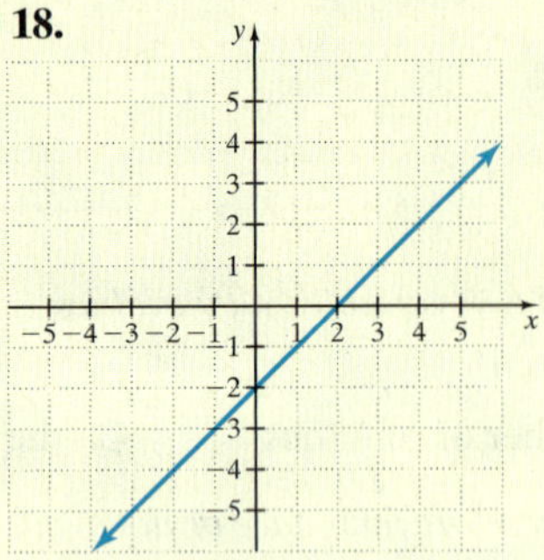

19. If $f(7) = 20$, write the corresponding ordered pair.

20. The table gives the number of Dish Network subscribers (in millions) for the years shown. (*Source:* Dish Network)

Year	Dish Network Subscribers (in millions)
2008	13.68
2009	14.10
2010	14.13
2011	13.97
2012	14.06
2013	14.06
2014	13.98

a. Write this data as a set of ordered pairs of the form (year, number of Dish Network subscribers in millions).

b. Create a scatter diagram of the data. Be sure to label the axes properly.

21. This graph approximates the number of movie ticket sales y (in millions) for the year x.

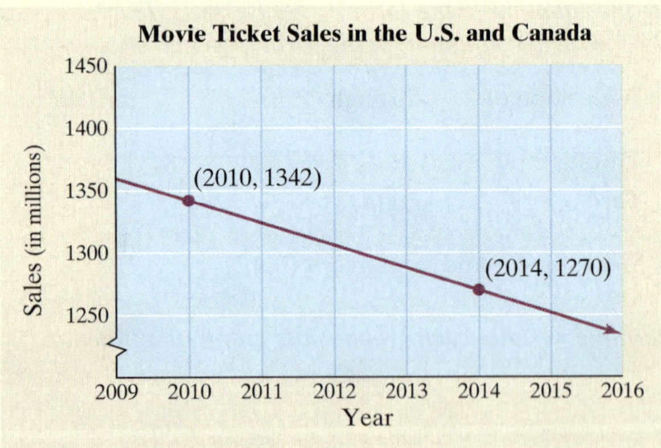

Data from National Association of Theater Owners

a. Find the slope of the line. Then write a sentence explaining the meaning of the slope as a rate of change. Don't forget to attach the proper units.

b. Write two ordered pairs of the form (years past 2000, number of tickets sold in millions).

c. Use the two ordered pairs from part (b) to write a linear equation. Write the equation in slope-intercept form.

d. Use the equation from part (c) to predict the number of movie tickets sold in 2020.

Is the ordered pair a solution of the given linear system?

22. $\begin{cases} 2x - 3y = 5 \\ 6x + y = 1 \end{cases}$; $(1, -1)$

Solve the system by graphing.

23. $\begin{cases} x - y = 2 \\ 3x - y = -2 \end{cases}$

Solve each system by the substitution method.

24. $\begin{cases} 3x - 2y = -14 \\ y = x + 5 \end{cases}$

25. $\begin{cases} \frac{1}{2}x + 2y = -\frac{15}{4} \\ 4x = -y \end{cases}$

Solve each system by the addition method.

26. $\begin{cases} x + y = 28 \\ x - y = 12 \end{cases}$

27. $\begin{cases} 4x - 6y = 7 \\ -2x + 3y = 0 \end{cases}$

Solve each system using the substitution method or the addition method.

28. $\begin{cases} 3x + y = 7 \\ 4x + 3y = 1 \end{cases}$

29. $\begin{cases} 3(2x + y) = 4x + 20 \\ x - 2y = 3 \end{cases}$

30. $\begin{cases} 8x - 4y = 12 \\ y = 2x - 3 \end{cases}$

Graph each inequality.

31. $y > -4x$

32. $2x - 3y > -6$

Graph the solutions of the following system of linear inequalities.

33. $\begin{cases} 2y - x \geq 1 \\ x + y \geq -4 \end{cases}$

Chapter 3 Cumulative Review

1. Insert $<$, $>$, or $=$ in the space between each pair of numbers to make each statement true.

   a. 2 3 b. 7 4 c. 72 27

2. Write the fraction $\frac{56}{64}$ in lowest terms.

3. Multiply $\frac{2}{15}$ and $\frac{5}{13}$. Simplify the product if possible.

4. Add: $\frac{10}{3} + \frac{5}{21}$

5. Simplify: $\frac{3 + |4 - 3| + 2^2}{6 - 3}$

6. Simplify: $16 - 3 \cdot 3 + 2^4$

7. Evaluate each expression.

   a. 3^2 b. $\left(\frac{1}{2}\right)^4$ c. -5^2

   d. $(-5)^2$ e. -5^3 f. $(-5)^3$

8. Simplify: $|9 + (-20)| + |-10|$

9. Simplify.

   a. $20 \div 2 \cdot 10$ b. $1 + 2(1 - 4)^2$

   c. $\frac{|-2|^3 + 1}{-7 - \sqrt{4}}$

10. Simplify: $-9 - (3 - 8)$

11. Evaluate each expression when $x = 4$ and $y = -3$.

    a. $3x - 7y$ b. $-2y^2$

    c. $\frac{\sqrt{x}}{y} - \frac{y}{x}$

12. Is -20 a solution of $\frac{x}{-10} = 2$?

13. Simplify each expression.

    a. $10 + (x + 12)$

    b. $-3(7x)$

14. Simplify: $(12 + x) - (4x - 7)$

15. Identify the numerical coefficient of each term.

    a. $-3y$ b. $22z^4$ c. y

    d. $-x$ e. $\frac{x}{7}$

16. Multiply: $-5(x - 7)$

17. Solve: $x - 7 = 10$ for x

18. Solve: $5(3 + z) - (8z + 9) = -4$

19. Solve: $\frac{5}{2}x = 15$

20. Solve: $\frac{x}{4} - 1 = -7$

21. If x is the first of three consecutive integers, express the sum of the three integers in terms of x. Simplify if possible.

22. Solve: $\frac{x}{3} - 2 = \frac{x}{3}$

23. Solve: $\frac{2(a + 3)}{3} = 6a + 2$

24. Solve $-3 < 4x - 1 \le 2$. Write the solution set in interval notation.

25. The 114th Congress began on January 3, 2015, and had a total of 435 Democrat and Republican representatives. There were 59 fewer Democratic representatives than Republican. Find the number of representatives from each party. (*Source:* congress.gov)

26. Solve $5(x + 4) \ge 4(2x + 3)$. Write the solution set in interval notation.

27. Solve $-5x + 7 < 2(x - 3)$. Write the solution in interval notation.

28. Graph: $y = -3x + 5$

29. Graph: $x \ge -1$ on a number line. Then write the solutions in interval notation.

30. Find the x- and y-intercepts of $2x + 4y = -8$.

31. Solve $-1 \le 2x - 3 < 5$. Graph the solution set and write it in interval notation.

32. Graph $x = 2$ on a rectangular coordinate system.

33. Determine whether each ordered pair is a solution of the equation $x - 2y = 6$.

    a. $(6, 0)$ b. $(0, 3)$

    c. $\left(1, -\frac{5}{2}\right)$

34. Find the slope of the line through $(0, 5)$ and $(-5, 4)$.

35. Determine whether each equation is a linear equation in two variables.

    a. $x - 1.5y = -1.6$ b. $y = -2x$

    c. $x + y^2 = 9$ d. $x = 5$

36. Find the slope of the line $x = -10$.

37. Find the slope of the line $y = -3$

38. Find the slope and y-intercept of the line whose equation is $2x - 5y = 10$.

39. Find an equation of the line with y-intercept $(0, -3)$ and slope of $\frac{1}{4}$.

40. Write an equation of the line through $(2, 3)$ and $(0, 0)$. Write the equation in standard form.

CHAPTER 4

Exponents and Polynomials

Recall from Chapter 1 that an exponent is a shorthand notation for repeated factors. This chapter explores additional concepts about exponents and exponential expressions. An especially useful type of exponential expression is a polynomial. Polynomials model many real-world phenomena. In this chapter, we focus on polynomials and operations on polynomials.

Can You Imagine a World Without the Internet?

In 1995, less than 1% of the world population was connected to the Internet. By 2015, that number had increased to 40%. Technology changes so fast that, if this trend continues, by the time you read this, far more than 40% of the world population will be connected to the Internet. The circle graph below shows Internet users by region of the world in 2015. In Section 4.2, Exercises 99 and 100, we explore more about the growth of Internet users.

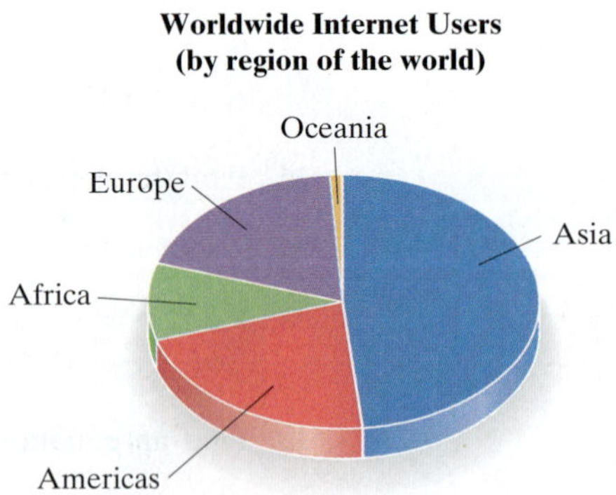

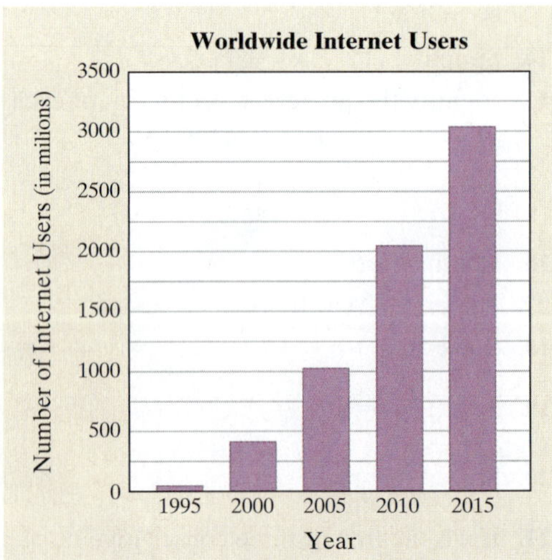

Data from International Telecommunication Union and United Nations Population Division

4.1 Exponents

OBJECTIVES

1 Evaluate Exponential Expressions.
2 Use the Product Rule for Exponents.
3 Use the Power Rule for Exponents.
4 Use the Power Rules for Products and Quotients.
5 Use the Quotient Rule for Exponents, and Define a Number Raised to the 0 Power.
6 Decide Which Rule(s) to Use to Simplify an Expression.

OBJECTIVE

1 Evaluating Exponential Expressions

As we reviewed in Section 1.3, an exponent is a shorthand notation for repeated factors. For example, $2 \cdot 2 \cdot 2 \cdot 2 \cdot 2$ can be written as 2^5. The expression 2^5 is called an **exponential expression.** It is also called the fifth **power** of 2, or we say that 2 is **raised** to the fifth power.

$$5^6 = \underbrace{5 \cdot 5 \cdot 5 \cdot 5 \cdot 5 \cdot 5}_{6 \text{ factors; each factor is } 5} \quad \text{and} \quad (-3)^4 = \underbrace{(-3) \cdot (-3) \cdot (-3) \cdot (-3)}_{4 \text{ factors; each factor is } -3}$$

The **base** of an exponential expression is the repeated factor. The **exponent** is the number of times that the base is used as a factor.

5^6 — 6 is the exponent, 5 is the base; $(-3)^4$ — 4 is the exponent, -3 is the base

EXAMPLE 1 Evaluate each expression.

a. 2^3 **b.** 3^1 **c.** $(-4)^2$ **d.** -4^2 **e.** $\left(\frac{1}{2}\right)^4$ **f.** $(0.5)^3$ **g.** $4 \cdot 3^2$

Solution

a. $2^3 = 2 \cdot 2 \cdot 2 = 8$

b. To raise 3 to the first power means to use 3 as a factor only once. Therefore, $3^1 = 3$. Also, when no exponent is shown, the exponent is assumed to be 1.

c. $(-4)^2 = (-4)(-4) = 16$ **d.** $-4^2 = -(4 \cdot 4) = -16$

e. $\left(\frac{1}{2}\right)^4 = \frac{1}{2} \cdot \frac{1}{2} \cdot \frac{1}{2} \cdot \frac{1}{2} = \frac{1}{16}$ **f.** $(0.5)^3 = (0.5)(0.5)(0.5) = 0.125$

g. $4 \cdot 3^2 = 4 \cdot 9 = 36$ □

PRACTICE 1 Evaluate each expression.

a. 3^3 **b.** 4^1 **c.** $(-8)^2$ **d.** -8^2

e. $\left(\frac{3}{4}\right)^3$ **f.** $(0.3)^4$ **g.** $3 \cdot 5^2$ ■

Notice how similar -4^2 is to $(-4)^2$ in the example above. The difference between the two is the parentheses. In $(-4)^2$, the parentheses tell us that the base, or repeated factor, is -4. In -4^2, only 4 is the base.

Helpful Hint

Be careful when identifying the base of an exponential expression. Pay close attention to the use of parentheses.

$(-3)^2$	-3^2	$2 \cdot 3^2$
The base is -3.	The base is 3.	The base is 3.
$(-3)^2 = (-3)(-3) = 9$	$-3^2 = -(3 \cdot 3) = -9$	$2 \cdot 3^2 = 2 \cdot 3 \cdot 3 = 18$

An exponent has the same meaning whether the base is a number or a variable. If x is a real number and n is a positive integer, then x^n is the product of n factors, each of which is x.

$$x^n = \underbrace{x \cdot x \cdot x \cdot x \cdot x \cdot \ldots \cdot x}_{n \text{ factors of } x}$$

EXAMPLE 2 Evaluate each expression for the given value of *x*.

a. $2x^3$; x is 5 **b.** $\frac{9}{x^2}$; x is -3

Solution **a.** If x is 5,
$$2x^3 = 2 \cdot (5)^3 = 2 \cdot (5 \cdot 5 \cdot 5) = 2 \cdot 125 = 250$$

b. If x is -3,
$$\frac{9}{x^2} = \frac{9}{(-3)^2} = \frac{9}{(-3)(-3)} = \frac{9}{9} = 1$$

□

PRACTICE 2 Evaluate each expression for the given value of *x*.

a. $3x^4$; x is 3 **b.** $\frac{6}{x^2}$; x is -4

■

OBJECTIVE 2

Using the Product Rule

Exponential expressions can be multiplied, divided, added, subtracted, and themselves raised to powers. By our definition of an exponent,

$$5^4 \cdot 5^3 = \underbrace{(5 \cdot 5 \cdot 5 \cdot 5)}_{4 \text{ factors of } 5} \cdot \underbrace{(5 \cdot 5 \cdot 5)}_{3 \text{ factors of } 5}$$
$$= \underbrace{5 \cdot 5 \cdot 5 \cdot 5 \cdot 5 \cdot 5 \cdot 5}_{7 \text{ factors of } 5}$$
$$= 5^7$$

Also,

$$x^2 \cdot x^3 = (x \cdot x) \cdot (x \cdot x \cdot x) = x \cdot x \cdot x \cdot x \cdot x = x^5$$

In both cases, notice that the result is exactly the same if the exponents are added.

$$5^4 \cdot 5^3 = 5^{4+3} = 5^7 \quad \text{and} \quad x^2 \cdot x^3 = x^{2+3} = x^5$$

This suggests the following rule.

Product Rule for Exponents

If m and n are positive integers and a is a real number, then

$$a^m \cdot a^n = a^{m+n}$$ ← Add exponents. (↑ Keep common base.)

For example, $3^5 \cdot 3^7 = 3^{5+7} = 3^{12}$ ← Add exponents. (↑ Keep common base.)

Helpful Hint

Don't forget that

$$3^5 \cdot 3^7 \neq 9^{12}$$ ← Add exponents. (↑ **Common base *not* kept.**)

$$3^5 \cdot 3^7 = \underbrace{3 \cdot 3 \cdot 3 \cdot 3 \cdot 3}_{5 \text{ factors of } 3} \cdot \underbrace{3 \cdot 3 \cdot 3 \cdot 3 \cdot 3 \cdot 3 \cdot 3}_{7 \text{ factors of } 3}$$
$$= 3^{12}$$ 12 factors of 3, *not* 9

In other words, to multiply two exponential expressions with the **same base,** we keep the base and add the exponents. We call this **simplifying** the exponential expression.

EXAMPLE 3 Use the product rule to simplify.

a. $4^2 \cdot 4^5$ **b.** $x^4 \cdot x^6$ **c.** $y^3 \cdot y$

d. $y^3 \cdot y^2 \cdot y^7$ **e.** $(-5)^7 \cdot (-5)^8$ **f.** $a^2 \cdot b^2$

Solution

a. $4^2 \cdot 4^5 = 4^{2+5} = 4^7$ ← Add exponents. ↑ **Keep** common base.

b. $x^4 \cdot x^6 = x^{4+6} = x^{10}$

c. $y^3 \cdot y = y^3 \cdot y^1 = y^{3+1} = y^4$

Helpful Hint
Don't forget that if no exponent is written, it is assumed to be 1.

d. $y^3 \cdot y^2 \cdot y^7 = y^{3+2+7} = y^{12}$

e. $(-5)^7 \cdot (-5)^8 = (-5)^{7+8} = (-5)^{15}$

f. $a^2 \cdot b^2$ Cannot be simplified because a and b are different bases. □

PRACTICE 3 Use the product rule to simplify.

a. $3^4 \cdot 3^6$ **b.** $y^3 \cdot y^2$

c. $z \cdot z^4$ **d.** $x^3 \cdot x^2 \cdot x^6$

e. $(-2)^5 \cdot (-2)^3$ **f.** $b^3 \cdot t^5$ ■

✔ CONCEPT CHECK

Where possible, use the product rule to simplify the expression.

a. $z^2 \cdot z^{14}$ **b.** $x^2 \cdot y^{14}$ **c.** $9^8 \cdot 9^3$ **d.** $9^8 \cdot 2^7$

EXAMPLE 4 Use the product rule to simplify $(2x^2)(-3x^5)$.

Solution Recall that $2x^2$ means $2 \cdot x^2$ and $-3x^5$ means $-3 \cdot x^5$.

$$(2x^2)(-3x^5) = 2 \cdot x^2 \cdot -3 \cdot x^5 \quad \text{Remove parentheses.}$$
$$= 2 \cdot -3 \cdot x^2 \cdot x^5 \quad \text{Group factors with common bases.}$$
$$= -6x^7 \quad \text{Simplify.}$$
□

PRACTICE 4 Use the product rule to simplify $(-5y^3)(-3y^4)$. ■

EXAMPLE 5 Simplify.

a. $(x^2y)(x^3y^2)$ **b.** $(-a^7b^4)(3ab^9)$

Solution

a. $(x^2y)(x^3y^2) = (x^2 \cdot x^3) \cdot (y^1 \cdot y^2)$ Group like bases and write y as y^1.

$= x^5 \cdot y^3$ or x^5y^3 Multiply.

b. $(-a^7b^4)(3ab^9) = (-1 \cdot 3) \cdot (a^7 \cdot a^1) \cdot (b^4 \cdot b^9)$

$= -3a^8b^{13}$ □

PRACTICE 5 Simplify.

a. $(y^7z^3)(y^5z)$ **b.** $(-m^4n^4)(7mn^{10})$ ■

Answers to Concept Check:

a. z^{16} **b.** cannot be simplified

c. 9^{11} **d.** cannot be simplified

Helpful Hint

These examples will remind you of the difference between adding and multiplying terms.

Addition

$5x^3 + 3x^3 = (5 + 3)x^3 = 8x^3$ By the distributive property.

$7x + 4x^2 = 7x + 4x^2$ Cannot be combined.

Multiplication

$(5x^3)(3x^3) = 5 \cdot 3 \cdot x^3 \cdot x^3 = 15x^{3+3} = 15x^6$ By the product rule.

$(7x)(4x^2) = 7 \cdot 4 \cdot x \cdot x^2 = 28x^{1+2} = 28x^3$ By the product rule.

OBJECTIVE

3 Using the Power Rule

Exponential expressions can themselves be raised to powers. Let's try to discover a rule that simplifies an expression like $(x^2)^3$. By definition,

$$(x^2)^3 = \underbrace{(x^2)(x^2)(x^2)}_{3 \text{ factors of } x^2}$$

which can be simplified by the product rule for exponents.

$$(x^2)^3 = (x^2)(x^2)(x^2) = x^{2+2+2} = x^6$$

Notice that the result is exactly the same if we multiply the exponents.

$$(x^2)^3 = x^{2 \cdot 3} = x^6$$

The following property states this result.

Power Rule for Exponents

If m and n are positive integers and a is a real number, then

$$(a^m)^n = a^{mn}$$ ← Multiply exponents. Keep common base.

For example, $(7^2)^5 = 7^{2 \cdot 5} = 7^{10}$ ← Multiply exponents. Keep common base.

To raise a power to a power, keep the base and multiply the exponents.

EXAMPLE 6 Use the power rule to simplify.

a. $(y^8)^2$ **b.** $(8^4)^5$ **c.** $[(-5)^3]^7$

Solution

a. $(y^8)^2 = y^{8 \cdot 2} = y^{16}$ **b.** $(8^4)^5 = 8^{4 \cdot 5} = 8^{20}$ **c.** $[(-5)^3]^7 = (-5)^{21}$ □

PRACTICE

6 Use the power rule to simplify.

a. $(z^3)^7$ **b.** $(4^9)^2$ **c.** $[(-2)^3]^5$ ■

Helpful Hint

Take a moment to make sure that you understand when to apply the product rule and when to apply the power rule.

Product Rule → Add Exponents	***Power Rule → Multiply Exponents***
$x^5 \cdot x^7 = x^{5+7} = x^{12}$	$(x^5)^7 = x^{5 \cdot 7} = x^{35}$
$y^6 \cdot y^2 = y^{6+2} = y^8$	$(y^6)^2 = y^{6 \cdot 2} = y^{12}$

OBJECTIVE

4 Using the Power Rules for Products and Quotients

When the base of an exponential expression is a product, the definition of x^n still applies. To simplify $(xy)^3$, for example,

$$\begin{aligned}(xy)^3 &= (xy)(xy)(xy) && (xy)^3 \text{ means 3 factors of } (xy).\\ &= x \cdot x \cdot x \cdot y \cdot y \cdot y && \text{Group factors with common bases.}\\ &= x^3y^3 && \text{Simplify.}\end{aligned}$$

Notice that to simplify the expression $(xy)^3$, we raise each factor within the parentheses to a power of 3.

$$(xy)^3 = x^3y^3$$

In general, we have the following rule.

Power of a Product Rule

If n is a positive integer and a and b are real numbers, then

$$(ab)^n = a^nb^n$$

For example, $(3x)^5 = 3^5x^5$.

In other words, to raise a product to a power, we raise each factor to the power.

EXAMPLE 7 Simplify each expression.

a. $(st)^4$ **b.** $(2a)^3$ **c.** $\left(\frac{1}{3}mn^3\right)^2$ **d.** $(-5x^2y^3z)^2$

Solution

a. $(st)^4 = s^4 \cdot t^4 = s^4t^4$ Use the power of a product rule.

b. $(2a)^3 = 2^3 \cdot a^3 = 8a^3$ Use the power of a product rule.

c. $\left(\frac{1}{3}mn^3\right)^2 = \left(\frac{1}{3}\right)^2 \cdot (m)^2 \cdot (n^3)^2 = \frac{1}{9}m^2n^6$ Use the power of a product rule.

d. $(-5x^2y^3z)^2 = (-5)^2 \cdot (x^2)^2 \cdot (y^3)^2 \cdot (z^1)^2$ Use the power of a product rule.

$= 25x^4y^6z^2$ Use the power rule for exponents.

PRACTICE

7 Simplify each expression.

a. $(pr)^5$ **b.** $(6b)^2$ **c.** $\left(\frac{1}{4}x^2y\right)^3$ **d.** $(-3a^3b^4c)^4$

Let's see what happens when we raise a quotient to a power. To simplify $\left(\frac{x}{y}\right)^3$, for example,

$$\begin{aligned}\left(\frac{x}{y}\right)^3 &= \left(\frac{x}{y}\right)\left(\frac{x}{y}\right)\left(\frac{x}{y}\right) && \left(\frac{x}{y}\right)^3 \text{ means 3 factors of } \left(\frac{x}{y}\right)\\ &= \frac{x \cdot x \cdot x}{y \cdot y \cdot y} && \text{Multiply fractions.}\\ &= \frac{x^3}{y^3} && \text{Simplify.}\end{aligned}$$

Notice that to simplify the expression $\left(\frac{x}{y}\right)^3$, we raise both the numerator and the denominator to a power of 3.

$$\left(\frac{x}{y}\right)^3 = \frac{x^3}{y^3}$$

In general, we have the following.

Power of a Quotient Rule

If n is a positive integer and a and c are real numbers, then

$$\left(\frac{a}{c}\right)^n = \frac{a^n}{c^n}, \quad c \neq 0$$

For example, $\left(\frac{y}{7}\right)^4 = \frac{y^4}{7^4}$.

In other words, to raise a quotient to a power, we raise both the numerator and the denominator to the power.

EXAMPLE 8 Simplify each expression.

a. $\left(\frac{m}{n}\right)^7$ **b.** $\left(\frac{x^3}{3y^5}\right)^4$

Solution

a. $\left(\frac{m}{n}\right)^7 = \frac{m^7}{n^7}, n \neq 0$ Use the power of a quotient rule.

b. $\left(\frac{x^3}{3y^5}\right)^4 = \frac{(x^3)^4}{3^4 \cdot (y^5)^4}, y \neq 0$ Use the power of a product and quotient rules.

$= \frac{x^{12}}{81y^{20}}$ Use the power rule for exponents. □

PRACTICE 8 Simplify each expression.

a. $\left(\frac{x}{y^2}\right)^5$ **b.** $\left(\frac{2a^4}{b^3}\right)^5$ ■

OBJECTIVE 5

Using the Quotient Rule and Defining the Zero Exponent

Another pattern for simplifying exponential expressions involves quotients.

To simplify an expression like $\frac{x^5}{x^3}$, in which the numerator and the denominator have a common base, we can apply the fundamental principle of fractions and divide the numerator and the denominator by the common base factors. Assume for the remainder of this section that denominators are not 0.

$$\begin{aligned}\frac{x^5}{x^3} &= \frac{x \cdot x \cdot x \cdot x \cdot x}{x \cdot x \cdot x} \\ &= \frac{x \cdot x \cdot x \cdot x \cdot x}{x \cdot x \cdot x} \\ &= x \cdot x \\ &= x^2\end{aligned}$$

Notice that the result is exactly the same if we subtract exponents of the common bases.

$$\frac{x^5}{x^3} = x^{5-3} = x^2$$

The quotient rule for exponents states this result in a general way.

Quotient Rule for Exponents

If m and n are positive integers and a is a real number, then

$$\frac{a^m}{a^n} = a^{m-n}$$

as long as a is not 0.

For example, $\frac{x^6}{x^2} = x^{6-2} = x^4$.

In other words, to divide one exponential expression by another with a common base, keep the base and subtract exponents.

EXAMPLE 9 Simplify each quotient.

a. $\frac{x^5}{x^2}$ **b.** $\frac{4^7}{4^3}$ **c.** $\frac{(-3)^5}{(-3)^2}$ **d.** $\frac{s^2}{t^3}$ **e.** $\frac{2x^5y^2}{xy}$

Solution

a. $\frac{x^5}{x^2} = x^{5-2} = x^3$ Use the quotient rule.

b. $\frac{4^7}{4^3} = 4^{7-3} = 4^4 = 256$ Use the quotient rule.

c. $\frac{(-3)^5}{(-3)^2} = (-3)^3 = -27$ Use the quotient rule.

d. $\frac{s^2}{t^3}$ Cannot be simplified because s and t are different bases.

e. Begin by grouping common bases.

$$\frac{2x^5y^2}{xy} = 2 \cdot \frac{x^5}{x^1} \cdot \frac{y^2}{y^1}$$

$$= 2 \cdot (x^{5-1}) \cdot (y^{2-1})$$ Use the quotient rule.

$$= 2x^4y^1 \quad \text{or} \quad 2x^4y$$

□

PRACTICE 9 Simplify each quotient.

a. $\frac{z^8}{z^4}$ **b.** $\frac{(-5)^5}{(-5)^3}$ **c.** $\frac{8^8}{8^6}$ **d.** $\frac{q^5}{t^2}$ **e.** $\frac{6x^3y^7}{xy^5}$

✔ CONCEPT CHECK

Suppose you are simplifying each expression. Tell whether you would *add* the exponents, *subtract* the exponents, *multiply* the exponents, *divide* the exponents, or *none of these*.

a. $(x^{63})^{21}$ **b.** $\frac{y^{15}}{y^3}$ **c.** $z^{16} + z^8$ **d.** $w^{45} \cdot w^9$

Let's now give meaning to an expression such as x^0. To do so, we will simplify $\frac{x^3}{x^3}$ in two ways and compare the results.

$\frac{x^3}{x^3} = x^{3-3} = x^0$ Apply the quotient rule.

$\frac{x^3}{x^3} = \frac{x \cdot x \cdot x}{x \cdot x \cdot x} = 1$ Apply the fundamental principle for fractions.

Answers to Concept Check:
a. multiply **b.** subtract
c. none of these **d.** add

Since $\frac{x^3}{x^3} = x^0$ and $\frac{x^3}{x^3} = 1$, we define that $x^0 = 1$ as long as x is not 0.

Zero Exponent

$a^0 = 1$, as long as a is not 0.

In other words, any base raised to the 0 power is 1 as long as the base is not 0.

EXAMPLE 10 Simplify each expression.

a. 3^0 **b.** $(5x^3y^2)^0$ **c.** $(-5)^0$ **d.** -5^0 **e.** $\left(\frac{3}{100}\right)^0$ **f.** $4x^0$

Solution

a. $3^0 = 1$

b. Assume that neither x nor y is zero.

$$(5x^3y^2)^0 = 1$$

c. $(-5)^0 = 1$

d. $-5^0 = -1 \cdot 5^0 = -1 \cdot 1 = -1$

e. $\left(\frac{3}{100}\right)^0 = 1$

f. $4x^0 = 4 \cdot x^0 = 4 \cdot 1 = 4$ □

PRACTICE 10 Simplify the following expressions.

a. -3^0 **b.** $(-3)^0$ **c.** 8^0 **d.** $(0.2)^0$

e. $(7a^2y^4)^0$ **f.** $7y^0$ ■

OBJECTIVE 6

Deciding Which Rule to Use

Let's practice deciding which rule(s) to use to simplify.

EXAMPLE 11 Simplify each expression.

a. $x^7 \cdot x^4$ **b.** $\left(\frac{t}{2}\right)^4$ **c.** $(9y^5)^2$

Solution

a. Here we have a product, so we use the product rule to simplify.

$$x^7 \cdot x^4 = x^{7+4} = x^{11}$$

b. This is a quotient raised to a power, so we use the power of a quotient rule.

$$\left(\frac{t}{2}\right)^4 = \frac{t^4}{2^4} = \frac{t^4}{16}$$

c. This is a product raised to a power, so we use the power of a product rule.

$$(9y^5)^2 = 9^2(y^5)^2 = 81y^{10}$$ □

PRACTICE 11 **a.** $\left(\frac{z}{12}\right)^2$ **b.** $(4x^6)^3$ **c.** $y^{10} \cdot y^3$ ■

EXAMPLE 12 Simplify each expression.

a. $4^2 - 4^0$ **b.** $(x^0)^3 + (2^0)^5$ **c.** $\left(\frac{3y^7}{6x^5}\right)^2$ **d.** $\frac{(2a^3b^4)^3}{-8a^9b^2}$

Solution

a. $4^2 - 4^0 = 16 - 1 = 15$ Remember that $4^0 = 1$.

b. $(x^0)^3 + (2^0)^5 = 1^3 + 1^5 = 1 + 1 = 2$

c. $\left(\frac{3y^7}{6x^5}\right)^2 = \frac{3^2(y^7)^2}{6^2(x^5)^2} = \frac{9 \cdot y^{14}}{36 \cdot x^{10}} = \frac{y^{14}}{4x^{10}}$

d. $\frac{(2a^3b^4)^3}{-8a^9b^2} = \frac{2^3(a^3)^3(b^4)^3}{-8a^9b^2} = \frac{8a^9b^{12}}{-8a^9b^2} = -1 \cdot (a^{9-9}) \cdot (b^{12-2})$

$= -1 \cdot a^0 \cdot b^{10} = -1 \cdot 1 \cdot b^{10} = -b^{10}$ □

PRACTICE 12 Simplify each expression.

a. $8^2 - 8^0$ **b.** $(z^0)^6 + (4^0)^5$ **c.** $\left(\frac{5x^3}{15y^4}\right)^2$ **d.** $\frac{(2z^8x^5)^4}{-16z^2x^{20}}$ ■

Vocabulary, Readiness & Video Check

Use the choices below to fill in each blank. Some choices may be used more than once.

0	base	add
1	exponent	multiply

1. Repeated multiplication of the same factor can be written using a(n) ________.

2. In 5^2, the 2 is called the ________ and the 5 is called the ________.

3. To simplify $x^2 \cdot x^7$, keep the base and ________ the exponents.

4. To simplify $(x^3)^6$, keep the base and ________ the exponents.

5. The understood exponent on the term y is ________.

6. If $x^{\square} = 1$, the exponent is ________.

Martin-Gay Interactive Videos

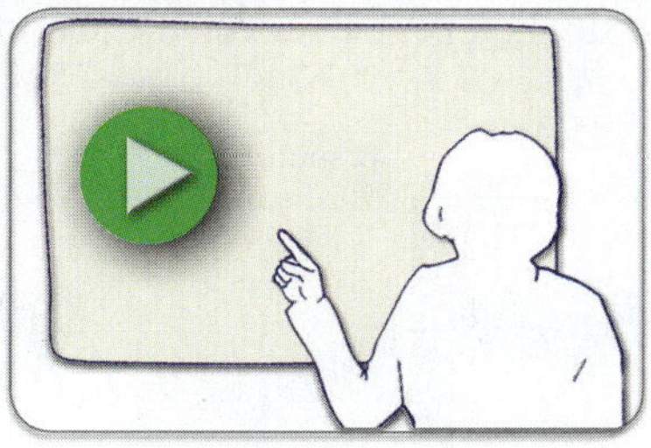

See Video 4.1

Watch the section lecture video and answer the following questions.

OBJECTIVE 1

7. Examples 3 and 4 illustrate how to find the base of an exponential expression both with and without parentheses. Explain how identifying the base of Example 7 is similar to identifying the base of Example 4.

OBJECTIVE 2

8. Why were the commutative and associative properties applied in Example 12? Were these properties used in another example?

OBJECTIVE 3

9. What point is made at the end of Example 15?

OBJECTIVE 4

10. Although it's not especially emphasized in Example 20, what is helpful to remind yourself about the −2 in the problem?

OBJECTIVE 5

11. In Example 24, which exponent rule is used to show that any nonzero base raised to zero is 1?

OBJECTIVE 6

12. When simplifying an exponential expression that's a fraction, will you always use the quotient rule? Refer to Example 30 for this objective to support your answer.

4.1 Exercise Set MyMathLab®

For each of the following expressions, state the exponent shown and its corresponding base.

1. 3^2 **2.** $(-3)^6$

3. -4^2 **4.** $5 \cdot 3^4$

5. $5x^2$ **6.** $(5x)^2$

Evaluate each expression. See Example 1.

7. 7^2 **8.** -3^2

9. $(-5)^1$ **10.** $(-3)^2$

11. -2^4 **12.** -4^3

13. $(-2)^4$ **14.** $(-4)^3$

15. $(0.1)^5$ **16.** $(0.2)^5$

17. $\left(\frac{1}{3}\right)^4$ **18.** $\left(-\frac{1}{9}\right)^2$

19. $7 \cdot 2^5$ **20.** $9 \cdot 1^7$

21. $-2 \cdot 5^3$ **22.** $-4 \cdot 3^3$

Evaluate each expression for the replacement values given. See Example 2.

23. $x^2; x = -2$ **24.** $x^3; x = -2$

25. $5x^3; x = 3$ **26.** $4x^2; x = -1$

27. $2xy^2; x = 3$ and $y = 5$ **28.** $-4x^2y^3; x = 2$ and $y = -1$

29. $\frac{2z^4}{5}; z = -2$ **30.** $\frac{10}{3y^3}; y = 5$

Use the product rule to simplify each expression. Write the results using exponents. See Examples 3 through 5

31. $x^2 \cdot x^5$ **32.** $y^2 \cdot y$

33. $(-3)^3 \cdot (-3)^9$ **34.** $(-5)^7 \cdot (-5)^6$

35. $(5y^4)(3y)$ **36.** $(-2z^3)(-2z^2)$

37. $(x^9y)(x^{10}y^5)$ **38.** $(a^2b)(a^{13}b^{17})$

39. $(-8mn^6)(9m^2n^2)$ **40.** $(-7a^3b^3)(7a^{19}b)$

41. $(4z^{10})(-6z^7)(z^3)$ **42.** $(12x^5)(-x^6)(x^4)$

△ **43.** The rectangle below has width $4x^2$ feet and length $5x^3$ feet. Find its area as an expression in x. $(A = l \cdot w)$

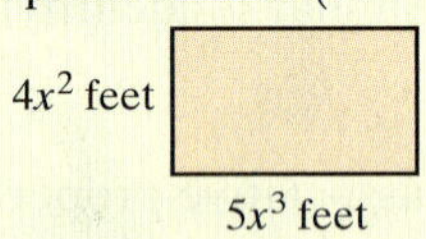

△ **44.** The parallelogram below has base length $9y^7$ meters and height $2y^{10}$ meters. Find its area as an expression in y. $(A = b \cdot h)$

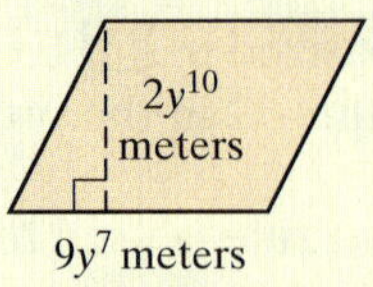

MIXED PRACTICE

Use the power rule and the power of a product or quotient rule to simplify each expression. See Examples 6 through 8.

45. $(x^9)^4$ **46.** $(y^7)^5$

47. $(pq)^8$ **48.** $(ab)^6$

49. $(2a^5)^3$ **50.** $(4x^6)^2$

51. $(x^2y^3)^5$ **52.** $(a^4b)^7$

53. $(-7a^2b^5c)^2$ **54.** $(-3x^7yz^2)^3$

55. $\left(\frac{r}{s}\right)^9$ **56.** $\left(\frac{q}{t}\right)^{11}$

57. $\left(\frac{mp}{n}\right)^5$ **58.** $\left(\frac{xy}{7}\right)^2$

59. $\left(\frac{-2xz}{y^5}\right)^2$ **60.** $\left(\frac{xy^4}{-3z^3}\right)^3$

△ **61.** The square shown has sides of length $8z^5$ decimeters. Find its area. $(A = s^2)$

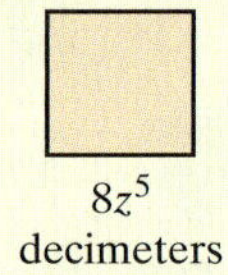

△ **62.** Given the circle below with radius $5y$ centimeters, find its area. Do not approximate π. $(A = \pi r^2)$

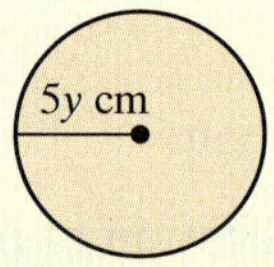

△ **63.** The vault below is in the shape of a cube. If each side is $3y^4$ feet, find its volume. $(V = s^3)$

△ **64.** The silo shown is in the shape of a cylinder. If its radius is $4x$ meters and its height is $5x^3$ meters, find its volume. Do not approximate π. $(V = \pi r^2 h)$

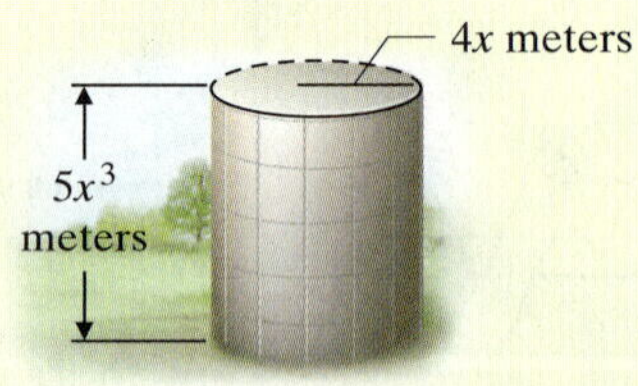

Use the quotient rule and simplify each expression. See Example 9.

65. $\frac{x^3}{x}$ **66.** $\frac{y^{10}}{y^9}$

67. $\frac{(-4)^6}{(-4)^3}$ **68.** $\frac{(-6)^{13}}{(-6)^{11}}$

69. $\frac{p^7q^{20}}{pq^{15}}$ **70.** $\frac{x^8y^6}{xy^5}$

71. $\frac{7x^2y^6}{14x^2y^3}$ **72.** $\frac{9a^4b^7}{27ab^2}$

Simplify each expression. See Example 10.

73. 7^0
74. 23^0
75. $(2x)^0$
76. $(4y)^0$
77. $-7x^0$
78. $-2x^0$
79. $5^0 + y^0$
80. $-3^0 + 4^0$

MIXED PRACTICE

Simplify each expression. See Examples 1 through 12.

81. -9^2
82. $(-9)^2$
83. $\left(\frac{1}{4}\right)^3$
84. $\left(\frac{2}{3}\right)^3$
85. b^4b^2
86. y^4y
87. $a^2a^3a^4$
88. $x^2x^{15}x^9$
89. $(2x^3)(-8x^4)$
90. $(3y^4)(-5y)$
91. $(a^7b^{12})(a^4b^8)$
92. $(y^2z^2)(y^{15}z^{13})$
93. $(-2mn^6)(-13m^8n)$
94. $(-3s^5t)(-7st^{10})$
95. $(z^4)^{10}$
96. $(t^5)^{11}$
97. $(4ab)^3$
98. $(2ab)^4$
99. $(-6xyz^3)^2$
100. $(-3xy^2a^3)^3$
101. $\frac{3x^5}{x}$
102. $\frac{5x^9}{x^3}$
103. $(9xy)^2$
104. $(2ab)^5$
105. $2^3 + 2^0$
106. $7^2 - 7^0$
107. $\left(\frac{3y^5}{6x^4}\right)^3$
108. $\left(\frac{2ab}{6yz}\right)^4$
109. $\frac{2x^3y^2z}{xyz}$
110. $\frac{x^{12}y^{13}}{x^5y^7}$
111. $(5^0)^3 + (y^0)^7$
112. $(9^0)^4 + (z^0)^5$
113. $\left(\frac{5x^9}{10y^{11}}\right)^2$
114. $\left(\frac{3a^4}{9b^5}\right)^2$
115. $\frac{(2a^5b^3)^4}{-16a^{20}b^7}$
116. $\frac{(2x^6y^2)^5}{-32x^{20}y^{10}}$

REVIEW AND PREVIEW

Use the histogram below to answer the exercises. See Section R.4.

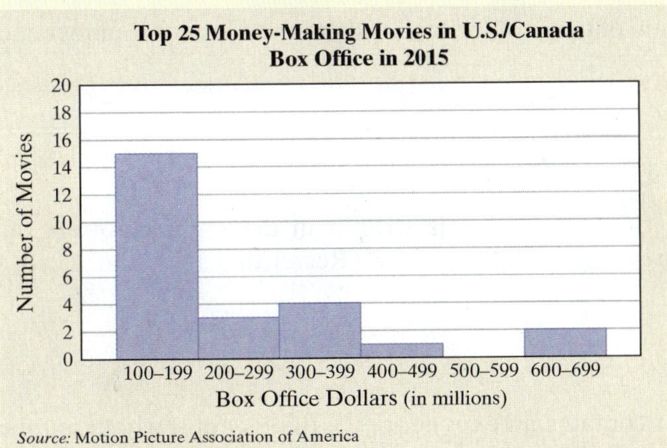

Source: Motion Picture Association of America

117. How many movies made \$200 million or more?
118. How many movies made \$299 million or less?
119. What fraction of the movies made \$100 to \$199 million?
120. What fraction of the movies made \$300 to \$499 million?
121. What percent of the movies made \$600 million or more?
122. What percent of the movies made \$499 million or less?

CONCEPT EXTENSIONS

Solve. See the Concept Checks in this section. For Exercises 123 through 126, match the expression with the operation needed to simplify each. A letter may be used more than once and a letter may not be used at all.

123. $(x^{14})^{23}$
124. $x^{14} \cdot x^{23}$
125. $x^{14} + x^{23}$
126. $\frac{x^{35}}{x^{17}}$

A. Add the exponents
B. Subtract the exponents
C. Multiply the exponents
D. Divide the exponents
E. None of these

Fill in the boxes so that each statement is true. (More than one answer is possible for each exercise.)

127. $x^{\square} \cdot x^{\square} = x^{12}$
128. $(x^{\square})^{\square} = x^{20}$
129. $\frac{y^{\square}}{y^{\square}} = y^7$
130. $(y^{\square})^{\square} \cdot (y^{\square})^{\square} = y^{30}$

△ 131. The formula $V = x^3$ can be used to find the volume V of a cube with side length x. Find the volume of a cube with side length 7 meters. (Volume is measured in cubic units.)

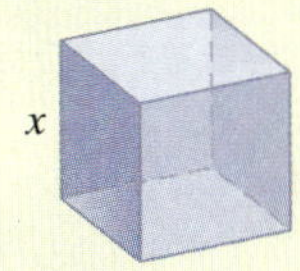

△ 132. The formula $S = 6x^2$ can be used to find the surface area S of a cube with side length x. Find the surface area of a cube with side length 5 meters. (Surface area is measured in square units.)

△ 133. To find the amount of water that a swimming pool in the shape of a cube can hold, do we use the formula for volume of the cube or surface area of the cube? (See Exercises 131 and 132.)

△ 134. To find the amount of material needed to cover an ottoman in the shape of a cube, do we use the formula for volume of the cube or surface area of the cube? (See Exercises 131 and 132.)

135. Explain why $(-5)^4 = 625$, while $-5^4 = -625$.
136. Explain why $5 \cdot 4^2 = 80$, while $(5 \cdot 4)^2 = 400$.
137. In your own words, explain why $5^0 = 1$.
138. In your own words, explain when $(-3)^n$ is positive and when it is negative.

Simplify each expression. Assume that variables represent positive integers.

139. $x^{5a}x^{4a}$
140. $b^{9a}b^{4a}$
141. $(a^b)^5$
142. $(2a^{4b})^4$
143. $\frac{x^{9a}}{x^{4a}}$
144. $\frac{y^{15b}}{y^{6b}}$

Solve. Round money amounts to 2 decimal places.

145. Suppose you borrow money for 6 months. If the interest is compounded monthly, the formula $A = P\left(1 + \frac{r}{12}\right)^6$ gives the total amount A to be repaid at the end of 6 months. For a loan of $P = \$1000$ and interest rate of 9% $(r = 0.09)$, how much money is needed to pay off the loan?

146. Suppose you borrow money for 3 years. If the interest is compounded quarterly, the formula $A = P\left(1 + \frac{r}{4}\right)^{12}$ gives the total amount A to be repaid at the end of 3 years. For a loan of \$10,000 and interest rate of 8% $(r = 0.08)$, how much money is needed to pay off the loan in 3 years?

4.2 Polynomial Functions and Adding and Subtracting Polynomials

OBJECTIVES

1 Define Polynomial, Monomial, Binomial, Trinomial, and Degree.

2 Define Polynomial Functions.

3 Simplify a Polynomial by Combining Like Terms.

4 Add and Subtract Polynomials.

OBJECTIVE

1 Defining Polynomial, Monomial, Binomial, Trinomial, and Degree

In this section, we introduce a special algebraic expression called a polynomial. Let's first review some definitions presented in Section 2.1.

Recall that a **term** is a number or the product of a number and variables raised to powers. The terms of the expression $4x^2 + 3x$ are $4x^2$ and $3x$.

The terms of the expression $9x^4 - 7x - 1$ are $9x^4$, $-7x$, and -1.

Expression	*Terms*
$4x^2 + 3x$	$4x^2, 3x$
$9x^4 - 7x - 1$	$9x^4, -7x, -1$
$7y^3$	$7y^3$
5	5

The **numerical coefficient** of a term, or simply the **coefficient**, is the numerical factor of each term. If no numerical factor appears in the term, then the coefficient is understood to be 1. If the term is a number only, it is called a **constant term** or simply a **constant**.

Term	*Coefficient*
x^5	1
$3x^2$	3
$-4x$	-4
$-x^2y$	-1
3 (constant)	3

Now we are ready to define a polynomial.

Polynomial

A **polynomial in** x is a finite sum of terms of the form ax^n, where a is a real number and n is a whole number.

For example,

$$x^5 - 3x^3 + 2x^2 - 5x + 1$$

is a polynomial. Notice that this polynomial is written in **descending powers** of x because the powers of x decrease from left to right. (Recall that the term 1 can be thought of as $1x^0$.)

On the other hand,

$$x^{-5} + 2x - 3$$

is **not** a polynomial because it contains an exponent, -5, that is not a whole number. (We study negative exponents in Section 4.5 of this chapter.)

Some polynomials are given special names.

Types of Polynomials

A **monomial** is a polynomial with exactly one term.

A **binomial** is a polynomial with exactly two terms.

A **trinomial** is a polynomial with exactly three terms.

The following are examples of monomials, binomials, and trinomials. Each of these examples is also a polynomial.

POLYNOMIALS

Monomials	*Binomials*	*Trinomials*	*More Than Three Terms*
ax^2	$x + y$	$x^2 + 4xy + y^2$	$5x^3 - 6x^2 + 3x - 6$
$-3z$	$3p + 2$	$x^5 + 7x^2 - x$	$-y^5 + y^4 - 3y^3 - y^2 + y$
4	$4x^2 - 7$	$-q^4 + q^3 - 2q$	$x^6 + x^4 - x^3 + 1$

Each term of a polynomial has a **degree.**

Degree of a Term

The degree of a term is the sum of the exponents on the variables contained in the term.

EXAMPLE 1 Find the degree of each term.

a. $3x^2$ **b.** -2^3x^5 **c.** y **d.** $12x^2yz^3$ **e.** 5

Solution

a. The exponent on x is 2, so the degree of the term is 2.

b. The exponent on x is 5, so the degree of the term is 5. (Recall that the degree is the sum of the exponents on only the *variables*.)

c. The degree of y, or y^1, is 1.

d. The degree is the sum of the exponents on the variables, or $2 + 1 + 3 = 6$.

e. The degree of 5, which can be written as $5x^0$, is 0. □

PRACTICE 1 Find the degree of each term.

a. $5y^3$ **b.** $10xy$ **c.** z **d.** $-3a^2b^5c$ **e.** 8 ■

From the preceding, we can say that **the degree of a constant is 0.**

Each polynomial also has a degree.

Degree of a Polynomial

The degree of a polynomial is the greatest degree of any term of the polynomial.

EXAMPLE 2 Find the degree of each polynomial and tell whether the polynomial is a monomial, binomial, trinomial, or none of these.

a. $-2t^2 + 3t + 6$ **b.** $15x - 10$ **c.** $7x + 3x^3 + 2x^2 - 1$

Solution

a. The degree of the trinomial $-2t^2 + 3t + 6$ is 2, the greatest degree of any of its terms.

b. The degree of the binomial $15x - 10$ or $15x^1 - 10$ is 1.

c. The degree of the polynomial $7x + 3x^3 + 2x^2 - 1$ is 3. □

PRACTICE 2 Find the degree of each polynomial and tell whether the polynomial is a monomial, binomial, trinomial, or none of these.

a. $5b^2 - 3b + 7$ **b.** $7t + 3$

c. $5x^2 + 3x - 6x^3 + 4$

EXAMPLE 3 Complete the table for the polynomial

$$7x^2y - 6xy + x^2 - 3y + 7$$

Use the table to give the degree of the polynomial.

Solution

> **Helpful Hint**
> This is the sum of the exponents on the variable factors.

Term	*Numerical Coefficient*	*Degree of Term*
$7x^2y$	7	3
$-6xy$	-6	2
x^2	1	2
$-3y$	-3	1
7	7	0

The degree of the polynomial is 3.

PRACTICE 3 Complete the table for the polynomial $-3x^3y^2 + 4xy^2 - y^2 + 3x - 2$.

Term	*Numerical Coefficient*	*Degree of Term*
$-3x^3y^2$		
$4xy^2$		
$-y^2$		
$3x$		
-2		

OBJECTIVE 2

Defining Polynomial Functions

At times, it is convenient to use function notation to represent polynomials. For example, we may write $P(x)$ to represent the polynomial $3x^2 - 2x - 5$. In symbols, this is

$$P(x) = 3x^2 - 2x - 5$$

This function is called a **polynomial function** because the expression $3x^2 - 2x - 5$ is a polynomial.

> **Helpful Hint**
> Recall that the symbol $P(x)$ **does not mean** P times x. It is a special symbol used to denote a function.

EXAMPLE 4 If $P(x) = 3x^2 - 2x - 5$, find the following.

a. $P(1)$ **b.** $P(-2)$

Solution

a. Substitute 1 for x in $P(x) = 3x^2 - 2x - 5$ and simplify.

$$P(x) = 3x^2 - 2x - 5$$
$$P(1) = 3(1)^2 - 2(1) - 5 = -4$$

Thus, $P(1) = -4$.

b. Substitute -2 for x in $P(x) = 3x^2 - 2x - 5$ and simplify.

$$P(x) = 3x^2 - 2x - 5$$
$$P(-2) = 3(-2)^2 - 2(-2) - 5 = 11$$

Thus, $P(-2) = 11$. □

PRACTICE 4 If $P(x) = -2x^2 - x + 7$, find

a. $P(1)$ **b.** $P(-4)$ ■

Many real-world phenomena are modeled by polynomial functions. If the polynomial function model is given, we can often find the solution of a problem by evaluating the function at a certain value.

EXAMPLE 5 **Finding the Height of a Dropped Object**

The Swiss Re Building, in London, is a unique building. Londoners often refer to it as the "pickle building." The building is 592.1 feet tall. An object is dropped from the highest point of this building. Neglecting air resistance, the height in feet of the object above ground at time t seconds is given by the polynomial function $P(t) = -16t^2 + 592.1$. Find the height of the object when $t = 1$ second, and when $t = 6$ seconds.

Solution To find each height, we find $P(1)$ and $P(6)$.

$$P(t) = -16t^2 + 592.1$$
$$P(1) = -16(1)^2 + 592.1 \quad \text{Replace } t \text{ with 1.}$$
$$P(1) = -16 + 592.1$$
$$P(1) = 576.1$$

The height of the object at 1 second is 576.1 feet.

$$P(t) = -16t^2 + 592.1$$
$$P(6) = -16(6)^2 + 592.1 \quad \text{Replace } t \text{ with 6.}$$
$$P(6) = -576 + 592.1$$
$$P(6) = 16.1$$

The height of the object at 6 seconds is 16.1 feet. □

PRACTICE 5 The cliff divers of Acapulco dive 130 feet into La Quebrada several times a day for the entertainment of the tourists. If a tourist is standing near the diving platform and drops his camera off the cliff, the height of the camera above the water at time t seconds is given by the polynomial function $P(t) = -16t^2 + 130$. Find the height of the camera when $t = 1$ second and when $t = 2$ seconds. ■

OBJECTIVE 3

Simplifying Polynomials by Combining Like Terms

Polynomials with like terms can be simplified by combining the like terms. Recall that like terms are terms that contain exactly the same variables raised to exactly the same powers.

Like Terms	*Unlike Terms*
$5x^2, -7x^2$	$3x, 3y$
$y, 2y$	$-2x^2, -5x$
$\frac{1}{2}a^2b, -a^2b$	$6st^2, 4s^2t$

Only like terms can be combined. We combine like terms by applying the distributive property.

EXAMPLE 6 Simplify each polynomial by combining any like terms.

a. $-3x + 7x$ **b.** $x + 3x^2$

c. $9x^3 + x^3$ **d.** $11x^2 + 5 + 2x^2 - 7$

e. $\frac{2}{5}x^4 + \frac{2}{3}x^3 - x^2 + \frac{1}{10}x^4 - \frac{1}{6}x^3$

Solution

a. $-3x + 7x = (-3 + 7)x = 4x$

b. $x + 3x^2$ These terms cannot be combined because x and $3x^2$ are not like terms.

c. $9x^3 + x^3 = 9x^3 + 1x^3 = 10x^3$

d. $11x^2 + 5 + 2x^2 - 7 = 11x^2 + 2x^2 + 5 - 7$

$= 13x^2 - 2$ Combine like terms.

e. $\frac{2}{5}x^4 + \frac{2}{3}x^3 - x^2 + \frac{1}{10}x^4 - \frac{1}{6}x^3$

$$= \left(\frac{2}{5} + \frac{1}{10}\right)x^4 + \left(\frac{2}{3} - \frac{1}{6}\right)x^3 - x^2$$

$$= \left(\frac{4}{10} + \frac{1}{10}\right)x^4 + \left(\frac{4}{6} - \frac{1}{6}\right)x^3 - x^2$$

$$= \frac{5}{10}x^4 + \frac{3}{6}x^3 - x^2$$

$$= \frac{1}{2}x^4 + \frac{1}{2}x^3 - x^2$$

□

PRACTICE 6 Simplify each polynomial by combining any like terms.

a. $-4y + 2y$ **b.** $z + 5z^3$ **c.** $15x^3 - x^3$

d. $7a^2 - 5 - 3a^2 - 7$ **e.** $\frac{3}{8}x^3 - x^2 + \frac{5}{6}x^4 + \frac{1}{12}x^3 - \frac{1}{2}x^4$

■

✔ **CONCEPT CHECK**

When combining like terms in the expression $5x - 8x^2 - 8x$, which of the following is the proper result?

a. $-11x^2$ b. $-8x^2 - 3x$ c. $-11x$ d. $-11x^4$

EXAMPLE 7 Combine like terms to simplify.

$$-9x^2 + 3xy - 5y^2 + 7yx$$

Solution

$$-9x^2 + 3xy - 5y^2 + 7yx = -9x^2 + (3 + 7)xy - 5y^2$$

$$= -9x^2 + 10xy - 5y^2$$

□

Helpful Hint

This term can be written as $7yx$ or $7xy$.

PRACTICE 7 Combine like terms to simplify: $9xy - 3x^2 - 4yx + 5y^2$. ■

Answer to Concept Check:
b

EXAMPLE 8 Write a polynomial that describes the total area of the squares and rectangles shown on page 245. Then simplify the polynomial.

Solution

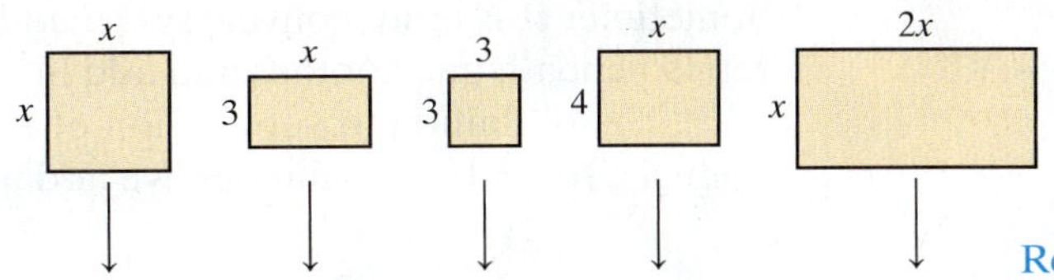

Area: $x \cdot x + 3 \cdot x + 3 \cdot 3 + 4 \cdot x + x \cdot 2x$ — Recall that the area of a rectangle is length times width.

$= x^2 + 3x + 9 + 4x + 2x^2$

$= 3x^2 + 7x + 9$ — Combine like terms. □

PRACTICE 8 Write a polynomial that describes the total area of the squares and rectangles shown below. Then simplify the polynomial.

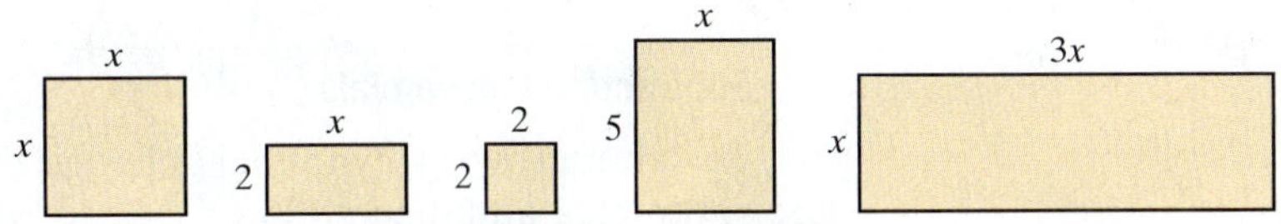

■

OBJECTIVE 4

Adding and Subtracting Polynomials

We now practice adding and subtracting polynomials.

> **Adding Polynomials**
>
> To add polynomials, combine all like terms.

EXAMPLE 9 Add.

a. $(7x^3y - xy^3 + 11) + (6x^3y - 4)$

b. $(3a^3 - b + 2a - 5) + (a + b + 5)$

Solution

a. To add, remove the parentheses and group like terms.

$(7x^3y - xy^3 + 11) + (6x^3y - 4)$

$= 7x^3y - xy^3 + 11 + 6x^3y - 4$

$= 7x^3y + 6x^3y - xy^3 + 11 - 4$ — Group like terms.

$= 13x^2y - xy^3 + 7$ — Combine like terms.

b. $(3a^3 - b + 2a - 5) + (a + b + 5)$

$= 3a^3 - b + 2a - 5 + a + b + 5$

$= 3a^3 - b + b + 2a + a - 5 + 5$ — Group like terms.

$= 3a^3 + 3a$ — Combine like terms. □

PRACTICE 9 Add.

a. $(4y^2 + x - 3y - 7) + (x + y^2 - 2)$

b. $(-8a^2b - ab^2 + 10) + (-2ab^2 - 10)$ ■

EXAMPLE 10 Add $(11x^3 - 12x^2 + x - 3)$ and $(x^3 - 10x + 5)$.

Solution

$(11x^3 - 12x^2 + x - 3) + (x^3 - 10x + 5)$

$= 11x^3 + x^3 - 12x^2 + x - 10x - 3 + 5$ — Group like terms.

$= 12x^3 - 12x^2 - 9x + 2$ — Combine like terms. □

PRACTICE 10 Add $(3x^2 - 9x + 11)$ and $(-3x^2 + 7x^3 + 3x - 4)$ ■

Sometimes it is more convenient to add polynomials vertically. To do this, line up like terms beneath one another and add like terms.

The definition of subtraction of real numbers can be extended to apply to polynomials. To subtract a number, we add its opposite.

$$a - b = a + (-b)$$

Likewise, to subtract a polynomial, we add its opposite. In other words, if P and Q are polynomials, then

$$P - Q = P + (-Q)$$

The polynomial $-Q$ is the **opposite, or additive inverse,** of the polynomial Q. We can find $-Q$ by writing the opposite of each term of Q.

Subtracting Polynomials

To subtract two polynomials, change the signs of the terms of the polynomial being subtracted and then add.

✔ CONCEPT CHECK

Which polynomial is the opposite of $16x^3 - 5x + 7$?

a. $-16x^3 - 5x + 7$ **b.** $-16x^3 + 5x - 7$ **c.** $16x^3 + 5x + 7$ **d.** $-16x^3 + 5x + 7$

EXAMPLE 11 Subtract: $(2x^3 + 8x^2 - 6x) - (2x^3 - x^2 + 1)$

Solution First, change the sign of each term of the second polynomial and then add.

Helpful Hint
Notice the sign of each term is changed.

$$\begin{aligned}(2x^3 + 8x^2 - 6x) - (2x^3 - x^2 + 1) &= (2x^3 + 8x^2 - 6x) + (-2x^3 + x^2 - 1)\\ &= 2x^3 - 2x^3 + 8x^2 + x^2 - 6x - 1\\ &= 9x^2 - 6x - 1 \quad \text{Combine like terms.}\end{aligned}$$

□

PRACTICE 11 Subtract: $(3x^3 - 5x^2 + 4x) - (x^3 - x^2 + 6)$ $2x^3 - 4x^2 + 4x - 6$ ■

EXAMPLE 12 Subtract $(5z - 7)$ from the sum of $(8z + 11)$ and $(9z - 2)$.

Solution Notice that $(5z - 7)$ is to be subtracted **from** a sum. The translation is

$$\begin{aligned}&[(8z + 11) + (9z - 2)] - (5z - 7)\\ &= 8z + 11 + 9z - 2 - 5z + 7 \quad \text{Remove grouping symbols.}\\ &= 8z + 9z - 5z + 11 - 2 + 7 \quad \text{Group like terms.}\\ &= 12z + 16 \quad \text{Combine like terms.}\end{aligned}$$

□

PRACTICE 12 Subtract $(3x + 5)$ from the sum of $(8x - 11)$ and $(2x + 5)$. ■

EXAMPLE 13 Add or subtract as indicated.

a. $(3x^2 - 6xy + 5y^2) + (-2x^2 + 8xy - y^2)$

b. $(9a^2b^2 + 6ab - 3ab^2) - (5b^2a + 2ab - 3 - 9b^2)$

Solution

a. $$\begin{aligned}&(3x^2 - 6xy + 5y^2) + (-2x^2 + 8xy - y^2)\\ &= 3x^2 - 6xy + 5y^2 - 2x^2 + 8xy - y^2\\ &= x^2 + 2xy + 4y^2 \quad \text{Combine like terms.}\end{aligned}$$

Answer to Concept Check:
b

b. $(9a^2b^2 + 6ab - 3ab^2) - (5b^2a + 2ab - 3 - 9b^2)$

$= 9a^2b^2 + 6ab - 3ab^2 - 5b^2a - 2ab + 3 + 9b^2$ Change the sign of each term of the polynomial being subtracted.

$= 9a^2b^2 + 4ab - 8ab^2 + 3 + 9b^2$ Combine like terms. □

PRACTICE 13 Add or subtract as indicated.

a. $(3a^2 - 4ab + 7b^2) + (-8a^2 + 3ab - b^2)$

b. $(5x^2y^2 - 6xy - 4xy^2) - (2x^2y^2 + 4xy - 5 + 6y^2)$

✔ CONCEPT CHECK

If possible, simplify each expression by performing the indicated operation.

a. $2y + y$ **b.** $2y \cdot y$ **c.** $-2y - y$ **d.** $(-2y)(-y)$ **e.** $2x + y$

To add or subtract polynomials vertically, just remember to line up like terms. For example, perform the subtraction $(10x^3y^2 - 7x^2y^2) - (4x^3y^2 - 3x^2y^2 + 2y^2)$ vertically.

Add the opposite of the second polynomial.

$$\begin{array}{r} 10x^3y^2 - 7x^2y^2 \\ \underline{-(4x^3y^2 - 3x^2y^2 + 2y^2)} \end{array} \quad \text{is equivalent to} \quad \begin{array}{r} 10x^3y^2 - 7x^2y^2 \\ \underline{-4x^3y^2 + 3x^2y^2 - 2y^2} \\ 6x^3y^2 - 4x^2y^2 - 2y^2 \end{array}$$

✔ CONCEPT CHECK

Why is the following subtraction incorrect?

$$\begin{aligned} &(7z - 5) - (3z - 4) \\ &= 7z - 5 - 3z - 4 \\ &= 4z - 9 \end{aligned}$$

Polynomial functions, like polynomials, can be added, subtracted, multiplied, and divided. For example, if

$$P(x) = x^2 + x + 1$$

then

$$2P(x) = 2(x^2 + x + 1) = 2x^2 + 2x + 2 \quad \text{Use the distributive property.}$$

Also, if $Q(x) = 5x^2 - 1$, then $P(x) + Q(x) = (x^2 + x + 1) + (5x^2 - 1)$
$= 6x^2 + x.$

A useful business and economics application of subtracting polynomial functions is finding the profit function $P(x)$ when given a revenue function $R(x)$ and a cost function $C(x)$. In business, it is true that

$$\text{profit} = \text{revenue} - \text{cost, or}$$
$$P(x) = R(x) - C(x)$$

For example, if the revenue function is $R(x) = 7x$ and the cost function is $C(x) = 2x + 5000$, then the profit function is

$$P(x) = R(x) - C(x)$$

or

$P(x) = 7x - (2x + 5000)$ Substitute $R(x) = 7x$ and $C(x) = 2x + 5000$.

$P(x) = 5x - 5000$

Answers to Concept Checks:
a. $3y$ **b.** $2y^2$ **c.** $-3y$ **d.** $2y^2$
e. cannot be simplified;
With parentheses removed, the expression should be $7z - 5 - 3z + 4 = 4z - 1$

Graphing Calculator Explorations

A graphing calculator may be used to visualize addition and subtraction of polynomials in one variable. For example, to visualize the following polynomial subtraction statement

$$(3x^2 - 6x + 9) - (x^2 - 5x + 6) = 2x^2 - x + 3$$

graph both

$$Y_1 = (3x^2 - 6x + 9) - (x^2 - 5x + 6) \quad \text{Left side of equation}$$

and

$$Y_2 = 2x^2 - x + 3 \quad \text{Right side of equation}$$

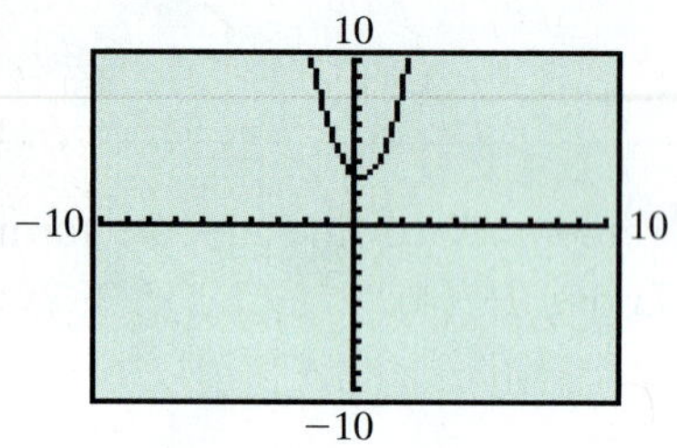

on the same screen and see that their graphs coincide. (*Note:* If the graphs do not coincide, we can be sure that a mistake has been made in combining polynomials or in calculator keystrokes. If the graphs appear to coincide, we cannot be sure that our work is correct. This is because it is possible for the graphs to differ so slightly that we do not notice it.)

The graphs of Y_1 and Y_2 are shown. The graphs appear to coincide, so the subtraction statement

$$(3x^2 - 6x + 9) - (x^2 - 5x + 6) = 2x^2 - x + 3$$

appears to be correct.

Perform the indicated operations. Then visualize by using the procedure described above.

1. $(2x^2 + 7x + 6) + (x^3 - 6x^2 - 14)$

2. $(-14x^3 - x + 2) + (-x^3 + 3x^2 + 4x)$

3. $(1.8x^2 - 6.8x - 1.7) - (3.9x^2 - 3.6x)$

4. $(-4.8x^2 + 12.5x - 7.8) - (3.1x^2 - 7.8x)$

5. $(1.29x - 5.68) + (7.69x^2 - 2.55x + 10.98)$

6. $(-0.98x^2 - 1.56x + 5.57) + (4.36x - 3.71)$

Vocabulary, Readiness & Video Check

Use the choices below to fill in each blank. Not all choices will be used.

least	monomial	trinomial	coefficient
greatest	binomial	constant	

1. A ___________ is a polynomial with exactly 2 terms.

2. A ___________ is a polynomial with exactly one term.

3. A ___________ is a polynomial with exactly three terms.

4. The numerical factor of a term is called the ___________.

5. A number term is also called a ___________.

6. The degree of a polynomial is the ___________ degree of any term of the polynomial.

Martin-Gay Interactive Videos

See Video 4.2

Watch the section lecture video and answer the following questions.

OBJECTIVE 1

7. For Example 2, why is the degree of each **term** found when the example asks for the degree of the **polynomial** only?

OBJECTIVE 2

8. From Example 3, how do you find the value of a polynomial function $Q(x)$ given a value for x?

OBJECTIVE 3

9. When combining any like terms in a polynomial, as in Examples 4–6, what are we doing to the polynomial?

OBJECTIVE 4

10. From Example 7, when we simply remove parentheses and combine the like terms of two polynomials, what operation do we perform? Is this true of Examples 9–11?

4.2 Exercise Set MyMathLab®

Find the degree of each of the following polynomials and determine whether it is a monomial, binomial, trinomial, or none of these. See Examples 1 through 3.

1. $x + 2$
2. $-6y^2 + 4$
3. $9m^3 - 5m^2 + 4m - 8$
4. $a + 5a^2 + 3a^3 - 4a^4$
5. $12x^4y - x^2y^2 - 12x^2y^4$
6. $7r^2s^2 + 2rs - 3rs^5$
7. $3 - 5x^8$
8. $5y^7 + 2$

In the second column, write the degree of the polynomial in the first column. See Examples 1 through 3.

	Polynomial	Degree
9.	$3xy^2 - 4$	
10.	$8x^2y^2$	
11.	$5a^2 - 2a + 1$	
12.	$4z^6 + 3z^2$	

If $P(x) = x^2 + x + 1$ and $Q(x) = 5x^2 - 1$, find the following. See Examples 4 and 5.

13. $P(7)$
14. $Q(4)$
15. $Q(-10)$
16. $P(-4)$
17. $P(0)$
18. $Q(0)$
19. $Q\left(\frac{1}{4}\right)$
20. $P\left(\frac{1}{2}\right)$

The CN Tower in Toronto, Ontario, is 1821 feet tall and is the world's tallest self-supporting structure. An object is dropped from the Skypod of the Tower, which is at 1150 feet. Neglecting air resistance, the height of the object at time t seconds is given by the polynomial function $P(t) = -16t^2 + 1150$. Find the height of the object at the given times.

	Time, t (in seconds)	Height $P(t) = -16t^2 + 1150$
21.	1	
22.	7	
23.	3	
24.	6	

25. The polynomial $-7.5x^2 + 103x + 2000$ models the yearly number of visitors (in thousands) x years after 2006 at Acadia National Park in Maine. Use this polynomial to estimate the number of visitors to the park in 2016.

26. The polynomial $-0.13x^2 + x + 827$ models the yearly number of visitors (in thousands) x years after 2006 at Canyon De Chelly National Monument in Arizona. Use this polynomial to estimate the number of visitors to the park in 2010.

Simplify each of the following by combining like terms. See Examples 6 and 7.

27. $14x^2 + 9x^2$
28. $18x^3 - 4x^3$
29. $15x^2 - 3x^2 - y$
30. $12k^3 - 9k^3 + 11$
31. $8s - 5s + 4s$
32. $5y + 7y - 6y$
33. $0.1y^2 - 1.2y^2 + 6.7 - 1.9$
34. $7.6y + 3.2y^2 - 8y - 2.5y^2$
35. $\frac{2}{5}x^2 - \frac{1}{3}x^3 + x^2 - \frac{1}{4}x^3 + 6$
36. $\frac{1}{6}x^4 - \frac{1}{7}x^2 + 5 - \frac{1}{2}x^4 - \frac{3}{7}x^2 + \frac{1}{3}$
37. $6a^2 - 4ab + 7b^2 - a^2 - 5ab + 9b^2$
38. $x^2y + xy - y + 10x^2y - 2y + xy$

Perform the indicated operations. See Examples 9 through 13.

39. $(-7x + 5) + (-3x^2 + 7x + 5)$
40. $(3x - 8) + (4x^2 - 3x + 3)$
41. $(2x^2 + 5) - (3x^2 - 9)$
42. $(5x^2 + 4) - (-2y^2 + 4)$
43. $3x - (5x - 9)$
44. $4 - (-y - 4)$
45. $(2x^2 + 3x - 9) - (-4x + 7)$
46. $(-7x^2 + 4x + 7) - (-8x + 2)$

47. $$\begin{array}{r} 3t^2 + 4 \\ \underline{+5t^2 - 8} \end{array}$$

48. $$\begin{array}{r} 7x^3 + 3 \\ \underline{+2x^3 - 1} \end{array}$$

49. $$\begin{array}{r} 4z^2 - 8z + 3 \\ \underline{-(6z^2 + 8z - 3)} \end{array}$$

50. $$\begin{array}{r} 5u^5 - 4u^2 + 3u - 7 \\ \underline{-(3u^5 + 6u^2 - 8u + 2)} \end{array}$$

51. $$\begin{array}{r} 5x^3 - 4x^2 + 6x - 2 \\ \underline{-(3x^3 - 2x^2 - x - 4)} \end{array}$$

52. $$\begin{array}{r} 7a^2 - 9a + 6 \\ \underline{-(11a^2 - 4a + 2)} \end{array}$$

MIXED PRACTICE

Perform the indicated operations.

53. $(-3y^2 - 4y) + (2y^2 + y - 1)$
54. $(7x^2 + 2x - 9) + (-3x^2 + 5)$
55. $(5x + 8) - (-2x^2 - 6x + 8)$
56. $(-6y^2 + 3y - 4) - (9y^2 - 3y)$
57. $(-8x^4 + 7x) + (-8x^4 + x + 9)$
58. $(6y^5 - 6y^3 + 4) + (-2y^5 - 8y^3 - 7)$
59. $(3x^2 + 5x - 8) + (5x^2 + 9x + 12) - (x^2 - 14)$
60. $(-a^2 + 1) - (a^2 - 3) + (5a^2 - 6a + 7)$

TRANSLATING

Perform each indicated operation.

61. Subtract $4x$ from $(7x - 3)$.
62. Subtract y from $(y^2 - 4y + 1)$.
63. Add $(4x^2 - 6x + 1)$ and $(3x^2 + 2x + 1)$.
64. Add $(-3x^2 - 5x + 2)$ and $(x^2 - 6x + 9)$.
65. Subtract $(19x^2 + 5)$ from $(81x^2 + 10)$.
66. Subtract $(2x + xy)$ from $(3x - 9xy)$.
67. Subtract $(5x + 7)$ from $(7x^2 + 3x + 9)$.
68. Subtract $(5y^2 + 8y + 2)$ from $(7y^2 + 9y - 8)$.
69. Subtract $(2x + 2)$ from the sum of $(8x + 1)$ and $(6x + 3)$.
70. Subtract $(-12x - 3)$ from the sum of $(-5x - 7)$ and $(12x + 3)$.
71. Subtract $(4y^2 - 6y - 3)$ from the sum of $(8y^2 + 7)$ and $(6y + 9)$.
72. Subtract $(4x^2 - 2x + 2)$ from the sum of $(x^2 + 7x + 1)$ and $(7x + 5)$.

Find the area of each figure. Write a polynomial that describes the total area of the rectangles and squares shown in Exercises 73–74. Then simplify the polynomial. See Example 8.

73.

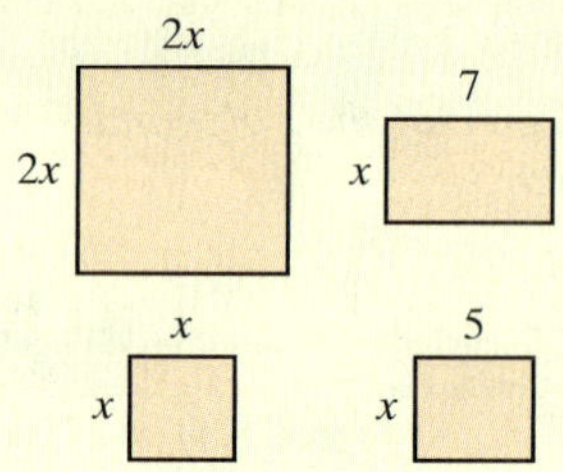

74.

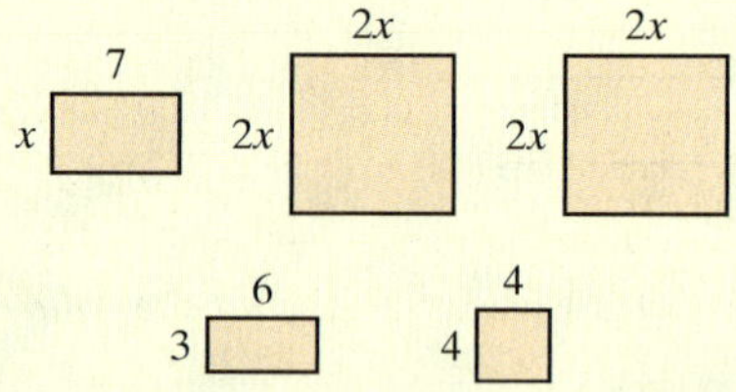

Add or subtract as indicated. See Examples 9 through 13.

75. $(9a + 6b - 5) + (-11a - 7b + 6)$
76. $(3x - 2 + 6y) + (7x - 2 - y)$
77. $(4x^2 + y^2 + 3) - (x^2 + y^2 - 2)$
78. $(7a^2 - 3b^2 + 10) - (-2a^2 + b^2 - 12)$
79. $(x^2 + 2xy - y^2) + (5x^2 - 4xy + 20y^2)$
80. $(a^2 - ab + 4b^2) + (6a^2 + 8ab - b^2)$
81. $(11r^2s + 16rs - 3 - 2r^2s^2) - (3sr^2 + 5 - 9r^2s^2)$
82. $(3x^2y - 6xy + x^2y^2 - 5) - (11x^2y^2 - 1 + 5yx^2)$

Simplify each polynomial by combining like terms.

83. $7.75x + 9.16x^2 - 1.27 - 14.58x^2 - 18.34$

84. $1.85x^2 - 3.76x + 9.25x^2 + 10.76 - 4.21x$

Perform each indicated operation.

85. $[(7.9y^4 - 6.8y^3 + 3.3y) + (6.1y^3 - 5)] - (4.2y^4 + 1.1y - 1)$

86. $[(1.2x^2 - 3x + 9.1) - (7.8x^2 - 3.1 + 8)] + (1.2x - 6)$

REVIEW AND PREVIEW

Rose gold is made by mixing gold with other metals, as shown in the circle graph below. Use this graph to answer the exercises. See Section R.4.

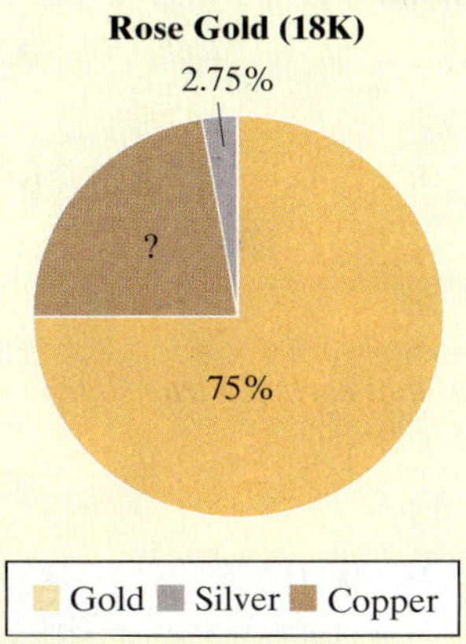

87. What percent of rose gold is copper?

88. What percent of rose gold is not gold?

89. What fraction of rose gold is gold?

90. How much gold is in 18 grams of rose gold?

91. How much silver is in 18 grams of rose gold?

92. How much copper is in 18 grams of rose gold?

CONCEPT EXTENSIONS

Recall that the perimeter of a figure is the sum of the lengths of its sides. For Exercises 93 through 96, find the perimeter of each polynomial.

△ **93.**

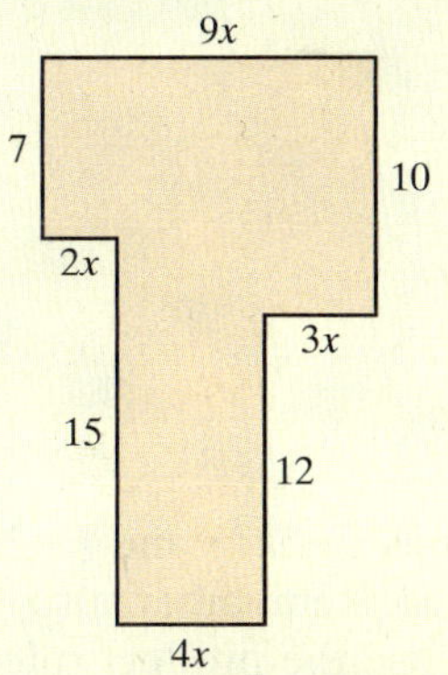

△ **94.**

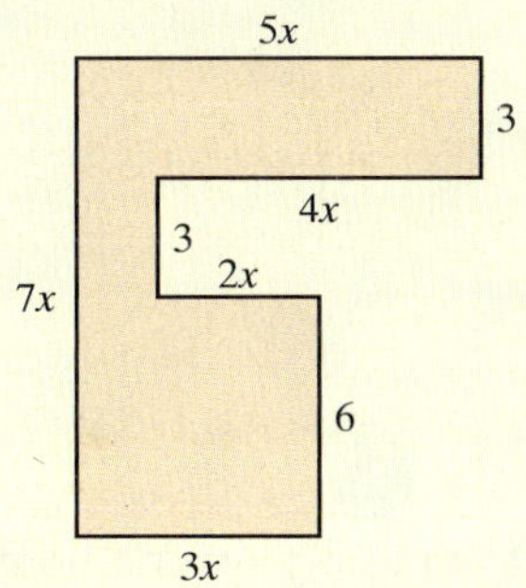

△ **95.**

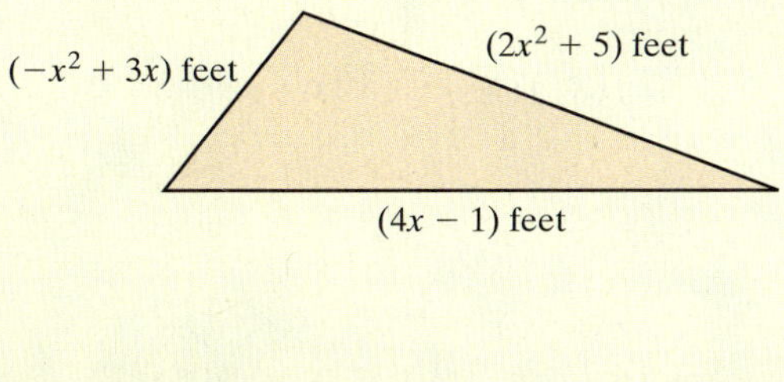

△ **96.**

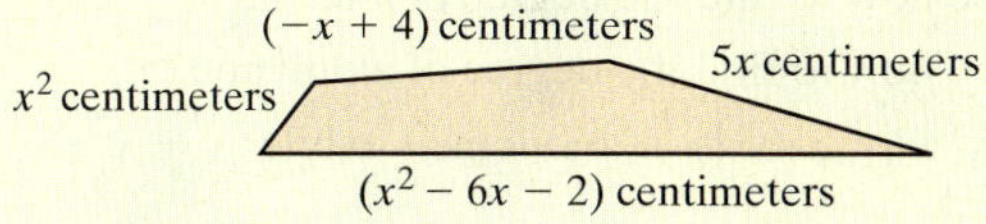

△ **97.** A wooden beam is $(4y^2 + 4y + 1)$ meters long. If a piece $(y^2 - 10)$ meters is cut, express the length of the remaining piece of beam as a polynomial in y.

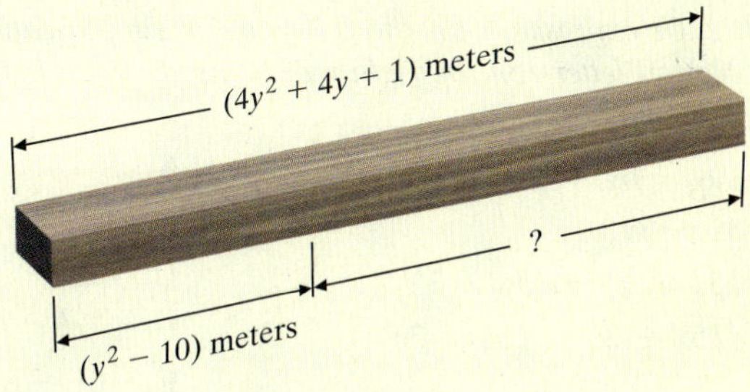

△ **98.** A piece of quarter-round molding is $(13x - 7)$ inches long. If a piece $(2x + 2)$ inches is removed, express the length of the remaining piece of molding as a polynomial in x.

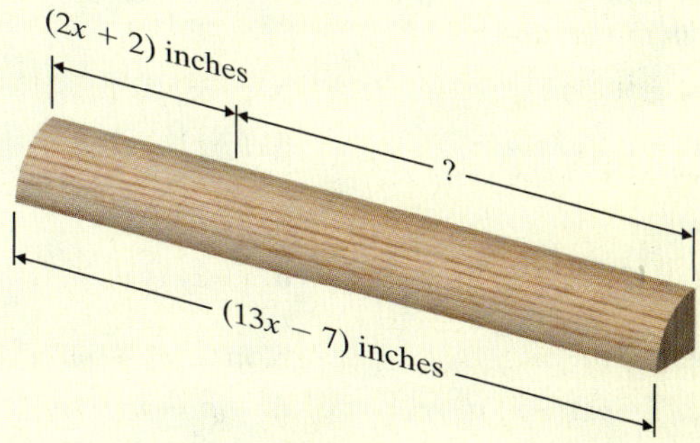

The number of worldwide Internet users (in millions) x years after the year 2000 can be modeled by the polynomial $4.8x^2 + 104x + 431$ for the years 1995 through 2015. Use this polynomial for Exercises 99 and 100.

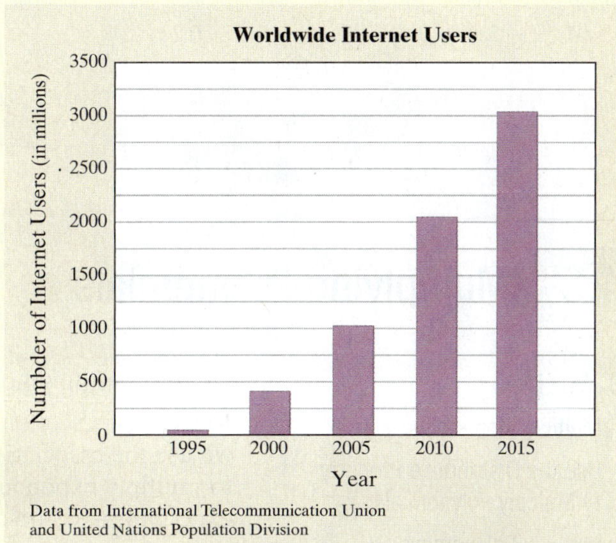

99. Estimate the number of Internet users in the world in 2015.

100. Use the given polynomial to predict the number of Internet users in the world in 2020.

101. Describe how to find the degree of a term.

102. Describe how to find the degree of a polynomial.

103. Explain why xyz is a monomial while $x + y + z$ is a trinomial.

104. Explain why the degree of the term $5y^3$ is 3 and the degree of the polynomial $2y + y + 2y$ is 1.

Match each expression on the left with its simplification on the right. Not all letters on the right will be used, and a letter may be used more than once.

105. $10y - 6y^2 - y$

106. $5x + 5x$

107. $(5x - 3) + (5x - 3)$

108. $(15x - 3) - (5x - 3)$

A. $3y$

B. $9y - 6y^2$

C. $10x$

D. $25x^2$

E. $10x - 6$

F. none of these

Simplify each expression by performing the indicated operation. Explain how you arrived at each answer. See the third Concept Check in this section.

109. **a.** $z + 3z$ **b.** $z \cdot 3z$ **c.** $-z - 3z$ **d.** $(-z)(-3z)$

110. **a.** $2y + y$ **b.** $2y \cdot y$ **c.** $-2y - y$ **d.** $(-2y)(-y)$

111. **a.** $m \cdot m \cdot m$ **b.** $m + m + m$ **c.** $(-m)(-m)(-m)$ **d.** $-m - m - m$

112. **a.** $x + x$ **b.** $x \cdot x$ **c.** $-x - x$ **d.** $(-x)(-x)$

Perform the indicated operations.

113. $(4x^{2a} - 3x^a + 0.5) - (x^{2a} - 5x^a - 0.2)$

114. $(9y^{5a} - 4y^{3a} + 1.5y) - (6y^{5a} - y^{3a} + 4.7y)$

115. $(8x^{2y} - 7x^y + 3) + (-4x^{2y} + 9x^y - 14)$

116. $(14z^{5x} + 3z^{2x} + z) - (2z^{5x} - 10z^{2x} + 3z)$

If $P(x) = 3x + 3$, $Q(x) = 4x^2 - 6x + 3$, and $R(x) = 5x^2 - 7$, find the following.

117. $P(x) + Q(x)$

118. $R(x) + P(x)$

119. $Q(x) - R(x)$

120. $P(x) - Q(x)$

121. $2[Q(x)] - R(x)$

122. $-5[P(x)] - Q(x)$

If $P(x)$ is the polynomial given, find **a.** $P(a)$, **b.** $P(-x)$, *and* **c.** $P(x + h)$.

123. $P(x) = 2x - 3$

124. $P(x) = 8x + 3$

125. $P(x) = 4x$

126. $P(x) = -4x$

Fill in the squares so that each is a true statement.

127. $3x^{\square} + 4x^2 = 7x^{\square}$

128. $9y^7 + 3y^{\square} = 12y^7$

129. $2x^{\square} + 3x^{\square} - 5x^{\square} + 4x^{\square} = 6x^4 - 2x^3$

130. $3y^{\square} + 7y^{\square} - 2y^{\square} - y^{\square} = 10y^5 - 3y^2$

Write a polynomial that describes the surface area of each figure. (Recall that the surface area of a solid is the sum of the areas of the faces or sides of the solid.)

△ **131.**

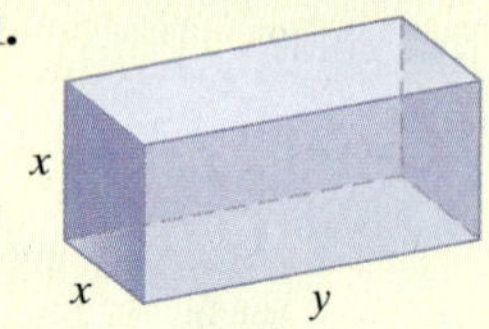

△ **132.**

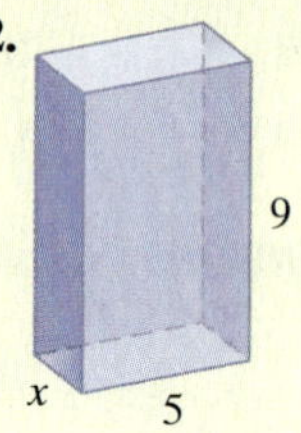

4.3 Multiplying Polynomials

OBJECTIVES

1 Multiply Monomials.

2 Use the Distributive Property to Multiply Polynomials.

3 Multiply Polynomials Vertically.

OBJECTIVE

1 Multiplying Monomials

Recall from Section 4.1 that to multiply two monomials such as $(-5x^3)$ and $(-2x^4)$, we use the associative and commutative properties and regroup. Remember, also, that to multiply exponential expressions with a common base, we use the product rule for exponents and add exponents.

$$(-5x^3)(-2x^4) = (-5)(-2)(x^3)(x^4) = 10x^7$$

EXAMPLES Multiply.

1. $6x \cdot 4x = (6 \cdot 4)(x \cdot x)$ Use the commutative and associative properties.
$= 24x^2$ Multiply.

2. $-7x^2 \cdot 0.2x^5 = (-7 \cdot 0.2)(x^2 \cdot x^5)$
$= -1.4x^7$

3. $\left(-\frac{1}{3}x^5\right)\left(-\frac{2}{9}x\right) = \left(-\frac{1}{3} \cdot -\frac{2}{9}\right) \cdot (x^5 \cdot x)$
$= \frac{2}{27}x^6$ □

PRACTICE 1–3 Multiply.

1. $5y \cdot 2y$ 2. $(5z^3) \cdot (-0.4z^5)$ 3. $\left(-\frac{1}{9}b^6\right)\left(-\frac{7}{8}b^3\right)$ ■

✔ CONCEPT CHECK

Simplify.

a. $3x \cdot 2x$ b. $3x + 2x$

OBJECTIVE 2

Using the Distributive Property to Multiply Polynomials

To multiply polynomials that are not monomials, use the distributive property.

EXAMPLE 4 Use the distributive property to find each product.

a. $5x(2x^3 + 6)$ b. $-3x^2(5x^2 + 6x - 1)$

Solution

a. $5x(2x^3 + 6) = 5x(2x^3) + 5x(6)$ Use the distributive property.
$= 10x^4 + 30x$ Multiply.

b. $-3x^2(5x^2 + 6x - 1)$
$= (-3x^2)(5x^2) + (-3x^2)(6x) + (-3x^2)(-1)$ Use the distributive property.
$= -15x^4 - 18x^3 + 3x^2$ Multiply. □

PRACTICE 4 Use the distributive property to find each product.

a. $3x(9x^5 + 11)$ b. $-6x^3(2x^2 - 9x + 2)$ ■

We also use the distributive property to multiply two binomials. To multiply $(x + 3)$ by $(x + 1)$, distribute the factor $(x + 3)$ first.

$(x + 3)(x + 1) = x(x + 1) + 3(x + 1)$ Distribute $(x + 3)$.
$= x(x) + x(1) + 3(x) + 3(1)$ Apply the distributive property a second time.
$= x^2 + x + 3x + 3$ Multiply.
$= x^2 + 4x + 3$ Combine like terms.

Answers to Concept Check:
a. $6x^2$ b. $5x$

This idea can be expanded so that we can multiply any two polynomials.

To Multiply Two Polynomials

Multiply each term of the first polynomial by each term of the second polynomial and then combine like terms.

EXAMPLE 5 Multiply: $(3x + 2)(2x - 5)$

Solution Multiply each term of the first binomial by each term of the second.

$$\begin{aligned}(3x + 2)(2x - 5) &= 3x(2x) + 3x(-5) + 2(2x) + 2(-5)\\ &= 6x^2 - 15x + 4x - 10 && \text{Multiply.}\\ &= 6x^2 - 11x - 10 && \text{Combine like terms.}\end{aligned}$$

□

PRACTICE 5 Multiply: $(5x - 2)(2x + 3)$ ■

EXAMPLE 6 Multiply: $(2x - y)^2$

Solution Recall that $a^2 = a \cdot a$, so $(2x - y)^2 = (2x - y)(2x - y)$. Multiply each term of the first polynomial by each term of the second.

$$\begin{aligned}(2x - y)(2x - y) &= 2x(2x) + 2x(-y) + (-y)(2x) + (-y)(-y)\\ &= 4x^2 - 2xy - 2xy + y^2 && \text{Multiply.}\\ &= 4x^2 - 4xy + y^2 && \text{Combine like terms.}\end{aligned}$$

□

PRACTICE 6 Multiply: $(5x - 3y)^2$ ■

✔CONCEPT CHECK

Square where indicated. Simplify if possible.

a. $(4a)^2 + (3b)^2$ **b.** $(4a + 3b)^2$

EXAMPLE 7 Multiply $(t + 2)$ by $(3t^2 - 4t + 2)$.

Solution Multiply each term of the first polynomial by each term of the second.

$$\begin{aligned}(t + 2)(3t^2 - 4t + 2) &= t(3t^2) + t(-4t) + t(2) + 2(3t^2) + 2(-4t) + 2(2)\\ &= 3t^3 - 4t^2 + 2t + 6t^2 - 8t + 4\\ &= 3t^3 + 2t^2 - 6t + 4 && \text{Combine like terms.}\end{aligned}$$

□

PRACTICE 7 Multiply $(y + 4)$ by $(2y^2 - 3y + 5)$. ■

EXAMPLE 8 Multiply: $(3a + b)^3$

Solution Write $(3a + b)^3$ as $(3a + b)(3a + b)(3a + b)$.

$$\begin{aligned}(3a + b)(3a + b)(3a + b) &= (9a^2 + 3ab + 3ab + b^2)(3a + b)\\ &= (9a^2 + 6ab + b^2)(3a + b)\\ &= 9a^2(3a + b) + 6ab(3a + b) + b^2(3a + b)\\ &= 27a^3 + 9a^2b + 18a^2b + 6ab^2 + 3ab^2 + b^3\\ &= 27a^3 + 27a^2b + 9ab^2 + b^3\end{aligned}$$

□

PRACTICE 8 Multiply: $(s + 2t)^3$ ■

Answers to Concept Check:

a. $16a^2 + 9b^2$

b. $16a^2 + 24ab + 9b^2$

OBJECTIVE 3

Multiplying Polynomials Vertically

Another convenient method for multiplying polynomials is to use a vertical format similar to the format used to multiply real numbers. We demonstrate this method by multiplying $(3y^2 - 4y + 1)$ by $(y + 2)$.

EXAMPLE 9 Multiply $(3y^2 - 4y + 1)$ by $(y + 2)$. Use a vertical format.

Solution

$$\begin{array}{rrrrl} & 3y^2 & -4y & +1 & \\ \times & & y & +2 & \text{1st, multiply } 3y^2 - 4y + 1 \text{ by } 2. \\ \hline & 6y^2 & -8y & +2 & \text{2nd, multiply } 3y^2 - 4y + 1 \text{ by } y. \\ 3y^3 & -4y^2 & +y & & \text{Line up like terms.} \\ \hline 3y^3 & +2y^2 & -7y & +2 & \text{3rd, combine like terms.} \end{array}$$

Helpful Hint

Make sure like terms are lined up.

Thus, $(3y^2 - 4y + 1)(y + 2) = 3y^3 + 2y^2 - 7y + 2$. □

PRACTICE 9 Multiply $(5x^2 - 3x + 5)$ by $(x - 4)$. Use a vertical format. ■

When multiplying vertically, be careful if a power is missing; you may want to leave space in the partial products and take care that like terms are lined up.

EXAMPLE 10 Multiply $(2x^3 - 3x + 4)$ by $(x^2 + 1)$. Use a vertical format.

Solution

$$\begin{array}{rrrrrl} & 2x^3 & & -3x & +4 & \\ \times & & & x^2 & +1 & \\ \hline & 2x^3 & & -3x & +4 & \text{Leave space for missing powers of } x. \\ 2x^5 & -3x^3 & +4x^2 & & & \leftarrow \text{Line up like terms.} \\ \hline 2x^5 & -x^3 & +4x^2 & -3x & +4 & \text{Combine like terms.} \end{array}$$

□

PRACTICE 10 Multiply $(x^3 - 2x^2 + 1)$ by $(x^2 + 2)$ using a vertical format. ■

EXAMPLE 11 Find the product of $(2x^2 - 3x + 4)$ and $(x^2 + 5x - 2)$ using a vertical format.

Solution First, we arrange the polynomials in a vertical format. Then we multiply each term of the second polynomial by each term of the first polynomial.

$$\begin{array}{rrrrrl} & & 2x^2 & -3x & +4 & \\ \times & & x^2 & +5x & -2 & \\ \hline & & -4x^2 & +6x & -8 & \text{Multiply } 2x^2 - 3x + 4 \text{ by } -2. \\ & 10x^3 & -15x^2 & +20x & & \text{Multiply } 2x^2 - 3x + 4 \text{ by } 5x. \\ 2x^4 & -3x^3 & +4x^2 & & & \text{Multiply } 2x^2 - 3x + 4 \text{ by } x^2. \\ \hline 2x^4 & +7x^3 & -15x^2 & +26x & -8 & \text{Combine like terms.} \end{array}$$

□

PRACTICE 11 Find the product of $(5x^2 + 2x - 2)$ and $(x^2 - x + 3)$ using a vertical format. ■

Vocabulary, Readiness & Video Check

Fill in each blank with the correct choice.

1. The expression $5x(3x + 2)$ equals $5x \cdot 3x + 5x \cdot 2$ by the ________________ property.
   a. commutative **b.** associative **c.** distributive
2. The expression $(x + 4)(7x - 1)$ equals $x(7x - 1) + 4(7x - 1)$ by the ________________ property.
   a. commutative **b.** associative **c.** distributive
3. The expression $(5y - 1)^2$ equals ________________.
   a. $2(5y - 1)$ **b.** $(5y - 1)(5y + 1)$ **c.** $(5y - 1)(5y - 1)$
4. The expression $9x \cdot 3x$ equals ________________.
   a. $27x$ **b.** $27x^2$ **c.** $12x$ **d.** $12x^2$

Martin-Gay Interactive Videos

See Video 4.3

Watch the section lecture video and answer the following questions.

OBJECTIVE 1

5. For Example 1, we use the product property to multiply the monomials. Is it possible to add the same two monomials? Why or why not?

OBJECTIVE 2

6. What property and what exponent rule is used in Examples 2–6?

OBJECTIVE 3

7. Would you say the vertical format used in Example 7 also applies the distributive property? Explain.

4.3 Exercise Set MyMathLab®

Multiply. See Examples 1 through 3.

1. $-4n^3 \cdot 7n^7$
2. $9t^6(-3t^5)$
3. $(-3.1x^3)(4x^9)$
4. $(-5.2x^4)(3x^4)$
5. $\left(-\frac{1}{3}y^2\right)\left(\frac{2}{5}y\right)$
6. $\left(-\frac{3}{4}y^7\right)\left(\frac{1}{7}y^4\right)$
7. $(2x)(-3x^2)(4x^5)$
8. $(x)(5x^4)(-6x^7)$

Multiply. See Example 4.

9. $3x(2x + 5)$
10. $2x(6x + 3)$
11. $-2a(a + 4)$
12. $-3a(2a + 7)$
13. $3x(2x^2 - 3x + 4)$
14. $4x(5x^2 - 6x - 10)$
15. $-2a^2(3a^2 - 2a + 3)$
16. $-4b^2(3b^3 - 12b^2 - 6)$
17. $-y(4x^3 - 7x^2y + xy^2 + 3y^3)$
18. $-x(6y^3 - 5xy^2 + x^2y - 5x^3)$
19. $\frac{1}{2}x^2(8x^2 - 6x + 1)$
20. $\frac{1}{3}y^2(9y^2 - 6y + 1)$

Multiply. See Examples 5 and 6.

21. $(x + 4)(x + 3)$
22. $(x + 2)(x + 9)$
23. $(a + 7)(a - 2)$
24. $(y - 10)(y + 11)$
25. $\left(x + \frac{2}{3}\right)\left(x - \frac{1}{3}\right)$
26. $\left(x + \frac{3}{5}\right)\left(x - \frac{2}{5}\right)$
27. $(3x^2 + 1)(4x^2 + 7)$
28. $(5x^2 + 2)(6x^2 + 2)$
29. $(2y - 4)^2$
30. $(6x - 7)^2$
31. $(4x - 3)(3x - 5)$
32. $(8x - 3)(2x - 4)$
33. $(3x^2 + 1)^2$
34. $(x^2 + 4)^2$
35. Perform the indicated operations.
    a. $4y^2(-y^2)$
    b. $4y^2 - y^2$
    c. Explain the difference between the two expressions.
36. Perform the indicated operations.
    a. $9x^2(-10x^2)$
    b. $9x^2 - 10x^2$
    c. Explain the difference between the two expressions.

Multiply. See Example 7.

37. $(x - 2)(x^2 - 3x + 7)$

38. $(x + 3)(x^2 + 5x - 8)$

39. $(x + 5)(x^3 - 3x + 4)$

40. $(a + 2)(a^3 - 3a^2 + 7)$

41. $(2a - 3)(5a^2 - 6a + 4)$

42. $(3 + b)(2 - 5b - 3b^2)$

Multiply. See Example 8.

43. $(x + 2)^3$

44. $(y - 1)^3$

45. $(2y - 3)^3$

46. $(3x + 4)^3$

Multiply vertically. See Examples 9 through 11.

47. $(2x - 11)(6x + 1)$

48. $(4x - 7)(5x + 1)$

49. $(5x + 1)(2x^2 + 4x - 1)$

50. $(4x - 5)(8x^2 + 2x - 4)$

51. $(x^2 + 5x - 7)(2x^2 - 7x - 9)$

52. $(3x^2 - x + 2)(x^2 + 2x + 1)$

MIXED PRACTICE

Multiply. See Examples 1 through 11.

53. $-1.2y(-7y^6)$

54. $-4.2x(-2x^5)$

55. $-3x(x^2 + 2x - 8)$

56. $-5x(x^2 - 3x + 10)$

57. $(x + 19)(2x + 1)$

58. $(3y + 4)(y + 11)$

59. $\left(x + \frac{1}{7}\right)\left(x - \frac{3}{7}\right)$

60. $\left(m + \frac{2}{9}\right)\left(m - \frac{1}{9}\right)$

61. $(3y + 5)^2$

62. $(7y + 2)^2$

63. $(a + 4)(a^2 - 6a + 6)$

64. $(t + 3)(t^2 - 5t + 5)$

65. $(2x - 5)^3$

66. $(3y - 1)^3$

67. $(4x + 5)(8x^2 + 2x - 4)$

68. $(5x + 4)(x^2 - x + 4)$

69. $(3x^2 + 2x - 4)(2x^2 - 4x + 3)$

70. $(a^2 + 3a - 2)(2a^2 - 5a - 1)$

Express as the product of polynomials. Then multiply.

△ **71.** Find the area of the rectangle.

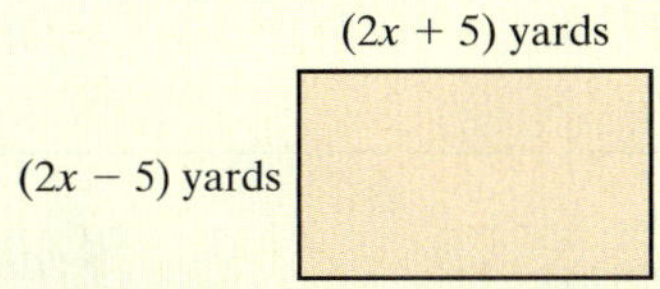

△ **72.** Find the area of the square field.

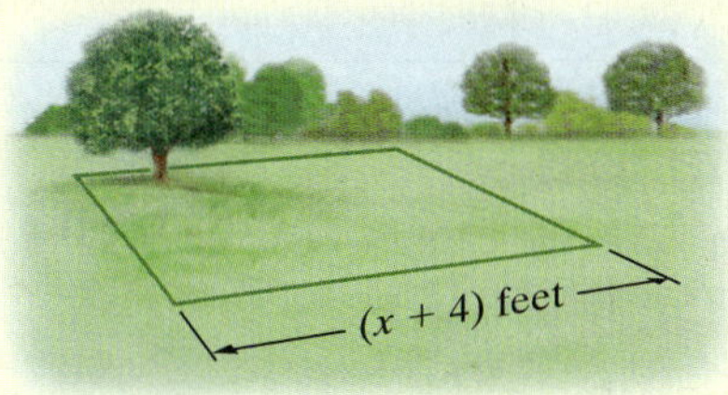

△ **73.** Find the area of the triangle.

(Triangle Area $= \frac{1}{2} \cdot \text{base} \cdot \text{height}$)

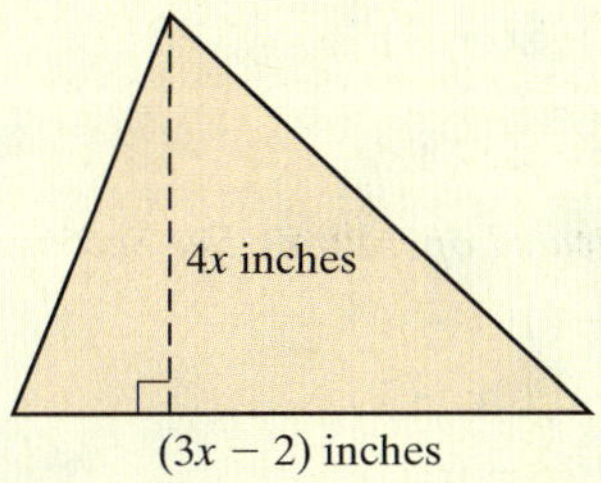

△ **74.** Find the volume of the cube.

(Volume = length · width · height)

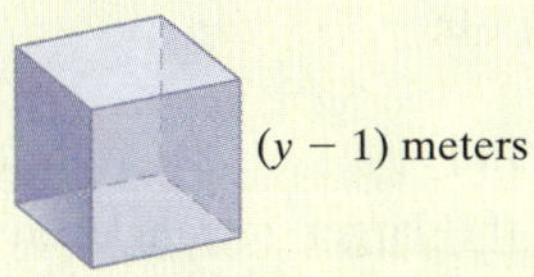

REVIEW AND PREVIEW

In this section, we review operations on monomials. Study the box below, then proceed. See Sections 2.1, 4.1, and 4.2. (Continued on next page)

Operations on Monomials	
Multiply	Review the product rule for exponents.
Divide	Review the quotient rule for exponents.
Add or Subtract	Remember, we may only combine like terms.

Perform the operations on the monomials if possible. The first two rows have been completed for you.

	Monomials	Add	Subtract	Multiply	Divide
	$6x, 3x$	$6x + 3x = 9x$	$6x - 3x = 3x$	$6x \cdot 3x = 18x^2$	$\frac{6x}{3x} = 2$
	$-12x^2, 2x$	$-12x^2 + 2x$, can't be simplified	$-12x^2 - 2x$, can't be simplified	$-12x^2 \cdot 2x = -24x^3$	$\frac{-12x^2}{2x} = -6x$
75.	$5a, 15a$				
76.	$4y^7, 4y^3$				
77.	$-3y^5, 9y^4$				
78.	$-14x^2, 2x^2$				

CONCEPT EXTENSIONS

79. Perform each indicated operation. Explain the difference between the two expressions.

a. $(3x + 5) + (3x + 7)$

b. $(3x + 5)(3x + 7)$

80. Perform each indicated operation. Explain the difference between the two expressions.

a. $(8x - 3) - (5x - 2)$

b. $(8x - 3)(5x - 2)$

MIXED PRACTICE

Perform the indicated operations. See Sections 4.2 and 4.3.

81. $(3x - 1) + (10x - 6)$

82. $(2x - 1) + (10x - 7)$

83. $(3x - 1)(10x - 6)$

84. $(2x - 1)(10x - 7)$

85. $(3x - 1) - (10x - 6)$

86. $(2x - 1) - (10x - 7)$

CONCEPT EXTENSIONS

△ **87.** The area of the larger rectangle on the right is $x(x + 3)$. Find another expression for this area by finding the sum of the areas of the two smaller rectangles.

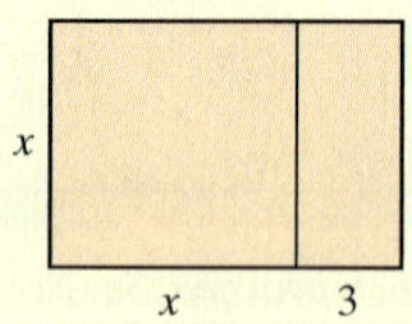

△ **88.** Write an expression for the area of the larger rectangle on the right in two different ways.

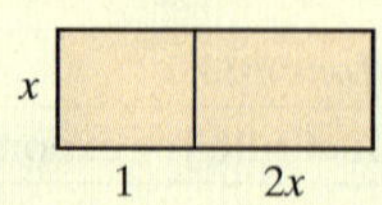

△ **89.** The area of the figure on the right is $(x + 2)(x + 3)$. Find another expression for this area by finding the sum of the areas of the four smaller rectangles.

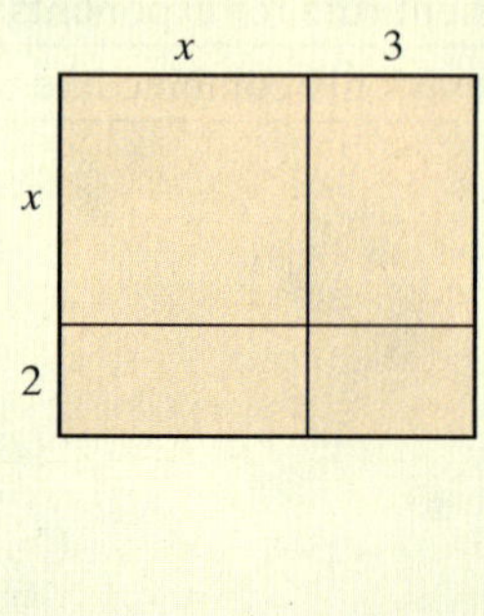

△ **90.** Write an expression for the area of the figure in two different ways.

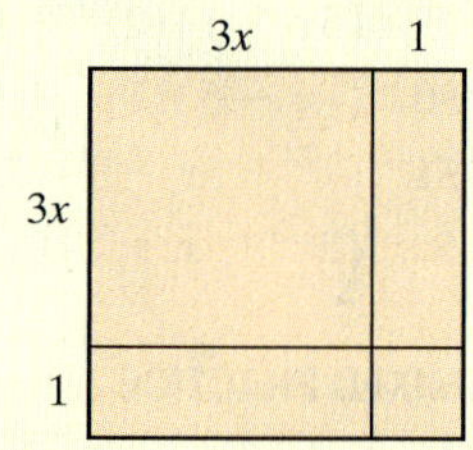

Simplify. See the Concept Checks in this section.

91. $5a + 6a$

92. $5a \cdot 6a$

Square where indicated. Simplify if possible.

93. $(5x)^2 + (2y)^2$

94. $(5x + 2y)^2$

95. Multiply each of the following polynomials.

a. $(a + b)(a - b)$

b. $(2x + 3y)(2x - 3y)$

c. $(4x + 7)(4x - 7)$

d. Can you make a general statement about all products of the form $(x + y)(x - y)$?

96. Evaluate each of the following.

a. $(2 + 3)^2; 2^2 + 3^2$

b. $(8 + 10)^2; 8^2 + 10^2$

c. Does $(a + b)^2 = a^2 + b^2$ no matter what the values of a and b are? Why or why not?

△ **97.** Write a polynomial that describes the area of the shaded region. (Find the area of the larger square minus the area of the smaller square.)

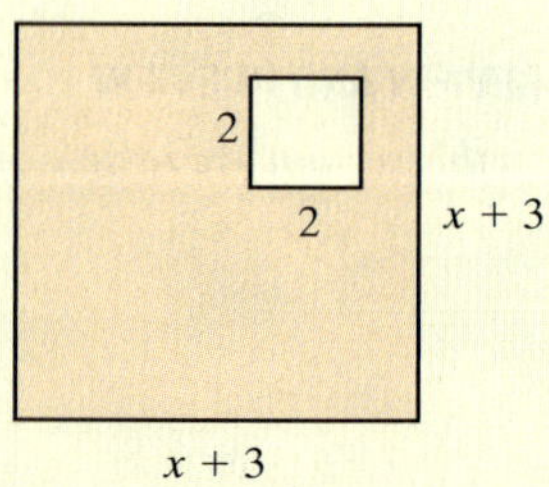

△ **98.** Write a polynomial that describes the area of the shaded region. (See Exercise 97.)

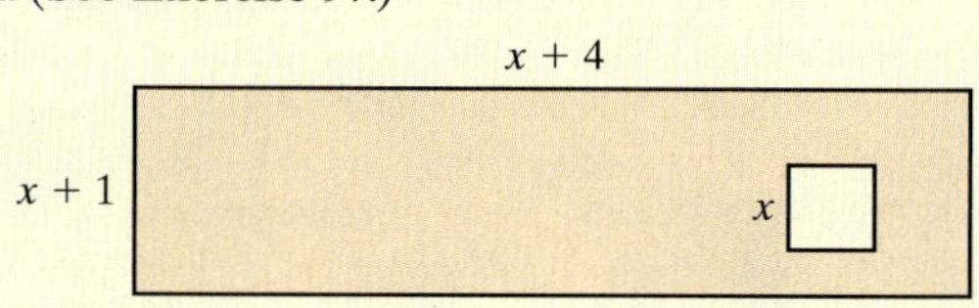

4.4 Special Products

OBJECTIVES

1 Multiply Two Binomials Using the FOIL Method.

2 Square a Binomial.

3 Multiply the Sum and Difference of Two Terms.

4 Use Special Products to Multiply Binomials.

OBJECTIVE

1 Using the FOIL Method

In this section, we multiply binomials using special products. First, a special order for multiplying binomials called the FOIL order or method is introduced. This method is demonstrated by multiplying $(3x + 1)$ by $(2x + 5)$ as shown below.

The FOIL Method

F stands for the product of the **First** terms. $(3x + 1)(2x + 5)$

$$(3x)(2x) = 6x^2 \quad \mathbf{F}$$

O stands for the product of the **Outer** terms. $(3x + 1)(2x + 5)$

$$(3x)(5) = 15x \quad \mathbf{O}$$

I stands for the product of the **Inner** terms. $(3x + 1)(2x + 5)$

$$(1)(2x) = 2x \quad \mathbf{I}$$

L stands for the product of the **Last** terms. $(3x + 1)(2x + 5)$

$$(1)(5) = 5 \quad \mathbf{L}$$

$$(3x + 1)(2x + 5) = \overset{F}{6x^2} + \overset{O}{15x} + \overset{I}{2x} + \overset{L}{5}$$

$$= 6x^2 + 17x + 5 \qquad \text{Combine like terms.}$$

✔ CONCEPT CHECK

Multiply $(3x + 1)(2x + 5)$ using methods from the last section. Show that the product is still $6x^2 + 17x + 5$.

EXAMPLE 1 Multiply $(x - 3)(x + 4)$ by the FOIL method.

Solution

$$(x - 3)(x + 4) = \overset{F}{(x)(x)} + \overset{O}{(x)(4)} + \overset{I}{(-3)(x)} + \overset{L}{(-3)(4)}$$

$$= x^2 + 4x - 3x - 12$$

$$= x^2 + x - 12 \qquad \text{Combine like terms.}$$

Helpful Hint

Remember that the FOIL order for multiplying can be used only for the product of two binomials.

PRACTICE

1 Multiply $(x + 2)(x - 5)$ by the FOIL method.

EXAMPLE 2 Multiply $(5x - 7)(x - 2)$ by the FOIL method.

Solution

$$(5x - 7)(x - 2) = \overset{F}{5x(x)} + \overset{O}{5x(-2)} + \overset{I}{(-7)(x)} + \overset{L}{(-7)(-2)}$$

$$= 5x^2 - 10x - 7x + 14$$

$$= 5x^2 - 17x + 14 \qquad \text{Combine like terms.}$$

PRACTICE

2 Multiply $(4x - 9)(x - 1)$ by the FOIL method.

Answer to Concept Check:
Multiply and simplify.
$3x(2x + 5) + 1(2x + 5)$

EXAMPLE 3 Multiply: $2(y + 6)(2y - 1)$

Solution

$$
\begin{aligned}
2(y+6)(2y-1) &= 2(\overset{F}{2y^2} - \overset{O}{1y} + \overset{I}{12y} - \overset{L}{6}) \\
&= 2(2y^2 + 11y - 6) \quad \text{Simplify inside parentheses.} \\
&= 4y^2 + 22y - 12 \quad \text{Now use the distributive property.}
\end{aligned}
$$

PRACTICE 3 Multiply: $3(x + 5)(3x - 1)$

OBJECTIVE 2

Squaring Binomials

Now, try squaring a binomial using the FOIL method.

EXAMPLE 4 Multiply: $(3y + 1)^2$

Solution

$$
\begin{aligned}
(3y+1)^2 &= (3y+1)(3y+1) \\
&= \overset{F}{(3y)(3y)} + \overset{O}{(3y)(1)} + \overset{I}{1(3y)} + \overset{L}{1(1)} \\
&= 9y^2 + 3y + 3y + 1 \\
&= 9y^2 + 6y + 1
\end{aligned}
$$

PRACTICE 4 Multiply: $(4x - 1)^2$

Notice the pattern that appears in Example 4.

$$(3y + 1)^2 = 9y^2 + 6y + 1$$

- $9y^2$ is the first term of the binomial squared. $(3y)^2 = 9y^2$.
- $6y$ is 2 times the product of both terms of the binomial. $(2)(3y)(1) = 6y$.
- 1 is the second term of the binomial squared. $(1)^2 = 1$.

This pattern leads to the following, which can be used when squaring a binomial. We call these **special products.**

> **Squaring a Binomial**
>
> A binomial squared is equal to the square of the first term plus or minus twice the product of both terms plus the square of the second term.
>
> $$(a + b)^2 = a^2 + 2ab + b^2$$
> $$(a - b)^2 = a^2 - 2ab + b^2$$

This product can be visualized geometrically.

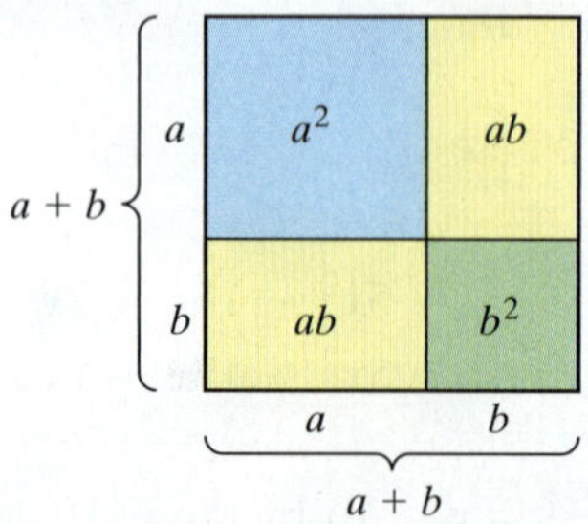

The area of the large square is side · side.

$$\text{Area} = (a + b)(a + b) = (a + b)^2$$

The area of the large square is also the sum of the areas of the smaller rectangles.

$$\text{Area} = a^2 + ab + ab + b^2 = a^2 + 2ab + b^2$$

Thus, $(a + b)^2 = a^2 + 2ab + b^2$.

EXAMPLE 5 Use a special product to square each binomial.

a. $(t + 2)^2$ **b.** $(p - q)^2$ **c.** $(2x + 5)^2$ **d.** $(x^2 - 7y)^2$

Solution

		first term squared	plus or minus	twice the product of the terms	plus	second term squared
a. $(t + 2)^2$	$=$	t^2	$+$	$2(t)(2)$	$+$	$2^2 = t^2 + 4t + 4$
b. $(p - q)^2$	$=$	p^2	$-$	$2(p)(q)$	$+$	$q^2 = p^2 - 2pq + q^2$
c. $(2x + 5)^2$	$=$	$(2x)^2$	$+$	$2(2x)(5)$	$+$	$5^2 = 4x^2 + 20x + 25$
d. $(x^2 - 7y)^2$	$=$	$(x^2)^2$	$-$	$2(x^2)(7y)$	$+$	$(7y^2) = x^4 - 14x^2y + 49y^2$

PRACTICE 5 Use a special product to square each binomial.

a. $(b + 3)^2$ **b.** $(x - y)^2$

c. $(3y + 2)^2$ **d.** $(a^2 - 5b)^2$

Helpful Hint

Notice that

$$(a + b)^2 \neq a^2 + b^2 \quad \text{The middle term } 2ab \text{ is missing.}$$
$$(a + b)^2 = (a + b)(a + b) = a^2 + 2ab + b^2$$

Likewise,

$$(a - b)^2 \neq a^2 - b^2$$
$$(a - b)^2 = (a - b)(a - b) = a^2 - 2ab + b^2$$

OBJECTIVE 3

Multiplying the Sum and Difference of Two Terms

Another special product is the product of the sum and difference of the same two terms, such as $(x + y)(x - y)$. Finding this product by the FOIL method, we see a pattern emerge.

$$\begin{aligned} & \qquad\quad\ \ \text{F} \quad\ \text{O} \quad\ \ \text{I} \quad\ \text{L} \\ (x + y)(x - y) &= x^2 - xy + xy - y^2 \\ &= x^2 - y^2 \end{aligned}$$

Notice that the middle two terms subtract out. This is because the **O**uter product is the opposite of the **I**nner product. Only the **difference of squares** remains.

Multiplying the Sum and Difference of Two Terms

The product of the sum and difference of two terms is the square of the first term minus the square of the second term.

$$(a + b)(a - b) = a^2 - b^2$$

EXAMPLE 6 Use a special product to multiply.

a. $4(x + 4)(x - 4)$ **b.** $(6t + 7)(6t - 7)$ **c.** $\left(x - \frac{1}{4}\right)\left(x + \frac{1}{4}\right)$

d. $(2p - q)(2p + q)$ **e.** $(3x^2 - 5y)(3x^2 + 5y)$

Solution

first term squared ↓ minus ↓ second term squared ↙

a. $4(x + 4)(x - 4) = 4(x^2 - 4^2) = 4(x^2 - 16) = 4x^2 - 64$

b. $(6t + 7)(6t - 7) = (6t)^2 - 7^2 = 36t^2 - 49$

c. $\left(x - \frac{1}{4}\right)\left(x + \frac{1}{4}\right) = x^2 - \left(\frac{1}{4}\right)^2 = x^2 - \frac{1}{16}$

d. $(2p - q)(2p + q) = (2p)^2 - q^2 = 4p^2 - q^2$

e. $(3x^2 - 5y)(3x^2 + 5y) = (3x^2)^2 - (5y)^2 = 9x^4 - 25y^2$ □

PRACTICE 6 Use a special product to multiply.

a. $3(x + 5)(x - 5)$ **b.** $(4b - 3)(4b + 3)$

c. $\left(x + \frac{2}{3}\right)\left(x - \frac{2}{3}\right)$ **d.** $(5s + t)(5s - t)$

e. $(2y - 3z^2)(2y + 3z^2)$ ■

✔CONCEPT CHECK

Match expression number 1 and number 2 to the equivalent expression or expressions in the list below.

1. $(a + b)^2$ **2.** $(a + b)(a - b)$

A. $(a + b)(a + b)$ **B.** $a^2 - b^2$ **C.** $a^2 + b^2$ **D.** $a^2 - 2ab + b^2$ **E.** $a^2 + 2ab + b^2$

OBJECTIVE 4

Using Special Products

Let's now practice multiplying polynomials in general. If possible, use a special product.

EXAMPLE 7 Use a special product to multiply, if possible.

a. $(x - 5)(3x + 4)$ **b.** $(7x + 4)^2$ **c.** $(y - 0.6)(y + 0.6)$

d. $\left(y^4 + \frac{2}{5}\right)\left(3y^2 - \frac{1}{5}\right)$ **e.** $(a - 3)(a^2 + 2a - 1)$

Solution

a. $(x - 5)(3x + 4) = 3x^2 + 4x - 15x - 20$ FOIL.

$= 3x^2 - 11x - 20$

b. $(7x + 4)^2 = (7x)^2 + 2(7x)(4) + 4^2$ Squaring a binomial.

$= 49x^2 + 56x + 16$

c. $(y - 0.6)(y + 0.6) = y^2 - (0.6)^2 = y^2 - 0.36$ Multiplying the sum and difference of 2 terms.

d. $\left(y^4 + \frac{2}{5}\right)\left(3y^2 - \frac{1}{5}\right) = 3y^6 - \frac{1}{5}y^4 + \frac{6}{5}y^2 - \frac{2}{25}$ FOIL.

e. I've inserted this product as a reminder that since it is not a binomial times a binomial, the FOIL order may not be used.

Answers to Concept Check:
1. A and E **2.** B

$$\begin{aligned}(a - 3)(a^2 + 2a - 1) &= a(a^2 + 2a - 1) - 3(a^2 + 2a - 1) \\ &= a^3 + 2a^2 - a - 3a^2 - 6a + 3 \\ &= a^3 - a^2 - 7a + 3\end{aligned}$$

Multiplying each term of the binomial by each term of the trinomial. □

PRACTICE 7 Use a special product to multiply, if possible.

a. $(4x + 3)(x - 6)$ **b.** $(7b - 2)^2$

c. $(x + 0.4)(x - 0.4)$ **d.** $\left(x^2 - \frac{3}{7}\right)\left(3x^4 + \frac{2}{7}\right)$

e. $(x + 1)(x^2 + 5x - 2)$

Helpful Hint

- When multiplying two binomials, you may always use the FOIL order or method.
- When multiplying any two polynomials, you may always use the distributive property to find the product.

Vocabulary, Readiness & Video Check

Answer each exercise true or false.

1. $(x + 4)^2 = x^2 + 16$

2. For $(x + 6)(2x - 1)$ the product of the first terms is $2x^2$.

3. $(x + 4)(x - 4) = x^2 + 16$

4. The product $(x - 1)(x^3 + 3x - 1)$ is a polynomial of degree 5.

Martin-Gay Interactive Videos

See Video 4.4

Watch the section lecture video and answer the following questions.

OBJECTIVE 1 **5.** From Examples 1–3, for what type of multiplication problem is the FOIL order of multiplication used?

OBJECTIVE 2 **6.** Name at least one other method you can use to multiply Example 4.

OBJECTIVE 3 **7.** From Example 5, why does multiplying the sum and difference of the same two terms always give you a binomial answer?

OBJECTIVE 4 **8.** Why was the FOIL method not used for Example 10?

4.4 Exercise Set MyMathLab®

Multiply using the FOIL method. See Examples 1 through 3.

1. $(x + 3)(x + 4)$

2. $(x + 5)(x - 1)$

3. $(x - 5)(x + 10)$

4. $(y - 12)(y + 4)$

5. $(5x - 6)(x + 2)$

6. $(3y - 5)(2y - 7)$

7. $5(y - 6)(4y - 1)$

8. $2(x - 11)(2x - 9)$

9. $(2x + 5)(3x - 1)$

10. $(6x + 2)(x - 2)$

11. $\left(x - \frac{1}{3}\right)\left(x + \frac{2}{3}\right)$

12. $\left(x - \frac{2}{5}\right)\left(x + \frac{1}{5}\right)$

Multiply. See Examples 4 and 5.

13. $(x + 2)^2$

14. $(x + 7)^2$

15. $(2x - 1)^2$

16. $(7x - 3)^2$

17. $(3a - 5)^2$

18. $(5a + 2)^2$

19. $(5x + 9)^2$

20. $(6s - 2)^2$

Multiply. See Example 6.

21. $(a - 7)(a + 7)$

22. $(b + 3)(b - 3)$

23. $(3x - 1)(3x + 1)$

24. $(4x - 5)(4x + 5)$

25. $\left(3x - \frac{1}{2}\right)\left(3x + \frac{1}{2}\right)$

26. $\left(10x + \frac{2}{7}\right)\left(10x - \frac{2}{7}\right)$
27. $(9x + y)(9x - y)$
28. $(2x - y)(2x + y)$
29. $(2x + 0.1)(2x - 0.1)$
30. $(5x - 1.3)(5x + 1.3)$

MIXED PRACTICE

Multiply. See Example 7.

31. $(a + 5)(a + 4)$
32. $(a - 5)(a - 7)$
33. $(a + 7)^2$
34. $(b - 2)^2$
35. $(4a + 1)(3a - 1)$
36. $(6a + 7)(6a + 5)$
37. $(x + 2)(x - 2)$
38. $(x - 10)(x + 10)$
39. $(3a + 1)^2$
40. $(4a - 2)^2$
41. $(x^2 + y)(4x - y^4)$
42. $(x^3 - 2)(5x + y)$
43. $(x + 3)(x^2 - 6x + 1)$
44. $(x - 2)(x^2 - 4x + 2)$
45. $(2a - 3)^2$
46. $(5b - 4x)^2$
47. $(5x - 6z)(5x + 6z)$
48. $(11x - 7y)(11x + 7y)$
49. $(x^5 - 3)(x^5 - 5)$
50. $(a^4 + 5)(a^4 + 6)$
51. $(x + 0.8)(x - 0.8)$
52. $(y - 0.9)(y + 0.9)$
53. $(a^3 + 11)(a^4 - 3)$
54. $(x^5 + 5)(x^2 - 8)$
55. $3(x - 2)^2$
56. $2(3b + 7)^2$
57. $(3b + 7)(2b - 5)$
58. $(3y - 13)(y - 3)$
59. $(7p - 8)(7p + 8)$
60. $(3s - 4)(3s + 4)$
61. $\left(\frac{1}{3}a^2 - 7\right)\left(\frac{1}{3}a^2 + 7\right)$
62. $\left(\frac{2}{3}a - b^2\right)\left(\frac{2}{3}a + b^2\right)$
63. $5x^2(3x^2 - x + 2)$
64. $4x^3(2x^2 + 5x - 1)$
65. $(2r - 3s)(2r + 3s)$
66. $(6r - 2x)(6r + 2x)$
67. $(3x - 7y)^2$
68. $(4s - 2y)^2$
69. $(4x + 5)(4x - 5)$
70. $(3x + 5)(3x - 5)$
71. $(8x + 4)^2$
72. $(3x + 2)^2$
73. $\left(a - \frac{1}{2}y\right)\left(a + \frac{1}{2}y\right)$
74. $\left(\frac{a}{2} + 4y\right)\left(\frac{a}{2} - 4y\right)$
75. $\left(\frac{1}{5}x - y\right)\left(\frac{1}{5}x + y\right)$
76. $\left(\frac{y}{6} - 8\right)\left(\frac{y}{6} + 8\right)$
77. $(a + 1)(3a^2 - a + 1)$
78. $(b + 3)(2b^2 + b - 3)$

Express each as a product of polynomials in x. Then multiply and simplify.

△ 79. Find the area of the square rug shown if its side is $(2x + 1)$ feet.

△ 80. Find the area of the rectangular canvas if its length is $(3x - 2)$ inches and its width is $(x - 4)$ inches.

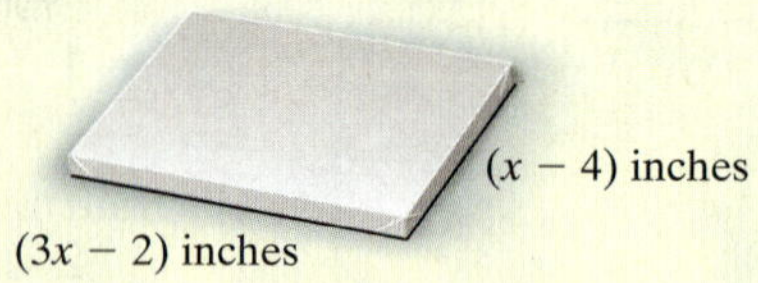

REVIEW AND PREVIEW

Identify the x- and y-intercepts. See Section 3.2.

81.

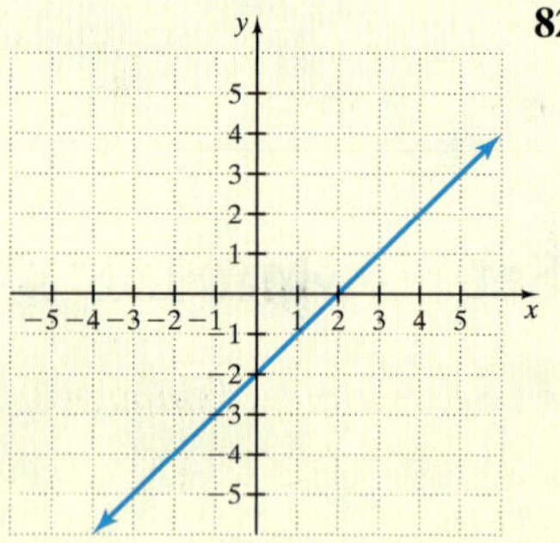

82.

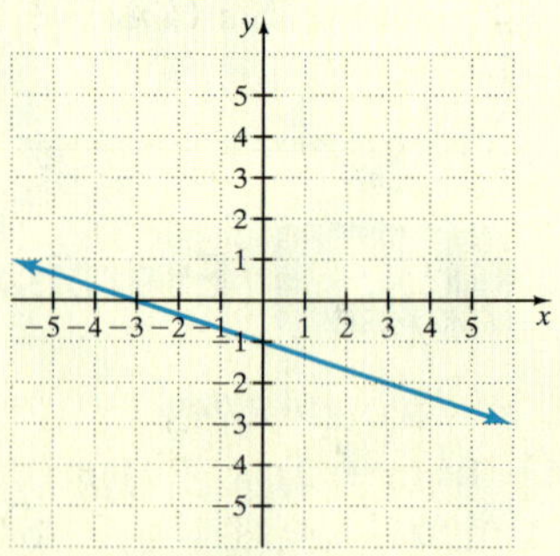

83.

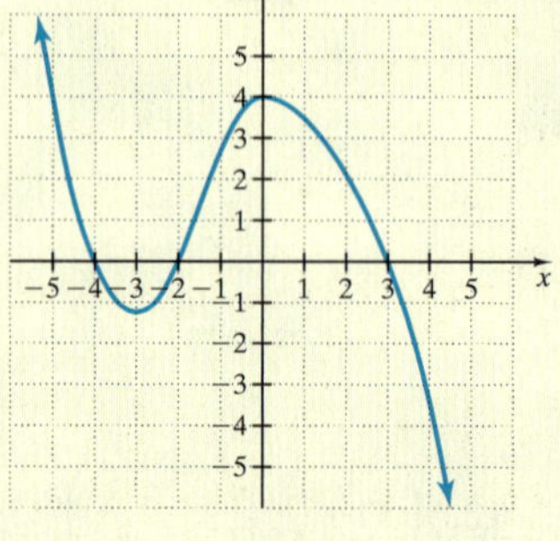

84.

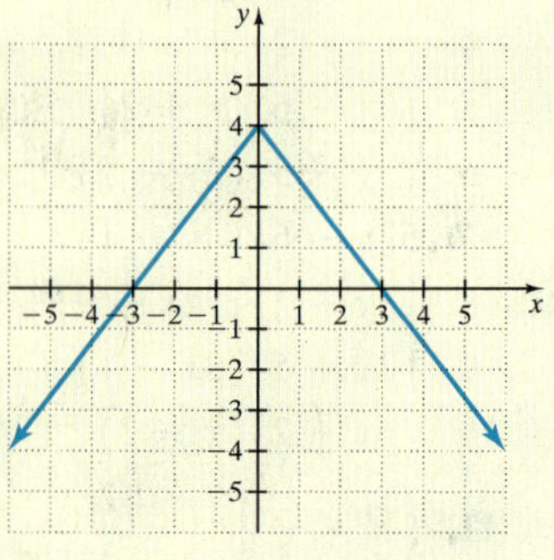

85.

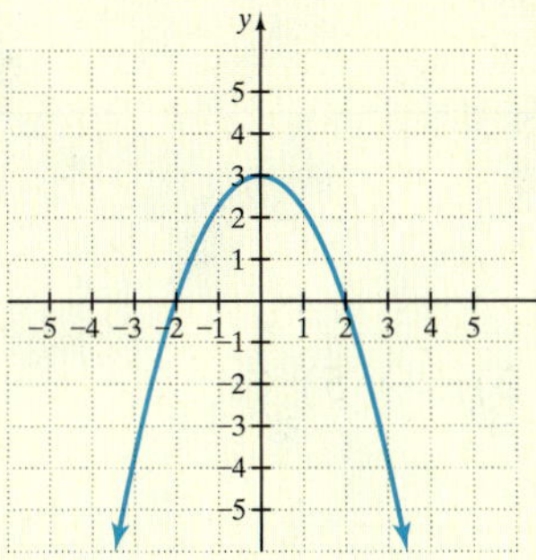

86.

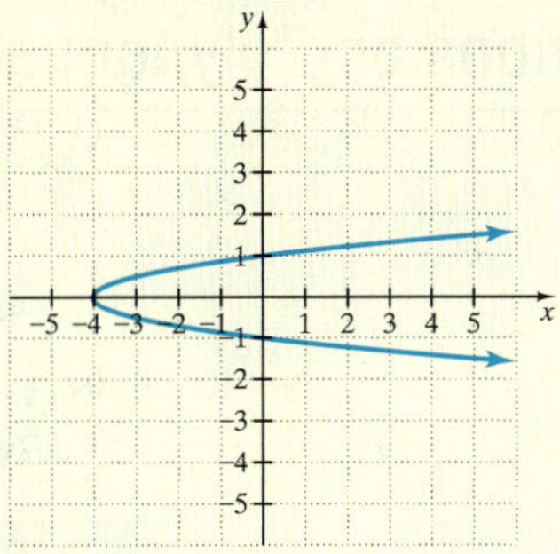

Find the slope of each line. See Section 3.2.

87.

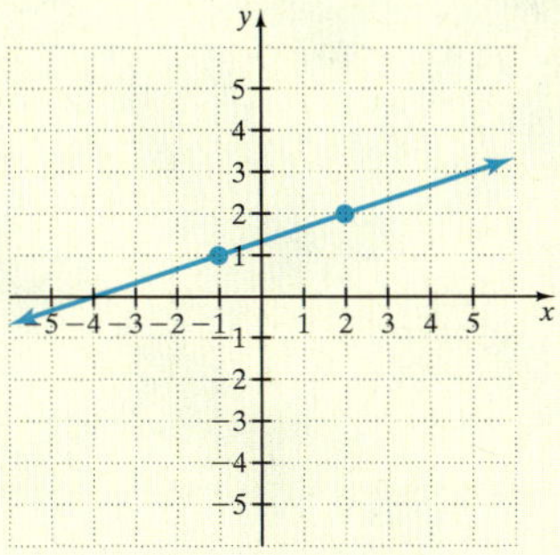

88.

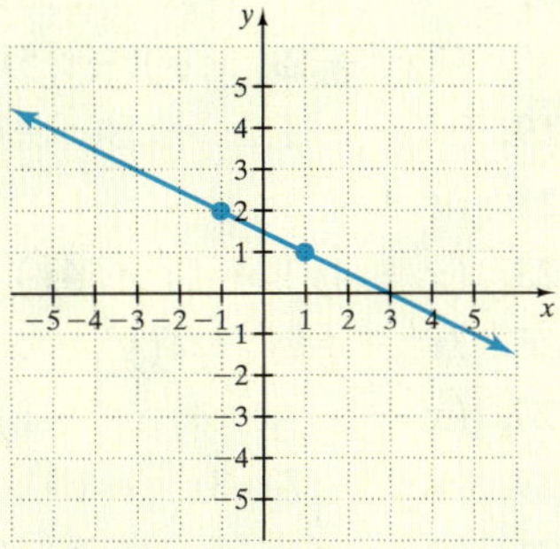

89.

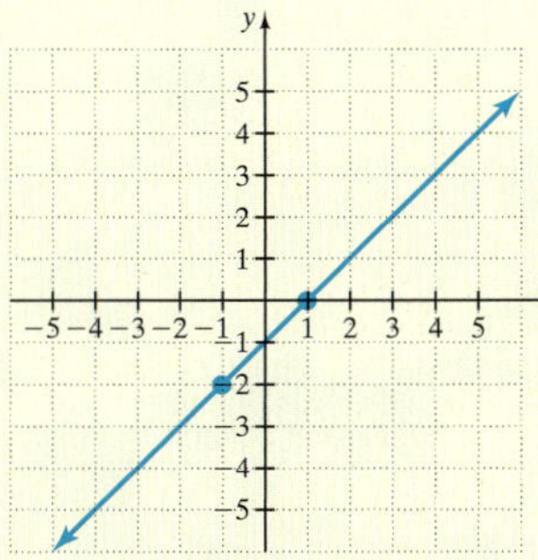

90.

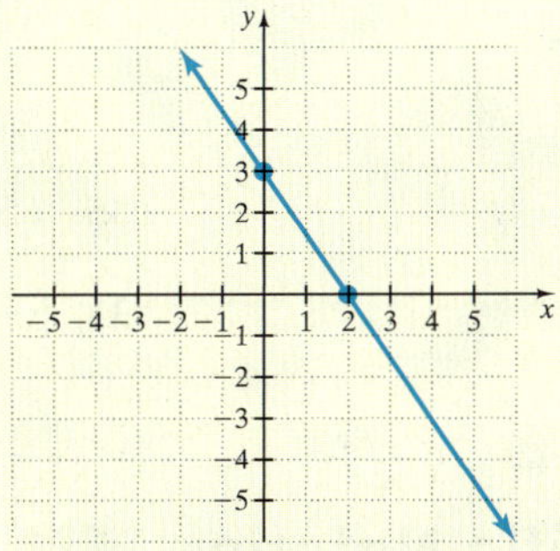

CONCEPT EXTENSIONS

Match each expression on the left to the equivalent expression on the right. See the second Concept Check in this section.

91. $(a - b)^2$

92. $(a - b)(a + b)$

93. $(a + b)^2$

94. $(a + b)^2(a - b)^2$

A. $a^2 - b^2$

B. $a^2 + b^2$

C. $a^2 - 2ab + b^2$

D. $a^2 + 2ab + b^2$

E. none of these

Fill in the squares so that a true statement forms.

95. $(x^{\square} + 7)(x^{\square} + 3) = x^4 + 10x^2 + 21$

96. $(5x^{\square} - 2)^2 = 25x^6 - 20x^3 + 4$

Find the area of the shaded figure. To do so, subtract the area of the smaller square(s) from the area of the larger geometric figure.

△ **97.**

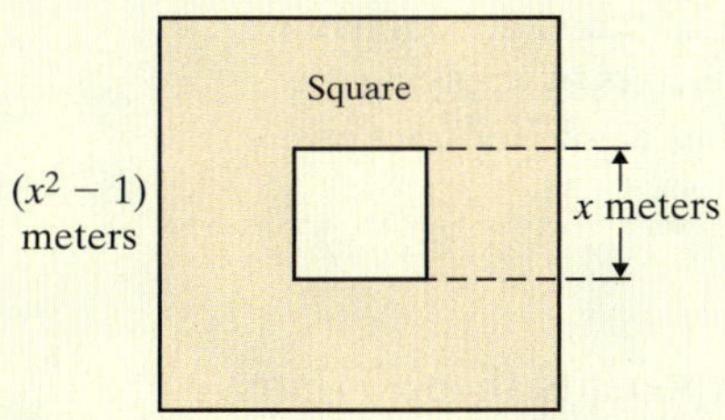

△ **98.**

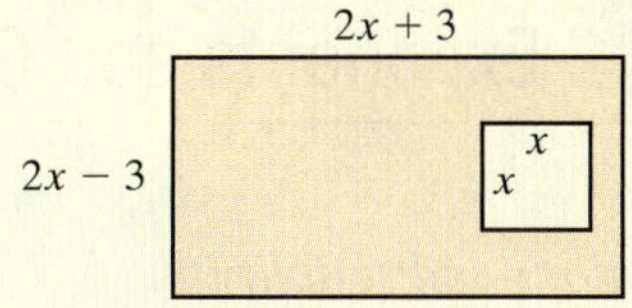

99.

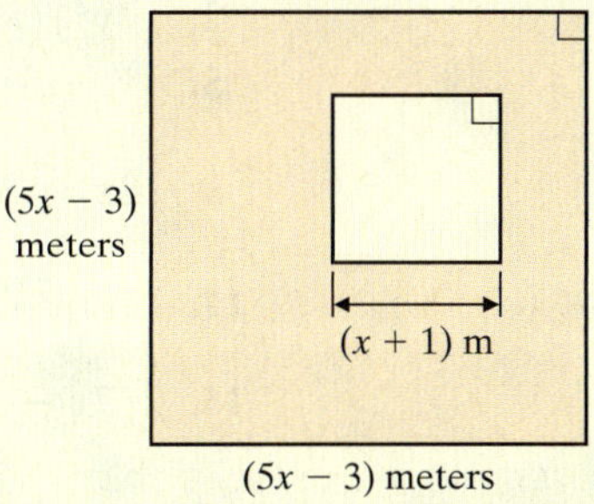

△ **100.**

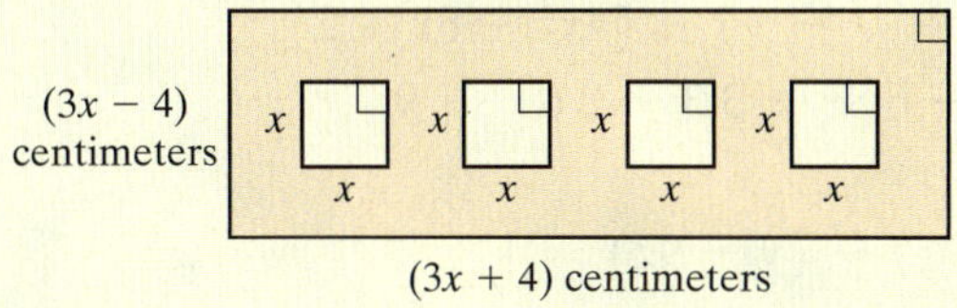

For Exercises 101 and 102, find the area of the shaded figure.

△ **101.**

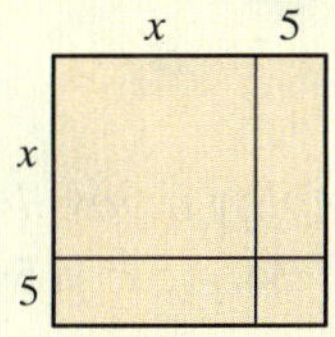

△ **102.**

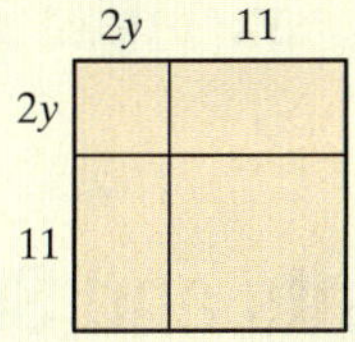

103. In your own words, describe the different methods that can be used to find the product $(2x - 5)(3x + 1)$.

104. In your own words, describe the different methods that can be used to find the product $(5x + 1)^2$.

105. Suppose that a classmate asked you why $(2x + 1)^2$ is **not** $(4x^2 + 1)$. Write down your response to this classmate.

106. Suppose that a classmate asked you why $(2x + 1)^2$ **is** $(4x^2 + 4x + 1)$. Write down your response to this classmate.

107. Using your own words, explain how to square a binomial such as $(a + b)^2$.

108. Explain how to find the product of two binomials using the FOIL method.

Find each product. For example,

$$[(a + b) - 2][(a + b) + 2] = (a + b)^2 - 2^2$$
$$= a^2 + 2ab + b^2 - 4$$

109. $[(x + y) - 3][(x + y) + 3]$

110. $[(a + c) - 5][(a + c) + 5]$

111. $[(a - 3) + b][(a - 3) - b]$

112. $[(x - 2) + y][(x - 2) - y]$

Integrated Review Exponents and Operations on Polynomials

Sections 4.1-4.4

Perform the indicated operations and simplify.

1. $(5x^2)(7x^3)$
2. $(4y^2)(8y^7)$
3. -4^2
4. $(-4)^2$
5. $(x-5)(2x+1)$
6. $(3x-2)(x+5)$
7. $(x-5)+(2x+1)$
8. $(3x-2)+(x+5)$
9. $\dfrac{7x^9y^{12}}{x^3y^{10}}$
10. $\dfrac{20a^2b^8}{14a^2b^2}$
11. $(12m^7n^6)^2$
12. $(4y^9z^{10})^3$
13. $3(4y-3)(4y+3)$
14. $2(7x-1)(7x+1)$
15. $(x^7y^5)^9$
16. $(3^1x^9)^3$
17. $(7x^2-2x+3)-(5x^2+9)$
18. $(10x^2+7x-9)-(4x^2-6x+2)$
19. $0.7y^2-1.2+1.8y^2-6y+1$
20. $7.8x^2-6.8x+3.3+0.6x^2-9$
21. $(x+4y)^2$
22. $(y-9z)^2$
23. $(x+4y)+(x+4y)$
24. $(y-9z)+(y-9z)$
25. $7x^2-6xy+4(y^2-xy)$
26. $5a^2-3ab+6(b^2-a^2)$
27. $(x-3)(x^2+5x-1)$
28. $(x+1)(x^2-3x-2)$
29. $(2x^3-7)(3x^2+10)$
30. $(5x^3-1)(4x^4+5)$
31. $(2x-7)(x^2-6x+1)$
32. $(5x-1)(x^2+2x-3)$

Perform the indicated operations and simplify if possible.

33. $5x^3+5y^3$
34. $(5x^3)(5y^3)$
35. $(5x^3)^3$
36. $\dfrac{5x^3}{5y^3}$
37. $x+x$
38. $x\cdot x$

4.5 Negative Exponents and Scientific Notation

OBJECTIVES

1. Simplify Expressions Containing Negative Exponents.
2. Use All the Rules and Definitions for Exponents to Simplify Exponential Expressions.
3. Write Numbers in Scientific Notation.
4. Convert Numbers from Scientific Notation to Standard Form.
5. Perform Operations on Numbers Written in Scientific Notation.

OBJECTIVE

1 Simplifying Expressions Containing Negative Exponents

Our work with exponential expressions so far has been limited to exponents that are positive integers or 0. Here we expand to give meaning to an expression like x^{-3}.

Suppose that we wish to simplify the expression $\frac{x^2}{x^5}$. If we use the quotient rule for exponents, we subtract exponents:

$$\frac{x^2}{x^5}=x^{2-5}=x^{-3},\quad x\neq 0$$

But what does x^{-3} mean? Let's simplify $\frac{x^2}{x^5}$ using the definition of x^n.

$$\frac{x^2}{x^5}=\frac{x\cdot x}{x\cdot x\cdot x\cdot x\cdot x}$$

$$=\frac{\overset{1}{\cancel{x}}\cdot\overset{1}{\cancel{x}}}{\underset{1}{\cancel{x}}\cdot\underset{1}{\cancel{x}}\cdot x\cdot x\cdot x}$$ Divide numerator and denominator by common factors by applying the fundamental principle for fractions.

$$=\frac{1}{x^3}$$

If the quotient rule is to hold true for negative exponents, then x^{-3} must equal $\frac{1}{x^3}$.

From this example, we state the definition for negative exponents.

Negative Exponents

If a is a real number other than 0 and n is an integer, then

$$a^{-n} = \frac{1}{a^n}$$

For example, $x^{-3} = \frac{1}{x^3}$.

In other words, another way to write a^{-n} is to take its reciprocal and change the sign of its exponent.

EXAMPLE 1 Simplify by writing each expression with positive exponents only.

a. 3^{-2} **b.** $2x^{-3}$ **c.** $2^{-1} + 4^{-1}$ **d.** $(-2)^{-4}$ **e.** y^{-4}

Solution

a. $3^{-2} = \frac{1}{3^2} = \frac{1}{9}$ Use the definition of negative exponents.

b. $2x^{-3} = 2 \cdot \frac{1}{x^3} = \frac{2}{x^3}$ Use the definition of negative exponents.

Helpful Hint

Don't forget that since there are no parentheses, only x is the base for the exponent -3.

c. $2^{-1} + 4^{-1} = \frac{1}{2} + \frac{1}{4} = \frac{2}{4} + \frac{1}{4} = \frac{3}{4}$

d. $(-2)^{-4} = \frac{1}{(-2)^4} = \frac{1}{(-2)(-2)(-2)(-2)} = \frac{1}{16}$

e. $y^{-4} = \frac{1}{y^4}$ □

PRACTICE 1 Simplify by writing each expression with positive exponents only.

a. 5^{-3} **b.** $3y^{-4}$ **c.** $3^{-1} + 2^{-1}$ **d.** $(-5)^{-2}$ **e.** x^{-5} ■

Helpful Hint

A negative exponent *does not affect* the sign of its base.

Remember: Another way to write a^{-n} is to take its reciprocal and change the sign of its exponent: $a^{-n} = \frac{1}{a^n}$. For example,

$$x^{-2} = \frac{1}{x^2}, \qquad 2^{-3} = \frac{1}{2^3} \text{ or } \frac{1}{8}$$

$$\frac{1}{y^{-4}} = \frac{1}{\frac{1}{y^4}} = y^4, \qquad \frac{1}{5^{-2}} = 5^2 \text{ or } 25$$

From the preceding Helpful Hint, we know that $x^{-2} = \frac{1}{x^2}$ and $\frac{1}{y^{-4}} = y^4$. We can use this to include another statement in our definition of negative exponents.

Negative Exponents

If a is a real number other than 0 and n is an integer, then

$$a^{-n} = \frac{1}{a^n} \quad \text{and} \quad \frac{1}{a^{-n}} = a^n$$

EXAMPLE 2 Simplify each expression. Write results using positive exponents only.

a. $\frac{1}{x^{-3}}$ **b.** $\frac{1}{3^{-4}}$ **c.** $\frac{p^{-4}}{q^{-9}}$ **d.** $\frac{5^{-3}}{2^{-5}}$

Solution

a. $\frac{1}{x^{-3}} = \frac{x^3}{1} = x^3$ **b.** $\frac{1}{3^{-4}} = \frac{3^4}{1} = 81$

c. $\frac{p^{-4}}{q^{-9}} = \frac{q^9}{p^4}$ **d.** $\frac{5^{-3}}{2^{-5}} = \frac{2^5}{5^3} = \frac{32}{125}$ □

PRACTICE 2 Simplify each expression. Write results using positive exponents only.

a. $\frac{1}{s^{-5}}$ **b.** $\frac{1}{2^{-3}}$ **c.** $\frac{x^{-7}}{y^{-5}}$ **d.** $\frac{4^{-3}}{3^{-2}}$ ■

EXAMPLE 3 Simplify each expression. Write answers with positive exponents.

a. $\frac{y}{y^{-2}}$ **b.** $\frac{3}{x^{-4}}$ **c.** $\frac{x^{-5}}{x^7}$ **d.** $\left(\frac{2}{3}\right)^{-3}$

Solution

a. $\frac{y}{y^{-2}} = \frac{y^1}{y^{-2}} = y^{1-(-2)} = y^3$ Remember that $\frac{a^m}{a^n} = a^{m-n}$.

b. $\frac{3}{x^{-4}} = 3 \cdot \frac{1}{x^{-4}} = 3 \cdot x^4$ or $3x^4$

c. $\frac{x^{-5}}{x^7} = x^{-5-7} = x^{-12} = \frac{1}{x^{12}}$

d. $\left(\frac{2}{3}\right)^{-3} = \frac{2^{-3}}{3^{-3}} = \frac{3^3}{2^3} = \frac{27}{8}$ □

PRACTICE 3 Simplify each expression. Write answers with positive exponents.

a. $\frac{x^{-3}}{x^2}$ **b.** $\frac{5}{y^{-7}}$ **c.** $\frac{z}{z^{-4}}$ **d.** $\left(\frac{5}{9}\right)^{-2}$ ■

OBJECTIVE 2

Simplifying Exponential Expressions

All the previously stated rules for exponents apply for negative exponents also. Here is a summary of the rules and definitions for exponents.

Summary of Exponent Rules

If m and n are integers and a, b, and c are real numbers, then:

Product rule for exponents: $a^m \cdot a^n = a^{m+n}$

Power rule for exponents: $(a^m)^n = a^{m \cdot n}$

Power of a product: $(ab)^n = a^n b^n$

Power of a quotient: $\left(\frac{a}{c}\right)^n = \frac{a^n}{c^n}, \quad c \neq 0$

Quotient rule for exponents: $\frac{a^m}{a^n} = a^{m-n}, \quad a \neq 0$

Zero exponent: $a^0 = 1, \quad a \neq 0$

Negative exponent: $a^{-n} = \frac{1}{a^n}, \quad a \neq 0$

EXAMPLE 4 Simplify the following expressions. Write each result using positive exponents only.

a. $(y^{-3}z^6)^{-6}$ **b.** $\dfrac{(2x^3)^4x}{x^7}$ **c.** $\left(\dfrac{3a^2}{b}\right)^{-3}$

Solution

a. $(y^{-3}z^6)^{-6} = y^{18} \cdot z^{-36} = \dfrac{y^{18}}{z^{36}}$

b. $\dfrac{(2x^3)^4x}{x^7} = \dfrac{2^4 \cdot x^{12} \cdot x}{x^7} = \dfrac{16 \cdot x^{12+1}}{x^7} = \dfrac{16x^{13}}{x^7} = 16x^{13-7} = 16x^6$ Use the power rule.

c. $\left(\dfrac{3a^2}{b}\right)^{-3} = \dfrac{3^{-3}(a^2)^{-3}}{b^{-3}}$ Raise each factor in the numerator and the denominator to the -3 power.

$= \dfrac{3^{-3}a^{-6}}{b^{-3}}$ Use the power rule.

$= \dfrac{b^3}{3^3a^6}$ Use the negative exponent rule.

$= \dfrac{b^3}{27a^6}$ Write 3^3 as 27. □

PRACTICE 4 Simplify the following expressions. Write each result using positive exponents only.

a. $(a^4b^{-3})^{-5}$ **b.** $\dfrac{x^2(x^5)^3}{x^7}$ **c.** $\left(\dfrac{5p^8}{q}\right)^{-2}$ ■

OBJECTIVE 3

Writing Numbers in Scientific Notation

Both very large and very small numbers frequently occur in many fields of science. For example, the distance between the Sun and the dwarf planet Pluto is approximately 5,906,000,000 kilometers, and the mass of a proton is approximately 0.000 000 000 000 000 000 000 001 65 gram. It can be tedious to write these numbers in this standard decimal notation, so **scientific notation** is used as a convenient shorthand for expressing very large and very small numbers.

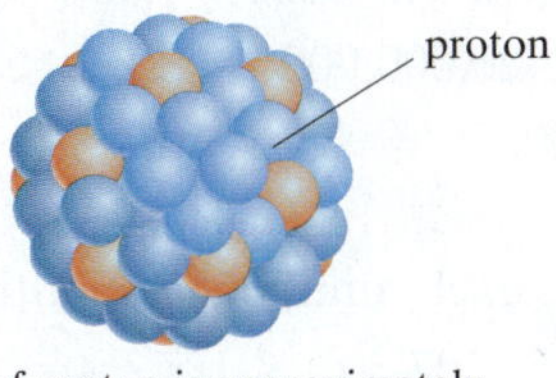

Mass of proton is approximately 0.000 000 000 000 000 000 000 001 65 gram

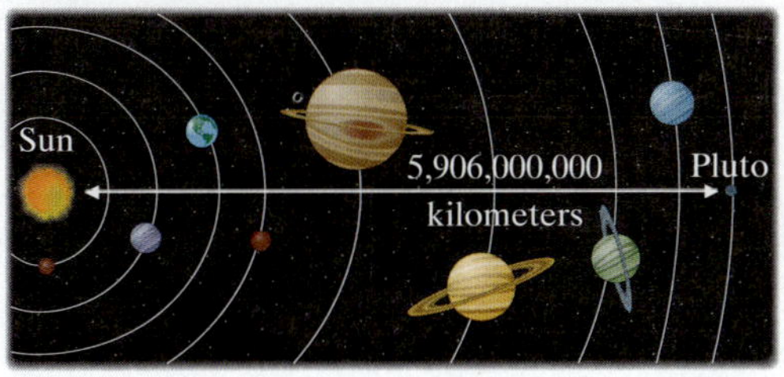

Helpful Hint

To help read small decimal numbers, we group the zeros to the right of the decimal point in groups of 3 zeros.

Scientific Notation

A positive number is written in scientific notation if it is written as the product of a number a, where $1 \le a < 10$, and an integer power r of 10:

$$a \times 10^r$$

Note: For negative numbers, the number a is such that $1 \le |a| < 10$.

The numbers below are written in scientific notation. The $\times$ sign for multiplication is used as part of the notation.

2.03×10^2	7.362×10^7	5.906×10^9	(Distance between the Sun and Pluto)
1×10^{-3}	8.1×10^{-5}	1.65×10^{-24}	(Mass of a proton)

The following steps are useful when writing numbers in scientific notation.

To Write a Number in Scientific Notation

Step 1. Move the decimal point in the original number to the left or right so that the new number has a value between 1 and 10 (including 1).

Step 2. Count the number of decimal places the decimal point is moved in Step 1. If the original number is 10 or greater, the count is positive. If the original number is less than 1, the count is negative.

Step 3. Multiply the new number in Step 1 by 10 raised to an exponent equal to the count found in Step 2.

EXAMPLE 5 Write each number in scientific notation.

a. 367,000,000 **b.** 0.000003 **c.** 20,520,000,000 **d.** 0.00085

Solution

a. Step 1. Move the decimal point until the number is between 1 and 10.

367,000,000.

8 places

Step 2. The decimal point is moved 8 places, and the original number is 10 or greater, so the count is positive 8.

Step 3. $367{,}000{,}000 = 3.67 \times 10^8$.

b. Step 1. Move the decimal point until the number is between 1 and 10.

0.000003

6 places

Step 2. The decimal point is moved 6 places, and the original number is less than 1, so the count is -6.

Step 3. $0.000003 = 3.0 \times 10^{-6}$

c. $20{,}520{,}000{,}000 = 2.052 \times 10^{10}$

d. $0.00085 = 8.5 \times 10^{-4}$ □

PRACTICE 5 Write each number in scientific notation.

a. 0.000007 **b.** 20,700,000

c. 0.0043 **d.** 812,000,000 ■

OBJECTIVE 4

Converting Numbers to Standard Form

A number written in scientific notation can be rewritten in standard form. For example, to write 8.63×10^3 in standard form, recall that $10^3 = 1000$.

$$8.63 \times 10^3 = 8.63(1000) = 8630$$

Notice that the exponent on the 10 is positive 3, and we moved the decimal point 3 places to the right.

To write 7.29×10^{-3} in standard form, recall that $10^{-3} = \frac{1}{10^3} = \frac{1}{1000}$.

$$7.29 \times 10^{-3} = 7.29\left(\frac{1}{1000}\right) = \frac{7.29}{1000} = 0.00729$$

The exponent on the 10 is negative 3, and we moved the decimal to the left 3 places.

In general, **to write a scientific notation number in standard form,** move the decimal point the same number of places as the exponent on 10. If the exponent is positive, move the decimal point to the right; if the exponent is negative, move the decimal point to the left.

EXAMPLE 6 Write each number in standard notation, without exponents.

a. 1.02×10^5 **b.** 7.358×10^{-3} **c.** 8.4×10^7 **d.** 3.007×10^{-5}

Solution

a. Move the decimal point 5 places to the right.

$$1.02 \times 10^5 = 102{,}000.$$

b. Move the decimal point 3 places to the left.

$$7.358 \times 10^{-3} = 0.007358$$

c. $8.4 \times 10^7 = 84{,}000{,}000.$ 7 places to the right

d. $3.007 \times 10^{-5} = 0.00003007$ 5 places to the left

PRACTICE 6 Write each number in standard notation, without exponents.

a. 3.67×10^{-4} **b.** 8.954×10^6

c. 2.009×10^{-5} **d.** 4.054×10^3

Large and small numbers not only occur in science, but in all types of displayed data. Scientific notation is just one convenient way to express these numbers. For example, study the graph below. The data labels represent billions of dollars. (A billion is 1,000,000,000 or 10^9.) In Example 7, we practice writing these numbers in standard form and scientific notation.

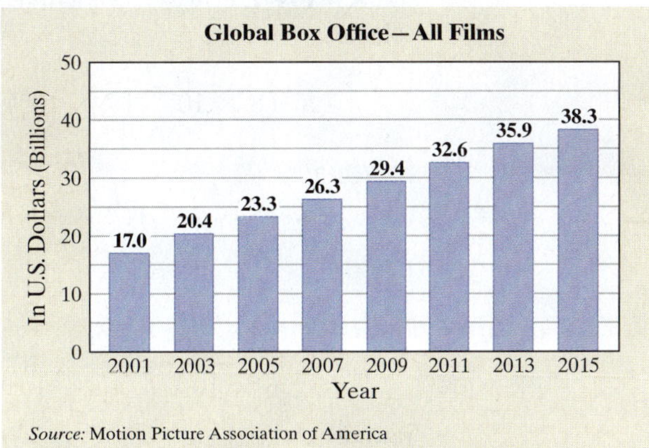

EXAMPLE 7 Global box office for all films released around the world reached \$38.3 billion in the year 2015.

a. Write this dollar amount in standard form.

b. Write this dollar amount in scientific notation.

(Continued on next page)

Solution

> **Helpful Hint**
> We read 38,300,000,000 as 38 billion, 300 million. It makes sense that this number equals 38.3 billion.

a. 38.3 billion $= 38.3 \times 1$ billion
$= 38.3 \times (1{,}000{,}000{,}000)$
$= 38{,}300{,}000{,}000$ in standard form.

b. 38.3 billion is not in scientific notation because 38.3 is not a number between 1 and 10.

38.3 billion $= 38.3 \times 1$ billion
$= (3.83 \times 10^1) \times (1 \times 10^9)$ Write 1 billion as 1×10^9 or just 10^9.
$= 3.83 \times 10^{10}$ Since $10^1 \times 10^9 = 10^{1+9} = 10^{10}$

To Check, write 3.83×10^{10} in standard form by moving the decimal point 10 places to the right.

Note: Another way to solve part b, is to use the standard form found in part a, and write this number in scientific notation. □

PRACTICE 7 Write the global box office dollar amount for 2001 in

a. standard form and

b. scientific notation ■

✔ CONCEPT CHECK

Which number in each pair is larger?

a. 7.8×10^3 or 2.1×10^5 **b.** 9.2×10^{-2} or 2.7×10^4 **c.** 5.6×10^{-4} or 6.3×10^{-5}

OBJECTIVE 5

Performing Operations with Scientific Notation

Performing operations on numbers written in scientific notation uses the rules and definitions for exponents.

EXAMPLE 8 Perform each indicated operation. Write each result in standard decimal notation.

a. $(8 \times 10^{-6})(7 \times 10^3)$ **b.** $\dfrac{12 \times 10^2}{6 \times 10^{-3}}$

Solution

a. $(8 \times 10^{-6})(7 \times 10^3) = (8 \cdot 7) \times (10^{-6} \cdot 10^3)$
$= 56 \times 10^{-3}$
$= 0.056$

b. $\dfrac{12 \times 10^2}{6 \times 10^{-3}} = \dfrac{12}{6} \times 10^{2-(-3)} = 2 \times 10^5 = 200{,}000$ □

PRACTICE 8 Perform each indicated operation. Write each result in standard decimal notation.

a. $(5 \times 10^{-4})(8 \times 10^6)$ **b.** $\dfrac{64 \times 10^3}{32 \times 10^{-7}}$ ■

Answers to Concept Check:
a. 2.1×10^5 **b.** 2.7×10^4
c. 5.6×10^{-4}

Calculator Explorations

Scientific Notation

To enter a number written in scientific notation on a scientific calculator, locate the scientific notation key, which may be marked EE or EXP. To enter 3.1×10^7, press 3.1 EE 7. The display should read 3.1 07.

Enter each number written in scientific notation on your calculator.

1. 5.31×10^3

2. -4.8×10^{14}

3. 6.6×10^{-9}

4. -9.9811×10^{-2}

Multiply each of the following on your calculator. Notice the form of the result.

5. $3{,}000{,}000 \times 5{,}000{,}000$

6. $230{,}000 \times 1000$

Multiply each of the following on your calculator. Write the product in scientific notation.

7. $(3.26 \times 10^6)(2.5 \times 10^{13})$

8. $(8.76 \times 10^{-4})(1.237 \times 10^9)$

Vocabulary, Readiness & Video Check

Fill in each blank with the correct choice.

1. The expression x^{-3} equals ______.

a. $-x^3$ **b.** $\frac{1}{x^3}$ **c.** $\frac{-1}{x^3}$ **d.** $\frac{1}{x^{-3}}$

2. The expression 5^{-4} equals ______.

a. -20 **b.** -625 **c.** $\frac{1}{20}$ **d.** $\frac{1}{625}$

3. The number 3.021×10^{-3} is written in ____________.

a. standard form **b.** expanded form
c. scientific notation

4. The number 0.0261 is written in ____________.

a. standard form **b.** expanded form
c. scientific notation

Martin-Gay Interactive Videos

See Video 4.5

Watch the section lecture video and answer the following questions.

OBJECTIVE 1 **5.** What important reminder is made at the end of Example 1?

OBJECTIVE 2 **6.** Name all the rules and definitions used to simplify Example 8.

OBJECTIVE 3 **7.** From Examples 9 and 10, explain how the movement of the decimal point in step 1 suggests the sign of the exponent on the number 10.

OBJECTIVE 4 **8.** From Example 11, what part of a number written in scientific notation is key in telling you how to write the number in standard form?

OBJECTIVE 5 **9.** For Example 13, what exponent rules were needed to evaluate?

4.5 Exercise Set MyMathLab®

Simplify each expression. Write each result using positive exponents only. See Examples 1 through 3.

1. 4^{-3}

2. 6^{-2}

3. $(-3)^{-4}$

4. $(-3)^{-5}$

5. $7x^{-3}$

6. $(7x)^{-3}$

7. $\left(\frac{1}{2}\right)^{-5}$

8. $\left(\frac{1}{8}\right)^{-2}$

9. $\left(-\frac{1}{4}\right)^{-3}$

10. $\left(-\frac{1}{8}\right)^{-2}$

11. $3^{-1} + 5^{-1}$

12. $4^{-1} + 4^{-2}$

13. $\frac{1}{p^{-3}}$ 14. $\frac{1}{q^{-5}}$ 15. $\frac{p^{-5}}{q^{-4}}$

16. $\frac{r^{-5}}{s^{-2}}$ 17. $\frac{x^{-2}}{x}$ 18. $\frac{y}{y^{-3}}$

19. $\frac{z^{-4}}{z^{-7}}$ 20. $\frac{x^{-4}}{x^{-1}}$ 21. $3^{-2} + 3^{-1}$

22. $4^{-2} - 4^{-3}$ 23. $\frac{-1}{p^{-4}}$ 24. $\frac{-1}{y^{-6}}$

25. $-2^0 - 3^0$ 26. $5^0 + (-5)^0$

MIXED PRACTICE

Simplify each expression. Write each result using positive exponents only. See Examples 1 through 4.

27. $\frac{x^2x^5}{x^3}$ 28. $\frac{y^4y^5}{y^6}$ 29. $\frac{p^2p}{p^{-1}}$

30. $\frac{y^3y}{y^{-2}}$ 31. $\frac{(m^5)^4m}{m^{10}}$ 32. $\frac{(x^2)^8x}{x^9}$

33. $\frac{r}{r^{-3}r^{-2}}$ 34. $\frac{p}{p^{-3}q^{-5}}$ 35. $(x^5y^3)^{-3}$

36. $(z^5x^5)^{-3}$ 37. $\frac{(x^2)^3}{x^{10}}$ 38. $\frac{(y^4)^2}{y^{12}}$

39. $\frac{(a^5)^2}{(a^3)^4}$ 40. $\frac{(x^2)^5}{(x^4)^3}$ 41. $\frac{8k^4}{2k}$

42. $\frac{27r^4}{3r^6}$ 43. $\frac{-6m^4}{-2m^3}$ 44. $\frac{15a^4}{-15a^5}$

45. $(a^{-5}b^2)^{-6}$ 46. $(4^{-1}x^5)^{-2}$

47. $\frac{4^2z^{-3}}{4^3z^{-5}}$ 48. $\frac{3^{-1}x^4}{3^3x^{-7}}$

49. $\frac{2^{-3}x^{-4}}{2^2x}$ 50. $\frac{5^{-1}z^7}{5^{-2}z^9}$

Write each number in scientific notation. See Examples 5 and 7.

51. 78,000 52. 9,300,000,000

53. 0.00000167 54. 0.00000017

55. 0.00635 56. 0.00194

57. 1,160,000 58. 700,000

Solve.

59. More than 2,000,000,000 pencils are manufactured in the United States annually. Write this number in scientific notation. (*Source:* AbsoluteTrivia.com)

60. The temperature at the interior of the Earth is 20,000,000 degrees Celsius. Write 20,000,000 in scientific notation.

61. As of this writing, the world's largest optical telescope is the Gran Telescopio Canaris, located in La Palma, Canary Islands, Spain. The elevation of this telescope is 2400 meters above sea level. Write 2400 in scientific notation.

62. In March 2004, the European Space Agency launched the Rosetta spacecraft, whose mission was to deliver the Philae lander to explore comet 67P/Churyumov-Gerasimenko. The lander finally arrived on the comet in late 2014. This comet is currently more than 320,000,000 miles from Earth. Write 320,000,000 in scientific notation. (*Source:* European Space Agency)

Write each number in standard notation. See Examples 6 and 7.

63. 8.673×10^{-10}

64. 9.056×10^{-4}

65. 3.3×10^{-2}

66. 4.8×10^{-6}

67. 2.032×10^{4}

68. 9.07×10^{10}

Solve.

69. Each second, the Sun converts 7.0×10^8 tons of hydrogen into helium and energy in the form of gamma rays. Write this number in standard notation. (*Source:* Students for the Exploration and Development of Space)

70. In chemistry, Avogadro's number is the number of atoms in one mole of an element. Avogadro's number is $6.02214199 \times 10^{23}$. Write this number in standard notation. (*Source:* National Institute of Standards and Technology)

71. The distance light travels in 1 year is 9.46×10^{12} kilometers. Write this number in standard notation.

72. The population of the world is 7.3×10^9. Write this number in standard notation. (*Source:* UN World Population Clock)

MIXED PRACTICE

See Examples 5 through 7. Below are some interesting facts about selected countries' external debts at a certain time. These are public and private debts owed to nonresidents of that country. If a number is written in standard form, write it in scientific notation. If a number is written in scientific notation, write it in standard form. (Source: CIA World Factbook)

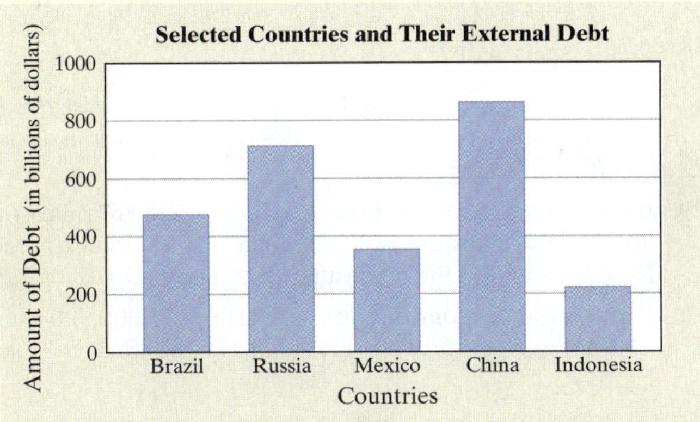

73. The external debt of Russia was $714,000,000,000.

74. The amount by which Russia's debt was greater than Mexico's debt was $359,000,000,000.

75. China's external debt was $\$8.63 \times 10^{11}$.

76. The external debt of the United States was $\$1.5 \times 10^{13}$.

77. The approximate population of the United States is 323.8 million. Write this number in

a. standard form

b. scientific notation

78. The approximate population of Canada is 36.3 million. Write this number in

a. standard form

b. scientific notation

Evaluate each expression using exponential rules. Write each result in standard notation. See Example 8.

79. $(1.2 \times 10^{-3})(3 \times 10^{-2})$ **80.** $(2.5 \times 10^{6})(2 \times 10^{-6})$

81. $(4 \times 10^{-10})(7 \times 10^{-9})$ **82.** $(5 \times 10^{6})(4 \times 10^{-8})$

83. $\dfrac{8 \times 10^{-1}}{16 \times 10^{5}}$ **84.** $\dfrac{25 \times 10^{-4}}{5 \times 10^{-9}}$

85. $\dfrac{1.4 \times 10^{-2}}{7 \times 10^{-8}}$ **86.** $\dfrac{0.4 \times 10^{5}}{0.2 \times 10^{11}}$

REVIEW AND PREVIEW

Use this histogram to answer the exercises. See Section R.4.

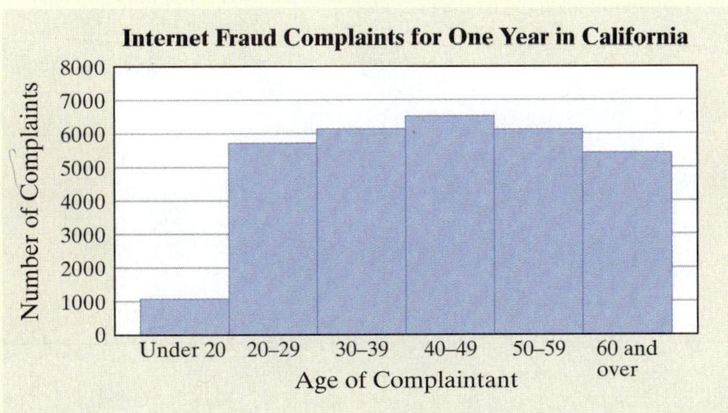

87. Which age range reported the fewest complaints?

88. Which age range reported the most complaints?

89. Estimate the number of complaints for the 40- to 49-year-olds.

90. Estimate the number of complaints for the 20- to 29-year-olds.

In this graph, there's a total of 30,923 complaint reports. Use this for Exercises 91 and 92 and round to the nearest tenth of a percent.

91. If 14,612 were from females, find the percent from females.

92. If 16,311 were from males, find the percent from males.

CONCEPT EXTENSIONS

△ **93.** Find the volume of the cube.

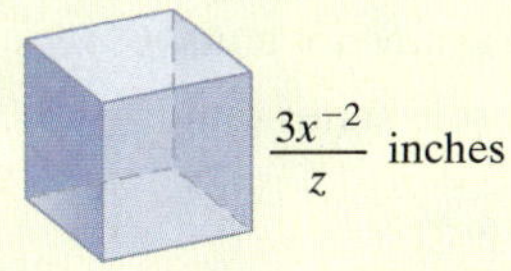

△ **94.** Find the area of the triangle.

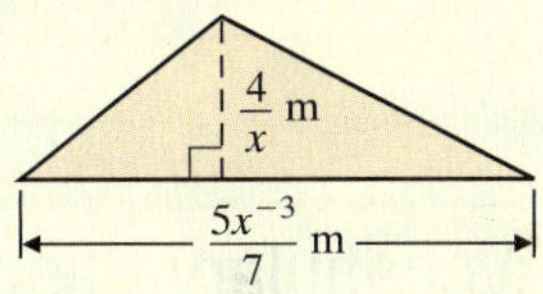

Simplify.

95. $(2a^3)^3a^4 + a^5a^8$ **96.** $(2a^3)^3a^{-3} + a^{11}a^{-5}$

Fill in the boxes so that each statement is true. (More than one answer may be possible for these exercises.)

97. $x^{\square} = \dfrac{1}{x^5}$ **98.** $7^{\square} = \dfrac{1}{49}$

99. $z^{\square} \cdot z^{\square} = z^{-10}$

100. $(x^{\square})^{\square} = x^{-15}$

101. Which is larger? See the Concept Check in this section.

a. 9.7×10^{-2} or 1.3×10^{1}

b. 8.6×10^{5} or 4.4×10^{7}

c. 6.1×10^{-2} or 5.6×10^{-4}

102. Determine whether each statement is true or false.

a. $5^{-1} < 5^{-2}$

b. $\left(\dfrac{1}{5}\right)^{-1} < \left(\dfrac{1}{5}\right)^{-2}$

c. $a^{-1} < a^{-2}$ for all nonzero numbers.

Simplify each expression. Write each result using positive exponents only.

103. $\dfrac{-24a^6b}{6ab^2}$ **104.** $\dfrac{-5x^4y^5}{15x^4y^2}$

105. $(-2x^3y^{-4})(3x^{-1}y)$ **106.** $(-5a^4b^{-7})(-a^{-4}b^3)$

107. $\left(\dfrac{x^{-2}y^4}{x^3y^7}\right)^2$ **108.** $\left(\dfrac{a^5b}{a^7b^{-2}}\right)^{-3}$

109. $\dfrac{7ab^{-4}}{7^{-1}a^{-3}b^2}$ **110.** $\dfrac{6^{-5}x^{-1}y^2}{6^{-2}x^{-4}y^4}$

111. $\left(\dfrac{a^{-5}b}{ab^3}\right)^{-4}$ **112.** $\left(\dfrac{r^{-2}s^{-3}}{r^{-4}s^{-3}}\right)^{-3}$

113. $\frac{(xy^3)^5}{(xy)^{-4}}$

114. $\frac{(rs)^{-3}}{(r^2s^3)^2}$

115. $\frac{(-2xy^{-3})^{-3}}{(xy^{-1})^{-1}}$

116. $\frac{(-3x^2y^2)^{-2}}{(xyz)^{-2}}$

117. $\frac{6x^2y^3}{-7xy^5}$

118. $\frac{-8xa^2b}{-5xa^5b}$

119. $\frac{(a^4b^{-7})^{-5}}{(5a^2b^{-1})^{-2}}$

120. $\frac{(a^6b^{-2})^4}{(4a^{-3}b^{-3})^3}$

121. It was stated earlier that for an integer n,

$$x^{-n} = \frac{1}{x^n}, \quad x \neq 0$$

Explain why x may not equal 0.

122. The quotient rule states that

$$\frac{a^m}{a^n} = a^{m-n}, a \neq 0.$$

Explain why a may not equal 0.

Simplify each expression. Assume that variables represent positive integers.

123. $a^{-4m} \cdot a^{5m}$

124. $(x^{-3s})^3$

125. $(3y^{2z})^3$

126. $a^{4m+1} \cdot a^4$

Simplify each expression. Write each result in standard notation.

127. $(2.63 \times 10^{12})(-1.5 \times 10^{-10})$

128. $(6.785 \times 10^{-4})(4.68 \times 10^{10})$

Light travels at a rate of 1.86×10^5 *miles per second. Use this information and the distance formula* $d = r \cdot t$ *to answer Exercises 129 and 130.*

129. If the distance from the moon to Earth is 238,857 miles, find how long it takes the reflected light of the moon to reach Earth. (Round to the nearest tenth of a second.)

130. If the distance from the Sun to Earth is 93,000,000 miles, find how long it takes the light of the Sun to reach Earth.

Chapter 4 Vocabulary Check

Fill in each blank with one of the words or phrases listed below.

term	coefficient	monomial	binomial	trinomial
polynomials	degree of a term	distributive	FOIL	degree of a polynomial

1. A ________________ is a number or the product of numbers and variables raised to powers.
2. The ________________ method may be used when multiplying two binomials.
3. A polynomial with exactly three terms is called a ________________.
4. The ________________ is the greatest degree of any term of the polynomial.
5. A polynomial with exactly two terms is called a ________________.
6. The ________________ of a term is its numerical factor.
7. The ________________ is the sum of the exponents on the variables in the term.
8. A polynomial with exactly one term is called a ________________.
9. Monomials, binomials, and trinomials are all examples of ________________.
10. The ________________ property is used to multiply $2x(x - 4)$.

Chapter 4 Highlights

DEFINITIONS AND CONCEPTS	EXAMPLES
Section 4.1 Exponents	
a^n means the product of n factors, each of which is a.	$3^2 = 3 \cdot 3 = 9$ $(-5)^3 = (-5)(-5)(-5) = -125$ $\left(\frac{1}{2}\right)^4 = \frac{1}{2} \cdot \frac{1}{2} \cdot \frac{1}{2} \cdot \frac{1}{2} = \frac{1}{16}$

DEFINITIONS AND CONCEPTS	EXAMPLES
Section 4.1 Exponents (Continued)	
If m and n are integers and no denominators are 0,	
Product Rule: $a^m \cdot a^n = a^{m+n}$	$x^2 \cdot x^7 = x^{2+7} = x^9$
Power Rule: $(a^m)^n = a^{mn}$	$(5^3)^8 = 5^{3\cdot 8} = 5^{24}$
Power of a Product Rule: $(ab)^n = a^n b^n$	$(7y)^4 = 7^4 y^4$
Power of a Quotient Rule: $\left(\frac{a}{b}\right)^n = \frac{a^n}{b^n}$	$\left(\frac{x}{8}\right)^3 = \frac{x^3}{8^3}$
Quotient Rule: $\frac{a^m}{a^n} = a^{m-n}$	$\frac{x^9}{x^4} = x^{9-4} = x^5$
Zero Exponent: $a^0 = 1, a \neq 0.$	$5^0 = 1, x^0 = 1, x \neq 0$
Section 4.2 Polynomial Functions and Adding and Subtracting Polynomials	
A **term** is a number or the product of numbers and variables raised to powers.	***Terms*** $-5x, 7a^2b, \frac{1}{4}y^4, 0.2$
The **numerical coefficient** or **coefficient** of a term is its numerical factor.	***Term*** — ***Coefficient*** $7x^2$ — 7 y — 1 $-a^2b$ — -1
A **polynomial** is a finite sum of terms in which all variables have exponents that are nonnegative integers and no variables appear in the denominator.	***Polynomials*** $1.3x^2$ (monomial) $-\frac{1}{3}y + 5$ (binomial) $6z^2 - 5z + 7$ (trinomial)
A function P is a **polynomial function** if $P(x)$ is a polynomial.	For the polynomial function $P(x) = -x^2 + 6x - 12$, find $P(-2)$ $P(-2) = -(-2)^2 + 6(-2) - 12 = -28.$
The **degree of a term** is the sum of the exponents on the variables in the term.	***Term*** — ***Degree*** $-5x^3$ — 3 3 (or $3x^0$) — 0 $2a^2b^2c$ — $2 + 2 + 1 = 5$
The **degree of a polynomial** is the greatest degree of any term of the polynomial.	***Polynomial*** — ***Degree*** $5x^2 - 3x + 2$ — 2 $7y + 8y^2z^3 - 12$ — $2 + 3 = 5$
To add polynomials, add or combine like terms.	Add: $(7x^2 - 3x + 2) + (-5x - 6) = 7x^2 - 3x + 2 - 5x - 6$ $= 7x^2 - 8x - 4$
To subtract two polynomials, change the signs of the terms of the second polynomial, then add.	Subtract: $(17y^2 - 2y + 1) - (-3y^3 + 5y - 6)$ $= (17y^2 - 2y + 1) + (3y^3 - 5y + 6)$ $= 17y^2 - 2y + 1 + 3y^3 - 5y + 6$ $= 3y^3 + 17y^2 - 7y + 7$

DEFINITIONS AND CONCEPTS	EXAMPLES
Section 4.3 Multiplying Polynomials	
To multiply two polynomials, multiply each term of one polynomial by each term of the other polynomial and then combine like terms.	Multiply: $(2x+1)(5x^2-6x+2)$ $= 2x(5x^2-6x+2) + 1(5x^2-6x+2)$ $= 10x^3 - 12x^2 + 4x + 5x^2 - 6x + 2$ $= 10x^3 - 7x^2 - 2x + 2$
Section 4.4 Special Products	
The **FOIL method** may be used when multiplying two binomials.	Multiply: $(5x-3)(2x+3)$ First, Last, Outer, Inner F O I L $(5x-3)(2x+3) = (5x)(2x) + (5x)(3) + (-3)(2x) + (-3)(3)$ $= 10x^2 + 15x - 6x - 9$ $= 10x^2 + 9x - 9$
Squaring a Binomial $(a+b)^2 = a^2 + 2ab + b^2$ $(a-b)^2 = a^2 - 2ab + b^2$	Square each binomial. $(x+5)^2 = x^2 + 2(x)(5) + 5^2$ $= x^2 + 10x + 25$ $(3x-2y)^2 = (3x)^2 - 2(3x)(2y) + (2y)^2$ $= 9x^2 - 12xy + 4y^2$
Multiplying the Sum and Difference of Two Terms $(a+b)(a-b) = a^2 - b^2$	Multiply: $(6y+5)(6y-5) = (6y)^2 - 5^2$ $= 36y^2 - 25$
Section 4.5 Negative Exponents and Scientific Notation	
If $a \neq 0$ and n is an integer, $a^{-n} = \frac{1}{a^n}$	$3^{-2} = \frac{1}{3^2} = \frac{1}{9};\ 5x^{-2} = \frac{5}{x^2}$
Rules for exponents are true for positive and negative integers.	Simplify: $\left(\frac{x^{-2}y}{x^5}\right)^{-2} = \frac{x^4y^{-2}}{x^{-10}}$ $= x^{4-(-10)}y^{-2}$ $= \frac{x^{14}}{y^2}$
A positive number is written in scientific notation if it is written as the product of a number a, $1 \le a < 10$, and an integer power r of 10. $a \times 10^r$	Write each number in scientific notation. $12{,}000 = 1.2 \times 10^4$ $0.00000568 = 5.68 \times 10^{-6}$

Chapter 4 Review

(4.1) *State the base and the exponent for each expression.*

1. 7^9

2. $(-5)^4$

3. -5^4

4. x^6

Evaluate each expression.

5. 8^3

6. $(-6)^2$

7. -6^2

8. $-4^3 - 4^0$

9. $(3b)^0$

10. $\frac{8b}{8b}$

Simplify each expression.

11. $y^2 \cdot y^7$

12. $x^9 \cdot x^5$

13. $(2x^5)(-3x^6)$

14. $(-5y^3)(4y^4)$

15. $(x^4)^2$

16. $(y^3)^5$

17. $(3y^6)^4$

18. $(2x^3)^3$

19. $\frac{x^9}{x^4}$

20. $\frac{z^{12}}{z^5}$

21. $\frac{a^5b^4}{ab}$

22. $\frac{x^4y^6}{xy}$

23. $\frac{3x^4y^{10}}{12xy^6}$

24. $\frac{2x^7y^8}{8xy^2}$

25. $5a^7(2a^4)^3$

26. $(2x)^2(9x)$

27. $(-5a)^0 + 7^0 + 8^0$

28. $8x^0 + 9^0$

Simplify the given expression and choose the correct result.

29. $\left(\frac{3x^4}{4y}\right)^3$

a. $\frac{27x^{64}}{64y^3}$ **b.** $\frac{27x^{12}}{64y^3}$ **c.** $\frac{9x^{12}}{12y^3}$ **d.** $\frac{3x^{12}}{4y^3}$

30. $\left(\frac{5a^6}{b^3}\right)^2$

a. $\frac{10a^{12}}{b^6}$ **b.** $\frac{25a^{36}}{b^9}$ **c.** $\frac{25a^{12}}{b^6}$ **d.** $25a^{12}b^6$

(4.2) *Find the degree of each term.*

31. $-5x^4y^3$

32. $10x^3y^2z$

33. $35a^5bc^2$

34. $95xyz$

Find the degree of each polynomial.

35. $y^5 + 7x - 8x^4$

36. $9y^2 + 30y + 25$

37. $-14x^2y - 28x^2y^3 - 42x^2y^2$

38. $6x^2y^2z^2 + 5x^2y^3 - 12xyz$

39. The Glass Bridge Skywalk is suspended 4000 feet over the Colorado River at the very edge of the Grand Canyon. Neglecting air resistance, the height of an object dropped from the Skywalk at time t seconds is given by the polynomial function $P(t) = -16t^2 + 4000$. Find the height of the object at the given times.

t	0 seconds	1 second	3 seconds	5 seconds
$P(t) = -16t^2 + 4000$				

△ **40.** The surface area of a box with a square base and a height of 5 units is given by the polynomial function $P(x) = 2x^2 + 20x$. Fill in the table below by evaluating the given values of x.

x	1	3	5.1	10
$P(x) = 2x^2 + 20x$				

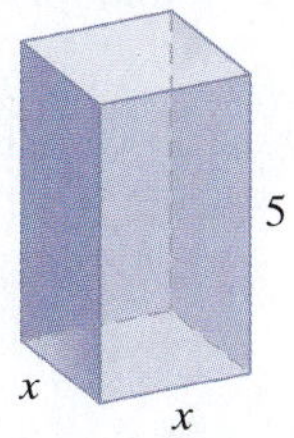

Combine like terms in each expression.

41. $7a^2 - 4a^2 - a^2$

42. $9y + y - 14y$

43. $6a^2 + 4a + 9a^2$

44. $21x^2 + 3x + x^2 + 6$

45. $4a^2b - 3b^2 - 8q^2 - 10a^2b + 7q^2$

46. $2s^{14} + 3s^{13} + 12s^{12} - s^{10}$

Add or subtract as indicated.

47. $(3x^2 + 2x + 6) + (5x^2 + x)$

48. $(2x^5 + 3x^4 + 4x^3 + 5x^2) + (4x^2 + 7x + 6)$

49. $(-5y^2 + 3) - (2y^2 + 4)$

50. $(3x^2 - 7xy + 7y^2) - (4x^2 - xy + 9y^2)$

51. $(-9x^2 + 6x + 2) + (4x^2 - x - 1)$

52. $(8x^6 - 5xy - 10y^2) - (7x^6 - 9xy - 12y^2)$

TRANSLATING

Perform the indicated operations.

53. Subtract $(3x - y)$ from $(7x - 14y)$.

54. Subtract $(4x^2 + 8x - 7)$ from the sum of $(x^2 + 7x + 9)$ and $(x^2 + 4)$.

(4.3) Multiply each expression.

55. $4(2a + 7)$

56. $9(6a - 3)$

57. $-7x(x^2 + 5)$

58. $-8y(4y^2 - 6)$

59. $(3a^3 - 4a + 1)(-2a)$

60. $(6b^3 - 4b + 2)(7b)$

61. $(2x + 2)(x - 7)$

62. $(2x - 5)(3x + 2)$

63. $(x - 9)^2$

64. $(x - 12)^2$

65. $(4a - 1)(a + 7)$

66. $(6a - 1)(7a + 3)$

67. $(5x + 2)^2$

68. $(3x + 5)^2$

69. $(x + 7)(x^3 + 4x - 5)$

70. $(x + 2)(x^5 + x + 1)$

71. $(x^2 + 2x + 4)(x^2 + 2x - 4)$

72. $(x^3 + 4x + 4)(x^3 + 4x - 4)$

73. $(x + 7)^3$

74. $(2x - 5)^3$

(4.4) Use special products to multiply each of the following.

75. $(x + 7)^2$

76. $(x - 5)^2$

77. $(3x - 7)^2$

78. $(4x + 2)^2$

79. $(5x - 9)^2$

80. $(5x + 1)(5x - 1)$

81. $(7x + 4)(7x - 4)$

82. $(a + 2b)(a - 2b)$

83. $(2x - 6)(2x + 6)$

84. $(4a^2 - 2b)(4a^2 + 2b)$

Express each as a product of polynomials in x. Then multiply and simplify.

△ **85.** Find the area of the square if its side is $(3x - 1)$ meters.

$(3x - 1)$ meters

△ **86.** Find the area of the rectangle.

$(x - 1)$ miles

$(5x + 2)$ miles

(4.5) Simplify each expression.

87. 7^{-2}

88. -7^{-2}

89. $2x^{-4}$

90. $(2x)^{-4}$

91. $\left(\frac{1}{5}\right)^{-3}$

92. $\left(\frac{-2}{3}\right)^{-2}$

93. $2^0 + 2^{-4}$

94. $6^{-1} - 7^{-1}$

Simplify each expression. Write each answer using positive exponents only.

95. $\frac{x^5}{x^{-3}}$

96. $\frac{z^4}{z^{-4}}$

97. $\frac{r^{-3}}{r^{-4}}$

98. $\frac{y^{-2}}{y^{-5}}$

99. $\left(\frac{bc^{-2}}{bc^{-3}}\right)^4$

100. $\left(\frac{x^{-3}y^{-4}}{x^{-2}y^{-5}}\right)^{-3}$

101. $\frac{x^{-4}y^{-6}}{x^2y^7}$

102. $\frac{a^5b^{-5}}{a^{-5}b^5}$

103. $a^{6m}a^{5m}$

104. $\frac{(x^{5+h})^3}{x^5}$

105. $(3xy^{2z})^3$

106. $a^{m+2}a^{m+3}$

Write each number in scientific notation.

107. 0.00027

108. 0.8868

109. 80,800,000

110. 868,000

111. Google.com is an Internet search engine that handles 2,500,000,000 searches every day. Write 2,500,000,000 in scientific notation. (*Source:* Google, Inc.)

112. The approximate diameter of the Milky Way galaxy is 150,000 light years. Write this number in scientific notation. (*Source:* NASA IMAGE/POETRY Education and Public Outreach Program)

Write each number in standard form.

113. 8.67×10^5

114. 3.86×10^{-3}

115. 8.6×10^{-4}

116. 8.936×10^5

117. The volume of the planet Jupiter is 1.43128×10^{15} cubic kilometers. Write this number in standard notation. (*Source:* National Space Science Data Center)

118. An angstrom is a unit of measure, equal to 1×10^{-10} meter, used for measuring wavelengths or the diameters of atoms. Write this number in standard notation. (*Source:* National Institute of Standards and Technology)

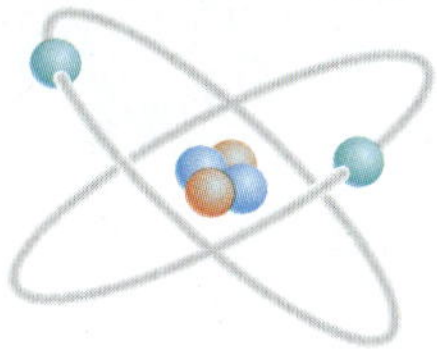

Simplify. Express each result in standard form.

119. $(8 \times 10^4)(2 \times 10^{-7})$

120. $\dfrac{8 \times 10^4}{2 \times 10^{-7}}$

MIXED REVIEW

Evaluate.

121. $\left(-\dfrac{1}{2}\right)^3$

Simplify each expression. Write each answer using positive exponents only.

122. $(4xy^2)(x^3y^5)$

123. $\dfrac{18x^9}{27x^3}$

124. $\left(\dfrac{3a^4}{b^2}\right)^3$

125. $(2x^{-4}y^3)^{-4}$

126. $\dfrac{a^{-3}b^6}{9^{-1}a^{-5}b^{-2}}$

Perform the indicated operations and simplify.

127. $(6x + 2) + (5x - 7)$

128. $(-y^2 - 4) + (3y^2 - 6)$

129. $(8y^2 - 3y + 1) - (3y^2 + 2)$

130. $(5x^2 + 2x - 6) - (-x - 4)$

131. $4x(7x^2 + 3)$

132. $(2x + 5)(3x - 2)$

133. $(x - 3)(x^2 + 4x - 6)$

134. $(7x - 2)(4x - 9)$

Use special products to multiply.

135. $(5x + 4)^2$

136. $(6x + 3)(6x - 3)$

Chapter 4 Getting Ready for the Test

MATCHING *Match the expression with the exponent operation needed to simplify. Letters may be used more than once or not at all.*

1. $x^2 \cdot x^5$

2. $(x^2)^5$

3. $x^2 + x^5$

4. $\dfrac{x^5}{x^2}$

A. multiply the exponents

B. divide the exponents

C. add the exponents

D. subtract the exponents

E. this expression will not simplify

MATCHING *Match the operation with the result when the operation is performed on the given terms. Letters may be used more than once or not at all.*

Given Terms: 20y and 4y

5. Add the terms

6. Subtract the terms

7. Multiply the terms

8. Divide the terms.

A. $80y$ **E.** $80y^2$

B. $24y^2$ **F.** $24y$

C. $16y$ **G.** $16y^2$

D. 16 **H.** $5y$

I. 5

9. MULTIPLE CHOICE *The expression 5^{-1} is equivalent to*

A. -5 **B.** 4 **C.** $\dfrac{1}{5}$ **D.** $-\dfrac{1}{5}$

10. MULTIPLE CHOICE *The expression 2^{-3} is equivalent to*

A. -6 **B.** -1 **C.** $-\dfrac{1}{6}$ **D.** $\dfrac{1}{8}$

MATCHING *Match each expression with its simplified form. Letters may be used more than once or not at all.*

11. $y + y + y$ **A.** $3y^3$ **E.** $-3y^3$

12. $y \cdot y \cdot y$ **B.** y^3 **F.** $-y^3$

13. $(-y)(-y)(-y)$ **C.** $3y$

14. $-y - y - y$ **D.** $-3y$

Chapter 4 Test

MyMathLab® YouTube

Evaluate each expression.

1. 2^5 **2.** $(-3)^4$ **3.** -3^4 **4.** 4^{-3}

Simplify each exponential expression. Write the result using only positive exponents.

5. $(3x^2)(-5x^9)$

6. $\frac{y^7}{y^2}$

7. $\frac{r^{-8}}{r^{-3}}$

8. $\left(\frac{x^2y^3}{x^3y^{-4}}\right)^2$

9. $\frac{6^2x^{-4}y^{-1}}{6^3x^{-3}y^7}$

Express each number in scientific notation.

10. 563,000 **11.** 0.0000863

Write each number in standard form.

12. 1.5×10^{-3} **13.** 6.23×10^4

14. Simplify. Write the answer in standard form.

$$(1.2 \times 10^5)(3 \times 10^{-7})$$

15. a. Complete the table for the polynomial $4xy^2 + 7xyz + x^3y - 2$.

Term	*Numerical Coefficient*	*Degree of Term*
$4xy^2$		
$7xyz$		
x^3y		
-2		

b. What is the degree of the polynomial?

16. Simplify by combining like terms.

$$5x^2 + 4xy - 7x^2 + 11 + 8xy$$

Perform each indicated operation.

17. $(8x^3 + 7x^2 + 4x - 7) + (8x^3 - 7x - 6)$

18. $5x^3 + x^2 + 5x - 2 - (8x^3 - 4x^2 + x - 7)$

19. Subtract $(4x + 2)$ from the sum of $(8x^2 + 7x + 5)$ and $(x^3 - 8)$.

Multiply.

20. $(3x + 7)(x^2 + 5x + 2)$

21. $3x^2(2x^2 - 3x + 7)$

22. $(x + 7)(3x - 5)$

23. $\left(3x - \frac{1}{5}\right)\left(3x + \frac{1}{5}\right)$

24. $(4x - 2)^2$

25. $(8x + 3)^2$

26. $(x^2 - 9b)(x^2 + 9b)$

Solve.

27. The height of the Bank of China in Hong Kong is 1001 feet. Neglecting air resistance, the height of an object dropped from this building at time t seconds is given by the polynomial function $P(t) = -16t^2 + 1001$. Find the height of the object at the given times below.

t	**0 seconds**	**1 second**	**3 seconds**	**5 seconds**
$P(t) = -16t^2 + 1001$				

△ **28.** Find the area of the top of the table. Express the area as a product, then multiply and simplify.

29. A pebble is hurled upward from the top of the Canada Trust Tower, which is 880 feet tall, with an initial velocity of 96 feet per second. Neglecting air resistance, the height $h(t)$ of the pebble after t seconds is given by the polynomial function

$$h(t) = -16t^2 + 96t + 880$$

a. Find the height of the pebble when $t = 1$.

b. Find the height of the pebble when $t = 5.1$.

Chapter 4 Cumulative Review

1. Tell whether each statement is true or false.
   a. $8 \geq 8$ b. $8 \leq 8$
   c. $23 \leq 0$ d. $23 \geq 0$
2. Find the absolute value of each number.
   a. $|-7.2|$ b. $|0|$ c. $\left|-\frac{1}{2}\right|$
3. Divide. Simplify all quotients if possible.
   a. $\frac{4}{5} \div \frac{5}{16}$ b. $\frac{7}{10} \div 14$ c. $\frac{3}{8} \div \frac{3}{10}$
4. Multiply. Write products in lowest terms.
   a. $\frac{3}{4} \cdot \frac{7}{21}$ b. $\frac{1}{2} \cdot 4\frac{5}{6}$
5. Evaluate the following.
   a. 3^2 b. 5^3 c. 2^4
   d. 7^1 e. $\left(\frac{3}{7}\right)^2$
6. Evaluate $\frac{2x - 7y}{x^2}$ for $x = 5$ and $y = 1$.
7. Decide whether 2 is a solution of $3x + 10 = 8x$.
8. Simplify: $8 + 3(2 \cdot 6 - 1)$
9. Subtract 7 from 4.
10. Is 1 a solution of $5x^2 + 2 = x - 8$?
11. Simplify: $\frac{8 + 2 \cdot 3}{2^2 - 1}$
12. Find the reciprocal of each number.
    a. 22 b. $\frac{3}{16}$
    c. -10 d. $-\frac{9}{13}$
13. Use an associative property to complete each statement.
    a. $5 + (4 + 6) =$ ______
    b. $(-1 \cdot 2) \cdot 5 =$ ______
14. Simplify: $\frac{4(-3) + (-8)}{5 + (-5)}$
15. Simplify each expression.
    a. $10 + (x + 12)$
    b. $-3(7x)$
16. Use the distributive property to write $-2(x + 3y - z)$ without parentheses.
17. Find each product by using the distributive property to remove parentheses.
    a. $5(3x + 2)$
    b. $-2(y + 0.3z - 1)$
    c. $-(9x + y - 2z + 6)$
18. Simplify: $2(6x - 1) - (x - 7)$
19. Solve $x - 7 = 10$ for x.
20. Write the phrase as an algebraic expression: double a number, subtracted from the sum of the number and seven
21. Solve: $\frac{5}{2}x = 15$
22. Solve: $2x + \frac{1}{8} = x - \frac{3}{8}$
23. Solve: $4(2x - 3) + 7 = 3x + 5$
24. Solve: $10 = 5j - 2$
25. Twice a number, added to seven, is the same as three subtracted from the number. Find the number.
26. Solve: $\frac{7x + 5}{3} = x + 3$
27. Graph $2 < x \leq 4$. Write the solutions in interval notation.
28. Graph $x < 5$. Then write the solutions in interval notation.
29. Complete the following ordered pair solutions for the equation $3x + y = 12$.
    a. (0,) b. (, 6)
    c. (−1,)
30. Recall that the grade of a road is its slope written as a percent. Find the grade of the road shown.

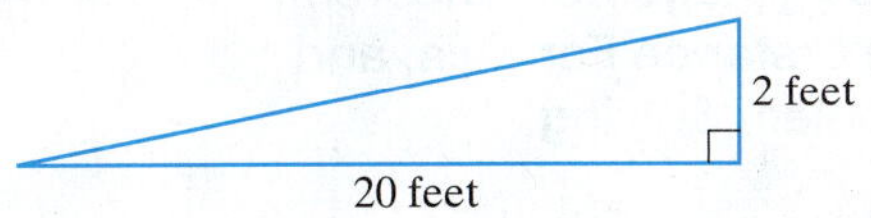

31. Graph the linear equation $x = 2$.
32. Evaluate.
    a. $(-5)^2$ b. -5^2
    c. $2 \cdot 5^2$
33. Find the slope of the line $y = -3$.
34. Simplify: $\frac{(z^2)^3 \cdot z^7}{z^9}$
35. Graph: $x + y < 7$
36. Subtract: $(5y^2 - 6) - (y^2 + 2)$
37. Use the product rule to simplify $(2x^2)(-3x^5)$.
38. Find the value of $-x^2$ when
    a. $x = 2$ b. $x = -2$
39. Add $(11x^3 - 12x^2 + x - 3)$ and $(x^3 - 10x + 5)$.
40. Multiply: $(10x^2 - 3)(10x^2 + 3)$
41. Multiply: $(2x - y)^2$
42. Multiply: $(10x^2 + 3)^2$
43. Evaluate.
    a. 3^{-2} b. $2^{-1} + 4^{-1}$
    c. $(-2)^{-4}$
44. Evaluate.
    a. 5^{-1} b. 7^{-2}

CHAPTER 5

Factoring Polynomials and Solving Quadratic Equations

CHECK YOUR PROGRESS

In Chapter 4, we learned how to multiply polynomials. This chapter deals with an operation that is the reverse process of multiplying, called *factoring*. Factoring is an important algebraic skill because this process allows us to write a sum as a product.

Also in this chapter, we use factoring to help us solve equations other than linear equations.

Why Are You in College?

There are probably as many answers as there are students. It may help you to know that college graduates have higher earnings and lower rates of unemployment. The double line graph below shows the increasing number of associate and bachelor degrees awarded over the years. It is also enlightening to know that an increasing number of high school graduates are looking to higher education.

In Exercise 110 of Section 5.1, we will explore how many students graduate from U.S. high schools each year, and how many of those may expect to go to college.

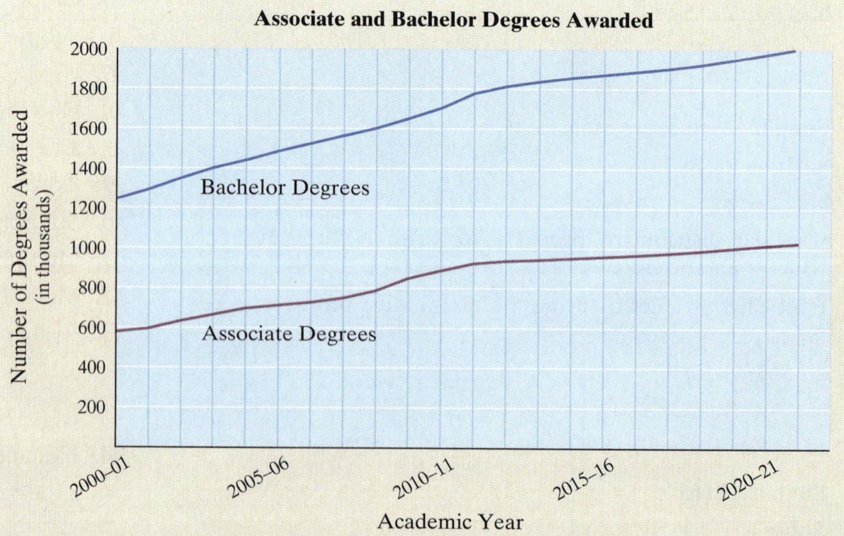

Source: National Center for Education Statistics (http://nces.ed.gov); U.S. Department of Education

Note: Some years are projected.

5.1 The Greatest Common Factor and Factoring by Grouping

OBJECTIVES

1. Find the Greatest Common Factor of a List of Integers.
2. Find the Greatest Common Factor of a List of Terms.
3. Factor Out the Greatest Common Factor from a Polynomial.
4. Factor a Polynomial by Grouping.

In the product $2 \cdot 3 = 6$, the numbers 2 and 3 are called **factors** of 6 and $2 \cdot 3$ is a **factored form** of 6. This is true of polynomials also. Since $(x+2)(x+3) = x^2 + 5x + 6$, $(x+2)$ and $(x+3)$ are factors of $x^2 + 5x + 6$, and $(x+2)(x+3)$ is a factored form of the polynomial.

a factored form of 6

$$\underbrace{2}_{\text{factor}} \cdot \underbrace{3}_{\text{factor}} = \underbrace{6}_{\text{product}}$$

a factored form of x^5

$$\underbrace{x^2}_{\text{factor}} \cdot \underbrace{x^3}_{\text{factor}} = \underbrace{x^5}_{\text{product}}$$

a factored form of $x^2 + 5x + 6$

$$\underbrace{(x+2)}_{\text{factor}}\underbrace{(x+3)}_{\text{factor}} = \underbrace{x^2 + 5x + 6}_{\text{product}}$$

The process of writing a polynomial as a product is called **factoring** the polynomial.

Do you see that factoring is the reverse process of multiplying?

$$x^2 + 5x + 6 = (x + 2)(x + 3)$$

(factoring: left to right; multiplying: right to left)

✔ CONCEPT CHECK

Multiply: $2(x - 4)$

What do you think the result of factoring $2x - 8$ would be? Why?

OBJECTIVE

1 Finding the Greatest Common Factor of a List of Integers

The first step in factoring a polynomial is to see whether the terms of the polynomial have a common factor. If there is one, we can write the polynomial as a product by **factoring out** the common factor. We will usually factor out the **greatest common factor (GCF).**

The GCF of a list of integers is the largest integer that is a factor of all the integers in the list. For example, the GCF of 12 and 20 is 4 because 4 is the largest integer that is a factor of both 12 and 20. With large integers, the GCF may not be found easily by inspection. When this happens, use the following steps.

Finding the GCF of a List of Integers

Step 1. Write each number as a product of prime numbers.

Step 2. Identify the common prime factors.

Step 3. The product of all common prime factors found in Step 2 is the greatest common factor. If there are no common prime factors, the greatest common factor is 1.

Recall from Section R.1 that a prime number is a whole number other than 1 whose only factors are 1 and itself.

Answers to Concept Check:
$2x - 8$; The result would be $2(x - 4)$ because factoring is the reverse process of multiplying.

EXAMPLE 1 Find the GCF of each list of numbers.

a. 28 and 40 **b.** 55 and 21 **c.** 15, 18, and 66

Solution

a. Write each number as a product of primes. (We have circled the common primes.)

$$28 = 2 \cdot 2 \cdot 7 = 2^2 \cdot 7$$
$$40 = 2 \cdot 2 \cdot 2 \cdot 5 = 2^3 \cdot 5$$

There are two common factors, each of which is 2, so the GCF is

$$\text{GCF} = 2 \cdot 2 = 4$$

b. $55 = 5 \cdot 11$

$21 = 3 \cdot 7$

There are no common prime factors; thus, the GCF is 1.

c. $15 = 3 \cdot 5$

$18 = 2 \cdot 3 \cdot 3 = 2 \cdot 3^2$

$66 = 2 \cdot 3 \cdot 11$

The only prime factor common to all three numbers is 3, so the GCF is

$$\text{GCF} = 3$$

□

PRACTICE 1 Find the GCF of each list of numbers.

a. 36 and 42 **b.** 35 and 44 **c.** 12, 16, and 40

■

OBJECTIVE

2 Finding the Greatest Common Factor of a List of Terms

The greatest common factor of a list of variables raised to powers is found in a similar way. For example, the GCF of x^2, x^3, and x^5 is x^2 because each term contains a factor of x^2 and no higher power of x is a factor of each term.

$$x^2 = x \cdot x$$
$$x^3 = x \cdot x \cdot x$$
$$x^5 = x \cdot x \cdot x \cdot x \cdot x$$

There are two common factors, each of which is x, so the GCF $= x \cdot x$ or x^2.

From this example, we see that **the GCF of a list of common variables raised to powers is the variable raised to the smallest exponent in the list.**

EXAMPLE 2 Find the GCF of each list of terms.

a. x^3, x^7, and x^5 **b.** y, y^4, and y^7

Solution

a. The GCF is x^3, since 3 is the smallest exponent to which x is raised.

b. The GCF is y^1 or y, since 1 is the smallest exponent on y.

□

PRACTICE 2 Find the GCF of each list of terms.

a. y^7, y^4, and y^6 **b.** x, x^4, and x^2

■

In general, the **greatest common factor (GCF) of a list of terms** is the product of the GCF of the numerical coefficients and the GCF of the variable factors.

$$20x^2y^2 = 2 \cdot 2 \cdot 5 \cdot x \cdot x \cdot y \cdot y$$
$$6xy^3 = 2 \cdot 3 \cdot x \cdot y \cdot y \cdot y$$
$$\text{GCF} = 2 \cdot x \cdot y \cdot y = 2xy^2$$

> **Helpful Hint**
>
> Remember that the GCF of a list of terms contains the smallest exponent on each common variable.
>
> The GCF of x^5y^6, x^2y^7, and x^3y^4 is x^2y^4. (Smallest exponent on x; Smallest exponent on y)

EXAMPLE 3 Find the GCF of each list of terms.

a. $6x^2$, $10x^3$, and $-8x$ **b.** $-18y^2$, $-63y^3$, and $27y^4$ **c.** a^3b^2, a^5b, and a^6b^2

Solution

a.
$$6x^2 = 2 \cdot 3 \cdot x^2$$
$$10x^3 = 2 \cdot 5 \cdot x^3$$
$$-8x = -1 \cdot 2 \cdot 2 \cdot 2 \cdot x^1$$
→ The GCF of x^2, x^3, and x^1 is x^1 or x.
$$\text{GCF} = 2 \cdot x^1 \quad \text{or} \quad 2x$$

b.
$$-18y^2 = -1 \cdot 2 \cdot 3 \cdot 3 \cdot y^2$$
$$-63y^3 = -1 \cdot 3 \cdot 3 \cdot 7 \cdot y^3$$
$$27y^4 = 3 \cdot 3 \cdot 3 \cdot y^4$$
→ The GCF of y^2, y^3, and y^4 is y^2.
$$\text{GCF} = 3 \cdot 3 \cdot y^2 \quad \text{or} \quad 9y^2$$

c. The GCF of a^3, a^5, and a^6 is a^3.
The GCF of b^2, b, and b^2 is b. Thus,
the GCF of a^3b^2, a^5b, and a^6b^2 is a^3b. □

PRACTICE 3 Find the GCF of each list of terms.

a. $5y^4$, $15y^2$, and $-20y^3$ **b.** $4x^2$, x^3, and $3x^8$ **c.** a^4b^2, a^3b^5, and a^2b^3 ■

OBJECTIVE

3 Factoring Out the Greatest Common Factor

The first step in factoring a polynomial is to find the GCF of its terms. Once we do so, we can write the polynomial as a product by **factoring out** the GCF.

The polynomial $8x + 14$, for example, contains two terms: $8x$ and 14. The GCF of these terms is 2. We factor out 2 from each term by writing each term as a product of 2 and the term's remaining factors.

$$8x + 14 = 2 \cdot 4x + 2 \cdot 7$$

Using the distributive property, we can write

$$8x + 14 = 2 \cdot 4x + 2 \cdot 7$$
$$= 2(4x + 7)$$

Thus, a factored form of $8x + 14$ is $2(4x + 7)$. We can check by multiplying:

$$2(4x + 7) = 2 \cdot 4x + 2 \cdot 7 = 8x + 14.$$

> **Helpful Hint**
>
> A factored form of $8x + 14$ is *not*
>
> $$2 \cdot 4x + 2 \cdot 7$$
>
> Although the *terms* have been factored (written as products), the *polynomial* $8x + 14$ has not been factored (written as a product).
>
> A factored form of $8x + 14$ is the *product* $2(4x + 7)$.

✔ **CONCEPT CHECK**

Which of the following is/are factored form(s) of $7t + 21$?

a. 7 b. $7 \cdot t + 7 \cdot 3$ c. $7(t + 3)$ d. $7(t + 21)$

EXAMPLE 4 Factor each polynomial by factoring out the GCF.

a. $6t + 18$ b. $y^5 - y^7$

Solution

a. The GCF of terms $6t$ and 18 is 6.

$$6t + 18 = 6 \cdot t + 6 \cdot 3$$
$$= 6(t + 3) \quad \text{Apply the distributive property.}$$

Our work can be checked by multiplying 6 and $(t + 3)$.

$$6(t + 3) = 6 \cdot t + 6 \cdot 3 = 6t + 18, \text{ the original polynomial.}$$

b. The GCF of y^5 and y^7 is y^5. Thus,

$$y^5 - y^7 = y^5(1) - y^5(y^2)$$
$$= y^5(1 - y^2)$$

Helpful Hint: Don't forget the 1.

□

PRACTICE 4 Factor each polynomial by factoring out the GCF.

a. $4t + 12$ b. $y^8 + y^4$

■

EXAMPLE 5 Factor: $-9a^5 + 18a^2 - 3a$

Solution

$$-9a^5 + 18a^2 - 3a = (3a)(-3a^4) + (3a)(6a) + (3a)(-1)$$
$$= 3a(-3a^4 + 6a - 1)$$

Helpful Hint: Don't forget the −1.

□

PRACTICE 5 Factor: $-8b^6 + 16b^4 - 8b^2$

■

In Example 5, we could have chosen to factor out a $-3a$ instead of $3a$. If we factor out a $-3a$, we have

$$-9a^5 + 18a^2 - 3a = (-3a)(3a^4) + (-3a)(-6a) + (-3a)(1)$$
$$= -3a(3a^4 - 6a + 1)$$

Helpful Hint: Notice the changes in signs when factoring out $-3a$.

EXAMPLES Factor.

6. $6a^4 - 12a = 6a(a^3 - 2)$

7. $\frac{3}{7}x^4 + \frac{1}{7}x^3 - \frac{5}{7}x^2 = \frac{1}{7}x^2(3x^2 + x - 5)$

8. $15p^2q^4 + 20p^3q^5 + 5p^3q^3 = 5p^2q^3(3q + 4pq^2 + p)$

□

PRACTICE 6–8 Factor.

6. $5x^4 - 20x$ 7. $\frac{5}{9}z^5 + \frac{1}{9}z^4 - \frac{2}{9}z^3$ 8. $8a^2b^4 - 20a^3b^3 + 12ab^3$

■

Answer to Concept Check: c

EXAMPLE 9 Factor: $5(x + 3) + y(x + 3)$

Solution The binomial $(x + 3)$ is the greatest common factor. Use the distributive property to factor out $(x + 3)$.

$$5(x + 3) + y(x + 3) = (x + 3)(5 + y)$$ □

PRACTICE 9 Factor: $8(y - 2) + x(y - 2)$ ■

EXAMPLE 10 Factor: $3m^2n(a + b) - (a + b)$

Solution The greatest common factor is $(a + b)$.

$$3m^2n(a + b) - 1(a + b) = (a + b)(3m^2n - 1)$$ □

PRACTICE 10 Factor: $7xy^3(p + q) - (p + q)$ ■

OBJECTIVE 4

Factoring by Grouping

Once the GCF is factored out, we can often continue to factor the polynomial, using a variety of techniques. We discuss here a technique for factoring polynomials called **factoring by grouping.**

EXAMPLE 11 Factor $xy + 2x + 3y + 6$ by grouping. Check by multiplying.

Solution The GCF of the first two terms is x, and the GCF of the last two terms is 3.

$$xy + 2x + 3y + 6 = (xy + 2x) + (3y + 6) \quad \text{Group terms.}$$
$$= x(y + 2) + 3(y + 2) \quad \text{Factor out GCF from each grouping.}$$

> **Helpful Hint**
> Notice that this form, $x(y + 2) + 3(y + 2)$, is *not* a factored form of the original polynomial. It is a sum, not a product.

Next we factor out the common binomial factor, $(y + 2)$.

$$x(y + 2) + 3(y + 2) = (y + 2)(x + 3)$$

Now the result is a factored form because it is a product. We were able to write the polynomial as a product because of the common binomial factor, $(y + 2)$, that appeared. If this does not happen, try rearranging the terms of the original polynomial.

Check: Multiply: $(y + 2)$ by $(x + 3)$

$$(y + 2)(x + 3) = xy + 2x + 3y + 6,$$

the original polynomial.
Thus, a factored form of $xy + 2x + 3y + 6$ is the product $(y + 2)(x + 3)$. □

PRACTICE 11 Factor $xy + 3y + 4x + 12$ by grouping. Check by multiplying. ■

You may want to try these steps when factoring by grouping.

To Factor a Four-Term Polynomial by Grouping

Step 1. Group the terms in two groups of two terms so that each group has a common factor.

Step 2. Factor out the GCF from each group.

Step 3. If there is now a common binomial factor in the groups, factor it out.

Step 4. If not, rearrange the terms and try these steps again.

EXAMPLES Factor by grouping.

12. $15x^3 - 10x^2 + 6x - 4$

$= (15x^3 - 10x^2) + (6x - 4)$ Group the terms.

$= 5x^2(3x - 2) + 2(3x - 2)$ Factor each group.

$= (3x - 2)(5x^2 + 2)$ Factor out the common factor, $(3x - 2)$.

13. $3x^2 + 4xy - 3x - 4y$

$= (3x^2 + 4xy) + (-3x - 4y)$

$= x(3x + 4y) - 1(3x + 4y)$ Factor each group. A -1 is factored from the second pair of terms so that there is a common factor, $(3x + 4y)$.

$= (3x + 4y)(x - 1)$ Factor out the common factor, $(3x + 4y)$.

14. $2a^2 + 5ab + 2a + 5b$

$= (2a^2 + 5ab) + (2a + 5b)$

$= a(2a + 5b) + 1(2a + 5b)$ Factor each group. An understood 1 is written before $(2a + 5b)$ to help remember that $(2a + 5b)$ is $1(2a + 5b)$.

$= (2a + 5b)(a + 1)$ Factor out the common factor, $(2a + 5b)$. □

Helpful Hint

Notice the factor of 1 is written when $(2a + 5b)$ is factored out.

PRACTICE 12–14

12. Factor $40x^3 - 24x^2 + 15x - 9$ by grouping.

13. Factor $2xy + 3y^2 - 2x - 3y$ by grouping.

14. Factor $7a^3 + 5a^2 + 7a + 5$ by grouping. ■

EXAMPLES Factor by grouping.

15. $3xy + 2 - 3x - 2y$

Notice that the first two terms have no common factor other than 1. However, if we rearrange these terms, a grouping emerges that does lead to a common factor.

$3xy + 2 - 3x - 2y$

$= (3xy - 3x) + (-2y + 2)$

$= 3x(y - 1) - 2(y - 1)$ Factor -2 from the second group so that there is a common factor, $(y - 1)$.

$= (y - 1)(3x - 2)$ Factor out the common factor, $(y - 1)$.

16. $5x - 10 + x^3 - x^2 = 5(x - 2) + x^2(x - 1)$

There is no common binomial factor that can now be factored out. No matter how we rearrange the terms, no grouping will lead to a common factor. Thus, this polynomial is not factorable by grouping. □

PRACTICE 15–16

15. Factor $4xy + 15 - 12x - 5y$ by grouping.

16. Factor $9y - 27 + y^3 - 4y^2$ by grouping. ■

Helpful Hint

One more reminder: When **factoring** a polynomial, make sure the polynomial is written as a **product.** For example, it is true that

$$3x^2 + 4xy - 3x - 4y = \underbrace{x(3x + 4y) - 1(3x + 4y)}_{\text{but is not a factored form}},$$

since it is a **sum (difference),** not a **product.** A factored form of $3x^2 + 4xy - 3x - 4y$ is the product $(3x + 4y)(x - 1)$.

Factoring out a greatest common factor first makes factoring by any method easier, as we see in the next example.

EXAMPLE 17 Factor: $4ax - 4ab - 2bx + 2b^2$

Solution First, factor out the common factor 2 from all four terms.

$4ax - 4ab - 2bx + 2b^2$

$= 2(2ax - 2ab - bx + b^2)$ Factor out 2 from all four terms.

$= 2[2a(x - b) - b(x - b)]$ Factor each pair of terms. A "$-b$" is factored from the second pair so that there is a common factor, $x - b$.

$= 2(x - b)(2a - b)$ Factor out the common binomial. □

Helpful Hint

Throughout this chapter, we will be factoring polynomials. Even when the instructions do not so state, it is always a good idea to check your answers by multiplying.

PRACTICE 17 Factor: $3xy - 3ay - 6ax + 6a^2$

Vocabulary, Readiness & Video Check

Use the choices below to fill in each blank. Some choices may be used more than once and some may not be used at all.

greatest common factor   factors   factoring   true   false   least   greatest

1. Since $5 \cdot 4 = 20$, the numbers 5 and 4 are called ________________ of 20.
2. The ________________ of a list of integers is the largest integer that is a factor of all the integers in the list.
3. The greatest common factor of a list of common variables raised to powers is the variable raised to the ________________ exponent in the list.
4. The process of writing a polynomial as a product is called ________________.
5. True or false: A factored form of $7x + 21 + xy + 3y$ is $7(x + 3) + y(x + 3)$. ________________
6. True or false: A factored form of $3x^3 + 6x + x^2 + 2$ is $3x(x^2 + 2)$. ________________

Martin-Gay Interactive Videos

See Video 5.1

Watch the section lecture video and answer the following questions.

OBJECTIVE 1
7. Based on Example 1, give a general definition for the greatest common factor (GCF) of a list of numbers.

OBJECTIVE 2
8. In Example 3, why are the numbers factored out, but not the variables?

OBJECTIVE 3
9. From Example 5, how can the number of terms in the other factor once you factor out the GCF help you determine if your factorization is correct?

OBJECTIVE 4
10. In Examples 7 and 8, what are you reminded to always do first when factoring a polynomial? Also, explain how a polynomial looks that suggests it might be factored by grouping.

5.1 Exercise Set MyMathLab®

Find the GCF for each list. See Examples 1 through 3.

1. 32, 36
2. 36, 90
3. 18, 42, 84
4. 30, 75, 135
5. 24, 14, 21
6. 15, 25, 27
7. y^2, y^4, y^7
8. x^3, x^2, x^5
9. z^7, z^9, z^{11}
10. y^8, y^{10}, y^{12}
11. $x^{10}y^2, xy^2, x^3y^3$
12. p^7q, p^8q^2, p^9q^3
13. $14x, 21$
14. $20y, 15$
15. $12y^4, 20y^3$
16. $32x^5, 18x^2$
17. $-10x^2, 15x^3$
18. $-21x^3, 14x$
19. $12x^3, -6x^4, 3x^5$
20. $15y^2, 5y^7, -20y^3$

21. $-18x^2y, 9x^3y^3, 36x^3y$ **22.** $7x^3y^3, -21x^2y^2, 14xy^4$

23. $20a^6b^2c^8, 50a^7b$ **24.** $40x^7y^2z, 64x^9y$

Factor out the GCF from each polynomial. See Examples 4 through 10.

25. $3a + 6$ **26.** $18a + 12$

27. $30x - 15$ **28.** $42x - 7$

29. $x^3 + 5x^2$ **30.** $y^5 + 6y^4$

31. $6y^4 + 2y^3$ **32.** $5x^2 + 10x^6$

33. $4x - 8y + 4$

34. $7x + 21y - 7$

35. $6x^3 - 9x^2 + 12x$

36. $12x^3 + 16x^2 - 8x$

37. $a^7b^6 - a^3b^2 + a^2b^5 - a^2b^2$

38. $x^9y^6 + x^3y^5 - x^4y^3 + x^3y^3$

39. $8x^5 + 16x^4 - 20x^3 + 12$

40. $9y^6 - 27y^4 + 18y^2 + 6$

41. $\frac{1}{3}x^4 + \frac{2}{3}x^3 - \frac{4}{3}x^5 + \frac{1}{3}x$

42. $\frac{2}{5}y^7 - \frac{4}{5}y^5 + \frac{3}{5}y^2 - \frac{2}{5}y$

43. $y(x^2 + 2) + 3(x^2 + 2)$

44. $x(y^2 + 1) - 3(y^2 + 1)$

45. $z(y + 4) - 3(y + 4)$

46. $8(x + 2) - y(x + 2)$

47. $r(z^2 - 6) + (z^2 - 6)$

48. $q(b^3 - 5) + (b^3 - 5)$

Factor a negative number or a GCF with a negative coefficient from each polynomial. See Example 5.

49. $-2x - 14$ **50.** $-7y - 21$

51. $-2x^5 + x^7$ **52.** $-5y^3 + y^6$

53. $-3a^4 + 9a^3 - 3a^2$

54. $-5m^6 + 10m^5 - 5m^3$

Factor each four-term polynomial by grouping. If this is not possible, write "not factorable by grouping." See Examples 11 through 17.

55. $x^3 + 2x^2 + 5x + 10$

56. $x^3 + 4x^2 + 3x + 12$

57. $5x + 15 + xy + 3y$

58. $xy + y + 2x + 2$

59. $6x^3 - 4x^2 + 15x - 10$

60. $16x^3 - 28x^2 + 12x - 21$

61. $5m^3 + 6mn + 5m^2 + 6n$

62. $8w^2 + 7wv + 8w + 7v$

63. $2y - 8 + xy - 4x$

64. $6x - 42 + xy - 7y$

65. $2x^3 - x^2 + 8x - 4$

66. $2x^3 - x^2 - 10x + 5$

67. $3x - 3 + x^3 - 4x^2$

68. $7x - 21 + x^3 - 2x^2$

69. $4x^2 - 8xy - 3x + 6y$

70. $5xy - 15x - 6y + 18$

71. $5q^2 - 4pq - 5q + 4p$

72. $6m^2 - 5mn - 6m + 5n$

73. $2x^4 + 5x^3 + 2x^2 + 5x$

74. $4y^4 + y^2 + 20y^3 + 5y$

75. $12x^2y - 42x^2 - 4y + 14$

76. $90 + 15y^2 - 18x - 3xy^2$

MIXED PRACTICE

Factor. See Examples 4 through 17.

77. $32xy - 18x^2$

78. $10xy - 15x^2$

79. $y(x + 2) - 3(x + 2)$

80. $z(y - 4) + 3(y - 4)$

81. $14x^3y + 7x^2y - 7xy$

82. $5x^3y - 15x^2y + 10xy$

83. $28x^3 - 7x^2 + 12x - 3$

84. $15x^3 + 5x^2 - 6x - 2$

85. $-40x^8y^6 - 16x^9y^5$

86. $-21x^3y - 49x^2y^2$

87. $6a^2 + 9ab^2 + 6ab + 9b^3$

88. $16x^2 + 4xy^2 + 8xy + 2y^3$

REVIEW AND PREVIEW

Multiply. See Section 4.3.

89. $(x + 2)(x + 5)$

90. $(y + 3)(y + 6)$

91. $(b + 1)(b - 4)$

92. $(x - 5)(x + 10)$

Fill in the chart by finding two numbers that have the given product and sum. The first column is filled in for you.

		93.	94.	95.	96.	97.	98.
Two Numbers	4, 7						
Their Product	28	12	20	8	16	−10	−24
Their Sum	11	8	9	−9	−10	3	−5

CONCEPT EXTENSIONS

See the Concept Checks in this section.

99. Which of the following is/are factored form(s) of $8a - 24$?

a. $8 \cdot a - 24$ **b.** $8(a - 3)$

c. $4(2a - 12)$ **d.** $8 \cdot a - 2 \cdot 12$

100. Which of the following is/are factored form(s) of $-2x + 14$?

a. $-2(x + 7)$ **b.** $-2 \cdot x + 14$

c. $-2(x - 14)$ **d.** $-2(x - 7)$

Determine whether the following expressions are factored.

101. $(a + 6)(a + 2)$

102. $(x + 5)(x + y)$

103. $5(2y + z) - b(2y + z)$

104. $3x(a + 2b) + 2(a + 2b)$

105. Construct a binomial whose greatest common factor is $5a^3$. (*Hint:* Multiply $5a^3$ by a binomial whose terms contain no common factor other than 1. $5a^3(\square + \square)$.)

106. Construct a trinomial whose greatest common factor is $2x^2$. See the hint for Exercise 105.

107. Explain how you can tell whether a polynomial is written in factored form.

108. Construct a four-term polynomial that can be factored by grouping.

109. The percent of total music industry revenues from streaming in the United States each year during 2007 through 2014 can be modeled by the polynomial $0.6x^2 - 0.6x + 3.6$, where x is the number of years since 2007. (*Source:* Recording Industry Association of America)

a. Find the percent of music industry revenue derived from streaming in 2013. To do so, let $x = 6$ and evaluate $0.6x^2 - 0.6x + 3.6$. Round to the nearest percent.

b. Use this expression to predict the percent revenue derived from streaming in 2018. Round to the nearest percent.

c. Factor the polynomial $0.6x^2 - 0.6x + 3.6$ by factoring 0.6 from each term.

110. The number (in thousands) of students who graduated from U.S. high schools, both public and private, each year during 2000 through 2013 can be modeled by $-3x^2 + 78x + 2904$, where x is the number of years since 2000. (*Source:* National Center for Educational Statistics)

a. Find the number of students who graduated from U.S. high schools in 2010. To do so, let $x = 10$ and evaluate $-3x^2 + 78x + 2904$.

b. Use this expression to predict the number of students who will graduate from U.S. high schools in 2018.

c. Factor the polynomial $-3x^2 + 78x + 2904$ by factoring -3 from each term.

d. For the year 2010, the National Center for Higher Education determined that 62.5% of U.S. high school graduates went on to higher education. Using your answer from part **a**, determine how many of those graduating in 2010 pursued higher education.

Write an expression for the area of each shaded region. Then write the expression as a factored polynomial.

△ **111.**

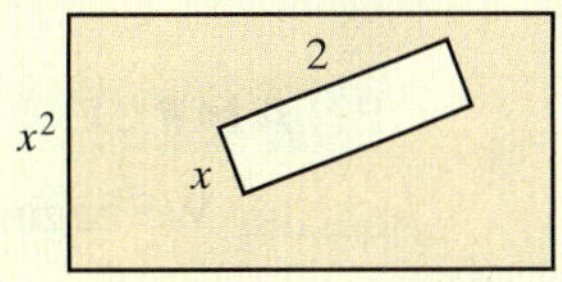

112.

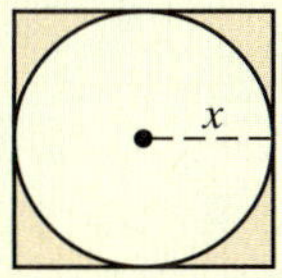

*Write an expression for the length of each rectangle. (**Hint:** Factor the area binomial and recall that Area = width · length.)*

△ **113.**

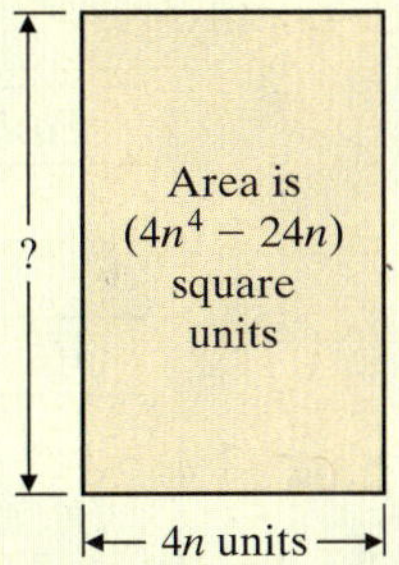

△ **114.**

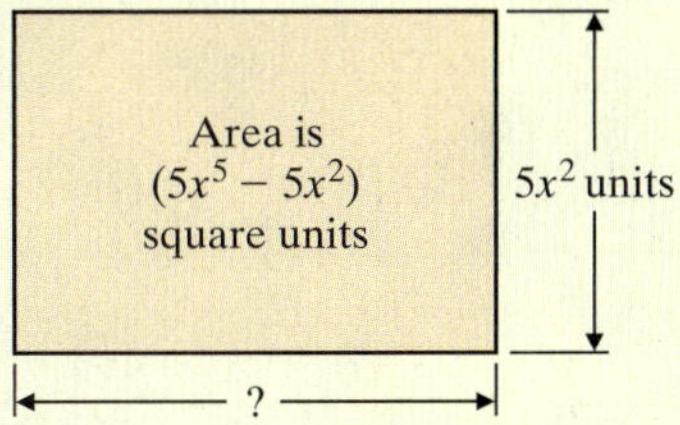

Factor each polynomial by grouping.

115. $x^{2n} + 2x^n + 3x^n + 6$
(*Hint:* Don't forget that $x^{2n} = x^n \cdot x^n$.)

116. $x^{2n} + 6x^n + 10x^n + 60$

117. $3x^{2n} + 21x^n - 5x^n - 35$

118. $12x^{2n} - 10x^n - 30x^n + 25$

5.2 Factoring Trinomials

OBJECTIVES

1 Factor Trinomials of the Form $x^2 + bx + c$.

2 Factor Trinomials of the Form $ax^2 + bx + c$.

a. Method 1—Trial and Check

b. Method 2—Grouping

OBJECTIVE 1

Factoring Trinomials of the Form $x^2 + bx + c$

In the previous section, we used factoring by grouping to factor four-term polynomials. In this section, we present techniques for factoring trinomials. Since $(x - 2)(x + 5) = x^2 + 3x - 10$, we say that $(x - 2)(x + 5)$ is a factored form of $x^2 + 3x - 10$. Taking a close look at how $(x - 2)$ and $(x + 5)$ are multiplied suggests a pattern for factoring trinomials of the form

$$x^2 + bx + c$$

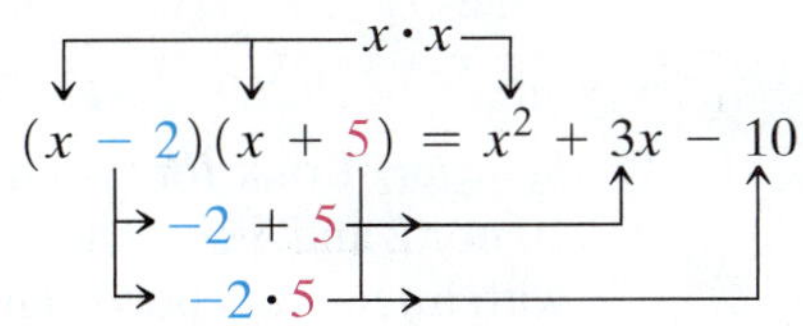

The pattern for factoring is summarized next.

Factoring a Trinomial of the Form $x^2 + bx + c$

The factored form of $x^2 + bx + c$ is

The product of these numbers is c.

$$x^2 + bx + c = (x + \square)(x + \square)$$

The sum of these numbers is b.

EXAMPLE 1 Factor: $x^2 + 7x + 12$

Solution We begin by writing the first terms of the binomial factors.

$$(x + \square)(x + \square)$$

Next we look for two numbers whose product is 12 and whose sum is 7. Since our numbers must have a positive product and a positive sum, we look at pairs of positive factors of 12 only.

Positive Factors of 12	*Sum of Factors*
1, 12	13
2, 6	8
3, 4	7

Correct sum, so the numbers are 3 and 4.

Thus, $x^2 + 7x + 12 = (x + 3)(x + 4)$

Check: $(x + 3)(x + 4) = x^2 + 4x + 3x + 12 = x^2 + 7x + 12.$ □

PRACTICE 1 Factor: $x^2 + 5x + 6$ ■

EXAMPLE 2 Factor: $x^2 + 4x - 12$

Solution $x^2 + 4x - 12 = (x + \square)(x + \square)$

We look for two numbers whose product is -12 and whose sum is 4. Since our numbers must have a negative product, we look at pairs of factors with opposite signs.

Factors of −12	*Sum of Factors*
−1, 12	11
1, −12	−11
−2, 6	4
2, −6	−4
−3, 4	1
3, −4	−1

Correct sum, so the numbers are −2 and 6.

Thus, $x^2 + 4x - 12 = (x - 2)(x + 6)$ □

PRACTICE 2 Factor: $x^2 + 5x - 14$ ■

EXAMPLE 3 Factor: $a^2 + 2a + 10$

Solution Look for two numbers whose product is 10 and whose sum is 2. Neither 1 and 10 nor 2 and 5 give the required sum, 2. We conclude that $a^2 + 2a + 10$ is not factorable with integers. A polynomial such as $a^2 + 2a + 10$ is called a **prime polynomial.** □

PRACTICE 3 Factor: $b^2 + 5b + 1$ ■

EXAMPLE 4 Factor: $x^4 + 5x^2 + 6$

Solution As usual, we begin by writing the first terms of the binomials. Since the greatest power of x in this polynomial is x^4, we write

$$(x^2 + \square)(x^2 + \square) \quad \text{since } x^2 \cdot x^2 = x^4$$

Now we look for two factors of 6 whose sum is 5. The numbers are 2 and 3. Thus,

$$x^4 + 5x^2 + 6 = (x^2 + 2)(x^2 + 3)$$ □

PRACTICE 4 Factor: $x^4 + 13x^2 + 12$ ■

If the terms of a polynomial are not written in descending powers of the variable, you may want to do so before factoring.

EXAMPLE 5 Factor: $40 - 13t + t^2$

Solution First, we rearrange terms so that the trinomial is written in descending powers of t.

$$40 - 13t + t^2 = t^2 - 13t + 40$$

Next, try to factor.

$$t^2 - 13t + 40 = (t + \square)(t + \square)$$

Now we look for two factors of 40 whose sum is -13. The numbers are -8 and -5. Thus,

$$t^2 - 13t + 40 = (t - 8)(t - 5)$$ □

PRACTICE 5 Factor: $48 - 14x + x^2$ ■

The following sign patterns may be useful when factoring trinomials.

Helpful Hint

A positive constant in a trinomial tells us to look for two numbers with the same sign. The sign of the coefficient of the middle term tells us whether the signs are both positive or both negative.

both positive ↓ same sign ↙

$$x^2 + 10x + 16 = (x + 2)(x + 8)$$

both negative ↓ same sign ↙

$$x^2 - 10x + 16 = (x - 2)(x - 8)$$

A negative constant in a trinomial tells us to look for two numbers with opposite signs.

opposite signs ↓

$$x^2 + 6x - 16 = (x + 8)(x - 2)$$

opposite signs ↓

$$x^2 - 6x - 16 = (x - 8)(x + 2)$$

Remember that the first step in factoring any polynomial is to factor out the greatest common factor (if there is one other than 1 or -1).

EXAMPLE 6 Factor: $2x^4 - 26x^3 + 84x^2$

Solution First we factor out the greatest common factor, $2x^2$, from each term.

$$2x^4 - 26x^3 + 84x^2 = 2x^2(x^2 - 13x + 42) \quad \text{Factor out common factor, } 2x^2.$$
$$= 2x^2(x - 6)(x - 7) \quad \text{Factor } x^2 - 13x + 42.$$ □

Helpful Hint

Remember to write the common factor $2x^2$ as part of the factored form.

PRACTICE 6 Factor: $3y^4 - 18y^3 - 21y^2$

OBJECTIVE

2 Factoring Trinomials of the Form $ax^2 + bx + c$

Next, we factor trinomials of the form $ax^2 + bx + c$, where the coefficient a of x^2 is not 1. Don't forget that the first step in factoring any polynomial is to factor out the greatest common factor of its terms. We will review two methods here. The first method we'll call trial and check.

OBJECTIVE

2a Method 1—Factoring $ax^2 + bx + c$ by Trial and Check

EXAMPLE 7 Factor: $3x^2 + 11x + 6$

Solution Since all numerical coefficients are positive, we use factors with positive numerical coefficients. We first find factors of $3x^2$.

$$\text{Factors of } 3x^2\text{:}\quad 3x^2 = 3x \cdot x$$

If factorable, the binomials will be of the form

$$3x^2 + 11x + 6 = (3x + \square)(x + \square)$$

Next we factor 6.

$$\text{Factors of 6:}\quad 6 = 1 \cdot 6,\quad 6 = 2 \cdot 3$$

Now we try combinations of factors of 6 until a middle term of $11x$ is obtained. Let's try 1 and 6 first.

$$(3x + 1)(x + 6) = 3x^2 + 19x + 6$$

$1x$
$+18x$
$19x$ ⟶ **Incorrect** middle term

Now let's next try 6 and 1.

$$(3x + 6)(x + 1)$$

Before multiplying, notice that the terms of the factor $3x + 6$ have a common factor of 3. The terms of the original trinomial $3x^2 + 11x + 6$ have no common factor other than 1, so the terms of its factors will also contain no common factor other than 1. This means that $(3x + 6)(x + 1)$ is not a factored form.

Next let's try 2 and 3 as last terms.

$$(3x + 2)(x + 3) = 3x^2 + 11x + 6$$

$2x$
$+9x$
$11x$ ⟶ **Correct** middle term

Thus a factored form of $3x^2 + 11x + 6$ is $(3x + 2)(x + 3)$. □

PRACTICE 7 Factor: $2x^2 + 11x + 15$

Below are the steps we are using to factor trinomials of the form $ax^2 + bx + c$.

Factoring a Trinomial of the Form $ax^2 + bx + c$

Step 1. Write all pairs of factors of ax^2.

Step 2. Write all pairs of factors of c, the constant term.

Step 3. Try various combinations of these factors until the correct middle term bx is found.

Step 4. If no combination exists, the polynomial is **prime.**

Helpful Hint

If the terms of a trinomial have no common factor (other than 1), then the terms of neither of its binomial factors will contain a common factor (other than 1).

✔ CONCEPT CHECK

Do the terms of $3x^2 + 29x + 18$ have a common factor? Without multiplying, decide which of the following factored forms could not be a factored form of $3x^2 + 29x + 18$.

a. $(3x + 18)(x + 1)$ **b.** $(3x + 2)(x + 9)$ **c.** $(3x + 6)(x + 3)$ **d.** $(3x + 9)(x + 2)$

EXAMPLE 8 Factor: $8x^2 - 22x + 5$

Solution Factors of $8x^2$: $8x^2 = 8x \cdot x$, $8x^2 = 4x \cdot 2x$
We'll try $8x$ and x.

$$8x^2 - 22x + 5 = (8x + \square)(x + \square)$$

Since the middle term, $-22x$, has a negative numerical coefficient, we factor 5 into negative factors.

Factors of 5: $5 = -1 \cdot -5$

Let's try -1 and -5.

$$(8x - 1)(x - 5) = 8x^2 - 41x + 5$$

$-1x$
$+(-40x)$
$-41x$ → Incorrect middle term

Now let's try -5 and -1.

$$(8x - 5)(x - 1) = 8x^2 - 13x + 5$$

$-5x$
$+(-8x)$
$-13x$ → Incorrect middle term

Don't give up yet! We can still try other factors of $8x^2$. Let's try $4x$ and $2x$ with -1 and -5.

$$(4x - 1)(2x - 5) = 8x^2 - 22x + 5$$

$-2x$
$+(-20x)$
$-22x$ → Correct middle term

A factored form of $8x^2 - 22x + 5$ is $(4x - 1)(2x - 5)$. □

Answers to Concept Check:
no; a, c, d

PRACTICE 8 Factor: $15x^2 - 22x + 8$ ■

EXAMPLE 9 Factor: $10x^2 - 13xy - 3y^2$

Solution Factors of $10x^2$: $10x^2 = 10x \cdot x$, $10x^2 = 2x \cdot 5x$

Factors of $-3y^2$: $-3y^2 = -3y \cdot y$, $-3y^2 = 3y \cdot -y$

We try some combinations of these factors:

Correct ↓ Correct ↓

$$(10x - 3y)(x + y) = 10x^2 + 7xy - 3y^2$$
$$(x + 3y)(10x - y) = 10x^2 + 29xy - 3y^2$$
$$(5x + 3y)(2x - y) = 10x^2 + xy - 3y^2$$
$$(2x - 3y)(5x + y) = 10x^2 - 13xy - 3y^2 \quad \text{Correct middle term}$$

A factored form of $10x^2 - 13xy - 3y^2$ is $(2x - 3y)(5x + y)$. □

PRACTICE 9 Factor: $21x^2 + 11xy - 2y^2$ ■

Helpful Hint

If a factoring attempt gives you a middle term whose numerical coefficient is the opposite of the desired numerical coefficient, try switching the signs of the last terms in the binomials.

Switched signs

$$(3x^2 + 8)(x^2 - 1) = 3x^4 + 5x^2 - 8 \quad \text{Middle term: } +5x^2$$
$$(3x^2 - 8)(x^2 + 1) = 3x^4 - 5x^2 - 8 \quad \text{Middle term: } -5x^2$$

Don't forget that the first step in factoring any polynomial is to look for a greatest common factor to factor out.

EXAMPLE 10 Factor: $24x^4 + 40x^3 + 6x^2$

Solution Notice that all three terms have a greatest common factor of $2x^2$. Thus we factor out $2x^2$ from all three terms first.

$$24x^4 + 40x^3 + 6x^2 = 2x^2(12x^2 + 20x + 3)$$

Next we factor $12x^2 + 20x + 3$.

Factors of $12x^2$: $12x^2 = 4x \cdot 3x$, $12x^2 = 12x \cdot x$, $12x^2 = 6x \cdot 2x$

Since all terms in the trinomial have positive numerical coefficients, we factor 3 using positive factors only.

Factors of 3: $3 = 1 \cdot 3$

We try some combinations of the factors.

$$2x^2(4x + 3)(3x + 1) = 2x^2(12x^2 + 13x + 3)$$
$$2x^2(12x + 1)(x + 3) = 2x^2(12x^2 + 37x + 3)$$
$$2x^2(2x + 3)(6x + 1) = 2x^2(12x^2 + 20x + 3) \quad \text{Correct middle term}$$

A factored form of $24x^4 + 40x^3 + 6x^2$ is $2x^2(2x + 3)(6x + 1)$. □

Helpful Hint

Don't forget to include the greatest common factor in the factored form.

PRACTICE 10 Factor: $3x^3 + 17x^2 + 10x$ ■

When the term containing the squared variable has a negative coefficient, you may want to first factor out a common factor of -1.

EXAMPLE 11 Factor: $-6x^2 - 13x + 5$

Solution We begin by factoring out a common factor of -1.

$$-6x^2 - 13x + 5 = -1(6x^2 + 13x - 5) \quad \text{Factor out } -1.$$
$$= -1(3x - 1)(2x + 5) \quad \text{Factor } 6x^2 + 13x - 5.$$

□

PRACTICE 11 Factor: $-8x^2 + 2x + 3$

Helpful Hint

If a trinomial has no common factor (other than 1), then none of its binomial factors will contain a common factor (other than 1).

EXAMPLE 12 Factor: $16x^2 + 24xy + 9y^2$

Solution No greatest common factor can be factored out of this trinomial.

Factors of $16x^2$: $16x \cdot x$, $8x \cdot 2x$, $4x \cdot 4x$

Factors of $9y^2$: $y \cdot 9y$, $3y \cdot 3y$

We try possible combinations until the correct factorization is found.

$$16x^2 + 24xy + 9y^2 = (4x + 3y)(4x + 3y) \quad \text{or} \quad (4x + 3y)^2$$

□

PRACTICE 12 Factor: $25x^2 + 20xy + 4y^2$

Note: The trinomial $16x^2 + 24xy + 9y^2$ in Example 12 is an example of a **perfect square trinomial** since its factors are two identical binomials.

OBJECTIVE

2b Method 2—Factoring $ax^2 + bx + c$ by Grouping

There is another method we can use when factoring trinomials of the form $ax^2 + bx + c$: Write the trinomial as a four-term polynomial and then factor by grouping.

To Factor Trinomials by Grouping

Step 1. Factor out the greatest common factor if there is one other than 1.

Step 2. For the resulting trinomial $ax^2 + bx + c$, find two numbers whose product is $a \cdot c$ and whose sum is b.

Step 3. Write the middle term, bx, using the factors found in Step 2.

Step 4. Factor by grouping.

EXAMPLE 13 Factor $3x^2 + 31x + 10$ by grouping.

Solution

Step 1. The terms of this trinomial contain no greatest common factor other than 1 (or -1).

Step 2. In $3x^2 + 31x + 10$, $a = 3$, $b = 31$, and $c = 10$.

(Continued on next page)

Let's find two numbers whose product is $a \cdot c$ or $3(10) = 30$ and whose sum is b or 31. The numbers are 1 and 30, as shown in the table below.

Factors of 30	*Sum of Factors*	
5, 6	11	
3, 10	13	
2, 15	17	
1, 30	31	Correct sum

Step 3. Write $31x$ as $1x + 30x$ so that $3x^2 + 31x + 10 = 3x^2 + 1x + 30x + 10$.

Step 4. Factor by grouping.

$$3x^2 + 1x + 30x + 10 = x(3x + 1) + 10(3x + 1)$$
$$= (3x + 1)(x + 10)$$
□

PRACTICE 13 Factor $5x^2 + 61x + 12$ by grouping. ■

EXAMPLE 14 Factor $8x^2 - 14x + 5$ by grouping.

Solution

Step 1. The terms of this trinomial contain no greatest common factor other than 1.

Step 2. This trinomial is of the form $ax^2 + bx + c$ with $a = 8$, $b = -14$, and $c = 5$. Find two numbers whose product is $a \cdot c$ or $8 \cdot 5 = 40$, and whose sum is b or -14.
The numbers are -4 and -10, as shown in the table below.

Factors of 40	*Sum of Factors*	
−40, −1	−41	
−20, −2	−22	
−10, −4	−14	Correct sum

Step 3. Write $-14x$ as $-4x - 10x$ so that

$$8x^2 - 14x + 5 = 8x^2 - 4x - 10x + 5$$

Step 4. Factor by grouping.

$$8x^2 - 4x - 10x + 5 = 4x(2x - 1) - 5(2x - 1)$$
$$= (2x - 1)(4x - 5)$$
□

PRACTICE 14 Factor $12x^2 - 19x + 5$ by grouping. ■

EXAMPLE 15 Factor $6x^2 - 2x - 20$ by grouping.

Solution

Step 1. First factor out the greatest common factor, 2.

$$6x^2 - 2x - 20 = 2(3x^2 - x - 10)$$

Step 2. Next, notice that $a = 3$, $b = -1$, and $c = -10$ in the resulting trinomial. Find two numbers whose product is $a \cdot c$ or $3(-10) = -30$ and whose sum is b, -1. The numbers are -6 and 5.

Step 3. $3x^2 - x - 10 = 3x^2 - 6x + 5x - 10$

Step 4. $3x^2 - 6x + 5x - 10 = 3x(x - 2) + 5(x - 2)$
$$= (x - 2)(3x + 5)$$

The factored form of $6x^2 - 2x - 20 = 2(x - 2)(3x + 5)$.

↑ Don't forget to include the GCF, 2. □

PRACTICE 15 Factor $30x^2 - 14x - 4$ by grouping.

EXAMPLE 16 Factor $4x^2 + 20x + 25$ by grouping.

Solution

Step 1. The terms of this trinomial contain no greatest common factor other than 1 (or -1).

Step 2. In $4x^2 + 20x + 25$, $a = 4$, $b = 20$, and $c = 25$. Find two numbers whose product is $a \cdot c$ or $4 \cdot 25 = 100$ and whose sum is 20. The numbers are 10 and 10.

Step 3. Write $20x$ as $10x + 10x$ so that

$$4x^2 + 20x + 25 = 4x^2 + 10x + 10x + 25$$

Step 4. Factor by grouping.

$$\begin{aligned} 4x^2 + 10x + 10x + 25 &= 2x(2x + 5) + 5(2x + 5) \\ &= (2x + 5)(2x + 5) \end{aligned}$$

The factored form of $4x^2 + 20x + 25$ is $(2x + 5)(2x + 5)$ or $(2x + 5)^2$. □

PRACTICE 16 Factor $16x^2 + 24x + 9$ by grouping.

Remember, a trinomial that is the square of a binomial, such as the trinomials in Examples 12 and 16, are called **perfect square trinomials.**

Vocabulary, Readiness & Video Check

For each trinomial $ax^2 + bx + c$, choose two numbers whose product is $a \cdot c$ and whose sum is b.

1. $x^2 + 6x + 8$
a. 4, 2 **b.** 7, 1 **c.** 6, 2 **d.** 6, 8

2. $x^2 + 11x + 24$
a. 6, 4 **b.** 24, 1 **c.** 8, 3 **d.** 2, 12

3. $2x^2 + 13x + 6$
a. 2, 6 **b.** 12, 1 **c.** 13, 1 **d.** 3, 4

4. $4x^2 + 8x + 3$
a. 4, 3 **b.** 4, 4 **c.** 12, 1 **d.** 2, 6

Martin-Gay Interactive Videos

See Video 5.2

Watch the section lecture video and answer the following questions.

OBJECTIVE 1

5. In Example 2, we know one of the factors of -24 is negative. How do we determine which one?

6. In Example 3, once the GCF is factored out how do we know both factors of $+8$ are negative?

OBJECTIVE 2a

7. From Example 4, explain in your own words how to factor a trinomial with a first-term coefficient $\neq 1$ by trial and check.

OBJECTIVE 2b

8. From Example 6 and the lecture before, why does writing the middle term of your trinomial as the sum of two terms suggest we'd need to factor by grouping?

5.2 Exercise Set MyMathLab®

Factor each trinomial. See Examples 1 through 6.

1. $x^2 + 9x + 18$
2. $x^2 + 9x + 20$
3. $x^2 - 12x + 32$
4. $x^2 - 12x + 27$
5. $x^2 + 10x - 24$
6. $x^2 + 3x - 54$
7. $x^2 - 2x - 24$
8. $x^2 - 9x - 36$
9. $3x^2 - 18x + 24$
10. $5x^2 - 45x + 70$
11. $4x^2z + 28xz + 40z$
12. $x^2y^2 + 4xy^2 + 3y^2$
13. $2x^2 - 24x - 64$
14. $3n^2 - 6n - 51$

Factor each trinomial. See Examples 7 through 16.

15. $5x^2 + 16x + 3$
16. $3x^2 + 8x + 4$
17. $2x^2 - 11x + 12$
18. $3x^2 - 19x + 20$
19. $2x^2 + 25x - 20$
20. $6x^2 + 13x + 8$
21. $4x^2 - 12x + 9$
22. $25x^2 - 30x + 9$
23. $12x^2 + 10x - 50$
24. $12y^2 - 48y + 45$
25. $3y^4 - y^3 - 10y^2$
26. $2x^2z + 5xz - 12z$
27. $6x^3 + 8x^2 + 24x$
28. $18y^3 + 6y^2 + 2y$
29. $2x^2 - 5xy - 3y^2$
30. $6x^2 + 11xy + 4y^2$
31. $28y^2 + 22y + 4$
32. $24y^3 - 2y^2 - y$
33. $2x^2 + 15x - 27$
34. $3x^2 + 14x + 15$
35. $x^4 + x^2 - 6$
36. $x^4 - x^2 - 20$

Factor each polynomial by grouping. See Examples 13 through 16.

37. $5x^4 - 3x^2 + 25x^2 - 15$
38. $2y^4 - 10y^2 + 7y^2 - 35$

Factor each trinomial by grouping. Exercises 39–42 are broken into parts to help you get started. See Examples 13 through 16.

39. $6x^2 + 11x + 3$
    a. Find two numbers whose product is $6 \cdot 3 = 18$ and whose sum is 11.
    b. Write $11x$ using the factors from part **a**.
    c. Factor by grouping.
40. $8x^2 + 14x + 3$
    a. Find two numbers whose product is $8 \cdot 3 = 24$ and whose sum is 14.
    b. Write $14x$ using the factors from part **a**.
    c. Factor by grouping.
41. $15x^2 - 23x + 4$
    a. Find two numbers whose product is $15 \cdot 4 = 60$ and whose sum is -23.
    b. Write $-23x$ using the factors from part **a**.
    c. Factor by grouping.
42. $6x^2 - 13x + 5$
    a. Find two numbers whose product is $6 \cdot 5 = 30$ and whose sum is -13.
    b. Write $-13x$ using the factors from part **a**.
    c. Factor by grouping.

MIXED PRACTICE

Factor each polynomial completely. See Examples 1 through 16.

43. $x^2 - 24x - 81$
44. $x^2 - 48x - 100$
45. $x^2 - 15x - 54$
46. $x^2 - 15x + 54$
47. $3x^2 - 6x + 3$
48. $8x^2 - 8x + 2$
49. $3x^2 - 5x - 2$
50. $5x^2 - 14x - 3$
51. $8x^2 - 26x + 15$
52. $12x^2 - 17x + 6$
53. $18x^4 + 21x^3 + 6x^2$
54. $20x^5 + 54x^4 + 10x^3$
55. $x^2 + 8xz + 7z^2$
56. $a^2 - 2ab - 15b^2$
57. $x^2 - x - 12$
58. $x^2 + 4x - 5$
59. $3a^2 + 12ab + 12b^2$
60. $2x^2 + 16xy + 32y^2$
61. $x^2 + 4x + 5$
62. $x^2 + 5x + 8$
63. $2x^2 + 19x + 39$
64. $3x^2 + 20x + 28$
65. $6x^2 - 49x + 30$
66. $4x^2 - 39x + 27$
67. $x^4 - 5x^2 - 6$
68. $x^4 - 5x^2 + 6$
69. $6x^3 - x^2 - x$
70. $12x^3 + x^2 - x$
71. $12a^2 - 29ab + 15b^2$
72. $16y^2 + 6yx - 27x^2$
73. $9x^2 + 30x + 25$
74. $4x^2 + 12x + 9$
75. $3x^2y - 11xy + 8y$
76. $5xy^2 - 9xy + 4x$
77. $2x^2 + 2x - 12$
78. $3x^2 + 6x - 45$
79. $x^2 - 5x - 14$
80. $x^2 - 8x + 7$
81. $2x^6 + 3x^3 - 9$
82. $3x^6 - 14x^3 + 8$
83. $72xy^4 - 24xy^2z + 2xz^2$
84. $36xy^2 - 48xyz^2 + 16xz^4$

85. $2x^3y + 2x^2y - 12xy$

86. $3x^2y^3 + 6x^2y^2 - 45x^2y$

87. $x^2 + 6xy + 5y^2$

88. $x^2 + 6xy + 8y^2$

REVIEW AND PREVIEW

The following graph shows the percent of online adults who participate in various social media sites. See Section R.4.

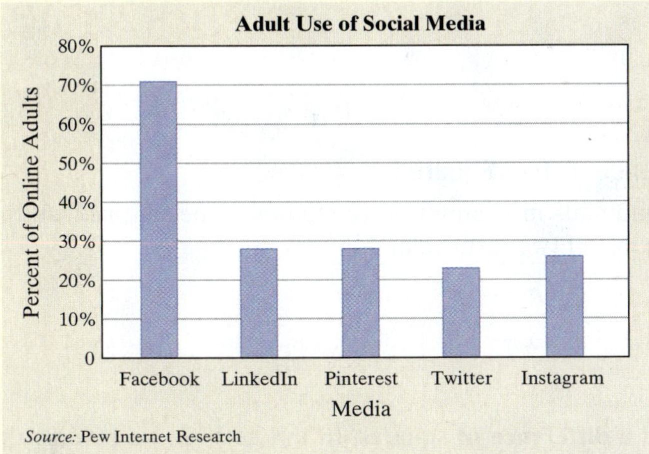

89. Which social medium has the highest percent of online adult users?

90. Which social medium has the lowest percent of online adult users?

91. Describe any trend you see.

92. Why don't the percents shown in the graph add up to 100%?

CONCEPT EXTENSIONS

See the Concept Check in this section.

93. Do the terms of $4x^2 + 19x + 12$ have a common factor (other than 1)?

94. Without multiplying, decide which of the following factored forms is not a factored form of $4x^2 + 19x + 12$.

a. $(2x + 4)(2x + 3)$

b. $(4x + 4)(x + 3)$

c. $(4x + 3)(x + 4)$

d. $(2x + 2)(2x + 6)$

95. Describe a perfect square trinomial.

96. Write the perfect square trinomial that factors as $(x + 3y)^2$.

Write the perimeter of each figure as a simplified polynomial. Then factor the polynomial.

97.

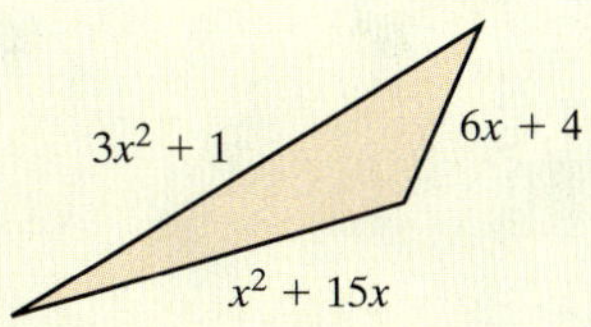

98.

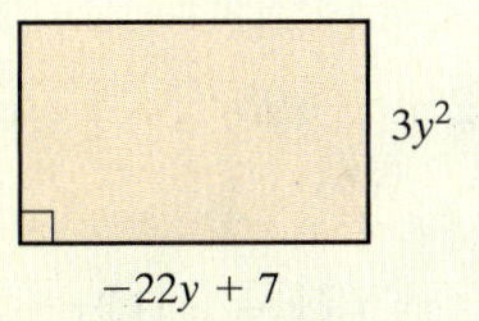

99. Find all positive and negative integers b such that $x^2 + bx + 6$ is factorable.

100. Find all positive and negative integers b such that $x^2 + bx - 10$ is factorable.

△ **101.** The volume $V(x)$ of a box in terms of its height x is given by the function $V(x) = x^3 + 2x^2 - 8x$. Factor this expression for $V(x)$.

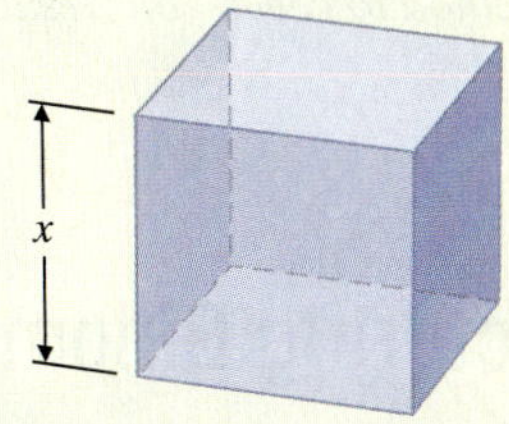

△ **102.** Based on your results from Exercise 101, find the length and width of the box if the height is 5 inches and the dimensions of the box are whole numbers.

103. Suppose that a movie is being filmed in New York City. An action shot requires an object to be thrown upward with an initial velocity of 80 feet per second off the top of 1 Madison Square Plaza, a height of 576 feet. The height $h(t)$ in feet of the object after t seconds is given by the function $h(t) = -16t^2 + 80t + 576$. (*Source: The World Almanac*)

a. Find the height of the object at $t = 0$ seconds, $t = 2$ seconds, $t = 4$ seconds, and $t = 6$ seconds.

b. Explain why the height of the object increases and then decreases as time passes.

c. Factor the polynomial $-16t^2 + 80t + 576$.

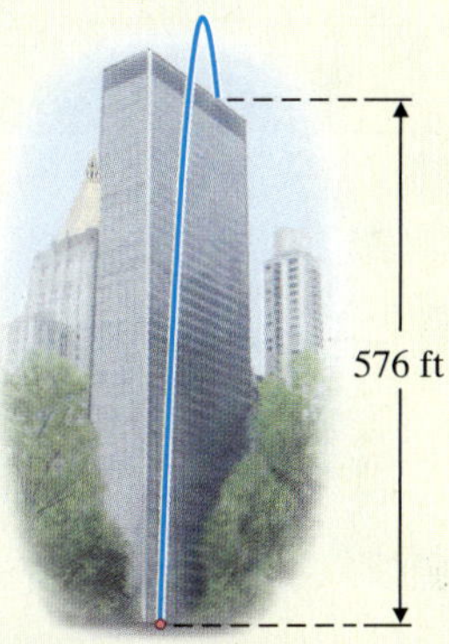

104. Suppose that an object is thrown upward with an initial velocity of 64 feet per second off the edge of a 960-foot cliff. The height $h(t)$ in feet of the object after t seconds is given by the function

$$h(t) = -16t^2 + 64t + 960$$

a. Find the height of the object at $t = 0$ seconds, $t = 3$ seconds, $t = 6$ seconds, and $t = 9$ seconds.

b. Explain why the height of the object increases and then decreases as time passes.

c. Factor the polynomial $-16t^2 + 64t + 960$.

Factor. Assume that variables used as exponents represent positive integers.

105. $x^{2n} + 10x^n + 16$

106. $x^{2n} - 7x^n + 12$

107. $x^{2n} - 3x^n - 18$

108. $x^{2n} + 7x^n - 18$

A graphing calculator may be used to check factoring of polynomials in one variable. For example, to see that

$$2x^3 - 9x^2 - 5x = x(2x + 1)(x - 5)$$

graph $Y_1 = 2x^3 - 9x^2 - 5x$ *and* $Y_2 = x(2x + 1)(x - 5)$*. Then trace along both graphs to see that they coincide. Factor the following and use this method to check your results.*

109. $x^4 + 6x^3 + 5x^2$

110. $x^3 + 6x^2 + 8x$

111. $30x^3 + 9x^2 - 3x$

112. $-6x^4 + 10x^3 - 4x^2$

5.3 Factoring Binomials

OBJECTIVES

1. Factor the Difference of Two Squares.
2. Factor the Sum or Difference of Two Cubes.

OBJECTIVE

1 Factoring the Difference of Two Squares

When learning to multiply binomials in Chapter 4, we studied a special product, the product of the sum and difference of two terms, *a* and *b*:

$$(a + b)(a - b) = a^2 - b^2$$

For example, the product of $x + 3$ and $x - 3$ is

$$(x + 3)(x - 3) = x^2 - 9$$

The binomial $x^2 - 9$ is called a **difference of squares**. In this section, we reverse the pattern for the product of a sum and difference to factor the binomial difference of squares.

Factoring the Difference of Two Squares

$$a^2 - b^2 = (a + b)(a - b)$$

Helpful Hint

Since multiplication is commutative, remember that the order of factors does not matter. In other words,

$$a^2 - b^2 = (a + b)(a - b) \text{ or } (a - b)(a + b)$$

EXAMPLE 1 Factor: $x^2 - 25$

Solution $x^2 - 25$ is the difference of two squares since $x^2 - 25 = x^2 - 5^2$. Therefore,

$$x^2 - 25 = x^2 - 5^2 = (x + 5)(x - 5)$$

Multiply to check. □

PRACTICE

1 Factor: $x^2 - 81$ ■

EXAMPLE 2 Factor each difference of squares.

a. $4x^2 - 1$ **b.** $25a^2 - 9b^2$ **c.** $y^2 - \frac{4}{9}$

Solution

a. $4x^2 - 1 = (2x)^2 - 1^2 = (2x + 1)(2x - 1)$

b. $25a^2 - 9b^2 = (5a)^2 - (3b)^2 = (5a + 3b)(5a - 3b)$

c. $y^2 - \frac{4}{9} = y^2 - \left(\frac{2}{3}\right)^2 = \left(y + \frac{2}{3}\right)\left(y - \frac{2}{3}\right)$ □

PRACTICE 2 Factor each difference of squares.

a. $9x^2 - 1$ **b.** $36a^2 - 49b^2$ **c.** $p^2 - \frac{25}{36}$

EXAMPLE 3 Factor: $x^4 - y^6$

Solution This is a difference of squares since $x^4 = (x^2)^2$ and $y^6 = (y^3)^2$. Thus,

$$x^4 - y^6 = (x^2)^2 - (y^3)^2 = (x^2 + y^3)(x^2 - y^3)$$

PRACTICE 3 Factor: $p^4 - q^{10}$

EXAMPLE 4 Factor each binomial.

a. $y^4 - 16$ **b.** $x^2 + 4$

Solution

a. $y^4 - 16 = (y^2)^2 - 4^2$

$= (y^2 + 4)(y^2 - 4)$ Factor the difference of two squares. This binomial can be factored further since it is the difference of two squares.

$= (y^2 + 4)(y + 2)(y - 2)$ Factor the difference of two squares.

b. $x^2 + 4$

Note that the binomial $x^2 + 4$ is the *sum* of two squares since we can write $x^2 + 4$ as $x^2 + 2^2$. We might try to factor using $(x + 2)(x + 2)$ or $(x - 2)(x - 2)$. But when we multiply to check, we find that neither factoring is correct.

$$(x + 2)(x + 2) = x^2 + 4x + 4$$
$$(x - 2)(x - 2) = x^2 - 4x + 4$$

In both cases, the product is a trinomial, not the required binomial. In fact, $x^2 + 4$ is a prime polynomial.

PRACTICE 4 Factor each binomial.

a. $z^4 - 81$ **b.** $m^2 + 49$

Helpful Hint

When factoring, don't forget:

- See whether the terms have a greatest common factor (GCF) (other than 1) that can be factored out.
- Other than the GCF, the **sum** of two squares cannot be factored using real numbers.
- Factor completely. Always check to see whether any factors can be factored further.

EXAMPLES Factor each binomial.

5. $4x^3 - 49x = x(4x^2 - 49)$ Factor out the GCF, x.

$= x[(2x)^2 - 7^2]$

$= x(2x + 7)(2x - 7)$ Factor the difference of two squares.

6. $162x^4 - 2 = 2(81x^4 - 1)$ Factor out the GCF, 2.

$= 2(9x^2 + 1)(9x^2 - 1)$ Factor the difference of two squares.

$= 2(9x^2 + 1)(3x + 1)(3x - 1)$ Factor the difference of two squares.

(Continued on next page)

PRACTICE
5–6 Factor each binomial.

5. $36y^3 - 25y$ **6.** $80y^4 - 5$

EXAMPLE 7 Factor: $-49x^2 + 16$

Solution Factor as is, or, if you like, rearrange terms.

Factor as is: $-49x^2 + 16 = -1(49x^2 - 16)$ Factor out -1.

$= -1(7x + 4)(7x - 4)$ Factor the difference of two squares.

Helpful Hint
When rearranging terms, keep in mind that the sign of a term is in front of the term.

Rewrite binomial: $-49x^2 + 16 = 16 - 49x^2 = 4^2 - (7x)^2$

$= (4 + 7x)(4 - 7x)$

Both factorizations are correct and are equal. To see this, factor -1 from $(4 - 7x)$ in the second factorization.

PRACTICE
7 Factor: $-9x^2 + 100$

OBJECTIVE
2 Factoring the Sum or Difference of Two Cubes

Although the sum of two squares usually does not factor, the sum or difference of two cubes can be factored and reveals factoring patterns. The pattern for the sum of cubes is illustrated by multiplying the binomial $x + y$ and the trinomial $x^2 - xy + y^2$.

$$\begin{array}{r} x^2 - xy + y^2 \\ x + y \\ \hline x^2y - xy^2 + y^3 \\ x^3 - x^2y + xy^2 \quad\quad \\ \hline x^3 \quad\quad\quad\quad + y^3 \end{array}$$

Thus, $(x + y)(x^2 - xy + y^2) = x^3 + y^3$ Sum of cubes

The pattern for the difference of two cubes is illustrated by multiplying the binomial $x - y$ by the trinomial $x^2 + xy + y^2$. The result is

$$(x - y)(x^2 + xy + y^2) = x^3 - y^3 \quad \text{Difference of cubes}$$

Factoring the Sum or Difference of Two Cubes

$$a^3 + b^3 = (a + b)(a^2 - ab + b^2)$$
$$a^3 - b^3 = (a - b)(a^2 + ab + b^2)$$

Recall that "factor" means "to write as a product." Above are patterns for writing sums and differences as products.

EXAMPLE 8 Factor: $x^3 + 8$

Solution First, write the binomial in the form $a^3 + b^3$.

$$x^3 + 8 = x^3 + 2^3 \quad \text{Write in the form } a^3 + b^3.$$

If we replace a with x and b with 2 in the formula above, we have

$$x^3 + 2^3 = (x + 2)[x^2 - (x)(2) + 2^2]$$
$$= (x + 2)(x^2 - 2x + 4)$$

PRACTICE
8 Factor: $x^3 + 64$

Helpful Hint

When factoring sums or differences of cubes, notice the sign patterns.

same sign

$$x^3 + y^3 = (x + y)(x^2 - xy + y^2)$$

opposite signs always positive

same sign

$$x^3 - y^3 = (x - y)(x^2 + xy + y^2)$$

opposite signs always positive

EXAMPLE 9 Factor: $y^3 - 27$

Solution

$$\begin{aligned} y^3 - 27 &= y^3 - 3^3 \quad \text{Write in the form } a^3 - b^3. \\ &= (y - 3)[y^2 + (y)(3) + 3^2] \\ &= (y - 3)(y^2 + 3y + 9) \end{aligned}$$

□

PRACTICE 9 Factor: $x^3 - 125$

EXAMPLE 10 Factor: $64x^3 + 1$

Solution

$$\begin{aligned} 64x^3 + 1 &= (4x)^3 + 1^3 \\ &= (4x + 1)[(4x)^2 - (4x)(1) + 1^2] \\ &= (4x + 1)(16x^2 - 4x + 1) \end{aligned}$$

□

PRACTICE 10 Factor: $27y^3 + 1$

EXAMPLE 11 Factor: $54a^3 - 16b^3$

Solution Remember to factor out the greatest common factor first before using other factoring methods.

$$\begin{aligned} 54a^3 - 16b^3 &= 2(27a^3 - 8b^3) \quad \text{Factor out the GCF, 2.} \\ &= 2[(3a)^3 - (2b)^3] \quad \text{Difference of two cubes} \\ &= 2(3a - 2b)[(3a)^2 + (3a)(2b) + (2b)^2] \\ &= 2(3a - 2b)(9a^2 + 6ab + 4b^2) \end{aligned}$$

□

PRACTICE 11 Factor: $32x^3 - 500y^3$

Graphing Calculator Explorations

Graphing

A graphing calculator is a convenient tool for evaluating an expression at a given replacement value. For example, let's evaluate $x^2 - 6x$ when $x = 2$. To do so, store the value 2 in the variable x and then enter and evaluate the algebraic expression.

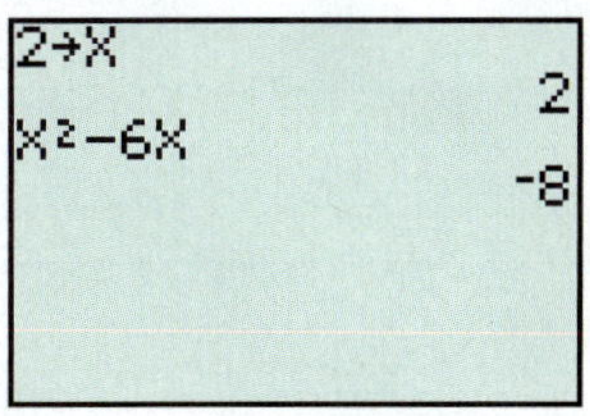

The value of $x^2 - 6x$ when $x = 2$ is -8. You may want to use this method for evaluating expressions as you explore the following.

We can use a graphing calculator to explore factoring patterns numerically. Use your calculator to evaluate $x^2 - 2x + 1$, $x^2 - 2x - 1$, and $(x - 1)^2$ for each value of x given in the table. What do you observe?

	$x^2 - 2x + 1$	$x^2 - 2x - 1$	$(x - 1)^2$
$x = 5$			
$x = -3$			
$x = 2.7$			
$x = -12.1$			
$x = 0$			

Notice in each case that $x^2 - 2x - 1 \neq (x - 1)^2$. Because for each x in the table the value of $x^2 - 2x + 1$ and the value of $(x - 1)^2$ are the same, we might guess that $x^2 - 2x + 1 = (x - 1)^2$. We can verify our guess algebraically with multiplication:

$$(x - 1)(x - 1) = x^2 - x - x + 1 = x^2 - 2x + 1$$

Vocabulary, Readiness & Video Check

Use the choices below to fill in each blank. Some choices may be used more than once and some choices may not be used at all.

true	difference of two squares	sum of two cubes
false	difference of two cubes	

1. The expression $x^3 - 27$ is called a ________________.

2. The expression $x^2 - 49$ is called a ________________.

3. The expression $z^3 + 1$ is called a ________________.

4. True or false: The binomial $y^2 + 9$ factors as $(y + 3)^2$. ________________

Write each term as a square.

5. $49x^2$

6. $25y^4$

Write each term as a cube.

7. $8y^3$

8. x^6

Martin-Gay Interactive Videos

See Video 5.3

Watch the section lecture video and answer the following questions.

OBJECTIVE 1

9. In Examples 1 and 2, what are two reasons the original binomial is rewritten so that each term is a square?

10. From Example 3, what is a prime polynomial?

OBJECTIVE 2

11. In Examples 6–8, what tips are given to remember how to factor the sum or difference of two cubes rather than memorizing the formulas?

5.3 Exercise Set MyMathLab®

Factor each binomial completely. See Examples 1 through 7.

1. $x^2 - 4$
2. $x^2 - 36$
3. $81p^2 - 1$
4. $49m^2 - 1$
5. $25y^2 - 9$
6. $49a^2 - 16$
7. $121m^2 - 100n^2$
8. $169a^2 - 49b^2$
9. $x^2y^2 - 1$
10. $a^2b^2 - 16$
11. $x^2 - \frac{1}{4}$
12. $y^2 - \frac{1}{16}$
13. $-4r^2 + 1$
14. $-9t^2 + 1$
15. $16r^2 + 1$
16. $49y^2 + 1$
17. $-36 + x^2$
18. $-1 + y^2$
19. $m^4 - 1$
20. $n^4 - 16$
21. $m^4 - n^{18}$
22. $n^4 - r^6$

Factor the sum or difference of two cubes. See Examples 8 through 11.

23. $x^3 + 125$
24. $p^3 + 1$
25. $8a^3 - 1$
26. $27y^3 - 1$
27. $m^3 + 27n^3$
28. $y^3 + 64z^3$
29. $5k^3 + 40$
30. $6r^3 + 162$
31. $x^3y^3 - 64$
32. $a^3b^3 - 8$
33. $250r^3 - 128t^3$
34. $24x^3 - 81y^3$

MIXED PRACTICE

Factor each binomial completely. See Examples 1 through 11.

35. $r^2 - 64$
36. $q^2 - 121$
37. $x^2 - 169y^2$
38. $x^2 - 225y^2$
39. $27 - t^3$
40. $125 - r^3$
41. $18r^2 - 8$
42. $32t^2 - 50$
43. $9xy^2 - 4x$
44. $36x^2y - 25y$
45. $8m^3 + 64$
46. $2x^3 + 54$
47. $xy^3 - 9xyz^2$
48. $x^3y - 4xy^3$
49. $36x^2 - 64y^2$
50. $225a^2 - 81b^2$
51. $144 - 81x^2$
52. $12x^2 - 27$
53. $x^3y^3 - z^6$
54. $a^3b^3 - c^9$
55. $49 - \frac{9}{25}m^2$
56. $100 - \frac{4}{81}n^2$
57. $t^3 + 343$
58. $s^3 + 216$
59. $n^3 + 49n$
60. $y^3 + 64y$
61. $x^6 - 81x^2$
62. $n^9 - n^5$
63. $64p^3q - 81pq^3$
64. $100x^3y - 49xy^3$
65. $27x^2y^3 + xy^2$
66. $8x^3y^3 + x^3y$
67. $125a^4 - 64ab^3$
68. $64m^4 - 27mn^3$
69. $16x^4 - 64x^2$
70. $25y^4 - 100y^2$

REVIEW AND PREVIEW

Solve each equation. See Section 2.3.

71. $x - 6 = 0$
72. $y + 5 = 0$
73. $2m + 4 = 0$
74. $3x - 9 = 0$
75. $5z - 1 = 0$
76. $4a + 2 = 0$

CONCEPT EXTENSIONS

Factor each expression completely.

77. $(x + 2)^2 - y^2$
78. $(y - 6)^2 - z^2$
79. $a^2(b - 4) - 16(b - 4)$
80. $m^2(n + 8) - 9(n + 8)$
81. $(x^2 + 6x + 9) - 4y^2$ (*Hint:* Factor the trinomial in parentheses first.)
82. $(x^2 + 2x + 1) - 36y^2$

83. $x^{2n} - 100$

84. $x^{2n} - 81$

85. What binomial multiplied by $(x - 6)$ gives the difference of two squares?

86. What binomial multiplied by $(5 + y)$ gives the difference of two squares?

87. In your own words, explain how to tell whether a binomial is a difference of squares. Then explain how to factor a difference of squares.

88. In your own words, explain how to tell whether a binomial is a sum of cubes. Then explain how to factor a sum of cubes.

89. The Toroweap Overlook, on the North Rim of the Grand Canyon, lies 3000 vertical feet above the Colorado River. The view is spectacular, and the sheer drop is dramatic. A film crew creating a documentary about the Grand Canyon has built a camera platform 136 feet above the Overlook. A camera filter comes loose and falls to the river below. The height $h(t)$ in feet of the filter above the river after t seconds is given by the function $h(t) = 3136 - 16t^2$.

a. Find the height of the filter above the river after 3 seconds.

b. Find the height of the filter above the river after 10 seconds.

c. To the nearest whole second, estimate when the filter lands in the river.

d. Factor $3136 - 16t^2$.

90. An object is dropped from the top of Pittsburgh's USX Tower, which is 841 feet tall. (*Source: World Almanac* research) The height $h(t)$ in feet of the object after t seconds is given by the function $h(t) = 841 - 16t^2$.

a. Find the height of the object after 2 seconds.

b. Find the height of the object after 5 seconds.

c. To the nearest whole second, estimate when the object hits the ground.

d. Factor $841 - 16t^2$.

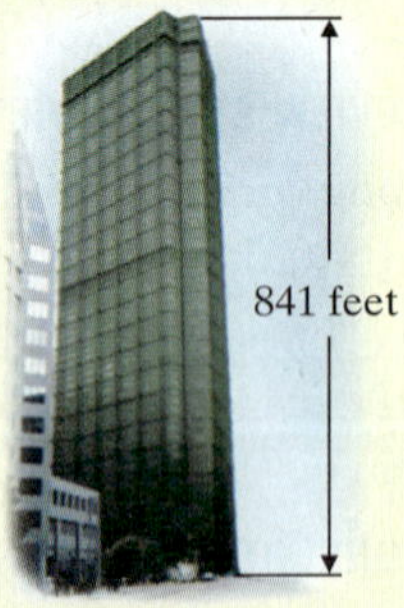

91. At this writing, the tallest completed building in the world is the Burj Khalifa, in Dubai, measuring a height of 2717 feet. (*Source:* Council on Tall Buildings and Urban Habitat) Suppose an action picture is being filmed there and a stunt man is making his way to the top of the spire. He sways in the wind and drops a clip from the height of 2704 feet. The height $h(t)$ in feet of the clip after t seconds is given by the function $h(t) = 2704 - 16t^2$.

a. Find the height of the clip after 3 seconds.

b. Find the height of the clip after 7 seconds.

c. To the nearest whole second, estimate when the clip will hit the ground.

d. Factor $2704 - 16t^2$.

92. A performer with the Moscow Circus is planning a stunt involving a free fall from the top of the Moscow State University building, which is 784 feet tall. (*Source:* Council on Tall Buildings and Urban Habitat) Neglecting air resistance, the performer's height $h(t)$ in feet above gigantic cushions positioned at ground level after t seconds is given by the function $h(t) = 784 - 16t^2$.

a. Find the performer's height after 2 seconds.

b. Find the performer's height after 5 seconds.

c. To the nearest whole second, estimate when the performer reaches the cushions positioned at ground level.

d. Factor $784 - 16t^2$.

Integrated Review Choosing a Factoring Strategy

Sections 5.1–5.3

The following steps may be helpful when factoring polynomials.

Factoring a Polynomial

Step 1. Are there any common factors? If so, factor out the GCF.

Step 2. How many terms are in the polynomial?

a. If there are **two** terms, decide if one of the following can be applied.

i. Difference of two squares: $a^2 - b^2 = (a + b)(a - b)$.

ii. Difference of two cubes: $a^3 - b^3 = (a - b)(a^2 + ab + b^2)$.

iii. Sum of two cubes: $a^3 + b^3 = (a + b)(a^2 - ab + b^2)$.

b. If there are **three** terms, factor using the methods presented in Section 5.2.

c. If there are **four** or more terms, try factoring by grouping.

Step 3. See if any factors in the factored polynomial can be factored further.

Step 4. Check by multiplying.

Study the next five examples to help you use the steps above.

EXAMPLE 1 Factor: $10t^2 - 17t + 3$

Solution

Step 1. The terms of this polynomial have no common factor (other than 1).

Step 2. There are three terms, so this polynomial is a trinomial. Factor using methods from Section 5.2.

$$\text{Factors of } 10t^2\text{:} \quad 10t^2 = 2t \cdot 5t, \qquad 10t^2 = t \cdot 10t$$

Since the middle term, $-17t$, has a negative numerical coefficient, find negative factors of 3.

$$\text{Factors of 3:} \quad 3 = -1 \cdot -3$$

Try different combinations of these factors. The correct combination is

$$(2t - 3)(5t - 1) = 10t^2 - 17t + 3$$

$-15t$

$-2t$

$-17t$ Correct middle term

Step 3. No factor can be factored further, so we have factored completely.

Step 4. To check, multiply $2t - 3$ and $5t - 1$.

$$(2t - 3)(5t - 1) = 10t^2 - 2t - 15t + 3 = 10t^2 - 17t + 3$$

A factored form of $10t^2 - 17t + 3$ is $(2t - 3)(5t - 1)$. □

PRACTICE 1 Factor: $6x^2 - 11x + 3$ ■

EXAMPLE 2 Factor: $2x^3 + 3x^2 - 2x - 3$

Solution

Step 1. There are no factors common to all terms.

Step 2. Try factoring by grouping since this polynomial has four terms.

$$2x^3 + 3x^2 - 2x - 3 = x^2(2x + 3) - 1(2x + 3)$$ Factor out the greatest common factor for each pair of terms.

$$= (2x + 3)(x^2 - 1)$$ Factor out $2x + 3$.

(Continued on next page)

Step 3. The binomial $x^2 - 1$ can be factored further. It is the difference of two squares.

$$= (2x + 3)(x + 1)(x - 1) \quad \text{Factor } x^2 - 1 \text{ as a difference of squares.}$$

Step 4. Check by finding the product of the three binomials. The polynomial factored completely is $(2x + 3)(x + 1)(x - 1)$. □

PRACTICE 2 Factor: $3x^3 + x^2 - 12x - 4$ ■

EXAMPLE 3 Factor: $12m^2 - 3n^2$

Solution

Step 1. The terms of this binomial contain a greatest common factor of 3.

$$12m^2 - 3n^2 = 3(4m^2 - n^2) \quad \text{Factor out the greatest common factor.}$$

Step 2. The binomial $4m^2 - n^2$ is a difference of squares.

$$= 3(2m + n)(2m - n) \quad \text{Factor the difference of squares.}$$

Step 3. No factor can be factored further.

Step 4. We check by multiplying.

$$3(2m + n)(2m - n) = 3(4m^2 - n^2) = 12m^2 - 3n^2$$

A factored form of $12m^2 - 3n^2$ is $3(2m + n)(2m - n)$. □

PRACTICE 3 Factor: $27x^2 - 3y^2$ ■

EXAMPLE 4 Factor: $x^3 + 27y^3$

Solution

Step 1. The terms of this binomial contain no common factor (other than 1).

Step 2. This binomial is the sum of two cubes.

$$\begin{aligned} x^3 + 27y^3 &= (x)^3 + (3y)^3 \\ &= (x + 3y)[x^2 - x(3y) + (3y)^2] \\ &= (x + 3y)(x^2 - 3xy + 9y^2) \end{aligned}$$

Step 3. No factor can be factored further.

Step 4. We check by multiplying.

$$\begin{aligned} (x + 3y)(x^2 - 3xy + 9y^2) &= x(x^2 - 3xy + 9y^2) + 3y(x^2 - 3xy + 9y^2) \\ &= x^3 - 3x^2y + 9xy^2 + 3x^2y - 9xy^2 + 27y^3 \\ &= x^3 + 27y^3 \end{aligned}$$

Thus, $x^3 + 27y^3$ factored completely is $(x + 3y)(x^2 - 3xy + 9y^2)$. □

PRACTICE 4 Factor: $8a^3 + b^3$ ■

EXAMPLE 5 Factor: $30a^2b^3 + 55a^2b^2 - 35a^2b$

Solution

Step 1. $30a^2b^3 + 55a^2b^2 - 35a^2b = 5a^2b(6b^2 + 11b - 7)$ Factor out the GCF.

Step 2. $= 5a^2b(2b - 1)(3b + 7)$ Factor the resulting trinomial.

Step 3. No factor can be factored further.

Step 4. Check by multiplying.

The trinomial factored completely is $5a^2b(2b - 1)(3b + 7)$. □

PRACTICE 5 Factor: $60x^3y^2 - 66x^2y^2 - 36xy^2$ ■

Factor the following completely.

1. $x^2 + 2xy + y^2$
2. $x^2 - 2xy + y^2$
3. $a^2 + 11a - 12$
4. $a^2 - 11a + 10$
5. $a^2 - a - 6$
6. $a^2 - 2a + 1$
7. $x^2 + 2x + 1$
8. $x^2 + x - 2$
9. $x^2 + 4x + 3$
10. $x^2 + x - 6$
11. $x^2 + 7x + 12$
12. $x^2 + x - 12$
13. $x^2 + 3x - 4$
14. $x^2 - 7x + 10$
15. $x^2 + 2x - 15$
16. $x^2 + 11x + 30$
17. $x^2 - x - 30$
18. $x^2 + 11x + 24$
19. $2x^2 - 98$
20. $3x^2 - 75$
21. $x^2 + 3x + xy + 3y$
22. $3y - 21 + xy - 7x$
23. $x^2 + 6x - 16$
24. $x^2 - 3x - 28$
25. $4x^3 + 20x^2 - 56x$
26. $6x^3 - 6x^2 - 120x$
27. $12x^2 + 34x + 24$
28. $8a^2 + 6ab - 5b^2$
29. $4a^2 - b^2$
30. $28 - 13x - 6x^2$
31. $20 - 3x - 2x^2$
32. $x^2 - 2x + 4$
33. $a^2 + a - 3$
34. $6y^2 + y - 15$
35. $4x^2 - x - 5$
36. $x^2y - y^3$
37. $4t^2 + 36$
38. $x^2 + x + xy + y$
39. $ax + 2x + a + 2$
40. $18x^3 - 63x^2 + 9x$
41. $12a^3 - 24a^2 + 4a$
42. $x^2 + 14x - 32$
43. $x^2 - 14x - 48$
44. $16a^2 - 56ab + 49b^2$
45. $25p^2 - 70pq + 49q^2$
46. $7x^2 + 24xy + 9y^2$
47. $125 - 8y^3$
48. $64x^3 + 27$
49. $-x^2 - x + 30$
50. $-x^2 + 6x - 8$
51. $14 + 5x - x^2$
52. $3 - 2x - x^2$
53. $3x^4y + 6x^3y - 72x^2y$
54. $2x^3y + 8x^2y^2 - 10xy^3$
55. $5x^3y^2 - 40x^2y^3 + 35xy^4$
56. $4x^4y - 8x^3y - 60x^2y$
57. $12x^3y + 243xy$
58. $6x^3y^2 + 8xy^2$
59. $4 - x^2$
60. $9 - y^2$
61. $3rs - s + 12r - 4$
62. $x^3 - 2x^2 + 3x - 6$
63. $4x^2 - 8xy - 3x + 6y$
64. $4x^2 - 2xy - 7yz + 14xz$
65. $6x^2 + 18xy + 12y^2$
66. $12x^2 + 46xy - 8y^2$
67. $xy^2 - 4x + 3y^2 - 12$
68. $x^2y^2 - 9x^2 + 3y^2 - 27$
69. $5(x + y) + x(x + y)$
70. $7(x - y) + y(x - y)$
71. $14t^2 - 9t + 1$
72. $3t^2 - 5t + 1$
73. $3x^2 + 2x - 5$
74. $7x^2 + 19x - 6$
75. $x^2 + 9xy - 36y^2$
76. $3x^2 + 10xy - 8y^2$
77. $1 - 8ab - 20a^2b^2$
78. $1 - 7ab - 60a^2b^2$
79. $9 - 10x^2 + x^4$
80. $36 - 13x^2 + x^4$
81. $x^4 - 14x^2 - 32$
82. $x^4 - 22x^2 - 75$
83. $x^2 - 23x + 120$
84. $y^2 + 22y + 96$
85. $6x^3 - 28x^2 + 16x$
86. $6y^3 - 8y^2 - 30y$
87. $27x^3 - 125y^3$
88. $216y^3 - z^3$
89. $x^3y^3 + 8z^3$
90. $27a^3b^3 + 8$
91. $2xy - 72x^3y$
92. $2x^3 - 18x$
93. $x^3 + 6x^2 - 4x - 24$
94. $x^3 - 2x^2 - 36x + 72$
95. $6a^3 + 10a^2$
96. $4n^2 - 6n$
97. $a^2(a + 2) + 2(a + 2)$
98. $a - b + x(a - b)$
99. $x^3 - 28 + 7x^2 - 4x$
100. $a^3 - 45 - 9a + 5a^2$

CONCEPT EXTENSIONS

Factor.

101. $(x - y)^2 - z^2$

102. $(x + 2y)^2 - 9$

103. $81 - (5x + 1)^2$

104. $b^2 - (4a + c)^2$

105. Explain why it makes good sense to factor out the GCF first, before using other methods of factoring.

106. The sum of two squares usually does not factor. Is the sum of two squares $9x^2 + 81y^2$ factorable?

107. Which of the following are equivalent to $(x + 10)(x - 7)$?

a. $(x - 7)(x + 10)$ **b.** $-1(x + 10)(x - 7)$

c. $-1(x + 10)(7 - x)$ **d.** $-1(-x - 10)(7 - x)$

108. Which of the following are equivalent to $(x - 2)(x - 5)$?

a. $-1(x + 2)(x + 5)$ **b.** $(x - 5)(x - 2)$

c. $(5 - x)(2 - x)$ **d.** $-1(x + 2)(x - 5)$

5.4 Solving Quadratic Equations by Factoring

OBJECTIVES

1. Solve Quadratic Equations by Factoring.
2. Solve Equations with Degree Greater Than 2 by Factoring.
3. Find the x-Intercepts of the Graph of a Quadratic Equation in Two Variables.

In this section, we introduce a new type of equation—the **quadratic equation.**

Quadratic Equation

A quadratic equation is one that can be written in the form

$$ax^2 + bx + c = 0$$

where a, b, and c are real numbers and $a \neq 0$.

Some examples of quadratic equations are shown below.

$$x^2 - 9x - 22 = 0 \qquad 4x^2 - 28 = -49 \qquad x(2x - 7) = 4$$

The form $ax^2 + bx + c = 0$ is called the **standard form** of a quadratic equation. The quadratic equation $x^2 - 9x - 22 = 0$ is the only equation above that is in standard form.

Quadratic equations model many real-life situations. For example, let's suppose we want to know how long before a person diving from a 144-foot cliff reaches the ocean. The answer to this question is found by solving the quadratic equation $-16t^2 + 144 = 0$. (See Example 4 in Section 5.7.)

OBJECTIVE

1 Solving Quadratic Equations by Factoring

Some quadratic equations can be solved by making use of factoring and the **zero factor property.**

Zero Factor Property

If a and b are real numbers and if $ab = 0$, then $a = 0$ or $b = 0$.

This property states that if the product of two numbers is 0 then at least one of the numbers must be 0.

EXAMPLE 1 Solve: $(x - 3)(x + 1) = 0$

Solution If this equation is to be a true statement, then either the factor $x - 3$ must be 0 or the factor $x + 1$ must be 0. In other words, either

$$x - 3 = 0 \quad \text{or} \quad x + 1 = 0$$

If we solve these two linear equations, we have

$$x = 3 \quad \text{or} \quad x = -1$$

Thus, 3 and -1 are both solutions of the equation $(x - 3)(x + 1) = 0$. To check, we replace x with 3 in the original equation. Then we replace x with -1 in the original equation.

Check: Let $x = 3$.

$$(x - 3)(x + 1) = 0$$
$$(3 - 3)(3 + 1) \stackrel{?}{=} 0 \quad \text{Replace } x \text{ with } 3.$$
$$0(4) = 0 \quad \text{True}$$

Let $x = -1$.

$$(x - 3)(x + 1) = 0$$
$$(-1 - 3)(-1 + 1) \stackrel{?}{=} 0 \quad \text{Replace } x \text{ with } -1.$$
$$(-4)(0) = 0 \quad \text{True}$$

The solutions are 3 and -1, or we say that the solution set is $\{-1, 3\}$. □

PRACTICE 1 Solve: $(x + 4)(x - 5) = 0$ ■

Helpful Hint

The zero factor property says that *if a product is 0, then a factor is 0.*

If $a \cdot b = 0$, then $a = 0$ or $b = 0$.

If $x(x + 5) = 0$, then $x = 0$ or $x + 5 = 0$.

If $(x + 7)(2x - 3) = 0$, then $x + 7 = 0$ or $2x - 3 = 0$.

Use this property only when the product is 0.

For example, if $a \cdot b = 8$, we do not know the value of a or b. The values may be $a = 2, b = 4$ or $a = 8, b = 1$, or any other two numbers whose product is 8.

EXAMPLE 2 Solve: $(x - 5)(2x + 7) = 0$

Solution The product is 0. By the zero factor property, this is true only when a factor is 0. To solve, we set each factor equal to 0 and solve the resulting linear equations.

$$(x - 5)(2x + 7) = 0$$
$$x - 5 = 0 \quad \text{or} \quad 2x + 7 = 0$$
$$x = 5 \quad \text{or} \quad 2x = -7$$
$$x = -\frac{7}{2}$$

Check: Let $x = 5$.

$$(x - 5)(2x + 7) = 0$$
$$(5 - 5)(2 \cdot 5 + 7) \stackrel{?}{=} 0 \quad \text{Replace } x \text{ with } 5.$$
$$0 \cdot 17 \stackrel{?}{=} 0$$
$$0 = 0 \quad \text{True}$$

Let $x = -\frac{7}{2}$.

$$(x - 5)(2x + 7) = 0$$
$$\left(-\frac{7}{2} - 5\right)\left(2\left(-\frac{7}{2}\right) + 7\right) \stackrel{?}{=} 0 \quad \text{Replace } x \text{ with } -\frac{7}{2}.$$
$$\left(-\frac{17}{2}\right)(-7 + 7) \stackrel{?}{=} 0$$
$$\left(-\frac{17}{2}\right) \cdot 0 \stackrel{?}{=} 0$$
$$0 = 0 \quad \text{True}$$

The solutions are 5 and $-\frac{7}{2}$. □

PRACTICE 2 Solve: $(x - 12)(4x + 3) = 0$ ■

EXAMPLE 3 Solve: $x(5x - 2) = 0$

Solution

$$x(5x - 2) = 0$$

$$x = 0 \quad \text{or} \quad 5x - 2 = 0 \quad \text{Use the zero factor property.}$$

$$5x = 2$$

$$x = \frac{2}{5}$$

Check: Let $x = 0$.

$$x(5x - 2) = 0$$

$$0(5 \cdot 0 - 2) \stackrel{?}{=} 0 \quad \text{Replace } x \text{ with } 0.$$

$$0(-2) \stackrel{?}{=} 0$$

$$0 = 0 \quad \text{True}$$

Let $x = \frac{2}{5}$.

$$x(5x - 2) = 0$$

$$\frac{2}{5}\left(5 \cdot \frac{2}{5} - 2\right) \stackrel{?}{=} 0 \quad \text{Replace } x \text{ with } \frac{2}{5}.$$

$$\frac{2}{5}(2 - 2) \stackrel{?}{=} 0$$

$$\frac{2}{5}(0) \stackrel{?}{=} 0$$

$$0 = 0 \quad \text{True}$$

The solutions are 0 and $\frac{2}{5}$. □

PRACTICE 3 Solve: $x(7x - 6) = 0$ ■

EXAMPLE 4 Solve: $x^2 - 9x - 22 = 0$

Solution One side of the equation is 0. However, to use the zero factor property, one side of the equation must be 0 *and* the other side must be written as a product (must be factored). Thus, we must first factor this polynomial.

$$x^2 - 9x - 22 = 0$$

$$(x - 11)(x + 2) = 0 \quad \text{Factor.}$$

Now we can apply the zero factor property.

$$x - 11 = 0 \quad \text{or} \quad x + 2 = 0$$

$$x = 11 \quad \text{or} \quad x = -2$$

Check: Let $x = 11$.

$$x^2 - 9x - 22 = 0$$

$$11^2 - 9 \cdot 11 - 22 \stackrel{?}{=} 0$$

$$121 - 99 - 22 \stackrel{?}{=} 0$$

$$22 - 22 \stackrel{?}{=} 0$$

$$0 = 0 \quad \text{True}$$

Let $x = -2$.

$$x^2 - 9x - 22 = 0$$

$$(-2)^2 - 9(-2) - 22 \stackrel{?}{=} 0$$

$$4 + 18 - 22 \stackrel{?}{=} 0$$

$$22 - 22 \stackrel{?}{=} 0$$

$$0 = 0 \quad \text{True}$$

The solutions are 11 and −2. □

PRACTICE 4 Solve: $x^2 - 8x - 48 = 0$ ■

EXAMPLE 5 Solve: $4x^2 - 28x = -49$

Solution First we rewrite the equation in standard form so that one side is 0. Then we factor the polynomial.

$$4x^2 - 28x = -49$$

$$4x^2 - 28x + 49 = 0 \quad \text{Write in standard form by adding 49 to both sides.}$$

$$(2x - 7)(2x - 7) = 0 \quad \text{Factor.}$$

Next we use the zero factor property and set each factor equal to 0. Since the factors are the same, the related equations will give the same solution.

$$2x - 7 = 0 \quad \text{or} \quad 2x - 7 = 0 \qquad \text{Set each factor equal to 0.}$$
$$2x = 7 \quad \text{or} \quad 2x = 7 \qquad \text{Solve.}$$
$$x = \frac{7}{2} \quad \text{or} \quad x = \frac{7}{2}$$

Check: Although $\frac{7}{2}$ occurs twice, there is a single solution. Check this solution in the original equation. The solution is $\frac{7}{2}$. □

PRACTICE 5 Solve: $9x^2 - 24x = -16$ ■

The following steps may be used to solve a quadratic equation by factoring.

To Solve Quadratic Equations by Factoring

Step 1. Write the equation in standard form so that one side of the equation is 0.

Step 2. Factor the quadratic expression completely.

Step 3. Set each factor containing a variable equal to 0.

Step 4. Solve the resulting equations.

Step 5. Check each solution in the original equation.

Since it is not always possible to factor a quadratic polynomial, not all quadratic equations can be solved by factoring. Other methods of solving quadratic equations are presented later in this chapter.

EXAMPLE 6 Solve: $x(2x - 7) = 4$

Solution First we write the equation in standard form; then we factor.

$$x(2x - 7) = 4$$
$$2x^2 - 7x = 4 \qquad \text{Multiply.}$$
$$2x^2 - 7x - 4 = 0 \qquad \text{Write in standard form.}$$
$$(2x + 1)(x - 4) = 0 \qquad \text{Factor.}$$
$$2x + 1 = 0 \quad \text{or} \quad x - 4 = 0 \qquad \text{Set each factor equal to zero.}$$
$$2x = -1 \quad \text{or} \quad x = 4 \qquad \text{Solve.}$$
$$x = -\frac{1}{2}$$

Check the solutions in the original equation. The solutions are $-\frac{1}{2}$ and 4. □

PRACTICE 6 Solve: $x(3x + 7) = 6$ ■

Helpful Hint

To solve the equation $x(2x - 7) = 4$, do **not** set each factor equal to 4. Remember that to apply the zero factor property, one side of the equation must be 0 and the other side of the equation must be in factored form.

Answer to Concept Check:
To use the zero factor property, one side of the equation must be 0, not 5. Correctly, $(x - 3)(x + 1) = 5$, $x^2 - 2x - 3 = 5$, $x^2 - 2x - 8 = 0$, $(x - 4)(x + 2) = 0$, $x - 4 = 0$ or $x + 2 = 0$, $x = 4$ or $x = -2$.

✔ CONCEPT CHECK

Explain the error and solve the equation correctly.

$$(x - 3)(x + 1) = 5$$
$$x - 3 = 5 \quad \text{or} \quad x + 1 = 5$$
$$x = 8 \quad \text{or} \quad x = 4$$

EXAMPLE 7 Solve: $-2x^2 - 4x + 30 = 0$

Solution The equation is in standard form, so we begin by factoring out the greatest common factor, -2.

$$-2x^2 - 4x + 30 = 0$$
$$-2(x^2 + 2x - 15) = 0 \quad \text{Factor out } -2.$$
$$-2(x + 5)(x - 3) = 0 \quad \text{Factor the quadratic.}$$

Next, set each factor **containing a variable** equal to 0.

$$x + 5 = 0 \quad \text{or} \quad x - 3 = 0 \quad \text{Set each factor containing a variable equal to 0.}$$
$$x = -5 \quad \text{or} \quad x = 3 \quad \text{Solve.}$$

Note: The factor -2 is a constant term containing no variables and can never equal 0. The solutions are -5 and 3. □

PRACTICE 7 Solve: $-3x^2 - 6x + 72 = 0$ ■

OBJECTIVE

2 Solving Equations with Degree Greater Than Two by Factoring

Some equations involving polynomials of degree higher than 2 may also be solved by factoring and then applying the zero factor property.

EXAMPLE 8 Solve: $3x^3 - 12x = 0$

Solution Factor the left side of the equation. Begin by factoring out the greatest common factor, $3x$.

$$3x^3 - 12x = 0$$
$$3x(x^2 - 4) = 0 \quad \text{Factor out the GCF, } 3x.$$
$$3x(x + 2)(x - 2) = 0 \quad \text{Factor } x^2 - 4 \text{, a difference of squares.}$$
$$3x = 0 \quad \text{or} \quad x + 2 = 0 \quad \text{or} \quad x - 2 = 0 \quad \text{Set each factor equal to 0.}$$
$$x = 0 \quad \text{or} \quad x = -2 \quad \text{or} \quad x = 2 \quad \text{Solve.}$$

Thus, the equation $3x^3 - 12x = 0$ has three solutions: 0, -2, and 2. To check, replace x with each solution in the original equation.

Let x = 0.	***Let x = −2.***	***Let x = 2.***
$3(0)^3 - 12(0) \stackrel{?}{=} 0$	$3(-2)^3 - 12(-2) \stackrel{?}{=} 0$	$3(2)^3 - 12(2) \stackrel{?}{=} 0$
$0 = 0$	$3(-8) + 24 \stackrel{?}{=} 0$	$3(8) - 24 \stackrel{?}{=} 0$
	$0 = 0$	$0 = 0$

Substituting 0, -2, or 2 into the original equation results each time in a true equation. The solutions are 0, -2, and 2. □

PRACTICE 8 Solve: $7x^3 - 63x = 0$ ■

EXAMPLE 9 Solve: $(5x - 1)(2x^2 + 15x + 18) = 0$

Solution

$$(5x - 1)(2x^2 + 15x + 18) = 0$$
$$(5x - 1)(2x + 3)(x + 6) = 0 \quad \text{Factor the trinomial.}$$
$$5x - 1 = 0 \quad \text{or} \quad 2x + 3 = 0 \quad \text{or} \quad x + 6 = 0 \quad \text{Set each factor equal to 0.}$$
$$5x = 1 \quad \text{or} \quad 2x = -3 \quad \text{or} \quad x = -6 \quad \text{Solve.}$$
$$x = \frac{1}{5} \quad \text{or} \quad x = -\frac{3}{2}$$

The solutions are $\frac{1}{5}$, $-\frac{3}{2}$, and -6. Check by replacing x with each solution in the original equation. The solutions are -6, $-\frac{3}{2}$, and $\frac{1}{5}$. □

PRACTICE 9 Solve: $(3x - 2)(2x^2 - 13x + 15) = 0$ ■

EXAMPLE 10 Solve: $2x^3 - 4x^2 - 30x = 0$

Solution Begin by factoring out the GCF, $2x$.

$$2x^3 - 4x^2 - 30x = 0$$

$$2x(x^2 - 2x - 15) = 0 \quad \text{Factor out the GCF, } 2x.$$

$$2x(x - 5)(x + 3) = 0 \quad \text{Factor the quadratic.}$$

$$2x = 0 \quad \text{or} \quad x - 5 = 0 \quad \text{or} \quad x + 3 = 0 \quad \text{Set each factor containing a variable equal to 0.}$$

$$x = 0 \quad \text{or} \quad x = 5 \quad \text{or} \quad x = -3 \quad \text{Solve.}$$

Check by replacing x with each solution in the cubic equation. The solutions are -3, 0, and 5. □

PRACTICE 10 Solve: $5x^3 + 5x^2 - 30x = 0$ ■

OBJECTIVE

3 Finding x-Intercepts of the Graph of a Quadratic Equation

In Chapter 3, we graphed linear equations in two variables, such as $y = 5x - 6$. Recall that to find the x-intercept of the graph of a linear equation, let $y = 0$ and solve for x. This is also how to find the x-intercepts of the graph of a **quadratic equation in two variables,** such as $y = x^2 - 5x + 4$.

EXAMPLE 11 Find the x-intercepts of the graph of $y = x^2 - 5x + 4$.

Solution Let $y = 0$ and solve for x.

$$y = x^2 - 5x + 4$$

$$0 = x^2 - 5x + 4 \quad \text{Let } y = 0.$$

$$0 = (x - 1)(x - 4) \quad \text{Factor.}$$

$$x - 1 = 0 \quad \text{or} \quad x - 4 = 0 \quad \text{Set each factor equal to 0.}$$

$$x = 1 \quad \text{or} \quad x = 4 \quad \text{Solve.}$$

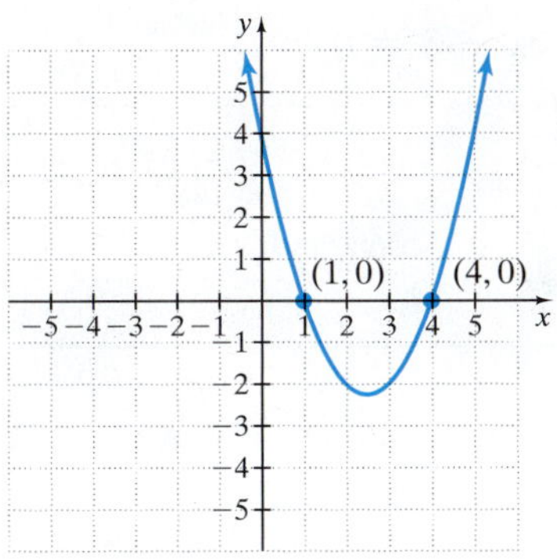

The x-intercepts of the graph of $y = x^2 - 5x + 4$ are $(1, 0)$ and $(4, 0)$.

The graph of $y = x^2 - 5x + 4$ is shown in the margin. □

PRACTICE 11 Find the x-intercepts of the graph of $y = x^2 - 6x + 8$. ■

In general, a quadratic equation in two variables is one that can be written in the form $y = ax^2 + bx + c$ where $a \neq 0$. The graph of such an equation is called a **parabola** and will open up or down depending on the sign of a.

Notice that the x-intercepts of the graph of $y = ax^2 + bx + c$ are the real number solutions of $0 = ax^2 + bx + c$. Also, the real number solutions of $0 = ax^2 + bx + c$ are the x-intercepts of the graph of $y = ax^2 + bx + c$. We study more about graphs of quadratic equations in two variables in Section 5.8.

Graph of $y = ax^2 + bx + c$
x-intercepts are solutions of $0 = ax^2 + bx + c$

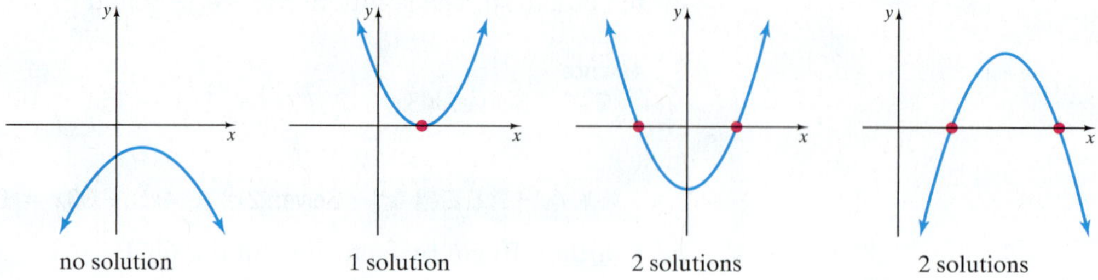

Graphing Calculator Explorations

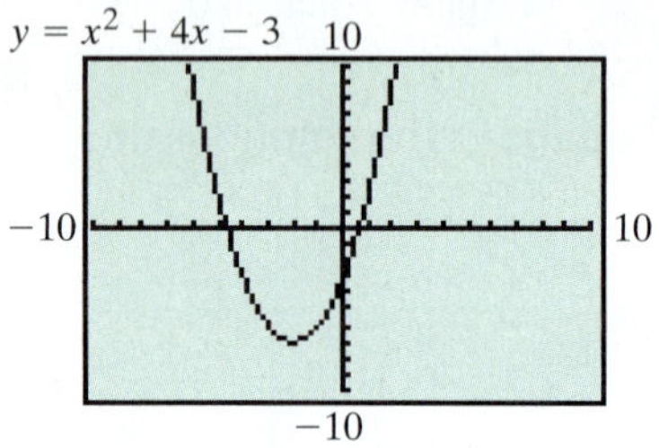

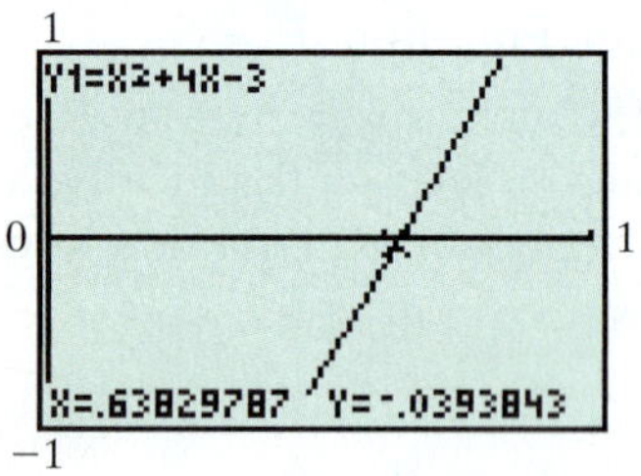

A grapher may be used to find solutions of a quadratic equation whether the related quadratic polynomial is factorable or not. For example, let's use a grapher to approximate the solutions of $0 = x^2 + 4x - 3$. To do so, graph $y_1 = x^2 + 4x - 3$. Recall that the x-intercepts of this graph are the solutions of $0 = x^2 + 4x - 3$.

Notice that the graph appears to have an x-intercept between -5 and -4 and one between 0 and 1. Many graphers contain a TRACE feature. This feature activates a graph cursor that can be used to *trace* along a graph while the corresponding x- and y-coordinates are shown on the screen. Use the TRACE feature to confirm that x-intercepts lie between -5 and -4 and between 0 and 1. To approximate the x-intercepts to the nearest tenth, use a ROOT or a ZOOM feature on your grapher or redefine the viewing window. (A ROOT feature calculates the x-intercept. A ZOOM feature magnifies the viewing window around a specific location such as the graph cursor.) If we redefine the window to $[0, 1]$ on the x-axis and $[-1, 1]$ on the y-axis, the graph to the left is generated.

By using the TRACE feature, we can conclude that one x-intercept is approximately 0.6 to the nearest tenth. By repeating these steps for the other x-intercept, we find that it is approximately -4.6.

Use a grapher to approximate the real number solutions to the nearest tenth. If an equation has no real number solution, state so.

1. $3x^2 - 4x - 6 = 0$

2. $x^2 - x - 9 = 0$

3. $2x^2 + x + 2 = 0$

4. $-4x^2 - 5x - 4 = 0$

5. $-x^2 + x + 5 = 0$

6. $10x^2 + 6x - 3 = 0$

Vocabulary, Readiness & Video Check

Use the choices below to fill in each blank. Not all choices will be used.

$-3, 5$	$a = 0$ or $b = 0$	0	linear
$3, -5$	quadratic	1	

1. An equation that can be written in the form $ax^2 + bx + c = 0$, with $a \neq 0$, is called a ______ equation.

2. If the product of two numbers is 0, then at least one of the numbers must be ______.

3. The solutions to $(x - 3)(x + 5) = 0$ are ______.

4. If $a \cdot b = 0$, then ______.

Martin-Gay Interactive Videos

See Video 5.4

Watch the section lecture video and answer the following questions.

OBJECTIVE 1

5. As shown in Examples 1–3, what two things have to be true in order to use the zero factor property?

OBJECTIVE 2

6. Example 4 implies that the zero factor property can be used with any number of factors on one side of the equation so long as the other side of the equation is zero. Why do you think this is true?

OBJECTIVE 3

7. From Example 5, how does finding the x-intercepts of the graph of a quadratic equation in two variables lead to solving a quadratic equation?

5.4 Exercise Set MyMathLab®

Solve each equation. See Examples 1 through 3.

1. $(x-6)(x-7)=0$
2. $(x-10)(x-5)=0$
3. $(x-2)(x+1)=0$
4. $(x+4)(x-10)=0$
5. $(x+9)(x+17)=0$
6. $(x+11)(x+1)=0$
7. $x(x+6)=0$
8. $x(x-7)=0$
9. $3x(x-8)=0$
10. $2x(x+12)=0$
11. $(2x+3)(4x-5)=0$
12. $(3x-2)(5x+1)=0$
13. $(2x-7)(7x+2)=0$
14. $(9x+1)(4x-3)=0$
15. $\left(x-\frac{1}{2}\right)\left(x+\frac{1}{3}\right)=0$
16. $\left(x+\frac{2}{9}\right)\left(x-\frac{1}{4}\right)=0$
17. $(x+0.2)(x+1.5)=0$
18. $(x+1.7)(x+2.3)=0$

Solve. See Examples 4 through 7.

19. $x^2-13x+36=0$
20. $x^2+2x-63=0$
21. $x^2+2x-8=0$
22. $x^2-5x+6=0$
23. $x^2-7x=0$
24. $x^2-3x=0$
25. $x^2-4x=32$
26. $x^2-5x=24$
27. $x^2=16$
28. $x^2=9$
29. $(x+4)(x-9)=4x$
30. $(x+3)(x+8)=x$
31. $x(3x-1)=14$
32. $x(4x-11)=3$
33. $-3x^2+75=0$
34. $-2y^2+72=0$
35. $24x^2+44x=8$
36. $6x^2+57x=30$

Solve each equation. See Examples 8 through 10.

37. $x^3-12x^2+32x=0$
38. $x^3-14x^2+49x=0$
39. $(4x-3)(16x^2-24x+9)=0$
40. $(2x+5)(4x^2+20x+25)=0$
41. $4x^3-x=0$
42. $4y^3-36y=0$
43. $32x^3-4x^2-6x=0$
44. $15x^3+24x^2-63x=0$

MIXED PRACTICE

Solve each equation. See Examples 1 through 10. (A few exercises are linear equations.)

45. $(x+3)(x-2)=0$
46. $(x-6)(x+7)=0$
47. $x^2+20x=0$
48. $x^2+15x=0$
49. $4(x-7)=6$
50. $5(3-4x)=9$
51. $4y^2-1=0$
52. $4y^2-81=0$
53. $(2x+3)(2x^2-5x-3)=0$
54. $(2x-9)(x^2+5x-36)=0$
55. $x^2-15=-2x$
56. $x^2-26=-11x$
57. $30x^2-11x-30=0$
58. $12x^2+7x-12=0$
59. $5x^2-6x-8=0$
60. $9x^2+7x=2$
61. $6y^2-22y-40=0$
62. $3x^2-6x-9=0$
63. $(y-2)(y+3)=6$
64. $(y-5)(y-2)=28$
65. $3x^3+19x^2-72x=0$
66. $36x^3+x^2-21x=0$
67. $x^2+14x+49=0$
68. $x^2+22x+121=0$
69. $12y=8y^2$
70. $9y=6y^2$
71. $7x^3-7x=0$
72. $3x^3-27x=0$
73. $3x^2+8x-11=13-6x$
74. $2x^2+12x-1=4+3x$
75. $3x^2-20x=-4x^2-7x-6$
76. $4x^2-20x=-5x^2-6x-5$

Find the x-intercepts of the graph of each equation. See Example 11.

77. $y = (3x + 4)(x - 1)$

78. $y = (5x - 3)(x - 4)$

79. $y = x^2 - 3x - 10$

80. $y = x^2 + 7x + 6$

81. $y = 2x^2 + 11x - 6$

82. $y = 4x^2 + 11x + 6$

For Exercises 83 through 88, match each equation with its graph. See Example 11.

83. $y = (x + 2)(x - 1)$ **84.** $y = (x - 5)(x + 2)$

85. $y = x(x + 3)$ **86.** $y = x(x - 4)$

87. $y = 2x^2 - 8$ **88.** $y = 2x^2 - 2$

A.

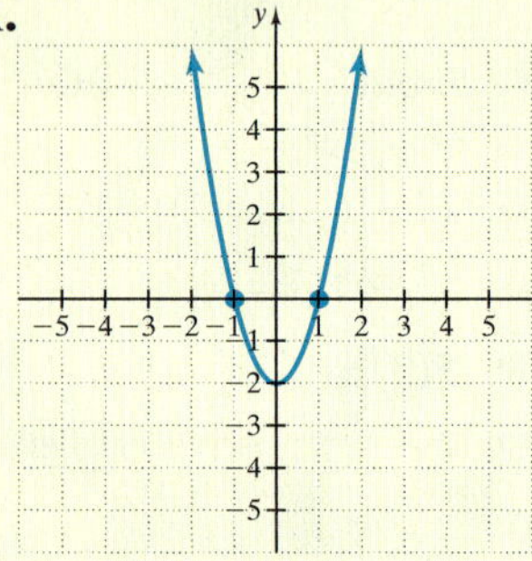

B.

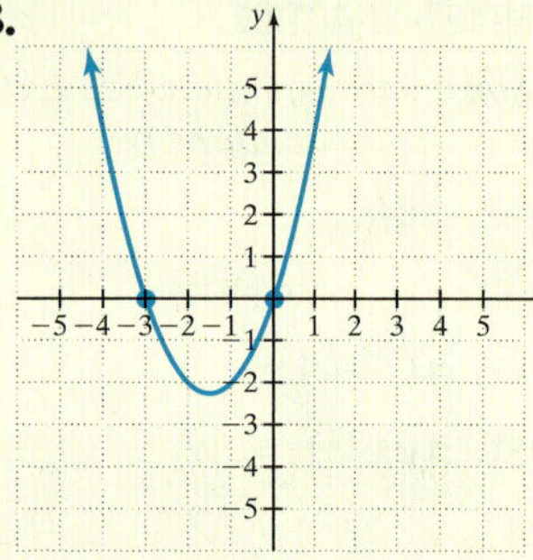

C.

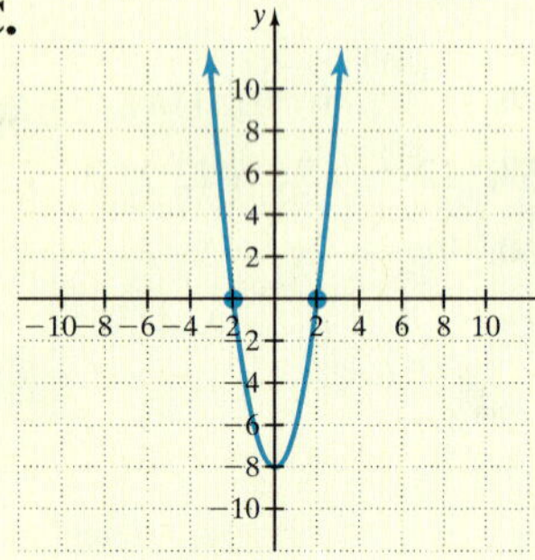

D.

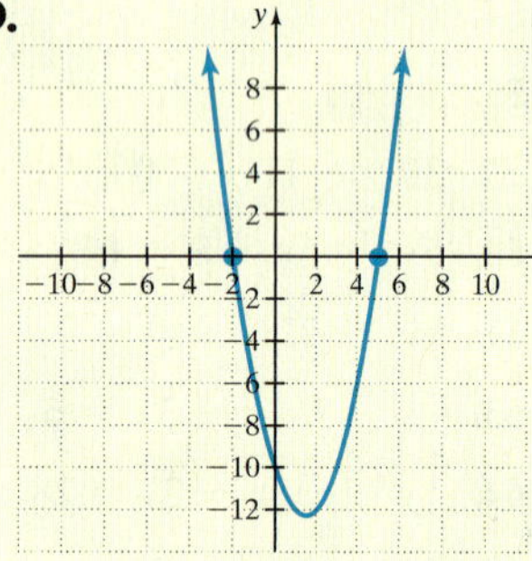

E.

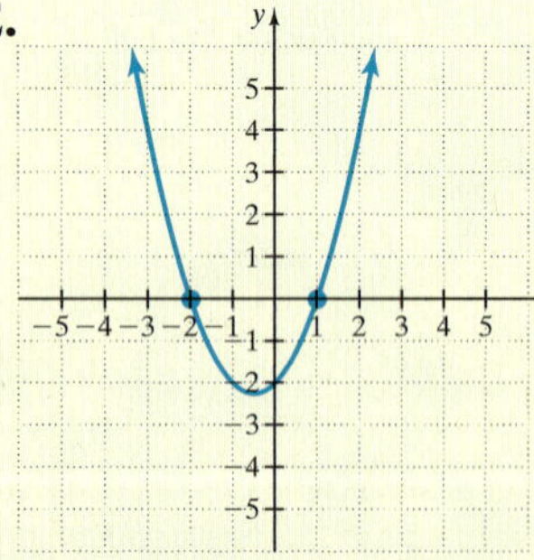

F.

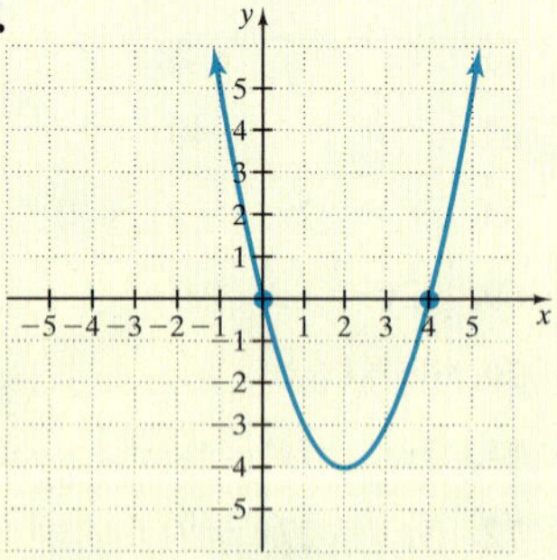

REVIEW AND PREVIEW

Perform the following operations. Write all results in lowest terms. See Section R.1.

89. $\frac{3}{5} + \frac{4}{9}$

90. $\frac{2}{3} + \frac{3}{7}$

91. $\frac{7}{10} - \frac{5}{12}$

92. $\frac{5}{9} - \frac{5}{12}$

93. $\frac{7}{8} \div \frac{7}{15}$

94. $\frac{5}{12} - \frac{3}{10}$

95. $\frac{4}{5} \cdot \frac{7}{8}$

96. $\frac{3}{7} \cdot \frac{12}{17}$

CONCEPT EXTENSIONS

For Exercises 97 and 98, see the Concept Check in this section.

97. Explain the error and solve correctly:

$x(x - 2) = 8$

$x = 8$ or $x - 2 = 8$

$x = 10$

98. Explain the error and solve correctly:

$(x - 4)(x + 2) = 0$

$x = -4$ or $x = 2$

99. Write a quadratic equation that has two solutions, 6 and −1. Leave the polynomial in the equation in factored form.

100. Write a quadratic equation that has two solutions, 0 and −2. Leave the polynomial in the equation in factored form.

101. Write a quadratic equation in standard form that has two solutions, 5 and 7.

102. Write an equation that has three solutions, 0, 1, and 2.

103. A compass is accidentally thrown upward and out of an air balloon at a height of 300 feet. The height, y, of the compass at time x in seconds is given by the equation

$$y = -16x^2 + 20x + 300$$

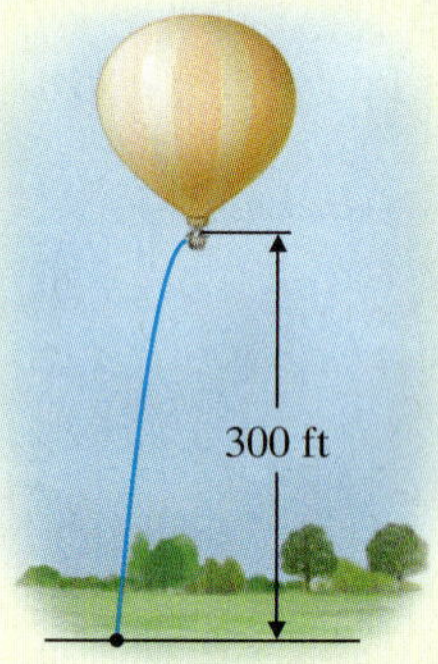

a. Find the height of the compass at the given times by filling in the table below.

time, x	0	1	2	3	4	5	6
height, y							

b. Use the table to determine when the compass strikes the ground.

c. Use the table to approximate the maximum height of the compass.

d. Plot the points (x, y) on a rectangular coordinate system and connect them with a smooth curve. Explain your results.

104. A rocket is fired upward from the ground with an initial velocity of 100 feet per second. The height, y, of the rocket at any time x is given by the equation

$$y = -16x^2 + 100x$$

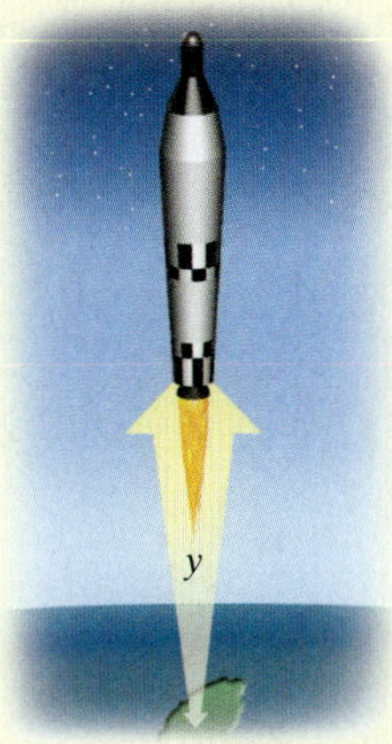

a. Find the height of the rocket at the given times by filling in the table below.

time, x	0	1	2	3	4	5	6	7
height, y								

b. Use the table to approximate when the rocket strikes the ground to the nearest second.

c. Use the table to approximate the maximum height of the rocket.

d. Plot the points (x, y) on a rectangular coordinate system and connect them with a smooth curve. Explain your results.

Solve each equation. First, multiply the binomials.

To solve $(x-6)(2x-3) = (x+2)(x+9)$, see below.

$$(x-6)(2x-3) = (x+2)(x+9)$$
$$2x^2 - 15x + 18 = x^2 + 11x + 18$$
$$x^2 - 26x = 0$$
$$x(x-26) = 0$$
$$x = 0 \quad \text{or} \quad x - 26 = 0$$
$$x = 26$$

105. $(x-3)(3x+4) = (x+2)(x-6)$

106. $(2x-3)(x+6) = (x-9)(x+2)$

107. $(2x-3)(x+8) = (x-6)(x+4)$

108. $(x+6)(x-6) = (2x-9)(x+4)$

5.5 Operations on Square Roots

OBJECTIVES

1. Find Square Roots.
2. Approximate Square Roots.
3. Use the Product Rule to Simplify Square Roots.
4. Use the Quotient Rule to Simplify Square Roots.
5. Add or Subtract Square Roots.
6. Multiply or Divide Square Roots.
7. Rationalize Denominators.

OBJECTIVE

1 Finding Square Roots

In this section, we review finding the **root** of a number by its reverse operation, raising a number to a power. We begin with squares and square roots.

The *square* of 5 is $5^2 = 25$.

The *square* of -5 is $(-5)^2 = 25$.

The *square* of $\frac{1}{2}$ is $\left(\frac{1}{2}\right)^2 = \frac{1}{4}$.

The reverse operation of squaring a number is finding the **square root** of a number. For example,

A *square root* of 25 is 5 because $5^2 = 25$.

A *square root* of 25 is also -5 because $(-5)^2 = 25$.

A *square root* of $\frac{1}{4}$ is $\frac{1}{2}$ because $\left(\frac{1}{2}\right)^2 = \frac{1}{4}$.

In general, a number b is a square root of a number a if $b^2 = a$.

Notice that both 5 and -5 are square roots of 25. The symbol $\sqrt{\ }$ denotes the **positive** or **principal square root** of a number. For example,

$\sqrt{25} = 5$ since $5^2 = 25$ and 5 is positive.

The symbol $-\sqrt{\ }$ denotes the **negative square root.** For example,

$$-\sqrt{25} = -5$$

The symbol $\sqrt{\ }$ is called a **radical** or **radical sign.** The expression within or under a radical sign is called the **radicand.** An expression containing a radical is called a **radical expression.**

$\sqrt{a}$ ← radical sign; a ← radicand

Square Root

If a is a positive number, then

$\sqrt{a}$ is the **positive square root** of a and

$-\sqrt{a}$ is the **negative square root** of a.

$$\sqrt{a} = b \quad \text{only if} \quad b^2 = a \text{ and } b > 0$$

Also, $\sqrt{0} = 0$.

EXAMPLE 1 Find each square root.

a. $\sqrt{36}$ **b.** $-\sqrt{16}$ **c.** $\sqrt{\frac{9}{100}}$ **d.** $\sqrt{0}$ **e.** $\sqrt{0.64}$

Solution

a. $\sqrt{36} = 6$ because $6^2 = 36$ and 6 is positive.

b. $-\sqrt{16} = -4$. The negative sign in front of the radical indicates the negative square root of 16.

c. $\sqrt{\frac{9}{100}} = \frac{3}{10}$ because $\left(\frac{3}{10}\right)^2 = \frac{9}{100}$ and $\frac{3}{10}$ is positive.

d. $\sqrt{0} = 0$ because $0^2 = 0$.

e. $\sqrt{0.64} = 0.8$ because $(0.8)^2 = 0.64$ and 0.8 is positive. □

PRACTICE 1 Find each square root.

a. $\sqrt{\frac{4}{81}}$ **b.** $-\sqrt{25}$ **c.** $\sqrt{144}$ **d.** $\sqrt{0.49}$ **e.** $-\sqrt{1}$ ■

Is the square root of a negative number a real number? For example, is $\sqrt{-4}$ a real number? To answer this question, we ask ourselves, is there a real number whose square is -4? Since there is no real number whose square is -4, we say that $\sqrt{-4}$ is not a real number. In general,

A square root of a negative number is not a real number.

Study the following table to make sure you understand the differences discussed earlier.

Number	*Square Roots of Number*	$\sqrt{\textit{number}}$	$-\sqrt{\textit{number}}$
25	$-5, 5$	$\sqrt{25} = 5$ only	$-\sqrt{25} = -5$
$\frac{1}{4}$	$-\frac{1}{2}, \frac{1}{2}$	$\sqrt{\frac{1}{4}} = \frac{1}{2}$ only	$-\sqrt{\frac{1}{4}} = -\frac{1}{2}$
-4	No real square roots.	$\sqrt{-4}$ is not a real number.	

OBJECTIVE 2

Approximating Square Roots

Recall that numbers such as 1, 4, 9, 25, and $\frac{4}{25}$ are called **perfect squares,** since $1^2 = 1$, $2^2 = 4$, $3^2 = 9$, $5^2 = 25$, and $\left(\frac{2}{5}\right)^2 = \frac{4}{25}$. Square roots of perfect square radicands simplify to rational numbers. What happens when we try to simplify a root such as $\sqrt{3}$? Since 3 is not a perfect square, $\sqrt{3}$ is not a rational number. It cannot be written as a quotient of integers. It is called an **irrational number** and we can find a decimal **approximation** of it. To find decimal approximations, use a calculator or an appendix.

EXAMPLE 2 Use a calculator or an appendix to approximate $\sqrt{3}$ to three decimal places.

Solution We may use an appendix or a calculator to approximate $\sqrt{3}$. To use a calculator, find the square root key $\boxed{\sqrt{\ }}$.

$$\sqrt{3} \approx 1.732050808$$

To three decimal places, $\sqrt{3} \approx 1.732$. □

PRACTICE

2 Use a calculator or an appendix to approximate $\sqrt{17}$ to three decimal places.

OBJECTIVE

3 Simplifying Square Roots Using the Product Rule

A square root is simplified when the radicand contains no perfect square factors (other than 1). For example, $\sqrt{20}$ is not simplified because $\sqrt{20} = \sqrt{4 \cdot 5}$ and 4 is a perfect square.

To begin simplifying square roots, we notice the following pattern.

$$\sqrt{9 \cdot 16} = \sqrt{144} = 12$$
$$\sqrt{9} \cdot \sqrt{16} = 3 \cdot 4 = 12$$

Since both expressions simplify to 12, we can write

$$\sqrt{9 \cdot 16} = \sqrt{9} \cdot \sqrt{16}$$

This suggests the following product rule for square roots.

Product Rule for Square Roots

If $\sqrt{a}$ and $\sqrt{b}$ are real numbers, then

$$\sqrt{a \cdot b} = \sqrt{a} \cdot \sqrt{b}$$

In other words, the square root of a product is equal to the product of the square roots.

To simplify $\sqrt{45}$, for example, we factor 45 so that one of its factors is a perfect square factor.

$$\begin{aligned} \sqrt{45} &= \sqrt{9 \cdot 5} && \text{Factor 45.} \\ &= \sqrt{9} \cdot \sqrt{5} && \text{Use the product rule.} \\ &= 3\sqrt{5} && \text{Write } \sqrt{9} \text{ as 3.} \end{aligned}$$

The notation $3\sqrt{5}$ means $3 \cdot \sqrt{5}$. Since the radicand 5 has no perfect square factor other than 1, the expression $3\sqrt{5}$ is in simplest form.

Helpful Hint

Remember, the notation $3\sqrt{5}$ means $3 \cdot \sqrt{5}$.

Helpful Hint

A radical expression in simplest form *does not mean* a decimal approximation. The simplest form of a radical expression is an exact form and may still contain a radical.

$$\underbrace{\sqrt{45} = 3\sqrt{5}}_{\text{exact}} \qquad \underbrace{\sqrt{45} \approx 6.71}_{\text{decimal approximation}}$$

EXAMPLE 3 Simplify.

a. $\sqrt{54}$ **b.** $\sqrt{12}$ **c.** $\sqrt{200}$ **d.** $\sqrt{35}$

Solution

a. Try to factor 54 so that at least one of the factors is a perfect square. Since 9 is a perfect square and $54 = 9 \cdot 6$,

$$\begin{aligned} \sqrt{54} &= \sqrt{9 \cdot 6} && \text{Factor 54 so that one factor is a perfect square.} \\ &= \sqrt{9} \cdot \sqrt{6} && \text{Apply the product rule.} \\ &= 3\sqrt{6} && \text{Write } \sqrt{9} \text{ as 3.} \end{aligned}$$

b.
$$\begin{aligned} \sqrt{12} &= \sqrt{4 \cdot 3} && \text{Factor 12 so that one factor is a perfect square.} \\ &= \sqrt{4} \cdot \sqrt{3} && \text{Apply the product rule.} \\ &= 2\sqrt{3} && \text{Write } \sqrt{4} \text{ as 2.} \end{aligned}$$

c.
$$\begin{aligned} \sqrt{200} &= \sqrt{100 \cdot 2} && \text{Factor 200 so that one factor is a perfect square.} \\ &= \sqrt{100} \cdot \sqrt{2} && \text{Apply the product rule.} \\ &= 10\sqrt{2} && \text{Write } \sqrt{100} \text{ as 10.} \end{aligned}$$

d. The radicand 35 contains no perfect square factors other than 1. Thus $\sqrt{35}$ is in simplest form. □

PRACTICE 3 Simplify.

a. $\sqrt{24}$ **b.** $\sqrt{60}$ **c.** $\sqrt{42}$ **d.** $\sqrt{300}$ ■

In Example 3(c), 100 is the largest perfect square factor of 200. What happens if we don't use the largest perfect square factor? Although using the largest perfect square factor saves time, the result is the same no matter what perfect square factor is used. For example, it is also true that $200 = 4 \cdot 50$. Then

$$\begin{aligned} \sqrt{200} &= \sqrt{4} \cdot \sqrt{50} \\ &= 2 \cdot \sqrt{50} \end{aligned}$$

Since $\sqrt{50}$ is not in simplest form, we continue.

$$\begin{aligned} \sqrt{200} &= 2 \cdot \sqrt{50} \\ &= 2 \cdot \sqrt{25 \cdot 2} \\ &= 2 \cdot \sqrt{25} \cdot \sqrt{2} \\ &= 2 \cdot 5 \cdot \sqrt{2} \\ &= 10\sqrt{2} \end{aligned}$$

EXAMPLE 4 Simplify: $3\sqrt{8}$

Solution Remember that $3\sqrt{8}$ means $3 \cdot \sqrt{8}$.

$$\begin{aligned} 3 \cdot \sqrt{8} &= 3 \cdot \sqrt{4 \cdot 2} && \text{Factor 8 so that one factor is a perfect square.} \\ &= 3 \cdot \sqrt{4} \cdot \sqrt{2} && \text{Use the product rule.} \\ &= 3 \cdot 2 \cdot \sqrt{2} && \text{Write } \sqrt{4} \text{ as 2.} \\ &= 6 \cdot \sqrt{2} \text{ or } 6\sqrt{2} && \text{Write } 3 \cdot 2 \text{ as 6.} \end{aligned}$$

□

PRACTICE 4 Simplify $5\sqrt{40}$. ■

OBJECTIVE 4

Simplifying Square Roots Using the Quotient Rule

Next, let's examine the square root of a quotient.

$$\sqrt{\frac{16}{4}} = \sqrt{4} = 2$$

Also,

$$\frac{\sqrt{16}}{\sqrt{4}} = \frac{4}{2} = 2$$

Since both expressions equal 2, we can write

$$\sqrt{\frac{16}{4}} = \frac{\sqrt{16}}{\sqrt{4}}$$

This suggests the following quotient rule.

Quotient Rule for Square Roots

If $\sqrt{a}$ and $\sqrt{b}$ are real numbers and $b \neq 0$, then

$$\sqrt{\frac{a}{b}} = \frac{\sqrt{a}}{\sqrt{b}}$$

In other words, the square root of a quotient is equal to the quotient of the square roots.

EXAMPLE 5 Simplify.

a. $\sqrt{\frac{25}{36}}$ **b.** $\sqrt{\frac{3}{64}}$ **c.** $\sqrt{\frac{40}{81}}$

Use the quotient rule.

Solution

a. $\sqrt{\frac{25}{36}} = \frac{\sqrt{25}}{\sqrt{36}} = \frac{5}{6}$

b. $\sqrt{\frac{3}{64}} = \frac{\sqrt{3}}{\sqrt{64}} = \frac{\sqrt{3}}{8}$

c. $\sqrt{\frac{40}{81}} = \frac{\sqrt{40}}{\sqrt{81}}$ Use the quotient rule.

$= \frac{\sqrt{4} \cdot \sqrt{10}}{9}$ Apply the product rule and write $\sqrt{81}$ as 9.

$= \frac{2\sqrt{10}}{9}$ Write $\sqrt{4}$ as 2. □

PRACTICE 5 Simplify.

a. $\sqrt{\frac{5}{49}}$ **b.** $\sqrt{\frac{9}{100}}$ **c.** $\sqrt{\frac{18}{25}}$ ■

OBJECTIVE 5

Adding and Subtracting Square Roots

To combine like terms, we use the distributive property.

$$5x + 3x = (5 + 3)x = 8x$$

The distributive property can also be applied to expressions containing radicals. For example,

$$5\sqrt{2} + 3\sqrt{2} = (5 + 3)\sqrt{2} = 8\sqrt{2}$$

Also,

$$9\sqrt{5} - 6\sqrt{5} = (9 - 6)\sqrt{5} = 3\sqrt{5}$$

Radical terms $5\sqrt{2}$ and $3\sqrt{2}$ are **like radicals,** as are $9\sqrt{5}$ and $6\sqrt{5}$.

Like Radicals

Like radicals are radical expressions that have the same index and the same radicand.

From the previous examples, we can see that **only like radicals can be combined** in this way. For example, the expression $2\sqrt{3} + 3\sqrt{2}$ cannot be simplified further since the radicals are not like radicals.

EXAMPLE 6 Simplify by combining like radical terms.

a. $4\sqrt{5} + 3\sqrt{5}$ **b.** $\sqrt{10} - 6\sqrt{10}$

Solution

a. $4\sqrt{5} + 3\sqrt{5} = (4 + 3)\sqrt{5} = 7\sqrt{5}$

b. $\sqrt{10} - 6\sqrt{10} = 1\sqrt{10} - 6\sqrt{10} = (1 - 6)\sqrt{10} = -5\sqrt{10}$ □

PRACTICE 6 Simplify by combining like radical terms.

a. $3\sqrt{2} + 5\sqrt{2}$ **b.** $\sqrt{6} - 8\sqrt{6}$ ■

✔ CONCEPT CHECK

Which is true?

a. $2 + 3\sqrt{5} = 5\sqrt{5}$
b. $2\sqrt{3} + 2\sqrt{7} = 2\sqrt{10}$
c. $\sqrt{3} + \sqrt{5} = \sqrt{8}$
d. $\sqrt{3} + \sqrt{3} = 3$
e. None of the above is true.

At first glance, it appears that the expression $\sqrt{50} + \sqrt{8}$ cannot be simplified further because the radicands are different. However, the product rule can be used to simplify each radical, and then further simplification might be possible.

EXAMPLE 7 Add or subtract by first simplifying each radical.

a. $\sqrt{50} + \sqrt{8}$ **b.** $7\sqrt{12} - \sqrt{75}$ **c.** $\sqrt{25} - \sqrt{27} - 2\sqrt{18} - \sqrt{16}$

Solution

a. First simplify each radical.

$\sqrt{50} + \sqrt{8} = \sqrt{25 \cdot 2} + \sqrt{4 \cdot 2}$ Factor radicands.

$= \sqrt{25} \cdot \sqrt{2} + \sqrt{4} \cdot \sqrt{2}$ Apply the product rule.

$= 5\sqrt{2} + 2\sqrt{2}$ Simplify $\sqrt{25}$ and $\sqrt{4}$.

$= 7\sqrt{2}$ Add like radicals.

b. $7\sqrt{12} - \sqrt{75} = 7\sqrt{4 \cdot 3} - \sqrt{25 \cdot 3}$ Factor radicands.

$= 7\sqrt{4} \cdot \sqrt{3} - \sqrt{25} \cdot \sqrt{3}$ Apply the product rule.

$= 7 \cdot 2\sqrt{3} - 5\sqrt{3}$ Simplify $\sqrt{4}$ and $\sqrt{25}$.

$= 14\sqrt{3} - 5\sqrt{3}$ Multiply.

$= 9\sqrt{3}$ Subtract like radicals.

c. $\sqrt{25} - \sqrt{27} - 2\sqrt{18} - \sqrt{16}$

$= 5 - \sqrt{9 \cdot 3} - 2\sqrt{9 \cdot 2} - 4$ Factor radicands.

$= 5 - \sqrt{9} \cdot \sqrt{3} - 2\sqrt{9} \cdot \sqrt{2} - 4$ Apply the product rule.

$= 5 - 3\sqrt{3} - 2 \cdot 3\sqrt{2} - 4$ Simplify.

$= 1 - 3\sqrt{3} - 6\sqrt{2}$ Write $5 - 4$ as 1 and $2 \cdot 3$ as 6. □

PRACTICE 7 Add or subtract by first simplifying each radical.

a. $\sqrt{45} + \sqrt{20}$ **b.** $\sqrt{36} + 3\sqrt{24} - \sqrt{40} - \sqrt{150}$

c. $\sqrt{98} - 5\sqrt{8}$ ■

Answer to Concept Check:
e (a, b, and c are not true since each left side cannot be simplified further. For d, $\sqrt{3} + \sqrt{3} = 2\sqrt{3}$.)

OBJECTIVE
6 Multiplying and Dividing Square Roots

We used the product and quotient rules for radicals to help us simplify radicals. We now use these same rules to simplify products and quotients of radicals.

For example, by the Product Rule,

$$\sqrt{3}\cdot\sqrt{2} = \sqrt{3\cdot 2} = \sqrt{6}$$

EXAMPLE 8 Multiply. Then simplify if possible.

a. $\sqrt{7}\cdot\sqrt{3}$ **b.** $\sqrt{3}\cdot\sqrt{3}$ **c.** $\sqrt{3}\cdot\sqrt{15}$ **d.** $2\sqrt{3}\cdot 5\sqrt{2}$

Solution

a. $\sqrt{7}\cdot\sqrt{3} = \sqrt{7\cdot 3} = \sqrt{21}$

b. $\sqrt{3}\cdot\sqrt{3} = \sqrt{3\cdot 3} = \sqrt{9} = 3$

c. $\sqrt{3}\cdot\sqrt{15} = \sqrt{45}$. Next, simplify $\sqrt{45}$.

$\sqrt{45} = \sqrt{9\cdot 5} = \sqrt{9}\cdot\sqrt{5} = 3\sqrt{5}$

d. $2\sqrt{3}\cdot 5\sqrt{2} = 2\cdot 5\sqrt{3\cdot 2} = 10\sqrt{6}$ □

PRACTICE
8 Multiply. Then simplify if possible.

a. $\sqrt{11}\cdot\sqrt{7}$ **b.** $9\sqrt{10}\cdot 8\sqrt{3}$ **c.** $\sqrt{5}\cdot\sqrt{10}$
d. $\sqrt{17}\cdot\sqrt{17}$ ■

From Example 8(b), we found that

$$\sqrt{3}\cdot\sqrt{3} = 3 \quad\text{or}\quad (\sqrt{3})^2 = 3$$

This is true in general.

If a is a positive number,

$$\sqrt{a}\cdot\sqrt{a} = a \quad\text{or}\quad (\sqrt{a})^2 = a$$

✔ **CONCEPT CHECK**

Identify the true statement(s).

a. $\sqrt{7}\cdot\sqrt{7} = 7$ **b.** $\sqrt{2}\cdot\sqrt{3} = 6$ **c.** $(\sqrt{131})^2 = 131$

EXAMPLE 9 Find: $(3\sqrt{2})^2$

Solution $(3\sqrt{2})^2 = 3^2\cdot(\sqrt{2})^2 = 9\cdot 2 = 18$ □

PRACTICE
9 Find $(2\sqrt{7})^2$. ■

When multiplying radical expressions containing more than one term, use the same techniques we use to multiply other algebraic expressions with more than one term.

EXAMPLE 10 Multiply. Then simplify if possible.

a. $\sqrt{5}(\sqrt{5}-\sqrt{2})$ **b.** $(\sqrt{5}-7)(\sqrt{5}+7)$

Solution

a. Using the distributive property, we have

$$\sqrt{5}(\sqrt{5}-\sqrt{2}) = \sqrt{5}\cdot\sqrt{5} - \sqrt{5}\cdot\sqrt{2}$$
$$= 5 - \sqrt{10}$$ Since $\sqrt{5}\cdot\sqrt{5} = 5$ and $\sqrt{5}\cdot\sqrt{2} = \sqrt{10}$

Answer to Concept Check: a, c

(Continued on next page)

b. Recall that $(a - b)(a + b) = a^2 - b^2$. Then

$$(\sqrt{5} - 7)(\sqrt{5} + 7) = (\sqrt{5})^2 - 7^2 = 5 - 49 = -44$$

□

PRACTICE

10 Multiply. Then simplify if possible.

a. $\sqrt{3}(\sqrt{3} - \sqrt{5})$ **b.** $(\sqrt{7} + 4)(\sqrt{7} - 4)$ ■

The special product formulas can be used to multiply expressions containing radicals.

To simplify quotients of radical expressions, we use the quotient rule.

EXAMPLE 11 Divide. Then simplify if possible.

a. $\dfrac{\sqrt{14}}{\sqrt{2}}$ **b.** $\dfrac{\sqrt{100}}{\sqrt{5}}$

Solution Use the quotient rule and then simplify the resulting radicand.

a. $\dfrac{\sqrt{14}}{\sqrt{2}} = \sqrt{\dfrac{14}{2}} = \sqrt{7}$

b. $\dfrac{\sqrt{100}}{\sqrt{5}} = \sqrt{\dfrac{100}{5}} = \sqrt{20} = \sqrt{4 \cdot 5} = \sqrt{4} \cdot \sqrt{5} = 2\sqrt{5}$ □

PRACTICE

11 Divide. Then simplify if possible.

a. $\dfrac{\sqrt{21}}{\sqrt{7}}$ **b.** $\dfrac{\sqrt{48}}{\sqrt{6}}$ ■

OBJECTIVE

7 Rationalizing Denominators

It is sometimes easier to work with radical expressions if the denominator does not contain a radical. To rewrite the expression so that the denominator does not contain a radical expression, we use the fact that we can multiply the numerator and the denominator of a fraction by the same nonzero number without changing the value of the expression. This is the same as multiplying the fraction by 1. For example, to get rid of the radical in the denominator of $\dfrac{\sqrt{5}}{\sqrt{2}}$, we multiply by 1 in the form of $\dfrac{\sqrt{2}}{\sqrt{2}}$. Then

$$\frac{\sqrt{5}}{\sqrt{2}} = \frac{\sqrt{5}}{\sqrt{2}} \cdot 1 = \frac{\sqrt{5}}{\sqrt{2}} \cdot \frac{\sqrt{2}}{\sqrt{2}} = \frac{\sqrt{5}}{\sqrt{2}} \cdot \frac{\sqrt{2}}{\sqrt{2}} = \frac{\sqrt{10}}{2}$$

This process is called **rationalizing** the denominator.

EXAMPLE 12 Rationalize each denominator.

a. $\dfrac{2}{\sqrt{7}}$ **b.** $\dfrac{\sqrt{5}}{\sqrt{12}}$

Solution

a. To rewrite $\dfrac{2}{\sqrt{7}}$ so that there is no radical in the denominator, we multiply by 1 in the form of $\dfrac{\sqrt{7}}{\sqrt{7}}$.

$$\frac{2}{\sqrt{7}} = \frac{2}{\sqrt{7}} \cdot \frac{\sqrt{7}}{\sqrt{7}} = \frac{2 \cdot \sqrt{7}}{\sqrt{7} \cdot \sqrt{7}} = \frac{2\sqrt{7}}{7}$$

b. We can multiply by $\frac{\sqrt{12}}{\sqrt{12}}$, but see what happens if we simplify first.

$$\frac{\sqrt{5}}{\sqrt{12}} = \frac{\sqrt{5}}{\sqrt{4 \cdot 3}} = \frac{\sqrt{5}}{2\sqrt{3}}$$

To rationalize the denominator now, we multiply by $\frac{\sqrt{3}}{\sqrt{3}}$.

$$\frac{\sqrt{5}}{2\sqrt{3}} = \frac{\sqrt{5}}{2\sqrt{3}} \cdot \frac{\sqrt{3}}{\sqrt{3}} = \frac{\sqrt{5} \cdot \sqrt{3}}{2\sqrt{3} \cdot \sqrt{3}} = \frac{\sqrt{15}}{2 \cdot 3} = \frac{\sqrt{15}}{6}$$ □

PRACTICE 12 Rationalize each denominator.

a. $\frac{4}{\sqrt{5}}$ **b.** $\frac{\sqrt{3}}{\sqrt{18}}$ ■

EXAMPLE 13 Simplify: $\frac{12 - \sqrt{18}}{9}$

Solution First simplify $\sqrt{18}$.

$$\frac{12 - \sqrt{18}}{9} = \frac{12 - \sqrt{9 \cdot 2}}{9} = \frac{12 - 3\sqrt{2}}{9}$$

Next, factor out a common factor of 3 from the terms in the numerator and the denominator and simplify.

$$\frac{12 - 3\sqrt{2}}{9} = \frac{3(4 - \sqrt{2})}{3 \cdot 3} = \frac{4 - \sqrt{2}}{3}$$ □

PRACTICE 13 Simplify $\frac{14 - \sqrt{28}}{6}$. ■

Calculator Explorations

To simplify or approximate square roots using a calculator, locate the key marked $\boxed{\sqrt{\ }}$. To simplify $\sqrt{25}$ using a scientific calculator, press $\boxed{25}$ $\boxed{\sqrt{\ }}$. The display should read $\boxed{\quad 5}$. To simplify $\sqrt{25}$ using a graphing calculator, press $\boxed{\sqrt{\ }}$ $\boxed{25}$ $\boxed{\text{ENTER}}$.

To approximate $\sqrt{30}$, press $\boxed{30}$ $\boxed{\sqrt{\ }}$ (or $\boxed{\sqrt{\ }}$ $\boxed{30}$ $\boxed{\text{ENTER}}$). The display should read $\boxed{5.4772256}$. This is an approximation for $\sqrt{30}$. A three-decimal-place approximation is

$$\sqrt{30} \approx 5.477$$

Is this answer reasonable? Since 30 is between perfect squares 25 and 36, $\sqrt{30}$ is between $\sqrt{25} = 5$ and $\sqrt{36} = 6$. The calculator result is then reasonable since 5.4772256 is between 5 and 6.

Use a calculator to approximate each expression to three decimal places. Decide whether each result is reasonable.

1. $\sqrt{7}$ **2.** $\sqrt{14}$ **3.** $\sqrt{11}$

4. $\sqrt{200}$ **5.** $\sqrt{82}$ **6.** $\sqrt{46}$

Vocabulary, Readiness & Video Check

Use the choices below to fill in each blank. Not all choices will be used.

principal	radical sign	radicand	$a \cdot b$
$\frac{a}{b}$	$\frac{\sqrt{a}}{\sqrt{b}}$	$\sqrt{a} \cdot \sqrt{b}$	

1. In the expression $\sqrt{16}$, the number 16 is called the __________, and $\sqrt{\ }$ is called the ____________.
2. The symbol $\sqrt{\ }$ is used to denote the positive, or __________, square root.
3. If $\sqrt{a}$ and $\sqrt{b}$ are real numbers, then $\sqrt{a \cdot b} =$ __________.
4. If $\sqrt{a}$ and $\sqrt{b}$ are real numbers, then $\sqrt{\frac{a}{b}} =$ __________.

Answer each exercise true or false.

5. $\sqrt{-16}$ simplifies to a real number. __________
6. $\sqrt{0} = 0$ and $\sqrt{1} = 1$. __________

Fill in each blank.

7. Radicals that have the same index and same radicand are called ____________.
8. The expressions $7\sqrt{2x}$ and $-\sqrt{2x}$ are called ____________.
9. $11\sqrt{2} + 6\sqrt{2} =$ __________.
   a. $66\sqrt{2}$ **b.** $17\sqrt{2}$ **c.** $17\sqrt{4}$
10. $\sqrt{5}$ is the same as ____________.
    a. $0\sqrt{5}$ **b.** $1\sqrt{5}$ **c.** $5\sqrt{5}$
11. $\sqrt{5} + \sqrt{5} =$ __________
    a. $\sqrt{10}$ **b.** 5 **c.** $2\sqrt{5}$
12. $9\sqrt{7} - \sqrt{7} =$ __________
    a. $8\sqrt{7}$ **b.** 9 **c.** 0
13. $\sqrt{7} \cdot \sqrt{3} =$ __________
14. $\sqrt{10} \cdot \sqrt{10} =$ __________
15. $\frac{\sqrt{15}}{\sqrt{3}} =$ __________
16. The process of eliminating the radical in the denominator of a radical expression is called ____________________.

Martin-Gay Interactive Videos

See Video 5.5

Watch the section lecture video and answer the following questions.

OBJECTIVE 1

17. Explain the differences between Examples 1 and 2, including how you know which one simplifies to a positive number and which one to a negative number.

OBJECTIVE 2

18. From Example 9, how do you determine if an approximate answer is reasonable?

OBJECTIVE 3

19. From Example 12, if you have trouble finding a perfect square factor in the radicand, what is recommended?

OBJECTIVE 4

20. Based on the lecture before Example 14, complete the following statement. In words, the quotient rule for square roots says that the square root of a quotient is equal to the square root of the _____ over the square root of the _____.

OBJECTIVE 5

21. From Examples 15–17, how is combining like radicals similar to combining like terms?

22. From Example 18, why should you always check to see if all radical terms in your expression are simplified before attempting to add or subtract the radicals?

OBJECTIVE 6

23. In Example 20, the product rule for radicals is applied twice, but in different ways. Explain.

24. From Example 22, when you're looking at a quotient of two radicals, what would make you think to apply the quotient rule in order to simplify?

OBJECTIVE 7

25. From the lecture before Example 23, what is the goal of rationalizing a denominator?

5.5 Exercise Set MyMathLab®

Find each square root. See Example 1.

1. $\sqrt{16}$
2. $\sqrt{49}$
3. $\sqrt{\frac{1}{25}}$
4. $\sqrt{\frac{1}{64}}$
5. $-\sqrt{100}$
6. $-\sqrt{36}$
7. $\sqrt{-4}$
8. $\sqrt{-25}$
9. $-\sqrt{121}$
10. $-\sqrt{49}$
11. $\sqrt{\frac{9}{25}}$
12. $\sqrt{\frac{25}{36}}$
13. $\sqrt{900}$
14. $\sqrt{400}$
15. $\sqrt{144}$
16. $\sqrt{169}$

Approximate each square root to three decimal places. See Example 2.

17. $\sqrt{7}$
18. $\sqrt{10}$
19. $\sqrt{37}$
20. $\sqrt{27}$
21. $\sqrt{136}$
22. $\sqrt{8}$

Use the product rule to simplify each radical. See Examples 3 and 4.

23. $\sqrt{20}$
24. $\sqrt{44}$
25. $\sqrt{50}$
26. $\sqrt{28}$
27. $\sqrt{33}$
28. $\sqrt{21}$
29. $\sqrt{98}$
30. $\sqrt{125}$
31. $\sqrt{180}$
32. $\sqrt{150}$
33. $3\sqrt{25}$
34. $9\sqrt{36}$
35. $7\sqrt{63}$
36. $11\sqrt{99}$
37. $-5\sqrt{27}$
38. $-6\sqrt{75}$

Use the quotient rule and the product rule to simplify each radical. See Example 5.

39. $\sqrt{\frac{8}{25}}$
40. $\sqrt{\frac{63}{16}}$
41. $\sqrt{\frac{27}{121}}$
42. $\sqrt{\frac{24}{169}}$
43. $\sqrt{\frac{9}{4}}$
44. $\sqrt{\frac{100}{49}}$
45. $-\sqrt{\frac{27}{144}}$
46. $-\sqrt{\frac{84}{121}}$

Add or subtract by first simplifying each radical (if needed) and then combining any like radical terms. See Examples 6 and 7.

47. $4\sqrt{3} - 8\sqrt{3}$
48. $2\sqrt{5} - 9\sqrt{5}$
49. $3\sqrt{6} + 8\sqrt{6} - 2\sqrt{6} - 5$
50. $12\sqrt{2} - 3\sqrt{2} + 8\sqrt{2} + 10$
51. $\sqrt{11} + \sqrt{11} + 11$
52. $\sqrt{13} + 13 + \sqrt{13}$
53. $\sqrt{12} + \sqrt{27}$
54. $\sqrt{50} + \sqrt{18}$
55. $\sqrt{45} + 3\sqrt{20}$
56. $\sqrt{28} + \sqrt{63}$
57. $2\sqrt{54} - \sqrt{20} + \sqrt{45} - \sqrt{24}$
58. $2\sqrt{8} - \sqrt{128} + \sqrt{48} + \sqrt{18}$
59. $\sqrt{35} - \sqrt{140}$
60. $\sqrt{6} - \sqrt{600}$
61. $\sqrt{\frac{5}{9}} + \sqrt{\frac{5}{81}}$
62. $\sqrt{\frac{3}{64}} + \sqrt{\frac{3}{16}}$
63. $\sqrt{\frac{3}{4}} - \sqrt{\frac{3}{64}}$
64. $\sqrt{\frac{7}{25}} - \sqrt{\frac{7}{100}}$

Multiply and simplify. See Examples 8 through 10.

65. $\sqrt{8}\cdot\sqrt{2}$
66. $\sqrt{3}\cdot\sqrt{12}$
67. $\sqrt{10}\cdot\sqrt{5}$
68. $\sqrt{2}\cdot\sqrt{14}$
69. $(\sqrt{6})^2$
70. $(\sqrt{10})^2$
71. $(2\sqrt{5})^2$
72. $(3\sqrt{10})^2$
73. $\sqrt{6}(\sqrt{5} + \sqrt{7})$
74. $\sqrt{10}(\sqrt{3} - \sqrt{7})$
75. $\sqrt{10}(\sqrt{2} + \sqrt{5})$
76. $\sqrt{6}(\sqrt{3} + \sqrt{2})$
77. $(\sqrt{3} + 6)(\sqrt{3} - 6)$
78. $(\sqrt{5} + 2)(\sqrt{5} - 2)$
79. $(2\sqrt{11} + 1)(\sqrt{11} - 6)$
80. $(5\sqrt{3} + 2)(\sqrt{3} - 1)$

Divide and simplify. See Example 11.

81. $\frac{\sqrt{32}}{\sqrt{2}}$ **82.** $\frac{\sqrt{40}}{\sqrt{10}}$

83. $\frac{\sqrt{21}}{\sqrt{3}}$ **84.** $\frac{\sqrt{55}}{\sqrt{5}}$

85. $\frac{\sqrt{90}}{\sqrt{5}}$ **86.** $\frac{\sqrt{96}}{\sqrt{8}}$

Rationalize each denominator and simplify. See Example 12.

87. $\frac{\sqrt{3}}{\sqrt{5}}$ **88.** $\frac{\sqrt{2}}{\sqrt{3}}$

89. $\frac{7}{\sqrt{2}}$ **90.** $\frac{8}{\sqrt{11}}$

91. $\sqrt{\frac{1}{8}}$ **92.** $\sqrt{\frac{1}{27}}$

93. $\sqrt{\frac{2}{15}}$ **94.** $\sqrt{\frac{11}{14}}$

95. $\sqrt{\frac{3}{20}}$ **96.** $\sqrt{\frac{3}{50}}$

Simplify the following. See Example 13.

97. $\frac{6 + 2\sqrt{3}}{2}$ **98.** $\frac{9 + 6\sqrt{2}}{3}$

99. $\frac{18 - 12\sqrt{5}}{6}$ **100.** $\frac{8 - 20\sqrt{3}}{4}$

101. $\frac{15\sqrt{3} + 5}{5}$ **102.** $\frac{8 + 16\sqrt{2}}{8}$

REVIEW AND PREVIEW

Solve each equation. See Sections 2.3 and 4.3.

103. $x + 5 = 7^2$

104. $2y - 1 = 3^2$

105. $4z^2 + 6z - 12 = (2z)^2$

106. $16x^2 + x + 9 = (4x)^2$

CONCEPT EXTENSIONS

The length of a side of a square is given by the expression $\sqrt{A}$, where A is the square's area. Use this expression for Exercises 107 through 110. Be sure to attach the appropriate units.

△ **107.** The area of a square is 49 square miles. Find the length of a side of the square.

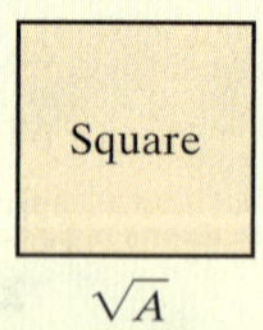

△ **108.** The area of a square is $\frac{1}{81}$ square meter. Find the length of a side of the square.

△ **109.** Sony currently makes the smallest portable mini disc player. It is approximately in the shape of a square with top area of 9.61 square inches. Find the length of a side. (*Source:* SONY)

△ **110.** A parking lot is in the shape of a square with area 2500 square yards. Find the length of a side.

111. Simplify $\sqrt{\sqrt{81}}$.

112. Simplify $\sqrt{\sqrt{10{,}000}}$.

For each square root below, give two whole numbers that the square root lies between. For example, since $\sqrt{11}$ is between $\sqrt{9}$ and $\sqrt{16}$, $\sqrt{11}$ is between 3 and 4.

113. $\sqrt{18}$ **114.** $\sqrt{28}$

115. $\sqrt{80}$ **116.** $\sqrt{98}$

117. In your own words, describe like radicals.

118. In the expression $\sqrt{5} + 2 - 3\sqrt{5}$, explain why 2 and -3 cannot be combined.

119. Find the perimeter of the rectangular picture frame.

120. Find the perimeter of the plot of land.

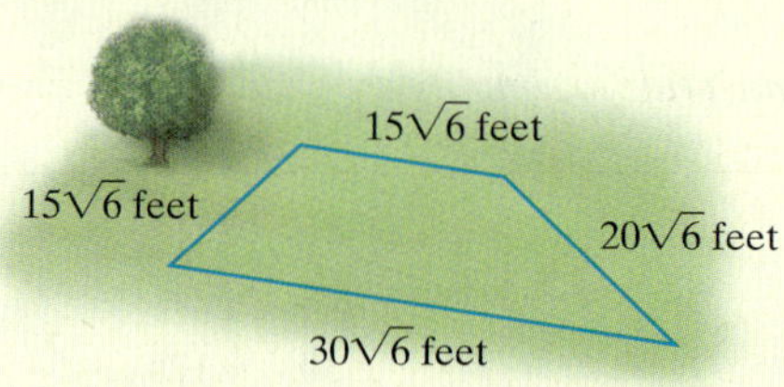

△ **121.** Find the area of a rectangle whose length is $13\sqrt{2}$ meters and width is $5\sqrt{6}$ meters.

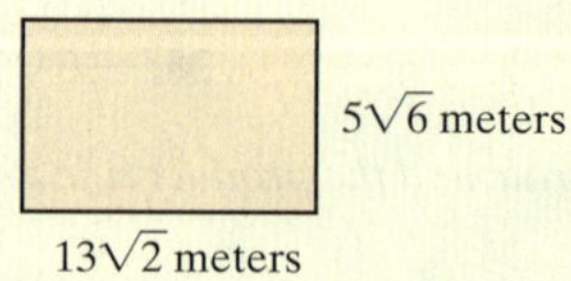

△ **122.** Find the volume of a box whose length is $\sqrt{3}$ feet, width is $\sqrt{2}$ feet, and height is $\sqrt{2}$ feet.

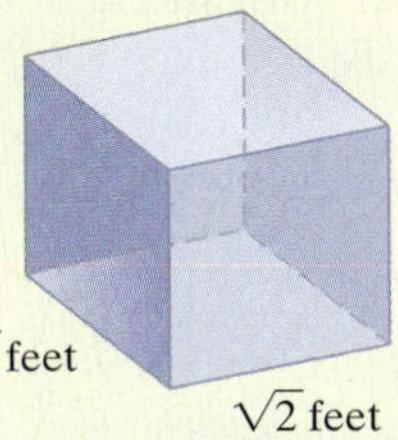

△ **123.** If a circle has area A, then the formula for the radius r of the circle is

$$r = \sqrt{\frac{A}{\pi}}$$

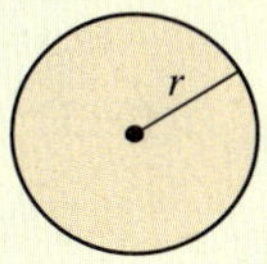

Simplify this expression by rationalizing the denominator.

124. The formula for calculating the period (time for one back-and-forth swing) of a pendulum is $T = 2\pi\sqrt{\frac{L}{g}}$, where T is the period of the swing, L is the length of the pendulum, and g is the acceleration of gravity. Suppose there is a Foucault's pendulum with length = 30 ft and $g = 32 \text{ ft/sec}^2$. Using $\pi \approx 3.14$, find the period of this pendulum. Round to the nearest tenth of a second.

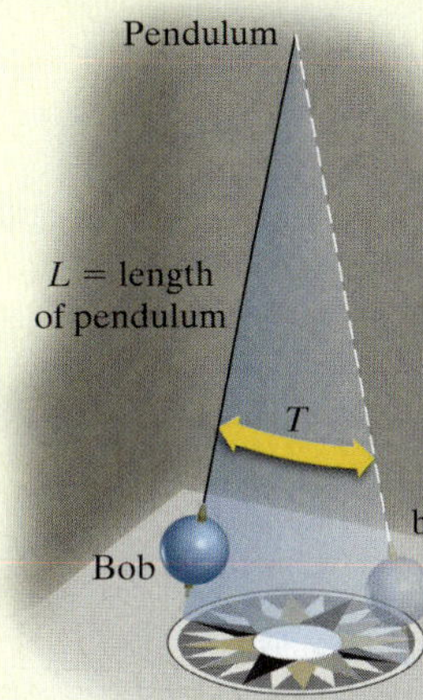

(*Note:* In 1851, French physicist Léon Foucault used a pendulum to demonstrate that the earth rotates on its axis.)

5.6 Solving Quadratic Equations by the Square Root Property and the Quadratic Formula

OBJECTIVES

1. Use the Square Root Property to Solve Quadratic Equations.
2. Use the Quadratic Formula to Solve Quadratic Equations

OBJECTIVE

1 Using the Square Root Property

Recall that a quadratic equation is an equation that can be written in the form

$$ax^2 + bx + c = 0$$

where a, b, and c are real numbers and $a \neq 0$.

Thus far, to solve quadratic equations by factoring, we used the **zero factor property:**

> **Zero Factor Property** If a and b are real numbers and if $ab = 0$, then $a = 0$ or $b = 0$.

Let's solve the quadratic equation $x^2 - 4 = 0$ using a method other than factoring. First, add 4 to both sides of the equation.

$$x^2 - 4 = 0$$
$$x^2 = 4 \quad \text{Add 4 to both sides.}$$

Now we see that the value for x must be a number whose square is 4. Therefore $x = \sqrt{4} = 2$ or $x = -\sqrt{4} = -2$. This reasoning is an example of the square root property.

> **Square Root Property**
> If $x^2 = a$ for $a \geq 0$, then
> $$x = \sqrt{a} \quad \text{or} \quad x = -\sqrt{a}$$

EXAMPLE 1 Use the square root property to solve $x^2 - 9 = 0$.

Solution First we solve for x^2 by adding 9 to both sides.

$$x^2 - 9 = 0$$
$$x^2 = 9 \quad \text{Add 9 to both sides.}$$

Next we use the square root property.

$$x = \sqrt{9} \quad \text{or} \quad x = -\sqrt{9}$$
$$x = 3 \quad \text{or} \quad x = -3$$

Check:

$$x^2 - 9 = 0 \quad \text{Original equation} \qquad x^2 - 9 = 0 \quad \text{Original equation}$$
$$3^2 - 9 \stackrel{?}{=} 0 \quad \text{Let } x = 3. \qquad (-3)^2 - 9 \stackrel{?}{=} 0 \quad \text{Let } x = -3.$$
$$0 = 0 \quad \text{True} \qquad 0 = 0 \quad \text{True}$$

The solutions are 3 and -3. □

PRACTICE

1 Use the square root property to solve $x^2 - 16 = 0$. ■

EXAMPLE 2 Use the square root property to solve $2x^2 = 7$.

Solution First we solve for x^2 by dividing both sides by 2. Then we use the square root property.

$$2x^2 = 7$$
$$x^2 = \frac{7}{2} \quad \text{Divide both sides by 2.}$$
$$x = \sqrt{\frac{7}{2}} \quad \text{or} \quad x = -\sqrt{\frac{7}{2}} \quad \text{Use the square root property.}$$

If the denominators are rationalized, we have

$$x = \frac{\sqrt{7} \cdot \sqrt{2}}{\sqrt{2} \cdot \sqrt{2}} \quad \text{or} \quad x = -\frac{\sqrt{7} \cdot \sqrt{2}}{\sqrt{2} \cdot \sqrt{2}} \quad \text{Rationalize the denominator.}$$
$$x = \frac{\sqrt{14}}{2} \quad \text{or} \quad x = -\frac{\sqrt{14}}{2} \quad \text{Simplify.}$$

Remember to check both solutions in the original equation. The solutions are $\frac{\sqrt{14}}{2}$ and $-\frac{\sqrt{14}}{2}$. □

PRACTICE

2 Use the square root property to solve $5x^2 = 13$. ■

EXAMPLE 3 Use the square root property to solve $(x - 3)^2 = 16$.

Solution Instead of x^2, here we have $(x - 3)^2$. But the square root property can still be used.

$$(x - 3)^2 = 16$$
$$x - 3 = \sqrt{16} \quad \text{or} \quad x - 3 = -\sqrt{16} \quad \text{Use the square root property.}$$
$$x - 3 = 4 \quad \text{or} \quad x - 3 = -4 \quad \text{Write } \sqrt{16} \text{ as 4 and } -\sqrt{16} \text{ as } -4.$$
$$x = 7 \quad \text{or} \quad x = -1 \quad \text{Solve.}$$

Check:

$(x-3)^2 = 16$	Original equation	$(x-3)^2 = 16$	Original equation
$(7-3)^2 \stackrel{?}{=} 16$	Let $x = 7$.	$(-1-3)^2 \stackrel{?}{=} 16$	Let $x = -1$.
$4^2 \stackrel{?}{=} 16$	Simplify.	$(-4)^2 \stackrel{?}{=} 16$	Simplify.
$16 = 16$	True	$16 = 16$	True

Both 7 and -1 are solutions. □

PRACTICE 3 Use the square root property to solve $(x-5)^2 = 36$. ■

EXAMPLE 4 Use the square root property to solve $(x+1)^2 = 8$.

Solution $(x+1)^2 = 8$

$x + 1 = \sqrt{8}$ or $x + 1 = -\sqrt{8}$ Use the square root property.

$x + 1 = 2\sqrt{2}$ or $x + 1 = -2\sqrt{2}$ Simplify the radical.

$x = -1 + 2\sqrt{2}$ or $x = -1 - 2\sqrt{2}$ Solve for x.

Check both solutions in the original equation. The solutions are $-1 + 2\sqrt{2}$ and $-1 - 2\sqrt{2}$. This can be written compactly as $-1 \pm 2\sqrt{2}$. The notation $\pm$ is read as "plus or minus." □

Helpful Hint

read "plus or minus"

The notation $-1 \pm \sqrt{5}$, for example, is just a shorthand notation for both $-1 + \sqrt{5}$ and $-1 - \sqrt{5}$.

PRACTICE 4 Use the square root property to solve $(x+2)^2 = 12$. ■

EXAMPLE 5 Use the square root property to solve $(x-1)^2 = -2$.

Solution This equation has no real solution because the square root of -2 is not a real number. □

PRACTICE 5 Use the square root property to solve $(x-8)^2 = -5$. ■

EXAMPLE 6 Use the square root property to solve $(5x-2)^2 = 10$.

Solution $(5x-2)^2 = 10$

$5x - 2 = \sqrt{10}$ or $5x - 2 = -\sqrt{10}$ Use the square root property.

$5x = 2 + \sqrt{10}$ or $5x = 2 - \sqrt{10}$ Add 2 to both sides.

$x = \dfrac{2+\sqrt{10}}{5}$ or $x = \dfrac{2-\sqrt{10}}{5}$ Divide both sides by 5.

Check both solutions in the original equation. The solutions are $\dfrac{2+\sqrt{10}}{5}$ and $\dfrac{2-\sqrt{10}}{5}$, which can be written as $\dfrac{2 \pm \sqrt{10}}{5}$. □

PRACTICE 6 Use the square root property to solve $(3x-5)^2 = 17$. ■

Helpful Hint

For some applications and graphing purposes, decimal approximations of exact solutions to quadratic equations may be desired.

Exact Solutions from Example 6		Decimal Approximations
$\dfrac{2+\sqrt{10}}{5}$	$\approx$	1.032
$\dfrac{2-\sqrt{10}}{5}$	$\approx$	−0.232

OBJECTIVE

2 Using the Quadratic Formula

We can use a technique called completing the square to develop a formula to find solutions of any quadratic equation. This formula is called the **quadratic formula** and is shown below. (*Note:* We do not develop this formula in this text.)

Quadratic Formula

If a, b, and c are real numbers and $a \neq 0$, a quadratic equation written in the form $ax^2 + bx + c = 0$ has solutions

$$x = \frac{-b \pm \sqrt{b^2 - 4ac}}{2a}$$

Helpful Hint

Don't forget that to identify a, b, and c correctly in the quadratic formula, you should write the equation in standard form.

Quadratic Equations in Standard Form

$5x^2 - 6x + 2 = 0$	$a = 5, b = -6, c = 2$
$4y^2 - 9 = 0$	$a = 4, b = 0, c = -9$
$x^2 + x = 0$	$a = 1, b = 1, c = 0$
$\sqrt{2}x^2 + \sqrt{5}x + \sqrt{3} = 0$	$a = \sqrt{2}, b = \sqrt{5}, c = \sqrt{3}$

EXAMPLE 7 Solve $3x^2 + x - 3 = 0$ using the quadratic formula.

Solution This equation is in standard form with $a = 3$, $b = 1$, and $c = -3$. By the quadratic formula, we have

$$x = \frac{-b \pm \sqrt{b^2 - 4ac}}{2a}$$

$$x = \frac{-1 \pm \sqrt{1^2 - 4 \cdot 3 \cdot (-3)}}{2 \cdot 3} \quad \text{Let } a = 3, b = 1, \text{ and } c = -3.$$

$$= \frac{-1 \pm \sqrt{1 + 36}}{6} \quad \text{Simplify.}$$

$$= \frac{-1 \pm \sqrt{37}}{6}$$

Check both solutions in the original equation. The solutions are $\frac{-1 + \sqrt{37}}{6}$ and $\frac{-1 - \sqrt{37}}{6}$. □

PRACTICE

7 Solve $5x^2 + x - 2 = 0$ using the quadratic formula. ■

EXAMPLE 8 Solve $2x^2 - 9x = 5$ using the quadratic formula.

Solution First we write the equation in standard form by subtracting 5 from both sides.

$$2x^2 - 9x = 5$$

$$2x^2 - 9x - 5 = 0$$

Next we note that $a = 2$, $b = -9$, and $c = -5$. We substitute these values into the quadratic formula.

$$x = \frac{-b \pm \sqrt{b^2 - 4ac}}{2a}$$

Helpful Hint

Notice that the fraction bar is under the entire numerator of $-b \pm \sqrt{b^2 - 4ac}$.

$$x = \frac{-(-9) \pm \sqrt{(-9)^2 - 4 \cdot 2 \cdot (-5)}}{2 \cdot 2}$$ Substitute in the formula.

$$= \frac{9 \pm \sqrt{81 + 40}}{4}$$ Simplify.

$$= \frac{9 \pm \sqrt{121}}{4} = \frac{9 \pm 11}{4}$$

Then,

$$x = \frac{9 - 11}{4} = -\frac{1}{2} \quad \text{or} \quad x = \frac{9 + 11}{4} = 5$$

Check $-\frac{1}{2}$ and 5 in the original equation. Both $-\frac{1}{2}$ and 5 are solutions. □

PRACTICE 8 Solve $3x^2 + 2x = 8$ using the quadratic formula.

The following steps may be useful when solving a quadratic equation by the quadratic formula.

Solving a Quadratic Equation by the Quadratic Formula

Step 1. Write the quadratic equation in standard form: $ax^2 + bx + c = 0$.

Step 2. If necessary, clear the equation of fractions to simplify calculations.

Step 3. Identify a, b, and c.

Step 4. Replace a, b, and c in the quadratic formula with the identified values and simplify.

✔ CONCEPT CHECK

For the quadratic equation $2x^2 - 5 = 7x$, if $a = 2$ and $c = -5$ in the quadratic formula, the value of b is which of the following?

a. $\frac{7}{2}$ **b.** 7 **c.** -5 **d.** -7

EXAMPLE 9 Solve $7x^2 = 1$ using the quadratic formula.

Solution First we write the equation in standard form by subtracting 1 from both sides.

$$7x^2 = 1$$
$$7x^2 - 1 = 0$$

Helpful Hint

$7x^2 - 1 = 0$ can be written as $7x^2 + 0x - 1 = 0$. This form helps you see that $b = 0$.

Next we replace a, b, and c with the identified values: $a = 7$, $b = 0$, and $c = -1$.

$$x = \frac{0 \pm \sqrt{0^2 - 4 \cdot 7 \cdot (-1)}}{2 \cdot 7}$$ Substitute in the formula.

$$= \frac{\pm\sqrt{28}}{14}$$ Simplify.

Answer to Concept Check:
d

(Continued on next page)

$$= \frac{\pm 2\sqrt{7}}{14}$$

$$= \pm\frac{\sqrt{7}}{7}$$

The solutions are $\frac{\sqrt{7}}{7}$ and $-\frac{\sqrt{7}}{7}$. □

PRACTICE 9 Solve $3x^2 = 5$ using the quadratic formula. ■

Notice that the equation in Example 9, $7x^2 = 1$, could have been easily solved by dividing both sides by 7 and then using the square root property. We solved the equation by the quadratic formula to show that this formula can be used to solve any quadratic equation.

EXAMPLE 10 Solve $x^2 = -x - 1$ using the quadratic formula.

Solution First we write the equation in standard form.

$$x^2 + x + 1 = 0$$

Next we replace a, b, and c in the quadratic formula with $a = 1$, $b = 1$, and $c = 1$.

$$x = \frac{-1 \pm \sqrt{1^2 - 4\cdot 1\cdot 1}}{2\cdot 1} \quad \text{Substitute in the formula.}$$

$$= \frac{-1 \pm \sqrt{-3}}{2} \quad \text{Simplify.}$$

There is no real number solution because $\sqrt{-3}$ is not a real number. □

PRACTICE 10 Solve $x^2 = 3x - 4$ using the quadratic formula. ■

EXAMPLE 11 Solve $\frac{1}{2}x^2 - x = 2$ using the quadratic formula.

Solution We write the equation in standard form and then clear the equation of fractions by multiplying both sides by the LCD, 2.

$$\frac{1}{2}x^2 - x = 2$$

$$\frac{1}{2}x^2 - x - 2 = 0 \quad \text{Write in standard form.}$$

$$x^2 - 2x - 4 = 0 \quad \text{Multiply both sides by 2.}$$

Here, $a = 1$, $b = -2$, and $c = -4$, so we substitute these values into the quadratic formula.

$$x = \frac{-(-2) \pm \sqrt{(-2)^2 - 4\cdot 1\cdot(-4)}}{2\cdot 1}$$

$$= \frac{2 \pm \sqrt{20}}{2} = \frac{2 \pm 2\sqrt{5}}{2} \quad \text{Simplify.}$$

$$= \frac{2(1 \pm \sqrt{5})}{2} = 1 \pm \sqrt{5} \quad \text{Factor and simplify.}$$

The solutions are $1 - \sqrt{5}$ and $1 + \sqrt{5}$. □

PRACTICE 11 Solve $\frac{1}{5}x^2 - x = 1$ using the quadratic formula. ■

Notice that in Example 11, although we cleared the equation of fractions, the coefficients $a = \frac{1}{2}$, $b = -1$, and $c = -2$ will give the same results.

Helpful Hint

When simplifying an expression such as

$$\frac{3 \pm 6\sqrt{2}}{6}$$

first factor out a common factor from the terms of the numerator and then simplify.

$$\frac{3 \pm 6\sqrt{2}}{6} = \frac{3(1 \pm 2\sqrt{2})}{2 \cdot 3} = \frac{1 \pm 2\sqrt{2}}{2}$$

Sometimes approximate solutions for quadratic equations are appropriate.

EXAMPLE 12 Approximate the exact solutions of the quadratic equation in Example 7. Round the approximations to the nearest tenth.

Solution From Example 7, we have exact solutions $\frac{-1 \pm \sqrt{37}}{6}$. Thus,

$$\frac{-1 + \sqrt{37}}{6} \approx 0.847127088 \approx 0.8 \text{ to the nearest tenth.}$$

$$\frac{-1 - \sqrt{37}}{6} \approx -1.180460422 \approx -1.2 \text{ to the nearest tenth.}$$

Thus approximate solutions to the quadratic equation in Example 7 are 0.8 and −1.2. □

PRACTICE 12 Approximate the exact solutions of the quadratic equation in Practice 7. Round the approximations to the nearest tenth. ■

Vocabulary, Readiness & Video Check

Fill in each blank.

1. The quadratic formula is ______________.

2. Identify the values of a, b, and c in the quadratic equation: $x^2 - 6 = 0$; $a =$ ______, $b =$ ______, $c =$ ______

Simplify the following.

3. $\frac{-1 \pm \sqrt{1^2 - 4(1)(-2)}}{2(1)}$

4. $\frac{-(-5) \pm \sqrt{(-5)^2 - 4(2)(3)}}{2(2)}$

5. $\frac{-5 \pm \sqrt{5^2 - 4(1)(2)}}{2(1)}$

6. $\frac{-7 \pm \sqrt{7^2 - 4(2)(1)}}{2(2)}$

Martin-Gay Interactive Videos

See Video 5.6

Watch the section lecture video and answer the following questions.

OBJECTIVE 1

7. As explained in Example 2, why is $a \geq 0$ in the statement of the square root property?

OBJECTIVE 2

8. Based on the lectures and Examples 6–8, answer the following.
a. Must a quadratic equation be written in standard form in order to use the quadratic formula? Why or why not?
b. Must fractions be cleared from the equation before using the quadratic formula? Why or why not?

9. From Example 9, how are approximate solutions found?

5.6 Exercise Set MyMathLab®

Use the square root property to solve each quadratic equation. See Examples 1 through 6.

1. $x^2 = 64$ **2.** $x^2 = 121$

3. $x^2 = \frac{1}{25}$ **4.** $x^2 = \frac{1}{16}$

5. $x^2 = -4$ **6.** $x^2 = -25$

7. $3x^2 = 13$ **8.** $5x^2 = 2$

9. $2x^2 - 10 = 0$ **10.** $7x^2 - 21 = 0$

11. $(x - 5)^2 = 49$ **12.** $(x + 2)^2 = 25$

13. $(x + 2)^2 = 7$ **14.** $(x - 7)^2 = 2$

15. $(p + 2)^2 = 10$ **16.** $(p - 7)^2 = 13$

17. $(3x - 7)^2 = 32$ **18.** $(5x - 11)^2 = 54$

19. $(2p - 5)^2 = 121$ **20.** $(3p - 1)^2 = 4$

Use the quadratic formula to solve each quadratic equation. See Examples 7 through 11.

21. $x^2 - 3x + 2 = 0$ **22.** $x^2 - 5x - 6 = 0$

23. $3k^2 + 7k + 1 = 0$ **24.** $7k^2 + 3k - 1 = 0$

25. $49x^2 - 4 = 0$ **26.** $25x^2 - 15 = 0$

27. $5z^2 - 4z + 3 = 0$ **28.** $3z^2 + 2z + 1 = 0$

29. $3 - x^2 = 4x$ **30.** $10 - x^2 = 2x$

31. $6x^2 + 9x = 2$ **32.** $3x^2 - 9x = 8$

33. $3p^2 - \frac{2}{3}p + 1 = 0$ **34.** $\frac{5}{2}p^2 - p + \frac{1}{2} = 0$

35. $5z^2 - 2z = \frac{1}{5}$ **36.** $9z^2 + 12z = -1$

Use the quadratic formula to solve each quadratic equation. Find the exact solutions; then approximate these solutions to the nearest tenth. See Example 12.

37. $x^2 + 6x + 1 = 0$ **38.** $x^2 + 4x + 2 = 0$

39. $x^2 = 9x + 4$ **40.** $x^2 = 7x + 5$

MIXED PRACTICE

An important skill in mathematics is learning when to use one technique in favor of another. The following steps may be used to solve a quadratic equation.

To Solve a Quadratic Equation

Step 1. If the equation is in the form $ax^2 = c$ or $(ax + b)^2 = c$, use the square root property and solve. If not, go to Step 2.

Step 2. Write the equation in standard form: $ax^2 + bx + c = 0$.

Step 3. Try to solve the equation by the factoring method. If not possible, go to Step 4.

Step 4. Solve the equation by the quadratic formula.

Choose and use a method to solve each equation.

41. $5x^2 - 11x + 2 = 0$ **42.** $5x^2 + 13x - 6 = 0$

43. $x^2 - 1 = 2x$ **44.** $x^2 + 7 = 6x$

45. $a^2 = 20$ **46.** $a^2 = 72$

47. $x^2 - x + 4 = 0$ **48.** $x^2 - 2x + 7 = 0$

49. $3x^2 - 12x + 12 = 0$ **50.** $5x^2 - 30x + 45 = 0$

51. $9 - 6p + p^2 = 0$ **52.** $49 - 28p + 4p^2 = 0$

53. $4y^2 - 16 = 0$ **54.** $3y^2 - 27 = 0$

55. $x^4 - 3x^3 + 2x^2 = 0$ **56.** $x^3 + 7x^2 + 12x = 0$

57. $(2z + 5)^2 = 25$ **58.** $(3z - 4)^2 = 16$

59. $30x = 25x^2 + 2$ **60.** $12x = 4x^2 + 4$

61. $\frac{2}{3}m^2 - \frac{1}{3}m - 1 = 0$ **62.** $\frac{5}{8}m^2 + m - \frac{1}{2} = 0$

63. $x^2 - \frac{1}{2}x - \frac{1}{5} = 0$ **64.** $x^2 + \frac{1}{2}x - \frac{1}{8} = 0$

65. $4x^2 - 27x + 35 = 0$ **66.** $9x^2 - 16x + 7 = 0$

67. $(7 - 5x)^2 = 18$ **68.** $(5 - 4x)^2 = 75$

69. $3z^2 - 7z = 12$

70. $6z^2 + 7z = 6$

71. $x = x^2 - 110$

72. $x = 56 - x^2$

73. $\frac{3}{4}x^2 - \frac{5}{2}x - 2 = 0$

74. $x^2 - \frac{6}{5}x - \frac{8}{5} = 0$

75. $x^2 - 0.6x + 0.05 = 0$

76. $x^2 - 0.1x - 0.06 = 0$

77. $10x^2 - 11x + 2 = 0$

78. $20x^2 - 11x + 1 = 0$

79. $\frac{1}{2}z^2 - 2z + \frac{3}{4} = 0$

80. $\frac{1}{5}z^2 - \frac{1}{2}z - 2 = 0$

REVIEW AND PREVIEW

Simplify each radical. See Section 5.5.

81. $\sqrt{48}$

82. $\sqrt{104}$

83. $\sqrt{50}$

84. $\sqrt{80}$

Solve the following. See Section 2.4.

△ **85.** The height of a triangle is 4 times the length of the base. The area of the triangle is 18 square feet. Find the height and base of the triangle. (*Note:* A $= \frac{1}{2} \cdot$ base $\cdot$ height)

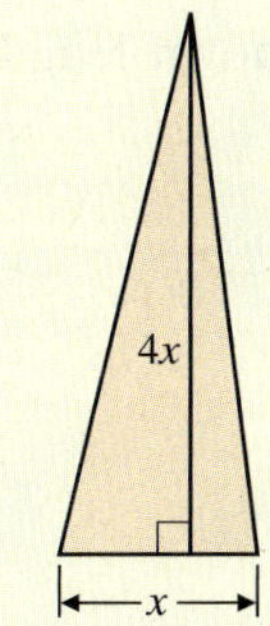

△ **86.** The length of a rectangle is 6 inches more than its width. The area of the rectangle is 391 square inches. Find the dimensions of the rectangle. (*Note:* A = length $\cdot$ width)

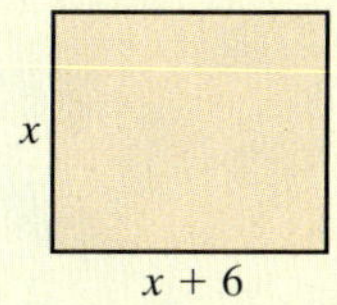

CONCEPT EXTENSIONS

Solve. See the Concept Check in this section. For the quadratic equation $5x^2 + 2 = x$, if $a = 5$ in the quadratic formula,

87. What is the value of b?

a. $\frac{1}{5}$ **b.** 0 **c.** -1 **d.** 1

88. What is the value of c?

a. 5 **b.** x **c.** -2 **d.** 2

For the quadratic equation $7y^2 = 3y$, if $b = 3$ in the quadratic formula,

89. What is the value of a?

a. 7 **b.** -7 **c.** 0 **d.** 1

90. What is the value of c?

a. 7 **b.** 3 **c.** 0 **d.** 1

Solve each equation using the quadratic formula.

91. $x^2 + 3\sqrt{2}x - 5 = 0$

92. $y^2 - 2\sqrt{5}y - 1 = 0$

93. Explain how to identify a, b, and c correctly when solving a quadratic equation by the quadratic formula.

94. Explain how the quadratic formula is developed and why it is useful.

Use the quadratic formula and a calculator to solve each equation. Round solutions to the nearest tenth.

95. $7.3z^2 + 5.4z - 1.1 = 0$

96. $1.2x^2 - 5.2x - 3.9 = 0$

5.7 The Pythagorean Theorem, the Distance Formula, and Problem-Solving

OBJECTIVES

1 Use the Pythagorean Formula to Solve Problems.

2 Use the Distance Formula.

3 Solve Problems Using Formulas Containing Radicals.

4 Solve Problems Modeled by Quadratic Equations

OBJECTIVE

1 Using the Pythagorean Formula

Applications of radicals can be found in geometry, finance, science, and other areas of technology. Our first application involves the Pythagorean theorem, which is a formula that relates the lengths of the three sides of a right triangle.

The Pythagorean Theorem

If a and b are lengths of the legs of a right triangle and c is the length of the hypotenuse, then $a^2 + b^2 = c^2$, or $(\text{leg})^2 + (\text{leg})^2 = (\text{hypotenuse})^2$

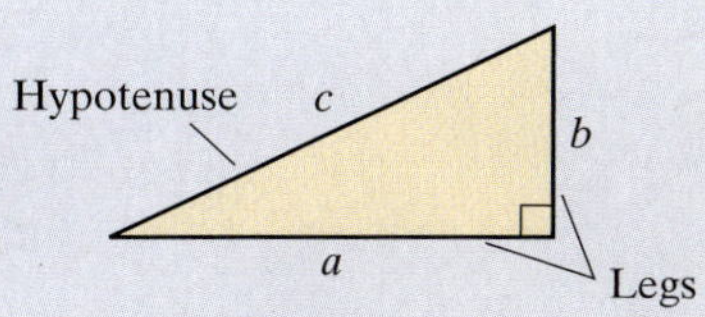

Helpful Hint

Don't forget that for this formula, c represents the hypotenuse.

That is, the square of the length of the hypotenuse is equal to the sum of the squares of the lengths of the legs.

EXAMPLE 1 Find the length of the hypotenuse of a right triangle whose legs are 6 inches and 8 inches long.

Solution Because this is a right triangle, we use the Pythagorean theorem. We let $a = 6$ inches and $b = 8$ inches. Length c must be the length of the hypotenuse.

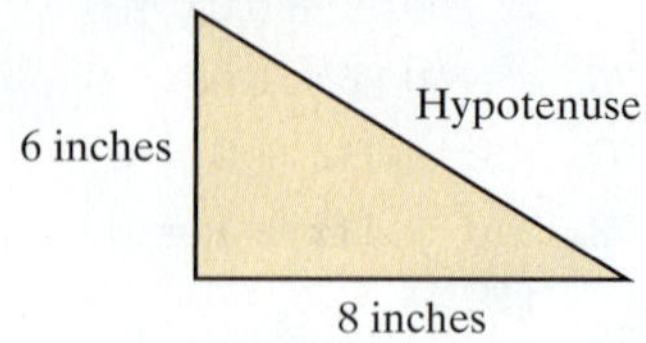

$$a^2 + b^2 = c^2 \quad \text{Use the Pythagorean theorem.}$$
$$6^2 + 8^2 = c^2 \quad \text{Substitute the lengths of the legs.}$$
$$36 + 64 = c^2 \quad \text{Simplify.}$$
$$100 = c^2$$

Since c represents a length, we know that c is positive and is the principal square root of 100.

$$100 = c^2$$
$$\sqrt{100} = c \quad \text{Use the definition of principal square root.}$$
$$10 = c \quad \text{Simplify.}$$

The hypotenuse has a length of 10 inches. □

PRACTICE

1 Find the length of the hypotenuse of a right triangle whose legs are 5 inches and 12 inches long. ■

EXAMPLE 2 Find the length of the leg of the right triangle shown. Give the exact length and a two-decimal-place approximation.

Solution We let $a = 2$ meters and b be the unknown length of the other leg. The hypotenuse is $c = 5$ meters.

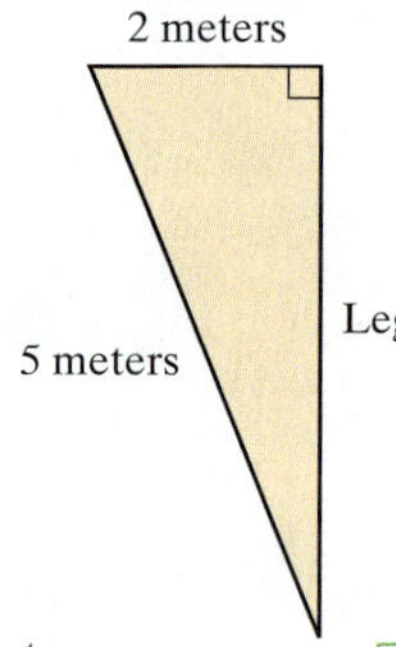

$$a^2 + b^2 = c^2 \quad \text{Use the Pythagorean theorem.}$$
$$2^2 + b^2 = 5^2 \quad \text{Let } a = 2 \text{ and } c = 5.$$
$$4 + b^2 = 25$$
$$b^2 = 21$$
$$b = \sqrt{21} \approx 4.58 \text{ meters}$$

The length of the leg is exactly $\sqrt{21}$ meters or approximately 4.58 meters. □

PRACTICE

2 Find the length of the leg of the right triangle shown. Give the exact length and a two-decimal-place approximation.

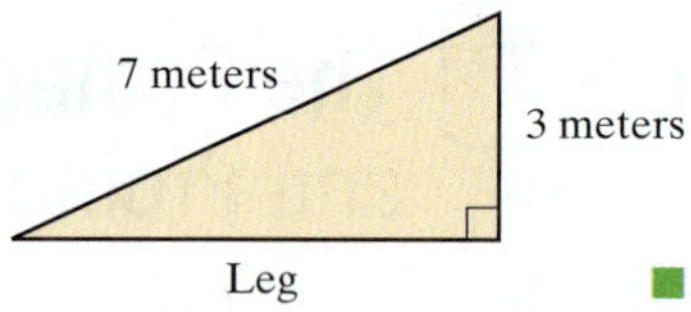

■

EXAMPLE 3 **Finding a Distance**

A surveyor must determine the distance across a lake at points P and Q as shown in the figure. To do this, she finds a third point R perpendicular to line PQ. If the length of $\overline{PR}$ is 320 feet and the length of $\overline{QR}$ is 240 feet, what is the distance across the lake? Approximate this distance to the nearest whole foot.

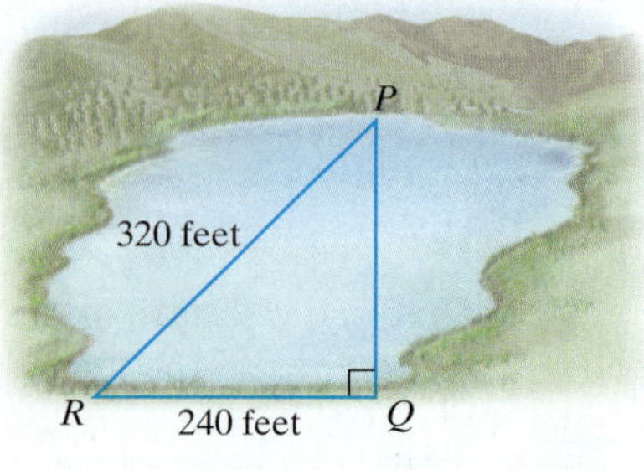

Solution

1. UNDERSTAND. Read and reread the problem. We will set up the problem using the Pythagorean theorem. By creating a line perpendicular to line PQ, the surveyor deliberately constructed a right triangle. The hypotenuse, $\overline{PR}$, has a length of 320 feet, so we let $c = 320$ in the Pythagorean theorem. The side $\overline{QR}$ is one of the legs, so we let $a = 240$ and $b =$ the unknown length.

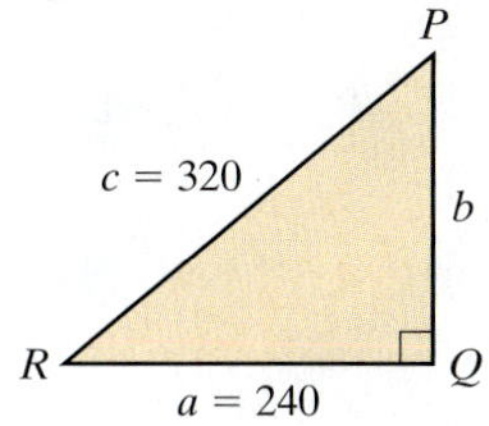

2. TRANSLATE.

$$a^2 + b^2 = c^2 \quad \text{Use the Pythagorean theorem.}$$
$$240^2 + b^2 = 320^2 \quad \text{Let } a = 240 \text{ and } c = 320.$$

3. SOLVE.

$$57{,}600 + b^2 = 102{,}400$$
$$b^2 = 44{,}800 \quad \text{Subtract 57,600 from both sides.}$$
$$b = \sqrt{44{,}800} \quad \text{Use the definition of principal square root.}$$

4. INTERPRET.

Check: See that $240^2 + (\sqrt{44{,}800})^2 = 320^2$.

State: The distance across the lake is **exactly** $\sqrt{44{,}800}$ feet. The surveyor can now use a calculator to find that $\sqrt{44{,}800}$ feet is **approximately** 211.6601 feet, so the distance across the lake is roughly 212 feet. □

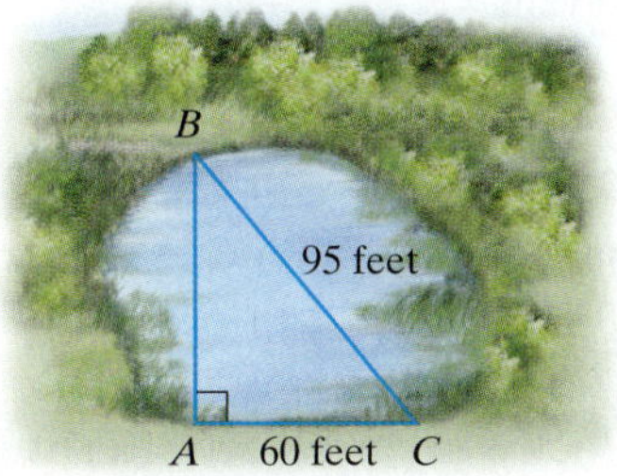

PRACTICE 3 Find the length of a bridge, to the nearest whole foot, to be constructed from point A to point B across Little Marsh. Use the distances shown in the diagram. ■

OBJECTIVE 2

Using the Distance Formula

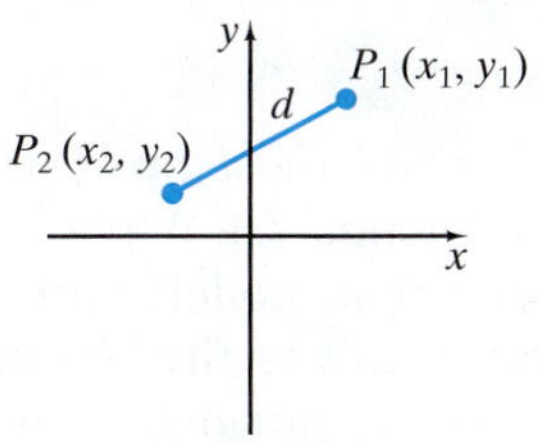

A second important application of radicals is in finding the distance between two points in the plane. By using the Pythagorean theorem, the following formula can be derived.

> **Distance Formula**
>
> The distance d between two points with coordinates (x_1, y_1) and (x_2, y_2) is given by
>
> $$d = \sqrt{(x_2 - x_1)^2 + (y_2 - y_1)^2}$$

EXAMPLE 4 Find the distance between $(-1, 9)$ and $(-3, -5)$.

Solution Use the distance formula with $(x_1, y_1) = (-1, 9)$ and $(x_2, y_2) = (-3, -5)$.

$$\begin{aligned} d &= \sqrt{(x_2 - x_1)^2 + (y_2 - y_1)^2} && \text{The distance formula.} \\ &= \sqrt{[-3 - (-1)]^2 + (-5 - 9)^2} && \text{Substitute known values.} \\ &= \sqrt{(-2)^2 + (-14)^2} && \text{Simplify.} \\ &= \sqrt{4 + 196} \\ &= \sqrt{200} = 10\sqrt{2} && \text{Simplify the radical.} \end{aligned}$$

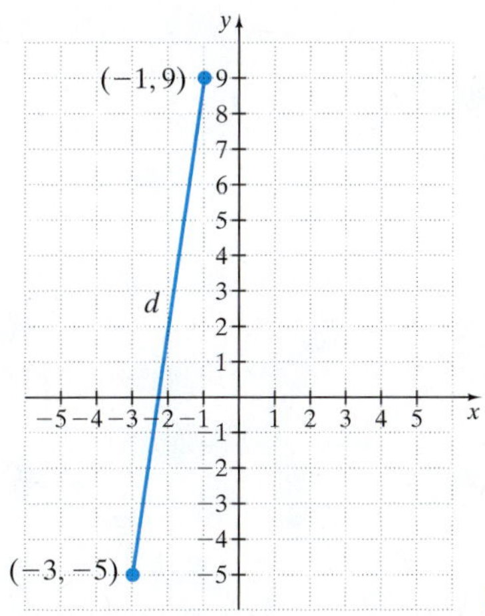

The distance is **exactly** $10\sqrt{2}$ units or **approximately** 14.1 units. □

(Continued on next page)

PRACTICE 4 Find the distance between $(-2, 5)$ and $(-4, -7)$.

OBJECTIVE 3 Using Formulas Containing Radicals

The Pythagorean theorem is an extremely important result in mathematics and should be memorized. But there are other applications involving formulas containing radicals that are not quite as well known, such as the velocity formula used in the next example.

EXAMPLE 5 Determining Velocity

A formula used to determine the velocity v, in feet per second, of an object (neglecting air resistance) after it has fallen a certain height is $v = \sqrt{2gh}$, where g is the acceleration due to gravity, and h is the height the object has fallen. On Earth, the acceleration g due to gravity is approximately 32 feet per second per second. Find the velocity of a person after falling 5 feet.

Solution We are told that $g = 32$ feet per second per second. To find the velocity v when $h = 5$ feet, we use the velocity formula.

$$v = \sqrt{2gh} \quad \text{Use the velocity formula.}$$
$$= \sqrt{2 \cdot 32 \cdot 5} \quad \text{Substitute known values.}$$
$$= \sqrt{320}$$
$$= 8\sqrt{5} \quad \text{Simplify the radicand.}$$

The velocity of the person after falling 5 feet is **exactly** $8\sqrt{5}$ feet per second, or **approximately** 17.9 feet per second.

PRACTICE 5 Use the formula in Example 5 to find the velocity of an object after it has fallen 12 feet.

OBJECTIVE 4 Solving Problems Modeled by Quadratic Equations

Some problems may be modeled by quadratic equations. To solve these problems, we use the same problem-solving steps that were introduced in Section 2.4. When solving these problems, keep in mind that a solution of an equation that models a problem may not be a solution to the problem. For example, a person's age or the length of a rectangle is always a positive number. Discard solutions that do not make sense as solutions of the problem.

EXAMPLE 6 Finding Free-Fall Time

Since the 1940s, one of the top tourist attractions in Acapulco, Mexico, is watching the La Quebrada cliff divers. The divers' platform is about 144 feet above the sea. These divers must time their descent just right, since they land in the crashing Pacific in an inlet that averages 11 feet deep. Neglecting air resistance, the height $h(t)$ in feet of a cliff diver above the ocean after t seconds is given by the quadratic function $h(t) = -16t^2 + 144$.

Find how long it takes the diver to reach the ocean. (*Source:* visitmexico.com)

Solution

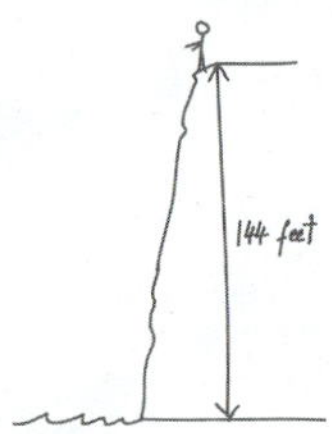

1. UNDERSTAND. Read and reread the problem. Then draw a picture of the problem.

   The function $h(t) = -16t^2 + 144$ models the height of the falling diver at time t. Familiarize yourself with this equation by finding the height of the diver at time $t = 1$ second and $t = 2$ seconds.

When $t = 1$ second, the height of the diver is $h(1) = -16(1)^2 + 144 = 128$ feet.
When $t = 2$ seconds, the height of the diver is $h(2) = -16(2)^2 + 144 = 80$ feet.

2. TRANSLATE. To find how long it takes the diver to reach the ocean, we want to know the value of t for which $h(t) = 0$.

3. SOLVE.

$$0 = -16t^2 + 144$$

$$0 = -16(t^2 - 9) \quad \text{Factor out } -16.$$

$$0 = -16(t - 3)(t + 3) \quad \text{Factor completely.}$$

$$t - 3 = 0 \text{ or } t + 3 = 0 \quad \text{Set each factor containing a variable equal to 0.}$$

$$t = 3 \text{ or } t = -3 \quad \text{Solve.}$$

4. INTERPRET. Since the time t cannot be negative, the proposed solution is 3 seconds.

Check: Verify that the height of the diver when t is 3 seconds is 0.

When $t = 3$ seconds, $h(3) = -16(3)^2 + 144 = -144 + 144 = 0$.

State: It takes the diver 3 seconds to reach the ocean. □

PRACTICE 6

Cliff divers also frequent the falls at Waimea Falls Park in Oahu, Hawaii. One of the popular diving spots is 64 feet high. Neglecting air resistance, the height of a diver above the pool after t seconds is $h(t) = -16t^2 + 64$. Find how long it takes a diver to reach the pool. ■

EXAMPLE 7 Finding an Unknown Number

The square of a number plus three times the number is 70. Find the number.

Solution

1. UNDERSTAND. Read and reread the problem. Suppose that the number is 5. The square of 5 is 5^2 or 25. Three times 5 is 15. Then $25 + 15 = 40$, not 70, so the number is not 5. Remember, the purpose of proposing a number, such as 5, is to understand the problem better. Now that we do, we will let $x =$ the number.

2. TRANSLATE.

the square of a number	plus	three times the number	is	70
↓	↓	↓	↓	↓
x^2	$+$	$3x$	$=$	70

3. SOLVE.

$$x^2 + 3x = 70$$

$$x^2 + 3x - 70 = 0 \quad \text{Subtract 70 from both sides.}$$

$$(x + 10)(x - 7) = 0 \quad \text{Factor.}$$

$$x + 10 = 0 \text{ or } x - 7 = 0 \quad \text{Set each factor equal to 0.}$$

$$x = -10 \quad x = 7 \quad \text{Solve.}$$

4. INTERPRET.

Check: The square of -10 is $(-10)^2$, or 100. Three times -10 is $3(-10)$ or -30. Then $100 + (-30) = 70$, the correct sum, so -10 checks.

The square of 7 is 7^2 or 49. Three times 7 is 3(7), or 21. Then $49 + 21 = 70$, the correct sum, so 7 checks.

State: There are two numbers. They are -10 and 7. □

PRACTICE 7

The square of a number minus eight times the number is equal to forty-eight. Find the number. ■

EXAMPLE 8 Finding the Dimensions of a Sail

The height of a triangular sail is 2 meters less than twice the length of the base. If the sail has an area of 30 square meters, find the length of its base and the height.

Solution

1. UNDERSTAND. Read and reread the problem. Since we are finding the length of the base and the height, we let

$$x = \text{the length of the base}$$

and since the height is 2 meters less than twice the base,

$$2x - 2 = \text{the height.}$$

An illustration is shown to the right.

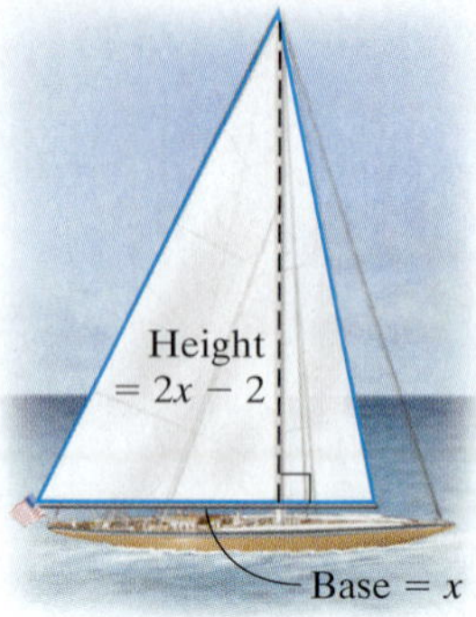

2. TRANSLATE. We are given that the area of the triangle is 30 square meters, so we use the formula for area of a triangle.

area of triangle	=	$\frac{1}{2}$	·	base	·	height
↓		↓		↓		↓
30	=	$\frac{1}{2}$	·	x	·	$(2x - 2)$

3. SOLVE. Now we solve the quadratic equation.

$$30 = \frac{1}{2}x(2x - 2)$$

$$30 = x^2 - x \quad \text{Multiply.}$$

$$x^2 - x - 30 = 0 \quad \text{Write in standard form.}$$

$$(x - 6)(x + 5) = 0 \quad \text{Factor.}$$

$$x - 6 = 0 \quad \text{or} \quad x + 5 = 0 \quad \text{Set each factor equal to 0.}$$

$$x = 6 \qquad x = -5$$

4. INTERPRET. Since x represents the length of the base, we discard the solution -5. The base of a triangle cannot be negative. The base is then 6 meters and the height is $2(6) - 2 = 10$ meters.

Check: To check this problem, we recall that $\frac{1}{2}\text{base} \cdot \text{height} = \text{area}$, or

$$\frac{1}{2}(6)(10) = 30. \quad \text{The required area}$$

State: The base of the triangular sail is 6 meters and the height is 10 meters. □

PRACTICE 8

An engineering team from Georgia Tech earned second place in a flight competition, with their triangular shaped paper hang glider. The base of their prize-winning entry was 1 foot less than three times the height. If the area of the triangular glider wing was 210 square feet, find the dimensions of the wing. (*Source: The Technique* [Georgia Tech's newspaper], April 18, 2003) ■

Study the following diagrams for a review of consecutive integers.

Examples

If x is the first integer, then consecutive integers are $x, x + 1, x + 2, \ldots$

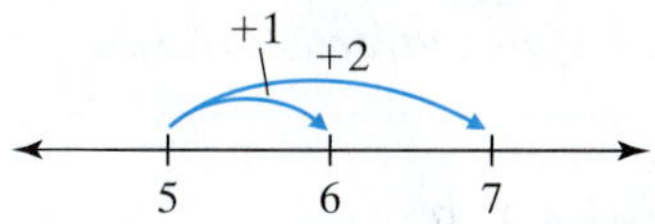

If x is the first even integer, then consecutive even integers are $x, x + 2, x + 4, \ldots$

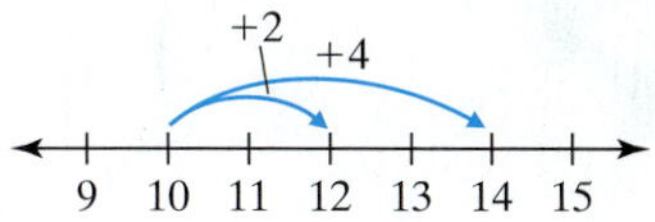

If x is the first odd integer, then consecutive odd integers are $x, x + 2, x + 4, \ldots$

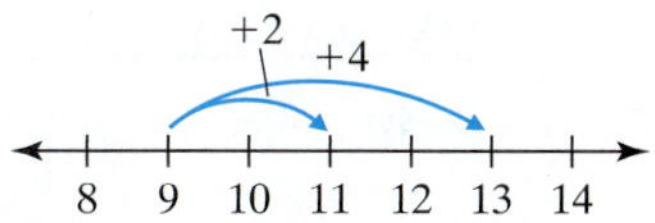

EXAMPLE 9 **Finding Consecutive Even Integers**

Find two consecutive even integers whose product is 34 more than their sum.

Solution

1. UNDERSTAND. Read and reread the problem. Let's just choose two consecutive even integers to help us better understand the problem. Let's choose 10 and 12. Their product is $10(12) = 120$ and their sum is $10 + 12 = 22$. The product is $120 - 22$, or 98, greater than the sum. Thus our guess is incorrect, but we have a better understanding of this example.

Let's let x and $x + 2$ be the consecutive even integers.

2. TRANSLATE.

Product of integers	is	34	more than	sum of integers
↓	↓	↓	↓	↓
$x(x + 2)$	$=$	34	$+$	$x + (x + 2)$

3. SOLVE. Now we solve the equation.

$$x(x + 2) = 34 + x + (x + 2)$$

$$x^2 + 2x = 34 + x + x + 2 \quad \text{Multiply.}$$

$$x^2 + 2x = 2x + 36 \quad \text{Combine like terms.}$$

$$x^2 - 36 = 0 \quad \text{Write in standard form.}$$

$$(x + 6)(x - 6) = 0 \quad \text{Factor.}$$

$$x + 6 = 0 \quad \text{or} \quad x - 6 = 0 \quad \text{Set each factor equal to 0.}$$

$$x = -6 \qquad x = 6 \quad \text{Solve.}$$

4. INTERPRET. If $x = -6$, then $x + 2 = -6 + 2$, or -4.

If $x = 6$, then $x + 2 = 6 + 2$, or 8.

Check: $-6, -4$

$$-6(-4) \stackrel{?}{=} 34 + (-6) + (-4)$$

$$24 \stackrel{?}{=} 34 + (-10)$$

$$24 = 24 \quad \text{True}$$

$6, 8$

$$6(8) \stackrel{?}{=} 34 + 6 + 8$$

$$48 \stackrel{?}{=} 34 + 14$$

$$48 = 48 \quad \text{True}$$

State: The two consecutive even integers are -6 and -4 or 6 and 8. □

PRACTICE 9 Find two consecutive integers whose product is 41 more than their sum. ■

Vocabulary, Readiness & Video Check

Martin-Gay Interactive Videos

See Video 5.7

Watch the section lecture video and answer the following questions.

OBJECTIVE 1

1. From Examples 1 and 2, when solving exercises using the Pythagorean theorem, what two things must you keep in mind?
2. What very important point is made as the final answer to Example 1 is being found?

OBJECTIVE 2

3. From Examples 3 and 4, the distance formula contains subtraction of the coordinates of two points, similar to the slope formula. In the distance formula, the subtractions are squared. In general, does order matter when subtractions are squared?

OBJECTIVE 3

4. In Examples 5 and 6, how do you know to give an estimated answer instead of an exact answer? In what form would the exact answer be given?

OBJECTIVE 4

5. In each of Examples 7–9, why aren't both solutions of the translated equation accepted as solutions to the application?

5.7 Exercise Set MyMathLab®

Use the Pythagorean theorem to find the unknown side of each right triangle. See Examples 1 and 2.

△ **1.**

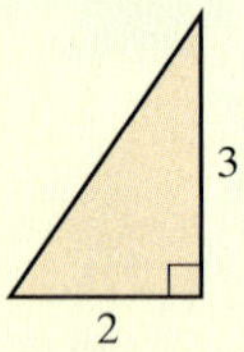

△ **2.**

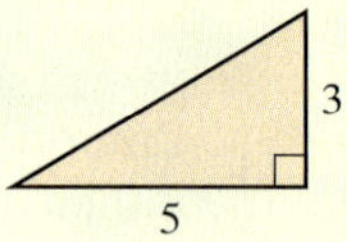

△ **3.**

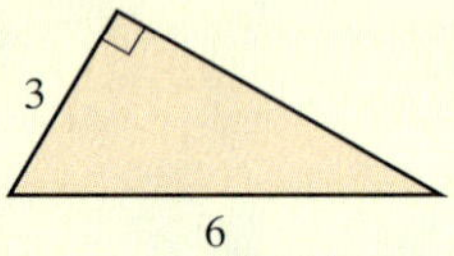

△ **4.**

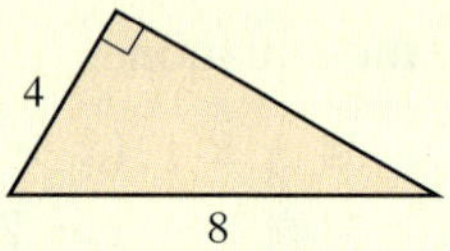

△ **5.**

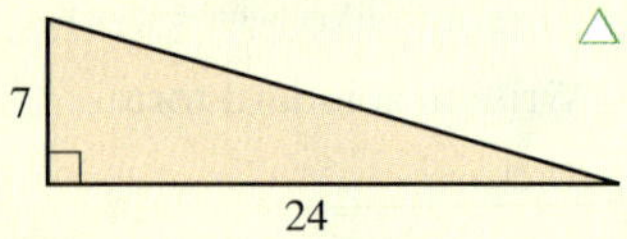

△ **6.**

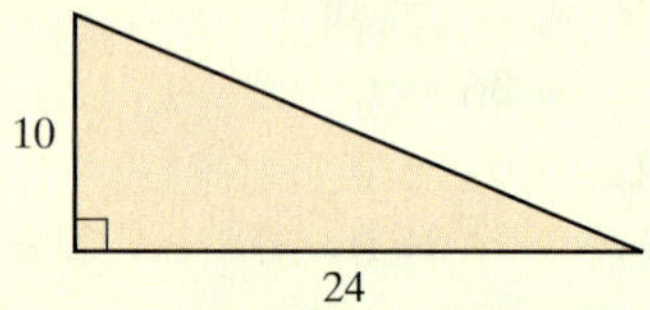

△ **7.**

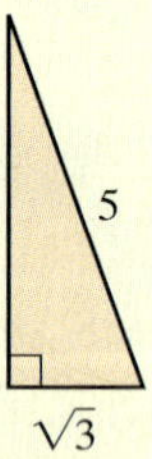

△ **8.**

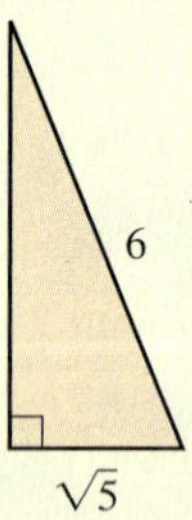

△ **9.**

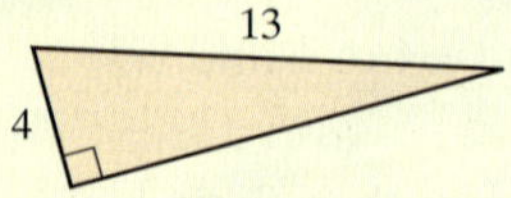

△ **10.**

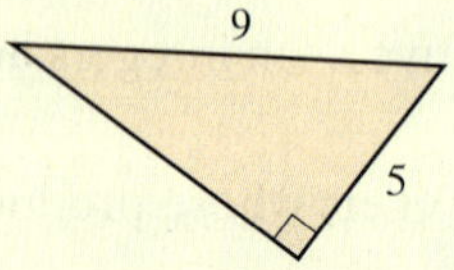

Find the length of the unknown side of each right triangle with sides a, b, and c, where c is the hypotenuse. See Examples 1 and 2.

△ **11.** $a = 4, b = 5$

△ **12.** $a = 2, b = 7$

△ **13.** $b = 2, c = 6$

△ **14.** $b = 1, c = 5$

△ **15.** $a = \sqrt{10}, c = 10$

△ **16.** $a = \sqrt{7}, c = \sqrt{35}$

Solve. See Examples 1, 2, 3, and 5.

△ **17.** Rod and Judy Pasch want to determine the distance at certain points across a pond on their property. They are able to measure the distances shown on the following diagram. Find how wide the pond is to the nearest tenth of a foot.

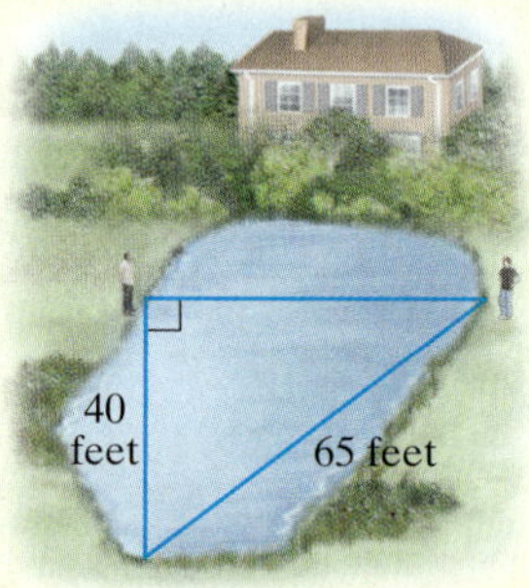

△ **18.** Use the formula from Example 5 and find the velocity of an object after it has fallen 20 feet.

△ **19.** A wire is used to anchor a 20-foot-tall pole. One end of the wire is attached to the top of the pole. The other end is fastened to a stake five feet away from the bottom of the pole. Find the length of the wire, to the nearest tenth of a foot.

△ **20.** Jim Spivey needs to connect two underground pipelines, which are offset by 3 feet, as pictured in the diagram. Neglecting the joints needed to join the pipes, find the length of the shortest possible connecting pipe rounded to the nearest hundredth of a foot.

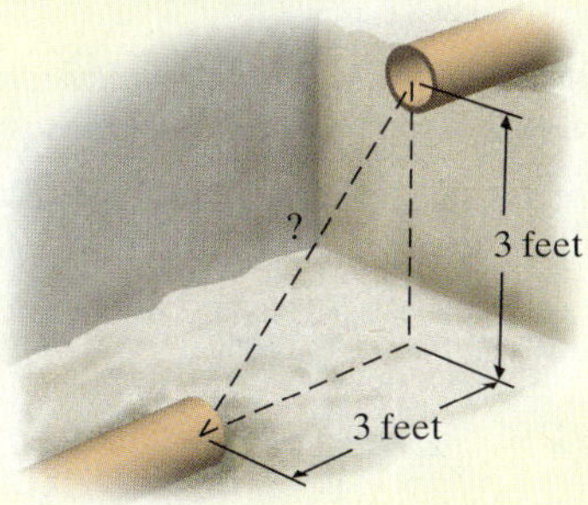

△ **21.** Robert Weisman needs to attach a diagonal brace to a rectangular frame to make it structurally sound. If the framework is 6 feet by 10 feet, find how long the brace needs to be to the nearest tenth of a foot.

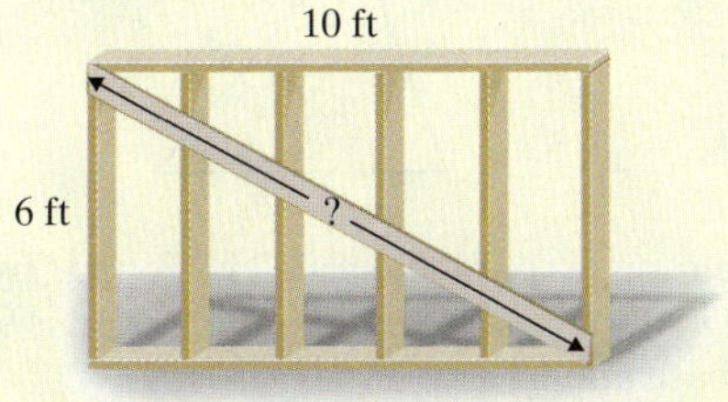

△ **22.** Elizabeth Kaster is flying a kite. She let out 80 feet of string and attached the string to a stake in the ground. The kite is now directly above her brother Mike, who is 32 feet away from Elizabeth. Find the height of the kite to the nearest foot.

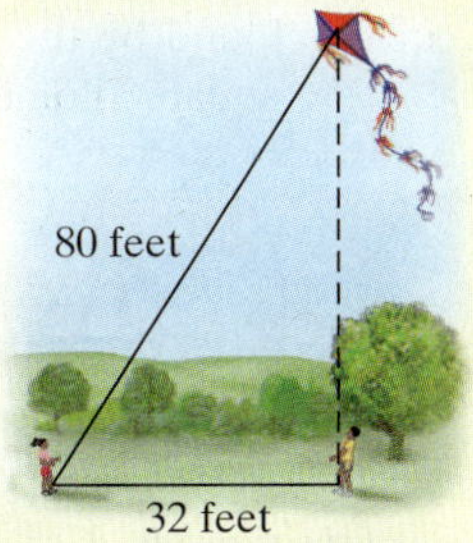

Use the distance formula to find the distance between the points given. See Example 4.

23. $(3, 6), (5, 11)$

24. $(2, 3), (9, 7)$

25. $(-3, 1), (5, -2)$

26. $(-2, 6), (3, -2)$

27. $(3, -2), (1, -8)$

28. $(-5, 8), (-2, 2)$

29. $\left(\frac{1}{2}, 2\right), (2, -1)$

30. $\left(\frac{1}{3}, 1\right), (1, -1)$

31. $(3, -2), (5, 7)$

32. $(-2, -3), (-1, 4)$

Solve each problem. See Example 5.

△ **33.** For a square-based pyramid, the formula $b = \sqrt{\frac{3V}{h}}$ describes the relationship between the length b of one side of the base, the volume V, and the height h. Find b if the volume of the pyramid is 24 cubic feet, and the pyramid is 2 feet high.

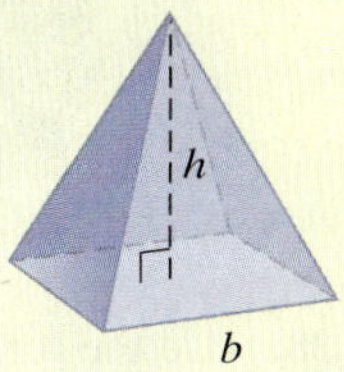

34. The formula $t = \frac{\sqrt{d}}{4}$ relates the distance d, in feet, that an object falls in t seconds, assuming that air resistance does not slow down the object. Find how long, to the nearest hundredth of a second, it takes an object to reach the ground from the top of the Willis Tower in Chicago, a distance of 1450 feet. (*Source: World Almanac and Book of Facts*)

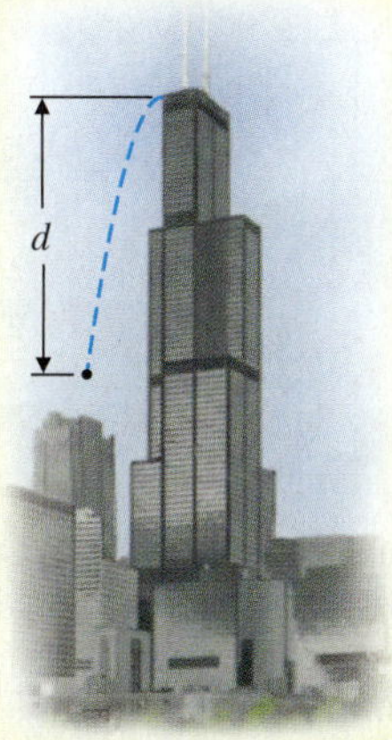

35. Police use the formula $s = \sqrt{30fd}$ to estimate the speed s of a car in miles per hour. In this formula, d represents the distance the car skidded in feet and f represents the coefficient of friction. The value of f depends on the type of road surface, and for wet concrete f is 0.35. Find how fast a car was moving if it skidded 280 feet on wet concrete, to the nearest mile per hour.

36. The coefficient of friction of a certain dry road is 0.95. Use the formula in Exercise 35 to find how fast a car was moving on this dry road if it skidded 126 feet. Round to the nearest mile per hour.

37. The formula $v = \sqrt{2.5r}$ can be used to estimate the maximum safe velocity, v, in miles per hour, at which a car can travel if it is driven along a curved road with a **radius of curvature,** r, in feet. To the nearest whole number, find the maximum safe speed if a cloverleaf exit on an expressway has a radius of curvature of 300 feet.

38. Use the formula from Exercise 37 to find the maximum safe speed if a cloverleaf exit on an expressway has a radius of curvature of 360 feet.

△ 39. The maximum distance d in kilometers that you can see from a height of h meters is given by $d = 3.5\sqrt{h}$. Find how far you can see from the top of the Bank One Tower in Indianapolis, a height of 285.4 meters. Round to the nearest tenth of a kilometer. (*Source: World Almanac and Book of Facts*)

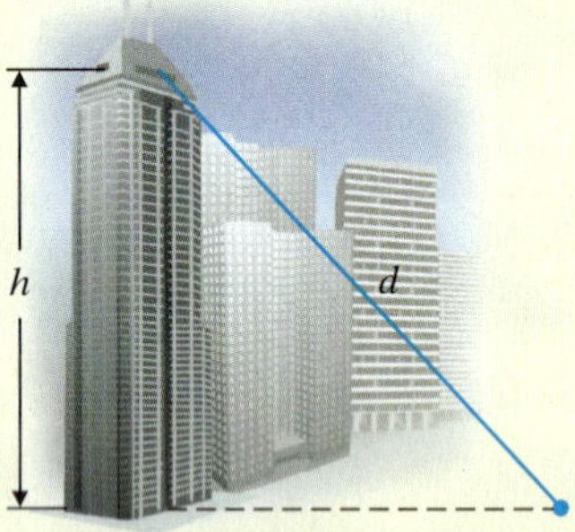

△ 40. Use the formula from Exercise 39 to find how far you can see from the top of the Chase Tower Building in Houston, Texas, a height of 305 meters. Round to the nearest tenth of a kilometer. (*Source:* Council on Tall Buildings and Urban Habitat)

41. Use the formula from Exercise 39 to find how far you can see from the top of the First Interstate Tower in Houston, Texas, a height of 295.7 meters. Round to the nearest tenth of a kilometer. (*Source:* Council on Tall Buildings and Urban Habitat)

42. Use the formula from Exercise 39 to find how far you can see from the top of the Gas Company Tower in Los Angeles, California, a height of 228.3 meters. Round to the nearest tenth of a kilometer. (*Source:* Council on Tall Buildings and Urban Habitat)

MIXED PRACTICE

Solve. See Examples 6 through 9.

TRANSLATING

For Exercises 43 through 48, represent each given condition using a single variable, x.

△ 43. The length and width of a rectangle whose length is 4 centimeters more than its width

△ 44. The length and width of a rectangle whose length is twice its width

45. Two consecutive odd integers

46. Two consecutive even integers

△ 47. The base and height of a triangle whose height is one more than four times its base

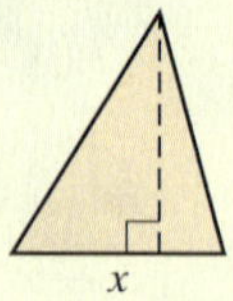

△ 48. The base and height of a trapezoid whose base is three less than five times its height

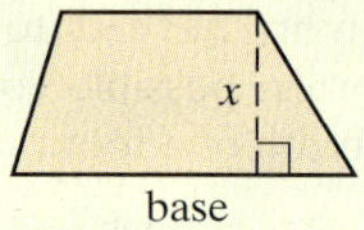

Use the information given to find the dimensions of each figure. For Exercises 49 and 50, recall that perimeter means distance around.

△ 49. The *perimeter* of the quadrilateral is 120 centimeters. Find the lengths of the sides.

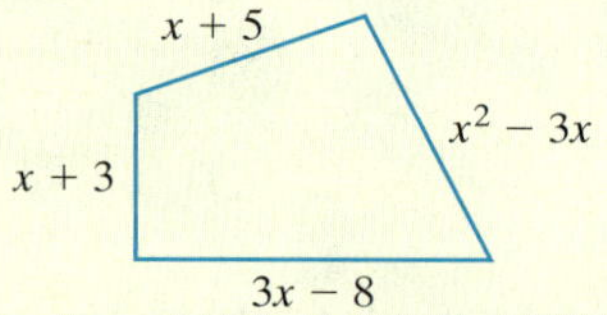

△ 50. The *perimeter* of the triangle is 85 feet. Find the lengths of its sides.

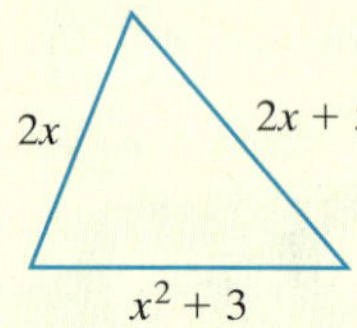

△ 51. The area of a parallelogram is found by the formula $A = b \cdot h$. If the *area* of the parallelogram is 96 square miles, find its base and height.

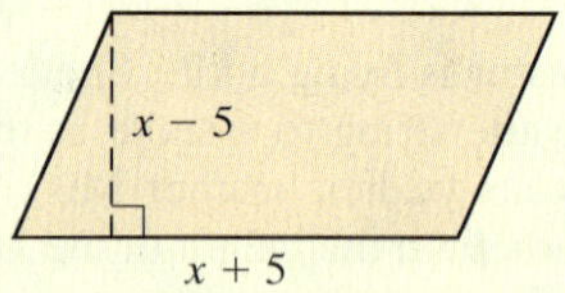

△ 52. The area of a circle is found by the formula $A = \pi r^2$. If the *area* of the circle is 25π square kilometers, find its radius.

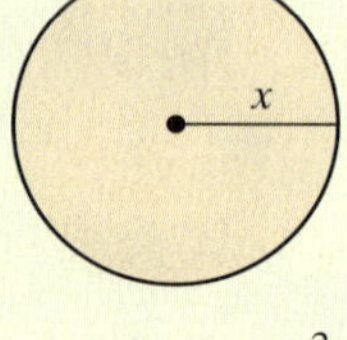

The formula for area of a square is $A = s^2$ where s is the length of a side. Use this formula for Exercises 53 through 56. For each exercise, give the exact answer and a two-decimal-place approximation.

△ 53. If the area of a square is 20 square inches, find the length of a side.

△ **54.** If the area of a square is 32 square meters, find the length of a side.

△ **55.** The "Water Cube" National Swimming Center was constructed in Beijing for the 2008 Summer Olympics. Its square base has an area of 31,329 square meters. Find the length of a side of this building. (*Source:* ARUP East Asia)

Note: The Beijing Water Cube was converted to an indoor Water Park and recently reopened.

△ **56.** The Washington Monument has a square base whose area is approximately 3039 square feet. Find the length of a side. (*Source: The World Almanac*)

Solve. See Example 6. For Exercises 57 through 60, use the formula $h = 16t^2$ and if necessary round answers to the nearest tenth of a second. (Recall that this formula does not take into account any air resistance.)

57. Each year, Brontallo, Switzerland hosts a cliff diving championship. The height of the dive is 85 feet into a shallow clear pool. Determine the time of a dive. (*Source: World High Diving Federation*)

58. In 1988, Eddie Turner saved Frank Fanan, who became unconscious after an injury while jumping out of an airplane. Fanan fell 11,136 feet before Turner pulled his ripcord. Determine the time of Fanan's unconscious free fall.

59. The Hualapai Indian Tribe allowed the Grand Canyon Skywalk to be built over the rim of the Grand Canyon on its tribal land. The skywalk extends 70 feet beyond the canyon's edge and is 4000 feet above the canyon floor. Determine the time it would take an object, dropped off the skywalk, to land at the bottom of the Grand Canyon. (*Source: Boston Globe*)

60. If a sandblaster drops his goggles from a bridge 400 feet from the water below, find how long it takes for the goggles to hit the water.

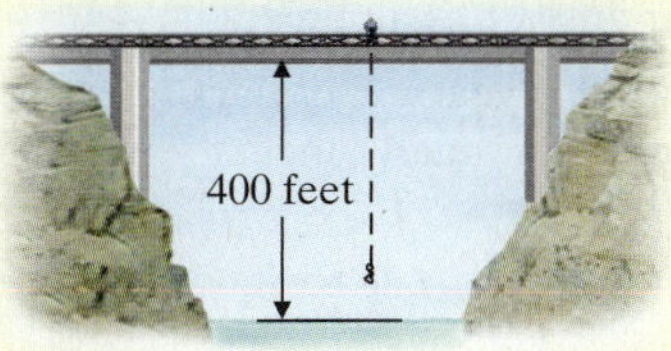

Solve.

△ **61.** The area of a circle is found by the formula $A = \pi r^2$. If the area A of a certain circle is 36π square inches, find its radius r.

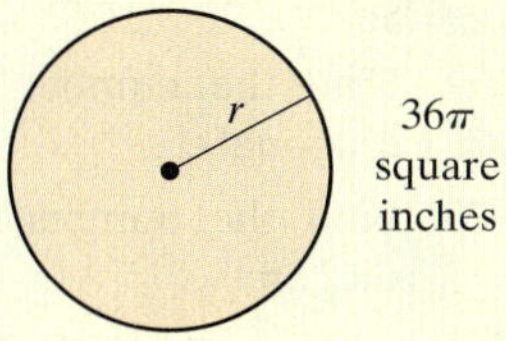

62. If the area of the circle to the right is 10π square units, find its exact radius. (See Exercise 61.)

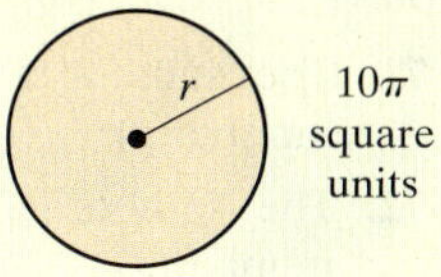

Solve.

63. An object is thrown upward from the top of an 80-foot building with an initial velocity of 64 feet per second. The height h of the object after t seconds is given by the quadratic equation $h = -16t^2 + 64t + 80$. When will the object hit the ground?

64. A hang glider pilot accidentally drops her compass from the top of a 400-foot cliff. The height h of the compass after t seconds is given by the quadratic equation $h = -16t^2 + 400$. When will the compass hit the ground?

△ **65.** The width of a rectangle is 7 centimeters less than twice its length. Its area is 30 square centimeters. Find the dimensions of the rectangle.

△ **66.** The length of a rectangle is 9 inches more than its width. Its area is 112 square inches. Find the dimensions of the rectangle.

The equation $D = \frac{1}{2}n(n - 3)$ gives the number of diagonals D for a polygon with n sides. For example, a polygon with 6 sides has $D = \frac{1}{2} \cdot 6(6 - 3)$ or $D = 9$ diagonals. (See if you can count all 9 diagonals. Some are shown in the figure.) Use this equation, $D = \frac{1}{2}n(n - 3)$, for Exercises 67 through 70.

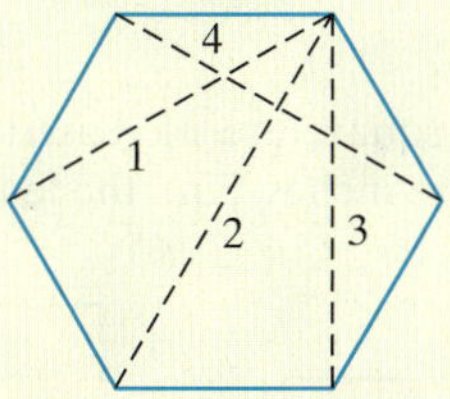

△ **67.** Find the number of diagonals for a polygon that has 12 sides.

68. Find the number of diagonals for a polygon that has 15 sides.

69. Find the number of sides n for a polygon that has 35 diagonals.

70. Find the number of sides n for a polygon that has 14 diagonals.

Solve.

71. The sum of a number and its square is 132. Find the number(s).

72. The sum of a number and its square is 182. Find the number(s).

73. The product of two consecutive room numbers is 210. Find the room numbers.

74. The product of two consecutive page numbers is 420. Find the page numbers.

75. A ladder is leaning against a building so that the distance from the ground to the top of the ladder is one foot less than the length of the ladder. Find the length of the ladder if the distance from the bottom of the ladder to the building is 5 feet.

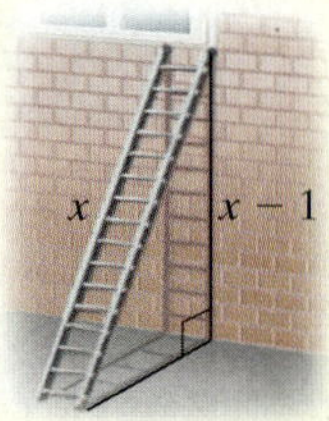

76. Use the given figure to find the length of the guy wire.

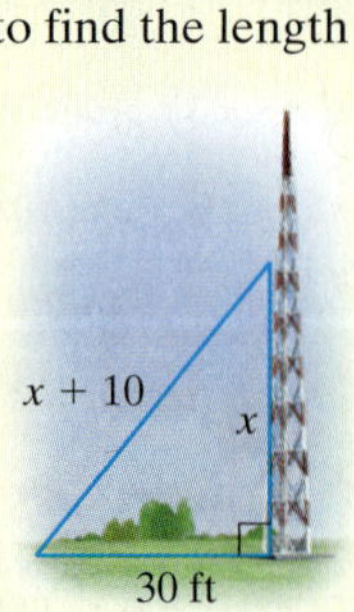

77. If the sides of a square are increased by 3 inches, the area becomes 64 square inches. Find the length of the sides of the original square.

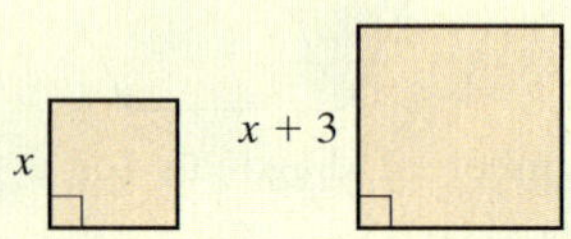

78. If the sides of a square are increased by 5 meters, the area becomes 100 square meters. Find the length of the sides of the original square.

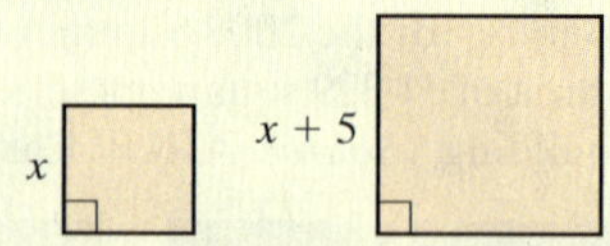

79. One leg of a right triangle is 4 millimeters longer than the smaller leg and the hypotenuse is 8 millimeters longer than the smaller leg. Find the lengths of the sides of the triangle.

80. One leg of a right triangle is 9 centimeters longer than the other leg and the hypotenuse is 45 centimeters. Find the lengths of the legs of the triangle.

81. The length of the base of a triangle is twice its height. If the area of the triangle is 100 square kilometers, find the height.

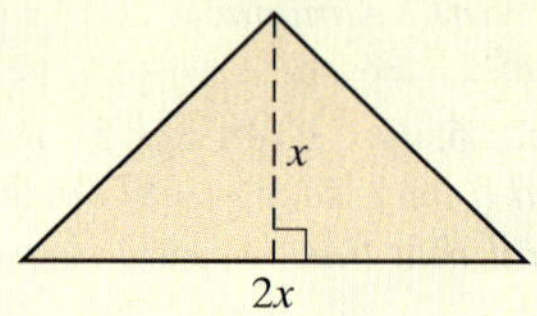

82. The height of a triangle is 2 millimeters less than the base. If the area is 60 square millimeters, find the base.

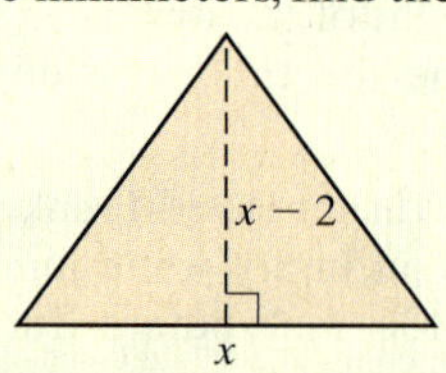

83. Find the length of the shorter leg of a right triangle if the longer leg is 12 feet more than the shorter leg and the hypotenuse is 12 feet less than twice the shorter leg.

84. Find the length of the shorter leg of a right triangle if the longer leg is 10 miles more than the shorter leg and the hypotenuse is 10 miles less than twice the shorter leg.

85. An object is dropped from 39 feet below the tip of the pinnacle atop one of the 1483-foot-tall Petronas Twin Towers in Kuala Lumpur, Malaysia. (*Source:* Council on Tall Buildings and Urban Habitat) The height h of the object after t seconds is given by the equation $h = -16t^2 + 1444$. Find how many seconds pass before the object reaches the ground.

86. An object is dropped from the top of 311 South Wacker Drive, a 961-foot-tall office building in Chicago. (*Source:* Council on Tall Buildings and Urban Habitat) The height h of the object after t seconds is given by the equation $h = -16t^2 + 961$. Find how many seconds pass before the object reaches the ground.

87. At the end of 2 years, P dollars invested at an interest rate r compounded annually increases to an amount, A dollars, given by

$$A = P(1 + r)^2$$

Find the interest rate if \$100 increased to \$144 in 2 years. Write your answer as a percent.

88. At the end of 2 years, P dollars invested at an interest rate r compounded annually increases to an amount, A dollars, given by

$$A = P(1 + r)^2$$

Find the interest rate if \$2000 increased to \$2420 in 2 years. Write your answer as a percent.

△ **89.** Find the dimensions of a rectangle whose width is 7 miles less than its length and whose area is 120 square miles.

△ **90.** Find the dimensions of a rectangle whose width is 2 inches less than half its length and whose area is 160 square inches.

91. If the cost, C, for manufacturing x units of a certain product is given by $C = x^2 - 15x + 50$, find the number of units manufactured at a cost of \$9500.

92. If a switchboard handles n telephones, the number C of telephone connections it can make simultaneously is given by the equation $C = \frac{n(n-1)}{2}$. Find how many telephones are handled by a switchboard making 120 telephone connections simultaneously.

REVIEW AND PREVIEW

The following double-line graph shows a comparison of the number of annual visitors (in millions) to Glacier National Park and Gettysburg National Military Park for the years shown. Use this graph to answer Exercises 93 through 98. See Section R.4.

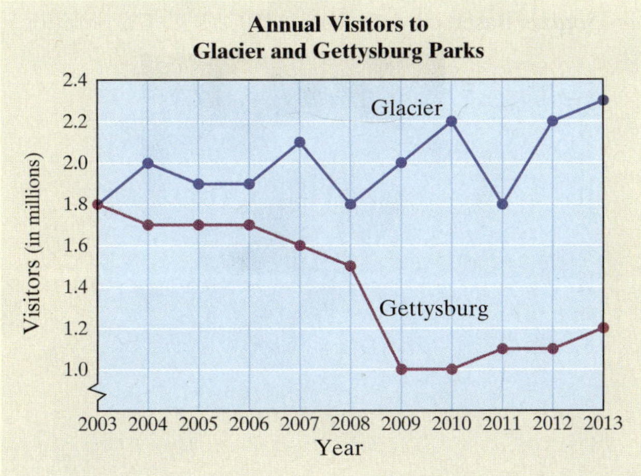

93. Approximate the number of visitors to Glacier National Park in 2011.

94. Approximate the number of visitors to Gettysburg National Military Park in 2011.

95. Approximate the number of visitors to Glacier National Park in 2013.

96. Approximate the number of visitors to Gettysburg National Military Park in 2013.

97. Determine the year that the colored lines in this graph intersect.

98. For what years on the graph is the number of visitors to Glacier Park greater than the number of visitors to Gettysburg Park?

CONCEPT EXTENSIONS

For each triangle, find the length of y, then x.

△ **99.**

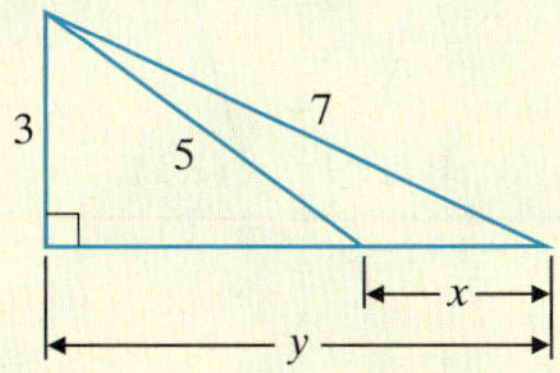

△ **100.**

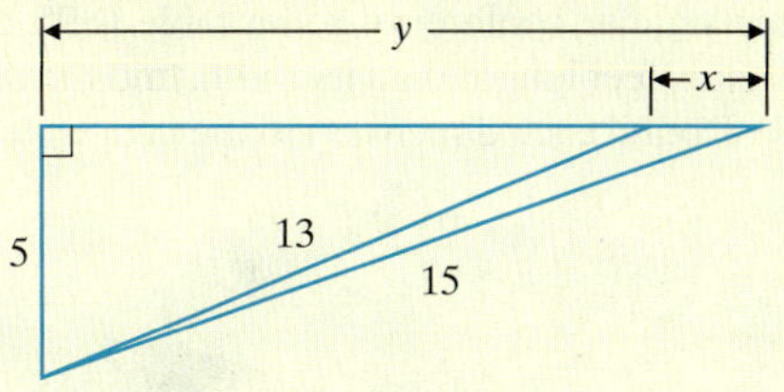

Solve.

△ **101.** Mike and Sandra Hallahan leave the seashore at the same time. Mike drives northward at a rate of 30 miles per hour while Sandra drives west at 60 mph. Find how far apart they are after 3 hours to the nearest mile.

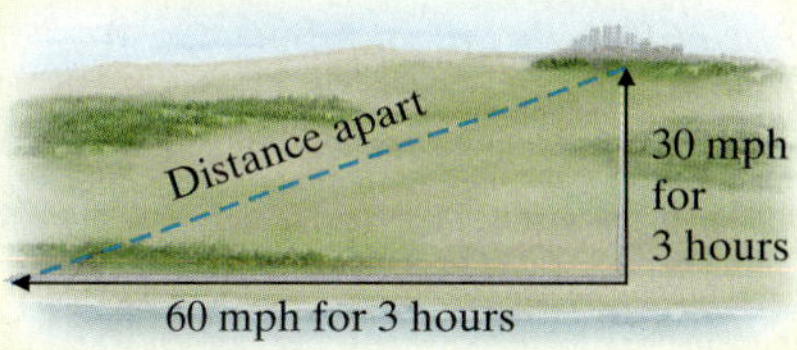

△ **102.** Railroad tracks are invariably made up of relatively short sections of rail connected by expansion joints. To see why this construction is necessary, consider a single rail 100 feet long (or 1200 inches). On an extremely hot day, suppose it expands 1 inch in the hot sun to a new length of 1201 inches. Theoretically, the track would bow upward as pictured.

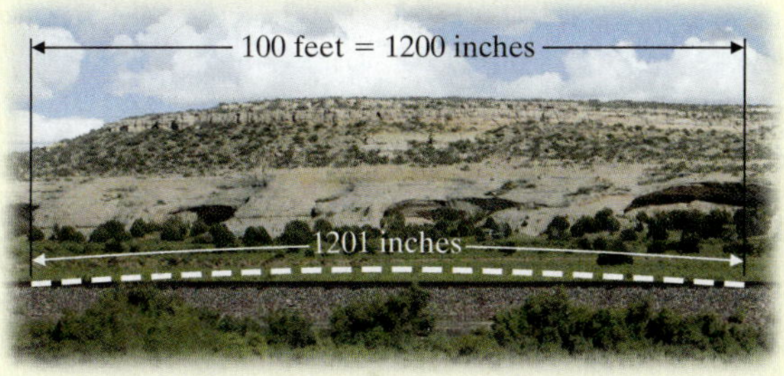

Let us approximate the bulge in the railroad this way.

a. Calculate the height h of the bulge.

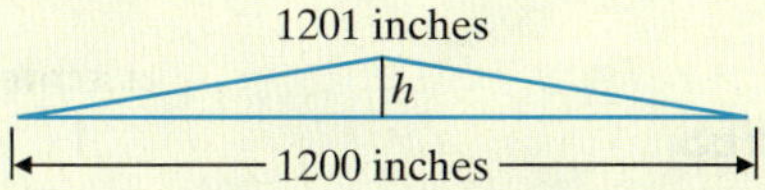

△ **b.** Based on the results of part **a.**, explain why railroads use short sections of rail connected by expansion joints.

△ **103.** In a recent year, Nestle created a chocolate bar that the company claimed weighed more than 2 tons. The rectangular bar had a base area of approximately 34.65 square feet, and its length was 0.6 foot shorter than three times its width. Find the length and width of the bar. (*Source:* Nestle)

△ **104.** The area of a rectangular conference room table is 95 square feet. If its length is six feet longer than its width, find the dimensions of the table. Round each dimension to the nearest tenth.

A rocket is launched from the top of an 80-foot cliff with an initial velocity of 120 feet per second. The height, h, of the rocket after t seconds is given by the equation $h = -16t^2 + 120t + 80$. *Use this for Exercises 105 and 106.*

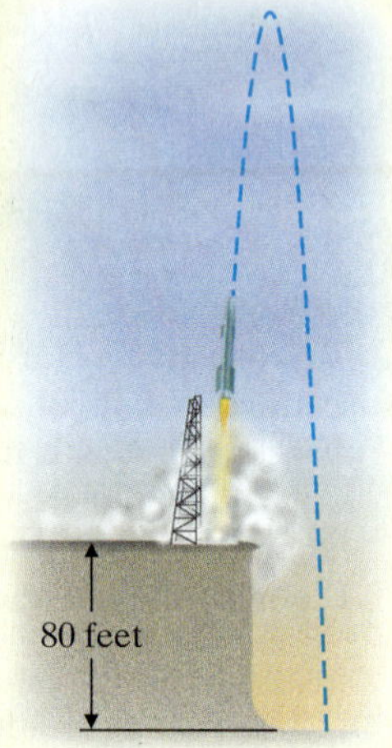

105. How long after the rocket is launched will it be 30 feet from the ground? Round to the nearest tenth of a second.

106. How long after the rocket is launched will it strike the ground? Round to the nearest tenth of a second. (*Hint:* The rocket will strike the ground when its height $h = 0$.)

107. The number of patients on the kidney waiting list from the beginning of year 2010 to 2014 can be modeled by the equation $y = 425x^2 + 1373x + 84{,}455$, where $x = 0$ represents the year 2010. Assume that this trend continued and predict the year in which 122,639 patients will be on the waiting list.

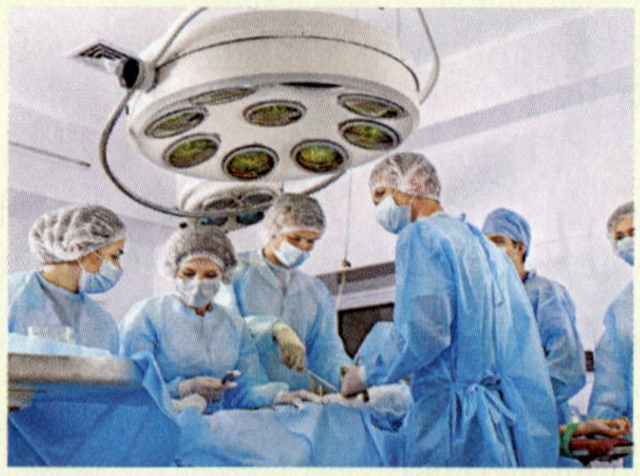

108. The revenues from product sales y (in millions of dollars) of ABIOMED, Inc., maker of the AbioCor artificial heart and Impella heart pumps, from 2010 through 2014 can be modeled by the equation $y = 2x^2 + 17x + 85.5$, where $x = 0$ represents the year 2010. Assume that this trend continued and predict the first year in which ABIOMED's revenues will be $250 million. Round up to the next whole number. (*Source:* Based on data from ABIOMED Corporation)

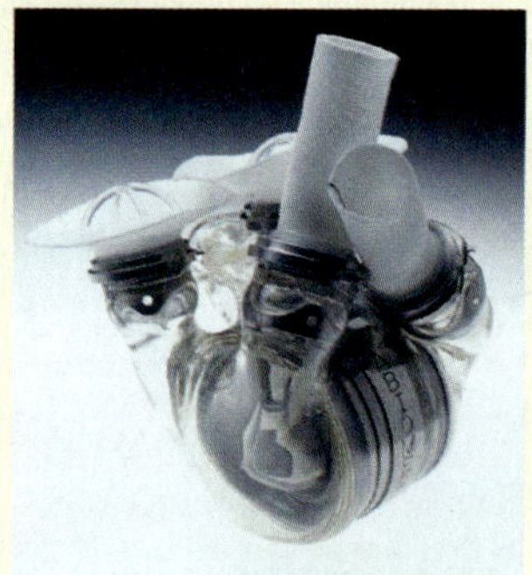

5.8 Extension—Graphing Quadratic Functions

OBJECTIVES

1. Graph Quadratic Equations of the Form $y = ax^2$.
2. Graph Quadratic Equations of the Form $y = ax^2 + bx + c$.
3. Use the Vertex Formula to Determine the Vertex of a Parabola.

Helpful Hint

Notice this graph to the right passes the vertical line test. It is the graph of a function.

OBJECTIVE

1 Graphing $y = ax^2$

Recall from Section 3.1 that the graph of a linear equation in two variables $Ax + By = C$ is a straight line. Also recall from Section 5.4 that the graph of a quadratic equation in two variables $y = ax^2 + bx + c$ is a parabola. In this section, we further investigate the graph of a quadratic equation.

To graph the quadratic equation $y = x^2$, select a few values for x and find the corresponding y-values. Make a table of values to keep track. Then plot the points corresponding to these solutions.

If $x = -3$, then $y = (-3)^2$, or 9.
If $x = -2$, then $y = (-2)^2$, or 4.
If $x = -1$, then $y = (-1)^2$, or 1.
If $x = 0$, then $y = 0^2$, or 0.
If $x = 1$, then $y = 1^2$, or 1.
If $x = 2$, then $y = 2^2$, or 4.
If $x = 3$, then $y = 3^2$, or 9.

$y = x^2$

x	y
−3	9
−2	4
−1	1
0	0
1	1
2	4
3	9

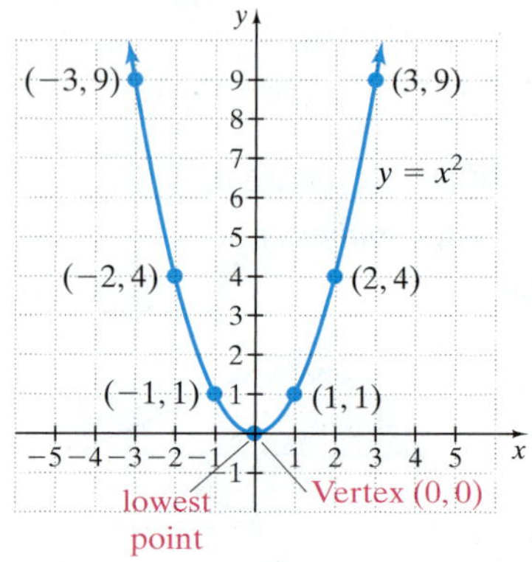

Clearly, these points are not on one straight line. As we saw earlier in this chapter, the graph of $y = x^2$ is a smooth curve through the plotted points. This curve is called a **parabola.** The lowest point on a parabola opening upward is called the **vertex.** The vertex is $(0, 0)$ for the parabola $y = x^2$. If we fold the graph paper along the y-axis, the two pieces of the parabola match perfectly. For this reason, we say the graph is **symmetric about the *y*-axis,** and we call the y-axis the **axis of symmetry.**

Notice that the parabola that corresponds to the equation $y = x^2$ opens upward. This happens when the coefficient of x^2 is positive. In the equation $y = x^2$, the coefficient of x^2 is 1. Example 1 shows the graph of a quadratic equation whose coefficient of x^2 is negative. All graphs in this section are graphs of functions. Notice they all pass the vertical line test.

EXAMPLE 1 Graph: $y = -2x^2$

Solution Select x-values and calculate the corresponding y-values. Plot the ordered pairs found. Then draw a smooth curve through those points. When the coefficient of x^2 is negative, the corresponding parabola opens downward. When a parabola opens downward, the vertex is the highest point of the parabola. The vertex of this parabola is $(0, 0)$ and the axis of symmetry is again the y-axis.

$y = -2x^2$

x	y
0	0
1	−2
2	−8
3	−18
−1	−2
−2	−8
−3	−18

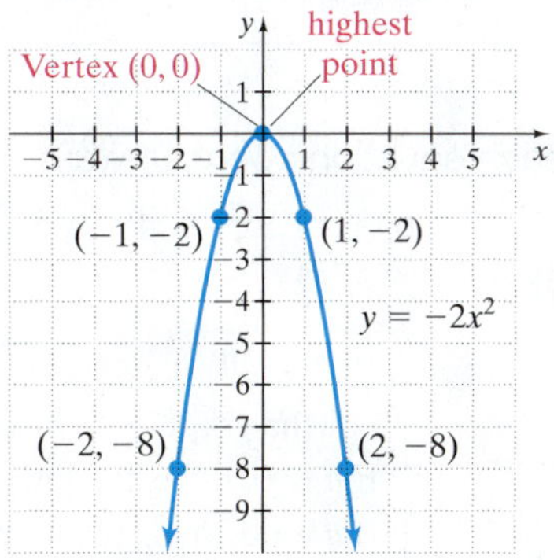

Helpful Hint
Notice this graph passes the vertical line test. It is the graph of a function.

PRACTICE 1 Graph: $y = -\frac{1}{2}x^2$

OBJECTIVE 2

Graphing $y = ax^2 + bx + c$

Just as for linear equations, we can use x- and y-intercepts to help graph quadratic equations. Recall from Chapter 3 that an x-intercept is the point where the graph intersects the x-axis. A y-intercept is the point where the graph intersects the y-axis. We find intercepts just as we did in Chapter 3.

Helpful Hint
Recall that:
To find x-intercepts, let $y = 0$ and solve for x.
To find y-intercepts, let $x = 0$ and solve for y.

EXAMPLE 2 Graph: $y = x^2 - 4$

Solution First, find intercepts. To find the y-intercept, let $x = 0$. Then

$$y = 0^2 - 4 = -4$$

To find x-intercepts, we let $y = 0$ and solve for x.

$$0 = x^2 - 4$$
$$0 = (x - 2)(x + 2) \quad \text{Factor } x^2 - 4.$$
$$x - 2 = 0 \quad \text{or} \quad x + 2 = 0 \quad \text{Set each factor equal to 0.}$$
$$x = 2 \quad \text{or} \quad x = -2 \quad \text{Solve.}$$

(Continued on next page)

Thus far, we have the y-intercept $(0, -4)$ and the x-intercepts $(2, 0)$ and $(-2, 0)$. Now we can select additional x-values, find the corresponding y-values, plot the points, and draw a smooth curve through the points.

$y = x^2 - 4$

x	y
0	−4
1	−3
2	0
3	5
−1	−3
−2	0
−3	5

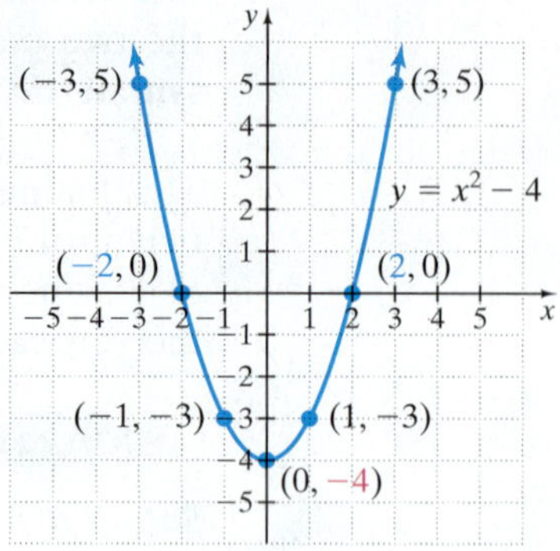

Notice that the vertex of this parabola is $(0, -4)$. □

PRACTICE 2 Graph: $y = x^2 + 1$ ■

✔CONCEPT CHECK

Tell whether the graph of each equation opens upward or downward.

a. $y = 2x^2$ **b.** $y = 3x^2 + 4x - 5$ **c.** $y = -5x^2 + 2$

Helpful Hint

For the graph of $y = ax^2 + bx + c$, where $a \neq 0$,
If a is positive, the parabola opens upward.
If a is negative, the parabola opens downward.
The parabola is the graph of a function and thus passes the vertical line test.

✔CONCEPT CHECK

For which of the following graphs of $y = ax^2 + bx + c$ would the value of a be negative?

a.

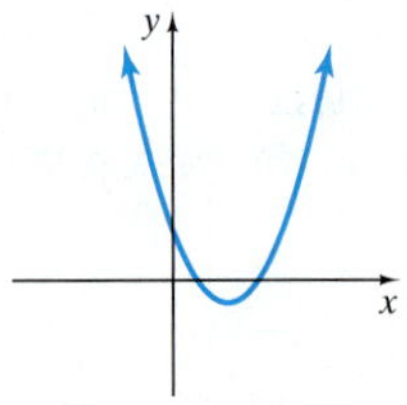

b.

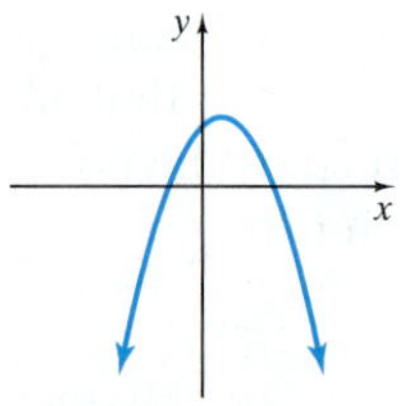

OBJECTIVE

3 Using the Vertex Formula

Thus far, we have accidentally stumbled upon the vertex of each parabola that we have graphed. However, our choice of values for x may not yield an ordered pair for the vertex of the parabola. It would be helpful if we could first find the vertex of a parabola, next determine whether the parabola opens upward or downward, and finally calculate additional points such as x- and y-intercepts as needed. In fact, there is a formula that may be used to find the vertex of a parabola.

One way to develop this formula is to notice that the x-value of the vertex of the parabolas that we are considering lies halfway between its x-intercepts. We can use this fact to find a formula for the vertex.

Answers to Concept Checks:
a. upward **b.** upward
c. downward;

b

Recall that the x-intercepts of a parabola may be found by solving the equation $0 = ax^2 + bx + c$. These solutions, by the quadratic formula, are

$$x = \frac{-b - \sqrt{b^2 - 4ac}}{2a}, \quad x = \frac{-b + \sqrt{b^2 - 4ac}}{2a}$$

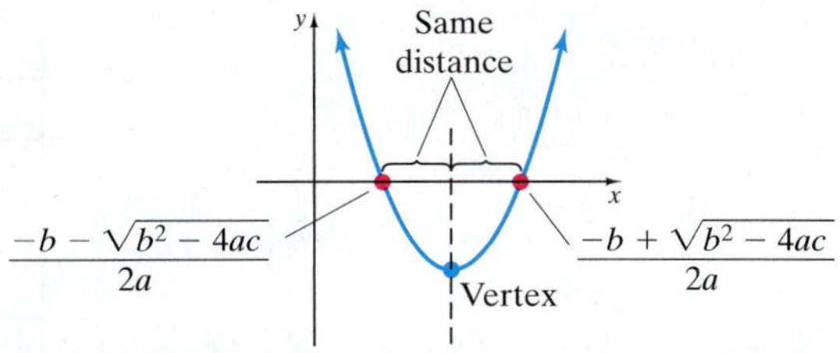

The x-coordinate of the vertex of a parabola is halfway between its x-intercepts, so the x-value of the vertex may be found by computing the average, or $\frac{1}{2}$ of the sum of the intercepts.

$$\begin{aligned} x &= \frac{1}{2}\left(\frac{-b - \sqrt{b^2 - 4ac}}{2a} + \frac{-b + \sqrt{b^2 - 4ac}}{2a}\right) \\ &= \frac{1}{2}\left(\frac{-b - \sqrt{b^2 - 4ac} - b + \sqrt{b^2 - 4ac}}{2a}\right) \\ &= \frac{1}{2}\left(\frac{-2b}{2a}\right) \\ &= \frac{-b}{2a} \end{aligned}$$

Vertex Formula

The vertex of the parabola $y = ax^2 + bx + c$ has x-coordinate

$$\frac{-b}{2a}$$

The corresponding y-coordinate of the vertex is found by substituting the x-coordinate into the equation and evaluating y.

EXAMPLE 3 Graph: $y = x^2 - 6x + 8$

Solution In the equation $y = x^2 - 6x + 8$, $a = 1$ and $b = -6$. The x-coordinate of the vertex is

$$x = \frac{-b}{2a} = \frac{-(-6)}{2 \cdot 1} = 3 \quad \text{Use the vertex formula, } x = \frac{-b}{2a}.$$

To find the corresponding y-coordinate, we let $x = 3$ in the original equation.

$$y = x^2 - 6x + 8 = 3^2 - 6 \cdot 3 + 8 = -1$$

The vertex is $(3, -1)$ and the parabola opens upward since a is positive. We now find and plot the intercepts.

To find the x-intercepts, we let $y = 0$ and solve for x.

$$\begin{aligned} 0 &= x^2 - 6x + 8 & & \\ 0 &= (x - 4)(x - 2) & & \text{Factor.} \\ x - 4 = 0 \quad &\text{or} \quad x - 2 = 0 & & \text{Set each factor equal to 0.} \\ x = 4 \quad &\text{or} \quad x = 2 & & \text{Solve.} \end{aligned}$$

The x-intercepts are $(4, 0)$ and $(2, 0)$.

(Continued on next page)

If we let $x = 0$ in the original equation, then $y = 8$ and the y-intercept is $(0, 8)$. Now we plot the vertex $(3, -1)$ and the intercepts $(4, 0)$, $(2, 0)$, and $(0, 8)$. Then we can sketch the parabola. These and two additional points are shown in the table.

$y = x^2 - 6x + 8$

	x	y
	3	−1
	4	0
	2	0
	0	8
Additional Points:	1	3
	5	3
	6	8

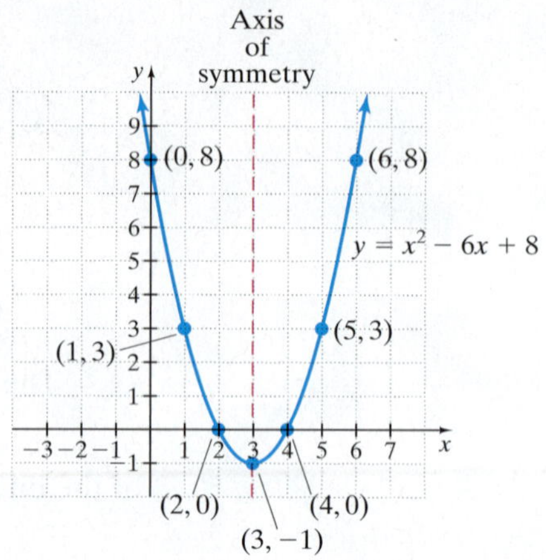

□

PRACTICE 3 Graph: $y = x^2 - 3x - 4$ ■

Study Example 3 and let's use it to write down a general procedure for graphing quadratic equations.

Graphing Parabolas Defined by $y = ax^2 + bx + c$

1. **Find the vertex by using the formula $x = \dfrac{-b}{2a}$.** Don't forget to find the y-value of the vertex.
2. **Find the intercepts.**
   - Let $x = 0$ and solve for y to find the y-intercept. There will be only one.
   - Let $y = 0$ and solve for x to find any x-intercepts. There may be 0, 1, or 2.
3. **Plot the vertex and the intercepts.**
4. **Find and plot additional points on the graph.** Then draw a smooth curve through the plotted points. Keep in mind if $a > 0$, the parabola opens up and, if $a < 0$, the parabola opens down.

EXAMPLE 4 Graph: $y = x^2 + 2x - 5$

Solution In the equation $y = x^2 + 2x - 5$, $a = 1$ and $b = 2$. Using the vertex formula, we find that the x-coordinate of the vertex is

$$x = \frac{-b}{2a} = \frac{-2}{2 \cdot 1} = -1$$

The y-coordinate of the vertex is

$$y = (-1)^2 + 2(-1) - 5 = -6$$

Thus the vertex is $(-1, -6)$.

To find the x-intercepts, we let $y = 0$.

$$0 = x^2 + 2x - 5$$

This cannot be solved by factoring, so we use the quadratic formula.

$$x = \frac{-2 \pm \sqrt{2^2 - 4(1)(-5)}}{2 \cdot 1} \qquad \text{Let } a = 1, b = 2, \text{ and } c = -5.$$

$$x = \frac{-2 \pm \sqrt{24}}{2}$$

$$x = \frac{-2 \pm 2\sqrt{6}}{2} \qquad \text{Simplify the radical.}$$

$$x = \frac{2(-1 \pm \sqrt{6})}{2} = -1 \pm \sqrt{6}$$

The x-intercepts are $(-1 + \sqrt{6}, 0)$ and $(-1 - \sqrt{6}, 0)$. We use a calculator to approximate these so that we can easily graph these intercepts.

$$-1 + \sqrt{6} \approx 1.4 \quad \text{and} \quad -1 - \sqrt{6} \approx -3.4$$

To find the y-intercept, we let $x = 0$ in the original equation and find that $y = -5$. Thus the y-intercept is $(0, -5)$.

$y = x^2 + 2x - 5$

x	y
-1	-6
$-1 + \sqrt{6}$	0
$-1 - \sqrt{6}$	0
0	-5
-2	-5

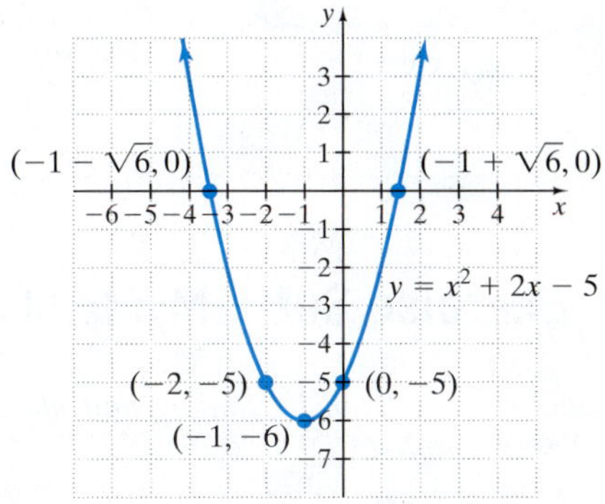

□

PRACTICE 4 Graph: $y = x^2 + 4x - 7$

Helpful Hint

Notice that the number of x-intercepts of the graph of the parabola $y = ax^2 + bx + c$ is the same as the number of real solutions of $0 = ax^2 + bx + c$.

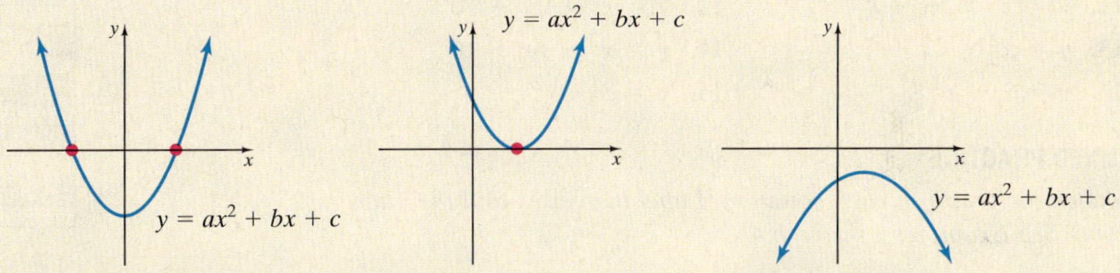

Two x-intercepts Two real solutions of $0 = ax^2 + bx + c$	One x-intercept One real solution of $0 = ax^2 + bx + c$	No x-intercepts No real solutions of $0 = ax^2 + bx + c$

Graphing Calculator Explorations

Recall that a graphing calculator may be used to solve quadratic equations. The x-intercepts of the graph of $y = ax^2 + bx + c$ are solutions of $0 = ax^2 + bx + c$. To solve $x^2 - 7x - 3 = 0$, for example, graph $y_1 = x^2 - 7x - 3$. The x-intercepts of the graph are the solutions of the equation.

Use a graphing calculator to solve each quadratic equation. Round solutions to two decimal places.

1. $x^2 - 7x - 3 = 0$

2. $2x^2 - 11x - 1 = 0$

3. $-1.7x^2 + 5.6x - 3.7 = 0$

4. $-5.8x^2 + 2.3x - 3.9 = 0$

5. $5.8x^2 - 2.6x - 1.9 = 0$

6. $7.5x^2 - 3.7x - 1.1 = 0$

Vocabulary, Readiness & Video Check

Martin-Gay Interactive Videos

See Video 5.8

Watch the section lecture video and answer the following questions.

OBJECTIVE 1
1. In Example 1, how are the vertex and axis of symmetry of a parabola explained?

OBJECTIVE 2
2. In Example 2, what important point was accidentally found? Why would it be useful to have an algebraic way to find this point?

OBJECTIVE 3
3. From Example 3, how can finding the vertex and noting whether the parabola opens up or down possibly help save you time and work? Explain.

5.8 Exercise Set MyMathLab®

Graph each quadratic equation by finding and plotting ordered pair solutions. See Example 1.

1. $y = 2x^2$
2. $y = 3x^2$
3. $y = -x^2$
4. $y = -4x^2$

Sketch the graph of each equation. Label the vertex and the intercepts. See Examples 2 through 4.

5. $y = x^2 - 1$
6. $y = x^2 - 16$
7. $y = x^2 + 4$
8. $y = x^2 + 9$
9. $y = -x^2 + 4x - 4$
10. $y = -x^2 - 2x - 1$
11. $y = x^2 + 5x + 4$
12. $y = x^2 + 7x + 10$
13. $y = x^2 - 4x + 5$
14. $y = x^2 - 6x + 10$
15. $y = 2 - x^2$
16. $y = 3 - x^2$

MIXED PRACTICE

Sketch the graph of each equation. Label the vertex and the intercepts. See Examples 1 through 4.

17. $y = \frac{1}{3}x^2$
18. $y = \frac{1}{2}x^2$
19. $y = x^2 + 6x$
20. $y = x^2 - 4x$
21. $y = x^2 + 2x - 8$
22. $y = x^2 - 2x - 3$
23. $y = -\frac{1}{2}x^2$
24. $y = -\frac{1}{3}x^2$
25. $y = 2x^2 - 11x + 5$
26. $y = 2x^2 + x - 3$
27. $y = -x^2 + 4x - 3$
28. $y = -x^2 + 6x - 8$
29. $y = x^2 + 2x - 2$
30. $y = x^2 - 4x - 3$
31. $y = x^2 - 3x + 1$
32. $y = x^2 - 2x - 5$

REVIEW AND PREVIEW

Find the slope of the line through each pair of points. Use the slope to determine whether the line is vertical, horizontal, or moves upward or downward from left to right. See Section 3.2.

33. $(-2, 5), (4, -3)$
34. $(0, 4), (2, 10)$
35. $(-3, -6), (1, 5)$
36. $(-2, 7), (3, -2)$
37. $(3, 7), (3, -2)$
38. $(0, -4), (2, -4)$

The following bar graph shows the capacity of the United States to generate electricity from the wind in the years shown. Use this graph for Exercises 39 and 40. See Section R.4.

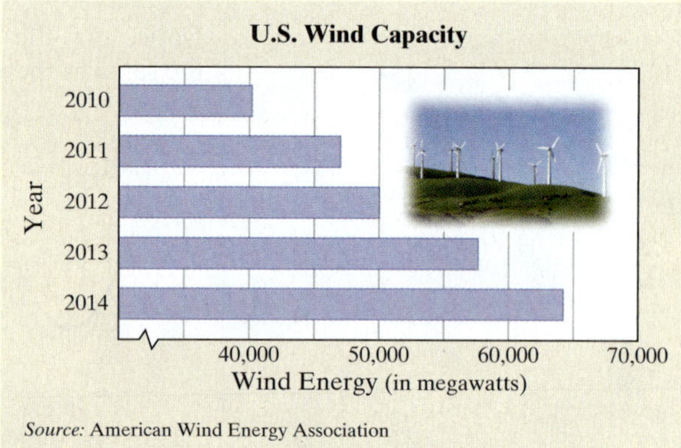

39. Find the approximate megawatt capacity in 2014.
40. Find the approximate megawatt capacity in 2013.

CONCEPT EXTENSIONS

The graph of a quadratic equation that takes the form $y = ax^2 + bx + c$ is the graph of a function. Write the domain and the range of each of the functions graphed.

41.

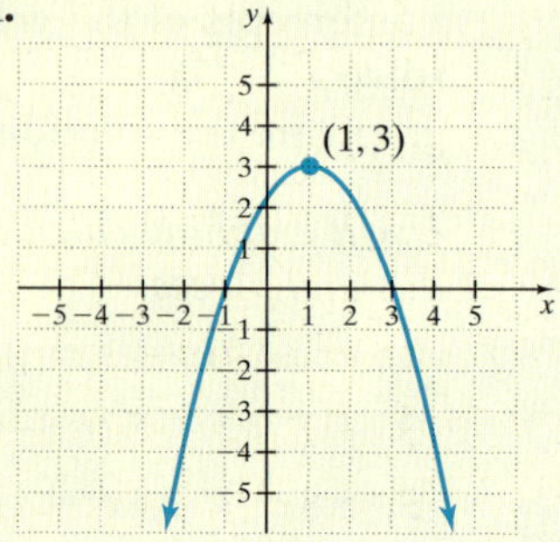

42.

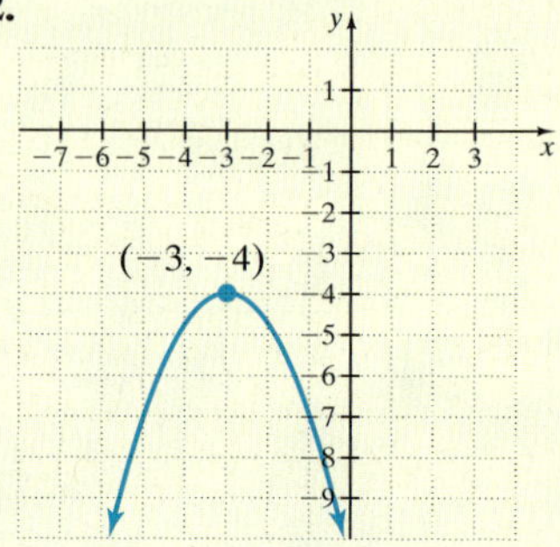

43.

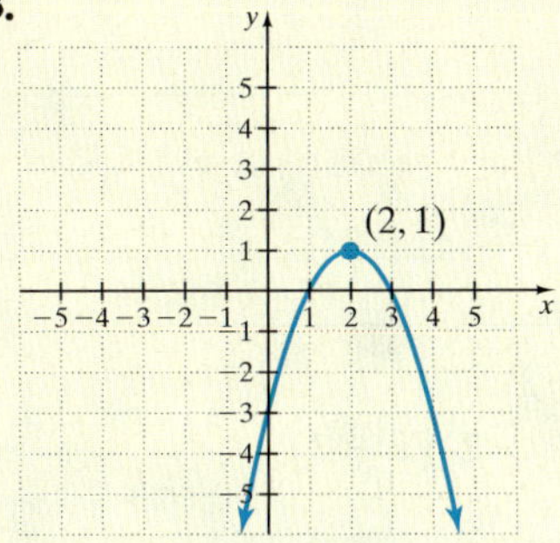

44.

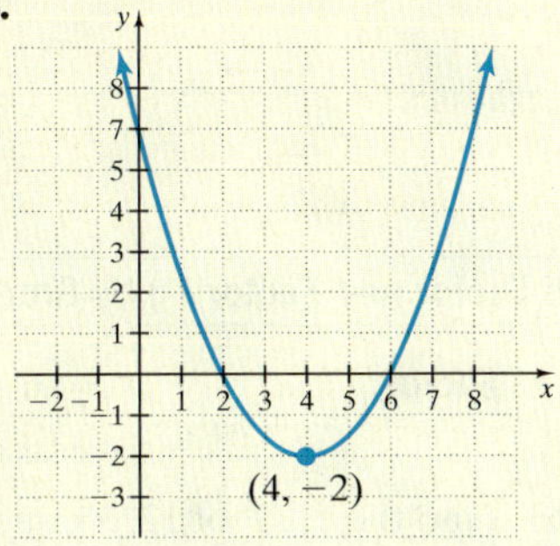

45. The height h of a fireball launched from a Roman candle with an initial velocity of 128 feet per second is given by the equation

$$h = -16t^2 + 128t$$

where t is time in seconds after launch.
Use the graph of this function to answer the questions.

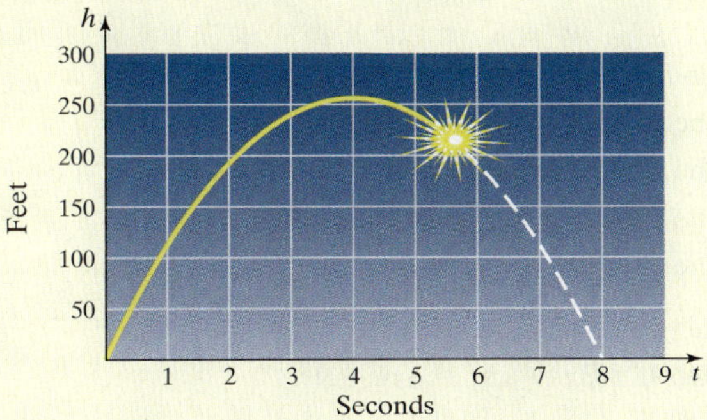

a. Estimate the maximum height of the fireball.

b. Estimate the time when the fireball is at its maximum height.

c. Estimate the time when the fireball returns to the ground.

46. Determine the maximum number and the minimum number of x-intercepts for a parabola. Explain your answer.

Match the values given with the correct graph of each quadratic equation of the form $y = a(x - h)^2 + k$. Here, (h, k) is the vertex.

47. $a > 0, h > 0, k > 0$

48. $a < 0, h > 0, k > 0$

49. $a > 0, h > 0, k < 0$

50. $a < 0, h > 0, k < 0$

A.

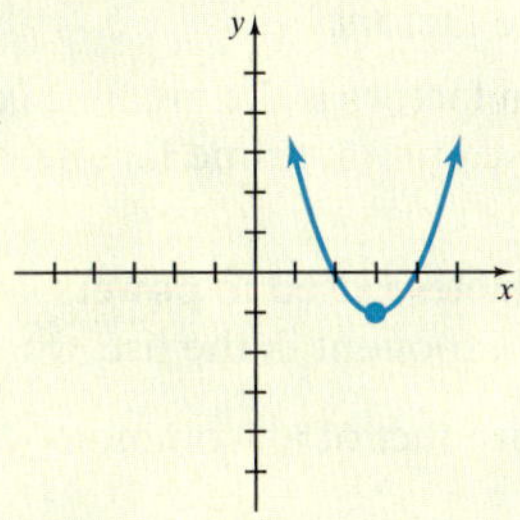

B.

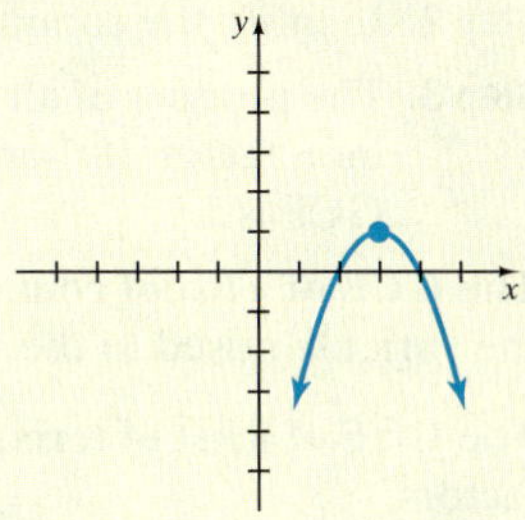

C.

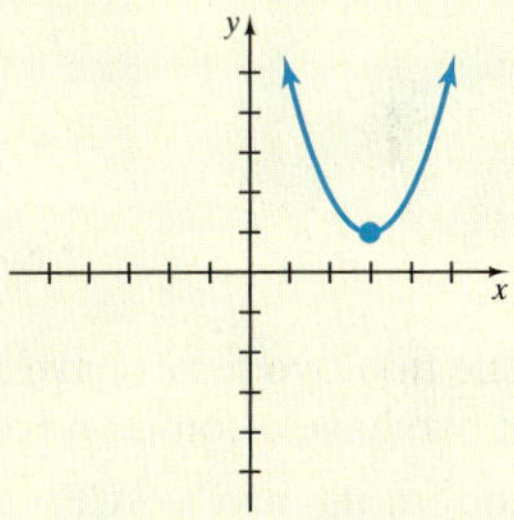

D.

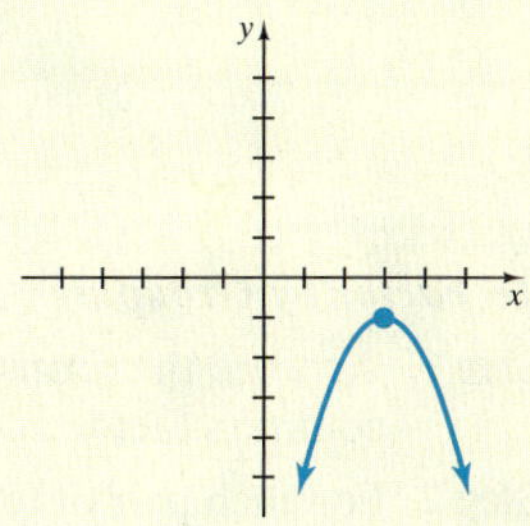

Chapter 5 Vocabulary Check

Fill in each blank with one of the words or phrases listed below. Not all choices will be used and some choices may be used more than once.

factoring	quadratic equation	sum of two cubes	rationalizing the denominator	quadratic
greatest common factor	difference of two squares	0	principal square root	radical
difference of two cubes	perfect square trinomial	1	radicand	like radicals
square root	complex			

1. An equation that can be written in the form $ax^2 + bx + c = 0$ (with a not 0) is called a(n) ________________.

2. ________________ is the process of writing an expression as a product.

3. The ________________ of a list of terms is the product of all common factors.

4. A trinomial that factors as the product of 2 identical binomials is called a(n) ________________.

5. The expression $a^2 - b^2$ is called a(n) ________________.

6. The expression $a^3 - b^3$ is called a(n) ________________.

7. The expression $a^3 + b^3$ is called a(n) ________________.

8. By the zero factor property, if the product of two numbers is 0, then at least one of the numbers must be ________________.

9. The expressions $5\sqrt{x}$ and $7\sqrt{x}$ are examples of ________________.

10. In the expression $\sqrt{45}$, the number 45 is the ________________, and $\sqrt{}$ is called the ________________ sign.

11. The ________________ of 25 is 5.

12. The process of eliminating the radical in the denominator of a radical expression is called ________________.

13. If $x^2 = a$, then $x = \sqrt{a}$ or $x = -\sqrt{a}$. This property is called the ________________ property.

14. The formula $x = \dfrac{-b \pm \sqrt{b^2 - 4ac}}{2a}$ is called the ________________ formula.

Chapter 5 Highlights

DEFINITIONS AND CONCEPTS	EXAMPLES
Section 5.1 The Greatest Common Factor and Factoring by Grouping	
Factoring is the process of writing an expression as a product.	Factor: $6 = 2 \cdot 3$ $x^2 + 5x + 6 = (x + 2)(x + 3)$
To Find the GCF of a List of Integers **Step 1.** Write each number as a product of primes. **Step 2.** Identify the common prime factors. **Step 3.** The product of all common factors is the greatest common factor. If there are no common prime factors, the GCF is 1.	Find the GCF of 12, 36, and 48. $12 = 2 \cdot 2 \cdot 3$ $36 = 2 \cdot 2 \cdot 3 \cdot 3$ $48 = 2 \cdot 2 \cdot 2 \cdot 2 \cdot 3$ $\text{GCF} = 2 \cdot 2 \cdot 3 = 12$
The GCF of a list of common variables raised to powers is the variable raised to the smallest exponent in the list.	The GCF of z^5, z^3, and z^{10} is z^3.
The GCF of a list of terms is the product of all common factors.	Find the GCF of $8x^2y$, $10x^3y^2$, and $26x^2y^3$. The GCF of 8, 10, and 26 is 2. The GCF of x^2, x^3, and x^2 is x^2. The GCF of y, y^2, and y^3 is y. The GCF of the terms is $2x^2y$.
To Factor by Grouping **Step 1.** Arrange the terms so that the first two terms have a common factor and the last two have a common factor. **Step 2.** For each pair of terms, factor out the pair's GCF. **Step 3.** If there is now a common binomial factor, factor it out. **Step 4.** If there is no common binomial factor, begin again, rearranging the terms differently. If no rearrangement leads to a common binomial factor, the polynomial cannot be factored by grouping.	Factor $10ax + 15a - 6xy - 9y$. **Step 1.** $10ax + 15a - 6xy - 9y$ **Step 2.** $5a(2x + 3) - 3y(2x + 3)$ **Step 3.** $(2x + 3)(5a - 3y)$
Section 5.2 Factoring Trinomials	
To Factor $x^2 + bx + c$ The product of these numbers is c. $x^2 + bx + c = (x + \square)(x + \square)$ The sum of these numbers is b.	Factor: $x^2 + 7x + 12$ $3 + 4 = 7$ $\quad 3 \cdot 4 = 12$ $x^2 + 7x + 12 = (x + 3)(x + 4)$
To Factor $ax^2 + bx + c$ Try various combinations of factors of ax^2 and c until a middle term of bx is obtained when checking.	Factor: $3x^2 + 14x - 5$ Factors of $3x^2$: $3x$, x Factors of -5: $-1, 5$ and $1, -5$. $(3x - 1)(x + 5)$ $-1x$ $+ 15x$ $14x$ **Correct** middle term
A **perfect square trinomial** is a trinomial that is the square of some binomial.	Perfect square trinomial = square of binomial $x^2 + 4x + 4 = (x + 2)^2$ $25x^2 - 10x + 1 = (5x - 1)^2$

DEFINITIONS AND CONCEPTS	EXAMPLES
Section 5.2 Factoring Trinomials (continued)	
To Factor $ax^2 + bx + c$ by Grouping	Factor: $3x^2 + 14x - 5$
Step 1. Find two numbers whose product is $a \cdot c$ and whose sum is b.	**Step 1.** Find two numbers whose product is $3 \cdot (-5)$ or -15 and whose sum is 14. They are 15 and -1.
Step 2. Rewrite bx, using the factors found in Step 1.	**Step 2.** $3x^2 + 14x - 5$ $= 3x^2 + 15x - 1x - 5$
Step 3. Factor by grouping.	**Step 3.** $= 3x(x + 5) - 1(x + 5)$ $= (x + 5)(3x - 1)$
Section 5.3 Factoring Binomials	
Difference of Squares $a^2 - b^2 = (a + b)(a - b)$	Factor. $x^2 - 9 = x^2 - 3^2 = (x + 3)(x - 3)$
Sum or Difference of Cubes $a^3 + b^3 = (a + b)(a^2 - ab + b^2)$ $a^3 - b^3 = (a - b)(a^2 + ab + b^2)$	$y^3 + 8 = y^3 + 2^3 = (y + 2)(y^2 - 2y + 4)$ $125z^3 - 1 = (5z)^3 - 1^3 = (5z - 1)(25z^2 + 5z + 1)$
Integrated Review—Choosing a Factoring Strategy	
To Factor a Polynomial	Factor: $2x^4 - 6x^2 - 8$
Step 1. Factor out the GCF.	**Step 1.** $2x^4 - 6x^2 - 8 = 2(x^4 - 3x^2 - 4)$
Step 2. a. If two terms **i.** $a^2 - b^2 = (a + b)(a - b)$ **ii.** $a^3 - b^3 = (a - b)(a^2 + ab + b^2)$ **iii.** $a^3 + b^3 = (a + b)(a^2 - ab + b^2)$	
b. If three terms Methods in Section 5.2. **c.** If four or more terms, try factoring by grouping.	**Step 2. b.** $= 2(x^2 + 1)(x^2 - 4)$
Step 3. See if any factors can be factored further.	**Step 3.** $= 2(x^2 + 1)(x + 2)(x - 2)$
Step 4. Check by multiplying.	**Step 4.** Check by multiplying. $2(x^2 + 1)(x + 2)(x - 2) = 2(x^2 + 1)(x^2 - 4)$ $= 2(x^4 - 3x^2 - 4)$ $= 2x^4 - 6x^2 - 8$
Section 5.4 Solving Quadratic Equations by Factoring	
A **quadratic equation** is an equation that can be written in the form $ax^2 + bx + c = 0$ with a not 0. The form $ax^2 + bx + c = 0$ is called the **standard form** of a quadratic equation.	***Quadratic Equation*** — ***Standard Form*** $x^2 = 16$ — $x^2 - 16 = 0$ $y = -2y^2 + 5$ — $2y^2 + y - 5 = 0$
Zero Factor Property If a and b are real numbers and if $ab = 0$, then $a = 0$ or $b = 0$.	If $(x + 3)(x - 1) = 0$, then $x + 3 = 0$ or $x - 1 = 0$
To Solve Quadratic Equations by Factoring	Solve: $3x^2 = 13x - 4$
Step 1. Write the equation in standard form: $ax^2 + bx + c = 0$.	**Step 1.** $3x^2 - 13x + 4 = 0$
Step 2. Factor the quadratic.	**Step 2.** $(3x - 1)(x - 4) = 0$
Step 3. Set each factor containing a variable equal to 0.	**Step 3.** $3x - 1 = 0$ or $x - 4 = 0$

(continued)

DEFINITIONS AND CONCEPTS	EXAMPLES
Section 5.4 Solving Quadratic Equations by Factoring (continued)	
Step 4. Solve the equations.	**Step 4.** $3x = 1$ or $x = 4$ $x = \frac{1}{3}$
Step 5. Check in the original equation.	**Step 5.** Check both $\frac{1}{3}$ and 4 in the original equation.
Section 5.5 Operations on Square Roots	
The **positive** or **principal square root** of a positive number a is written as $\sqrt{a}$. The **negative square root** of a is written as $-\sqrt{a}$. $\sqrt{a} = b$ only if $b^2 = a$ and $b > 0$.	$\sqrt{25} = 5$ $\quad\sqrt{100} = 10$ $-\sqrt{9} = -3$ $\quad\sqrt{\frac{4}{49}} = \frac{2}{7}$
A square root of a negative number is not a real number.	$\sqrt{-4}$ is not a real number.
In $\sqrt{a}$, the symbol $\sqrt{}$ is called a **radical,** and the expression within the radical is called the **radicand.**	radical ↓ $\sqrt{a}$ ↖ radicand
Product Rule for Square Roots If $\sqrt{a}$ and $\sqrt{b}$ are real numbers, then $\sqrt{a}\cdot\sqrt{b} = \sqrt{a\cdot b}$.	$\sqrt{2}\cdot\sqrt{3} = \sqrt{6}$
A square root is in **simplified form** if the radicand contains no perfect square factors other than 1. To simplify a square root, factor the radicand so that one of its factors is a perfect square factor.	$\sqrt{45} = \sqrt{9\cdot 5}$ $= \sqrt{9}\cdot\sqrt{5}$ $= 3\sqrt{5}$ in simplest form.
Quotient Rule for Square Roots If $\sqrt{a}$ and $\sqrt{b}$ are real numbers and $b \neq 0$, then $\sqrt{\frac{a}{b}} = \frac{\sqrt{a}}{\sqrt{b}}$	$\sqrt{\frac{18}{25}} = \frac{\sqrt{9\cdot 2}}{\sqrt{25}} = \frac{\sqrt{9}\cdot\sqrt{2}}{5} = \frac{3\sqrt{2}}{5}$
Like radicals are radical expressions that have the same index and the same radicand.	***Like Radicals*** $5\sqrt{2}, -7\sqrt{2}, \sqrt{2}$
To combine like radicals, use the distributive property.	$2\sqrt{7} - 13\sqrt{7} = (2 - 13)\sqrt{7} = -11\sqrt{7}$
The product and quotient rules for radicals may be used to simplify products and quotients of radicals.	Perform the indicated operations and simplify. $\sqrt{3}\cdot\sqrt{11} = \sqrt{33}$ $(2\sqrt{5})^2 = 2^2\cdot(\sqrt{5})^2 = 4\cdot 5 = 20$ Multiply. $(\sqrt{3} + 1)(\sqrt{5} - \sqrt{3})$ $= \sqrt{15} - \sqrt{9} + \sqrt{5} - \sqrt{3}$ $= \sqrt{15} - 3 + \sqrt{5} - \sqrt{3}$
The process of eliminating the radical in the denominator of a radical expression is called **rationalizing the denominator.**	Rationalize the denominator. $\frac{5}{\sqrt{11}} = \frac{5\cdot\sqrt{11}}{\sqrt{11}\cdot\sqrt{11}} = \frac{5\sqrt{11}}{11}$
Section 5.6 Solving Quadratic Equations by the Square Root Property and the Quadratic Formula	
Square Root Property If $x^2 = a$ for $a \geq 0$, then $x = \pm\sqrt{a}$	Solve the equation. $(x - 1)^2 = 15$ $x - 1 = \pm\sqrt{15}$ $x = 1 \pm \sqrt{15}$

DEFINITIONS AND CONCEPTS | **EXAMPLES**

Section 5.6 Solving Quadratic Equations by the Square Root Property and the Quadratic Formula (continued)

Quadratic Formula

If a, b, and c are real numbers and $a \neq 0$, the quadratic equation $ax^2 + bx + c = 0$ has solutions

$$x = \frac{-b \pm \sqrt{b^2 - 4ac}}{2a}$$

Identify a, b, and c in the quadratic equation

$$4x^2 - 6x = 5$$

First, subtract 5 from both sides.

$$4x^2 - 6x - 5 = 0$$
$$a = 4, b = -6, \text{ and } c = -5$$

To Solve a Quadratic Equation by the Quadratic Formula

Step 1. Write the equation in standard form: $ax^2 + bx + c = 0$.

Step 2. If necessary, clear the equation of fractions.

Step 3. Identify a, b, and c.

Step 4. Replace a, b, and c in the quadratic formula by known values and simplify.

Solve: $3x^2 - 2x - 2 = 0$

In this equation, $a = 3$, $b = -2$, and $c = -2$.

$$x = \frac{-(-2) \pm \sqrt{(-2)^2 - 4(3)(-2)}}{2 \cdot 3}$$
$$= \frac{2 \pm \sqrt{4 - (-24)}}{6}$$
$$= \frac{2 \pm \sqrt{28}}{6} = \frac{2 \pm \sqrt{4 \cdot 7}}{6} = \frac{2 \pm 2\sqrt{7}}{6}$$
$$= \frac{2(1 \pm \sqrt{7})}{2 \cdot 3} = \frac{1 \pm \sqrt{7}}{3}$$

Section 5.7 The Pythagorean Theorem, the Distance Formula, and Problem Solving

Problem-Solving Steps

1. UNDERSTAND. Read and reread the problem.

A gutter is mounted on the eaves of a house 15 feet above the ground. A garden is adjacent to the house so that the closest a ladder can be placed to the house is 6 feet. How long a ladder is needed for installing the gutter? Let $x =$ the length of the ladder.

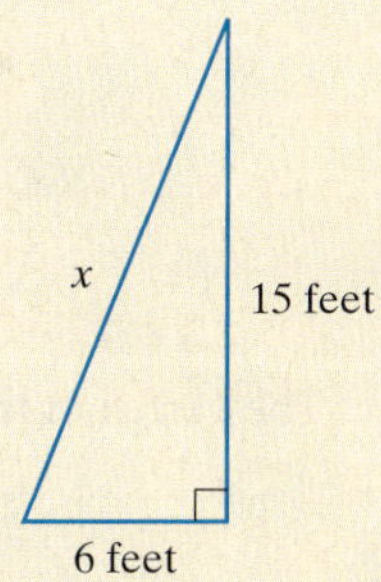

2. TRANSLATE.

Here, we use the Pythagorean theorem. The unknown length x is the hypotenuse.
In words:

$$(\text{leg})^2 + (\text{leg})^2 = (\text{hypotenuse})^2$$

Translate:

3. SOLVE.

$$6^2 + 15^2 = x^2$$
$$36 + 225 = x^2$$
$$261 = x^2$$
$$\sqrt{261} = x \quad \text{or} \quad x = 3\sqrt{29}$$

4. INTERPRET.

Check and state. The ladder needs to be $3\sqrt{29}$ feet or approximately 16.2 feet long.

Problem-Solving Steps

A garden is in the shape of a rectangle whose length is two feet more than its width. If the area of the garden is 35 square feet, find its dimensions.

1. UNDERSTAND the problem.

1. Read and reread the problem. Guess a solution and check your guess.

   Let x be the width of the rectangular garden. Then $x + 2$ is the length.

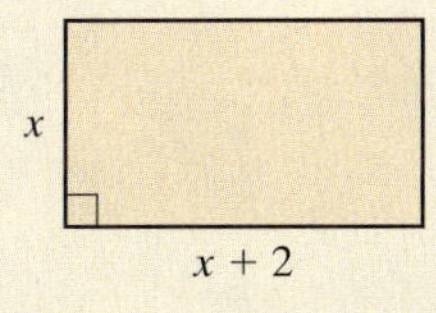

2. TRANSLATE.

2. In words: length $\cdot$ width $=$ area

   Translate: $(x + 2) \cdot x = 35$

(continued)

DEFINITIONS AND CONCEPTS	EXAMPLES

Section 5.7 The Pythagorean Theorem, the Distance Formula, and Problem Solving (continued)

3. SOLVE.

3.
$$(x + 2)x = 35$$
$$x^2 + 2x - 35 = 0$$
$$(x - 5)(x + 7) = 0$$
$$x - 5 = 0 \quad \text{or} \quad x + 7 = 0$$
$$x = 5 \quad \text{or} \quad x = -7$$

4. INTERPRET.

4. Discard the solution of -7 since x represents width.

Check: If x is 5 feet then $x + 2 = 5 + 2 = 7$ feet. The area of a rectangle whose width is 5 feet and whose length is 7 feet is (5 feet)(7 feet) or 35 square feet.

State: The garden is 5 feet by 7 feet.

Section 5.8 Extension—Graphing Quadratic Functions

The graph of a quadratic equation $y = ax^2 + bx + c$, $a \neq 0$, is called a **parabola.** The lowest point on a parabola opening upward or the highest point on a parabola opening downward is called the **vertex.** The vertical line through the vertex is the **axis of symmetry.**

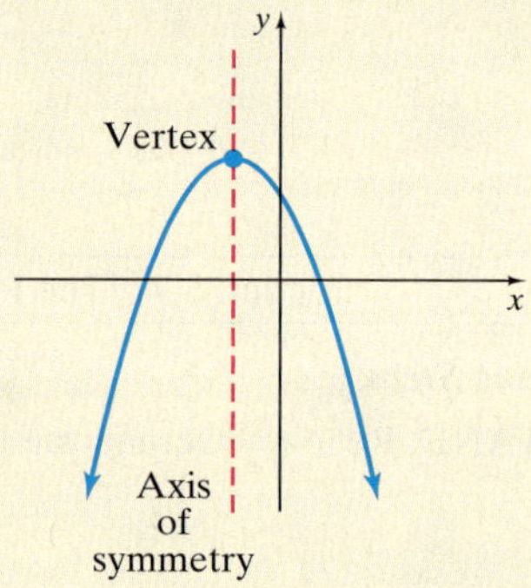

The vertex of the parabola $y = ax^2 + bx + c$ has x-value $\dfrac{-b}{2a}$.

Graph: $y = 2x^2 - 6x + 4$

The x-value of the vertex is

$$x = \frac{-b}{2a} = \frac{-(-6)}{2(2)} = \frac{6}{4} = \frac{3}{2}$$

The y-value is

$$y = 2\left(\frac{3}{2}\right)^2 - 6\left(\frac{3}{2}\right) + 4 = -\frac{1}{2}$$

The vertex is $\left(\frac{3}{2}, -\frac{1}{2}\right)$.

Find the intercepts.

- Let $x = 0$ and solve for y to find the y-intercept.
- Let $y = 0$ and solve for x to find any x-intercept.

The y-intercept is

$$y = 2 \cdot 0^2 - 6 \cdot 0 + 4 = 4$$

The x-intercepts are found by solving

$$0 = 2x^2 - 6x + 4$$
$$0 = 2(x^2 - 3x + 2)$$
$$0 = 2(x - 2)(x - 1)$$
$$x - 2 = 0 \quad \text{or} \quad x - 1 = 0$$
$$x = 2 \quad \text{or} \quad x = 1$$

Find more ordered pair solutions as needed.

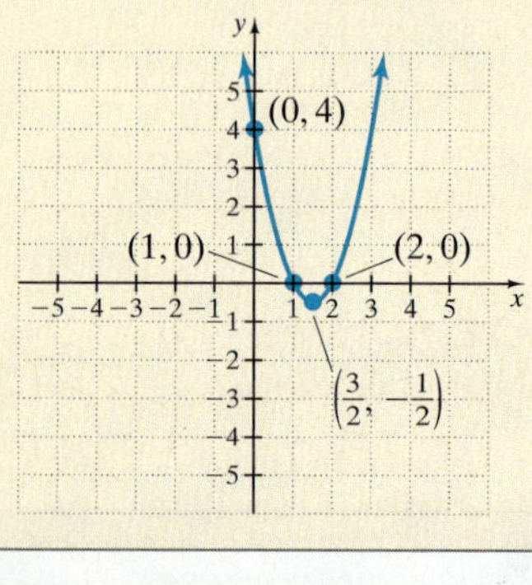

Chapter 5 Review

(5.1) *Complete the factoring.*

1. $6x^2 - 15x = 3x(\quad)$

2. $2x^3y + 6x^2y^2 + 8xy^3 = 2xy(\quad)$

Factor the GCF from each polynomial.

3. $20x^2 + 12x$

4. $6x^2y^2 - 3xy^3$

5. $3x(2x + 3) - 5(2x + 3)$

6. $5x(x + 1) - (x + 1)$

Factor each polynomial by grouping.

7. $3x^2 - 3x + 2x - 2$

8. $3a^2 + 9ab + 3b^2 + ab$

9. $10a^2 + 5ab + 7b^2 + 14ab$

10. $6x^2 + 10x - 3x - 5$

(5.2) *Factor each trinomial.*

11. $x^2 + 6x + 8$

12. $x^2 - 11x + 24$

13. $x^2 + x + 2$

14. $x^2 - x + 2$

15. $x^2 + 4xy - 12y^2$

16. $x^2 + 8xy + 15y^2$

17. $10a^3 - 110a^2 + 100a$

18. $5y^3 - 50y^2 + 120y$

19. $2x^2 + 13x + 6$

20. $4x^2 + 4x - 3$

21. $6x^2 + 5xy - 4y^2$

22. $18x^2 - 9xy - 20y^2$

23. $10y^3 + 25y^2 - 60y$

24. $60y^3 - 39y^2 + 6y$

25. $18x^2 - 60x + 50$

26. $4x^2 - 28xy + 49y^2$

(5.3) *Factor each binomial.*

27. $4x^2 - 9$

28. $9t^2 - 25s^2$

29. $16x^2 + y^2$

30. $x^3 - 8y^3$

31. $8x^3 + 27$

32. $2x^3 + 8x$

33. $54 - 2x^3y^3$

34. $9x^2 - 4y^2$

35. $16x^4 - 1$

36. $x^4 + 16$

(5.4) *Solve the following equations.*

37. $(x + 6)(x - 2) = 0$

38. $3x(x + 1)(7x - 2) = 0$

39. $4(5x + 1)(x + 3) = 0$

40. $x^2 + 8x + 7 = 0$

41. $x^2 - 2x - 24 = 0$

42. $x^2 + 10x = -25$

43. $56x^2 - 5x = 6$

44. $20x^2 - 7x = 6$

(5.5) *Find each square root.*

45. $\sqrt{81}$

46. $-\sqrt{49}$

47. $-\sqrt{\dfrac{9}{64}}$

48. $\sqrt{\dfrac{36}{81}}$

Simplify each expression.

49. $\sqrt{54}$

50. $\sqrt{88}$

51. $\sqrt{\dfrac{18}{25}}$

52. $\sqrt{\dfrac{75}{64}}$

Add or subtract by combining like radicals.

53. $5\sqrt{7} + \sqrt{7}$

54. $5\sqrt{2} - 5\sqrt{3}$

55. $\sqrt{50} - \sqrt{8}$

56. $\sqrt{\dfrac{11}{25}} + \sqrt{\dfrac{11}{16}}$

Find the product and simplify if possible.

57. $\sqrt{3}(2\sqrt{6} - 3\sqrt{12})$

58. $4\sqrt{5}(2\sqrt{10} - 5\sqrt{5})$

59. $(\sqrt{3} + 2)(\sqrt{6} - 5)$

60. $(2\sqrt{5} + 1)(4\sqrt{5} - 3)$

Divide and simplify if possible.

61. $\dfrac{\sqrt{27}}{\sqrt{3}}$

62. $\dfrac{\sqrt{20}}{\sqrt{5}}$

63. $\dfrac{\sqrt{160}}{\sqrt{8}}$

64. $\dfrac{\sqrt{96}}{\sqrt{3}}$

Rationalize each denominator and simplify.

65. $\dfrac{\sqrt{2}}{\sqrt{11}}$

66. $\dfrac{\sqrt{3}}{\sqrt{13}}$

67. $\sqrt{\dfrac{5}{6}}$

68. $\sqrt{\dfrac{7}{10}}$

(5.6) *Use the square root property to solve each quadratic equation.*

69. $x^2 = 36$

70. $x^2 = 81$

71. $k^2 = 50$

72. $k^2 = 45$

73. $(4p + 5)^2 = 41$

74. $(3p + 7)^2 = 37$

Solve. For Exercises 75 and 76, use the formula $h = 16t^2$, where h is the height in feet at time t seconds.

75. If Kara Washington dives from a height of 100 feet, how long before she hits the water?

76. How long does a 5-mile free fall take? Round your result to the nearest tenth of a second.

(*Hint:* 1 mi = 5280 ft)

Use the quadratic formula to solve each quadratic equation.

77. $7x^2 = 35$

78. $11x^2 = 33$

79. $x^2 - 10x + 7 = 0$

80. $x^2 + 4x - 7 = 0$

81. $3x^2 + x - 1 = 0$

82. $x^2 + 3x - 1 = 0$

83. $2x^2 + x + 5 = 0$

84. $7x^2 - 3x + 1 = 0$

For the Exercise numbers given, approximate the exact solutions to the nearest tenth.

85. Exercise 81

86. Exercise 82

(5.7) *Use the Pythagorean theorem to find the length of each unknown side. Give the exact answer and a two-decimal-place approximation.*

△ **87.**

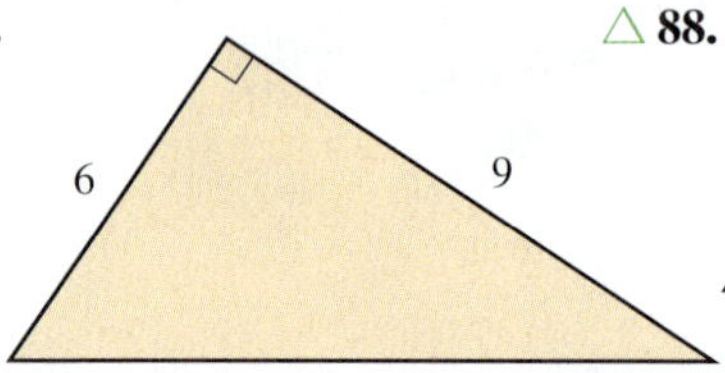

△ **88.**

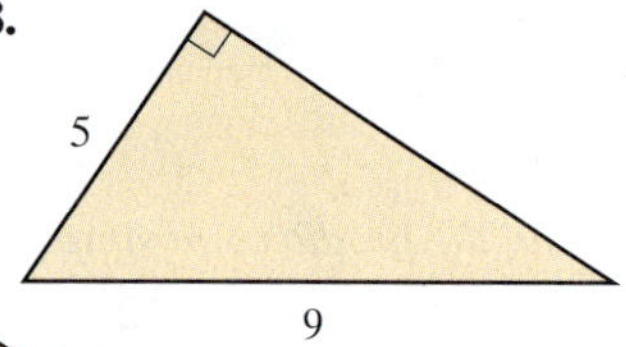

Solve. Give the exact answer and a two-decimal-place approximation.

△ **89.** Romeo is standing 20 feet away from the wall below Juliet's balcony during a school play. Juliet is on the balcony, 12 feet above the ground. Find how far apart Romeo and Juliet are.

△ **90.** The diagonal of a rectangle is 10 inches long. If the width of the rectangle is 5 inches, find the length of the rectangle.

Use the distance formula to find the distance between the points.

91. $(6, -2)$ and $(-3, 5)$

92. $(2, 8)$ and $(-6, 10)$

Use the formula $r = \sqrt{\frac{S}{4\pi}}$, *where r = the radius of a sphere and S = the surface area of the sphere, for Exercises 93 and 94.*

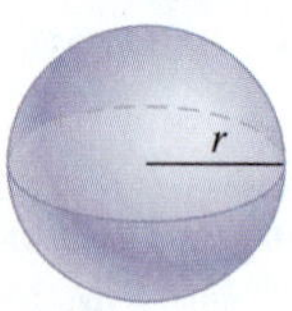

△ **93.** Find the radius of a sphere to the nearest tenth of an inch if the area is 72 square inches.

△ **94.** Find the exact surface area of a sphere if its radius is 6 inches. (Do not approximate π.)

Use the given information to choose the correct dimensions.

△ **95.** The perimeter of a rectangle is 24 inches. The length is twice the width. Find the dimensions of the rectangle.

a. 5 inches by 7 inches **b.** 5 inches by 10 inches
c. 4 inches by 8 inches **d.** 2 inches by 10 inches

△ **96.** The area of a rectangle is 80 square meters. The length is one more than three times the width. Find the dimensions of the rectangle.

a. 8 meters by 10 meters **b.** 4 meters by 13 meters
c. 4 meters by 20 meters **d.** 5 meters by 16 meters

Use the given information to find the dimensions of each figure.

△ **97.** The *area* of the square is 81 square units. Find the length of a side.

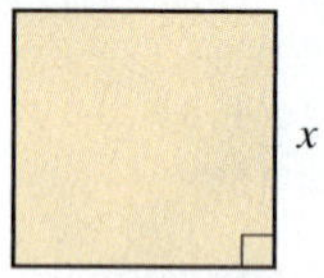

△ **98.** The *perimeter* of the quadrilateral is 47 units. Find the lengths of the sides.

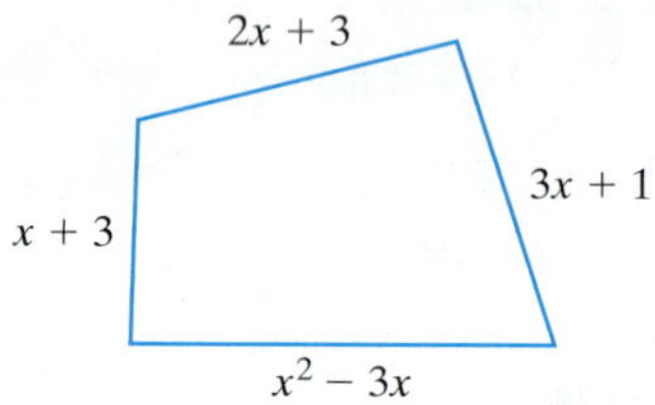

△ **99.** A flag for a local organization is in the shape of a rectangle whose length is 15 inches less than twice its width. If the area of the flag is 500 square inches, find its dimensions.

△ **100.** The base of a triangular sail is four times its height. If the area of the triangle is 162 square yards, find the base.

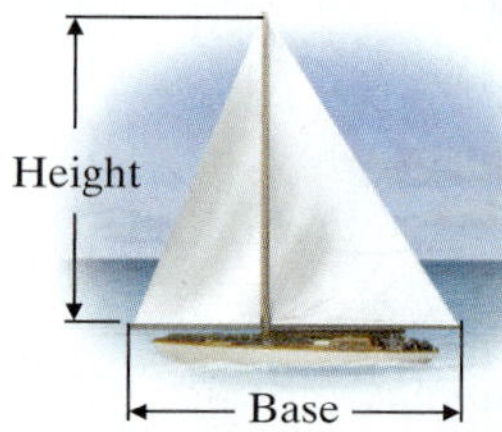

101. Find two consecutive positive integers whose product is 380.

102. Find two consecutive positive even integers whose product is 440.

103. A rocket is fired from the ground with an initial velocity of 440 feet per second. Its height h after t seconds is given by the equation

$$h = -16t^2 + 440t$$

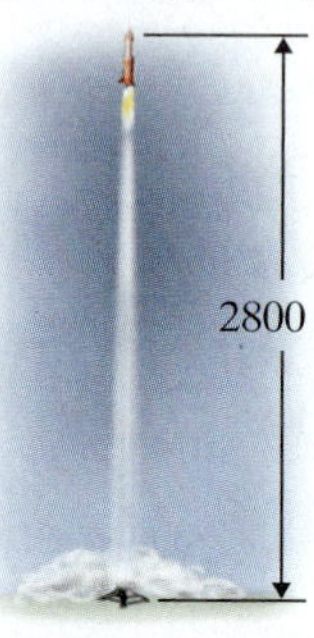

a. Find how many seconds pass before the rocket reaches a height of 2800 feet. Explain why two answers are obtained.

b. Find how many seconds pass before the rocket reaches the ground again.

104. An architect's squaring instrument is in the shape of a right triangle. Find the length of the longer leg of the right triangle if the hypotenuse is 8 centimeters longer than the longer leg and the shorter leg is 8 centimeters shorter than the longer leg.

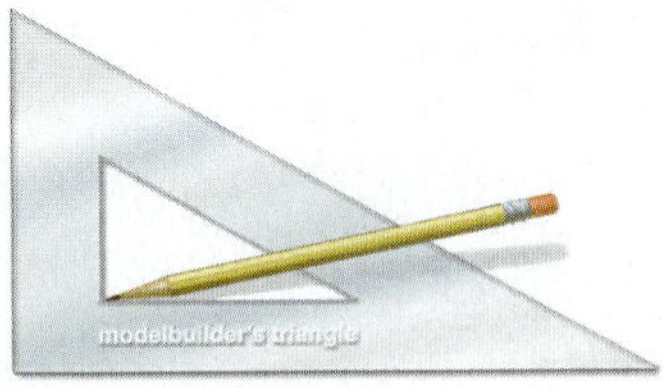

(5.8) *Graph each quadratic equation and find and plot any intercepts.*

105. $y = 5x^2$

106. $y = -\frac{1}{2}x^2$

Graph each quadratic equation. Label the vertex and the intercepts with their coordinates.

107. $y = x^2 - 25$

108. $y = x^2 - 36$

109. $y = x^2 + 3$

110. $y = x^2 + 8$

111. $y = -4x^2 + 8$

112. $y = -3x^2 + 9$

113. $y = x^2 + 3x - 10$

114. $y = x^2 + 3x - 4$

115. $y = -x^2 - 5x - 6$

116. $y = -x^2 + 4x + 8$

117. $y = 2x^2 - 11x - 6$

118. $y = 3x^2 - x - 2$

MIXED REVIEW

Factor completely.

119. $7x - 63$

120. $11x(4x - 3) - 6(4x - 3)$

121. $m^2 - \frac{4}{25}$

122. $3x^3 - 4x^2 + 6x - 8$

123. $xy + 2x - y - 2$

124. $2x^2 + 2x - 24$

125. $3x^3 - 30x^2 + 27x$

126. $4x^2 - 81$

127. $2x^2 - 18$

128. $125x^3 + 27$

Write the perimeter of each figure as a simplified polynomial. Then factor each polynomial.

△ **129.**

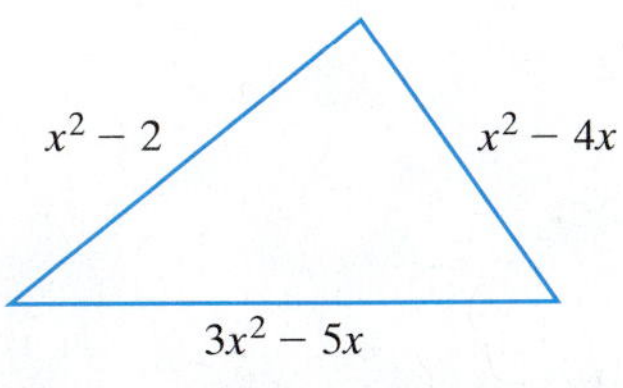

△ **130.**

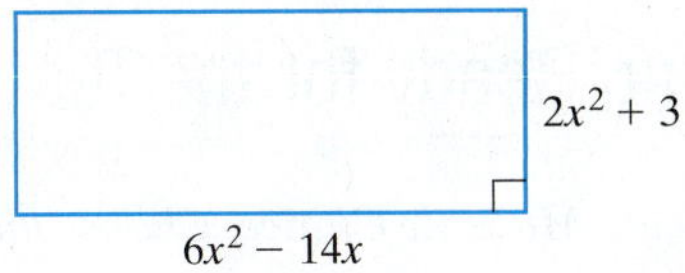

Solve.

131. $2x^2 - x - 28 = 0$

132. $x(x - 5) = -6$

Add or subtract by simplifying and then combining like terms.

133. $\sqrt{12} + \sqrt{75}$

134. $\sqrt{\frac{3}{16}} - \sqrt{\frac{3}{4}}$

Multiply and simplify if possible.

135. $\sqrt{3}(\sqrt{9} - \sqrt{2})$

136. $(\sqrt{2} + 4)(\sqrt{5} - 1)$

Rationalize each denominator and simplify.

137. $\sqrt{\frac{2}{7}}$

138. $\frac{3}{\sqrt{2}}$

Match each quadratic equation with its graph.

139. $y = 2x^2$

140. $y = -x^2$

141. $y = x^2 + 4x + 4$

142. $y = x^2 + 5x + 4$

A.

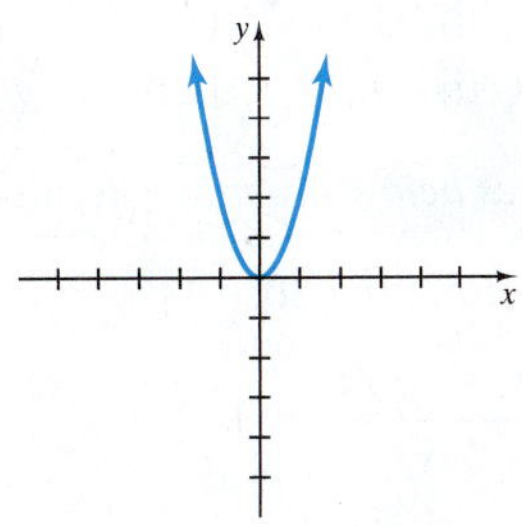

B.

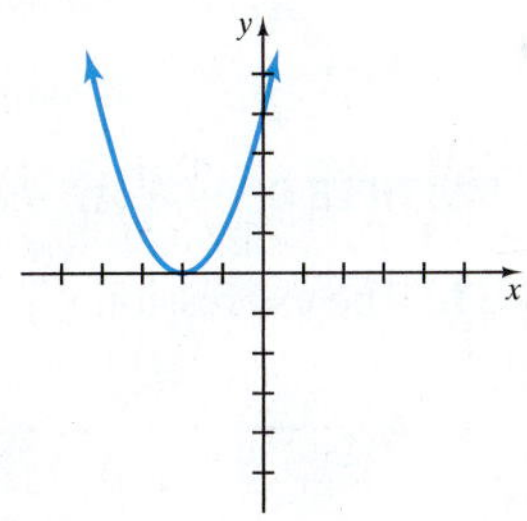

C.

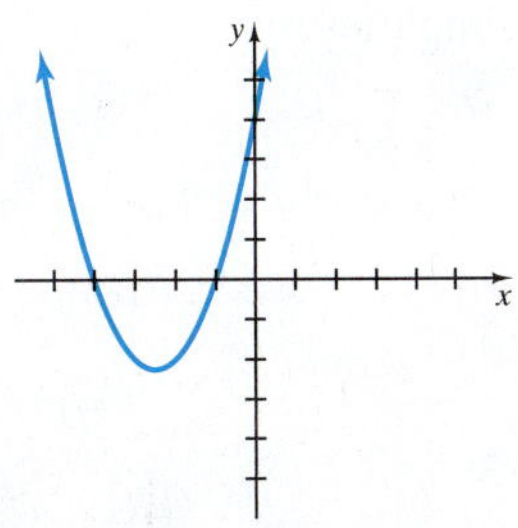

D.

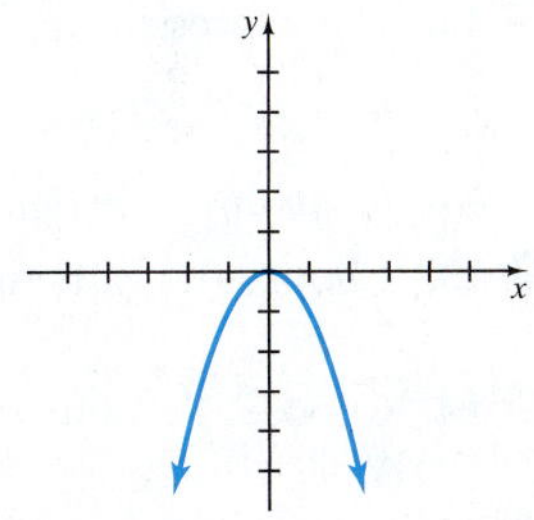

Use the square root property to solve each quadratic equation.

143. $y^2 = 75$

144. $(x - 7)^2 = 64$

Use the quadratic formula to solve each quadratic equation.

145. $4x^2 - 3x - 2 = 0$

146. $5x^2 + x - 2 = 0$

Chapter 5 Getting Ready for the Test

*All the exercises below are **Multiple Choice**. Choose the correct letter. Also, letters may be used more than once.*

1. The greatest common factor of the terms of $10x^4 - 70x^3 + 2x^2 - 14x$ is
   A. $2x^2$ **B.** $2x$ **C.** $7x^2$ **D.** $7x$
2. Choose the expression that is NOT a factored form of $9y^3 - 18y^2$.
   A. $9(y^3 - 2y^2)$ **B.** $9y(y^2 - 2y)$ **C.** $9y^2(y - 2)$ **D.** $9 \cdot y^3 - 18 \cdot y^2$

Identify each expression as:
A. A factored expression or **B.** Not a factored expression

3. $(x - 1)(x + 5)$
4. $z(z + 12)(z - 12)$
5. $y(x - 6) + 1(x - 6)$
6. $m \cdot m - 5 \cdot 5$

MULTIPLE CHOICE *Choose the correct letter.*

7. $7\sqrt{3} - \sqrt{3} =$
   A. 7 **B.** 6 **C.** $6\sqrt{3}$ **D.** cannot be simplified
8. $7\sqrt{3} \cdot \sqrt{2} =$
   A. 35 **B.** $7\sqrt{6}$ **C.** 42 **D.** $8\sqrt{6}$ **E.** cannot be simplified
9. $(4\sqrt{5})^2 =$
   A. 40 **B.** 80 **C.** $8\sqrt{5}$ **D.** $16\sqrt{5}$ **E.** cannot be simplified
10. To rationalize the denominator of $\frac{\sqrt{5}}{\sqrt{2}}$, we multiply by:
    A. $\frac{\sqrt{5}}{\sqrt{2}}$ **B.** $\frac{\sqrt{10}}{\sqrt{10}}$ **C.** $\frac{\sqrt{2}}{\sqrt{2}}$ **D.** $\frac{\sqrt{5}}{\sqrt{5}}$

MULTIPLE CHOICE *All the exercises below are multiple choice. Choose the correct letter.*

11. The expression $\frac{12 \pm 3\sqrt{7}}{9}$ simplifies to:
    A. $\frac{\pm 15\sqrt{7}}{9}$ **B.** $\frac{4 \pm 3\sqrt{7}}{3}$ **C.** $\frac{4 \pm \sqrt{7}}{3}$ **D.** $4 \pm \sqrt{7}$
12. The expression $\frac{5 \pm 10\sqrt{2}}{5}$ simplifies to:
    A. $1 \pm 2\sqrt{2}$ **B.** $\pm 10\sqrt{2}$ **C.** $\pm 3\sqrt{2}$ **D.** $1 \pm 10\sqrt{2}$

For the quadratic equation $7x^2 = 3 - x$, if $a = 7$ in the quadratic formula, answer Exercises 13 and 14.

13. Choose the value of c.
    A. 3 **B.** -3 **C.** 1 **D.** -1 **E.** x
14. Choose the value of b.
    A. 3 **B.** -3 **C.** 1 **D.** -1 **E.** x
15. To solve $x(x + 2) = 15$, which is an incorrect next step?
    A. $x^2 + 2x = 15$ **B.** $x(x + 2) - 15 = 0$ **C.** $x = 15$ and $x + 2 = 15$

Chapter 5 Test

MyMathLab® YouTube

Factor each polynomial completely. If a polynomial cannot be factored, write "prime."

1. $x^2 + 11x + 28$
2. $49 - m^2$
3. $y^2 + 22y + 121$
4. $y^2 - 8y - 48$
5. $x^2 + x - 10$
6. $9x^3 + 39x^2 + 12x$
7. $3a^2 + 3ab - 7a - 7b$
8. $3x^2 - 5x + 2$
9. $x^2 + 14xy + 24y^2$
10. $180 - 5x^2$
11. $6t^2 - t - 5$
12. $xy^2 - 7y^2 - 4x + 28$
13. $64x^3 - 1$
14. $8y^3 - 64$

Solve each equation.

15. $(x - 3)(x + 9) = 0$

16. $x^2 + 5x = 14$

17. $x(x + 6) = 7$

18. $5t^3 - 45t = 0$

Simplify each radical expression.

19. $\sqrt{16}$

20. $\sqrt{54}$

21. $\sqrt{92}$

Perform each indicated operation. Assume that all variables represent positive numbers.

22. $\sqrt{13} + \sqrt{13} - 4\sqrt{13}$

23. $\sqrt{12} - 2\sqrt{75}$

24. $\sqrt{\frac{3}{4}} + \sqrt{\frac{3}{25}}$

25. $\sqrt{7} \cdot \sqrt{14}$

26. $\sqrt{2}(\sqrt{6} - \sqrt{5})$

27. $\frac{\sqrt{50}}{\sqrt{10}}$

Rationalize the denominator.

28. $\sqrt{\frac{2}{3}}$

Solve using the square root property.

29. $5k^2 = 80$

30. $(3m - 5)^2 = 8$

Solve by the most appropriate method.

31. $(3x - 5)(x + 2) = -6$

32. $(3x - 1)^2 = 16$

33. $3x^2 - 7x - 2 = 0$

34. $3x^2 - 7x + 2 = 0$

35. $2x^2 - 6x + 1 = 0$

36. $9x^3 = x$

Solve.

37. Find the length of the unknown leg of a right triangle if the other leg is 8 inches long and the hypotenuse is 12 inches long.

38. Find the distance between $(-3, 6)$ and $(-2, 8)$.

Solve.

39. The highest dive from a diving board by a woman was made by Lucy Wardle Streeter of the United States. She dove from a height of 120.75 feet at Ocean Park, Hong Kong, in 1986. To the nearest tenth of a second, how long did the dive take? Use the formula $h = 16t^2$.

40. An object is dropped from the top of the Woolworth Building on Broadway in New York City. The height h of the object after t seconds is given by the equation

$$h = -16t^2 + 784$$

Find how many seconds pass before the object reaches the ground.

Graph each quadratic equation. Label the vertex and the intercepts with their coordinates.

41. $y = -5x^2$

42. $y = x^2 - 4$

43. $y = x^2 - 7x + 10$

44. $y = 2x^2 + 4x - 1$

Chapter 5 Cumulative Review

1. Translate each sentence into a mathematical statement.
   a. Nine is less than or equal to eleven.
   b. Eight is greater than one.
   c. Three is not equal to four.

2. Insert < or > in the space to make each statement true.
   a. $|-5| \quad |-3|$
   b. $|0| \quad |-2|$

3. Simplify each fraction (write it in lowest terms).
   a. $\frac{42}{49}$ b. $\frac{11}{27}$ c. $\frac{88}{20}$

4. Evaluate $\frac{x}{y} + 5x$ if $x = 20$ and $y = 10$.

5. Simplify $\frac{8 + 2 \cdot 3}{2^2 - 1}$.

6. Evaluate $\frac{x}{y} + 5x$ if $x = -20$ and $y = 10$.

7. Simplify each expression by combining like terms.
   a. $7x - 3x$
   b. $10y^2 + y^2$
   c. $8x^2 + 2x - 3x$
   d. $9n^2 - 5n^2 + n^2$

8. Solve: $0.8y + 0.2(y - 1) = 1.8$

Solve.

9. $-3x = 33$

10. $-\frac{2}{3}x = -22$

11. $8(2 - t) = -5t$

12. $-z = \frac{7z + 3}{5}$

13. Balsa wood sticks are commonly used for building models (for example, bridge models). A 48-inch balsa wood stick is to be cut into two pieces so that the longer piece is 3 times the shorter. Find the length of each piece.

14. Solve $3x + 9 \leq 5(x - 1)$. Write the solution set using interval notation.

15. Graph the linear equation $y = -\frac{1}{3}x + 2$.

16. Is the ordered pair $(-1, 2)$ a solution of $-7x - 8y = -9$?

17. Find the slope and the y-intercept of the line whose equation is $3x - 4y = 4$.

18. Find the slope of the line through $(5, -6)$ and $(5, 2)$.

19. Evaluate each expression for the given value of x.
   a. $2x^3$; x is 5
   b. $\frac{9}{x^2}$; x is -3

20. Find the slope and y-intercept of the line whose equation is $7x - 3y = 2$.

21. Find the degree of each term.

a. $3x^2$ **b.** $12x^2yz^3$

c. 5

22. Find an equation of the vertical line through $(0, 7)$.

23. Subtract $(2x^3 + 8x^2 - 6x) - (2x^3 - x^2 + 1)$.

24. Find an equation of the line with slope 4 and y-intercept $\left(0, \frac{1}{2}\right)$. Write the equation in standard form.

25. Multiply $(3x + 2)(2x - 5)$.

26. Write an equation of the line through $(-4, 0)$ and $(6, -1)$. Write the equation in standard form.

27. Multiply $(3y + 1)^2$.

28. Solve the system. $\begin{cases} -x + 3y = 18 \\ -3x + 2y = 19 \end{cases}$

29. Simplify by writing each expression with positive exponents only.

a. 3^{-2} **b.** $2x^{-3}$ **c.** $2^{-1} + 4^{-1}$

d. $(-2)^{-4}$ **e.** y^{-4}

30. Simplify: $\dfrac{(5a^7)^2}{a^5}$

31. Write each number in scientific notation.

a. 367,000,000

b. 0.000003

c. 20,520,000,000

d. 0.00085

32. Multiply $(3x - 7y)^2$.

33. Find the GCF of each list of terms.

a. x^3, x^7, and x^5

b. y, y^4, and y^7

Factor.

34. $z^3 + 7z + z^2 + 7$

35. $x^2 + 7x + 12$

36. $2x^3 + 2x^2 - 84x$

37. $8x^2 - 22x + 5$

38. $-4x^2 - 23x + 6$

39. $25a^2 - 9b^2$

40. $9xy^2 - 16x$

41. Solve: $(x - 3)(x + 1) = 0$

42. Solve $x^2 - 13x = -36$.

CHAPTER 6

Geometry and Measurement

The word *geometry* is formed from the Greek words *geo*, meaning "Earth," and *metron*, meaning "measure." Geometry literally means to measure the Earth. In this chapter we learn about various geometric figures and their properties, such as perimeter, area, and volume. Knowledge of geometry can help us solve practical problems in all types of real-life situations.

New Horizons is NASA's robotic spacecraft mission, shown above. The spacecraft is about the size and shape of a grand piano with a satellite dish attached. It was launched in January 2006 and is now heading toward Pluto. How do we receive the images of Pluto on Earth when this spacecraft is so far into deep space, and how long before we receive the images?

The Deep Space Network (DSN) is a worldwide network of large antennas and communication facilities. When a mission is in deep space, fewer sites are needed for sending and receiving transmissions; thus, the DSN has only three sites—one in California (Goldstone), one in Spain (Madrid), and one in Australia (Canberra). The diagram below shows an overview of Earth from the vantage point of the North Pole and the location of these three sites.

We study some geometry of the DSN in Section 6.1, Exercises 65 and 66, and Section 6.6, Exercises 23–26.

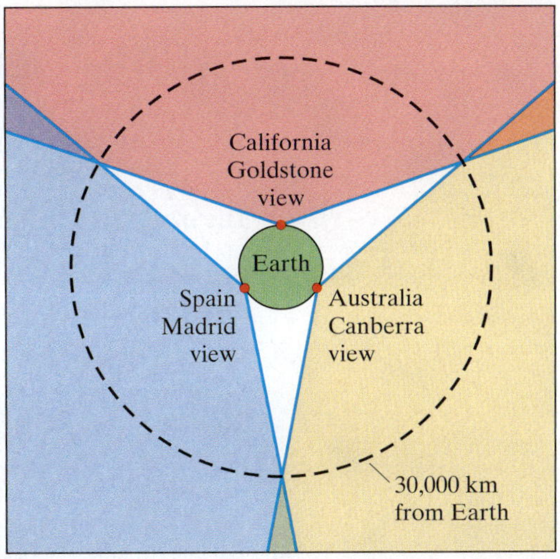

6.1 Lines and Angles

OBJECTIVES

1 Identify Lines, Line Segments, Rays, and Angles.
2 Classify Angles as Acute, Right, Obtuse, or Straight.
3 Identify Complementary and Supplementary Angles.
4 Find Measures of Angles.

OBJECTIVE

1 Identifying Lines, Line Segments, Rays, and Angles

Let's begin with a review of two important concepts—space and plane.

Space extends in all directions indefinitely. Examples of objects in space are houses, grains of salt, bushes, your *Path to College Mathematics* textbook, and you.

A **plane** is a flat surface that extends indefinitely. Surfaces like a plane are a classroom floor or a blackboard or whiteboard.

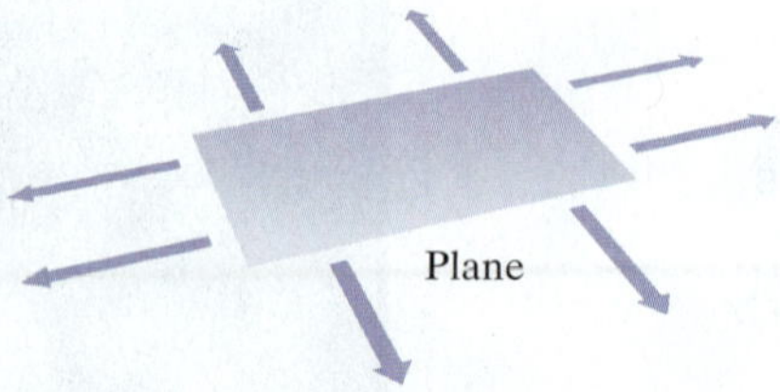

The most basic concept of geometry is the idea of a point in space. A **point** has no length, no width, and no height, but it does have location. We represent a point by a dot, and we usually label points with capital letters.

Point *P*

A **line** is a set of points extending indefinitely in two directions. A line has no width or height, but it does have length. We can name a line by any two of its points or by a single lowercase letter. A **line segment** is a piece of a line with two endpoints.*

Line *AB*, $\overleftrightarrow{AB}$, or line *l**

Line segment *AB* or $\overline{AB}$

A **ray** is a part of a line with one endpoint. A ray extends indefinitely in one direction. An **angle** is made up of two rays that share the same endpoint. The common endpoint is called the **vertex.**

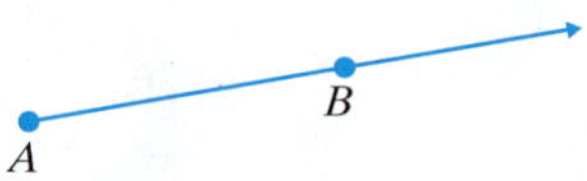

Ray *AB* or $\overrightarrow{AB}$

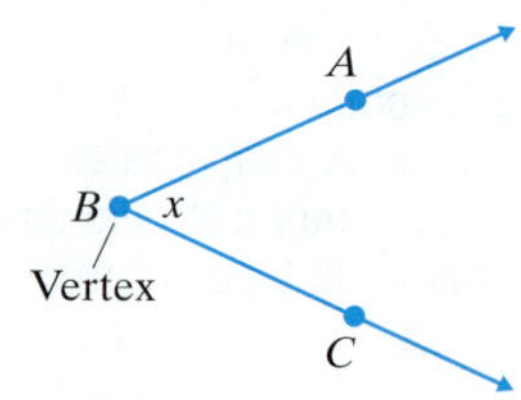

The angle in the figure above can be named

$$\angle ABC \quad \angle CBA \quad \angle B \quad \text{or} \quad \angle x$$

↑ ↑ The vertex is the middle point.

Rays *BA* and *BC* are **sides** of the angle.

*Although line *l* is also line *BA* or $\overleftrightarrow{BA}$, we will use only one order of points to name a line or line segment.

> **Helpful Hint**
>
> **Naming an Angle**
>
> When there is no confusion as to what angle is being named, you may use the vertex alone.
>
> 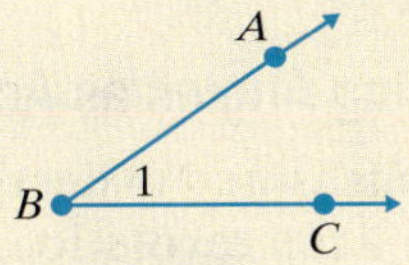
>
>
> Name of $\angle B$ is all right.
> There is no confusion. $\angle B$ means $\angle 1$.
>
> 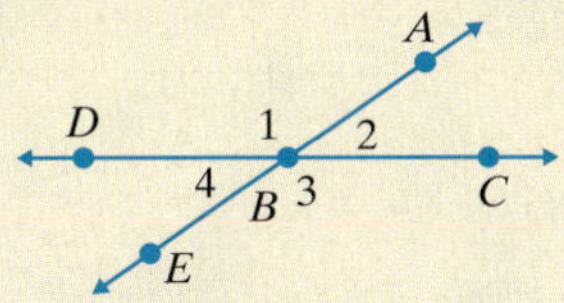
>
>
> Name of $\angle B$ is *not* all right.
> There is confusion. Does $\angle B$ mean $\angle 1$, $\angle 2$, $\angle 3$, or $\angle 4$?

EXAMPLE 1 Identify each figure as a line, a ray, a line segment, or an angle. Then name the figure using the given points.

a.

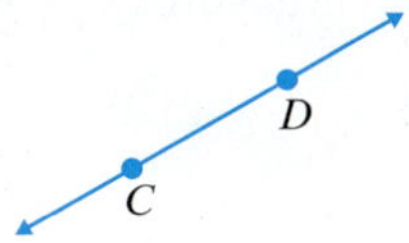

b.

c.

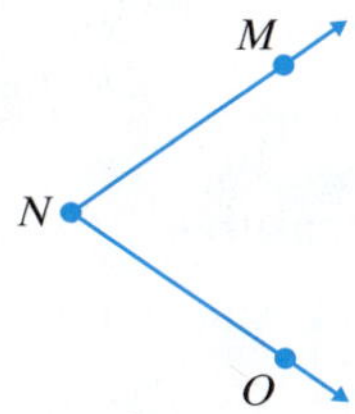

d.

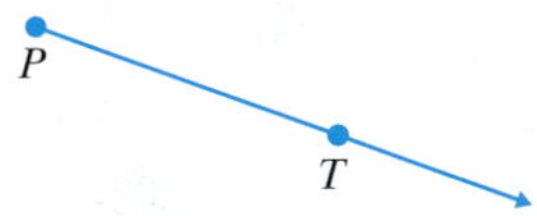

Solution

Figure (a) extends indefinitely in two directions. It is line CD or $\overleftrightarrow{CD}$.
Figure (b) has two endpoints. It is line segment EF or $\overline{EF}$.
Figure (c) has two rays with a common endpoint. It is $\angle MNO$, $\angle ONM$, or $\angle N$.
Figure (d) is part of a line with one endpoint. It is ray PT or $\overrightarrow{PT}$. □

PRACTICE 1 Identify each figure as a line, a ray, a line segment, or an angle. Then name the figure using the given points.

a.

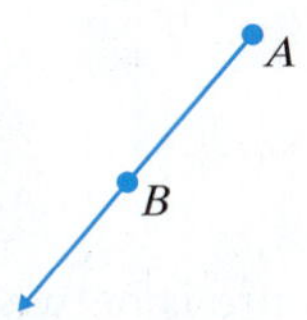

b.

c.

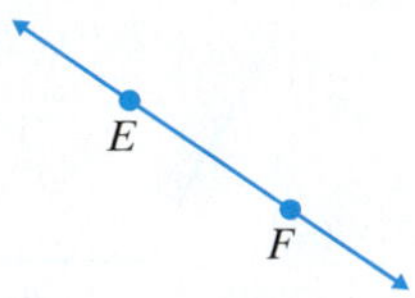

d.

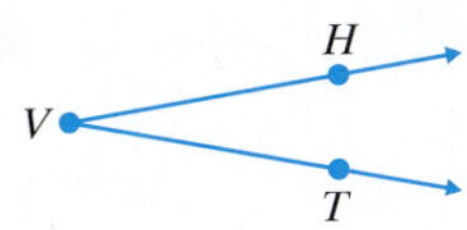

■

EXAMPLE 2 List other ways to name $\angle y$.

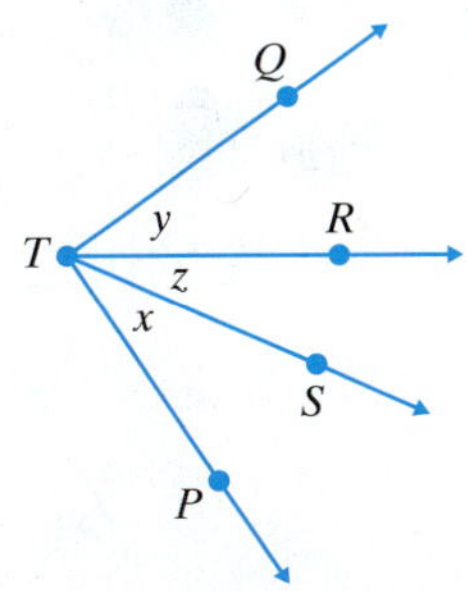

(Continued on next page)

Solution Two other ways to name $\angle y$ are $\angle QTR$ and $\angle RTQ$. We may *not* use the vertex alone to name this angle because several different angles have T as their vertex. □

PRACTICE 2 Use the figure in Example 2 to list other ways to name $\angle z$. ■

OBJECTIVE 2 Classifying Angles as Acute, Right, Obtuse, or Straight

An angle can be measured in **degrees.** The symbol for degrees is a small, raised circle, °. There are 360° in a full revolution, or a full circle.

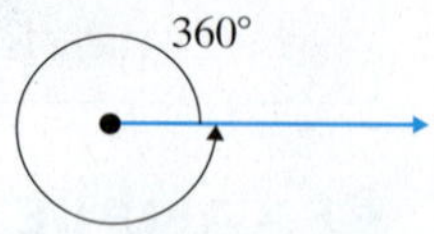

$\frac{1}{2}$ of a revolution measures $\frac{1}{2}(360°) = 180°$. An angle that measures 180° is called a **straight angle.**

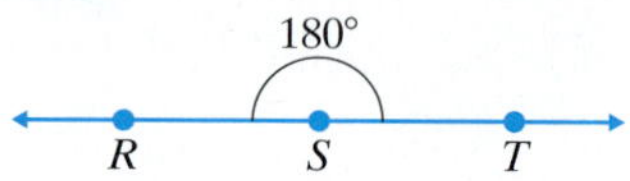

$\angle RST$ is a straight angle.

$\frac{1}{4}$ of a revolution measures $\frac{1}{4}(360°) = 90°$. An angle that measures 90° is called a **right angle.** The symbol ∟ is used to denote a right angle.

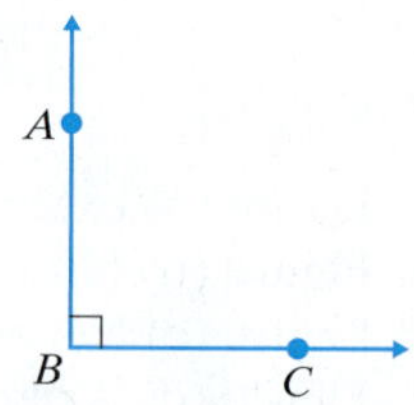

$\angle ABC$ is a right angle.

An angle whose measure is between 0° and 90° is called an **acute angle.**

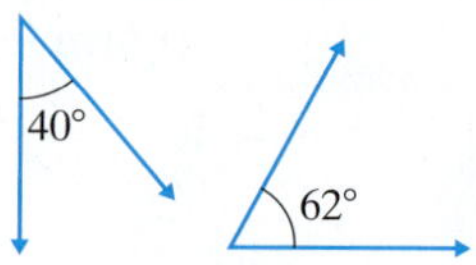

Acute angles

An angle whose measure is between 90° and 180° is called an **obtuse angle.**

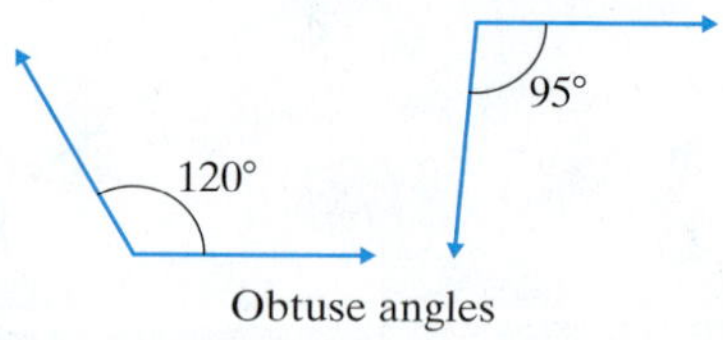

Obtuse angles

EXAMPLE 3 Classify each angle as acute, right, obtuse, or straight.

a.

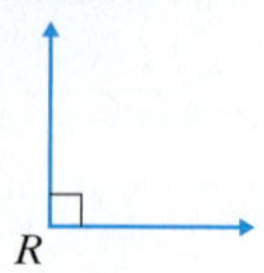

b.

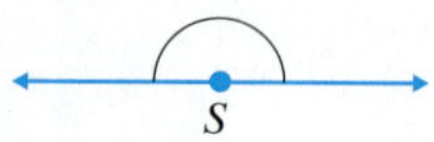

c.

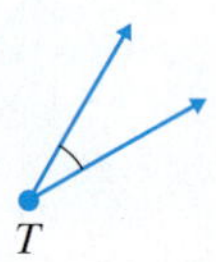

d.

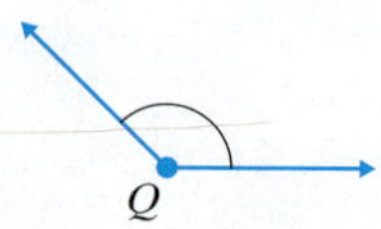

Solution

a. $\angle R$ is a right angle, denoted by ∟. It measures $90°$.

b. $\angle S$ is a straight angle. It measures $180°$.

c. $\angle T$ is an acute angle. It measures between $0°$ and $90°$.

d. $\angle Q$ is an obtuse angle. It measures between $90°$ and $180°$. □

PRACTICE 3 Classify each angle as acute, right, obtuse, or straight.

a.

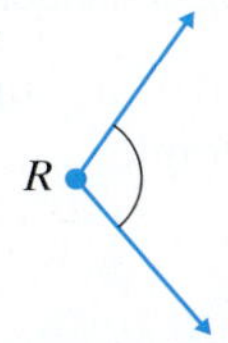

b.

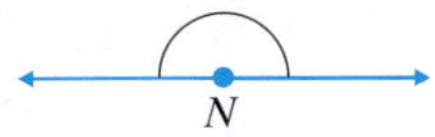

c.

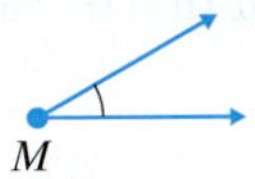

d.

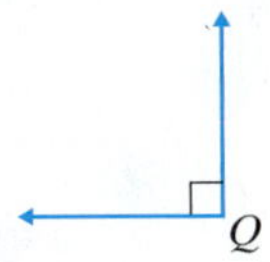

■

Let's look at $\angle B$ below, whose measure is $62°$.

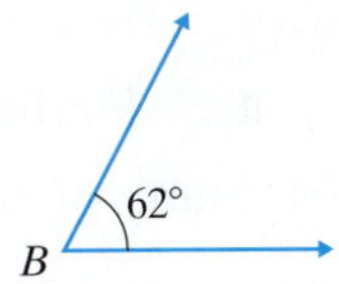

There is a shorthand notation for writing the measure of this angle. To write "The measure of $\angle B$ is 62°," we can write,

$$m\angle B = 62°.$$

By the way, note that $\angle B$ is an acute angle because $m\angle B$ is between $0°$ and $90°$.

OBJECTIVE 3

Identifying Complementary and Supplementary Angles

Two angles that have a sum of $90°$ are called **complementary angles.** We say that each angle is the **complement** of the other.

$\angle R$ and $\angle S$ are complementary angles because

$$m\angle R + m\angle S = 60° + 30° = 90°$$

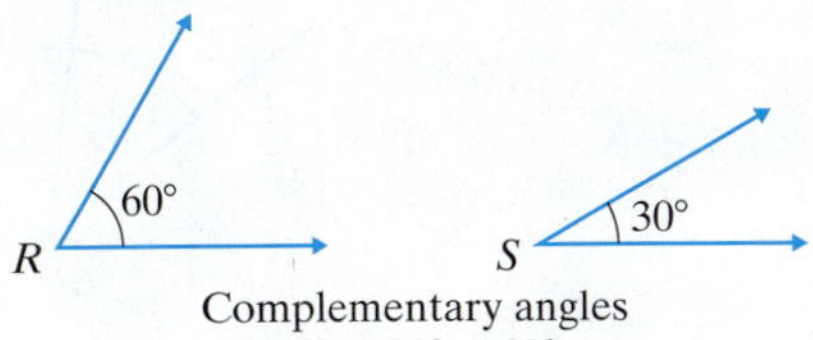

Complementary angles
$60° + 30° = 90°$

Two angles that have a sum of $180°$ are called **supplementary angles.** We say that each angle is the **supplement** of the other.

$\angle M$ and $\angle N$ are supplementary angles because

$$m\angle M + m\angle N = 125° + 55° = 180°$$

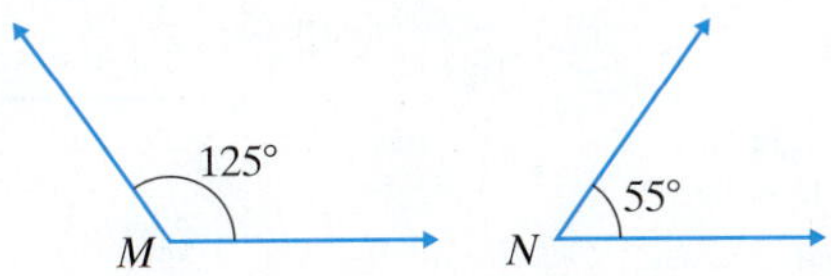

Supplementary angles
$125° + 55° = 180°$

EXAMPLE 4 Find the complement of a 48° angle.

Solution Two angles that have a sum of 90° are complementary. This means that the complement of an angle that measures 48° is an angle that measures $90° - 48° = 42°$. □

PRACTICE 4 Find the complement of a 29° angle. ■

EXAMPLE 5 Find the supplement of a 107° angle.

Solution Two angles that have a sum of 180° are supplementary. This means that the supplement of an angle that measures 107° is an angle that measures $180° - 107° = 73°$. □

PRACTICE 5 Find the supplement of a 67° angle. ■

✓ CONCEPT CHECK
True or false? The supplement of a 48° angle is 42°. Explain.

OBJECTIVE 4
Finding Measures of Angles

Measures of angles can be added or subtracted to find measures of related angles.

EXAMPLE 6 Find the measure of $\angle x$. Then classify $\angle x$ as an acute, obtuse, or right angle.

Solution
$$m\angle x = m\angle QTS - m\angle RTS$$
$$= 87° - 52°$$
$$= 35°$$

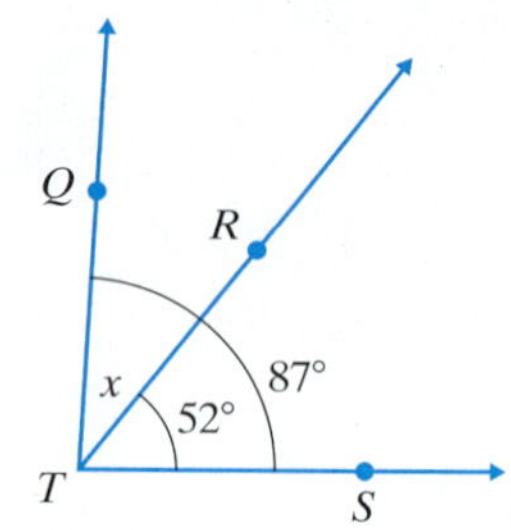

Thus, the measure of $\angle x$ $(m\angle x)$ is 35°.

Since $\angle x$ measures between 0° and 90°, it is an acute angle. □

PRACTICE 6

a. Find the measure of $\angle y$.

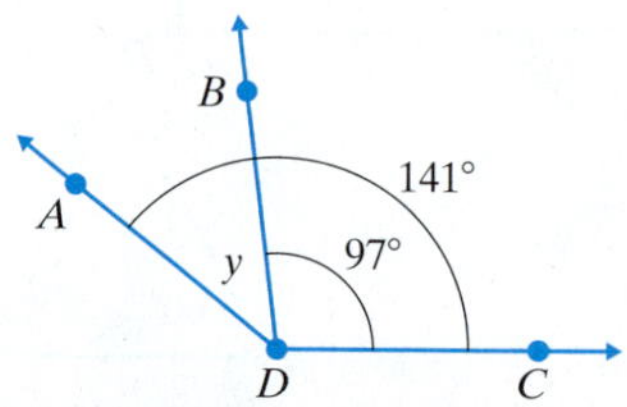

b. Find the measure of $\angle x$.

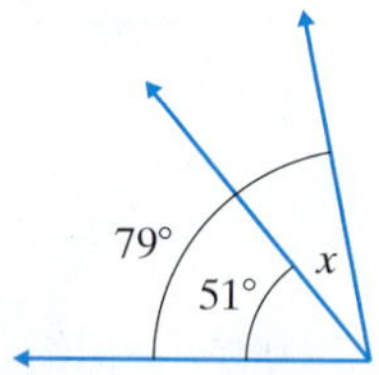

c. Classify $\angle x$ and $\angle y$ as acute, obtuse, or right angles. ■

Two lines in a plane can be either parallel or intersecting. **Parallel lines** never meet. **Intersecting lines** meet at a point. The symbol $\|$ is used to indicate "is parallel to." For example, in the figure, $p \| q$.

Parallel lines

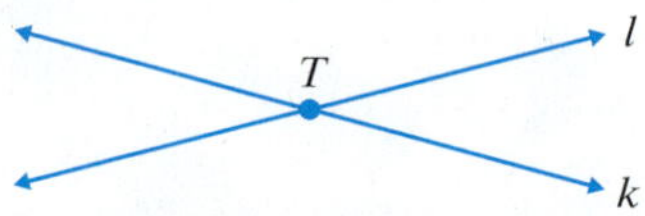

Intersecting lines

Answer to Concept Check:
false; the *complement* of a 48° angle is 42°; the *supplement* of a 48° angle is 132°

Some intersecting lines are perpendicular. Two lines are **perpendicular** if they form right angles when they intersect. The symbol $\perp$ is used to denote "is perpendicular to." For example, in the figure below, $n \perp m$.

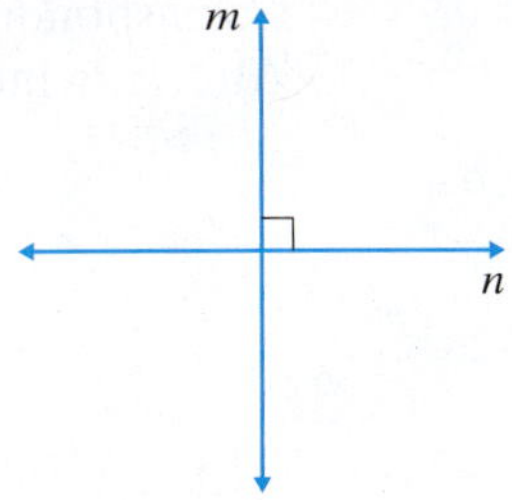

Perpendicular lines

When two lines intersect, four angles are formed. Two angles that are opposite each other are called **vertical angles.** Vertical angles have the same measure.

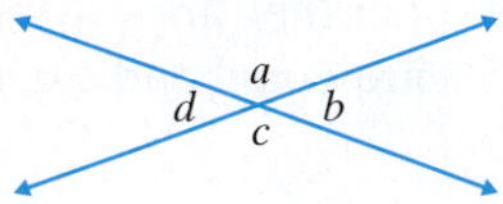

Vertical angles:
$\angle a$ and $\angle c$
$\angle d$ and $\angle b$

Adjacent angles:
$\angle a$ and $\angle b$
$\angle b$ and $\angle c$
$\angle c$ and $\angle d$
$\angle d$ and $\angle a$

Two angles that share a common side are called **adjacent angles.** Adjacent angles formed by intersecting lines are supplementary. That is, they have a sum of 180°.

Here are a few real-life examples of the lines we just discussed.

Parallel lines

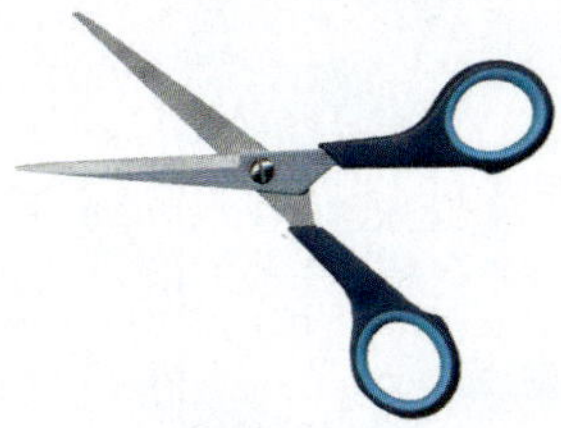

Vertical angles

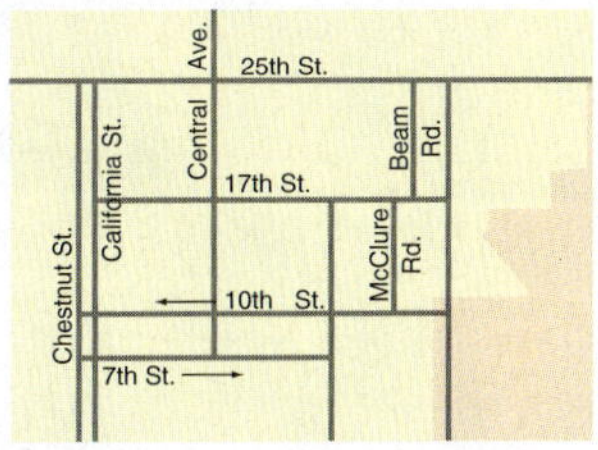

Perpendicular lines

EXAMPLE 7 Find the measure of $\angle x$, $\angle y$, and $\angle z$ if the measure of $\angle t$ is 42°.

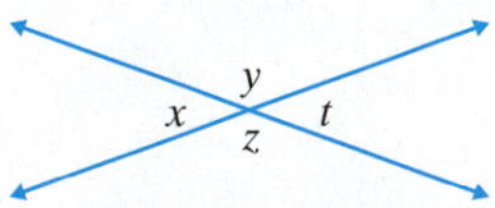

Solution Since $\angle t$ and $\angle x$ are vertical angles, they have the same measure, so $\angle x$ measures 42°.

Since $\angle t$ and $\angle y$ are adjacent angles, their measures have a sum of 180°. So $\angle y$ measures $180° - 42° = 138°$.

Since $\angle y$ and $\angle z$ are vertical angles, they have the same measure. So $\angle z$ measures 138°. □

PRACTICE 7 Find the measure of $\angle a$, $\angle b$, and $\angle c$.

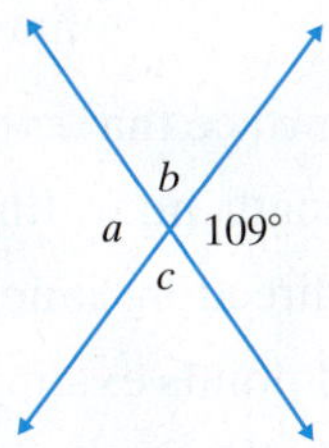

A line that intersects two or more lines at different points is called a **transversal.** Line l is a transversal that intersects lines m and n. The eight angles formed have special names. Some of these names are:

Corresponding angles: $\angle a$ and $\angle e$, $\angle c$ and $\angle g$, $\angle b$ and $\angle f$, $\angle d$ and $\angle h$
Alternate interior angles: $\angle c$ and $\angle f$, $\angle d$ and $\angle e$

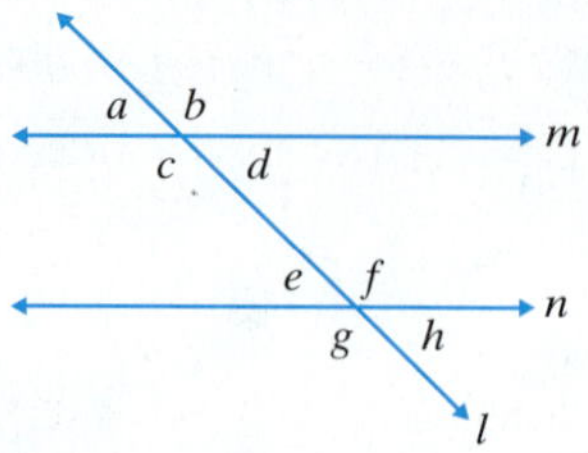

When two lines cut by a transversal are *parallel,* the following statement is true:

Parallel Lines Cut by a Transversal

If two parallel lines are cut by a transversal, then the measures of **corresponding angles are equal** and the measures of the **alternate interior angles are equal.**

EXAMPLE 8 Given that $m \| n$ and that the measure of $\angle w$ is 100°, find the measures of $\angle x$, $\angle y$, and $\angle z$.

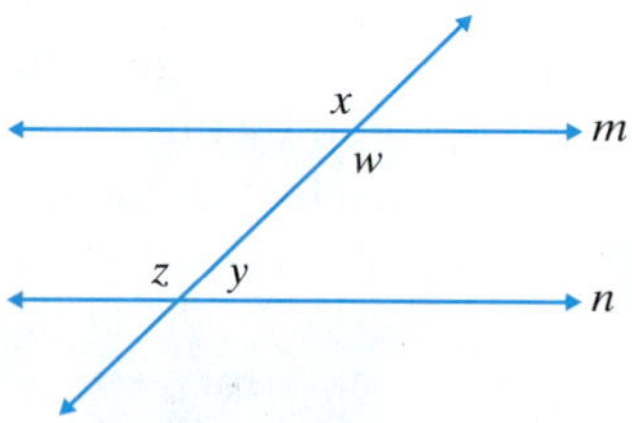

Solution

$m\angle x = 100°$ — $\angle x$ and $\angle w$ are vertical angles.

$m\angle z = 100°$ — $\angle x$ and $\angle z$ are corresponding angles.

$m\angle y = 180° - 100° = 80°$ — $\angle z$ and $\angle y$ are supplementary angles. □

PRACTICE 8 Given that $m \| n$ and that the measure of $\angle w = 45°$, find the measures of all the angles shown.

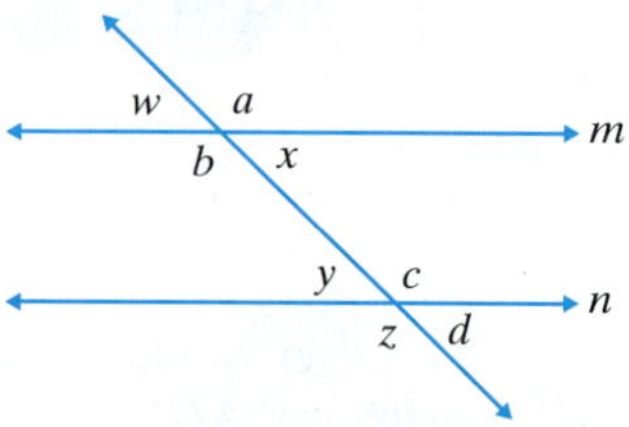

Vocabulary, Readiness & Video Check

Use the choices below to fill in each blank.

acute	straight	degrees	adjacent	parallel	intersecting
obtuse	space	plane	point	vertical	vertex
right	angle	ray	line	perpendicular	transversal

1. A(n) ______________ is a flat surface that extends indefinitely.
2. A(n) ______________ has no length, no width, and no height.
3. ______________ extends in all directions indefinitely.
4. A(n) ______________ is a set of points extending indefinitely in two directions.
5. A(n) ______________ is part of a line with one endpoint.

6. A(n) ______________ is made up of two rays that share a common endpoint. The common endpoint is called the ______________.
7. A(n) ______________ angle measures 180°.
8. A(n) ______________ angle measures 90°.
9. A(n) ______________ angle measures between 0° and 90°.
10. A(n) ______________ angle measures between 90° and 180°.
11. ______________ lines never meet and ______________ lines meet at a point.
12. Two intersecting lines are ______________ if they form right angles when they intersect.
13. An angle can be measured in ______________.
14. A line that intersects two or more lines at different points is called a(n) ______________.
15. When two lines intersect, four angles are formed. The two angles opposite one another are called ______________ angles.
16. Two angles that share a common side are called ______________ angles.

Martin-Gay Interactive Videos

See Video 6.1

Watch the section lecture video and answer the following questions.

OBJECTIVE 1
17. In the lecture after Example 2, what are the four ways we can name the angle shown?

OBJECTIVE 2
18. In the lecture before Example 3, what type of angle forms a line? What is its measure?

OBJECTIVE 3
19. What calculation is used to find the answer to Example 6?

OBJECTIVE 4
20. In the lecture before Example 7, two lines in a plane that aren't parallel must what?

6.1 Exercise Set MyMathLab®

Identify each figure as a line, a ray, a line segment, or an angle. Then name the figure using the given points. See Examples 1 and 2.

1.

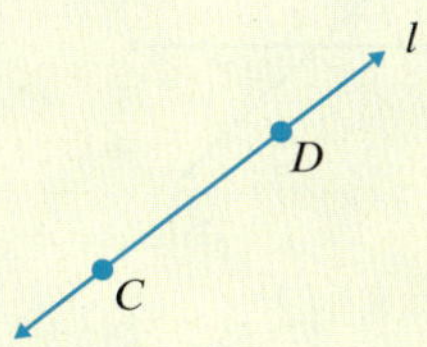

2.

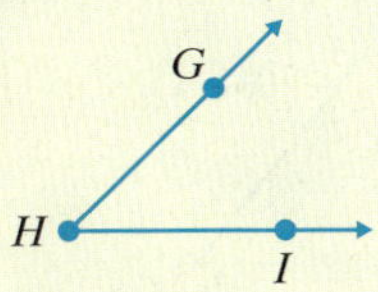

3.

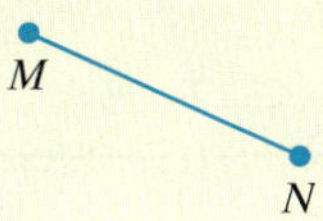

4.

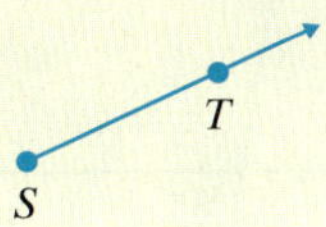

5.

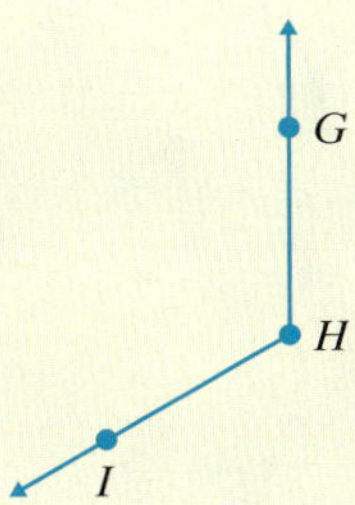

6.

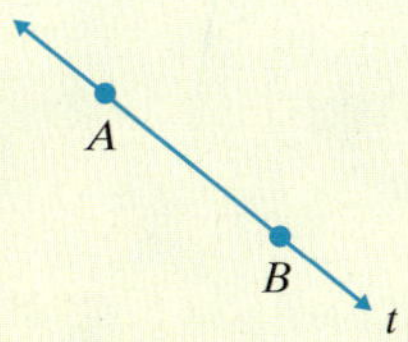

7.

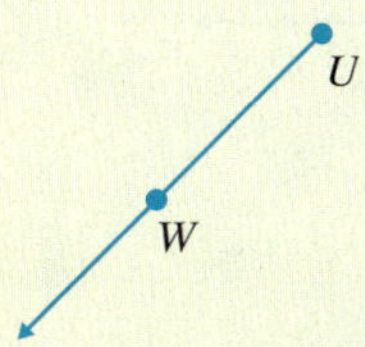

8.

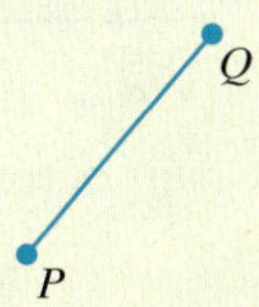

List two other ways to name each angle. See Example 2.

9. $\angle x$

10. $\angle w$

11. $\angle z$

12. $\angle y$

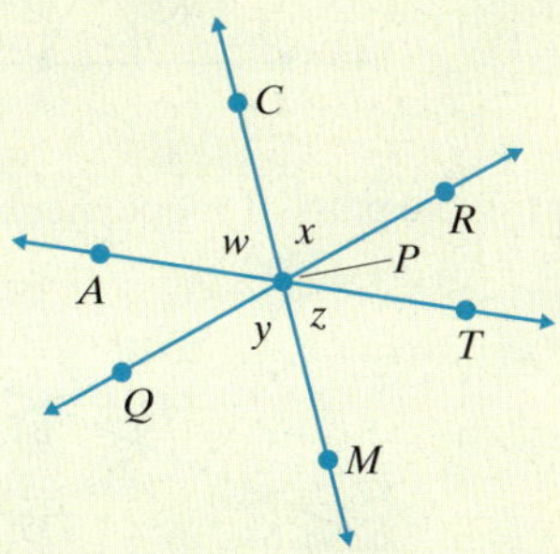

Classify each angle as acute, right, obtuse, or straight. See Example 3.

13.

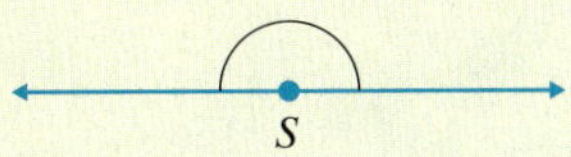

14.

15.

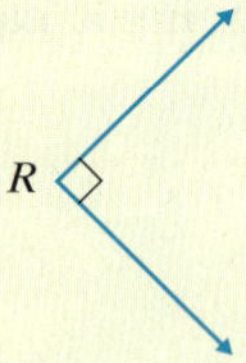

16.

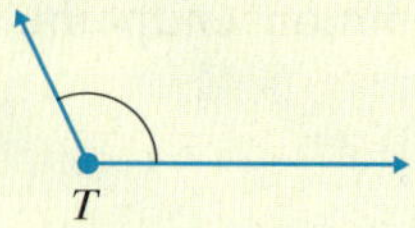

17.

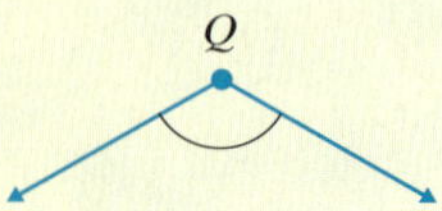

18.

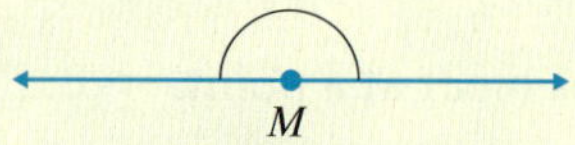

19.

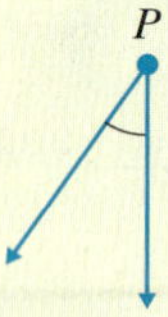

20.

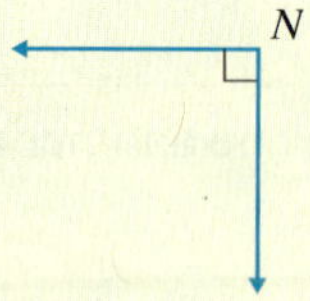

Find each complementary or supplementary angle as indicated. See Examples 4 and 5.

21. Find the complement of a 23° angle.

22. Find the complement of a 77° angle.

23. Find the supplement of a 17° angle.

24. Find the supplement of a 77° angle.

25. Find the complement of a 58° angle.

26. Find the complement of a 22° angle.

27. Find the supplement of a 150° angle.

28. Find the supplement of a 130° angle.

29. Identify the pairs of complementary angles.

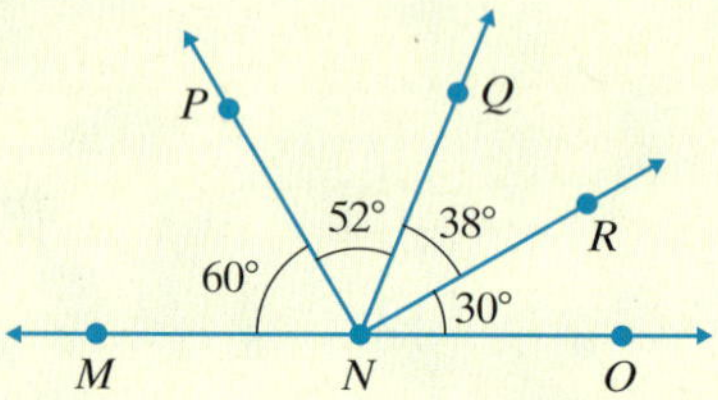

30. Identify the pairs of complementary angles.

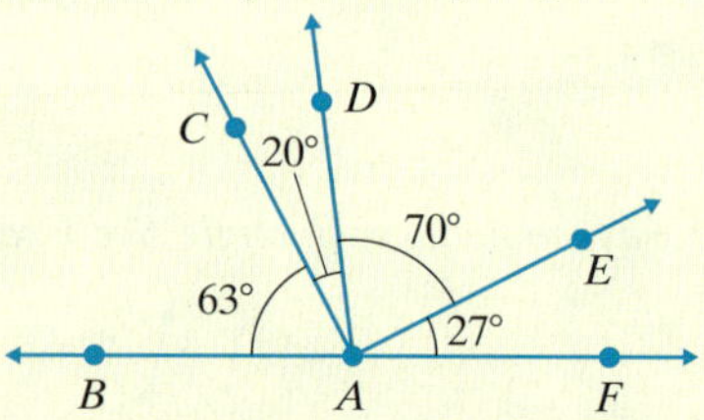

31. Identify the pairs of supplementary angles.

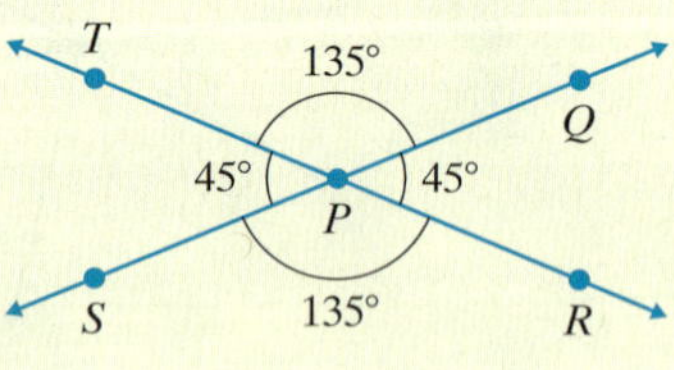

32. Identify the pairs of supplementary angles.

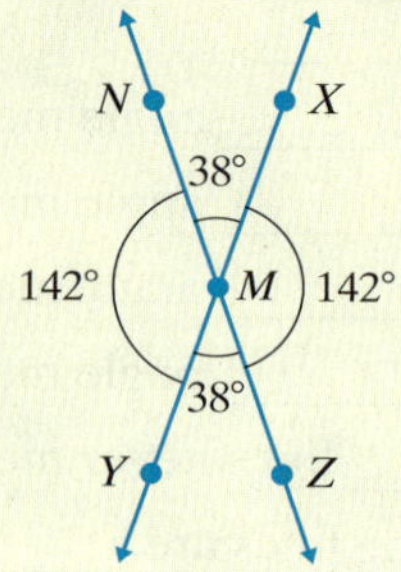

Find the measure of ∠x in each figure. See Example 6.

33.

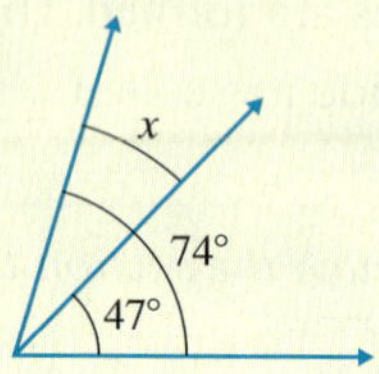

34.

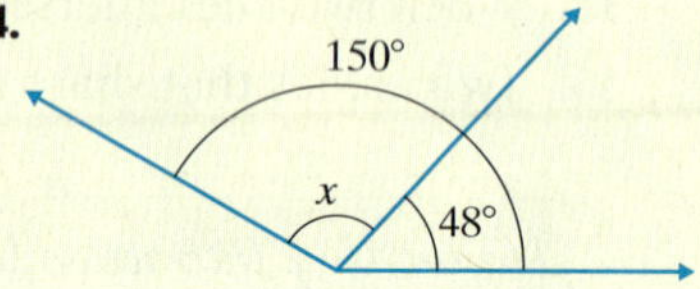

35.

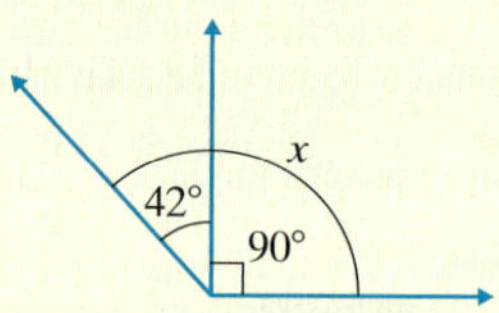

36.

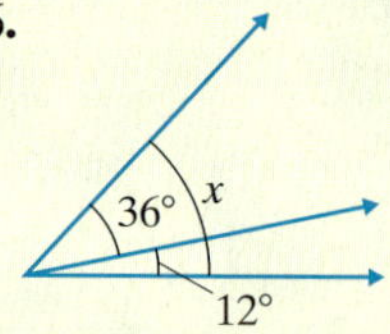

Find the measures of angles x, y, and z in each figure. See Examples 7 and 8.

37.

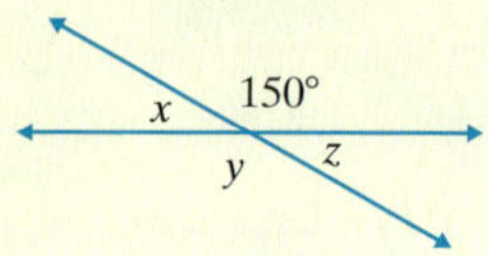

38.

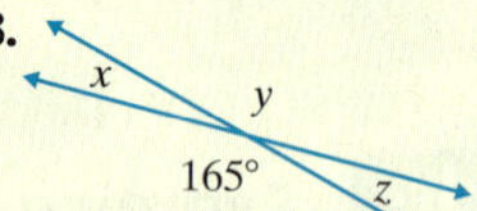

39.

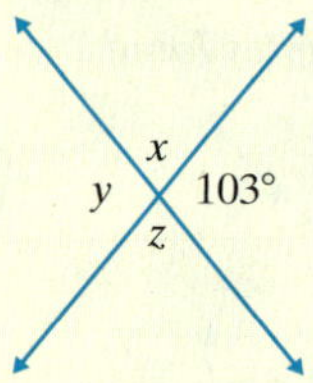

40.

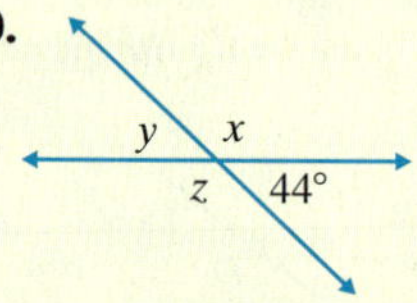

41. $m \parallel n$

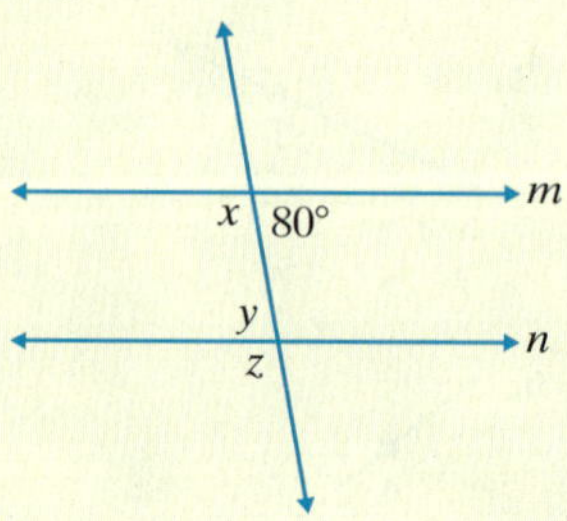

42. $m \parallel n$

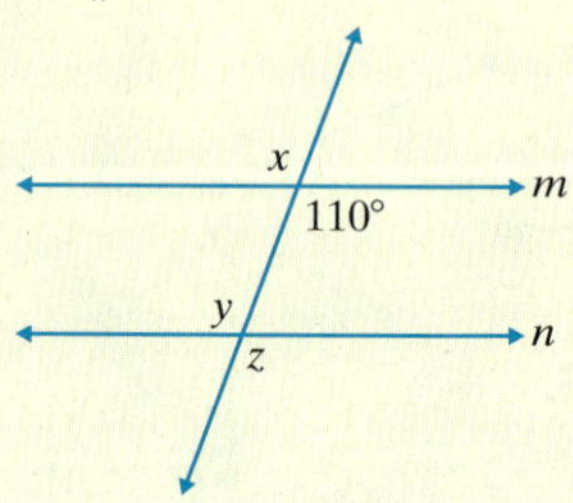

43. $m \parallel n$

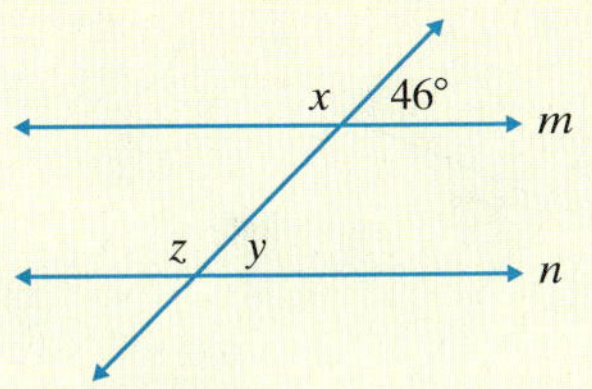

44. $m \parallel n$

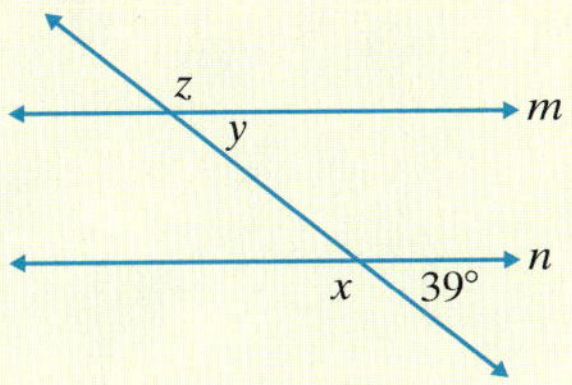

MIXED PRACTICE

Find two other ways of naming each angle. See Example 2.

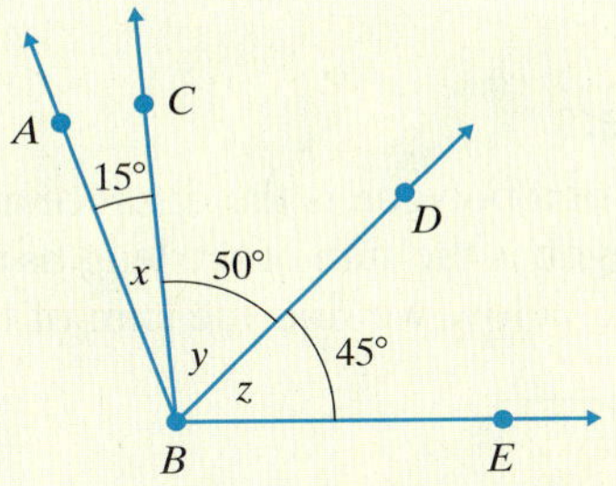

45. $\angle x$

46. $\angle y$

47. $\angle z$

48. $\angle ABE$ (just name one other way)

Find the measure of each angle in the figure above. See Example 6.

49. $\angle ABC$ **50.** $\angle EBD$ **51.** $\angle CBD$

52. $\angle CBA$ **53.** $\angle DBA$ **54.** $\angle EBC$

55. $\angle CBE$ **56.** $\angle ABE$

REVIEW AND PREVIEW

Perform each indicated operation. See Section R.1.

57. $\frac{7}{8} + \frac{1}{4}$ **58.** $\frac{7}{8} - \frac{1}{4}$ **59.** $\frac{7}{8} \cdot \frac{1}{4}$

60. $\frac{7}{8} \div \frac{1}{4}$ **61.** $3\frac{1}{3} - 2\frac{1}{2}$ **62.** $3\frac{1}{3} + 2\frac{1}{2}$

63. $3\frac{1}{3} \div 2\frac{1}{2}$ **64.** $3\frac{1}{3} \cdot 2\frac{1}{2}$

CONCEPT EXTENSIONS

Use this North Pole overhead view of the three sites of the Deep Space Network to answer Exercises 65 and 66. (See the Chapter Opener.)

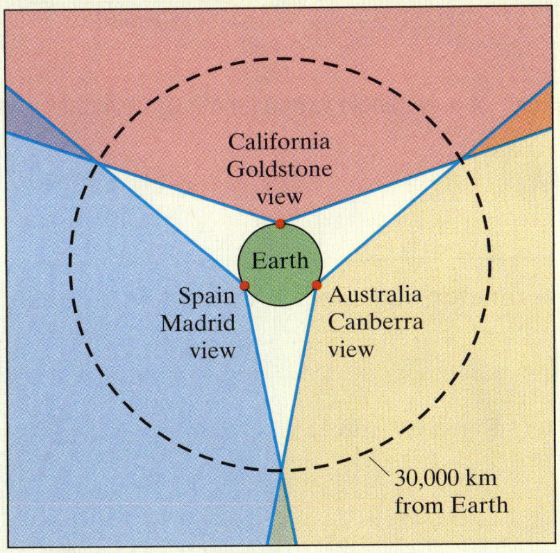

65. How many degrees are there around the Earth at the equator?

66. If the three sites of the Deep Space Network (red dots shown) are about the same number of degrees apart, how many degrees apart are they?

67. The angle between the two walls of the Vietnam Veterans Memorial in Washington, D.C., is 125.2°. Find the supplement of this angle. (*Source:* National Park Service)

68. The faces of Khafre's Pyramid at Giza, Egypt, are inclined at an angle of 53.13°. Find the complement of this angle. (*Source:* PBS *NOVA* Online)

Answer true or false for Exercises 69 through 72. See the Concept Check in this section. If false, explain why.

69. The complement of a 100° angle is an 80° angle.

70. It is possible to find the complement of a 120° angle.

71. It is possible to find the supplement of a 120° angle.

72. The supplement of a 5° angle is a 175° angle.

73. If lines m and n are parallel, find the measures of angles a through e.

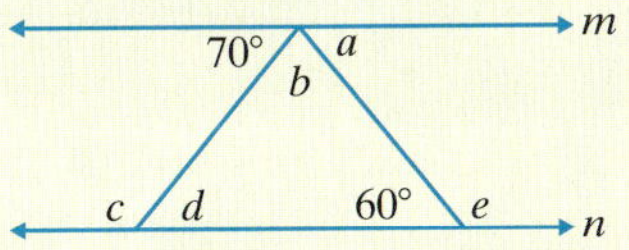

74. Below is a rectangle. List which segments, if extended, would be parallel lines.

75. Can two supplementary angles both be acute? Explain why or why not.

76. In your own words, describe how to find the complement and the supplement of a given angle.

77. Find two complementary angles with the same measure.

78. Is the figure below possible? Why or why not?

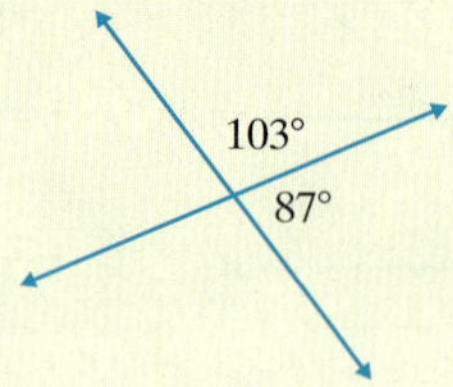

6.2 Perimeter, Area, and Volume

OBJECTIVES

1. Use Formulas to Find Perimeters.
2. Use Formulas to Find Circumferences.
3. Find the Areas of Plane Figures.
4. Find the Volumes of Solids.

OBJECTIVE

1 Using Formulas to Find Perimeters

Recall from Section R.1 that the perimeter of a polygon is the distance around the polygon. This means that the perimeter of a polygon is the sum of the lengths of its sides.

To find the perimeter of the rectangle below, we find the sum of the length of its sides.

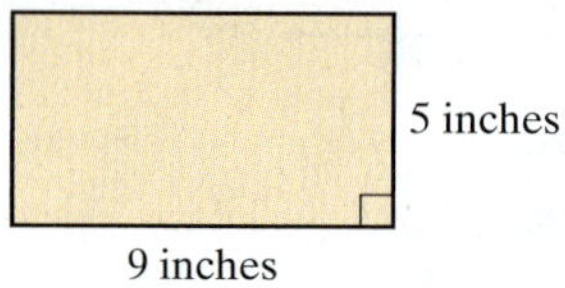

$$\begin{aligned}\text{perimeter} &= 9\text{ inches} + 9\text{ inches} + 5\text{ inches} + 5\text{ inches}\\ &= 2\cdot(\underset{\text{length}}{\underset{\uparrow}{9\text{ inches}}}) + 2\cdot(\underset{\text{width}}{\underset{\uparrow}{5\text{ inches}}})\\ &= 28\text{ inches}\end{aligned}$$

In general, we can say that the perimeter of a rectangle is always

$$2\cdot\text{length} + 2\cdot\text{width}$$

As we have just seen, the perimeters of some special figures such as rectangles form patterns. These patterns are given as **formulas.** The formula for the perimeter of a rectangle is shown next:

Perimeter of a Rectangle

Perimeter = 2 · **l**ength + 2 · **w**idth

In symbols, this can be written as

$$P = 2l + 2w$$

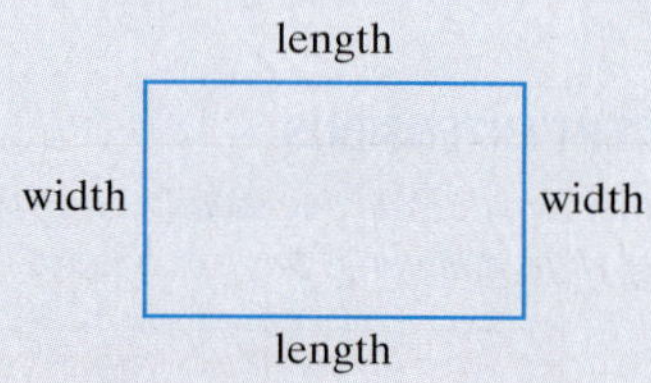

EXAMPLE 1 Find the perimeter of a rectangle with a length of 11 inches and a width of 3 inches.

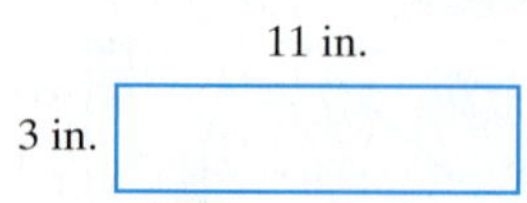

Solution We use the formula for perimeter and replace the letters by their known lengths.

$$\begin{aligned}P &= 2l + 2w\\ &= 2\cdot 11\text{ in.} + 2\cdot 3\text{ in.} \quad \text{Replace } l \text{ with 11 in. and } w \text{ with 3 in.}\\ &= 22\text{ in.} + 6\text{ in.}\\ &= 28\text{ in.}\end{aligned}$$

The perimeter is 28 inches. □

PRACTICE 1

a. Find the perimeter of the rectangle.

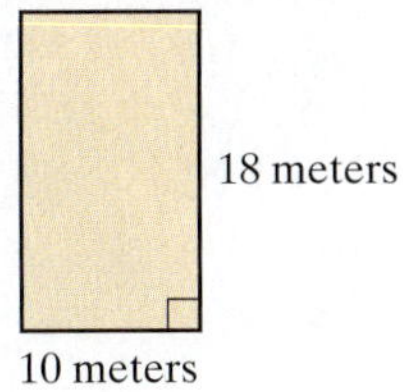

b. Find the perimeter of the rectangular lot shown below:

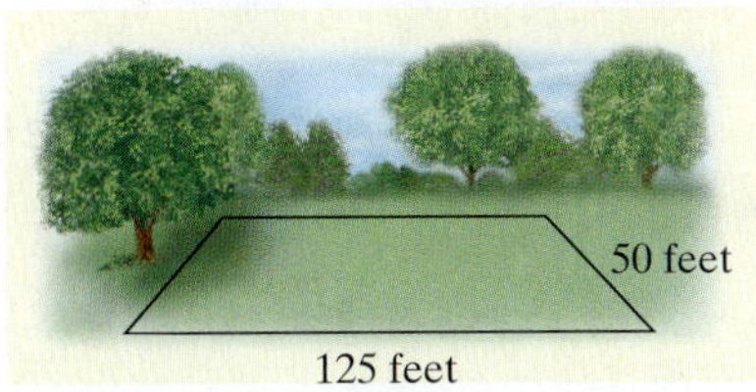

Recall that a square is a special rectangle with all four sides the same length. The formula for the perimeter of a square is shown next:

Perimeter of a Square

Perimeter = **s**ide + **s**ide + **s**ide + **s**ide

= 4 · **s**ide

In symbols,

$P = 4s$

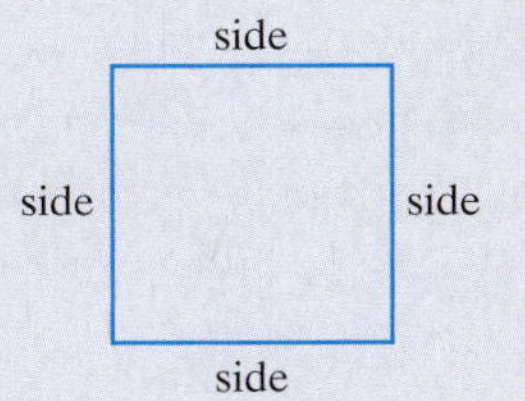

EXAMPLE 2 **Finding the Perimeter of a Field**

How much fencing is needed to enclose a square field 50 yards on a side?

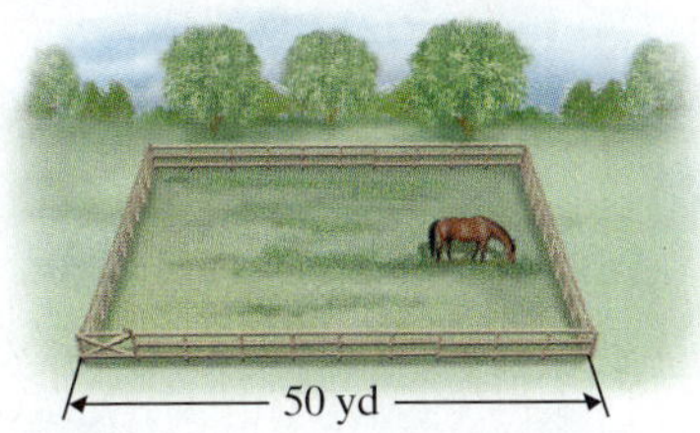

Solution To find the amount of fencing needed, we find the distance around, or perimeter. The formula for the perimeter of a square is $P = 4s$. We use this formula and replace s by 50 yards.

$$\begin{aligned} P &= 4s \\ &= 4 \cdot 50 \text{ yd} \\ &= 200 \text{ yd} \end{aligned}$$

The amount of fencing needed is 200 yards.

PRACTICE 2

Find the perimeter of a square tabletop if each side is 4 feet long.

The formula for the perimeter of a triangle with sides of lengths a, b, and c is given next:

Perimeter of a Triangle

Perimeter = side **a** + side **b** + side **c**

In symbols,

$P = a + b + c$

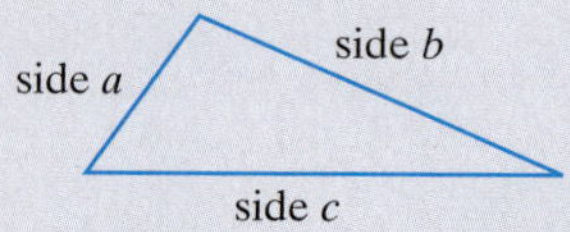

EXAMPLE 3 Find the perimeter of a triangle if the sides are 3 inches, 7 inches, and 6 inches.

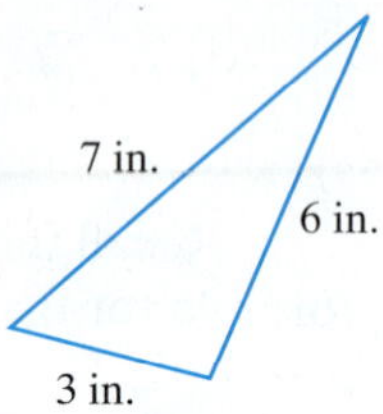

Solution The formula for the perimeter is $P = a + b + c$, where a, b, and c are the lengths of the sides. Thus,

$$\begin{aligned} P &= a + b + c \\ &= 3 \text{ in.} + 7 \text{ in.} + 6 \text{ in.} \\ &= 16 \text{ in.} \end{aligned}$$

The perimeter of the triangle is 16 inches. □

PRACTICE 3 Find the perimeter of a triangle if the sides are 6 centimeters, 10 centimeters, and 8 centimeters in length. ■

The method for finding the perimeter of any polygon is given next:

Perimeter of a Polygon

The perimeter of a polygon is the sum of the lengths of its sides.

EXAMPLE 4 Find the perimeter of the trapezoid shown below:

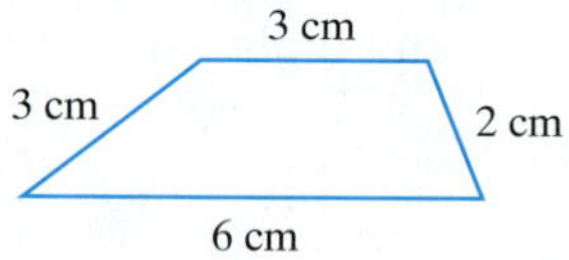

Solution To find the perimeter, we find the sum of the lengths of its sides.

$$\text{perimeter} = 3 \text{ cm} + 2 \text{ cm} + 6 \text{ cm} + 3 \text{ cm} = 14 \text{ cm}$$

The perimeter is 14 centimeters. □

PRACTICE 4 Find the perimeter of the trapezoid shown.

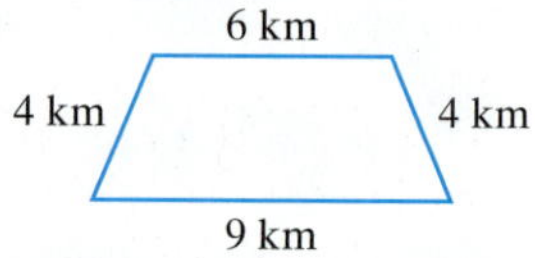

■

EXAMPLE 5 **Finding the Perimeter of a Room**

Find the perimeter of the room shown below:

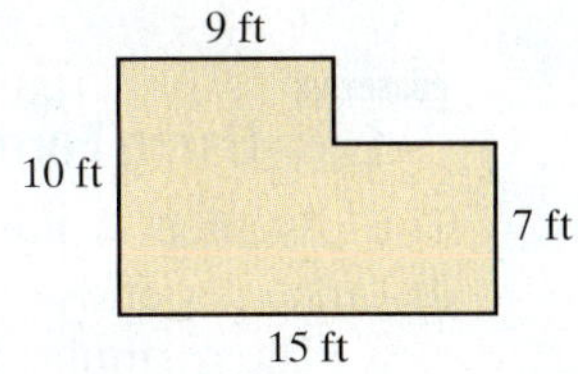

Solution To find the perimeter of the room, we first need to find the lengths of all sides of the room.

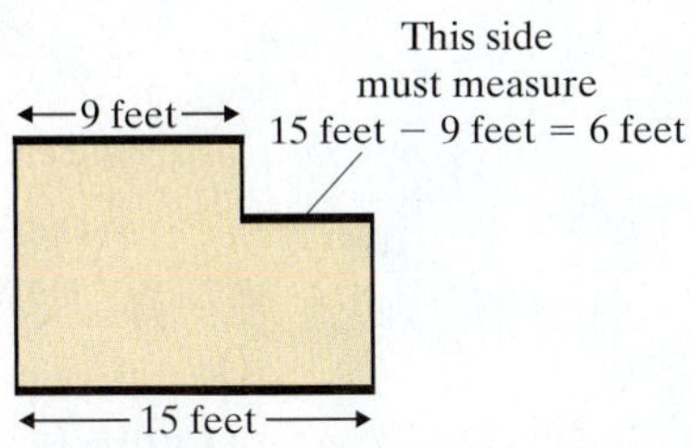

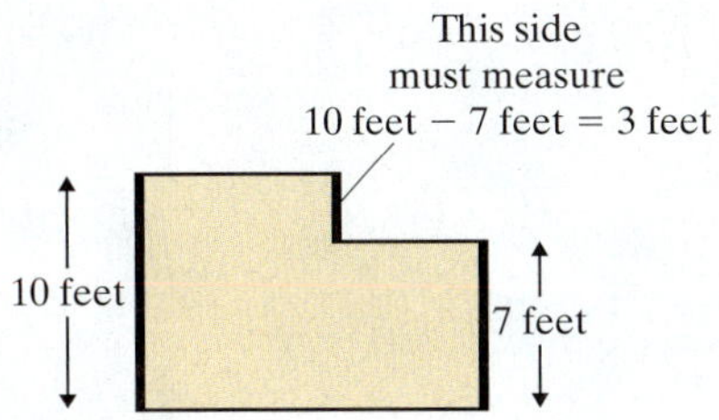

Now that we know the measures of all sides of the room, we can add the measures to find the perimeter.

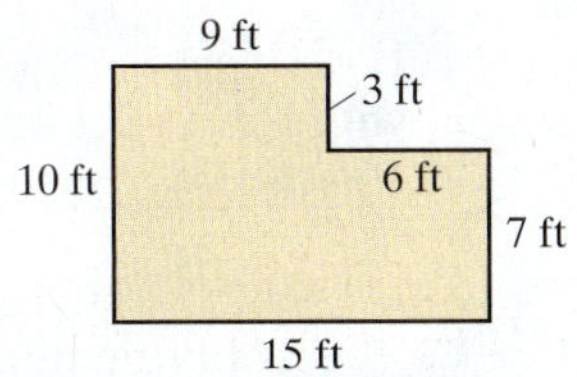

$$\begin{aligned}\text{perimeter} &= 10\text{ ft} + 9\text{ ft} + 3\text{ ft} + 6\text{ ft} + 7\text{ ft} + 15\text{ ft}\\ &= 50\text{ ft}\end{aligned}$$

The perimeter of the room is 50 feet. □

PRACTICE 5 Find the perimeter of the room shown.

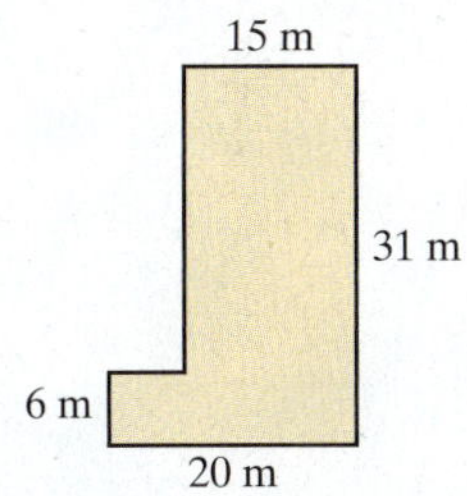

■

EXAMPLE 6 **Calculating the Cost of Wallpaper Border**

A rectangular room measures 10 feet by 12 feet. Find the cost to hang a wallpaper border on the walls close to the ceiling if the cost of the wallpaper border is $1.09 per foot.

Solution First we find the perimeter of the room.

$$\begin{aligned}P &= 2l + 2w\\ &= 2\cdot 12\text{ ft} + 2\cdot 10\text{ ft} \quad \text{Replace } l \text{ with 12 feet and } w \text{ with 10 feet.}\\ &= 24\text{ ft} + 20\text{ ft}\\ &= 44\text{ ft}\end{aligned}$$

The cost of the wallpaper is

$$\text{cost} = \$1.09\cdot 44\text{ ft} = 47.96$$

The cost of the wallpaper is $47.96. □

(Continued on next page)

PRACTICE 6 A rectangular lot measures 60 feet by 120 feet. Find the cost to install fencing around the lot if the cost of fencing is $1.90 per foot.

OBJECTIVE 2

Using Formulas to Find Circumferences

The distance around a circle is given a special name, called the **circumference.** This distance depends on the radius or the diameter of the circle.

The formulas for circumference are shown next:

Circumference of a Circle

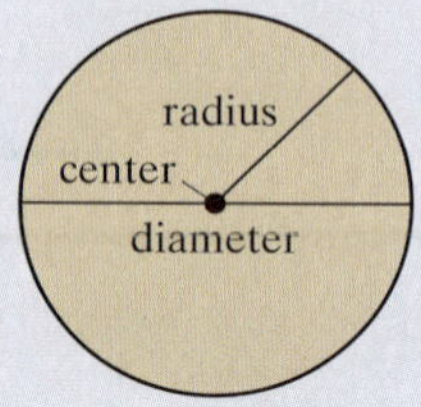

Circumference = $2 \cdot \pi \cdot$ **r**adius or Circumference = $\pi \cdot$ **d**iameter

In symbols,

$$C = 2\pi r \quad \text{or} \quad C = \pi d$$

where $\pi \approx 3.14$ or $\pi \approx \frac{22}{7}$ or use a calculator approximation for π.

To better understand circumference and π (pi), try the following experiment. Take any can and measure its circumference and its diameter.

The can in the figure above has a circumference of 23.5 centimeters and a diameter of 7.5 centimeters. Now divide the circumference by the diameter.

$$\frac{\text{circumference}}{\text{diameter}} = \frac{23.5 \text{ cm}}{7.5 \text{ cm}} \approx 3.13$$

Try this with other sizes of cylinders and circles—you should always get a number close to 3.1. The exact ratio of circumference to diameter is π. (Recall that $\pi \approx 3.14$ or $\approx \frac{22}{7}$.)

EXAMPLE 7 **Finding Circumference of a Circular Spa**

A homeowner plans to install a border of new tiling around the circumference of her circular spa. If her spa has a diameter of 14 feet, find its exact circumference. Then use a calculator approximation for π to approximate the circumference. Round to two decimal places.

Solution Because we are given the diameter, we use the formula $C = \pi d$.

$$\begin{aligned} C &= \pi d \\ &= \pi \cdot 14 \text{ ft} \quad \text{Replace } d \text{ with 14 feet.} \\ &= 14\pi \text{ ft} \end{aligned}$$

The circumference of the spa is *exactly* 14π feet. By using a calculator approximation for π, we find that the circumference is *approximately* 43.98 feet.

$$14\pi \text{ ft} \approx 43.98 \text{ ft}$$

□

PRACTICE 7 An irrigation device waters a circular region with a diameter of 20 yards. Find the exact circumference of the watered region, then use a calculator approximation for π. Round to two decimal places.

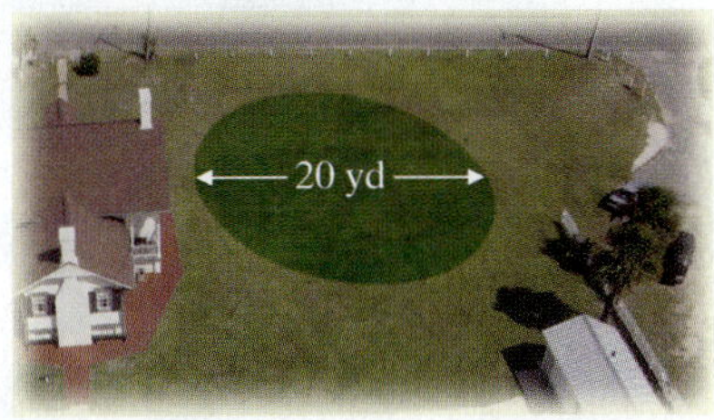

✓ CONCEPT CHECK
The distance around which figure is greater: a square with side length 5 inches or a circle with radius 3 inches?

OBJECTIVE 3

Finding Areas of Plane Figures

Recall that area measures the number of square units that cover the surface of a plane figure; that is, a figure that lies in a plane. Thus far, we know how to find the areas of a rectangle, a square, and a triangle. These formulas, as well as formulas for finding the areas of other common geometric figures, are given next.

Area Formulas of Common Geometric Figures

Geometric Figure | **Area Formula**

RECTANGLE

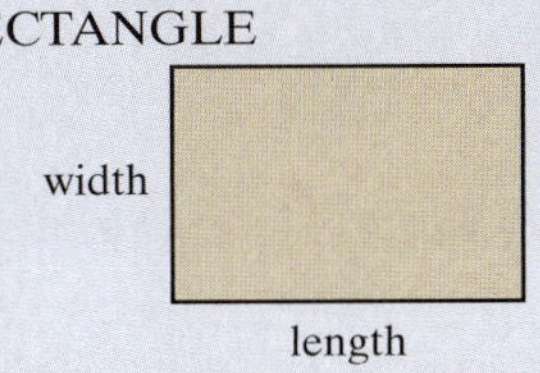

Area of a rectangle:

Area = **l**ength · **w**idth

$A = lw$

SQUARE

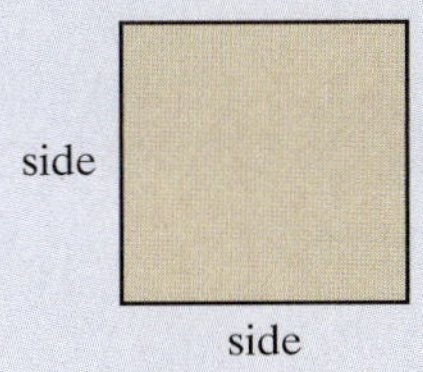

Area of a square:

Area = **s**ide · **s**ide

$A = s \cdot s = s^2$

Answer to Concept Check:
a square with side length 5 in.

(Continued on next page)

Geometric Figure	**Area Formula**
TRIANGLE 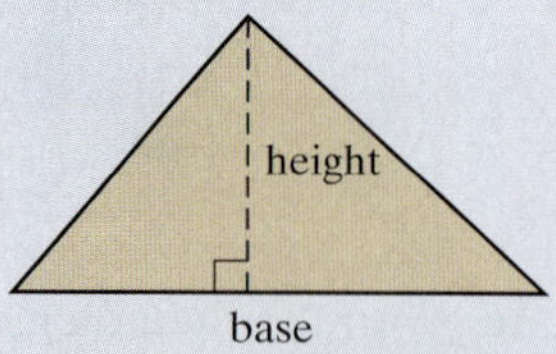	Area of a triangle: **A**rea $= \frac{1}{2} \cdot$ **b**ase $\cdot$ **h**eight; $A = \frac{1}{2}bh$
PARALLELOGRAM 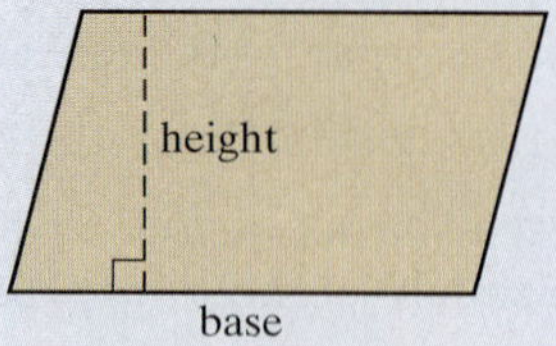	Area of a parallelogram: **A**rea $=$ **b**ase $\cdot$ **h**eight; $A = bh$
TRAPEZOID 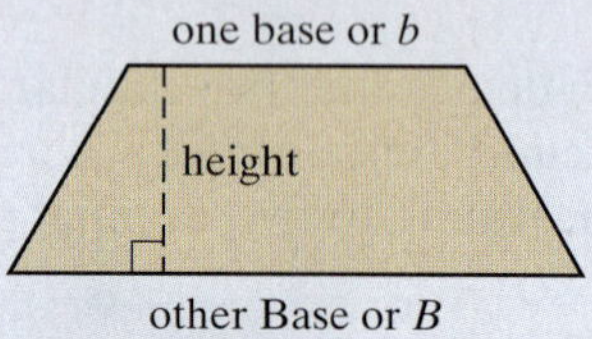	Area of a trapezoid: **A**rea $= \frac{1}{2} \cdot$ (one **b**ase + other **B**ase) $\cdot$ **h**eight; $A = \frac{1}{2}(b + B)h$
CIRCLE 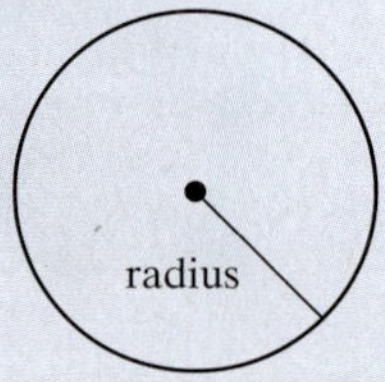	Area of a circle: **A**rea $= \pi \cdot ($**r**adius$)^2$; $A = \pi r^2$ (A fraction approximation for π is $\frac{22}{7}$.) (A decimal approximation for π is 3.14.)

Figure labels — triangle: height, base; parallelogram: height, base; trapezoid: one base or b, height, other Base or B; circle: radius.

Helpful Hint

Note: Throughout this chapter, exact answers and approximations are asked for with formulas containing π. Approximate answers may vary depending on whether $\frac{22}{7}$, 3.14, or a calculator approximation is used for π.

Use these formulas for the following examples.

Helpful Hint

Area is always measured in square units.

EXAMPLE 8 Find the area of the triangle.

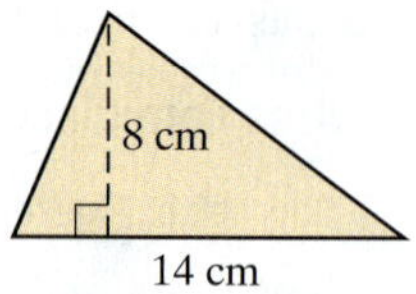

Solution

$$A = \frac{1}{2}bh$$

$$= \frac{1}{2} \cdot 14 \text{ cm} \cdot 8 \text{ cm} \quad \text{Replace } b \text{, base, with 14 cm and } h \text{, height, with 8 cm.}$$

$$= \frac{\overset{1}{\cancel{2}} \cdot 7 \cdot 8}{\underset{1}{\cancel{2}}} \text{ square centimeters} \quad \text{Write 14 as } 2 \cdot 7.$$

$$= 56 \text{ square centimeters}$$

The area is 56 square centimeters. □

PRACTICE 8 Find the area of the triangle.

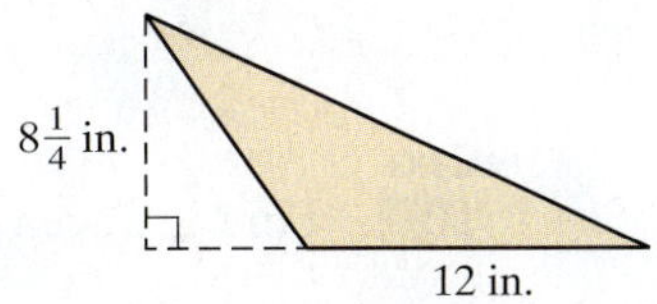

EXAMPLE 9 Find the area of the parallelogram.

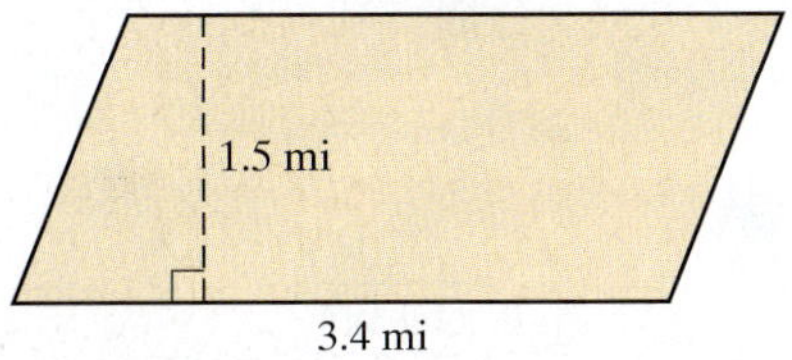

Solution

$$A = bh$$
$$= 3.4 \text{ miles} \cdot 1.5 \text{ miles} \quad \text{Replace } b\text{, base, with 3.4 miles and } h\text{, height, with 1.5 miles.}$$
$$= 5.1 \text{ square miles}$$

The area is 5.1 square miles.

PRACTICE 9 Find the area of the trapezoid.

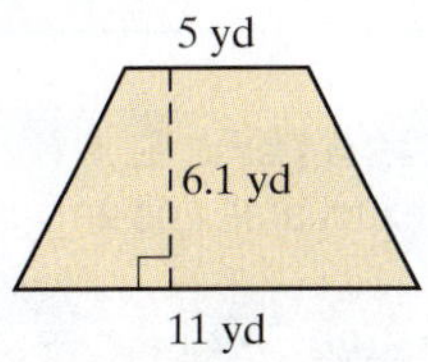

Helpful Hint

When finding the area of figures, check to make sure that all measurements are in the same units before calculations are made.

EXAMPLE 10 Find the area of the figure.

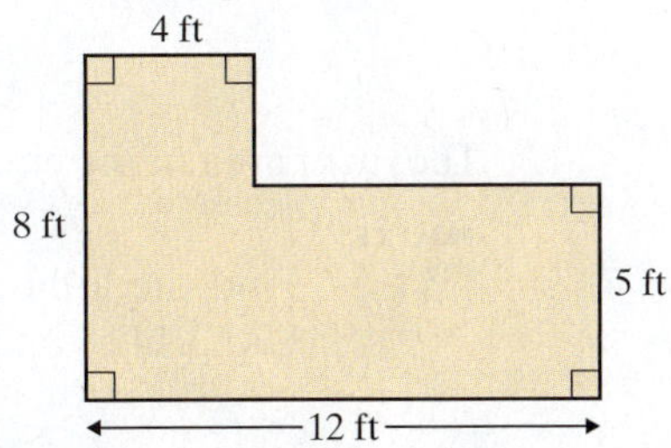

Solution Split the figure into two rectangles. To find the area of the figure, we find the sum of the areas of the two rectangles.

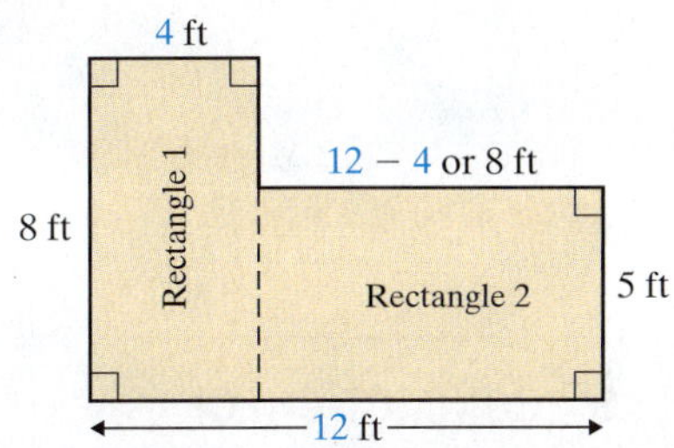

$$\text{area of Rectangle 1} = lw$$
$$= 8 \text{ feet} \cdot 4 \text{ feet}$$
$$= 32 \text{ square feet}$$

Notice that the length of Rectangle 2 is 12 feet − 4 feet or 8 feet.

$$\text{area of Rectangle 2} = lw$$
$$= 8 \text{ feet} \cdot 5 \text{ feet}$$
$$= 40 \text{ square feet}$$

(Continued on next page)

$$\text{area of the figure} = \text{area of Rectangle 1} + \text{area of Rectangle 2}$$
$$= 32 \text{ square feet} + 40 \text{ square feet}$$
$$= 72 \text{ square feet}$$

PRACTICE 10 Find the area of the figure.

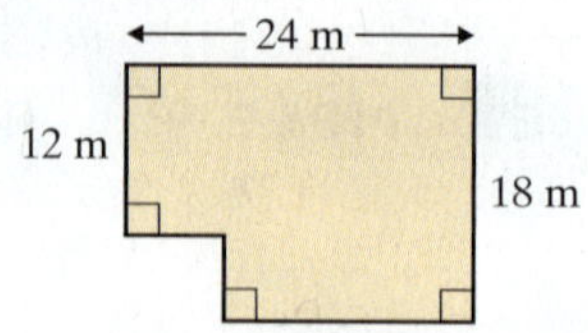

Helpful Hint

The figure in Example 10 could also be split into two rectangles as shown.

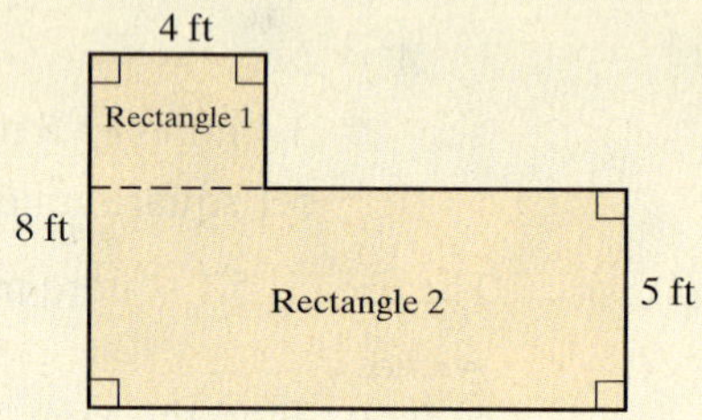

EXAMPLE 11 Find the area of a circle with a radius of 3 feet. Find the exact area and an approximation. Use 3.14 as an approximation for π.

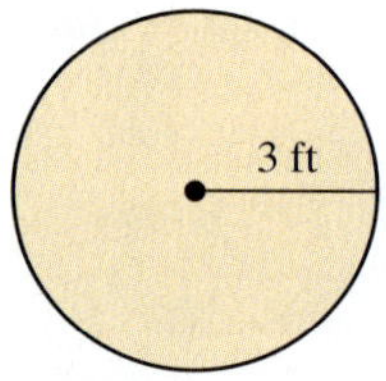

Solution We let $r = 3$ feet and use the formula

$$A = \pi r^2$$
$$= \pi \cdot (3 \text{ feet})^2 \quad \text{Replace } r \text{ with 3 feet.}$$
$$= 9 \cdot \pi \text{ square feet} \quad \text{Replace } (3 \text{ feet})^2 \text{ with 9 sq ft.}$$

To approximate this area, we substitute 3.14 for π.

$$9 \cdot \pi \text{ square feet} \approx 9 \cdot 3.14 \text{ square feet}$$
$$= 28.26 \text{ square feet}$$

The *exact* area of the circle is 9π square feet, which is *approximately* 28.26 square feet.

PRACTICE 11 Find the area of the given circle. Find the exact area and an approximation. Use 3.14 as an approximation for π.

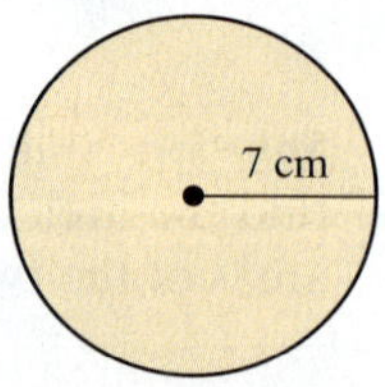

✔ CONCEPT CHECK

Use estimation to decide which figure would have a larger area: a circle of diameter 10 in. or a square 10 in. long on each side.

Answer to Concept Check:
A square 10 in. long on each side would have a larger area.

OBJECTIVE

4 Finding Volumes of Solids

A **convex solid** is a set of points, S, not all in one plane, such that for any two points A and B in S, all points between A and B are also in S. In this objective, we will find the volume of special types of solids called polyhedrons. A solid formed by the intersection of a finite number of planes is called a **polyhedron.** The box to the right is an example of a polyhedron.

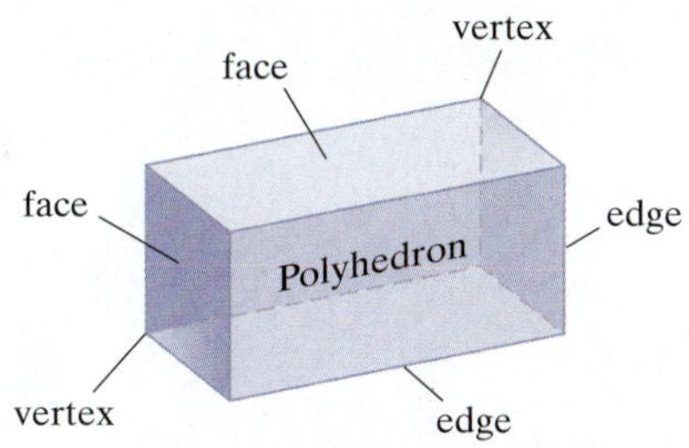

Each of the plane regions of a polyhedron is called a **face** of the polyhedron. If the intersection of two faces is a line segment, this line segment is an **edge** of the polyhedron. The intersections of the edges are the **vertices** of the polyhedron.

Volume is a measure of the space of a region. The volume of a box or can, for example, is the amount of space inside. Volume can be used to describe the amount of juice in a pitcher or the amount of concrete needed to pour a foundation for a house.

The volume of a solid is the number of **cubic units** in the solid. A cubic centimeter and a cubic inch are illustrated.

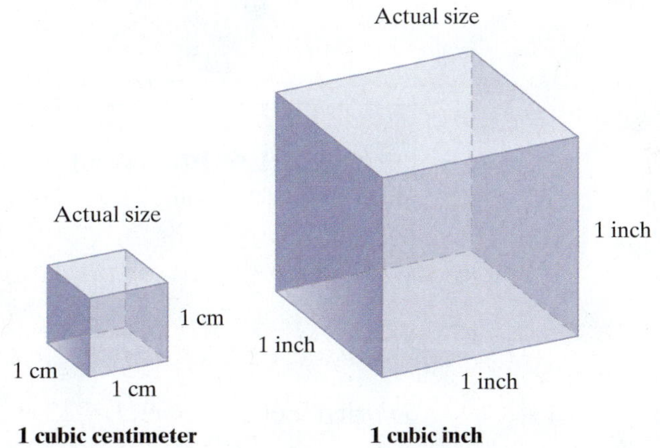

Formulas for finding the volumes, V, of some common solids are given next. (Note: Spheres, circular cylinders, and cones are not polyhedrons, but they are solids and we will calculate volumes of these solids.)

Volume Formulas of Common Solids

Solid	*Formulas*
RECTANGULAR SOLID 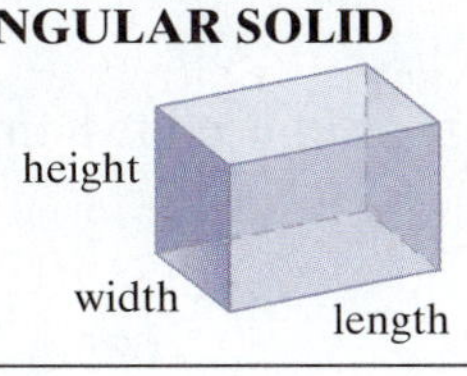	$V = lwh$ where h = height, w = width, l = length
CUBE 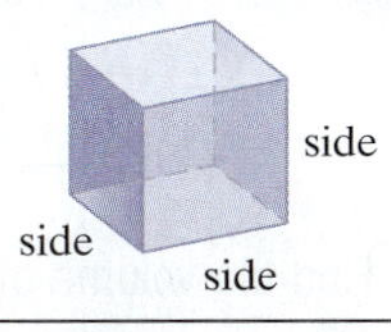	$V = s^3$ where s = side
SPHERE 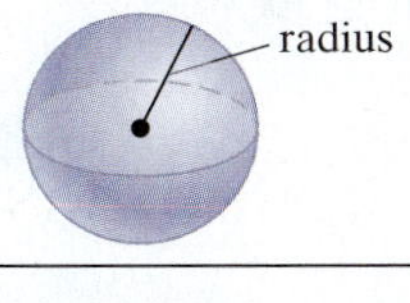	$V = \frac{4}{3}\pi r^3$ where r = radius

(Continued on next page)

Solid	*Formulas*
CIRCULAR CYLINDER 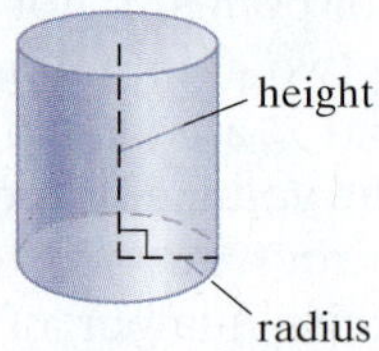	$V = \pi r^2 h$ where h = height, r = radius
CONE 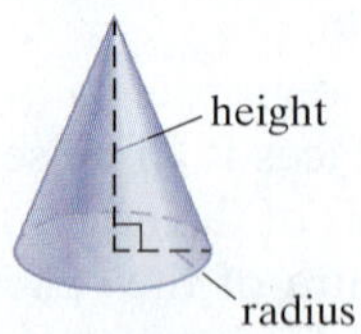	$V = \frac{1}{3}\pi r^2 h$ where h = height, r = radius
SQUARE-BASED PYRAMID 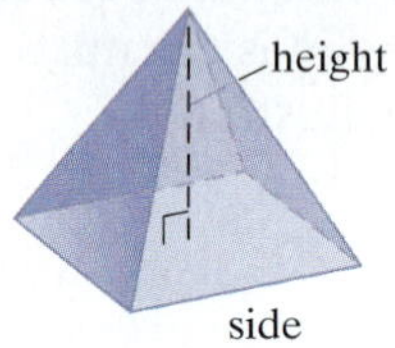	$V = \frac{1}{3}s^2 h$ where h = height, s = side

EXAMPLE 12 Find the volume of a rectangular box that is 12 inches long, 6 inches wide, and 3 inches high.

Solution Let $h = 3$ in., $l = 12$ in., and $w = 6$ in.

$$V = lwh$$

$$V = 12 \text{ inches} \cdot 6 \text{ inches} \cdot 3 \text{ inches} = 216 \text{ cubic inches}$$

The volume of the rectangular box is 216 cubic inches. □

Answer to Concept Check:
Volume $= lwh$
$= 14 \text{ cm} \cdot 8 \text{ cm} \cdot 5 \text{ cm}$
$= 560$ cu cm

PRACTICE 12 Find the volume of a rectangular box that is 7 feet long, 3 feet wide, and 4 feet high. ■

✔ CONCEPT CHECK

Juan is calculating the volume of the following rectangular solid. Find the error in his calculation.

Volume $= l + w + h$
$= 14 \text{ cm} + 8 \text{ cm} + 5 \text{ cm}$
$= 27$ cu cm

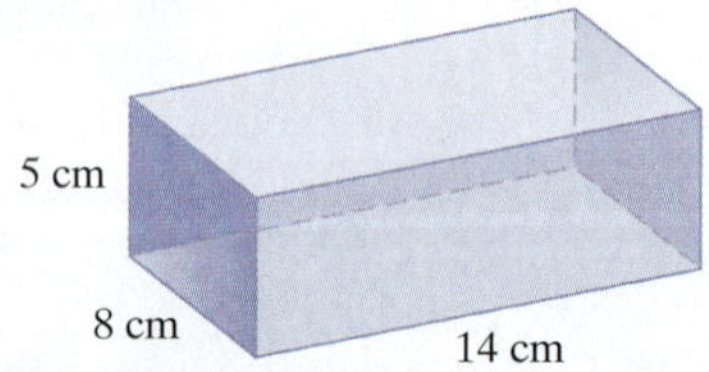

EXAMPLE 13 Find the volume of a ball of radius 2 inches. Give the exact volume, and then use the approximation $\frac{22}{7}$ for π.

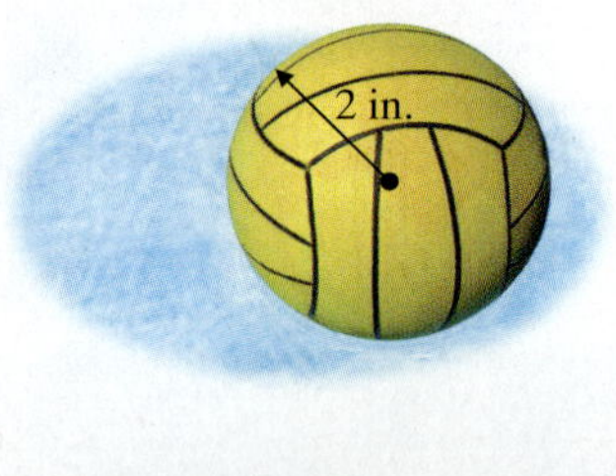

Solution

$$V = \frac{4}{3}\pi r^3 \quad \text{Formula for volume of a sphere}$$

$$V = \frac{4}{3}\cdot\pi(2\text{ in.})^3 \quad \text{Let } r = 2 \text{ inches.}$$

$$= \frac{32}{3}\pi \text{ cu in.} \quad \text{Exact volume}$$

$$\approx \frac{32}{3}\cdot\frac{22}{7} \text{ cu in.} \quad \text{Approximate } \pi \text{ with } \frac{22}{7}.$$

$$= \frac{704}{21} \text{ or } 33\frac{11}{21} \text{ cu in.} \quad \text{Approximate volume}$$

The volume of the sphere is exactly $\frac{32}{3}\pi$ cubic inches or approximately $33\frac{11}{21}$ cubic inches. □

PRACTICE 13 Find the volume of a ball of radius $\frac{1}{2}$ centimeter. Give the exact volume. Then use $\frac{22}{7}$ for π and approximate the value. ■

Vocabulary, Readiness & Video Check

Use the choices below to fill in each blank.

circumference	radius	area	cubic	π	$\frac{22}{7}$
diameter	perimeter	volume	square	3.14	

1. The ______________ of a polygon is the sum of the lengths of its sides.
2. The distance around a circle is called the ______________.
3. The exact ratio of circumference to diameter is ______________.
4. The diameter of a circle is double its ______________.
5. Both ______________ and ______________ are approximations for π.
6. The radius of a circle is half its ______________.
7. The measure of the amount of space inside a solid is its ______________.
8. ______________ measures the amount of surface enclosed by a region.
9. Volume is measured in ______________ units.
10. Area is measured in ______________ units.

Martin-Gay Interactive Videos

See Video 6.2

Watch the section lecture video and answer the following questions.

OBJECTIVE 1
11. In Example 1, how can the perimeter be found if we forget the formula?

OBJECTIVE 2
12. From the lecture before Example 6, circumference is a special name for what?

OBJECTIVE 3
13. Explain why we need to use a formula twice to solve Example 9.

OBJECTIVE 4
14. In Examples 13 and 14, explain the difference in the two volume answers found for each.

6.2 Exercise Set MyMathLab®

Find the perimeter of each figure. See Examples 1 through 5.

1.

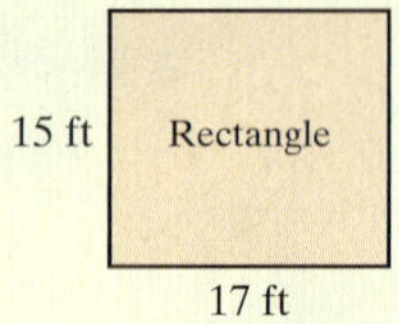

2.

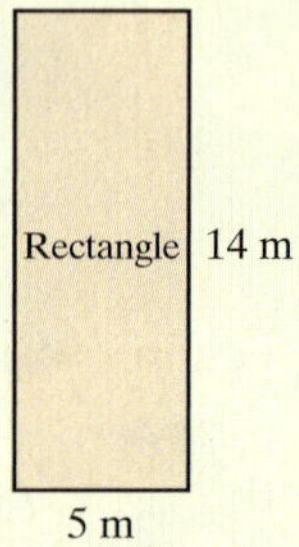

3.

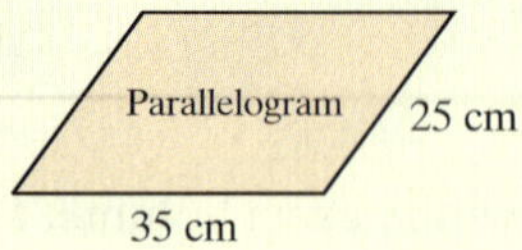

4.

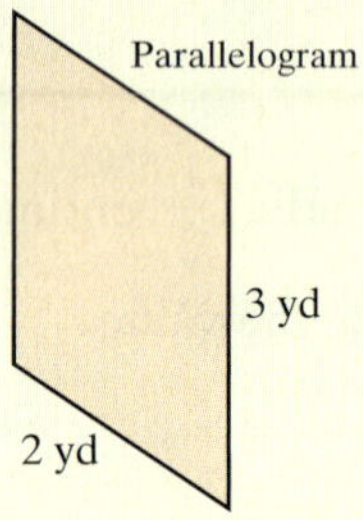

5.

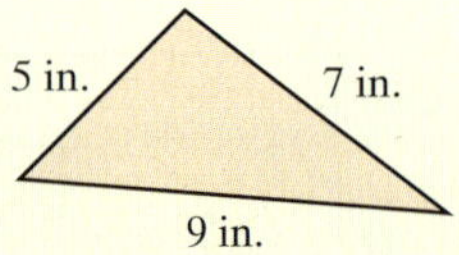

6.

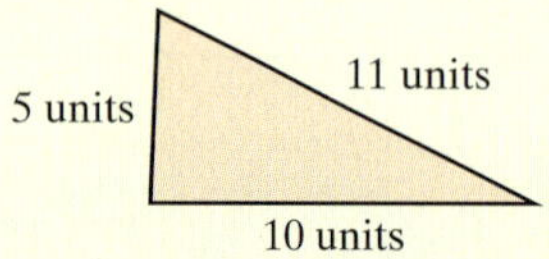

7.

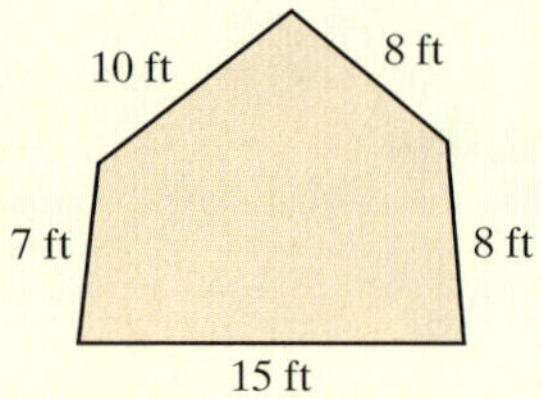

8.

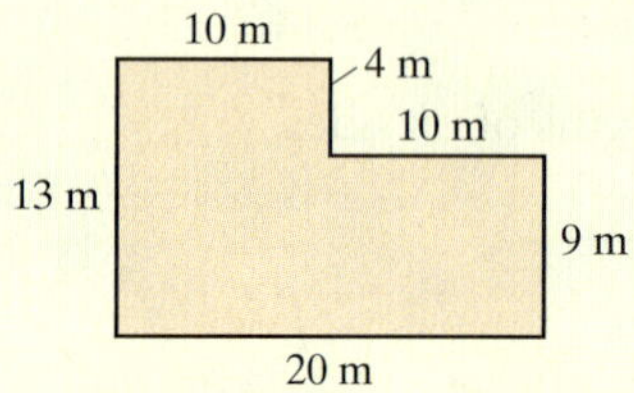

Find the perimeter of each regular polygon. (The sides of a regular polygon have the same length.)

9.

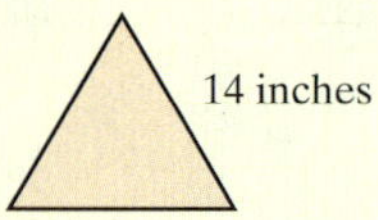

10.

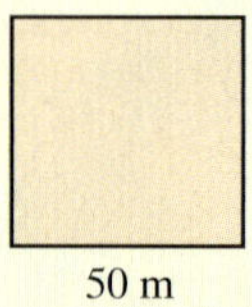

11.

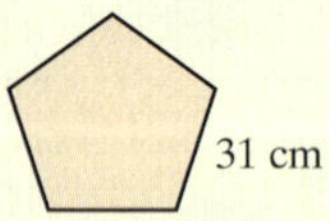

12.

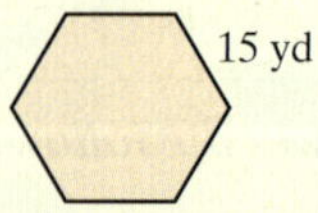

Solve. See Examples 1 through 6.

13. A polygon has sides of length 5 feet, 3 feet, 2 feet, 7 feet, and 4 feet. Find its perimeter.

14. A triangle has sides of length 8 inches, 12 inches, and 10 inches. Find its perimeter.

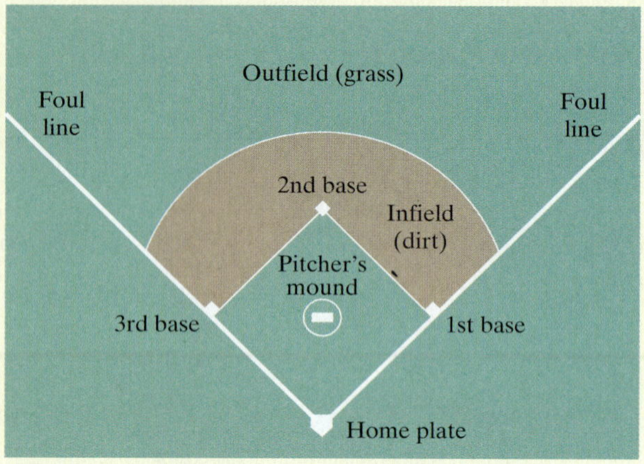

15. A line-marking machine lays down lime powder to mark both foul lines on a baseball field. If each foul line for this field measures 312 feet, how many feet of lime powder will be deposited?

16. A baseball diamond has 4 sides, with each side length 90 feet. If a baseball player hits a home run, how far does the player run (home plate, around the bases, then back to home plate)?

17. If a football field is 53 yards wide and 120 yards long, what is the perimeter?

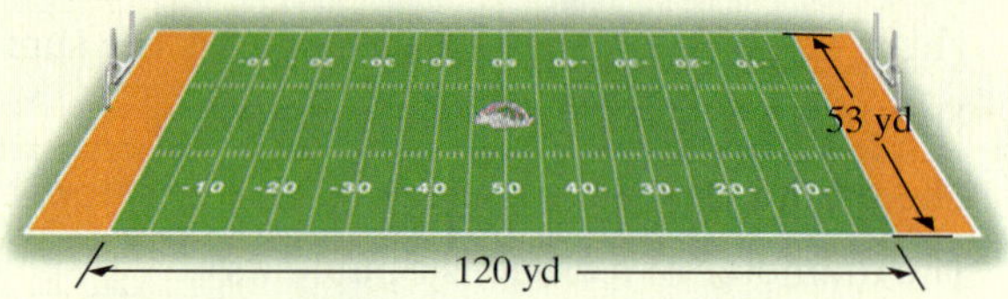

18. A stop sign has eight equal sides of length 12 inches. Find its perimeter.

19. A metal strip is being installed around a workbench that is 8 feet long and 3 feet wide. Find how much stripping is needed for this project.

20. Find how much fencing is needed to enclose a rectangular garden 70 feet by 21 feet.

21. If the stripping in Exercise **19** costs \$2.50 per foot, find the total cost of the stripping.

22. If the fencing in Exercise **20** costs \$2 per foot, find the total cost of the fencing.

23. A regular octagon has a side length of 9 inches.

a. How many sides does an octagon have?

b. Find its perimeter.

24. A regular pentagon has a side length of 14 meters.

a. How many sides does a pentagon have?

b. Find its perimeter.

25. Find the perimeter of the top of a square compact disc case if the length of one side is 7 inches.

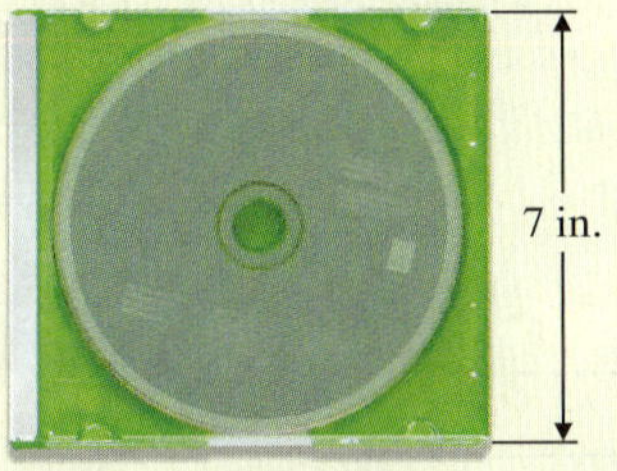

26. Find the perimeter of a square ceramic tile with a side of length 3 inches.

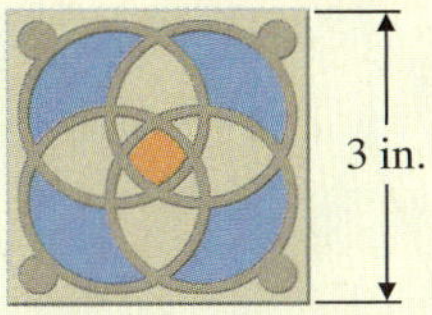

27. A rectangular room measures 10 feet by 11 feet. Find the cost of installing a strip of wallpaper around the room if the wallpaper costs \$0.86 per foot.

28. A rectangular house measures 85 feet by 70 feet. Find the cost of installing gutters around the house if the cost is \$2.36 per foot.

Find the perimeter of each figure. See Example 5.

29.

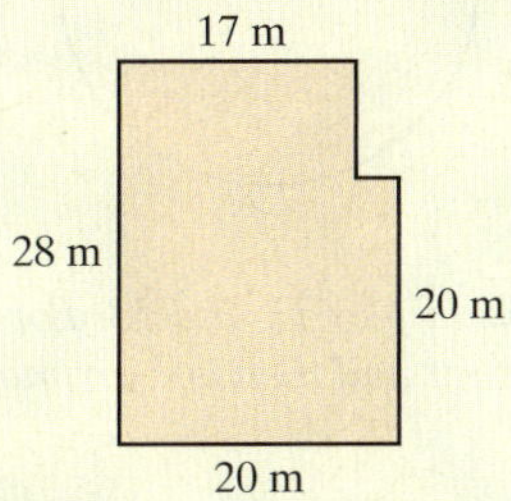

30.

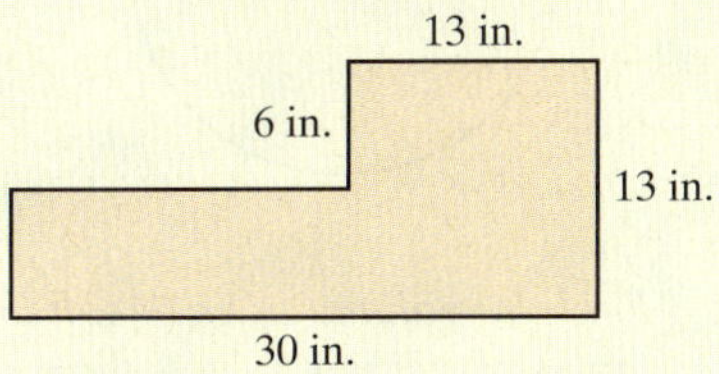

31.

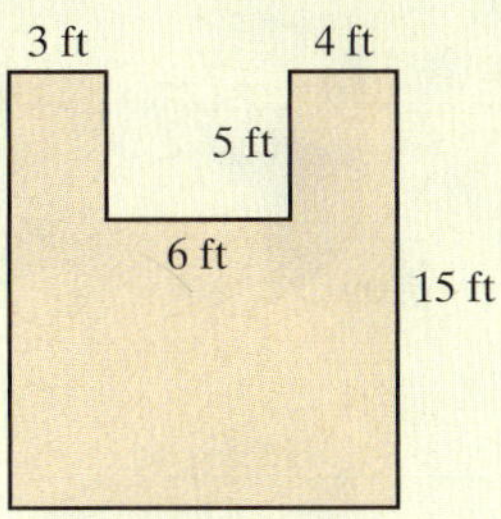

32.

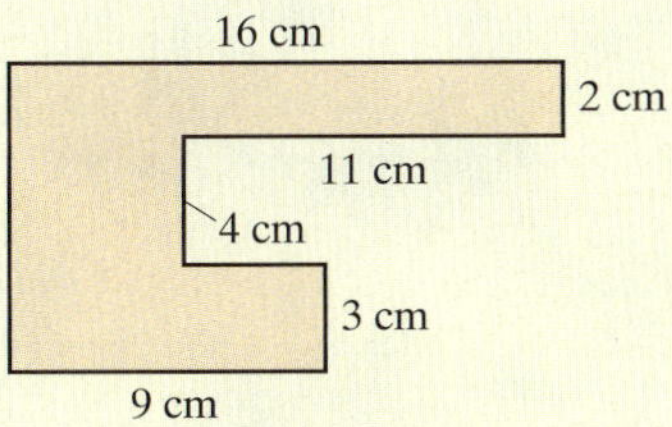

33.

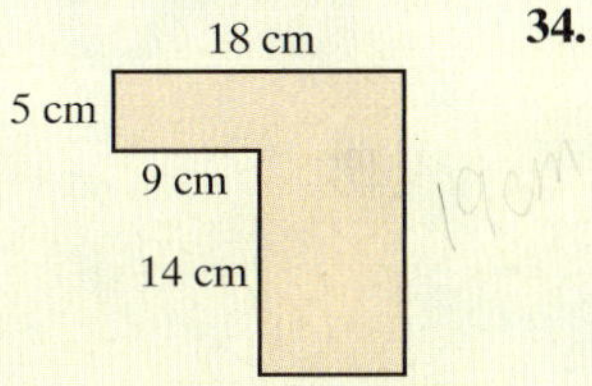

34.

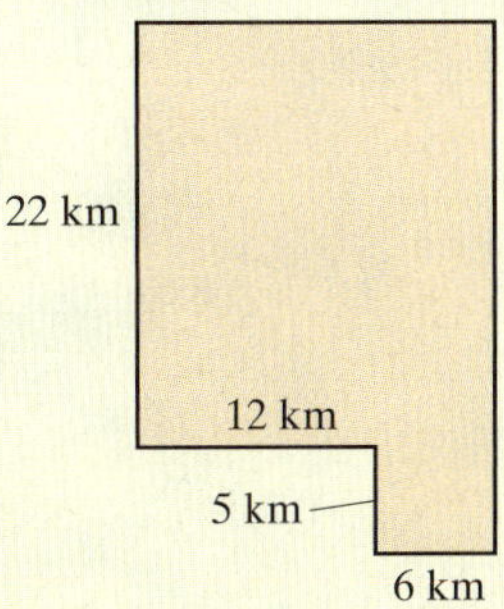

Find the circumference of each circle. Give the exact circumference and then an approximation. For Exercises 35–43, use $\pi \approx 3.14$. *See Example 7.*

35.

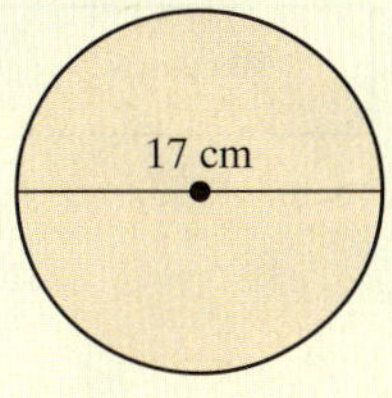

36.

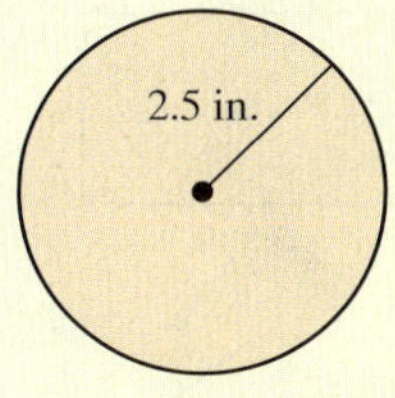

37.

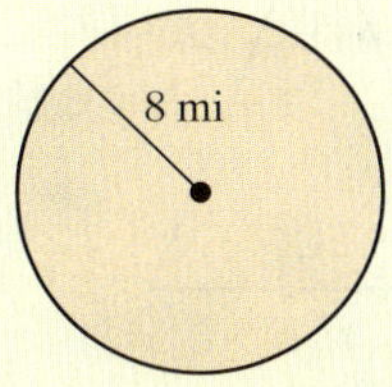

38.

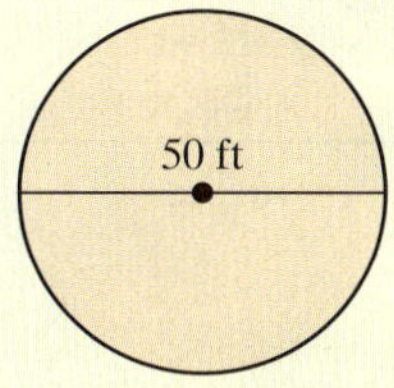

39.

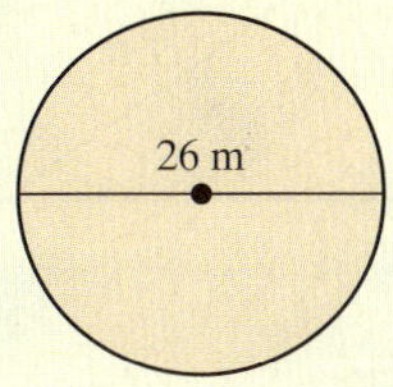

40.

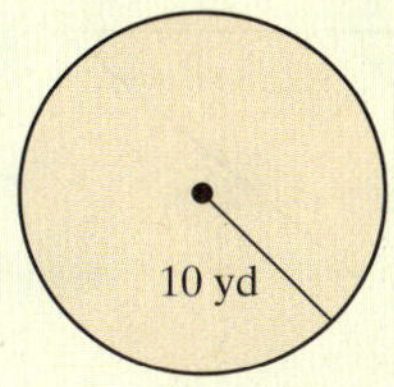

41. Wyley Robinson just bought a trampoline for his children to use. The trampoline has a diameter of 15 feet. If Wyley wishes to buy netting to go around the outside of the trampoline, how many feet of netting does he need?

42. The largest round barn in the world is located at the Marshfield Fairgrounds in Wisconsin. The barn has a diameter of 150 ft. What is the circumference of the barn? (*Source: The Milwaukee Journal Sentinel*)

43. Meteor Crater, near Winslow, Arizona, is 4000 feet in diameter. Approximate the distance around the crater. (*Source: The Handy Science Answer Book*)

44. The *Pearl of Lao-tze* has a diameter of $5\frac{1}{2}$ inches. Approximate the distance around the pearl. Use $\frac{22}{7}$ for π. (*Source: The Guinness World Records*)

Find the area of each geometric figure. If the figure is a circle, give the exact area and then approximate π and round to two decimal places if no other instructions are given. See Examples 8 through 11.

45.

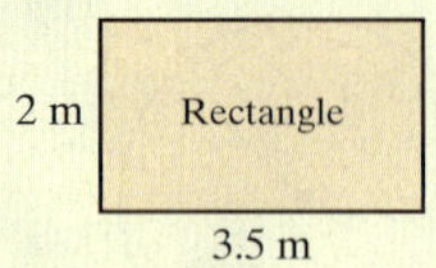

46.

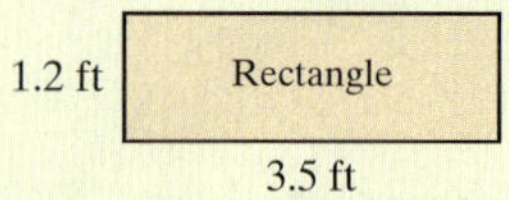

47.

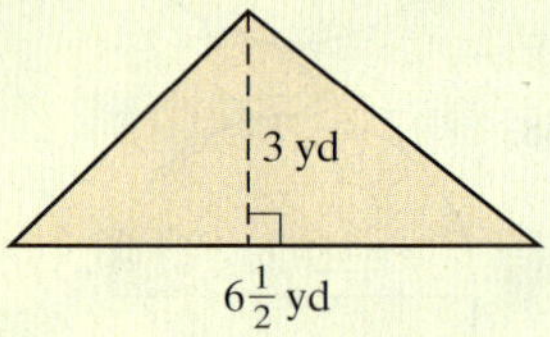

48.

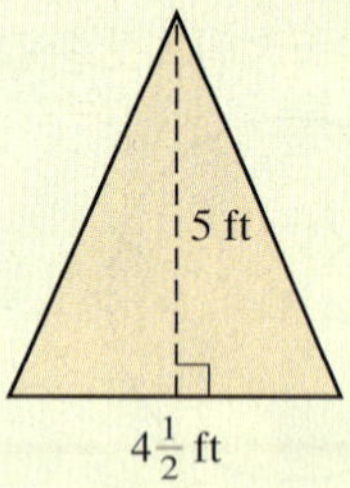

49.

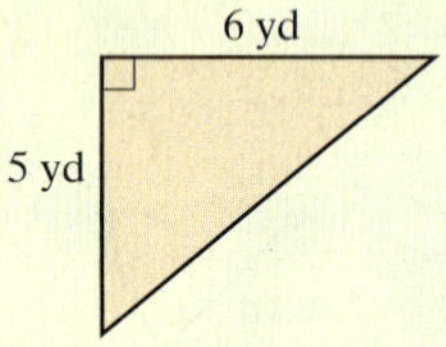

50.

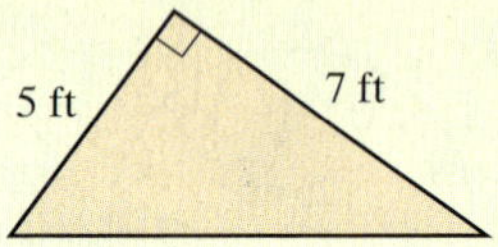

51. Use 3.14 for π.

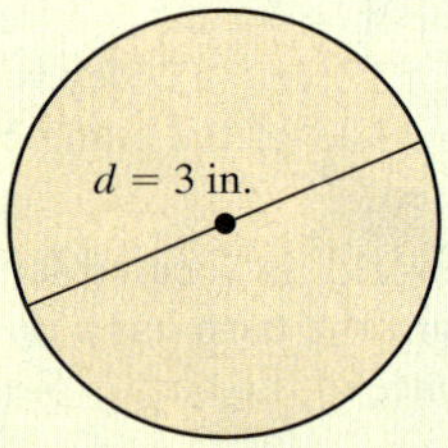

52. Use $\frac{22}{7}$ for π.

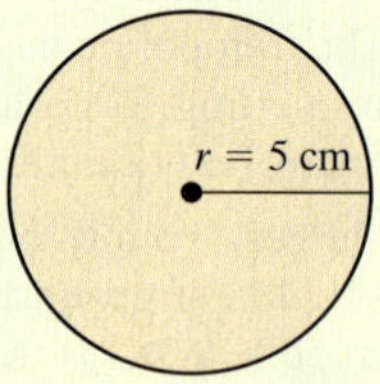

53.

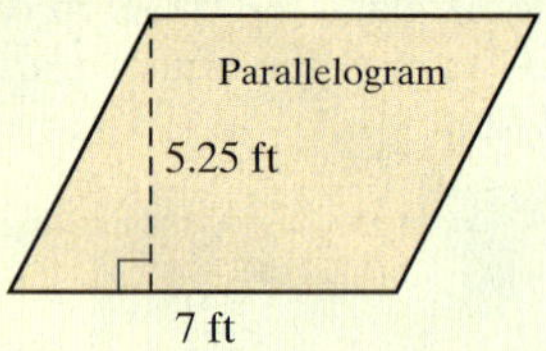

54.

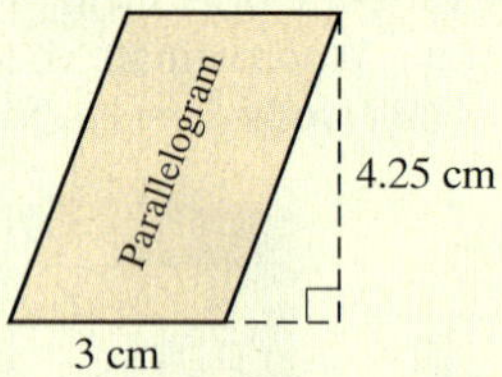

55.

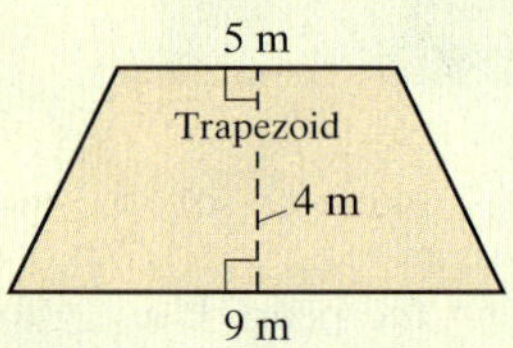

56.

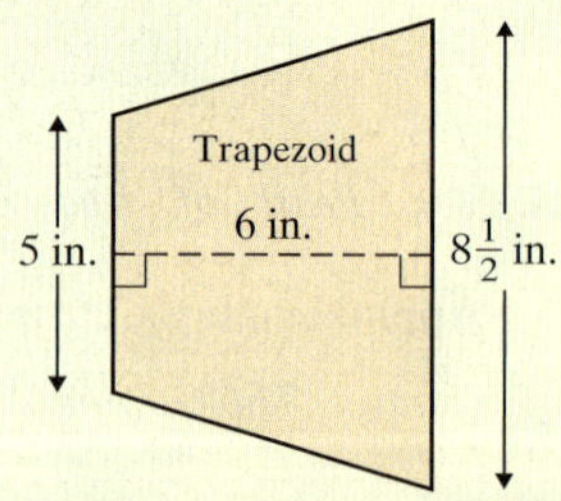

57.

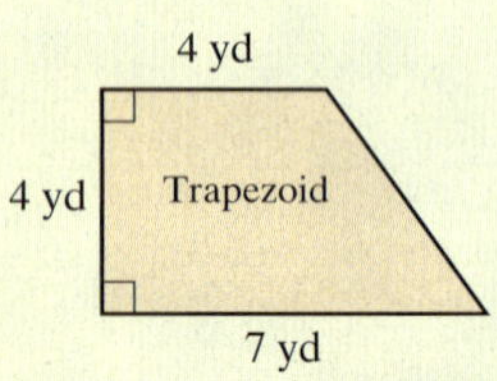

58.

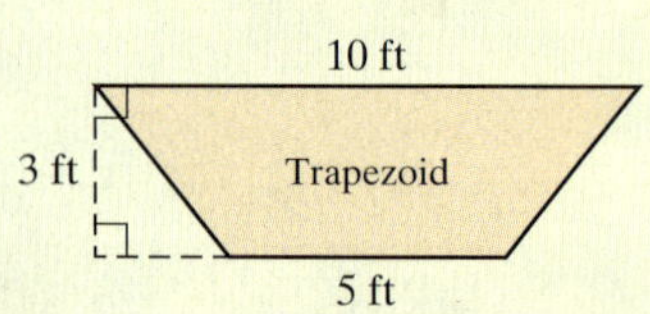

59.

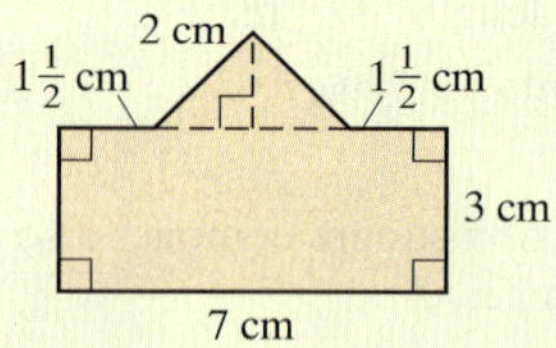

60.

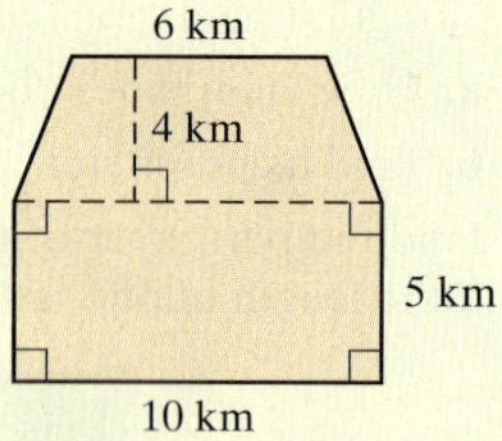

61.

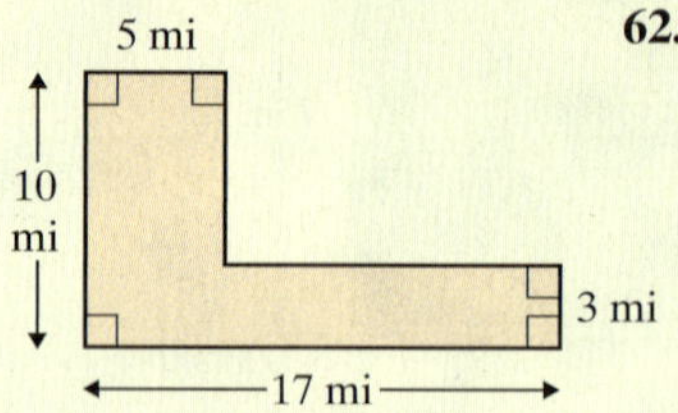

62.

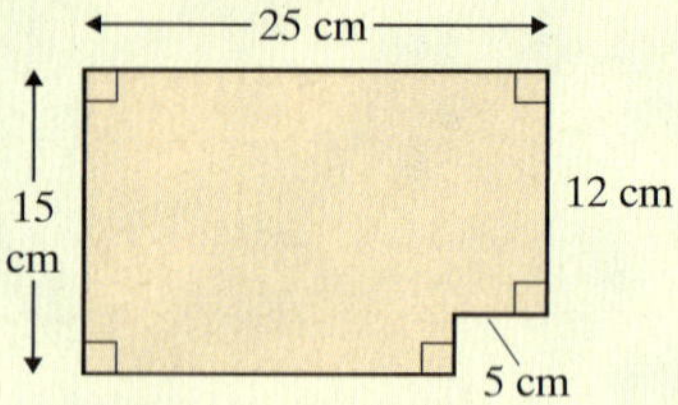

63.

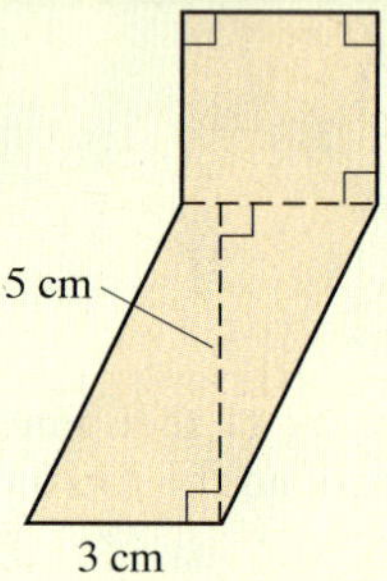

64.

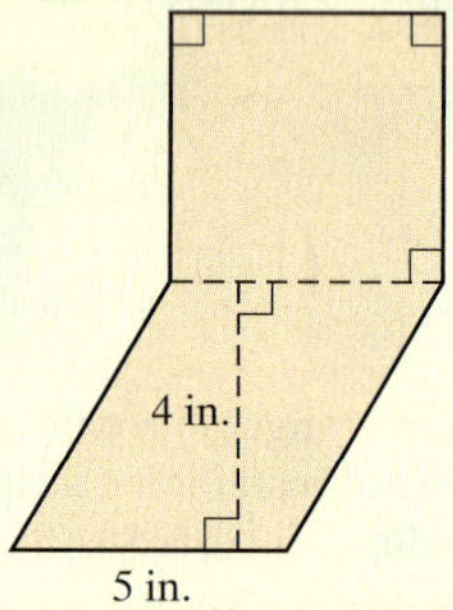

65.

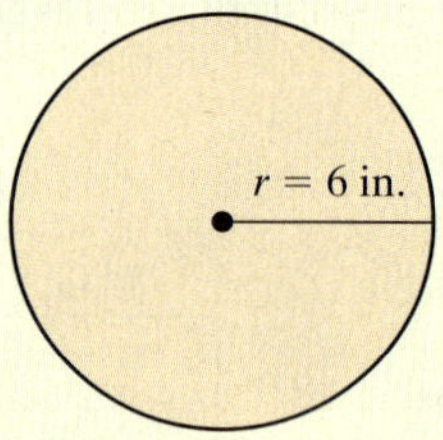

66.

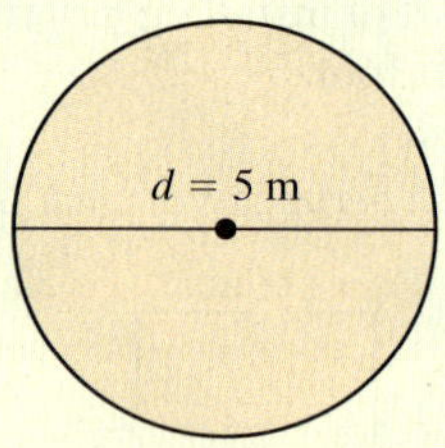

Find the volume of each solid. See Examples 12 and 13. For formulas containing π, give the exact answer and then approximate π and round to 2 decimal places.

67.

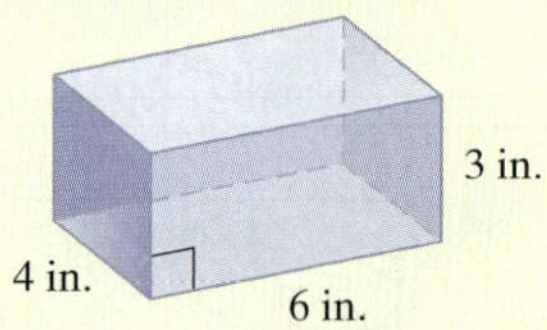

68.

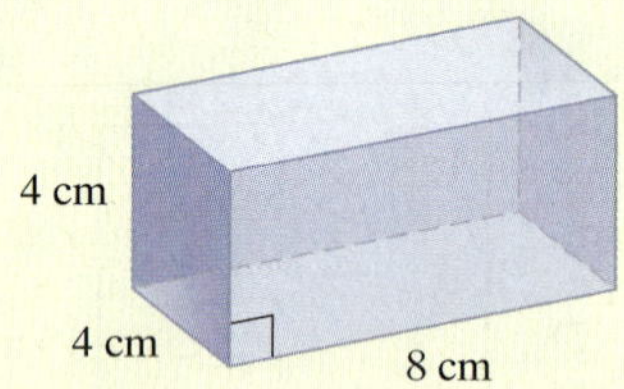

69.

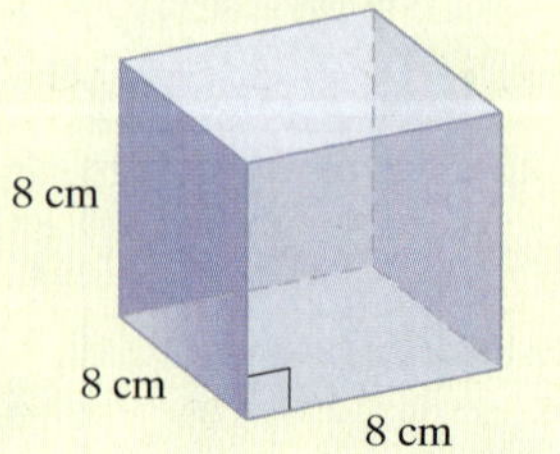

70.

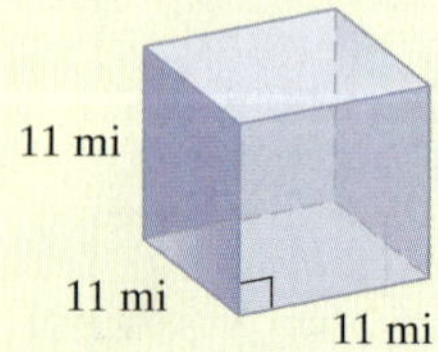

71.

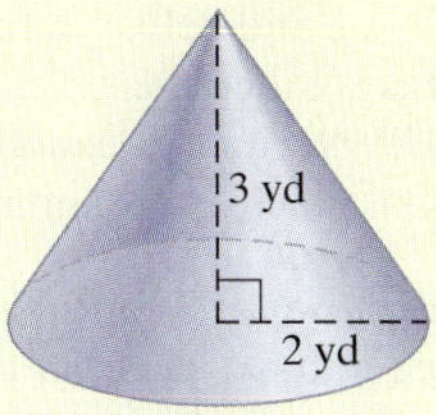

72.

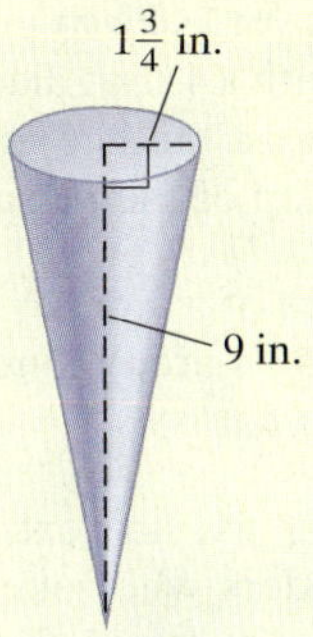

73.

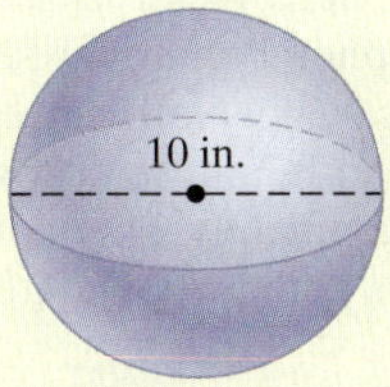

74.

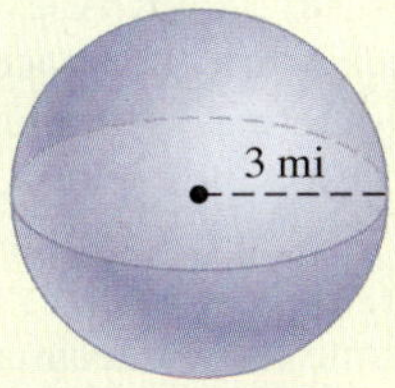

75.

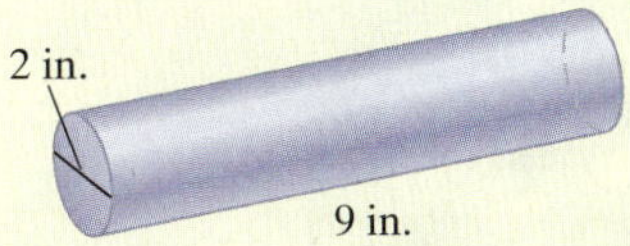

76.

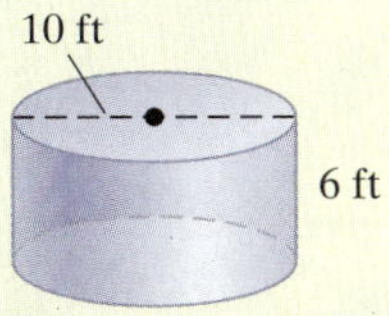

77.

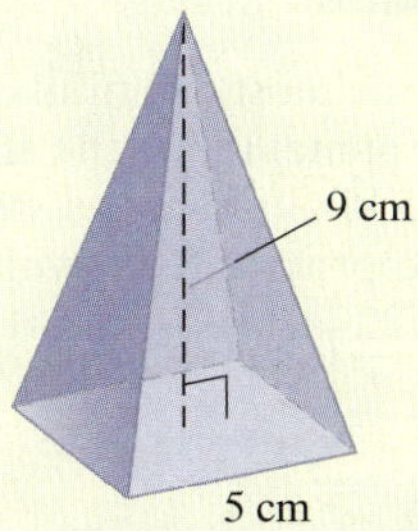

78.

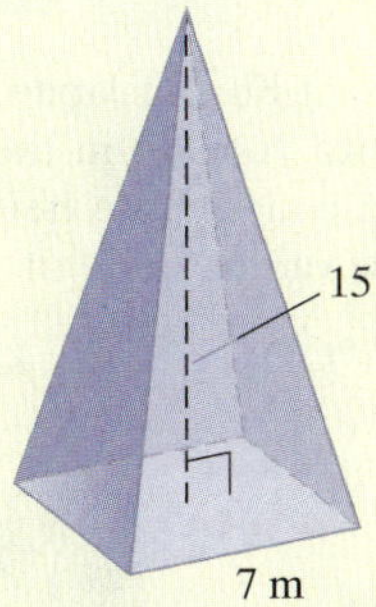

MIXED PRACTICE

Solve. See Examples 8 through 13.

79. Find the volume of a cube with edges of $1\frac{1}{3}$ inches.

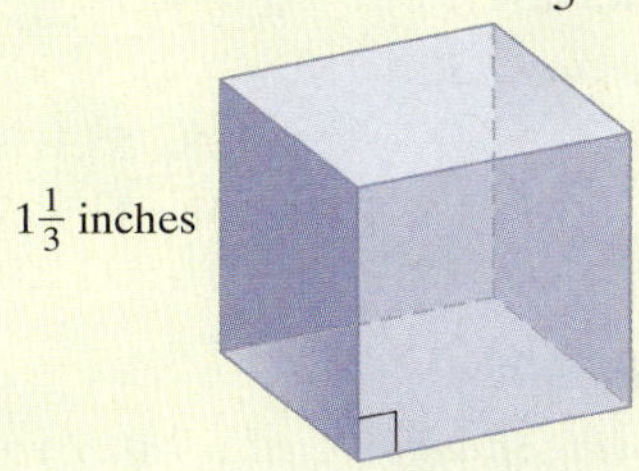

80. A water storage tank is in the shape of a cone with the pointed end down. If the radius is 14 ft and the depth of the tank is 15 ft, approximate the volume of the tank in cubic feet. Use $\frac{22}{7}$ for π.

81. Find the volume of a rectangular box 2 ft by 1.4 ft by 3 ft.

82. Find the volume of a box in the shape of a cube that is 5 ft on each side.

83. The largest American flag measures 505 feet by 225 feet. It's the U.S. "Super flag" owned by "Ski" Demski of Long Beach, California. Find its area. (*Source: Guinness World Records*)

84. The largest indoor illuminated sign is a billboard at Dubai Airport. It measures 28 meters by 6.2 meters. Find its area. (*Source: The Guinness Book of World Records*)

85. A drapery panel measures 6 ft by 7 ft. Find how many square feet of material are needed for *four* panels.

86. A page in a book measures 27.6 cm by 21.5 cm. Find its area.

87. A paperweight is in the shape of a square-based pyramid 20 centimeters tall. If an edge of the base is 12 centimeters, find the volume of the paperweight.

88. A birdbath is made in the shape of a hemisphere (half-sphere). If its radius is 10 inches, approximate its volume. Use $\frac{22}{7}$ for π.

89. The outlined part of the roof shown is in the shape of a trapezoid and needs to be shingled. The number of shingles to buy depends on the area.

a. Use the dimensions given to find the area of the outlined part of the roof to the nearest whole square foot.

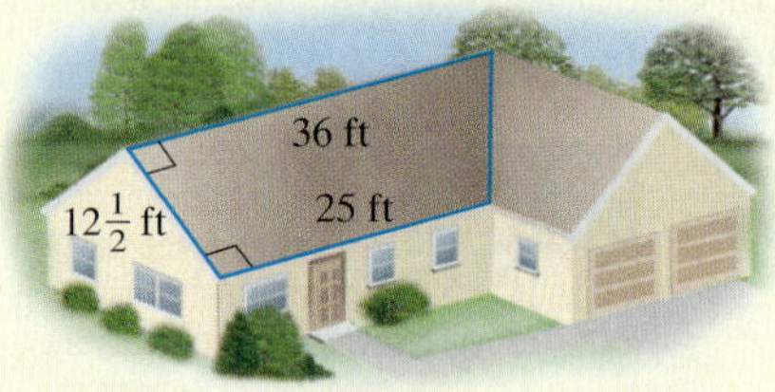

b. Shingles are packaged in a unit called a "square." If a "square" covers 100 square feet, how many whole squares need to be purchased to shingle this part of the roof?

90. The entire side of the building shaded in the drawing is to be bricked. The number of bricks to buy depends on the area.

a. Find the area.

b. If the side area of each brick (including mortar room) is $\frac{1}{6}$ square ft, find the number of bricks that are needed to brick the end of the building.

91. Find the exact volume of a waffle ice cream cone with a 3-in. diameter and a height of 7 inches.

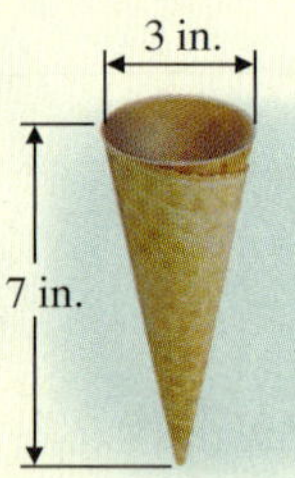

92. A snow globe has a diameter of 6 inches. Find its exact volume. Then approximate its volume using 3.14 for π.

93. Paul Revere's Pizza in the USA will bake and deliver a round pizza with a 4-foot diameter. This pizza is called the "Ultimate Party Pizza" and its current price is $99.99. Find the exact area of the top of the pizza and a 2 decimal place approximation.

94. The face of a circular watch has a diameter of 2 centimeters. What is its area? Find the exact area and a 2 decimal place approximation.

95. Zorbing is an extreme sport invented by two New Zealanders who joke that they were looking for a way to walk on water. A Zorb is a large sphere inside a second sphere with the space between the spheres pumped full of air. There is a tunnel-like opening so a person can crawl into the inner sphere. You are strapped in and sent down a Zorbing hill. A standard Zorb is approximately 3 m in diameter. Find the exact volume of a Zorb, and approximate the volume to two decimal places.

96. Mount Fuji, in Japan, is considered the most beautiful composite volcano in the world. The mountain is in the shape of a cone whose height is about 3.5 kilometers and whose base radius is about 3 kilometers. Approximate the volume of Mt. Fuji in cubic kilometers. Use $\frac{22}{7}$ for π.

REVIEW AND PREVIEW

Evaluate. See Section 1.3.

97. $1^2 + 2^2$

98. $5^2 + 3^2$

99. $4^2 + 2^2$

100. $1^2 + 6^2$

CONCEPT EXTENSIONS

Given the following situations, tell whether you are more likely to be concerned with area or perimeter.

101. ordering fencing to fence a yard

102. ordering grass seed to plant in a yard

103. buying carpet to install in a room

104. buying gutters to install on a house

105. ordering paint to paint a wall

106. ordering baseboards to install in a room

107. buying a wallpaper border to go on the walls around a room

108. buying fertilizer for your yard

Solve.

109. A pizza restaurant recently advertised two specials. The first special was a 12-inch pizza for \$10. The second special was two 8-inch pizzas for \$9. Determine the better buy. (*Hint:* First compare the areas of the two specials and then find a price per square inch for both specials.)

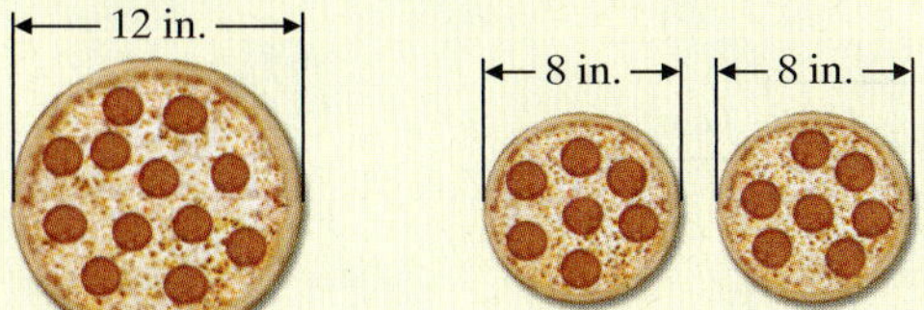

110. Find the approximate area of the state of Utah.

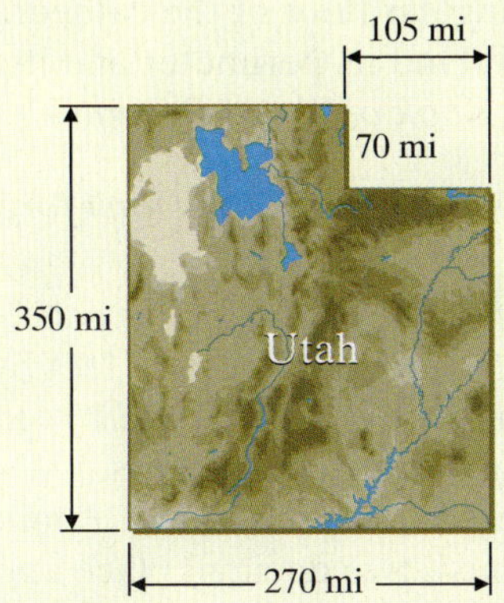

Solve Exercises 111 and 112. See a Concept Check in this section. Choose the figure that has the greater distance around.

111. **a.** A square with side length 3 inches

b. A circle with diameter 4 inches

112. **a.** A circle with diameter 7 inches

b. A square with side length 7 inches

113. **a.** Find the circumference of each circle. Approximate the circumference by using 3.14 for π.

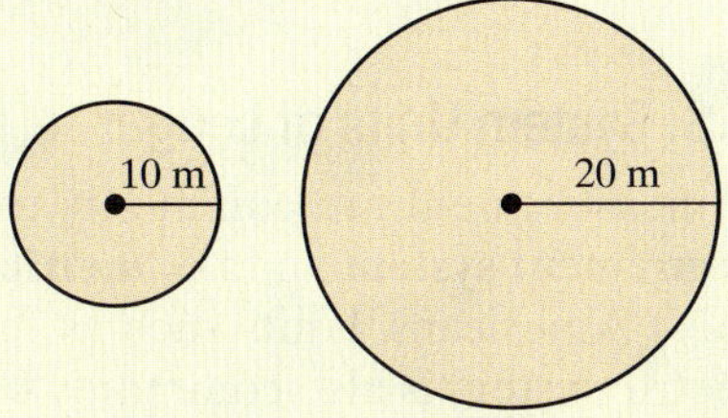

b. If the radius of a circle is doubled, is its corresponding circumference doubled?

114. **a.** Find the circumference of each circle. Approximate the circumference by using 3.14 for π.

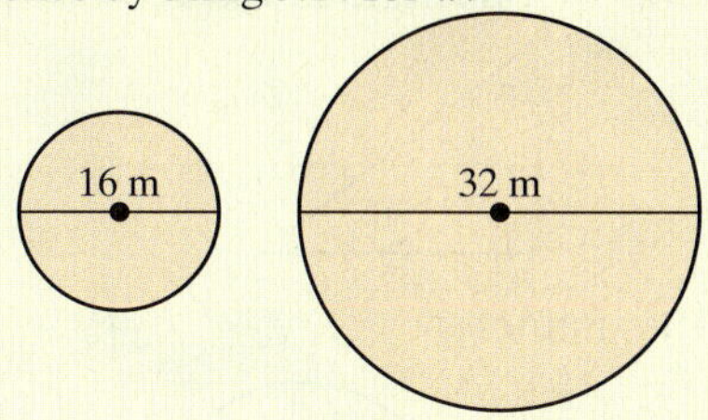

b. If the diameter of a circle is doubled, is its corresponding circumference doubled?

115. The Hayden Planetarium, at the Museum of Natural History in New York City, boasts a dome that has a diameter of 20 m. The dome is a hemisphere, or half a sphere. What is the volume enclosed by the dome at the Hayden Planetarium? Round to the nearest hundredth. (*Source:* Hayden Planetarium)

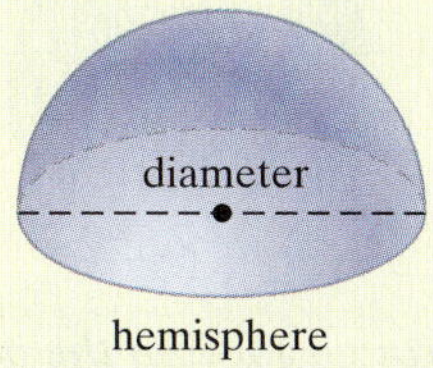

116. The Adler Museum in Chicago has a planetarium, its StarRider Theater, that has a diameter of 55 feet. Find the volume of its hemispheric (half a sphere) dome. Round to the nearest hundredth. (*Source:* The Adler Museum)

117. Can you compute the volume of a rectangle? Why or why not?

118. In your own words, explain why perimeter is measured in units and area is measured in square units.

119. Do two rectangles with the same perimeter have the same area? To see, find the perimeter and the area of each rectangle.

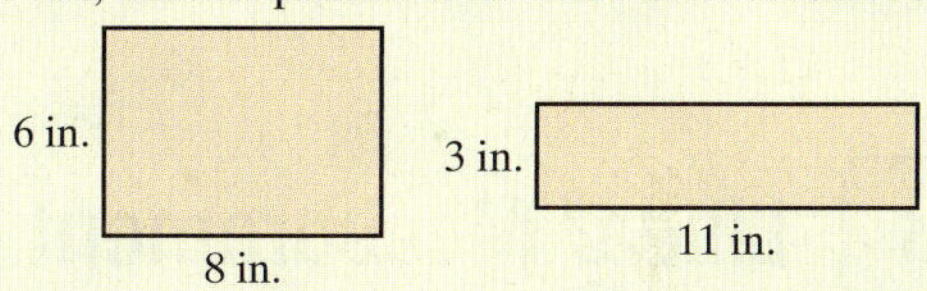

120. Do two rectangular solids with the same volume have the same dimensions? To see, find the volume of each rectangular solid.

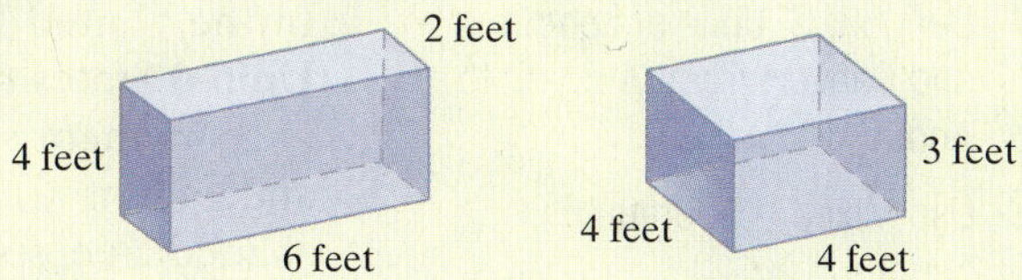

Integrated Review Geometry Concepts

Sections 6.1–6.2

1. Find the supplement and the complement of a 27° angle.

Find the measures of angles x, y, and z in each figure in Exercises 2 and 3.

2.

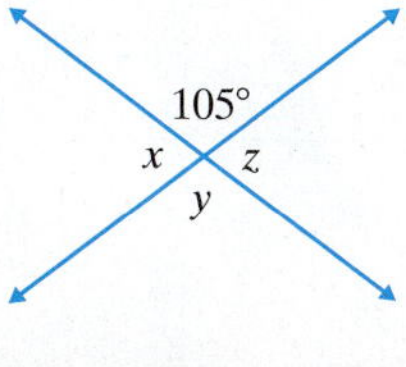

3. $m \parallel n$

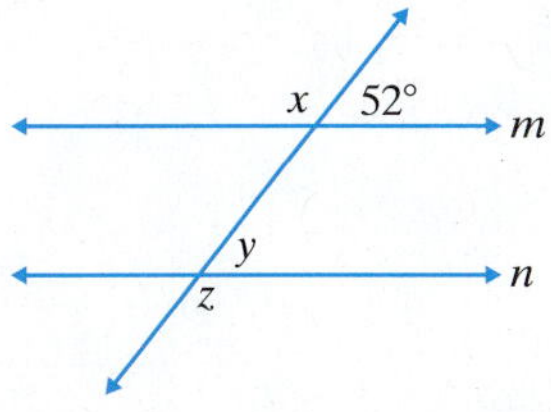

4. Find the measure of $\angle x$. (*Hint:* The sum of the angle measures of a triangle is 180°.)

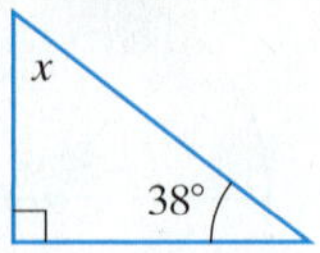

5. Find the diameter.

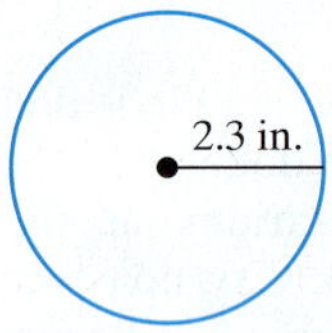

6. Find the radius.

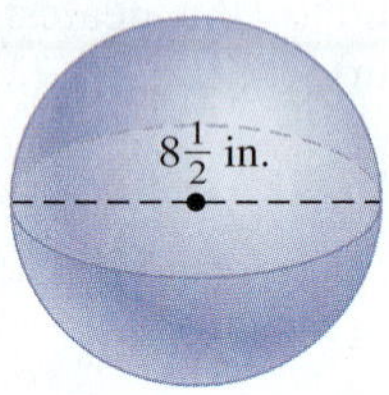

For Exercises 7 through 11, find the perimeter (or circumference) and area of each figure. For the circle, give the exact circumference and area. Then use $\pi \approx 3.14$ to approximate each. Don't forget to attach correct units.

7.

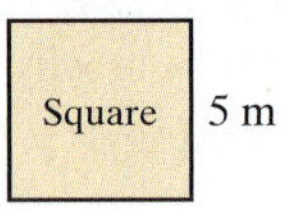

8.

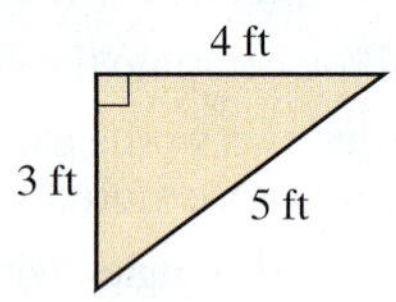

9.

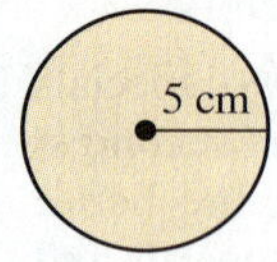

10.

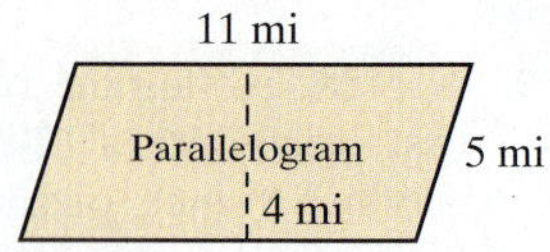

11.

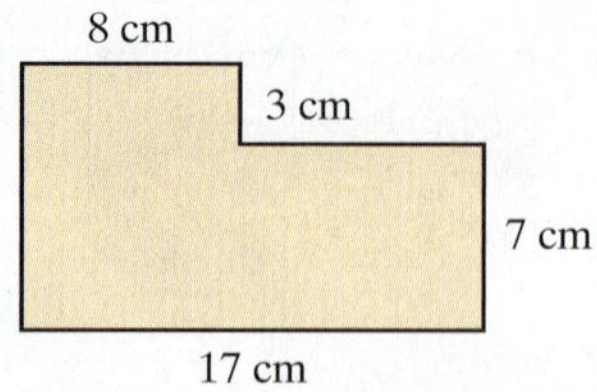

12. The smallest cathedral is in Highlandville, Missouri. The rectangular floor of the cathedral measures 14 feet by 17 feet. Find its perimeter and its area. (Source: *The Guinness Book of World Records*)

Find the volume of each solid. Don't forget to attach correct units.

13. A cube with edges of 4 inches each

14. A rectangular box 2 feet by 3 feet by 5.1 feet

15. A pyramid with a square base 10 centimeters on a side and a height of 12 centimeters

16. A sphere with a diameter of 3 miles. Give the exact volume and a two decimal place approximation.

6.3 Linear Measurement

OBJECTIVES

1. Define U.S. Units of Length and Convert from One Unit to Another.
2. Use Mixed U.S. Units of Length.
3. Perform Arithmetic Operations on U.S. Units of Length.
4. Define Metric Units of Length and Convert from One Unit to Another.
5. Perform Arithmetic Operations on Metric Units of Length.

OBJECTIVE

1 Defining and Converting U.S. System Units of Length

In the United States, two systems of measurement are commonly used. They are the **United States (U.S.), or English, measurement system** and the **metric system.** The U.S. measurement system is familiar to most Americans. Units such as feet, miles, ounces, and gallons are used. However, the metric system is also commonly used in fields such as medicine, sports, international marketing, and certain physical sciences. We are accustomed to buying 2-liter bottles of soft drinks, watching televised coverage of the 100-meter dash at the Olympic Games, or taking a 200-milligram dose of pain reliever.

The U.S. system of measurement uses the **inch, foot, yard,** and **mile** to measure **length.** The following is a summary of equivalencies between units of length:

U.S. Units of Length

12 inches (in.) = 1 foot (ft)
3 feet = 1 yard (yd)
36 inches = 1 yard
5280 feet = 1 mile (mi)

To convert from one unit of length to another, we will use **unit fractions.** We define a unit fraction to be a fraction that is equivalent to 1. Examples of unit fractions are as follows:

Unit Fractions

$$\frac{12 \text{ in.}}{1 \text{ ft}} = 1 \text{ or } \frac{1 \text{ ft}}{12 \text{ in.}} = 1 \text{ (since } 12 \text{ in.} = 1 \text{ ft)}$$

$$\frac{3 \text{ ft}}{1 \text{ yd}} = 1 \text{ or } \frac{1 \text{ yd}}{3 \text{ ft}} = 1 \text{ (since } 3 \text{ ft} = 1 \text{ yd)}$$

$$\frac{5280 \text{ ft}}{1 \text{ mi}} = 1 \text{ or } \frac{1 \text{ mi}}{5280 \text{ ft}} = 1 \text{ (since } 5280 \text{ ft} = 1 \text{ mi)}$$

Remember that multiplying a number by 1 does not change the value of the number.

EXAMPLE 1 Convert 8 feet to inches.

Solution We multiply 8 feet by a unit fraction that uses the equality 12 inches = 1 foot. The unit fraction should be in the form $\frac{\text{units to convert to}}{\text{original units}}$ or, in this case, $\frac{12 \text{ inches}}{1 \text{ foot}}$. We do this so that like units will divide out to 1, as shown.

$$\begin{aligned} 8 \text{ ft} &= \frac{8 \text{ ft}}{1} \cdot 1 && \text{Multiply by 1 in the form of } \frac{12 \text{ in.}}{1 \text{ ft}}. \\ &= \frac{8 \cancel{\text{ft}}}{1} \cdot \frac{12 \text{ in.}}{1 \cancel{\text{ft}}} \\ &= 8 \cdot 12 \text{ in.} \\ &= 96 \text{ in.} && \text{Multiply.} \end{aligned}$$

Thus, 8 ft = 96 in., as shown in the diagram:

8 feet = 96 inches

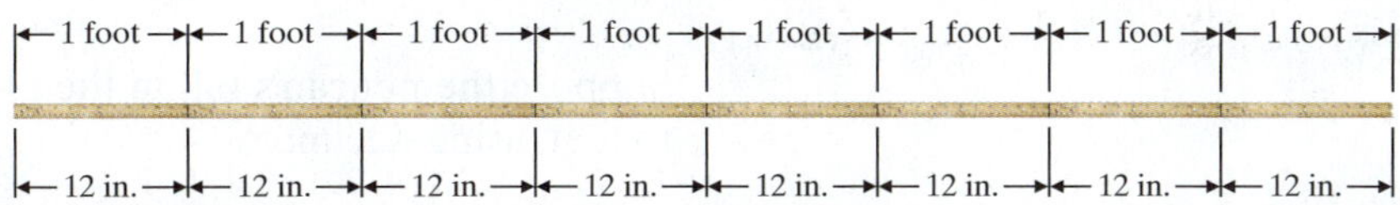

PRACTICE 1 Convert 6 feet to inches.

EXAMPLE 2 Convert 7 feet to yards.

Solution We multiply by a unit fraction that compares 1 yard to 3 feet.

$$\begin{aligned} 7 \text{ ft} &= \frac{7 \text{ ft}}{1} \cdot 1 \\ &= \frac{7 \cancel{\text{ft}}}{1} \cdot \frac{1 \text{ yd}}{3 \cancel{\text{ft}}} && \begin{array}{l} \leftarrow \text{Units to convert to} \\ \leftarrow \text{Original units} \end{array} \\ &= \frac{7}{3} \text{ yd} \\ &= 2\frac{1}{3} \text{ yd} && \text{Divide.} \end{aligned}$$

(Continued on next page)

> **Helpful Hint**
>
> When converting from one unit to another, select a unit fraction with the properties below:
>
> $$\frac{\text{units you are converting to}}{\text{original units}}$$
>
> By using this unit fraction, the original units will divide out, as wanted.

Thus, $7 \text{ ft} = 2\frac{1}{3} \text{ yd}$, as shown in the diagram.

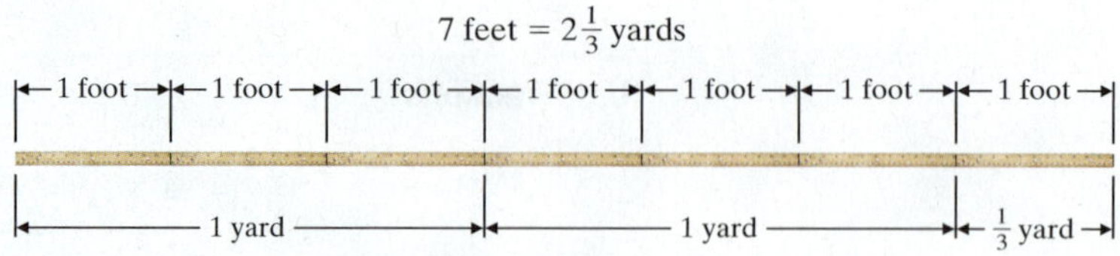

PRACTICE 2 Convert 8 yards to feet.

EXAMPLE 3 Finding the Length of a Pelican's Bill

The Australian pelican has the longest bill, measuring from 13 to 18.5 inches long. The pelican in the photo has a 15-inch bill. Convert 15 inches to feet, using decimals in your final answer.

Solution

$$15 \text{ in.} = \frac{15 \text{ in.}}{1} \cdot \frac{1 \text{ ft}}{12 \text{ in.}} \quad \begin{matrix}\leftarrow \text{Units to convert to} \\ \leftarrow \text{Original units}\end{matrix}$$

$$= \frac{15}{12} \text{ ft}$$

$$= \frac{5}{4} \text{ ft} \qquad \text{Simplify } \frac{15}{12}.$$

$$= 1.25 \text{ ft} \qquad \text{Divide.}$$

Thus, 15 in. = 1.25 ft, as shown in the diagram.

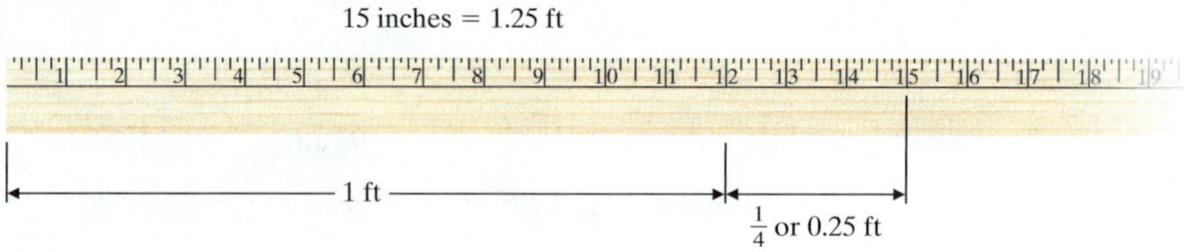

PRACTICE 3 Suppose the pelican's bill in the photo measures 18 inches. Convert 18 inches to feet, using decimals.

OBJECTIVE 2

Using Mixed U.S. System Units of Length

Sometimes it is more meaningful to express a measurement of length with mixed units such as 1 ft and 5 in. We usually condense this and write 1 ft 5 in.

In Example 2, we found that 7 feet is the same as $2\frac{1}{3}$ yards. The measurement can also be written as a mixture of yards and feet. That is,

$$7 \text{ ft} = \_\_\_\_ \text{ yd } \_\_\_\_ \text{ ft}$$

Because 3 ft = 1 yd, we divide 3 into 7 to see how many whole yards are in 7 feet. The quotient is the number of yards, and the remainder is the number of feet.

$$\begin{array}{r} 2 \text{ yd } 1 \text{ ft} \\ 3\overline{)7} \\ \underline{-6} \\ 1 \end{array}$$

Thus, 7 ft = 2 yd 1 ft, as seen in the diagram:

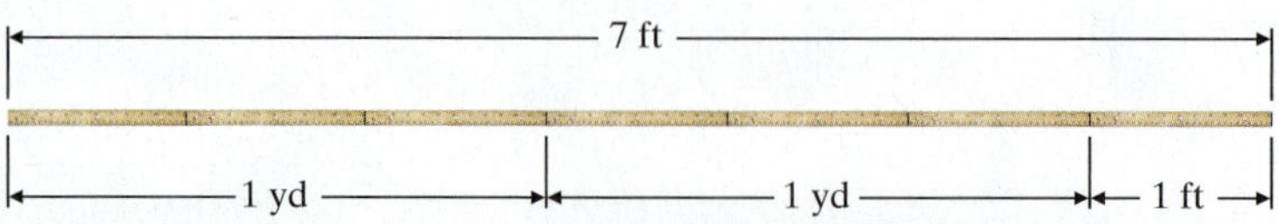

EXAMPLE 4 Convert: 134 in. = ____ ft ____ in.

Solution Because 12 in. = 1 ft, we divide 12 into 134. The quotient is the number of feet. The remainder is the number of inches. To see why we divide 12 into 134, notice that

$$134 \text{ in.} = \frac{134 \text{ in.}}{1} \cdot \frac{1 \text{ ft}}{12 \text{ in.}} = \frac{134}{12} \text{ ft}$$

$$\begin{array}{r} 11 \text{ ft } 2 \text{ in.} \\ 12\overline{)134} \\ -12 \\ \hline 14 \\ -12 \\ \hline 2 \end{array}$$

Thus, 134 in. = 11 ft 2 in. □

PRACTICE 4 Convert: 68 in. = ____ft ____in. ■

EXAMPLE 5 Convert 3 feet 7 inches to inches.

Solution First, we convert 3 feet to inches. Then we add 7 inches.

$$3 \text{ ft} = \frac{3 \text{ ft}}{1} \cdot \frac{12 \text{ in.}}{1 \text{ ft}} = 36 \text{ in.}$$

Then

$$3 \text{ ft } 7 \text{ in.} = 36 \text{ in.} + 7 \text{ in.} = 43 \text{ in.}$$ □

PRACTICE 5 Convert 5 yards 2 feet to feet. ■

OBJECTIVE 3

Performing Operations on U.S. System Units of Length

Finding sums or differences of measurements often involves converting units, as shown in the next example. Just remember that, as usual, only like units can be added or subtracted.

EXAMPLE 6 Add 3 ft 2 in. and 5 ft 11 in.

Solution To add, we line up the similar units.

$$\begin{array}{r} 3 \text{ ft } \ \ 2 \text{ in.} \\ +\ 5 \text{ ft } 11 \text{ in.} \\ \hline 8 \text{ ft } 13 \text{ in.} \end{array}$$

Since 13 inches is the same as 1 ft 1 in., we have

$$\begin{aligned} 8 \text{ ft } 13 \text{ in.} &= 8 \text{ ft} + 1 \text{ ft } 1 \text{ in.} \\ &= 9 \text{ ft } 1 \text{ in.} \end{aligned}$$ □

Answer to Concept Check:
round each to the nearest yard:
7 yd + 4 yd = 11 yd

PRACTICE 6 Add 4 ft 8 in. to 8 ft 11 in. ■

✔ **CONCEPT CHECK**

How could you estimate the following sum?

$$\begin{array}{r} 7 \text{ yd } \ \ 4 \text{ in.} \\ +\ 3 \text{ yd } 27 \text{ in.} \\ \hline \end{array}$$

EXAMPLE 7 Multiply 8 ft 9 in. by 3.

Solution By the distributive property, we multiply 8 ft by 3 and 9 in. by 3.

$$\begin{array}{r} 8 \text{ ft} \;\; 9 \text{ in.} \\ \times \qquad\quad 3 \\ \hline 24 \text{ ft } 27 \text{ in.} \end{array}$$

Since 27 in. is the same as 2 ft 3 in., we simplify the product as

$$\begin{aligned} 24 \text{ ft } 27 \text{ in.} &= 24 \text{ ft} + 2 \text{ ft } 3 \text{ in.} \\ &= 26 \text{ ft } 3 \text{ in.} \end{aligned}$$

□

PRACTICE 7 Multiply 4 ft 7 in. by 4. ■

We divide in a similar manner as above.

EXAMPLE 8 Finding the Length of a Piece of Rope

A rope of length 6 yd 1 ft has 2 yd 2 ft cut from one end. Find the length of the remaining rope.

Solution Subtract 2 yd 2 ft from 6 yd 1 ft.

beginning length	→	6 yd 1 ft
− amount cut	→	− 2 yd 2 ft
remaining length		

We cannot subtract 2 ft from 1 ft, so we borrow 1 yd from the 6 yd. One yard is converted to 3 ft and combined with the 1 ft already there.

Borrow 1 yd = 3 ft

5 yd + (1 yd) (3 ft)

$$\begin{array}{r} \not{6}\,\text{yd } 1 \text{ ft} \\ -2 \text{ yd } 2 \text{ ft} \\ \hline \end{array} \; = \; \begin{array}{r} 5 \text{ yd } 4 \text{ ft} \\ -2 \text{ yd } 2 \text{ ft} \\ \hline 3 \text{ yd } 2 \text{ ft} \end{array}$$

The remaining rope is 3 yd 2 ft long. □

PRACTICE 8 A carpenter cuts 1 ft 9 in. from a board of length 5 ft 8 in. Find the remaining length of the board. ■

OBJECTIVE 4 Defining and Converting Metric System Units of Length

The basic unit of length in the metric system is the **meter.** A meter is slightly longer than a yard. It is approximately 39.37 inches long. Recall that a yard is 36 inches long.

1 yard = 36 inches

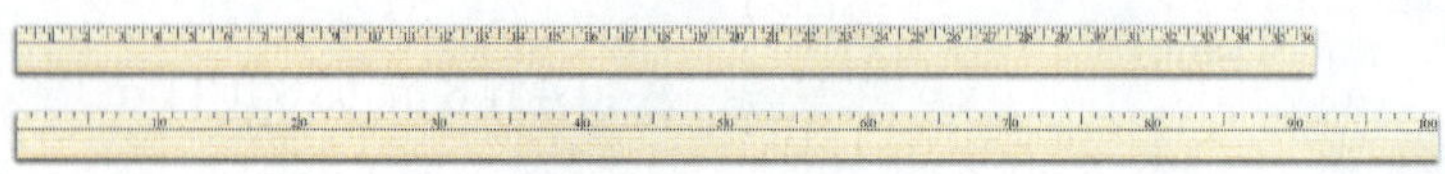

1 meter ≈ 39.37 inches

All units of length in the metric system are based on the meter. The following is a summary of the prefixes used in the metric system. Also shown are equivalencies between units of length. Like the decimal system, the metric system uses powers of 10 to define units.

Metric Units of Length
1 **kilo**meter (km) = 1000 meters (m)
1 **hecto**meter (hm) = 100 m
1 **deka**meter (dam) = 10 m
1 meter (m) = 1 m
1 **deci**meter (dm) = 1/10 m or 0.1 m
1 **centi**meter (cm) = 1/100 m or 0.01 m
1 **milli**meter (mm) = 1/1000 m or 0.001 m

The figure below will help you with decimeters, centimeters, and millimeters.

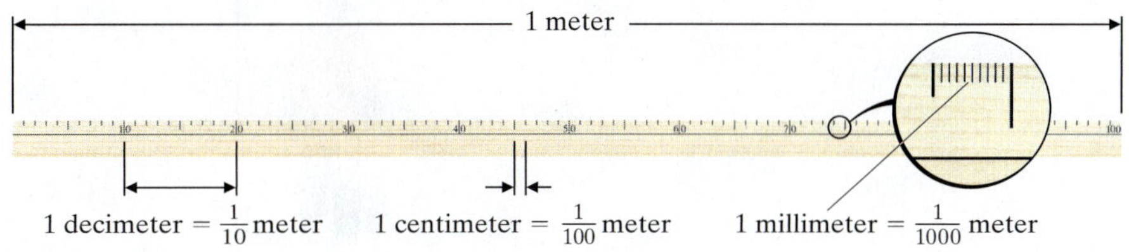

Helpful Hint

Study the figure above for other equivalencies between metric units of length.

10 decimeters = 1 meter 10 millimeters = 1 centimeter
100 centimeters = 1 meter 10 centimeters = 1 decimeter
1000 millimeters = 1 meter

These same prefixes are used in the metric system for mass and capacity. The most commonly used measurements of length in the metric system are the **meter, millimeter, centimeter,** and **kilometer.**

✓ CONCEPT CHECK

Is this statement reasonable? "The screen of a home television set has a 30-meter diagonal." Why or why not?

Being comfortable with the metric units of length means gaining a "feeling" for metric lengths, just as you have a "feeling" for the lengths of an inch, a foot, and a mile. To help you accomplish this, study the following examples:

- A millimeter is about the thickness of a large paper clip.

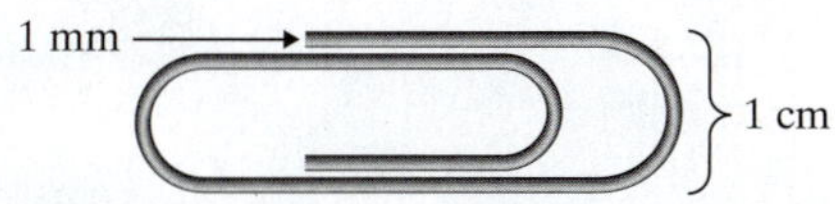

- A centimeter is about the width of a large paper clip.

- A meter is slightly longer than a yard.
- A kilometer is about two-thirds of a mile.

Answer to Concept Check:
no; answers may vary

- The width of this book is approximately 21.5 centimeters.
- The distance between New York City and Philadelphia is about 160 kilometers.

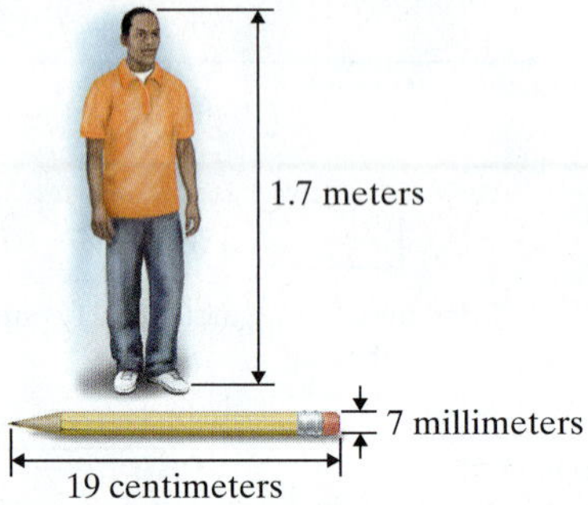

$2\frac{1}{2}$ centimeters is about 1 inch.

As with the U.S. system of measurement, unit fractions may be used to convert from one unit of length to another. For example, let's convert 1200 meters to kilometers. To do so, we will multiply by 1 in the form of the unit fraction

$$\frac{1\text{ km}}{1000\text{ m}} \begin{matrix} \leftarrow \text{Units to convert to} \\ \leftarrow \text{Original units} \end{matrix}$$

$$1200\text{ m} = \frac{1200\text{ m}}{1} \cdot 1 = \frac{1200\text{ m}}{1} \cdot \overbrace{\frac{1\text{ km}}{1000\text{ m}}}^{\text{Unit fraction}} = \frac{1200\text{ km}}{1000} = 1.2\text{ km}$$

The metric system does, however, have a distinct advantage over the U.S. system of measurement: the ease of converting from one unit of length to another. Since all units of length are powers of 10 of the meter, converting from one unit of length to another is as simple as moving the decimal point. Listing units of length in order from largest to smallest helps to keep track of how many places to move the decimal point when converting.

Let's again convert 1200 meters to kilometers. This time, to convert from meters to kilometers, we move along the chart shown, 3 units to the left, from meters to kilometers. This means that we move the decimal point 3 places to the left.

km hm dam **m** dm cm mm

3 units to the left

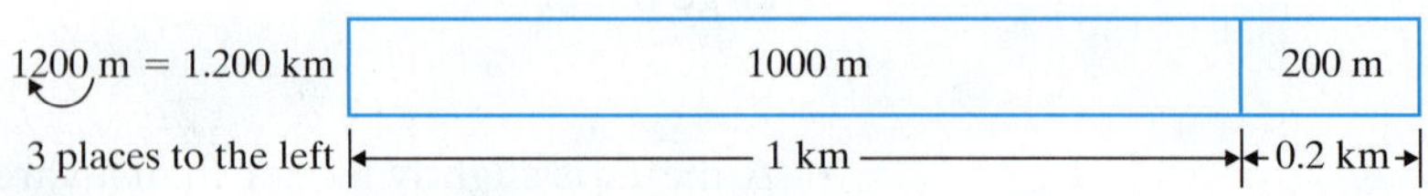

Thus, 1200 m = 1.2 km, as shown in the diagram.

EXAMPLE 9 Convert 2.3 m to centimeters.

Solution First we will convert by using a unit fraction.

$$2.3 \text{ m} = \frac{2.3 \cancel{\text{m}}}{1} \cdot \underbrace{\frac{100 \text{ cm}}{1 \cancel{\text{m}}}}_{\text{Unit fraction}} = 230 \text{ cm}$$

Now we will convert by listing the units of length in order from left to right and moving from meters to centimeters.

km hm dam m dm cm mm

2 units to the right

$2.30 \text{ m} = 230. \text{ cm}$

2 places to the right

With either method, we get 230 cm. □

PRACTICE 9 Convert 2.5 m to millimeters. ■

EXAMPLE 10 Convert 450,000 mm to meters.

Solution We list the units of length in order from left to right and move from millimeters to meters.

km hm dam m dm cm mm

3 units to the left

Thus, move the decimal point 3 places to the left.

$450{,}000 \text{ mm} = 450.000 \text{ m} \quad \text{or} \quad 450 \text{ m}$ □

PRACTICE 10 Convert 3500 m to kilometers. ■

✔CONCEPT CHECK

What is wrong with the following conversion of 150 cm to meters?

~~150.00 cm = 15,000 m~~

OBJECTIVE 5

Performing Operations on Metric System Units of Length

To add, subtract, multiply, or divide with metric measurements of length, we write all numbers using the same unit of length and then add, subtract, multiply, or divide as with decimals.

EXAMPLE 11 Subtract 430 m from 1.3 km.

Solution First we convert both measurements to kilometers or both to meters.

430 m = 0.43 km or 1.3 km = 1300 m

$$\begin{array}{r} 1.30 \text{ km} \\ -0.43 \text{ km} \\ \hline 0.87 \text{ km} \end{array} \qquad \begin{array}{r} 1300 \text{ m} \\ -430 \text{ m} \\ \hline 870 \text{ m} \end{array}$$

The difference is 0.87 km or 870 m. □

PRACTICE 11 Subtract 640 m from 2.1 km. ■

Answer to Concept Check:
decimal point should be moved two places to the left: 1.5 m

EXAMPLE 12 Multiply 5.7 mm by 4.

Solution Here we simply multiply the two numbers. Note that the unit of measurement remains the same.

$$\begin{array}{r} 5.7 \text{ mm} \\ \times \quad 4 \\ \hline 22.8 \text{ mm} \end{array}$$

□

PRACTICE 12 Multiply 18.3 hm by 5. ■

EXAMPLE 13 **Finding a Person's Height**

Fritz Martinson was 1.2 meters tall on his last birthday. Since then, he has grown 14 centimeters. Find his current height in meters.

Solution

$$\begin{array}{rcr} \text{original height} & \rightarrow & 1.20 \text{ m} \\ \underline{+ \text{ height grown}} & \rightarrow & \underline{+ \ 0.14 \text{ m}} \\ \text{current height} & & 1.34 \text{ m} \end{array} \quad (\text{Since } 14 \text{ cm} = 0.14 \text{ m})$$

Fritz is now 1.34 meters tall. □

PRACTICE 13 Doris Blackwell is knitting a scarf that is currently 0.8 meter long. If she knits an additional 45 centimeters, how long will the scarf be? ■

Vocabulary, Readiness & Video Check

Use the choices below to fill in each blank. Some choices may be used more than once.

inches   yard   unit fraction
feet   meter

1. The basic unit of length in the metric system is the ____________.
2. The expression $\frac{1 \text{ foot}}{12 \text{ inches}}$ is an example of a(n) ____________.
3. A meter is slightly longer than a(n) ____________.
4. One foot equals 12 ____________.
5. One yard equals 3 ____________.
6. One yard equals 36 ____________.
7. One mile equals 5280 ____________.

Martin-Gay Interactive Videos

See Video 6.3

Watch the section lecture video and answer the following questions.

OBJECTIVE 1
8. In Example 3, what units are used in the denominator of the unit fraction and why was this decided?

OBJECTIVE 2
9. In Example 4, how is a mixed unit similar to a mixed number? Use examples in your answer.

OBJECTIVE 3
10. In Example 5, why is the sum of the addition problem not the final answer? Reference the sum in your answer.

OBJECTIVE 4
11. In the lecture before Example 6, why is it easier to convert metric units than U.S. units?

OBJECTIVE 5
12. What two answers did we get for Example 8? Explain why both answers are correct.

6.3 Exercise Set MyMathLab®

Convert each measurement as indicated. See Examples 1 through 3.

1. 60 in. to feet
2. 84 in. to feet
3. 12 yd to feet
4. 18 yd to feet
5. 42,240 ft to miles
6. 36,960 ft to miles
7. $8\frac{1}{2}$ ft to inches
8. $12\frac{1}{2}$ ft to inches
9. 10 ft to yards
10. 25 ft to yards
11. 6.4 mi to feet
12. 3.8 mi to feet
13. 162 in. to yd (Write answer as a decimal.)
14. 7216 yd to mi (Write answer as a decimal.)
15. 3 in. to ft (Write answer as a decimal.)
16. 129 in. to ft (Write answer as a decimal.)

Convert each measurement as indicated. See Examples 4 and 5.

17. 40 ft = _____ yd _____ ft
18. 100 ft = _____ yd _____ ft
19. 85 in. = _____ ft _____ in.
20. 59 in. = _____ ft _____ in.
21. 10,000 ft = _____ mi _____ ft
22. 25,000 ft = _____ mi _____ ft
23. 5 ft 2 in. = _____ in.
24. 4 ft 11 in. = _____ in.
25. 8 yd 2 ft = _____ ft
26. 4 yd 1 ft = _____ ft
27. 2 yd 1 ft = _____ in.
28. 1 yd 2 ft = _____ in.

Perform each indicated operation. Simplify the result if possible. See Examples 6 through 8.

29. 3 ft 10 in. + 7 ft 4 in.
30. 12 ft 7 in. + 9 ft 11 in.
31. 12 yd 2 ft + 9 yd 2 ft
32. 16 yd 2 ft + 8 yd 2 ft
33. 22 ft 8 in. − 16 ft 3 in.
34. 15 ft 5 in. − 8 ft 2 in.
35. 18 ft 3 in. − 10 ft 9 in.
36. 14 ft 8 in. − 3 ft 11 in.
37. 28 ft 8 in. ÷ 2
38. 34 ft 6 in. ÷ 2
39. 16 yd 2 ft × 5
40. 15 yd 1 ft × 8

Convert as indicated. See Examples 9 and 10.

41. 60 m to centimeters
42. 46 m to centimeters
43. 40 mm to centimeters
44. 14 mm to centimeters
45. 500 m to kilometers
46. 400 m to kilometers
47. 1700 mm to meters
48. 6400 mm to meters
49. 1500 cm to meters
50. 6400 cm to meters
51. 0.42 km to centimeters
52. 0.95 km to centimeters
53. 7 km to meters
54. 5 km to meters
55. 8.3 cm to millimeters
56. 4.6 cm to millimeters
57. 20.1 mm to decimeters
58. 140.2 mm to decimeters
59. 0.04 m to millimeters
60. 0.2 m to millimeters

Perform each indicated operation. Remember to insert units when writing your answers. See Examples 11 through 13.

61. 8.6 m + 0.34 m
62. 14.1 cm + 3.96 cm
63. 2.9 m + 40 mm
64. 30 cm + 8.9 m
65. 24.8 mm − 1.19 cm
66. 45.3 m − 2.16 dam
67. 15 km − 2360 m
68. 14 cm − 15 mm
69. 18.3 m × 3
70. 14.1 m × 4
71. 6.2 km ÷ 4
72. 9.6 m ÷ 5

MIXED PRACTICE

Solve. Remember to insert units when writing your answers. For Exercises 73 through 82, complete the charts. See Examples 1 through 13.

	Yards	Feet	Inches
73. Chrysler Building in New York City		1046	
74. 4-story building			792
75. Python length		35	
76. Ostrich height			108

		Meters	*Millimeters*	*Kilometers*	*Centimeters*
77.	Length of elephant	5			
78.	Height of grizzly bear	3			
79.	Tennis ball diameter				6.5
80.	Golf ball diameter				4.6
81.	Distance from London to Paris			342	
82.	Distance from Houston to Dallas			396	

83. The National Zoo maintains a small patch of bamboo, which it grows as a food supply for its pandas. Two weeks ago, the bamboo was 6 ft 10 in. tall. Since then, the bamboo has grown 3 ft 8 in. How tall is the bamboo now?

84. While exploring in the Marianas Trench, a submarine probe was lowered to a point 1 mile 1400 feet below the ocean's surface. Later it was lowered an additional 1 mile 4000 feet below this point. How far was the probe below the surface of the Pacific?

85. At its deepest point, the Grand Canyon of the Colorado River in Arizona is about 6000 ft. The Grand Canyon of the Yellowstone River, which is in Yellowstone National Park in Wyoming, is at most 300 yards deep. How much deeper is the Grand Canyon of the Colorado River than the Grand Canyon of the Yellowstone River? (*Note:* Notice that the dimensions are different.) (*Source:* National Park Service)

86. The Grand Canyon of the Gunnison River, in Colorado, is often called the Black Canyon of the Gunnison because it is so steep that light rarely penetrates the depth of the canyon. The Black Canyon of the Gunnison is only 1150 ft wide at its narrowest point. At its narrowest, the Grand Canyon of the Yellowstone is $\frac{1}{2}$ mile wide. Find the difference in width between the Grand Canyon of the Yellowstone and the Black Canyon of the Gunnison. (See the note for Exercise 85) (*Source:* National Park Service)

87. The tallest man in the world is recorded as Robert Pershing Wadlow of Alton, Illinois. Born in 1918, he measured 8 ft 11 in. at his tallest. The shortest man in the world is Chandra Bahadur Dangi of Nepal, who measures 21.5 in. How many times taller than Chandra is Robert? Round to one decimal place. (*Source: Guinness World Records*)

88. A 3.4-m rope is attached to a 5.8-m rope. However, when the ropes are tied, 8 cm of length is lost to form the knot. What is the length of the tied ropes?

89. The ice on a pond is 5.33 cm thick. For safe skating, the owner of the pond insists that it be 80 mm thick. How much thicker must the ice be before skating is allowed?

90. The sediment on the bottom of the Towamencin Creek is normally 14 cm thick, but a recent flood washed away 22 mm of sediment. How thick is it now?

91. The Amana Corporation stacks up its microwave ovens in a distribution warehouse. Each stack is 1 ft 9 in. wide. How far from the wall would 9 of these stacks extend?

92. The highway commission is installing concrete sound barriers along a highway. Each barrier is 1 yd 2 ft long. Find the total length of 25 barriers placed end to end.

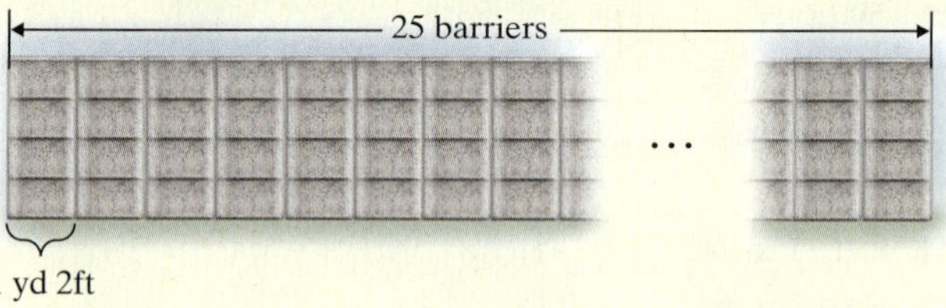

93. A logging firm needs to cut a 67-m-long redwood log into 20 equal pieces before loading it onto a truck for shipment. How long will each piece be?

94. A 112.5-foot-tall dead pinewood tree is removed by starting at the top and cutting off 9-foot-long sections. How many whole sections are removed?

95. The longest truck in the world is operated by Gould Transport in Australia, and is the 182-ft Road Train. How many *yards* long are 2 of these trucks? (*Source: Guinness World Records*)

96. The large Coca-Cola sign formerly in the Tokyo Ginza shopping district was in the shape of a rectangle whose length was 31 yards and whose width was 49 feet. Find the area of a rectangle with those dimensions in square feet. (*Source:* Coca-Cola Company) (*Hint:* Recall that the area of a rectangle is the product length times width.)

REVIEW AND PREVIEW

Write each decimal as a fraction and each fraction as a decimal. See Section R.3.

97. 0.21

98. 0.86

99. $\frac{13}{100}$

100. $\frac{47}{100}$

101. $\frac{1}{4}$

102. $\frac{3}{20}$

CONCEPT EXTENSIONS

Determine whether the measurement in each statement is reasonable.

103. The width of a twin-size bed is 20 meters.

104. A window measures 1 meter by 0.5 meter.

105. A drinking glass is made of glass 2 millimeters thick.

106. A paper clip is 4 kilometers long.

107. The distance across the Colorado River is 50 kilometers.

108. A model's hair is 30 centimeters long.

Estimate each sum or difference. See the first Concept Check in this section.

109.
$$\begin{array}{r} 5 \text{ yd } 2 \text{ in.} \\ +\ 7 \text{ yd } 30 \text{ in.} \\ \hline \end{array}$$

110.
$$\begin{array}{r} 45 \text{ ft } 1 \text{ in.} \\ -\ 10 \text{ ft } 11 \text{ in.} \\ \hline \end{array}$$

111. Using a unit other than the foot, write a length that is equivalent to 4 feet. (*Hint:* There are many possibilities.)

112. Using a unit other than the meter, write a length that is equivalent to 7 meters. (*Hint:* There are many possibilities.)

113. To convert from meters to centimeters, the decimal point is moved two places to the right. Explain how this relates to the fact that the prefix *centi* means $\frac{1}{100}$.

114. Explain why conversions in the metric system are easier to make than conversions in the U.S. system of measurement.

115. An advertisement sign outside Fenway Park in Boston measures 18.3 m by 18.3 m. What is the area of this sign?

6.4 Weight and Mass

OBJECTIVES

1. Define U.S. Units of Weight and Convert from One Unit to Another.
2. Perform Arithmetic Operations on U.S. Units of Weight.
3. Define Metric Units of Mass and Convert from One Unit to Another.
4. Perform Arithmetic Operations on Metric Units of Mass.

OBJECTIVE

1 Defining and Converting U.S. System Units of Weight

Whenever we talk about how heavy an object is, we are concerned with the object's **weight.** We discuss weight when we refer to a 12-ounce box of Rice Krispies, a 15-pound tabby cat, or a barge hauling 24 tons of garbage.

12 ounces

15 pounds

24 tons of garbage

The most common units of weight in the U.S. measurement system are the **ounce,** the **pound,** and the **ton.** The following is a summary of equivalencies between units of weight:

U.S. Units of Weight	Unit Fraction
16 ounces (oz) = 1 pound (lb)	$\frac{16 \text{ oz}}{1 \text{ lb}} = \frac{1 \text{ lb}}{16 \text{ oz}} = 1$
2000 pounds = 1 ton	$\frac{2000 \text{ lb}}{1 \text{ ton}} = \frac{1 \text{ ton}}{2000 \text{ lb}} = 1$

✔ CONCEPT CHECK

If you were describing the weight of a fully loaded semitrailer, which type of unit would you use: ounce, pound, or ton? Why?

Unit fractions that equal 1 are used to convert between units of weight in the U.S. system. When converting using unit fractions, recall that the numerator of a unit fraction should contain the units we are converting to and the denominator should contain the original units.

EXAMPLE 1 Convert 9000 pounds to tons.

Solution We multiply 9000 lb by a unit fraction that uses the equality 2000 pounds = 1 ton.

Remember, the unit fraction should be $\frac{\text{units to convert to}}{\text{original units}}$ or $\frac{1 \text{ ton}}{2000 \text{ lb}}$.

$$9000 \text{ lb} = \frac{9000 \text{ lb}}{1} \cdot 1 = \frac{9000 \cancel{\text{lb}}}{1} \cdot \frac{1 \text{ ton}}{2000 \cancel{\text{lb}}} = \frac{9000 \text{ tons}}{2000} = \frac{9}{2} \text{ tons or } 4\frac{1}{2} \text{ tons}$$

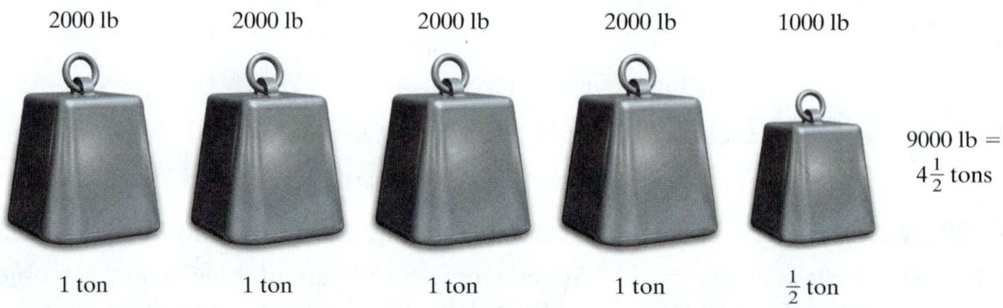

PRACTICE 1 Convert 6500 pounds to tons.

EXAMPLE 2 Convert 3 pounds to ounces.

Solution We multiply by the unit fraction $\frac{16 \text{ oz}}{1 \text{ lb}}$ to convert from pounds to ounces.

$$3 \text{ lb} = \frac{3 \text{ lb}}{1} \cdot 1 = \frac{3 \cancel{\text{lb}}}{1} \cdot \frac{16 \text{ oz}}{1 \cancel{\text{lb}}} = 3 \cdot 16 \text{ oz} = 48 \text{ oz}$$

Answer to Concept Check:
ton

PRACTICE 2 Convert 72 ounces to pounds.

As with length, it is sometimes useful to simplify a measurement of weight by writing it in terms of mixed units.

EXAMPLE 3 Convert: 33 ounces = ____ lb ____ oz

Solution Because 16 oz = 1 lb, divide 16 into 33 to see how many pounds are in 33 ounces. The quotient is the number of pounds, and the remainder is the number of ounces. To see why we divide 16 into 33, notice that

$$33 \text{ oz} = 33 \cancel{\text{oz}} \cdot \frac{1 \text{ lb}}{16 \cancel{\text{oz}}} = \frac{33}{16} \text{ lb}$$

$$\begin{array}{r} 2 \text{ lb } 1 \text{ oz} \\ 16\overline{)33} \\ -32 \\ \hline 1 \end{array}$$

Thus, 33 ounces is the same as 2 lb 1 oz.

PRACTICE 3 Convert: 47 ounces = ____ lb ____ oz

OBJECTIVE 2

Performing Operations on U.S. System Units of Weight

Performing arithmetic operations on units of weight works the same way as performing arithmetic operations on units of length.

EXAMPLE 4 Subtract 3 tons 1350 lb from 8 tons 1000 lb.

Solution To subtract, we line up similar units.

$$\begin{array}{r} 8 \text{ tons } 1000 \text{ lb} \\ -3 \text{ tons } 1350 \text{ lb} \\ \hline \end{array}$$

Since we cannot subtract 1350 lb from 1000 lb, we borrow 1 ton from the 8 tons. To do so, we write 1 ton as 2000 lb and combine it with the 1000 lb.

$$\begin{array}{rcr} & 7 \text{ tons} + 1 \text{ ton } 2000 \text{ lb} & \\ \cancel{8} \text{ tons } 1000 \text{ lb} & = & 7 \text{ tons } 3000 \text{ lb} \\ -3 \text{ tons } 1350 \text{ lb} & = & -3 \text{ tons } 1350 \text{ lb} \\ \hline & & 4 \text{ tons } 1650 \text{ lb} \end{array}$$

To check, see that the sum of 4 tons 1650 lb and 3 tons 1350 lb is 8 tons 1000 lb.

PRACTICE 4 Subtract 5 tons 1200 lb from 8 tons 100 lb.

EXAMPLE 5 Divide 9 lb 6 oz by 2.

Solution We divide each of the units by 2.

$$
\begin{array}{r@{\quad}r}
 & 4\text{ lb} \quad 11\text{ oz} \\
2\overline{)} & 9\text{ lb} \quad\;\; 6\text{ oz} \\
 & -8 \qquad\qquad\;\; \\
 & 1\text{ lb} = 16\text{ oz} \\
 & 22\text{ oz}
\end{array}
\quad \text{Divide 2 into 22 oz to get 11 oz.}
$$

To check, multiply 4 pounds 11 ounces by 2. The result is 9 pounds 6 ounces. □

PRACTICE 5 Divide 5 lb 8 oz by 4. ■

EXAMPLE 6 **Finding the Weight of a Child**

Bryan weighed 8 lb 8 oz at birth. By the time he was 1 year old, he had gained 11 lb 14 oz. Find his weight at age 1 year.

Solution

birth weight	→	8 lb 8 oz
+ weight gained	→	+ 11 lb 14 oz
total weight	→	19 lb 22 oz

Since 22 oz equals 1 lb 6 oz,

$$
\begin{aligned}
19\text{ lb } 22\text{ oz} &= 19\text{ lb} + 1\text{ lb } 6\text{ oz} \\
&= 20\text{ lb } 6\text{ oz}
\end{aligned}
$$

Bryan weighed 20 lb 6 oz on his first birthday. □

PRACTICE 6 A 5-lb 14-oz batch of cookies is packed into a 6-oz container before it is mailed. Find the total weight. ■

OBJECTIVE 3

Defining and Converting Metric System Units of Mass

In scientific and technical areas, a careful distinction is made between **weight** and **mass.** **Weight** is really a measure of the pull of gravity. The farther from Earth an object gets, the less it weighs. However, **mass** is a measure of the amount of substance in the object and does not change. Astronauts orbiting Earth weigh much less than they weigh on Earth, but they have the same mass in orbit as they do on Earth. Here on Earth, weight and mass are the same, so either term may be used.

The basic unit of mass in the metric system is the **gram.** It is defined as the mass of water contained in a cube 1 centimeter (cm) on each side.

Actual size

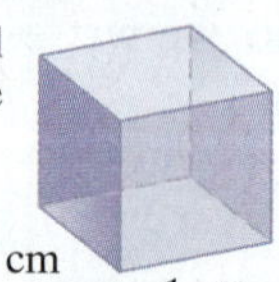

1 cm

1 cm

1 cm

1 cubic centimeter

The following examples may help you get a feeling for metric masses:

- A tablet contains 200 milligrams of ibuprofen.

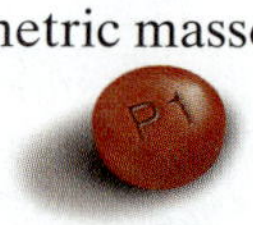

- A large paper clip weighs approximately 1 gram.

- A box of crackers weighs 453 grams.

A kilogram is slightly over 2 pounds. An adult woman may weigh 60 kilograms.

The prefixes for units of mass in the metric system are the same as for units of length, as shown in the following table:

Metric Units of Mass
1 **kilo**gram (kg) = 1000 grams (g)
1 **hecto**gram (hg) = 100 g
1 **deka**gram (dag) = 10 g
1 gram (g) = 1 g
1 **deci**gram (dg) = 1/10 g or 0.1 g
1 **centi**gram (cg) = 1/100 g or 0.01 g
1 **milli**gram (mg) = 1/1000 g or 0.001 g

✔ CONCEPT CHECK
True or false? A decigram is larger than a dekagram. Explain.

The **milligram,** the **gram,** and the **kilogram** are the three most commonly used units of mass in the metric system.

As with lengths, all units of mass are powers of 10 of the gram, so converting from one unit of mass to another only involves moving the decimal point. To convert from one unit of mass to another in the metric system, list the units of mass in order from largest to smallest.

Let's convert 4300 milligrams to grams. To convert from milligrams to grams, we move along the list 3 units to the left.

kg hg dag **g** dg cg **mg**

3 units to the left

This means that we move the decimal point 3 places to the left to convert from milligrams to grams.

$$4300 \text{ mg} = 4.3 \text{ g}$$

Don't forget, the same conversion can be done with unit fractions.

$$4300 \text{ mg} = \frac{4300 \text{ mg}}{1} \cdot 1 = \frac{4300 \cancel{\text{mg}}}{1} \cdot \frac{0.001 \text{ g}}{1 \cancel{\text{mg}}} \quad \text{Can also use } \frac{1 \text{ g}}{1000 \text{ mg}}.$$

$$= 4300 \cdot 0.001 \text{ g}$$

$$= 4.3 \text{ g} \quad \text{To multiply by 0.001, move the decimal point 3 places to the left.}$$

Answer to Concept Check:
false

To see that this is reasonable, study the diagram:

1000 mg 1000 mg 1000 mg 1000 mg 300 mg

1 g 1 g 1 g 1 g 0.3 g

4300 mg = 4.3 g

Thus, 4300 mg = 4.3 g

EXAMPLE 7 Convert 3.2 kg to grams.

Solution First we convert by using a unit fraction.

$$3.2 \text{ kg} = 3.2 \text{ kg} \cdot 1 = 3.2 \cancel{\text{kg}} \cdot \overbrace{\frac{1000 \text{ g}}{1 \cancel{\text{kg}}}}^{\text{Unit fraction}} = 3200 \text{ g}$$

Now let's list the units of mass in order from left to right and move from kilograms to grams.

kg hg dag g dg cg mg

3 units to the right

$$3.200\text{ kg} = 3200.\text{ g}$$

3 places to the right

1 kg 1 kg 1 kg 0.2 kg

1000 g 1000 g 1000 g 200 g

3.2 kg = 3200 g

□

PRACTICE 7 Convert 3.41 g to milligrams. ■

EXAMPLE 8 Convert 2.35 cg to grams.

Solution We list the units of mass in a chart and move from centigrams to grams.

kg hg dag g dg cg mg

2 units to the left

$$02.35\text{ cg} = 0.0235\text{ g}$$

2 places to the left

□

PRACTICE 8 Convert 56.2 cg to grams. ■

OBJECTIVE

4 Performing Operations on Metric System Units of Mass

Arithmetic operations can be performed with metric units of mass just as we performed operations with metric units of length. We convert each number to the same unit of mass and add, subtract, multiply, or divide as with decimals.

EXAMPLE 9 Subtract 5.4 dg from 1.6 g.

Solution We convert both numbers to decigrams or to grams before subtracting.

$$5.4\,\text{dg} = 0.54\,\text{g} \qquad \text{or} \qquad 1.6\,\text{g} = 16\,\text{dg}$$

$$\begin{array}{r} 1.60\,\text{g} \\ -0.54\,\text{g} \\ \hline 1.06\,\text{g} \end{array} \qquad\qquad \begin{array}{r} 16.0\,\text{dg} \\ -5.4\,\text{dg} \\ \hline 10.6\,\text{dg} \end{array}$$

The difference is 1.06 g or 10.6 dg. □

PRACTICE 9 Subtract 3.1 dg from 2.5 g. ■

EXAMPLE 10 **Calculating Allowable Weight in an Elevator**

An elevator has a weight limit of 1400 kg. A sign posted in the elevator indicates that the maximum capacity of the elevator is 17 persons. What is the average allowable weight for each passenger, rounded to the nearest kilogram?

Solution To solve, notice that the total weight of 1400 kilograms ÷ 17 = average weight.

$$\begin{array}{r} 82.3\,\text{kg} \approx 82\,\text{kg} \\ 17\overline{)1400.0\,\text{kg}} \\ -136\phantom{.0\,\text{kg}} \\ \hline 40\phantom{\,\text{kg}} \\ -34\phantom{\,\text{kg}} \\ \hline 60\phantom{\,\text{kg}} \\ -51\phantom{\,\text{kg}} \\ \hline 9\phantom{\,\text{kg}} \end{array}$$

Each passenger can weigh an average of 82 kg. (Recall that a kilogram is slightly over 2 pounds, so 82 kilograms is over 164 pounds.) □

PRACTICE 10 Twenty-four bags of cement weigh a total of 550 kg. Find the average weight of 1 bag, rounded to the nearest kilogram.

■

Vocabulary, Readiness & Video Check

Use the choices below to fill in each blank.

mass weight gram

1. __________ is a measure of the amount of substance in an object. This measure does not change.
2. __________ is the measure of the pull of gravity.
3. The basic unit of mass in the metric system is the __________.

Fill in these blanks with the correct number. Choices for these blanks are not shown in the list of terms above.

4. One pound equals __________ ounces.
5. One ton equals __________ pounds.

Martin-Gay Interactive Videos

See Video 6.4

Watch the section lecture video and answer the following questions.

OBJECTIVE 1

6. In Example 2, what units are used in the numerator of the unit fraction and why was this decided?

OBJECTIVE 2

7. In Example 4, explain the first step taken to solve the problem.

OBJECTIVE 3

8. In Example 5, how many places is the decimal moved and in what direction? What is the final conversion?

OBJECTIVE 4

9. What is the answer to Example 7 in decigrams?

6.4 Exercise Set MyMathLab®

Convert as indicated. See Examples 1 through 3.

1. 2 pounds to ounces

2. 5 pounds to ounces

3. 5 tons to pounds

4. 7 tons to pounds

5. 18,000 pounds to tons

6. 28,000 pounds to tons

7. 60 ounces to pounds

8. 90 ounces to pounds

9. 3500 pounds to tons

10. 11,000 pounds to tons

11. 12.75 pounds to ounces

12. 9.5 pounds to ounces

13. 4.9 tons to pounds

14. 8.3 tons to pounds

15. $4\frac{3}{4}$ pounds to ounces

16. $9\frac{1}{8}$ pounds to ounces

17. 2950 pounds to the nearest tenth of a ton

18. 51 ounces to the nearest tenth of a pound

19. $\frac{4}{5}$ oz to pounds

20. $\frac{1}{4}$ oz to pounds

21. $5\frac{3}{4}$ lb to ounces

22. $2\frac{1}{4}$ lb to ounces

23. 10 lb 1 oz to ounces

24. 7 lb 6 oz to ounces

25. 89 oz = _____ lb _____ oz

26. 100 oz = _____ lb _____ oz

Perform each indicated operation. See Examples 4 through 6.

27. 34 lb 12 oz + 18 lb 14 oz

28. 6 lb 10 oz + 10 lb 8 oz

29. 3 tons 1820 lb + 4 tons 930 lb

30. 1 ton 1140 lb + 5 tons 1200 lb

31. 5 tons 1050 lb − 2 tons 875 lb

32. 4 tons 850 lb − 1 ton 260 lb

33. 12 lb 4 oz − 3 lb 9 oz

34. 45 lb 6 oz − 26 lb 10 oz

35. 5 lb 3 oz × 6

36. 2 lb 5 oz × 5

37. 6 tons 1500 lb ÷ 5

38. 5 tons 400 lb ÷ 4

Convert as indicated. See Examples 7 and 8.

39. 500 g to kilograms

40. 820 g to kilograms

41. 4 g to milligrams

42. 9 g to milligrams

43. 25 kg to grams

44. 18 kg to grams

45. 48 mg to grams

46. 112 mg to grams

47. 6.3 g to kilograms

48. 4.9 g to kilograms

49. 15.14 g to milligrams

50. 16.23 g to milligrams

51. 6.25 kg to grams

52. 3.16 kg to grams

53. 35 hg to centigrams

54. 4.26 cg to dekagrams

Perform each indicated operation. Remember to insert units when writing your answers. See Examples 9 and 10.

55. 3.8 mg + 9.7 mg

56. 41.6 g + 9.8 g

57. 205 mg + 5.61 g

58. 2.1 g + 153 mg

59. 9 g − 7150 mg

60. 6.13 g − 418 mg

61. 1.61 kg − 250 g

62. 4 kg − 2410 g

63. 5.2 kg × 2.6

64. 4.8 kg × 9.3

65. 17 kg ÷ 8

66. 8.25 g ÷ 6

MIXED PRACTICE

Solve. Remember to insert units when writing your answers. For Exercises 67 through 74, complete the chart. See Examples 1 through 10.

	Object	*Tons*	*Pounds*	*Ounces*
67.	Statue of Liberty—weight of copper sheeting	100		
68.	Statue of Liberty—weight of steel	125		
69.	 A 12-inch cube of osmium (heaviest metal)		1345	
70.	 A 12-inch cube of lithium (lightest metal)		32	

	Object	*Grams*	*Kilograms*	*Milligrams*	*Centigrams*
71.	Capsule of amoxicillin (antibiotic)			500	
72.	Tablet of Topamax (epilepsy and migraine uses)			25	
73.	A six-year-old boy		21		
74.	A golf ball	45			

75. A can of 7-Up weighs 336 grams. Find the weight in kilograms of 24 cans.

76. Guy Green normally weighs 73 kg, but he lost 2800 grams after being sick with the flu. Find Guy's new weight.

77. Sudafed is a decongestant that comes in two strengths. Regular strength contains 60 mg of medication. Extra strength contains 0.09 g of medication. How much extra medication is in the extra-strength tablet?

78. A small can of Planters sunflower seeds weighs 177 g. If each can contains 6 servings, find the weight of one serving.

79. Doris Johnson has two open containers of rice. If she combines 1 lb 10 oz from one container with 3 lb 14 oz from the other container, how much total rice does she have?

80. Dru Mizel maintains the records of the amount of coal delivered to his department in the steel mill. In January, 3 tons 1500 lb were delivered. In February, 2 tons 1200 lb were delivered. Find the total amount delivered in these two months.

81. Carla Hamtini was amazed when she grew a 28-lb 10-oz zucchini in her garden, but later she learned that the heaviest zucchini ever grown weighed 64 lb 8 oz in Llanharry, Wales, by B. Lavery in 1990. How far below the record weight was Carla's zucchini? (*Source: Guinness World Records*)

82. The heaviest baby born in good health weighed an incredible 22 lb 8 oz. He was born in Italy in September 1955. How much heavier is this than a 7-lb 12-oz baby? (*Source: Guinness World Records*)

83. The smallest baby born in good health weighed only 8.6 ounces, less than a can of soda. She was born in Chicago in December 2004. How much lighter was she than an average baby, who weighs about 7 lb 8 ounces?

84. A large bottle of Hire's Root Beer weighs 1900 grams. If a carton contains 6 large bottles of root beer, find the weight in kilograms of 5 cartons.

85. Three milligrams of preservatives are added to a 0.5-kg box of dried fruit. How many milligrams of preservatives are in 3 cartons of dried fruit if each carton contains 16 boxes?

86. One box of Swiss Miss Cocoa Mix weighs 0.385 kg, but 39 grams of this weight is the packaging. Find the actual weight of the cocoa in 8 boxes.

87. A carton of 12 boxes of Quaker Oats Oatmeal weighs 6.432 kg. Each box includes 26 grams of packaging material. What is the actual weight of the oatmeal in the carton?

88. The supermarket prepares hamburger in 85-gram market packages. When Leo Gonzalas gets home, he divides the package in half before refrigerating the meat. How much will each package weigh?

89. The Shop 'n Bag supermarket chain ships hamburger meat by placing 10 packages of hamburger in a box, with each package weighing 3 lb 4 oz. How much will 4 boxes of hamburger weigh?

90. The Quaker Oats Company ships its 1-lb 2-oz boxes of oatmeal in cartons containing 12 boxes of oatmeal. How much will 3 such cartons weigh?

91. A carton of Del Monte Pineapple weighs 55 lb 4 oz, but 2 lb 8 oz of this weight is due to packaging. Find the actual weight of the pineapple in 4 cartons.

92. The Hormel Corporation ships cartons of canned ham weighing 43 lb 2 oz each. Of this weight, 3 lb 4 oz is due to packaging. Find the actual weight of the ham found in 3 cartons.

REVIEW AND PREVIEW

Write each fraction as a decimal. See Section R.3.

93. $\frac{4}{25}$

94. $\frac{3}{5}$

95. $\frac{7}{8}$

96. $\frac{3}{16}$

CONCEPT EXTENSIONS

Determine whether the measurement in each statement is reasonable.

97. The doctor prescribed a pill containing 2 kg of medication.

98. A full-grown cat weighs approximately 15 g.

99. A bag of flour weighs 4.5 kg.

100. A staple weighs 15 mg.

101. A professor weighs less than 150 g.

102. A car weighs 2000 mg.

103. Use a unit other than centigram and write a mass that is equivalent to 25 centigrams. (*Hint:* There are many possibilities.)

104. Use a unit other than pound and write a weight that is equivalent to 4000 pounds. (*Hint:* There are many possibilities.)

True or false? See the second Concept Check in this section.

105. A kilogram is larger than a gram.

106. A decigram is larger than a milligram.

107. Why is the decimal point moved to the right when grams are converted to milligrams?

108. To change 8 pounds to ounces, multiply by 16. Why is this the correct procedure?

6.5 Capacity

OBJECTIVES

1 Define U.S. Units of Capacity and Convert from One Unit to Another.

2 Perform Arithmetic Operations on U.S. Units of Capacity.

3 Define Metric Units of Capacity and Convert from One Unit to Another.

4 Perform Arithmetic Operations on Metric Units of Capacity.

OBJECTIVE

1 Defining and Converting U.S. System Units of Capacity

Units of **capacity** are generally used to measure liquids. The number of gallons of gasoline needed to fill a gas tank in a car, the number of cups of water needed in a bread recipe, and the number of quarts of milk sold each day at a supermarket are all examples of using units of capacity. The following summary shows equivalencies between units of capacity:

U.S. Units of Capacity

8 fluid ounces (fl oz) = 1 cup (c)

2 cups = 1 pint (pt)

2 pints = 1 quart (qt)

4 quarts = 1 gallon (gal)

Just as with units of length and weight, we can form unit fractions to convert between different units of capacity. For instance,

$$\frac{2\text{ c}}{1\text{ pt}} = \frac{1\text{ pt}}{2\text{ c}} = 1 \quad \text{and} \quad \frac{2\text{ pt}}{1\text{ qt}} = \frac{1\text{ qt}}{2\text{ pt}} = 1$$

EXAMPLE 1 Convert 9 quarts to gallons.

Solution We multiply by the unit fraction $\frac{1 \text{ gal}}{4 \text{ qt}}$.

$$\begin{aligned} 9 \text{ qt} &= \frac{9 \text{ qt}}{1} \cdot 1 \\ &= \frac{9 \cancel{\text{qt}}}{1} \cdot \frac{1 \text{ gal}}{4 \cancel{\text{qt}}} \\ &= \frac{9 \text{ gal}}{4} \\ &= 2\frac{1}{4} \text{ gal} \end{aligned}$$

Thus, 9 quarts is the same as $2\frac{1}{4}$ gallons, as shown in the diagram:

□

PRACTICE

1 Convert 43 pints to quarts. ■

EXAMPLE 2 Convert 14 cups to quarts.

Solution Our equivalency table contains no direct conversion from cups to quarts. However, from this table we know that

$$1 \text{ qt} = 2 \text{ pt} = \frac{2 \text{ pt}}{1} \cdot 1 = \frac{2 \cancel{\text{pt}}}{1} \cdot \frac{2 \text{ c}}{1 \cancel{\text{pt}}} = 4 \text{ c}$$

so 1 qt = 4 c. Now we have the unit fraction $\frac{1 \text{ qt}}{4 \text{ c}}$. Thus,

$$14 \text{ c} = \frac{14 \text{ c}}{1} \cdot 1 = \frac{14 \cancel{\text{c}}}{1} \cdot \frac{1 \text{ qt}}{4 \cancel{\text{c}}} = \frac{14 \text{ qt}}{4} = \frac{7}{2} \text{ qt} \quad \text{or} \quad 3\frac{1}{2} \text{ qt}$$

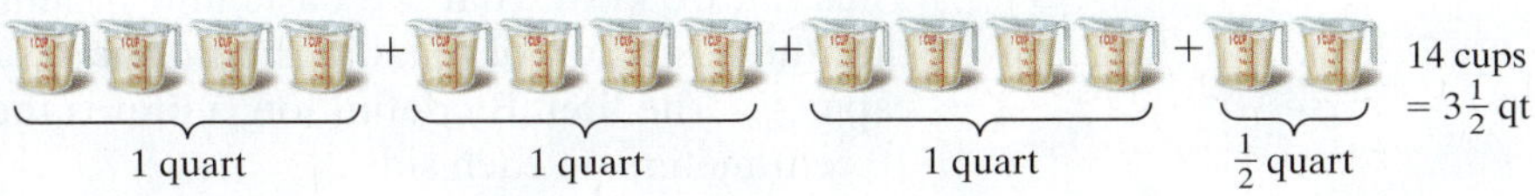

□

PRACTICE

2 Convert 26 quarts to cups. ■

> ✔ **CONCEPT CHECK**
> If 50 cups is converted to quarts, will the equivalent number of quarts be less than or greater than 50? Explain.

OBJECTIVE

2 Performing Operations on U.S. System Units of Capacity

As is true of units of length and weight, units of capacity can be added, subtracted, multiplied, and divided.

Answer to Concept Check:
less than 50

EXAMPLE 3 Subtract 3 qt from 5 gal 2 qt.

Solution To subtract, we line up similar units.

$$\begin{array}{r} 5 \text{ gal } 2 \text{ qt} \\ - \quad\quad 3 \text{ qt} \\ \hline \end{array}$$

We cannot subtract 3 qt from 2 qt. We need to borrow 1 gallon from the 5 gallons, convert it to 4 quarts, and then combine it with the 2 quarts.

$$\begin{array}{rcr} \overbrace{5 \text{ gal}}^{4 \text{ gal } + \text{ (1 gal) } 4 \text{ qt}}\ 2 \text{ qt} & = & 4 \text{ gal } 6 \text{ qt} \\ - \quad\quad 3 \text{ qt} & = & - \quad\quad 3 \text{ qt} \\ \hline & & 4 \text{ gal } 3 \text{ qt} \end{array}$$

To check, see that the sum of 4 gal 3 qt and 3 qt is 5 gal 2 qt. □

PRACTICE 3 Subtract 2 qt from 1 gal 1 qt. ■

EXAMPLE 4 **Finding the Amount of Water in an Aquarium**

An aquarium contains 6 gal 3 qt of water. If 2 gal 2 qt of water is added, what is the total amount of water in the aquarium?

Solution

$$\begin{array}{rcr} \text{beginning water} & \rightarrow & 6 \text{ gal } 3 \text{ qt} \\ + \quad \text{water added} & \rightarrow & + 2 \text{ gal } 2 \text{ qt} \\ \hline \text{total water} & \rightarrow & 8 \text{ gal } 5 \text{ qt} \end{array}$$

Since 5 qt = 1 gal 1 qt, we have

$$= \underbrace{8 \text{ gal}}_{8 \text{ gal}} + \underbrace{1 \text{ gal } 1 \text{ qt}}_{5 \text{ qt}}$$
$$= 9 \text{ gal } 1 \text{ qt}$$

The total amount of water is 9 gal 1 qt. □

PRACTICE 4 A large oil drum contains 15 gal 3 qt of oil. How much will be in the drum if an additional 4 gal 3 qt of oil is poured into it? ■

OBJECTIVE

3 Defining and Converting Metric System Units of Capacity

Thus far, we know that the basic unit of length in the metric system is the meter and that the basic unit of mass in the metric system is the gram. What is the basic unit of capacity? The **liter.** By definition, a **liter** is the capacity or volume of a cube measuring 10 centimeters on each side.

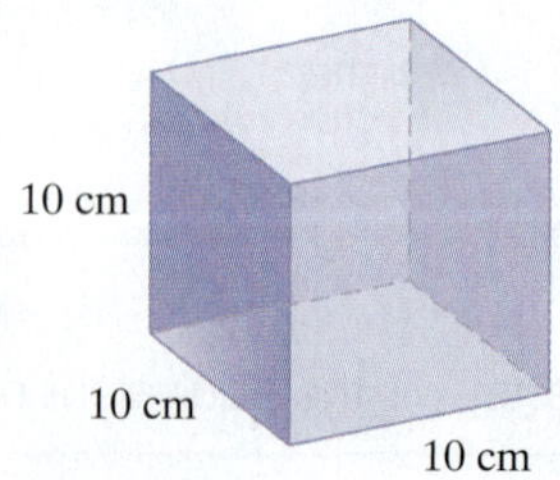

The following examples may help you get a feeling for metric capacities:

- One liter of liquid is slightly more than one quart.

1 liter 1 quart

- Many soft drinks are packaged in 2-liter bottles.

2 liters

The metric system was designed to be a consistent system. Once again, the prefixes for metric units of capacity are the same as for metric units of length and mass, as summarized in the following table:

Metric Units of Capacity
1 **kilo**liter (kl) = 1000 liters (L)
1 **hecto**liter (hl) = 100 L
1 **deka**liter (dal) = 10 L
1 liter (L) = 1 L
1 **deci**liter (dl) = 1/10 L or 0.1 L
1 **centi**liter (cl) = 1/100 L or 0.01 L
1 **milli**liter (ml) = 1/1000 L or 0.001 L

The **milliliter** and the **liter** are the two most commonly used metric units of capacity.

Converting from one unit of capacity to another involves multiplying by powers of 10 or moving the decimal point to the left or to the right. Listing units of capacity in order from largest to smallest helps to keep track of how many places to move the decimal point when converting.

Let's convert 2.6 liters to milliliters. To convert from liters to milliliters, we move along the chart 3 units to the right.

kl hl dal **L** dl cl **ml**

3 units to the right

This means that we move the decimal point 3 places to the right to convert from liters to milliliters.

$$2.600\text{ L} = 2600.\text{ml}$$

This same conversion can be done with unit fractions.

$$\begin{aligned} 2.6\text{ L} &= \frac{2.6\text{ L}}{1} \cdot 1 \\ &= \frac{2.6\text{ \cancel{L}}}{1} \cdot \frac{1000\text{ ml}}{1\text{ \cancel{L}}} \\ &= 2.6 \cdot 1000\text{ ml} \\ &= 2600\text{ ml} \end{aligned}$$

To multiply by 1000, move the decimal point 3 places to the right.

To visualize the result, study the diagram below:

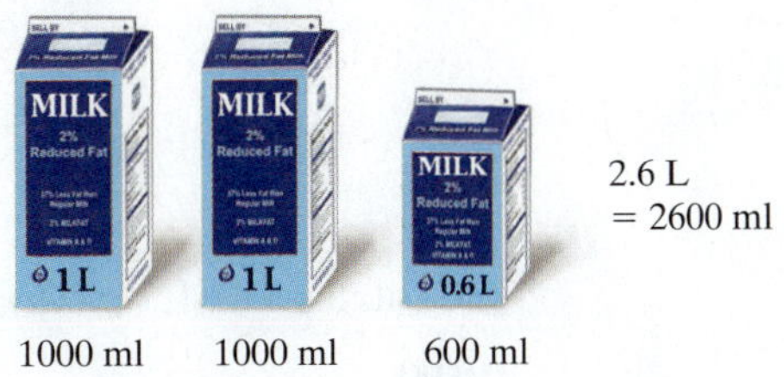

Thus, 2.6 L = 2600 ml.

EXAMPLE 5 Convert 3210 ml to liters.

Solution Let's use the unit fraction method first.

$$3210 \text{ ml} = \frac{3210 \text{ ml}}{1} \cdot 1 = 3210 \cancel{\text{ml}} \cdot \overbrace{\frac{1 \text{ L}}{1000 \cancel{\text{ml}}}}^{\text{Unit fraction}} = 3.21 \text{ L}$$

Now let's list the unit measures in order from left to right and move from milliliters to liters.

kl hl dal L dl cl ml

3 units to the left

3210 ml = 3.210 L, the same results as before and shown below in the diagram.

3 places to the left

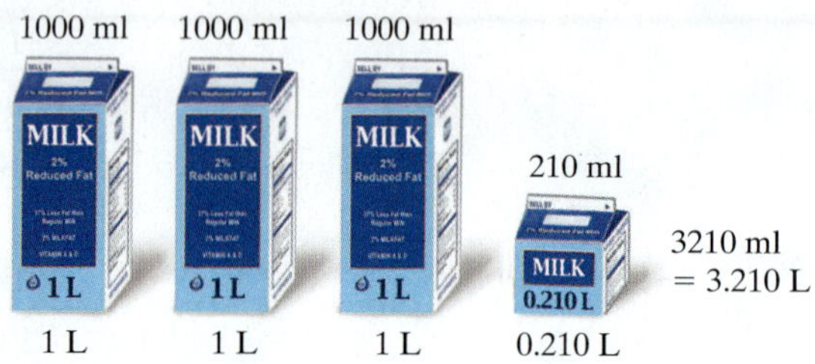

PRACTICE 5 Convert 2100 ml to liters.

EXAMPLE 6 Convert 0.185 dl to milliliters.

Solution We list the unit measures in order from left to right and move from deciliters to milliliters.

kl hl dal L dl cl ml

2 units to the right

0.185 dl = 18.5 ml

2 places to the right

PRACTICE 6 Convert 2.13 dal to liters.

OBJECTIVE

4 Performing Operations on Metric System Units of Capacity

As was true for length and weight, arithmetic operations involving metric units of capacity can also be performed. Make sure that the metric units of capacity are the same before adding, subtracting, multiplying, or dividing.

EXAMPLE 7 Add 2400 ml to 8.9 L.

Solution We must convert both to liters or both to milliliters before adding the capacities together.

2400 ml = 2.4 L or 8.9 L = 8900 ml

$$\begin{array}{r} 2.4 \text{ L} \\ +\ 8.9 \text{ L} \\ \hline 11.3 \text{ L} \end{array} \qquad \begin{array}{r} 2400 \text{ ml} \\ +\ 8900 \text{ ml} \\ \hline 11{,}300 \text{ ml} \end{array}$$

The total is 11.3 L or 11,300 ml. They both represent the same capacity.

PRACTICE 7 Add 1250 ml to 2.9 L.

✓ CONCEPT CHECK
How could you estimate the following operation? Subtract 950 ml from 7.5 L.

EXAMPLE 8 **Finding the Amount of Medication a Person Has Received**

A patient hooked up to an IV unit in the hospital is to receive 12.5 ml of medication every hour. How much medication does the patient receive in 3.5 hours?

Solution We multiply 12.5 ml by 3.5.

medication per hour	→	12.5 ml
× hours	→	× 3.5
total medication		625
		3750
		43.75 ml

The patient receives 43.75 ml of medication. □

Answer to Concept Check:
950 ml = 0.95 L; round 0.95 to 1;
7.5 − 1 = 6.5 L

PRACTICE 8 If 28.6 L of water can be pumped every minute, how much water can be pumped in 85 minutes? ■

Vocabulary, Readiness & Video Check

Use the choices below to fill in each blank. Some choices may be used more than once.

cups	pints	liter
quarts	fluid ounces	capacity

1. Units of ______________ are generally used to measure liquids.

2. The basic unit of capacity in the metric system is the ______________.

3. One cup equals 8 ______________.

4. One quart equals 2 ______________.

5. One pint equals 2 ______________.

6. One quart equals 4 ______________.

7. One gallon equals 4 ______________.

Martin-Gay Interactive Videos

See Video 6.5

Watch the section lecture video and answer the following questions.

OBJECTIVE 1
8. Complete this statement based on Example 1: When using a unit fraction, we are not changing the ______, we are changing the ______.

OBJECTIVE 2
9. In Example 4, explain the first step taken to solve the exercise.

OBJECTIVE 3
10. In Example 5, how many places is the decimal moved and in what direction? What is the final conversion?

OBJECTIVE 4
11. What is the answer to Example 7 in dekaliters?

6.5 Exercise Set MyMathLab®

Convert each measurement as indicated. See Examples 1 and 2.

1. 32 fluid ounces to cups
2. 16 quarts to gallons
3. 8 quarts to pints
4. 9 pints to quarts
5. 14 quarts to gallons
6. 11 cups to pints
7. 80 fluid ounces to pints
8. 18 pints to gallons
9. 2 quarts to cups
10. 3 pints to fluid ounces
11. 120 fluid ounces to quarts
12. 20 cups to gallons
13. 42 cups to quarts
14. 7 quarts to cups
15. $4\frac{1}{2}$ pints to cups
16. $6\frac{1}{2}$ gallons to quarts
17. 5 gal 3 qt to quarts
18. 4 gal 1 qt to quarts
19. $\frac{1}{2}$ cup to pints
20. $\frac{1}{2}$ pint to quarts
21. 58 qt = _____ gal _____ qt
22. 70 qt = _____ gal _____ qt
23. 39 pt = _____ gal _____ qt _____ pt
24. 29 pt = _____ gal _____ qt _____ pt
25. $2\frac{3}{4}$ gallons to pints
26. $3\frac{1}{4}$ quarts to cups

Perform each indicated operation. See Examples 3 and 4.

27. 5 gal 3 qt + 7 gal 3 qt
28. 2 gal 2 qt + 9 gal 3 qt
29. 1 c 5 fl oz + 2 c 7 fl oz
30. 2 c 3 fl oz + 2 c 6 fl oz
31. 3 gal − 1 gal 3 qt
32. 2 pt − 1 pt 1 c
33. 3 gal 1 qt − 1 qt 1 pt
34. 3 qt 1 c − 1 c 4 fl oz
35. 8 gal 2 qt × 2
36. 6 gal 1 pt × 2
37. 9 gal 2 qt ÷ 2
38. 5 gal 6 fl oz ÷ 2

Convert as indicated. See Examples 5 and 6.

39. 5 L to milliliters
40. 8 L to milliliters
41. 0.16 L to kiloliters
42. 0.127 L to kiloliters
43. 5600 ml to liters
44. 1500 ml to liters
45. 3.2 L to centiliters
46. 1.7 L to centiliters
47. 410 L to kiloliters
48. 250 L to kiloliters
49. 64 ml to liters
50. 39 ml to liters
51. 0.16 kl to liters
52. 0.48 kl to liters
53. 3.6 L to milliliters
54. 1.9 L to milliliters

Perform each indicated operation. Remember to insert units when writing your answers. See Examples 7 and 8.

55. 3.4 L + 15.9 L
56. 18.5 L + 4.6 L
57. 2700 ml + 1.8 L
58. 4.6 L + 1600 ml
59. 8.6 L − 190 ml
60. 4.8 L − 283 ml
61. 17,500 ml − 0.9 L
62. 6850 ml − 0.3 L
63. 480 ml × 8
64. 290 ml × 6
65. 81.2 L ÷ 0.5
66. 5.4 L ÷ 3.6

MIXED PRACTICE

Solve. Remember to insert units when writing your answers. For Exercises 67 through 70, complete the chart.

	Capacity	Cups	Gallons	Quarts	Pints
67.	An average-size bath of water		21		
68.	A dairy cow's daily milk yield				38
69.	Your kidneys filter about this amount of blood every minute	4			
70.	The amount of water needed in a punch recipe	2			

71. Mike Schaferkotter drank 410 ml of Mountain Dew from a 2-liter bottle. How much Mountain Dew remains in the bottle?
72. The Werners' Volvo has a 54.5-L gas tank. Only 3.8 liters of gasoline still remain in the tank. How much is needed to fill it?

73. Margie Phitts added 354 ml of Prestone dry gas to the 18.6 L of gasoline in her car's tank. Find the total amount of gasoline in the tank.

74. Chris Peckaitis wishes to share a 2-L bottle of Coca-Cola equally with 7 of his friends. How much will each person get?

75. A garden tool engine requires a 30-to-1 gas-to-oil mixture. This means that $\frac{1}{30}$ of a gallon of oil should be mixed with 1 gallon of gas. Convert $\frac{1}{30}$ gallon to fluid ounces. Round to the nearest tenth.

76. Henning's Supermarket sells homemade soup in 1 qt 1 pt containers. How much soup is contained in three such containers?

77. Can 5 pt 1 c of fruit punch and 2 pt 1 c of ginger ale be poured into a 1-gal container without it overflowing?

78. Three cups of prepared Jell-O are poured into 6 dessert dishes. How many fluid ounces of Jell-O are in each dish?

79. Stanley Fisher paid $14 to fill his car with 44.3 liters of gasoline. Find the price per liter of gasoline to the nearest thousandth of a dollar.

80. A student carelessly misread the scale on a cylinder in the chemistry lab and added 40 cl of water to a mixture instead of 40 ml. Find the excess amount of water.

REVIEW AND PREVIEW

Write each fraction in simplest form. See Section R.1.

81. $\frac{20}{25}$ **82.** $\frac{75}{100}$ **83.** $\frac{27}{45}$

84. $\frac{56}{60}$ **85.** $\frac{72}{80}$ **86.** $\frac{18}{20}$

CONCEPT EXTENSIONS

Determine whether the measurement in each statement is reasonable.

87. Clair took a dose of 2 L of cough medicine to cure her cough.

88. John drank 250 ml of milk for lunch.

89. Jeannie likes to relax in a tub filled with 3000 ml of hot water.

90. Sarah pumped 20 L of gasoline into her car yesterday.

Solve. See the Concept Checks in this section.

91. If 70 pints are converted to gallons, will the equivalent number of gallons be less than or greater than 70? Explain why.

92. If 30 gallons are converted to quarts, will the equivalent number of quarts be less than or greater than 30? Explain why.

93. Explain how to estimate the following operation: Add 986 ml to 6.9 L.

94. Explain how to borrow in order to subtract 1 gal 2 qt from 3 gal 1 qt.

95. Find the number of fluid ounces in 1 gallon.

96. Find the number of fluid ounces in 1.5 gallons.

A cubic centimeter (cc) is the amount of space that a volume of 1 ml occupies. Because of this, we will say that 1 cc = 1 ml.

A common syringe is one with a capacity of 3 cc. Use the diagram and give the measurement indicated by each arrow.

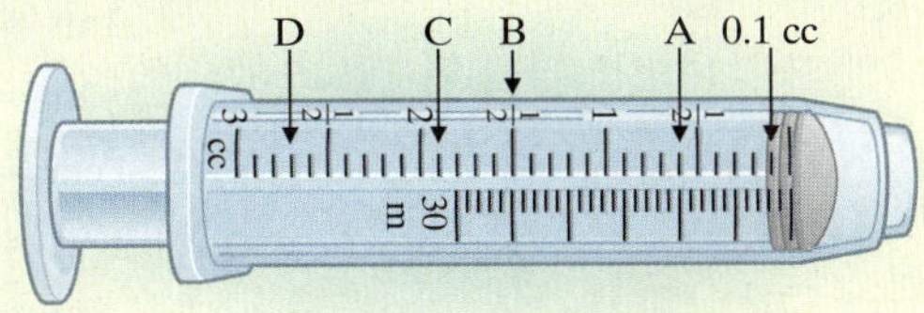

97. B **98.** A

99. D **100.** C

In order to measure small dosages, such as for insulin, u-100 syringes are used. For these syringes, 1 cc has been divided into 100 equal units (u). Use the diagram and give the measurement indicated by each arrow in units (u) and then in cubic centimeters. Use 100 u = 1 cc.

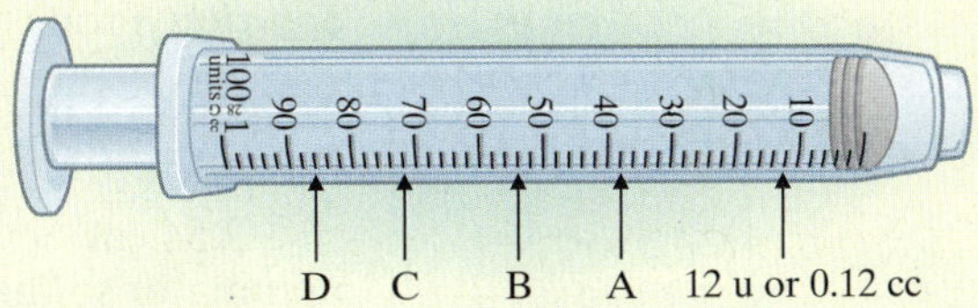

101. B **102.** A

103. D **104.** C

6.6 Temperature and Conversions Between the U.S. and Metric Systems

OBJECTIVES

1 Convert Between the U.S. and Metric Systems.

2 Convert Temperatures from Degrees Celsius to Degrees Fahrenheit.

3 Convert Temperatures from Degrees Fahrenheit to Degrees Celsius.

OBJECTIVE

1 Converting Between the U.S. and Metric Systems

The metric system probably had its beginnings in France in the 1600s, but it was the Metric Act of 1866 that made the use of this system legal (although not mandatory) in the United States. Other laws have followed that allow for a slow, but deliberate, transfer to the modernized metric system. In April 2001, for example, the U.S. Stock Exchanges completed their change to decimal trading instead of fractions. By the end of 2009, all products sold in Europe (with some exceptions) were required to have only metric units on their labels. (*Source:* U.S. Metric Association and National Institute of Standards and Technology)

You may be surprised at the number of everyday items we use that are already manufactured in metric units. We easily recognize 1 L and 2 L soda bottles, but what about the following?

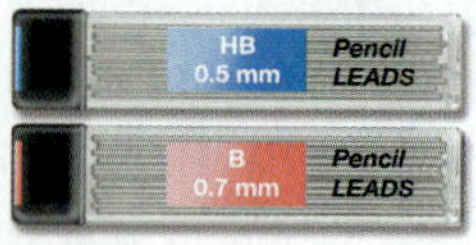

- Pencil leads (0.5 mm or 0.7 mm)
- Camera film (35 mm)
- Sporting events (5-km or 10-km races)
- Medicines (500-mg capsules)

- Labels on retail goods (dual-labeled since 1994)

1 yard

1 meter

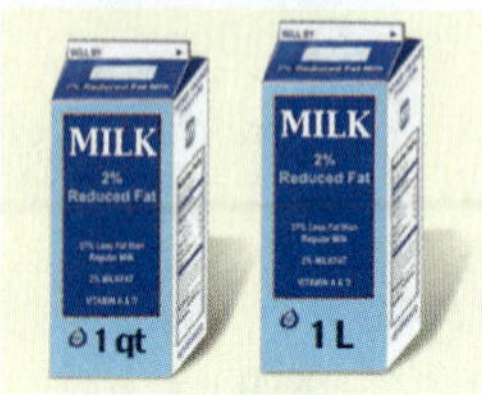

1 quart 1 liter

1 pound 1 kilogram

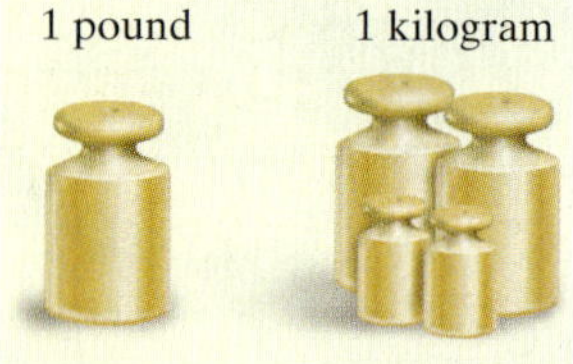

Since the United States has not completely converted to the metric system, we need to practice converting from one system to the other. Below is a table of mostly approximate conversions.

Length:		***Capacity:***		***Weight (mass):***	
Metric	U.S. System	Metric	U.S. System	Metric	U.S. System
1 m ≈	1.09 yd	1 L ≈	1.06 qt	1 kg ≈	2.20 lb
1 m ≈	3.28 ft	1 L ≈	0.26 gal	1 g ≈	0.04 oz
1 km ≈	0.62 mi	3.79 L ≈	1 gal	0.45 kg ≈	1 lb
2.54 cm =	1 in.	0.95 L ≈	1 qt	28.35 g ≈	1 oz
0.30 m ≈	1 ft	29.57 ml ≈	1 fl oz		
1.61 km ≈	1 mi				

There are many ways to perform these metric-to-U.S. conversions. We will do so by using unit fractions.

EXAMPLE 1 **Compact Discs**

Standard-sized compact discs are 12 centimeters in diameter. Convert this length to inches. Round the result to two decimal places.

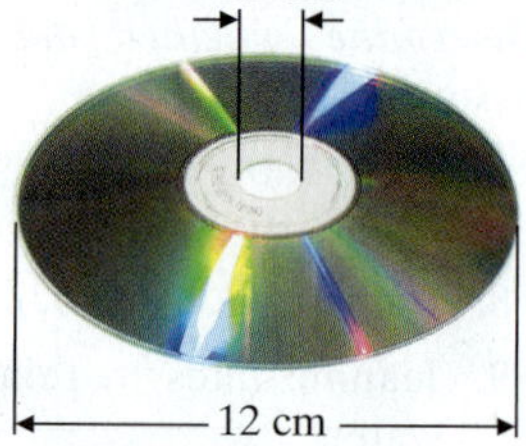

Solution From our length conversion table, we know that 2.54 cm = 1 in. This fact gives us two unit fractions: $\frac{2.54\text{ cm}}{1\text{ in.}}$ and $\frac{1\text{ in.}}{2.54\text{ cm}}$. We use the unit fraction with cm in the denominator so that these units divide out.

$$12\text{ cm} = \frac{12\text{ cm}}{1}\cdot 1 = \frac{12\ \cancel{\text{cm}}}{1}\cdot\overbrace{\frac{1\text{ in.}}{2.54\ \cancel{\text{cm}}}}^{\text{Unit fraction}} \begin{array}{l}\leftarrow \text{ Units to convert to}\\ \leftarrow \text{ Original units}\end{array}$$

$$= \frac{12\text{ in.}}{2.54}$$

$$\approx 4.72\text{ in.} \quad \text{Divide.}$$

Thus, the diameter of a standard compact disc is exactly 12 cm or approximately 4.72 inches. For a dimension this size, you can use a ruler to check. Another method is to approximate. Our result, 4.72 in., is close to 5 inches. Since 1 in. is about 2.5 cm, then 5 in. is about 5(2.5 cm) = 12.5 cm, which is close to 12 cm. □

PRACTICE

1 The center hole of a standard-sized compact disc is 1.5 centimeters in diameter. Convert this length to inches. Round the result to 2 decimal places. ■

EXAMPLE 2 **Liver**

The liver is your largest internal organ. It weighs about 3.5 pounds in a grown man. Convert this weight to kilograms. Round to the nearest tenth. (*Source: Some Body!* by Dr. Pete Rowan)

Solution

$$3.5 \text{ lb} \approx \frac{3.5 \cancel{\text{lb}}}{1} \cdot \overbrace{\frac{0.45 \text{ kg}}{1 \cancel{\text{lb}}}}^{\text{Unit fraction}} = 3.5(0.45 \text{ kg}) \approx 1.6 \text{ kg}$$

Thus 3.5 pounds is approximately 1.6 kilograms. From the table of conversions, we know that 1 kg ≈ 2.2 lb. So that means 0.5 kg ≈ 1.1 lb and after adding, we have 1.5 kg ≈ 3.3 lb. Our result is reasonable. □

PRACTICE 2 A full-grown human heart weighs about 8 ounces. Convert this weight to grams. If necessary, round your result to the nearest tenth of a gram. ■

EXAMPLE 3 **Postage Stamp**

Australia converted to the metric system in 1973. In that year, four postage stamps were issued to publicize this conversion. One such stamp is shown. Let's check the mathematics on the stamp by converting 7 fluid ounces to milliliters. Round to the nearest hundred.

Solution $$7 \text{ fl oz} \approx \frac{7 \cancel{\text{fl oz}}}{1} \cdot \overbrace{\frac{29.57 \text{ ml}}{1 \cancel{\text{fl oz}}}}^{\text{Unit fraction}} = 7(29.57 \text{ ml}) = 206.99 \text{ ml}$$

Rounded to the nearest hundred, 7 fl oz ≈ 200 ml. □

PRACTICE 3 Convert 237 ml to fluid ounces. Round to the nearest whole fluid ounce. ■

Now that we have practiced converting between two measurement systems, let's practice converting between two temperature scales.

Temperature When Gabriel Fahrenheit and Anders Celsius independently established units for temperature scales, each based his unit on the heat of water the moment it boils compared to the moment it freezes. One degree Celsius is $\frac{1}{100}$ of the difference in heat. One degree Fahrenheit is $\frac{1}{180}$ of the difference in heat. Celsius arbitrarily labeled the temperature at the freezing point at 0°C, making the boiling point 100°C; Fahrenheit labeled the freezing point 32°F, making the boiling point 212°F. Water boils at 212°F or 100°C.

By comparing the two scales in the figure, we see that a 20°C day is as warm as a 68°F day. Similarly, a sweltering 104°F day in the Mojave desert corresponds to a 40°C day.

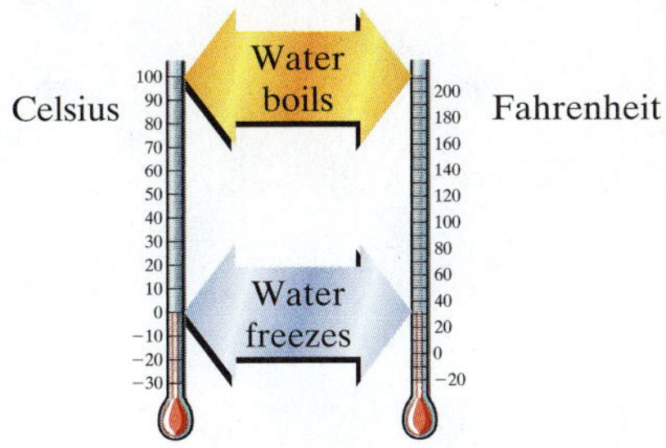

✓ CONCEPT CHECK

Which of the following statements is correct? Explain.

a. 6°C is below the freezing point of water.

b. 6°F is below the freezing point of water.

Answer to Concept Check:
b

OBJECTIVE

2 Converting Degrees Celsius to Degrees Fahrenheit

To convert from Celsius temperatures to Fahrenheit temperatures, see the box below. In this box, we use the symbol F to represent degrees Fahrenheit and the symbol C to represent degrees Celsius.

Converting Celsius to Fahrenheit

$$F = \frac{9}{5}C + 32 \quad \text{or} \quad F = 1.8C + 32$$

(To convert to Fahrenheit temperature, multiply the Celsius temperature by $\frac{9}{5}$ or 1.8, and then add 32.)

EXAMPLE 4 Convert 15°C to degrees Fahrenheit.

Solution

$$\begin{aligned} F &= \frac{9}{5}C + 32 \\ &= \frac{9}{5} \cdot 15 + 32 && \text{Replace C with 15.} \\ &= 27 + 32 && \text{Simplify.} \\ &= 59 && \text{Add.} \end{aligned}$$

Thus, 15°C is equivalent to 59°F. □

PRACTICE

4 Convert 60°C to degrees Fahrenheit. ■

EXAMPLE 5 Convert 29°C to degrees Fahrenheit.

Solution

$$\begin{aligned} F &= 1.8\,C + 32 \\ &= 1.8 \cdot 29 + 32 && \text{Replace C with 29.} \\ &= 52.2 + 32 && \text{Multiply 1.8 by 29.} \\ &= 84.2 && \text{Add.} \end{aligned}$$

Therefore, 29°C is the same as 84.2°F. □

PRACTICE

5 Convert 32°C to degrees Fahrenheit. ■

OBJECTIVE

3 Converting Degrees Fahrenheit to Degrees Celsius

To convert from Fahrenheit temperatures to Celsius temperatures, see the box below. The symbol C represents degrees Celsius and the symbol F represents degrees Fahrenheit.

Converting Fahrenheit to Celsius

$$C = \frac{5}{9}(F - 32)$$

(To convert to Celsius temperature, subtract 32 from the Fahrenheit temperature, and then multiply by $\frac{5}{9}$.)

EXAMPLE 6 Convert 59°F to degrees Celsius.

Solution We evaluate the formula $C = \frac{5}{9}(F - 32)$ when F is 59.

$$\begin{aligned} C &= \frac{5}{9}(F - 32) \\ &= \frac{5}{9}\cdot(59 - 32) && \text{Replace F with 59.} \\ &= \frac{5}{9}\cdot(27) && \text{Subtract inside parentheses.} \\ &= 15 && \text{Multiply.} \end{aligned}$$

Therefore, 59°F is the same temperature as 15°C. □

PRACTICE 6 Convert 68°F to degrees Celsius. ■

EXAMPLE 7 Convert 114°F to degrees Celsius. If necessary, round to the nearest tenth of a degree.

Solution

$$\begin{aligned} C &= \frac{5}{9}(F - 32) \\ &= \frac{5}{9}(114 - 32) && \text{Replace F with 114.} \\ &= \frac{5}{9}\cdot(82) && \text{Subtract inside parentheses.} \\ &\approx 45.6 && \text{Multiply.} \end{aligned}$$

Therefore, 114°F is approximately 45.6°C. □

PRACTICE 7 Convert 113°F to degrees Celsius. If necessary, round to the nearest tenth of a degree. ■

EXAMPLE 8 **Body Temperature**

Normal body temperature is 98.6°F. What is this temperature in degrees Celsius?

Solution We evaluate the formula $C = \frac{5}{9}(F - 32)$ when F is 98.6.

$$\begin{aligned} C &= \frac{5}{9}(F - 32) \\ &= \frac{5}{9}(98.6 - 32) && \text{Replace F with 98.6.} \\ &= \frac{5}{9}\cdot(66.6) && \text{Subtract inside parentheses.} \\ &= 37 && \text{Multiply.} \end{aligned}$$

Therefore, normal body temperature is 37°C. □

PRACTICE 8 During a bout with the flu, Albert's temperature reaches 102.8°F. What is his temperature measured in degrees Celsius? Round to the nearest tenth of a degree. ■

Answer to Concept Check:
She used the conversion for Celsius to Fahrenheit instead of Fahrenheit to Celsius.

✔ CONCEPT CHECK

Clarissa must convert 40°F to degrees Celsius. What is wrong with her work shown below?

$$\begin{aligned} F &= 1.8\cdot C + 32 \\ F &= 1.8\cdot 40 + 32 \\ F &= 72 + 32 \\ F &= 104 \end{aligned}$$

Vocabulary, Readiness & Video Check

Martin-Gay Interactive Videos

See Video 6.6

Watch the section lecture video and answer the following questions.

OBJECTIVE 1

1. Write two conversions that may be used to solve Example 2.

OBJECTIVE 2

2. Why isn't 0.1125 kg the final answer to Example 3?
3. Which version of the formula is used to solve Example 4? What is the replacement value for C?

OBJECTIVE 3

4. In Example 5, what is the replacement value for F? What is the final conversion?

6.6 Exercise Set MyMathLab®

Note: Because approximations are used, your answers may vary slightly from the answers given in the back of the book.

Convert as indicated. If necessary, round answers to two decimal places. See Examples 1 through 3.

1. 756 milliliters to fluid ounces
2. 18 liters to quarts
3. 86 inches to centimeters
4. 86 miles to kilometers
5. 1000 grams to ounces
6. 100 kilograms to pounds
7. 93 kilometers to miles
8. 9.8 meters to feet
9. 14.5 liters to gallons
10. 150 milliliters to fluid ounces
11. 30 pounds to kilograms
12. 15 ounces to grams

Fill in the chart. Give exact answers or round to one decimal place. See Examples 1 through 3.

		Meters	Yards	Centimeters	Feet	Inches
13.	The height of a woman				5	
14.	Statue of Liberty length of nose	1.37				
15.	Leaning Tower of Pisa		60			
16.	Blue whale		36			

Solve. If necessary, round answers to two decimal places. See Examples 1 through 3.

17. The balance beam for female gymnasts is 10 centimeters wide. Convert this width to inches.

18. In men's gymnastics, the rings are 250 centimeters from the floor. Convert this height to inches, then to feet.

19. In many states, the maximum speed limit for recreational vehicles is 50 miles per hour. Convert this to kilometers per hour.

20. In some states, the speed limit is 70 miles per hour. Convert this to kilometers per hour.

21. Ibuprofen comes in 200-milligram tablets. Convert this to ounces. (Round your answer to this exercise to 3 decimal places.)

22. Vitamin C tablets come in 500-milligram caplets. Convert this to ounces.

The 70-meter-diameter antenna is the largest and most sensitive Deep Space Network antenna. See the Chapter Opener and answer Exercises 23 through 26.

70-Meter Antenna

23. Convert 70 meters to feet.

24. The Deep Space Network sites also have a 26-meter antenna. Convert 26 meters to feet.

25. The 70-meter-diameter antenna can track a spacecraft traveling more than 16 billion kilometers from Earth. Convert this distance to miles.

26. The dish reflector and the mount atop the concrete pedestal of the 70-meter antenna weigh nearly 2.7 million kilograms. Convert this number to tons.

27. A stone is a unit in the British customary system. Use the conversion 14 pounds = 1 stone to check the equivalencies in this 1973 Australian stamp. Is 100 kilograms approximately 15 stone 10 pounds?

28. Convert 5 feet 11 inches to centimeters and check the conversion on this 1973 Australian stamp. Is it correct?

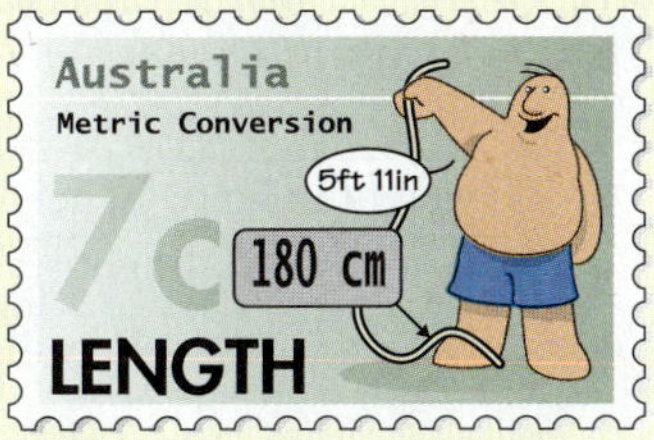

29. The Monarch butterfly migrates annually between the northern United States and central Mexico. The trip is about 4500 km long. Convert this to miles.

30. There is a species of African termite that builds nests up to 18 ft high. Convert this to meters.

31. A $3\frac{1}{2}$-inch diskette is not really $3\frac{1}{2}$ inches. To find its actual width, convert this measurement to centimeters, then to millimeters. Round the result to the nearest ten.

32. The average two-year-old is 84 centimeters tall. Convert this to feet and inches.

33. For an average adult, the weight of the right lung is greater than the weight of the left lung. If the right lung weighs 1.5 pounds and the left lung weighs 1.25 pounds, find the difference in grams. (*Source: Some Body!*)

34. The skin of an average adult weighs 9 pounds and is the heaviest organ. Find the weight in grams. (*Source: Some Body!*)

35. A fast sneeze has been clocked at about 167 kilometers per hour. Convert this to miles per hour. Round to the nearest whole.

36. A Boeing 747 has a cruising speed of about 980 kilometers per hour. Convert this to miles per hour. Round to the nearest whole.

37. The General Sherman giant sequoia tree has a diameter of about 8 meters at its base. Convert this to feet. (*Source: Fantastic Book of Comparisons*)

38. The largest crater on the near side of the moon is Billy Crater. It has a diameter of 303 kilometers. Convert this to miles. (*Source: Fantastic Book of Comparisons*)

39. The total length of the track on a CD is about 4.5 kilometers. Convert this to miles. Round to the nearest whole mile.

40. The distance between Mackinaw City, Michigan, and Cheyenne, Wyoming, is 2079 kilometers. Convert this to miles. Round to the nearest whole mile.

41. A doctor orders a dosage of 5 ml of medicine every 4 hours for 1 week. How many fluid ounces of medicine should be purchased? Round up to the next whole fluid ounce.

42. A doctor orders a dosage of 12 ml of medicine every 6 hours for 10 days. How many fluid ounces of medicine should be purchased? Round up to the next whole fluid ounce.

Without actually converting, choose the most reasonable answer.

43. This math book has a height of about _____.

a. 28 mm **b.** 28 cm
c. 28 m **d.** 28 km

44. A mile is _____ a kilometer.

a. shorter than
b. longer than
c. the same length as

45. A liter has _____ capacity than a quart.

a. less
b. greater
c. the same

46. A foot is _____ a meter.

a. shorter than

b. longer than

c. the same length as

47. A kilogram weighs _____ a pound.

a. the same as

b. less than

c. greater than

48. A football field is 100 yards, which is about _____.

a. 9 m **b.** 90 m

c. 900 m **d.** 9000 m

49. An $8\frac{1}{2}$-ounce glass of water has a capacity of about _____.

a. 250 L **b.** 25 L

c. 2.5 L **d.** 250 ml

50. A 5-gallon gasoline can has a capacity of about _____.

a. 19 L **b.** 1.9 L

c. 19 ml **d.** 1.9 ml

51. The weight of an average man is about _____.

a. 700 kg **b.** 7 kg

c. 0.7 kg **d.** 70 kg

52. The weight of a pill is about _____.

a. 200 kg **b.** 20 kg

c. 2 kg **d.** 200 mg

MIXED PRACTICE

Convert as indicated. When necessary, round to the nearest tenth of a degree. See Examples 4 through 8.

53. 77°F to degrees Celsius

54. 86°F to degrees Celsius

55. 104°F to degrees Celsius

56. 140°F to degrees Celsius

57. 50°C to degrees Fahrenheit

58. 80°C to degrees Fahrenheit

59. 115°C to degrees Fahrenheit

60. 225°C to degrees Fahrenheit

61. 20°F to degrees Celsius

62. 26°F to degrees Celsius

63. 142.1°F to degrees Celsius

64. 43.4°F to degrees Celsius

65. 92°C to degrees Fahrenheit

66. 75°C to degrees Fahrenheit

67. 12.4°C to degrees Fahrenheit

68. 48.6°C to degrees Fahrenheit

69. The hottest temperature ever recorded in the United States, in Death Valley, was 134°F. Convert this temperature to degrees Celsius. (*Source:* National Climatic Data Center)

70. The hottest temperature ever recorded in the United States in January was 95°F in Los Angeles. Convert this temperature to degrees Celsius. (*Source:* National Climatic Data Center)

71. A weather forecaster in Caracas predicts a high temperature of 27°C. Find this measurement in degrees Fahrenheit.

72. While driving to work, Alan Olda notices a temperature of 18°C flash on the local bank's temperature display. Find the corresponding temperature in degrees Fahrenheit.

73. At Mack Trucks' headquarters, the room temperature is to be set at 70°F, but the thermostat is calibrated in degrees Celsius. Find the temperature to be set.

74. The computer room at Merck, Sharp, and Dohm is normally cooled to 66°F. Find the corresponding temperature in degrees Celsius.

75. In a European cookbook, a recipe requires the ingredients for caramels to be heated to 118°C, but the cook has access only to a Fahrenheit thermometer. Find the temperature in degrees Fahrenheit that should be used to make the caramels.

76. The ingredients for divinity should be heated to 127°C, but the candy thermometer that Myung Kim has is calibrated to degrees Fahrenheit. Find how hot he should heat the ingredients.

77. The temperature of Earth's core is estimated to be 4000°C. Find the corresponding temperature in degrees Fahrenheit.

78. In 2012, the average temperature of Earth's surface was 58.3°F. Convert this temperature to degrees Celsius. (*Source:* NASA)

REVIEW AND PREVIEW

Perform the indicated operations. *See Section 1.3.*

79. $6 \cdot 4 + 5 \div 1$

80. $10 \div 2 + 9(8)$

81. $3[(1 + 5) \cdot (8 - 6)]$

82. $5[(18 - 8) - 9]$

CONCEPT EXTENSIONS

Determine whether the measurement in each statement is reasonable.

83. A 72°F room feels comfortable.

84. Water heated to 110°F will boil.

85. Josiah has a fever if a thermometer shows his temperature to be 40°F.

86. An air temperature of 20°F on a Vermont ski slope can be expected in the winter.

87. When the temperature is 30°C outside, an overcoat is needed.

88. An air-conditioned room at 60°C feels quite chilly.

89. Barbara has a fever when a thermometer records her temperature at 40°C.

90. Water cooled to 32°C will freeze.

Body surface area (BSA) is often used to calculate dosages for some drugs. BSA is calculated in square meters using a person's weight and height.

$$\text{BSA} = \sqrt{\frac{(\text{weight in kg}) \times (\text{height in cm})}{3600}}$$

For Exercises 91 through 96, calculate the BSA for each person. Round to the nearest hundredth. You will need to use the square root key on your calculator.

91. An adult whose height is 182 cm and weight is 90 kg

92. An adult whose height is 157 cm and weight is 63 kg

93. A child whose height is 40 in. and weight is 50 kg (*Hint:* Don't forget to first convert inches to centimeters.)

94. A child whose height is 26 in. and weight is 13 kg (*Hint:* Don't forget to first convert inches to centimeters.)

95. An adult whose height is 60 in. and weight is 150 lb

96. An adult whose height is 69 in. and weight is 172 lb

97. In February 2010, at the Brookhaven National Laboratory in Long Island, NY, the highest temperature produced in a laboratory was achieved. This temperature was 7,200,000,000°F. Convert this temperature to degrees Celsius. Round your answer to the nearest million degrees. (*Source: Guinness World Records*)

98. The hottest-burning substance known is carbon subnitride. Its flame at one atmospheric pressure reaches 9010°F. Convert this temperature to degrees Celsius. (*Source: Guinness World Records*)

99. In your own words, describe how to convert from degrees Celsius to degrees Fahrenheit.

100. In your own words, describe how to convert from degrees Fahrenheit to degrees Celsius.

Chapter 6 Vocabulary Check

Fill in each blank with one of the words or phrases listed below.

transversal	line segment	obtuse	straight	adjacent	right
acute	perimeter	vertical	supplementary	ray	angle
vertex	mass	unit fractions	gram	weight	meter
volume	area	line	complementary	liter	

1. ____________________ is a measure of the pull of gravity.

2. ____________________ is a measure of the amount of substance in an object. This measure does not change.

3. The basic unit of length in the metric system is the ____________________.

4. To convert from one unit of length to another, ____________________ may be used.

5. The ____________________ is the basic unit of mass in the metric system.

6. The ____________________ is the basic unit of capacity in the metric system.

7. A(n) ____________________ is a piece of a line with two endpoints.

8. Two angles that have a sum of 90° are called ____________________ angles.

9. A(n) ____________________ is a set of points extending indefinitely in two directions.

10. The ____________________ of a polygon is the distance around the polygon.

11. A(n) ____________________ is made up of two rays that share the same endpoint. The common endpoint is called the ____________________.

12. ____________________ measures the amount of surface of a region.

13. A(n) ____________________ is a part of a line with one endpoint. A ray extends indefinitely in one direction.

14. A line that intersects two or more lines at different points is called a(n) ____________________.

15. An angle that measures 180° is called a(n) ____________________ angle.

16. The measure of the space of a solid is called its ____________________.

17. When two lines intersect, four angles are formed. Two of these angles that are opposite each other are called ____________________ angles.

18. Two of the angles from Exercise **17** that share a common side are called ____________________ angles.

19. An angle whose measure is between 90° and 180° is called a(n) ____________________ angle.

20. An angle that measures 90° is called a(n) ____________________ angle.

21. An angle whose measure is between 0° and 90° is called a(n) ____________________ angle.

22. Two angles that have a sum of 180° are called ____________________ angles.

Chapter 6 Highlights

DEFINITIONS AND CONCEPTS | **EXAMPLES**

Section 6.1 Lines and Angles

A **line** is a set of points extending indefinitely in two directions. A line has no width or height, but it does have length. We name a line by any two of its points.

Line AB or $\overleftrightarrow{AB}$

A **line segment** is a piece of a line with two endpoints.

Line segment AB or $\overline{AB}$

A **ray** is a part of a line with one endpoint. A ray extends indefinitely in one direction.

Ray AB or $\overrightarrow{AB}$

An **angle** is made up of two rays that share the same endpoint. The common endpoint is called the **vertex.**

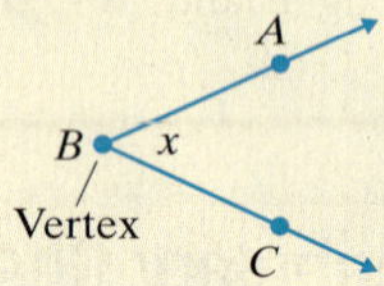

Section 6.2 Perimeter, Area, and Volume

Perimeter Formulas

Rectangle: $P = 2l + 2w$

Square: $P = 4s$

Triangle: $P = a + b + c$

Circumference of a Circle: $C = 2\pi r$ or $C = \pi d$

where $\pi \approx 3.14$ or $\pi \approx \frac{22}{7}$

Find the perimeter of the rectangle.

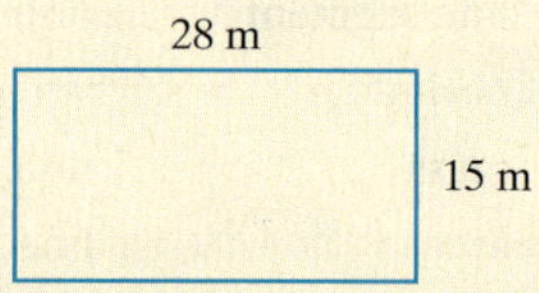

$$\begin{aligned} P &= 2l + 2w \\ &= 2 \cdot 28 \text{ meters} + 2 \cdot 15 \text{ meters} \\ &= 56 \text{ meters} + 30 \text{ meters} \\ &= 86 \text{ meters} \end{aligned}$$

The perimeter is 86 meters.

Area Formulas

Rectangle: $A = lw$

Square: $A = s^2$

Triangle: $A = \frac{1}{2}bh$

Parallelogram: $A = bh$

Trapezoid: $A = \frac{1}{2}(b + B)h$

Circle: $A = \pi r^2$

Find the area of the square.

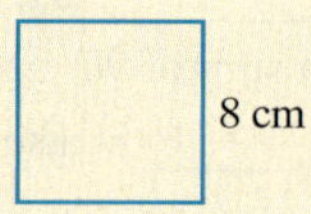

$$\begin{aligned} A &= s^2 \\ &= (8 \text{ centimeters})^2 \\ &= 64 \text{ square centimeters} \end{aligned}$$

The area of the square is 64 square centimeters.

Volume Formulas

Rectangular Solid: $V = lwh$

Cube: $V = s^3$

Sphere: $V = \frac{4}{3}\pi r^3$

Right Circular Cylinder: $V = \pi r^2 h$

Cone: $V = \frac{1}{3}\pi r^2 h$

Square-Based Pyramid: $V = \frac{1}{3}s^2 h$

Find the volume of the sphere. Use $\frac{22}{7}$ for π.

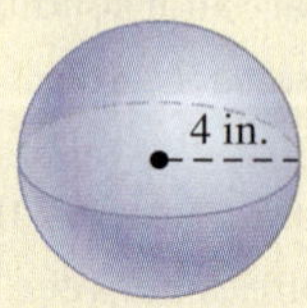

$$\begin{aligned} V &= \frac{4}{3}\pi r^3 \\ &\approx \frac{4}{3} \cdot \frac{22}{7} \cdot (4 \text{ inches})^3 \\ &= \frac{4 \cdot 22 \cdot 64}{3 \cdot 7} \text{ cubic inches} \\ &= \frac{5632}{21} \text{ or } 268\frac{4}{21} \text{ cubic inches} \end{aligned}$$

DEFINITIONS AND CONCEPTS | **EXAMPLES**

Section 6.3 Linear Measurement

To convert from one unit of length to another, multiply by a **unit fraction** in the form

$$\frac{\text{units to convert to}}{\text{original units}}$$

Length: U.S. System of Measurement

$$12 \text{ inches (in.)} = 1 \text{ foot (ft)}$$
$$3 \text{ feet} = 1 \text{ yard (yd)}$$
$$5280 \text{ feet} = 1 \text{ mile (mi)}$$

Length: Metric System of Measurement

The basic unit of length in the metric system is the **meter.** A meter is slightly longer than a yard.

Metric Units of Length
1 **kilo**meter (km) = 1000 meters (m)
1 **hecto**meter (hm) = 100 m
1 **deka**meter (dam) = 10 m
1 meter (m) = 1 m
1 **deci**meter (dm) = 1/10 m or 0.1 m
1 **centi**meter (cm) = 1/100 m or 0.01 m
1 **milli**meter (mm) = 1/1000 m or 0.001 m

$$\frac{12 \text{ inches}}{1 \text{ foot}}, \frac{1 \text{ foot}}{12 \text{ inches}}, \frac{3 \text{ feet}}{1 \text{ yard}}$$

Convert 6 feet to inches.

$$6 \text{ ft} = \frac{6 \text{ ft}}{1} \cdot 1$$
$$= \frac{6 \cancel{\text{ft}}}{1} \cdot \frac{12 \text{ in.}}{1 \cancel{\text{ft}}} \begin{matrix} \leftarrow \text{units to convert to} \\ \leftarrow \text{original units} \end{matrix}$$
$$= 6 \cdot 12 \text{ in.}$$
$$= 72 \text{ in.}$$

Convert 3650 centimeters to meters.

$$3650 \text{ cm} = 3650 \text{ cm} \cdot 1$$
$$= \frac{3650 \cancel{\text{cm}}}{1} \cdot \frac{0.01 \text{ m}}{1 \cancel{\text{cm}}} = 36.5 \text{ m}$$

or

km hm dam m dm cm mm

2 units to the left

3650 cm = 36.5 m

2 places to the left

Section 6.4 Weight and Mass

Weight is really a measure of the pull of gravity. **Mass** is a measure of the amount of substance in an object and does not change.

Weight: U.S. System of Measurement

$$16 \text{ ounces (oz)} = 1 \text{ pound (lb)}$$
$$2000 \text{ pounds} = 1 \text{ ton}$$

Mass: Metric System of Measurement

The **gram** is the basic unit of mass in the metric system. It is the mass of water contained in a cube 1 centimeter on each side. A paper clip weighs about 1 gram.

Metric Units of Mass
1 **kilo**gram (kg) = 1000 grams (g)
1 **hecto**gram (hg) = 100 g
1 **deka**gram (dag) = 10 g
1 gram (g) = 1 g
1 **deci**gram (dg) = 1/10 g or 0.1 g
1 **centi**gram (cg) = 1/100 g or 0.01 g
1 **milli**gram (mg) = 1/1000 g or 0.001 g

Convert 5 pounds to ounces.

$$5 \text{ lb} = 5 \text{ lb} \cdot 1 = \frac{5 \cancel{\text{lb}}}{1} \cdot \frac{16 \text{ oz}}{1 \cancel{\text{lb}}} = 80 \text{ oz}$$

Convert 260 grams to kilograms.

$$260 \text{ g} = \frac{260 \text{ g}}{1} \cdot 1 = \frac{260 \cancel{\text{g}}}{1} \cdot \frac{1 \text{ kg}}{1000 \cancel{\text{g}}} = 0.26 \text{ kg}$$

or

kg hg dag g dg cg mg

3 units to the left

260 g = 0.26 kg

3 places to the left

DEFINITIONS AND CONCEPTS	EXAMPLES
Section 6.5 Capacity	
Capacity: U.S. System of Measurement 8 fluid ounces (fl oz) = 1 cup (c) 2 cups = 1 pint (pt) 2 pints = 1 quart (qt) 4 quarts = 1 gallon (gal)	Convert 5 pints to gallons. $1 \text{ gal} = 4 \text{ qt} = 8 \text{ pt}$ $5 \text{ pt} = 5 \text{ pt} \cdot 1 = \frac{5 \cancel{\text{pt}}}{1} \cdot \frac{1 \text{ gal}}{8 \cancel{\text{pt}}} = \frac{5}{8} \text{ gal}$
Capacity: Metric System of Measurement The **liter** is the basic unit of capacity in the metric system. It is the capacity or volume of a cube measuring 10 centimeters on each side. A liter of liquid is slightly more than 1 quart. *Metric Units of Capacity* 1 **kilo**liter (kl) = 1000 liters (L) 1 **hecto**liter (hl) = 100 L 1 **deka**liter (dal) = 10 L 1 liter (L) = 1 L 1 **deci**liter (dl) = 1/10 L or 0.1 L 1 **centi**liter (cl) = 1/100 L or 0.01 L 1 **milli**liter (ml) = 1/1000 L or 0.001 L	Convert 1.5 liters to milliliters. $1.5 \text{ L} = \frac{1.5 \text{ L}}{1} \cdot 1 = \frac{1.5 \cancel{\text{L}}}{1} \cdot \frac{1000 \text{ ml}}{1 \cancel{\text{L}}} = 1500 \text{ ml}$ or kl hl dal L dl cl ml 3 units to the right 1.500 L = 1500 ml 3 places to the right
Section 6.6 Temperature and Conversions Between the U.S. and Metric Systems	
To convert between systems, use approximate unit fractions. See page 432.	Convert 7 feet to meters. $7 \text{ ft} \approx \frac{7 \cancel{\text{ft}}}{1} \cdot \frac{0.30 \text{ m}}{1 \cancel{\text{ft}}} = 2.1 \text{ m}$ Convert 8 liters to quarts. $8 \text{ L} \approx \frac{8 \cancel{\text{L}}}{1} \cdot \frac{1.06 \text{ qt}}{1 \cancel{\text{L}}} = 8.48 \text{ qt}$ Convert 363 grams to ounces. $363 \text{ g} \approx \frac{363 \cancel{\text{g}}}{1} \cdot \frac{0.04 \text{ oz}}{1 \cancel{\text{g}}} = 14.52 \text{ oz}$
Celsius to Fahrenheit $F = \frac{9}{5}C + 32$ or $F = 1.8C + 32$	Convert 35°C to degrees Fahrenheit. $F = \frac{9}{5} \cdot 35 + 32 = 63 + 32 = 95$ 35°C = 95°F
Fahrenheit to Celsius $C = \frac{5}{9}(F - 32)$	Convert 50°F to degrees Celsius. $C = \frac{5}{9} \cdot (50 - 32) = \frac{5}{9} \cdot (18) = 10$ 50°F = 10°C

Chapter 6 Review

(6.1) *Classify each angle as acute, right, obtuse, or straight.*

1.

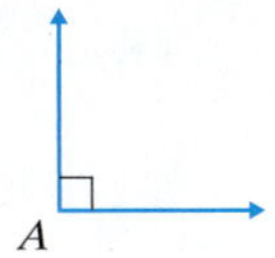

2.

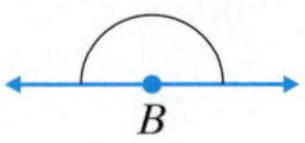

3.

4.

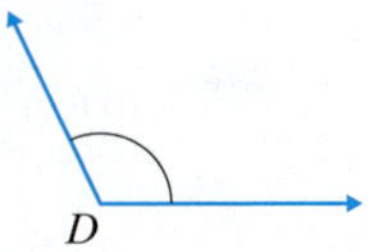

5. Find the complement of a 25° angle.

6. Find the supplement of a 105° angle.

Find the measure of angle x in each figure.

7.

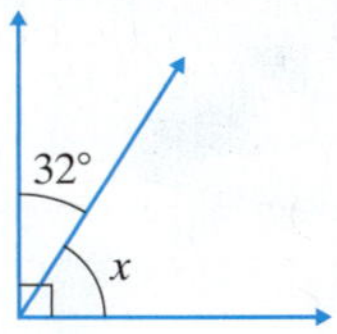

8.

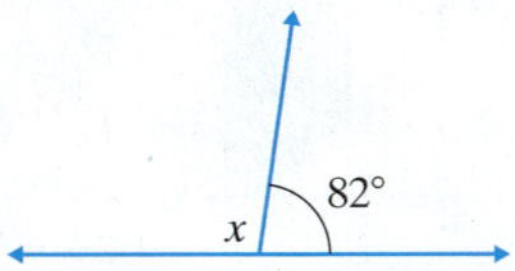

9.

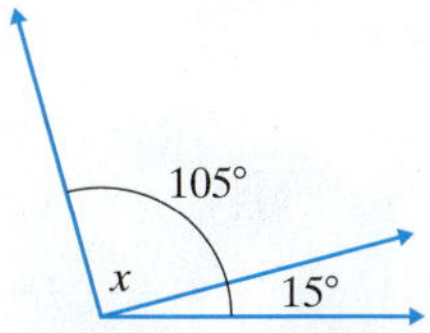

10.

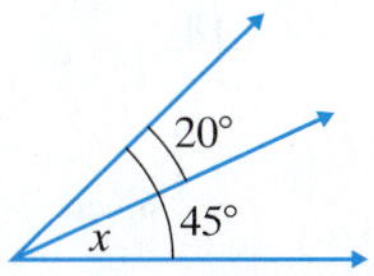

11. Identify the pairs of supplementary angles.

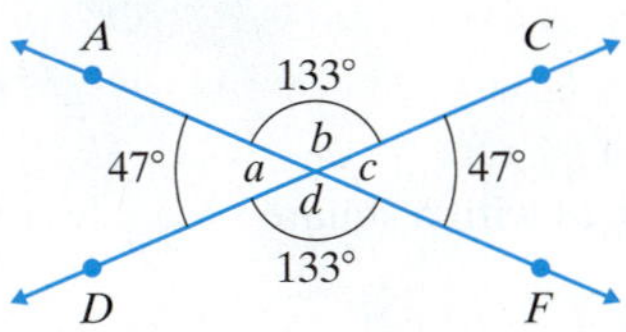

12. Identify the pairs of complementary angles.

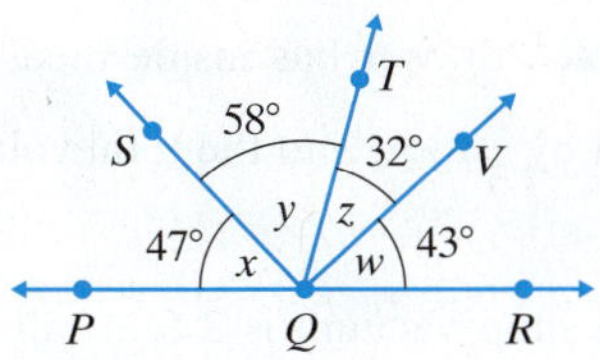

Find the measures of angles x, y, and z in each figure.

13.

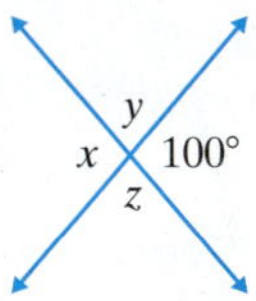

14.

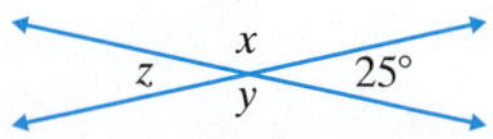

15. Given that $m \parallel n$.

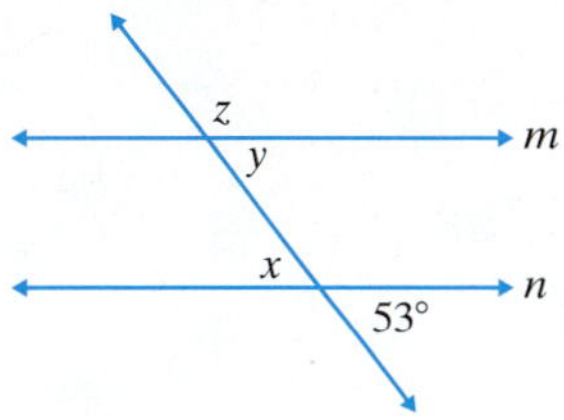

16. Given that $m \parallel n$.

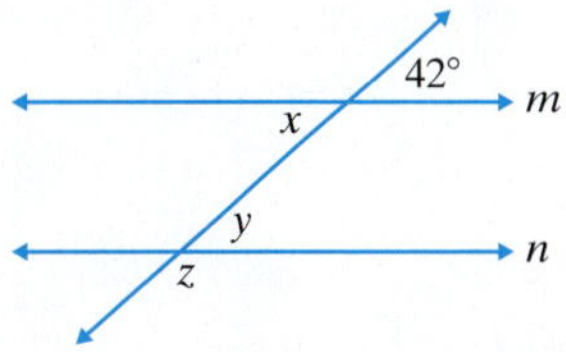

(6.2) *Find the perimeter of each figure.*

17.

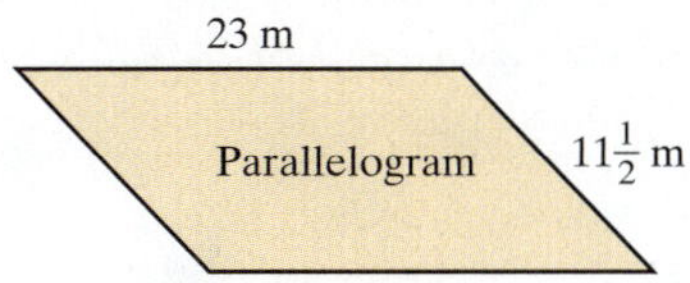

18.

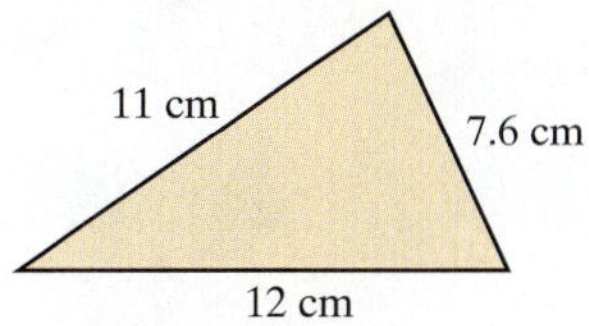

19.

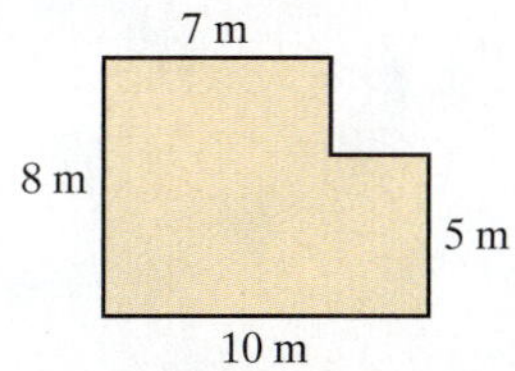

20.

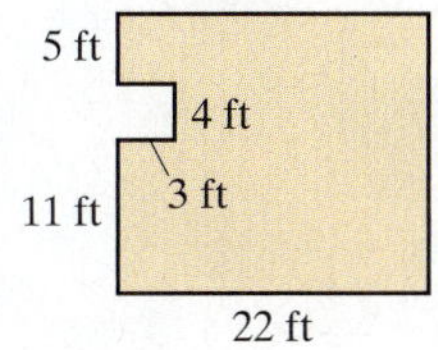

Solve.

21. Find the perimeter of a rectangular sign that measures 6 feet by 10 feet.

22. Find the perimeter of a town square that measures 110 feet on a side.

Find the circumference of each circle. Use $\pi \approx 3.14$.

23.

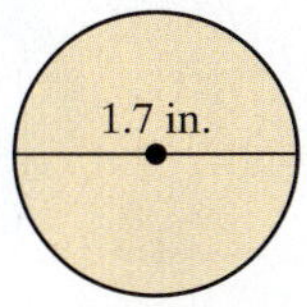

24.

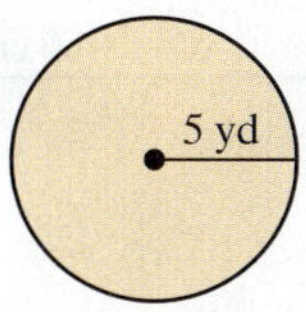

Find the area of each figure. For the circles, find the exact area and then use $\pi \approx 3.14$ to approximate the area.

25.

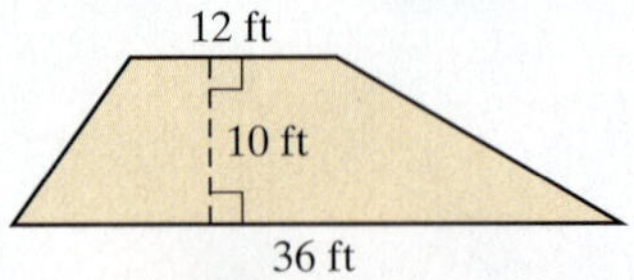

26.

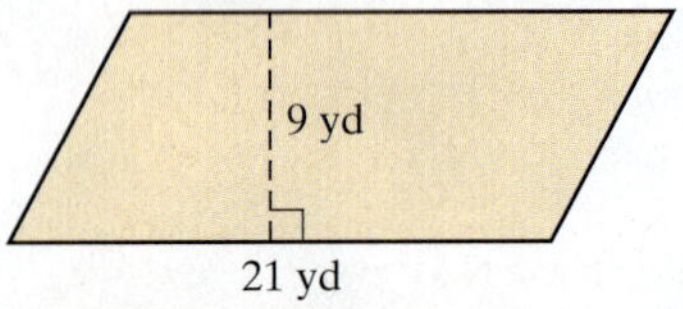

27.

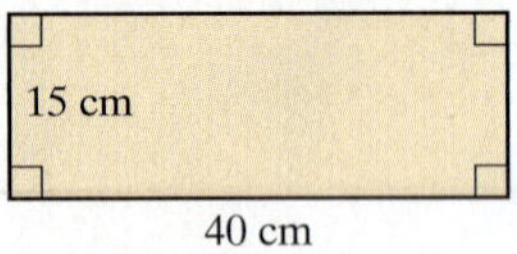

28.

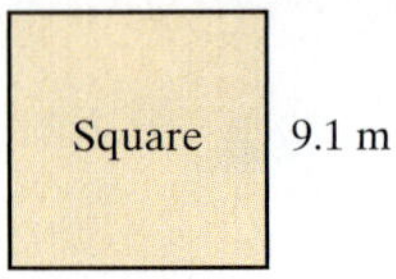

29.

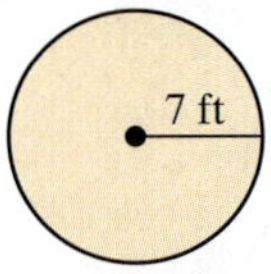

30.

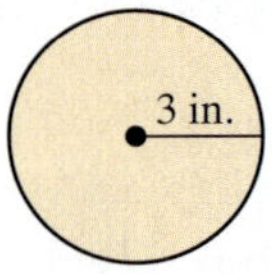

31.

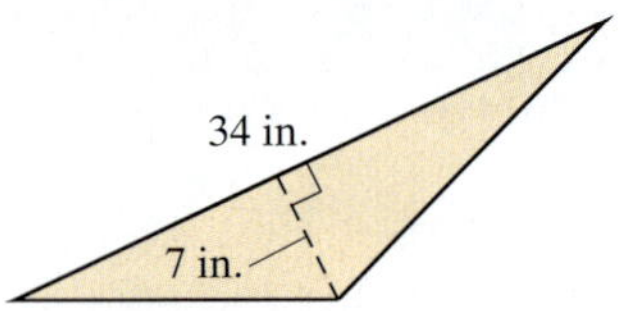

32.

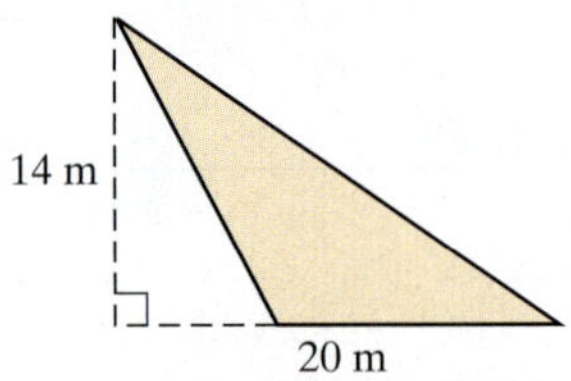

33.

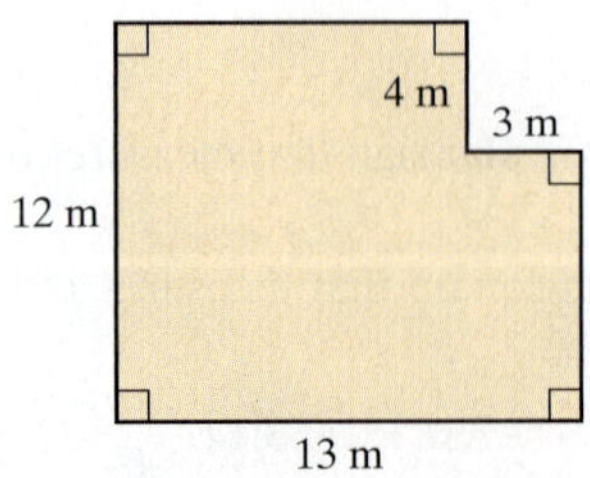

34.

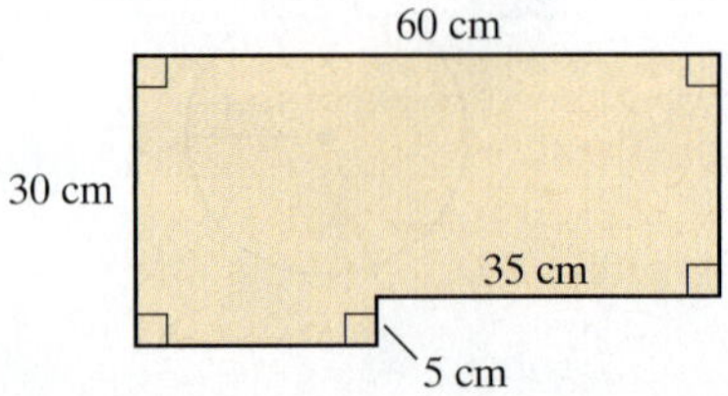

35. The amount of sealer necessary to seal a driveway depends on the area. Find the area of a rectangular driveway 36 feet by 12 feet.

36. Find how much carpet is necessary to cover the floor of the room shown.

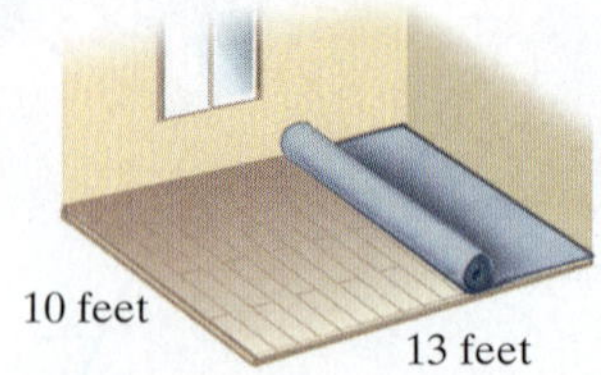

Find the volume of the solids in Exercises 37 and 38. For Exercises 39 and 40, give the exact volume and an approximation.

37.

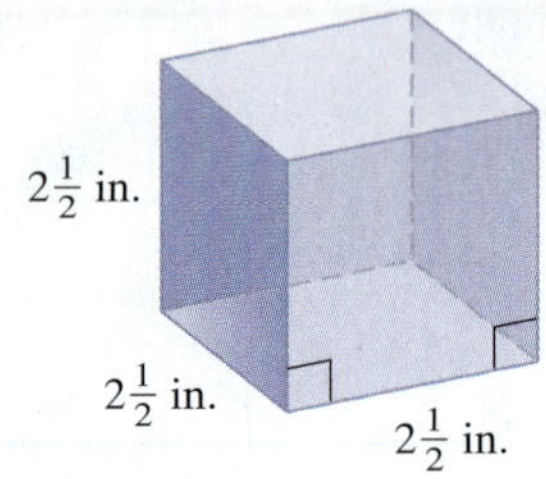

38.

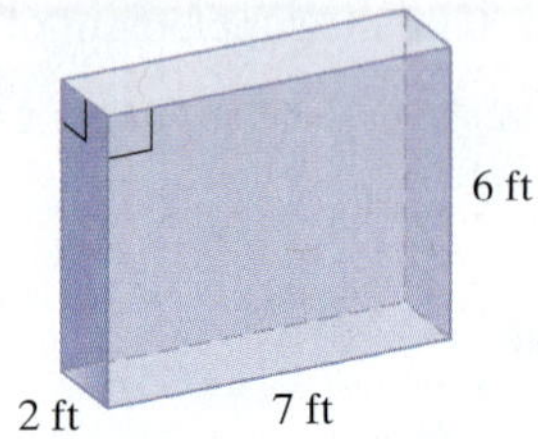

39. Use $\pi \approx 3.14$.

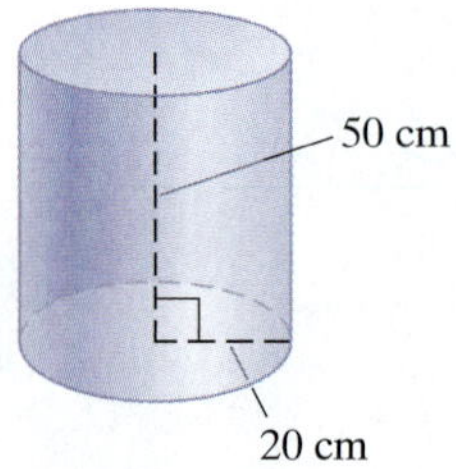

40. Use $\pi \approx \frac{22}{7}$.

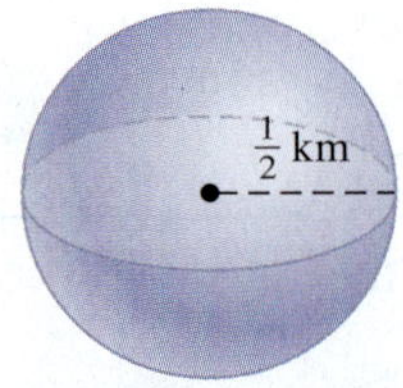

41. Find the volume of a pyramid with a square base 2 feet on a side and a height of 2 feet.

42. Approximate the volume of a tin can 8 inches high and 3.5 inches in radius. Use 3.14 for π.

43. A chest has 3 drawers. If each drawer has inside measurements of $2\frac{1}{2}$ feet by $1\frac{1}{2}$ feet by $\frac{2}{3}$ foot, find the total volume of the 3 drawers.

44. A cylindrical canister for a shop vacuum is 2 feet tall and 1 foot in *diameter*. Find its exact volume.

(6.3) *Convert.*

45. 108 in. to feet

46. 72 ft to yards

47. 1.5 mi to feet

48. $\frac{1}{2}$ yd to inches

49. 52 ft = _____ yd _____ ft

50. 46 in. = _____ ft _____ in.

51. 42 m to centimeters

52. 82 cm to millimeters

53. 12.18 mm to meters

54. 2.31 m to kilometers

Perform each indicated operation.

55. 4 yd 2 ft + 16 yd 2 ft

56. 7 ft 4 in. ÷ 2

57. 8 cm + 15 mm

58. 4 m − 126 cm

Solve.

59. A bolt of cloth contains 333 yd 1 ft of cotton ticking. Find the amount of material that remains after 163 yd 2 ft is removed from the bolt.

60. The student activities club is sponsoring a walk for hunger, and all students who participate will receive a sash with the name of the school to wear on the walk. If each sash requires 5 ft 2 in. of material and there are 50 students participating in the walk, how much material will the student activities club need?

61. The trip from El Paso, TX, to Ontario, CA, is about 1235 km each way. Four friends agree to share the driving equally. How far must each drive on this round-trip vacation?

△ **62.** The college has ordered that NO SMOKING signs be placed above the doorway of each classroom. Each sign is 0.8 m long and 30 cm wide. Find the area of each sign. (*Hint:* Recall that the area of a rectangle = width · length.)

(6.4) *Convert.*

63. 66 oz to pounds

64. 2.3 tons to pounds

65. 52 oz = _____ lb _____ oz

66. 10,300 lb = _____ tons _____ lb

67. 27 mg to grams

68. 40 kg to grams

69. 2.1 hg to dekagrams

70. 0.03 mg to decigrams

Perform each indicated operation.

71. 6 lb 5 oz − 2 lb 12 oz

72. 8 lb 6 oz × 4

73. 4.3 mg × 5

74. 4.8 kg − 4200 g

Solve.

75. Donshay Berry ordered 1 lb 12 oz of soft-center candies and 2 lb 8 oz of chewy-center candies for his party. Find the total weight of the candy ordered.

76. Four local townships jointly purchase 38 tons 300 lb of cinders to spread on their roads during an ice storm. Determine the weight of the cinders each township receives if they share the purchase equally.

(6.5) *Convert.*

77. 28 pints to quarts

78. 40 fluid ounces to cups

79. 3 qt 1 pt to pints

80. 18 quarts to cups

81. 9 pt = _____ qt _____ pt

82. 15 qt = _____ gal _____ qt

83. 3.8 L to milliliters

84. 14 hl to kiloliters

85. 30.6 L to centiliters

86. 2.45 ml to liters

Perform each indicated operation.

87. 1 qt 1 pt + 3 qt 1 pt

88. 3 gal 2 qt × 2

89. 0.946 L − 210 ml

90. 6.1 L + 9400 ml

Solve.

91. Carlos Perez prepared 4 gal 2 qt of iced tea for a block party. During the first 30 minutes of the party, 1 gal 3 qt of the tea is consumed. How much iced tea remains?

92. A recipe for soup stock calls for 1 c 4 fl oz of beef broth. How much should be used if the recipe is cut in half?

93. Each bottle of Kiwi liquid shoe polish holds 85 ml of the polish. Find the number of liters of shoe polish contained in 8 boxes if each box contains 16 bottles.

94. Ivan Miller wants to pour three separate containers of saline solution into a single vat with a capacity of 10 liters. Will 6 liters of solution in the first container combined with 1300 milliliters in the second container and 2.6 liters in the third container fit into the larger vat?

(6.6) *Note: Because approximations are used in this section, your answers may vary slightly from the answers given in the back of the book.*

Convert as indicated. If necessary, round to two decimal places.

95. 7 meters to feet

96. 11.5 yards to meters

97. 17.5 liters to gallons

98. 7.8 liters to quarts

99. 15 ounces to grams

100. 23 pounds to kilograms

101. A compact disc is 1.2 mm thick. Find the height (in inches) of 50 discs.

102. If a person weighs 82 kilograms, how many pounds is this?

Convert. Round to the nearest tenth of a degree, if necessary.

103. 42°C to degrees Fahrenheit

104. 160°C to degrees Fahrenheit

105. 41.3°F to degrees Celsius

106. 80°F to degrees Celsius

Solve. Round to the nearest tenth of a degree, if necessary.

107. A sharp dip in the jet stream caused the temperature in New Orleans to drop to 35°F. Find the corresponding temperature in degrees Celsius.

108. A recipe for meat loaf calls for a 165°C oven. Find the setting used if the oven has a Fahrenheit thermometer.

MIXED REVIEW

Find the following.

109. Find the supplement of a 72° angle.

110. Find the complement of a 1° angle.

Find the measure of angle x in each figure.

111.

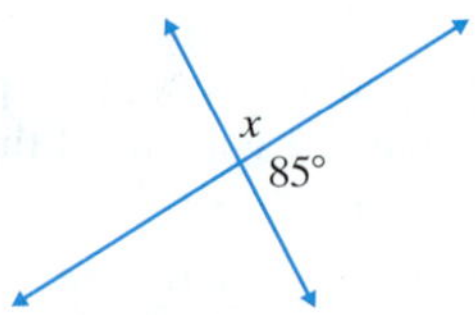

112.

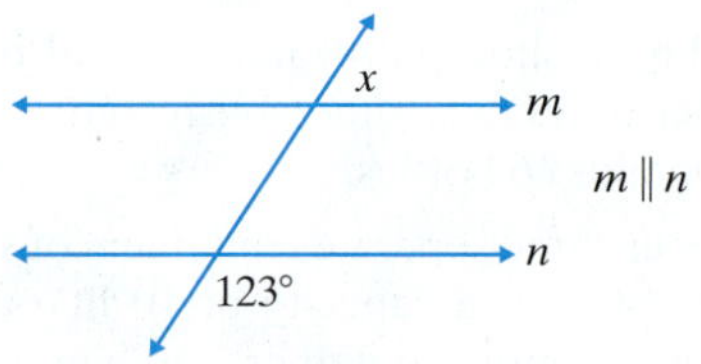

Find the perimeter of each figure.

113.

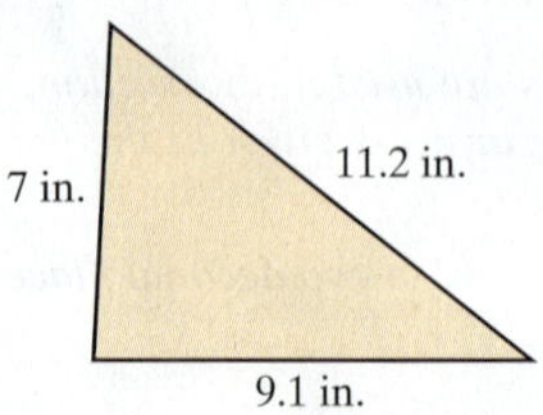

114.

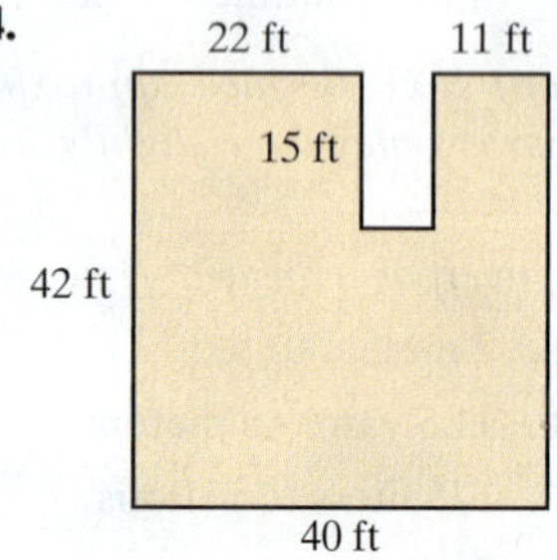

Find the area of each figure. For the circle, find the exact area and then use $\pi \approx 3.14$ to approximate the area.

115.

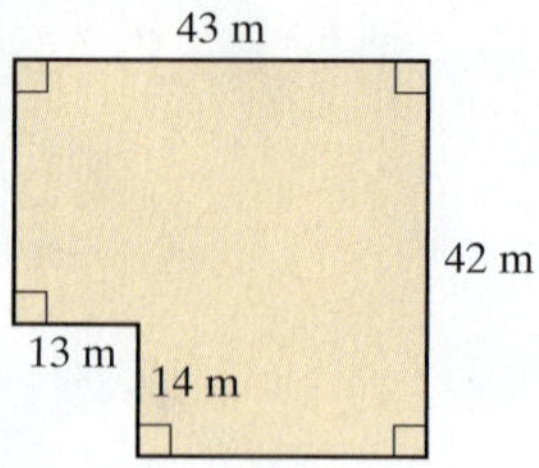

116.

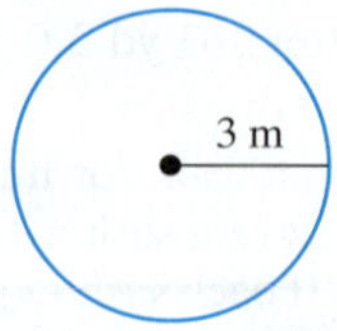

Find the volume of each solid.

117. Give an approximation using $\frac{22}{7}$ for π.

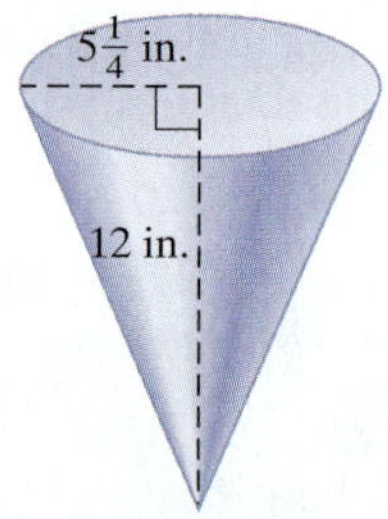

118.

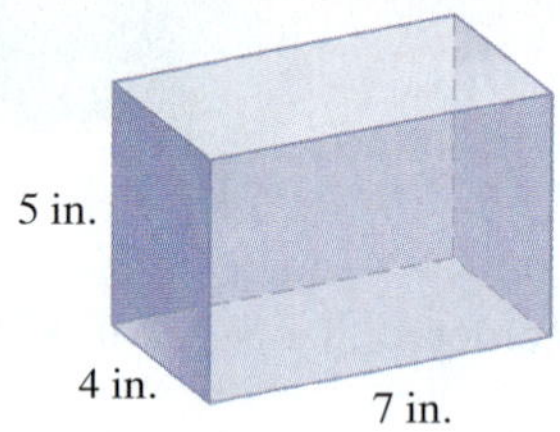

Convert the following.

119. 6.25 ft to inches

120. 8200 lb = _____ tons _____ lb

121. 5 m to centimeters

122. 286 mm to kilometers

123. 1400 mg to grams

124. 6.75 gallons to quarts

125. 86°C to degrees Fahrenheit

126. 51.8°F to degrees Celsius

Perform the indicated operations and simplify.

127. 9.3 km − 183 m

128. 35 L + 700 ml

129. 3 gal 3 qt + 4 gal 2 qt

130. 3.2 kg × 4

Chapter 6 Getting Ready for the Test

True or False. For all exercises, choose A. *True or* B. *False.*

1. The complement of a 70° angle is a 110° angle.
2. The supplement of a 50° angle is a 130° angle.
3. Area is measured in square units.
4. Perimeter is measured in cubic units.
5. A meter is slightly longer than a yard.
6. A yard equals 30 inches.
7. A ton equals 2000 pounds.
8. A millimeter is longer than a meter.
9. A quart equals 4 pints.
10. A gallon equals 4 quarts.
11. The freezing temperature of water is 32°F .
12. The boiling temperature of water is 212°C.

Chapter 6 Test

MyMathLab® YouTube

1. Find the complement of a 78° angle.
2. Find the supplement of a 124° angle.
3. Find the measure of $\angle x$.

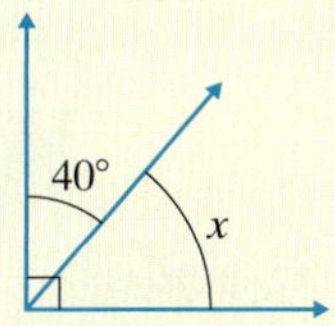

Find the measures of x, y, and z in each figure.

4.

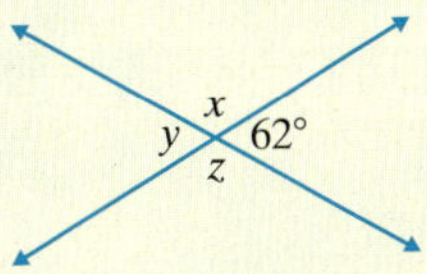

5. Given that $m \parallel n$.

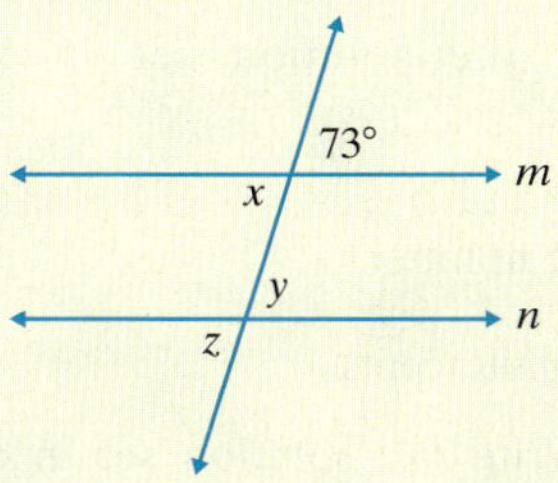

Find the unknown diameter or radius as indicated.

6.

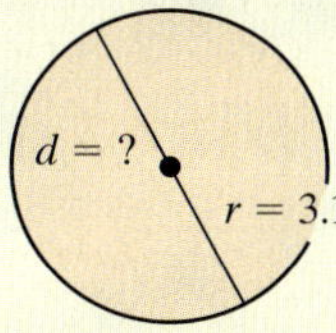

7.

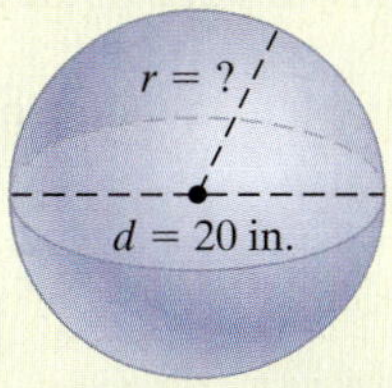

Find the perimeter (or circumference) and area of each figure. For the circle, give the exact value and an approximation using $\pi \approx 3.14$.

8.

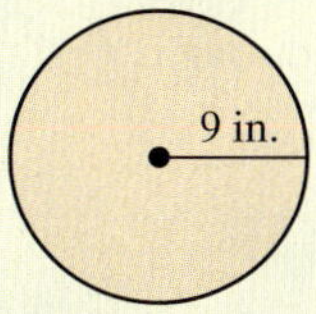

9.

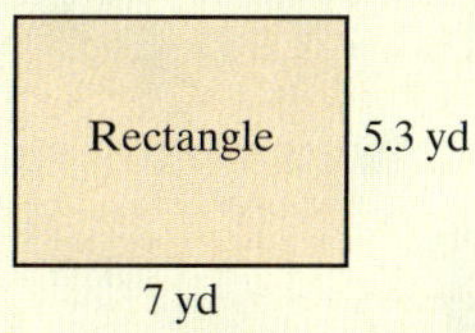

10.

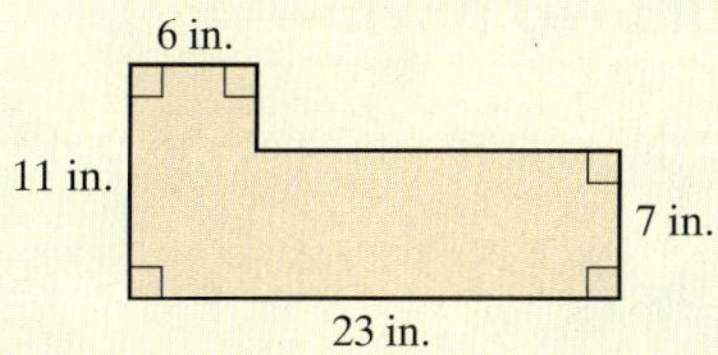

Find the volume of each solid. For the cylinder, use $\pi \approx \frac{22}{7}$.

11.

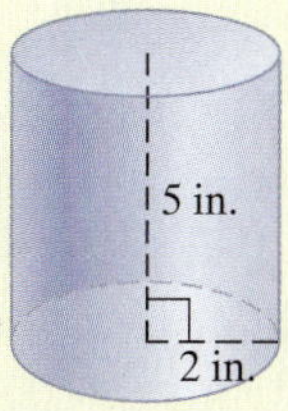

12.

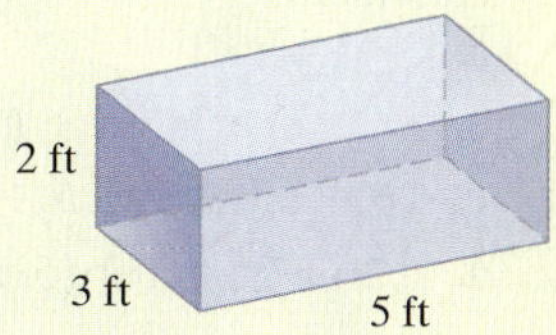

Solve.

13. Find the perimeter of a square photo with a side length of 4 inches.

14. How much soil is needed to fill a rectangular hole 3 feet by 3 feet by 2 feet?

15. Find how much baseboard is needed to go around a rectangular room that measures 18 feet by 13 feet. If baseboard costs \$1.87 per foot, also calculate the total cost needed for materials.

Convert.

16. 280 in. = _____ ft _____ in.

17. $2\frac{1}{2}$ gal to quarts

18. 30 oz to pounds

19. 2.8 tons to pounds

20. 38 pt to gallons

21. 40 mg to grams

22. 2.4 kg to grams

23. 3.6 cm to millimeters

24. 4.3 dg to grams

25. 0.83 L to milliliters

Perform each indicated operation.

26. 3 qt 1 pt + 2 qt 1 pt

27. 8 lb 6 oz − 4 lb 9 oz

28. 2 ft 9 in. × 3

29. 5 gal 2 qt ÷ 2

30. 8 cm − 14 mm

31. 1.8 km + 456 m

Convert. Round to the nearest tenth of a degree, if necessary.

32. 84°F to degrees Celsius

33. 12.6°C to degrees Fahrenheit

34. The sugar maples in front of Bette MacMillan's house are 8.4 meters tall. Because they interfere with the phone lines, the telephone company plans to remove the top third of the trees. How tall will the maples be after they are shortened?

35. A total of 15 gal 1 qt of oil has been removed from a 20-gallon drum. How much oil still remains in the container?

36. The engineer in charge of bridge construction said that the span of a certain bridge would be 88 m. But the actual construction required it to be 340 cm longer. Find the span of the bridge, in meters.

37. If 2 ft 9 in. of material is used to manufacture one scarf, how much material is needed for 6 scarves?

Chapter 6 Cumulative Review

1. Insert <, >, or = in the appropriate space to make each statement true.

a. $-1 \quad 0$ **b.** $7 \quad \frac{14}{2}$ **c.** $-5 \quad -6$

2. Evaluate.

a. 5^2 **b.** 2^5

3. Name the property or properties illustrated by each true statement.

a. $3 \cdot y = y \cdot 3$

b. $(x + 7) + 9 = x + (7 + 9)$

c. $(b + 0) + 3 = b + 3$

d. $0.2 \cdot (z \cdot 5) = 0.2 \cdot (5 \cdot z)$

e. $-2 \cdot \left(-\frac{1}{2}\right) = 1$

f. $-2 + 2 = 0$

g. $-6 \cdot (y \cdot 2) = (-6 \cdot 2) \cdot y$

4. Evaluate $y^2 - 3x$ for $x = 8$ and $y = 5$.

5. Subtract $4x - 2$ from $2x - 3$.

6. Simplify: $7 - 12 + (-5) - 2 + (-2)$

7. Solve: $7 = -5(2a - 1) - (-11a + 6)$.

8. Evaluate $2y^2 - x^2$ for $x = -7$ and $y = -3$.

9. Solve: $\frac{5}{2}x = 15$

10. Simplify: $0.4y - 6.7 + y - 0.3 - 2.6y$

11. Solve: $\frac{x}{2} - 1 = \frac{2}{3}x - 3$

12. Solve: $7(x - 2) - 6(x + 1) = 20$

13. Twice the sum of a number and 4 is the same as four times the number, decreased by 12. Find the number.

14. Solve: $5(y - 5) = 5y + 10$

Factor.

15. $2x^3 + 3x^2 - 2x - 3$

16. $3x^3 + 5x^2 - 12x - 20$

17. Solve $-2x \le -4$. Write the solution set in interval notation.

18. Five times the sum of a number and −1 is the same as 6 times the number. Find the number.

19. Graph $y = -\frac{1}{2}x$ by plotting intercepts.

20. Solve $3x + 7 \ge x - 9$. Write the solution set in interval notation.

21. Find the slope of the line through $(-1, 5)$ and $(2, -3)$.

22. Complete the table of values for $x - 3y = 3$.

x	y
	−1
3	
	2

23. Find the slope of the line whose equation is $y = \frac{3}{4}x + 6$.

24. Find the slope of a line passing through $(-1, 3)$ and $(2, -8)$.

25. Find the slope and the y-intercept of the line whose equation is $3x - 4y = 4$.

26. Find the slope and y-intercept of the line whose equation is $y = 7x$.

27. Find an equation of the line with slope -2 that passes through $(-1, 5)$. Write the equation in standard form: $Ax + By = C$.

28. Determine whether the lines are parallel, perpendicular, or neither.

$$y = 4x - 5$$
$$-4x + y = 7$$

29. Find an equation of the vertical line through $(-1, 5)$.

30. Write an equation of the line with slope -5, through $(-2, 3)$.

31. Find the domain and the range of the relation $\{(0, 2), (3, 3), (-1, 0), (3, -2)\}$.

32. If $f(x) = 5x^2 - 6$, find $f(0)$ and $f(-2)$.

33. Determine whether each relation is also a function.
a. $\{(-1, 1), (2, 3), (7, 3), (8, 6)\}$
b. $\{(0, -2), (1, 5), (0, 3), (7, 7)\}$

34. Determine whether each graph is the graph of a function.

a. **b.** **c.**

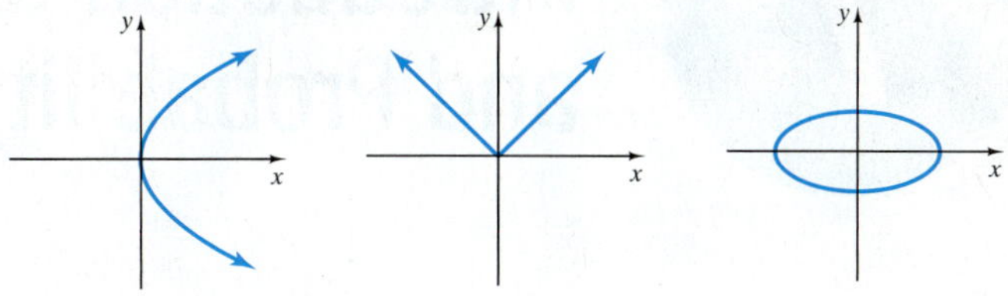

Solve each system.

35. $\begin{cases} x - 5y = -12 \\ -x + y = 4 \end{cases}$

36. $\begin{cases} 3x - 4y = 10 \\ y = 2x \end{cases}$

37. $\begin{cases} 3x - 2y = 10 \\ 4x - 3y = 15 \end{cases}$

38. $\begin{cases} x = 5y - 3 \\ x = 8y + 4 \end{cases}$

39. Graph: $2x - y \geq 3$

40. Graph: $x > 1$

CHAPTER 7

Introduction to Counting and Probability

The last section in this chapter is an introduction to probability, or the likelihood that an event will occur. For example, the probability that heads will show when a coin is tossed is $\frac{1}{2}$. Why? We know that the denominator, 2 is the number of possible outcomes (heads or tails) and we know that the numerator, 1, is the only outcome we are interested in (heads). Before we study probability, the earlier sections are devoted to counting methods so that we can count the number of possible outcomes of an experiment.

Platonic Solids

***Triangular Pyramid or* Tetrahedron**	**Cube *or Hexahedron***	**Octahedron**	**Dodecahedron**	**Icosahedron**
Four Faces (equilateral triangles)	Six Faces (cubes)	Eight Faces (equilateral triangles)	Twelve Faces (regular pentagons)	Twenty Faces (equilateral triangles)
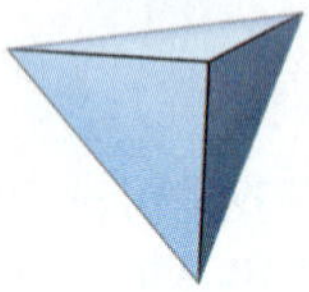	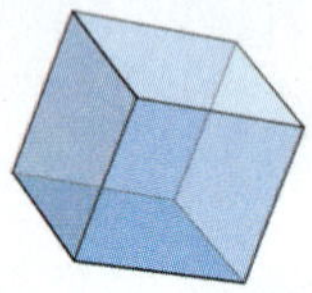	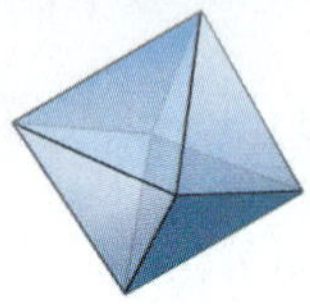	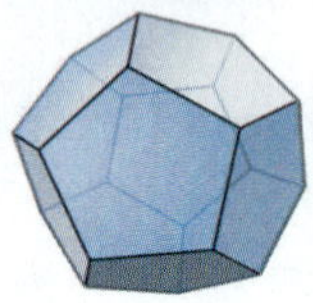	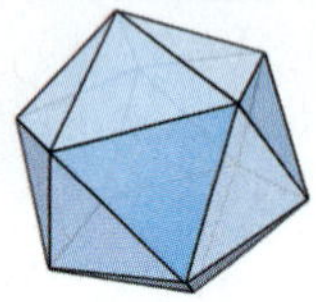

What do platonic solids have to do with probability?

A platonic solid is a solid that has congruent regular polygonal faces with the same number of faces meeting at each vertex. Although the 6-sided cube is the most common die used, each of the solids, or dice above are uniform fair dice. In other words all the dice (solids) shown above have faces with equal probability of outcome.

Below, we show the solids as dice with numbered faces. Throughout this chapter, we have probability examples of dice, mostly the 6-sided cube that is used quite often.

Four faces

Six faces

Eight faces

Twelve faces

Twenty faces

Note: The four-sided die does not have a face upward when at rest, but will have a vertex upward. Thus, the same number is printed about the three faces close to the vertex it forms. That is how a "roll" is read, and the diagram shown reads a roll of 4.

Source: Britannica.com, the evansgroupllc.com, neatorama.com

7.1 Counting and the Fundamental Counting Principle

OBJECTIVES

1 Use a Tree Diagram to Count Outcomes.

2 Use the Fundamental Counting Principle to Count Outcomes.

OBJECTIVE

1 Using a Tree Diagram

In our daily conversations, we often talk about the likelihood or **probability** of a given result occurring. For example:

The *chance* of thundershowers is 70 percent.

What are the *odds* that the New Orleans Saints will go to the Super Bowl?

What is the *probability* that you will finish cleaning your room today?

Each of these chance happenings for which the outcome is uncertain—thundershowers, the New Orleans Saints playing in the Super Bowl, and finishing cleaning your room today—is called an **experiment.** The possible results of an experiment are called **outcomes.**

In Sections 7.1, 7.2, and 7.3, we learn ways to count the number of possible outcomes. In Section 7.4, we then calculate probability.

Let's begin with counting outcomes. For example, flipping a coin is an experiment, and the 2 possible outcomes are heads (H) or tails (T).

Heads (H)

Tails (T)

What if we toss a coin twice? We can still easily list the possible outcomes. They are:

(H, H), (H, T), (T, T), (T, H)   There are 4 possible outcomes.

As we increase the number of tosses or parts, it is a good idea to devise a method or a system so that we don't accidentally miss an outcome. One systematic way of listing outcomes is by using a **tree diagram,** as shown in Example 1.

EXAMPLE 1 Draw a tree diagram to determine the possible outcomes for the experiment of tossing a coin 3 times. Then use the diagram to give the number of possible outcomes.

Solution This experiment has 3 parts, which correspond to the 3 tosses of the coin. Read the tree diagram below from left to right. The tree splits to show possibilities for the first toss. From there, the tree splits again to show the possibilities for the second toss, and so on.

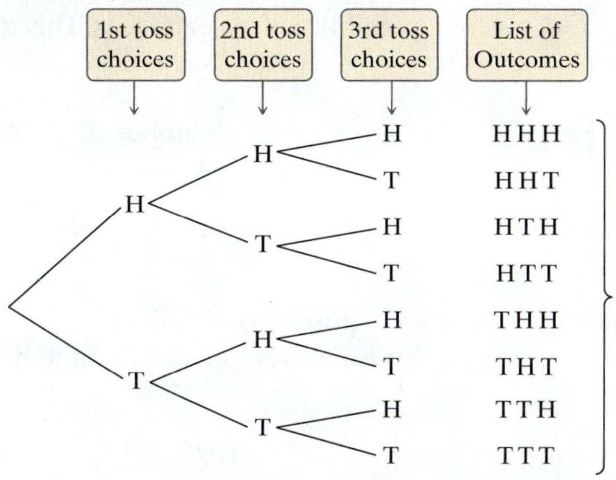

There are 8 possible outcomes, and they are listed to the left.

There are 8 possible outcomes for the experiment of tossing a coin three times. □

(Continued on next page)

PRACTICE 1 Draw a tree diagram for the experiment of tossing a coin 4 times. Then use the diagram to find the number of possible outcomes. ■

EXAMPLE 2 Draw a tree diagram for an experiment consisting of rolling a die and then tossing a coin. Then use the diagram to find the number of possible outcomes.

Die (plural is dice)

Solution Recall that a die has six sides and that each side represents a number, 1 through 6.

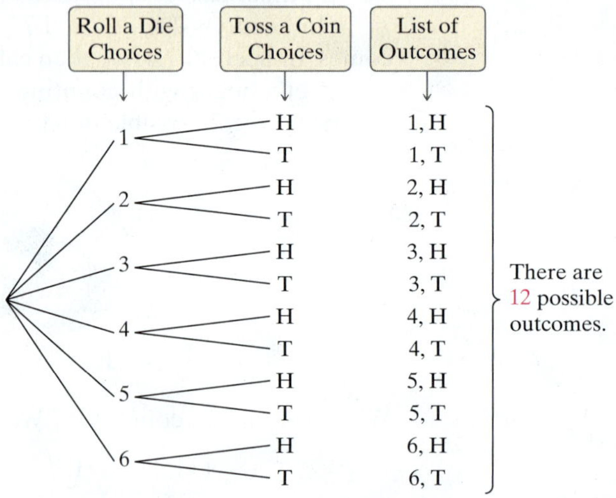

There are 12 possible outcomes for rolling a die and then tossing a coin. □

PRACTICE 2 Draw a tree diagram for an experiment consisting of tossing a coin and then rolling a die. Then use the diagram to find the number of possible outcomes. ■

OBJECTIVE 2

Using the Fundamental Counting Principle

Thus far, we have found the number of possible outcomes of an experiment by listing the outcomes. Suppose that we don't need a listing of the actual outcomes, but we do need to know the *number* of possible outcomes.

From the experiment in Example 1 of tossing a coin 3 times, notice that there were 2 possibilities for each toss. This means that the number of possible outcomes is:

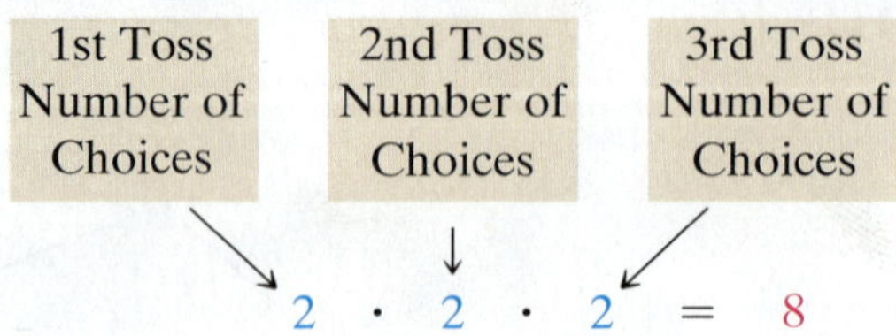

There are 8 possible outcomes.

This is an example of the Fundamental Counting Principle.

> **Helpful Hint**
> Notice that the Fundamental Counting Principle can be extended to as many parts as there are in the experiment.

The Fundamental Counting Principle

Suppose an experiment consists of two separate parts. If one part has m choices and the other part has n choices, then the total number of outcomes for this 2-part experiment is

$$m \cdot n$$

Note: This principle can be extended to three or more parts.

EXAMPLE 3 An alarm system is installed in your home. The installer tells you to choose any 4-digit number for your access code as long as the last digit is not a 0. How many access codes are possible?

Solution Think of this experiment as having 4 parts, corresponding to the 4 digits. Select a first digit, then a second digit, then a third digit, and finally a fourth digit. There are 10 ways to select a first digit (count the digits from 0 to 9), 10 ways to select a second digit, 10 ways to select a third digit, then 9 ways to select a fourth digit (count the digits from 1 to 9). By the Fundamental Counting Principle, the number of access code possibilities is

1st Digit Number of Choices		2nd Digit Number of Choices		3rd Digit Number of Choices		4th Digit Number of Choices	
↓		↓		↓		↓	
10	·	10	·	10	·	9	= 9000

> **Helpful Hint**
> Recall that the last digit can not be 0, so there are 9 choices.

Thus, there are 9000 possible access codes. □

PRACTICE

3 An alarm system is installed at work, and each employee has his or her own access code. The installer allows employees to choose any 3-digit code as long as the first digit is not 0. How many access codes are possible? ■

EXAMPLE 4 From 1969 to 1980, California automobile license plates were formed by 3 digits followed by 3 letters. How many distinct license plates could be manufactured?

Solution Using the Fundamental Counting Principle and the fact that there are 10 digits and 26 letters of the alphabet, the number of distinct license plates is

1st Digit Number of Choices		2nd Digit Number of Choices		3rd Digit Number of Choices		1st Letter Number of Choices		2nd Letter Number of Choices		3rd Letter Number of Choices	
↓		↓		↓		↓		↓		↓	
10	·	10	·	10	·	26	·	26	·	26	= 17,576,000

Thus, 17,576,000 distinct license plates can be manufactured. □

(Continued on next page)

PRACTICE 4 In June 2012 the state of Texas introduced a new, larger license plate called the "Texas Classic." It consists of 3 letters followed by 4 digits. How many distinct license plates can be manufactured?

Note: Around 1980, California was running out of distinct license plates of the form in Example 4. Current California license plates now have a digit, then 3 letters, then 3 digits. (See Exercise 25.)

EXAMPLE 5 Dr. Thomas Hall and wife Diane are traveling to a medical convention. He packs 3 suits, 5 shirts, and 7 pairs of socks for this trip. How many possible outfits can he assemble using all three items of clothing?

Solution This experiment has 3 parts.

Suit Number of Choices		Shirt Number of Choices		Socks Number of Choices	
↓		↓		↓	
3	·	5	·	7	= 105

Thus, 105 possible outfits can be assembled. □

PRACTICE 5 If the salesperson in Example 5 packs 2 suits, 8 shirts, and 8 pairs of socks, how many possible outfits can be assembled using all three items of clothing?

EXAMPLE 6 A multiple-choice test has 5 questions, and each question has 4 choices, with one correct choice per question. If no question is left blank, how many ways can this test be answered?

Solution Since there are 5 questions, this experiment has 5 parts.

Question 1 Number of Choices		Question 2 Number of Choices		Question 3 Number of Choices		Question 4 Number of Choices		Question 5 Number of Choices	
↓		↓		↓		↓		↓	
4	·	4	·	4	·	4	·	4	$= 4^5 = 1024$

There are 1024 ways that this test can be answered. □

PRACTICE 6 Suppose a test similar to the one in Example 6 has 7 questions, and each question has 3 choices with one correct choice per question. Find how many ways this test can be answered if no question is left blank.

Vocabulary, Readiness & Video Check

Use the choices below to fill in each blank. Choices may be used more than once or not at all.

probability outcome experiment tree diagram

1. A possible result of an experiment is called a(n) ____________.

2. A(n) ____________ shows each outcome of an experiment as a separate branch.

3. A chance happening for which the outcome is uncertain is called a(n) ____________.

Martin-Gay Interactive Videos

See Video 7.1

Watch the section lecture video and answer the following questions.

OBJECTIVE 1

4. In Example 1, how is the possible number of outcomes to the experiment determined from the tree diagram?

OBJECTIVE 2

5. After Example 1, we discussed the different information that can be found using a tree diagram vs. using the Fundamental Counting Principle. What information can be found using a tree diagram, but not the Fundamental Counting Principle?

7.1 Exercise Set MyMathLab®

For Exercises 1 through 10, draw a tree diagram for each experiment. Then use the diagram to find the number of possible outcomes. See Examples 1 and 2.

1. Choose a letter in the word MATH, then a number (1, 2, or 3)

2. Choose a number (1 or 2) and then a vowel (a, e, i, o, u)

Use Spinners A and B for Exercises 3 through 6.

Spinner A

Spinner B

3. Spin Spinner A once

4. Spin Spinner B once

5. Spin Spinner A and then Spinner B

6. Spin Spinner B and then Spinner A

7. Find the number of lunch specials possible from a restaurant. The lunch special consists of 1 sandwich, 1 side, and 1 drink.

Sandwiches choices: turkey, ham, egg salad

Side choices: french fries, salad, fruit

Drink choices: soda or tea

8. The same restaurant as in Exercise 7 offers an early dinner special. This meal consists of 1 entree, 1 side, and 1 appetizer. Find the number of possible early dinner specials.

Entree choices: steak, chicken, fish

Side choices: baked potato or french fries

Appetizer choices: soup or salad

9. Find the number of 3-digit numbers that can be written using the digits {4, 5, 6}:

a. If the digits can be repeated.

b. If the digits cannot be repeated.

10. Find the number of 2-digit numbers that can be written using the digits {2, 3, 7, 8}:

a. If the digits can be repeated.

b. If the digits cannot be repeated.

Use the Fundamental Counting Principle to find the number of possible outcomes for each experiment. See Examples 3 through 6.

11. A single coin is flipped 6 times.

12. A single coin is flipped 7 times.

13. A multiple-choice quiz consists of 3 questions with 4 choices for each question. (Each question has one answer.)

14. A multiple choice quiz consists of 6 questions with 3 choices for each question. (Each question has one answer.)

15. A single 8-sided die is rolled three times.

16. A single 20-sided die is rolled twice.

Solve.

17. A pizza shop offers the following toppings:

Sauces	***Meats***	***Vegetables***	***Cheeses***
red	sausage	olives	feta
white	Canadian bacon	peppers	mozzarella
pesto	pepperoni	onion	
		artichokes	
		pineapple	

If one item is chosen from each group, how many different pizzas can be made?

18. A sandwich shop offers the following options:

Breads	***Meats***	***Toppings***
wheat	ham	lettuce
whole-grain	turkey	tomato
rye	roast beef	onion
croissant		cheese

If one item is chosen from each group, how many different sandwiches can be made?

Suppose a 6-sided die is tossed twice. (This is a cube with each side numbered 1 through 6 with dots.) Draw a tree diagram for each exercise to show the outcomes for Exercises 19 through 22. Then list the outcomes. (Note: To save time, complete only those branches that make each exercise possible.)

19. The sum of the tosses is 7.

20. The sum of the tosses is 5.

21. The sum of the tosses is greater than 9.

22. The sum of the tosses is less than 4.

Solve.

23. Suppose you pack 2 suits, 4 shirts, and 5 ties for a business trip. How many different outfits can you form using all three items of clothing?

24. Suppose you pack 3 suits and 5 shirts for a business trip. How many different outfits can you form using both items of clothing?

25. Current license plates in California now consist of a digit, then 3 letters, then 3 digits. Find the number of distinct arrangements possible given that the first digit is not a 0.

26. Some specialty plates in Colorado consist of 3 letters, then 2 digits. Find the number of different ways to choose 3 letters, then 2 digits.

Although there are exceptions, the call signs for radio stations west of the Mississippi River start with a K and those east start with a W. Earlier radio stations had 3-letter call signs, and in 1922 4-letter call stations were established. (Today, a few older stations continue with a 3-letter call sign while a majority have a 4-letter call sign.) (Use this for Exercises 27 and 28.)

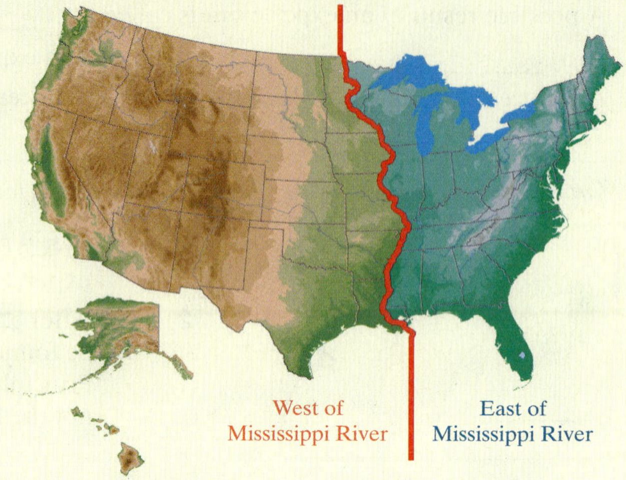

27. Find the number of 3-letter call signs possible given that the first letter must be a K or W.

28. Find the number of 4-letter call signs possible given that the first letter must be a K or W.

Solve.

29. How many different 5-digit access codes are available given that the first digit may not be a 0 or a 9?

30. How many different 4-digit access codes are available given that the first digit must be 6 or greater?

31. How many 3-digit numbers are in our number system? (The first digit may not be 0.)

32. How many 2-digit numbers are in our number system? (The first digit may not be 0.)

33. You are unsure of 5 questions on a true/false exam. How many different ways can you answer these questions?

34. You are unsure of 3 questions on a multiple-choice exam. If each question has choices A, B, C, or D, how many different ways can you answer these questions?

REVIEW AND PREVIEW

Current and predicted users of social network sites are shown below. Use this graph for Exercises 35 through 38. See Section R.4.

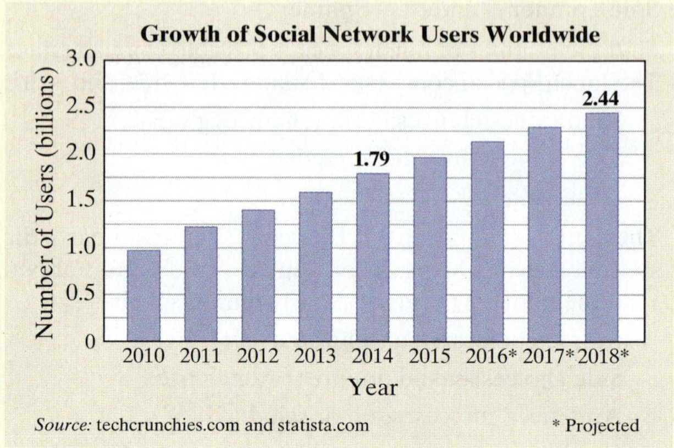

35. Estimate the number of social network users in 2016.

36. Estimate the number of social network users predicted in 2017.

37. Use the label on the 2018 bar and give the projected number of social network users in standard form.

38. Use the label on the 2014 bar and give the number of social network users in standard form.

CONCEPT EXTENSIONS

For security purposes, a company is starting to issue identification badges. Each badge will begin with the letter S or T followed by 2 digits. Use this for Exercises 39 and 40.

39. The company presently has 320 employees who will need badges.

a. How many different badges are possible?

b. Will this ID code system provide enough different codes for the employees?

40. If your answer to Exercise 39 part **b** is no, suggest an ID code change so that all current and new employees up to a total of 700 will have a different code.

41. Now that you know about counting problems and some applications of counting, write an application of your own.

42. In your own words, explain the Fundamental Counting Principle.

7.2 Permutations

OBJECTIVES

1. Use the Fundamental Counting Principle to Count Permutations.
2. Define and Evaluate Factorials and Factorial Expressions.
3. Use the Permutation Formula to Count Permutations.

OBJECTIVE 1 Using the Fundamental Counting Principle to Count Permutations

In Section 7.1, we introduced tree diagrams as an organized way to list and count the outcomes of an experiment. We also introduced the Fundamental Counting Principle as a convenient way to calculate the *number* of outcomes of an experiment.

You may not have noticed, but for the most part, repetitions (of digits and letters) were allowed in Section 7.1. Let's now look at a different counting experiment—one in which repetitions are not allowed.

Suppose there are five posters that you are to arrange hanging on a wall from left to right. Can you see that once the first poster is chosen, there are no longer five posters to choose from, but only four? Let's use the Fundamental Counting Principle to see how many arrangements are possible.

1st Poster Number of Choices		2nd Poster Number of Choices		3rd Poster Number of Choices		4th Poster Number of Choices		5th Poster Number of Choices	
↓		↓		↓		↓		↓	
5	·	4	·	3	·	2	·	1	= 120

Here, there are 120 possible arrangements.

Although we are still using the Fundamental Counting Principle, this is the first time we've had to think about no repetitions. (We certainly cannot hang the same poster twice.)

Let's try a few examples. Compare the following example with Example 3 in Section 7.1.

EXAMPLE 1 An alarm system is installed in your home. The installer tells you to choose any 4-digit number for your code as long as no digit is used twice.

Solution We use the Fundamental Counting Principle—keeping in mind the rules given to use by the installer.

Remember: There are 10 digits (choices) from 0 to 9.

1st Digit Number of Choices	2nd Digit Number of Choices	3rd Digit Number of Choices	4th Digit Number of Choices	
↓	↓	↓	↓	
10 ·	9 ·	8 ·	7	= 5040

There are 5040 possible access codes. □

PRACTICE

1 Determine the number of security access codes possible for a 5-digit access code with no digit used twice. ■

Notice that Example 1 has to do with no repetitions. Let's now look at an example where order matters (as well as repetition).

EXAMPLE 2 An executive committee of 6 people needs to elect among themselves a president, a vice president, and a secretary—in that order. How many ways can this be done given that no person may serve in more than one of these positions?

Solution In this example, although repetition matters, notice that order matters as well. Keep this in mind as we use the Fundamental Counting Principle.

President Number of Choices	Vice President Number of Choices	Secretary Number of Choices	
↓	↓	↓	
6 ·	5 ·	4	= 120

There are 120 possible election outcomes. □

PRACTICE

2 A college club consists of 52 members. How many different ways can the members elect a president, vice president, secretary, and treasurer—in that order—given that no person may serve in more than one of these positions? ■

OBJECTIVE

2 Evaluating Factorials and Factorial Expressions

Notice that the solutions of Examples 1 and 2 are found using products of decreasing natural numbers.

$$\text{Example 1: } 10 \cdot 9 \cdot 8 \cdot 7$$
$$\text{Example 2: } 6 \cdot 5 \cdot 4$$

Before we continue, let's define a shorthand notation for products of decreasing natural numbers. This new notation is called a **factorial.**

The **factorial of n,** written $n!$ (read "n factorial"), is the product of the first n consecutive natural numbers.

Factorial of n: $n!$

If n is a natural number, then

$$n! = n(n-1)(n-2)(n-3)\cdot\cdots\cdot 3\cdot 2\cdot 1.$$

The factorial of 0, written 0!, is defined to be 1.

$$0! = 1$$

Helpful Hint
Remember that $0! = 1$

For example, $3! = 3\cdot 2\cdot 1 = 6$, $5! = 5\cdot 4\cdot 3\cdot 2\cdot 1 = 120$, and $0! = 1$.

EXAMPLE 3 Evaluate each expression.

a. $\frac{5!}{6!}$ **b.** $\frac{10!}{7!3!}$ **c.** $\frac{3!}{2!1!}$ **d.** $\frac{7!}{7!0!}$

Solution

a. $\frac{5!}{6!} = \frac{5 \cdot 4 \cdot 3 \cdot 2 \cdot 1}{6 \cdot 5 \cdot 4 \cdot 3 \cdot 2 \cdot 1} = \frac{1}{6}$

b. $\frac{10!}{7!3!} = \frac{10 \cdot 9 \cdot 8 \cdot 7!}{7! \cdot 3 \cdot 2 \cdot 1} = \frac{10 \cdot 9 \cdot 8}{3 \cdot 2 \cdot 1} = 10 \cdot 3 \cdot 4 = 120$

c. $\frac{3!}{2!1!} = \frac{3 \cdot 2 \cdot 1}{2 \cdot 1 \cdot 1} = 3$

d. $\frac{7!}{7!0!} = \frac{7!}{7! \cdot 1} = 1$ □

PRACTICE 3 Evaluate each expression.

a. $\frac{6!}{7!}$ **b.** $\frac{8!}{6!2!}$ **c.** $\frac{5!}{4!1!}$ **d.** $\frac{9!}{9!0!}$ ■

> **Helpful Hint**
> We can use a calculator with a factorial key to evaluate a factorial. A calculator uses scientific notation for large results.

OBJECTIVE 3

Using the Permutation Formula

Example 2 in this section had to do with a committee of 6 people electing 3 positions—president, vice president, and secretary—with no one in more than one position. Another way of stating Example 2 is that we were finding the number of ordered arrangements of 6 items (people) taken 3 at a time.

In this context, **ordered arrangements** are also called **permutations.**

Thus, to solve Example 2, we were finding the number of permutations of 6 items taken 3 at a time.

In general, the notation $_nP_r$ means **the number of permutations of n items taken r at a time.**

> **Permutation Formula**
> The number of permutations, or ordered arrangements, of n distinct items taken r at a time, where $r \le n$, is:
> $$_nP_r = \frac{n!}{(n-r)!}$$

> **Helpful Hint**
> A counting exercise is a permutation if
> - Repetitions are not allowed and
> - Order is important.

If we refer once again to Example 2, the number of permutations of 6 items taken 3 at a time is $_6P_3$.

By our formula,

$$_6P_3 = \frac{6!}{(6-3)!} = \frac{6!}{3!} = \frac{6 \cdot 5 \cdot 4 \cdot 3 \cdot 2 \cdot 1}{3 \cdot 2 \cdot 1} = \frac{6 \cdot 5 \cdot 4}{1} = 120.$$

This checks with our answer to Example 2, which is 120.

Let's practice evaluating some permutations.

EXAMPLE 4 Evaluate each permutation.

a. ${}_7P_3$ **b.** ${}_9P_4$ **c.** ${}_5P_5$

Solution

$$\textbf{a. } {}_7P_3 = \frac{7!}{(7-3)!} = \frac{7!}{4!} = \frac{7\cdot6\cdot5\cdot4\cdot3\cdot2\cdot1}{4\cdot3\cdot2\cdot1} = 7\cdot6\cdot5 = 210$$

$$\textbf{b. } {}_9P_4 = \frac{9!}{(9-4)!} = \frac{9!}{5!} = \frac{9\cdot8\cdot7\cdot6\cdot5!}{5!} = 9\cdot8\cdot7\cdot6 = 3024$$

$$\textbf{c. } {}_5P_5 = \frac{5!}{(5-5)!} = \frac{5!}{0!} = \frac{5!}{1} = 5\cdot4\cdot3\cdot2\cdot1 = 120 \quad (\text{Remember: } 0! = 1)$$

□

PRACTICE 4 Evaluate each permutation.

a. ${}_{10}P_2$ **b.** ${}_{11}P_6$ **c.** ${}_6P_6$ ■

Although we will not use the following formula in our solutions, there is an alternate permutation formula shown next.

Alternate Permutation Formula

The number of permutations, or ordered arrangements of n distinct items chosen r at a time, where $r \le n$ is

$${}_nP_r = n\cdot(n-1)\cdot(n-2)\cdot(n-3)\cdots(n-r+1)$$

EXAMPLE 5 How many different ways can we arrange the letters in the word "stop"?

Solution Notice that we are interested in an ordered arrangement and all the items (letters) are distinct. Thus, we have a permutation of 4 letters chosen 4 at a time, or ${}_4P_4$.

$${}_4P_4 = \frac{4!}{(4-4)!} = \frac{4!}{0!} = \frac{4!}{1} = 24$$

The number of arrangements is 24. □

PRACTICE 5 How many different ways can we arrange the letters in the word "tie"? ■

EXAMPLE 6 A combination lock contains the numbers 1 through 30 and requires the correct 3-number code in order to open it. How many 3-number codes are possible if no number is used more than once?

Solution This is an ordered arrangement of distinct numbers 1 through 30 chosen 3 at a time and no number is to be used more than once. Thus, we have a permutation of 30 numbers chosen 3 at a time, or ${}_{30}P_3$.

$${}_{30}P_3 = \frac{30!}{(30-3)!} = \frac{30!}{27!} = \frac{30\cdot29\cdot28\cdot27!}{27!} = 30\cdot29\cdot28 = 24{,}360$$

There are 24,360 possible 3-number codes. □

Helpful Hint

The number of outcomes for permutations may also be calculated using the Fundamental Counting Principle.

PRACTICE 6 Suppose the combination lock in Example 6 only contains the numbers 1 through 20. Now how many 3-number codes are possible? ■

Graphing Calculator Explorations

Graphing calculators have a menu item for calculating permutations. For TI graphing calculators, this menu item appears as [nPr].

To find this feature, press the [MATH] key and go to the [PRB] drop-down menu.

To calculate ${}_7P_2$, for example, press [7] [nPr] [2] [ENTER].

The result is 42.

Note: Some scientific calculators have this feature as well.

Vocabulary, Readiness & Video Check

Use the choices below to fill in each blank. Choices may be used more than once or not at all.

permutation	factorial of n	$\frac{n!}{r!}$	$\frac{n!}{(n-r)!}$
the Fundamental Counting Principle	0	1	$\frac{n!}{r!(n-r)!}$

1. By definition, $0! =$ ____________________.
2. The ____________________, written as $n!$, is the product of the first n consecutive natural numbers.
3. Another name for an ordered arrangement of items is a(n) ____________________.
4. A formula for calculating a permutation of n items taken r at a time is ____________________.
5. Permutations can also be counted using ____________________.
6. The permutation ${}_nP_0$ always equals ____________________.

Martin-Gay Interactive Videos

See Video 7.2

Watch the section lecture video and answer the following questions.

OBJECTIVE 1

7. When using the Fundamental Counting Principle to solve Example 1, describe the pattern of the factors.

OBJECTIVE 2

8. In the definition of factorial of n, $n!$, how is $0!$ defined?

OBJECTIVE 3

9. After Example 5, what is the phrase used to describe a permutation?

7.2 Exercise Set MyMathLab®

Evaluate. See Examples 3 through 6.

1. $\frac{7!}{3!}$
2. $\frac{9!}{6!}$
3. $\frac{5!}{5!}$
4. $\frac{9!}{9!}$
5. $\frac{10!}{7!2!}$
6. $\frac{9!}{5!3!}$
7. $(8-6)!$
8. $(12-7)!$
9. $(11-5)!$
10. $(13-10)!$
11. $\frac{10!}{(10-2)!}$
12. $\frac{12!}{(12-5)!}$
13. $\frac{201!}{(201-4)!}$
14. $\frac{190!}{(190-3)!}$

15. ${}_9P_2$

16. ${}_{12}P_2$

17. ${}_4P_1$

18. ${}_5P_1$

19. ${}_{10}P_{10}$

20. ${}_8P_8$

21. ${}_6P_5$

22. ${}_7P_6$

23. Find the number of permutations of 12 objects taken 4 at a time.

24. Find the number of permutations of 100 objects taken 4 at a time.

25. Find the number of permutations of 200 objects taken 3 at a time.

26. Find the number of permutations of 11 objects taken 7 at a time.

Solve. See Examples 1, 2, and 4 through 6.

27. In how many different ways can we arrange the letters in the word "first"

28. In how many different ways can we arrange the letters in the word "second"

29. A baseball team consists of 9 players. How many different ways can a batting order be assigned?

30. A charity committee of 6 persons is to be lined up from left to right for a newspaper photo. How many different ways can these persons be arranged?

31. How many different ways can a 9-member committee choose a chairperson and a secretary given that no person can hold both positions?

32. How many different ways can a 20-member honor society choose a president, vice president, and treasurer given that no person may hold more than one position?

33. In how many ways can 7 people line up at a theatre box office?

34. In how many ways can 8 people line up at a bus stop?

35. A 10 K (10-kilometer) race has 102 runners. If first-, second-, and third-place medals are to be awarded, how many different outcomes are possible?

36. A half-marathon (13.1-mile) race has 47 runners. If first- and second-place medals are to be awarded, how many different outcomes are possible?

37. Find the number of distinct arrangements of 4 of the letters of the word "flower." No letter may be used more than once.

38. Find the number of distinct arrangements of 3 of the letters of the word "tiger" if no letter may be used more than once.

MIXED PRACTICE

Solve. Some of these exercises may be solved using the Fundamental Counting Principle or this principle along with the permutation formula.

39. Four children, Ali, Charles, Leo, and Emilio, are having their photo taken. How many different ways can the children be arranged given that Ali or Charles will be seated first on the left and Charles and Emilio can't sit together because they always start laughing? (*Hint:* Use a tree diagram, keeping in mind the rules given.)

40. Four adults, Chris, Tran, Juanita, and Mike, are attending a Broadway play. How many different ways can the adults sit in a row given that Chris and Juanita want to sit by each other? (See the hint from Exercise 39.)

41. There are 5 seats in an SUV vehicle. How many ways can 5 people be seated if only 2 can drive?

42. There are 7 seats in a van. How many ways can 7 people be seated if 4 of the 7 can drive?

For Exercises 43 through 48, use the word "garnish." How many distinct ways can the letters of this word be arranged if

43. there are no restrictions?

44. the first letter must be an *n*?

45. the first letter must not be an *n*?

46. the first letter is an *r* and the last letter is an *h*?

47. the first two letters are *sr*?

48. the last letter is a vowel?

For Exercises 49 through 54, use the word "friend." How many distinct ways can the letters of this word be arranged if

49. there are no restrictions?

50. the first letter must be a *d*?

51. the first letter must not be a *d*?

52. the first letter is an *r* and the last letter is an *n*?

53. the first two letters are *ne*?

54. the last letter is a vowel?

REVIEW AND PREVIEW

Use the bar graph to solve Exercises 55 through 60. See Section R.4.

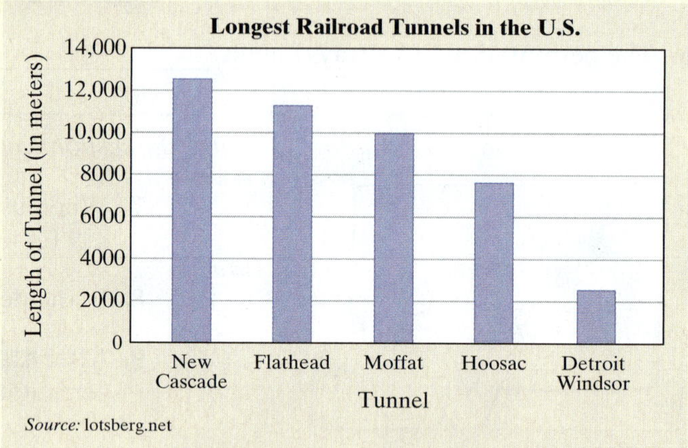

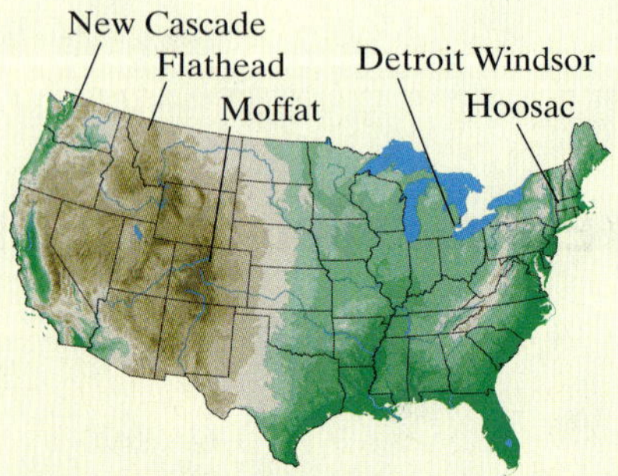

55. Name the longest railway tunnel in the United States.

56. Estimate the length of the Moffat tunnel.

57. Use Exercise 56 and the conversion $1\,\text{mi} \approx 1609\,\text{m}$ to convert the length of the Moffat tunnel to miles. Round to the nearest hundredth.

58. Name the tunnel(s) shorter than 8000 meters.

59. If the New Cascade tunnel is approximately 7.8 mi, give this tunnel length in meters. Use the conversion in Exercise 57.

60. Is your answer to Exercise 59 reasonable when compared with the bar height on the graph?

CONCEPT EXTENSIONS

61. Is it possible to evaluate ${}_6P_5$? Explain why or why not.

62. Is it possible to evaluate ${}_5P_6$? Explain why or why not.

63. How many distinct ways can the letters of the word "lottery" be arranged? (*Hint:* This word has two Ts, so some arrangements will not be distinct and should not be counted twice.

64. How many distinct ways can the letters of the word "middle" be arranged? (See the hint for Exercise 63.)

65. Explain why ${}_5P_5$ can be calculated but is not really useful.

66. Explain why ${}_5P_0$ can be calculated but is not really useful.

67. If ${}_nP_2 = 56$, find the value of n.

68. If ${}_nP_2 = 90$, find the value of n.

7.3 Combinations

OBJECTIVES

1. Determine Whether a Counting Problem Is a Permutation or a Combination.
2. Use the Combination Formula to Count Combinations.

OBJECTIVE

1 Determining Whether a Permutation or a Combination

In the previous section, we introduced methods for counting permutations, or counting ordered arrangements of distinct items, with no repetitions. In this section, we begin by introducing *combinations*. A counting problem is a **combination** when the order of items makes no difference and there are still no repetitions.

Permutation Example: How many ways can a president, vice president, and secretary be selected from a group of 7 people?

Here, **order does matter** because a different title goes with each person.

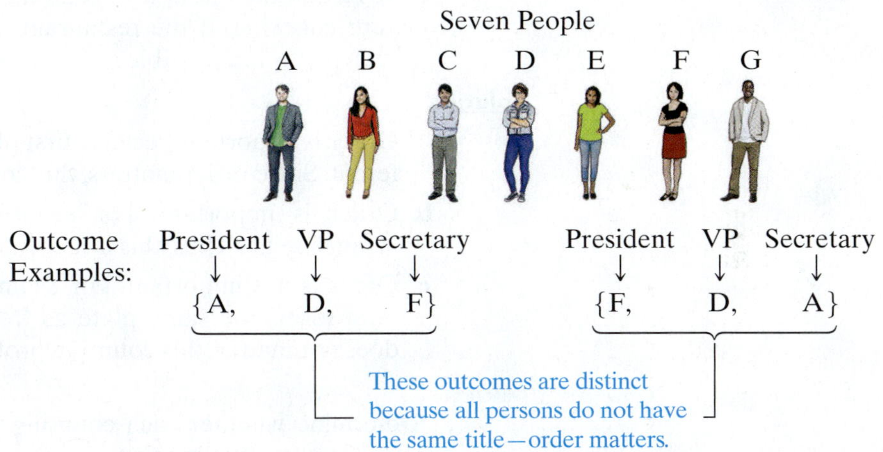

Combination Example: How many ways can a committee of 3 be selected from a group of 7 people?

Here, **order does not matter**. Why? A committee of persons {A, D, F} is the same as a committee of persons {F, D, A}.

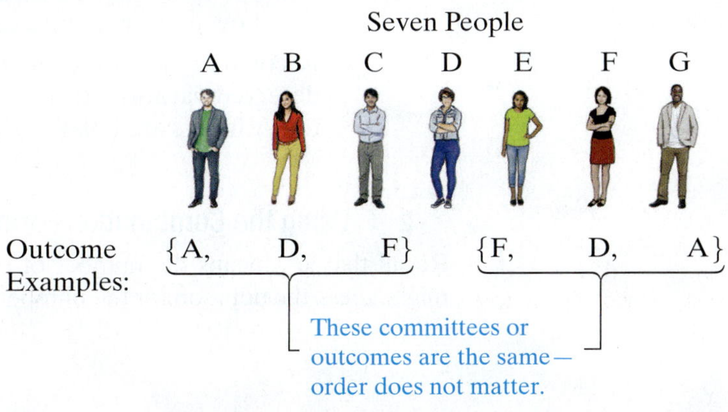

Helpful Hint

The table below shows the simularity and difference between permutations and combinations.

Permutations	*Combinations*
• Repetitions **are not** allowed. • Order **does** matter.	• Repetitions **are not** allowed. • Order **does not** matter.

EXAMPLE 1 Determine whether each counting problem is a permutation or a combination. Do not actually solve.

a. A 5-kilometer race is held with first-, second-, and third-place medals given. If 38 people show up for the race, how many outcomes are possible for the three medals?

b. How many distinct ways can the letters in the word "tropic" be arranged?

c. A restaurant offers a vegetarian plate consisting of 4 vegetable sides of your choosing. If the restaurant has 11 vegetable side options, how many such plates are possible?

Solution

a. Order is important because first place, second place, and third place are different. Since order matters, this counting problem is a permutation.

b. Order is important. For example, "pictro" and "pictor" are different arrangements. Thus, this is a permutation.

c. Order is not important. For example, a choice of corn, french fries, beans, okra forms the same plate as french fries, okra, beans, corn. Since order does not matter, this counting problem is a combination. □

PRACTICE 1 Determine whether each counting problem is a permutation or a combination. Do not actually solve.

a. How many distinct 3-digit numbers may be formed using the digits {7, 2, 3, 8, 4} if no digit is used more than once?

b. The board of directors for a hospital decides to form a committee to study whether a new MRI machine should be purchased. If the board has 8 members, how many committees of 3 members can be formed?

c. A company lottery is held, with 4 winners to be chosen to go on a well-deserved vacation. If the company has 21 employees, how many groups of four winners are possible? ■

OBJECTIVE 2

Using the Combination Formula

Recall that ${}_nP_r$ means the number of permutations of n items taken r at a time. As you might guess, the notation for the number of combinations of n items taken r at a time is ${}_nC_r$.

Let's see if we can discover a formula for a combination, ${}_nC_r$, by using the formula for a permutation, ${}_nP_r$. To do so, remember that

Order matters for permutations and

Order does not matter for combinations.

Suppose that 5 people are going to play tennis in 2-person teams.

Players: **A**very, **B**ill, **C**arol, **D**ennis, **E**li, or persons $\underbrace{\text{A, B, C, D, E.}}_{\text{1st letter of each name}}$

Note: Since order is not important, remember that a 2-person team of {A, B} is the same as the team {B, A}.

Thus, the list of possible teams, or combinations, is:

$$\underbrace{\{A, B\}, \{A, C\}, \{A, D\}, \{A, E\}, \{B, C\}, \{B, D\}, \{B, E\}, \{C, D\}, \{C, E\}, \{D, E\}}_{10 \text{ possible teams}}$$

or

$${}_5C_2 = 10.$$

The only difference in ${}_5P_2$ and ${}_5C_2$ is that in ${}_5C_2$, we eliminate the rearrangements. For this example of 2-player teams, there are $2! = 2$ rearrangements. (Remember that {A, B} is the same as {B, A}.)

Thus, for this example,

$${}_5C_2 = \frac{{}_5P_2}{2!} \text{ or } \frac{5!}{2!(5-2)!}$$

In general, we have ${}_nC_r = \dfrac{{}_nP_r}{r!} = \dfrac{n!}{r!(n-r)!}$, stated next.

Combination Formula

The number of combinations of n distinct items taken r at a time where $r \leq n$ is:

$${}_nC_r = \frac{n!}{r!(n-r)!}$$

Let's practice evaluating some combinations.

EXAMPLE 2 Evaluate each combination.

a. ${}_5C_3$ **b.** ${}_{12}C_9$ **c.** ${}_9C_9$

Solution

$$\textbf{a. } {}_5C_3 = \frac{5!}{3!(5-3)!} = \frac{5!}{3!2!} = \frac{5\cdot4\cdot3\cdot2\cdot1}{\underbrace{3\cdot2\cdot1}_{3!}\cdot\underbrace{2\cdot1}_{2!}} = \frac{5\cdot4}{2\cdot1} = 10$$

$$\textbf{b. } {}_{12}C_9 = \frac{12!}{9!(12-9)!} = \frac{12!}{9!3!} = \frac{12\cdot11\cdot10\cdot9!}{9!\cdot3\cdot2\cdot1} = \frac{12\cdot11\cdot10}{3\cdot2\cdot1} = 220$$

$$\textbf{c. } {}_9C_9 = \frac{9!}{9!(9-9)!} = \frac{9!}{9!0!} = \frac{9!}{9!\cdot1} = 1$$

$0! = 1$

□

PRACTICE 2 Evaluate each combination.

a. ${}_8C_6$ **b.** ${}_{15}C_{12}$ **c.** ${}_{21}C_{21}$

■

EXAMPLE 3 An ice cream shop offers a dish called "the monster." "The monster" consists of a bowl of 9 scoops of different flavors of ice cream. If the ice cream shop has 20 flavors of ice cream, how many different monster bowls are possible?

Solution Notice that no repetition of flavors is allowed and the order does not matter, so this is a combination of 20 items taken 9 at a time.

$$_{20}C_9 = \frac{20!}{9!(20-9)!} = \frac{20!}{9!11!} = \frac{20\cdot19\cdot18\cdot17\cdot16\cdot15\cdot14\cdot13\cdot12\cdot11!}{9\cdot8\cdot7\cdot6\cdot5\cdot4\cdot3\cdot2\cdot1\cdot11!}$$
$$= 167{,}960$$

There are 167,960 possible monster bowls. □

PRACTICE
3 Suppose the ice cream shop in Example 3 only has 15 flavors of ice cream on a particular day. How many monster bowls are possible? ■

EXAMPLE 4 Celeste goes to the bookstore and finds 12 paperback books that she would like to purchase. She decides to only purchase 5. How many different groups of 5 books are possible?

Solution There is no repetition since she is not buying the same book more than once. Also, the order of the 5 books does not matter, so this is a combination of 12 items taken 5 at a time.

$$_{12}C_5 = \frac{12!}{5!(12-5)!} = \frac{12!}{5!7!} = \frac{12\cdot11\cdot10\cdot9\cdot8\cdot7!}{5\cdot4\cdot3\cdot2\cdot1\cdot7!} = \frac{12\cdot11\cdot10\cdot9\cdot8}{5\cdot4\cdot3\cdot2\cdot1} = 792$$ □

PRACTICE
4 The U.S. Senate consists of 100 senators (2 from each of 50 states). A committee of 6 senators is to be formed. How many ways is this possible? ■

Helpful Hint

Although permutations may also be counted using the Fundamental Counting Principle, this principle may not be used to count combinations.

Graphing Calculator Explorations

Graphing calculators have a menu item for calculating combinations. For TI graphing calculators, this menu item appears as [nCr]. To find this feature, press the [MATH] key and go to the [PRB] drop-down menu.

To calculate ${}_7C_2$, for example, use the key strokes [7] [nCr] [2] [ENTER]. The result is 21.

Note: Some scientific calculators have this feature as well.

Vocabulary, Readiness & Video Check

Use the choices below to fill in each blank. Choices may be used more than once or not at all.

permutation	combination	0	$\dfrac{n!}{r!}$	$\dfrac{n!}{(n-r)!}$	$\dfrac{n!}{r!(n-r)!}$
the Fundamental Counting Principle	1				

1. By definition, $0! =$ ____________.

2. Order does not matter for a(n) ____________.

3. Order does matter for a(n) ____________.

4. A formula for calculating a combination of n items taken r at a time is ____________.

5. The combination ${}_nC_0$ always equals ____________.

6. The combination ${}_nC_n$ always equals ____________.

Martin-Gay Interactive Videos

See Video 7.3

Watch the section lecture video and answer the following questions.

OBJECTIVE 1

7. Before Examples 1 and 2, the major difference between permutations and combinations is discussed. What is that major difference?

OBJECTIVE 2

8. Why is Example 6 an example of a combination?

7.3 Exercise Set MyMathLab®

Evaluate each combination. See Examples 2 through 4.

1. ${}_7C_6$ **2.** ${}_9C_8$ **3.** ${}_{13}C_3$

4. ${}_{14}C_4$ **5.** ${}_{30}C_{30}$ **6.** ${}_{28}C_{28}$

7. ${}_5C_1$ **8.** ${}_7C_1$ **9.** ${}_{120}C_0$

10. ${}_{117}C_0$ **11.** ${}_9C_6$ **12.** ${}_9C_3$

13. ${}_{26}C_{10}$ **14.** ${}_{28}C_5$

MIXED PRACTICE

Evaluate each expression. Note that we have a mixture of factorials, permutations, and combinations.

15. ${}_8C_6 \cdot {}_{10}C_9$ **16.** ${}_7C_4 \cdot {}_{11}C_{10}$

17. $\dfrac{{}_5P_3}{{}_5C_4}$ **18.** $\dfrac{{}_7P_2}{{}_3C_2}$

19. ${}_6C_3 \cdot {}_4P_2$ **20.** ${}_8C_5 \cdot {}_5P_2$

21. ${}_8P_3 - {}_8C_3$ **22.** ${}_7P_3 - {}_7C_3$

Identify each exercise in Exercises 23 through 26 as either a ***permutation*** *(order matters) or* ***combination*** *(order does not matter) counting problem. Do not actually solve these exercises. See Example 1.*

23. A charity race is held and the first 5 people who cross the finish line receive the same prize and are not recognized as first, second, and so on. How many outcomes are possible if 95 people entered this race?

24. A charity race is held and first, second, third, fourth, and fifth places are awarded with different prizes. How many outcomes are possible if 95 people entered this race?

25. A club with 30 members is electing a president, vice president, and treasurer. How many outcomes are possible for this election?

26. A club with 30 members is selecting a committee of 3 members. How many outcomes are possible for this committee?

Solve each combination exercise. See Examples 3 and 4.

27. Tryouts are starting for the flag team at a local high school. If 37 students try out for 15 positions, how many different selections are possible?

28. Two members of a 16-member committee are to be selected to travel to a national meeting. How many selections of 2 members are possible?

29. A standard deck of cards contains 52 different cards if we consider card name and suit. (For example, there is only 1 queen of hearts. See a diagram on page 472.) If order does not matter, how many outcomes of 5 cards are possible?

30. For the standard deck of 52 different cards in Exercise 29, how many outcomes of 2 cards are possible if order does not matter?

31. Thirty dancers arrive to try out for the chorus line in a new Broadway play. If only 12 are needed, how many selections of 12 dancers from the 30 are possible?

32. A new department at a company is to be formed. The CEO decided that 7 people will initially be hired at the same position and salary. If 40 people are interviewed, how many selections of 7 people from the 40 are possible?

33. Louisiana Lottery has an Easy 5 game in which one picks 5 different numbers from 1 to 37. If the order of the 5 numbers, does not matter, how many different selections are possible?

34. Illinois has a lottery game called LOTTO in which one picks 6 different numbers between 1 and 52. If the order of the 6 numbers does not matter, how many different selections are possible?

MIXED PRACTICE

Solve. Use any method needed from Sections 7.1 through 7.3.

35. How many ways can 6 framed photographs be arranged from left to right on a fireplace mantel?

36. How many ways can 4 framed photographs be arranged from left to right on a wall?

37. How many company ID cards may be made using 1 letter followed by 3 digits where repetition of digits is allowed?

38. Parking spaces at a large hospital are about to be identified by floor (1, 2, 3, 4 only), then by what part of a floor (A, B, C only), then by space (3 digits and any digit may be a 0). How many spaces may be numbered using this system?

39. How many ways can a faculty advocate committee of 6 be formed from the university senate of 20 faculty?

40. How many ways can a PTO committee of 4 be formed from 10 volunteers?

Use the digits 0, 1, 2, 3, 4, 5, 6, 7, 8, 9 to find how many numbers can be formed for Exercises 41 through 44. For all exercises, the first digit may not be 0.

41. Form a 5-digit number so that no digit may be repeated.

42. Form a 5-digit number. Digits may be repeated.

43. Form a 3-digit odd number. Digits may be repeated.

44. Form a 3-digit even number. Digits may not be repeated.

REVIEW AND PREVIEW

A biology experiment consists of growing a plant from a seed. The results are shown on the histogram below. Use this to answer Exercises 45 through 54. See Section R.4.

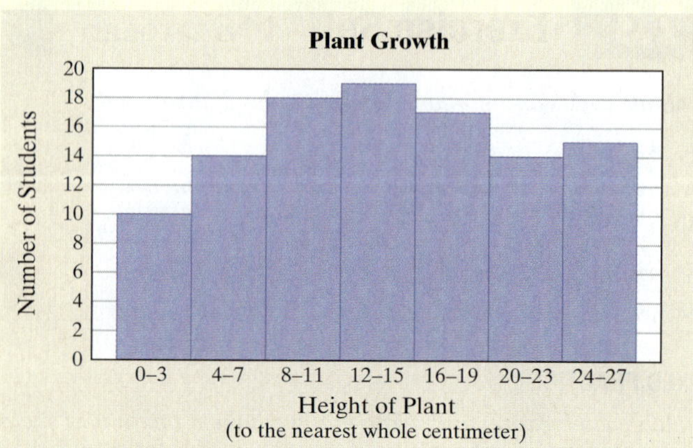

45. How many students grew plants to a height of 24–27 centimeters?

46. How many students grew plants to a height of 16–19 centimeters?
47. The greatest number of students had plants at what range of heights?
48. The fewest number of students had plants at what range of heights?
49. How many students had plant heights of 11 cm or less?
50. How many students had plant heights of 16 cm or greater?
51. How many students were in this plant experiment?
52. How many students had a plant of height 8–19 cm?
53. Did any bars in this histogram have the same heights?
54. If the answer to Exercise 53 is yes, explain what this means in terms of number of students and plant heights.

CONCEPT EXTENSIONS

The notation $\binom{n}{r}$ means $\frac{n!}{r!(n-r)!}$. Thus, $\binom{n}{r} = {}_nC_r$.

For example,

$$\binom{5}{3} = \frac{5!}{3!(5-3)!} = \frac{5!}{3!2!} = \frac{5\cdot4\cdot3\cdot2\cdot1}{(3\cdot2\cdot1)\cdot(2\cdot1)} = 10.$$

Evaluate the following.

55. $\binom{9}{5}$

56. $\binom{4}{3}$

57. $\binom{8}{2}$

58. $\binom{12}{11}$

59. Show that $\binom{n}{n} = 1$ for any natural number n.
60. Show that $\binom{n}{0} = 1$ for any natural number n.
61. Write down the formula for calculating ${}_7C_5$.
62. Write down the formula for calculating ${}_7C_2$.
63. Write down the formula for ${}_nC_r$ and for ${}_nC_{n-r}$.
64. Use the results of Exercises 61 through 63 to make a statement about ${}_nC_r$ and ${}_nC_{n-r}$.
65. Find the value of ${}_nC_n$. In your own words, explain why your answer makes sense.
66. Find the value of ${}_nC_1$. In your own words, explain why your answer makes sense.
67. Find the value of ${}_nC_0$. In your own words, explain why your answer makes sense.

Integrated Review Counting

Sections 7.1–7.3

Evaluate.

1. $\frac{9!}{8!}$

2. $\frac{400!}{399!}$

3. $5!2!$

4. $7!3!$

5. $\frac{4!}{0!}$

6. $\frac{286!}{286!}$

7. ${}_{13}P_3$

8. ${}_{12}P_2$

9. ${}_{13}C_3$

10. ${}_{12}C_2$

Solve using a method of your choice.

11. The menu of the day offers a choice of soup or salad, then 3 entrees, 5 side dishes, and 3 desserts. If 1 item is chosen for each of the 4 categories, how many outcomes are possible?
12. How many 7-digit numbers can be formed if the first digit may not be 0?
13. How many committees of 4 may be chosen from a group of 10 people?
14. Colorado Lottery has a game called Lotto where you pick 6 numbers from 1 to 42. If no repetitions are allowed and the order you pick does not matter, how many outcomes are possible?

15. A race has 23 runners. How many first-, second-, and third-place medals may be awarded?

7.4 Introduction to Probability

OBJECTIVES

1. Calculate Theoretical Probability.
2. Calculate Empirical Probability.

OBJECTIVE

1 Calculating Theoretical Probability

Now that we understand simple counting methods, we will use these methods to introduce **probability,** or the likelihood that an event will occur. Let's begin with a few more vocabulary words.

The set S of all possible outcomes of an experiment is called the **sample space** of the experiment. For the experiment of flipping a coin once, the sample space S is

$$S = \{\text{heads, tails}\} \text{ or } S = \{H, T\}.$$

If a fair die is tossed once, the sample space S is

$$S = \{1, 2, 3, 4, 5, 6\}.$$

A subset of a sample space is called an **event.** We find the probability of an event by dividing two numbers. Let's look at a simple example to make sure we understand our vocabulary words—sample space, event, and probability.

EXAMPLE 1 If a coin is tossed, find the probability that heads occurs.

Solution For now, let's assume that heads and tails are equally likely, or that we have a "fair" coin. Here, the sample space is $S = \{H, T\}$. The experiment is tossing a coin and the outcome is uncertain. The event that we are interested in is the occurrence of heads on the single toss of a coin. In symbols, we can write this event (a subset of the sample space) as $E = \{H\}$. Since *one* of the *two* possible outcomes is heads, the probability of heads is 1 divided by 2, or $\frac{1}{2}$. □

PRACTICE

1 If a fair coin is tossed, find the probability that tails occurs. ■

> **Helpful Hint**
> Notice that
> $P(\text{heads}) + P(\text{not heads}) = 1.$
> Also,
> $P(\text{tails}) + P(\text{not tails}) = 1.$

Notice that the only two possibilities from the single toss of a coin is heads or tails and the sum of their probabilities is $\frac{1}{2} + \frac{1}{2} = 1$. This is true in general.

Example 1 is an example of a special type of probability called **theoretical probability.** For theoretical probability, the sample space only contains equally likely outcomes. Since all outcomes are equally likely, to calculate this probability, we divide the number of favorable outcomes (ones that lead to the desired event) by the total number of outcomes in the sample space.

Theoretical Probability of an Event

If all the outcomes of a sample space S are equally likely, then the theoretical probability of event E, denoted by $P(E)$ is given by

$$P(E) = \frac{\text{number of outcomes in event } E}{\text{total number of outcomes}} = \frac{n(E)}{n(S)}$$

Note: If a probability is asked for, we will assume it to be theoretical probability.

> **Helpful Hint**
>
> Let's make sure we understand the notation used in the box on the previous page.
>
> $P(E)$ is the probability that event E occurs.
> $n(S)$ is the *number* of outcomes in sample space S.
> $n(E)$ is the *number* of outcomes in set S that belong to event E.

EXAMPLE 2 If a fair die is rolled, find the probability that the result is a 2.

Solution Here, we have that $S = \underbrace{\{1, 2, 3, 4, 5, 6\}}_{\text{6 outcomes}}$ and $E = \underbrace{\{2\}}_{\text{1 outcome}}$. Thus,

$$P(E) = \frac{n(E)}{n(S)} = \frac{1}{6} \begin{array}{l} \leftarrow \textbf{Number} \text{ of outcomes in } E \\ \leftarrow \textbf{Number} \text{ of outcomes in } S \end{array}$$

Possible outcomes for rolling a die

Desired event for Example 2

PRACTICE 2 If a fair die is rolled, find the probability that the result is a 6.

Now let's look at two special cases of probability. We will continue to use the experiment of rolling a die.

EXAMPLE 3 If a fair die is rolled, find the probability of rolling

a. a 1, 2, 3, 4, 5, or 6 **b.** a number greater than 6

Solution

a. Here, we have that $S = \underbrace{\{1, 2, 3, 4, 5, 6\}}_{\text{6 outcomes}}$ and $E = \underbrace{\{1, 2, 3, 4, 5, 6\}}_{\text{6 outcomes}}$. Thus,

$$P(E) = \frac{n(E)}{n(S)} = \frac{6}{6} = 1$$

b. Since there is no way to roll a number greater than 6, can you see that $E = \{\ \}$, the empty set. It contains no element, so the numerator in the probability fraction is 0 and we have that

$$P(E) = \frac{n(E)}{n(S)} = \frac{0}{6} = 0$$

PRACTICE 3 Three marbles are placed in a bag: 1 red, 1 green, and 1 blue. If you choose a marble without looking, and each are equally likely to be chosen, find the probability of

a. choosing a yellow marble

b. choosing a red, green, or blue marble

Example 3 along with the probability definition tells us a few important facts about probability:

Important Facts about Probability

- The probability of an event that is certain to occur is 1.
- The probability of an event that is impossible to occur is 0.
- The probability of any event occurring is at least 0 and at most 1.

Helpful Hint

Study the box above. It is also true that the closer a probability is to 1, the more likely the event is to occur.

The closer a probability is to 0, the less likely the event is to occur.

Our next example has to do with a standard deck of cards, so these 52 cards are reviewed next.

Example of 52 playing cards (13 of each suit: clubs, diamonds, hearts, and spades)

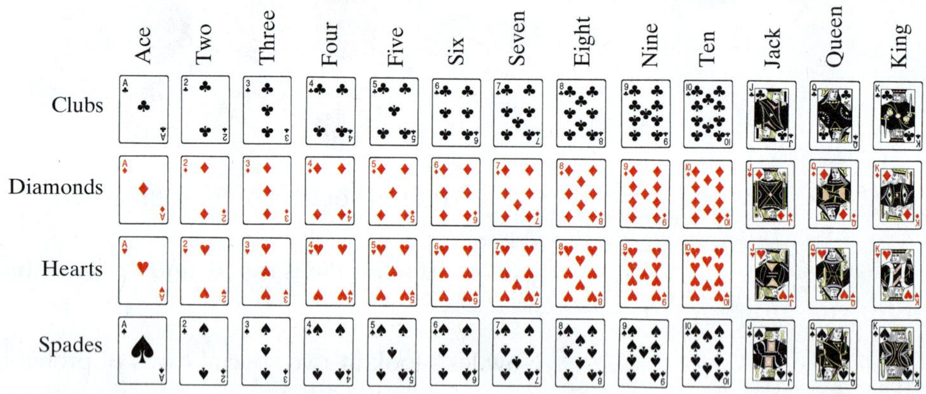

EXAMPLE 4 One card is chosen from a deck of 52. Find the probability that each is chosen:

a. the jack of diamonds **b.** a club **c.** a queen

Solution Since there are 52 different cards, the number of outcomes in our sample space, $n(S) = 52$ for parts a, b, and c.

a. There is only 1 card that leads to the event of choosing the jack of diamonds, thus $n(E) = 1$.

$$P(\text{jack of diamonds}) = \frac{n(E)}{n(S)} = \frac{1}{52}$$

The probability of choosing the jack of diamonds is $\frac{1}{52}$.

b. How many cards are clubs? Study the diagram of cards and see that there are 13 cards that have clubs on them, so here, $n(E) = 13$. Thus,

$$P(\text{a club}) = \frac{n(E)}{n(S)} = \frac{13}{52} = \frac{1}{4}$$

The probability of choosing a club is $\frac{1}{4}$.

c. How many queens are in a deck of cards? There are 4 queens in the diagram of cards, so $n(E) = 4$. Thus,

$$P(\text{a queen}) = \frac{n(E)}{n(S)} = \frac{4}{52} = \frac{1}{13}$$

The probability of choosing a queen is $\frac{1}{13}$. □

PRACTICE 4 One card is dealt from a deck of 52. Find the probability that each is dealt:

a. a diamond **b.** a king **c.** the 8 of hearts ■

Notice that we can use our various methods of counting from earlier in this chapter to find total numbers of outcomes for different situations.

OBJECTIVE 2 Calculating Empirical Probability

Now let's study a type of probability that is based on actual observed results of an experiment. We call this empirical probability.

Empirical Probability Formula

The empirical probability of event E, denoted by $P(E)$, is the number of times event E occurs divided by the total number of trials of the experiment.

$$P(E) = \frac{\text{number times event } E \text{ occurs}}{\text{total number of trials of experiment}}$$

Helpful Hint

Remember: Theoretical probability is based on our assumption that each outcome is equally likely.

Empirical probability is based on observed outcomes of actual experiments.

To understand the difference between theoretical and empirical probability, let's calculate the empirical probability of rolling a die and the result is 2.

Note: We calculated this theoretical probability and found that if a fair die is rolled,

$$P(\text{rolling a 2}) = \frac{1}{6}(\approx 0.167) \quad \text{(See Example 2.)}$$

EXAMPLE 5 Find the empirical probability of rolling a fair die and the result is 2.

Solution To find empirical probability, we need the results of the experiment of rolling a die.

Below are the results when I rolled a single die 100 times. (Yes, I rolled a die 100 times—*Elayn Martin-Gay*)

Outcomes of Rolling a Die 100 times

Roll Result	*Number of Times Rolled*
1	11
2	20
3	19
4	18
5	15
6	17
	100 Total rolls

(Continued on next page)

Using our empirical probability formula, we have

$$P(\text{rolling a } 2) = \frac{\text{number of times 2 is rolled}}{\text{total number of times die is rolled}} = \frac{20}{100} = \frac{1}{5} = 0.2$$

□

PRACTICE 5 For Practice 2, we found that the theoretical probability of rolling a 6 on a die is $\frac{1}{6}(\approx 0.167)$. Use the table in Example 5 to find the empirical probability of rolling a 6. ■

Suppose another person is asked to roll a die 100 times and record your results. Do you think the results will be the same as mine? Probably not. What does this mean? It means that empirical probability will change depending on many factors—especially the number of times an experiment is performed or, in this case, the number of times a die is rolled. This leads us to the Law of Large Numbers.

Law of Large Numbers

As the number of trials of an experiment increases, the proportion with which an event is observed gets closer and closer to the theoretical probability of that event.

For our example of rolling a die, the more times we perform that experiment, the closer each event (rolling a certain number) approaches the theoretical probability of that event.

Technology Note

Although rolling a die 100 times and recording the results can be educational, the same cannot be said of rolling a die 1000 times. There are many random number–generating programs that will simulate this quickly and produce a graph of the results.

EXAMPLE 6 The table below shows real data for the 204.6 million U.S. citizens age 25 years or older.

If an American 25 years old or older is selected, find the probability that the person has an associate's degree. Round the probability to 2 decimal places.

U.S. Educational Attainment for Persons 25 Years or Older (in millions)

Not a High School Graduate	*High School Graduate*	*Some College*	*Associate's Degree*	*Bachelor's Degree*	*Advanced Degree*	*Total*
25.4	62.2	34.2	19.6	40.5	22.7	204.6 million

Source: U.S. Census Bureau

Solution Since this is observed data, this is an empirical probability example. The event is that the person has an associate's degree.

$$P(\text{American 25 years old or older has an associate's degree})$$

$$= \frac{\text{number of persons with an associate's degree}}{\text{total number of persons} \geq \text{25 years old}} = \frac{19.6}{204.6} \approx 0.10$$

The empirical probability that a selected American age 25 years or older has an associate's degree is approximately 0.10. □

PRACTICE 6 Use the table in Example 6 to find the probability that a person age 25 years or older selected from the U.S. population has a bachelor's degree. Round to 2 decimal places. ■

Let's review some rules for probability.

Rules for Probability—Theoretical or Empirical

- The probability of an event, $P(E)$, is 0, 1, or between 0 and 1.

$$0 \le P(E) \le 1.$$

- The probability of an outcome that is impossible is 0.
- The probability of an outcome that is certain to happen is 1.
- The sum of the probability that an event E will occur and the probability that E will not occur is 1.

$$P(E) + P(\text{not } E) = 1$$

(In other words, an event E will either occur or it will not occur.)

Vocabulary, Readiness & Video Check

Use the choices below to fill in each blank. Choices may be used more than once or not at all.

event	0	$\dfrac{\text{number of outcomes in event } E}{\text{total number of outcomes}}$	$\dfrac{\text{number of times event } E \text{ occured}}{\text{total number of trials of experiment}}$
sample space	1	probability	

1. ______________________ is the likelihood that an event will occur.

2. The probability of an event that is impossible to happen is ______________________.

3. The probability of an event that is certain to happen is ______________________.

4. The set of all possible outcomes of an experiment is called the ______________________.

5. A subset of the sample space is called a(n) ______________________.

6. The empirical probability of event E is ______________________.

7. The theoretical probability of event E is ______________________.

Martin-Gay Interactive Videos

See Video 7.4

Watch the section lecture video and answer the following questions.

OBJECTIVE 1

8. For Example 1, why is the answer $\frac{1}{6}$ and not $\frac{5}{6}$?

OBJECTIVE 2

9. After Example 9, empirical probability and the empirical probability formula is introduced. What is the difference between theoretical probability and empirical probability?

7.4 Exercise Set MyMathLab®

If a single die is tossed once, find the probability of each event. See Examples 1 through 3.

1. A 5
2. A 9
3. A 1 or a 6
4. A 2 or a 3
5. An even number
6. An odd number
7. A number greater than 2
8. A number less than 6

The spinner shown is spun once. Find the probability that the spinner stops on each. See Examples 1 through 4.

9. the 2
10. the 9
11. the 3 or the 4
12. the 10 or the 11
13. a blue
14. a yellow
15. a red or a yellow
16. a blue or a red
17. an odd number
18. an even number
19. the 4, 7, or 8
20. the 1, 5, or 12
21. a red or a yellow or a blue
22. a natural number from 1 to 12

One card is dealt from a standard deck of 52. Find the probability that each is dealt. See Example 4.

23. the king of hearts
24. the 10 of spades
25. a king
26. a 10
27. a heart
28. a club
29. a jack or a queen
30. a king or an ace
31. a diamond or a club
32. a diamond, a club, or a heart

A single die is rolled twice and the equally likely 36 outcomes are shown.

First Die \ Second Die	⚀	⚁	⚂	⚃	⚄	⚅
⚀	(1, 1)	(1, 2)	(1, 3)	(1, 4)	(1, 5)	(1, 6)
⚁	(2, 1)	(2, 2)	(2, 3)	(2, 4)	(2, 5)	(2, 6)
⚂	(3, 1)	(3, 2)	(3, 3)	(3, 4)	(3, 5)	(3, 6)
⚃	(4, 1)	(4, 2)	(4, 3)	(4, 4)	(4, 5)	(4, 6)
⚄	(5, 1)	(5, 2)	(5, 3)	(5, 4)	(5, 5)	(5, 6)
⚅	(6, 1)	(6, 2)	(6, 3)	(6, 4)	(6, 5)	(6, 6)

Find the probability of each event. See Examples 1 through 4.

33. a 1, then a 5
34. a 6, then a 2
35. two of the same number
36. two numbers that are not the same
37. two even numbers
38. two odd numbers
39. two numbers whose sum is 4
40. two numbers whose sum is 5
41. two numbers whose sum is 2
42. two numbers whose sum is 12
43. two numbers whose sum is 15
44. two numbers whose sum is ≤ 12

A new drug is being tested that is supposed to lower blood pressure. This drug was given to 200 people and the results are shown below. See Example 6.

Lower Blood Pressure	Higher Blood Pressure	Blood Pressure Not Changed
152	38	10

45. If a person is testing this drug, what is the probability that his or her blood pressure will be higher?
46. If a person is testing this drug, what is the probability that his or her blood pressure will be lower?
47. If a person is testing this drug, what is the probability that his or her blood pressure will not change?
48. What is the sum of the answers to Exercises 45, 46, and 47? In your own words, explain why.

Suppose that we roll a die 30 times with the following results:

2	2	1	5	4	4	2	3	1	3
2	2	6	5	3	1	1	6	6	5
4	5	2	3	2	5	6	3	5	6

Organize these data in a frequency distribution by filling in each blank with the number of times that number occurs.

	Number	Frequency
49.	1	
50.	2	
51.	3	
52.	4	
53.	5	
54.	6	

Use the data from Exercises 49 through 54 to find the empirical probability of each. Write each as a fraction, then as a decimal rounded to 2 decimal places. See Examples 5 and 6.

55. rolling a 1

56. rolling a 2

57. rolling a 3

58. rolling a 4

59. rolling a 5

60. rolling a 6

REVIEW AND PREVIEW

Use this bar graph to answer Exercises 61 through 66. See Section R.4.

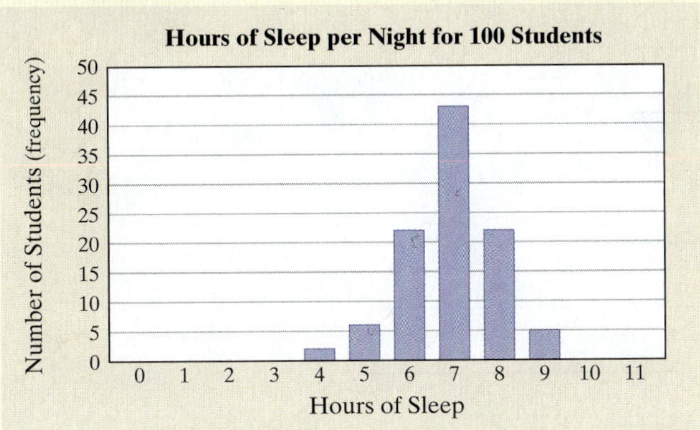

61. How many hours per night did most students sleep?

62. How many hours per night did the fewest students sleep?

63. How many students slept 5, 6, or 7 hours per night?

64. How many students slept 8 hours or more per night?

65. How many students slept less than 4 hours per night?

66. How many students slept greater than 9 hours per night?

Let's recall the experiment of tossing a coin once. Thus,

Sample space: $S = \{H, T\}$ and $P(\text{heads}) = \frac{1}{2}$; $P(\text{tails}) = \frac{1}{2}$

Perform your own experiment of tossing a coin.

67. Toss a coin 20 times. Record the frequency of heads and tails and write each empirical probability as a fraction.

Outcomes	*Frequency*	*Empirical Probability*
H		
T		

68. Toss a coin 50 times and use the same directions as for Exercise 67.

Outcomes	*Frequency*	*Empirical Probability*
H		
T		

69. Find the sum of your empirical probabilities for Exercise 67.

70. Find the sum of your empirical probabilities for Exercise 68.

71. Use the results of Exercises 69 and 70 and write what happened.

72. Will a theoretical probability vary? Explain why or why not.

73. Will an empirical probability vary? Explain why or why not.

Chapter 7 Vocabulary Check

Use the choices below to fill in each blank. Choices may be used more than once or not at all.

probability	outcome	$\dfrac{\text{number of outcomes in event } E}{\text{total number of outcomes}}$	$\dfrac{n!}{(n-r)!}$
event	0	$\dfrac{n!}{r!}$	$\dfrac{n!}{r!(n-r)!}$
sample space	1	factorial of n	
permutation	combination	tree diagram	
the Fundamental Counting Principle	experiment	$\dfrac{\text{number of times event } E \text{ occured}}{\text{total number of trials of experiment}}$	

1. A possible result of an experiment is called a(n) ______________.

2. A(n) ______________ shows each outcome of an experiment as a separate branch.

3. A chance happening for which the outcome is uncertain is called a(n) ______________.

4. By definition, $0! =$ ______________.

5. The ______________, written as $n!$, is the product of the first n consecutive natural numbers.

6. Another name for an ordered arrangement of items is a(n) ______________.

7. A formula for calculating a permutation of n items taken r at a time is ______________________.

8. Permutations can also be counted using ______________________.

9. The permutation ${}_nP_0$ always equals ______________________.

10. Order does not matter for a(n) ______________________.

11. Order does matter for a(n) ______________________.

12. A formula for calculating a combination of n items taken r at a time is ______________________.

13. The combination ${}_nC_0$ always equals ______________________.

14. The combination ${}_nC_n$ always equals ______________________.

15. ______________________ is the likelihood that an event will occur.

16. The probability of an event that is impossible to happen is ______________________.

17. The probability of an event that is certain to happen is ______________________.

18. The set of all possible outcomes of an experiment is call the ______________________.

19. A subset of the sample space is called a(n) ______________________.

20. The empirical probability of event E is ______________________.

21. The theoretical probability of event E is ______________________.

Chapter 7 Highlights

DEFINITIONS AND CONCEPTS | **EXAMPLES**

Section 7.1 Counting and the Fundamental Counting Principle

A chance happening is called an **experiment.**

Tossing a coin once is an experiment.

The possible results of an experiment are called **outcomes.**

The outcomes of this experiment are heads (H) or tails (T).

The Fundamental Counting Principle

Suppose an experiment consists of separate parts. If one part has m choices and another part has n choices, then the total number of outcomes for this 2-part experiment is

$$m \cdot n$$

Note: This principle can be extended to three or more parts.

If a 6-sided die is rolled 3 times, how many outcomes are possible?

By the Fundamental Counting Principle,

1st roll number of choices		2nd roll number of choices		3rd roll number of choices	
↓		↓		↓	
6	·	6	·	6	= 216

One way to picture the outcomes of an experiment is to draw a **tree diagram.**

Draw a tree diagram for the experiment of tossing a coin, then choosing a primary color—red, blue, or yellow.

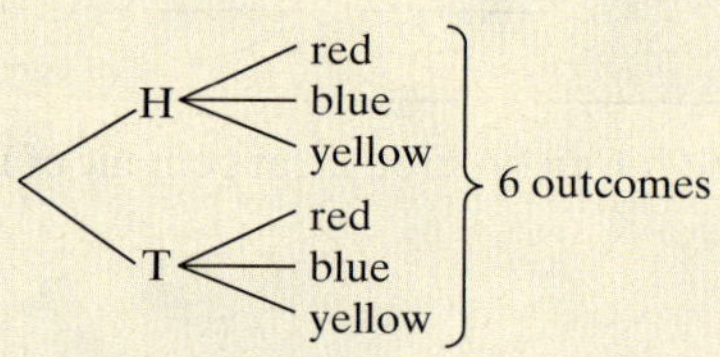

DEFINITIONS AND CONCEPTS	EXAMPLES

Section 7.2 Permutations

Factorial of n: $n!$

If n is a natural number, then

$$n! = n(n-1)(n-2)(n-3)\cdots 3\cdot 2\cdot 1.$$

The factorial of 0, written 0!, is defined to be 1.

$$0! = 1$$

Evaluate.

$$4! = 4\cdot 3\cdot 2\cdot 1 = 24$$
$$0! = 1$$

Ordered arrangements are also called **permutations**.

$_nP_r$ means **the number of permutations of n items taken r at a time.**

The number of ways that 6 photos can be arranged on a wall is a permutation of 6 items (photos) taken 6 at a time, or $_6P_6$.

Permutation Formula

The number of permutations or ordered arrangements of n distinct items taken r at a time where $r \le n$ is:

$$_nP_r = \frac{n!}{(n-r)!}$$

$$_6P_6 = \frac{6!}{(6-6)!} = \frac{6!}{0!} = \frac{6\cdot 5\cdot 4\cdot 3\cdot 2\cdot 1}{1} = 720$$

Section 7.3 Combinations

A counting problem is a **combination** when the order of items makes no difference, and there are no repetitions.

Permutations	***Combinations***
• Repetitions **are not** allowed. • Order **does** matter.	• Repetitions **are not** allowed. • Order **does not** matter.

The number of ways that 6 photos can be chosen from a collection of 10 photos is a combination of 10 items (photos) taken 6 at a time, or $_{10}C_6$.

Combination Formula

The number of combinations of n distinct items taken r at a time, where $r \le n$ is:

$$_nC_r = \frac{n!}{r!(n-r)!}$$

$$_{10}C_6 = \frac{10!}{6!(10-6)!} = \frac{10!}{6!4!} = \frac{10\cdot 9\cdot 8\cdot 7\cdot 6!}{6!4\cdot 3\cdot 2\cdot 1} = \frac{10\cdot 9\cdot 8\cdot 7}{4\cdot 3\cdot 2\cdot 1} = 210$$

Section 7.4 Introduction to Probability

Probability is the likelihood that some result will occur.

If a coin is tossed twice, what is the probability of the result being heads, heads?

The set S of all possible outcomes of an experiment is called the **sample space** of the experiment.

A subset of a sample space is called an **event.**

Sample space, $S = \{(H, T), (T, H), (H, H), (T, T)\}$

Event, $E = \{(H, H)\}$

Theoretical Probability of an Event

If all the outcomes of a sample space S are equally likely, then the theoretical probability of event E, denoted by P(E) is given by

$$P(E) = \frac{\text{number of outcomes in Event } E}{\text{total number of outcomes}} = \frac{n(E)}{n(S)}$$

Note: If a probability is asked for, we will assume it to be a theoretical probability.

Given the above, find the probability of heads, heads.

$$P(E) = \frac{n(E)}{n(S)} = \frac{1}{4}$$

The probability of (H, H) is $\frac{1}{4}$.

DEFINITIONS AND CONCEPTS	EXAMPLES

Section 7.4 Introduction to Probability (continued)

Theoretical probability is based on our assumption that each outcome is equally likely. **Empirical probability** is based on observed outcomes of actual experiments.

Empirical Probability Formula

The empirical probability of event E, denoted by $P(E)$, is the number of times event E occurs divided by the total number of trials of the experiment.

$$P(E) = \frac{\text{number times Event } E \text{ occurs}}{\text{total number of trials of experiment}}$$

Suppose a single coin is tossed 50 times and the results are:

H	T
29	21

Find the empirical probability of event "tails."

$$P(T) = \frac{21}{50} = 0.42$$

What is the theoretical probability of event "tails"?

$$P(T) = \frac{1}{2} = 0.5$$

Thus, theoretical $P(T) = 0.5$ and empirical $P(T) = 0.42$.

Rules for Probability — Theoretical or Empirical

- The probability of an event, $P(E)$ is 0, 1, or between 0 and 1

$$0 \le P(E) \le 1$$

- The probability of an outcome that is impossible is 0.
- The probability of an outcome that is certain to happen is 1.
- $P(E) + P(\text{not } E) = 1$

A coin is tossed once. Notice that

$$P(H) + P(T) = \frac{1}{2} + \frac{1}{2} = 1$$

The probability of rolling a die once and rolling a 7 is 0.

The probability of rolling a die once and rolling a 1, 2, 3, 4, 5, or 6 is 1.

Chapter 7 Review

(7.1) Draw a tree diagram for each experiment. Then use the diagram to find the number of possible outcomes.

Spinner A

Spinner B

1. Spinning Spinner B twice
2. Spinning Spinner A twice
3. Tossing a coin and then spinning Spinner B
4. Spinning Spinner A and then tossing a coin

Solve.

5. Find the number of outcomes if a 12-sided die is rolled twice.

6. Find the number of outcomes if a 12-sided die is rolled three times.
7. A multiple-choice exam has 5 questions whose answers you are unsure of. If each question has choices A, B, C, or D, how many different ways can you answer these questions?
8. A true/false exam has 7 questions whose answers you are unsure of. How many different ways can you answer these questions?
9. How many different 6-digit access codes are available given that the first digit may not be a 0 and the last digit may not be a 9?
10. How many different 5-digit access codes are available given that the first digit must be a 6, 7, 8, or 9?
11. A company is issuing identification badges. Each badge will begin with a letter followed by 2 digits. How many possible different badges are available?

12. Suppose the badges in Exercise 11 will begin with the letter A, B, or C followed by 3 digits. How many different badges are now possible?

(7.2) *Evaluate.*

13. 9! **14.** 8!

15. 0! **16.** 1!

17. $\frac{12!}{11!}$ **18.** $\frac{200!}{199!}$

19. $(10 - 8)!$ **20.** $(23 - 20)!$

21. ${}_6P_4$ **22.** ${}_9P_8$

23. ${}_{12}P_{12}$ **24.** ${}_{12}P_0$

25. Find the number of permutations of 10 objects taken 5 at a time.

26. Find the number of permutations of 8 objects taken 4 at a time.

27. In how many ways can we arrange the letters of the word "travel"?

28. In how many ways can we arrange the letters of the word "clear"?

29. How many different ways can an 8-member committee choose a chair person and a treasurer given that no person can hold both positions?

30. How many different ways can a 10-member committee choose a president, secretary, and treasurer given that no person can hold more than one position?

(7.3) *Evaluate*

31. ${}_{12}C_{11}$ **32.** ${}_{15}C_{14}$

33. ${}_{12}C_1$ **34.** ${}_{15}C_1$

35. ${}_{12}C_5$ **36.** ${}_{15}C_5$

37. ${}_8C_8$ **38.** ${}_9C_9$

39. ${}_{10}C_8$ **40.** ${}_{11}C_8$

41. ${}_{14}C_0$ **42.** ${}_{17}C_0$

Solve.

43. Fifteen college graduates apply for 3 jobs that are essentially the same. How many selections of 3 graduates from the 15 are possible?

44. Tryouts are held for the dancing team of a community college. If 42 students try out for 10 positions, how many different selections are possible?

45. The state of Colorado has a lottery game called Lotto. For this game, you pick 6 numbers from 1 to 42. If the order of the numbers does not matter, how many different selections are possible?

46. The state of Oregon has a lottery game called Pick 4. For this game, you pick 4 numbers (digits) from 0 to 9. If order does not matter, how many different selections are possible?

(7.4) *Suppose the spinner shown is spun once. Find the probability of each event.*

47. The result of the spin is 2.

48. The result of the spin is 3.

49. The result of the spin is 1, 2, or 3.

50. The result of the spin is not 3.

51. The result of the spin is an odd number.

52. The result of the spin is an even number.

If a single choice is made from the bag of marbles shown, find the probability of each event.

53. A red marble is chosen.

54. A blue marble is chosen.

55. A yellow marble is chosen.

56. A green marble is chosen.

57. A green or red marble is chosen.

58. A blue or yellow marble is chosen.

One card is dealt from a deck of 52. Find the probability that each is dealt.

Example of 52 playing cards; 13 of each suit:clubs, diamonds, hearts, and spades

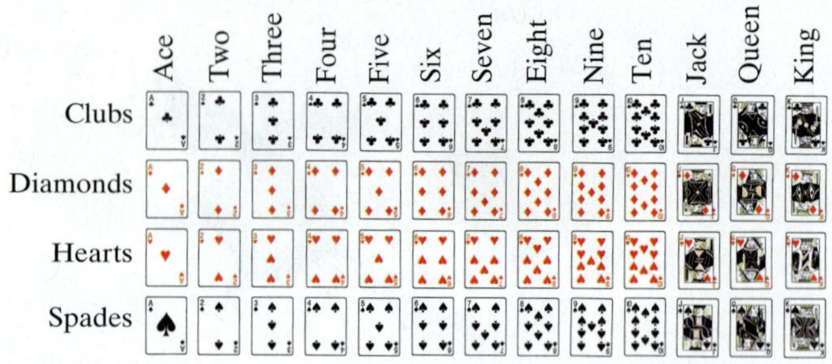

59. the 8 of hearts

60. the 4 of diamonds

61. an 8

62. a 4

63. a diamond

64. a club

65. an ace or a king

66. a face card (a jack, a queen, or a king)

A die is rolled once. Find the theoretical probability of each. Write the probability as a decimal rounded to 3 places if needed.

67. rolling a 5

68. rolling a 2

69. rolling an even number

70. rolling an odd number

A die is actually rolled 50 times, and the results are in the table. Find each empirical probabily. Write the probability as a decimal.

Number Rolled	*Frequency*
1	8
2	10
3	5
4	10
5	8
6	9
	50 Total

71. rolling a 5

72. rolling a 2

73. rolling an even number

74. rolling an odd number

Chapter 7 Getting Ready for the Test

MULTIPLE CHOICE *Choices A, B, C, and D are given. For Exercises 1 through 12, select the evaluation of each expression. Choices A, B, C, and D may not be used or may be used more than once.*

A. 0 **B.** 1 **C.** 3 **D.** 6

1. 0!

2. 1!

3. $(6 - 6)!$

4. 3!

5. $\frac{3!}{2!}$

6. ${}_6C_6$

7. ${}_6C_5$

8. ${}_6C_0$

9. ${}_3P_3$

10. ${}_3P_1$

11. ${}_3P_0$

12. $\frac{3 - 3}{(3 - 3)!}$

MATCHING *Match each concept in the first column with all statements in the second column that apply to the concept.*

13. combinations

14. permutations

A. repetitions are not allowed

B. order does matter

C. order does not matter

Chapter 7 Test

MyMathLab® YouTube

Evaluate.

1. 6!

2. 0!

3. $(7 - 4)!$

4. $7! - 4!$

5. ${}_9P_6$

6. ${}_9C_6$

7. Draw a tree diagram for the experiment of tossing a coin, then choosing a vowel: *a, e, i, o, u*. Then use the diagram to find the number of outcomes.

8. Find the number of 3-digit numbers that can be written using the digits {2, 4, 6, 8}

a. if the digits can be repeated.

b. if the digits cannot be repeated.

9. Find the number of outfits that can be chosen from 6 shirts, 2 suits, and 5 ties.

Exercises 10 and 11 have to do with a committee of 7 people.

10. In how many ways can 2 members be selected to travel to a national meeting?

11. In how many ways can a president and a secretary be selected if no member may hold both positions?

12. License plates in Tennessee consist of 3 letters followed by 3 digits. How many of these arrangements are possible?

13. South Dakota has a lottery game called Dakota Cash in which one selects 5 numbers from 1 to 35. If the order of the five numbers does not matter, how many selections are possible?

14. How many 3 digit *even* numbers are there? *Note:* The first digit may not be 0.

The spinner shown is spun once. Find the probability that the spinner stops on each.

15. a purple sector

16. an orange sector

17. the number 7

18. an odd number

A single die is rolled twice, and the equally likely 36 outcomes are shown.

First Die \ Second Die	⚀	⚁	⚂	⚃	⚄	⚅
⚀	(1, 1)	(1, 2)	(1, 3)	(1, 4)	(1, 5)	(1, 6)
⚁	(2, 1)	(2, 2)	(2, 3)	(2, 4)	(2, 5)	(2, 6)
⚂	(3, 1)	(3, 2)	(3, 3)	(3, 4)	(3, 5)	(3, 6)
⚃	(4, 1)	(4, 2)	(4, 3)	(4, 4)	(4, 5)	(4, 6)
⚄	(5, 1)	(5, 2)	(5, 3)	(5, 4)	(5, 5)	(5, 6)
⚅	(6, 1)	(6, 2)	(6, 3)	(6, 4)	(6, 5)	(6, 6)

Find the probability of each event.

19. a 2, then a 1

20. two of the same numbers

21. two numbers that are not the same

22. two numbers whose sum is 6

23. two numbers whose sum is 7

24. two numbers whose sum is 13

25. two numbers whose sum is ≤ 12

One card is dealt from a standard deck of 52. Find the probability that each is dealt.

Example of 52 playing cards; 13 of each suit:clubs, diamonds, hearts, and spades

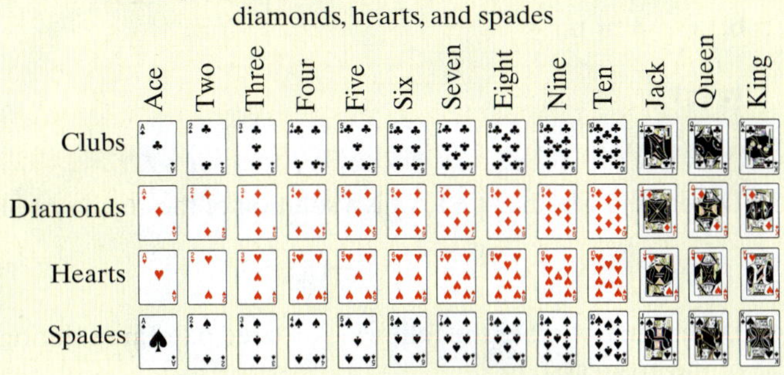

26. the ace of diamonds

27. a heart

28. an ace

29. a queen or a king

An experiment was performed by actually tossing a coin 100 times and the results are below.

	Frequency
H	59
T	41
	100 Total

30. Find the empirical probability of tossing a head.

31. What is the theoretical probability of tossing a head?

Chapter 7 Cumulative Review

1. Simplify.

$$\frac{-5\sqrt{30-5}+(-2)^2}{4^2+|7-10|}$$

2. Simplify each expression.

a. $\dfrac{4(-3)-(-6)}{-8+4}$

b. $\dfrac{3+(-3)(-2)^3}{-1-(-4)}$

3. Solve: $0.25x + 0.10(x - 3) = 0.05(22)$

4. Solve: $6y - 11 + 4 + 2y = 8 + 15y - 8y$

5. Complete the table for the equation $y = \frac{1}{2}x - 5$.

x	y
−2	
0	
	0

6. Complete the table for the equation $2x + y = 6$.

x	y
0	
	−2
3	

7. Find an equation of the line with y-intercept $(0, -3)$ and slope of $\frac{1}{4}$.

8. Find an equation of the line with y-intercept $(0, 4)$ and slope of -2.

9. Find an equation of the line parallel to the line $y = 5$ and passing through $(-2, -3)$.

10. Find an equation of the line perpendicular to $y = 4$ and passing through $(1, 5)$.

11. Decide whether the equation describes a function.

a. $y = x$

b. $y = 2x + 1$

c. $y = 5$

d. $x = -1$

12. Decide whether the equation describes a function.

a. $2x + 3 = y$

b. $x + 4 = 0$

c. $\frac{1}{2}y = 2x$

d. $y = 0$

13. Determine whether $(-1, 1)$ is a solution of the given system.

$$\begin{cases} -x + y = 2 \\ 2x - y = -3 \end{cases}$$

14. Determine whether the following ordered pairs are solutions of the given system.

$$\begin{cases} 2x + y = 4 \\ x + y = 2 \end{cases}$$

a. $(1, 1)$ **b.** $(2, 0)$

15. Solve the system.

$$\begin{cases} 7x - 3y = -14 \\ -3x + y = 6 \end{cases}$$

16. Solve the system.

$$\begin{cases} 3y = x + 10 \\ 2x + 5y = 24 \end{cases}$$

17. Solve the system.

$$\begin{cases} 3x + \dfrac{y}{2} = 2 \\ 6x + y = 5 \end{cases}$$

18. Solve the system.

$$\begin{cases} \dfrac{x}{2} + y = \dfrac{5}{6} \\ 2x - y = \dfrac{5}{6} \end{cases}$$

19. The Swiss Re Building, in London, is a unique building. Londoners often refer to it as the "pickle building." The building is 592.1 feet tall. An object is dropped from the highest point of this building. Neglecting air resistance, the height in feet of the object above ground at time t seconds is given by the polynomial function $P(t) = -16t^2 + 592.1$. Find the height of the object when $t = 1$ second, and when $t = 6$ seconds.

20. Find the height of the object in Exercise 19 when $t = 3$ seconds and when $t = 5$ seconds.

21. Graph the solution of the system.

$$\begin{cases} 3x \geq y \\ x + 2y \leq 8 \end{cases}$$

22. Graph the solution of the system.

$$\begin{cases} x + y \leq 1 \\ 2x - y \geq 2 \end{cases}$$

23. Combine like terms to simplify $-9x^2 + 3xy - 5y^2 + 7yx$.

24. Combine like terms to simplify.

$4a^2 + 3a - 2a^2 + 7a - 5$

25. Factor: $2x^4 - 26x^3 + 84x^2$

26. Factor: $3x^2 + 15x + 18$

27. Simplify.

a. $\sqrt{54}$ **b.** $\sqrt{12}$

c. $\sqrt{200}$ **d.** $\sqrt{35}$

28. Simplify.

a. $\sqrt{40}$ **b.** $\sqrt{500}$

c. $\sqrt{63}$ **d.** $\sqrt{169}$

29. Multiply. Then simplify if possible.

a. $\sqrt{5}(\sqrt{5} - \sqrt{2})$

b. $(\sqrt{5} - 7)(\sqrt{5} + 7)$

30. Multiply. Then simplify if possible.

a. $(\sqrt{6} + 2)^2$

b. $(\sqrt{2} + 5)(\sqrt{2} - 5)$

31. Find the length of the hypotenuse of a right triangle whose legs are 6 inches and 8 inches long.

32. Find the length of the unknown leg of a right triangle whose other leg is 9 and whose hypotenuse is 13.

CHAPTER 8

Introduction to Statistics

In this chapter, we introduce statistics, or the analysis and interpretation of data. First, we study a type of descriptive statistics called measures of central tendency. Here, we learn ways to best describe a list of numerical data items with a single number. Next, we study ways to measure the dispersion of data items, or the degree in which the data values are scattered. Finally, this chapter ends with a study of data items that—when graphed—form a bell curve, or a normal distribution. These data items have much in common.

A drawing of a quincunx by Sir Francis Galton on the left and a working model on the right

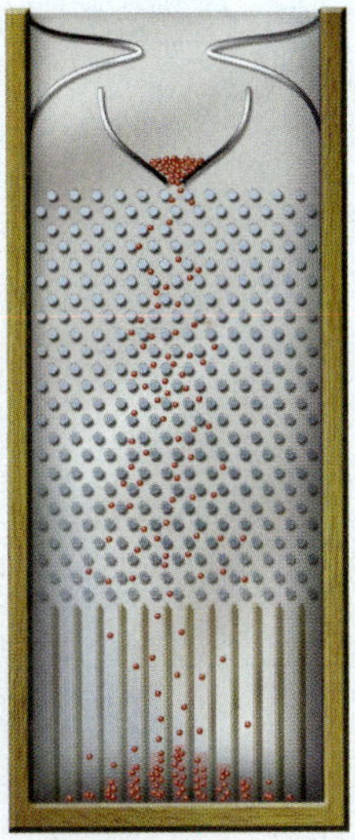

What is a Quincunx?

The quincunx is also known as the bean machine or Galton box. This device was invented by *Sir Francis Galton*, an English statistician who lived from 1822 to 1911. This device is meant to be placed vertically. Small balls are dropped from the top which then bounce randomly off the pins and then drop into a column and collect at the bottom. The pattern the collected balls form is a bell curve, or normal distribution curve.

Below is a bell curve, or a normal distribution curve showing standard deviations and the 68-95-99.7 rule. We study all of these concepts throughout this chapter.

(Notes:

*The quincunx is also known as the geometric pattern of 5 points arranged as on a die, ⁙

*Many games under many names have developed from the idea of balls randomly bouncing off pins. Among these games are pinball, Plinko, Pachinko, and Peggle.)

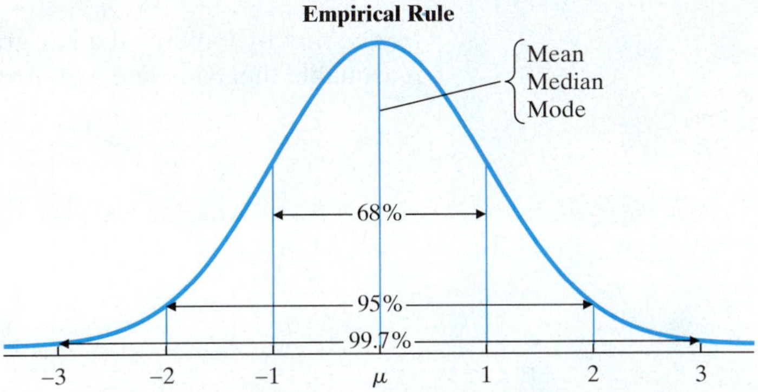

8.1 Measures of Central Tendency

OBJECTIVES

1 Determine the Average or Arithmetic Mean of a Data Set.

2 Determine the Median of a Data Set.

3 Determine the Mode of a Data Set.

We are certainly familiar with the word "average." The next four examples show real-life averages. For example,

- Adults employed in the United States report working an **average** of 47 hours per week, according to a Gallup poll.
- The U.S. miles per gallon **average** for light vehicles is 25.5, according to autonews.com (*Automotive News*).

As our accumulation of data increases, our ability to gather, store, and present these tremendous amounts of data increases. Sometimes it is desirable to be able to describe a set of data by a single "middle" number or a measure of central tendency. Three of the most common measures of central tendency are the mean (or average), the median, and the mode.

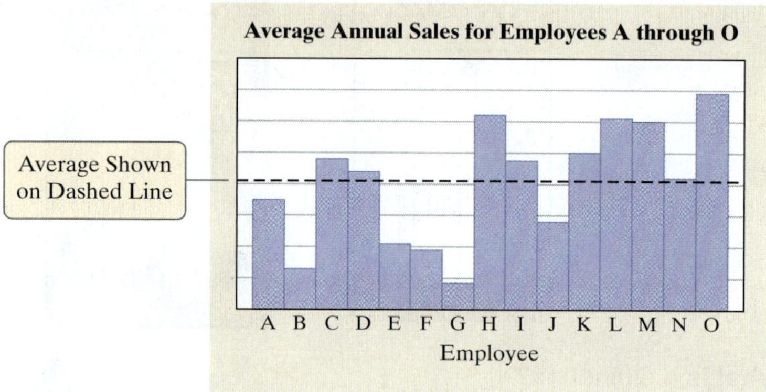

The Average Retirement Age in the U.S.

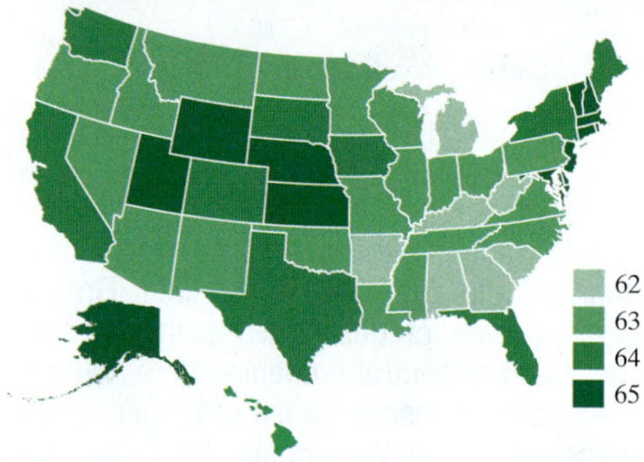

Based on U.S. Census Bureau labor force participation data.

OBJECTIVE

1 Finding the Average or Mean of a Data Set

Although we are probably familiar with finding an **average,** or **mean,** let's introduce a few new symbols. Keep in mind, though, that the way we calculate mean is the way we are familiar with: mean $= \dfrac{\text{sum of items}}{\text{number of items}}$.

For example: To find the mean of four test scores—86, 82, 93, and 75—we find the sum of the grades and then divide by the number of grades, 4.

$$\text{mean} = \frac{86 + 82 + 93 + 75}{4} = \frac{336}{4} = 84$$

Notice that by looking at a bar graph of the scores with a dashed line at the mean, it is reasonable that 84 is one *measure of central tendency*.

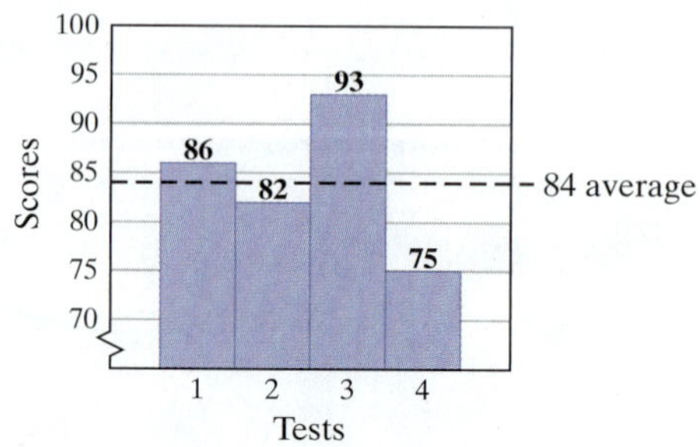

Now we introduce new vocabulary words. We now use the notation

$\bar{x}$ (read "x bar") to denote the mean of a sample of data.

Finding sums is very common in statistics, so we use the Greek letter

$$\Sigma \text{(sigma) to mean "sum of".}$$

Since finding the mean requires us to find the sum of n data items, say $x_1, x_2, x_3, \ldots, x_n$, the notation

$$\Sigma x = x_1 + x_2 + \cdots + x_n.$$

Let's use our new notation to write a formula for mean.

Mean

The mean of n data items, $x_1, x_2, x_3, \ldots, x_n$, is given by:

$$\bar{x} = \frac{\Sigma x}{n}$$

EXAMPLE 1 **Finding the Mean Time in an Experiment**

Seven students in a psychology class conducted an experiment on mazes. Each student was given a pencil and asked to successfully complete the same maze. The timed results are below:

Student	A	B	C	D	E	F	G
Time (Seconds)	13.2	11.8	10.7	16.2	15.9	13.8	18.5

a. Which student completed the maze in the shortest time? Which student completed the maze in the longest time?

b. Find the mean time.

c. How many students took longer than the mean time? How many students took shorter than the mean time?

Solution

a. Student C completed the maze in 10.7 seconds, the shortest time. Student G completed the maze in 18.5 seconds, the longest time.

b. To find the mean (or average), we use our formula with $n = 7$, the number of items.

$$\bar{x} = \frac{\Sigma x}{n} = \frac{13.2 + 11.8 + 10.7 + 16.2 + 15.9 + 13.8 + 18.5}{7}$$

$$= \frac{100.1}{7} = 14.3$$

c. Three students, Students D, E, and G, had times longer than the mean time. Four students, Students A, B, C, and F, had times shorter than the mean time. □

PRACTICE 1 Find the mean of the following test scores: 87, 75, 96, 91, and 78.

✓CONCEPT CHECK

Estimate the mean of the following set of data:

5, 10, 10, 10, 10, 15

The mean has one main disadvantage. This measure of central tendency can be greatly affected by *outliers*. (Outliers are values that are especially large or small when compared with the rest of the data set.) Let's see an example of this next.

Answer to Concept Check:
10

EXAMPLE 2 The table lists the rounded salary of 10 staff numbers.

Staff	A	B	C	D	E	F	G	H	I	J
Salary (in thousands)	\$32	\$34	\$46	\$38	\$42	\$95	\$102	\$50	\$42	\$41

a. Find the mean of all 10 staff members.

b. Find the mean of all staff members except F and G.

Solution

a. $\bar{x} = \dfrac{\Sigma x}{n} = \dfrac{32 + 34 + 46 + 38 + 42 + 95 + 102 + 50 + 42 + 41}{10} = \dfrac{522}{10}$

$= 52.2$

The mean salary is \$52.2 thousand or \$52,200.

b. $\bar{x} = \dfrac{\Sigma x}{n} = \dfrac{32 + 34 + 46 + 38 + 42 + 50 + 42 + 41}{8} = \dfrac{325}{8}$

$= 40.625$

Now, the mean salary is \$40.625 thousand or \$40,625. □

PRACTICE 2 Use the table in Example 2 and find the mean salary of all staff members except G. Round thousands of dollars to 2 decimal places.

The mean in part **a.** does not appear to be a measure of central tendency because this mean, \$52.2 thousand, is greater than all salaries except 2 of the 10. Also, notice the difference in the means for parts **a.** and **b.** By removing the 2 outliers, the mean was greatly reduced.

Although the mean was calculated correctly each time, parts **a.** and **b.** of Example 2 show one disadvantage of the mean. That is, a few numerical outliers can greatly affect the mean.

Helpful Hint

Remember an important disadvantage of the mean.
If our data set has a few outliers, the mean may not be the best measure of central tendency.

Often in college, the calculation of a **grade point average** (GPA) is a **weighted mean** and is calculated as shown in Example 3.

EXAMPLE 3 **Calculating Grade Point Average (GPA)**

The following grades were earned by a student during one semester. Find the student's grade point average.

Course	*Grade*	*Credit Hours*
College mathematics	A	3
Biology	B	3
English	A	3
PE	C	1
Social studies	D	2

Solution To calculate the grade point average, we need to know the point values for the different possible grades. The point values of grades commonly used in colleges and universities are given below:

A: 4, B: 3, C: 2, D: 1, F: 0

Now, to find the grade point average, we multiply the number of credit hours for each course by the point value of each grade. The grade point average is the sum of these products divided by the sum of the credit hours.

Course	*Grade*	*Point Value of Grade*	*Credit Hours*	*Point Value of Credit Hours*
College mathematics	A	4	3	12
Biology	B	3	3	9
English	A	4	3	12
PE	C	2	1	2
Social studies	D	1	2	2
		Totals:	12	37

$$\text{grade point average} = \frac{37}{12} \approx 3.08 \text{ rounded to two decimal places}$$

This student earned a grade point average of 3.08. □

PRACTICE 3 Find the grade point average if the following grades were earned in one semester.

Grade	*Credit Hours*
A	2
B	4
C	5
D	2
A	2

OBJECTIVE 2

Finding the Median of a Data Set

You may have noticed that a very low number or a very high number can affect the mean of a list of numbers. Because of this, you may sometimes want to use another measure of central tendency. A second measure of central tendency is called the **median.** The median of a list of numbers is not affected by a low or high number in the list.

The **median** of a set of numbers in numerical order is the middle number.

If the number of items is odd, the median is the middle number.

If the number of items is even, the median is the *mean* of the two middle numbers.

For a long list of data items, this formula will give you the position of the median.

Position of the Median

For n data items in order from smallest to largest, the median is the item in the $\frac{n+1}{2}$ position.

Helpful Hint

The formula above, $\frac{n+1}{2}$ does not give the *value* of the median, just the **position of the median.**

EXAMPLE 4 Find the median of the following list of numbers:

25, 54, 56, 57, 60, 71, 98

Solution Because this list is in numerical order, and there is an odd number of items in the list, the median is the middle number, 57. □

PRACTICE 4 Find the median of the list of numbers: 5, 11, 14, 23, 24, 35, 38, 41, 43

EXAMPLE 5 Find the median of the following list of scores: 67, 91, 75, 86, 55, 91

Solution First we list the scores in numerical order and then we find the middle number.

55, 67, 75, 86, 91, 91

Since there is an even number of scores, there are two middle numbers, 75 and 86. The median is the mean of the two middle numbers.

$$\text{median} = \frac{75 + 86}{2} = 80.5$$

The median is 80.5. □

Helpful Hint
Don't forget to write the numbers in order from smallest to largest before finding the median.

PRACTICE 5 Find the median of the list of scores:

36, 91, 78, 65, 95, 95, 88, 71 ■

OBJECTIVE

3 Finding the Mode of a Data Set

The last common measure of central tendency is called the **mode.**

The **mode** of a set of numbers is the number that occurs most often. (It is possible for a set of numbers to have more than one mode or to have no mode.)

EXAMPLE 6 Find the mode of the list of numbers:

11, 14, 14, 16, 31, 56, 65, 77, 77, 78, 79

Solution There are two numbers that occur the most often. They are 14 and 77. This list of numbers has two modes, 14 and 77. □

PRACTICE 6 Find the mode of the list of numbers:

14, 10, 10, 13, 15, 15, 15, 17, 18, 18, 20 ■

EXAMPLE 7 Find the median and the mode of the following set of numbers. These numbers were high temperatures for 14 consecutive days in a city in Montana.

76, 80, 85, 86, 89, 87, 82, 77, 76, 79, 82, 89, 89, 92

Solution First we write the numbers in numerical order.

76, 76, 77, 79, 80, 82, 82, 85, 86, 87, 89, 89, 89, 92

Since there is an even number of items, the median is the mean of the two middle numbers, 82 and 85.

$$\text{median} = \frac{82 + 85}{2} = 83.5$$

The mode is 89, since 89 occurs most often. □

PRACTICE 7 Find the median and the mode of the list of numbers:

26, 31, 15, 15, 26, 30, 16, 18, 15, 35 ■

✔ **CONCEPT CHECK**
True or false? Every set of numbers *must* have a mean, median, and mode. Explain your answer.

Answer to Concept Check:
false; a set of numbers may have no mode

Helpful Hint

Don't forget that it is possible for a list of numbers to have no mode. For example, the list

2, 4, 5, 6, 8, 9

has no mode. There is no number or numbers that occur more often than the others.

Recall that in Section R.4, we organized data with frequency distribution tables.

EXAMPLE 8 Find the (a) mean, (b) median, and (c) mode from this frequency distribution table of retirement ages. If needed, round answers to 1 decimal place.

Age	*Frequency*
60	3
61	1
62	1
63	2
64	2
65	2

Solution Study the table for a moment. From the frequency column, we see that there are 11 data items $(3 + 1 + 1 + 2 + 2 + 2)$.

a. To find the mean, we use our mean formula:

$$\bar{x} = \frac{\Sigma x}{n} = \frac{3 \cdot 60 + 61 + 62 + 2 \cdot 63 + 2 \cdot 64 + 2 \cdot 65}{11}$$

$$= \frac{690}{11} \approx 62.7$$

b. Since there are 11 data items and the items are arranged in numerical order in the table, we find $\frac{n+1}{2}$ to locate the middle item. This is $\frac{11+1}{2} = \frac{12}{2} = 6$, or the sixth item.

The median is the sixth number, or 63.

c. The mode has the greatest frequency, so the mode is 60. □

PRACTICE 8 One state with a young retirement age is Michigan. The table below is from a poll of retirement ages from that state.

Age	*Frequency*
50	1
59	3
60	3
62	5
63	2
67	1

Find the (a) mean, (b) median, and (c) mode of these data. If needed, round answers to 1 decimal place. ■

Vocabulary, Readiness & Video Check

Use the choices below to fill in each blank. Choices may be used more than once or not at all.

mean	mode	grade point average	x
median	average	$\bar{x}$	

1. Another word for "mean" is ________________.

2. The number that occurs most often in a set of numbers is called the ________________.

3. The formula for ________________ of a set of number items is $\frac{\text{sum of items}}{\text{number of items}}$ or $\frac{\Sigma x}{n}$.

4. The ________________ of a set of numbers is the middle number. If the number of numbers is even, it is the ________________ of the two middle numbers.

5. An example of weighted mean is a calculation of ________________.

6. We denote mean by the symbol ________________.

Martin-Gay Interactive Videos

See Video 8.1

Watch the section lecture video and answer the following questions.

OBJECTIVE 1

7. Why is the ≈ symbol used in Example 1?

8. For Example 2, we used another symbol for mean. What is that symbol and what is the new way to write the formula for mean?

OBJECTIVE 2

9. From Example 4, what is always the first step when finding the median of a set of data numbers?

OBJECTIVE 3

10. From Example 5, why do you think it is helpful to have data numbers in numerical order when finding the mode?

8.1 Exercise Set MyMathLab®

MIXED PRACTICE

For each set of numbers, find the mean, median, and mode. If necessary, round the mean to one decimal place. See Examples 1, 2 and 4 through 8.

1. 15, 23, 24, 18, 25

2. 45, 36, 28, 46, 52

3. 7.6, 8.2, 8.2, 9.6, 5.7, 9.1

4. 4.9, 7.1, 6.8, 6.8, 5.3, 4.9

5. 0.5, 0.2, 0.2, 0.6, 0.3, 1.3, 0.8, 0.1, 0.5

6. 0.6, 0.6, 0.8, 0.4, 0.5, 0.3, 0.7, 0.8, 0.1

7. 231, 543, 601, 293, 588, 109, 334, 268

8. 451, 356, 478, 776, 892, 500, 467, 780

The ten tallest buildings in the world, completed as of 2016, are listed in the following table. Use this table to answer Exercises 9 through 14. If necessary, round results to one decimal place. See Examples 1, 2, 4, 5, 7, and 8.

Building	*Height (in feet)*
Burj Khalifa, Dubai	2717
Shanghai Tower, China	2073
Makkah Royal Clock Tower Hotel, Saudi Arabia	1972
Ping An Finance Centre, China	1965
Lotte World Tower, South Korea	1823
Freedom Tower, New York City	1776
CTF Finance Centre, China	1740
Taipei 101	1667
Shanghai World Financial Center	1614
International Commerce Centre, Hong Kong	1588

(*Source:* skyscrapercenter.com and Internet research)

9. Find the mean height of the six tallest buildings.

10. Find the median height of the six tallest buildings.

11. Find the median height of the ten tallest buildings.

12. Find the mean height of the ten tallest buildings.

13. Given the building heights, explain how you know, without calculating, that the answer to Exercise **10** is greater than the answer to Exercise **11**.

14. Given the building heights, explain how you know, without calculating, that the answer to Exercise **12** is less than the answer to Exercise **9**.

For Exercises 15 through 18, the grades are given for a student for a particular semester. Find the grade point average. If necessary, round the grade point average to the nearest hundredth. See Example 3.

15.

Grade	*Credit Hours*
B	3
C	3
A	4
C	4

16.

Grade	*Credit Hours*
D	1
F	1
C	4
B	5

17.

Grade	*Credit Hours*
A	3
A	3
A	4
B	3
C	1

18.

Grade	*Credit Hours*
B	2
B	2
C	3
A	3
B	3

MIXED PRACTICE

During an experiment, the following times (in seconds) were recorded:

7.8, 6.9, 7.5, 4.7, 6.9, 7.0.

19. Find the mean.

20. Find the median.

21. Find the mode.

In a mathematics class, the following test scores were recorded for a student:

93, 85, 89, 79, 88, 91.

22. Find the mean.

23. Find the median.

24. Find the mode.

The following pulse rates were recorded for 1 minute for each of 15 students:

78, 80, 66, 68, 71, 64, 82, 71, 70, 65, 70, 75, 77, 86, 72.

25. Find the mean.

26. Find the median.

27. Find the mode.

28. How many pulse rates were higher than the mean?

29. How many pulse rates were lower than the mean?

30. Explain how to find the position of the median.

Below are lengths for the six longest rivers in the world.

Name	*Length (miles)*
Nile	4160
Amazon	4000
Yangtze	3915
Mississippi-Missouri	3709
Ob-Irtysh	3459
Huang Ho	3395

Find the mean and the median for each of the following.

31. the six longest rivers

32. the three longest rivers

Use each frequency distribution table to find the **a.** *mean,* **b.** *median, and* **c.** *mode. If needed, round the mean to 1 decimal place.*

33.

Data Item	*Frequency*
5	1
6	1
7	2
8	5
9	6
10	2

34.

Data Item	*Frequency*
3	2
4	1
5	4
6	7
7	2
8	1

35.

Data Item	*Frequency*
2	5
3	7
4	4
5	7
6	8
7	8
8	8
9	6
10	5

36.

Data Item	*Frequency*
4	3
5	8
6	5
7	8
8	2

*Use each graph of data items to find the **a.** mean, **b.** median, and **c.** mode. If needed, round the mean to one decimal place.*

37.

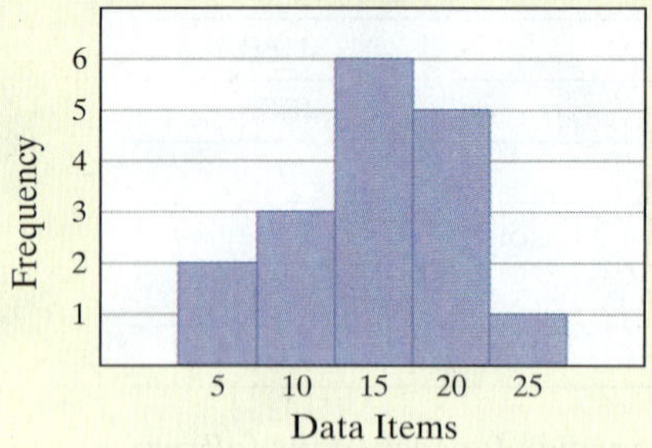

38.

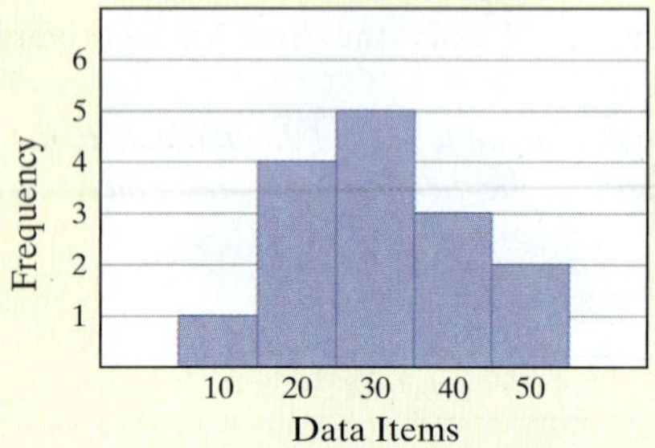

39.

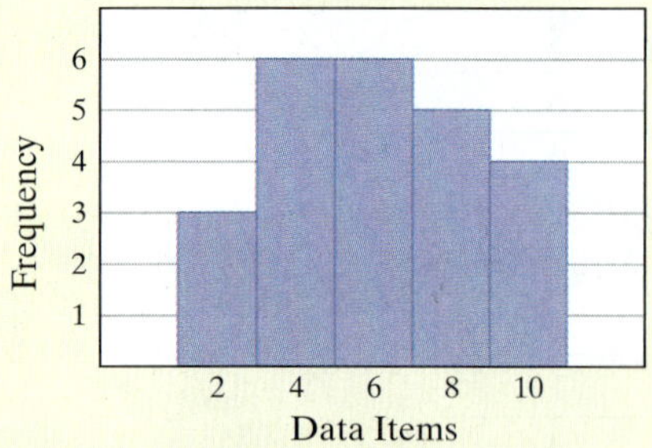

40.

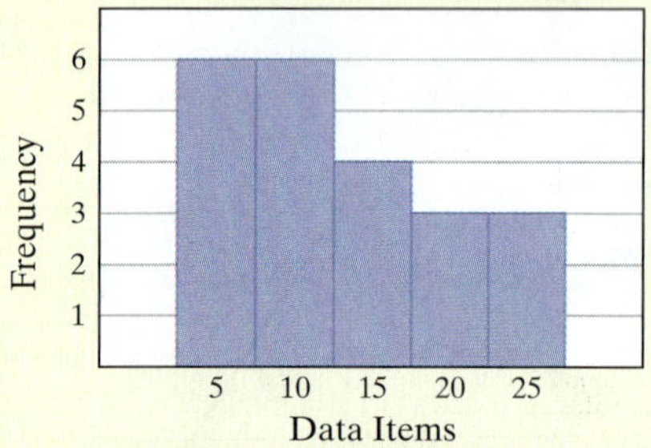

REVIEW AND PREVIEW

*When a set of data contains a large number of items, we may arrange the items into groups or **classes**. The follow items are test scores for 30 students in a prealgebra class.*

35	85	72	59	99	28	45	75	77	95
38	21	60	60	92	82	85	78	81	59
52	35	80	75	88	90	72	82	88	86

Use the table below to find the number of items in each of the following group. Then use this information to find the empirical probability for each class. Write the empirical probability as a fraction. See Section 7.4.

Class	*Frequency*	*Empirical Probability*
20–29	**41.** ________	**49.** ________
30–39	**42.** ________	**50.** ________
40–49	**43.** ________	**51.** ________
50–59	**44.** ________	**52.** ________
60–69	**45.** ________	**53.** ________
70–79	**46.** ________	**54.** ________
80–89	**47.** ________	**55.** ________
90–100	**48.** ________	**56.** ________

57. Find the sum of the empirical probabilities.

58. Count the number of data items then find the frequency total.

CONCEPT EXTENSIONS

Find the missing numbers in each set of numbers.

59. 16, 18, ________, ________, ________. The mode is 21. The median is 20.

60. ________, ________, ________, 40, ________. The mode is 35. The median is 37. The mean is 38.

61. Write a list of numbers for which you feel the median would be a better measure of central tendency than the mean.

62. Without making any computations, decide whether the median of the following list of numbers is a whole number. Explain your reasoning.

36, 77, 29, 58, 43

8.2 Measures of Dispersion

OBJECTIVES

1 Calculate the Range of a Data Set.

2 Calculate the Standard Deviation of a Data Set.

In Section 8.1, we studied three measures of central tendency—the mean, the median, and the mode. All three of these can be used to calculate a number that might best describe a data set.

In this section, we study two ways to describe the dispersion of a data set. What is dispersion? In statistics, **dispersion** is a way to describe the degree to which the data values are scattered. In the case of standard deviation, we measure the degree in which the data values are scattered about the mean.

OBJECTIVE

1 Finding the Range of a Data Set

A quick and simplè measure of dispersion of a data set is range.

Range

The range of a data set is the difference between the largest data value and the smallest data value.

$$\text{range} = \text{largest data value} - \text{smallest data value}$$

EXAMPLE 1 The following pulse rates (for 1 minute) were recorded for a group of 15 students. Find the range.

$$78, 80, 66, 68, 71, 64, 82, 71, 70, 65, 70, 75, 77, 86, 72.$$

Solution

$$\begin{aligned}\text{range} &= \text{largest data value} - \text{smallest data value}\\ &= 86 - 64\\ &= 22\end{aligned}$$

The range of this data set is 22. □

PRACTICE

1 The table lists the rounded salary of 10 staff members. Find the range.

Staff	A	B	C	D	E	F	G	H	J
Salary (in thousands)	\$32	\$34	\$46	\$38	\$42	\$95	\$102	\$50	\$42

OBJECTIVE

2 Finding the Standard Deviation

Above, we calculated range, or the degree in which data values are scattered. Now, we calculate another measure of despersion called **standard deviation,** which is the degree in which data values are scattered from the mean, or $\bar{x}$.

Important Notation Note

In statistics, it is important to recognize the difference between sample and population. Although we will not expand on this discussion, sample and population have different meanings, and this is important in the study of data. Although **sample** mean, $\bar{x}$, and **population** mean, μ (the Greek letter read "mew"), are calculated in the same way, this is not true of **sample** standard deviation and **population** standard deviation.

In this section, we calculate **sample** standard deviation by steps and by a graphing calculator program or computer program if allowed. When using a program, notice that most of the time both standard deviations are calculated—**sample** standard deviation, denoted by s, and **population** standard deviation, denoted by lowercase σ (a Greek lowercase sigma).

Helpful Hint

When we refer to standard deviation, we mean sample standard deviation, s.

Since a standard deviation has to do with deviations from the mean, let's start with a given data set, find the mean, then find deviations from the mean.

EXAMPLE 2 Find each deviation from the mean for the data set:

261, 278, 200, 142, 169

Solution Let's first find the mean, where $n = 5$, the number of data items.

$$\bar{x} = \frac{\Sigma x}{n} = \frac{261 + 278 + 200 + 142 + 169}{5} = \frac{1050}{5} = 210$$

Thus, the mean, $\bar{x} = 210$

Now let's calculate how much each data item deviates from the mean.

Data Item	*Deviation: Data Item − Mean*
261	$261 - 210 = 51$
278	$278 - 210 = 68$
200	$200 - 210 = -10$
142	$142 - 210 = -68$
169	$169 - 210 = -41$

□

PRACTICE 2 Find each deviation from the mean for the data set:

48, 30, 40, 42, 16, 45, 45.

■

For Example 2, notice that the sum of the deviations is 0.

$$51 + 68 + (-10) + (-68) + (-41) = 0.$$

In fact, this is always true. Since mean is a measure of central tendency, the deviations from the mean should sum to 0. Since we have a sum of 0 each time, calculating standard deviation can't simply be the mean (average) of the deviations.

Helpful Hint

For each data set, the sum of the deviations (data item − mean) is always 0.

Although there are many graphing calculator and computer programs that calculate standard deviation, it is a good idea to calculate a few of these standard deviations step by step. This way, we have a better understanding and appreciation of what standard deviation means.

Below is the formula for sample standard deviation. We historically use the lowercase letter s to denote sample standard deviation. (With normal distributions, we will use the Greek lowercase sigma, σ, to stand for population standard deviation.)

Calculating the Sample Standard Deviation of a Data Set

Suppose a data set of n items, $x_1, x_2, x_3, \ldots, x_n$, has a mean $\bar{x}$. Then the standard deviation s of this data set is given by the formula

$$s = \sqrt{\frac{\Sigma(x - \bar{x})^2}{n - 1}}$$

Steps for Calculating s:

1. Find $\bar{x}$, the mean of the data items.
2. Find the deviations from the mean.
3. Square each deviation.
4. Find the sum of the squared deviations.
5. Take the square root of the quotient, $\dfrac{\text{sum in Step 4}}{n - 1}$.

Helpful Hint

Notice that the steps on the previous page are simply a step-by-step method for the calculations in the standard deviation formula.

Let's continue with the table from Example 2 and calculate the (sample) standard deviation.

EXAMPLE 3 Find the standard deviation for the data set in Example 2:

261, 278, 200, 142, 169

Round the standard deviation to 2 decimal places.

Solution

Step 1. $\bar{x} = 210$; **Step 2.**; **Step 3.**

Data Item	*Deviation: Data Item-Mean*	*Square Each Deviation*
261	$261 - 210 = 51$	$(51)^2 = 2601$
278	$278 - 210 = 68$	$(68)^2 = 4624$
200	$200 - 210 = -10$	$(-10)^2 = 100$
142	$142 - 210 = -68$	$(-68)^2 = 4624$
169	$169 - 210 = -41$	$(-41)^2 = 1681$

Helpful Hint

Check to see that the sum of the deviations (data item − mean) is 0.

Step 4. Sum of squared deviations: 13,630

Step 5. Take the square root of $\frac{\text{sum in Step 4}}{5 - 1}$:

$$s = \sqrt{\frac{13{,}630}{4}} \approx 58.37$$

The standard deviation is approximately 58.37. □

PRACTICE 3 Find the standard deviation for the data set in Practice 2:

48, 30, 40, 42, 16, 45, 45.

Round the standard deviation to two decimal places. ■

In the next section, we will study the normal distribution and how it relates to standard deviation. For now, know that the larger the standard deviation, the larger the dispersion (or the spread of data).

Graphing Calculator Explorations

We can use a graphing calculator to calculate sample standard deviation. For example, let's calculate the sample standard deviation, s, of the data set

38, 21, 26, 15.

First enter the data set as list L_1. To do so, press [STAT] [ENTER]. Then press [38] [ENTER] to enter 38 as the first number in L_1. Next press [21] [ENTER] to enter 21 as the second number in L_1. Continue in the same manner and enter 26 and 15.

To find sample standard deviation s, press [STAT] then right arrow key [▶], to go to the CALC drop down menu.

After that, simply press [ENTER] since we want the first item in the CALC drop down menu, which is 1-Var Stats. The phrase 1-Var Stats enters on the main screen, so we press [ENTER] again and here's what appears.

1-Var Stats	most of these items we know
$\bar{x} = 25$	mean
$\Sigma x = 100$	sum of data items
$\Sigma x^2 = 2786$	sum of squared data items
$Sx = 9.763879011$	sample SD
$\sigma x = 8.455767263$	(identified in the next section)
$\downarrow n = 4$	number of data items

Use a graphing calculator to calculate the sample standard deviation of each data set. Round each to 2 decimal places.

1. 86, 43, 95, 90, 98

2. 66, 48, 87, 87, 95

3. 1.7, 1.6, 2.8, 2.6, 1.6, 2

4. 2.5, 1.9, 3.1, 2.8, 2.6, 2.9

5. 9, 9, 9, 8, 8, 10, 10, 8, 10

6. 10, 10, 13, 13, 13, 16, 16

Vocabulary, Readiness & Video Check

Use the choices below to fill in each blank. Choices may be used more than once or not at all.

sample standard deviation	0	dispersion	$\sqrt{\frac{\Sigma(x-\bar{x})^2}{n}}$
range	1	$\sqrt{\frac{\Sigma(x-\bar{x})^2}{n-1}}$	

1. ____________________ = largest data value − smallest data value

2. For a single data set, the sum of the deviations from the mean is always ____________________.

3. The formula for ____________________ is $s =$ ____________________.

4. The degree in which the data values are scattered is called ____________________.

Martin-Gay Interactive Videos

See Video 8.2

Watch the section lecture video and answer the following questions.

OBJECTIVE 1

5. Based on the results of Example 2, give an example of a data set of four data items whose range is 0.

OBJECTIVE 2

6. When working Example 3, we learned that the sum of the deviations from the mean is always what number?

8.2 Exercise Set MyMathLab®

Find the range for each data set. See Example 1.

1. 14, 16, 8, 10, 20

2. 25, 15, 11, 40, 37

3. 206, 206, 555, 556

4. 129, 188, 188, 276

5. 9, 9, 9, 9, 11

6. 7, 7, 7, 7, 10

7.

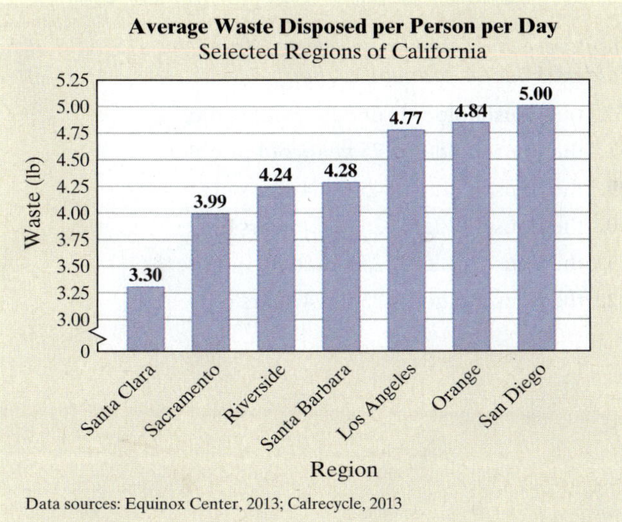

8.

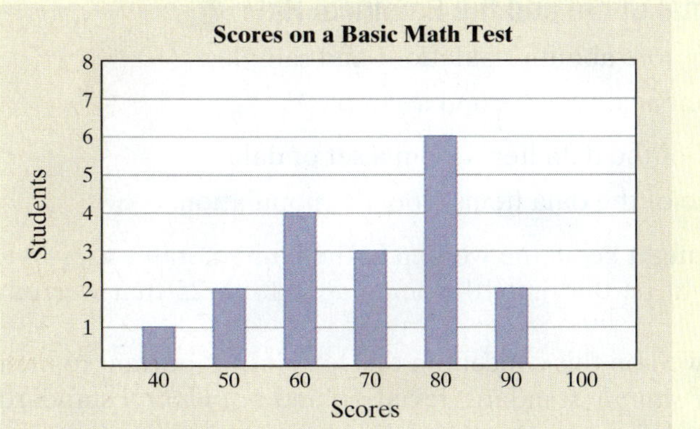

For the data sets in Exercises 9 through 16.

a. Find the mean, $\bar{x}$, of the data.
b. Find deviation from the mean for each data item.
c. Find the sum of the deviations. See Example 2.

9. 17, 10, 18, 18, 12

10. 25, 22, 22, 19, 17

11. 260, 256, 190, 286

12. 500, 462, 528, 526

13. 3.7, 4.0, 3.3, 2.8, 4.8, 3.0

14. 2.0, 3.0, 2.5, 3.4, 1.8, 1.1

15. 9, 5, 7, 4, 10

16. 2, 10, 5, 7, 6

For Exercises 17 through 24, find the standard deviation of the data set. Round to 2 decimal places. See Example 3.

17. 12, 15, 17, 12

18. 14, 18, 20, 16

19. 5, 10, 10, 15

20. 11, 18, 18, 25

21. 10, 11, 12, 13, 14

22. 15, 16, 17, 18, 19

23. 27, 34, 40, 46, 53

24. 16, 21, 26, 31, 36

MIXED PRACTICE

For each graph,

a. list the data items, then
b. calculate the mean,
c. calculate the standard deviation, and
d. calculate the range.

If needed, round standard deviations to 1 decimal place.

25.

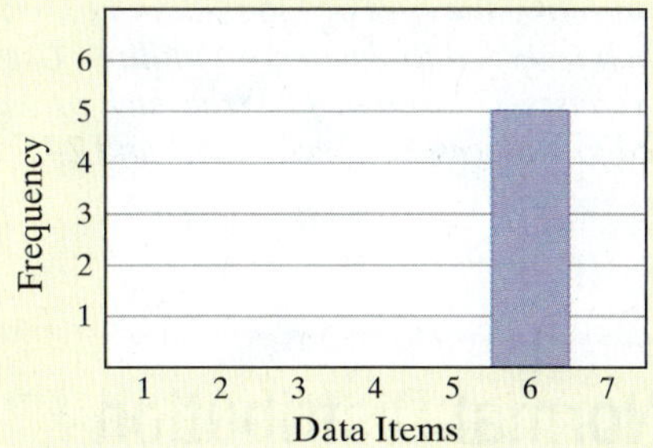

26.

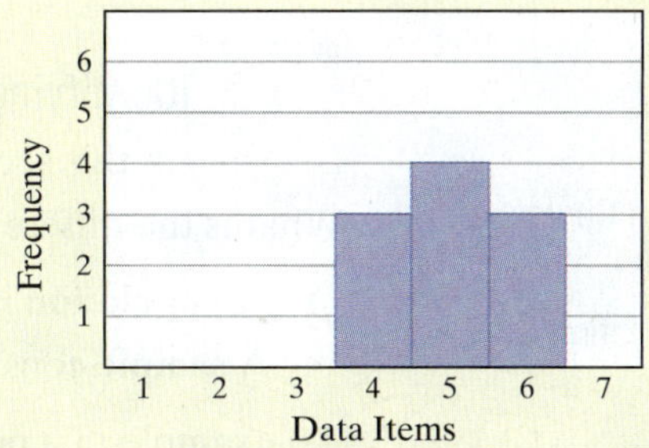

27.

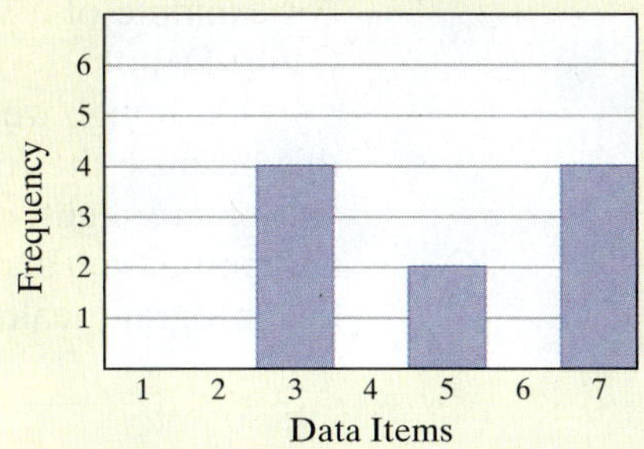

28.

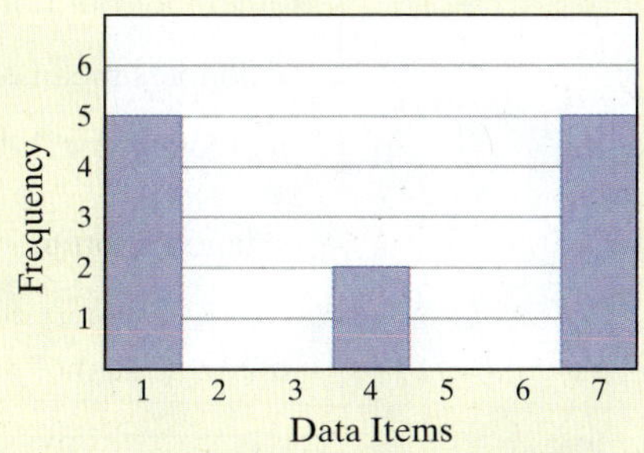

For Exercises 29 through 36, study the graphs and answers from Exercises 25 through 28.

29. Explain how the mean can be determined from each graph.

30. Explain how standard deviation is a good measure of dispersion.

31. Explain how range can be determined from each graph.

32. Do you think range is a good measure of dispersion for these graphs? Explain why or why not.

33. Use the graphs and give the mode for Exercises 25 through 28.

34. Use the graphs and give the median for Exercises 25 through 28.

35. Compare the medians found in Exercise 34 and the means, $\bar{x}$. Do you notice any similarities? If so, explain.

36. Compare the modes found in Exercise 33 and the means, $\bar{x}$. Do you notice any similarities? If so, explain.

REVIEW AND PREVIEW

The following histogram shows the ages of 115 million householders for a recent year, each rounded to the nearest million. Use this histogram to answer Exercises 37 through 42. Write each probability as a decimal rounded to 2 places. See Section R.4 and 7.4.

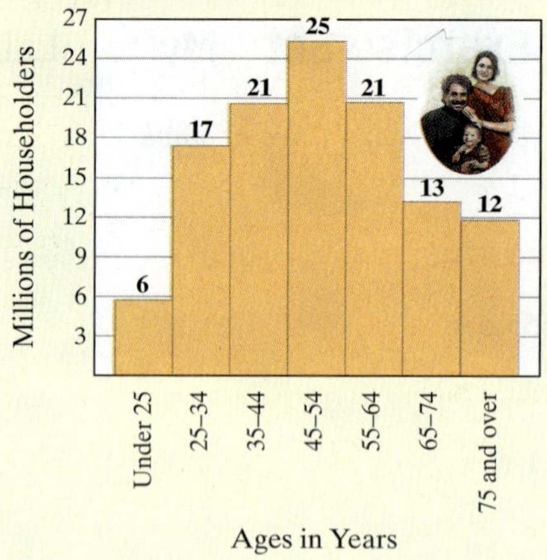

Source: U.S. Bureau of the Census, *Current Population Reports*

Suppose a householder is selected from the 115 million householders. Find the probability that:

37. the householder is under 25 years old.

38. the householder is 75 years old or older.

39. the householder is 45 to 64 years old.

40. the householder is 25 to 54 years old.

41. the householder is 25 years old or older.

42. the householder is 25 to 74 years old.

8.3 The Normal Distribution

OBJECTIVES

1 Identify Properties of the Normal Distribution Curve and the Empirical Rule.

2 Calculate the Population Standard Deviation of a Data Set.

OBJECTIVE

1 Identifying the Normal Curve and the Empirical Rule

Let's begin this section with a note about populations and samples.

What is the difference between a population and a sample?

- A **population** includes all of the data items from a set of data.
- A **sample** consists of some of the data items from the population.

An example of a population might be all the women in the United States ages 18 to 25. An example of a sample might be the first 10 women ages 18 to 25 that walked into Office Depot.

Although we will not dwell on this concept in this text, it is important to mention this because the formulas for *sample* standard deviation and *population* standard deviation are slightly different. (Sample standard deviation has $n - 1$ in the denominator and population standard deviation has simply n in the denominator.) Also, most statistical programs calculate both, and we don't want to introduce confusion.

Helpful Hint

Remember our notations for *samples* and *populations*.

$\bar{x}$ denotes mean of a **sample**.
μ (read "mew") denotes mean of a **population**.
} These are calculated the same.

s denotes **sample** standard deviation: $s = \sqrt{\dfrac{\Sigma(x - \bar{x})^2}{n - 1}}$

σ (read "sigma") denotes **population** standard deviation: $\sigma = \sqrt{\dfrac{\Sigma(x - \bar{x})^2}{n}}$

} These are calculated differently.

Some population examples, when graphed, have what we call a bell-shaped curve, or a **normal distribution.** For example, histograms of populations of IQ scores, SAT scores, heights of men or women, and birth weights of babies, for example, do appear to have this bell-shape or normal distribution.

Below is a histogram of SAT scores. Notice this shape appears to be bell-shaped, and we can approximate this graph with a normal distribution curve, shown in red.

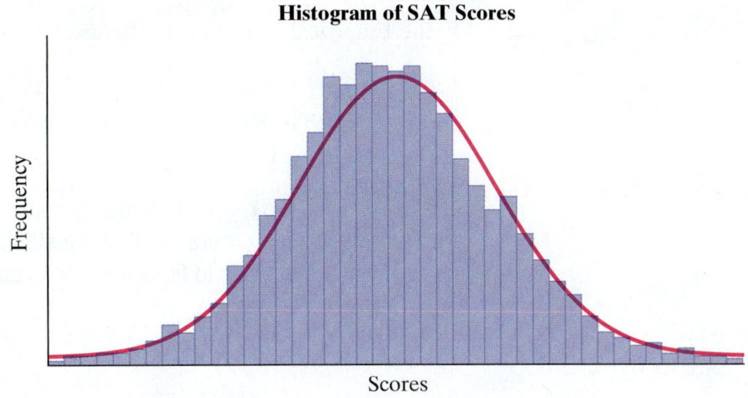

Let's now review some properties that are true of bell-shaped, or normal distribution, curve.

Study the graphs below. They are all examples of normal curves.

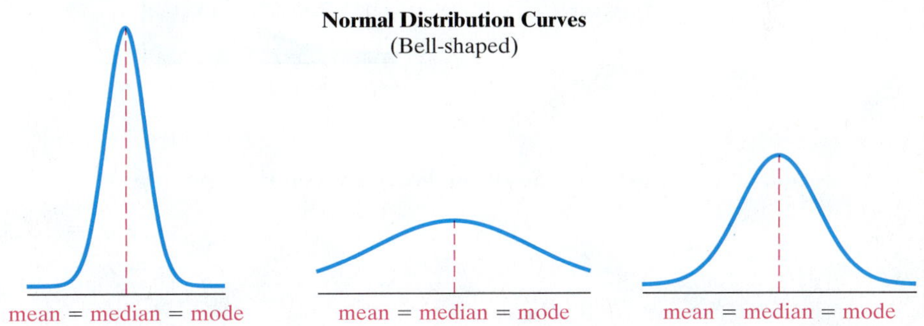

Each normal distribution curve has:

- mean-median-mode.
- symmetry about the dashed red line for mean.
- half the values less than the mean and half greater than the mean.

Much of statistics and probability is associated with the normal curve. In this text, we will concentrate on the mean and standard deviation for a normal curve and the information we then know about data items.

The empirical rule (for a normal curve) allows us to estimate how many data items occur within regions of a normal curve. This is given by the Empirical Rule graph below.

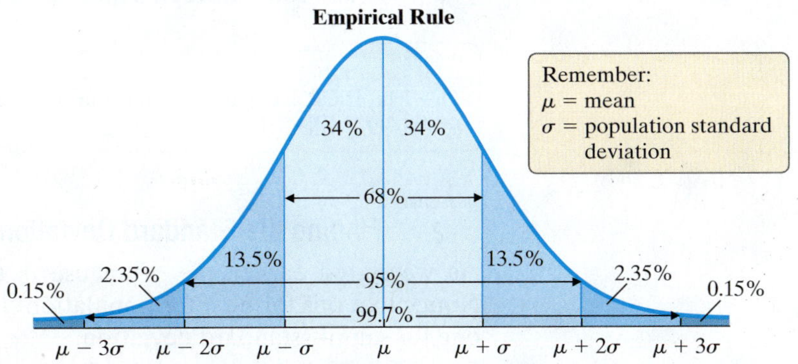

In normal distributions, the Empirical Rule tells us that:

- About 68% of all data values will fall within ± 1 standard deviation of the mean.
- About 95% of all data values will fall within ± 2 standard deviations of the mean.
- About 99.7% of all data values will fall within ± 3 standard deviations of the mean.

Helpful Hint

The Empirical Rule can also be used on sample data. Just use

$\underbrace{\bar{x} \text{ instead of } \mu}_{\text{(calculated the same)}}$ and $\underbrace{s \text{ instead of } \sigma}_{\text{(calculated a little differently)}}$

EXAMPLE 1 A company reports that for one year the ACT scores were normally distributed with a mean score of 18 and a standard deviation of 6. (These numbers are rounded to the nearest whole.) Use this to fill in the blanks on the graph below.

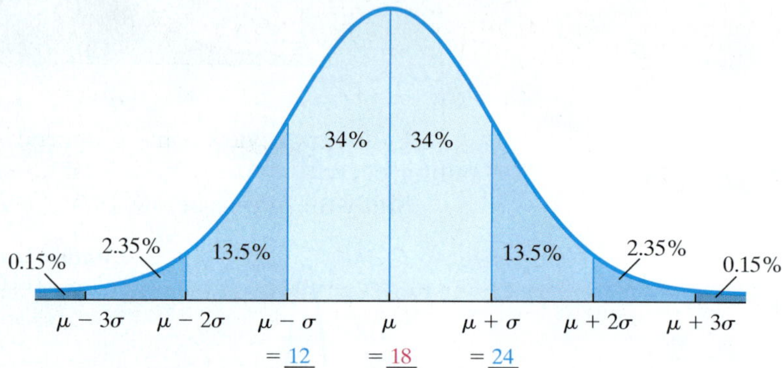

Solution We are given that mean $\mu = 18$ and standard deviation $\sigma = 6$. Thus:

$$\mu + \sigma = 18 + 6 = 24$$
$$\mu - \sigma = 18 - 6 = 12$$ □

PRACTICE 1 Use Example 1 and find $\mu + 2\sigma$ and $\mu - 2\sigma$. ■

EXAMPLE 2 Use the results of Example 1 and give the percent of students who scored between 12 and 24 on the ACT exam.

Solution Read the percents from the graph with Example 1. We have

$$34\% + 34\% = 68\%.$$

Approximately 68% of students scored between 12 and 24 on the ACT exam. □

PRACTICE 2 Use the results of Example 1 and Practice 1 to give the percent of students who score between 6 and 30 on the ACT exam. ■

Helpful Hint

The Empirical Rule for normal curves is also called the 68%–95%–99.7% rule (or the 68–95–99.7 rule).

OBJECTIVE

2 Finding the Standard Deviation of a Population

As we stated earlier, we often use σ to represent population standard deviation. Notice that this formula for population standard deviation is slightly different than the one for sample standard deviation.

Calculating the Population Standard Deviation of a Data Set

Suppose a data set of n items, $x_1, x_2, x_3, \ldots, x_n$, has a mean $\bar{x}$. Then the population standard deviation of this data set is given by the formula

$$\sigma = \sqrt{\frac{\Sigma(x - \bar{x})^2}{n}}$$

Steps for Calculating:

1. Find $\bar{x}$, the mean of the data items.
2. Find the deviations from the mean.
3. Square each deviation.
4. Find the sum of the squared deviations.
5. Take the square root of the quotient $\frac{\text{sum in Step 4}}{n}$.

EXAMPLE 3 Given the data, find (a) the mean and (b) the population standard deviation.

population data: 7, 8, 12, 12, 16

Solution Here, $n = 5$.

a. $\text{mean} = \bar{x} = \frac{\Sigma x}{n} = \frac{7 + 8 + 12 + 12 + 16}{5} = \frac{55}{5} = 11$

b. Let's use the steps above.

Helpful Hint
The steps above are similar to the steps in Section 8.2, except that we divide by n.

Step 1. $\bar{x} = 11$; **Step 2.** ↓ **Step 3.** ↓

Data Item	*Deviation: Data Item − Mean*	*Square Each Deviation*
7	$7 - 11 = -4$	$(-4)^2 = 16$
8	$8 - 11 = -3$	$(-3)^2 = 9$
12	$12 - 11 = 1$	$1^2 = 1$
12	$12 - 11 = 1$	$1^2 = 1$
16	$16 - 11 = 5$	$5^2 = 25$

Step 4. Sum = 52

Step 5. Take the square root of $\frac{\text{sum in Step 4}}{n}$.

$$\sigma = \sqrt{\frac{52}{5}} \approx 3.2$$

The mean is 11, and the population standard deviation is 3.2. □

PRACTICE 3 Given the data, find (a) the mean and (b) the population standard deviation. If needed, round answers to 1 decimal place.

population data: 8, 14, 11, 17, 20 ■

Graphing Calculator Explorations

We can use a graphing calculator to calculate population standard deviation. For example, let's calculate the population standard deviation, σ, of the data set

38, 21, 26, 15.

First enter the data set as list L_1. To do so, press [STAT] [ENTER]. Then press [38] [ENTER] to enter 38 as the first number in L_1. Next press [21] [ENTER] to enter 21 as the second number in L_1. Continue in the same manner and enter 26 and 15.

To find population standard deviation, σ, press STAT then right arrow key ▶, to go to the CALC drop down menu. After that, simply press ENTER since we want the first item in the CALC drop down menu, which is 1-Var Stats. The phrase 1-Var Stats enters on the main screen, so we press ENTER again and here's what appears.

1-Var Stats

$\bar{x} = 25$	mean
$\Sigma x = 100$	sum of data items
$\Sigma x^2 = 2786$	sum of squared data items
$Sx = 9.763879011$	sample SD
$\sigma x = 8.455767263$	population SD
$\downarrow n = 4$	number of data items

Use a graphing calculator to calculate the population standard deviation of each data set. Compare this SD, σ, with the sample SD, s, found in Section 8.2. Round each to 2 decimal places.

1. 86, 43, 95, 90, 98

2. 66, 48, 87, 87, 95

3. 1.7, 1.6, 2.8, 2.6, 1.6, 2

4. 2.5, 1.9, 3.1, 2.8, 2.6, 2.9

5. 9, 9, 9, 8, 8, 10, 10, 8, 10

6. 10, 10, 13, 13, 13, 16, 16

Vocabulary, Readiness & Video Check

Use the choices below to fill in each blank. Choices may be used more than once or not at all.

sample standard deviation	population standard deviation	Empirical
mode	median	mean
normal distribution curve	sample	population

1. A(n) ______________ looks bell-shaped.

2. A(n) ______________ includes all of the data items from a set of data.

3. The formula for ______________ is $\sqrt{\dfrac{\Sigma(x-\bar{x})^2}{n-1}}$.

4. The formula for ______________ is $\sqrt{\dfrac{\Sigma(x-\bar{x})^2}{n}}$.

5. A(n) ______________ consists of some of the data items from the population.

6. For a normal distribution curve, mean = ______________ = ______________.

7. Another name for the 68–95–99.7 rule is the ______________ Rule.

Martin-Gay Interactive Videos

See Video 8.3

Watch the section lecture video and answer the following questions.

OBJECTIVE 1

8. Before Example 1, we learned that another name for the Empirical Rule is what name?

OBJECTIVE 2

9. After working Example 11, what is the difference between the formulas for sample standard deviation and population standard deviation?

8.3 Exercise Set MyMathLab®

Let the scores for Tests A and B be normally distributed.

Test A has a mean of 50 and a standard deviation of 15.

Test B has a mean of 200 and a standard deviation of 40.

Use this information to answer Exercises 1 through 12. See Examples 1 and 2.

1. Find the Test A score that is 1 standard deviation (SD) below the mean.

2. Find the Test A score that is 1 SD above the mean.

3. Find the Test B score that is 1 SD above the mean.

4. Find the Test B score that is 1 SD below the mean.

5. Find the Test A score that is 2 SD above the mean.

6. Find the Test A score that is 2 SD below the mean.

7. Find the Test B score that is 3 SD below the mean.

8. Find the Test B score that is 3 SD above the mean.

9. Find the Test A score that is 1.5 SD above the mean.

10. Find the Test A score that is 2.5 SD above the mean.

11. Find the Test B score that is 2.5 SD below the mean.

12. Find the Test B score that is 1.5 SD below the mean.

The College Board reports that the mean of the mathematics portion of the SAT is 514 with a SD (standard deviation) of 118. Also, the mean of the critical reading portion is 496 with a SD of 115. Use this to fill in the blanks for Exercises 13 and 14. See Examples 1 and 2.

13.

Math Portion of SAT

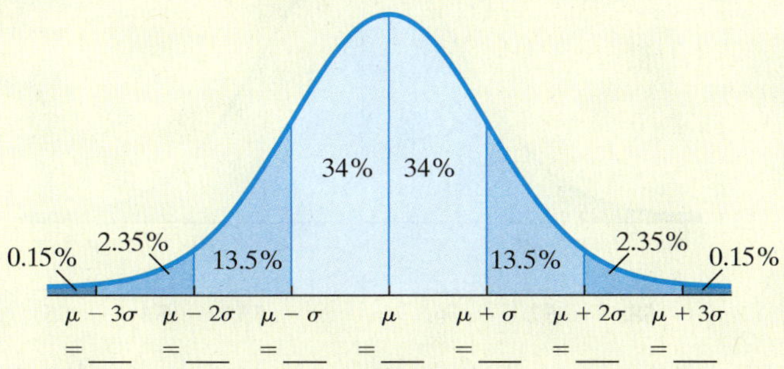

14.

Critical Reading Portion of SAT

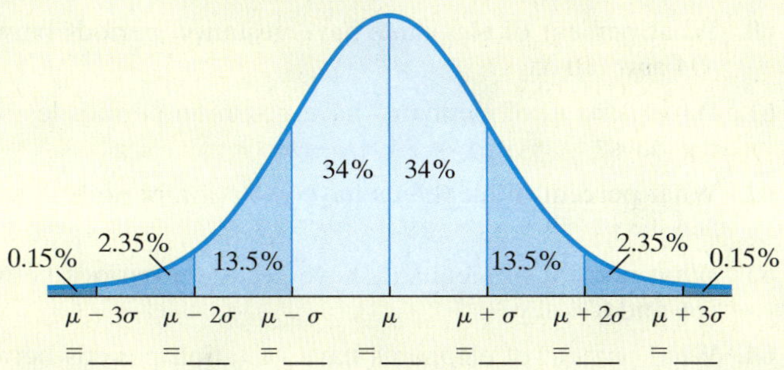

Use the filled-in graphs from Exercises 13 and 14 to answer Exercises 15 through 24. See Examples 1 and 2.

15. Find the math score that is 2 SD above the mean.

16. Find the math score that is 1 SD below the mean.

17. Find the reading score that is 1 SD below the mean.

18. Find the reading score that is 2 SD above the mean.

19. What percent of people scored between 396 and 632 on the math portion?

20. What percent of people scored between 266 and 726 on the reading portion?

21. What percent of people scored between 496 and 726 on the reading portion?

22. What percent of people scored between 278 and 514 on the math portion?

23. What percent of people scored greater than 632 on the math portion?

24. What percent of people scored less than 266 on the math portion?

A commuter tallied the time to work for 200 days and found that it resembled a normal curve. The mean time was 42 minutes, and the standard deviation was 3 minutes. Use this graph for Exercises 25 through 36. See Examples 1 and 2.

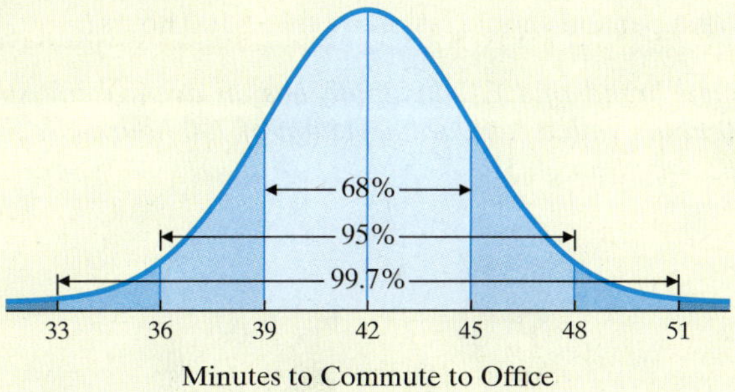

For the 200-work-day experiment, find the percent of time commuting to work:

25. between 39 and 45 minutes

26. between 36 and 48 minutes

27. between 39 and 42 minutes

28. between 42 and 45 minutes

29. between 42 and 48 minutes

30. between 36 and 42 minutes

31. between 39 and 48 minutes

32. between 36 and 45 minutes

33. more than 48 minutes

34. more than 51 minutes

35. less than 33 minutes

36. less than 36 minutes

A certain brand of a standard lightbulb has an average life span of 2500 hours with a standard deviation of 300 hours. Use this for Exercises 37 through 44.

37. Find the life span for 1 SD above the mean.

38. Find the life span for 1 SD below the mean.

39. What percent of these lightbulbs should last more than 2500 hours?

40. What percent of these lightbulb should last less than 2500 hours?

41. What percent of these lightbulbs should last more than 2800 hours?

42. What percent of these lightbulbs should last more than 2200 hours?

43. What percent should last less than 3100 hours?

44. What percent should last more than 3100 hours?

A certain brand of LED lightbulb has an average life span of 10,000 hours with a standard deviation of 790 hours. Use this for Exercises 45 through 52.

45. Find the life span for 2 standard deviations below the mean.

46. Find the life span for 2 standard deviations above the mean.

47. What percent of these lightbulbs should last more than 10,000 hours?

48. What percent of these lightbulbs should last less than 10,000 hours?

49. What percent of these lightbulbs should last more than 10,790 hours?

50. What percent of these lightbulbs should last less than 9210 hours?

51. What percent of these lightbulbs should last between 9210 and 10,790 hours?

52. What percent of these lightbulbs should last more than 10,790 hours or less than 9210 hours?

The mammal with the longest gestation period is an elephant. (Gestation period is defined as the time from conception to birth.) Suppose one species of Asian elephant has a mean gestation period of 602 days with a standard deviation of 36 days. Use this to answer Exercises 53 through 64.

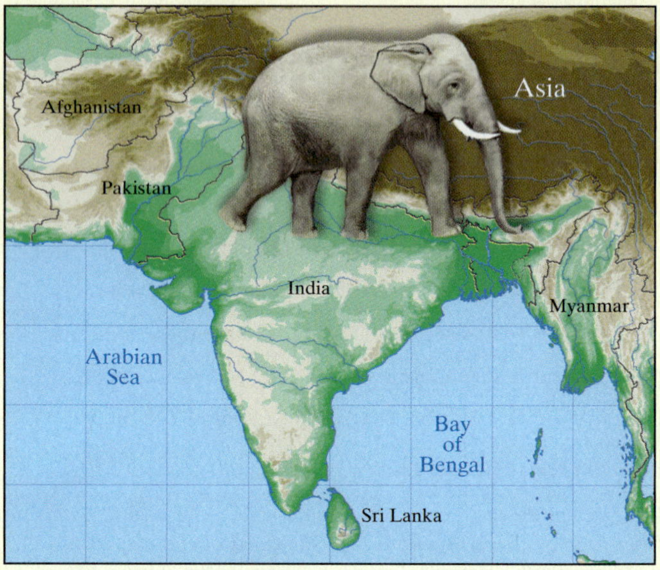

For Exercises 53 through 58, fill in the blanks below the graph.

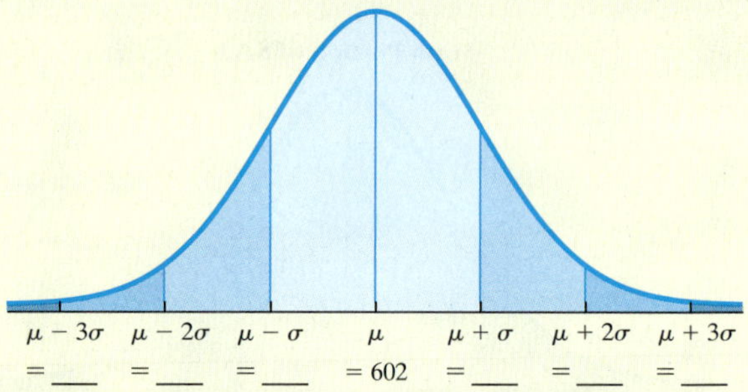

58. **57.** **56.** **53.** **54.** **55.**

59. What percent of elephants have gestation periods between 530 and 674 days?

60. What percent of elephants have gestation periods between 494 and 710 days?

61. What percent of elephants have gestation periods less than 638 days?

62. What percent of elephants have gestation periods less than 566 days?

63. What percent of elephants have gestation periods between 566 and 674 days?

64. What percent of elephants have gestation periods between 494 and 638 days?

REVIEW AND PREVIEW

Use the bar graph to answer Exercises 65 through 68. See Sections R.4 and 8.1.

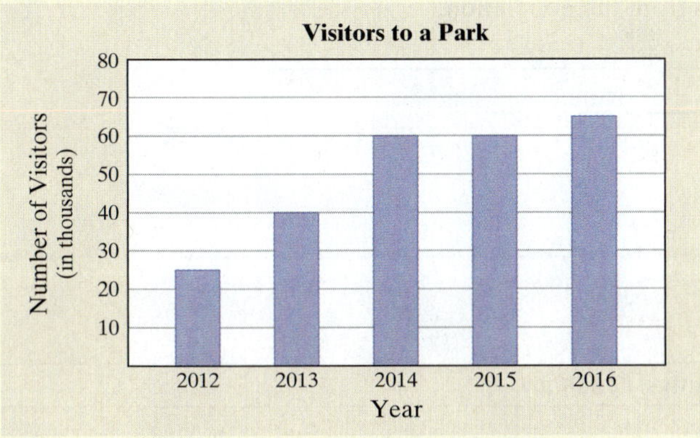

65. Find the mean number of visitors to the park for the years given.

66. Find the mode number of visitors to the park for the years given. What year did the mode occur?

67. Name the year(s) that the number of visitors was fewer than the mean.

68. Name the year(s) that the number of visitors was more than the mean.

CONCEPT EXTENSIONS

For Exercises 69 through 72, use the given data set and calculate:

a. sample standard deviation and

b. population standard deviation

Round standard deviations to 1 decimal place. See Example 3.

69. 45, 73, 87, 95, 90

70. 32, 48, 50, 30, 45

71. 10, 15, 12, 19, 11, 15, 16

72. 28, 19, 25, 22, 27, 23

Use this graph for Exercises 73 through 77. No calculations should be needed.

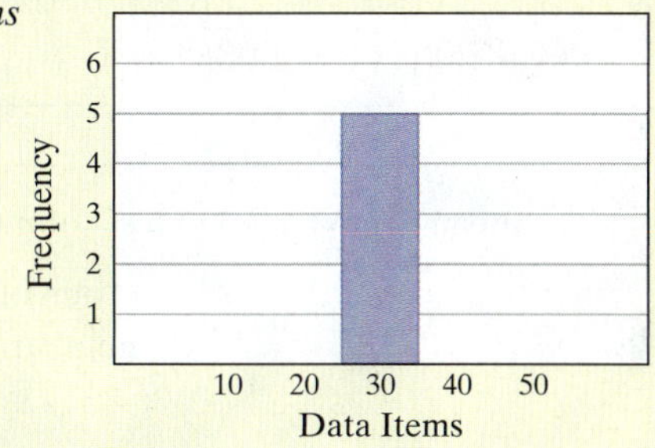

Find the

73. mean

74. median

75. mode

76. range

77. standard deviation

78. Describe a normal distribution curve. Include an example of data that are normally distributed.

79. Describe the Empirical Rule for a normal distribution curve.

Chapter 8 Vocabulary Check

Use the choices below to fill in each blank. Choices may be used more than once or not at all.

sample standard deviation	0	dispersion	x
range	1	grade point average	mode
sample standard deviation	population standard deviation	mean	median
normal distribution curve	$\sqrt{\dfrac{\Sigma(x-\overline{x})^2}{n-1}}$	$\sqrt{\dfrac{\Sigma(x-\overline{x})^2}{n}}$	$\overline{x}$
sample	population	Empirical	

1. Another word for average is ______________.

2. The ______________ of a set of numbers is the middle number. If the number of numbers is even, it is the ______________ of the two middle numbers.

3. The number that occurs most often in a set of number is called the ______________.

4. The formula for ______________ is $\dfrac{\Sigma x}{n}$.

5. We denote mean by the symbol ______________.

6. An example of weighted mean is a calculation of ______________.

7. ______________ = largest data value − smallest data value

8. For a single data set, the sum of the deviations from the mean is always ______________.

9. The formula for ______________ is $\sqrt{\dfrac{\Sigma(x-\overline{x})^2}{n-1}}$.

10. The degree in which the data values are scattered is called ______________.

11. A(n) ______________ looks bell-shaped.

12. A(n) ______________________ includes all of the data items from a set of data.

13. The formula for ______________________ is $\sqrt{\frac{\Sigma(x-\bar{x})^2}{n}}$.

14. A(n) ______________________ consists of some of the data items from the population.

15. For a normal distribution curve, mean = ______________________ = ______________________.

16. Another name for the 68–95–99.7 rule is the ______________________ Rule.

Chapter 8 Highlights

DEFINITIONS AND CONCEPTS	EXAMPLES
Section 8.1 Measures of Central Tendency	
The **mean** (or **average**) of a set of number items is $$\text{mean} = \frac{\text{sum of items}}{\text{number of items}}$$ Also, another notation for mean:	Find the mean, median, and mode of the following set of numbers: 33, 35, 35, 43, 68, 68 $$\text{mean} = \frac{33+35+35+43+68+68}{6} = 47$$ $$\bar{x} = \frac{\Sigma x}{n} = \frac{33+35+35+43+68+68}{6} = 47$$
Mean The mean of n data items, $x_1, x_2, x_3, \cdots, x_n$ is given by: $$\bar{x} = \frac{\Sigma x}{n}$$	
The **median** of a set of numbers in numerical order is the middle number. If the number of items is even, the median is the mean of the two middle numbers.	The median is the mean of the two middle numbers, 35 and 43 $$\text{median} = \frac{35+43}{2} = 39$$
The **mode** of a set of numbers is the number that occurs most often. (A set of numbers may have no mode or more than one mode.)	There are two modes because there are two numbers that occur twice: 35 and 68
Section 8.2 Measures of Dispersion	
Range The range of a data set is the difference between the largest data value and the smallest data value. $$\text{range} = \text{largest data value} - \text{smallest data value}$$	The range of data set 5, 7, 9, 11, 21, 21 is: $$\text{range} = 21 - 5 = 16$$
Calculating the Sample Standard Deviation of a Data Set Suppose a data set of n items, $x_1, x_2, x_3, \cdots, x_n$ has a mean $\bar{x}$. Then the standard deviation of this data set is given by the formula $$s = \sqrt{\frac{\Sigma(x-\bar{x})^2}{n-1}}$$ Steps for calculating: **1.** Find $\bar{x}$, the mean of the data items. **2.** Find the deviations from the mean. **3.** Square each deviation. **4.** Find the sum of the squared deviations. **5.** Take the square root of the quotient $\frac{\text{sum in Step 4}}{n-1}$.	Calculate the standard deviation of 38, 21, 26, 15. **Step 1.** $\bar{x} = 25$; **Step 2.** ↓ **Step 3.** ↓ (see table below) **Step 4.** Sum of deviations: 286 **Step 5.** $s = \sqrt{\frac{286}{4-1}} \approx 9.76$

Data Item	*Deviation* *Data Item − mean*	*Square each deviation*
38	$38 - 25 = 13$	$13^2 = 169$
21	$21 - 25 = -4$	$(-4)^2 = 16$
26	$26 - 25 = 1$	$1^2 = 1$
15	$15 - 25 = -10$	$(-10)^2 = 100$

DEFINITIONS AND CONCEPTS | **EXAMPLES**

Section 8.3 The Normal Distribution

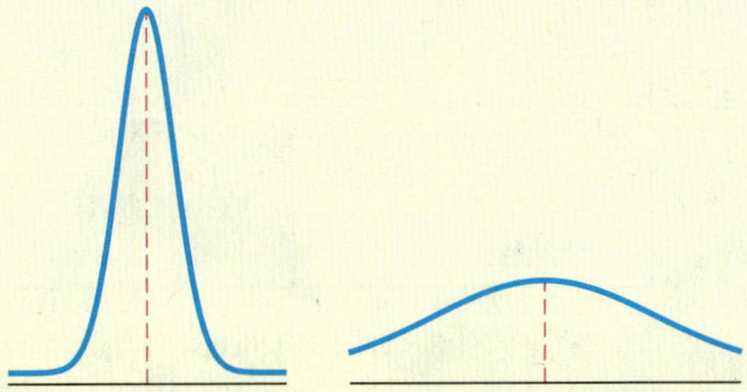

For Each Normal Distribution Curve:

- mean = median = mode
- Symmetry about the dashed red line for mean
- Half the values are less than the mean and half are greater than the mean

Suppose a set of data is normally distributed.

If the mean of the data is 105, then median = 105 and mode = 105.

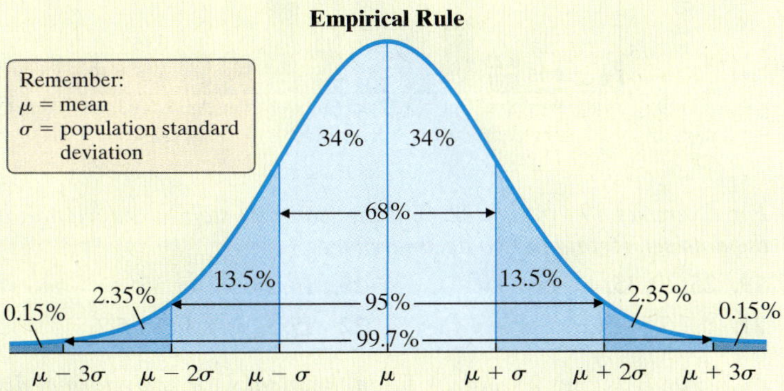

A brand of light bulbs has a life span of 2000 hours with a standard deviation of 78 hours. 95% of the light bulbs last between how many hours?

$$\mu + 2\sigma = 2000 + 2(78) = 2156$$

$$\mu - 2\sigma = 2000 - 2(78) = 1844$$

Thus, 95% of the light bulbs last between 1844 and 2156 hours.

Calculating the Population Standard Deviation of a Data Set

Suppose a data set of n items, $x_1, x_2, x_3, \cdots, x_n$ has a mean $\bar{x}$. Then the population standard deviation of this data set is given by the formula

$$\sigma = \sqrt{\frac{\Sigma(x - \bar{x})^2}{n}}$$

For the data set

38, 21, 26, 15,

the sample standard deviation,

$s \approx 9.76$. (See Highlights Section 8.2.)

The population standard deviation,

$\sigma \approx 8.46$.

Chapter 8 Review

(8.1) *Find the mean, median, and any mode(s) for each list of numbers. If necessary, round to the nearest tenth.*

1. 13, 23, 33, 14, 6

2. 45, 86, 21, 60, 86, 64, 45

3. 14,000, 20,000, 12,000, 20,000, 36,000, 45,000

4. 560, 620, 123, 400, 410, 300, 400, 780, 430, 450

For Exercises 5 and 6, the grades are given for a student for a particular semester. Find each grade point average. If necessary, round the grade point average to the nearest hundredth.

5.

Grade	*Credit Hours*
A	3
A	3
C	2
B	3
C	1

6.

Grade	*Credit Hours*
B	3
B	4
C	2
D	2
B	3

Use each frequency distribution table to find the **a.** *mean,* **b.** *median, and* **c.** *mode. If needed, round the mean to 1 decimal place.*

7.

Data Item	Frequency
60	2
61	10
62	5
63	11
64	3

8.

Data Item	Frequency
60	5
61	7
62	8
63	10
64	15

9.

Data Item	Frequency
11	5
12	5
13	3
14	1
16	3
17	1

10.

Data Item	Frequency
15	3
16	2
17	1
18	6
19	4
20	6

Use each graph of data items to find the **a.** *mean,* **b.** *median, and* **c.** *mode. If needed round the mean to 1 decimal place.*

11.

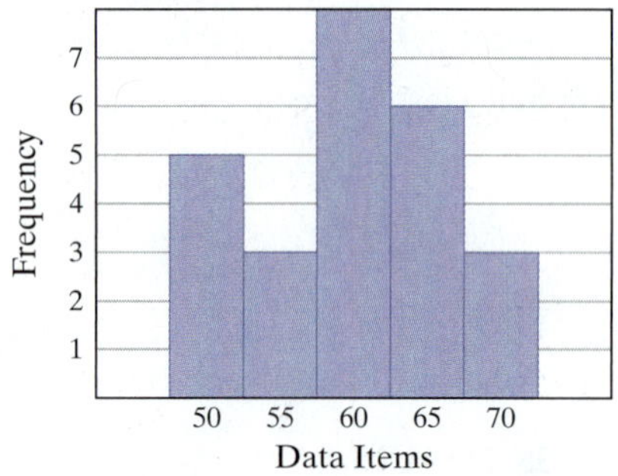

12.

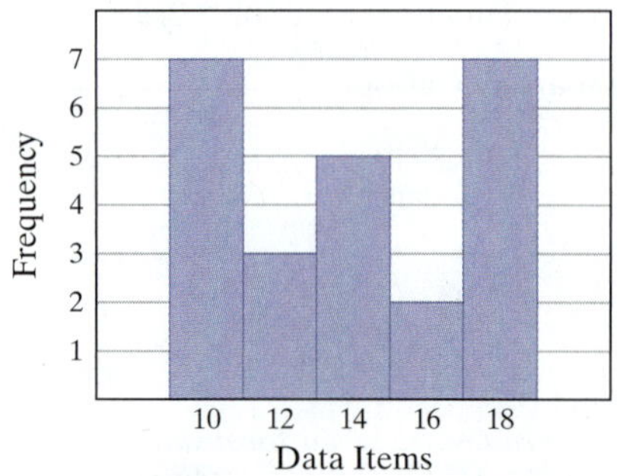

(8.2) *Find the range for each data set.*

13. 4, 4, 4, 3, 3, 5, 6

14. 1, 1, 1, 8, 8

15. 70, 75, 95, 60, 88

16. 80, 87, 97, 99, 85

17.

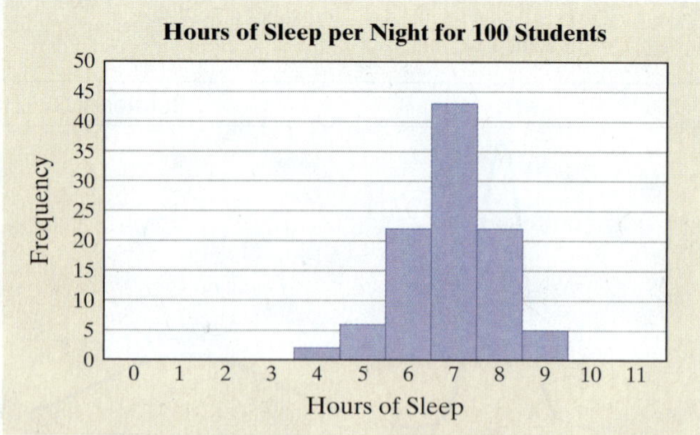

18.

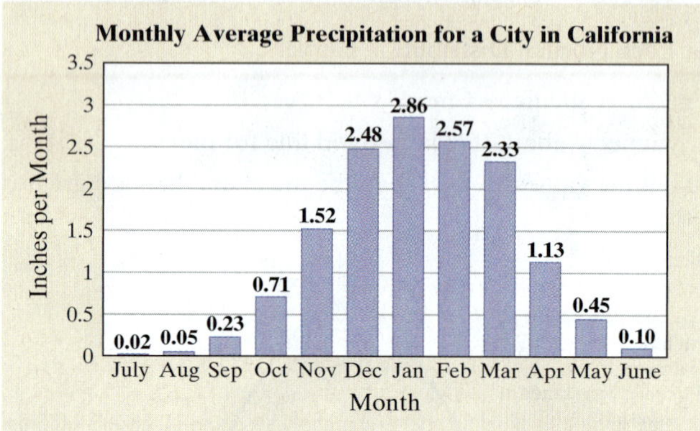

For Exercises 19 through 22, find the (sample) standard deviation of the data set. Round to two decimal places.

19. 25, 30, 35, 40, 45

20. 15, 25, 35, 45, 55

21. 8, 7, 7, 5, 10, 8

22. 15, 3, 10, 8, 9, 12, 5, 6

(8.3) *For Exercises 23 and 24, fill in the blanks for each normal distribution graph. Remember that μ stands for mean and σ stands for population standard deviation.*

23. Professor A: Test scores: average grade 84; standard deviation 2

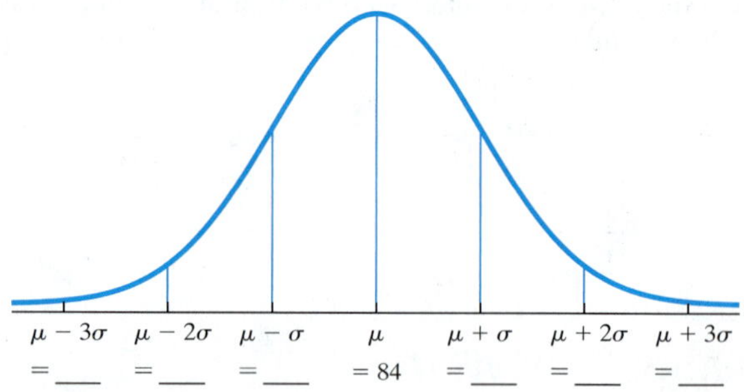

24. Professor B: Test scores: average grade 84, standard deviation 5

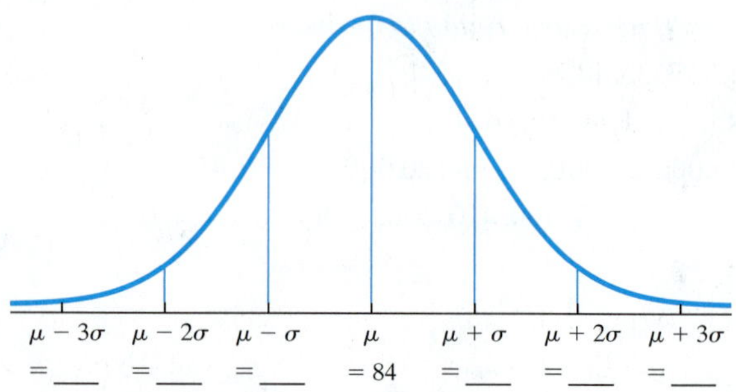

Use the normal distribution graph with mean and standard deviation for a set of tests to answer Exercises 25 through 32.

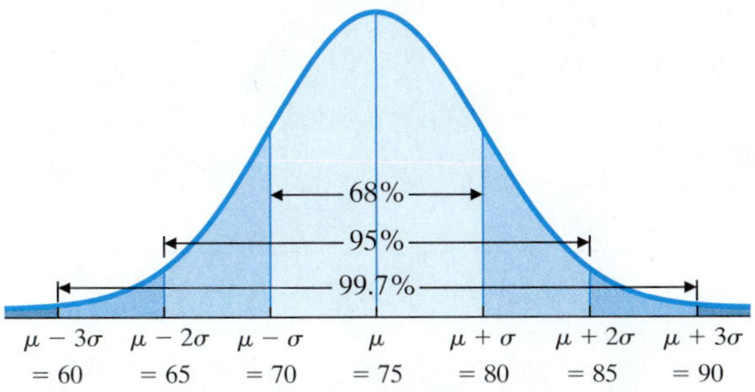

25. What is the mean for the data set?

26. What is the standard deviation for the data set?

27. What percent scored between 70 and 80?

28. What percent scored between 65 and 85?

29. What percent scored between 80 and 85?

30. What percent scored between 70 and 75?

31. What percent scored greater than 80?

32. What percent scored less than 65?

MIXED PRACTICE

*For each graph, calculate parts **a.–e.** Round the mean and the standard deviation to two decimal places.*

a. mean **b.** median **c.** mode

d. range **e.** (sample) standard deviation

33.

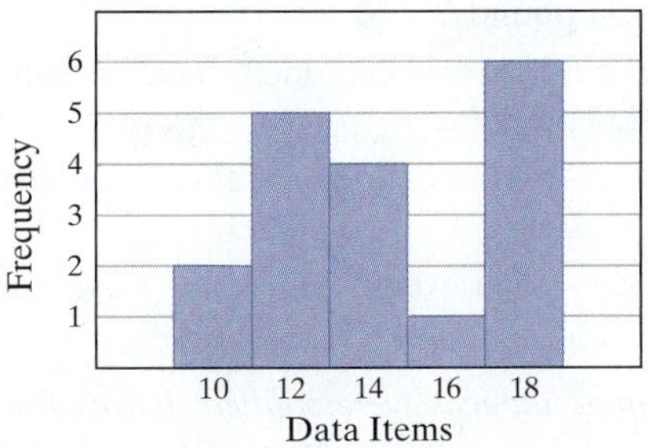

34.

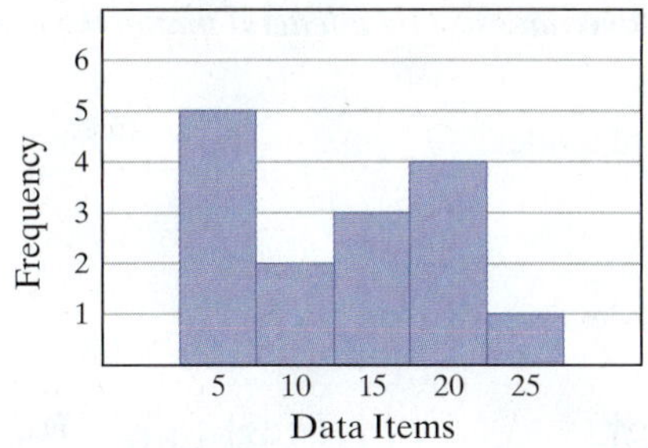

*Find the **a.** range and **b.** (sample) standard deviation for the data items in each exercise.*

35. Exercise 11

36. Exercise 12

A certain brand of fluorescent lightbulb has an average life span of 8000 hours with a standard deviation of 430 hours. Use this for Exercises 37 through 44.

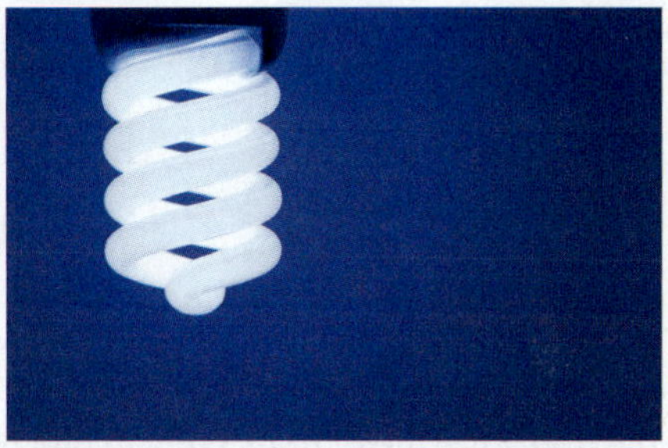

37. Find the life span for 1 standard deviation below the mean.

38. Find the life span for 1 standard deviation above the mean.

39. What percent of these lightbulbs should last more than 8000 hours?

40. What percent of these lightbulbs should last less than 8000 hours?

41. What percent of these lightbulbs should last more than 8860 hours?

42. What percent of these lightbulbs should last more than 7570 hours?

43. What percent of these lightbulbs should last between 7570 and 8430 hours?

44. What percent of these lightbulbs should last between 6710 and 9290 hours?

lc

Chapter 8 Getting Ready for the Test

MULTIPLE CHOICE *For Exercises 1 through 6, use the data sets and choices below to answer.*

A. data set: 10, 10, 10, 10, 10

B. data set: 6, 8, 10, 12, 14

C. data set: 8, 9, 10, 11, 12

D. they are the same

1. Which data set has the greatest range?

2. Which data set has a mode?

3. Which data set has the greatest median?

4. Which data set has the greatest mean?

5. Without calculating, which data set has a standard deviation of 0?

6. Without calculating, which data set has the greatest standard deviation?

7. A data set has a mean of 50 and a standard deviation of 4. Choose the value that is two standard deviations from the mean.

A. 34 **B.** 38 **C.** 54 **D.** 46 **E.** 42

MATCHING *Choose* **two** *data sets (two letters) that make each statement true. Data sets may be used more than once or not at all*

8. equal means
9. equal number of modes
10. equal ranges
11. equal medians
12. equal standard deviations

A. 10, 20, 30, 30, 40, 50
B. 9, 11, 11, 30, 48
C. 20, 25, 30, 35, 40
D. 50, 55, 60, 65, 70

Chapter 8 Test

MyMathLab® YouTube

If no other instructions, round values in this test to one decimal place.

Find the mean, median, and mode of each list of numbers.

1. 26, 32, 42, 43, 49

2. 8, 10, 16, 16, 14, 12, 12, 13

Find the grade point average. Round to the nearest hundredth.

3.

Grade	Credit Hours
A	3
B	3
C	3
B	4
A	1

4. Use the frequency distribution table to find the following:
a. mean **b.** median
c. mode **d.** range

Data Item	Frequency
90	3
91	1
92	2
93	8
94	8

5. Use the graph of data items to find the following:
a. mean **b.** median
c. mode **d.** range

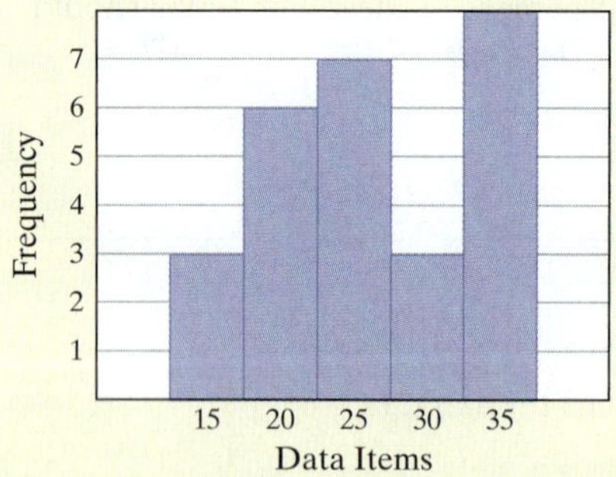

Given the data items:

10, 18, 13, 16, 13

6. Find the range.

7. Find the standard deviation rounded to 2 decimal places.

The mean weight of adult American men is 172 pounds with a standard deviation of 31 pounds. If the weights are normally distributed, use this data to solve Exercises 8 through 13.

8. Label the normal curve below with the values for μ, $\mu \pm \sigma$, $\mu \pm 2\sigma$, and $\mu \pm 3\sigma$ given below.

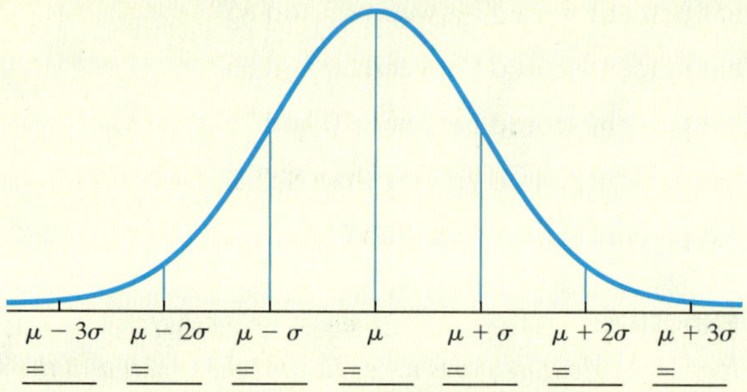

9. What percent of adult American men have a weight between 141 and 203 pounds?

10. What percent of adult American men have a weight between 110 and 234 pounds?

11. What percent of adult American men have a weight between 172 and 234 pounds?

12. 16% of the heaviest men weigh more than how many pounds?

13. 2.5% of the lightest men weigh less than how many pounds?

A large number of female alpacas were studied. It was found that this species of alpaca has a mean gestation period of 360 days with a standard deviation of 15 days. (Gestation period is defined as the time from conception to birth.) Use this to answer Exercises 14 through 19.

14. Label the normal curve below with the values for $\mu, \mu \pm \sigma, \mu \pm 2\sigma$, and $\mu \pm 3\sigma$ given below.

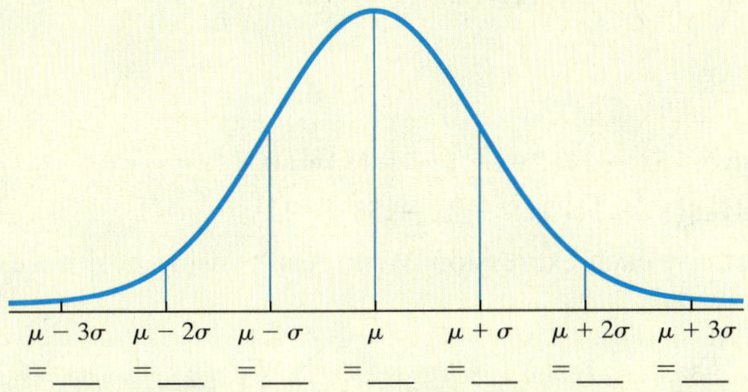

15. What percent of alpacas have gestation periods between 330 and 390 days?

16. What percent of alpacas have gestation periods between 315 and 405 days?

17. What percent of alpacas have gestation periods less than 375 days?

18. What percent of alpacas have gestation periods less than 345 days?

19. What percent of alpacas have gestation periods between 330 and 375 days?

Chapter 8 Cumulative Review

1. Simplify each expression by combining like terms.

a. $2x + 3x + 5 + 2$

b. $-5a - 3 + a + 2$

c. $4y - 3y^2$

d. $2.3x + 5x - 6$

e. $-\frac{1}{2}b + b$

2. Simplify each expression by combining like terms.

a. $4x - 3 + 7 - 5x$

b. $-6y + 3y - 8 + 8y$

c. $2 + 8.1a + a - 6$

d. $2x^2 - 2x$

3. Identify the x- and y-intercepts.

a.

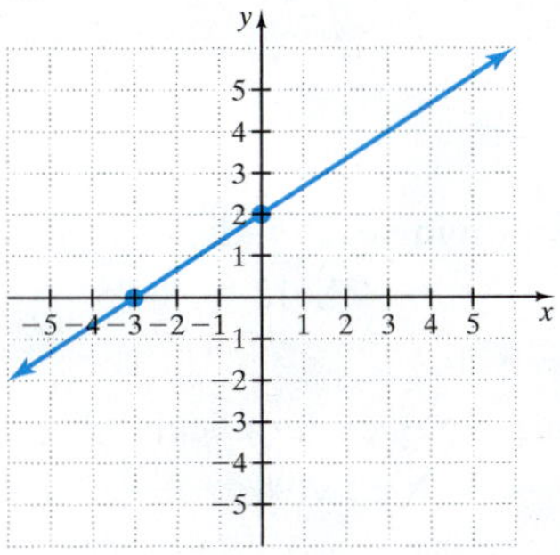

b.

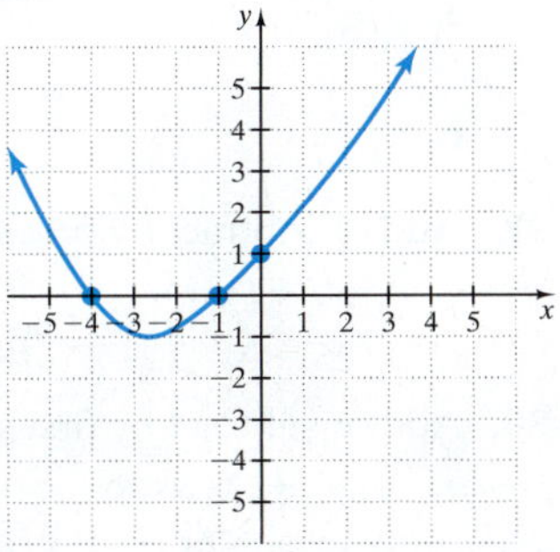

c.

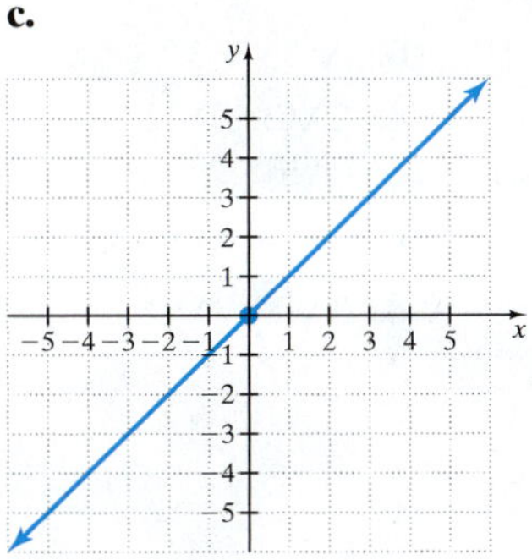

d.

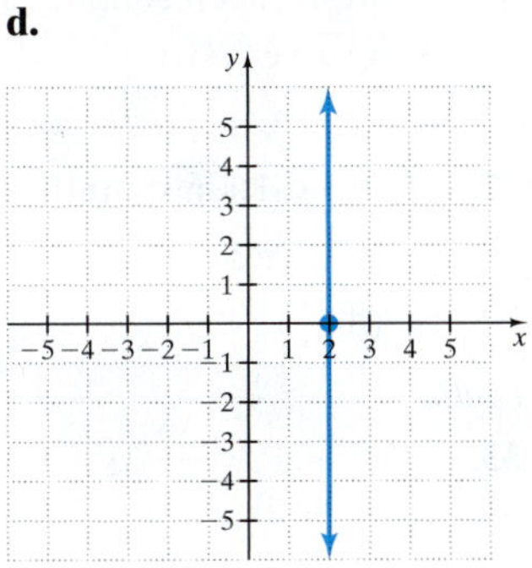

e.

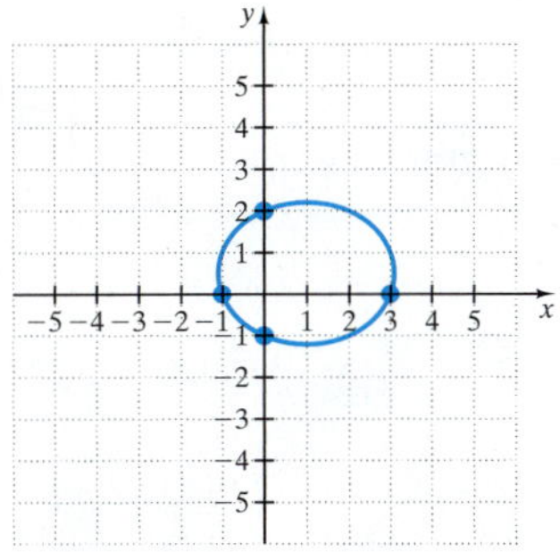

4. Identify the x- and y-intercepts.

a.

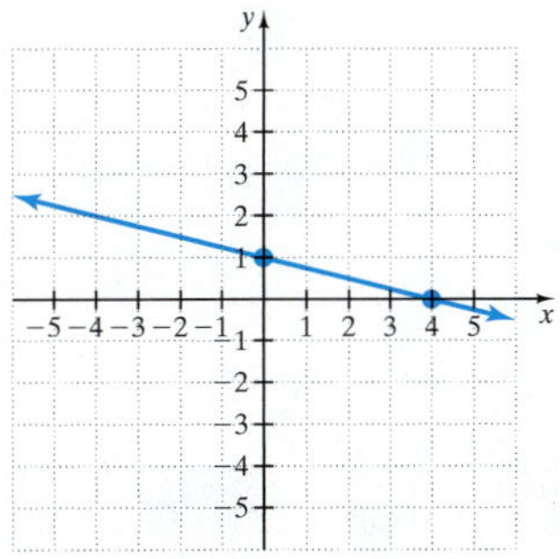

b.

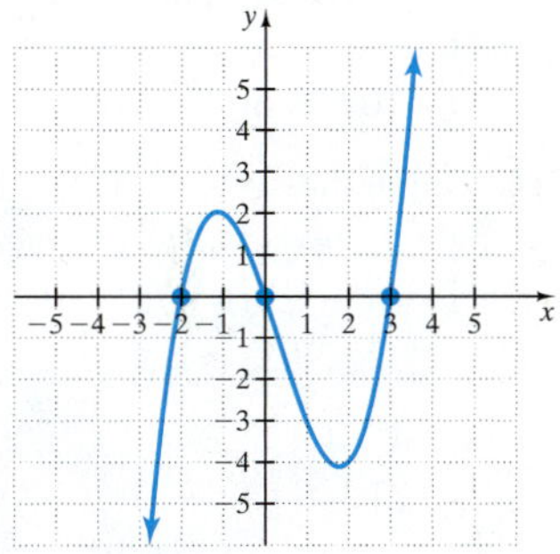

c.

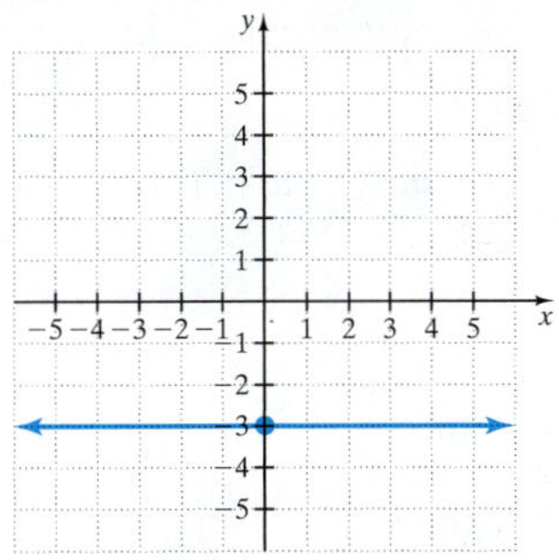

d.

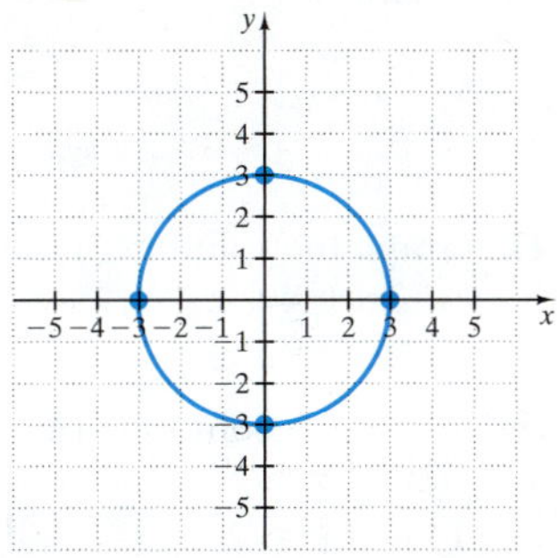

TRANSLATING

5. Write each sentence as an equation or inequality. Let x represent the unknown number.
a. The quotient of 15 and a number is 4.
b. Three subtracted from 12 is a number.
c. Four times a number, added to 17, is not equal to 21.
d. Triple a number is less than 48.

6. Write each sentence as an equation. Let x represent the unknown number.
a. The difference of 12 and a number is -45.
b. The product of 12 and a number is -45.
c. A number less 10 is twice the number.

7. Graph $x - 3y = 6$ by finding and plotting intercepts.

8. Find the slope of the line whose equation is $7x + 2y = 9$.

9. Solve the following system of equations by graphing.
$$\begin{cases} x - 2y = 4 \\ x = 2y \end{cases}$$

10. Solve the following system by graphing.
$$\begin{cases} y = x + 2 \\ 2x + y = 5 \end{cases}$$

11. Solve the system.
$$\begin{cases} 7x - 3y = -14 \\ -3x + y = 6 \end{cases}$$

12. Solve the system.
$$\begin{cases} 5x + y = 3 \\ y = -5x \end{cases}$$

13. Solve the system.
$$\begin{cases} -5x - 3y = 9 \\ 10x + 6y = -18 \end{cases}$$

14. Solve the system.
$$\begin{cases} -2x + y = 7 \\ 6x - 3y = -21 \end{cases}$$

15. Use the product rule to simplify each expression.
a. $4^2 \cdot 4^5$ **b.** $x^4 \cdot x^6$
c. $y^3 \cdot y$ **d.** $y^3 \cdot y^2 \cdot y^7$
e. $(-5)^7 \cdot (-5)^8$ **f.** $a^2 \cdot b^2$

16. Simplify.
a. $\dfrac{x^9}{x^7}$ **b.** $\dfrac{x^{19}y^5}{xy}$
c. $(x^5y^2)^3$ **d.** $(-3a^2b)(5a^3b)$

17. Graph the solution of the system.
$$\begin{cases} -3x + 4y < 12 \\ x \geq 2 \end{cases}$$

18. Graph the solution of the system.
$$\begin{cases} 2x - y \leq 6 \\ y \geq 2 \end{cases}$$

19. Simplify each expression.
a. $x^7 \cdot x^4$ **b.** $\left(\dfrac{t}{2}\right)^4$
c. $(9y^5)^2$

20. Simplify.
a. $\left(\dfrac{-6x}{y^3}\right)^3$ **b.** $\dfrac{(2b^2)^5}{a^2b^7}$
c. $\dfrac{(3y)^2}{y^2}$ **d.** $\dfrac{(x^2y^4)^2}{xy^3}$

21. Solve: $(5x - 1)(2x^2 + 15x + 18) = 0$.

22. Solve: $(x + 1)(2x^2 - 3x - 5) = 0$.

23. Simplify each expression. Write results using positive exponents only.
a. $\dfrac{1}{x^{-3}}$ **b.** $\dfrac{1}{3^{-4}}$
c. $\dfrac{p^{-4}}{q^{-9}}$ **d.** $\dfrac{5^{-3}}{2^{-5}}$

24. Simplify. Write results with positive exponents.
a. 5^{-3} **b.** $\dfrac{9}{x^{-7}}$ **c.** $\dfrac{11^{-1}}{7^{-2}}$

25. Find each root.
a. $\sqrt{36}$ **b.** $-\sqrt{16}$ **c.** $\sqrt{\dfrac{9}{100}}$
d. $\sqrt{0}$ **e.** $\sqrt{0.64}$

26. Find each root.
a. $\sqrt{9}$ **b.** $-\sqrt{\dfrac{1}{16}}$
c. $-\sqrt{25}$ **d.** $\sqrt{1}$

27. Simplify the following expressions.
a. $\sqrt{\dfrac{25}{36}}$ **b.** $\sqrt{\dfrac{3}{64}}$
c. $\sqrt{\dfrac{40}{81}}$

28. Simplify the following expressions.
a. $\sqrt{\dfrac{4}{25}}$ **b.** $\sqrt{\dfrac{16}{121}}$
c. $\sqrt{\dfrac{2}{49}}$

29. Add or subtract by first simplifying each radical.
a. $\sqrt{50} + \sqrt{8}$ **b.** $7\sqrt{12} - \sqrt{75}$
c. $\sqrt{25} - \sqrt{27} - 2\sqrt{18} - \sqrt{16}$

30. Add or subtract by first simplifying each radical.
a. $\sqrt{80} + \sqrt{20}$ **b.** $2\sqrt{98} - 2\sqrt{18}$
c. $\sqrt{32} + \sqrt{121} - \sqrt{12}$

31. Multiply. Then simplify if possible.
a. $\sqrt{7} \cdot \sqrt{3}$ **b.** $\sqrt{3} \cdot \sqrt{3}$
c. $\sqrt{3} \cdot \sqrt{15}$ **d.** $2\sqrt{3} \cdot 5\sqrt{2}$

32. Multiply. Then simplify if possible.
a. $\sqrt{2} \cdot \sqrt{5}$ **b.** $\sqrt{56} \cdot \sqrt{7}$
c. $(4\sqrt{3})^2$ **d.** $3\sqrt{8} \cdot 7\sqrt{2}$

Factor.

33. $-9a^5 + 18a^2 - 3a$

34. $7x^6 - 7x^5 + 7x^4$

35. $x^3 + 8$

36. $x^2 - 4$

37. A surveyor must determine the distance across a lake at points P and Q. To do this, she finds a third point R perpendicular to line PQ. If the length of $\overline{PR}$ is 320 feet and the length of $\overline{QR}$ is 240 feet, what is the distance across the lake? Approximate this distance to the nearest whole foot.

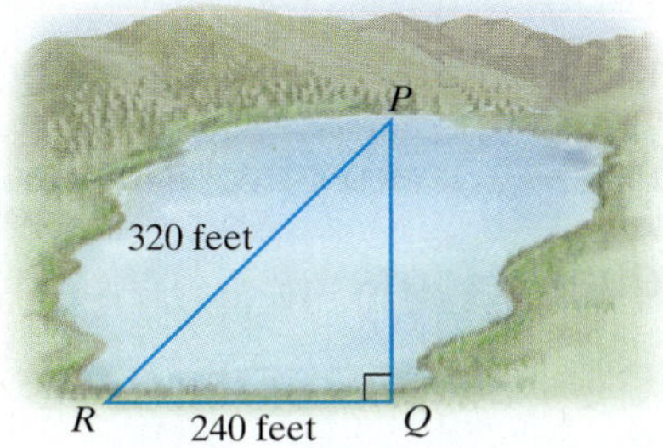

38. Find the distance between $(-7, 4)$ and $(2, 5)$.

39. Use the square root property to solve $2x^2 = 7$.

40. Use the square root property to solve $(x - 4)^2 = 3$.

41. Solve $2x^2 - 9x = 5$ using the quadratic formula.

42. Solve $2x^2 + 5x = 7$ using the quadratic formula.

43. Graph $y = x^2 - 4$.

44. Graph $y = x^2 + 2x + 3$.

Appendices

A Tables

A.1 Table of Percent, Decimal, and Fraction Equivalents

A.2 Table on Finding Common Percents of a Number

A.3 Table of Squares and Square Roots

B Inductive and Deductive Reasoning

C Ratios and Rates

D Proportion and Problem Solving

E Variation and Problem Solving

Student Resources

Study Skills Builders

Answers to Selected Exercises

Appendix A

Tables

A.1 Table of Percent, Decimal, and Fraction Equivalents

Percent	***Decimal***	***Fraction***
1%	0.01	$\frac{1}{100}$
5%	0.05	$\frac{1}{20}$
10%	0.1	$\frac{1}{10}$
12.5% or $12\frac{1}{2}$%	0.125	$\frac{1}{8}$
$16.\overline{6}$% or $16\frac{2}{3}$%	$0.1\overline{6}$	$\frac{1}{6}$
20%	0.2	$\frac{1}{5}$
25%	0.25	$\frac{1}{4}$
30%	0.3	$\frac{3}{10}$
$33.\overline{3}$% or $33\frac{1}{3}$%	$0.\overline{3}$	$\frac{1}{3}$
37.5% or $37\frac{1}{2}$%	0.375	$\frac{3}{8}$
40%	0.4	$\frac{2}{5}$
50%	0.5	$\frac{1}{2}$
60%	0.6	$\frac{3}{5}$
62.5% or $62\frac{1}{2}$%	0.625	$\frac{5}{8}$
$66.\overline{6}$% or $66\frac{2}{3}$%	$0.\overline{6}$	$\frac{2}{3}$
70%	0.7	$\frac{7}{10}$
75%	0.75	$\frac{3}{4}$
80%	0.8	$\frac{4}{5}$
$83.\overline{3}$% or $83\frac{1}{3}$%	$0.8\overline{3}$	$\frac{5}{6}$
87.5% or $87\frac{1}{2}$%	0.875	$\frac{7}{8}$
90%	0.9	$\frac{9}{10}$
100%	1.0	1
110%	1.1	$1\frac{1}{10}$
125%	1.25	$1\frac{1}{4}$
$133.\overline{3}$% or $133\frac{1}{3}$%	$1.\overline{3}$	$1\frac{1}{3}$
150%	1.5	$1\frac{1}{2}$
$166.\overline{6}$% or $166\frac{2}{3}$%	$1.\overline{6}$	$1\frac{2}{3}$
175%	1.75	$1\frac{3}{4}$
200%	2.0	2

A.2 Table on Finding Common Percents of a Number

*Common Percent Equivalences**	*Shortcut Method for Finding Percent*	*Examples*
1% = 0.01 (or $\frac{1}{100}$)	To find 1% of a number, multiply by 0.01. To do so, move the decimal point 2 places to the left.	1% of 210 is 2.10 or 2.1. 1% of 1500 is 15. 1% of 8.6 is 0.086.
10% = 0.1 (or $\frac{1}{10}$)	To find 10% of a number, multiply by 0.1, or move the decimal point of the number 1 place to the left.	10% of 140 is 14. 10% of 30 is 3. 10% of 17.6 is 1.76.
25% = $\frac{1}{4}$	To find 25% of a number, find $\frac{1}{4}$ of the number, or divide the number by 4.	25% of 20 is $\frac{20}{4}$ or 5. 25% of 8 is 2. 25% of 10 is $\frac{10}{4}$ or $2\frac{1}{2}$.
50% = $\frac{1}{2}$	To find 50% of a number, find $\frac{1}{2}$ of the number, or divide the number by 2.	50% of 64 is $\frac{64}{2}$ or 32. 50% of 1000 is 500. 50% of 9 is $\frac{9}{2}$ or $4\frac{1}{2}$.
100% = 1	To find 100% of a number, multiply the number by 1. In other words, 100% of a number is the number.	100% of 98 is 98. 100% of 1407 is 1407. 100% of 18.4 is 18.4.
200% = 2	To find 200% of a number, multiply the number by 2.	200% of 31 is $31 \cdot 2$ or 62. 200% of 750 is 1500. 200% of 6.5 is 13.

*See Appendix A.1.

A.3 Table of Squares and Square Roots (Note: Square roots are rounded to 3 decimal places)

n	n^2	$\sqrt{n}$	n	n^2	$\sqrt{n}$
1	1	1.000	51	2601	7.141
2	4	1.414	52	2704	7.211
3	9	1.732	53	2809	7.280
4	16	2.000	54	2916	7.348
5	25	2.236	55	3025	7.416
6	36	2.449	56	3136	7.483
7	49	2.646	57	3249	7.550
8	64	2.828	58	3364	7.616
9	81	3.000	59	3481	7.681
10	100	3.162	60	3600	7.746
11	121	3.317	61	3721	7.810
12	144	3.464	62	3844	7.874
13	169	3.606	63	3969	7.937
14	196	3.742	64	4096	8.000
15	225	3.873	65	4225	8.062
16	256	4.000	66	4356	8.124
17	289	4.123	67	4489	8.185
18	324	4.243	68	4624	8.246
19	361	4.359	69	4761	8.307
20	400	4.472	70	4900	8.367
21	441	4.583	71	5041	8.426
22	484	4.690	72	5184	8.485
23	529	4.796	73	5329	8.544
24	576	4.899	74	5476	8.602
25	625	5.000	75	5625	8.660
26	676	5.099	76	5776	8.718
27	729	5.196	77	5929	8.775
28	784	5.292	78	6084	8.832
29	841	5.385	79	6241	8.888
30	900	5.477	80	6400	8.944
31	961	5.568	81	6561	9.000
32	1024	5.657	82	6724	9.055
33	1089	5.745	83	6889	9.110
34	1156	5.831	84	7056	9.165
35	1225	5.916	85	7225	9.220
36	1296	6.000	86	7396	9.274
37	1369	6.083	87	7569	9.327
38	1444	6.164	88	7744	9.381
39	1521	6.245	89	7921	9.434
40	1600	6.325	90	8100	9.487
41	1681	6.403	91	8281	9.539
42	1764	6.481	92	8464	9.592
43	1849	6.557	93	8649	9.644
44	1936	6.633	94	8836	9.695
45	2025	6.708	95	9025	9.747
46	2116	6.782	96	9216	9.798
47	2209	6.856	97	9409	9.849
48	2304	6.928	98	9604	9.899
49	2401	7.000	99	9801	9.950
50	2500	7.071	100	10,000	10.000

Appendix B

Inductive and Deductive Reasoning

OBJECTIVES

1 Understand and Use Inductive Reasoning.

2 Understand and Use Deductive Reasoning.

Logic and logical reasoning have applications in many fields, including science, law, psychology and mathematics. For example, computers must have logic built into their circuits in order to process information correctly. We begin our study of logic by examining inductive reasoning.

OBJECTIVE 1 Using Inductive Reasoning

Inductive Reasoning

This is the process of forming a general conclusion based on observing a number of specific examples or outcomes.

$$\text{Specific Observations} \xrightarrow{\text{to a}} \text{general conclusion}$$

EXAMPLE 1 Find the next number in the sequence, or listing of numbers.

$$1, 5, 25, 125$$

Solution Each number after the first is obtained by multiplying the previous number by 5. If we assume that this pattern continues, the next number is $125 \times 5 = 625$.

$$1, \quad 5, \quad 25, \quad 125, \quad 625$$
$$1 \cdot 5 \quad 5 \cdot 5 \quad 25 \cdot 5 \quad 125 \cdot 5$$

□

PRACTICE 1 Find the next number in the sequence

$$2, 6, 18, 54$$

■

EXAMPLE 2 Find the next two numbers in the given sequence.

$$1, 1, 2, 3, 5, 8$$

Solution Each number after the first two numbers is obtained by adding the two previous numbers in the list. Notice that $1 + 1 = 2, 1 + 2 = 3, 2 + 3 = 5$, and so on. If this pattern is to continue, the next number in the sequence is $5 + 8 = 13$, and the next number is $8 + 13 = 21$. □

PRACTICE 2 Find the next number in the sequence

$$2, 4, 6, 10, 16$$

■

The sequence described in Example 2 is called the *Fibonacci sequence*. There are many examples of this sequence found in nature. This sequence also has many applications in science, business, economics, operations research, archeology, fine arts, architecture, and poetry.

EXAMPLE 3 Find the next letter in each sequence.

a. J, F, M, A, M **b.** A, E, D, H, G

Solution

a. Each letter is the first letter of some of the months of the year, in order: January, February, March, April, May. The next month is June, so the letter is J.

b. Let's look for a pattern by corresponding each letter of the alphabet to a number, according to order.

A → 1 + 4 = 5 → E (Add 4)
E → 5 − 1 = 4 → D (Subtract 1)
D → 4 + 4 = 8 → H (Add 4)
H → 8 − 1 = 7 → G (Subtract 1)

$$1 + 4 = 5 \quad 5 - 1 = 4 \quad 4 + 4 = 8 \quad 8 - 1 = 7$$

From this pattern, the next letter is

G → 7 + 4 = 11 → K

$$7 + 4 = 11$$

PRACTICE 3 Find the next letter in each sequence.

a. S, M, T, W, T
b. A, F, D, I, G

EXAMPLE 4 Use inductive reasoning to find the next shape in the sequence.

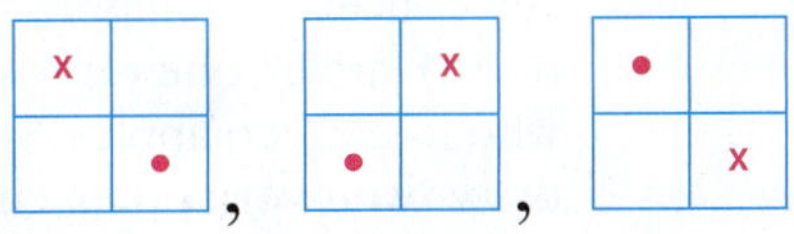

Solution Notice that the "x" rotates clockwise in each square. Also, there is a dot always diagonally across from each x. Thus, we might reason that the next shape is the following.

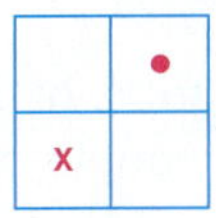

PRACTICE 4 Use inductive reasoning to find the next possible shape in the sequence.

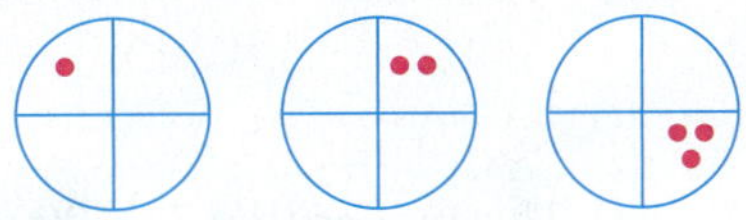

Next, let's study deductive reasoning. We begin with a definition.

OBJECTIVE 2

Using Deductive Reasoning

Deductive Reasoning
This is the process of forming a specific conclusion based on accepted assumptions.

General Assumptions $\xrightarrow{\text{to a}}$ specific conclusion

In short, with inductive reasoning, we reason from specific examples observed to a general conclusion; with deductive reasoning, we reason logically from general statements or assumptions to a specific conclusion.

Diagrams, called *Venn diagrams*, can help us reason deductively. Let's discuss the diagram below that has to do with pet ownership.

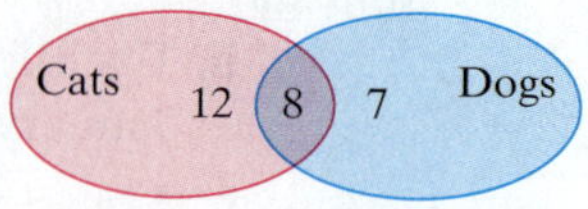

Since 8 people are in both regions, these 8 people have both cats and dogs as pets. See if you understand each answer below.

How many people have cats? $12 + 8 = 20$
How many people have cats and no dogs? 12
How many people have dogs? $8 + 7 = 15$
How many people have dogs and no cats? 7

EXAMPLE 5 The results of a survey of 50 people are as follows.

27 people like red apples.
25 people like green apples.
20 people like both red and green apples.

How many people like neither red nor green apples?

Solution We draw a Venn diagram to organize the information in the survey. This survey concerns red apples and green apples which indicates that our diagram will consist of two circles, one representing people who like red apples and one representing people who like green apples. Since there are 20 people who like both red and green apples, we draw two overlapping circles and place 20 in the overlapping section.

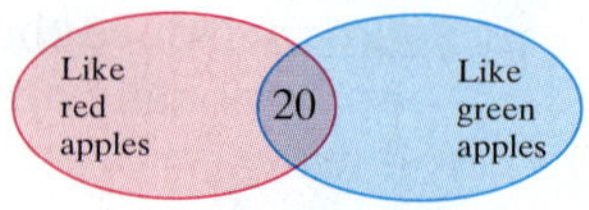

A total of 27 people liked red apples. This means that $27 - 20 = 7$ is the number of people who liked red apples only. Since a total of 25 people liked green apples, $25 - 20 = 5$ people liked green apples only.

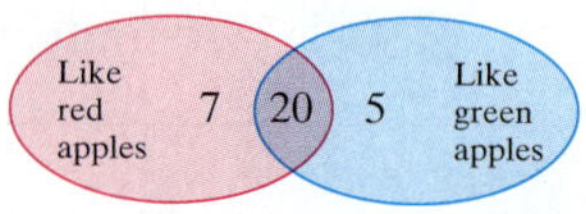

This means that $7 + 20 + 5 = 32$ people like red or green apples. Since 50 people were polled, $50 - 32 = 18$ people liked neither red nor green apples. □

PRACTICE

5 The results of a survey of 30 people are as follows.

20 people like potato chips.
17 people like tortilla chips.
13 people like both potato and tortilla chips.

How many people like neither potato nor tortilla chips? ■

Sometimes a grid may be useful in organizing information.

EXAMPLE 6 Max, Jim, Michael, and Dong Ming have careers as an accountant, a mathematician, a computer programmer, and a manager, not necessarily in the given order. Use the statements below and determine which person is the mathematician.

1. Jim and Max went to lunch with the mathematician.
2. The accountant and the computer programmer taught Jim in college.
3. Max and Dong Ming went to Florida with the computer programmer.

Solution Draw a grid to organize the information. Then record the information given in each statement. From statement 1, we know that neither Jim nor Max is the mathematician. To indicate this on the grid, place X's in the appropriate places. (See below on the left.)

	math.	comp. prog.	acct.	manager
Jim	X			
Max	X			
Michael				
Dong Ming				

	math.	comp. prog.	acct.	manager
Jim	X	X	X	
Max	X			
Michael				
Dong Ming				

From statement 2, we know Jim is not the accountant and he is not the computer programmer, so place an X in the grid corresponding to Jim/accountant and an X in the grid corresponding to Jim/computer programmer. (See above on the right.)

At this point, notice that Jim must be the manager. Place a check mark in the grid corresponding to Jim/manager. Since Jim is the manager, no one else is and we can place X's in the rest of the manager column to indicate the no one else is the manager. (See below on the left.)

	math.	comp. prog.	acct.	manager
Jim	X	X	X	✓
Max	X			X
Michael				X
Dong Ming				X

	math.	comp. prog.	acct.	manager
Jim	X	X	X	✓
Max	X	X		X
Michael				X
Dong Ming		X		X

From statement 3, we know neither Max nor Dong Ming is the computer programmer. Record this information in the grid. (See above on the right.) Notice that Max must be the accountant. Place a check mark in the grid under Max/accountant and mark the rest of that column with X's. (See below on the left.) **Now Dong Ming must be the mathematician** and Michael must be the computer programmer. (See below on the right.)

	math.	comp. prog.	acct.	manager
Jim	X	X	X	✓
Max	X	X	✓	X
Michael			X	X
Dong Ming		X	X	X

	math.	comp. prog.	acct.	manager
Jim	X	X	X	✓
Max	X	X	✓	X
Michael	X	✓	X	X
Dong Ming	✓	X	X	X

□

PRACTICE 6 Persons A, B, C, and D sit in the front row of their mathematics class. Use the statements below to determine their order.

1. Person D is sitting between persons A and B.
2. Person C is sitting next to person B only.
3. Person C is not on the far left.

■

B Exercise Set MyMathLab®

Determine whether each is an example of inductive or deductive reasoning. See Examples 1 through 6.

1. My coat is red. My neighbor's coat is red. Therefore, all coats are red.
2. All typewriters type the letter b. I have a typewriter. Therefore, my typewriter will type the letter b.
3. Rabbits do not lay eggs. Therefore, my pet rabbit will not lay an egg.
4. The last two times Ken flew on a commercial jet, his luggage was lost. Ken reasons that the next time he flies on a commercial jet, his luggage will be also lost.
5. A scientist holds a piece of salt over a burning candle and notices that it burns with a yellow flame. She does this again with another piece of salt and notices that is also burns with a yellow flame. She therefore reasons that all salt burns with a yellow flame.
6. Sherlock Holmes knows that the murderer was either the butler, the maid, or the cook. The night of the murder, the cook catered a party in a nearby town and has plenty of witnesses who saw him. The butler was in the hospital with pneumonia. Therefore, Detective Holmes reasoned that the maid did it.

Use inductive reasoning to determine the next number or figure in each sequence. See Examples 1 through 4.

7. 1, 3, 9, 27
8. 2, 4, 6, 8
9. A, Z, B, Y
10. A, C, E, G
11. 1, 4, 9, 16, 25
12. 2, 4, 8, 16
13. O, T, T, F, F, S, S
14. S, S, M, T, W, T
15. 9, 99, 999, 9999
16. 1, 10, 100, 1000
17. 2, 7, 4, 8, 6
18. 3, 1, 4, 2, 5
19. 9, 12, 20, 33, 51
20. 4, 13, 29, 52, 82
21. 1, 11, 38, 84, 151
22. 2, 12, 24, 41, 66

23. 

24. 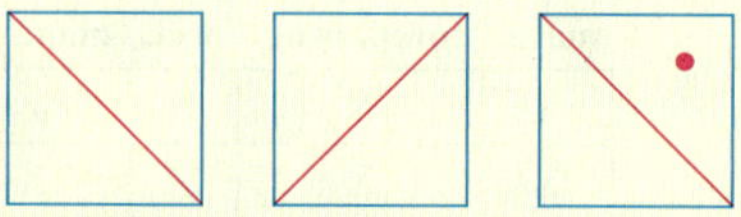

25. 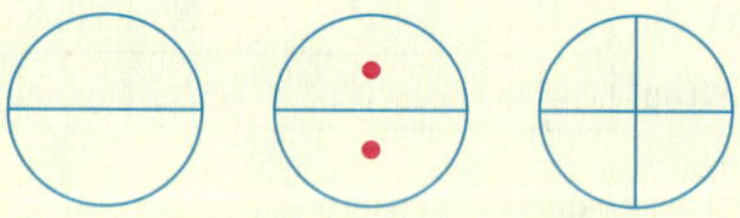

26. 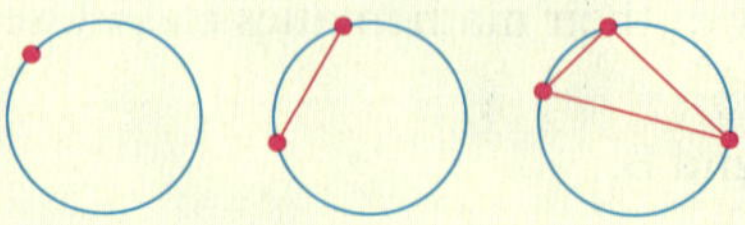

27. 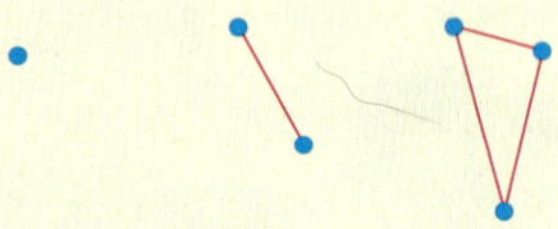

28. 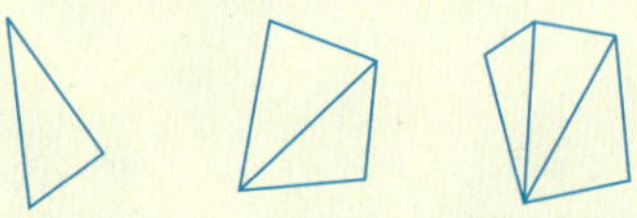

29. 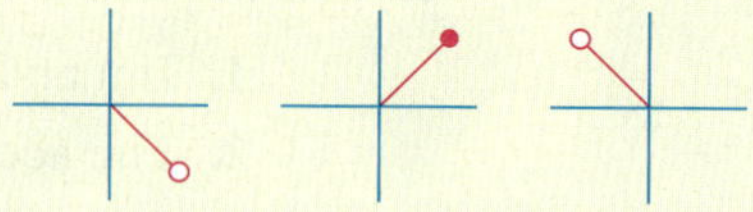

30. 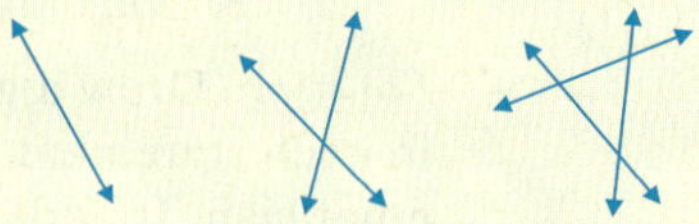

31. 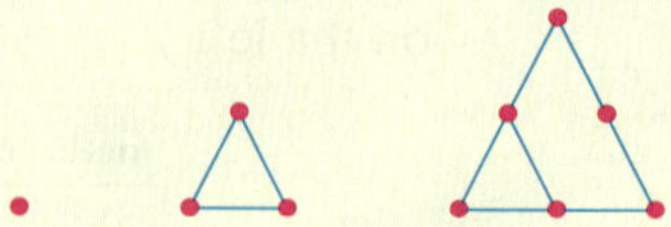

32. 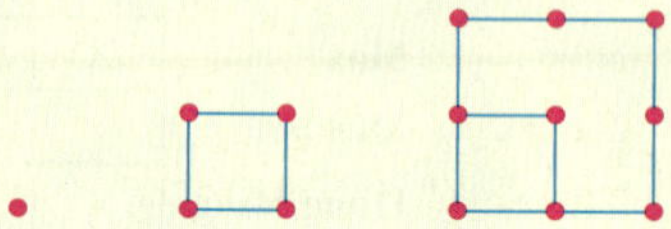

33. 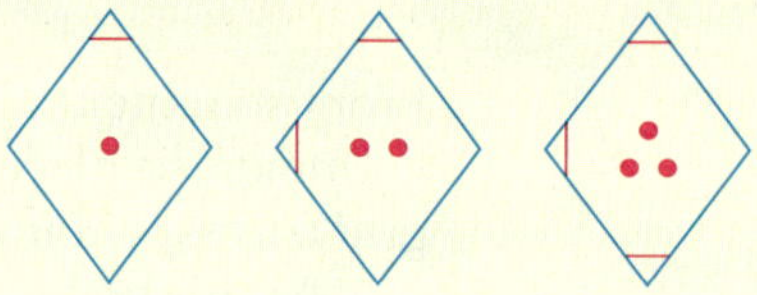

34. 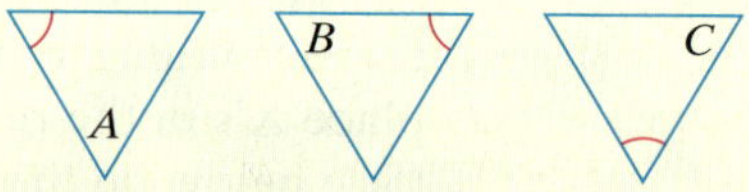

35. Give an example occurring outside the classroom where you used inductive reasoning.
36. Give an example occurring outside the classroom where you used deductive reasoning.

Draw a Venn diagram illustrating each of the following statements. See Example 5. Draw a region representing birds and let a point represent Wendy.

37. Wendy is a bird.

38. Wendy is not a bird.

Draw a Venn diagram illustrating each of the following statements. See Example 5.

39. No bicycles are tricycles.

40. All whales are mammals.

41. Some roosters crow.

42. No trees have wheels.

43. All politicians make promises.

44. Some dryers are heated by electricity.

Use diagrams and deductive reasoning to solve each problem. See Example 6.

45. Four men in a wheelchair race finished 1st, 2nd, 3rd, and 4th.

Mark beat Bob.
Joey finished between Mark and Bob.
Sal beat Mark.

Who finished 3rd in the race?

46. Four women in a marathon finished 1st, 2nd, 3rd, and 4th.

Wynonna beat Mary.
Sally beat Wynonna.
June did not beat Mary.

Who finished 1st in the race?

47. 75 people in a health club were surveyed and the following information was gathered.

55 people drink water after exercising.
40 people drink Gatorade after exercising.
35 people drink water and Gatorade after exercising.

How many people drink neither water nor Gatorade after exercising?

48. In a poll of 200 people, the following information was gathered.

80 people listen to country music.
100 people listen to rock-n-roll music.
50 people listen to both country and rock-n-roll music.

How many people listen to country music only?

49. Aaron, Marty, Donna, and Maria have careers as a civil engineer, an electrical engineer, a mechanical engineer, and a nautical engineer, not necessarily in the given order. Given the information below, determine who is the civil engineer.

a. Aaron and Donna wrote a research paper with the electrical engineer.
b. Marty and Maria carpool with the civil engineer.
c. Aaron works on the same floor as the civil engineer and the nautical engineer.

50. Celeste, Bryan, Clay, and Eric are majoring in business, elementary education, art, and computer science but not necessarily in that order. Given the information below, determine who is majoring in art.

a. Celeste, Bryan, and the elementary education major are in a class together.
b. Bryan and the art major had lunch with the business major.
c. Clay and the elementary education major are cousins.
d. Celeste and the art major carpool together.

51. John, Leon, Alberto, and Julio sit in a theater according to the following.

Alberto is on the far left.
If Leon is on the far right, then John is not sitting next to him.

If Leon is on the far right, who is sitting next to him on his left?

52. Ralph, Anoa, Tumulish, and Pier sit on a bench during graduation ceremonies according to the following.

Tumulish is sitting on the far left.
If Anoa is sitting on the far right, then Ralph is not sitting next to her.

If Anoa is indeed sitting on the far right, list their seating order starting with the person sitting on the far left.

53. 100 people were surveyed and the results are as follows.

65 people said they drink Coke.
40 people said they drink Pepsi.
10 people said they drink both Coke and Pepsi.

How many people drink neither Coke nor Pepsi?

54. A poll was taken in Gotham City one day about the type of transportation used to commute to work. One hundred fifty-five people were polled.

70 people ride the bus.
95 people ride the subway.
30 people ride both the bus and the subway.

How many people ride neither the bus nor the subway?

Appendix C

Ratios and Rates

OBJECTIVES

1. Write Ratios as Fractions.
2. Write Rates as Fractions.
3. Find Unit Rates.
4. Find Unit Prices.

OBJECTIVE 1

Writing Ratios as Fractions

A **ratio** is the quotient of two quantities. A ratio, in fact, is no different from a fraction, except that a ratio is sometimes written using notation other than fractional notation. For example, the ratio of 1 to 2 can be written as

$$1 \text{ to } 2 \quad \text{or} \quad \frac{1}{2} \quad \text{or} \quad 1:2$$

fractional notation ($\frac{1}{2}$) colon notation ($1:2$)

These ratios are all read as, "the ratio of 1 to 2."

CONCEPT CHECK

How should each ratio be read aloud?

a. $\frac{8}{5}$ **b.** $\frac{5}{8}$

In this section, we write ratios using fractional notation. If the fraction happens to be an improper fraction, do not write the fraction as a mixed number. Why? The mixed number form is not a ratio or quotient of two quantities.

Writing a Ratio as a Fraction

The order of the quantities is important when writing ratios. To write a ratio as a fraction, write the *first number* of the ratio as the *numerator* of the fraction and the *second number* as the *denominator*.

Helpful Hint

The ratio of 6 to 11 is $\frac{6}{11}$, *not* $\frac{11}{6}$.

EXAMPLE 1 Write the ratio of 21 to 29 using fractional notation.

Solution The ratio is $\frac{21}{29}$. □

PRACTICE 1

Write the ratio of 19 to 30 using fractional notation. ■

To simplify a ratio, we just write the fraction in simplest form. Common factors as well as common units can be divided out.

Answers to Concept Check:
a. "the ratio of eight to five"
b. "the ratio of five to eight"

EXAMPLE 2 Write the ratio of \$15 to \$10 as a fraction in simplest form.

Solution

$$\frac{\$15}{\$10} = \frac{15}{10} = \frac{3 \cdot \overset{1}{\cancel{5}}}{2 \cdot \underset{1}{\cancel{5}}} = \frac{3}{2}$$ □

PRACTICE 2 Write the ratio of \$16 to \$12 as a fraction in simplest form. ■

> **Helpful Hint**
>
> In the previous example, although $\frac{3}{2} = 1\frac{1}{2}$, a ratio is a quotient of *two* quantities. For that reason, ratios are not written as mixed numbers.

If a ratio contains decimal numbers or mixed numbers, we simplify by writing the ratio as a ratio of whole numbers.

EXAMPLE 3 Write the ratio of 2.5 to 3.15 as a fraction in simplest form.

Solution The ratio in fraction form is

$$\frac{2.5}{3.15}$$

Now let's clear the ratio of decimals.

$$\frac{2.5}{3.15} = \frac{2.5}{3.15} \cdot 1 = \frac{2.5}{3.15} \cdot \frac{100}{100} = \frac{2.5 \cdot 100}{3.15 \cdot 100} = \frac{250}{315} = \frac{50}{63} \quad \text{Simplest form}$$ □

PRACTICE 3 Write the ratio of 1.68 to 4.8 as a fraction in simplest form. ■

EXAMPLE 4 Write the ratio of $2\frac{5}{8}$ to $8\frac{3}{4}$ as a fraction in simplest form.

Solution The ratio in fraction form is $\dfrac{2\frac{5}{8}}{8\frac{3}{4}}$.

To simplify, remember that the fraction bar means division.

$$\frac{2\frac{5}{8}}{8\frac{3}{4}} = 2\frac{5}{8} \div 8\frac{3}{4} = \frac{21}{8} \div \frac{35}{4} = \frac{21}{8} \cdot \frac{4}{35} = \frac{3 \cdot \overset{1}{\cancel{7}} \cdot \overset{1}{\cancel{4}}}{2 \cdot \underset{1}{\cancel{4}} \cdot 5 \cdot \underset{1}{\cancel{7}}} = \frac{3}{10} \quad \text{Simplest form}$$ □

PRACTICE 4 Write the ratio of $2\frac{2}{3}$ to $1\frac{13}{15}$ as a fraction in simplest form. ■

EXAMPLE 5 **Writing a Ratio from a Circle Graph**

The circle graph at the right shows the part of a car's total mileage that falls into a particular category. Write the ratio of family business miles to total miles as a fraction in simplest form.

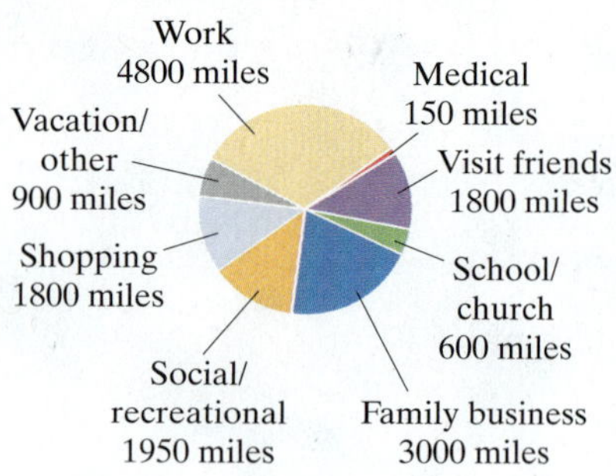

Sources: The American Automobile Manufacturers Association and The National Automobile Dealers Association.

Solution

$$\frac{\text{family business miles}}{\text{total miles}} = \frac{3000 \text{ miles}}{15{,}000 \text{ miles}} = \frac{3000}{15{,}000} = \frac{\overset{1}{\cancel{3000}}}{5 \cdot \underset{1}{\cancel{3000}}} = \frac{1}{5}$$

□

PRACTICE 5 Use the circle graph for Example 5 to write the ratio of work miles to total miles as a fraction in simplest form. ■

EXAMPLE 6 Given the rectangle shown:

a. Find the ratio of its width (shorter side) to its length (longer side).

b. Find the ratio of its length to its perimeter.

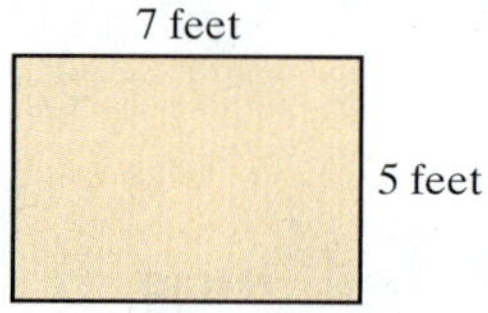

Solution

a. The ratio of its width to its length is

$$\frac{\text{width}}{\text{length}} = \frac{5 \cancel{\text{ feet}}}{7 \cancel{\text{ feet}}} = \frac{5}{7}$$

b. Recall that the perimeter of a rectangle is the distance around the rectangle: $7 + 5 + 7 + 5 = 24$ feet. The ratio of its length to its perimeter is

$$\frac{\text{length}}{\text{perimeter}} = \frac{7 \cancel{\text{ feet}}}{24 \cancel{\text{ feet}}} = \frac{7}{24}$$

□

PRACTICE 6 Given the triangle shown:

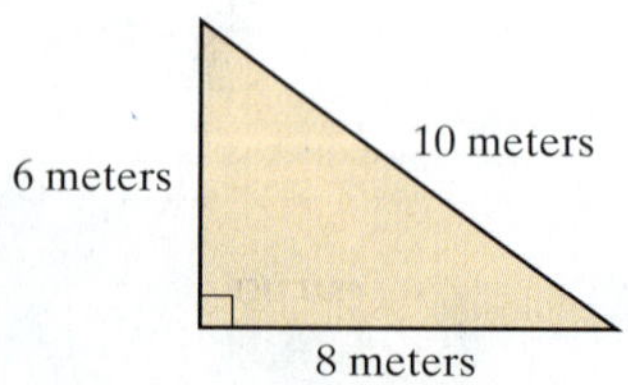

a. Find the ratio of the length of the shortest side to the length of the longest side.

b. Find the ratio of the length of the longest side to the perimeter of the triangle. ■

OBJECTIVE 2

Writing Rates as Fractions

A special type of ratio is a rate. **Rates** are used to compare *different* kinds of quantities. For example, suppose that a recreational runner can run 3 miles in 33 minutes. If we write this rate as a fraction, we have

$$\frac{3 \text{ miles}}{33 \text{ minutes}} = \frac{1 \text{ mile}}{11 \text{ minutes}} \quad \text{In simplest form}$$

Helpful Hint

When comparing quantities with different units, write the units as part of the comparison. Units do not divide out unless they are the same.

Same Units: $\frac{3 \cancel{\text{inches}}}{12 \cancel{\text{inches}}} = \frac{1}{4}$ Units are the same and divide out.

Different Units: $\frac{2 \text{ miles}}{20 \text{ minutes}} = \frac{1 \text{ mile}}{10 \text{ minutes}}$ Units are still written.

EXAMPLES Write each rate as a fraction in simplest form.

7. \$2160 for 12 weeks is $\frac{2160 \text{ dollars}}{12 \text{ weeks}} = \frac{180 \text{ dollars}}{1 \text{ week}}$

8. 360 miles on 16 gallons of gasoline is $\frac{360 \text{ miles}}{16 \text{ gallons}} = \frac{45 \text{ miles}}{2 \text{ gallons}}$ □

PRACTICE 7–8 Write each rate as a fraction in simplest form.

7. \$1350 for 6 weeks

8. 295 miles on 15 gallons of gasoline ■

✔ CONCEPT CHECK

True or false? $\frac{16 \text{ gallons}}{4 \text{ gallons}}$ is a rate. Explain.

OBJECTIVE 3

Finding Unit Rates

A **unit rate** is a rate with a denominator of 1. A familiar example of a unit rate is 55 mph, read as "55 **miles per hour.**" This means 55 miles per 1 hour or

$$\frac{55 \text{ miles}}{1 \text{ hour}} \quad \text{Denominator of 1}$$

Helpful Hint

In this context, the word *per* translates to "division."

Writing a Rate as a Unit Rate

To write a rate as a unit rate, divide the numerator of the rate by the denominator.

EXAMPLE 9 Write as a unit rate: \$31,500 every 7 months

Solution

$$\frac{31{,}500 \text{ dollars}}{7 \text{ months}} \qquad 7\overline{)31{,}500} = 4500$$

(Continued on next page)

Answer to Concept Check:
false; a rate compares different kinds of quantities

The unit rate is

$$\frac{4500 \text{ dollars}}{1 \text{ month}} \text{ or } 4500 \text{ dollars/month}$$ Read as, "4500 dollars per month." □

PRACTICE 9 Write as a unit rate: 3200 feet every 8 seconds ■

EXAMPLE 10 Write as a unit rate: 337.5 miles every 15 gallons of gas

Solution

$$\frac{337.5 \text{ miles}}{15 \text{ gallons}} \qquad 15\overline{)337.5} = 22.5$$

The unit rate is

$$\frac{22.5 \text{ miles}}{1 \text{ gallon}} \text{ or } 22.5 \text{ miles/gallon}$$ Read as, "22.5 miles per gallon." □

PRACTICE 10 Write as a unit rate: 78 bushels of fruit from 12 trees ■

OBJECTIVE 4

Finding Unit Prices

Rates are used extensively in sports, business, medicine, and science. One of the most common uses of rates is in consumer economics. When a unit rate is "money per item," it is also called a **unit price.**

$$\text{unit price} = \frac{\text{price}}{\text{number of units}}$$

EXAMPLE 11 **Finding Unit Price**

A store charges $3.36 for a 16-ounce jar of picante sauce. What is the unit price in dollars per ounce?

Solution

$$\text{unit price} = \frac{\$3.36}{16 \text{ ounces}} = \frac{\$0.21}{1 \text{ ounce}} \quad \text{or} \quad \$0.21 \text{ per ounce}$$ □

PRACTICE 11 An automobile rental agency charges $170 for 5 days for a certain model car. What is the unit price in dollars per day? ■

EXAMPLE 12 **Finding the Best Buy**

Approximate each unit price to decide which is the better buy: 4 bars of soap for $0.99 or 5 bars of soap for $1.19.

Solution

$$\text{unit price} = \frac{\text{price}}{\text{no. of units}} = \frac{\$0.99}{4 \text{ bars}} \approx \$0.25 \text{ per bar of soap} \qquad 4\overline{)0.990} = 0.247 \approx 0.25$$ ("is approximately")

$$\text{unit price} = \frac{\text{price}}{\text{no. of units}} = \frac{\$1.19}{5 \text{ bars}} \approx \$0.24 \text{ per bar of soap} \qquad 5\overline{)1.190} = 0.238 \approx 0.24$$

Since the 5-bar package has a cheaper price per bar, it is the better buy. □

PRACTICE 12 Approximate each unit price to decide which is the better buy for a bag of nacho chips: 11 ounces for \$3.99 or 16 ounces for \$5.99.

Vocabulary, Readiness & Video Check

Use the choices below to fill in each blank. Not all choices will be used.

rate	different	denominator	unit
numerator	unit price	true	
division	ratio	false	

1. A rate with a denominator of 1 is called a(n) ____________ rate.
2. When a rate is written as money per item, a unit rate is called a(n) ____________.
3. The word *per* translates to "____________."
4. Rates are used to compare ____________ types of quantities.
5. To write a rate as a unit rate, divide the ____________ of the rate by the ____________.
6. The quotient of two quantities is called a(n) ____________.
7. Answer true or false: The ratio $\frac{7}{5}$ means the same as the ratio $\frac{5}{7}$. ____________

Martin-Gay Interactive Videos

See Video App C

Watch the section lecture video and answer the following questions.

OBJECTIVE 1
8. How is the ratio in Example 2 rewritten as an equivalent ratio containing no decimals?

OBJECTIVE 2
9. Why can't we divide out the units in Example 5 as we did in Example 4?

OBJECTIVE 3
10. Why did we divide the first quantity of the rate in Example 8 by the second quantity?

OBJECTIVE 4
11. From Example 9, unit prices can be especially helpful when?

C Exercise Set MyMathLab®

Write each ratio as a ratio of whole numbers using fractional notation. Write the fraction in simplest form. See Examples 1 through 6.

1. 16 to 24
2. 25 to 150
3. 7.7 to 10
4. 8.1 to 10
5. 4.63 to 8.21
6. 9.61 to 7.62
7. 6 ounces to 16 ounces
8. 35 meters to 100 meters
9. \$32 to \$100
10. \$46 to \$102
11. 24 days to 14 days
12. 120 miles to 80 miles
13. $3\frac{1}{2}$ to $12\frac{1}{4}$
14. $3\frac{1}{3}$ to $4\frac{1}{6}$
15. $7\frac{3}{5}$ hours to $1\frac{9}{10}$ hours
16. $25\frac{1}{2}$ days to $2\frac{5}{6}$ days

Find the ratio described in each exercise as a fraction in simplest form. See Examples 5 and 6.

17.

Average Weight of Mature Whales	
Blue Whale	***Fin Whale***
145 tons	50 tons

Use the table to find the ratio of the weight of an average mature Fin Whale to the weight of an average mature Blue Whale.

18.

Countries with Small Land Areas	
Tuvalu	***San Marino***
10 sq mi	24 sq mi

(*Source: World Almanac*)

Use the table to find the ratio of the land area of Tuvalu to the land area of San Marino.

△ **19.** Find the ratio of the width of a regulation size basketball court to its perimeter.

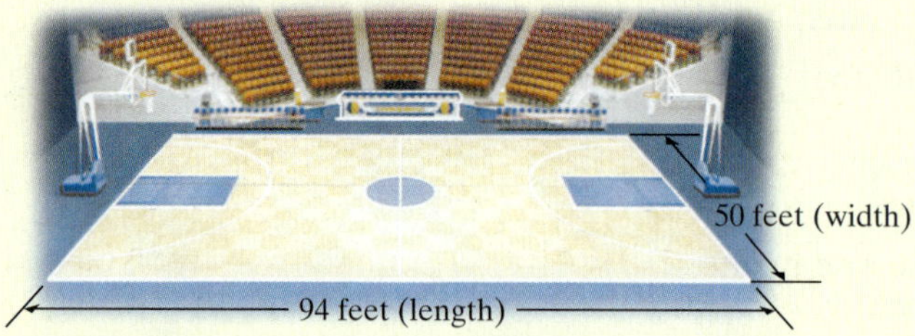

△ **20.** Find the ratio of the width to the perimeter shown of the swimming pool.

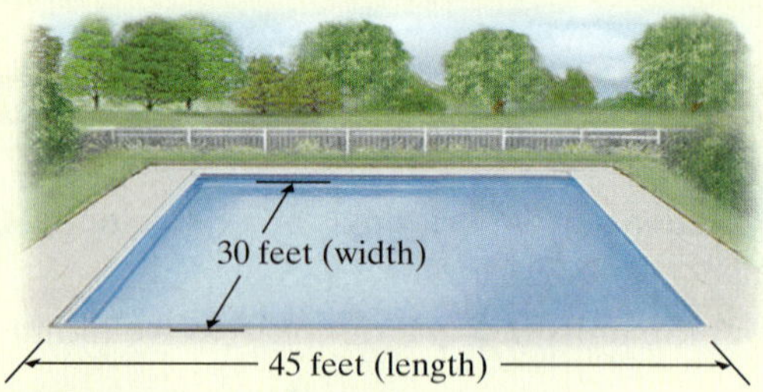

At the Hidalgo County School Board meeting one night, there were 125 women and 100 men present.

21. Find the ratio of women to men.

22. Find the ratio of men to the total number of people present.

Blood contains three types of cells: red blood cells, white blood cells, and platelets. For approximately every 600 red blood cells in healthy humans, there are 40 platelets and 1 white blood cell. Use this information for Exercises 23 and 24. (Source: American Red Cross Biomedical Services*)*

23. Write the ratio of red blood cells to platelet cells.

24. Write the ratio of white blood cells to red blood cells.

Exercises 25 through 28 have to do with disk storage.

25. A standard DVD has a diameter of 12 centimeters while a miniDVD has a diameter of 8 centimeters. Write the ratio of the miniDVD diameter to the standard DVD diameter.

26. LaserDiscs have diameters of 12 centimeters and 20 centimeters. Write the ratio of the smaller diameter to the larger diameter.

27. Hard drives come in many diameters. Find the ratio of a $2\frac{1}{2}''$ diameter to a $5\frac{1}{4}''$ diameter hard drive.

28. Floppy disks come in many diameters. Find the ratio of a $3\frac{1}{2}''$ diameter to a $5\frac{1}{4}''$ diameter floppy disk.

△ **29.** Find the ratio of the longest side to the perimeter of the right-triangular-shaped billboard.

△ **30.** Find the ratio of the base to the perimeter of the triangular mainsail.

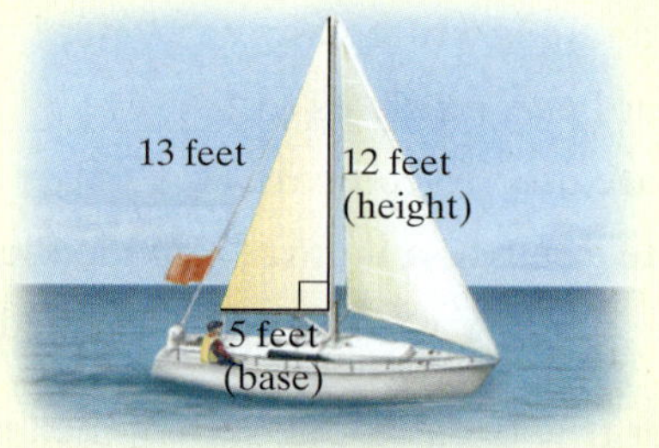

31. Citizens of Mexico consume the most Coca-Cola beverages per capita of any country in the world. Mexicans drink an average of the equivalent of 573 8-oz beverages per year each. In contrast, the average American consumes about the equivalent of 423 8-oz beverages per year. Find the ratio of the average amount of Coca-Cola beverages drunk by Mexicans to the average amount of Coca-Cola beverages drunk by Americans. (*Source:* Coca-Cola Company)

32. A large order of McDonald's french fries has 500 calories. Of this total, 220 calories are from fat. Find the ratio of the calories from fat to total calories in a large order of McDonald's french fries. (*Source:* McDonald's Corporation)

Write each rate as a fraction in simplest form. See Examples 7 and 8.

33. 5 shrubs every 15 feet

34. 14 lab tables for 28 students

35. 15 returns for 100 sales

36. 8 phone lines for 36 employees

37. 6 laser printers for 28 computers

38. 4 inches of rain in 18 hours

39. 18 gallons of pesticide for 4 acres of crops

40. 150 graduate students for 8 advisors

Write each rate as a unit rate. See Examples 9 and 10.

41. 330 calories in a 3-ounce serving

42. 275 miles in 11 hours

43. 375 riders in 5 subway cars

44. 18 signs in 6 blocks

45. A hummingbird moves its wings at a rate of 5400 wingbeats a minute. Write this rate in wingbeats per second.

46. A bat moves its wings at a rate of 1200 wingbeats a minute. Write this rate in wingbeats per second.

47. A $1,000,000 lottery winning paid over 20 years

48. 400,000 library books for 8000 students

49. The state of Delaware has 631,500 registered voters for two senators. (*Source:* Delaware.gov)

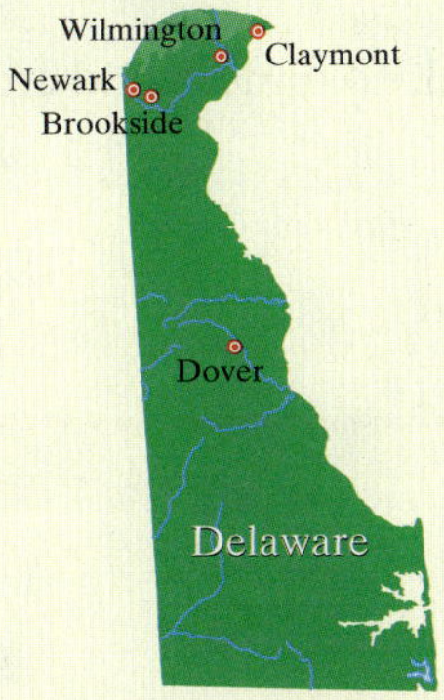

50. The 2020 projected population of Louisiana is approximately 4,588,800 residents for 64 parishes. (*Note:* Louisiana is the only U.S. state with parishes instead of counties.) (*Source:* Louisiana.gov)

51. 12,000 good assembly line products to 40 defective products

52. 5,000,000 lottery tickets for 4 lottery winners

53. The combined salary for the 20 highest-paid players of the 2012 World Series Champion San Francisco Giants was approximately $118,494,000. (*Source:* ESPN)

54. The top-grossing concert tour was the 2009–2011 U2 360° tour, which grossed over $735,900 thousand for 110 shows worldwide. (*Source:* Pollstar)

55. Charlie Catlett can assemble 250 computer boards in an 8-hour shift while Suellen Catlett can assemble 402 computer boards in a 12-hour shift.

a. Find the unit rate of Charlie.
b. Find the unit rate of Suellen.
c. Who can assemble computer boards faster, Charlie or Suellen?

56. Jerry Stein laid 713 bricks in 46 minutes while his associate, Bobby Burns, laid 396 bricks in 30 minutes.

a. Find the unit rate of Jerry.
b. Find the unit rate of Bobby.
c. Who is the faster bricklayer?

For Exercises 57 and 58, round the rates to the nearest tenth.

57. One student drove 400 miles in his car on 14.5 gallons of gasoline. His sister drove 270 miles in her truck on 9.25 gallons of gasoline.

a. Find the unit rate of the car.
b. Find the unit rate of the truck.
c. Which vehicle gets better gas mileage?

58. Charlotte Leal is a grocery scanner who can scan an average of 100 items in 3.5 minutes while her cousin Leo can scan 148 items in 5.5 minutes.

a. Find the unit rate of Charlotte.
b. Find the unit rate of Leo.
c. Who is the faster scanner?

Find each unit price. See Example 11.

59. $57.50 for 5 compact discs

60. $0.87 for 3 apples

61. $1.19 for 7 bananas

62. $73.50 for 6 lawn chairs

Find each unit price and decide which is the better buy. Round to three decimal places. Assume that we are comparing different sizes of the same brand. See Example 12.

63. Crackers:
$3.29 for 8 ounces
$4.79 for 12 ounces

64. Pickles:
$2.79 for 32 ounces
$1.49 for 18 ounces

65. Frozen orange juice:
$1.89 for 16 ounces
$0.69 for 6 ounces

66. Eggs:
$1.56 for a dozen
$3.69 for a flat $\left(2\frac{1}{2}\text{dozen}\right)$

67. Soy sauce:
12 ounces for $2.29
8 ounces for $1.49

68. Shampoo:
20 ounces for $1.89
32 ounces for $3.19

69. Napkins:
100 for $0.59
180 for $0.93

70. Crackers:
20 ounces for $2.39
8 ounces for $0.99

Appendix D

Proportion and Problem Solving

OBJECTIVES

1 Solve Proportions.

2 Use Proportions to Solve Problems.

OBJECTIVE

1 Solving Proportions

A **ratio** is the quotient of two numbers or two quantities. For example, the ratio of 2 to 5 can be written as $\frac{2}{5}$, the quotient of 2 and 5.

If two ratios are equal, we say the ratios are **in proportion** to each other. A **proportion** is a mathematical statement that two ratios are equal.

For example, the equation $\frac{1}{2} = \frac{4}{8}$ is a proportion, as is $\frac{x}{5} = \frac{8}{10}$, because both sides of the equations are ratios. When we want to emphasize the equation as a proportion, we

read the proportion $\frac{1}{2} = \frac{4}{8}$ as "one is to two as four is to eight"

In a proportion, cross products are equal. To understand cross products, let's start with the proportion

$$\frac{a}{b} = \frac{c}{d}$$

and multiply both sides by the LCD, bd.

$$bd\left(\frac{a}{b}\right) = bd\left(\frac{c}{d}\right) \quad \text{Multiply both sides by the LCD, } bd.$$

$$\underbrace{ad}_{\text{Cross product}} = \underbrace{bc}_{\text{Cross product}} \quad \text{Simplify.}$$

Notice why ad and bc are called cross products.

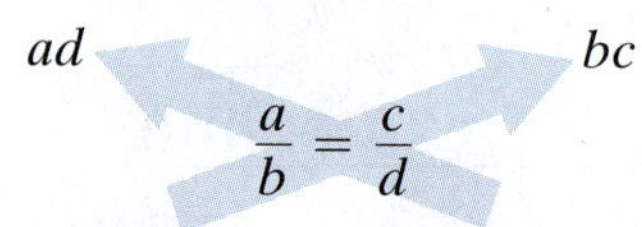

Cross Products

If $\frac{a}{b} = \frac{c}{d}$, then $ad = bc$.

For example, if

$$\frac{1}{2} = \frac{4}{8}, \text{ then } 1 \cdot 8 = 2 \cdot 4 \text{ or}$$

$$8 = 8$$

Notice that a proportion contains four numbers (or expressions). If any three numbers are known, we can solve and find the fourth number.

EXAMPLE 1 Solve for x: $\frac{45}{x} = \frac{5}{7}$

Solution This equation is a proportion. Since this is a proportion, we may set cross products equal.

$$\frac{45}{x} = \frac{5}{7}$$

$$45 \cdot 7 = x \cdot 5 \quad \text{Set cross products equal.}$$
$$315 = 5x \quad \text{Multiply.}$$
$$\frac{315}{5} = \frac{5x}{5} \quad \text{Divide both sides by 5.}$$
$$63 = x \quad \text{Simplify.}$$

Check: To check, substitute 63 for x in the original proportion. The solution is 63. □

PRACTICE 1 Solve for x: $\frac{36}{x} = \frac{4}{11}$ ■

EXAMPLE 2 Solve for x: $\frac{x-5}{3} = \frac{x+2}{5}$

Solution

$$\frac{x-5}{3} = \frac{x+2}{5}$$

$$5(x-5) = 3(x+2) \quad \text{Set cross products equal.}$$
$$5x - 25 = 3x + 6 \quad \text{Multiply.}$$
$$5x = 3x + 31 \quad \text{Add 25 to both sides.}$$
$$2x = 31 \quad \text{Subtract } 3x \text{ from both sides.}$$
$$\frac{2x}{2} = \frac{31}{2} \quad \text{Divide both sides by 2.}$$
$$x = \frac{31}{2}$$

Check: Verify that $\frac{31}{2}$ is the solution. □

PRACTICE 2 Solve for x: $\frac{3x+2}{9} = \frac{x-1}{2}$ ■

OBJECTIVE 2

Using Proportions to Solve Problems

Proportions can be used to model and solve many real-life problems. When using proportions in this way, it is important to judge whether the solution is reasonable. Doing so helps us decide if the proportion has been formed correctly. We use the same problem-solving steps that were introduced in Section 2.4.

EXAMPLE 3 **Calculating the Cost of a Redbox Rental**

Not everyone is streaming movies. Many people still rent movies and video games from Redbox. If renting 3 movies costs $4.50, how much should 5 movies cost?

Solution

1. UNDERSTAND. Read and reread the problem. We know that the cost of renting 5 movies is more than the cost of renting three movies, or $4.50, and less than the cost of 6 rentals, which is double the cost of 3 rentals, or $2(\$4.50) = \$9.00$. Let's suppose that 5 rentals cost $7.00. To check, we see if 3 rentals is to 5 rentals as the *price* of three rentals is to the *price* of 5 rentals. In other words, we see if

$$\frac{3 \text{ rentals}}{5 \text{ rentals}} = \frac{\textit{price} \text{ of } 3 \text{ rentals}}{\textit{price} \text{ of } 5 \text{ rentals}}$$

or

$$\frac{3}{5} = \frac{4.50}{7}$$

$$3(7) = 5(4.50) \quad \text{Set cross products equal.}$$

or

$$21 = 22.5 \quad \text{Not a true statement.}$$

Thus, $7 is not correct, but we now have a better understanding of the problem.
Let x = price of renting 5 videos.

2. TRANSLATE.

$$\frac{3 \text{ rentals}}{5 \text{ rentals}} = \frac{\textit{price} \text{ of } 3 \text{ rentals}}{\textit{price} \text{ of } 5 \text{ rentals}}$$

$$\frac{3}{5} = \frac{4.50}{x}$$

3. SOLVE.

$$\frac{3}{5} = \frac{4.50}{x}$$

$$3x = 5(4.50) \quad \text{Set cross products equal.}$$

$$3x = 22.50$$

$$x = 7.50 \quad \text{Divide both sides by 3.}$$

4. INTERPRET.

Check: Verify that 3 rentals is to 5 rentals as $4.50 is to $7.50. Also, notice that our solution is a reasonable one as discussed in Step 1.

State: Five rentals from Redbox cost $7.50. □

PRACTICE 3 Four 2-liter bottles of Diet Pepsi cost $5.16. How much will seven 2-liter bottles cost? ■

Helpful Hint

The proportion $\frac{5 \text{ rentals}}{3 \text{ rentals}} = \frac{\textit{price} \text{ of } 5 \text{ rentals}}{\textit{price} \text{ of } 3 \text{ rentals}}$ could also have been used to solve Example 3. Notice that the cross products are the same.

Similar triangles have the same shape but not necessarily the same size. In similar triangles, the measures of corresponding angles are equal, and corresponding sides are in proportion.

If triangle ABC and triangle XYZ shown are similar, then we know that the measure of angle A = the measure of angle X, the measure of angle B = the measure of angle Y, and the measure of angle C = the measure of angle Z. We also know that corresponding sides are in proportion: $\frac{a}{x} = \frac{b}{y} = \frac{c}{z}$.

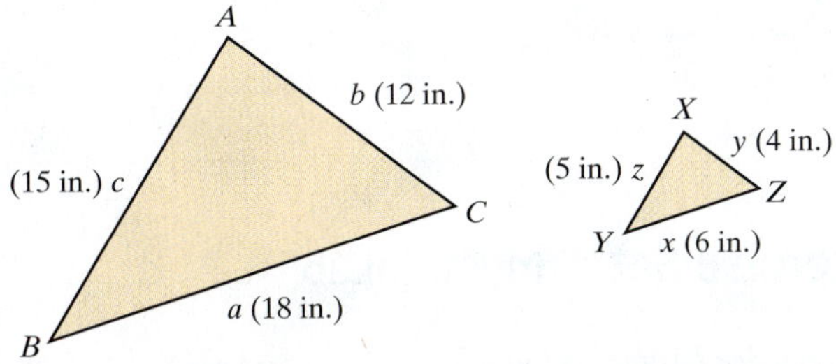

In this section, we will position similar triangles so that they have the same orientation.

To show that corresponding sides are in proportion for the triangles above, we write the ratios of the corresponding sides.

$$\frac{a}{x} = \frac{18}{6} = 3 \quad \frac{b}{y} = \frac{12}{4} = 3 \quad \frac{c}{z} = \frac{15}{5} = 3$$

EXAMPLE 4 **Finding the Length of a Side of a Triangle**

If the following two triangles are similar, find the unknown length x.

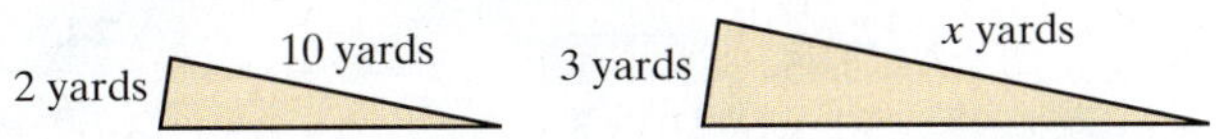

Solution

1. UNDERSTAND. Read the problem and study the figure.
2. TRANSLATE. Since the triangles are similar, their corresponding sides are in proportion and we have

$$\frac{2}{3} = \frac{10}{x}$$

3. SOLVE. To solve, we cross multiply.

$$2x = 30$$

$$x = 15 \quad \text{Divide both sides by 2.}$$

4. INTERPRET.

Check: To check, replace x with 15 in the original proportion and see that a true statement results.

State: The unknown length is 15 yards. □

PRACTICE 4 If the following two triangles are similar, find x.

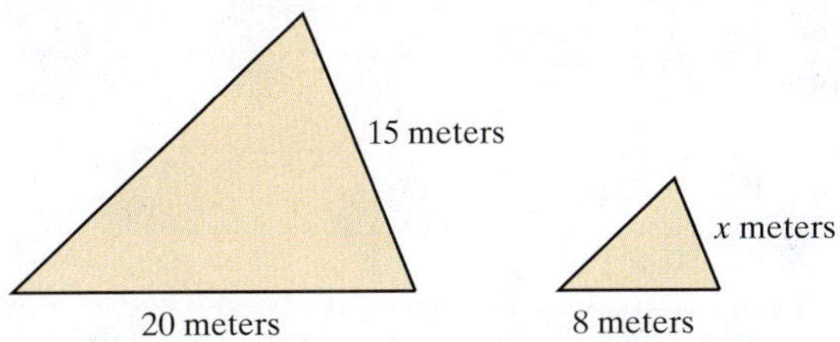

Martin-Gay Interactive Videos

See Video App D

Watch the section lecture video and answer the following questions.

OBJECTIVE 1

1. Based on Examples 1 and 2, can proportions only be solved by using cross products? Explain.

OBJECTIVE 2

2. In Example 3 we are told there are many ways to set up a correct proportion. Why does this fact make it even more important to check that your solution is reasonable?

D Exercise Set MyMathLab®

Solve each proportion. See Examples 1 and 2.

1. $\frac{2}{3} = \frac{x}{6}$

2. $\frac{x}{2} = \frac{16}{6}$

3. $\frac{x}{10} = \frac{5}{9}$

4. $\frac{9}{4x} = \frac{6}{2}$

5. $\frac{x+1}{2x+3} = \frac{2}{3}$

6. $\frac{x+1}{x+2} = \frac{5}{3}$

7. $\frac{9}{5} = \frac{12}{3x+2}$

8. $\frac{6}{11} = \frac{27}{3x-2}$

9. $\frac{7x}{18} = \frac{5}{3}$

10. $\frac{6x}{5} = \frac{19}{10}$

11. $\frac{11}{7} = \frac{4}{x+1}$

12. $\frac{12}{5} = \frac{7}{x-1}$

13. $\frac{x-3}{2x+1} = \frac{4}{9}$

14. $\frac{x+1}{3x-4} = \frac{3}{8}$

15. $\frac{2x+1}{4} = \frac{6x-1}{5}$

16. $\frac{11x+2}{3} = \frac{10x-5}{2}$

Solve. See Example 3.

17. The ratio of the weight of an object on Earth to the weight of the same object on Pluto is 100 to 3. If an elephant weighs 4100 pounds on Earth, find the elephant's weight on Pluto.

18. If a 170-pound person weighs approximately 65 pounds on Mars, about how much does a 9000-pound satellite weigh? Round your answer to the nearest pound.

19. There are 110 calories per 177.4 grams of Frosted Flakes cereal. Find out how many calories are in 212.5 grams of this cereal. Round to the nearest whole calorie.

20. On an architect's blueprint, 1 inch corresponds to 4 feet. Find the length of a wall represented by a line that is $3\frac{7}{8}$ inches long on the blueprint.

Find the unknown length x or y in the following pairs of similar triangles. See Example 4.

△ 21.

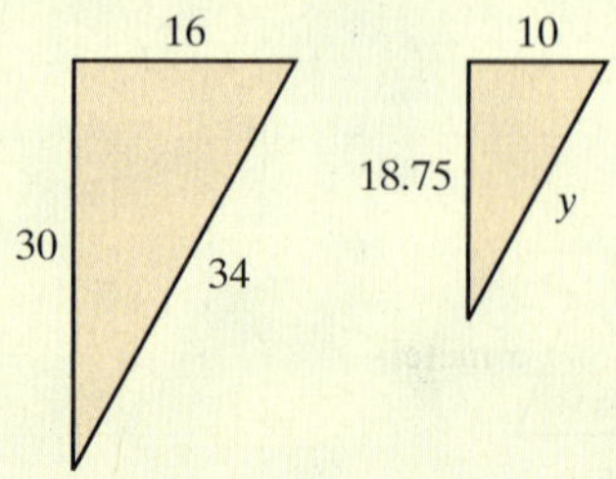

△ 22.

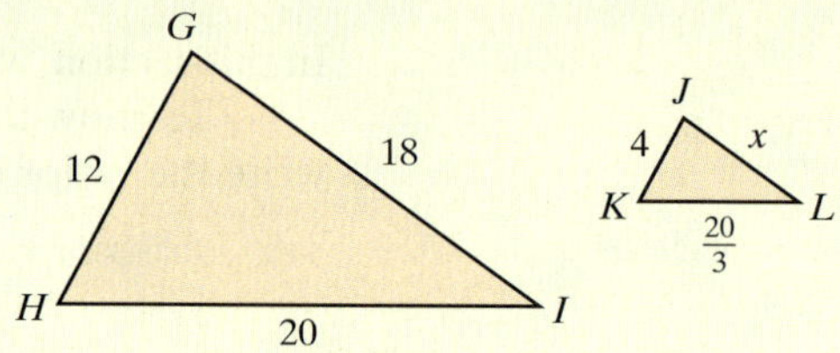

△ 23.

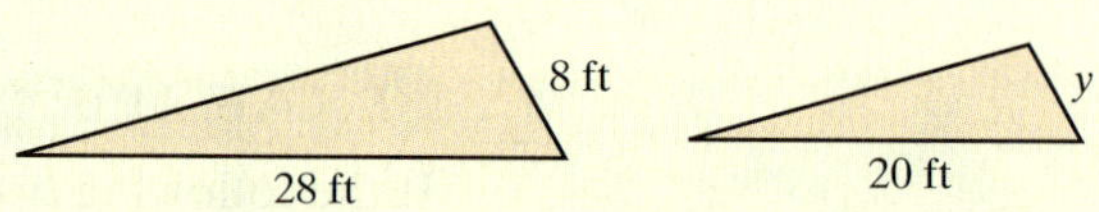

△ 24.

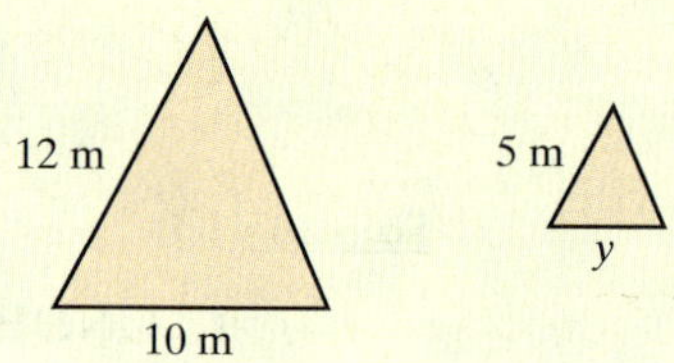

Solve. A self-tanning lotion advertises that a 3-oz bottle will provide four applications.

25. Jen Haddad found a great deal on a 14-oz bottle of the self-tanning lotion she had been using. Based on the advertising claims, how many applications of the self-tanner should Jen expect? Round down to the next whole number.

26. The Community College thespians need fake tans for a play they are doing. If the play has a cast of 35, how many ounces of self-tanning lotion should the cast purchase? Round up to the next whole number of ounces.

The school's computer lab goes through 5 reams of printer paper every 3 weeks.

27. Find out how long a case of printer paper is likely to last (a case of paper holds 8 reams of paper). Round to the nearest week.

28. How many cases of printer paper should be purchased to last the entire semester of 15 weeks? Round up to the next case.

A recipe for pancakes calls for 2 cups flour and $1\frac{1}{2}$ cups milk to make a serving for four people.

29. Ming has plenty of flour, but only 4 cups milk. How many servings can he make?

30. The swim team has a weekly breakfast after early practice. How much flour will it take to make pancakes for 18 swimmers?

Solve.

31. A 16-oz grande Tazo Black Iced Tea at Starbucks has 80 calories. How many calories are there in a 24-oz venti Tazo Black Iced Tea? (*Source:* Starbucks Coffee Company)

32. A 16-oz nonfat Caramel Macchiato at Starbucks has 220 calories. How many calories are in a 12-oz nonfat Caramel Macchiato? (*Source:* Starbucks Coffee Company)

33. Mosquitos are annoying insects. To eliminate mosquito larvae, a certain granular substance can be applied to standing water in a ratio of 1 tsp per 25 sq ft of standing water.

a. At this rate, find how many teaspoons of granules must be used for 450 square feet.

b. If 3 tsp = 1 tbsp, how many tablespoons of granules must be used?

34. Another type of mosquito control is liquid, where 3 oz of pesticide is mixed with 100 oz of water. This mixture is sprayed on roadsides to control mosquito breeding grounds hidden by tall grass.

a. If one mixture of water with this pesticide can treat 150 feet of roadway, how many ounces of pesticide are needed to treat one mile? (*Hint:* 1. mile = 5280 feet)

b. If 8 liquid ounces equals one cup, write your answer to part **a** in cups. Round to the nearest cup.

35. The daily supply of oxygen for one person is provided by 625 square feet of lawn. A total of 3750 square feet of lawn would provide the daily supply of oxygen for how many people? (*Source:* Professional Lawn Care Association of America)

36. In the United States, approximately 71 million of the 200 million cars and light trucks in service have driver-side air bags. In a parking lot containing 800 cars and light trucks, how many would be expected to have driver-side air bags? (*Source:* Insurance Institute for Highway Safety)

Given that the following pairs of triangles are similar, find each unknown length.

△ **37.**

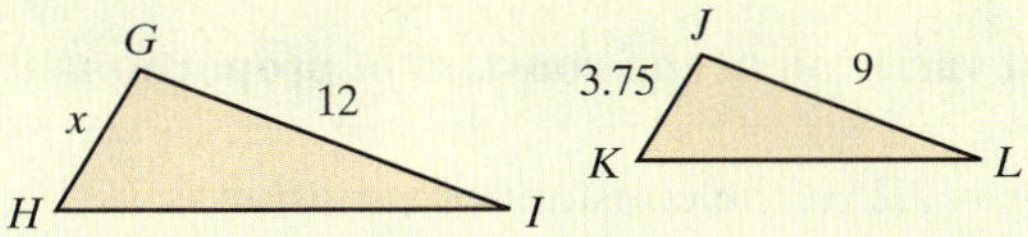

△ **38.**

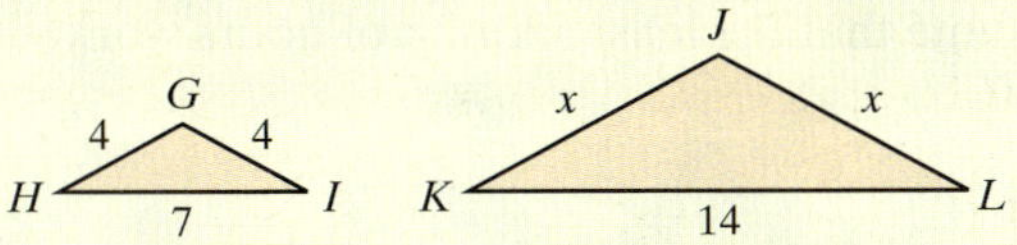

△ **39.**

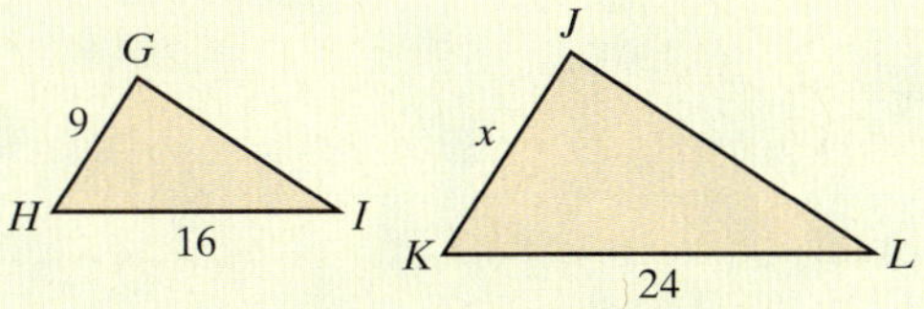

△ **40.**

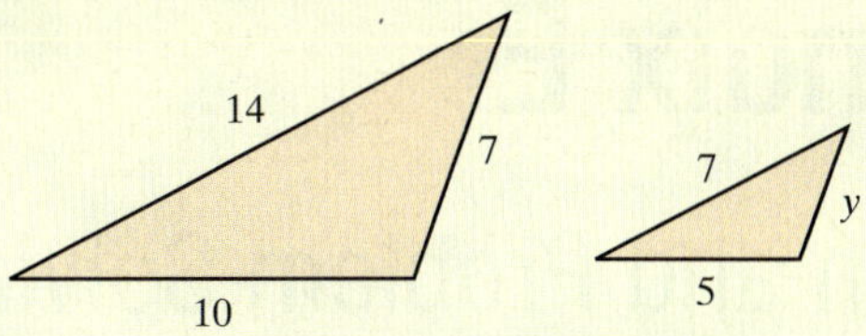

41. A student would like to estimate the height of the Statue of Liberty in New York City's harbor. The length of the Statue of Liberty's right arm is 42 feet. The student's right arm is 2 feet long and her height is $5\frac{1}{3}$ feet. Use this information to estimate the height of the Statue of Liberty. How close is your estimate to the statue's actual height of 111 feet, 1 inch from heel to top of head? (*Source:* National Park Service)

42. The length of the Statue of Liberty's index finger is 8 feet while the height to the top of the head is about 111 feet. Suppose your measurements are proportionally the same as this statue and your height is 5 feet.

a. Use this information to find the proposed length of your index finger. Give the exact measurement and then a decimal rounded to the nearest hundredth.

b. Measure your index finger and write it as a decimal in feet rounded to the nearest hundredth. How close is the length of your index finger to the answer to **a**? Explain why.

43. There are 72 milligrams of cholesterol in a 3.5 ounce serving of lobster. How much cholesterol is in 5 ounces of lobster? Round to the nearest tenth of a milligram. (*Source:* The National Institute of Health)

44. There are 76 milligrams of cholesterol in a 3-ounce serving of skinless chicken. How much cholesterol is in 8 ounces of chicken? (*Source:* USDA)

45. Medication is prescribed in 7 out of every 10 hospital emergency room visits that involve an injury. If a large urban hospital had 620 emergency room visits involving an injury in the past month, how many of these visits would you expect included a prescription for medication? (*Source:* National Center for Health Statistics)

46. Currently in the American population of people aged 65 years old and older, there are 145 women for every 100 men. In a nursing home with 280 male residents over the age of 65, how many female residents over the age of 65 would be expected? (*Source:* U.S. Bureau of the Census)

47. One out of three American adults has worked in the restaurant industry at some point during his or her life. In an office of 84 workers, how many of these people would you expect to have worked in the restaurant industry at some point? (*Source:* National Restaurant Association)

48. One pound of firmly packed brown sugar yields $2\frac{1}{4}$ cups. How many pounds of brown sugar will be required in a recipe that calls for 6 cups of firmly packed brown sugar? (*Source:* Based on data from *Family Circle* magazine)

Appendix E

Variation and Problem Solving

OBJECTIVES

1. Solve Problems Involving Direct Variation.
2. Solve Problems Involving Inverse Variation.
3. Other Types of Direct and Inverse Variation.
4. Variation and Problem Solving.

In Chapter 3, we studied linear equations in two variables. Recall that such an equation can be written in the form $Ax + By = C$, where A and B are not both 0.

Also recall that the graph of a linear equation in two variables is a line. In this section, we begin by looking at a particular family of linear equations—those that can be written in the form

$$y = kx,$$

where k is a constant. This family of equations is called *direct variation*.

OBJECTIVE

1 Solving Direct Variation Problems

Let's suppose that you are earning \$7.25 per hour at a part-time job. The amount of money you earn depends on the number of hours you work. This is illustrated by the following table:

Hours Worked	0	1	2	3	4
Money Earned (before deductions)	0	7.25	14.50	21.75	29.00

and so on

In general, to calculate your earnings (before deductions) multiply the constant \$7.25 by the number of hours you work. If we let y represent the amount of money earned and x represent the number of hours worked, we get the direct variation equation

$$y = 7.25 \cdot x$$

earnings = \$7.25 · hours worked

Notice that in this direct variation equation, as the number of hours increases, the pay increases as well.

> **Direct Variation**
> **y varies directly as x**, or **y is directly proportional to x**, if there is a nonzero constant k such that
> $$y = kx$$
> The number k is called the **constant of variation** or the **constant of proportionality.**

In our direct variation example $y = 7.25x$, the constant of variation is 7.25.

Let's use the previous table to graph $y = 7.25x$. We begin our graph at the ordered pair solution $(0, 0)$. Why? We assume that the least amount of hours worked is 0. If 0 hours are worked, then the pay is \$0.

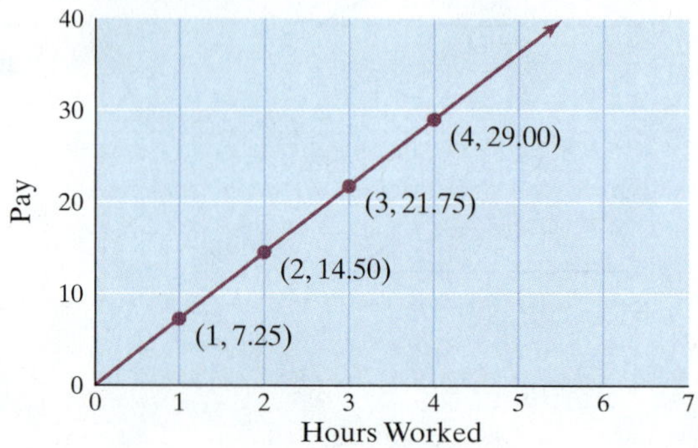

As illustrated in this graph, a direct variation equation $y = kx$ is linear. Also notice that $y = 7.25x$ is a function since its graph passes the vertical line test.

EXAMPLE 1 Write a direct variation equation of the form $y = kx$ that satisfies the ordered pairs in the table below.

x	2	9	1.5	−1
y	6	27	4.5	−3

Solution We are given that there is a direct variation relationship between x and y. This means that

$$y = kx.$$

By studying the given values, you may be able to calculate k mentally. If not, to find k, we simply substitute one given ordered pair into this equation and solve for k. We'll use the given pair $(2, 6)$.

$$y = kx$$
$$6 = k \cdot 2$$
$$\frac{6}{2} = \frac{k \cdot 2}{2}$$
$$3 = k \quad \text{Solve for } k.$$

Since $k = 3$, we have the equation $y = 3x$.
To check, see that each given y is 3 times the given x. □

PRACTICE
1 Write a direct variation of the form $y = kx$ that satisfies the ordered pairs in the table below.

x	2	8	−4	1.3
y	10	40	−20	6.5

■

Let's try another type of direct variation example.

EXAMPLE 2 Suppose that y varies directly as x. If y is 17 when x is 34, find the constant of variation and the direct variation equation. Then find y when x is 12.

Solution Let's use the same method as in Example 1 to find x. Since we are told that y varies directly as x, we know the relationship is of the form

$$y = kx.$$

Let $y = 17$ and $x = 34$ and solve for k.

$$17 = k \cdot 34$$
$$\frac{17}{34} = \frac{k \cdot 34}{34}$$
$$\frac{1}{2} = k \quad \text{Solve for } k.$$

Thus, the constant of variation is $\frac{1}{2}$ and the equation is $y = \frac{1}{2}x$.

To find y when $x = 12$, use $y = \frac{1}{2}x$ and replace x with 12.

$$y = \frac{1}{2}x$$
$$y = \frac{1}{2} \cdot 12 \quad \text{Replace } x \text{ with 12.}$$
$$y = 6$$

Thus, when x is 12, y is 6. □

(Continued on next page)

PRACTICE 2 If y varies directly as x and y is 12 when x is 48, find the constant of variation and the direct variation equation. Then find y when x is 20. ■

Let's review a few facts about linear equations of the form $y = kx$.

Direct Variation: $y = kx$

- There is a direct variation relationship between x and y.
- The graph is a line.
- The line will always go through the origin $(0, 0)$. Why? Let $x = 0$. Then $y = k \cdot 0$ or $y = 0$.
- The slope of the graph of $y = kx$ is k, the constant of variation. Why? Remember that the slope of an equation of the form $y = mx + b$ is m, the coefficient of x.
- The equation $y = kx$ describes a function. Each x has a unique y and its graph passes the vertical line test.

EXAMPLE 3 The line is the graph of a direct variation equation. Find the constant of variation and the direct variation equation.

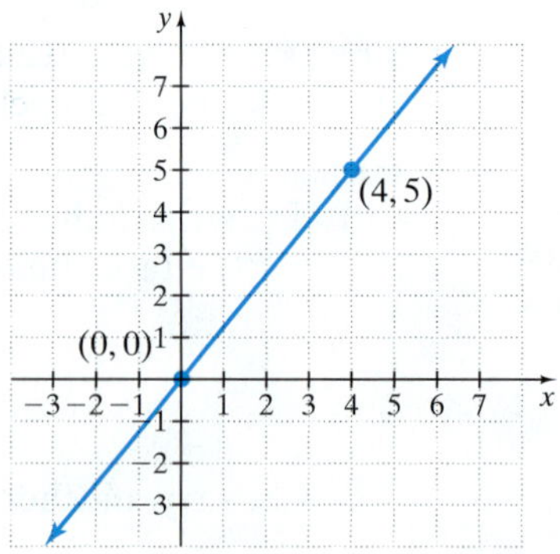

Solution Recall that k, the constant of variation, is the same as the slope of the line. Thus, to find k, we use the slope formula and find slope.

Using the given points $(0, 0)$, and $(4, 5)$, we have

$$\text{slope} = \frac{5 - 0}{4 - 0} = \frac{5}{4}.$$

Thus, $k = \frac{5}{4}$ and the variation equation is $y = \frac{5}{4}x$. □

PRACTICE 3 Find the constant of variation and the direct variation equation for the line below.

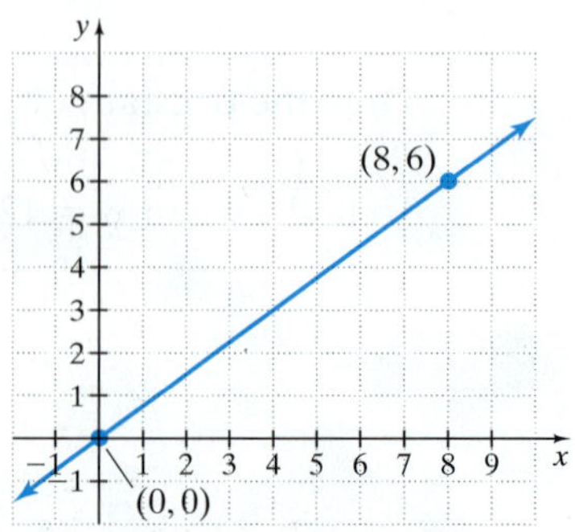

■

OBJECTIVE

2 Solving Inverse Variation Problems

In this section, we will introduce another type of variation, called inverse variation.

Let's suppose you need to drive a distance of 40 miles. You know that the faster you drive the distance, the sooner you arrive at your destination. Recall that there is a mathematical relationship between distance, rate, and time. It is $d = r \cdot t$. In our example, distance is a constant 40 miles, so we have $40 = r \cdot t$ or $t = \frac{40}{r}$.

For example, if you drive 10 mph, the time to drive the 40 miles is

$$t = \frac{40}{r} = \frac{40}{10} = 4 \text{ hours}$$

If you drive 20 mph, the time is

$$t = \frac{40}{r} = \frac{40}{20} = 2 \text{ hours}$$

Again, notice that as speed increases, time decreases. Below are some ordered pair solutions of $t = \frac{40}{r}$ and its graph.

Rate (mph)	r	5	10	20	40	60	80
Time (hr)	t	8	4	2	1	$\frac{2}{3}$	$\frac{1}{2}$

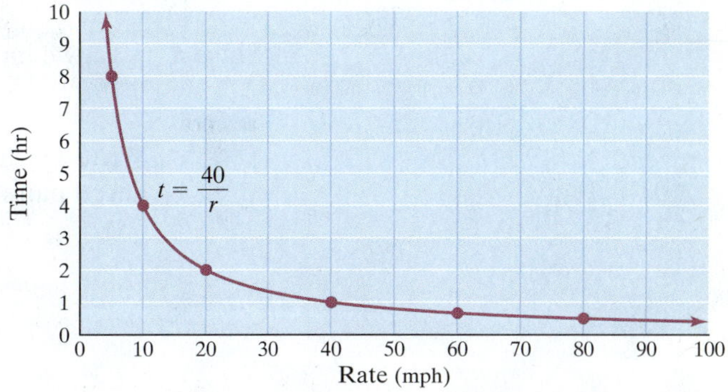

Notice that the graph of this variation is not a line, but it passes the vertical line test, so $t = \frac{40}{r}$ does describe a function. This is an example of inverse variation.

Inverse Variation

***y* varies inversely as *x*,** or ***y* is inversely proportional to *x*,** if there is a nonzero constant k such that

$$y = \frac{k}{x}$$

The number k is called the **constant of variation** or the **constant of proportionality.**

In our inverse variation example, $t = \frac{40}{r}$ or $y = \frac{40}{x}$, the constant of variation is 40.

We can immediately see differences and similarities in direct variation and inverse variation.

Direct variation	$y = kx$	linear equation	both functions
Inverse variation	$y = \frac{k}{x}$	not a linear equation	

Remember that $y = \frac{k}{x}$ is not a linear equation. Also notice that because x is in the denominator, x can be any value except 0.

We can still derive an inverse variation equation from a table of values.

EXAMPLE 4 Write an inverse variation equation of the form $y = \frac{k}{x}$ that satisfies the ordered pairs in the table below.

x	2	4	$\frac{1}{2}$
y	6	3	24

Solution Since there is an inverse variation relationship between x and y, we know that $y = \frac{k}{x}$. To find k, choose one given ordered pair and substitute the values into the equation. We'll use $(2, 6)$.

$$y = \frac{k}{x}$$

$$6 = \frac{k}{2}$$

$$2 \cdot 6 = 2 \cdot \frac{k}{2} \quad \text{Multiply both sides by 2.}$$

$$12 = k \quad \text{Solve for } k.$$

Since $k = 12$, we have the equation $y = \frac{12}{x}$. □

PRACTICE 4 Write an inverse variation equation of the form $y = \frac{k}{x}$ that satisfies the ordered pairs in the table below.

x	2	-1	$\frac{1}{3}$
y	4	-8	24

■

Helpful Hint

Multiply both sides of the inverse variation relationship equation $y = \frac{k}{x}$ by x (as long as x is not 0), and we have $xy = k$. This means that if y varies inversely as x, their product is always the constant of variation k. For an example of this, check the table from Example 4.

x	2	4	$\frac{1}{2}$
y	6	3	24

$$2 \cdot 6 = 12 \quad 4 \cdot 3 = 12 \quad \frac{1}{2} \cdot 24 = 12$$

EXAMPLE 5 Suppose that y varies inversely as x. If $y = 0.02$ when $x = 75$, find the constant of variation and the inverse variation equation. Then find y when x is 30.

Solution Since y varies inversely as x, the constant of variation may be found by simply finding the product of the given x and y.

$$k = xy = 75(0.02) = 1.5$$

To check, we will use the inverse variation equation

$$y = \frac{k}{x}.$$

Let $y = 0.02$ and $x = 75$ and solve for k.

$$0.02 = \frac{k}{75}$$

$$75(0.02) = 75 \cdot \frac{k}{75} \quad \text{Multiply both sides by 75.}$$

$$1.5 = k \quad \text{Solve for } k.$$

Thus, the constant of variation is 1.5 and the equation is $y = \frac{1.5}{x}$.

To find y when $x = 30$, use $y = \frac{1.5}{x}$ and replace x with 30.

$$y = \frac{1.5}{x}$$

$$y = \frac{1.5}{30} \quad \text{Replace } x \text{ with 30.}$$

$$y = 0.05$$

Thus, when x is 30, y is 0.05. □

PRACTICE 5 If y varies inversely as x and y is 0.05 when x is 42, find the constant of variation and the inverse variation equation. Then find y when x is 70.

OBJECTIVE 3

Solving Other Types of Direct and Inverse Variation Problems

It is possible for y to vary directly or inversely as powers of x.

Direct and Inverse Variation as *n*th Powers of *x*

y* varies directly as a power of *x if there is a nonzero constant k and a natural number n such that

$$y = kx^n$$

y* varies inversely as a power of *x if there is a nonzero constant k and a natural number n such that

$$y = \frac{k}{x^n}$$

EXAMPLE 6 The surface area of a cube A varies directly as the square of a length of its side s. If A is 54 when s is 3, find A when $s = 4.2$.

Solution Since the surface area A varies directly as the square of side s, we have

$$A = ks^2.$$

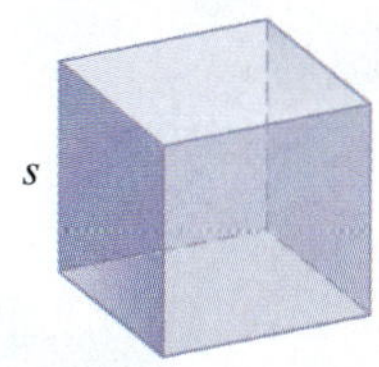

To find k, let $A = 54$ and $s = 3$.

$$A = k \cdot s^2$$

$$54 = k \cdot 3^2 \quad \text{Let } A = 54 \text{ and } s = 3.$$

$$54 = 9k \quad 3^2 = 9.$$

$$6 = k \quad \text{Divide by 9.}$$

(Continued on next page)

The formula for surface area of a cube is then

$$A = 6s^2 \text{ where } s \text{ is the length of a side.}$$

To find the surface area when $s = 4.2$, substitute.

$$A = 6s^2$$
$$A = 6 \cdot (4.2)^2$$
$$A = 105.84$$

The surface area of a cube whose side measures 4.2 units is 105.84 square units. □

PRACTICE 6 The area of an isosceles right triangle A varies directly as the square of one of its legs x. If A is 32 when x is 8, find A when $x = 3.6$. ■

OBJECTIVE 4

Solving Applications of Variation

There are many real-life applications of direct and inverse variation.

EXAMPLE 7 The weight of a body w varies inversely with the square of its distance from the center of Earth d. If a person weighs 160 pounds on the surface of Earth, what is the person's weight 200 miles above the surface? (Assume that the radius of Earth is 4000 miles.)

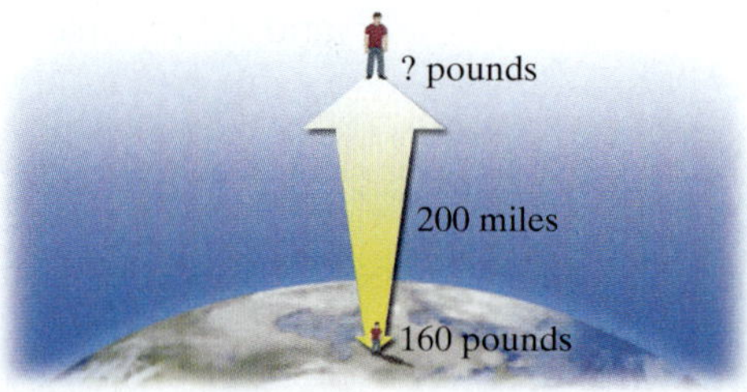

Solution

1. UNDERSTAND. Make sure you read and reread the problem.
2. TRANSLATE. Since we are told that weight w varies inversely with the square of its distance from the center of Earth, d, we have

$$w = \frac{k}{d^2}.$$

3. SOLVE. To solve the problem, we first find k. To do so, use the fact that the person weighs 160 pounds on Earth's surface, which is a distance of 4000 miles from Earth's center.

$$w = \frac{k}{d^2}$$
$$160 = \frac{k}{(4000)^2}$$
$$2{,}560{,}000{,}000 = k$$

Thus, we have $w = \frac{2{,}560{,}000{,}000}{d^2}$

Since we want to know the person's weight 200 miles above Earth's surface, we let $d = 4200$ and find w.

$$w = \frac{2{,}560{,}000{,}000}{d^2}$$
$$w = \frac{2{,}560{,}000{,}000}{(4200)^2} \quad \text{A person 200 miles above Earth's surface is 4200 miles from Earth's center.}$$
$$w \approx 145 \quad \text{Simplify.}$$

4. INTERPRET.

Check: Our answer is reasonable since the farther a person is from Earth, the less the person weighs.

State: Thus, 200 miles above the surface of Earth, a 160-pound person weighs approximately 145 pounds. □

PRACTICE 7 Robert Boyle investigated the relationship between volume of a gas and its pressure. He developed Boyle's law, which states that the volume of a gas varies inversely with pressure if the temperature is held constant. If 50 ml of oxygen is at a pressure of 20 atmospheres, what will the volume of the oxygen be at a pressure of 40 atmospheres? ■

Vocabulary, Readiness & Video Check

State whether each equation represents direct or inverse variation.

1. $y = \frac{k}{x}$, where k is a constant. ________

2. $y = kx$, where k is a constant. ________

3. $y = 5x$ ________

4. $y = \frac{5}{x}$ ________

5. $y = \frac{7}{x^2}$ ________

6. $y = 6.5x^4$ ________

7. $y = \frac{11}{x}$ ________

8. $y = 18x$ ________

9. $y = 12x^2$ ________

10. $y = \frac{20}{x^3}$ ________

Martin-Gay Interactive Videos

See Video App E

Watch the section lecture video and answer the following questions.

OBJECTIVE 1 **11.** Based on the lecture before Example 1, what kind of equation is a direct variation equation? What does k represent in this equation?

OBJECTIVE 2 **12.** In Example 4, why is it not necessary to replace the given values of x and y into the inverse variation equation in order to find k?

OBJECTIVE 3 **13.** From Examples 5–7, does a power on x change the basic direct and inverse variation formula relationships?

OBJECTIVE 4 **14.** In Example 8, why is it reasonable to expect our answer to be a greater distance than the original distance given in the problem?

E Exercise Set MyMathLab®

Write a direct variation equation, $y = kx$, that satisfies the ordered pairs in each table. See Example 1.

1.

x	0	6	10
y	0	3	5

2.

x	0	2	−1	3
y	0	14	−7	21

3.

x	−2	2	4	5
y	−12	12	24	30

4.

x	3	9	−2	12
y	1	3	$-\frac{2}{3}$	4

Write a direct variation equation, $y = kx$, that describes each graph. See Example 3.

5.

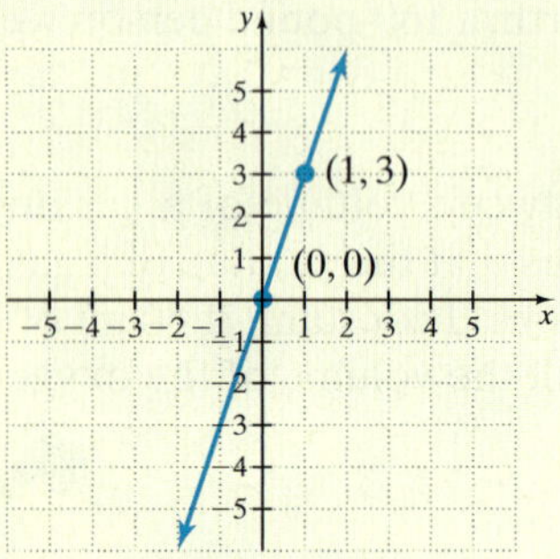

6.

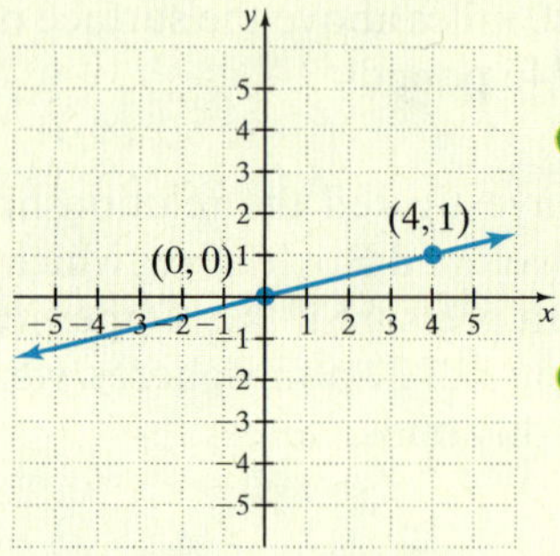

7.

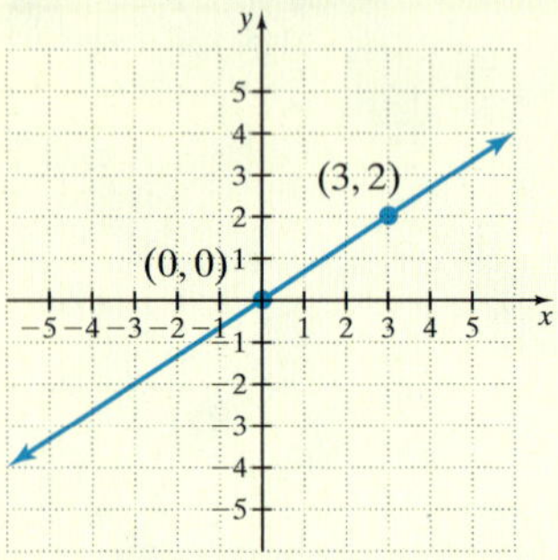

8.

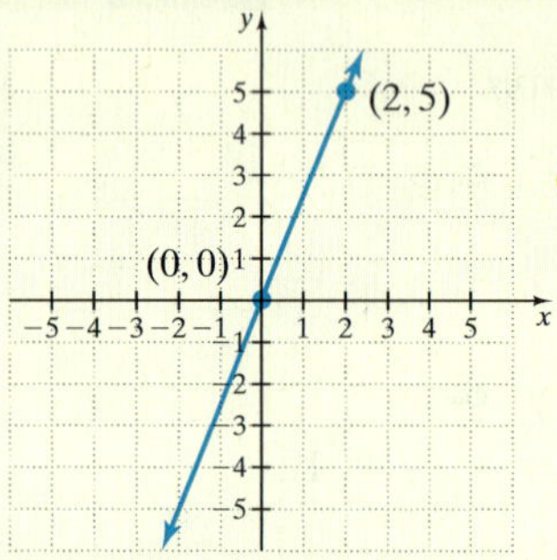

Write an inverse variation equation, $y = \frac{k}{x}$, that satisfies the ordered pairs in each table. See Example 4.

9.

x	1	−7	3.5	−2
y	7	−1	2	−3.5

10.

x	2	−11	4	−4
y	11	−2	5.5	−5.5

11.

x	10	$\frac{1}{2}$	$-\frac{3}{2}$
y	0.05	1	$-\frac{1}{3}$

12.

x	4	$\frac{1}{5}$	−8
y	0.1	2	−0.05

MIXED PRACTICE

Write an equation to describe each variation. Use k for the constant of proportionality. See Examples 1 through 6.

13. y varies directly as x

14. a varies directly as b

15. h varies inversely as t

16. s varies inversely as t

17. z varies directly as x^2

18. p varies inversely as x^2

19. y varies inversely as z^3

20. x varies directly as y^4

21. x varies inversely as $\sqrt{y}$

22. y varies directly as d^2

Solve. See Examples 2, 5, and 6.

23. y varies directly as x. If $y = 20$ when $x = 5$, find y when x is 10.

24. y varies directly as x. If $y = 27$ when $x = 3$, find y when x is 2.

25. y varies inversely as x. If $y = 5$ when $x = 60$, find y when x is 100.

26. y varies inversely as x. If $y = 200$ when $x = 5$, find y when x is 4.

27. z varies directly as x^2. If $z = 96$ when $x = 4$, find z when $x = 3$.

28. s varies directly as t^3. If $s = 270$ when $t = 3$, find s when $t = 1$.

29. a varies inversely as b^3. If $a = \frac{3}{2}$ when $b = 2$, find a when b is 3.

30. p varies inversely as q^2. If $p = \frac{5}{16}$ when $q = 8$, find p when $q = \frac{1}{2}$.

Solve. If needed, round money amounts to 2 decimal places. See Examples 1 through 7.

31. Your paycheck (before deductions) varies directly as the number of hours you work. If your paycheck is \$112.50 for 18 hours, find your pay for 10 hours.

32. If your paycheck (before deductions) is \$244.50 for 30 hours, find your pay for 34 hours. (See Exercise 31.)

33. The cost of manufacturing a certain type of headphone varies inversely as the number of headphones increases. If 5000 headphones can be manufactured for \$9.00 each, find the cost per headphone to manufacture 7500 headphones.

34. The cost of manufacturing a certain composition notebook varies inversely as the number of notebooks increases. If 10,000 notebooks can be manufactured for \$0.50 each, find the cost per notebook to manufacture 18,000 notebooks.

35. The distance a spring stretches varies directly with the weight attached to the spring. If a 60-pound weight stretches the spring 4 inches, find the distance that an 80-pound weight stretches the spring.

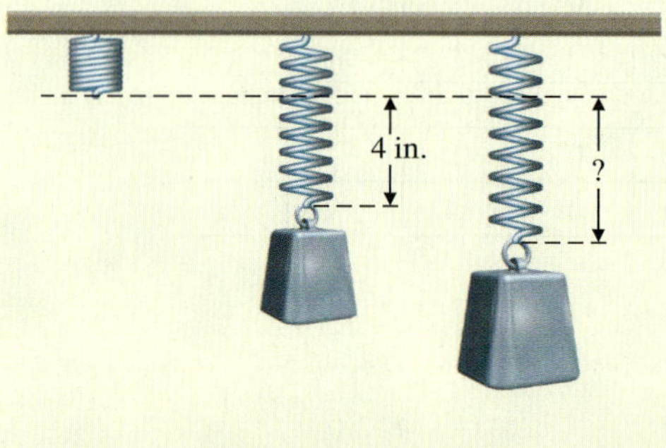

If a 30-pound weight stretches a spring 10 inches, find the distance a 20-pound weight stretches the spring. (See Exercise 35.)

36. The weight of an object varies inversely as the square of its distance from the *center* of Earth. If a person weighs 180 pounds on Earth's surface, what is his weight 10 miles above the surface of Earth? Round to the nearest tenth of a pound. (Assume that Earth's radius is 4000 miles.)

37. For a constant distance, the rate of travel varies inversely as the time traveled. If a family travels 55 mph and arrives at a destination in 4 hours, how long will the return trip take traveling at 60 mph?

38. The distance d that an object falls is directly proportional to the square of the time of the fall, t. A person who is parachuting for the first time is told to wait 10 seconds before opening the parachute. If the person falls 64 feet in 2 seconds, find how far he falls in 10 seconds.

39. The distance needed for a car to stop, d, is directly proportional to the square of its rate of travel, r. Under certain driving conditions, a car traveling 60 mph needs 300 feet to stop. With these same driving conditions, how long does it take a car to stop if the car is traveling 30 mph when the brakes are applied?

Contents of Student Resources

Student Resources

Study Skills Builders

Attitude and Study Tips

Study Skills Builder 1

Have You Decided to Complete This Course Successfully?

Ask yourself if one of your current goals is to complete this course successfully.

If it is not a goal of yours, ask yourself why. One common reason is fear of failure. Amazingly enough, fear of failure alone can be strong enough to keep many of us from doing our best in any endeavor.

Another common reason is that you simply haven't taken the time to think about or write down your goals for this course. To help accomplish this, answer the questions below.

Exercises

1. Write down your goal(s) for this course.
2. Now list steps you will take to make sure your goal(s) in Exercise 1 are accomplished.
3. Rate your commitment to this course with a number between 1 and 5. Use the diagram below to help.

High Commitment		Average Commitment		Not Commited at All
5	4	3	2	1

4. If you have rated your personal commitment level (from the exercise above) as a 1, 2, or 3, list the reasons why this is so. Then determine whether it is possible to increase your commitment level to a 4 or 5.

Good luck, and don't forget that a positive attitude will make a big difference.

Study Skills Builder 2

Tips for Studying for an Exam

To prepare for an exam, try the following study techniques:

- Start the study process days before your exam.
- Make sure that you are up-to-date on your assignments.
- If there is a topic that you are unsure of, use one of the many resources that are available to you. For example,
  - See your instructor.
  - View a lecture video on the topic.
  - Visit a learning resource center on campus.
  - Read the textbook material and examples on the topic.
- Reread your notes and carefully review the Chapter Highlights at the end of any chapter.
- Work the review exercises at the end of the chapter.
- Find a quiet place to take the Chapter Test found at the end of the chapter. Do not use any resources when taking this sample test. This way, you will have a clear indication of how prepared you are for your exam. Check your answers and use the Chapter Test Prep Videos to make sure that you correct any missed exercises.

Good luck, and keep a positive attitude.

Exercises

Let's see how you did on your last exam.

1. How many days before your last exam did you start studying for that exam?
2. Were you up-to-date on your assignments at that time or did you need to catch up on assignments?
3. List the most helpful text supplement (if you used one).
4. List the most helpful campus resource (if you used one).
5. List your process for preparing for a mathematics test.
6. Was this process helpful? In other words, were you satisfied with your performance on your exam?
7. If not, what changes can you make in your process that will make it more helpful to you?

Study Skills Builder 3

What to Do the Day of an Exam

Your first exam may be soon. On the day of an exam, don't forget to try the following:

- Allow yourself plenty of time to arrive.
- Read the directions on the test carefully.
- Read each problem carefully as you take your test. Make sure that you answer the question asked.
- Watch your time and pace yourself so that you may attempt each problem on your test.
- Check your work and answers.
- ***Do not turn your test in early.*** If you have extra time, spend it double-checking your work.

Good luck!

Exercises

Answer the following questions based on your most recent mathematics exam, whenever that was.

1. How soon before class did you arrive?

2. Did you read the directions on the test carefully?

3. Did you make sure you answered the question asked for each problem on the exam?

4. Were you able to attempt each problem on your exam?

5. If your answer to Exercise 4 is no, list reasons why.

6. Did you have extra time on your exam?

7. If your answer to Exercise 6 is yes, describe how you spent that extra time.

Study Skills Builder 4

Are You Satisfied with Your Performance on a Particular Quiz or Exam?

If not, don't forget to analyze your quiz or exam and look for common errors. Were most of your errors a result of:

- *Carelessness?* Did you turn in your quiz or exam before the allotted time expired? If so, resolve to use any extra time to check your work.
- *Running out of time?* Answer the questions you are sure of first. Then attempt the questions you are unsure of, and delay checking your work until all questions have been answered.
- *Not understanding a concept?* If so, review that concept and correct your work so that you make sure you understand it before the next quiz or the final exam.
- *Test conditions?* When studying for a quiz or exam, make sure you place yourself in conditions similar to test conditions. For example, before your next quiz or exam, take a sample test without the aid of your notes or text.

(For a sample test, see your instructor or use the Chapter Test at the end of each chapter.)

Exercises

1. Have you corrected all your previous quizzes and exams?

2. List any errors you have found common to two or more of your graded papers.

3. Is one of your common errors not understanding a concept? If so, are you making sure you understand all the concepts for the next quiz or exam?

4. Is one of your common errors making careless mistakes? If so, are you now taking all the time allotted to check over your work so that you can minimize the number of careless mistakes?

5. Are you satisfied with your grades thus far on quizzes and tests?

6. If your answer to Exercise 5 is no, are there any more suggestions you can make to your instructor or yourself to help? If so, list them here and share these with your instructor.

Study Skills Builder 5

How Are You Doing?

If you haven't done so yet, take a few moments and think about how you are doing in this course. Are you working toward your goal of successfully completing this course? Is your performance on homework, quizzes, and tests satisfactory? If not, you might want to see your instructor to see if he/she has any suggestions on how you can improve your performance. Reread Section 1.1 for ideas on places to get help with your mathematics course.

Exercises

Answer the following.

1. List any textbook supplements you are using to help you through this course.
2. List any campus resources you are using to help you through this course.
3. Write a short paragraph describing how you are doing in your mathematics course.
4. If improvement is needed, list ways that you can work toward improving your situation as described in Exercise 3.

Study Skills Builder 6

Are You Preparing for Your Final Exam?

To prepare for your final exam, try the following study techniques:

- Review the material that you will be responsible for on your exam. This includes material from your textbook, your notebook, and any handouts from your instructor.
- Review any formulas that you may need to memorize.
- Check to see if your instructor or mathematics department will be conducting a final exam review.
- Check with your instructor to see whether final exams from previous semesters/quarters are available to students for review.
- Use your previously taken exams as a practice final exam. To do so, rewrite the test questions in mixed order on blank sheets of paper. This will help you prepare for exam conditions.
- If you are unsure of a few concepts, see your instructor or visit a learning lab for assistance. Also, view the video segment of any troublesome sections.
- If you need further exercises to work, try the Cumulative Reviews at the end of the chapters.

Once again, good luck! I hope you are enjoying this textbook and your mathematics course.

Organizing Your Work

Study Skills Builder 7

Learning New Terms

Many of the terms used in this text may be new to you. It will be helpful to make a list of new mathematical terms and symbols as you encounter them and to review them frequently. Placing these new terms (including page references) on 3×5 index cards might help you later when you're preparing for a quiz.

Exercises

1. Name one way you might place a word and its definition on a 3×5 card.
2. How do new terms stand out in this text so that they can be found?

Study Skills Builder 8

Are You Organized?

Have you ever had trouble finding a completed assignment? When it's time to study for a test, are your notes neat and organized? Have you ever had trouble reading your own mathematics handwriting? (Be honest—I have.)

When any of these things happens, it's time to get organized. Here are a few suggestions:

- Write your notes and complete your homework assignments in a notebook with pockets (spiral or ring binder).
- Take class notes in this notebook, and then follow the notes with your completed homework assignment.
- When you receive graded papers or handouts, place them in the notebook pocket so that you will not lose them.
- Mark (possibly with an exclamation point) any note(s) that seem extra important to you.
- Mark (possibly with a question mark) any notes or homework that you are having trouble with.
- See your instructor or a math tutor to help you with the concepts or exercises that you are having trouble understanding.
- If you are having trouble reading your own handwriting, *slow down* and write your mathematics work clearly!

Exercises

1. Have you been completing your assignments on time?
2. Have you been correcting any exercises you may be having difficulty with?
3. If you are having trouble with a mathematical concept or correcting any homework exercises, have you visited your instructor, a tutor, or your campus math lab?
4. Are you taking lecture notes in your mathematics course? (By the way, these notes should include worked-out examples solved by your instructor.)
5. Is your mathematics course material (handouts, graded papers, lecture notes) organized?
6. If your answer to Exercise 5 is no, take a moment and review your course material. List at least two ways that you might better organize it.

Study Skills Builder 9

Organizing a Notebook

It's never too late to get organized. If you need ideas about organizing a notebook for your mathematics course, try some of these:

- Use a spiral or ring binder notebook with pockets and use it for mathematics only.
- Start each page by writing the book's section number you are working on at the top.
- When your instructor is lecturing, take notes. *Always* include any examples your instructor works for you.
- Place your worked-out homework exercises in your notebook immediately after the lecture notes from that section. This way, a section's worth of material is together.
- Homework exercises: Attempt and check all assigned homework.
- Place graded quizzes in the pockets of your notebook or a special section of your binder.

Exercises

Check your notebook organization by answering the following questions.

1. Do you have a spiral or ring binder notebook for your mathematics course only?
2. Have you ever had to flip through several sheets of notes and work in your mathematics notebook to determine what section's work you are in?
3. Are you now writing the textbook's section number at the top of each notebook page?
4. Have you ever lost or had trouble finding a graded quiz or test?
5. Are you now placing all your graded work in a dedicated place in your notebook?
6. Are you attempting all of your homework and placing all of your work in your notebook?
7. Are you checking and correcting your homework in your notebook? If not, why not?
8. Are you writing in your notebook the examples your instructor works for you in class?

Study Skills Builder 10

How Are Your Homework Assignments Going?

It is very important in mathematics to keep up with homework. Why? Many concepts build on each other. Often your understanding of a day's concepts depends on an understanding of the previous day's material.

Remember that completing your homework assignment involves a lot more than attempting a few of the problems assigned.

To complete a homework assignment, remember these four things:

- Attempt all of it.
- Check it.
- Correct it.
- If needed, ask questions about it.

Exercises

Take a moment and review your completed homework assignments. Answer the questions below based on this review.

1. Approximate the fraction of your homework you have attempted.
2. Approximate the fraction of your homework you have checked (if possible).
3. If you are able to check your homework, have you corrected it when errors have been found?
4. When working homework, if you do not understand a concept, what do you do?

MyMathLab and MathXL

Study Skills Builder 11

Tips for Turning In Your Homework on Time

It is very important to keep up with your mathematics homework assignments. Why? Many concepts in mathematics build upon each other.

Remember these four tips to help ensure your work is completed on time:

- Know the assignments and due dates set by your instructor.
- Do not wait until the last minute to submit your homework.
- Set a goal to submit your homework 6–8 hours before the scheduled due date in case you have unexpected technology trouble.
- Schedule enough time to complete each assignment.

Following these tips will also help you avoid potentially losing points for late or missed assignments.

Exercises

Take a moment to consider your work on your homework assignments to date and answer the following questions:

1. What percentage of your assignments have you turned in on time?
2. Why might it be a good idea to submit your homework 6–8 hours before the scheduled deadline?
3. If you have missed submitting any homework by the due date, list some of the reasons why this occurred.
4. What steps do you plan to take in the future to ensure your homework is submitted on time?

Study Skills Builder 12

Tips for Doing Your Homework Online

Practice is one of the main keys to success in any mathematics course. Did you know that MyMathLab/MathXL provides you with **immediate feedback** for each exercise? If you are incorrect, you are given hints to work the exercise correctly. You have **unlimited practice opportunities** and can rework any exercises you have trouble with until you master them, and submit homework assignments unlimited times before the deadline.

Remember these success tips when doing your homework online:

- Attempt all assigned exercises.
- Write down (neatly) your step-by-step work for each exercise before entering your answer.
- Use the immediate feedback provided by the program to help you check and correct your work for each exercise.
- Rework any exercises you have trouble with until you master them.
- Work through your homework assignment as many times as necessary until you are satisfied.

Exercises

Take a moment to think about your homework assignments to date and answer the following:

1. Have you attempted all assigned exercises?

2. Of the exercises attempted, have you also written out your work before entering your answer—so that you can check it?

3. Are you familiar with how to enter answers using the MathXL player so that you avoid answer entry type errors?

4. List some ways the immediate feedback and practice supports have helped you with your homework. If you have not used these supports, how do you plan to use them with the success tips above on your next assignment?

Study Skills Builder 13

Organizing Your Work

Have you ever used any readily available paper (such as the back of a flyer, another course assignment, Post-it® Notes, etc.) to work out homework exercises before entering the answer in MathXL? To save time, have you ever entered answers directly into MathXL without working the exercises on paper? When it's time to study, have you ever been unable to find your completed work or read and follow your own mathematics handwriting?

When any of these things happen, it's time to get organized. Here are some suggestions:

- Write your step-by-step work for each homework exercise (neatly) on lined, loose-leaf paper and keep this in a 3-ring binder.
- Refer to your step-by-step work when you receive feedback that your answer is incorrect in MathXL. Double-check against the steps and hints provided by the program and correct your work accordingly.
- Keep your written homework with your class notes for that section.
- Identify any exercises you are having trouble with and ask questions about them.
- Keep all graded quizzes and tests in this binder as well to study later.

If you follow the suggestions above, you and your instructor or tutor will be able to follow your steps and correct any mistakes. You will have a written copy of your work to refer to later to ask questions and study for tests.

Exercises

1. Why is it important that you write out your step-by-step work on homework exercises and keep a hard copy of all work submitted online?

2. If you have gotten an incorrect answer, are you able to follow your steps and find your error?

3. If you were asked today to review your previous homework assignments and first test, could you find them? If not, list some ways you might better organize your work.

Study Skills Builder 14

Getting Help with Your Homework Assignments

There are many helpful resources available to you through MathXL to help you work through any homework exercises you may have trouble with. It is important that you know what these resources are and know when and how to use them.

Let's review these features found in the homework exercises:

- **Help Me Solve This**—provides step-by-step help for the exercise you are working. You must work an additional exercise of the same type (without this help) before you can get credit for having worked it correctly.
- **View an Example**—allows you to view a correctly worked exercise similar to the one you are having trouble with. You can go back to your original exercise and work it on your own.
- **E-book**—allows you to read examples from your text and find similar exercises.
- **Video**—your text author, Elayn Martin-Gay, works an exercise similar to the one you need help with. **Not all exercises have an accompanying video clip.
- **Ask My Instructor**—allows you to email your instructor for help with an exercise.

Exercises

1. How does the "Help Me Solve This" feature work?
2. If the "View an Example" feature is used, is it necessary to work an additional problem before continuing the assignment?
3. When might be a good time to use the "Video" feature? Do all exercises have an accompanying video clip?
4. Which of the features above have you used? List those you found the most helpful to you.
5. If you haven't used the features discussed, list those you plan to try on your next homework assignment.

Study Skills Builder 15

Tips for Preparing for an Exam

Did you know that you can rework your previous homework assignments in MyMathLab and MathXL? This is a great way to prepare for tests. To do this, open a previous homework assignment and click "similar exercise." This will generate new exercises similar to the homework you have submitted. You can then rework the exercises and assignments until you feel confident that you understand them.

To prepare for an exam, follow these tips:

- Review your written work for your previous homework assignments along with your class notes.
- Identify any exercises or topics that you have questions on or have difficulty understanding.
- Rework your previous assignments in MyMathLab and MathXL until you fully understand them and can do them without help.
- Get help for any topics you feel unsure of or for which you have questions.

Exercises

1. Are your current homework assignments up to date and is your written work for them organized in a binder or notebook? If the answer is no, it's time to get organized. For tips on this, see Study Skills Builder 13—Organizing Your Work.
2. How many days in advance of an exam do you usually start studying?
3. List some ways you think that practicing previous homework assignments can help you prepare for your test.
4. List two or three resources you can use to get help for any topics you are unsure of or have questions on.

Good luck!

Study Skills Builder 16

How Well Do You Know the Resources Available to You in MyMathLab?

There are many helpful resources available to you in MyMathLab. Let's take a moment to locate and explore a few of them now. Go into your MyMathLab course, and visit the multimedia library, tools for success, and E-book.

Let's see what you found.

Exercises

1. List the resources available to you in the Multimedia Library.
2. List the resources available to you in the Tools for Success folder.
3. Where did you find the English/Spanish Audio Glossary?
4. Can you view videos from the E-book?
5. Did you find any resources you did not know about? If so, which ones?
6. Which resources have you used most often or found most helpful?

Additional Help Inside and Outside Your Textbook

Study Skills Builder 17

How Well Do You Know Your Textbook?

The questions below will help determine whether you are familiar with your textbook. For additional information, see Section 1.1 in this text.

1. What does the ▶ icon mean?
2. What does the ╲ icon mean?
3. What does the △ icon mean?
4. Where can you find a review for each chapter? What answers to this review can be found in the back of your text?
5. Each chapter contains an overview of the chapter along with examples. What is this feature called?
6. Each chapter contains a review of vocabulary. What is this feature called?
7. There are practice exercises that are contained in this text. What are they and how can they be used?
8. This text contains a student section in the back entitled Student Resources. List the contents of this section and how they might be helpful.
9. What exercise answers are available in this text? Where are they located?

Study Skills Builder 18

Are You Familiar with Your Textbook Supplements?

Below is a review of some of the student supplements available for additional study. Check to see if you are using the ones most helpful to you.

- Interactive DVD Lecture Series. These are keyed to each section of the text. The material is presented by me, Elayn Martin-Gay, and I have placed a ▶ by the exercises in the text that I have worked on the video.
- Getting Ready for the Test Videos. These videos provide video clip solutions to the exercises in Getting Ready for the Test. These exercises are found before each Chapter Test in your text.
- Chapter Test Prep Videos. These videos provide video clip solutions to the Chapter Test exercises in this text. You will find this extremely useful when studying for tests or exams.
- The *Student Solutions Manual*. This contains worked-out solutions to odd-numbered exercises as well as Practice Exercises and every exercise in the Integrated Reviews, Chapter Reviews, Chapter Tests, and Cumulative Reviews.
- Pearson Tutor Center. Mathematics questions may be phoned, faxed, or e-mailed to this center.
- MyMathLab is a text-specific online course. MathXL is an online homework, tutorial, and assessment system. Take a moment and determine whether these are available to you.

As usual, your instructor is your best source of information.

Exercises

Let's see how you are doing with textbook supplements.

1. Name one way the Lecture Videos can be helpful to you.
2. Name one way the Chapter Test Prep Video can help you prepare for a chapter test.
3. List any textbook supplements that you have found useful.
4. Have you located and visited a learning resource lab located on your campus?
5. List the textbook supplements that are currently housed in your campus's learning resource lab.

Study Skills Builder 19

Are You Getting All the Mathematics Help That You Need?

Remember that, in addition to your instructor, there are many places to get help with your mathematics course. For example:

- This text has an accompanying video lesson for every section, and the Chapter Test Prep videos provide worked-out solutions to every Chapter Test exercise.
- The back of the book contains answers to odd-numbered exercises.
- A *Student Solutions Manual* is available that contains worked-out solutions to odd-numbered exercises as well as solutions to Practice Exercises and every exercise in the Integrated Reviews, Chapter Reviews, Chapter Tests, and Cumulative Reviews.
- Don't forget to check with your instructor for other local resources available to you, such as a tutoring center.

Exercises

1. List items you find helpful in the text and all student supplements to this text.
2. List all the campus help that is available to you for this course.
3. List any help (besides the textbook) from Exercises 1 and 2 above that you are using.
4. List any help (besides the textbook) that you feel you should try.
5. Write a goal for yourself that includes trying everything you listed in Exercise 4 during the next week.

Answers to Selected Exercises

CHAPTER R PREALGEBRA REVIEW

Section R.1

Practice Exercises

1. a. $2 \cdot 2 \cdot 3 \cdot 3$ **b.** $2 \cdot 2 \cdot 2 \cdot 5 \cdot 5$ **2. a.** $\frac{7}{8}$ **b.** $\frac{16}{3}$ **c.** $\frac{7}{25}$ **3.** $\frac{7}{24}$ **4. a.** $\frac{27}{16}$ **b.** $\frac{1}{36}$ **c.** $\frac{5}{2}$ **5. a.** 1 **b.** $\frac{6}{5}$ **c.** $\frac{4}{5}$ **d.** $\frac{1}{2}$ **6.** $\frac{14}{21}$ **7. a.** $\frac{46}{77}$ **b.** $\frac{1}{14}$ **c.** $\frac{1}{2}$ **8.** $\frac{341}{15}$ or $22\frac{11}{15}$ **9.** $40\frac{5}{6}$

Vocabulary, Readiness & Video Check R.1

1. fraction **3.** product **5.** factors, product **7.** equivalent **9.** The division operation changes to multiplication and the second fraction $\frac{1}{20}$ changes to its reciprocal $\frac{20}{1}$. **11.** The number $4\frac{7}{6}$ is not in proper mixed number form as the fraction part, $\frac{7}{6}$, should not be an improper fraction.

Exercise Set R.1

1. $\frac{3}{8}$ **3.** $\frac{5}{7}$ **5.** $3 \cdot 11$ **7.** $2 \cdot 7 \cdot 7$ **9.** $2 \cdot 2 \cdot 5$ **11.** $3 \cdot 5 \cdot 5$ **13.** $3 \cdot 3 \cdot 5$ **15.** $\frac{1}{2}$ **17.** $\frac{2}{3}$ **19.** $\frac{3}{7}$ **21.** $\frac{3}{5}$ **23.** $\frac{30}{61}$ **25.** $\frac{3}{8}$ **27.** $\frac{1}{2}$ **29.** $\frac{6}{7}$ **31.** 15 **33.** $\frac{1}{6}$ **35.** $\frac{25}{27}$ **37.** $\frac{11}{20}$ sq mi **39.** $\frac{7}{36}$ sq ft **41.** $\frac{3}{5}$ **43.** 1 **45.** $\frac{1}{3}$ **47.** $\frac{9}{35}$ **49.** $\frac{21}{30}$ **51.** $\frac{4}{18}$ **53.** $\frac{16}{20}$ **55.** $\frac{23}{21}$ **57.** $\frac{11}{60}$ **59.** $\frac{5}{66}$ **61.** $\frac{7}{5}$ **63.** $\frac{1}{5}$ **65.** $\frac{3}{8}$ **67.** $\frac{1}{9}$ **69.** $18\frac{20}{27}$ **71.** $2\frac{28}{29}$ **73.** $48\frac{1}{15}$ **75.** $7\frac{1}{12}$ **77.** $\frac{5}{7}$ **79.** $\frac{65}{21}$ **81.** $\frac{2}{5}$ **83.** $\frac{10}{9}$ **85.** $5\frac{2}{3}$ or $\frac{17}{3}$ **87.** 37 **89.** $\frac{5}{66}$ **91.** $\frac{1}{5}$ **93.** $5\frac{1}{6}$ **95.** $\frac{17}{18}$ **97.** $55\frac{1}{4}$ ft **99.** answers may vary **101.** $3\frac{3}{8}$ mi **103.** $\frac{3}{40}$ **105.** $\frac{37}{40}$ **107.** incorrect; $\frac{12}{24} = \frac{2 \cdot 2 \cdot 3}{2 \cdot 2 \cdot 2 \cdot 3} = \frac{1}{2}$ **109.** incorrect; $\frac{2}{7} + \frac{9}{7} = \frac{11}{7}$

Section R.2

Practice Exercises

1. $\frac{27}{100}$ **2.** $\frac{51}{10}$ **3.** $\frac{7685}{1000}$ **4. a.** 27.978 **b.** 12.82 **5. a.** 57.248 **b.** 89.81 **6. a.** 1.426 **b.** 0.0378 **7. a.** 43.5 **b.** 2600 **8.** 12.92 **9.** 245.3 **10.** 0.4 **11.** $0.8\overline{3}$ **12.** 0.111

Vocabulary, Readiness & Video Check R.2

1. decimal **3.** vertically **5.** Reading a decimal correctly gives us the correct place value, which tells us the denominator of our equivalent fraction. **7.** When rounding whole numbers, digits to the right of the rounding place are replaced by zeros; when rounding decimal numbers to the right of the decimal point, digits to the right of the rounding place are not replaced by zeros.

Exercise Set R.2

1. $\frac{6}{10}$ **3.** $\frac{186}{100}$ **5.** $\frac{114}{1000}$ **7.** $\frac{1231}{10}$ **9.** 6.83 **11.** 27.0578 **13.** 6.5 **15.** 15.22 **17.** 0.12 **19.** 0.2646 **21.** 1.68 **23.** 5.8 **25.** 56.431 **27.** 67.5 **29.** 70 **31.** 598.23 **33.** 43.274 **35.** 840 **37.** 84.97593 **39.** 0.6 **41.** 0.23 **43.** 0.595 **45.** 98,207.2 **47.** 12.35 **49.** 0.75 **51.** $0.\overline{3} \approx 0.33$ **53.** 0.4375 **55.** $0.\overline{54} \approx 0.55$ **57.** $4.8\overline{3} \approx 4.83$ **59. a.** tenths **b.** thousandths **c.** ones **61.** answers may vary **63.** 182.6 lb **65.** $\frac{3}{50}$ **67.** 0.06 **69.** answers may vary

Section R.3

Practice Exercises

1. 27% **2.** 31% **3.** 0.49 **4.** 0.031 **5.** 1.75 **6.** 0.0046 **7.** 6 **8.** $\frac{1}{2}$ **9.** $\frac{23}{1000}$ **10.** $\frac{3}{2}$ or $1\frac{1}{2}$ **11.** $\frac{2}{3}$ **12.** $\frac{3}{25}$ **13.** 14% **14.** 175% **15.** 5.7% **16.** 50% **17.** 12% **18.** $22\frac{1}{2}\%$ **19.** 550% **20.** 17.65% **21.** 0.275, $\frac{11}{40}$ **22.** 175%

Vocabulary, Readiness & Video Check R.3

1. Percent **3.** percent **5.** 0.01 **7.** Percent means "per 100." **9.** 1

Exercise Set R.3

1. 96% **3.** football; 37% **5.** 50% **7.** 0.41 **9.** 0.06 **11.** 1.00 or 1 **13.** 0.736 **15.** 0.028 **17.** 0.006 **19.** 3.00 or 3 **21.** 0.3258 **23.** $\frac{2}{25}$ **25.** $\frac{1}{25}$ **27.** $\frac{9}{200}$ **29.** $\frac{7}{4}$ or $1\frac{3}{4}$ **31.** $\frac{1}{16}$ **33.** $\frac{31}{300}$ **35.** $\frac{179}{800}$ **37.** 22% **39.** 0.6% **41.** 530%

43. 5.6% **45.** 22.28% **47.** 300% **49.** 70% **51.** 70% **53.** 80% **55.** 68% **57.** $37\frac{1}{2}\%$ **59.** $33\frac{1}{3}\%$ **61.** 450% **63.** 190% **65.** 81.82% **67.** 26.67% **69.** 0.6; $\frac{3}{5}$; $23\frac{1}{2}\%$; $\frac{47}{200}$; 80%; 0.8; 0.3333; $\frac{1}{3}$; 87.5%; 0.875; 0.075; $\frac{3}{40}$ **71.** 2; 2; 280%; $2\frac{4}{5}$; 7.05; $7\frac{1}{20}$; 454%; 4.54 **73.** 0.38; $\frac{19}{50}$ **75.** 0.358; $\frac{179}{500}$ **77.** 0.91; $\frac{91}{100}$ **79.** 0.005; $\frac{1}{200}$ **81.** 0.142; $\frac{71}{500}$ **83.** 78.1% **85.** 0.7% **87.** 0.47 **89. a.** 52.9% **b.** 52.86% **91.** b, d **93.** 4% **95.** 75% **97.** greater **99.** 0.266; 26.6% **101.** biomedical engineers **103.** 0.39 **105.** answers may vary

Section R.4

Practice Exercises

1. a. 400 million people **b.** 175 million people **2. a.** 78 **b.** arachnids **3.**

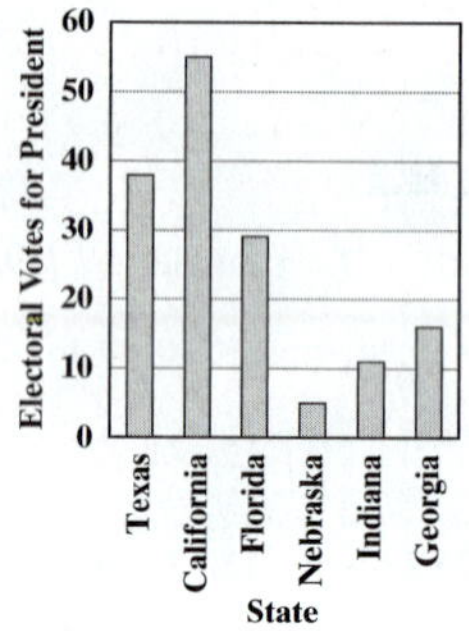

4. 12 **5.** 16

6.

Tally	Class Frequency (Number of Months)
|||	3
||||	4
||	2
|	1
|	1
|	1

7.

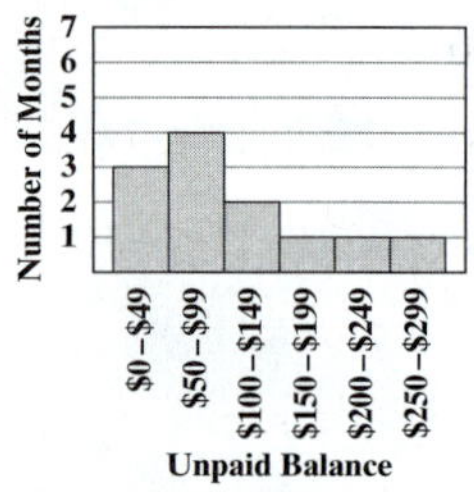

8. a. January **b.** December **c.** June, July, and August **9.** $\frac{2}{25}$ **10.** 38% **11.** 18,900,000

Vocabulary, Readiness & Video Check R.4

1. bar **3.** line **5.** circle **7.** 100 **9.** Count the number of symbols and multiply this number by how much each symbol stands for (from the key). **11.** bar graph **13.** 100%

Exercise Set R.4

1. Kansas **3.** 5.5 million or 5,500,000 acres **5.** Texas **7.** Montana (or Oklahoma) and Washington **9.** 48,000 **11.** 2009 **13.** 18,000 **15.** 60,000 wildfires/year **17.** September **19.** 77 **21.** $\frac{2}{77}$ **23.** Tokyo, Japan; about 34.7 million or 34,700,000 **25.** New York; 21.6 million or 21,600,000 **27.** approximately 2 million **29.**

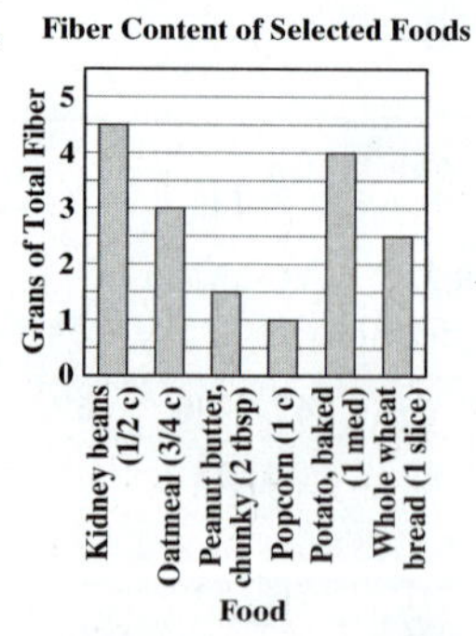

31.

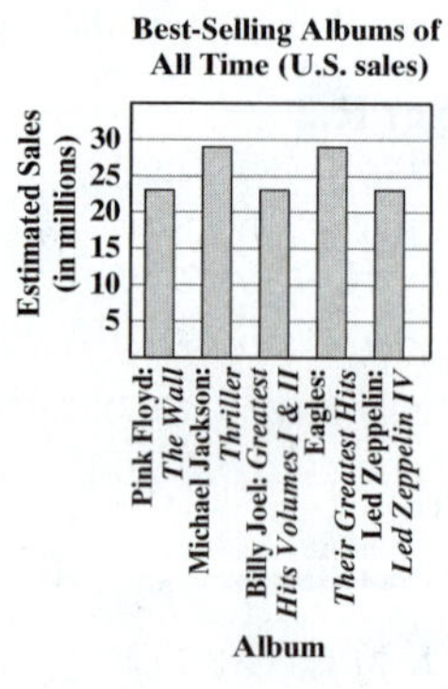

33. 15 adults **35.** 61 adults **37.** 24 adults **39.** 12 adults **41.** $\frac{9}{100}$ **43.** 45–54 **45.** 21 million householders **47.** 44 million householders **49.** 4 million householders **51.** |; 1 **53.** 𝍸 |||; 8 **55.** 𝍸 |; 6 **57.** 𝍸 |; 6 **59.** ||; 2 **61.**

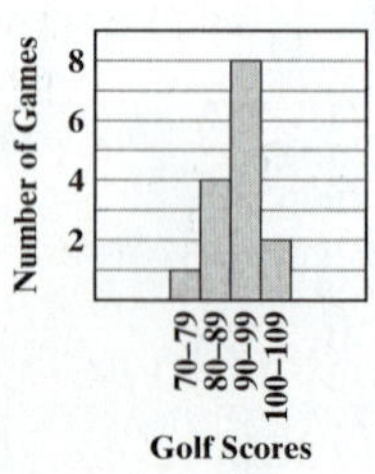

63. 7.9 goals/game **65.** 2003 **67.** increase **69.** 2001, 2007, 2013, 2015 **71.** parent or guardian's home **73.** $\frac{9}{35}$ **75.** $\frac{9}{16}$ **77.** Asia **79.** 37% **81.** 17,100,000 sq mi **83.** 2,850,000 sq mi **85.** 55% **87.** nonfiction **89.** 31,400 books **91.** 27,632 books **93.** 25,120 books **95.** 83°F **97.** Sunday; 68°F **99.** Tuesday; 13°F **101.** answers may vary **103.** answers may vary **105.** 129,600,002 sq km **107.** 55,542,858 sq km **109.** 672 respondents **111.** 2408 respondents **113.** $\frac{12}{31}$ **115.** no; answers may vary

Chapter R Vocabulary Check

1. factor **2.** composite number **3.** equivalent **4.** prime number **5.** simplified **6.** mixed number **7.** of **8.** Percent **9.** 100% **10.** 0.01 **11.** $\frac{1}{100}$ **12.** bar **13.** pictograph **14.** line **15.** circle **16.** histogram; class interval; class frequency

Chapter R Review

1. $2\cdot2\cdot3\cdot3$ **3.** $\frac{12}{25}$ **5.** $\frac{13}{10}$ **7.** $9\frac{3}{8}$ **9.** 15 **11.** $\frac{7}{12}$ **13.** $A = 1\frac{1}{6}$ sq m; $P = 4\frac{5}{12}$ m **15.** $14\frac{1}{8}$ lb **17.** $18\frac{7}{16}$ lb **19.** Baby E **21.** $\frac{181}{100}$ **23.** 95.118 **25.** 13.38 **27.** 91.2 **29.** 28.6 **31.** 0.77 **33.** 0.5 **35.** $0.\overline{36} \approx 0.364$ **37.** 37% **39.** 0.26 **41.** 0.035 **43.** 2.75 **45.** 0.4785 **47.** 160% **49.** 7.6% **51.** 71% **53.** 600% **55.** $\frac{7}{100}$ **57.** $\frac{1}{4}$ **59.** $\frac{51}{500}$ **61.** $\frac{1}{3}$ **63.** 40% **65.** $58\frac{1}{3}\%$ **67.** 125% **69.** 6.25% **71.** 2,250,000 homes **73.** South **75.** South, West **77.** 30% **79.** 1990, 2000, 2010, 2014 **81.** 962 (exact number) **83.** 930 **85.** 25 **87.** 1 employee **89.** 18 employees **91.** $\cancel{||||}$; 5 **93.** ||||; 4 **95.** mortgage payment **97.** \$1225 **99.** $\frac{39}{160}$ **101.** 31 **103.** 5

Chapter R Test

1. $\frac{53}{40}$ **2.** $\frac{18}{49}$ **3.** $\frac{1}{20}$ **4.** $\frac{29}{36}$ **5.** $4\frac{8}{9}$ **6.** $2\frac{5}{22}$ **7.** 55 **8.** $13\frac{13}{20}$ **9.** 45.11 **10.** 65.88 **11.** 12.688 **12.** 320 **13.** 23.73 **14.** 0.875 **15.** $\frac{1}{8}$ sq ft **16.** $\frac{63}{64}$ sq cm **17.** 0.85 **18.** 0.006 **19.** 5.6% **20.** 610% **21.** $\frac{6}{5}$ or $1\frac{1}{5}$ **22.** $\frac{77}{200}$ **23.** 37.5% **24.** 175% **25.** 20% **26.** $\frac{16}{25}$ **27.** \$225 **28.** 3rd week; \$350 **29.** \$1100 **30.** June, August, September **31.** February; 3 cm **32.** March and November **33.** $\frac{17}{40}$ **34.** $\frac{31}{22}$ **35.** 45 million **36.** 12 million **37.** 9 students **38.** 11 students **39.** |; 1; |||; 3; ||||; 4; $\cancel{||||}$; 5; $\cancel{||||}$ |||; 8; ||||; 4 **40.**

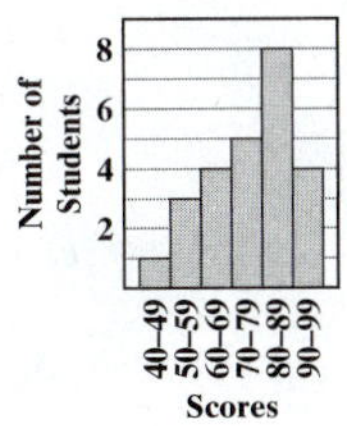

CHAPTER 1 REVIEW OF REAL NUMBERS

Section 1.2

Practice Exercises

1. a. < **b.** > **c.** < **2. a.** True **b.** False **c.** True **d.** True **3. a.** $3 < 8$ **b.** $15 \ge 9$ **c.** $6 \ne 7$ **4.** -10 **5. a.** 25 **b.** 25 **c.** 25, -15, -99 **d.** 25, $\frac{7}{3}$, -15, $-\frac{3}{4}$, -3.7, 8.8, -99 **e.** $\sqrt{5}$ **f.** all numbers in the given set **6. a.** < **b.** > **c.** = **7. a.** 8 **b.** 9 **c.** 2.5 **d.** $\frac{5}{11}$ **e.** $\sqrt{3}$ **8. a.** = **b.** > **c.** < **d.** > **e.** < **9. a.** -5.4 **b.** $\frac{3}{5}$ **c.** -18

Vocabulary, Readiness & Video Check 1.2

1. whole **3.** inequality **5.** real **7.** irrational **9.** $-a$ **11.** irrational **13.** To form a true statement: $0 < 7$. **15.** 0 belongs to the whole numbers, the integers, the rational numbers, and the real numbers; because 0 is a rational number, it cannot also be an irrational number. **17.** The absolute value of a number is its *distance* from 0 on a number line, regardless of direction—distance must be positive or zero. The opposite of a number is the *number* that lies the same distance from 0 on the number line as the original number but on the order side of 0—the opposite of a number can be negative.

Exercise Set 1.2

1. > **3.** = **5.** < **7.** < **9.** $32 < 212$ **11.** $30 \le 45$ **13.** true **15.** false **17.** false **19.** true **21.** false **23.** $8 < 12$ **25.** $5 \ge 4$ **27.** $15 \ne -2$ **29.** 14,494; -282 **31.** $-28{,}000$ **33.** 350; -126 **35.** whole numbers, integers, rational numbers, real numbers **37.** integers, rational numbers, real numbers **39.** natural numbers, whole numbers, integers, rational numbers, real numbers **41.** rational numbers, real numbers **43.** irrational numbers, real numbers **45.** false **47.** true **49.** true **51.** true **53.** false **55.** > **57.** > **59.** < **61.** < **63.** > **65.** = **67.** < **69.** < **71.** 6.2 **73.** $-\frac{4}{7}$ **75.** $\frac{2}{3}$ **77.** 0 **79.** 40 million > 16 million **81.** 40 million pounds less, or -40 million **83.** 1966–1975 and 2006–2015 **85.** 1946–1955, 1966–1975, 1976–1985, 2006–2015 **87.** $38 < 49$ **89.** $-0.04 > -26.7$ **91.** Sun **93.** Sun **95.** $20 \le 25$ **97.** $6 > 0$ **99.** $-12 < -10$ **101.** answers may vary

Section 1.3

Practice Exercises

1. a. 1 **b.** 25 **c.** $\frac{1}{100}$ **d.** 9 **e.** $\frac{8}{125}$ **2. a.** 33 **b.** 11 **c.** $\frac{32}{9}$ **d.** 36 **e.** $\frac{3}{16}$ **3.** $\frac{31}{11}$ **4.** 4 **5.** $\frac{9}{22}$ **6. a.** 9 **b.** $\frac{8}{15}$

c. $\frac{19}{10}$ **d.** 33 **7.** no **8. a.** $6x$ **b.** $x - 8$ **c.** $x \cdot 9$ or $9x$ **d.** $2x + 3$ **e.** $7 + x$ **9. a.** $x + 7 = 13$ **b.** $x - 2 = 11$ **c.** $2x + 9 \neq 25$ **d.** $5(11) \geq x$

Graphing Calculator Explorations 1.3

1. 625 **3.** 59,049 **5.** 30 **7.** 9857 **9.** 2376

Vocabulary, Readiness & Video Check 1.3

1. base; exponent **3.** variable **5.** equation **7.** solving **9.** The replacement value for z is not used because it's not needed—there is no variable z in the given algebraic expression. **11.** We translate phrases to mathematical expressions and sentences to mathematical equations.

Exercise Set 1.3

1. 243 **3.** 27 **5.** 1 **7.** 5 **9.** 49 **11.** $\frac{16}{81}$ **13.** $\frac{1}{125}$ **15.** 1.44 **17.** 0.000064 **19.** 17 **21.** 20 **23.** 10 **25.** 21 **27.** 45 **29.** 0 **31.** 30 **33.** 2 **35.** $\frac{7}{18}$ **37.** $\frac{27}{10}$ **39.** $\frac{7}{5}$ **41.** 32 **43.** $\frac{23}{27}$ **45. a.** 64 **b.** 43 **c.** 19 **d.** 22 **47.** 9 **49.** 1 **51.** 1 **53.** 11 **55.** 8 **57.** 45 **59.** 27 **61.** 132 **63.** $\frac{37}{18}$ **65.** 16; 64; 144; 256 **67.** yes **69.** no **71.** no **73.** yes **75.** no **77.** $x + 15$ **79.** $x - 5$ **81.** $\frac{x}{4}$ **83.** $3x + 22$ **85.** $1 + 2 = 9 \div 3$ **87.** $3 \neq 4 \div 2$ **89.** $5 + x = 20$ **91.** $7.6x = 17$ **93.** $13 - 3x = 13$ **95.** multiply **97.** subtract **99.** no; answers may vary **101.** 14 in., 12 sq in. **103.** 14 in., 9.01 sq in. **105.** Rectangles with the same perimeter can have different areas. **107.** $(20 - 4) \cdot 4 \div 2$ **109. a.** expression **b.** equation **c.** equation **d.** expression **e.** expression **111.** answers may vary **113.** answers may vary, for example, $2(5) + 1$ **115.** 12,000 sq ft **117.** 51 mph

Integrated Review

1. 16 **2.** 27 **3.** 0 **4.** 11 **5.** 21 **6.** 18 **7.** undefined **8.** 44 **9.** 3 **10.** 24 **11.** 35 **12.** 20 **13.** negative **14.** negative **15.** positive **16.** 0 **17.** positive **18.** 0 **19.** positive **20.** positive **21.** $-\frac{1}{7}; \frac{1}{7}$ **22.** $\frac{12}{5}; \frac{12}{5}$ **23.** 3; 3 **24.** $-\frac{9}{11}; \frac{9}{11}$

Section 1.4

Practice Exercises

1. a. −8 **b.** −3 **c.** 5 **d.** −8.1 **e.** $\frac{1}{15}$ **f.** $-\frac{7}{22}$ **2. a.** −8 **b.** −3 **c.** −12 **d.** 7.7 **e.** $-\frac{22}{21}$ **f.** 1.8 **g.** −7 **3. a.** 12 **b.** −4 **4. a.** −15 **b.** $\frac{1}{2}$ **c.** −10.2 **d.** 0 **e.** $-\frac{2}{13}$ **f.** 36 **g.** −11.5 **5. a.** −2 **b.** 5 **c.** $-\frac{1}{6}$ **d.** −6 **e.** $\frac{1}{9}$ **f.** 0 **6. a.** 8 **b.** $\frac{1}{9}$ **c.** −144 **d.** 144 **e.** −64 **f.** −64 **7. a.** 7 **b.** $\frac{1}{4}$ **c.** −8 **d.** not a real number **e.** 10 **8. a.** 2 **b.** 27 **c.** −29 **9.** 19 **10.** $-\frac{17}{44}$ **11. a.** 67 **b.** −100 **c.** $-\frac{39}{80}$ **12.** 23; 50; 77

Vocabulary, Readiness & Video Check 1.4

1. 0 **3.** reciprocal **5.** exponent **7.** square root **9.** addition **11.** The parentheses, or lack of them, determine the base of the expression. In Example 7, -7^2, the base is 7 and only 7 is squared. In Example 8, $(-7)^2$, the base is −7 and all of −7 is squared. **13.** It allows each expression to evaluate to a single number.

Exercise Set 1.4

1. 5 **3.** −24 **5.** 1 **7.** −12 **9.** −5 **11.** 0 **13.** −10 **15.** −5 **17.** 19 **19.** 5 **21.** −11 **23.** −4 **25.** $\frac{4}{3}$ **27.** −2 **29.** $-\frac{1}{2}$ **31.** 13 **33.** −1 **35.** −23 **37.** −60 **39.** −2 **41.** 50 **43.** 0 **45.** −3 **47.** 3 **49.** 0 **51.** $-\frac{1}{6}$ **53.** 0.56 **55.** −7 **57.** −8 **59.** −49 **61.** 36 **63.** −8 **65.** $-\frac{1}{27}$ **67.** 7 **69.** $-\frac{2}{3}$ **71.** 8 **73.** 4 **75.** not a real number **77.** 48 **79.** −1 **81.** −9 **83.** 17 **85.** −4 **87.** −2 **89.** −3 **91.** 2 **93.** $\frac{6}{5}$ **95.** $\frac{3}{2}$ **97.** −2.1 **99.** $-\frac{20}{33}$ **101.** −21 **103.** 41 **105.** −134 **107.** 3 **109.** 0 **111.** $-\frac{23}{18}$ **113.** 5 **115. a.** 18; 22; 28; 208 **b.** increase; answers may vary **117. a.** 600; 150; 105 **b.** decrease; answers may vary **119.** −162°F **121.** 130° **123.** 30° **125.** answers may very **127.** −3 **129.** −22 **131.** negative **133.** positive **135.** true **137.** false **139.** not possible **141.** negative

Section 1.5

Practice Exercises

1. a. $8 \cdot x$ **b.** $17 + x$ **2. a.** $2 + (9 + 7)$ **b.** $(-4 \cdot 2) \cdot 7$ **3. a.** $x + 14$ **b.** $-30x$ **4. a.** $5x - 5y$ **b.** $-24 - 12t$ **c.** $6x - 8y - 2z$ **d.** $-3 + y$ **e.** $-x + 7 - 2s$ **f.** $x + 11$ **5. a.** $5(w + 3)$ **b.** $9(w + z)$ **6. a.** commutative property of multiplication **b.** associative property of addition **c.** identity element for addition **d.** multiplicative inverse property **e.** commutative property of addition **f.** additive inverse property **g.** commutative and associative properties of multiplication

Vocabulary, Readiness & Video Check 1.5

1. commutative property of addition **3.** distributive property **5.** associative property of addition **7.** opposites or additive inverses **9.** 2 is outside the parentheses, so the point is made that you should only distribute the −9 to the terms within the parentheses and not also to the 2.

Exercise Set 1.5

1. $16 + x$ **3.** $y \cdot (-4)$ **5.** yx **7.** $13 + 2x$ **9.** $x \cdot (yz)$ **11.** $(2 + a) + b$ **13.** $(4a) \cdot b$ **15.** $a + (b + c)$ **17.** $17 + b$ **19.** $24y$ **21.** y **23.** $26 + a$ **25.** $-72x$ **27.** s **29.** $2 + x$ **31.** $4x + 4y$ **33.** $9x - 54$ **35.** $6x + 10$ **37.** $28x - 21$ **39.** $18 + 3x$ **41.** $-2y + 2z$ **43.** $-21y - 35$ **45.** $5x + 20m + 10$ **47.** $-4 + 8m - 4n$ **49.** $-5x - 2$ **51.** $-r + 3 + 7p$ **53.** $3x + 4$ **55.** $-x + 3y$ **57.** $6r + 8$ **59.** $-36x - 70$ **61.** $-16x - 25$ **63.** $4(1 + y)$ **65.** $11(x + y)$ **67.** $-1(5 + x)$ **69.** $30(a + b)$ **71.** commutative property of multiplication **73.** associative property of addition **75.** distributive property **77.** associative property of multiplication **79.** identity element for addition **81.** distributive property **83.** commutative and associative properties of multiplication **85.** $-8; \frac{1}{8}$ **87.** $-x; \frac{1}{x}$ **89.** $2x; -2x$ **91.** false **93.** no **95.** yes **97.** yes **99.** no **101. a.** commutative property of addition **b.** commutative property of addition **c.** associative property of addition **103.** answers may vary **105.** answers may vary

Chapter 1 Vocabulary Check

1. inequality symbols **2.** equation **3.** absolute value **4.** variable **5.** opposites **6.** solution **7.** reciprocals **8.** base; exponent **9.** grouping symbols **10.** set

Chapter 1 Review

1. $<$ **3.** $>$ **5.** $<$ **7.** $=$ **9.** $>$ **11.** $4 \geq -3$ **13.** $0.03 < 0.3$ **15. a.** 1, 3 **b.** 0, 1, 3 **c.** −6, 0, 1, 3 **d.** $-6, 0, 1, 1\frac{1}{2}, 3, 9.62$ **e.** π **f.** $-6, 0, 1, 1\frac{1}{2}, 3, \pi, 9.62$ **17.** Friday **19.** 9 **21.** −2 **23.** c **25.** $\frac{4}{49}$ **27.** 37 **29.** $\frac{18}{7}$ **31.** $20 - 12 = 2 \cdot 4$ **33.** 18 **35.** 5 **37.** 63° **39.** yes **41.** −11 **43.** $-\frac{3}{16}$ **45.** −13.9 **47.** −14 **49.** 5 **51.** −48 **53.** 3 **55.** undefined **57.** $-7 \cdot x$ or $-7x$ **59.** $-20 - x$ **61.** −12 **63.** 9 **65.** 3 **67.** $-\frac{5}{4}$ **69.** $\frac{5}{8}$ **71.** −1 **73.** 1 **75.** −4 **77.** −10 **79.** $\frac{1}{5}$ **81.** −5 **83.** 5 **85.** 6.28; 62.8; 628 **87.** commutative property of addition **89.** distributive property **91.** associative property of addition **93.** distributive property **95.** multiplicative inverse property **97.** $5y - 10$ **99.** $-7 + x - 4z$ **101.** $-12z - 27$ **103.** $<$ **105.** −15.3 **107.** −80 **109.** $-\frac{1}{4}$ **111.** 16 **113.** −5 **115.** $-\frac{5}{6}$ **117.** $\frac{3}{4}; -\frac{4}{3}$ **119.** 2.2 **121.** 2.3 **123.** 1.4

Chapter 1 Getting Ready for the Test

1. A **2.** C **3.** B **4.** A **5.** A **6.** B **7.** B **8.** A **9.** A **10.** B **11.** C **12.** D **13.** B **14.** C **15.** B **16.** C **17.** C **18.** A **19.** C **20.** B **21.** B and D **22.** A and B

Chapter 1 Test

1. $|-7| > 5$ **2.** $9 + 5 \geq 4$ **3.** −5 **4.** −11 **5.** −3 **6.** −39 **7.** 12 **8.** −2 **9.** undefined **10.** −8 **11.** −3 **12.** −56 **13.** −225 **14.** −32 **15.** −48 **16.** 3 **17.** $-\frac{3}{2}$ **18.** $>$ **19.** $>$ **20.** $<$ **21.** $=$ **22.** $2221 < 10{,}993$ **23. a.** 1, 7 **b.** 0, 1, 7 **c.** −5, −1, 0, 1, 7 **d.** $-5, -1, 0, \frac{1}{4}, 1, 7, 11.6$ **e.** $\sqrt{7}, 3\pi$ **f.** $-5, -1, 0, \frac{1}{4}, 1, 7, 11.6, \sqrt{7}, 3\pi$ **24.** 40 **25.** 12 **26.** 22 **27.** −1 **28.** associative property of addition **29.** commutative property of multiplication **30.** distributive property **31.** multiplicative inverse property **32.** 9 **33.** −3 **34.** second down **35.** yes **36.** 17° **37.** \$650 million **38.** \$420

CHAPTER 2 EQUATIONS, INEQUALITIES, AND PROBLEM SOLVING

Section 2.1
Practice Exercises

1. a. 1 **b.** −7 **c.** $-\frac{1}{5}$ **d.** 43 **e.** −1 **2. a.** like terms **b.** unlike terms **c.** like terms **d.** like terms **3. a.** $8y$ **b.** $5x^2$ **c.** $5x + 5x^2$ **d.** $21y^2$ **4. a.** $11y - 5$ **b.** $5x - 6$ **c.** $-\frac{1}{4}t$ **d.** $12.2y + 13$ **e.** $5z - 3z^4$ **5. a.** $6x - 21$ **b.** $-5x + 2.5z + 25$ **c.** $-2x + y - z + 2$ **6. a.** $36x + 10$ **b.** $-11x + 1$ **c.** $-30x - 17$ **7.** $-5x + 4$ **8. a.** $2x + 3$ **b.** $x - 1$ **c.** $2x + 10$ **d.** $\frac{13}{2}x$

Vocabulary, Readiness & Video Check 2.1

1. expression; term **3.** numerical coefficient **5.** numerical coefficient **7.** Although these terms have exactly the same variables, the exponents on each are not exactly the same—the exponents on x differ in each term. **9.** −1

Exercise Set 2.1

1. −7 **3.** 1 **5.** 17 **7.** like **9.** unlike **11.** like **13.** $15y$ **15.** $13w$ **17.** $-7b - 9$ **19.** $-m - 6$ **21.** −8 **23.** $7.2x - 5.2$ **25.** $4x - 3$ **27.** $5x^2$ **29.** $1.3x + 3.5$ **31.** $5y - 20$ **33.** $-2x - 4$ **35.** $7d - 11$ **37.** $-10x + 15y - 30$ **39.** $-3x + 2y - 1$ **41.** $2x + 14$ **43.** $10x - 3$ **45.** $-4x - 9$ **47.** $-4m - 3$ **49.** $k - 6$ **51.** $-15x + 18$ **53.** 16 **55.** $x + 5$ **57.** $x + 2$ **59.** $2k + 10$ **61.** $-3x + 5$ **63.** −11 **65.** $3y + \frac{5}{6}$ **67.** $-22 + 24x$ **69.** $0.9m + 1$ **71.** $10 - 6x - 9y$ **73.** $-x - 38$ **75.** $5x - 7$ **77.** $2x - 4$ **79.** $2x + 7$ **81.** $\frac{3}{4}x + 12$ **83.** $12x - 2$ **85.** $8(x + 6)$ or $8x + 48$ **87.** $x - 10$ **89.** $7x - 7$ **91.** 51–54 in. **93.** 21 students

95. 48 students **97.** $(18x - 2)$ ft **99.** balanced **101.** balanced **103.** answers may vary **105.** $(15x + 23)$ in. **107.** answers may vary **109.** $5b^2c^3 + b^3c^2$ **111.** $5x^2 + 9x$ **113.** $-7x^2y$

Section 2.2

Practice Exercises

1. -8 **2.** -1.8 **3.** -10 **4.** 18 **5.** 20 **6.** -12 **7.** 65 **8.** -3 **9. a.** 7 **b.** $9 - x$ **c.** $(9 - x)$ ft **10.** $3x + 6$

Vocabulary, Readiness & Video Check 2.2

1. equation; expression **3.** solution **5.** addition **7.** multiplication **9.** true **11.** both sides **13.** addition property; multiplication property; answers may vary

Exercise Set 2.2

1. 3 **3.** -2 **5.** -14 **7.** 0.5 **9.** -3 **11.** -0.7 **13.** 3 **15.** 11 **17.** 0 **19.** -3 **21.** 16 **23.** -4 **25.** 0 **27.** 12 **29.** 10 **31.** -12 **33.** 3 **35.** -2 **37.** 0 **39.** answers may vary **41.** 10 **43.** -20 **45.** 0 **47.** -5 **49.** 0 **51.** $-\frac{3}{2}$ **53.** -21 **55.** $\frac{11}{2}$ **57.** 1 **59.** $-\frac{1}{4}$ **61.** 12 **63.** -30 **65.** $\frac{9}{10}$ **67.** -30 **69.** 2 **71.** -2 **73.** 23 **75.** $20 - p$ **77.** $(10 - x)$ ft **79.** $(180 - x)°$ **81.** $(n + 284)$ votes **83.** $(m - 60)$ ft **85.** $(n - 28{,}000)$ students **87.** $7x$ sq mi **89.** $2x + 2$ **91.** $2x + 2$ **93.** $5x + 20$ **95.** $7x - 12$ **97.** 1 **99.** 86% **101.** 8.46 million sq mi **103.** $(173 - 3x)°$ **105.** answers may vary **107.** 4; 4 **109.** answers may vary **111.** answers may vary **113.** -48 **115.** $\frac{700}{3}$ mg **117.** solution **119.** -2.95 **121.** 0.02

Section 2.3

Practice Exercises

1. 3 **2.** $\frac{21}{13}$ **3.** -15 **4.** 3 **5.** 0 **6.** no solution **7.** all real numbers

Calculator Explorations 2.3

1. solution **3.** not a solution **5.** solution

Vocabulary, Readiness & Video Check 2.3

1. equation **3.** expression **5.** expression **7.** equation **9.** 3; distributive property, addition property of equality, multiplication property of equality **11.** The number of decimal places in each number helps you determine the least power of 10 you can multiply through by so you are no longer dealing with decimals.

Exercise Set 2.3

1. -6 **3.** 3 **5.** 1 **7.** $\frac{3}{2}$ **9.** 0 **11.** -1 **13.** 4 **15.** -4 **17.** -3 **19.** 2 **21.** 50 **23.** 1 **25.** $\frac{7}{3}$ **27.** 0.2 **29.** all real numbers **31.** no solution **33.** no solution **35.** all real numbers **37.** 18 **39.** $\frac{19}{9}$ **41.** $\frac{14}{3}$ **43.** 13 **45.** 4 **47.** all real numbers **49.** $-\frac{3}{5}$ **51.** -5 **53.** 10 **55.** no solution **57.** 3 **59.** -17 **61.** -4 **63.** 3 **65.** all real numbers **67.** 80–89 in. **69.** 29 trees **71.** $\frac{5}{22}$ **73.** $(6x - 8)$ m **75. a.** all real numbers **b.** answers may vary **c.** answers may vary **77.** A **79.** B **81.** C **83.** answers may vary **85. a.** $x + x + x + 2x + 2x = 28$ **b.** $x = 4$ **c.** $x = 4$ cm; $2x = 8$ cm **87.** answers may vary **89.** 15.3 **91.** -0.2 **93.** $-\frac{7}{8}$ **95.** no solution

Integrated Review

1. 6 **2.** -17 **3.** 12 **4.** -26 **5.** -3 **6.** -1 **7.** $\frac{27}{2}$ **8.** $\frac{25}{2}$ **9.** 8 **10.** -64 **11.** 2 **12.** -3 **13.** no solution **14.** no solution **15.** -2 **16.** -2 **17.** $-\frac{5}{6}$ **18.** $\frac{1}{6}$ **19.** 1 **20.** 6 **21.** 4 **22.** 1 **23.** $\frac{9}{5}$ **24.** $-\frac{6}{5}$ **25.** all real numbers **26.** all real numbers **27.** 0 **28.** -1.6 **29.** $\frac{4}{19}$ **30.** $-\frac{5}{19}$ **31.** $\frac{7}{2}$ **32.** $-\frac{1}{4}$ **33.** 2 **34.** 2 **35.** no solution **36.** no solution **37.** $\frac{7}{6}$ **38.** $\frac{1}{15}$

Section 2.4

Practice Exercises

1. 9 **2.** 2 **3.** 9 in. and 36 in. **4.** 28 Republican and 21 Democratic governors **5.** 25°, 75°, 80° **6.** 46, 48, 50

Vocabulary, Readiness & Video Check 2.4

1. $2x$; $2x - 31$ **3.** $x + 5$; $2(x + 5)$ **5.** $20 - y$; $\frac{20 - y}{3}$ or $(20 - y) \div 3$ **7.** in the statement of the application **9.** That the 3 angle measures are consecutive even integers and that they sum to 180°.

Exercise Set 2.4

1. $6x + 1 = 5x; -1$ **3.** $3x - 6 = 2x + 8; 14$ **5.** $2(x - 8) = 3(x + 3); -25$ **7.** $2(-2 + x) = x - \frac{1}{2}; \frac{7}{2}$ **9.** 3 in.; 6 in.; 16 in. **11.** 1st piece: 5 in.; 2nd piece: 10 in.; 3rd piece: 25 in. **13.** 16,146; 27,119; In 2014, 16,146 screens were 3D. **15.** 1st angle: 37.5°; 2nd angle: 37.5°; 3rd angle: 105° **17.** $3x + 3$ **19.** $x + 2$; $x + 4$; $2x + 4$ **21.** $x + 1; x + 2; x + 3; 4x + 6$ **23.** $x + 2; x + 4; 2x + 6$ **25.** 234, 235 **27.** Belgium: 32; France: 33; Spain: 34 **29.** Sahara: 3,500,000 sq mi; Gobi: 500,000 sq mi **31.** 5 ft, 12 ft **33.** $\frac{5}{4}$ **35.** Botswana: 32,000,000 carats; Angola: 8,000,000 carats **37.** 58°, 60°, 62° **39.** Netherlands: 24; Canada: 25; Norway: 26 **41.** −16 **43.** 43°, 137° **45.** 1 **47.** $\frac{3}{2}$ **49.** Maglev: 267 mph; Harmony: 236 mph **51.** $\frac{5}{2}$ **53.** California: 58; Montana: 56 **55.** 111° **57.** 1st piece: 3 ft; 2nd piece: 12 ft; 3rd piece: 15 ft **59.** Call of Duty: Advanced Warfare **61.** Destiny: 3.8 million; Madden NFL 15: 2.9 million **63.** answers may vary **65.** 42% **67.** 91.1 million cats **69.** 15 ft by 24 ft **71.** 5400 chirps per hour; 129,600 chirps per day; 47,304,000 chirps per year **73.** answers may vary **75.** answers may vary **77.** c

Section 2.5

Practice Exercises

1. 62.5% **2.** 360 **3.** discount: $408; new price: $72 **4.** tax: $12.35; new price: $142.34 **5.** 80.8% **6.** 14,732 screens **7.** $292.50 **8.** $40 **9.** $1540 **10.** $2898.19 **11.** $7720.73

Vocabulary, Readiness & Video Check 2.5

1. no **3.** yes **5.** simple **7.** Compound **9.** Total amount **11. a.** equals; = **b.** multiplication; · **c.** Drop the percent symbol and move the decimal point two places to the left. **13.** You must first find the actual amount of increase in price by subtracting the original price from the new price. **15.** semiannually; 2

Exercise Set 2.5

1. 11.2 **3.** 55% **5.** 180 **7.** $8.90; $80.10 **9.** $98.25; $98.25 **11.** $3255; $18,445 **13.** $27.45; $332.45 **15.** $3.08; $59.08 **17.** 30; 60% **19.** 52; 80% **21.** 2; 25% **23.** 120; 75% **25.** $104 **27.** $42,500 **29.** $73.60; $233.60 **31.** $750; $5750 **33.** $33.75; $408.75 **35.** $700; $3200 **37.** $936.31 **39.** $9674.77 **41.** $634.49 **43.** 4.6 **45.** 50 **47.** 30% **49.** discount: $1480; new price: $17,020 **51.** $46.58 **53.** 9.3% **55.** 30% **57.** 45 min **59.** 32.1% **61.** 63%; 5%; 24%; 2%; 101% due to rounding **63.** decrease: $64; sale price: $192 **65.** 7.2% **67.** 251 million **69.** 300% **71.** 16% **73.** 120 employees **75.** $101,562.50; $264,062.50 **77.** $5562.50 **79.** $46,815.37 **81.** $2327.14 **83.** $58,163.65 **85.** $2938.66 **87.** $2971.89 **89.** 854 thousand Scoville units **91.** > **93.** = **95.** > **97.** answers may vary **99.** $4.00; $6.00; $8.00 **101.** $7.20; 10.80; $14.40 **103.** a discount of 60% is better; answers may vary **105.** 9.6% **107.** 26.9%; yes **109.** 17.1%

Section 2.6

Practice Exercises

1. 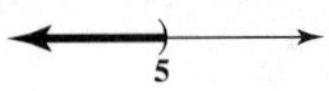 $(-\infty, 5)$ **2.** 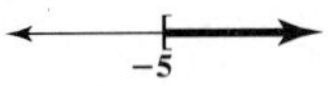 $[-5, \infty)$ **3.** 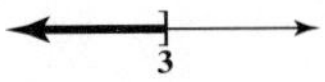 $(-\infty, 3]$ **4.** 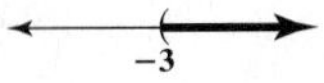 $(-3, \infty)$ **5.** 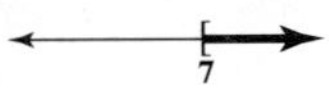 $[7, \infty)$ **6.** 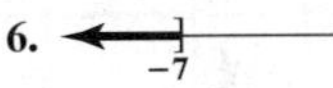 $(-\infty, -7]$ **7.** 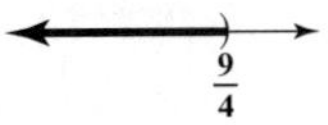 $\left(-\infty, \frac{9}{4}\right)$ **8.** 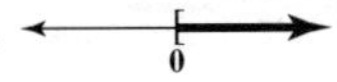 $[0, \infty)$ **9.** 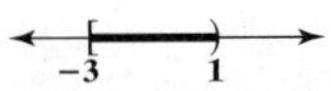 $[-3, 1)$ **10.**  $(-2, 2]$ **11.** 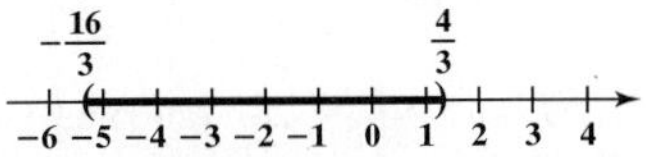 $\left(-\frac{16}{3}, \frac{4}{3}\right)$ **12.** all numbers less than 10 **13.** Kasonga can afford at most 3 classes.

Vocabulary, Readiness & Video Check 2.6

1. expression **3.** inequality **5.** −5 **7.** The graph of Example 1 is shaded from $-\infty$ to and including −1, as indicated by a bracket. To write interval notation, you write down what is shaded for the inequality from left to right. A parenthesis is always used with $-\infty$, so from the graph, the interval notation is $(-\infty, -1]$. **9.** You would divide the left, middle, and right by −3 instead of 3, which would reverse the directions of both inequality symbols.

Exercise Set 2.6

1. 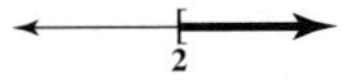 $x \geq 2$ **3.** 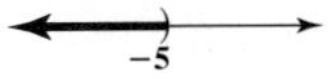 $x < -5$ **5.** 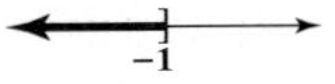 $(-\infty, -1]$ **7.** 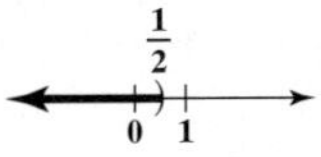 $\left(-\infty, \frac{1}{2}\right)$ **9.** 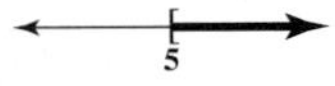 $[5, \infty)$ **11.** $x < -3$, 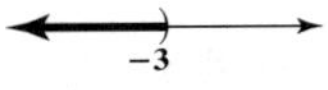 $(-\infty, -3)$ **13.** $x \geq -5$, 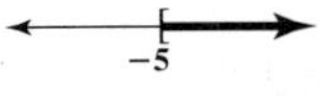 $[-5, \infty)$ **15.** $x \geq -2$ 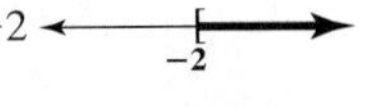 $[-2, \infty)$ **17.** $x > -3$, 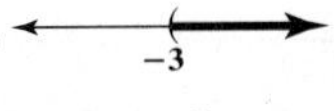 $(-3, \infty)$ **19.** $x \leq 1$, 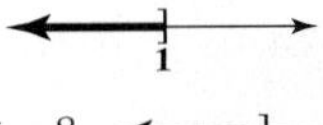 $(-\infty, 1]$ **21.** $x > -5$, 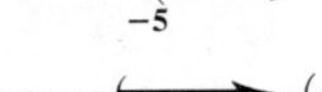 $(-5, \infty)$ **23.** $x \leq -2$, 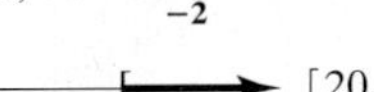 $(-\infty, -2]$ **25.** $x \leq -8$, 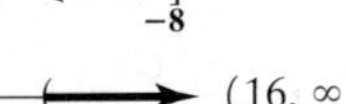 $(-\infty, -8]$ **27.** $x > 4$, $(4, \infty)$ **29.** $x \geq 20$, 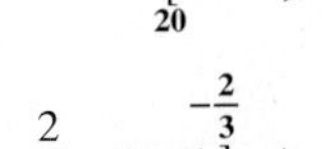 $[20, \infty)$ **31.** $x > 16$,  $(16, \infty)$ **33.** $x > -3$, 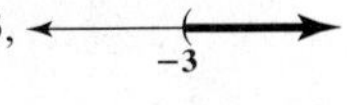 $(-3, \infty)$ **35.** $x \leq -\frac{2}{3}$, 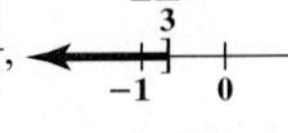 $\left(-\infty, -\frac{2}{3}\right]$ **37.** $x > \frac{8}{3}$, 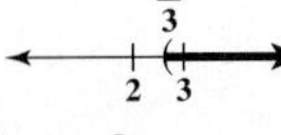 $\left(\frac{8}{3}, \infty\right)$ **39.** $x > -13$, 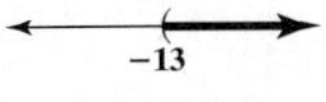 $(-13, \infty)$ **41.** $x < 0$, 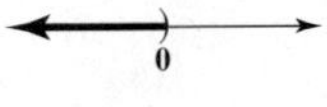 $(-\infty, 0)$ **43.** $x \leq 0$, 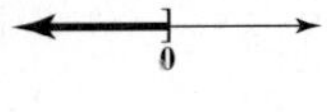 $(-\infty, 0]$

45. $x > 3$, $(3, \infty)$ **47.** $x > \frac{8}{3}$, $\left(\frac{8}{3}, \infty\right)$ **49.** $(-1, 3)$ **51.** $[0, 2)$

53. $(-1, 2)$ **55.** $[4, 5]$ **57.** $(1, 5]$ **59.** $(1, 4)$

61. $\left(0, \frac{14}{3}\right]$ **63.** all numbers greater than -10 **65.** 35 cm **67.** at least 193 **69.** 86 people **71.** at least 35 min

73. $-3 < x < 3$ **75.** Serena Williams **77.** \$8 million **79.** \$0.966 or \$966,000 per tournament **81.** $>$ **83.** $\geq$ **85.** when multiplying or dividing by a negative number **87.** final exam score ≥ 78.5 **89.** answers may vary **91.** answers may vary **93.** $0.924 \leq d \leq 0.987$

95. $(1, \infty)$ **97.** $\left(-\infty, \frac{5}{8}\right)$

Chapter 2 Vocabulary Check

1. like terms **2.** unlike terms **3.** linear equation in one variable **4.** linear inequality in one variable **5.** compound inequalities **6.** 0 **7.** numerical coefficient **8.** equivalent equations **9.** all real numbers **10.** no solution **11.** the same **12.** reversed **13.** of **14.** is **15.** Percent **16.** Compound interest **17.** Percent of decrease **18.** Percent of increase

Chapter 2 Review

1. $6x$ **3.** $4x - 2$ **5.** $3n - 18$ **7.** $-6x + 7$ **9.** $3x - 7$ **11.** 4 **13.** 6 **15.** 0 **17.** -23 **19.** 5; 5 **21.** b **23.** b **25.** -12 **27.** 0 **29.** 0.75 **31.** -6 **33.** -1 **35.** $-\frac{1}{5}$ **37.** $3x + 3$ **39.** -4 **41.** 2 **43.** no solution **45.** $\frac{3}{4}$ **47.** 20 **49.** $\frac{23}{7}$ **51.** 102 **53.** 6665.5 in. **55.** Target Canada: 133; Target U.S. 1793 **57.** 3 **59.** 20% **61.** 110 **63.** markup: \$209; new price: \$2109 **65.** 18% **67.** 966 customers **69.** 32% **71.** \$273.75 **73.** discount: \$900; sale price: \$2100 **75.** \$160 **77.** \$30,104.61 **79.** \$180.61 **81.** $(0, \infty)$ **83.** $[0.5, 1.5)$ **85.** $(-\infty, -4)$ **87.** $(-\infty, 4]$

89. $\left(-\frac{1}{2}, \frac{3}{4}\right)$ **91.** $\left(-\infty, \frac{19}{3}\right]$ **93.** \$2500 **95.** 4 **97.** $-\frac{3}{2}$ **99.** all real numbers **101.** -13

103. 160 **105.** \$617.70 **107.** \$11,687.50 **109.** $(9, \infty)$ **111.** $(-\infty, 0]$

Chapter 2 Getting Ready for the Test

1. C **2.** A **3.** D **4.** B **5.** B **6.** C **7.** B **8.** A **9.** C **10.** C **11.** B **12.** D

Chapter 2 Test

1. $y - 10$ **2.** $5.9x + 1.2$ **3.** $-2x + 10$ **4.** $10y + 1$ **5.** -5 **6.** 8 **7.** $\frac{7}{10}$ **8.** 0 **9.** 27 **10.** 3 **11.** 0.25 **12.** $\frac{25}{7}$

13. no solution **14.** $(-\infty, -2]$ **15.** $(-\infty, 4)$ **16.** $\left(-1, \frac{7}{3}\right)$

17. $\left(\frac{2}{5}, \infty\right)$ **18.** 21 **19.** 62 ft by 64 ft **20.** 401, 802 **21.** 552 tornadoes **22.** 40% **23.** 16% **24.** \$383.21

25. 5% **26.** discount: \$18; sale price: \$102 **27.** \$647.50 **28.** \$2005.63 **29.** \$427

Chapter 2 Cumulative Review

1. a. 11, 112 **b.** 0, 11, 112 **c.** $-3, -2, 0, 11, 112$ **d.** $-3, -2, -1.5, 0, \frac{1}{4}, 11, 112$ **e.** $\sqrt{2}$ **f.** all numbers in the given set; Sec. 1.2, Ex. 5

3. a. 4 **b.** 5 **c.** 0 **d.** $\frac{1}{2}$ **e.** 5.6; Sec. 1.2, Ex. 7 **5. a.** $2 \cdot 2 \cdot 2 \cdot 5$ **b.** $3 \cdot 3 \cdot 7$; Sec. R.1, Ex. 1 **7.** $\frac{8}{20}$; Sec. R.1, Ex. 6

9. 66; Sec. 1.3, Ex. 4 **11.** 2 is a solution; Sec. 1.3, Ex. 7 **13.** -4; Sec. 1.4, Ex. 1.a **15.** 5; Sec. 1.4, Ex. 1.c **17. a.** -6 **b.** -7 **c.** 20.5 **d.** $-\frac{7}{6}$; Sec. 1.4, Ex. 2 **19. a.** 6 **b.** -7; Sec. 1.4, Ex. 3 **21. a.** 130° **b.** 30°; Sec. 1.4, Exercises 121, 123 **23. a.** 8 **b.** -0.36 **c.** $-\frac{2}{11}$; Sec. 1.4, Ex. 4 **25. a.** -5 **b.** 3 **c.** $\frac{1}{4}$ **d.** $-\frac{1}{8}$; Sec. 1.4, Ex. 5 **27. a.** $5 + x$ **b.** $x \cdot 3$; Sec. 1.5, Ex. 1 **29. a.** $8(2 + x)$ **b.** $7(s + t)$; Sec. 1.5, Ex. 5 **31.** $-2x - 1$; Sec. 2.1, Ex. 7 **33.** -1.6; Sec. 2.2, Ex. 2 **35.** 8; Sec. 2.2, Ex. 4 **37.** 140; Sec. 2.2, Ex. 7

39. 2; Sec. 2.3, Ex. 1 **41.** 10; Sec. 2.4, Ex. 2 **43.** $(-\infty, -10]$; Sec. 2.6, Ex. 2

CHAPTER 3 GRAPHING AND SYSTEMS IN TWO VARIABLES

Section 3.1

Practice Exercises

1. 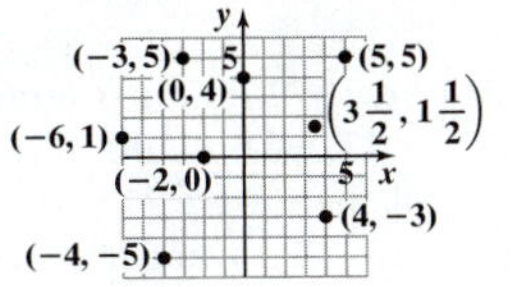

a. Quadrant IV; **b.** Quadrant II; **c.** y-axis; **d.** Quadrant II; **e.** x-axis; **f.** Quadrant I; **g.** Quadrant I; **h.** Quadrant III

2. a. (2008, 79), (2009, 79), (2010, 72), (2011, 74), (2012, 68), (2013, 48), (2014, 63), (2015, 68) **b.** 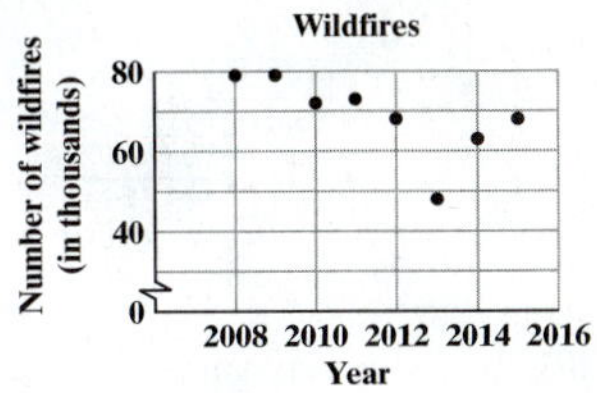

3. a. yes **b.** yes **c.** no **4. a.** $(0, -8)$ **b.** $(6, 4)$ **c.** $(-3, -14)$ **5.**

	x	y
a.	−10	−4
b.	0	−2
c.	10	0

6. a. yes **b.** no **c.** yes **d.** yes

7. 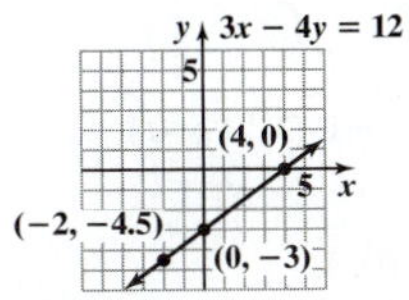

8. 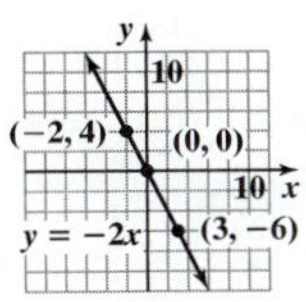

9. 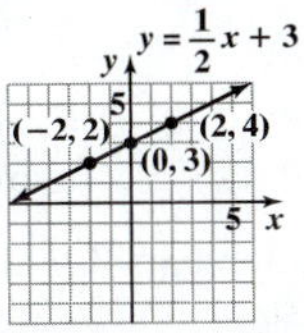

10. a. 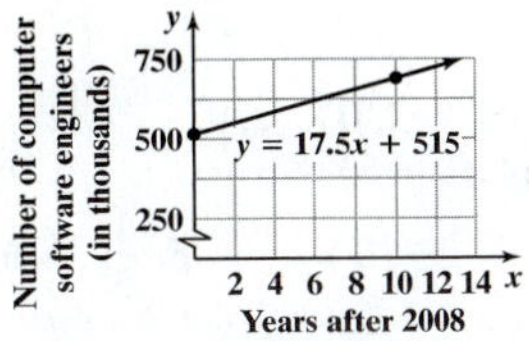

b. We predict 725 thousand computer software application engineers in the year 2020.

Calculator Explorations 3.1

1. 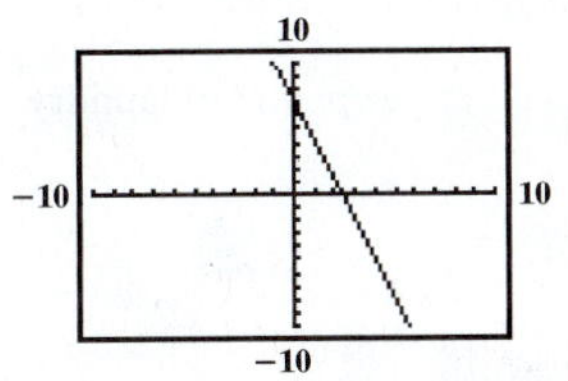 **3.** 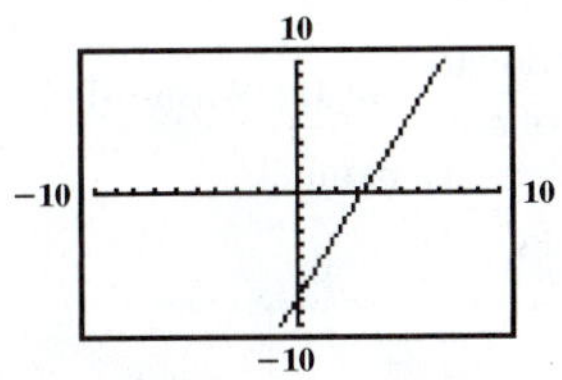 **5.** 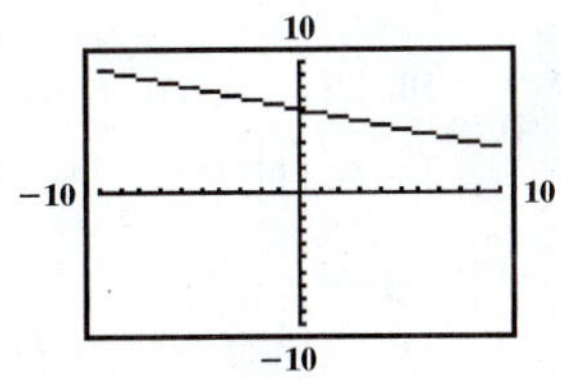

Vocabulary, Readiness & Video Check 3.1

1. x-axis; y-axis **3.** quadrants; four **5.** one **7.** linear **9.** origin; left or right; up or down **11.** (7, 0) and (0, 7); since one of these points is a solution and one is not, it shows that it is very important to remember that the first number is the x-value and the second number is the y-value and not to mix them up. **13.** In the definition, x and y both have an understood power of 1. Example 14 shows an equation where y has a power of 2, so it is not a linear equation in two variables. **15.** An infinite number of points make up the line and each point corresponds to an ordered pair that is a solution of the linear equation in two variables.

Exercise Set 3.1

1. 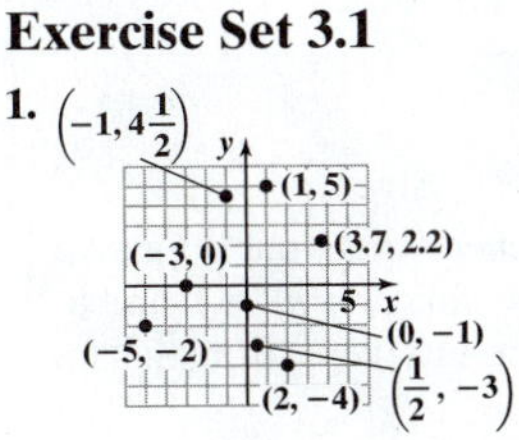

(1, 5) and (3.7, 2.2) are in quadrant I, $\left(-1, 4\frac{1}{2}\right)$ is in quadrant II, $(-5, -2)$ is in quadrant III, $(2, -4)$ and $\left(\frac{1}{2}, -3\right)$ are in quadrant IV, $(-3, 0)$ lies on the x-axis, $(0, -1)$ lies on the y-axis **3.** $(0, 0)$ **5.** $(3, 2)$ **7.** $(-2, -2)$ **9.** $(2, -1)$ **11.** $(0, -3)$ **13.** $(1, 3)$ **15.** $(-3, -1)$ **17. a.** (2011, 32.6), (2012, 34.7), (2013, 35.9), (2014, 36.4), (2015, 38.3) **b.** In the year 2015, the worldwide box office was $38.3 billion.

c. 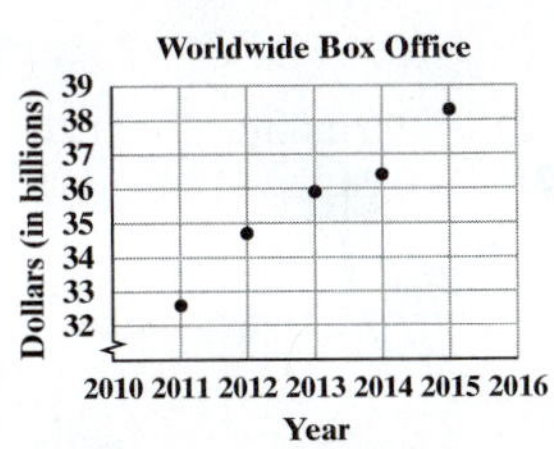

d. The worldwide box office increased every year. **19. a.** (2313, 2), (2085, 1), (2711, 21), (2869, 39), (2920, 42), (4038, 99), (1783, 0), (2493, 9) **b.** 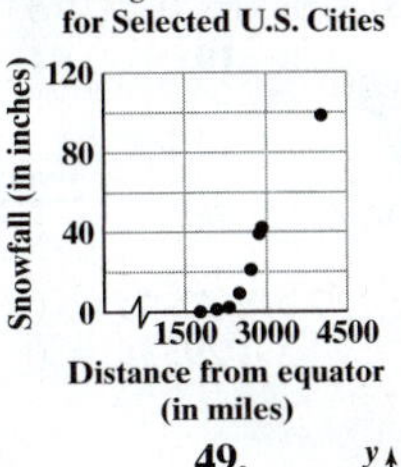

c. The farther from the equator, the more snowfall. **21.** yes; no; yes **23.** yes; yes **25.** $(-4, -2)$, $(4, 0)$ **27.** $(-8, -5)$, $(16, 1)$ **29.** 0; 7; $-\frac{2}{7}$ **31.** 0; −3; 2 **33.** −12; 5; −6 **35.** $\frac{5}{7}$; $\frac{5}{2}$; −1 **37.** yes **39.** yes **41.** no

43. 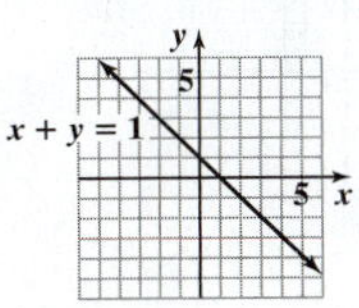

45. 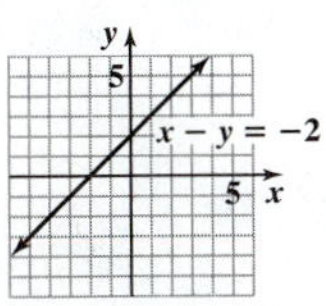

47. 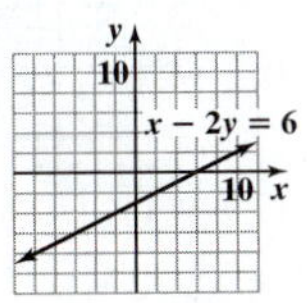

49. 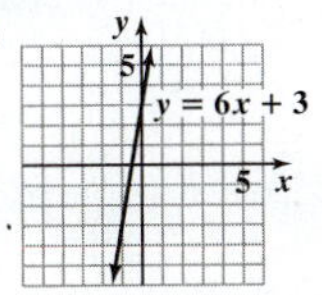

51. 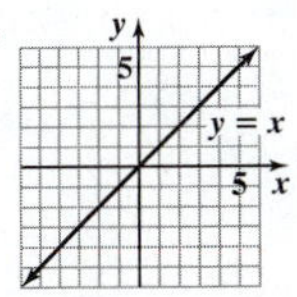

53. 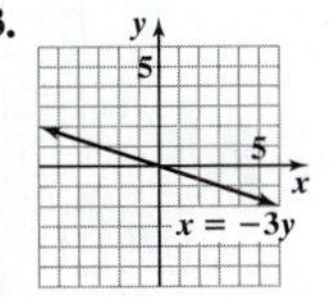

55. 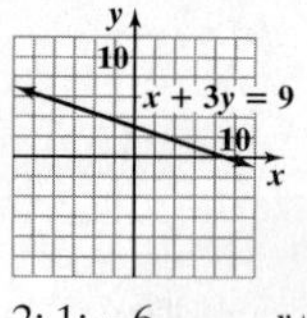

57. 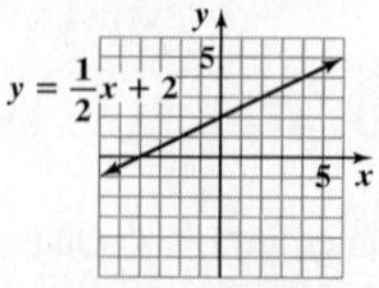

59. 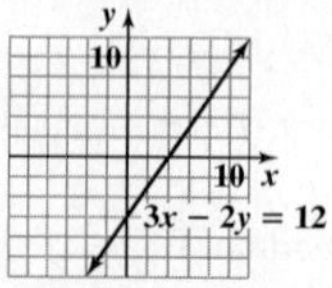

61. 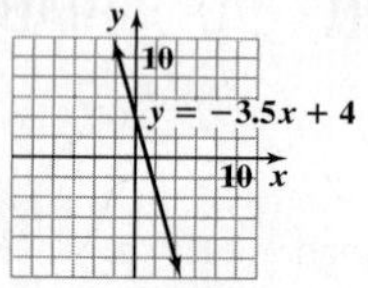

63. 0; −5; −2 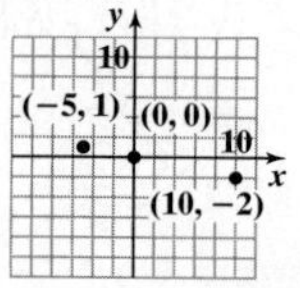

65. 2; 1; −6 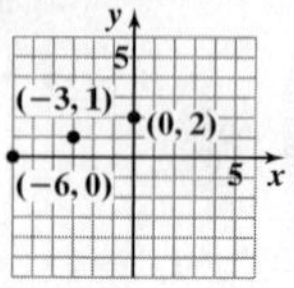

67. a. 13,000; 21,000; 29,000 **b.** 45 desks **69. a.** 7.85; 8.03; 8.21 **b.** year 3; 2013 **c.** 2034 **d.** In 2011, the average cinema admission price was $7.94. **71.** In 2014, there were 4203 Walmart stores in the U.S. **73.** year 8: 100 stores; year 9: 105 stores; year 10: 100 stores **75.** The *y*-values are all 0. **77. a.** (8, 31) **b.** In 2008, there were 31 million joggers. **c.** 40 million joggers **79. a.** (12, 216.4) **b.** In 2012, there were 216.4 million people with driver's licenses. **c.** 234 million **81.** China **83.** France, U.S., Spain, China **85.** 53 million **87.** false **89.** true **91.** negative; negative **93.** positive; negative **95.** 0; 0 **97.** y **99.** no; answers may vary

101. answers may vary **103.** 0; 1; 1; 4; 4 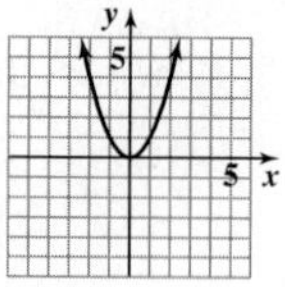

Section 3.2

Practice Exercises

1. a. x-intercept: $(-4, 0)$ y-intercept: $(0, -6)$ **b.** x-intercepts: $(-2, 0)$, $(2, 0)$ y-intercept: $(0, -3)$ **c.** x-intercept: (0, 0) y-intercept: (0, 0) **d.** x-intercept: none y-intercept: (0, 3) **e.** x-intercepts: $(-1, 0)$, $(5, 0)$ y-intercepts: $(0, 2)$, $(0, -2)$

2. 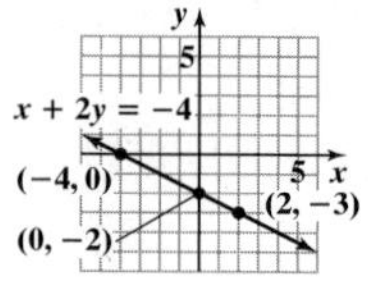

3. 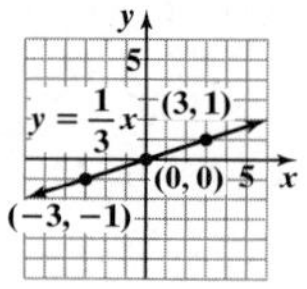

4. −1 **5.** $\frac{1}{3}$ **6.** $m = \frac{2}{3}$; y-intercept: $(0, -2)$ **7.** $m = -\frac{5}{2}$; y-intercept: (0, 4)

8. 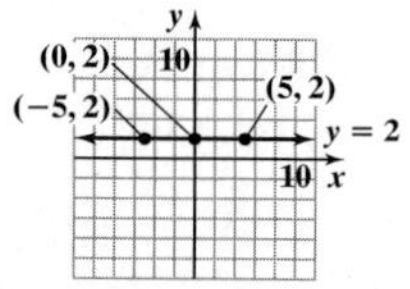

9. 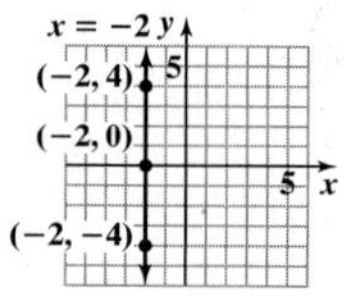

10. 25% **11. a.** $m = \frac{0.75 \text{ dollar}}{1 \text{ pound}}$ **b.** The Wash-n-Fold charges $0.75 per pound of laundry. **12. a.** perpendicular **b.** neither **c.** parallel

Calculator Explorations 3.2

1. 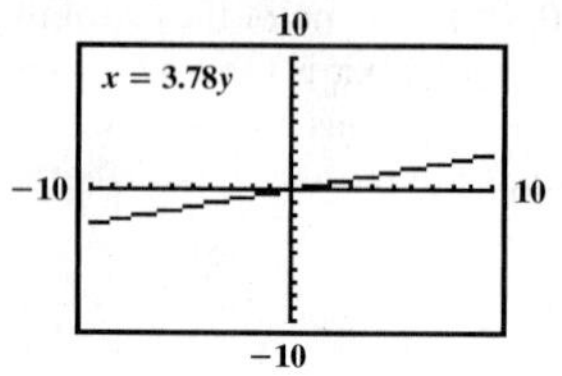

3. 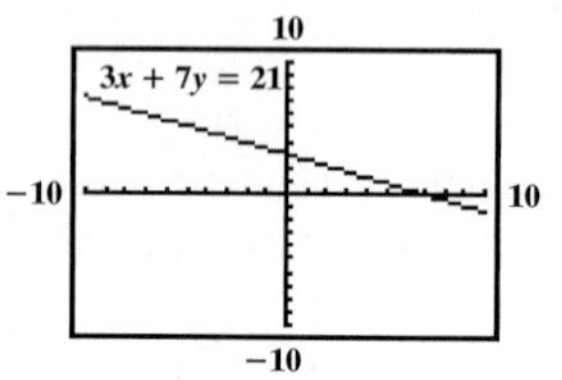

5. 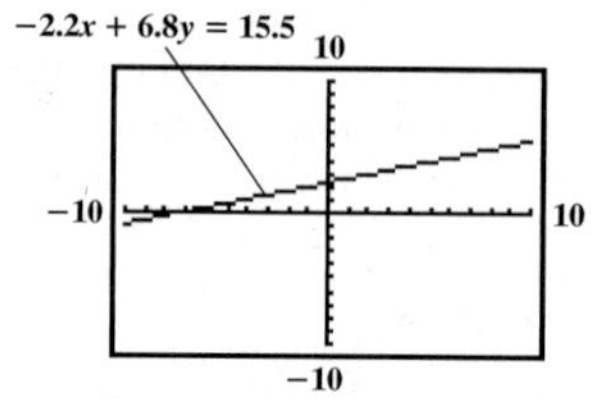

Vocabulary, Readiness & Video Check 3.2

1. horizontal **3.** y-intercept **5.** y; x **7.** slope **9.** 0 **11.** positive **13.** y; x **15.** Using a third point as a check that your points lie along a straight line is always good practice. **17.** solve the equation for y; the slope is the coefficient of x **19.** Zero slope indicates $m = 0$ and a horizontal line; undefined slope indicates m is undefined and a vertical line; no slope refers to an undefined slope. **21.** slope-intercept form; this form makes the slope easy to see, and you need to compare slopes to determine if two lines are parallel or perpendicular

Exercise Set 3.2

1. $(-1, 0)$; $(0, 1)$ **3.** $(-2, 0)$; $(2, 0)$; $(0, -2)$ **5.** $(-2, 0)$; $(1, 0)$; $(3, 0)$; $(0, 3)$ **7.** $(-1, 0)$; $(1, 0)$; $(0, 1)$; $(0, -2)$ **9.** infinite **11.** 0

13. 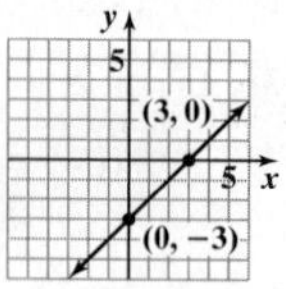

15. 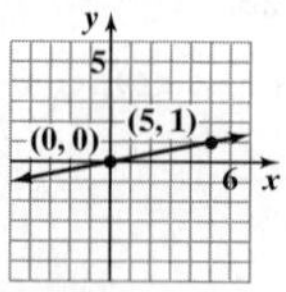

17. 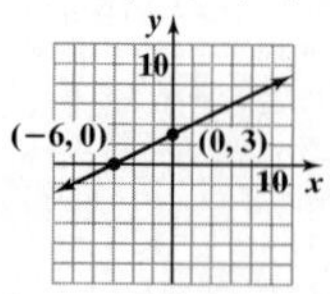

19. 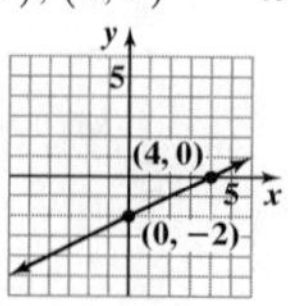

21. 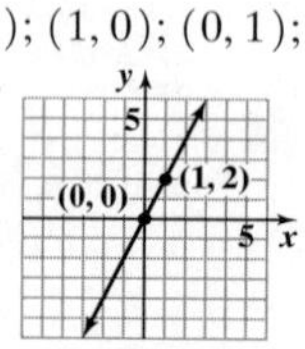

23. 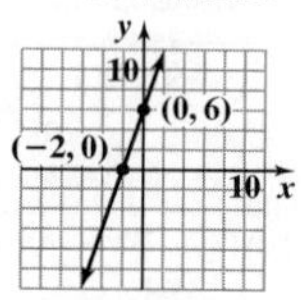

25. undefined slope 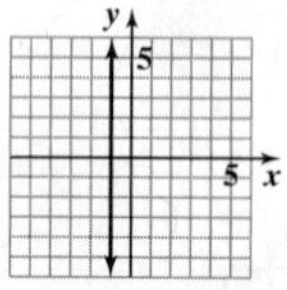

27. $m = 0$ 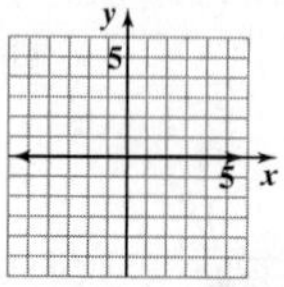

29. $m = 0$ 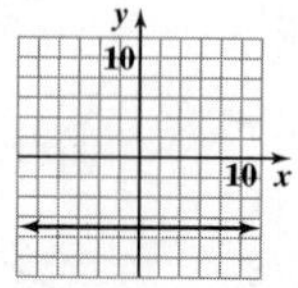

31. undefined slope 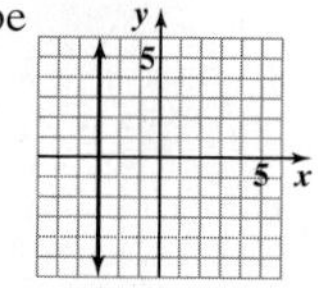

33. -1 **35.** undefined **37.** $-\frac{2}{3}$ **39.** 0 **41.** $m = -\frac{4}{3}$ **43.** undefined slope **45.** $m = \frac{5}{2}$ **47.** negative **49.** undefined **51.** D **53.** B **55.** E **57.** $m = 5$ **59.** $m = -0.3$ **61.** $m = -2$ **63.** $m = \frac{2}{3}$ **65.** undefined slope **67.** $m = 0$ **69.** $m = -\frac{3}{4}$ **71.** $\frac{3}{5}$ **73.** 12.5% **75.** 40% **77.** 37%; 35% **79.** $m = 1.4$ or $\frac{1.4}{1}$; every 1 year, there are 1.4 million more U.S. households with computers. **81.** C **83.** E **85.** B **87. a.** 1 **b.** -1 **89. a.** $\frac{9}{11}$ **b.** $-\frac{11}{9}$ **91.** neither **93.** parallel **95.** perpendicular **97.** 15.8 **99.** from 2008 to 2010 **101.** 2010 to 2014 **103.** false **105.** true **107. a.** (4.5, 0) **b.** 4.5 years after 2009, there should be no more analog screens. **c.** answers may vary **109.** $x = 6$

Integrated Review

1. $m = 2$ **2.** $m = 0$ **3.** $m = -\frac{2}{3}$ **4.** undefined slope **5.**

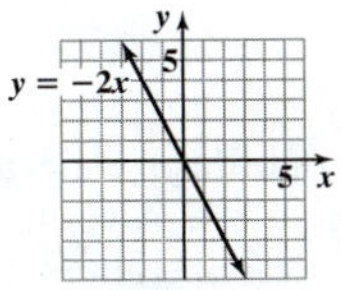

6.

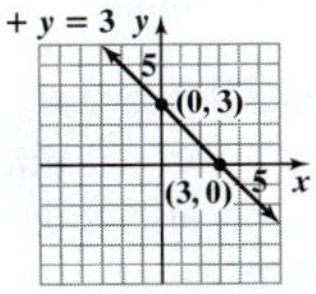

7.

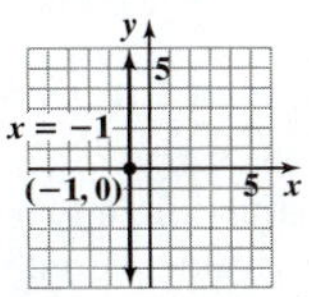

8.

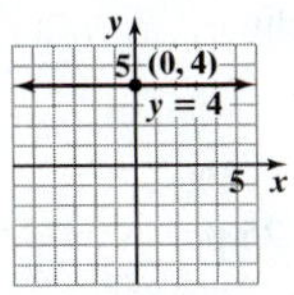

9.

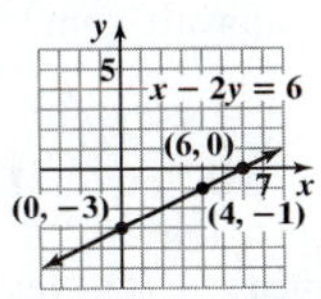

10.

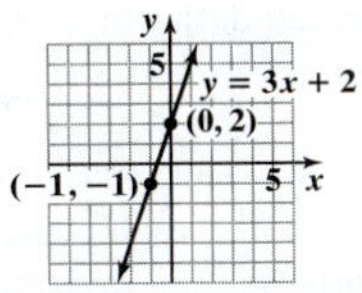

11.

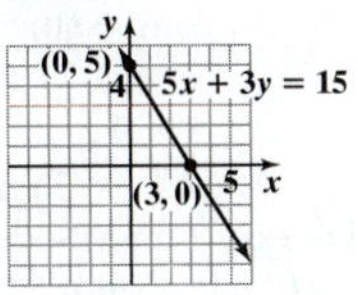

12.

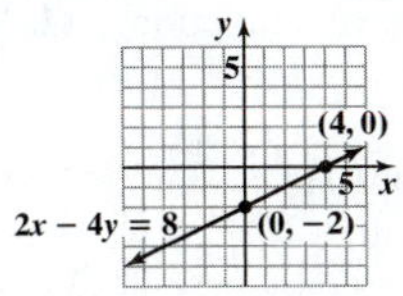

13. a. (0, 490) **b.** In 2002, there were 490 thousand bridges on public roads. **c.** 4.09 or $\frac{4.09}{1}$ **d.** For the years 2002 through 2013, the number of bridges on public roads increased at a rate of 4.09 thousand per 1 year. **14. a.** (4, 43.9) **b.** In 2013, the revenue for online advertising was \$43.9 billion. **15.** parallel **16.** neither

Section 3.3
Practice Exercises

1.

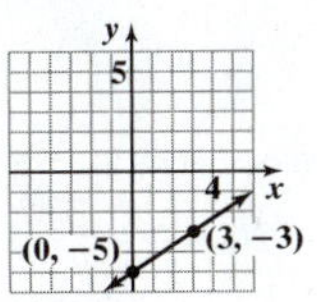

2.

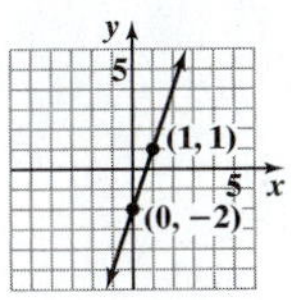

3. $y = \frac{1}{2}x + 7$ **4.** $y = 4x - 5$; $4x - y = 5$ **5.** $y = -\frac{5}{4}x + \frac{19}{4}$ **6.** $x = 3$ **7.** $y = 3$ **8. a.** $y = -1500x + 195{,}000$ **b.** \$105,000

Calculator Explorations 3.3

1.

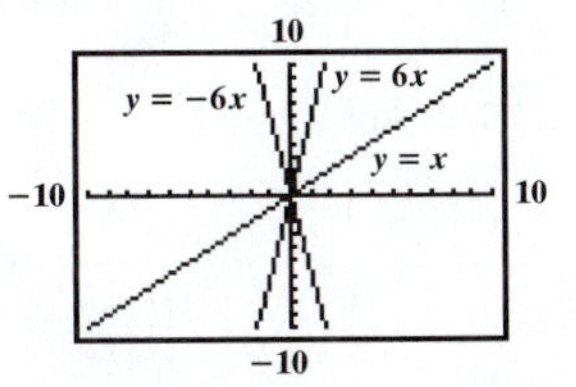

3.

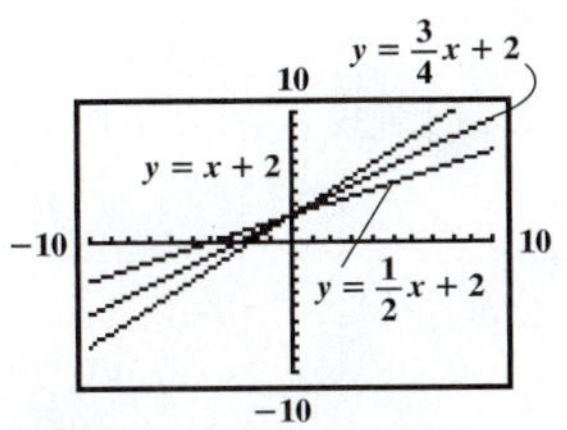

5.

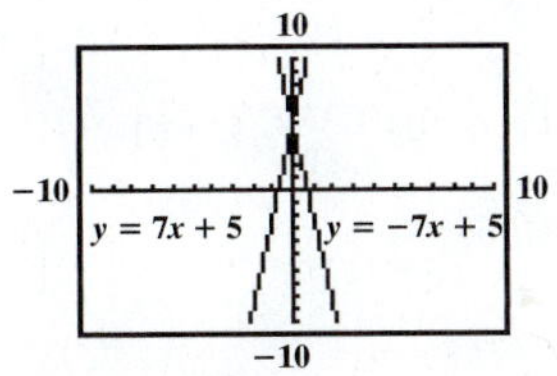

Vocabulary, Readiness & Video Check 3.3

1. slope-intercept; m; b **3.** y-intercept; fraction **5.** Write the equation with x- and y-terms on one side of the equal sign and a constant on the other side. **7.** Example 6: $y = -3$; Example 7: $x = -2$

Exercise Set 3.3

1.

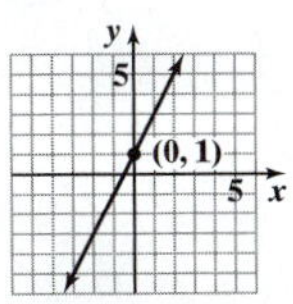

3.

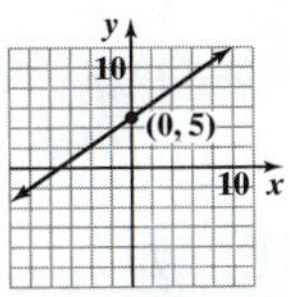

5.

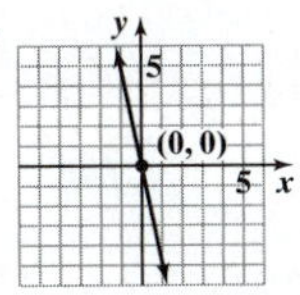

7.

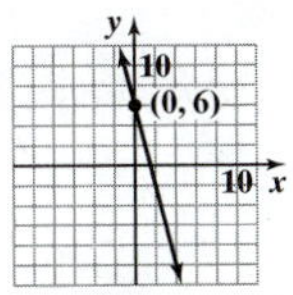

9.

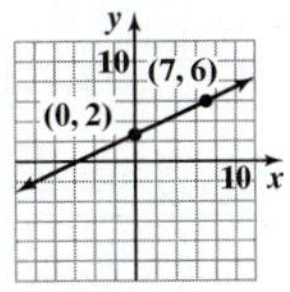

11.

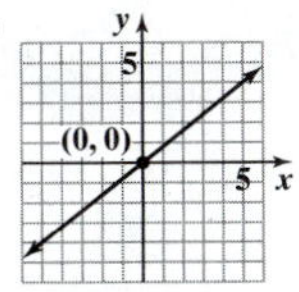

13. $y = 5x + 3$ **15.** $y = -4x - \frac{1}{6}$ **17.** $y = \frac{2}{3}x$ **19.** $y = -8$ **21.** $y = -\frac{1}{5}x + \frac{1}{9}$ **23.** $y = 6x - 10$ **25.** $y = -8x - 13$ **27.** $y = \frac{3}{2}x - \frac{27}{2}$ **29.** $y = -\frac{1}{2}x - \frac{3}{2}$ **31.** $y = 2x - 4$ **33.** $y = 8x + 11$ **35.** $y = \frac{4}{3}x + \frac{1}{3}$ **37.** $y = -\frac{8}{13}x$ **39.** $x = 0$ **41.** $y = 3$ **43.** $x = -\frac{7}{3}$ **45.** $y = 2$ **47.** $y = 5$ **49.** $x = 6$ **51.** $y = -\frac{1}{2}x + \frac{5}{3}$ **53.** $y = -x + 17$ **55.** $x = -\frac{3}{4}$ **57.** $y = x + 16$

59. $y = -5x + 7$ **61.** $y = 7$ **63.** $y = \frac{3}{2}x$ **65.** $y = -3$ **67.** $y = -\frac{4}{7}x - \frac{18}{7}$ **69. a.** $s = 32t$ **b.** 128 ft/sec **71. a.** $y = 2x + 66$ **b.** 80 thousand cars **73. a.** $y = 0.56x + 85$ **b.** 91.7 persons per sq mi **75. a.** $y = -1.1x + 55$ **b.** 38.5 million **77. a.** $S = -1000p + 13{,}000$ **b.** 9500 Fun Noodles **79.** 12–17 **81.** 5.9 million **83.** 5.4 million **85.** no **87.** point-slope **89.** slope-intercept **91.** horizontal **93.** answers may vary **95. a.** $3x - y = -5$ **b.** $x + 3y = 5$ **97. a.** $3x + 2y = -1$ **b.** $2x - 3y = 21$

Section 3.4
Practice Exercises

1. domain: $\{0, 1, 5\}$; range: $\{-2, 0, 3, 4\}$ **2. a.** function **b.** not a function **3. a.** not a function **b.** function **4. a.** function **b.** function **c.** function **d.** not a function **5. a.** function **b.** function **c.** function **d.** not a function **6. a.** 69°F **b.** February **c.** yes **7. a.** $h(2) = 9$; $(2, 9)$ **b.** $h(-5) = 30$; $(-5, 30)$ **c.** $h(0) = 5$; $(0, 5)$ **8. a.** domain: $(-\infty, \infty)$ **b.** domain: $(-\infty, 0) \cup (0, \infty)$ **9. a.** domain: $[-4, 6]$; range: $[-2, 3]$ **b.** domain: $(-\infty, \infty)$; range: $(-\infty, 3]$

Vocabulary, Readiness & Video Check 3.4

1. relation **3.** range **5.** vertical **7.** A relation is a set of ordered pairs and an equation in two variables defines a set of ordered pairs. Therefore, an equation in two variables can also define a relation. **9.** A vertical line represents one x-value paired with many y-values. A function only allows an x-value paired with exactly one y-value, so if a vertical line intersects a graph more than once, there's an x-value paired with more than one y-value, and we don't have a function. **11.** The r-value is positive since the graphed data points are moving upward from left to right.

Exercise Set 3.4

1. $\{-7, 0, 2, 10\}$; $\{-7, 0, 4, 10\}$ **3.** $\{0, 1, 5\}$; $\{-2\}$ **5.** yes **7.** no **9.** no **11.** yes **13.** yes **15.** no **17.** yes **19.** yes **21.** yes **23.** no **25.** no **27.** 9:30 p.m. **29.** January 1 and December 1 **31.** yes; it passes the vertical line test **33.** $4.25 per hour **35.** 2009 **37.** yes; answers may vary **39.** $1.80 **41.** more than 5 ounces and less than or equal to 6 ounces **43.** yes; answers may vary **45.** $-9, -5, 1$ **47.** $6, 2, 11$ **49.** $-6, 0, 9$ **51.** $2, 0, 3$ **53.** $5, 0, -20$ **55.** $5, 3, 35$ **57.** $(3, 6)$ **59.** $\left(0, -\frac{1}{2}\right)$ **61.** $(-2, 9)$ **63.** $(-\infty, \infty)$ **65.** all real number except -5 or $(-\infty, -5) \cup (-5, \infty)$ **67.** $(-\infty, \infty)$ **69.** domain: $(-\infty, \infty)$; range: $[-4, \infty)$ **71.** domain: $(-\infty, \infty)$; range: $(-\infty, \infty)$ **73.** domain: $(-\infty, \infty)$; range: $\{2\}$ **75.** -1 **77.** -1 **79.** $-1, 5$ **81.** c **83.** a **85.** positive **87.** $44.75 billion **89.** $(-2, 1)$ **91.** $(-3, -1)$ **93.** $f(-5) = 12$ **95.** $(3, -4)$ **97.** $f(5) = 0$ **99. a.** 166.38 cm **b.** 148.25 cm **101.** answers may vary **103.** $f(x) = x + 7$ **105. a.** 11 **b.** $2a + 7$ **107. a.** 16 **b.** $a^2 + 7$

Section 3.5
Practice Exercises

1. a. yes **b.** no **2.** solution: $(1, 5)$

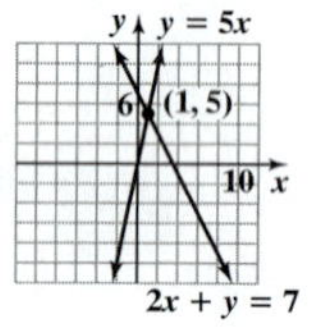

3. no solution

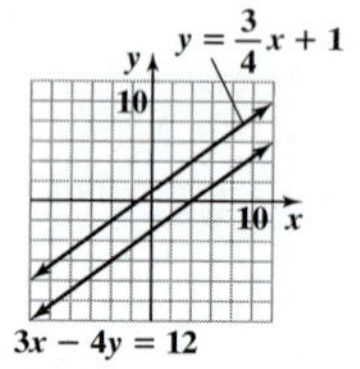

4. $\{(x, y) \mid 3x - 2y = 4\}$ or $\{(x, y) \mid -9x + 6y = -12\}$

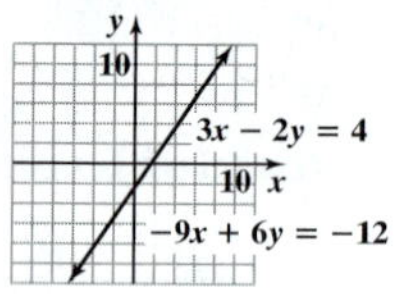

5. $\left(-\frac{1}{2}, 5\right)$ **6.** $(-3, 2)$ **7.** $(2, 1)$ **8.** $(-2, 0)$ **9.** $\varnothing$ or $\{\ \}$ **10.** $\{(x, y) \mid -3x + 2y = -1\}$ or $\{(x, y) \mid 9x - 6y = 3\}$ **11.** 1500 packages

Graphing Calculator Explorations 3.5

1. $(2.11, 0.17)$ **3.** $(0.57, -1.97)$

Vocabulary, Readiness & Video Check 3.5

1. B **3.** A **5.** The ordered pair must be a solution of *both* equations of the system in order to be a solution of the system. **7.** Solve one equation for a variable. Next be sure to substitute this expression for the variable into the *other* equation. **9.** The break-even point occurs when revenue equals cost—money has not been lost or made; set the revenue function equal to the cost function and solve for the variable.

Exercise Set 3.5

1. yes **3.** no **5.** yes **7.** no **9.** $(2, -1)$

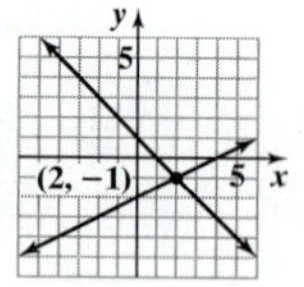

11. $(1, 2)$

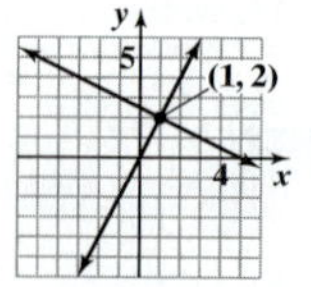

13. $\varnothing$

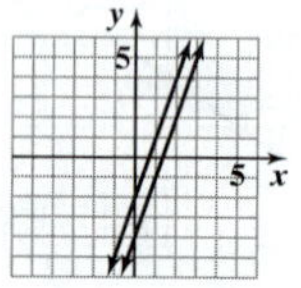

15. $(2, 8)$ **17.** $(0, -9)$ **19.** $(1, -1)$ **21.** $(-5, 3)$ **23.** $\left(\frac{5}{2}, \frac{5}{4}\right)$ **25.** $(1, -2)$ **27.** $(8, 2)$ **29.** $(7, 2)$ **31.** $\varnothing$ **33.** $\{(x, y) \mid 3x + y = 1\}$ **35.** $\left(\frac{3}{2}, 1\right)$ **37.** $(2, -1)$ **39.** $(-5, 3)$ **41.** $\{(x, y) \mid 2x + 6y = 8\}$ **43.** $\varnothing$ **45.** $\left(\frac{1}{2}, \frac{1}{5}\right)$ **47.** $(9, 9)$ **49.** $\{(x, y) \mid x = 3y + 2\}$ **51.** $\left(-\frac{1}{4}, \frac{1}{2}\right)$ **53.** $(3, 2)$ **55.** $(7, -3)$ **57.** $\varnothing$ **59.** $(3, 4)$ **61.** $(-2, 1)$ **63.** $(1.2, -3.6)$ **65.** 10 and 8 **67.** 625 units **69.** 3000 units **71.** 1280 units **73. a.** $R(x) = 450x$ **b.** $C(x) = 200x + 6000$ **c.** 24 desks **75.** 5% **77.** $\frac{1}{10}$ **79.** 1 g **81.** 15 g **83.** no solution **85.** infinite number of solutions **87.** no; answers may vary

89. 5000 DVDs; $21 **91.** supply greater than demand **93.** (1875, 4687.5) **95.** makes money **97.** for x-values greater than 1875

99. answers may vary; One possibility: $\begin{cases} -2x + y = 1 \\ x - 2y = -8 \end{cases}$ **101. a.** Mobile phone subscriptions are increasing, while landline subscriptions are decreasing. **b.** (4, 60) **c.** In the year 2004, the number of mobile phone subscriptions was about the same as the number of landline subscriptions.

103. $\left(\frac{1}{4}, 8\right)$ **105.** $\left(\frac{1}{3}, \frac{1}{2}\right)$ **107.** $\varnothing$ **109.** $\left(-\frac{1}{4}, \frac{1}{3}\right)$

Section 3.6

Practice Exercises

1. a. no **b.** yes

2. 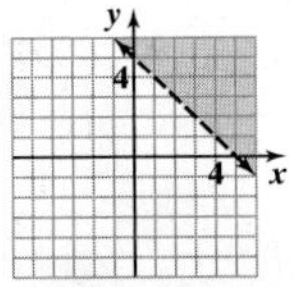

3. 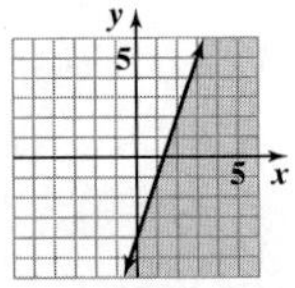

4. 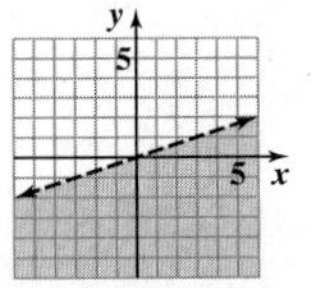

5. 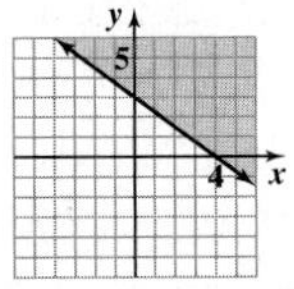

6. 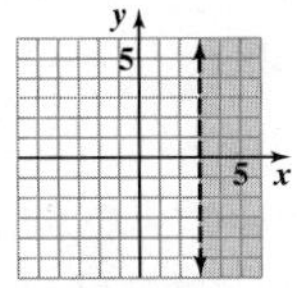

7. 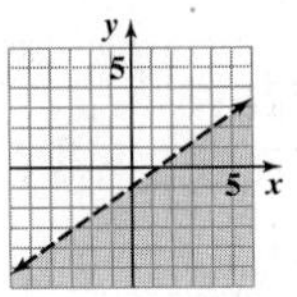

8. 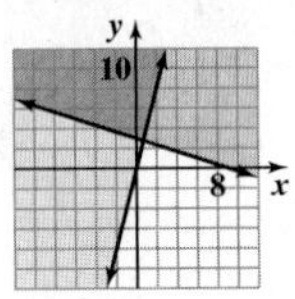

9. 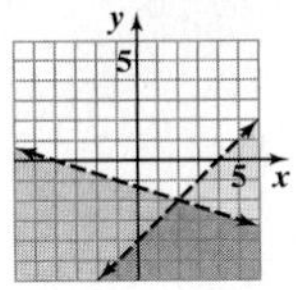

10. 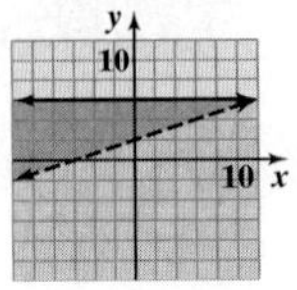

Vocabulary, Readiness & Video Check 3.6

1. linear inequality in two variables **3.** false **5.** true **7.** An ordered pair is a solution of an inequality if replacing the variables with the coordinates of the ordered pair results in a true statement. **9.** No; we can choose any test point except a point on the second inequality's own boundary line.

Exercise Set 3.6

1. no; no **3.** no; yes

5. 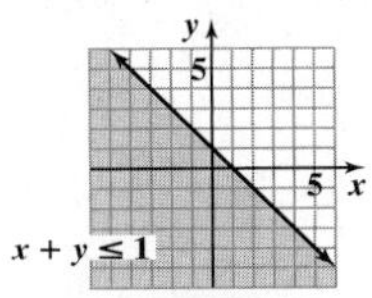

7. 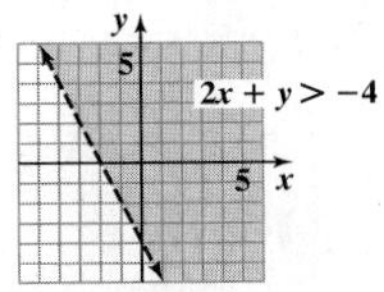

9. 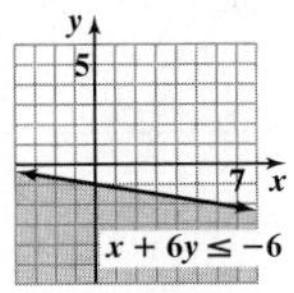

11. 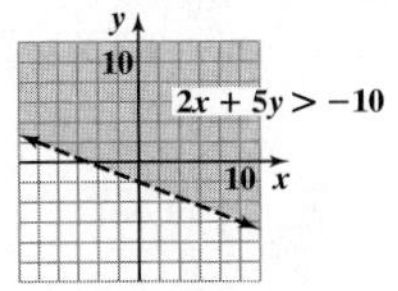

13. 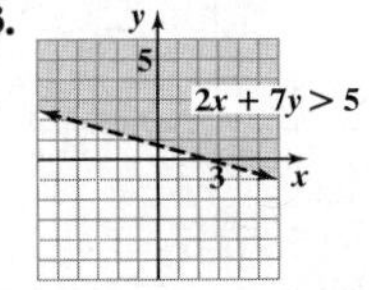

15. 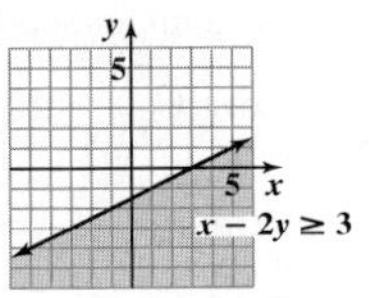

17. 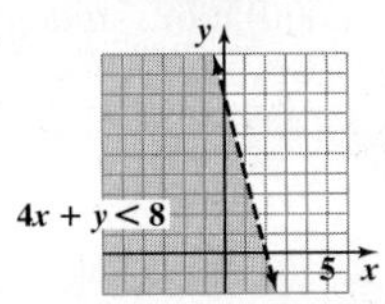

19. 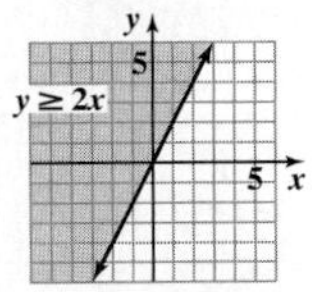

21. 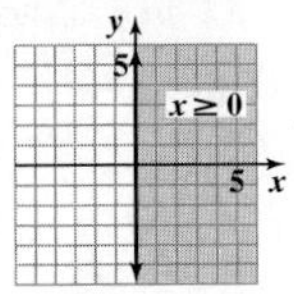

23. 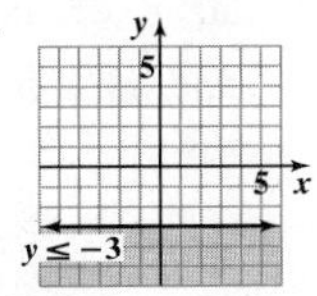

25. 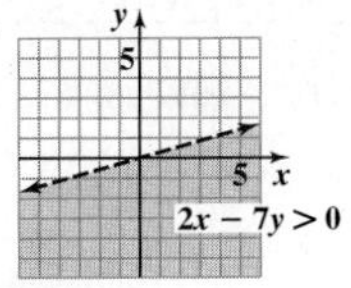

27. E **29.** C **31.** F

33. 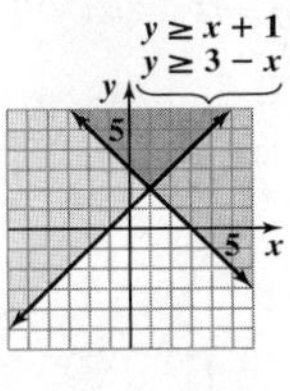

35. 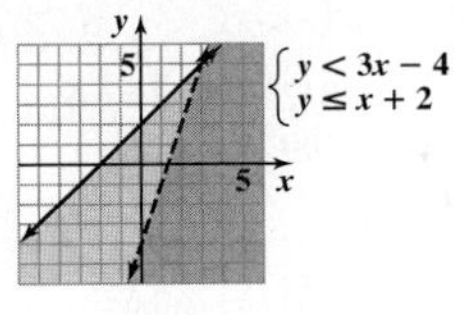

37. 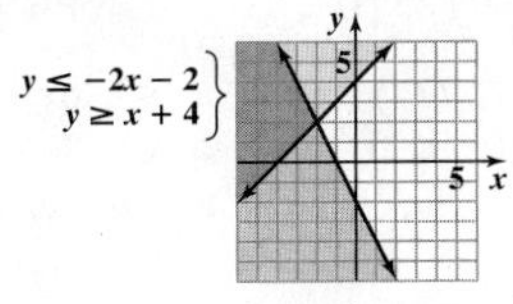

39. 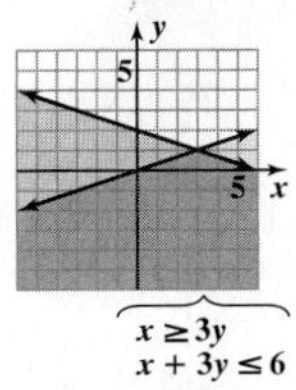

41. 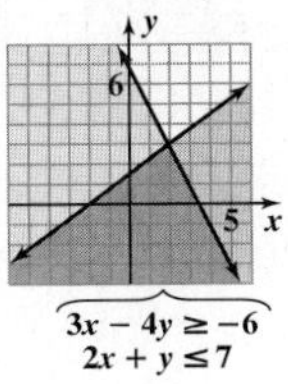

43. 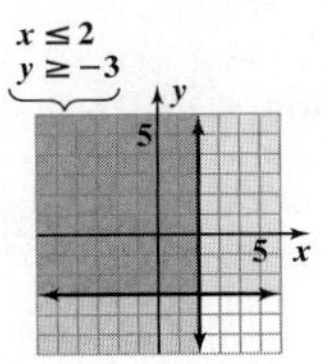

45. 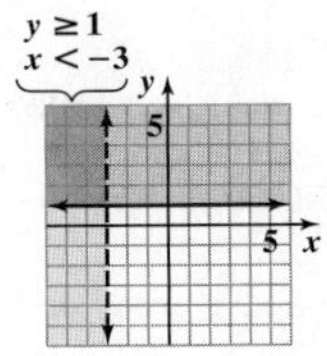

47. 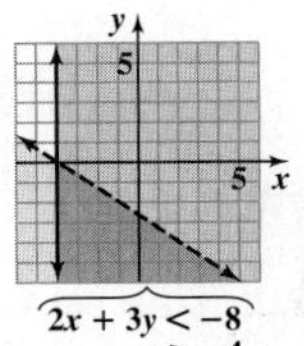

49. 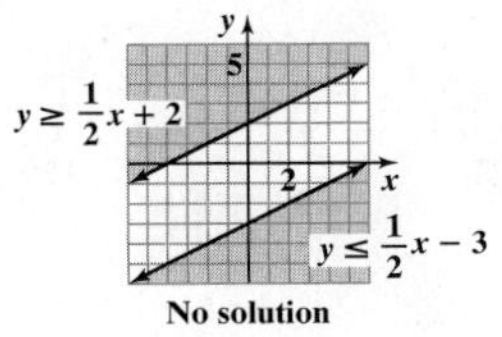

No solution

51. C **53.** D **55.** 81–90 **57.** 25 **59.** 8.3% **61.** 51.0% **63.** yes **65.** yes **67.** answers may vary

69. 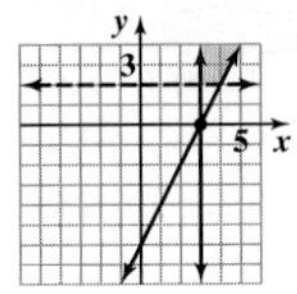

71. answers may vary

73. 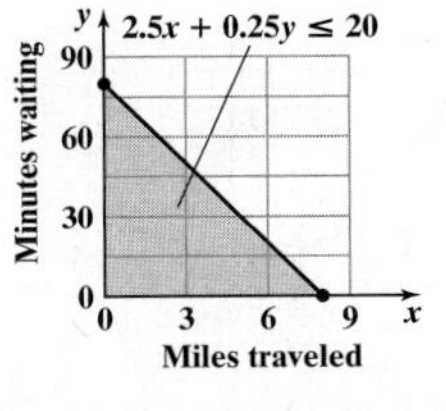

Chapter 3 Vocabulary Check

1. solution **2.** y-axis **3.** linear **4.** x-intercept **5.** standard **6.** y-intercept **7.** slope-intercept **8.** point-slope **9.** y **10.** x-axis **11.** x **12.** slope **13.** function **14.** domain **15.** range **16.** relation **17.** dependent **18.** system of linear equations **19.** consistent **20.** solution **21.** addition; substitution **22.** inconsistent **23.** independent **24.** system of linear inequalities

Chapter 3 Review

1.
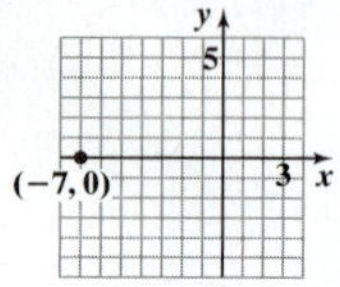

3.
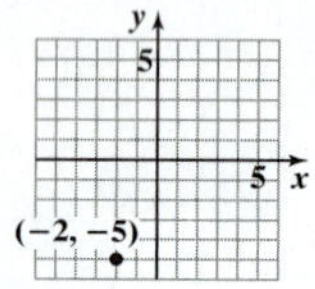

5.
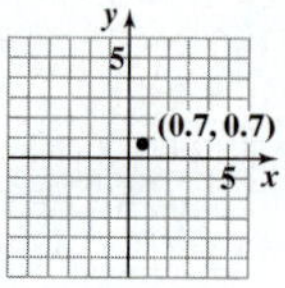

7. a. (5.00, 50), (8.50, 100), (20.00, 250), (27.00, 500) **b.**
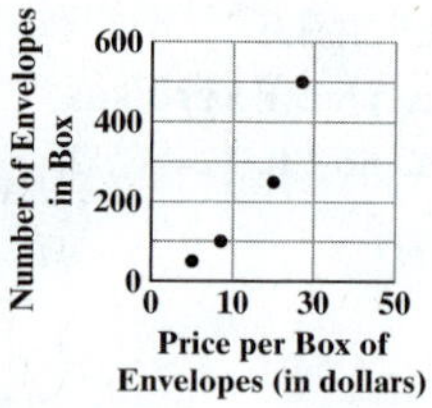

9. no; yes **11.** yes; yes **13.** (7, 44) **15.** $(-3, 0)$; $(1, 3)$; $(9, 9)$
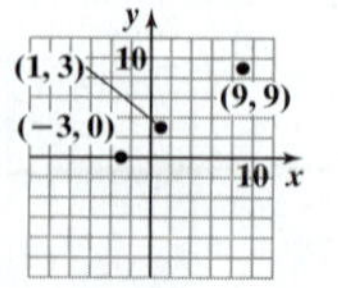

17. 2005; 2500; 7000 **19.**
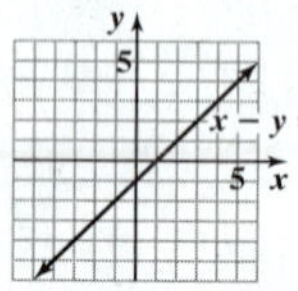

21.
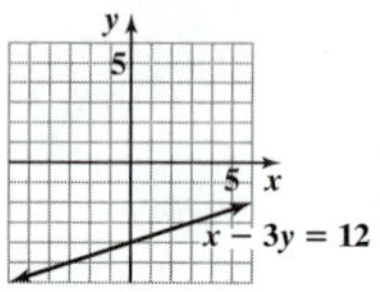

23.
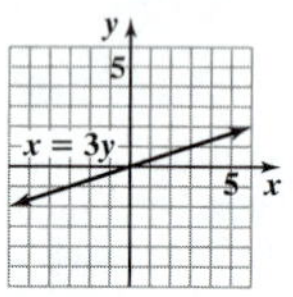

25.
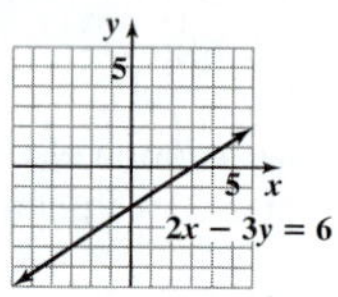

27. $(4, 0)$, $(0, -2)$ **29.** $(-2, 0)$, $(2, 0)$, $(0, 2)$, $(0, -2)$

31.
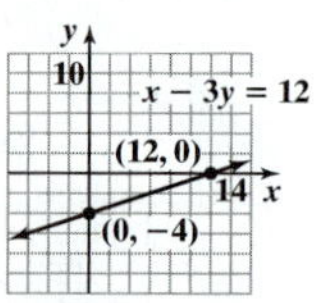

33.
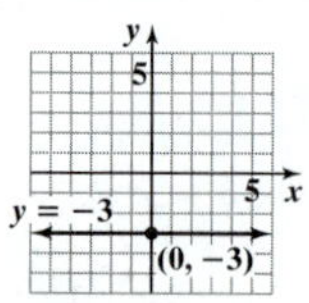

35.
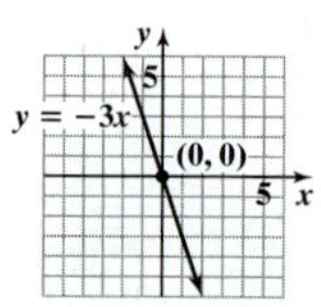

37.
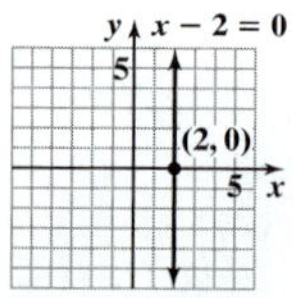

39. $m = -\frac{3}{4}$ **41.** D **43.** C **45.** $\frac{3}{4}$ **47.** 4

49. 3 **51.** 0 **53.** perpendicular **55.** neither **57.** $m = 24$ or $\frac{24}{1}$; Every 1 year, 24 thousand (24,000) more students graduate with an associate's degree. **59.** $m = -3$; $(0, 7)$ **61.** $m = 0$; $(0, 2)$ **63.**
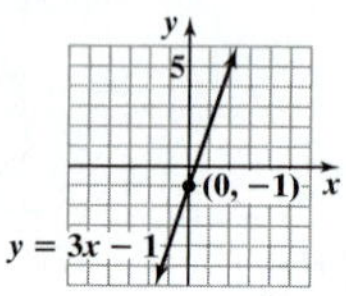

65.
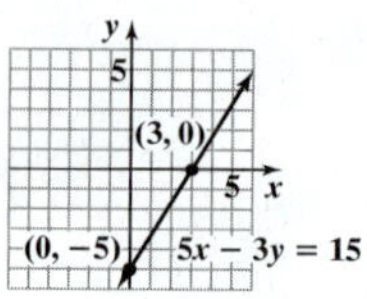

67. $y = -5x + \frac{1}{2}$ **69.** C **71.** B

73. (0, 1859) **75.** $3x + y = -5$ **77.** $y = -3$ **79.** $6x + y = 11$ **81.** $x + y = 6$ **83.** $x = 5$ **85.** $x = 6$ **87.** no **89.** yes **91.** no **93.** no **95. a.** 6 **b.** 10 **c.** 5 **97. a.** 45 **b.** −35 **c.** 0 **99.** $(-\infty, \infty)$ **101.** domain: $[-3, 5]$ range: $[-4, 2]$ **103.** domain: $\{3\}$; range: $(-\infty, \infty)$ **105. a.** no **b.** yes **c.** no **107.**
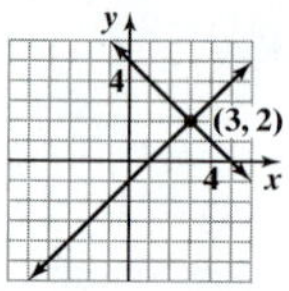

109.
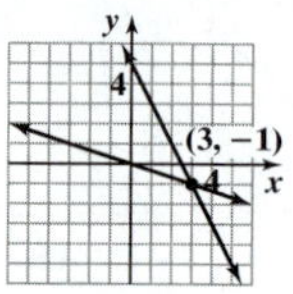

111. no solution
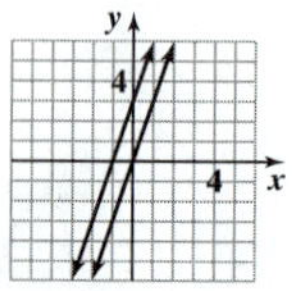

113. $(-1, 4)$ **115.** $(3, -2)$ **117.** $\{(x, y) \mid 4y = 2x + 6\}$ or $\{(x, y) \mid x - 2y = -3\}$ **119.** no solution **121.** $(-6, 2)$ **123.** $(3, 7)$

125.
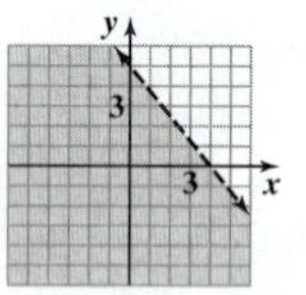

127.
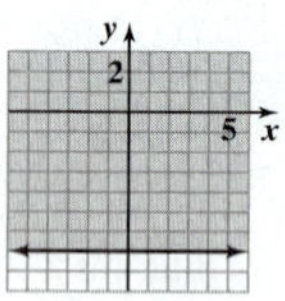

129.
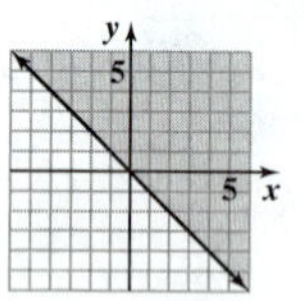

131.
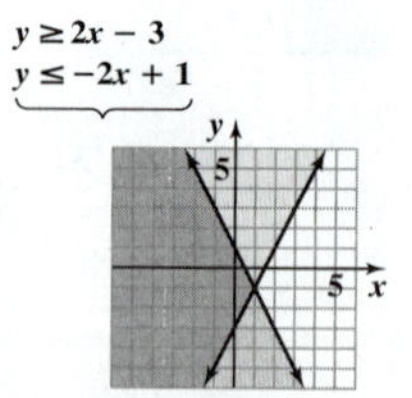

133.
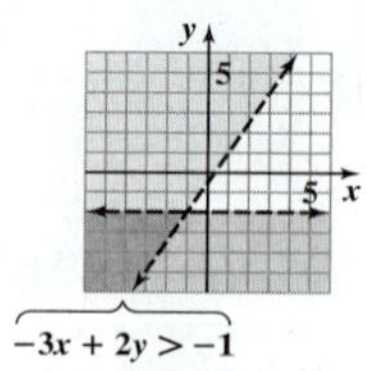

135. 7; −1; −3 **137.** $(3, 0)$; $(0, -2)$ **139.**
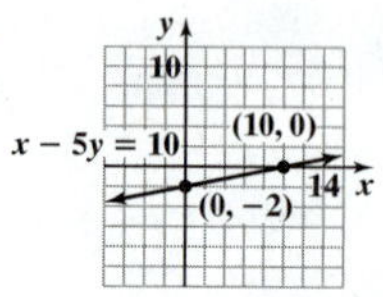

141.
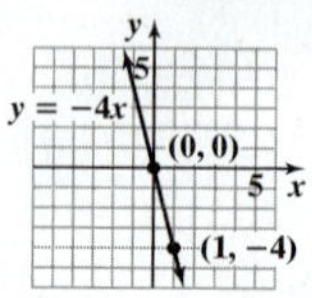

143.
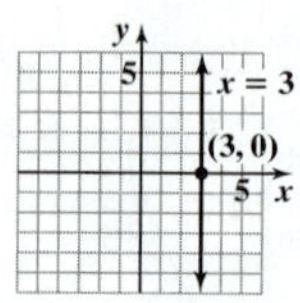

145. $m = -1$ **147.** $m = 2$

149. $m = \frac{2}{3}$; $(0, -5)$ **151.** $y = -5x + 8$ **153.** $y = -4x - 3$ **155.** $(3, 2)$ **157.** $\left(\frac{3}{2}, -3\right)$ **159.** $(-5, 2)$ **161.**
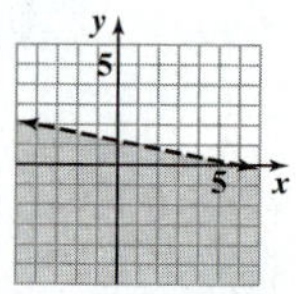

Chapter 3 Getting Ready for the Test

1. C **2.** A **3.** B **4.** B **5.** B **6.** C **7.** C **8.** B **9.** D **10.** A or C; A or C **11.** E **12.** C **13.** B **14.** D **15.** A **16.** B **17.** D **18.** C

Chapter 3 Test

1. 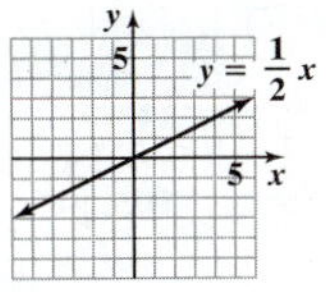

2. 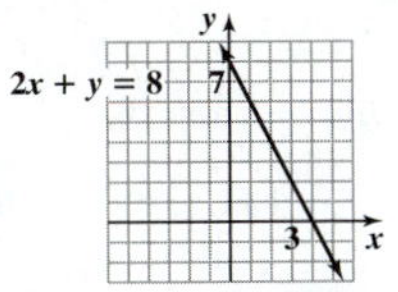

3. 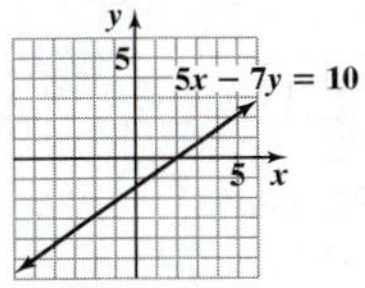

4. 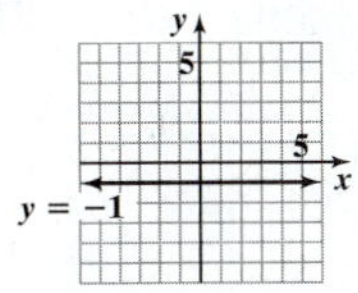

5. $\frac{2}{5}$ **6.** 0 **7.** −1 **8.** undefined

9. $m = \frac{7}{3}$; $\left(0, -\frac{2}{3}\right)$ **10.** $x + 4y = 10$ **11.** $7x + 6y = 0$ **12.** $y = -8x + 11$ **13.** $y = \frac{1}{8}x + 12$ **14.** yes **15.** no **16. a.** 0 **b.** 0 **c.** 60 **17. a.** x-intercepts: (0, 0), (4, 0); y-intercept: (0, 0) **b.** domain: $(-\infty, \infty)$; range: $(-\infty, 4]$ **18. a.** x-intercept: (2, 0); y-intercept; (0, −2) **b.** domain: $(-\infty, \infty)$; range: $(-\infty, \infty)$ **19.** (7, 20) **20. a.** (2008, 13.68), (2009, 14.1), (2010, 14.13), (2011, 13.97), (2012, 14.06), (2013, 14.06), (2014, 13.98)

b. 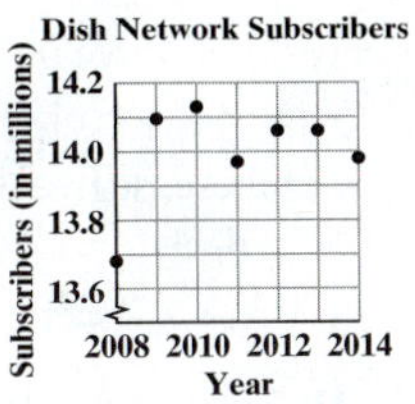

21. a. $m = -18$ or $\frac{-18}{1}$; For every 1 year 18 million fewer tickets are sold. **b.** (10, 1342), (14, 1270) **c.** $y = -18x + 1522$ **d.** In 2020, we predict that 1162 million movie tickets will be sold in the U.S. and Canada. **22.** no

23. 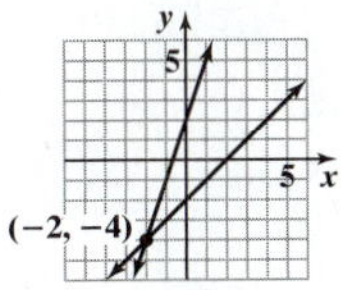

24. (−4, 1) **25.** $\left(\frac{1}{2}, -2\right)$ **26.** (20, 8) **27.** no solution **28.** (4, −5) **29.** (7, 2)

30. $\{(x, y) \mid 8x - 4y = 12\}$ or $\{(x, y) \mid y = 2x - 3\}$ **31.** 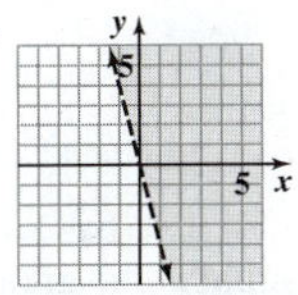

32. 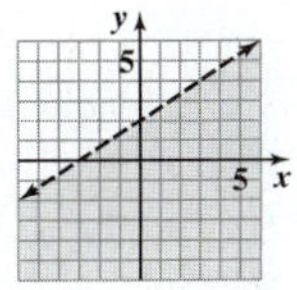

33. 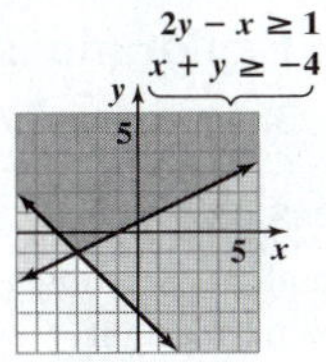

Chapter 3 Cumulative Review

1. a. < **b.** > **c.** >; Sec. 1.2, Ex. 1 **3.** $\frac{2}{39}$; Sec. R.1, Ex. 3 **5.** $\frac{8}{3}$; Sec. 1.3, Ex. 3 **7. a.** 9 **b.** $\frac{1}{16}$ **c.** −25 **d.** 25 **e.** −125 **f.** −125; Sec. 1.4, Ex. 6 **9. a.** 100 **b.** 19 **c.** −1; Sec. 1.4, Ex. 8 **11. a.** 33 **b.** −18 **c.** $\frac{1}{12}$; Sec. 1.4, Ex. 11 **13. a.** $22 + x$ **b.** $-21x$; Sec. 1.5, Ex. 3 **15. a.** −3 **b.** 22 **c.** 1 **d.** −1 **e.** $\frac{1}{7}$; Sec. 2.1, Ex. 1 **17.** 17; Sec. 2.2, Ex. 1 **19.** 6; Sec. 2.2, Ex. 5 **21.** $3x + 3$; Sec. 2.2, Ex. 10 **23.** 0; Sec. 2.3, Ex. 4 **25.** 247 Republicans, 188 Democrats; Sec. 2.4, Ex. 4 **27.** $\left(\frac{13}{7}, \infty\right)$; Sec. 2.6, Ex. 7 **29.** 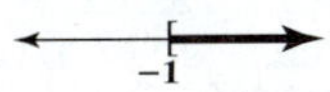

$[-1, \infty)$; Sec. 2.6, Ex. 1

31. 

$[1, 4)$; Sec. 2.6, Ex. 10 **33. a.** solution **b.** not a solution **c.** solution; Sec. 3.1, Ex. 3 **35. a.** yes **b.** yes **c.** no **d.** yes; Sec. 3.1, Ex. 6 **37.** 0; Sec. 3.2, Ex. 8 **39.** $y = \frac{1}{4}x - 3$; Sec. 3.3, Ex. 3

CHAPTER 4 EXPONENTS AND POLYNOMIALS

Section 4.1

Practice Exercises

1. a. 27 **b.** 4 **c.** 64 **d.** −64 **e.** $\frac{27}{64}$ **f.** 0.0081 **g.** 75 **2. a.** 243 **b.** $\frac{3}{8}$ **3. a.** 3^{10} **b.** y^5 **c.** z^5 **d.** x^{11} **e.** $(-2)^8$ **f.** $b^3 \cdot t^5$ **4.** $15y^7$ **5. a.** $y^{12}z^4$ **b.** $-7m^5n^{14}$ **6. a.** z^{21} **b.** 4^{18} **c.** $(-2)^{15}$ **7. a.** p^5r^5 **b.** $36b^2$ **c.** $\frac{1}{64}x^6y^3$ **d.** $81a^{12}b^{16}c^4$ **8. a.** $\frac{x^5}{y^{10}}$ **b.** $\frac{32a^{20}}{b^{15}}$ **9. a.** z^4 **b.** 25 **c.** 64 **d.** $\frac{q^5}{t^2}$ **e.** $6x^2y^2$ **10. a.** −1 **b.** 1 **c.** 1 **d.** 1 **e.** 1 **f.** 7 **11. a.** $\frac{z^2}{144}$ **b.** $64x^{18}$ **c.** y^{13} **12. a.** 63 **b.** 2 **c.** $\frac{x^6}{9y^8}$ **d.** $-z^{30}$

Vocabulary, Readiness & Video Check 4.1

1. exponent **3.** add **5.** 1 **7.** Example 4 can be written as $-4^2 = -1 \cdot 4^2$, which is similar to Example 7, $4 \cdot 3^2$, and shows why the negative sign should not be considered part of the base when there are no parentheses. **9.** Be careful not to confuse the power rule with the product rule. The power rule involves a power raised to a power (exponents are multiplied), and the product rule involves a product (exponents are added). **11.** the quotient rule

Exercise Set 4.1

1. exponent: 2; base: 3 **3.** exponent: 2; base: 4 **5.** exponent: 2; base: x **7.** 49 **9.** -5 **11.** -16 **13.** 16 **15.** 0.00001 **17.** $\frac{1}{81}$ **19.** 224 **21.** -250 **23.** 4 **25.** 135 **27.** 150 **29.** $\frac{32}{5}$ **31.** x^7 **33.** $(-3)^{12}$ **35.** $15y^5$ **37.** $x^{19}y^6$ **39.** $-72m^3n^8$ **41.** $-24z^{20}$ **43.** $20x^5$ sq ft **45.** x^{36} **47.** p^8q^8 **49.** $8a^{15}$ **51.** $x^{10}y^{15}$ **53.** $49a^4b^{10}c^2$ **55.** $\frac{r^9}{s^9}$ **57.** $\frac{m^5p^5}{n^5}$ **59.** $\frac{4x^2z^2}{y^{10}}$ **61.** $64z^{10}$ sq dm **63.** $27y^{12}$ cu ft **65.** x^2 **67.** -64 **69.** p^6q^5 **71.** $\frac{y^3}{2}$ **73.** 1 **75.** 1 **77.** -7 **79.** 2 **81.** -81 **83.** $\frac{1}{64}$ **85.** b^6 **87.** a^9 **89.** $-16x^7$ **91.** $a^{11}b^{20}$ **93.** $26m^9n^7$ **95.** z^{40} **97.** $64a^3b^3$ **99.** $36x^2y^2z^6$ **101.** $3x^4$ **103.** $81x^2y^2$ **105.** 9 **107.** $\frac{y^{15}}{8x^{12}}$ **109.** $2x^2y$ **111.** 2 **113.** $\frac{x^{18}}{4y^{22}}$ **115.** $-b^5$ **117.** 10 movies **119.** $\frac{3}{5}$ **121.** 8% **123.** C **125.** E **127.** answers may vary **129.** answers may vary **131.** 343 cu m **133.** volume **135.** answers may vary **137.** answers may vary **139.** x^{9a} **141.** a^{5b} **143.** x^{5a} **145.** $1045.85

Section 4.2
Practice Exercises

1. a. degree 3 **b.** degree 2 **c.** degree 1 **d.** degree 8 **e.** degree 0 **2. a.** trinomial, degree 2 **b.** binomial, degree 1 **c.** none of these, degree 3

3.

Term	Numerical Coefficient	Degree of Term
$-3x^3y^2$	-3	5
$4xy^2$	4	3
$-y^2$	-1	2
$3x$	3	1
-2	-2	0

4. a. 4 **b.** -21 **5.** 114 ft; 66 ft **6. a.** $-2y$ **b.** $z + 5z^3$ **c.** $14x^3$ **d.** $4a^2 - 12$ **e.** $\frac{1}{3}x^4 + \frac{11}{24}x^3 - x^2$ **7.** $-3x^2 + 5xy + 5y^2$ **8.** $x^2 + 2x + 4 + 5x + 3x^2; 4x^2 + 7x + 4$ **9. a.** $5y^2 - 3y + 2x - 9$ **b.** $-8a^2b - 3ab^2$ **10.** $7x^3 - 6x + 7$ **11.** $2x^3 - 4x^2 + 4x - 6$ **12.** $7x - 11$ **13. a.** $-5a^2 - ab + 6b^2$ **b.** $3x^2y^2 - 10xy - 4xy^2 + 5 - 6y^2$

Graphing Calculator Explorations 4.2

1. $x^3 - 4x^2 + 7x - 8$ **3.** $-2.1x^2 - 3.2x - 1.7$ **5.** $7.69x^2 - 1.26x + 5.3$

Vocabulary, Readiness & Video Check 4.2

1. binomial **3.** trinomial **5.** constant **7.** The degree of the polynomial is the greatest degree of any of its terms, so we need to find the degree of each term first. **9.** simplifying it

Exercise Set 4.2

1. 1; binomial **3.** 3; none of these **5.** 6; trinomial **7.** 8; binomial **9.** 3 **11.** 2 **13.** 57 **15.** 499 **17.** 1 **19.** $-\frac{11}{16}$ **21.** 1134 ft **23.** 1006 ft **25.** 2280 thousand **27.** $23x^2$ **29.** $12x^2 - y$ **31.** $7s$ **33.** $-1.1y^2 + 4.8$ **35.** $-\frac{7}{12}x^3 + \frac{7}{5}x^2 + 6$ **37.** $5a^2 - 9ab + 16b^2$ **39.** $-3x^2 + 10$ **41.** $-x^2 + 14$ **43.** $-2x + 9$ **45.** $2x^2 + 7x - 16$ **47.** $8t^2 - 4$ **49.** $-2z^2 - 16z + 6$ **51.** $2x^3 - 2x^2 + 7x + 2$ **53.** $-y^2 - 3y - 1$ **55.** $2x^2 + 11x$ **57.** $-16x^4 + 8x + 9$ **59.** $7x^2 + 14x + 18$ **61.** $3x - 3$ **63.** $7x^2 - 4x + 2$ **65.** $62x^2 + 5$ **67.** $7x^2 - 2x + 2$ **69.** $12x + 2$ **71.** $4y^2 + 12y + 19$ **73.** $4x^2 + 7x + x^2 + 5x; 5x^2 + 12x$ **75.** $-2a - b + 1$ **77.** $3x^2 + 5$ **79.** $6x^2 - 2xy + 19y^2$ **81.** $8r^2s + 16rs - 8 + 7r^2s^2$ **83.** $-5.42x^2 + 7.75x - 19.61$ **85.** $3.7y^4 - 0.7y^3 + 2.2y - 4$ **87.** 22.25% **89.** $\frac{3}{4}$ **91.** 0.495 g **93.** $18x + 44$ **95.** $(x^2 + 7x + 4)$ ft **97.** $(3y^2 + 4y + 11)$ m **99.** 3071 million users **101.** answers may vary **103.** answers may vary **105.** B **107.** E **109. a.** $4z$ **b.** $3z^2$ **c.** $-4z$ **d.** $3z^2$; answers may vary **111. a.** m^3 **b.** $3m$ **c.** $-m^3$ **d.** $-3m$; answers may vary **113.** $3x^{2a} + 2x^a + 0.7$ **115.** $4x^{2y} + 2x^y - 11$ **117.** $4x^2 - 3x + 6$ **119.** $-x^2 - 6x + 10$ **121.** $3x^2 - 12x + 13$ **123. a.** $2a - 3$ **b.** $-2x - 3$ **c.** $2x + 2h - 3$ **125. a.** $4a$ **b.** $-4x$ **c.** $4x + 4h$ **127.** 2; 2 **129.** 4; 3; 3; 4 **131.** $2x^2 + 4xy$

Section 4.3
Practice Exercises

1. $10y^2$ **2.** $-2z^8$ **3.** $\frac{7}{72}b^9$ **4. a.** $27x^6 + 33x$ **b.** $-12x^5 + 54x^4 - 12x^3$ **5.** $10x^2 + 11x - 6$ **6.** $25x^2 - 30xy + 9y^2$ **7.** $2y^3 + 5y^2 - 7y + 20$ **8.** $s^3 + 6s^2t + 12st^2 + 8t^3$ **9.** $5x^3 - 23x^2 + 17x - 20$ **10.** $x^5 - 2x^4 + 2x^3 - 3x^2 + 2$ **11.** $5x^4 - 3x^3 + 11x^2 + 8x - 6$

Vocabulary, Readiness & Video Check 4.3

1. c. distributive **3. c.** $(5y - 1)(5y - 1)$ **5.** No. The monomials are unlike terms. **7.** Yes. The parentheses have been removed for the vertical format, but every term in the first polynomial is still distributed to every term in the second polynomial.

Exercise Set 4.3

1. $-28n^{10}$ **3.** $-12.4x^{12}$ **5.** $-\frac{2}{15}y^3$ **7.** $-24x^8$ **9.** $6x^2 + 15x$ **11.** $-2a^2 - 8a$ **13.** $6x^3 - 9x^2 + 12x$ **15.** $-6a^4 + 4a^3 - 6a^2$ **17.** $-4x^3y + 7x^2y^2 - xy^3 - 3y^4$ **19.** $4x^4 - 3x^3 + \frac{1}{2}x^2$ **21.** $x^2 + 7x + 12$ **23.** $a^2 + 5a - 14$ **25.** $x^2 + \frac{1}{3}x - \frac{2}{9}$ **27.** $12x^4 + 25x^2 + 7$ **29.** $4y^2 - 16y + 16$ **31.** $12x^2 - 29x + 15$ **33.** $9x^4 + 6x^2 + 1$ **35. a.** $-4y^4$ **b.** $3y^2$ **c.** answers may vary **37.** $x^3 - 5x^2 + 13x - 14$ **39.** $x^4 + 5x^3 - 3x^2 - 11x + 20$ **41.** $10a^3 - 27a^2 + 26a - 12$ **43.** $x^3 + 6x^2 + 12x + 8$ **45.** $8y^3 - 36y^2 + 54y - 27$ **47.** $12x^2 - 64x - 11$ **49.** $10x^3 + 22x^2 - x - 1$ **51.** $2x^4 + 3x^3 - 58x^2 + 4x + 63$ **53.** $8.4y^7$ **55.** $-3x^3 - 6x^2 + 24x$ **57.** $2x^2 + 39x + 19$ **59.** $x^2 - \frac{2}{7}x - \frac{3}{49}$ **61.** $9y^2 + 30y + 25$ **63.** $a^3 - 2a^2 - 18a + 24$ **65.** $8x^3 - 60x^2 + 150x - 125$ **67.** $32x^3 + 48x^2 - 6x - 20$ **69.** $6x^4 - 8x^3 - 7x^2 + 22x - 12$ **71.** $(4x^2 - 25)$ sq yd **73.** $(6x^2 - 4x)$ sq in. **75.** $5a + 15a = 20a; 5a - 15a = -10a; 5a \cdot 15a = 75a^2; \frac{5a}{15a} = \frac{1}{3}$ **77.** $-3y^5 + 9y^4$, can't be simplified; $-3y^5 - 9y^4$, can't be simplified; $-3y^5 \cdot 9y^4 = -27y^9; \frac{-3y^5}{9y^4} = -\frac{y}{3}$ **79. a.** $6x + 12$ **b.** $9x^2 + 36x + 35$; answers may vary **81.** $13x - 7$ **83.** $30x^2 - 28x + 6$ **85.** $-7x + 5$ **87.** $x^2 + 3x$ **89.** $x^2 + 5x + 6$ **91.** $11a$ **93.** $25x^2 + 4y^2$ **95. a.** $a^2 - b^2$ **b.** $4x^2 - 9y^2$ **c.** $16x^2 - 49$ **d.** answers may vary **97.** $(x^2 + 6x + 5)$ sq units

Section 4.4

Practice Exercises

1. $x^2 - 3x - 10$ **2.** $4x^2 - 13x + 9$ **3.** $9x^2 + 42x - 15$ **4.** $16x^2 - 8x + 1$ **5. a.** $b^2 + 6b + 9$ **b.** $x^2 - 2xy + y^2$ **c.** $9y^2 + 12y + 4$ **d.** $a^4 - 10a^2b + 25b^2$ **6. a.** $3x^2 - 75$ **b.** $16b^2 - 9$ **c.** $x^2 - \frac{4}{9}$ **d.** $25s^2 - t^2$ **e.** $4y^2 - 9z^4$ **7. a.** $4x^2 - 21x - 18$ **b.** $49b^2 - 28b + 4$ **c.** $x^2 - 0.16$ **d.** $3x^6 - \frac{9}{7}x^4 + \frac{2}{7}x^2 - \frac{6}{49}$ **e.** $x^3 + 6x^2 + 3x - 2$

Vocabulary, Readiness & Video Check 4.4

1. false **3.** false **5.** a binomial times a binomial **7.** Multiplying gives you four terms, and the two like terms will always subtract out.

Exercise Set 4.4

1. $x^2 + 7x + 12$ **3.** $x^2 + 5x - 50$ **5.** $5x^2 + 4x - 12$ **7.** $20y^2 - 125y + 30$ **9.** $6x^2 + 13x - 5$ **11.** $x^2 + \frac{1}{3}x - \frac{2}{9}$ **13.** $x^2 + 4x + 4$ **15.** $4x^2 - 4x + 1$ **17.** $9a^2 - 30a + 25$ **19.** $25x^2 + 90x + 81$ **21.** $a^2 - 49$ **23.** $9x^2 - 1$ **25.** $9x^2 - \frac{1}{4}$ **27.** $81x^2 - y^2$ **29.** $4x^2 - 0.01$ **31.** $a^2 + 9a + 20$ **33.** $a^2 + 14a + 49$ **35.** $12a^2 - a - 1$ **37.** $x^2 - 4$ **39.** $9a^2 + 6a + 1$ **41.** $4x^3 - x^2y^4 + 4xy - y^5$ **43.** $x^3 - 3x^2 - 17x + 3$ **45.** $4a^2 - 12a + 9$ **47.** $25x^2 - 36z^2$ **49.** $x^{10} - 8x^5 + 15$ **51.** $x^2 - 0.64$ **53.** $a^7 + 11a^4 - 3a^3 - 33$ **55.** $3x^2 - 12x + 12$ **57.** $6b^2 - b - 35$ **59.** $49p^2 - 64$ **61.** $\frac{1}{9}a^4 - 49$ **63.** $15x^4 - 5x^3 + 10x^2$ **65.** $4r^2 - 9s^2$ **67.** $9x^2 - 42xy + 49y^2$ **69.** $16x^2 - 25$ **71.** $64x^2 + 64x + 16$ **73.** $a^2 - \frac{1}{4}y^2$ **75.** $\frac{1}{25}x^2 - y^2$ **77.** $3a^3 + 2a^2 + 1$ **79.** $(2x + 1)(2x + 1)$ sq ft or $(4x^2 + 4x + 1)$ sq ft **81.** $(2, 0), (0, -2)$ **83.** $(-4, 0), (-2, 0), (3, 0), (0, 4)$ **85.** $(-2, 0), (2, 0), (0, 3)$ **87.** $\frac{1}{3}$ **89.** 1 **91.** C **93.** D **95.** 2; 2 **97.** $(x^4 - 3x^2 + 1)$ sq m **99.** $(24x^2 - 32x + 8)$ sq m **101.** $(x^2 + 10x + 25)$ sq units **103.** answers may vary **105.** answers may vary **107.** answers may vary **109.** $x^2 + 2xy + y^2 - 9$ **111.** $a^2 - 6a + 9 - b^2$

Integrated Review

1. $35x^5$ **2.** $32y^9$ **3.** -16 **4.** 16 **5.** $2x^2 - 9x - 5$ **6.** $3x^2 + 13x - 10$ **7.** $3x - 4$ **8.** $4x + 3$ **9.** $7x^6y^2$ **10.** $\frac{10b^6}{7}$ **11.** $144m^{14}n^{12}$ **12.** $64y^{27}z^{30}$ **13.** $48y^2 - 27$ **14.** $98x^2 - 2$ **15.** $x^{63}y^{45}$ **16.** $27x^{27}$ **17.** $2x^2 - 2x - 6$ **18.** $6x^2 + 13x - 11$ **19.** $2.5y^2 - 6y - 0.2$ **20.** $8.4x^2 - 6.8x - 5.7$ **21.** $x^2 + 8xy + 16y^2$ **22.** $y^2 - 18yz + 81z^2$ **23.** $2x + 8y$ **24.** $2y - 18z$ **25.** $7x^2 - 10xy + 4y^2$ **26.** $-a^2 - 3ab + 6b^2$ **27.** $x^3 + 2x^2 - 16x + 3$ **28.** $x^3 - 2x^2 - 5x - 2$ **29.** $6x^5 + 20x^3 - 21x^2 - 70$ **30.** $20x^7 - 4x^4 + 25x^3 - 5$ **31.** $2x^3 - 19x^2 + 44x - 7$ **32.** $5x^3 + 9x^2 - 17x + 3$ **33.** cannot simplify **34.** $25x^3y^3$ **35.** $125x^9$ **36.** $\frac{x^3}{y^3}$ **37.** $2x$ **38.** x^2

Section 4.5

Practice Exercises

1. a. $\frac{1}{125}$ **b.** $\frac{3}{y^4}$ **c.** $\frac{5}{6}$ **d.** $\frac{1}{25}$ **e.** $\frac{1}{x^5}$ **2. a.** s^5 **b.** 8 **c.** $\frac{y^5}{x^7}$ **d.** $\frac{9}{64}$ **3. a.** $\frac{1}{x^5}$ **b.** $5y^7$ **c.** z^5 **d.** $\frac{81}{25}$ **4. a.** $\frac{b^{15}}{a^{20}}$ **b.** x^{10} **c.** $\frac{q^2}{25p^{16}}$ **5. a.** 7×10^{-6} **b.** 2.07×10^7 **c.** 4.3×10^{-3} **d.** 8.12×10^8 **6. a.** 0.000367 **b.** 8,954,000 **c.** 0.00002009 **d.** 4054 **7. a.** 17,000,000,000 **b.** 1.7×10^{10} **8. a.** 4000 **b.** 20,000,000,000

Calculator Explorations 4.5

1. 5.31 EE 3 **3.** 6.6 EE −9 **5.** 1.5×10^{13} **7.** 8.15×10^{19}

Vocabulary, Readiness & Video Check 4.5

1. b. $\frac{1}{x^3}$ **3. c.** scientific notation **5.** A negative exponent has nothing to do with the sign of the simplified result. **7.** When you move the decimal point to the left, the sign of the exponent will be positive; when you move the decimal point to the right, the sign of the exponent will be negative. **9.** the quotient rule

Exercise Set 4.5

1. $\frac{1}{64}$ **3.** $\frac{1}{81}$ **5.** $\frac{7}{x^3}$ **7.** 32 **9.** -64 **11.** $\frac{8}{15}$ **13.** p^3 **15.** $\frac{q^4}{p^5}$ **17.** $\frac{1}{x^3}$ **19.** z^3 **21.** $\frac{4}{9}$ **23.** $-p^4$ **25.** -2 **27.** x^4 **29.** p^4 **31.** m^{11} **33.** r^6 **35.** $\frac{1}{x^{15}y^9}$ **37.** $\frac{1}{x^4}$ **39.** $\frac{1}{a^2}$ **41.** $4k^3$ **43.** $3m$ **45.** $\frac{a^{30}}{b^{12}}$ **47.** $\frac{z^2}{4}$ **49.** $\frac{1}{32x^5}$ **51.** 7.8×10^4 **53.** 1.67×10^{-6} **55.** 6.35×10^{-3} **57.** 1.16×10^6 **59.** 2×10^9 **61.** 2.4×10^3 **63.** 0.0000000008673 **65.** 0.033 **67.** 20,320 **69.** 700,000,000 **71.** 9,460,000,000,000 **73.** $\$7.14 \times 10^{11}$ **75.** $863,000,000,000 **77. a.** 323,800,000 **b.** 3.238×10^8 **79.** 0.000036 **81.** 0.0000000000000000028 **83.** 0.0000005 **85.** 200,000 **87.** Under 20 **89.** 6500 **91.** 47.3% **93.** $\frac{27}{x^6z^3}$ cu in. **95.** $9a^{13}$ **97.** -5 **99.** answers may vary **101. a.** 1.3×10^1 **b.** 4.4×10^7 **c.** 6.1×10^{-2} **103.** $-\frac{4a^5}{b}$ **105.** $-\frac{6x^2}{y^3}$ **107.** $\frac{1}{x^{10}y^6}$ **109.** $\frac{49a^4}{b^6}$ **111.** $a^{24}b^8$ **113.** x^9y^{19} **115.** $-\frac{y^8}{8x^2}$ **117.** $-\frac{6x}{7y^2}$ **119.** $\frac{25b^{33}}{a^{16}}$ **121.** answers may vary **123.** a^m **125.** $27y^{6z}$ **127.** -394.5 **129.** 1.3 sec

Chapter 4 Vocabulary Check

1. term **2.** FOIL **3.** trinomial **4.** degree of a polynomial **5.** binomial **6.** coefficient **7.** degree of a term **8.** monomial **9.** polynomials **10.** distributive

Chapter 4 Review

1. base: 7; exponent: 9 **3.** base: 5; exponent: 4 **5.** 512 **7.** -36 **9.** 1 **11.** y^9 **13.** $-6x^{11}$ **15.** x^8 **17.** $81y^{24}$ **19.** x^5 **21.** a^4b^3 **23.** $\frac{x^3y^4}{4}$ **25.** $40a^{19}$ **27.** 3 **29.** b **31.** 7 **33.** 8 **35.** 5 **37.** 5 **39.** 4000 ft; 3984 ft; 3856 ft; 3600 ft **41.** $2a^2$ **43.** $15a^2 + 4a$ **45.** $-6a^2b - 3b^2 - q^2$ **47.** $8x^2 + 3x + 6$ **49.** $-7y^2 - 1$ **51.** $-5x^2 + 5x + 1$ **53.** $4x - 13y$ **55.** $8a + 28$ **57.** $-7x^3 - 35x$ **59.** $-6a^4 + 8a^2 - 2a$ **61.** $2x^2 - 12x - 14$ **63.** $x^2 - 18x + 81$ **65.** $4a^2 + 27a - 7$ **67.** $25x^2 + 20x + 4$ **69.** $x^4 + 7x^3 + 4x^2 + 23x - 35$ **71.** $x^4 + 4x^3 + 4x^2 - 16$ **73.** $x^3 + 21x^2 + 147x + 343$ **75.** $x^2 + 14x + 49$ **77.** $9x^2 - 42x + 49$ **79.** $25x^2 - 90x + 81$ **81.** $49x^2 - 16$ **83.** $4x^2 - 36$ **85.** $(9x^2 - 6x + 1)$ sq m **87.** $\frac{1}{49}$ **89.** $\frac{2}{x^4}$ **91.** 125 **93.** $\frac{17}{16}$ **95.** x^8 **97.** r **99.** c^4 **101.** $\frac{1}{x^6y^{13}}$ **103.** a^{11m} **105.** $27x^3y^{6z}$ **107.** 2.7×10^{-4} **109.** 8.08×10^7 **111.** 2.5×10^9 **113.** 867,000 **115.** 0.00086 **117.** 1,431,280,000,000,000 **119.** 0.016 **121.** $-\frac{1}{8}$ **123.** $\frac{2x^6}{3}$ **125.** $\frac{x^{16}}{16y^{12}}$ **127.** $11x - 5$ **129.** $5y^2 - 3y - 1$ **131.** $28x^3 + 12x$ **133.** $x^3 + x^2 - 18x + 18$ **135.** $25x^2 + 40x + 16$

Chapter 4 Getting Ready for the Test

1. C **2.** A **3.** E **4.** D **5.** F **6.** C **7.** E **8.** I **9.** C **10.** D **11.** C **12.** B **13.** F **14.** D

Chapter 4 Test

1. 32 **2.** 81 **3.** -81 **4.** $\frac{1}{64}$ **5.** $-15x^{11}$ **6.** y^5 **7.** $\frac{1}{r^5}$ **8.** $\frac{y^{14}}{x^2}$ **9.** $\frac{1}{6xy^8}$ **10.** 5.63×10^5 **11.** 8.63×10^{-5} **12.** 0.0015 **13.** 62,300 **14.** 0.036 **15. a.** 4, 3; 7, 3; 1, 4; -2, 0 **b.** 4 **16.** $-2x^2 + 12xy + 11$ **17.** $16x^3 + 7x^2 - 3x - 13$ **18.** $-3x^3 + 5x^2 + 4x + 5$ **19.** $x^3 + 8x^2 + 3x - 5$ **20.** $3x^3 + 22x^2 + 41x + 14$ **21.** $6x^4 - 9x^3 + 21x^2$ **22.** $3x^2 + 16x - 35$ **23.** $9x^2 - \frac{1}{25}$ **24.** $16x^2 - 16x + 4$ **25.** $64x^2 + 48x + 9$ **26.** $x^4 - 81b^2$ **27.** 1001 ft; 985 ft; 857 ft; 601 ft **28.** $(4x^2 - 9)$ sq in. **29. a.** 960 ft **b.** 953.44 ft

Chapter 4 Cumulative Review

1. a. true **b.** true **c.** false **d.** true; Sec. 1.2, Ex. 2 **3. a.** $\frac{64}{25}$ **b.** $\frac{1}{20}$ **c.** $\frac{5}{4}$; Sec. R.1, Ex. 4 **5. a.** 9 **b.** 125 **c.** 16 **d.** 7 **e.** $\frac{9}{49}$; Sec. 1.3, Ex. 1 **7.** yes; Sec. 1.3, Ex. 7 **9.** -3; Sec. 1.4, Ex. 2g **11.** $\frac{14}{3}$; Sec. 1.3, Ex. 5 **13. a.** $(5 + 4) + 6$ **b.** $-1 \cdot (2 \cdot 5)$; Sec. 1.5, Ex. 2 **15. a.** $22 + x$ **b.** $-21x$; Sec. 1.5, Ex. 3 **17. a.** $15x + 10$ **b.** $-2y - 0.6z + 2$ **c.** $-9x - y + 2z - 6$; Sec. 2.1, Ex. 5 **19.** 17; Sec. 2.2, Ex. 1 **21.** 6; Sec. 2.2, Ex. 5 **23.** 2; Sec. 2.3, Ex. 1 **25.** -10; Sec. 2.4, Ex. 1 **27.** $(2, 4]$ 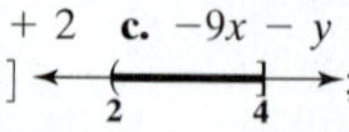 ; Sec. 2.6, Ex. 9 **29. a.** $(0, 12)$ **b.** $(2, 6)$ **c.** $(-1, 15)$; Sec. 3.1, Ex. 4

31. 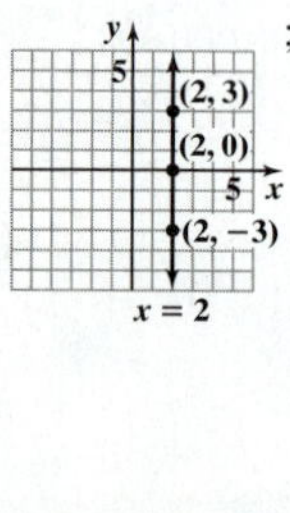

; Sec. 3.2, Ex. 9 **33.** $m = 0$; Sec. 3.2, Ex. 8 **35.** 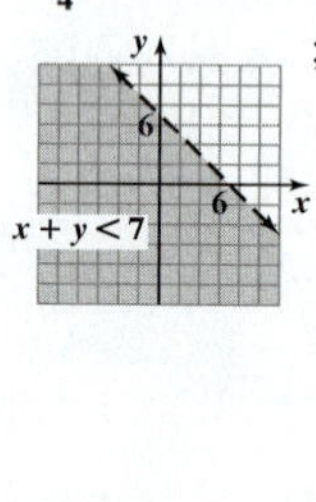

; Sec. 3.6, Ex. 2 **37.** $-6x^7$; Sec. 4.1, Ex. 4 **39.** $12x^3 - 12x^2 - 9x + 2$; Sec. 4.2, Ex. 10 **41.** $4x^2 - 4xy + y^2$; Sec. 4.3, Ex. 6 **43. a.** $\frac{1}{9}$ **b.** $\frac{3}{4}$ **c.** $\frac{1}{16}$; Sec. 4.5, parts of Ex. 1

CHAPTER 5 FACTORING POLYNOMIALS AND SOLVING QUADRATIC EQUATIONS

Section 5.1

Practice Exercises

1. a. 6 **b.** 1 **c.** 4 **2. a.** y^4 **b.** x **3. a.** $5y^2$ **b.** x^2 **c.** a^2b^2 **4. a.** $4(t + 3)$ **b.** $y^4(y^4 + 1)$ **5.** $8b^2(-b^4 + 2b^2 - 1)$ or $-8b^2(b^4 - 2b^2 + 1)$ **6.** $5x(x^3 - 4)$ **7.** $\frac{1}{9}z^3(5z^2 + z - 2)$ **8.** $4ab^3(2ab - 5a^2 + 3)$ **9.** $(y - 2)(8 + x)$ **10.** $(p + q)(7xy^3 - 1)$ **11.** $(x + 3)(y + 4)$ **12.** $(5x - 3)(8x^2 + 3)$ **13.** $(2x + 3y)(y - 1)$ **14.** $(7a + 5)(a^2 + 1)$ **15.** $(y - 3)(4x - 5)$ **16.** cannot be factored by grouping **17.** $3(x - a)(y - 2a)$

Vocabulary, Readiness & Video Check 5.1

1. factors **3.** least **5.** false **7.** The GCF of a list of numbers is the largest number that is a factor of all numbers in the list. **9.** When factoring out a GCF, the number of terms in the other factor should have the same number of terms as your original polynomial.

Exercise Set 5.1

1. 4 **3.** 6 **5.** 1 **7.** y^2 **9.** z^7 **11.** xy^2 **13.** 7 **15.** $4y^3$ **17.** $5x^2$ **19.** $3x^3$ **21.** $9x^2y$ **23.** $10a^6b$ **25.** $3(a + 2)$ **27.** $15(2x - 1)$ **29.** $x^2(x + 5)$ **31.** $2y^3(3y + 1)$ **33.** $4(x - 2y + 1)$ **35.** $3x(2x^2 - 3x + 4)$ **37.** $a^2b^2(a^5b^4 - a + b^3 - 1)$ **39.** $4(2x^5 + 4x^4 - 5x^3 + 3)$ **41.** $\frac{1}{3}x(x^3 + 2x^2 - 4x^4 + 1)$ **43.** $(x^2 + 2)(y + 3)$ **45.** $(y + 4)(z - 3)$ **47.** $(z^2 - 6)(r + 1)$ **49.** $-2(x + 7)$ **51.** $-x^5(2 - x^2)$ **53.** $-3a^2(a^2 - 3a + 1)$ **55.** $(x + 2)(x^2 + 5)$ **57.** $(x + 3)(5 + y)$ **59.** $(3x - 2)(2x^2 + 5)$ **61.** $(5m^2 + 6n)(m + 1)$ **63.** $(y - 4)(2 + x)$ **65.** $(2x - 1)(x^2 + 4)$ **67.** not factorable by grouping **69.** $(x - 2y)(4x - 3)$ **71.** $(5q - 4p)(q - 1)$ **73.** $x(x^2 + 1)(2x + 5)$ **75.** $2(2y - 7)(3x^2 - 1)$ **77.** $2x(16y - 9x)$ **79.** $(x + 2)(y - 3)$ **81.** $7xy(2x^2 + x - 1)$ **83.** $(4x - 1)(7x^2 + 3)$ **85.** $-8x^8y^5(5y + 2x)$ **87.** $3(2a + 3b^2)(a + b)$ **89.** $x^2 + 7x + 10$ **91.** $b^2 - 3b - 4$ **93.** 2, 6 **95.** $-1, -8$ **97.** $-2, 5$ **99.** b **101.** factored **103.** not factored **105.** answers may vary **107.** answers may vary **109. a.** 22% **b.** 70% **c.** $0.6(x^2 - x + 6)$ **111.** $12x^3 - 2x; 2x(6x^2 - 1)$ **113.** $(n^3 - 6)$units **115.** $(x^n + 2)(x^n + 3)$ **117.** $(3x^n - 5)(x^n + 7)$

Section 5.2

Practice Exercises

1. $(x + 2)(x + 3)$ **2.** $(x + 7)(x - 2)$ **3.** prime polynomial **4.** $(x^2 + 12)(x^2 + 1)$ **5.** $(x - 6)(x - 8)$ **6.** $3y^2(y - 7)(y + 1)$ **7.** $(2x + 5)(x + 3)$ **8.** $(5x - 4)(3x - 2)$ **9.** $(7x - y)(3x + 2y)$ **10.** $x(3x + 2)(x + 5)$ **11.** $-1(4x - 3)(2x + 1)$ **12.** $(5x + 2y)^2$ **13.** $(5x + 1)(x + 12)$ **14.** $(4x - 5)(3x - 1)$ **15.** $2(5x + 1)(3x - 2)$ **16.** $(4x + 3)^2$

Vocabulary, Readiness & Video Check 5.2

1. a **3.** b **5.** Check by multiplying. If you get a middle term of $2x$, not $-2x$, switch the signs of your factors of -24. **7.** Write down the factors of the first and last terms. Try various combinations of these factors and look at the sum of the outer and inner products to see if you get the middle term of the trinomial. If not, try another combination of factors.

Exercise Set 5.2

1. $(x + 3)(x + 6)$ **3.** $(x - 8)(x - 4)$ **5.** $(x + 12)(x - 2)$ **7.** $(x - 6)(x + 4)$ **9.** $3(x - 2)(x - 4)$ **11.** $4z(x + 2)(x + 5)$ **13.** $2(x^2 - 12x - 32)$ **15.** $(5x + 1)(x + 3)$ **17.** $(2x - 3)(x - 4)$ **19.** prime polynomial **21.** $(2x - 3)^2$ **23.** $2(3x - 5)(2x + 5)$ **25.** $y^2(3y + 5)(y - 2)$ **27.** $2x(3x^2 + 4x + 12)$ **29.** $(2x + y)(x - 3y)$ **31.** $2(7y + 2)(2y + 1)$ **33.** $(2x - 3)(x + 9)$ **35.** $(x^2 + 3)(x^2 - 2)$ **37.** $(5x^2 - 3)(x^2 + 5)$ **39. a.** 9, 2 **b.** $9x + 2x$ **c.** $(3x + 1)(2x + 3)$ **41. a.** $-20, -3$ **b.** $-20x - 3x$ **c.** $(3x - 4)(5x - 1)$ **43.** $(x - 27)(x + 3)$ **45.** $(x - 18)(x + 3)$ **47.** $3(x - 1)^2$ **49.** $(3x + 1)(x - 2)$ **51.** $(4x - 3)(2x - 5)$ **53.** $3x^2(2x + 1)(3x + 2)$ **55.** $(x + 7z)(x + z)$ **57.** $(x - 4)(x + 3)$ **59.** $3(a + 2b)^2$ **61.** prime polynomial **63.** $(2x + 13)(x + 3)$ **65.** $(3x - 2)(2x - 15)$ **67.** $(x^2 - 6)(x^2 + 1)$ **69.** $x(3x + 1)(2x - 1)$ **71.** $(4a - 3b)(3a - 5b)$ **73.** $(3x + 5)^2$ **75.** $y(3x - 8)(x - 1)$ **77.** $2(x + 3)(x - 2)$ **79.** $(x + 2)(x - 7)$ **81.** $(2x^3 - 3)(x^3 + 3)$ **83.** $2x(6y^2 - z)^2$ **85.** $2xy(x + 3)(x - 2)$ **87.** $(x + 5y)(x + y)$ **89.** Facebook **91.** answers may vary **93.** no **95.** answers may vary **97.** $4x^2 + 21x + 5$; $(4x + 1)(x + 5)$ **99.** $\pm 5, \pm 7$ **101.** $x(x + 4)(x - 2)$ **103. a.** 576 ft; 672 ft; 640 ft; 480 ft **b.** answers may vary **c.** $-16(t + 4)(t - 9)$ **105.** $(x^n + 2)(x^n + 8)$ **107.** $(x^n - 6)(x^n + 3)$ **109.** $x^2(x + 5)(x + 1)$ **111.** $3x(5x - 1)(2x + 1)$

Section 5.3

Practice Exercises

1. $(x + 9)(x - 9)$ **2. a.** $(3x - 1)(3x + 1)$ **b.** $(6a - 7b)(6a + 7b)$ **c.** $\left(p + \frac{5}{6}\right)\left(p - \frac{5}{6}\right)$ **3.** $(p^2 - q^5)(p^2 + q^5)$ **4. a.** $(z^2 + 9)(z + 3)(z - 3)$ **b.** prime polynomial **5.** $y(6y + 5)(6y - 5)$ **6.** $5(4y^2 + 1)(2y + 1)(2y - 1)$ **7.** $-1(3x + 10)(3x - 10)$ or $(10 + 3x)(10 - 3x)$ **8.** $(x + 4)(x^2 - 4x + 16)$ **9.** $(x - 5)(x^2 + 5x + 25)$ **10.** $(3y + 1)(9y^2 - 3y + 1)$ **11.** $4(2x - 5y)(4x^2 + 10xy + 25y^2)$

Graphing Calculator Explorations 5.3

	$x^2 - 2x + 1$	$x^2 - 2x - 1$	$(x - 1)^2$
$x = 5$	16	14	16
$x = -3$	16	14	16
$x = 2.7$	2.89	0.89	2.89
$x = -12.1$	171.61	169.61	171.61
$x = 0$	1	−1	1

Vocabulary, Readiness & Video Check 5.3

1. difference of two cubes **3.** sum of two cubes **5.** $(7x)^2$ **7.** $(2y)^3$ **9.** In order to recognize the binomial as a difference of squares and also to identify the terms to use in the special factoring formula. **11.** First rewrite the original binomial with terms writtten as cubes. Answers will then vary depending on your interpretation.

Exercise Set 5.3

1. $(x + 2)(x - 2)$ **3.** $(9p + 1)(9p - 1)$ **5.** $(5y - 3)(5y + 3)$ **7.** $(11m + 10n)(11m - 10n)$ **9.** $(xy - 1)(xy + 1)$ **11.** $\left(x - \frac{1}{2}\right)\left(x + \frac{1}{2}\right)$ **13.** $-1(2r + 1)(2r - 1)$ or $(1 - 2r)(1 + 2r)$ **15.** prime **17.** $(x - 6)(x + 6)$ or $-1(6 + x)(6 - x)$ **19.** $(m^2 + 1)(m + 1)(m - 1)$ **21.** $(m^2 + n^9)(m^2 - n^9)$ **23.** $(x + 5)(x^2 - 5x + 25)$ **25.** $(2a - 1)(4a^2 + 2a + 1)$ **27.** $(m + 3n)(m^2 - 3mn + 9n^2)$ **29.** $5(k + 2)(k^2 - 2k + 4)$ **31.** $(xy - 4)(x^2y^2 + 4xy + 16)$ **33.** $2(5r - 4t)(25r^2 + 20rt + 16t^2)$ **35.** $(r + 8)(r - 8)$ **37.** $(x + 13y)(x - 13y)$ **39.** $(3 - t)(9 + 3t + t^2)$ **41.** $2(3r + 2)(3r - 2)$ **43.** $x(3y + 2)(3y - 2)$ **45.** $8(m + 2)(m^2 - 2m + 4)$ **47.** $xy(y - 3z)(y + 3z)$ **49.** $4(3x - 4y)(3x + 4y)$ **51.** $9(4 - 3x)(4 + 3x)$ **53.** $(xy - z^2)(x^2y^2 + xyz^2 + z^4)$ **55.** $\left(7 - \frac{3}{5}m\right)\left(7 + \frac{3}{5}m\right)$ **57.** $(t + 7)(t^2 - 7t + 49)$ **59.** $n(n^2 + 49)$ **61.** $x^2(x^2 + 9)(x + 3)(x - 3)$ **63.** $pq(8p + 9q)(8p - 9q)$ **65.** $xy^2(27xy + 1)$ **67.** $a(5a - 4b)(25a^2 + 20ab + 16b^2)$ **69.** $16x^2(x + 2)(x - 2)$ **71.** 6 **73.** −2 **75.** $\frac{1}{5}$ **77.** $(x + 2 + y)(x + 2 - y)$ **79.** $(a + 4)(a - 4)(b - 4)$ **81.** $(x + 3 + 2y)(x + 3 - 2y)$ **83.** $(x^n + 10)(x^n - 10)$ **85.** $(x + 6)$ **87.** answers may vary **89. a.** 2992 ft **b.** 1536 ft **c.** 14 sec **d.** $16(14 - t)(14 + t)$ **91. a.** 2560 ft **b.** 1920 ft **c.** 13 sec **d.** $16(13 + t)(13 - t)$

Integrated Review

Practice Exercises

1. $(3x - 1)(2x - 3)$ **2.** $(3x + 1)(x - 2)(x + 2)$ **3.** $3(3x - y)(3x + y)$ **4.** $(2a + b)(4a^2 - 2ab + b^2)$ **5.** $6xy^2(5x + 2)(2x - 3)$

Exercise Set

1. $(x + y)^2$ **2.** $(x - y)^2$ **3.** $(a + 12)(a - 1)$ **4.** $(a - 10)(a - 1)$ **5.** $(a + 2)(a - 3)$ **6.** $(a - 1)^2$ **7.** $(x + 1)^2$ **8.** $(x + 2)(x - 1)$ **9.** $(x + 1)(x + 3)$ **10.** $(x + 3)(x - 2)$ **11.** $(x + 3)(x + 4)$ **12.** $(x + 4)(x - 3)$ **13.** $(x + 4)(x - 1)$ **14.** $(x - 5)(x - 2)$ **15.** $(x + 5)(x - 3)$ **16.** $(x + 6)(x + 5)$ **17.** $(x - 6)(x + 5)$ **18.** $(x + 8)(x + 3)$ **19.** $2(x + 7)(x - 7)$ **20.** $3(x + 5)(x - 5)$ **21.** $(x + 3)(x + y)$ **22.** $(y - 7)(3 + x)$ **23.** $(x + 8)(x - 2)$ **24.** $(x - 7)(x + 4)$ **25.** $4x(x + 7)(x - 2)$ **26.** $6x(x - 5)(x + 4)$ **27.** $2(3x + 4)(2x + 3)$ **28.** $(2a - b)(4a + 5b)$ **29.** $(2a + b)(2a - b)$ **30.** $(4 - 3x)(7 + 2x)$ **31.** $(5 - 2x)(4 + x)$ **32.** prime **33.** prime **34.** $(3y + 5)(2y - 3)$ **35.** $(4x - 5)(x + 1)$ **36.** $y(x + y)(x - y)$ **37.** $4(t^2 + 9)$ **38.** $(x + 1)(x + y)$ **39.** $(x + 1)(a + 2)$ **40.** $9x(2x^2 - 7x + 1)$ **41.** $4a(3a^2 - 6a + 1)$ **42.** $(x + 16)(x - 2)$ **43.** prime **44.** $(4a - 7b)^2$ **45.** $(5p - 7q)^2$ **46.** $(7x + 3y)(x + 3y)$ **47.** $(5 - 2y)(25 + 10y + 4y^2)$ **48.** $(4x + 3)(16x^2 - 12x + 9)$ **49.** $-(x - 5)(x + 6)$ **50.** $-(x - 2)(x - 4)$ **51.** $(7 - x)(2 + x)$ **52.** $(3 + x)(1 - x)$ **53.** $3x^2y(x + 6)(x - 4)$ **54.** $2xy(x + 5y)(x - y)$ **55.** $5xy^2(x - 7y)(x - y)$ **56.** $4x^2y(x - 5)(x + 3)$ **57.** $3xy(4x^2 + 81)$ **58.** $2xy^2(3x^2 + 4)$ **59.** $(2 + x)(2 - x)$ **60.** $(3 + y)(3 - y)$ **61.** $(s + 4)(3r - 1)$ **62.** $(x - 2)(x^2 + 3)$ **63.** $(4x - 3)(x - 2y)$ **64.** $(2x - y)(2x + 7z)$ **65.** $6(x + 2y)(x + y)$ **66.** $2(x + 4y)(6x - y)$ **67.** $(x + 3)(y + 2)(y - 2)$ **68.** $(y + 3)(y - 3)(x^2 + 3)$ **69.** $(5 + x)(x + y)$ **70.** $(x - y)(7 + y)$ **71.** $(7t - 1)(2t - 1)$ **72.** prime **73.** $(3x + 5)(x - 1)$ **74.** $(7x - 2)(x + 3)$ **75.** $(x + 12y)(x - 3y)$ **76.** $(3x - 2y)(x + 4y)$ **77.** $(1 - 10ab)(1 + 2ab)$ **78.** $(1 + 5ab)(1 - 12ab)$ **79.** $(3 + x)(3 - x)(1 + x)(1 - x)$ **80.** $(3 + x)(3 - x)(2 + x)(2 - x)$ **81.** $(x + 4)(x - 4)(x^2 + 2)$ **82.** $(x + 5)(x - 5)(x^2 + 3)$ **83.** $(x - 15)(x - 8)$ **84.** $(y + 16)(y + 6)$ **85.** $2x(3x - 2)(x - 4)$ **86.** $2y(3y + 5)(y - 3)$ **87.** $(3x - 5y)(9x^2 + 15xy + 25y^2)$ **88.** $(6y - z)(36y^2 + 6yz + z^2)$ **89.** $(xy + 2z)(x^2y^2 - 2xyz + 4z^2)$ **90.** $(3ab + 2)(9a^2b^2 - 6ab + 4)$ **91.** $2xy(1 + 6x)(1 - 6x)$ **92.** $2x(x + 3)(x - 3)$ **93.** $(x + 2)(x - 2)(x + 6)$ **94.** $(x - 2)(x + 6)(x - 6)$ **95.** $2a^2(3a + 5)$ **96.** $2n(2n - 3)$ **97.** $(a^2 + 2)(a + 2)$ **98.** $(a - b)(1 + x)$ **99.** $(x + 2)(x - 2)(x + 7)$ **100.** $(a + 3)(a - 3)(a + 5)$ **101.** $(x - y + z)(x - y - z)$ **102.** $(x + 2y + 3)(x + 2y - 3)$ **103.** $(9 + 5x + 1)(9 - 5x - 1)$ **104.** $(b + 4a + c)(b - 4a - c)$ **105.** answers may vary **106.** yes; $9(x^2 + 9y^2)$ **107.** a, c **108.** b, c

Section 5.4

Practice Exercises

1. −4, 5 **2.** $-\frac{3}{4}$, 12 **3.** 0, $\frac{6}{7}$ **4.** −4, 12 **5.** $\frac{4}{3}$ **6.** −3, $\frac{2}{3}$ **7.** −6, 4 **8.** −3, 0, 3 **9.** $\frac{2}{3}, \frac{3}{2}$, 5 **10.** −3, 0, 2 **11.** (2, 0) and (4, 0)

Calculator Explorations 5.4

1. $-0.9, 2.2$ **3.** no real solution **5.** $-1.8, 2.8$

Vocabulary, Readiness & Video Check 5.4

1. quadratic **3.** $3, -5$ **5.** One side of the equation must be a factored polynomial and the other side must be zero. **7.** To find the x-intercepts of any graph in two variables, we let $y = 0$. Doing this with our quadratic equation gives us a quadratic equation with 0 on one side, which we can try to solve by factoring.

Exercise Set 5.4

1. $6, 7$ **3.** $2, -1$ **5.** $-9, -17$ **7.** $0, -6$ **9.** $0, 8$ **11.** $-\frac{3}{2}, \frac{5}{4}$ **13.** $\frac{7}{2}, -\frac{2}{7}$ **15.** $\frac{1}{2}, -\frac{1}{3}$ **17.** $-0.2, -1.5$ **19.** $9, 4$ **21.** $-4, 2$ **23.** $0, 7$ **25.** $8, -4$ **27.** $4, -4$ **29.** $-3, 12$ **31.** $\frac{7}{3}, -2$ **33.** $-5, 5$ **35.** $-2, \frac{1}{6}$ **37.** $0, 4, 8$ **39.** $\frac{3}{4}$ **41.** $-\frac{1}{2}, 0, \frac{1}{2}$ **43.** $-\frac{3}{8}, 0, \frac{1}{2}$ **45.** $-3, 2$ **47.** $-20, 0$ **49.** $\frac{17}{2}$ **51.** $-\frac{1}{2}, \frac{1}{2}$ **53.** $-\frac{3}{2}, -\frac{1}{2}, 3$ **55.** $-5, 3$ **57.** $-\frac{5}{6}, \frac{6}{5}$ **59.** $2, -\frac{4}{5}$ **61.** $-\frac{4}{3}, 5$ **63.** $-4, 3$ **65.** $\frac{8}{3}, -9, 0$ **67.** -7 **69.** $0, \frac{3}{2}$ **71.** $0, 1, -1$ **73.** $-6, \frac{4}{3}$ **75.** $\frac{6}{7}, 1$ **77.** $\left(-\frac{4}{3}, 0\right), (1, 0)$ **79.** $(-2, 0), (5, 0)$ **81.** $(-6, 0), \left(\frac{1}{2}, 0\right)$ **83.** E **85.** B **87.** C **89.** $\frac{47}{45}$ **91.** $\frac{17}{60}$ **93.** $\frac{15}{8}$ **95.** $\frac{7}{10}$ **97.** didn't write equation in standard form; should be $x = 4$ or $x = -2$ **99.** answers may vary; for example $(x - 6)(x + 1) = 0$ **101.** answers may vary; for example, $x^2 - 12x + 35 = 0$ **103. a.** 300; 304; 276; 216; 124; 0; -156 **b.** 5 sec **c.** 304 ft **d.**

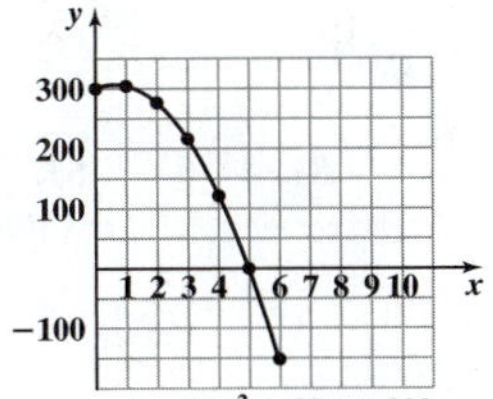

$y = -16x^2 + 20x + 300$

105. $0, \frac{1}{2}$ **107.** $0, -15$

Section 5.5
Practice Exercises

1. a. $\frac{2}{9}$ **b.** -5 **c.** 12 **d.** 0.7 **e.** -1 **2.** 4.123 **3. a.** $2\sqrt{6}$ **b.** $2\sqrt{15}$ **c.** $\sqrt{42}$ **d.** $10\sqrt{3}$ **4.** $10\sqrt{10}$ **5. a.** $\frac{\sqrt{5}}{7}$ **b.** $\frac{3}{10}$ **c.** $\frac{3\sqrt{2}}{5}$ **6. a.** $8\sqrt{2}$ **b.** $-7\sqrt{6}$ **7. a.** $5\sqrt{5}$ **b.** $6 - 2\sqrt{10} + \sqrt{6}$ **c.** $-3\sqrt{2}$ **8. a.** $\sqrt{77}$ **b.** $72\sqrt{30}$ **c.** $5\sqrt{2}$ **d.** 17 **9.** 28 **10. a.** $3 - \sqrt{15}$ **b.** -9 **11. a.** $\sqrt{3}$ **b.** $2\sqrt{2}$ **12. a.** $\frac{4\sqrt{5}}{5}$ **b.** $\frac{\sqrt{6}}{6}$ **13.** $\frac{7 - \sqrt{7}}{3}$

Calculator Explorations 5.5

1. 2.646; yes **3.** 3.317; yes **5.** 9.055; yes

Vocabulary, Readiness & Video Check 5.5

1. radicand; radical sign **3.** $\sqrt{a} \cdot \sqrt{b}$ **5.** false **7.** like radicals **9. b.** $17\sqrt{2}$ **11. c.** $2\sqrt{5}$ **13.** $\sqrt{21}$ **15.** $\sqrt{\frac{15}{3}}$ or $\sqrt{5}$ **17.** The radical sign, $\sqrt{\ }$, indicates a positive square root only. A negative sign before the radical sign, $-\sqrt{\ }$, indicates a negative square root. **19.** Factor until you have a product of primes. A repeated prime factor means a perfect square—if more than one factor is repeated you can multiply all the repeated factors together to get one larger perfect square factor. **21.** Both like terms and like radicals are combined using the distributive property; also only like (vs. unlike) terms can be combined, as with like radicals (same radicand). **23.** In the example, the product rule is first used to multiply the radicals and then later used to simplify the radicals. **25.** To write an equivalent expression without a radical in the denominator.

Exercise Set 5.5

1. 4 **3.** $\frac{1}{5}$ **5.** -10 **7.** not a real number **9.** -11 **11.** $\frac{3}{5}$ **13.** 30 **15.** 12 **17.** 2.646 **19.** 6.083 **21.** 11.662 **23.** $2\sqrt{5}$ **25.** $5\sqrt{2}$ **27.** $\sqrt{33}$ **29.** $7\sqrt{2}$ **31.** $6\sqrt{5}$ **33.** 15 **35.** $21\sqrt{7}$ **37.** $-15\sqrt{3}$ **39.** $\frac{2\sqrt{2}}{5}$ **41.** $\frac{3\sqrt{3}}{11}$ **43.** $\frac{3}{2}$ **45.** $-\frac{\sqrt{3}}{4}$ **47.** $-4\sqrt{3}$ **49.** $9\sqrt{6} - 5$ **51.** $2\sqrt{11} + 11$ **53.** $5\sqrt{3}$ **55.** $9\sqrt{5}$ **57.** $4\sqrt{6} + \sqrt{5}$ **59.** $-\sqrt{35}$ **61.** $\frac{4\sqrt{5}}{9}$ **63.** $\frac{3\sqrt{3}}{8}$ **65.** 4 **67.** $5\sqrt{2}$ **69.** 6 **71.** 20 **73.** $\sqrt{30} + \sqrt{42}$ **75.** $2\sqrt{5} + 5\sqrt{2}$ **77.** -33 **79.** $16 - 11\sqrt{11}$ **81.** 4 **83.** $\sqrt{7}$ **85.** $3\sqrt{2}$ **87.** $\frac{\sqrt{15}}{5}$ **89.** $\frac{7\sqrt{2}}{2}$ **91.** $\frac{\sqrt{2}}{4}$ **93.** $\frac{\sqrt{30}}{15}$ **95.** $\frac{\sqrt{15}}{10}$ **97.** $3 + \sqrt{3}$ **99.** $3 - 2\sqrt{5}$ **101.** $3\sqrt{3} + 1$ **103.** 44 **105.** 2 **107.** 7 mi **109.** 3.1 in. **111.** 3 **113.** 4, 5 **115.** 8, 9 **117.** answers may vary **119.** $8\sqrt{5}$ in. **121.** $130\sqrt{3}$ sq m **123.** $\frac{\sqrt{A\pi}}{\pi}$

Section 5.6

Practice Exercises

1. $-4, 4$ **2.** $-\frac{\sqrt{65}}{5}, \frac{\sqrt{65}}{5}$ **3.** $-1, 11$ **4.** $-2 \pm 2\sqrt{3}$ **5.** no real solution **6.** $\frac{5 \pm \sqrt{17}}{3}$ **7.** $\frac{-1 \pm \sqrt{41}}{10}$ **8.** $\frac{4}{3}, -2$ **9.** $\pm\frac{\sqrt{15}}{3}$ **10.** no real solution **11.** $\frac{5 \pm 3\sqrt{5}}{2}$ **12.** $\frac{-1 + \sqrt{41}}{10} \approx 0.5; \frac{-1 - \sqrt{41}}{10} \approx -0.7$

Vocabulary, Readiness & Video Check 5.6

1. $x = \frac{-b \pm \sqrt{b^2 - 4ac}}{2a}$ **3.** $-2, 1$ **5.** $\frac{-5 \pm \sqrt{17}}{2}$ **7.** To solve, a becomes the radicand of a square root and the square root of a negative number is not a real number. **9.** The exact solution values are keyed into a calculator, and the decimal approximations are then rounded to the requested place value.

Exercise Set 5.6

1. ± 8 **3.** $\pm\frac{1}{5}$ **5.** no real solution **7.** $\pm\frac{\sqrt{39}}{3}$ **9.** $\pm\sqrt{5}$ **11.** $12, -2$ **13.** $-2 \pm \sqrt{7}$ **15.** $-2 \pm \sqrt{10}$ **17.** $\frac{7 \pm 4\sqrt{2}}{3}$ **19.** $8, -3$ **21.** $1, 2$ **23.** $\frac{-7 \pm \sqrt{37}}{6}$ **25.** $\pm\frac{2}{7}$ **27.** no real solution **29.** $-2 \pm \sqrt{7}$ **31.** $\frac{-9 \pm \sqrt{129}}{12}$ **33.** no real solution **35.** $\frac{1 \pm \sqrt{2}}{5}$ **37.** $-3 \pm 2\sqrt{2}; -5.8, -0.2$ **39.** $\frac{9 \pm \sqrt{97}}{2}; 9.4, -0.4$ **41.** $2, \frac{1}{5}$ **43.** $1 \pm \sqrt{2}$ **45.** $\pm 2\sqrt{5}$ **47.** no real solution **49.** 2 **51.** 3 **53.** ± 2 **55.** $0, 1, 2$ **57.** $0, -5$ **59.** $\frac{3 \pm \sqrt{7}}{5}$ **61.** $\frac{3}{2}, -1$ **63.** $\frac{5 \pm \sqrt{105}}{20}$ **65.** $5, \frac{7}{4}$ **67.** $\frac{-7 \pm 3\sqrt{2}}{-5}$ or $\frac{7 \pm 3\sqrt{2}}{5}$ **69.** $\frac{7 \pm \sqrt{193}}{6}$ **71.** $11, -10$ **73.** $4, -\frac{2}{3}$ **75.** $0.5, 0.1$ **77.** $\frac{11 \pm \sqrt{41}}{20}$ **79.** $\frac{4 \pm \sqrt{10}}{2}$ **81.** $4\sqrt{3}$ **83.** $5\sqrt{2}$ **85.** base: 3 ft; height: 12 ft **87.** c **89.** b **91.** $\frac{-3\sqrt{2} \pm \sqrt{38}}{2}$ **93.** answers may vary **95.** $-0.9, 0.2$

Section 5.7

Practice Exercises

1. 13 in. **2.** $2\sqrt{10}$ m ≈ 6.32 m **3.** $\sqrt{5425}$ ft ≈ 74 ft **4.** $2\sqrt{37}$ units **5.** $16\sqrt{3}$ ft/sec ≈ 27.71 ft/sec **6.** 2 sec **7.** There are two numbers. They are -4 and 12. **8.** base: 35 ft; height: 12 ft **9.** 7 and 8 or -6 and -5

Vocabulary, Readiness & Video Check 5.7

1. The Pythagorean theorem applies to right triangles only, and in the formula $a^2 + b^2 = c^2$, c is the length of the hypotenuse. **3.** No, order does not matter when subtractions are squared. For example, $(3 - 5)^2 = (5 - 3)^2$. It is still a good habit to always keep order correct when subtracting. **5.** In applications, the context of the problem needs to be considered. Each exercise resulted in both a positive and a negative solution, and a negative solution is not appropriate for any of the problems.

Exercise Set 5.7

1. $\sqrt{13}$ **3.** $3\sqrt{3}$ **5.** 25 **7.** $\sqrt{22}$ **9.** $3\sqrt{17}$ **11.** $\sqrt{41}$ **13.** $4\sqrt{2}$ **15.** $3\sqrt{10}$ **17.** 51.2 ft **19.** 20.6 ft **21.** 11.7 ft **23.** $\sqrt{29}$ **25.** $\sqrt{73}$ **27.** $2\sqrt{10}$ **29.** $\frac{3\sqrt{5}}{2}$ **31.** $\sqrt{85}$ **33.** 6 ft **35.** 54 mph **37.** 27 mph **39.** 59.1 km **41.** 60.2 km **43.** width $= x$; length $= x + 4$ **45.** x and $x + 2$ if x is an odd integer **47.** base $= x$; height $= 4x + 1$ **49.** 15 cm, 13 cm, 70 cm, 22 cm **51.** base = 16 mi; height = 6 mi **53.** $2\sqrt{5}$ in. ≈ 4.47 in. **55.** 177 m **57.** 2.7 sec **59.** 15.8 sec **61.** 6 in. **63.** 5 sec **65.** width = 5 cm; length = 6 cm **67.** 54 diagonals **69.** 10 sides **71.** -12 or 11 **73.** 14, 15 **75.** 13 feet **77.** 5 in. **79.** 12 mm, 16 mm, 20 mm **81.** 10 km **83.** 36 ft **85.** 9.5 sec **87.** 20% **89.** length: 15 mi; width: 8 mi **91.** 105 units **93.** 1.8 million **95.** 2.3 million **97.** 2003 **99.** $y = 2\sqrt{10}, x = 2\sqrt{10} - 4$ **101.** 201 mi **103.** width: 3.5 ft; length: 9.9 ft **105.** 7.9 sec **107.** 2018

Section 5.8

Practice Exercises

1.

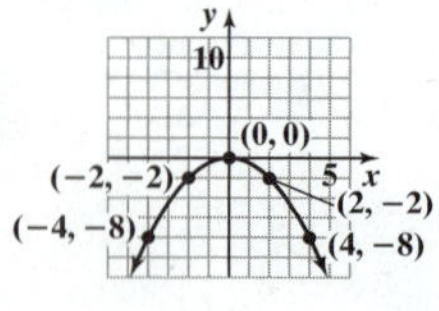

2.

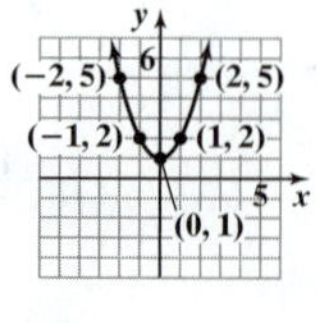

3.

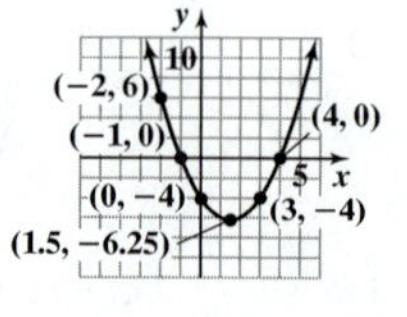

4.

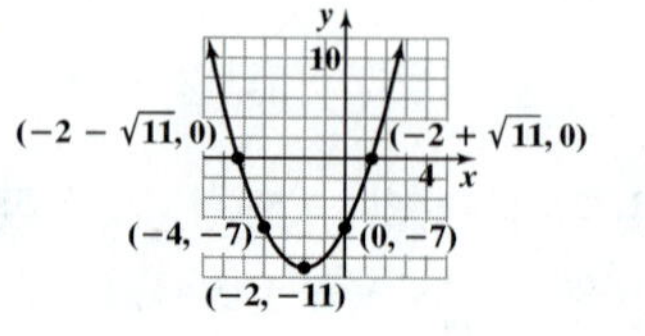

Calculator Explorations 5.8

1. $x = -0.41, 7.41$ **3.** $x = 0.91, 2.38$ **5.** $x = -0.39, 0.84$

Vocabulary, Readiness & Video Check 5.8

1. If a parabola opens upward, the lowest point is called the vertex; if a parabola opens downward, the highest point is called the vertex. If a graph can be folded along a line such that the two sides coincide or form mirror images of each other, we say the graph is symmetric with respect to that line and that line is an axis of symmetry. **3.** For example, if the vertex is in quadrant III or IV and the parabola opens downward, then there won't be any x-intercepts, and there's no need to let $y = 0$ and solve the equation for x.

Exercise Set 5.8

1.
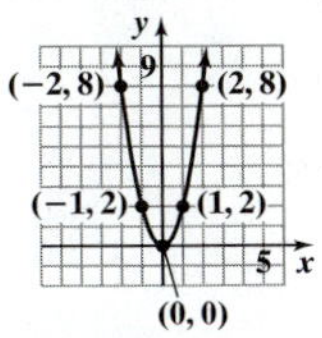

3.
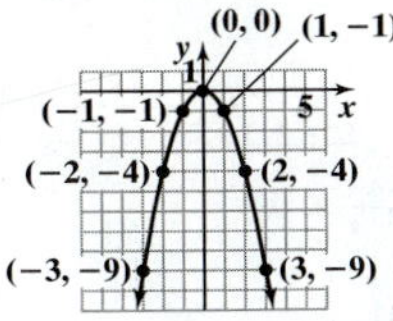

5.
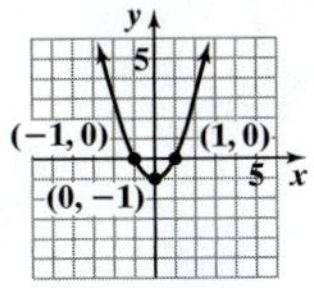

7.
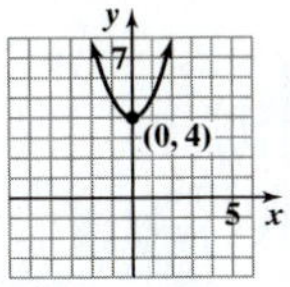

9.
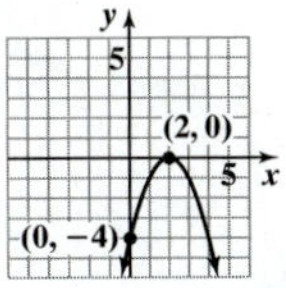

11.
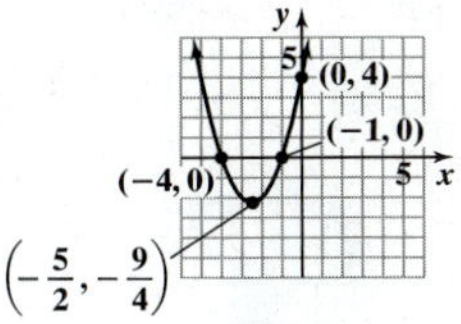

13.
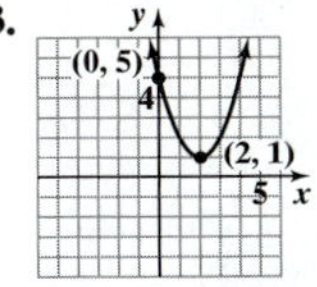

15.
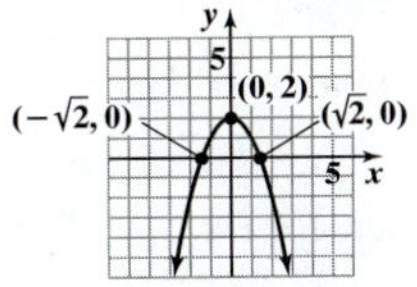

17.
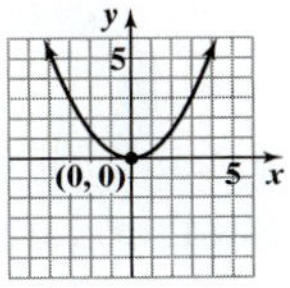

19.
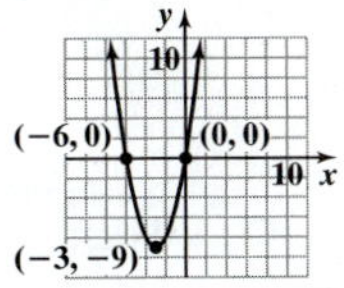

21.
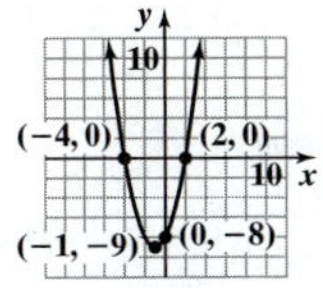

23.
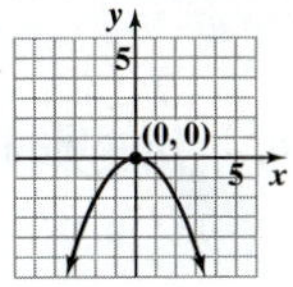

25.
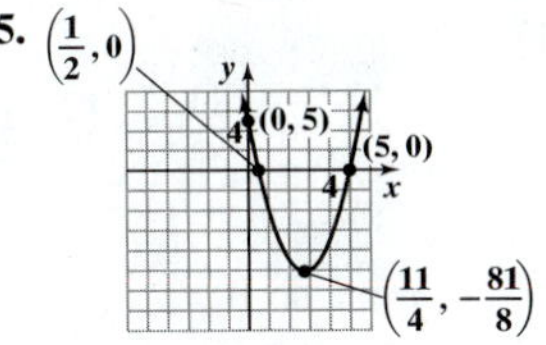

27.
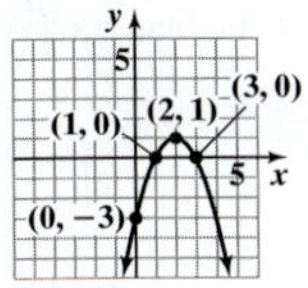

29.
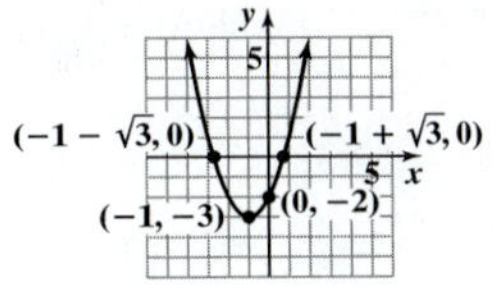

31.
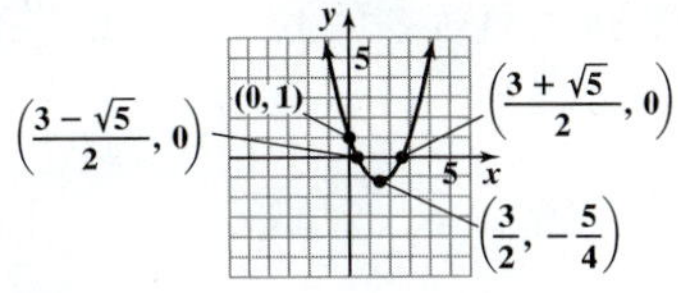

33. $-\frac{4}{3}$; downward **35.** $\frac{11}{4}$; upward **37.** undefined slope; vertical **39.** 64,000 megawatts **41.** domain: $(-\infty, \infty)$; range: $(-\infty, 3]$ **43.** domain: $(-\infty, \infty)$; range: $(-\infty, 1]$ **45. a.** 256 ft **b.** $t = 4$ sec **c.** $t = 8$ sec **47.** C **49.** A

Chapter 5 Vocabulary Check

1. quadratic equation **2.** Factoring **3.** greatest common factor **4.** perfect square trinomial **5.** difference of two squares **6.** difference of two cubes **7.** sum of two cubes **8.** 0 **9.** like radicals **10.** radicand; radical **11.** principal square root **12.** rationalizing the denominator **13.** square root **14.** quadratic

Chapter 5 Review

1. $2x - 5$ **3.** $4x(5x + 3)$ **5.** $(2x + 3)(3x - 5)$ **7.** $(x - 1)(3x + 2)$ **9.** $(2a + b)(5a + 7b)$ **11.** $(x + 4)(x + 2)$ **13.** prime **15.** $(x + 6y)(x - 2y)$ **17.** $10a(a - 1)(a - 10)$ **19.** $(2x + 1)(x + 6)$ **21.** $(3x + 4y)(2x - y)$ **23.** $5y(2y - 3)(y + 4)$ **25.** $2(3x - 5)^2$ **27.** $(2x + 3)(2x - 3)$ **29.** prime **31.** $(2x + 3)(4x^2 - 6x + 9)$ **33.** $2(3 - xy)(9 + 3xy + x^2y^2)$ **35.** $(4x^2 + 1)(2x + 1)(2x - 1)$ **37.** $-6, 2$ **39.** $-\frac{1}{5}, -3$ **41.** $-4, 6$ **43.** $-\frac{2}{7}, \frac{3}{8}$ **45.** 9 **47.** $-\frac{3}{8}$ **49.** $3\sqrt{6}$ **51.** $\frac{3\sqrt{2}}{5}$ **53.** $6\sqrt{7}$ **55.** $3\sqrt{2}$ **57.** $6\sqrt{2} - 18$ **59.** $3\sqrt{2} - 5\sqrt{3} + 2\sqrt{6} - 10$ **61.** 3 **63.** $2\sqrt{5}$ **65.** $\frac{\sqrt{22}}{11}$ **67.** $\frac{\sqrt{30}}{6}$ **69.** ± 6 **71.** $\pm 5\sqrt{2}$ **73.** $\frac{-5 \pm \sqrt{41}}{4}$ **75.** 2.5 sec **77.** $\pm\sqrt{5}$ **79.** $5 \pm 3\sqrt{2}$ **81.** $\frac{-1 \pm \sqrt{13}}{6}$ **83.** no real solution **85.** 0.4, −0.8 **87.** $3\sqrt{13}$;10.82 **89.** $4\sqrt{34}$ ft; 23.32 ft **91.** $\sqrt{130}$ units **93.** 2.4 in. **95.** c **97.** 9 units **99.** width: 20 in.; length: 25 in. **101.** 19 and 20 **103. a.** 17.5 sec and 10 sec; The rocket reaches a height of 2800 ft on its way up and on its way back down. **b.** 27.5 sec

105.
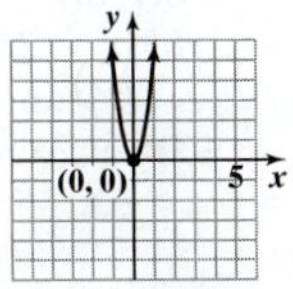

107.
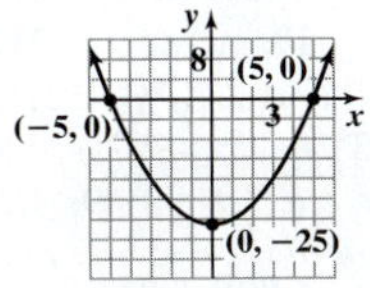

109.
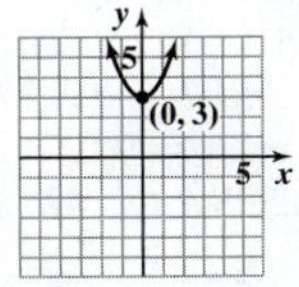

111.
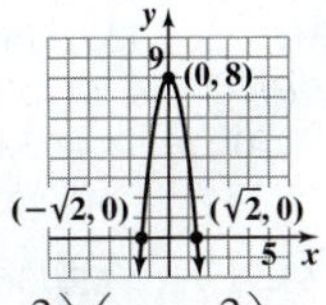

113.
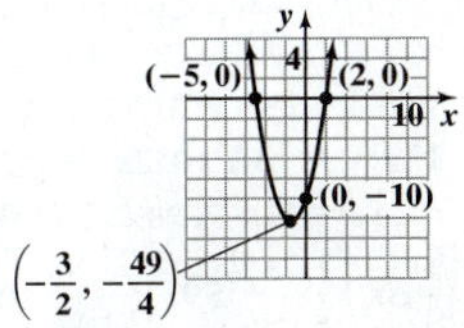

115.
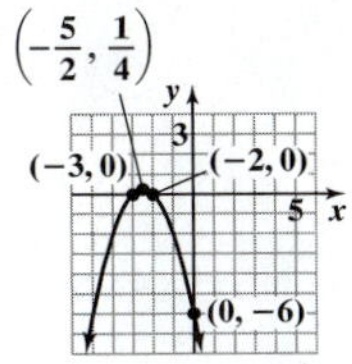

117.
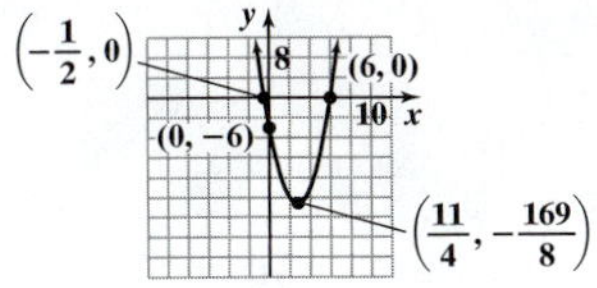

119. $7(x - 9)$ **121.** $\left(m + \frac{2}{5}\right)\left(m - \frac{2}{5}\right)$ **123.** $(y + 2)(x - 1)$ **125.** $3x(x - 9)(x - 1)$ **127.** $2(x + 3)(x - 3)$ **129.** $5x^2 - 9x - 2$; $(5x + 1)(x - 2)$ **131.** $-\frac{7}{2}, 4$ **133.** $7\sqrt{3}$ **135.** $3\sqrt{3} - \sqrt{6}$ **137.** $\frac{\sqrt{14}}{7}$ **139.** A **141.** B **143.** $\pm 5\sqrt{3}$ **145.** $\frac{3 \pm \sqrt{41}}{8}$

Chapter 5 Getting Ready for the Test

1. B **2.** D **3.** A **4.** A **5.** B **6.** B **7.** C **8.** B **9.** B **10.** C **11.** C **12.** A **13.** B **14.** C **15.** C

Chapter 5 Test

1. $(x + 7)(x + 4)$ **2.** $(7 - m)(7 + m)$ **3.** $(y + 11)^2$ **4.** $(y - 12)(y + 4)$ **5.** prime **6.** $3x(3x + 1)(x + 4)$ **7.** $(3a - 7)(a + b)$ **8.** $(3x - 2)(x - 1)$ **9.** $(x + 12y)(x + 2y)$ **10.** $5(6 + x)(6 - x)$ **11.** $(6t + 5)(t - 1)$ **12.** $(y + 2)(y - 2)(x - 7)$ **13.** $(4x - 1)(16x^2 + 4x + 1)$ **14.** $8(y - 2)(y^2 + 2y + 4)$ **15.** $-9, 3$ **16.** $-7, 2$ **17.** $-7, 1$ **18.** $0, 3, -3$ **19.** 4 **20.** $3\sqrt{6}$ **21.** $2\sqrt{23}$ **22.** $-2\sqrt{13}$ **23.** $-8\sqrt{3}$ **24.** $\frac{7\sqrt{3}}{10}$ **25.** $7\sqrt{2}$ **26.** $2\sqrt{3} - \sqrt{10}$ **27.** $\sqrt{5}$ **28.** $\frac{\sqrt{6}}{3}$ **29.** ± 4 **30.** $\frac{5 \pm 2\sqrt{2}}{3}$

31. $-\frac{4}{3}, 1$ **32.** $-1, \frac{5}{3}$ **33.** $\frac{7 \pm \sqrt{73}}{6}$ **34.** $\frac{1}{3}, 2$ **35.** $\frac{3 \pm \sqrt{7}}{2}$ **36.** $0, \pm\frac{1}{3}$ **37.** $4\sqrt{5}$ in. **38.** $\sqrt{5}$ units **39.** 2.7 sec **40.** 7 sec

41.

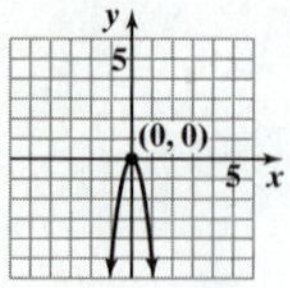

42.

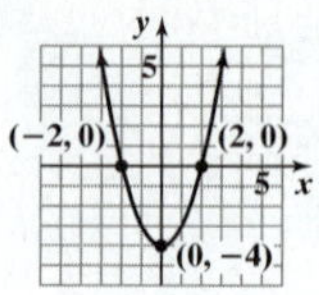

43.

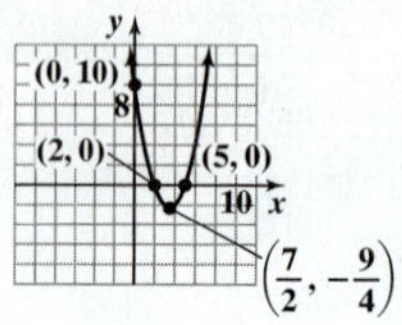

44.

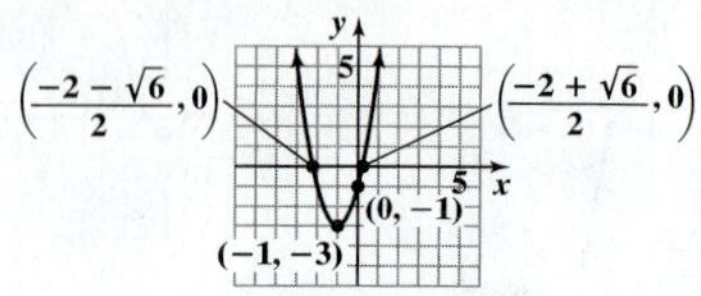

Chapter 5 Cumulative Review

1. a. $9 \le 11$ **b.** $8 > 1$ **c.** $3 \ne 4$; Sec. 1.2, Ex. 3 **3. a.** $\frac{6}{7}$ **b.** $\frac{11}{27}$ **c.** $\frac{22}{5}$; Sec. R.1, Ex. 2 **5.** $\frac{14}{3}$; Sec. 1.3, Ex. 5 **7. a.** $4x$ **b.** $11y^2$ **c.** $8x^2 - x$; **d.** $5n^2$; Sec. 2.1, Ex. 3 **9.** -11; Sec. 2.2, Ex. 6 **11.** $\frac{16}{3}$; Sec. 2.3, Ex. 2 **13.** shorter: 12 in.; longer: 36 in.; Sec. 2.4, Ex. 3 **15.**

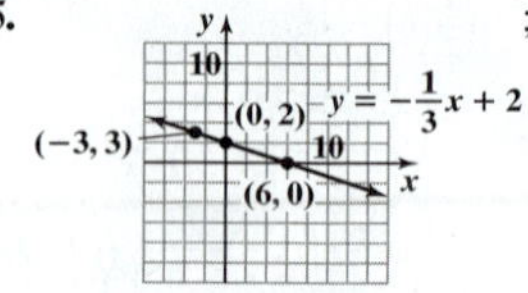

; Sec. 3.1, Ex. 9 **17.** $m = \frac{3}{4}$; y-intercept: $(0, -1)$; Sec. 3.2, Ex. 7 **19. a.** 250 **b.** 1; Sec. 4.1, Ex. 2 **21. a.** 2 **b.** 6 **c.** 0; Sec. 4.2, Ex. 1 **23.** $9x^2 - 6x - 1$; Sec. 4.2, Ex. 11 **25.** $6x^2 - 11x - 10$; Sec. 4.3, Ex. 5 **27.** $9y^2 + 6y + 1$; Sec. 4.4, Ex. 4 **29. a.** $\frac{1}{9}$ **b.** $\frac{2}{x^3}$ **c.** $\frac{3}{4}$ **d.** $\frac{1}{16}$ **e.** $\frac{1}{y^4}$; Sec. 4.5, Ex. 1 **31. a.** 3.67×10^8 **b.** 3.0×10^{-6} **c.** 2.052×10^{10} **d.** 8.5×10^{-4}; Sec. 4.5, Ex. 5 **33. a.** x^3 **b.** y; Sec. 5.1, Ex. 2 **35.** $(x + 3)(x + 4)$; Sec. 5.2, Ex. 1 **37.** $(4x - 1)(2x - 5)$; Sec. 5.2, Ex. 8 **39.** $(5a + 3b)(5a - 3b)$; Sec. 5.3, Ex. 2b **41.** $3, -1$; Sec. 5.4, Ex. 1

CHAPTER 6 GEOMETRY AND MEASUREMENT

Section 6.1

Practice Exercises

1. a. ray; ray AB or $\overrightarrow{AB}$ **b.** line segment; line segment RS or $\overline{RS}$ **c.** line; line EF or $\overleftrightarrow{EF}$ **d.** angle; $\angle TVH$ or $\angle HVT$ or $\angle V$ **2.** $\angle RTS$, $\angle STR$ **3. a.** obtuse **b.** straight **c.** acute **d.** right **4.** 61° **5.** 113° **6. a.** 44° **b.** 28° **c.** acute **7.** $m\angle a = 109°$; $m\angle b = 71°$; $m\angle c = 71°$ **8.** $m\angle x = 45°$; $m\angle y = 45°$; $m\angle z = 135°$; $m\angle a = 135°$; $m\angle b = 135°$; $m\angle c = 135°$; $m\angle d = 45°$

Vocabulary, Readiness & Video Check 6.1

1. plane **3.** Space **5.** ray **7.** straight **9.** acute **11.** Parallel; intersecting **13.** degrees **15.** vertical **17.** $\angle WUV$, $\angle VUW$, $\angle U$, $\angle x$ **19.** $180° - 17° = 163°$

Exercise Set 6.1

1. line; line CD or line l or $\overleftrightarrow{CD}$ **3.** line segment; line segment MN or $\overline{MN}$ **5.** angle; $\angle GHI$ or $\angle IHG$ or $\angle H$ **7.** ray; ray UW or $\overrightarrow{UW}$ **9.** $\angle CPR$, $\angle RPC$ **11.** $\angle TPM$, $\angle MPT$ **13.** straight **15.** right **17.** obtuse **19.** acute **21.** 67° **23.** 163° **25.** 32° **27.** 30° **29.** $\angle MNP$ and $\angle RNO$; $\angle PNQ$ and $\angle QNR$ **31.** $\angle SPT$ and $\angle TPQ$; $\angle SPR$ and $\angle RPQ$; $\angle SPT$ and $\angle SPR$; $\angle TPQ$ and $\angle QPR$ **33.** 27° **35.** 132° **37.** $m\angle x = 30°$; $m\angle y = 150°$; $m\angle z = 30°$ **39.** $m\angle x = 77°$; $m\angle y = 103°$; $m\angle z = 77°$ **41.** $m\angle x = 100°$; $m\angle y = 80°$; $m\angle z = 100°$ **43.** $m\angle x = 134°$; $m\angle y = 46°$; $m\angle z = 134°$ **45.** $\angle ABC$ or $\angle CBA$ **47.** $\angle DBE$ or $\angle EBD$ **49.** 15° **51.** 50° **53.** 65° **55.** 95° **57.** $\frac{9}{8}$ or $1\frac{1}{8}$ **59.** $\frac{7}{32}$ **61.** $\frac{5}{6}$ **63.** $1\frac{1}{3}$ **65.** 360° **67.** 54.8° **69.** false; answers may vary **71.** true **73.** $m\angle a = 60°$; $m\angle b = 50°$; $m\angle c = 110°$; $m\angle d = 70°$; $m\angle e = 120°$ **75.** no; answers may vary **77.** 45°; 45°

Section 6.2

Practice Exercises

1. a. 56 m **b.** 350 ft **2.** 16 ft **3.** 24 cm **4.** 23 km **5.** 102 m **6.** $684 **7.** exactly 20π yd $\approx$ 62.83 yd **8.** $49\frac{1}{2}$ sq in. **9.** 48.8 sq yd **10.** 396 sq m **11.** 49π sq cm $\approx$ 153.86 sq cm **12.** 84 cu ft **13.** $\frac{1}{6}\pi$ cu cm $\approx \frac{11}{21}$ cu cm

Vocabulary, Readiness & Video Check 6.2

1. perimeter **3.** π **5.** $\frac{22}{7}$ (or 3.14); 3.14 $\left(\text{or } \frac{22}{7}\right)$ **7.** volume **9.** cubic **11.** Opposite sides of a rectangle have the same measure, so we can just find the sum of the measures of all four sides. **13.** We don't have a formula for an L-shaped figure, so we divide it into two rectangles, use the formula to find the area of each, and then add these two areas.

Exercise Set 6.2

1. 64 ft **3.** 120 cm **5.** 21 in. **7.** 48 ft **9.** 42 in. **11.** 155 cm **13.** 21 ft **15.** 624 ft **17.** 346 yd **19.** 22 ft **21.** $55 **23. a.** 8 **b.** 72 in. **25.** 28 in. **27.** $36.12 **29.** 96 m **31.** 66 ft **33.** 74 cm **35.** 17π cm; 53.38 cm **37.** 16π mi; 50.24 mi **39.** 26π m; 81.64 m **41.** 15π ft; 47.1 ft **43.** 12,560 ft **45.** 7 sq m **47.** $9\frac{3}{4}$ sq yd **49.** 15 sq yd **51.** 2.25π sq in. $\approx$ 7.065 sq in.

53. 36.75 sq ft **55.** 28 sq m **57.** 22 sq yd **59.** 25 sq cm **61.** 86 sq mi **63.** 24 sq cm **65.** 36π sq in. ≈ 113.10 sq in. **67.** 72 cu in. **69.** 512 cu cm **71.** 4π cu yd ≈ 12.57 cu yd **73.** $\frac{500}{3}\pi$ cu in. ≈ 523.60 cu in. **75.** 9π cu in. ≈ 28.27 cu in. **77.** 75 cu cm **79.** $2\frac{10}{27}$ cu in. **81.** 8.4 cu ft **83.** 113,625 sq ft **85.** 168 sq ft **87.** 960 cu cm **89. a.** 381 sq ft **b.** 4 squares **91.** 5.25π cu in. **93.** 4π sq ft ≈ 12.57 sq ft **95.** 4.5π cu m; 14.14 cu m **97.** 5 **99.** 20 **101.** perimeter **103.** area **105.** area **107.** perimeter **109.** 12-in. pizza **111.** b **113. a.** 62.8 m; 125.6 m **b.** yes **115.** 2094.40 cu m **117.** no; answers may vary **119.** no; answers may vary

Integrated Review

1. 153°; 63° **2.** $m\angle x = 75°; m\angle y = 105°; m\angle z = 75°$ **3.** $m\angle x = 128°; m\angle y = 52°; m\angle z = 128°$ **4.** $m\angle x = 52°$ **5.** 4.6 in. **6.** $4\frac{1}{4}$ in. **7.** 20 m; 25 sq m **8.** 12 ft; 6 sq ft **9.** 10π cm ≈ 31.4 cm; 25π sq cm ≈ 78.5 sq cm **10.** 32 mi; 44 sq mi **11.** 54 cm; 143 sq cm **12.** 62 ft; 238 sq ft **13.** 64 cu in. **14.** 30.6 cu ft **15.** 400 cu cm **16.** $4\frac{1}{2}\pi$ cu mi ≈ 14.14 cu mi

Section 6.3

Practice Exercises

1. 72 in. **2.** 24 ft **3.** 1.5 ft **4.** 5 ft 8 in. **5.** 17 ft **6.** 13 ft 7 in. **7.** 18 ft 4 in. **8.** 3 ft 11 in. **9.** 2500 mm **10.** 3.5 km **11.** 1.46 km or 1460 m **12.** 91.5 hm **13.** 125 cm or 1.25 m

Vocabulary, Readiness & Video Check 6.3

1. meter **3.** yard **5.** feet **7.** feet **9.** Both mean addition; $5\frac{2}{5} = 5 + \frac{2}{5}$ and 5 ft 2 in. = 5 ft + 2 in. **11.** Since the metric system is based on base 10, we just need to move the decimal point to convert from one unit to another.

Exercise Set 6.3

1. 5 ft **3.** 36 ft **5.** 8 mi **7.** 102 in. **9.** $3\frac{1}{3}$ yd **11.** 33,792 ft **13.** 4.5 yd **15.** 0.25 ft **17.** 13 yd 1 ft **19.** 7 ft 1 in. **21.** 1 mi 4720 ft **23.** 62 in. **25.** 26 ft **27.** 84 in. **29.** 11 ft 2 in. **31.** 22 yd 1 ft **33.** 6 ft 5 in. **35.** 7 ft 6 in. **37.** 14 ft 4 in. **39.** 83 yd 1 ft **41.** 6000 cm **43.** 4 cm **45.** 0.5 km **47.** 1.7 m **49.** 15 m **51.** 42,000 cm **53.** 7000 m **55.** 83 mm **57.** 0.201 dm **59.** 40 mm **61.** 8.94 m **63.** 2.94 m or 2940 mm **65.** 1.29 cm or 12.9 mm **67.** 12.64 km or 12,640 m **69.** 54.9 m **71.** 1.55 km **73.** $348\frac{2}{3}$; 12,552 **75.** $11\frac{2}{3}$; 420 **77.** 5000; 0.005; 500 **79.** 0.065; 65; 0.000065 **81.** 342,000; 342,000,000; 34,200,000 **83.** 10 ft 6 in. **85.** 5100 ft **87.** 5.0 times **89.** 26.7 mm **91.** 15 ft 9 in. **93.** 3.35 m **95.** $121\frac{1}{3}$ yd **97.** $\frac{21}{100}$ **99.** 0.13 **101.** 0.25 **103.** no **105.** yes **107.** no **109.** Estimate: 13 yd **111.** answers may vary; for example, $1\frac{1}{3}$ yd or 48 in. **113.** answers may vary **115.** 334.89 sq m

Section 6.4

Practice Exercises

1. $3\frac{1}{4}$ tons **2.** $4\frac{1}{2}$ lb **3.** 2 lb 16 oz **4.** 2 tons 900 lb **5.** 1 lb 6 oz **6.** 6 lb 4 oz **7.** 3410 mg **8.** 0.562 g **9.** 2.19 g or 21.9 dg **10.** 23 kg

Vocabulary, Readiness & Video Check 6.4

1. Mass **3.** gram **5.** 2000 **7.** We can't subtract 9 oz from 4 oz, so we borrow 1 lb(= 16 oz) from 12 lb to add to the 4 oz; 12 lb 4 oz becomes 11 lb 20 oz. **9.** 18.50 dg

Exercise Set 6.4

1. 32 oz **3.** 10,000 lb **5.** 9 tons **7.** $3\frac{3}{4}$ lb **9.** $1\frac{3}{4}$ tons **11.** 204 oz **13.** 9800 lb **15.** 76 oz **17.** 1.5 tons **19.** $\frac{1}{20}$ lb **21.** 92 oz **23.** 161 oz **25.** 5 lb 9 oz **27.** 53 lb 10 oz **29.** 8 tons 750 lb **31.** 3 tons 175 lb **33.** 8 lb 11 oz **35.** 31 lb 2 oz **37.** 1 ton 700 lb **39.** 0.5 kg **41.** 4000 mg **43.** 25,000 g **45.** 0.048 g **47.** 0.0063 kg **49.** 15,140 mg **51.** 6250 g **53.** 350,000 cg **55.** 13.5 mg **57.** 5.815 g or 5815 mg **59.** 1850 mg or 1.85 g **61.** 1360 g or 1.36 kg **63.** 13.52 kg **65.** 2.125 kg **67.** 200,000; 3,200,000 **69.** $\frac{269}{400}$ or 0.6725; 21,520 **71.** 0.5; 0.0005; 50 **73.** 21,000; 21,000,000; 2,100,000 **75.** 8.064 kg **77.** 30 mg **79.** 5 lb 8 oz **81.** 35 lb 14 oz **83.** 6 lb 15.4 oz **85.** 144 mg **87.** 6.12 kg **89.** 130 lb **91.** 211 lb **93.** 0.16 **95.** 0.875 **97.** no **99.** yes **101.** no **103.** answers may vary; for example, 250 mg or 0.25 g **105.** true **107.** answers may vary

Section 6.5

Practice Exercises

1. $21\frac{1}{2}$ qt **2.** 104 c **3.** 3 qt **4.** 200 gal 2 qt **5.** 2.1 L **6.** 21.3 L **7.** 4150 ml or 4.15 L **8.** 2431 L

Vocabulary, Readiness & Video Check 6.5

1. capacity **3.** fluid ounces **5.** cups **7.** quarts **9.** We can't subtract 3 qt from 0 qt, so we borrow 1 gal(= 4 qt) from 3 gal to get 2 gal 4 qt. **11.** 0.45 dal

Exercise Set 6.5

1. 4 c **3.** 16 pt **5.** $3\frac{1}{2}$ gal **7.** 5 pt **9.** 8 c **11.** $3\frac{3}{4}$ qt **13.** $10\frac{1}{2}$ qt **15.** 9 c **17.** 23 qt **19.** $\frac{1}{4}$ pt **21.** 14 gal 2 qt **23.** 4 gal 3 qt 1 pt **25.** 22 pt **27.** 13 gal 2 qt **29.** 4 c 4 fl oz **31.** 1 gal 1 qt **33.** 2 gal 3 qt 1 pt **35.** 17 gal **37.** 4 gal 3 qt **39.** 5000 ml

41. 0.00016 kl **43.** 5.6 L **45.** 320 cl **47.** 0.41 kl **49.** 0.064 L **51.** 160 L **53.** 3600 ml **55.** 19.3 L **57.** 4.5 L or 4500 ml **59.** 8410 ml or 8.41 L **61.** 16,600 ml or 16.6 L **63.** 3840 ml **65.** 162.4 L **67.** 336; 84; 168 **69.** $\frac{1}{4}$; 1; 2 **71.** 1.59 L **73.** 18.954 L **75.** 4.3 fl oz **77.** yes **79.** \$0.316 **81.** $\frac{4}{5}$ **83.** $\frac{3}{5}$ **85.** $\frac{9}{10}$ **87.** no **89.** no **91.** less than; answers may vary **93.** answers may vary **95.** 128 fl oz **97.** 1.5 cc **99.** 2.7 cc **101.** 54 u or 0.54 cc **103.** 86 u or 0.86 cc

Section 6.6

Practice Exercises

1. 0.59 in. **2.** 226.8 g **3.** 8 fl oz **4.** 140°F **5.** 89.6°F **6.** 20°C **7.** 45°C **8.** 39.3°C

Vocabulary, Readiness & Video Check 6.6

1. 1 L ≈ 0.26 gal or 3.79 L ≈ 1 gal **3.** F = 1.8C + 32; 27

Exercise Set 6.6

1. 25.57 fl oz **3.** 218.44 cm **5.** 40 oz **7.** 57.66 mi **9.** 3.77 gal **11.** 13.5 kg **13.** 1.5; $1\frac{2}{3}$; 150; 60 **15.** 55; 5500; 180; 2160 **17.** 3.94 in. **19.** 80.5 kph **21.** 0.008 oz **23.** 229.6 ft **25.** 9.92 billion mi **27.** yes **29.** 2790 mi **31.** 90 mm **33.** 112.5 g **35.** 104 mph **37.** 26.24 ft **39.** 3 mi **41.** 8 fl oz **43.** b **45.** b **47.** c **49.** d **51.** d **53.** 25°C **55.** 40°C **57.** 122°F **59.** 239°F **61.** −6.7°C **63.** 61.2°C **65.** 197.6°F **67.** 54.3°F **69.** 56.7°C **71.** 80.6°F **73.** 21.1°C **75.** 244.4°F **77.** 7232°F **79.** 29 **81.** 36 **83.** yes **85.** no **87.** no **89.** yes **91.** 2.13 sq m **93.** 1.19 sq m **95.** 1.69 sq m **97.** 4,000,000,000°C **99.** answers may vary

Chapter 6 Vocabulary Check

1. Weight **2.** Mass **3.** meter **4.** unit fractions **5.** gram **6.** liter **7.** line segment **8.** complementary **9.** line **10.** perimeter **11.** angle; vertex **12.** Area **13.** ray **14.** transversal **15.** straight **16.** volume **17.** vertical **18.** adjacent **19.** obtuse **20.** right **21.** acute **22.** supplementary

Chapter 6 Review

1. right **3.** acute **5.** 65° **7.** 58° **9.** 90° **11.** $\angle a$ and $\angle b$; $\angle b$ and $\angle c$; $\angle c$ and $\angle d$; $\angle d$ and $\angle a$ **13.** $m\angle x = 100°; m\angle y = 80°; m\angle z = 80°$ **15.** $m\angle x = 53°; m\angle y = 53°; m\angle z = 127°$ **17.** 69 m **19.** 36 m **21.** 32 ft **23.** 5.338 in. **25.** 240 sq ft **27.** 600 sq cm **29.** 49π sq ft ≈ 153.86 sq ft **31.** 119 sq in. **33.** 144 sq m **35.** 432 sq ft **37.** $15\frac{5}{8}$ cu in. **39.** $20{,}000\pi$ cu cm ≈ 62,800 cu cm **41.** $2\frac{2}{3}$ cu ft **43.** $7\frac{1}{2}$ cu ft **45.** 9 ft **47.** 7920 ft **49.** 17 yd 1 ft **51.** 4200 cm **53.** 0.01218 m **55.** 21 yd 1 ft **57.** 9.5 cm or 95 mm **59.** 169 yd 2 ft **61.** 617.5 km **63.** $4\frac{1}{8}$ lb **65.** 3 lb 4 oz **67.** 0.027 g **69.** 21 dag **71.** 3 lb 9 oz **73.** 21.5 mg **75.** 4 lb 4 oz **77.** 14 qt **79.** 7 pt **81.** 4 qt 1 pt **83.** 3800 ml **85.** 3060 cl **87.** 1 gal 1 qt **89.** 736 ml or 0.736 L **91.** 2 gal 3 qt **93.** 10.88 L **95.** 22.96 ft **97.** 4.55 gal **99.** 425.25 g **101.** 2.36 in. **103.** 107.6°F **105.** 5.2°C **107.** 1.7°C **109.** 108° **111.** 95° **113.** 27.3 in. **115.** 1624 sq m **117.** $346\frac{1}{2}$ cu in. **119.** 75 in. **121.** 500 cm **123.** 1.4 g **125.** 186.8°F **127.** 9117 m or 9.117 km **129.** 8 gal 1 qt

Chapter 6 Getting Ready for the Test

1. B **2.** A **3.** A **4.** B **5.** A **6.** B **7.** A **8.** B **9.** B **10.** A **11.** A **12.** B

Chapter 6 Test

1. 12° **2.** 56° **3.** 50° **4.** $m\angle x = 118°; m\angle y = 62°; m\angle z = 118°$ **5.** $m\angle x = 73°; m\angle y = 73°; m\angle z = 73°$ **6.** 6.2 m **7.** 10 in. **8.** circumference = 18π in. ≈ 56.52 in.; area = 81π sq in. ≈ 254.34 sq in. **9.** perimeter = 24.6 yd; area = 37.1 sq yd **10.** perimeter = 68 in.; area = 185 sq in. **11.** $62\frac{6}{7}$ cu in. **12.** 30 cu ft **13.** 16 in. **14.** 18 cu ft **15.** 62 ft; \$115.94 **16.** 23 ft 4 in. **17.** 10 qt **18.** $1\frac{7}{8}$ lb **19.** 5600 lb **20.** $4\frac{3}{4}$ gal **21.** 0.04 g **22.** 2400 g **23.** 36 mm **24.** 0.43 g **25.** 830 ml **26.** 1 gal 2 qt **27.** 3 lb 13 oz **28.** 8 ft 3 in. **29.** 2 gal 3 qt **30.** 66 mm or 6.6 cm **31.** 2.256 km or 2256 m **32.** 28.9°C **33.** 54.7°F **34.** 5.6 m **35.** 4 gal 3 qt **36.** 91.4 m **37.** 16 ft 6 in.

Chapter 6 Cumulative Review

1. a. < **b.** = **c.** >; Sec. 1.2, Ex. 6 **3. a.** commutative property of multiplication **b.** associative property of addition **c.** identity element for addition **d.** commutative property of multiplication **e.** multiplicative inverse property **f.** additive inverse property **g.** commutative and associative properties of multiplication; Sec. 1.5, Ex. 6 **5.** $-2x - 1$; Sec. 2.1, Ex. 7 **7.** 8; Sec. 2.2, Ex. 4 **9.** 6; Sec. 2.2, Ex. 5 **11.** 12; Sec. 2.3, Ex. 3 **13.** 10; Sec. 2.4, Ex. 2 **15.** $(2x + 3)(x + 1)(x - 1)$; Ch. 5 Int. Rev. Ex. 2 **17.** $[2, \infty)$; Sec. 2.6, Ex. 3 **19.** 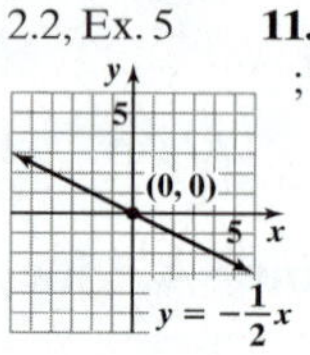

; Sec. 3.2, Ex. 3 **21.** $-\frac{8}{3}$; Sec. 3.2, Ex. 4 **23.** $\frac{3}{4}$; Sec. 3.2, Ex. 6 **25.** slope: $\frac{3}{4}$; *y*-intercept: $(0, -1)$; Sec. 3.2, Ex. 7 **27.** $2x + y = 3$; Sec. 3.3, Ex. 4

29. $x = -1$; Sec. 3.3, Ex. 6 **31.** domain: $\{-1, 0, 3\}$; range: $\{-2, 0, 2, 3\}$; Sec. 3.4, Ex. 1 **33. a.** function **b.** not a function; Sec. 3.4, Ex. 2
35. $(-2, 2)$; Sec. 3.5, Ex. 7 **37.** $(0, -5)$; Sec. 3.5, Ex. 8 **39.** 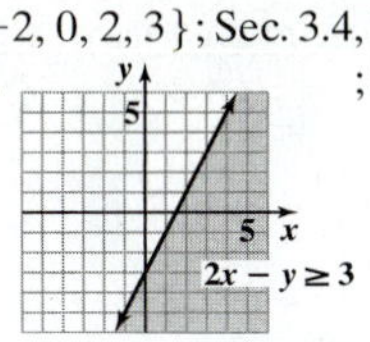

; Sec. 3.6, Ex. 3

CHAPTER 7 INTRODUCTION TO COUNTING AND PROBABILITY

Section 7.1
Practice Exercises

1. 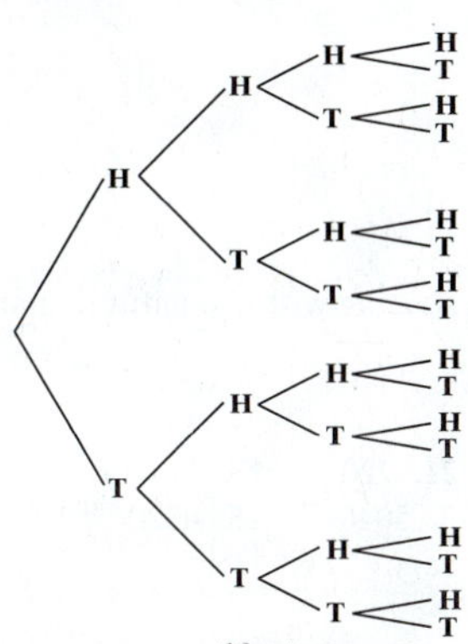

16 outcomes

2. 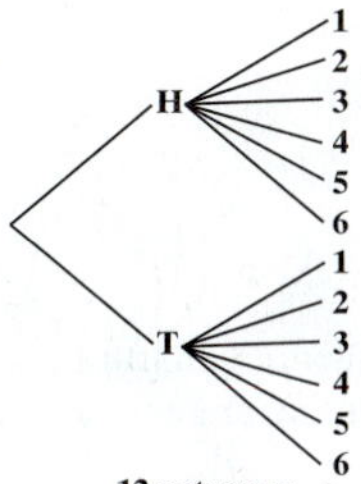

12 outcomes

3. 900 **4.** 175,760,000 **5.** 128 **6.** 2187

Vocabulary, Readiness & Video Check 7.1

1. outcome **3.** experiment **5.** A tree diagram can be used to list outcomes and determine the number of outcomes. The Fundamental Counting Principle can be used to determine the number of outcomes only.

Exercise Set 7.1

1. 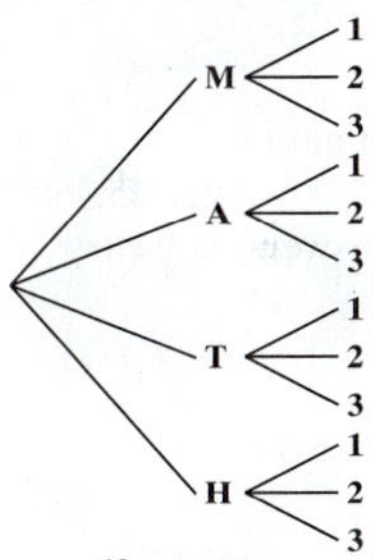

12 outcomes

3. 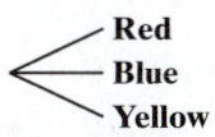

3 outcomes

5. 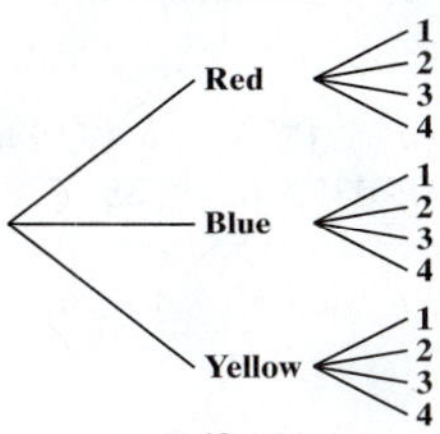

12 outcomes

7. 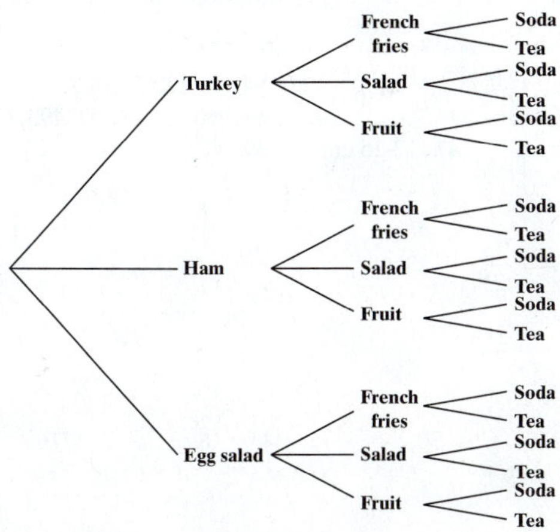

18 outcomes

9. a. 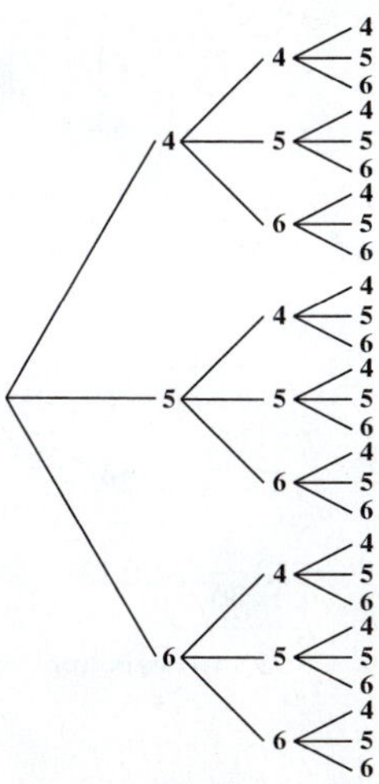

27 outcomes

b. 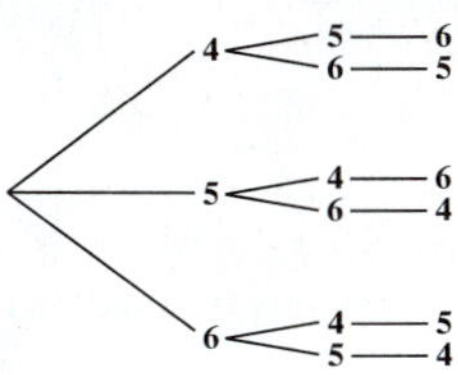

6 outcomes

11. 64 **13.** 64 **15.** 512 **17.** 90 **19.** 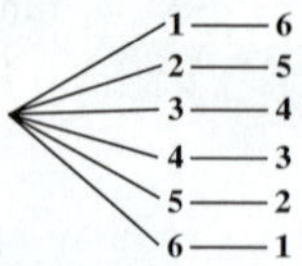 (1, 6), (2, 5), (3, 4), (4, 3), (5, 2), (6, 1) **21.** 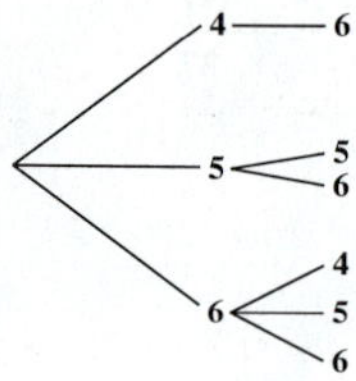 (4, 6), (5, 5), (5, 6), (6, 4), (6, 5), (6, 6)

23. 40 **25.** 158,184,000 **27.** 1352 **29.** 80,000 **31.** 900 **33.** 32 **35.** 2.15 billion **37.** 2,440,000,000 **39. a.** 200 **b.** no
41. answers may vary

Section 7.2

Practice Exercises

1. 30,240 **2.** 6,497,400 **3. a.** $\frac{1}{7}$ **b.** 28 **c.** 5 **d.** 1 **4. a.** 90 **b.** 332,640 **c.** 720 **5.** 6 **6.** 6840

Vocabulary, Readiness & Video Check 7.2

1. 1 **3.** permutation **5.** the Fundamental Counting Principle **7.** The factors were descending natural numbers, starting with the natural number 5. In other words, the product was $5 \cdot 4 \cdot 3 \cdot 2 \cdot 1$. **9.** A permutation is also called an ordered arrangement.

Exercise Set 7.2

1. 840 **3.** 1 **5.** 360 **7.** 2 **9.** 720 **11.** 90 **13.** 1,583,960,400 **15.** 72 **17.** 4 **19.** 3,628,800 **21.** 720 **23.** 11,880
25. 7,880,400 **27.** 120 **29.** 362,880 **31.** 72 **33.** 5040 **35.** 1,030,200 **37.** 360 **39.** 6 **41.** 48 **43.** 5040 **45.** 4320
47. 120 **49.** 720 **51.** 600 **53.** 24 **55.** New Cascade **57.** 6.22 mi **59.** 12,550.2 m **61.** answers may vary **63.** 2520
65. answers may vary **67.** $n = 8$

Section 7.3

Practice Exercises

1. a. permutation **b.** combination **c.** combination **2. a.** 28 **b.** 455 **c.** 1 **3.** 5005 **4.** 1,192,052,400

Vocabulary, Readiness & Video Check 7.3

1. 1 **3.** permutation **5.** 1 **7.** For permutations, order does matter. For combinations, order does not matter.

Exercise Set 7.3

1. 7 **3.** 286 **5.** 1 **7.** 5 **9.** 1 **11.** 84 **13.** 5,311,735 **15.** 280 **17.** 12 **19.** 240 **21.** 280 **23.** combination
25. permutation **27.** 9,364,199,760 **29.** 2,598,960 **31.** 86,493,225 **33.** 435,897 **35.** 720 **37.** 26,000 **39.** 38,760 **41.** 45,360
43. 450 **45.** 15 students **47.** 12-15 cm **49.** 42 students **51.** 107 students **53.** yes **55.** 126 **57.** 28 **59.** answers may vary

61. ${}_7C_5 = \frac{7!}{5!(7-5)!} = \frac{7!}{5!2!}$ **63.** ${}_nC_r = \frac{n!}{r!(n-r)!}$; ${}_nC_{n-r} = \frac{n!}{(n-r)!(n-(n-r))!} = \frac{n!}{(n-r)!r!}$ **65.** 1; answers may vary
67. 1; answers may vary

Integrated Review

1. 9 **2.** 400 **3.** 240 **4.** 30,240 **5.** 24 **6.** 1 **7.** 1716 **8.** 132 **9.** 286 **10.** 66 **11.** 90 **12.** 9,000,000 **13.** 210
14. 5,245,786 **15.** 10,626

Section 7.4

Practice Exercises

1. $\frac{1}{2}$ **2.** $\frac{1}{6}$ **3. a.** 0 **b.** 1 **4. a.** $\frac{1}{4}$ **b.** $\frac{1}{13}$ **c.** $\frac{1}{52}$ **5.** 0.17 **6.** 0.20

Vocabulary, Readiness & Video Check 7.4

1. Probability **3.** 1 **5.** event **7.** $\frac{\text{number of outcomes in event } E}{\text{total number of outcomes}}$ **9.** Empirical probability is based on observed outcomes from the trials of an experiment. Theoretical probability is not.

Exercise Set 7.4

1. $\frac{1}{6}$ **3.** $\frac{1}{3}$ **5.** $\frac{1}{2}$ **7.** $\frac{2}{3}$ **9.** $\frac{1}{12}$ **11.** $\frac{1}{6}$ **13.** $\frac{1}{3}$ **15.** $\frac{2}{3}$ **17.** $\frac{1}{2}$ **19.** $\frac{1}{4}$ **21.** 1 **23.** $\frac{1}{52}$ **25.** $\frac{1}{13}$ **27.** $\frac{1}{4}$ **29.** $\frac{2}{13}$
31. $\frac{1}{2}$ **33.** $\frac{1}{36}$ **35.** $\frac{1}{6}$ **37.** $\frac{1}{4}$ **39.** $\frac{1}{12}$ **41.** $\frac{1}{36}$ **43.** 0 **45.** $\frac{19}{100}$ **47.** $\frac{1}{20}$ **49.** 4 **51.** 5 **53.** 6 **55.** $\frac{4}{30} \approx 0.13$
57. $\frac{5}{30} \approx 0.17$ **59.** $\frac{6}{30} = 0.2$ **61.** 7 hours **63.** 71 students **65.** 0 students **67.** answers may vary **69.** 1 **71.** answers may vary
73. answers may vary

Chapter 7 Vocabulary Check

1. outcome **2.** tree diagram **3.** experiment **4.** 1 **5.** factorial of n **6.** permutation **7.** $\frac{n!}{(n-r)!}$ **8.** the Fundamental Counting Principle **9.** 1 **10.** combination **11.** permutation **12.** $\frac{n!}{r!(n-r)!}$ **13.** 1 **14.** 1 **15.** Probability **16.** 0 **17.** 1 **18.** sample space **19.** event **20.** $\frac{\text{number of times event } E \text{ occurred}}{\text{total number of trials of experiment}}$ **21.** $\frac{\text{number of outcomes in event } E}{\text{total number of outcomes}}$

Chapter 7 Review

1.

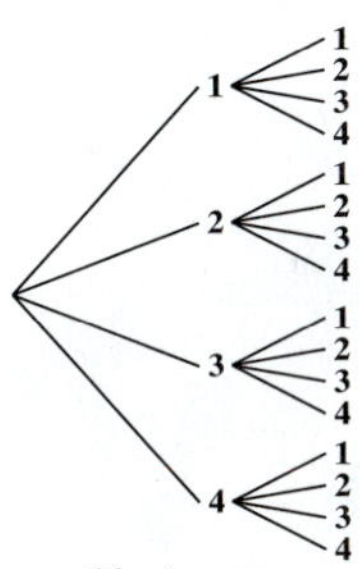

16 outcomes

3.

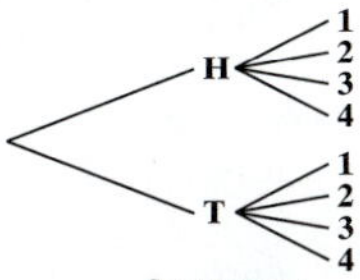

8 outcomes

5. 144 **7.** 1024 **9.** 810,000 **11.** 2600 **13.** 362,880 **15.** 1 **17.** 12 **19.** 2 **21.** 360 **23.** 479,001,600 **25.** 30,240 **27.** 720 **29.** 56 **31.** 12 **33.** 12 **35.** 792 **37.** 1 **39.** 45 **41.** 1 **43.** 455 **45.** 5,245,786 **47.** $\frac{1}{3}$ **49.** 1 **51.** $\frac{2}{3}$ **53.** $\frac{1}{7}$ **55.** $\frac{2}{7}$ **57.** $\frac{4}{7}$ **59.** $\frac{1}{52}$ **61.** $\frac{1}{13}$ **63.** $\frac{1}{4}$ **65.** $\frac{2}{13}$ **67.** 0.167 **69.** 0.5 **71.** 0.16 **73.** 0.58

Chapter 7 Getting Ready for the Test

1. B **2.** B **3.** B **4.** D **5.** C **6.** B **7.** D **8.** B **9.** D **10.** C **11.** B **12.** A **13.** A and C **14.** A and B

Chapter 7 Test

1. 720 **2.** 1 **3.** 6 **4.** 5016 **5.** 60,480 **6.** 84 **7.**

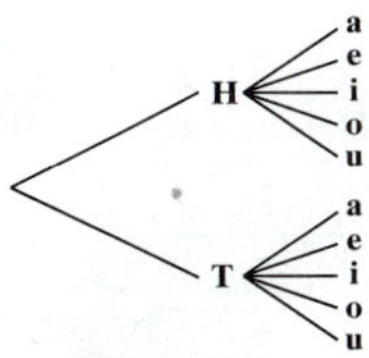

10 outcomes

8. a. 64 **b.** 24 **9.** 60 **10.** 21 **11.** 42 **12.** 17,576,000 **13.** 324,632 **14.** 450 **15.** $\frac{3}{8}$ **16.** $\frac{1}{4}$ **17.** $\frac{1}{8}$ **18.** $\frac{1}{2}$ **19.** $\frac{1}{36}$ **20.** $\frac{1}{6}$ **21.** $\frac{5}{6}$ **22.** $\frac{5}{36}$ **23.** $\frac{1}{6}$ **24.** 0 **25.** 1 **26.** $\frac{1}{52}$ **27.** $\frac{1}{4}$ **28.** $\frac{1}{13}$ **29.** $\frac{2}{13}$ **30.** $\frac{59}{100}$ or 0.59 **31.** $\frac{1}{2}$ or 0.5

Chapter 7 Cumulative Review

1. $-\frac{21}{19}$ Sec. 1.4, Ex. 10 **3.** 4; Sec. 2.3, Ex. 5 **5.**

x	y
−2	−6
0	−5
10	0

; Sec. 3.1, Ex. 5 **7.** $y = \frac{1}{4}x - 3$; Sec. 3.3, Ex. 3 **9.** $y = -3$; Sec. 3.3, Ex. 7 **11. a.** function **b.** function **c.** function **d.** not a function; Sec. 3.4, Ex. 5 **13.** yes; Sec. 3.5, Ex. 1a **15.** $(-2, 0)$; Sec. 3.5, Ex. 6 **17.** no solution or { }; Sec. 3.5, Ex. 9 **19.** 576.1 ft; 16.1 ft; Sec. 4.2, Ex. 5

21.

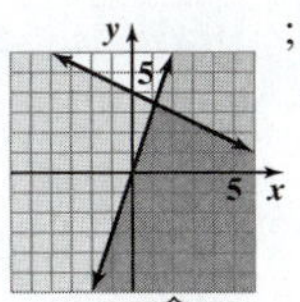

$3x \ge y$
$x + 2y \le 8$

; Sec. 3.6, Ex. 8 **23.** $-9x^2 + 10xy - 5y^2$; Sec. 4.2, Ex. 7 **25.** $2x^2(x-6)(x-7)$; Sec. 5.2, Ex. 6 **27. a.** $3\sqrt{6}$ **b.** $2\sqrt{3}$ **c.** $10\sqrt{2}$ **d.** $\sqrt{35}$; Sec. 5.5, Ex. 3 **29. a.** $5-\sqrt{10}$ **b.** −44; Sec. 5.5, Ex. 10 **31.** 10 in.; Sec. 5.7, Ex. 1

CHAPTER 8 INTRODUCTION TO STATISTICS

Section 8.1

Practice Exercises

1. 85.4 **2.** $46.67 thousand or $46,670 **3.** 2.67 **4.** 24 **5.** 83 **6.** 15 **7.** median: 22; mode: 15 **8. a.** 60.7 **b.** 62 **c.** 62

Vocabulary, Readiness & Video Check 8.1

1. average **3.** mean (or average) **5.** grade point average **7.** The answer is not exact; it's an approximation since we rounded to the nearest tenth. **9.** Place the data numbers in numerical order (or verify that they already are).

Exercise Set 8.1

1. mean: 21; median: 23; no mode **3.** mean: 8.1; median: 8.2; mode: 8.2 **5.** mean: 0.5; median: 0.5; mode: 0.2 and 0.5 **7.** mean: 370.9; median: 313.5; no mode **9.** 2054.3 ft **11.** 1799.5 ft **13.** answers may vary **15.** 2.79 **17.** 3.64 **19.** 6.8 **21.** 6.9 **23.** 88.5 **25.** 73 **27.** 70 and 71 **29.** 9 rates **31.** mean: 3773 mi; median: 3812 mi **33. a.** 8.2 **b.** 8 **c.** 9 **35. a.** 6.1 **b.** 6 **c.** 6, 7, 8 **37. a.** 15 **b.** 15 **c.** 15 **39. a.** 6.1 **b.** 6 **c.** 4, 6 **41.** 2 **43.** 1 **45.** 2 **47.** 9 **49.** $\frac{1}{15}$ **51.** $\frac{1}{30}$ **53.** $\frac{1}{15}$ **55.** $\frac{3}{10}$ **57.** 1 **59.** 20, 21, 21, **61.** answers may vary

Section 8.2

Practice Exercises

1. $70 thousand or $70,000 **2.**

Data Item	*Deviation*
48	10
30	−8
40	2
42	4
16	−22
45	7
45	7

3. $s \approx 11.30$

Graphing Calculator Explorations 8.2

1. $s \approx 22.50$ **3.** $s \approx 0.53$ **5.** $s \approx 0.87$

Vocabulary, Readiness & Video Check 8.2

1. range **3.** sample standard deviation; $\sqrt{\frac{\Sigma(x - \bar{x})^2}{n - 1}}$ **5.** Answers may vary. For example, 6, 6, 6, 6.

Exercise Set 8.2

1. 12 **3.** 350 **5.** 2 **7.** 1.7 **9. a.** 15 **b.**

Data Item	*Deviation*
17	2
10	−5
18	3
18	3
12	−3

c. 0 **11. a.** 248 **b.**

Data Item	*Deviation*
260	12
256	8
190	−58
286	38

c. 0

13. a. 3.6 **b.**

Data Item	*Deviation*
3.7	0.1
4.0	0.4
3.3	−0.3
2.8	−0.8
4.8	1.2
3.0	−0.6

c. 0 **15. a.** 7 **b.**

Data Item	*Deviation*
9	2
5	−2
7	0
4	−3
10	3

c. 0 **17.** $s \approx 2.45$ **19.** $s \approx 4.08$ **21.** $s \approx 1.58$

23. $s \approx 10.12$ **25. a.** 6, 6, 6, 6, 6 **b.** $\bar{x} = 6$ **c.** $s = 0$ **d.** range = 0 **27. a.** 3, 3, 3, 3, 5, 5, 7, 7, 7, 7 **b.** $\bar{x} = 5$ **c.** $s \approx 1.9$ **d.** range = 4 **29.** answers may vary **31.** answers may vary **33. 25.** mode = 6; **26.** mode = 5; **27.** mode = 3,7; **28.** mode = 1, 7 **35.** answers may vary **37.** 0.05 **39.** 0.40 **41.** 0.95

Section 8.3

Practice Exercises

1. $\mu + 2\sigma = 30$; $\mu - 2\sigma = 6$ **2.** 95% **3.** $\bar{x} = 14$; $\sigma = 4.2$

Graphing Calculator Explorations 8.3

1. $s \approx 22.50$; $\sigma \approx 20.13$ **3.** $s \approx 0.53$; $\sigma \approx 0.48$ **5.** $s \approx 0.87$; $\sigma \approx 0.82$

Vocabulary, Readiness & Video Check 8.3

1. normal distribution curve **3.** sample standard deviation **5.** sample **7.** Empirical **9.** The denominator of the radical for sample standard deviation is $n - 1$. The denominator of the radical for population standard deviation is n.

Exercise Set 8.3

1. 35 **3.** 240 **5.** 80 **7.** 80 **9.** 72.5 **11.** 100 **13.** 160; 278; 396; 514; 632; 750; 868 **15.** 750 **17.** 381 **19.** 68% **21.** 47.5% **23.** 16% **25.** 68% **27.** 34% **29.** 47.5% **31.** 81.5% **33.** 2.5% **35.** 0.15% **37.** 2800 hr **39.** 50% **41.** 16% **43.** 97.5%

45. 8420 hr **47.** 50% **49.** 16% **51.** 68% **53.** 638 **55.** 710 **57.** 530 **59.** 95% **61.** 84% **63.** 81.5% **65.** 50,000 **67.** 2012, 2013 **69.** $s \approx 20.2; \sigma \approx 18.0$ **71.** $s \approx 3.2; \sigma \approx 2.9$ **73.** 30 **75.** 30 **77.** 0 **79.** answers may vary

Chapter 8 Vocabulary Check

1. mean **2.** median; mean **3.** mode **4.** mean **5.** $\bar{x}$ **6.** grade point average **7.** range **8.** 0 **9.** sample standard deviation **10.** dispersion **11.** normal distribution curve **12.** population **13.** population standard deviation **14.** sample **15.** median or mode; median or mode **16.** Empirical

Chapter 8 Review

1. mean: 17.8; median: 14; no mode **3.** mean: 24,500; median: 20,000; mode: 20,000 **5.** 3.25 **7. a.** 62.1 **b.** 62 **c.** 63 **9. a.** 12.9 **b.** 12 **c.** 11 and 12 **11. a.** 59.8 **b.** 60 **c.** 60 **13.** 3 **15.** 35 **17.** 5 **19.** $s \approx 7.91$ **21.** $s \approx 1.64$ **23.** 78; 80; 82; 86; 88; 90 **25.** 75 **27.** 68% **29.** 13.5% **31.** 16% **33. a.** 14.44 **b.** 14 **c.** 18 **d.** 8 **e.** 2.96 **35. a.** 20 **b.** 6.53 **37.** 7570 hr **39.** 50% **41.** 2.5% **43.** 68%

Chapter 8 Getting Ready for the Test

1. B **2.** A **3.** D **4.** D **5.** A **6.** B **7.** E **8.** A and C **9.** A and B **10.** C and D **11.** A and C **12.** C and D

Chapter 8 Test

1. mean: 38.4; median: 42; no mode **2.** mean: 12.625; median: 12.5; mode: 12 and 16 **3.** 3.07 **4. a.** 92.8 **b.** 93 **c.** 93 and 94 **d.** 4 **5. a.** 26.3 **b.** 25 **c.** 35 **d.** 20 **6.** 8 **7.** 3.08 **8.** $\mu = 172; \mu + \sigma = 203; \mu + 2\sigma = 234; \mu + 3\sigma = 265; \mu - \sigma = 141; \mu - 2\sigma = 110; \mu - 3\sigma = 79$ **9.** 68% **10.** 95% **11.** 47.5% **12.** 203 lb **13.** 110 lb **14.** $\mu = 360; \mu + \sigma = 375; \mu + 2\sigma = 390; \mu + 3\sigma = 405; \mu - \sigma = 345; \mu - 2\sigma = 330; \mu - 3\sigma = 315$ **15.** 95% **16.** 99.7% **17.** 84% **18.** 16% **19.** 81.5%

Chapter 8 Cumulative Review

1. a. $5x + 7$ **b.** $-4a - 1$ **c.** $4y - 3y^2$ **d.** $7.3x - 6$ **e.** $\frac{1}{2}b$; Sec. 2.1, Ex. 4 **3. a.** x-int: $(-3, 0)$; y-int: $(0, 2)$ **b.** x-int: $(-4, 0), (-1, 0)$; y-int: $(0, 1)$ **c.** x-int and y-int: $(0, 0)$ **d.** x-int: $(2, 0)$; y-int: none **e.** x-int: $(-1, 0), (3, 0)$; y-int: $(0, -1), (0, 2)$; Sec. 3.2, Ex. 1 **5. a.** $\frac{15}{x} = 4$ **b.** $12 - 3 = x$ **c.** $4x + 17 \neq 21$ **d.** $3x < 48$; Sec. 1.3, Ex. 9 **7.**

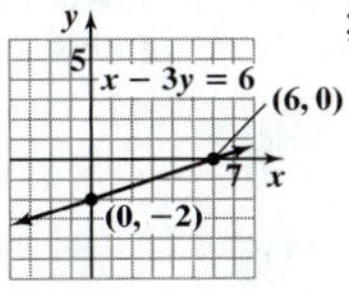

; Sec. 3.2, Ex. 2 **9.**

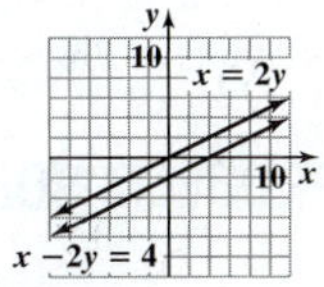

no solution; Sec. 3.5, Ex. 3

11. $(-2, 0)$; Sec. 3.5, Ex. 6 **13.** $\{(x, y) \mid -5x - 3y = 9\}$ or $\{(x, y) \mid 10x + 6y = -18\}$; Sec. 3.5, Ex. 10 **15. a.** 4^7 **b.** x^{10} **c.** y^4 **d.** y^{12} **e.** $(-5)^{15}$ **f.** a^2b^2; Sec. 4.1, Ex. 3 **17.**

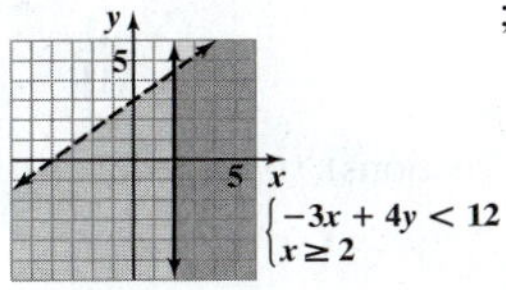

; Sec. 3.6, Ex. 10 **19. a.** x^{11} **b.** $\frac{t^4}{16}$ **c.** $81y^{10}$; Sec. 4.1, Ex. 11 **21.** $-6, -\frac{3}{2}, \frac{1}{5}$; Sec. 5.4, Ex. 9 **23. a.** x^3 **b.** 81 **c.** $\frac{q^9}{p^4}$ **d.** $\frac{32}{125}$; Sec. 4.5, Ex. 2 **25. a.** 6 **b.** -4 **c.** $\frac{3}{10}$ **d.** 0 **e.** 0.8; Sec. 5.5, Ex. 1 **27. a.** $\frac{5}{6}$ **b.** $\frac{\sqrt{3}}{8}$ **c.** $\frac{2\sqrt{10}}{9}$; Sec. 5.5, Ex. 5 **29. a.** $7\sqrt{2}$ **b.** $9\sqrt{3}$ **c.** $1 - 3\sqrt{3} - 6\sqrt{2}$; Sec. 5.5, Ex. 7 **31. a.** $\sqrt{21}$ **b.** 3 **c.** $3\sqrt{5}$ **d.** $10\sqrt{6}$; Sec. 5.5, Ex. 8 **33.** $-3a(3a^4 - 6a + 1)$; Sec. 5.1, Ex. 5 **35.** $(x + 2)(x^2 - 2x + 4)$; Sec. 5.3, Ex. 8 **37.** $\sqrt{44{,}800}$ ft $\approx$ 212 ft; Sec. 5.7, Ex. 3 **39.** $\frac{\sqrt{14}}{2}, -\frac{\sqrt{14}}{2}$; Sec. 5.6, Ex. 2 **41.** $-\frac{1}{2}, 5$; Sec. 5.6, Ex. 8 **43.**

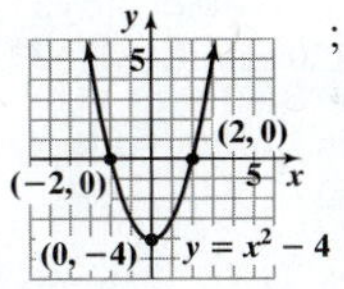

; Sec. 5.8, Ex. 2

APPENDIX B INDUCTIVE AND DEDUCTIVE REASONING

Practice Exercises

1. 162 **2.** 26 **3. a.** F(each letter is the first letter of the days of the week) **b.** L **4.**

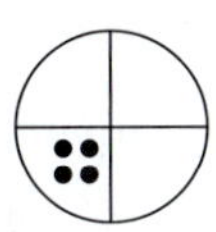

5. 6 people **6.** A, D, B, C

Exercise Set Appendix B

1. inductive **3.** deductive **5.** inductive **7.** 81 **9.** C **11.** 36 **13.** E **15.** 99999 **17.** 9 **19.** 74 **21.** 241

23.

25.

27.

29.

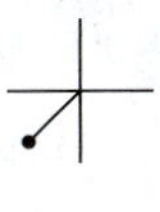

31.

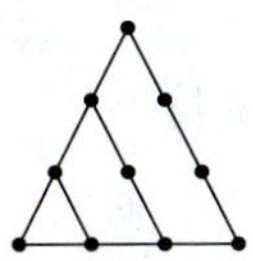

33.

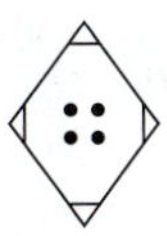

35. answers may vary **37.**

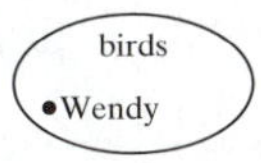

39. 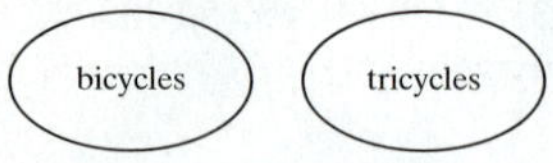

41. 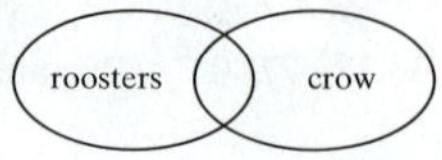

43. 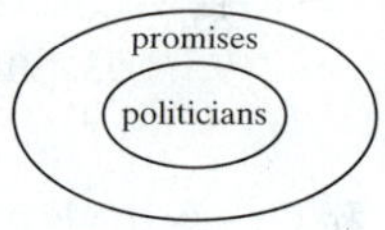

45. Joey **47.** 15 people **49.** Donna **51.** Julio **53.** 5 people

APPENDIX C RATIOS AND RATES

Practice Exercises

1. $\frac{19}{30}$ **2.** $\frac{4}{3}$ **3.** $\frac{7}{20}$ **4.** $\frac{10}{7}$ **5.** $\frac{8}{25}$ **6. a.** $\frac{3}{5}$ **b.** $\frac{5}{12}$ **7.** $\frac{\$225}{1\text{ wk}}$ **8.** $\frac{59\text{ mi}}{3\text{ gal}}$ **9.** $\frac{400\text{ ft}}{1\text{ sec}}$ or 400 ft/sec **10.** $\frac{6.5\text{ bushels}}{1\text{ tree}}$ or 6.5 bushels/tree **11.** \$34 per day **12.** 11-oz bag

Vocabulary, Readiness & Video Check Appendix C

1. unit **3.** division **5.** numerator; denominator **7.** false **9.** The units are different in Example 5 (shrubs and feet); they were the same in Example 4 (days). **11.** When shopping for the best buy

Exercise Set Appendix C

1. $\frac{2}{3}$ **3.** $\frac{77}{100}$ **5.** $\frac{463}{821}$ **7.** $\frac{3}{8}$ **9.** $\frac{8}{25}$ **11.** $\frac{12}{7}$ **13.** $\frac{2}{7}$ **15.** $\frac{4}{1}$ **17.** $\frac{10}{29}$ **19.** $\frac{25}{144}$ **21.** $\frac{5}{4}$ **23.** $\frac{15}{1}$ **25.** $\frac{2}{3}$ **27.** $\frac{10}{21}$ **29.** $\frac{17}{40}$ **31.** $\frac{191}{141}$ **33.** $\frac{1\text{ shrub}}{3\text{ ft}}$ **35.** $\frac{3\text{ returns}}{20\text{ sales}}$ **37.** $\frac{3\text{ laser printers}}{14\text{ computers}}$ **39.** $\frac{9\text{ gal}}{2\text{ acres}}$ **41.** 110 cal/oz **43.** 75 riders/car **45.** 90 wingbeats/sec **47.** \$50,000/yr **49.** 315,750 voters/senator **51.** 300 good/defective **53.** \$5,924,700/player **55. a.** 31.25 computer boards/hr **b.** 33.5 computer boards/hr **c.** Suellen **57. a.** $\approx$27.6 miles/gal **b.** $\approx$29.2 miles/gal **c.** the truck **59.** \$11.50 per compact disc **61.** \$0.17 per banana **63.** 8 oz: \$0.411 per oz; 12 oz: \$0.399 per oz; 12 oz **65.** 16 oz: \$0.118 per oz; 6 oz: \$0.115 per oz; 6 oz **67.** 12 oz: \$0.191 per oz; 8 oz: \$0.186 per oz; 8 oz **69.** 100: \$0.006 per napkin; 180: \$0.005 per napkin; 180 napkins

APPENDIX D PROPORTION AND PROBLEM SOLVING

Practice Exercises

1. 99 **2.** $\frac{13}{3}$ **3.** \$9.03 **4.** 6 **5.** 15

Vocabulary, Readiness & Video Check Appendix D

1. No. Proportions are actually equations containing rational expressions (algebraic fractions), so they can also be solved by using the steps to solve those equations.

Exercise Set Appendix D

1. 4 **3.** $\frac{50}{9}$ **5.** -3 **7.** $\frac{14}{9}$ **9.** $\frac{30}{7}$ **11.** $\frac{17}{11}$ **13.** 31 **15.** $\frac{9}{14}$ **17.** 123 lb **19.** 132 cal **21.** $y = 21.25$ **23.** $y = 5\frac{5}{7}$ ft **25.** 18 applications **27.** 5 weeks **29.** $10\frac{2}{3}$ servings **31.** 120 calories **33. a.** 18 tsp **b.** 6 tbsp **35.** 6 people **37.** $x = 5$ **39.** $x = 13.5$ **41.** 112 ft; 11-in. difference **43.** 102.9 mg **45.** 434 emergency room visits **47.** 28 workers

APPENDIX E VARIATION AND PROBLEM SOLVING

Practice Exercises

1. $y = 5x$ **2.** $k = \frac{1}{4}; y = \frac{1}{4}x$; when x is 20, $y = 5$. **3.** $k = \frac{3}{4}; y = \frac{3}{4}x$ **4.** $y = \frac{8}{x}$ **5.** $k = 2.1; y = \frac{2.1}{x}$; when x is 70, $y = 0.03$. **6.** 6.48 sq units **7.** 25 ml

Vocabulary, Readiness & Video Check Appendix E

1. inverse **3.** direct **5.** inverse **7.** inverse **9.** direct **11.** linear; slope **13.** No. The direct relationship is the power of x times a constant, and the inverse relationship is the reciprocal of the power of x times a constant.

Exercise Set Appendix E

1. $y = \frac{1}{2}x$ **3.** $y = 6x$ **5.** $y = 3x$ **7.** $y = \frac{2}{3}x$ **9.** $y = \frac{7}{x}$ **11.** $y = \frac{0.5}{x}$ **13.** $y = kx$ **15.** $h = \frac{k}{t}$ **17.** $z = kx^2$ **19.** $y = \frac{k}{z^3}$ **21.** $x = \frac{k}{\sqrt{y}}$ **23.** $y = 40$ **25.** $y = 3$ **27.** $z = 54$ **29.** $a = \frac{4}{9}$ **31.** \$62.50 **33.** \$6 **35.** $5\frac{1}{3}$ in. **37.** 179.1 lb **39.** 1600 ft

Index